Values of Some Physical Constants

	CGS	SI*
Avogadro's number, N_{av}	6.02×10^{23} molecules/mole	6.02×10^{26} molecules/kmole
Electronic charge, e	1.60×10^{-20} emu	1.60×10^{-19} C
	4.8×10^{-10} esu	
Electronic mass, m	9.11×10^{-28} g	9.11×10^{-31} Kg
Planck's constant, h	6.63×10^{-27} erg · sec	6.63×10^{-34} J-sec
Velocity of light, c	3.00×10^{10} cm/sec	3.00×10^{8} m/sec
Boltzmann's constant, k	1.38×10^{-16} erg/°K	1.38×10^{-23} J/°K
	(8.63×10^{-5} eV/°K)	
Gas constant, R	8.31×10^{7} erg/mole °K	8.31 J/mole, °K
	(1.98 cal/mole °K)	

*The International System of Units

Some Useful Conversion Factors

1 A = 10^{-8} cm
1 Atm = 14.7 psi = 1.01 Bars
1 cal = 4.19 J
1 eV = 1.6×10^{-12} erg = 1.6×10^{-19} J
1 eV/molecule = 23,100 cal/mole
1 erg = 10^{-7} J = 2.39×10^{-8} cal
1 gauss = 10^{-4} Weber/m^2
1 J = 2.78×10^{7} kW · hr
°K = °C + 273
1 kg/mm^2 = 1422 psi
1 psi = 6.89×10^{4} dynes/cm^2
1 Mil = 10^{-3} in = 25.4 μm
1 μm = 3.94×10^{-5} in
1 oersted = $\frac{1}{4x} \times 10^{3}$ amp/m
1 lb = 0.454 kg
1 W = 1 J/sec = 10^{7} erg/sec

Some Useful Mathematical Relations

Natural logarithm base $e = 2.718$
$e^{x} = 10^{(x/2.3)}$
$e^{-x} = 1/e^{x}$
$\ln e^{x} = x$
$\ln (A\, e^{x}) = \ln A + x$
$\ln x = 2.3 \log_{10} x$
$e^{x}e^{y} = e^{(x+y)}$

The Principles of Materials Selection for Engineering Design

by

Pat L. Mangonon, Ph.D., P.E., FASM
Florida Institute of Technology
Melbourne, Florida 32901

Prentice Hall, Upper Saddle River, New Jersey 07458

Library of Congress Cataloging-in-Publication Data

Mangonon, P. L.
The principles of materials selection for engineering design by
P. L. Mangonon
p. cm.
Includes bibliographical references and index.
ISBN 0–13–242595–5
1. Materials. 2. Engineering design. I. Title.
TA403.6.M29 1999
620.1′1—dc21 97–39039
CIP

Acquisitions Editor: *Bill Stenquist*
Editorial/Production Supervision: *Sharyn Vitrano*
Editor-in-Chief: *Marcia Horton*
Managing Editor: *Bayani Mendoza DeLeon*
Copy Editor: *Barbara Danziger*
Cover Director: *Jayne Conte*
Director of Production and Manufacturing: *David W. Riccardi*
Manufacturing Buyer: *Pat Brown*
Editorial Assistant: *Meg Weist*

The cover photo illustrating materials selection for performance, safety, and economy is courtesy of NASA's Public Affairs Office at the KSC.

Printed in the United States of America

10 9 8 7 6 5 4 3 2 1

ISBN 0-13-242595-5

Prentice-Hall International (UK) Limited, *London*
Prentice-Hall of Australia Pty. Limited, *Sydney*
Prentice-Hall Canada Inc., *Toronto*
Prentice-Hall Hispanoamericana, S.A., *Mexico*
Prentice-Hall of India Private Limited, *New Delhi*
Prentice-Hall of Japan, Inc., *Tokyo*
Simon & Schuster Asia Pte. Ltd., *Singapore*
Editora Prentice-Hall do Brasil, Ltda., *Rio de Janeiro*

Contents

Preface xii

Part I Structures and Properties 1

1. Atomic Structure and Bonding in Solids 2

1.1 Introduction 2
1.2 The Unbonded Atom 2
1.2.1 Structure 2
1.2.2 Atomic Number 3
1.2.3 Atomic Mass 3
1.2.4 Fundamental Particles 4*
1.3 Arrangement of Electrons in Atoms—Electron Configuration 5
1.3.1 Energy Levels of Electron and Principal Quantum Number 5
1.3.2 Subshells and Other Quantum Numbers—Quantum States of Electrons 6
1.3.3 Electron Configuration 6
1.4 The Periodic Arrangement (Table) of Materials and Their Chemical Characteristics 10
1.5 Characteristic X-ray Emission and Chemical Analysis 13
1.6 Bonding of Atoms to Form Molecules, Liquids, or Solids 15
1.6.1 Bonding Forces and Energy 15
1.6.2 The Principal Bonding Mechanisms and Properties Derived from Them 16
1.6.3 Mixed Ionic and Covalent Bonding—Electronegativity 20
1.6.4 Secondary Bonding —The van der Waals Bond 21
1.7 The Sizes of Atoms and Ions 22
Summary 26
References 26
Terms and Concepts 26
Questions and Exercise Problems 27

2. Crystalline Structures—Ideal and Actual 29

2.1 Introduction 29
2.2 Crystalline and Amorphous (Noncrystalline) Materials 29
2.3 The Unit Cell, Crystal Systems, and Point or Space Lattices 30

2.4 The Common Metallic Structures 32
2.4.1 The Body Centered Cubic (BCC) Structure 32
2.4.2 The Face Centered Cubic (FCC) Structure 33
2.4.3 The Hexagonal Close-Packed (HCP) Structure 34
2.4.4 Interstices or Holes in Crystal Structures 35
2.5 Covalent Structure—The Diamond Cubic Structure 38
2.6 Ionic Crystal Structures—Ceramic Materials 40
2.6.1 The CsCl Structure 41
2.6.2 The NaCl Structure 41
2.6.3 The Zinc Blende Structure 42
2.6.4 The Silica Structure 43
2.7 Crystallinity in Polymers 44
2.8 Allotropy or Polymorphism 45
2.9 Packing Fraction and Density 47
2.10 Directions and Planes in Crystals—Miller Indices 49
2.10.1 Directions 49
2.10.2 Planes 52
2.10.3 HCP Planes and Directions 57
2.11 Slip Systems in Metals and Deformability 60
2.11.1 Slip Planes 60
2.11.2 Slip Directions 61
2.11.3 Number of Slip Systems and Deformability 61
2.12 Imperfections in Crystalline Structures—Actual Structures of Materials 63
2.12.1 Surface Structures 63
2.12.2 Imperfections within a Grain or Single Crystal 66
2.13 Crystal Structure and Lattice Parameter Determination —X-ray Diffraction* 70
2.13.1 Constructive and Destructive Interference 70
2.13.2 Bragg's Law 71
2.13.3 Interplanar Spacing, d 72
2.13.4 Diffraction Angle or Direction 72
2.13.5 Intensity of Diffracted Beam 72
2.13.6 Structure Factors for Cubic Structures 73
2.13.7 Determining the Crystal Structure and the Lattice Parameter 74
Summary 78
References 78
Terms and Concepts 78
Questions and Exercise Problems 79

3. Electrical Properties of Materials 82

3.1 Introduction 82
3.2 Metals and Nonmetals (Insulators) 82
3.2.1 Relationship to Bonding and Energy Bands 82
3.2.2 Nonmetal to Metal Transition and Vice Versa 85*
3.3 Electrical Conductivity 89
3.4 Electronic Conduction in Metals and Alloys 91
3.5 Superconductivity and Its Characteristics* 96
3.5.1 Critical Properties 97
3.5.2 Motion and Interaction of Electrons in LTS Materials 100
3.5.3 Theory for HTS Materials 101
3.6 Work Function and Contact Potential 101
3.7 Semiconductivity 102
3.7.1 Intrinsic Semiconductors 102
3.7.2 Density of States and the Fermi-Dirac Function 103
3.7.3 Intrinsic N_{i_e} and N_{i_h} 105
3.7.4 Extrinsic Semiconductors 108
3.7.5 Conductivities of Extrinsic Semiconductors 110
3.7.6 Compound Semiconductors 112
3.7.7 Interdependency of N_e and N_h 114
3.7.8 Semiconductor Devices Based on p-n Junctions 114
Summary 118
References 119
Terms and Concepts 119
Questions and Exercise Problems 120

4. Materials Properties for Design of Structures and Components 122

4.1 Introduction 122
4.2 Physical Properties—Microstructure Insensitive 123
4.2.1 Density 123
4.2.2 The Melting Point 123
4.2.3 The Glass Transition Temperature 124
4.2.4 Coefficient of Linear Thermal Expansion 124
4.2.5 Thermal Conductivity 125

4.3 Concepts of Stress, Strain, and Young's Modulus 127
4.3.1 Concept of Stress 127
4.3.2 Shape Changes and Concept of Strain 128
4.3.3 Young's Modulus and Interatomic Bonding [2] 131*
4.4 Mechanical Properties of Materials and Test Methods 133
4.4.1 Strength of Materials 134
4.4.2 Strengthening of Metallic Materials 134*
4.4.2a Solid Solution Strengthening 134
4.4.2b Grain Size Strengthening 135
4.4.2c Dislocation Strengthening 136
4.4.2d Dispersed Phase or Particle Strengthening 136
4.4.3 Ductility 139
4.4.4 The Tensile Test—ASTM E-8 and Properties Derived From It. 140
4.4.4a The Engineering Stress-Strain Curve 140
4.4.4b Uniform and Nonuniform Deformation of Test Specimen 143
4.4.4c Measures of Ductility 144
4.4.4d The True Stress—True Strain Curve 144
4.4.5 Tensile Test for Plastics—ASTM D 68 146
4.4.6 Factor of Safety and Allowable Design Stress 150
4.4.7 Fracture Mechanics and Fracture Toughness [5] 151
4.4.8 Fracture Toughness Testing 153
4.4.9 Design Using Fracture Toughness 154
4.4.10 Fatigue Properties and Fatigue Testing 156
4.4.10a The Constant Stress Amplitude Test 156
4.4.10b The Constant Strain Amplitude Test 158
4.4.11 Creep and Creep Testing 164
4.4.11a The Mechanisms of Creep 165
4.4.12 Design Involving Creep—Time-Temperature Parameters 167
4.4.12a The Shelby-Dorn Parameter 168
4.4.12b The Larson-Miller Parameter 169
4.4.13 Impact Properties and Tests 171
4.4.14 Hardness and Hardness Tests 173
Summary 175
References 176
Terms and Concepts 177
Questions and Exercise Problems 178

Part II Processing 183

5. Equilibrium Phase Diagrams 184

5.1 Introduction 184
5.2 Basic Concepts and Terms 185
5.3 Complete Solubility in the Solid State 186
5.4 The Gibbs Phase Rule 188
5.5 Lever Rule 189
5.6 Three-Phase Reactions 191
5.7 The Eutectic Systems 192
5.7.1 Complete Insolubility in the Solid State 192
5.7.2 Partial Solubility in One Component 193
5.7.3 Partial Solubilities in Both Components 193
5.7.4 Development and Control of Microstructures in Eutectic Systems 195
5.8 The Iron–Carbon System 199
5.8.1 The Fe-Fe_3C Metastable System 200
5.8.1a The Eutectoid System in the Fe-Fe_3C Diagram 202
5.8.2 The Stable Fe-C (Graphite) System 206
5.9 Intermetallic Compounds and Secondary Solid Solutions 207
5.10 More Complicated Phase Diagrams—Other Three-Phase Invariant Reactions 208
5.11 Determination of the Liquidus and Solidus Curves of the Phase Diagram 211
5.11.1 Cooling and Liquidus Curves 211
5.11.2 Heating Curves and Solidus Curves 213
Summary 215
References 215
Terms and Concepts 215
Questions and Exercise Problems 216

6. Diffusion in Solids 219

6.1 Introduction 219
6.2 Macroscopic Movements of Atoms—Fick's Laws of Diffusion 220
6.2.1 Fick's First Law 220
6.2.2 The Diffusion Coefficient 221
6.2.3 Fick's Second Law 221
6.2.4 Applications of Fick's Second Law in Processing 222
6.2.4a Homogenization [Ref. 1] 222
6.2.4b Carburizing and Nitriding Processes [Ref. 2] 223

6.2.4c Decarburization [Ref. 1] 227
6.2.4d Doping of Pure Silicon 231
6.2.5 Processing of Microelectronic Circuits 233
6.3 Microscopic Atomic Movements—Mechanisms Of Diffusion* 236
6.3.1 Nature of Diffusion Coefficient, D 236
6.3.2 Mechanisms of Diffusion, Activation Energy, and Γ 237
6.3.3 Diffusion along Defects and on Surfaces 238
Summary 239
References 239
Terms and Concepts 240
Questions and Exercise Problems 240

7. Melting, Solidification, And Casting 243

7.1 Introduction 243
7.2 Processing of Melts 244
7.2.1 Primary Processing 244
7.2.2 Secondary Processing or Ladle Metallurgy 246
7.2.3 Remelting 249
7.3 Characteristics of the Liquid State 250
7.4 Solidification—A Nucleation and Growth Process 251
7.4.1 Homogeneous Nucleation 252
7.4.2 Heterogeneous Nucleation 254
7.4.3 Growth of Nuclei 256
7.5 Casting and Control of Cast Structure 258
7.5.1 Possible Macroscopic Defects in Castings and Ingots 258
7.5.2 Cast Structure and Segregation 260
7.5.3 Continuous Casting 263
7.6 Metallic Glasses or Amorphous Metals [6c] 264
7.7 Single Crystals 265
7.8 Homogenization 266
Summary 266
References 267
Terms and Concepts 267
Questions and Exercise Problems 268

8. Plastic Deformation and Annealing 270

8.1 Introduction 270
8.2 Deformation Processes 271
8.3 Plastic Deformation of Single Crystals 273
8.3.1 Existence and Evidence of Dislocations 273*
8.3.2 Schmid's Critical Resolved Shear Stress (CRSS) for Slip 275
8.3.3 Rotation of Crystal Planes and Effects of Constraints During Deformation 277
8.4 Cold and Warm Working 278
8.4.1 Strengthening with Cold-Work 279
8.4.2 Specifying the Cold-Work Condition 281
8.4.3 Deformation (Crystallographic) Texture 285
8.4.4 Sheet Properties for Good Formability 286
8.4.4a Drawing and Properties for Good Drawability 286
8.4.4b Stretching and Properties for Stretchability 289
8.4.4c Actual Forming Operations 291
8.4.5 Residual Stresses 291
8.4.6 Plastic Work and Stored Energy of Cold-Work 292
8.4.6a Plastic Work with a Tensile Specimen 292
8.4.6b Extrusion and Wire Drawing 295
8.5 Annealing—Heating to Soften Materials 296
8.5.1 Recovery—The First Stage 297
8.5.2 Recrystallization or Primary Recrystallization—The Second Stage 298
8.5.2a Kinetics of Recrystallization 299
8.5.2b Recrystallization or Annealing Texture 300
8.5.3 Grain Growth—The Third Stage 301
8.6 Hot Working 302
8.6.1 Dynamic Recovery and Recrystallization 303
8.6.2 Static Recrystallization 304
8.6.3 Controlled Thermal Mechanical Processing—A Process Design 306
8.7 Superplasticity and Superplastic Forming 308
8.8 Forming of Ceramics, Glass, Plastics/Reinforced Plastics (Composites) 310
Summary 311
References 312
Terms and Concepts 312
Questions and Exercise Problems 313

9. Thermal Processes—Heat-Treating 316

9.1 Introduction 316
9.2 Nonhardening Thermal Processes 316
9.2.1 Thermal Stress-Relief of Residual Stresses 316
9.2.2 Annealing 319
9.2.3 Normalizing 321

9.3 Hardening (Heat Treating) Processes—General Principles 323
9.3.1 Transformation C-Curves and Their Implications 325
9.3.2 The Martensite Transformation—A Displacive Transformation 328
9.4 Precipitation Hardening 331
9.4.1 Aluminum Alloys 332
9.5 Heat-Treating of Titanium Alloys [6] 335
9.6 Hardening Due to Spinodal and Order Transformations 339
9.7 Heat-Treating of Steels 342
9.7.1 Nonequilibrium Processing Diagrams 343
9.7.1a Isothermal Transformation (I-T) Diagrams (C-Curves) of Steels 343
9.7.1b Continuous Cooling Transformation (CCT) Diagrams 345
9.7.2 Through Hardening 347
9.7.2a Hardness of As-Quenched Martensite 347
9.7.2b Tempering and Predicting Tempering Condition 351
9.7.2c Hardenability and the Jominy Test 355
9.7.2d Hardenability Bands 358
9.7.2e Severity of Quench 358
9.7.2f Ideal and Actual Critical Diameters 365
9.7.2g Special Processes Using the I-T Diagrams 367
9.7.3 Precautions in Processing and Use of Through Hardened Steels 369*
9.7.3a Homogeneous and Retained Austenite 371
9.7.3b Embrittling Phenomena 373
9.7.4 Surface Hardening Processes 374
9.7.4a Processes with No Change in Surface Composition 374
9.7.4b Processes with Change in Surface Composition 377
Summary 384
References 385
Terms and Concepts 386
Questions and Exercise Problems 387

Part III Performance and Materials Selection for Engineering Design 389

10. Corrosion and Corrosion Control 390

10.1 Introduction 390
10.2 Forms of Corrosion and Environments 391
10.3 Electrochemical Aqueous Corrosion 391
10.3.1 Electrochemical Reactions at the Anode and at the Cathode 392
10.3.2 Cell Voltage and Electrode Potential 393
10.3.3 Thermodynamics and Cell Voltage 394
10.3.4 Dependence of Equilibrium Cell Voltages and Electrode Potentials on Concentration— Nernst Equation 396
10.3.5 Types of Electrochemical Corrosion Cells 397
10.3.5a Galvanic Cells 397
10.3.5b Electrolyte Concentration Cells 397
10.3.5c Differential Temperature Cells 401
10.3.5d Differential Cold-Work or Stress Cells 401
10.3.6 The Pourbaix Diagram 404
10.4 Amount and Rate of Corrosion 409
10.5 Corrosion Current and Change in Electrode Potentials → Polarization 412
10.6 Some Factors Affecting Aqueous Corrosion Rates 414
10.6.1 Presence of Oxygen in the Electrolyte—Aerated Solutions 414
10.6.2 Hard versus Soft Water 414
10.6.3 Presence of More Noble Ions in Water 415
10.7 Effect of Flowing Electrolyte 415
10.8 Effect of Corrosion on Mechanical Properties 416
10.9 Designing to Control Corrosion 418
10.9.1 Elimination of the Electrodes 418
10.9.2 Eliminate or Avoid Contact of the Electrodes 420
10.9.3 Eliminate or Control the Electrolyte 420
10.9.4 Cathodic Protection 421
10.9.5 Anodic Protection 422
10.9.6 Design Details and Assembly of Components to Minimize Corrosion 424
Summary 426
References 426
Terms and Concepts 426
Questions and Exercise Problems 427

11. Design, Selection, and Failure of Materials 429

11.1 Introduction 429
11.2 General Methodology of Design 429
11.2.1 The Phases of Design 430
11.2.2 Design Activities 431
11.2.3 Materials Selection 431

11.3 Factors in Materials Selection 431
11.3.1 Interrelated Constraints 431
11.3.1a Physical Factors 433
11.3.1b Mechanical Factors 433
11.3.1c Processing and Fabricability 433
11.3.1d Life of Component Factors 433
11.3.1e Cost and Availability 433
11.3.1f Codes, Statutory, and Other Factors 433
11.3.2 Criteria and Tools for Materials Selection [2] 434
11.3.3 Performance and Efficiency of Materials 437
11.3.4 Materials Charts [3,5] 439
11.3.5 Shape Factors [4,5] 445
11.3.6 Efficiency of Shape Sections 448
11.3.7 Influence of Internal Regular Shaped Features 450
11.4 Practical Issues in Engineering Design 451
11.4.1 Risk Issues 451
11.4.2 Performance and Cost 452
11.5 Materials and Component Failures—As Sources of Engineering Experience [10] 452
11.5.1 Deficiencies in Design 452
11.5.2 Deficiencies in Materials Selection 453
11.5.3 Imperfections in the Material 455
11.5.4 Failures Arising from Processing and Fabrication 455
11.5.5 Failures Due to Errors in Handling and Assembly 456
11.5.6 Failures Due to Service Conditions 456
11.6 Information Sources and Specifications 456
Summary 457
References 458
Terms and Concepts 458
Questions and Exercise Problems 459

12. Selection of Ferrous Materials 460

12.1 Introduction 460
12.2 Classification, Designations, and Specifications for Steels 461
12.2.1 Classification According to Composition 462
12.2.2 Classification According to Strength 464
12.2.3 Classification According to Product Shape, Finish Processing, and Quality Descriptors 469
12.3 Procurement and Specifications 477
12.4 Selection of Carbon and Low-Alloy Structural Steels [4a] 479
12.4.1 Weldability, Formability, and Machinability of Carbon and Low-Alloy Steels 480
12.5 Heat-treatable Carbon and Low-Alloy Steels 486
12.5.1 Carbon Steels 489
12.5.2 Alloy Steels 491
12.6 Selection of Tool Steels [5,6] 498
12.6.1 High-Speed Steels 498
12.6.2 Hot-Work Tool Steels—Group H 499
12.6.3 Cold-Work Tool Steels 501
12.6.4 Shock Resisting Steels—Group S 501
12.6.5 Low-Alloy Special-Purpose Steels—Group L 501
12.6.6 Mold Tool Steels—Group P 501
12.6.7 Water Hardening Tool Steels—Group W 502
12.6.8 Guide to Selection and Summary of Properties of Tool Steels 502
12.7 Selection of Stainless Steels [7] 502
12.7.1 Designations, Classes, and Properties of Stainless Steels (SS) 502
12.7.2 Selection of SS 512
12.8 Selection of Cast Irons 512
12.8.1 Types and Characteristics of Cast Irons [8] 516
12.8.2 Specifications, Grades, and Mechanical Properties of Cast Irons 522
12.8.2a Gray Cast Iron 522
12.8.2b Ductile Cast Iron 525
12.8.2c Compacted Graphite (CG) Cast Iron 526
12.8.2d Malleable Cast Iron 528
12.8.3 Design of Iron Castings [8] 531
12.8.4 Special Design Concepts in Using Gray Iron 534
Summary 535
References 536
Terms and Concepts 537
Questions and Exercise Problems 537

13. Selection of Nonferrous Metals 539

13.1 Introduction 539
13.2 Aluminum and Aluminum Alloys [1,2] 540
13.2.1 Alloy Designations 540
13.2.1a Wrought Aluminum and Aluminum Alloys 540
13.2.1b Cast Aluminum and Aluminum Alloys 542

13.2.2 Temper Designations 542
13.2.2a Basic Temper Designations 542
13.2.2b Designation for Degree of Cold-Work or Strain-Hardening 544
13.2.2c Designations for Heat-Treated Conditions 545
13.2.2d Additional T Temper Variations 546
13.2.2e Annealed Products 546
13.2.3 Selection and Applications of Wrought Aluminum and Aluminum Alloys [2] 547
13.2.3a Nonheat-Treatable Alloys 547
13.2.3b Heat-Treatable Alloys 554
13.2.4 Selection of Aluminum and Aluminum Alloy Castings [2] 557
13.3 Titanium and Titanium Alloys [5] 563
13.3.1 Commercially Pure and Modified Titanium 563
13.3.2 Titanium Alloys 568
13.3.2a Classifications and Designations of Alloys 568
13.3.2b Room Temperature Mechanical Properties 571
13.3.2c Selection of Wrought Titanium Alloys 572
13.3.2d Titanium Castings 573
13.4 Magnesium and Magnesium Alloys [6] 580
13.4.1 Magnesium Alloys and Their Designations 580
13.4.2 Applications of Magnesium 582
13.5 Copper and Copper Alloys 583
13.5.1 Alloy and Temper Designations of Copper and Copper Alloys 583
13.5.2 Properties and Selection Criteria of Copper and Copper Alloys 585
13.5.3 Applications of Copper and Copper Alloys 591
13.6 Nickel and Nickel Alloys [6] 592
13.6.1 Alloys—Their Designations and Characteristics 593
Summary 601
References 601
Terms and Concepts 602
Questions 602

14. Inorganic Materials—Ceramics and Glasses 604

14.1 Introduction 604
14.2 Ceramics 605
14.2.1 Crystal Structures 605
14.2.2 Ceramic Phase Diagrams 608
14.2.3 Processing of Ceramics 611
14.2.3a Batching and Preparation of Powders 611
14.2.3b Forming 612
14.2.3c Drying and Firing 614
14.2.3d Shaping and Surface Finishing 618
14.2.4 Properties of Structural Ceramics 620
14.2.4a Porosity or Void Content 621
14.2.4b Uniaxial Tensile Strength 621
14.2.4c Uniaxial Compression Strength 626
14.2.4d Modulus of Elasticity 627
14.2.5 Fracture Toughness of Ceramics and Zirconia 628*
14.2.5a ZrO_2 in Another Ceramic Matrix 628
14.2.5b Partially Stabilized Zirconia (PSZ) 629
14.2.5c Tetragonal Zirconia Polycrystals (TZP) 630
14.3 Glasses 631
14.3.1 Formation andPrimary Processing of Glass 631
14.3.1a Container Manufacture 632
14.3.1b Flat Glass—Sheet and Plate 633
14.3.1c Tubings 634
14.3.1d Glass Fibers 634
14.3.2 Silicates and Glass Structures 634
14.3.2a Island Structures 635
14.3.2b Isolated Group Structures 636
14.3.2c Chain Structures 636
14.3.2d Sheet Structures 636
14.3.2e Framework or Network Structures 636
14.3.2f Glass or Amorphous Structures 637
14.3.3 Physical and Chemical Properties of Glasses 639
14.3.3a Viscosity—Melting and Glass Transition Temperatures 639
14.3.3b Density 641
14.3.3c Thermal Expansion 641
14.3.3d Thermal Conductivity and Diffusivity 642
14.3.3e Chemical Durability and Weathering of Glasses 642
14.3.3f Young's Modulus, E, and Poisson's Ratio 642
14.3.4 Mechanical Properties 643
14.4 Glass-Ceramics and Secondary Processing 644
14.4.1 Glass-Ceramics 644
14.4.1a Nucleation 644
14.4.1b Crystal Growth 645
14.4.2 Other Secondary Processing 647
14.4.2a Tempering Process 648
14.4.2b Ion-Exchange 649
14.5 Design and Selection of Ceramics and Glasses 650
14.5.1 Design Methodology 650
14.5.2 Selection of Structural Ceramics 652
14.5.3 Selection of Refractories 655

Summary 662
References 663
Terms and Concepts 664
Questions and Exercise Problems 665

15. Plastics and Reinforced Plastics 666

15.1 Introduction 666
15.2 Polymerization Processes 667
15.2.1 Addition Polymerization 667
15.2.2 Condensation Polymerization 668
15.2.3 Combination Polymerization 668
15.3 Structural Features of Polymers and Basic Properties 669
15.3.1 Molecular Weight (MW) 669
15.3.2 Molecular Weight Distribution 671
15.3.3 Linear and Branched Molecules 671
15.3.4 Cross-Linked Molecules 672
15.3.5 Crystalline and Amorphous Polymers 673
15.3.6 Intramolecular Chemical Composition 675
15.3.7 Co-polymerization 675
15.3.8 Non-Carbon Backboned Polymers 678
15.4 Types and Classification of Plastics 678
15.4.1 Classification of Thermoplastics 681
15.4.1a Commodity Thermoplastics 681
15.4.1b Engineering Thermoplastics 684
15.4.2 Thermosets 690
15.5 Compounding of Polymers 696
15.5.1 Alloying and Blending 696
15.5.2 Additives and Fillers [1] 697
15.6 Composites and Reinforced Plastics 701
15.6.1 Classification of Composites 701
15.6.2 Reinforced Plastics 703
15.6.2a Fibers 703
15.6.2b Matrices 708
15.7 Plastics Properties Used in Design 709
15.7.1 Density 709
15.7.2 Water Absorption and Water Transmission 709
15.7.3 Thermal Properties 709
15.7.4 Mechanical Properties 710
15.7.4a Short-Term Properties 712
15.7.4b Long-Term Properties 716
15.8 Properties of Reinforced Plastics and Composite Materials 719
15.8.1 Types of Composites 720
15.8.2 Properties of Fiber-Reinforced Plastics 720
15.8.2a Stressing Unidirectional Fiber Composite in Longitudinal Direction 721
15.8.2b Stressing the Composite along the Transverse Fiber Direction 722
15.8.2c Critical Fiber Volume Fraction 722
15.9 Laminates 725
15.9.1 Fiber Orientation 725
15.9.2 Laminate Codes [5] 726
15.9.3 Properties of Laminates 727
15.10 Fabrication of Plastics 727
15.10.1 Extrusion 729
15.10.2 Injection Molding 730
15.10.3 Blow Molding 730
15.10.4 Forming Processes 731
15.10.5 Other Processes 733
15.11 Fabrication of Reinforced Plastics 734
15.11.1 Contact Molding Methods 734
15.11.2 Matched Mold Methods 735
15.11.3 Other Methods 735
15.12 Designing with Plastics and Reinforced Plastics 736
15.12.1 Design of Continuous Fiber-Reinforced Plastics/Composites—Engineered Materials 738
15.13 Selection of Plastics 740
15.13.1 Guides to Selecting Plastics 740
15.13.2 Matrix Selection for Reinforced Plastics/Composites 740
15.13.3 Composite Selection Methodology 744
15.14 ASTM D 4000 Line Callout for Plastic Materials 749
Summary 750
References 751
Terms and Concepts 751
Questions and Exercise Problems 752

16. Materials Selection Case Studies 754

16.1 Introduction 754
16.2 Steels 754
16.2.1 Structural Steels 755
16.2.1a Statically Loaded Structures [1a] 762
16.2.1b Dynamically Loaded Structures [1a,c] 763
16.2.1c Use of Microalloyed Steels in Automobiles and Off-highway Vehicles and Equipment 764
16.2.1d Benefits in Adopting High-performance Steels—an Example 764
16.2.2 Stainless Steel as a Viable Structural Material [1g] 766
16.2.2a Case Study: Stainless Steels for Reinforcement of Concrete in Highway Bridges 766
16.2.3 Selection of Heat Treated Steel [4] 768

16.3 Aluminum Alloys 771
16.3.1 Justification of Aluminum Use in Skins of Aircraft [5] 771
16.3.1a Case Study 1—Selection and Development of an Aluminum Alloy for the Concorde [5] 774
16.3.1b Case Study 2—Materials Selection for NASA's Space Shuttle [6] 775
16.4 Materials Selection for the Automotive Industry 781
16.4.1 Structural Ceramics for Engine Components 782
16.4.1a The Swirl Chamber [9b,10] 782
16.4.1b Turbocharger Rotor [9b,c,d] 785
16.4.1c Selection of the Ceramic Material 786
16.4.2 Application of Concurrent Engineering in Product Development [11] 787
16.5 Selection Based on Life-Cycle Analysis (LCA) [12,13] 790
16.5.1 Material Selection for Beverage Containers [12] 793
16.5.2 Material Selection for Automotive Fenders [13] 795
16.6 Selection of Glass-Reinforced Plastics [14] 797
16.7 Application of Fracture Mechanics to Materials Selection 797
References 802

Appendix 803

Index 818

Preface

The need for writing this textbook was prompted by the recent change in the educational guidelines from the Accreditation Board for Engineering and Technology (ABET). To give students a broader design experience, ABET wants to see some design content in courses other than the major design course. For a materials course geared for a Materials Science and Engineering major, the design content can be directed to the development of the optimum macro- or microstructures via controlled processing to obtain the optimum properties of the material. However, for a course geared for other engineering disciplines, the design content can only be directed to the optimum use of materials in the design of safe and efficient structures and components at a minimum cost. Therefore, the purpose of this textbook is to help the latter students make sound, intelligent selections of materials in the design of safe and efficient structures or components. It is intended to be the only course in materials engineering that most nonmaterials engineering students need to take in their curricula.

In order to be able to make a judicious and discerning selection of materials, an engineer or designer must know the attributes of materials that are important for the design of the structure or component. These attributes encompass the physical, chemical, and mechanical properties of the material. The physical properties, essentially the microstructure insensitive properties, such as density, modulus, and the coefficient of linear thermal expansion, are unaffected by the processing and the microstructure of the material. On the other hand, the mechanical properties, such as the strength, fracture toughness, and fatigue, are affected by the processing and the resultant microstructure of the material. It is important, therefore, for engineers and designers to understand how processing affects these properties and to realize also that for a chosen material, the properties may change during the fabrication of the structure or component. Thus, the performance of the material in the fabricated structure can no longer be based on the values given in handbooks or by the materials' suppliers. In addition, the values in the handbook have very little significance in corrosive environments where chemical and mechanical properties interact. It is equally important to understand that the handbook properties were obtained using test specimens that may be thinner than those used in the structure. Thus, the prudent selection of materials must be a holistic and concurrent consideration according to the following: how the properties of a chosen material arise; how the properties of the material are affected by the composition, processing, and fabrication; how the structure or component is going to be fabricated from the material; and how the structure or component will be affected by the environmental conditions that they are exposed to. This approach will indicate that there is no single attribute that will satisfy all the design requirements and constraints. Rather the design will involve the combination of various attributes of a material that will yield the optimum performance in the structure or component.

In view of the foregoing, the objective of this book is to provide the underlying principles on how the properties of materials arise; how these properties are affected by the microstructures via composition, processing, and fabrication; how the properties are affected by corrosion; and how the performance of a material in a structure or component is determined by the mode of loading and by its shape. A unique feature of the book is that it contains an entire chapter on the methodology of design and how materials selection should be simultaneously chosen at every stage of the design process—the concept of concurrent engineering. The materials selection should be effected through consideration not only of their properties, their processing, and their fabrication, but also their recyclability, recovery, and disposal after their use—the concept of life-cycle analysis.

The book is organized according to its objective and consists of three parts: Part I, Structures and Properties; Part II, Processing; and Part III, Performance and Materials Selection for Engineering Design. The principles are conveyed in Parts I and II as well as in Chap. 10, Corrosion and Corrosion Control and Chap. 11, Design, Selection, and Failure of Materials. Then, chapters on Ferrous Materials, Nonferrous Metals, Ceramics and Glasses, and Plastics and Reinforced Plastics follow. The last chapter presents actual materials selection cases to fortify the principles. It includes an example of the application of concurrent engineering and the use of life-cycle analysis during concurrent design in materials selection. The book concludes with an example that illustrates a catastrophic failure due to the selection of a wrong material.

Part I indicates how the properties arise from the atomic structures and the microstructures of materials. It consists of four basic interrelated chapters—Chap. 1, Atomic Structure and Bonding in Solids; Chap. 2, Crystalline Structures—Ideal and Actual; Chap. 3, Electrical Properties of Materials; and Chap. 4, Material Properties for Design of Structures and Components.

Chapter 1 discusses the electronic structure (configuration) of an atom, how this leads naturally to the Periodic Table and the chemical characteristics of materials, and how this relates to the type of bonding and bonding energies that may result in large aggregates of atoms such as those in solids.

Chapter 2 defines crystalline and noncrystalline (amorphous) materials according to the arrangement of atoms or molecules in a solid. In addition to the discussion of the basic crystallographic terms, the densities and deformabilities of materials are shown to be related to the crystal structures of materials. The imperfections in actual crystals that may influence the mechanical properties of materials are introduced and the basic rudiments of X-ray diffraction of cubic crystals are discussed to familiarize the students with how crystal structures and lattice parameters are determined.

Chapter 3 describes how electrons are localized or delocalized from their respective atoms after bonding in nonmetals or metals. The localization or delocalization depends on the nonavailability or availability of quantum states for electrons to occupy when an applied electric field is imposed on the material. Metals are characterized by freed or delocalized electrons that move from an orbital of one atom to orbitals of other atoms due to the proximity of atoms that allow the orbitals to overlap. This concept of electrical conductivity was tested in many instances to transform ordinarily nonconductive materials at 0 K to be conductive or have "metallic character" by doping (adding impurities) with a critical concentration. For the same reason, a metallic material can be made nonconductive by separating the atoms to prevent the overlap of orbitals. These are the bases of current discussions on metal-to-nonmetal transitions. Superconductivity and semiconductivity are discussed and their importance indicated. The importance of the p-n junctions of extrinsic semiconductors as the basis of integrated circuits is pointed out.

In Chap. 4, the materials properties are related to the bonding energies of atoms, their crystal structures, and the imperfections of actual crystals. There is a presentation of the importance of each property and how each property is used in design. The melting point, the coefficient of linear thermal expansion, and the elastic moduli are shown to be related to the bonding energies. The strength of materials is related to the imperfections, while fracture toughness is the resistance to crack propagation. The tensile test is discussed to show: (1) how the tensile properties are obtained; (2) the concepts of elastic and plastic strains, strain hardening or cold-working, and recovered elastic strains; (3) the difference between ductile and brittle materials; (4) the difference between the true stress-true strain curve and the engineering stress-engineering strain curve; and (5) the importance of the strain-hardening exponent in forming operations. In design, the allowable stress-design method is stated. To prevent catastrophic failure in static loading, the design must incorporate the fracture toughness of the material and the fracture mechanics approach. In dynamic loading, the fatigue properties of materials obtained with the strain-life tests can be used to predict crack initiation life. With the fracture toughness and the crack propagation rate of the material, the total fatigue life of a structure before fracture can be predicted. For creep, parameters are used to predict the rupture life for a certain stress or the maximum stress for a certain rupture life at various temperatures of application. Charpy V-Notch impact properties are empirically correlated to fracture toughness for some steels, while the tensile strength of steel is correlated to hardness.

In Part II, the actual chapters on processing are: Chap. 7, Melting, Solidification, and Casting; Chap. 8, Plastic Deformation and Annealing; and Chap. 9, Thermal Processes—Heat Treating. These chapters illustrate how the microstructures change or develop during the processing; therefore, processing can be used to control the microstructures and the properties.

Chapter 5, Equilibrium Phase Diagram, and Chap. 6, Diffusion in Solids, lay the foundation to help understand the processes. The equilibrium phase diagram is indicated to be a guide in the processing of materials. Basic terms and concepts relating to the phase diagram are introduced. Binary (two-component) diagrams are used to show (1) complete solubility, (2) insolubility, (3) partial solubility, and (4) compound formation between two elements. Counterparts of these situations in binary liquid solutions are indicated to help understand them in the solid state. The importance of the eutectic and the eutectoid systems, as well as the experimental methods to determine the liquidus and the solidus in simple binary systems, are discussed.

Chapter 6 exposes the students to rate processes involving atomic movements (diffusion) that must be given time to produce the desired effects. Thermal processing involves diffusion and the macroscopic effects can be expressed by Fick's Laws (in particular, the second law). Examples of these processes are homogenization, carburizing of steels, and the doping of silicon. The microscopic aspects indicate how defects in the crystalline structure help in the diffusion process and give physical significance to the diffusion coefficient and the activation energy. Because of the "open" spaces in them, imperfections such as dislocations, grain boundaries, and the surface provide faster diffusion paths than vacancy (volume) diffusion within the crystal.

The sequence of the chapters on processing follows the actual processing of materials. The materials are first melted, refined, solidified, and cast. For wrought products, the cast ingots or slabs or blooms are deformed usually by hot-rolling, then cold-rolling, and then formed into the final product. In between the deformation processes, there could be a need for annealing and recrystallization. Finally, the materials are thermally processed before assembly or fabrication. Discussion in Chap. 7 indicates that the selection of materials starts with the knowledge of how the material is processed in the liquid state. It centers on the importance in getting the desired composition, the elimination of undesired impurities called inclusions, and the careful control during casting that such inclusions do not reappear. Refining of the melt is done primarily in the melting furnace but considerable refining may be further achieved in the ladle before the melt is poured and cast. Using principles acquired from the phase diagram, it is pointed out that during solidification of the melt, conpositional segregation results. In addition, the presence of gases in the melt and the volume contraction from liquid to solid during solidification can potentially produce macroscopic defects. Therefore, control during melting and casting is paramount in order to obtain the desired microstructures. Continuous casting technology reduces segregation and defect formation. Critical applications, especially for fatigue and fracture toughness, require extra-clean (void of inclusions) materials and these may require remelting in a vacuum that may involve zone-refining types of processes similar to those used with semi-conductor silicon.

In Chap. 8, deformation processes are classified according to primary and secondary deformation, and as hot- or cold-working processes. Primary processes are usually done by the metal supplier and usually by hot-working; while secondary processes are those that are done at the fabricating or consumers' plants and may be either hot-worked or cold-worked. Cold-working reemphasizes the importance of the strain hardening discussed in Chap. 4, that is, that cold-worked materials are strengthened—such strengthening can be taken advantage of especially in the final product. Specifications and temper designations for desired cold-worked conditions are discussed in the procurement of materials. These designations are discussed more fully in Chap. 13 for aluminum and copper alloys but are discussed here to illustrate the just-in-time approach. Annealing is introduced to indicate how to resoften a material if further cold-work is required to attain desired thickness or dimensions. The first stage of annealing is recovery and involves minor changes in the dislocation rearrangement but not in the shape of the "pancaked" cold-worked grains. The softening of the materials is due to recrystallization, which is the complete replacement of the cold-worked pancaked grains with new equi-axed grains. There is a discussion on the laws on recrystallization and how to control the grain sizes of materials. Hot-working is done at temperatures where recrystallization can occur. In the hot-working of materials, the laws of recrystallization can be applied to inhibit recrystallization, or to induce the recrystallization or the transformation to other phases with very fine grain sizes. This processing control produces high-strength materials and it is used particularly in the processing of microalloyed high-strength low-alloy steels. For aluminum and titanium, fine grain sizes are required for superplastic forming. Other conditions are presented for typical alloys and for those exhibiting superplasticity.

In Chap. 9, thermal processes are classified as nonhardening and hardening. The nonhardening processes include the recrystallization and stress-relief or recovery-annealing processes for all materials discussed in Chap. 8. For steels, nonhardening processes include: (1) annealing that may be full annealing, intercritical annealing, and subcritical annealing; (2) normalizing; and (3) spheroidization. The hardening processes are commonly referred to in the industry as "heat-treating" that has the specific purpose to harden or to strengthen the material. The generic principles of the formation of the C-curves (isothermal transformation diagrams) and heat treating that is applicable to all systems are discussed. The C-curve is the result of the interaction of thermodynamic (free energy of transformation) and kinetic (diffusion) factors. The heat-treating principles consist of a solution anneal of a two- or multiphase alloy to a single phase field in the phase diagram, quenching it rapidly to retain the solid solution condition at high temperatures to produce a supersaturated solid solution at a much lower temperature, and then the controlled reforming or precipitation of transition phases hardens the material. The specific precipitation hardening of aluminum alloys and the heat treating of titanium and copper alloys are discussed.

The primary hardening process for steels is due to the martensite formation, which can be either through the entire section or on the surface of the steel only—if through the section, the steels are called through-hardened steels and if on the surface, they are called surface-hardened steels. The hardness of the as-quenched martensite depends only on the carbon content, and because of its hardness, the as-quenched martensite is brittle. Thus, hardened steels are not used in the as-quenched condition and need to be tempered to give them some ductility and toughness. The through-hardened steels are, therefore, in the quenched and tempered condition and the degree of tempering (varying temperature) will depend on the desired properties. For very high-alloy steels such as the tool steels, there is a secondary hardening (precipitation hardening) that arises during tempering due to the precipitation of alloy carbides. The secondary hardening maintains the hardness of the tool steels that is needed in the cutting/machining to produce a good machined surface.

The ease of forming martensite is called the hardenability of the steel and is determined by the Jominy test. The result of the latter is shown as a hardenability or Jominy curve. The through hardening of a steel depends on its hardenability, on its section size, and on the severity of quench. Actual selection of heat-treated steels depends on the hardness and the strength required at a particular location in the section. The principle of Jominy-equivalent cooling is used, which says that the hardness of a steel at a particular point in the Jominy curve can be attained in a part or section of the same steel if the latter can be cooled with the same cooling rate corresponding to that at the point in the Jominy curve. Ideal critical diameters can be calculated from compositions of steels and can roughly indicate whether certain compositions can be through-hardened with the severity of quench to achieve the desired hardening. Precautions and the embrittling phenomena in heat treating and surface hardening processes are discussed. The latter includes the carburizing process as stated in Chap. 6.

Part III starts with Chap. 10, Corrosion and Corrosion Control, to discuss the performance or nonperformance of materials when exposed to different environments. This chapter stresses the inevitability of degradation (corrosion) when materials are exposed to service; it is, therefore, important to take measures to control it starting from the design stage. It discusses the electrochemical principles involved, the types of electrochemical cells, and the factors that affect aqueous corrosion. It also indicates the influence of corrosion on the mechanical properties of materials and how to design to control corrosion. Chapter 11 presents the methodology of design and the different factors considered in the selection of materials. It indicates that materials selection must be done concurrently in every phase of the design process to make sure that a material selected has the required properties, that it can be processed in the most efficient and economical manner, and that it can meet environmental concerns for its recyclability, recovery, or disposal after its initial use. It shows how the performance of a material in the structure is measured by a performance or efficiency index. This index is a combination of materials properties that depends on the mode of loading and the failure criterion of the structure. It shows how I-beams and honeycomb-shaped structures are used in design to increase the performance or efficiency of materials. As a source of engineering experience for use in design, the common reasons why materials and components fail are discussed.

Chapter 12 discusses the attributes of ferrous materials—the different types of steels and cast irons. Chapter 13 delves into the attributes of the most common nonferrous metals and alloys used for structures or components—aluminum and aluminum alloys, titanium and titanium alloys, magnesium and magnesium alloys, copper and copper alloys, and nickel and nickel alloys. Chapter 14 discusses the attributes and processing of ceramics and glasses. It shows how the Weibull statistics are used in evaluating the tensile properties and in design using brittle materials. Chapter 15 presents the processing and the properties of plastics/polymers and reinforced plastics (composites) as well as how to design with plastics and reinforced plastics (composite materials).

Chapter 16 contains actual case studies of materials selection. It has case studies on the selection of steels, aluminum, ceramics, and plastics. Case studies for steels involve: (1) conventional and high-performance structural steels in statically (buildings, storage tanks, pipes) and dynamically (bridge) loaded structures, and in off-highway vehicles; (2) construction materials for the oil and related industries; (3) stainless steels as construction materials such as rebars; and (4) heat-treated steels. The justification of aluminum alloys for aircraft structures is presented and the selections of aluminum alloys for the Concorde and NASA's space shuttle are documented. A case study on the selection of structural ceramics is included in the section on materials selection in the automotive industry. This section also contains an example of how concurrent engineering is used in product development. Case studies on materials selection based on life-cycle analysis are given for a beverage container and an automotive fender. Two actual cases in the selection of plastics and reinforced plastics are presented. A final case study using fracture mechanics indicates how an incorrect selection of a material for a motor case being designed for NASA led to a catastrophic failure. One of the lessons learned from this case study and reinforced by the case on concurrent engineering is that the properties of the material in the fabricated condition are the properties that need to be considered! This is especially important when welding is used for fabrication and when tensile residual stresses are present after fabrication.

The author wishes to thank the numerous publishers and other sources for granting permission for copyrighted photographs, microstructures, figures, and tables used in the textbook. Special thanks to the ASM International, Materials Park, OH for a lot of materials from the ASM Handbooks and other ASM copyrighted sources. In the years it took to complete and publish the text, there was a conscious effort to acknowledge the work of others. Where the author's record and memory have failed, he asks forgiveness.

The author wishes also to thank the mechanical, aerospace, and chemical engineering students at Florida Institute of Technology who provided ideas and for their patience in the formative years of the text.

The author also thanks Mr. William (Bill) Stenquist, Prentice Hall Executive Editor, for embracing the overall concept of the text and for his support and encouragement throughout the preparation of the manuscript. Thanks for the critical comments of various reviewers and especially those of Professor Norman E. Dowling of the Virginia Polytechnic Institute and State University. Thanks also to the various professionals who contributed to the production of the text, expecially to Mrs. Sharyn Vitrano.

Finally, the author wishes to thank his family, especially his wife Aurora, sons Patrick and Michael and their wives, for their support, encouragement, and patience during the preparations of the text. I ultimately thank God for giving me the strength, the perseverance, the health, and the mind to complete this enormous task.

Pat L. Mangonon

Part I

Structures and Properties

1

Atomic Structure and Bonding in Solids

1.1 Introduction

The atom is the smallest entity that has a direct influence on the properties of materials. We shall review in this chapter how the atomic structure influences these properties. We shall learn how the arrangement of the electrons in the unbonded atom leads to the Periodic Table from which we can immediately infer some of the chemical and bonding characteristics of the elements. These characteristics determine the type and strength of the bonding the element may have with other atoms (elements). The type and the strength of the bond determine the physical and mechanical properties of the solid material that forms.

1.2 The Unbonded Atom

1.2.1 Structure

The structure of an atom is now well established. According to wave mechanics, it consists of a positively charged nucleus surrounded by and electrostatically bound by Coulombic attraction (of positive and negative charges) with a smeared negatively charged "cloud" of electrons, as shown in Fig. 1-1. The electron is a particle, but its motion through the potential (Coulombic) field around the relatively stationary nucleus can be represented by a wave function, ψ—the principle of the dual behavior of an electron. The position of the electron cannot be definitely specified at one time, but it is indicated by the probability density as defined in Eq. 1-1.

$$|\psi|^2 dV \equiv \text{Probability Density} \qquad (1\text{-}1)$$

The probability density is the probability of finding the electron in a small volume dV located at a point (x, y, z) in space. Thus, there is a probability that the electron can be found at any point in space and this probability is represented by the smeared orbit or cloud in Fig. 1-1. The most probable location of the electron in space is that which has the highest probability density in Eq. 1-1 and is represented in Fig. 1-1 as the darkest region. The diameter of this orbit gives an indication of the atomic size.

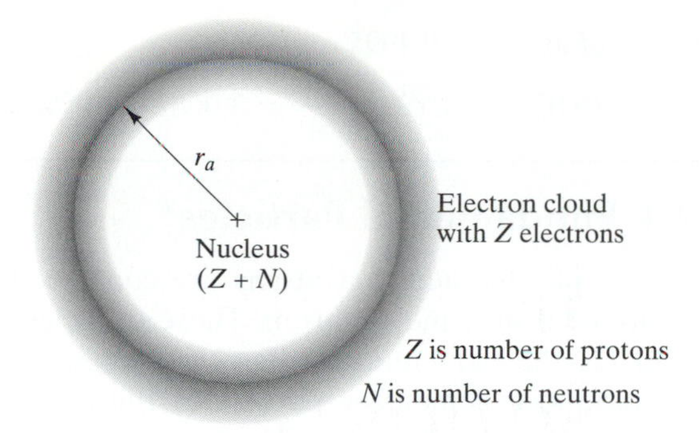

Fig. 1-1 Schematic of atomic structure with a spherical cloud of Z electrons moving around a relatively stationary nucleus.

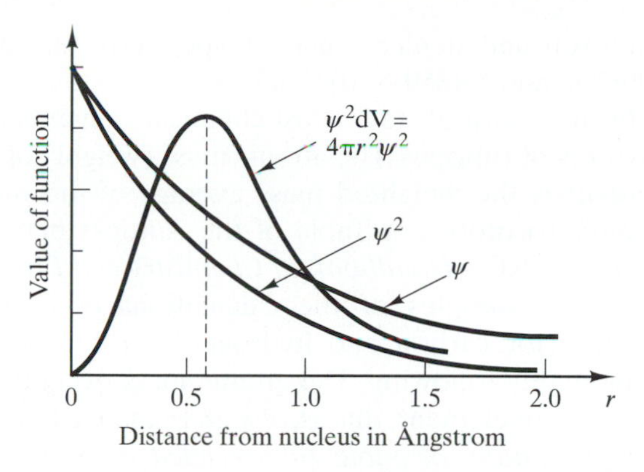

Fig. 1-2 Wave functions and the probability density of the electron in hydrogen atom **[Ref. 3]**.

The atomic size for hydrogen, the simplest atom, with one positive charge and one electron, was calculated and the results are shown schematically in Fig. 1-2. This figure shows the ψ, ψ^2, and $\psi^2 dV = 4\pi r^2 \psi^2$ functions for the electron motion around the nucleus as a function of radius, r, from the nucleus. The most probable location of the electron in the "smeared" orbit, that is, the location with the highest probability density ($\psi^2 dV = 4\pi r^2 \psi^2$), is seen to be about 0.6 Å (Ångstrom) away from the nucleus (indicated by the arrow in Fig. 1-1 and represented with a darkened orbit). Thus, the diameter of the orbit and, therefore, the atomic size is about $1 \text{ Å} = 10^{-10}$ m; this is about 4 to 5 orders of magnitude greater than the sizes of the nucleus (10^{-15} to 10^{-14} m) and the electrons (10^{-15} m).

1.2.2 Atomic Number

The nucleus contains the elementary particles called *nucleons.* These are the positively charged *protons* and the electrically uncharged *neutrons.* The charge of a proton ($e = 1.602 \times 10^{-19}$ Coulomb) is equal to the charge of the electron. The number of protons, Z, is called the *atomic number* and will have a total positive charge of Ze. For electrical neutrality, this total positive charge is balanced by Z-negatively-charged electrons, each having a charge of e. The bonding force between the positive nucleus and the negative orbital electrons in the atom is Coulombic, that is, the electrostatic attraction of oppositely charged particles. An atom with a specific atomic number, Z, is an element in the Periodic Table. The arrangement of the elements in the Periodic Table is explained by modern quantum mechanical principles and is such that elements in the same column (group) exhibit similar chemical characteristics. These chemical characteristics form the basis of bonding with other atoms of the same or different elements to form small or macro-molecules of materials. The types and strengths of the bonds that are produced determine the properties of materials and illustrate the relation of atomic structure to materials properties.

1.2.3 Atomic Mass

The mass of the electron, at rest, is about 9.1094×10^{-28} grams, while each of the nucleons (each proton and each neutron) is on the average 1,837 times heavier. Thus, the mass of an atom is concentrated in the nucleus. If *N is the number of neutrons in the nucleus,* $(Z + N)$ *is sometimes referred to as the atomic mass number,* which is roughly equal to the atomic mass or atomic weight of the element. Atoms with the same atomic number, Z, but with different N values are called *isotopes.*

It is conventional to indicate atomic masses in terms of a *unified atomic mass unit, u.* This is defined as $\frac{1}{12}$ (the mass of ^{12}C) $= 1.6605 \times 10^{-24}$ grams. (The symbol ^{12}C signifies the carbon atom isotope with a mass of 12.000 grams per mol.) The reciprocal of u is *the Avogadro's number,* $N_A = 6.0223 \times 10^{-23}$, the number of atoms in 1 mol. The masses of a proton, a neutron, and an electron are, respectively, 1.0073 u, 1.0087 u, and 5.4859×10^{-4} u.

In their natural state, most chemical elements are mixtures of isotopes. The atomic mass (weight) of an element is the weighted mass average of the most abundant isotopes. A Table of the isotopes may be found in CRC's *Handbook of Chemistry and Physics* and two examples of the calculations of atomic weights (for carbon and hydrogen) are illustrated immediately following. The atomic mass (weight) in grams of an element, that is, *one gram atomic mass is one gram-mole or mole (abbreviated as mol) and contains N_A (6.0223×10^{23}) atoms.*

EXAMPLE 1-1

The most abundant isotopes of the element carbon are the ^{12}C and the ^{13}C. The percent abundance of these isotopes are 98.90 percent and 1.10 percent, respectively, and their atomic masses are 12.000000 and 13.003355 grams respectively. Calculate the atomic mass of carbon.

SOLUTION: The weighted average of the masses of the isotopes is

$$\text{Atomic mass} = 0.9890 \times 12.000000 + 0.0110 \times 13.003355 = 12.0110.$$

EXAMPLE 1-2

The atomic mass of ^1H is 1.0078 and that of ^2H is 2.0140. The natural abundance of ^1H is 99.985 percent and that of ^2H is 0.015 percent. Calculate the atomic mass of hydrogen.

SOLUTION: The weighted average of the masses of hydrogen isotopes is

$$\text{Atomic mass} = 0.99985 \times 1.0078 + 0.00015 \times 2.0140 = 1.0079$$

EXAMPLE 1-3

Calculate the number of moles and the number of atoms of carbon with a mass of 6 grams.

SOLUTION:

$$\text{number of moles} = \frac{6\text{ g}}{12.0110\,\frac{\text{g}}{\text{mol}}} = 0.4995\text{ mol}$$

$$\text{number of atoms} = 0.4995\text{ mol} \times (6.0223 \times 10^{23})\,\frac{\text{atoms}}{\text{mol}} = 3.0081 \times 10^{23}\text{atoms}$$

1.2.4 Fundamental Particles*

We have just learned that atoms are composed of protons, neutrons, and electrons. These particles are considered the elementary particles of atoms. At one time, the most elementary particle of the elements was thought to be the atom. Now, we see that the atoms are composed of protons, neutrons, and electrons. *The field of high energy and particle physics has identified even more fundamental particles.* The term fundamental particle(s) refer to a particle(s) that is (are) truly a fundamental constituent of matter and is (are) not a compound(s) in any sense. The current belief is that the fundamental particles are the *quarks* and the *leptons*. In addition to protons and neutrons being compounds of quarks, there are other strong interacting particles, called *hadrons,* that are formed from quarks. Quarks have yet to be isolated, however, and at this time it is justified to call the hadrons (includes protons and neutrons) as elementary particles. A lepton is a particle that interacts through weak electromagnetic, and gravitational forces, but does not interact through the strong (nuclear) force. Leptons are found to be less than 10^{-18} m in size (less than 10^{-3} the size of the electron and less than 10^{-8} the size of the atom) and as such are considered point particles.

Atomic nuclei are believed to be compounds of quarks. It is believed that the *initial stage of matter formation* is the combination of the quarks. Quarks may associate themselves *two by two* to form *pions* or *three by three* to form the *nucleons*. This process is made possible by the action of an exchange particle called the *gluon*. The next stage is the fusion of the nucleons into atomic nuclei. It is believed that the particles responsible for fusing the neutrons and protons are *mesons,* whose masses are of the order of 200–300 times that of an electron. The mesons are of different kinds and have charges of $+e$ or $-e$, and some may be neutral. They produce fields that are not electrostatic in nature, and it is these meson fields that bind the protons and neutrons together in the atomic nuclei.

1.3 Arrangement of Electrons in Atoms—Electron Configuration

The arrangement of electrons around the nucleus of an atom is based upon the occupancy of the available discrete quantum state(s) with the lowest energy level(s). These quantum states and their energy levels are established by wave mechanics and their occupancy results in the Periodic Table. We shall now briefly indicate how these quantum states were established and indicate their energy levels.

1.3.1 Energy Levels of Electron and Principal Quantum Number

To determine the energy levels of the electron, its motion as a particle around the nucleus was modeled by the motion of a particle in a box, as shown in Fig. 1-3. Because the electron is bound to the nucleus via Coulombic attraction (electrostatic attraction of positive and negative charges), it cannot escape this potential field. This is represented as barrier potential V in Fig. 1-3. The energy of the moving electron within the atom is designated as E. As indicated in the last section, the motion of the electron particle in a potential field was represented as a wave motion in the form of the Schrödinger equation. For our purpose, it is sufficient to use the one-dimensional form of this equation, which is given as Eq. 1-2.

$$\frac{d^2\psi}{dx^2} + (E - V)\psi = 0 \tag{1-2}$$

The potential V is constant and to facilitate the solution of Eq. 1-2, it is assumed to be zero. The solutions of Eq. 1-2 are obtained by using the boundary conditions:

$$(1)\ \psi = 0 \text{ at } x = 0 \qquad \text{and} \qquad (2)\ \psi = 0 \text{ at } x = L \tag{1-3}$$

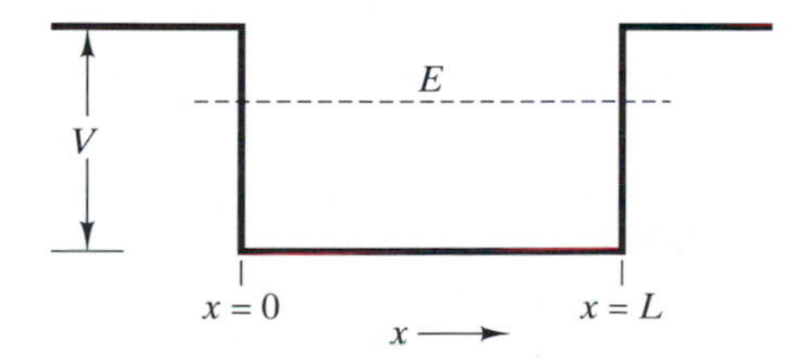

Fig. 1-3 Particle in a box—model of electron motion in atom.

The simplest wave function is

$$\psi = A \sin\left(\frac{2\pi x}{\lambda}\right) \tag{1-4}$$

where λ is the wavelength. This wavelength is related to the energy of the particle in motion by the principle of dual behavior through the following equation,

$$\lambda = \frac{h}{p} = \frac{h}{mv} \tag{1-5}$$

or

$$\lambda^2 = \frac{h^2}{2mE} \tag{1-5a}$$

where m is the mass of the electron, v is its velocity, $E = \frac{1}{2}mv^2$, and h is Planck's constant. (Planck's constant in physics is the proportionality constant that relates the discrete energy of any particle to its vibration frequency.)

If we differentiate Eq. 1-4 twice and use Eq. 1-5a, the Schrödinger equation becomes

$$\frac{d^2\psi}{dx^2} + \frac{8\pi^2 mE}{h^2}\psi = 0 \tag{1-6}$$

Using the boundary conditions in Eq. 1-3, the solutions to Eq. 1-6 are

$$\lambda = \frac{2L}{n}, \qquad \text{where } n = 1, 2, 3, \ldots, \text{etc.} \tag{1-7}$$

These wavelengths represent the discrete wavelengths that the electron can possess in a stationary state. Through Eq. 1-5a, the electron can only have discrete energy levels,

$$E = \frac{n^2h^2}{8mL^2} \tag{1-8}$$

The integer n is called the principal quantum number and determines the discrete quantum states that are allowed for the electron. Each principal quantum number, n, is considered a "shell" and we see from

Eq. 1-8 that the energies of all electrons with the same n are equal. The shells corresponding to $n = 1$, $n = 2$, $n = 3$, $n = 4, \ldots$ are designated the K shell, L shell, M shell, N shell, …, respectively.

1.3.2 Subshells and Other Quantum Numbers—Quantum States of Electrons

For each of the shells mentioned above, that is, for each n, there are subshells to reflect slight differences of discrete energy levels electrons can occupy within the shell. These differences are due to three other quantum numbers: (1) the orbital angular momentum number, l; (2) the magnetic quantum number, m_l; and (3) the spin quantum number, m_s. These numbers can have the following values:

$$\left.\begin{aligned} l &= 0, 1, \ldots, (n-1) \\ m_l &= -l, -1, 0, +1, +l \\ m_s &= \pm 1/2 \end{aligned}\right\} \quad (1\text{-}9)$$

The number of quantum states for each n value is the total number of m_l. This is shown in Table 1-1 for $n = 1, 2, 3$, and 4. It turns out that the total number of m_l for each n is equal to n^2; this can be verified by adding all the m_l's under the column "Quantum States" in Table 1-1 for each n value. The letters *s, p, d,* and *f* are the letter designations for $l = 0, 1, 2$, and 3, respectively. Each of the quantum states can be occupied by two electrons with opposite spin numbers—the *Pauli Exclusion Principle*. Thus, each electron occupies an energy state that has a unique set of the four quantum numbers, n, l, m_l, and m_s.

1.3.3 Electron Configuration

In the ground state, the Z electrons of an element with atomic number Z will occupy the available quantum states with the lowest energy while simultaneously adhering to the Pauli Exclusion Principle. The relative energy levels of the quantum states for the various n values in Table 1-1 are shown in Fig. 1-4. Each of the []'s in Fig. 1-4 is a quantum state that is capable of accommodating two electrons with opposite spin numbers, that is, $+1/2$ and $-1/2$. The quantum states are obtained by attaching the quantum number to the states directly above it. We see that the 4*s*-state has a lower energy than the 3*d*-states and therefore gets to be filled first before the 3*d*-states. This order of occupancy of the quantum states leads naturally to the positions of the elements or materials in the Periodic Table that enable us to discern quickly the materials behavior or the most probable bonding mechanism and to predict the materials properties.

We indicate the occupancy of the quantum states by a superscript on them. The highest superscript for a given state is twice the number of []s in Fig. 1-4.

Table 1-1 Quantum states for quantum numbers 1 through 4.

n		l					
Number	Letter	Number	Letter	m_l	m_s	Quantum states	Max. no. of electrons
1	K	0	s	0	$\pm 1/2$	1 − 1s	2
2	L	0	s	0	$\pm 1/2$	1 − 2s	8
		1	p	−1, 0, 1	$\pm 1/2$	3 − 2p	
3	M	0	s	0	$\pm 1/2$	1 − 3s	
		1	p	−1, 0, 1	$\pm 1/2$	3 − 3p	18
		2	d	−2, −1, 0,1, 2	$\pm 1/2$	5 − 3d	
4	N	0	s	0	$\pm 1/2$	1 − 4s	
		1	p	−1, 0, 1	$\pm 1/2$	3 − 4p	32
		2	d	−2, −1, 0,1, 2	$\pm 1/2$	5 − 4d	
		3	f	−3, −2, −1, 0, 1, 2, 3	$\pm 1/2$	7 − 4f	

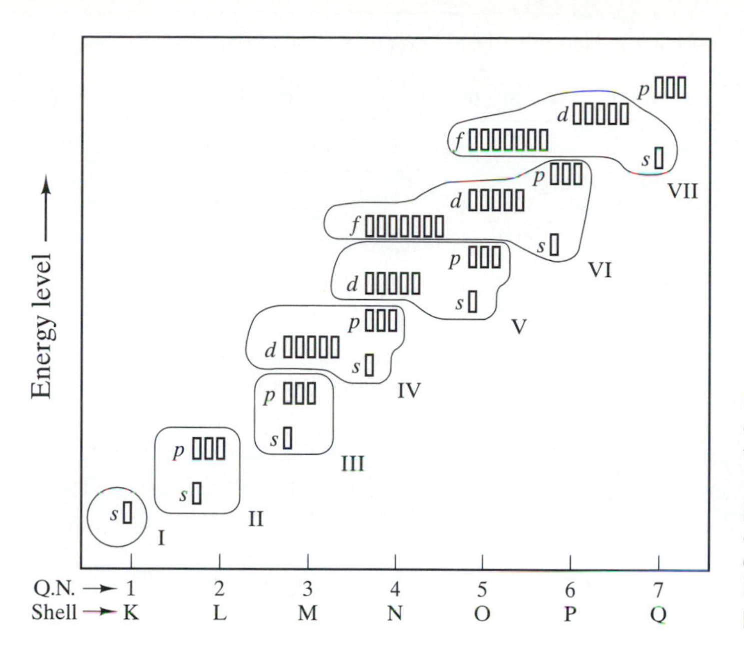

Fig. 1-4 Schematic of relative energies of quantum states of the shells and subshells. The Roman numerals indicate the periods in the Periodic Table and the quantum states within each of the envelopes indicate how many elements are in the period.

Thus, the *s*, *p*, *d*, and *f* states can have 2, 6, 10, and 14 as highest superscripts corresponding to the maximum electrons, respectively, that each state can accommodate. A number less than the highest possible superscript indicates that the corresponding state(s) is (are) partially filled. When we distribute the *Z* electrons of an atom to the lowest energy quantum states, we obtain what is called the *electron configuration* of an atom in the ground state. The sum of the superscripts in this representation equals *Z* (the atomic number). Thus, hydrogen (Z = 1) has an electron configuration (EC) of $1s^1$, helium (Z = 2) has an EC of $1s^2$, and sodium (Z = 11) has an EC of $1s^2 2s^2 2p^6 3s^1$. We may also use Fig. 1-5 to indicate the electron configuration of an element. In this figure, the electrons sequentially fill the states from the bottom arrow and from tail to head of the succeeding arrows. The Periodic Table ends with partially filled 6*d* states. A summary of the EC of all the elements is given in Table 1-2.

What we just learned is the regular occupancy of the quantum states by electrons in the ground state of an atom. There are some exceptions to this regular occupancy and we shall now discuss these with the help of Table 1-2, the summary of the electron configuration of the unbonded, neutral atoms in the ground state. This table indicates by an asterisk (*) some of the elements that exhibit irregularity in their electron

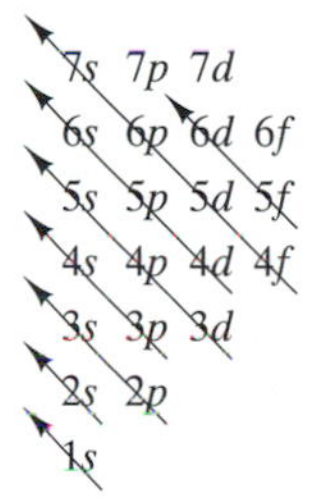

Fig. 1-5 Another way to indicate the electron configuration of an element.

configuration. We shall discuss this irregularity using the box representation of the quantum states and the occupying electrons represented with up and down arrows signifying $+\frac{1}{2}$ and $-\frac{1}{2}$ spin quantum number, respectively. We start with carbon that is supposed to have the electron configuration, $1s^2 2s^2 2p^2$. The total number of *s* and *p* quantum states for $n = 2$ is four and the predicted regular arrangement of the electrons in these states is shown in Fig. 1-6a which shows that one of the *p*-states is unoccupied. We note that there are two electrons in the *s*-state with opposite spins. These electrons have both negative charges and exert a repulsive force (positive energy) on each other that causes the energy of the entire atomic system to increase. The energy of the entire atomic system can be lowered, and thus the system becomes more stable when the two electrons in the *s*-state are separated and one of these electrons is transferred to the empty *p*-state. The resulting box representation,

Table 1-2 Electron configuration of neutral atoms in ground state [**Ref. 1**].

		K	L		M			N				O				P			Q
Atomic	**n→**	**1**	**2**		**3**			**4**				**5**				**6**			**7**
no.	**Element**	**s**	**s**	**p**	**s**	**p**	**d**	**s**	**p**	**d**	**f**	**s**	**p**	**d**	**f**	**s**	**p**	**d**	**s**
1	H	1																	
2	He	2																	
3	Li	2	1																
4	Be	2	2																
5	B	2	2	1															
6	C	2	1*	3															
7	N	2	2	3															
8	O	2	2	4															
9	F	2	2	5															
10	Ne	2	2	6															
11	Na	2	2	6	1														
12	Mg	2	2	6	2														
13	Al	2	2	6	2	1													
14	Si	2	2	6	1*	3													
15	P	2	2	6	2	3													
16	S	2	2	6	2	4													
17	Cl	2	2	6	2	5													
18	Ar	2	2	6	2	6													
19	K	2	2	6	2	6		1											
20	Ca	2	2	6	2	6		2											
21	Sc	2	2	6	2	6	1	2											
22	Ti	2	2	6	2	6	2	2											
23	V	2	2	6	2	6	3	2											
24	Cr	2	2	6	2	6	5*	1											
25	Mn	2	2	6	2	6	5	2											
26	Fe	2	2	6	2	6	6	2											
27	Co	2	2	6	2	6	7	2											
28	Ni	2	2	6	2	6	8	2											
29	Cu	2	2	6	2	6	10*	1											
30	Zn	2	2	6	2	6	10	2											
31	Ga	2	2	6	2	6	10	2	1										
32	Ge	2	2	6	2	6	10	1*	3										
33	As	2	2	6	2	6	10	2	3										
34	Se	2	2	6	2	6	10	2	4										
35	Br	2	2	6	2	6	10	2	5										
36	Kr	2	2	6	2	6	10	2	6										
37	Rb	2	2	6	2	6	10	2	6			1							
38	Sr	2	2	6	2	6	10	2	6			2							
39	Y	2	2	6	2	6	10	2	6	1		2							
40	Zr	2	2	6	2	6	10	2	6	2		2							
41	Nb	2	2	6	2	6	10	2	6	4*		1							
42	Mo	2	2	6	2	6	10	2	6	5		1							
43	Tc	2	2	6	2	6	10	2	6	5		2							
44	Ru	2	2	6	2	6	10	2	6	7		1							
45	Rh	2	2	6	2	6	10	2	6	8		1							
46	Pd	2	2	6	2	6	10	2	6	10*									
47	Ag	2	2	6	2	6	10	2	6	10		1							
48	Cd	2	2	6	2	6	10	2	6	10		2							
49	In	2	2	6	2	6	10	2	6	10		2	1						
50	Sn	2	2	6	2	6	10	2	6	10		2	2						
51	Sb	2	2	6	2	6	10	2	6	10		2	3						
52	Te	2	2	6	2	6	10	2	6	10		2	4						
53	I	2	2	6	2	6	10	2	6	10		2	5						

(*Continued*)

Table 1-2 (*Continued*)

Atomic	n→	K	L		M			N				O				P			Q
		1	2		3			4				5				6			7
no.	Element	s	s	p	s	p	d	s	p	d	f	s	p	d	f	s	p	d	s
54	Xe	2	2	6	2	6	10	2	6	10		2	6						
55	Cs	2	2	6	2	6	10	2	6	10		2	6			1			
56	Ba	2	2	6	2	6	10	2	6	10		2	6			2			
57	La	2	2	6	2	6	10	2	6	10		2	6	1		2			
58	Ce	2	2	6	2	6	10	2	6	10	1*	2	6	1		2			
59	Pr	2	2	6	2	6	10	2	6	10	3	2	6			2			
60	Nd	2	2	6	2	6	10	2	6	10	4	2	6			2			
61	Pm	2	2	6	2	6	10	2	6	10	5	2	6			2			
62	Sm	2	2	6	2	6	10	2	6	10	6	2	6			2			
63	Eu	2	2	6	2	6	10	2	6	10	7	2	6			2			
64	Gd	2	2	6	2	6	10	2	6	10	7	2	6	1		2			
65	Tb	2	2	6	2	6	10	2	6	10	9*	2	6			2			
66	Dy	2	2	6	2	6	10	2	6	10	10	2	6			2			
67	Ho	2	2	6	2	6	10	2	6	10	11	2	6			2			
68	Er	2	2	6	2	6	10	2	6	10	12	2	6			2			
69	Tm	2	2	6	2	6	10	2	6	10	13	2	6			2			
70	Yb	2	2	6	2	6	10	2	6	10	14	2	6			2			
71	Lu	2	2	6	2	6	10	2	6	10	14	2	6	1		2			
72	Hf	2	2	6	2	6	10	2	6	10	14	2	6	2		2			
73	Ta	2	2	6	2	6	10	2	6	10	14	2	6	3		2			
74	W	2	2	6	2	6	10	2	6	10	14	2	6	4		2			
75	Re	2	2	6	2	6	10	2	6	10	14	2	6	5		2			
76	Os	2	2	6	2	6	10	2	6	10	14	2	6	6		2			
77	Ir	2	2	6	2	6	10	2	6	10	14	2	6	7		2			
78	Pt	2	2	6	2	6	10	2	6	10	14	2	6	9		1			
79	Au	2	2	6	2	6	10	2	6	10	14	2	6	10		1			
80	Hg	2	2	6	2	6	10	2	6	10	14	2	6	10		2			
81	Tl	2	2	6	2	6	10	2	6	10	14	2	6	10		2	1		
82	Pb	2	2	6	2	6	10	2	6	10	14	2	6	10		2	2		
83	Bi	2	2	6	2	6	10	2	6	10	14	2	6	10		2	3		
84	Po	2	2	6	2	6	10	2	6	10	14	2	6	10		2	4		
85	At	2	2	6	2	6	10	2	6	10	14	2	6	10		2	5		
86	Rn	2	2	6	2	6	10	2	6	10	14	2	6	10		2	6		
87	Fr	2	2	6	2	6	10	2	6	10	14	2	6	10		2	6		1
88	Ra	2	2	6	2	6	10	2	6	10	14	2	6	10		2	6		2
89	Ac	2	2	6	2	6	10	2	6	10	14	2	6	10		2	6	1	2
90	Th	2	2	6	2	6	10	2	6	10	14	2	6	10		2	6	2	2
91	Pa	2	2	6	2	6	10	2	6	10	14	2	6	10	2*	2	6	1	2
92	U	2	2	6	2	6	10	2	6	10	14	2	6	10	3	2	6	1	2
93	Np	2	2	6	2	6	10	2	6	10	14	2	6	10	4	2	6	1	2
94	Pu	2	2	6	2	6	10	2	6	10	14	2	6	10	6*	2	6		2
95	Am	2	2	6	2	6	10	2	6	10	14	2	6	10	7	2	6		2
96	Cm	2	2	6	2	6	10	2	6	10	14	2	6	10	7*	2	6	1	2
97	Bk	2	2	6	2	6	10	2	6	10	14	2	6	10	9	2	6		2
98	Cf	2	2	6	2	6	10	2	6	10	14	2	6	10	10	2	6		2
99	Es	2	2	6	2	6	10	2	6	10	14	2	6	10	11	2	6		2
100	Fm	2	2	6	2	6	10	2	6	10	14	2	6	10	12	2	6		2
101	Md	2	2	6	2	6	10	2	6	10	14	2	6	10	13	2	6		2
102	No	2	2	6	2	6	10	2	6	10	14	2	6	10	14	2	6		2
103	Lr	2	2	6	2	6	10	2	6	10	14	2	6	10	14	2	6	1	2
104	Rf	2	2	6	2	6	10	2	6	10	14	2	6	10	14	2	6	2	2

*Note Irregularity.

$[\uparrow\downarrow]_{2s}\{[\uparrow][\uparrow][\ ]\}_{2p} \rightarrow\rightarrow\rightarrow [\uparrow]_{2s}\{[\uparrow][\uparrow][\uparrow]\}_{2p}$

$2s^2 2p^2$ (a) $\qquad$ $2s^1 2p^3$ (b)

Fig. 1-6 Hybridization of $2s^2 2p^2$ to $2s^1 2p^3$.

$[\uparrow\downarrow]_{4s}\{[\uparrow][\uparrow][\uparrow][\uparrow][\ ]\}_{3d} \rightarrow\rightarrow\rightarrow [\uparrow]_{4s}\{[\uparrow][\uparrow][\uparrow][\uparrow][\uparrow]\}_{3d}$

$4s^2 3d^4$ (a) $\qquad$ $4s^1 3d^5$ (b)

Fig. 1-7 Transformation of $4s^2 3d^4$ to $4s^1 3d^5$ in chromium to lower energy.

indicated in Fig. 1-6b, shows one electron in each available state and that all the electrons have the same spin quantum number. The electron configuration changed from $2s^2 2p^2$ to $2s^1 2p^3$. This transformation is called the hybridization of the s- and the p-states and the hybridized state is sometimes indicated simply as sp^3. Silicon and germanium have the same $s^2 p^2$ configuration at the end that will also hybridize to the sp^3 configuration as in Fig. 1-6. The hybridization leads to all the states being equal and during bonding, leads to tetrahedral bonding.

The other example is chromium with Z = 24. The expected electron configuration for chromium is $1s^2 2s^2 2p^6 3s^2 3p^6 4s^2 3d^4$. Again, this configuration has two electrons in the $4s$ state and 4 electrons in the $3d$ states. The chromium atom will be more stable if one of the $4s$ electrons is moved to the $3d$ state, as indicated in Fig. 1-7. Thus, the electron configuration will change from $4s^2 3d^4$ to $4s^1 3d^5$ and all the electrons in the six states have the same spin quantum number.

EXAMPLE 1-4

Without looking at the electron configuration table (Table 1-2),

(a) write the most appropriate electron configuration of the element with Z = 14, and
(b) indicate the set of quantum numbers (energy state) for each of the electrons beyond the closed configuration.

SOLUTION:

(a) To obtain the electron configuration, we fill sequentially the quantum states in Fig. 1-4 and use the Pauli Exclusion Principle for each state. The electrons in each state are indicated by the superscript and the sum of the superscripts must equal Z. Thus, the sequential energy states, filled or partially filled, are

$$1s^2\, 2s^2\, 2p^6\, 3s^2\, 3p^2$$

The partially filled states are the $3p$ (three states—a maximum of six electrons, as in $2p^6$). We note as indicated earlier that the energy of an atom with $s^2 p^2$ configuration can be lowered with $s^1 p^3$ configuration. Therefore, the most appropriate configuration for this element is

$$1s^2\, 2s^2\, 2p^6\, 3s^1\, 3p^3$$

with all four electrons having the same spin quantum number. With the four electrons in the $3s$ and $3p$ states, this element belongs to Group 14 (formerly Group IV) and has the same $s^1 p^3$ hybridization as carbon and thus will have tetrahedral covalent bonding. The element is silicon that will covalently bond as carbon.

(b) The four valence electrons are in the hybridized states $3s^1 p^3$. From these states we can deduce the set of quantum numbers for each electron. We start by recognizing that the number before the letters is the principal quantum number n. The letters represent the orbital quantum number l; s for $l = 0$ and p for $l = 1$ from which we can obtain the magnetic quantum number m_l. The spin quantum number is the same for all the four electrons either $+1/2$ or $-1/2$. The result is

Electron	n	l	m_l	m_s
$3s^1$	3	0	0	$+1/2$
$3p^3$	3	1	−1	$+1/2$
	3	1	0	$+1/2$
	3	1	+1	$+1/2$

1.4 The Periodic Arrangement (Table) of Materials and Their Chemical Characteristics

As indicated earlier, the occupancy of the available lowest-energy quantum states and the adherence to the Pauli Exclusion Principle by the Z electrons of an atom result in a periodic arrangement (table) of the materials (elements) from which we can discern quickly their chemical characteristics. In the Periodic Table, shown in Fig. 1-8, there are rows and columns

New IUPAC notation
Old IUPAC notation

IUPAC = International Union of Pure and Applied Chemistry

Key to Chart

26	← Atomic Number
Fe	← Element Symbol
55.847	← Atomic Weight

1 IA	2 IIA	3 IIIA	4 IVA	5 VA	6 VIA	7 VIIA	8	9 VIIIA	10	11 IB	12 IIB	13 IIIB	14 IVB	15 VB	16 VIB	17 VIIB	18 VIIIB
1 **H** 1.00794																	2 **He** 4.00260
3 **Li** 6.941	4 **Be** 9.01218											5 **B** 10.81	6 **C** 12.011	7 **N** 14.0067	8 **O** 15.9994	9 **F** 18.998403	10 **Ne** 20.1797
11 **Na** 22.98977	12 **Mg** 24.305											13 **Al** 26.98154	14 **Si** 28.0855	15 **P** 30.97376	16 **S** 32.066	17 **Cl** 35.453	18 **Ar** 39.948
19 **K** 39.0983	20 **Ca** 40.078	21 **Sc** 44.9559	22 **Ti** 47.88	23 **V** 50.9415	24 **Cr** 51.996	25 **Mn** 54.9380	26 **Fe** 55.847	27 **Co** 58.9332	28 **Ni** 58.69	29 **Cu** 63.546	30 **Zn** 65.39	31 **Ga** 69.72	32 **Ge** 72.61	33 **As** 74.9216	34 **Se** 78.96	35 **Br** 79.904	36 **Kr** 83.80
37 **Rb** 85.4678	38 **Sr** 87.62	39 **Y** 88.9059	40 **Zr** 91.224	41 **Nb** 92.9064	42 **Mo** 95.94	43 **Tc** (98)	44 **Ru** 101.07	45 **Rh** 102.9055	46 **Pd** 106.42	47 **Ag** 107.8682	48 **Cd** 112.41	49 **In** 114.82	50 **Sn** 118.710	51 **Sb** 121.757	52 **Te** 127.60	53 **I** 126.9045	54 **Xe** 131.29
55 **Cs** 132.9054	56 **Ba** 137.33	* 57 **La** 138.9055	72 **Hf** 178.49	73 **Ta** 180.9479	74 **W** 183.85	75 **Re** 186.207	76 **Os** 190.2	77 **Ir** 192.22	78 **Pt** 195.08	79 **Au** 196.9665	80 **Hg** 200.59	81 **Tl** 204.383	82 **Pb** 207.2	83 **Bi** 208.9804	84 **Po** (209)	85 **At** (210)	86 **Rn** (222)
87 **Fr** (223)	88 **Ra** 226.0254	† 89 **Ac** 227.0278	104 **Rf** (261)	105 **Db** (262)	106 **Sg** (263)	107 **Bh** (262)	108 **Hs** (265)	109 **Mt** (266)	110 (269)	111 (272)	112 (277)						

*Lanthanide series	58 **Ce** 140.12	59 **Pr** 140.9077	60 **Nd** 144.24	61 **Pm** (145)	62 **Sm** 150.36	63 **Eu** 151.96	64 **Gd** 157.25	65 **Tb** 158.9254	66 **Dy** 162.50	67 **Ho** 164.9304	68 **Er** 167.26	69 **Tm** 168.9342	70 **Yb** 173.04	71 **Lu** 174.967
†Actinide series	90 **Th** 232.0381	91 **Pa** 231.0359	92 **U** 238.0289	93 **Np** 237.048	94 **Pu** (244)	95 **Am** (243)	96 **Cm** (247)	97 **Bk** (247)	98 **Cf** (251)	99 **Es** (252)	100 **Fm** (257)	101 **Md** (258)	102 **No** (259)	103 **Lr** (260)

Fig. 1-8 Periodic Table of the elements **[Based on Periodic Table in Ref. 1]**.

that are called periods and groups, respectively. The period starts when the s-state begins to be occupied, and except for the first period, ends when the $3p$-states are all occupied. Because the $2s$-state begins to be occupied after Z = 2 (helium), the material with Z = 3 (lithium) starts the second period. For the same reason, when the $4s$-state begins to be occupied, the fourth period starts in spite of the unfilled $3d$-states, as seen in Fig. 1-4. Materials (elements) belonging to a period are those occupying the quantum states within an envelope as shown in Fig. 1-4. The roman numerals marking the envelopes in Fig. 1-4 signify the periods. With each quantum state accommodating two electrons (two elements), the number of elements in the period is twice the number within the envelopes. Thus, there are two elements (H and He) in the first period, eight elements each in the second and third periods, 18 elements each in the fourth and fifth periods, 32 elements in the sixth period, and 18 in the partially filled seventh period.

As a result of the convention adopted in the last paragraph, elements with s^1 electron configuration (EC) are in one column or group. Likewise, the elements with s^2p^6 EC are in one group that includes helium (with s^2 EC), neon, argon, krypton, xenon, and radon. The s^2p^6 EC signifies a closed shell that means that elements with this EC have no reactivity or tendency to react at all, and consequently their valence is zero. Therefore, their atoms remain unbonded and independent and they are predominantly gases.

There is a recent change in numbering of the groups in the periodic table adopted by the International Union of Pure and Applied Chemistry (IUPAC). This is shown in Fig. 1-8. The old system used Roman numerals while the new system uses ordinary numbers. In the old system, the elements with s^1 were designated Group I and those with the closed shell s^2p^6 were designated Group VIII because of the sum of the superscripts. Elements with EC $s^2p^0, s^2p^1, \ldots, s^2p^5$ are designated Group II, III, ..., and VII, respectively.

Elements in Groups I and II (Groups 1 and 2) on the left side of the table, and Groups VI and VII (Groups 16 and 17) on the right side of the table will tend to react by giving up and accepting electrons, respectively, in order to attain the s^2p^6 EC in their outer shell. Because of this, they are referred to as *electropositive* and *electronegative* materials because they attain positive and negative valences, respectively, in ionic bonding. The valence reflects the number of electrons given up (+) or accepted (−). The ease with which the electrons are given up or accepted is a measure of the reactivity of the material. Thus, lithium and sodium easily give up their electrons and react violently in water and acids. Flourine is the most electronegative element that explains its reactivity with almost anything. For example, silicate glasses can contain any acids but are attacked by hydrofluoric acid (HF). Atoms from elements in the old Group III, IV, and V tend to react by partially sharing electrons; in particular, the atoms from Group IV combine with covalent bonding.

Materials that easily give up electrons are also called *metallic* and those that accept are called *nonmetallic*. Except for the elements to the right of the heavy-lined portion of the periodic table and shaded, the rest of the elements exhibit metallic characteristics that include the transition metals in Groups 3 to 12 in Fig. 1-8. The metallic character of a material is based on the ease of conducting current; a nonconducting material is nonmetallic. The transition metals have either filled or partially filled d-states in their EC. They acquire oxidation states or valences that reflect the ease of giving up either the s or the d electrons from their EC. The remaining outer shell does not necessarily attain the closed shell configuration as in the elements of Groups 1 and 2.

EXAMPLE 1-5

Write the electron configuration of (a) the Sr^{2+} ion, (b) the Br^{1-} ion, and (c) the V^{2+} and the V^{5+} ions.

SOLUTION:

(a) Strontium (Sr) has an atomic number of 38, is in the fifth row, second column in the Periodic Table. With this, we deduce that the valence electrons have the $5s^2$ configuration. The Sr^{2+} ion is obtained by giving up the two valence electrons and is thus left with 36 electrons and the closed shell configuration of krypton (Kr),

$$[\mathrm{Kr}] \qquad 1s^2\,2s^2\,2p^6\,3s^2\,3p^6\,4s^2\,3d^{10}\,4p^6$$

(b) Bromine (Br) has an atomic number of 35, is in the fourth row, Group 17 (formerly VII) column. This says that the ground state electron configuration is

$4s^2 4p^5$. The atom can accept one electron and become Br^{1-} with a closed shell configuration of 36 electrons, that is, that of krypton. Therefore we see that

$$[Sr^{2+}] = [Kr] = [Br^{1-}]$$

where the [ ] signifies the electron configuration.

(c) Vanadium (V) has an atomic number of 23 and is one of the transition elements, that is, those with partially filled *d*-quantum states. The ground state electron configuration is $[Ar]4s^2\,3d^3$. In transition elements, the energy levels of the *s* and the *d* electrons are very close to each other. Thus, the atom can give up the *s* electrons as easily as the *d* electrons. It is customary to designate the *s* electrons to be given up first. A closed shell configuration is not necessarily achieved. Thus,

$$[V^{2+}] = [Ar]3d^3 \quad \text{and} \quad [V^{5+}] = [Ar]$$

1.5 Characteristic X-ray Emission and Chemical Analysis

We have seen that electrons in each of the atoms occupy specific discrete energy states. We shall now see the evidence of these in the *characteristic X-ray emissions* that are considered the *"fingerprints" of the element* and are the basis for chemical analysis of materials. To start, we recall that the energy states are due to the Coulombic (electrostatic) interactions between the electrons and the nucleus (protons). The Coulombic force between a positive and a negative particle is

$$F = \frac{e^2}{4\,\pi\,\epsilon_o r^2} \qquad (1\text{-}10)$$

where *e* is the charge of each of the particles, *r* is the distance between the opposite charges, and ϵ_o is the permittivity constant equal to 8.85×10^{-12}, F per m^{-1}. This is an attractive force and is positive. The energy of attraction is the integral of *Fdr* from ∞ to *r*, that is,

$$E = \int_{\infty}^{r} F\,dr = \frac{e^2}{4\pi\epsilon_o}\int_{\infty}^{r}\frac{dr}{r^2} = -\frac{e^2}{4\,\pi\epsilon_o r} \qquad (1\text{-}11)$$

Equation 1-10 shows that the force is zero when $r = \infty$. It follows that the energy of attraction is also zero when the electron is at ∞. The resultant negative attractive energy at the electron's most probable location *r* arises from the fact that work has to be done to move the electron from ∞ to *r*. A larger negative energy means a larger attractive energy and a more stable state. We must note that Eq. 1-11 is for one proton and one electron—the hydrogen atom. When we increase the atomic number Z to 2, a nucleus with two positive charges binds the electrons more strongly and the ψ functions (stationary states) of the electrons change. Every time we change the atomic number (add positive and negative charges), the energy levels of the stationary states change. Thus, we see that the levels of the energy states are unique for each atom.

Figure 1-9 schematically illustrates the energy levels of the various shells for a particular atom. When one of the electrons in the K-shell is removed, an energy state is vacated. Electrons from the nearest and next nearest shells, that is, from the L and the M shells, can move to the vacant K-shell energy state because it is a more stable state. The transitions are from higher energy states to a lower energy state and result in discrete (quanta) losses in energy of the electrons. The energy losses are characteristic of the atom and are emitted as electromagnetic radiation called X rays. The X rays also have a wavelike character and their frequencies and wavelengths are related to the energy loss, ΔE, by

$$\Delta E = h\nu = \frac{hc}{\lambda} = eV \qquad (1\text{-}12)$$

where *h* is Planck's constant, ν is the frequency, λ is the wavelength of the radiation, *c* is the velocity of light, *e* is the electron charge, and *V* is the energy expressed in electron volts (*eV*). When the constants are substituted, the following equation results,

$$\lambda = \frac{1.2397}{V} \qquad (1\text{-}13)$$

where λ is in nanometers (nm) and *V* is in kilovolts (keV). Equation 1-13 implies that an emitted X ray has a definite λ and a definite *V*. For this reason, analysts refer to X rays as definite lines in the spectrum. Actually, there is some uncertainty in the energy (*V*) of the X rays but the uncertainty is a very small fraction of the energy and thus it is justified to call them lines. The fundamental lines that elements

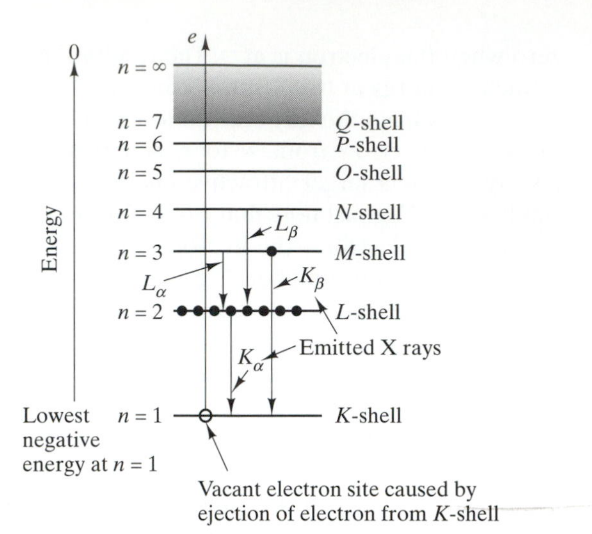

Fig. 1-9 Emission of characteristic X rays, K_α, K_β, L_α, and L_β, when electrons move from a high energy to a lower energy state.

Table 1-3 K_α and L_α energies (in KeV) of selected elements.

Atomic number, Z	Element	K_α (KeV)	L_α (KeV)
13	Al	1.487	0.452
22	Ti	4.511	0.573
24	Cr	5.415	0.573
26	Fe	6.404	0.705
28	Ni	7.478	0.852
29	Cu	8.048	0.930
30	Zn	8.639	1.012
41	Nb	16.615	2.166
42	Mo	17.479	2.293
74	W	59.32	8.398
79	Au	66.99	9.713

are identified with are the K X rays. As indicated in Fig. 1-9, the K_α X ray is formed from the transition of an electron from the L-shell to the vacant K-shell energy state, while the K_β X ray results from the transition from the M-shell to the vacant K-shell. In case of doubt, the other X rays such as the L_α and the L_β X rays are used by the analyst to either prove or disprove the presence of an element. The L_α line arises from the transition of an electron from the M to the L-shell, while the L_β forms with an electron transition from the N to the L-shell. Energies of K_α and L_α lines of selected elements are listed in Table 1-3.

We shall now discuss how X rays are emitted from elements when they are bombarded by accelerated high-energy particles such as electrons themselves that we shall call *beam electrons*. As the beam electrons hit the material of interest, they get scattered and rescattered, both elastically and inelastically, continuously lose energy, and decelerate. This phenomenon is exhibited as a continuous emitted radiation from the beam electrons. If the beam electron is completely stopped on its impact with the material, all its energy is emitted as radiation, and the wavelength of this radiation is the short wavelength, λ_{SWL}. Thus, the continuous spectrum starts at the λ_{SWL} as shown in Fig. 1-10. When the energy of the beam electrons increases, the λ_{SWL} shifts to shorter wavelengths. If the beam electrons have sufficient energy to eject a K-shell electron from the material, then the transition of electrons from the L and the M-shell of the material, as described above, will form the characteristic K_α or K_β lines. These are the bases for indicating the presence of elements in a material—a qualitative analysis.

So far we have discussed using the X rays as a means of identifying elements. We shall see in the next chapter that they are also used for diffraction to identify the arrangement of atoms (crystal structure) in a material. For diffraction, we need a single wavelength (monochromatic) radiation and the most common radiation used is the characteristic K_α. We can obtain a monochromatic K_α by filtering all the continuous spectrum and any other lines in the spectrum and allowing only the K_α to pass through the filter.

EXAMPLE 1-6

Calculate the wavelengths of the Fe K_α and L_α from the data in Table 1-3.

SOLUTION: We use Eq. 1-12 and noting from Table 1-3 that the energies of the Fe K and L lines are 6.404 Ke*V* and 0.705 Ke*V*, respectively.

$$\lambda \text{ for } K_\alpha = \frac{1.2397}{6.404} = 0.1936 \text{ nm } (1.936 \text{ Å})$$

$$\lambda \text{ for } L_\alpha = \frac{1.2397}{0.705} = 1.7584 \ (17.584 \text{ Å})$$

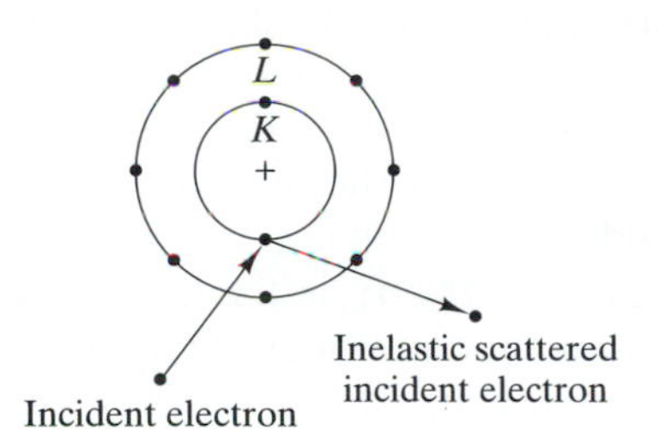

(a) Continuous spectrum

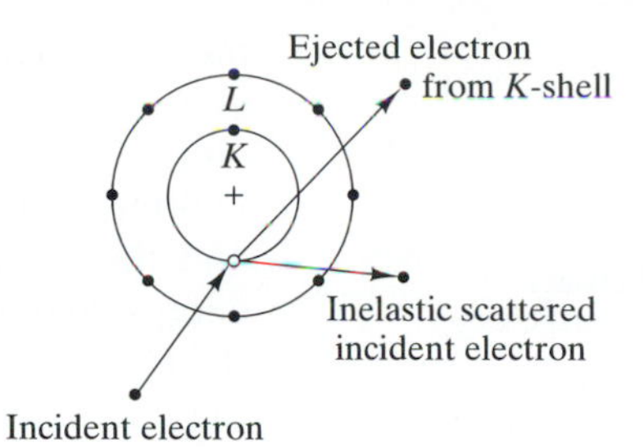

(b) Critical voltage (energy) of incident electron to eject K-shell electron

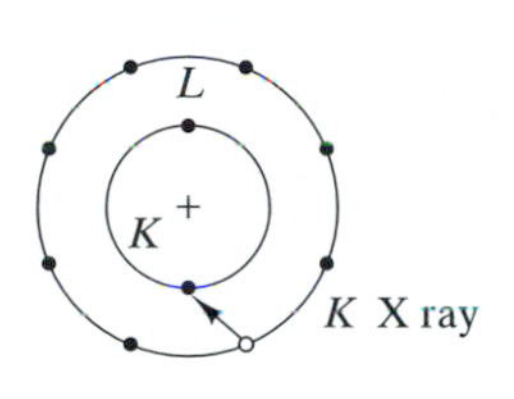

(c) Emission of *K* X ray

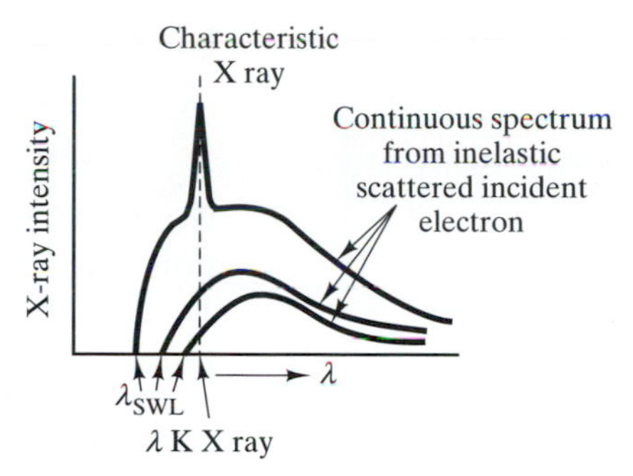

(d) Characteristic X ray and continuous spectra

Fig. 1-10 Schematic of (a) inelastic scattering of an incident electron; loss in energy of incident beam shows as continuous X-ray spectrum; (b) ejection of K-shell electron when incident electron has critical voltage, (c) emission of K X ray; and (d) characteristic and continuous spectra—characteristic from target material and continuous from inelastic electrons.

1.6 Bonding of Atoms to Form Molecules, Liquids, or Solids

1.6.1 Bonding Forces and Energy

When bonding occurs between atoms, there is chemical affinity between the atoms. When there is *no chemical affinity, the atoms do not bond,* each of the atoms are well separated from each other, and the elements are gases at ordinary temperatures and pressures. The examples of these are, of course, the Group 18 elements (He, Ne, Ar, Kr, Xe, and Ra) which were discussed in Sect. 1.4. These elements have closed shell configurations and, therefore, have no resultant affinity because all the *s*- and the *p*-states are all occupied. This state of affairs may also be referred to as *saturation bonding,* that is, all the available energy quantum states for electrons in the atom are occupied. When there is *chemical affinity,* there exists a long-range *positive electrostatic (Coulombic) attractive force,* F_A, that brings atoms very close together. This is shown in Fig. 1-11a where the attraction is between the nucleus of one atom (N_1) and the e_2 electrons of the other atom N_2. When the atoms are at close range, a *negative repulsive force,* F_R, sets in that is due to the mutual repulsion of the electrons, e_1 and e_2. The net total bonding force, F_B, is the sum of the attractive and the repulsive forces, that is,

$$F_B = F_A + F_R \tag{1-14}$$

An equilibrium spacing exists between a pair of atoms in which the *repulsive force balances the attractive force,* that is, the net total bonding force, F_B, is zero.

It is customary to use energy, instead of force, in discussing bonding between atoms. The attractive energy is negative, long-range, and is still that expressed by Eq. 1-11, except that the Coulombic attractive force is no longer limited to the electron-proton within one atom, but as indicated in Fig. 1-11a. The repulsive energy is short-range and arises when the electron clouds of the two atoms begin to overlap. This short-ranged positive repulsive energy may be described by

$$E_R = \frac{C}{r^n} \tag{1-15}$$

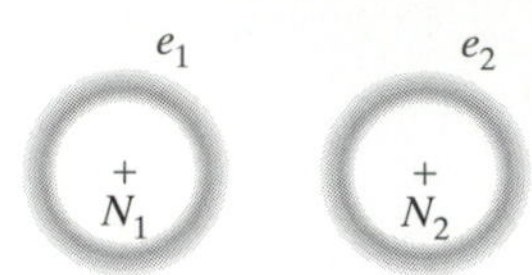

Long-range attractive force between positive (nuclei) and negative (electrons) charges

Short-range repulsive force between charges of the same sign (electrons–electrons)

(a)

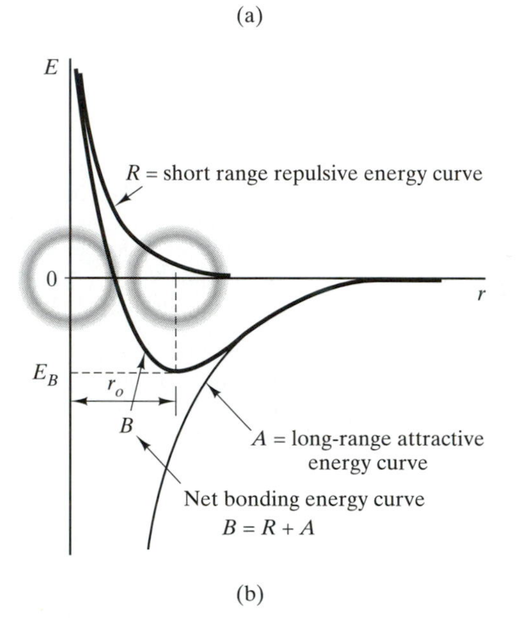

(b)

Fig. 1-11 Schematic of (a) two separated atoms indicating attractive and repulsive forces, and (b) the bonding energy curve.

where C is a constant and the exponent n depends on the atoms. When we schematically depict the attractive and the repulsive energies, they will look like the curves A and R, respectively, in Fig. 1-11b. The attractive energy will be zero at infinity and will essentially be unchanged until the atoms are within one or two atomic diameters of each other when it begins to increase (becomes more negative). The repulsive energy is shown to set in only when the electron "clouds" are very close to each other. The total energy is the sum of the attractive and repulsive energies and the resulting curve is the bonding energy curve, B. The equilibrium distance, r_o, is the distance between the centers of the nuclei of the two atoms when the attractive force is balanced by the repulsive force. It is schematically shown as that when the "clouds" are about to overlap and is commonly used as a measure of the atomic diameter. The energy, E_o, at r_o is the bonding energy of the pair of atoms. This bonding energy gives an indication of the melting point, the coefficient of thermal expansion, and the Young's modulus. The higher the bonding energy, that is, the more negative E_o is, the higher are the melting point and the modulus of elasticity, while the coefficient of thermal expansion decreases. The quantitative relationships between E_o and these properties are discussed in Chap. 4.

1.6.2 The Principal Bonding Mechanisms and Properties Derived from Them

The three principal bonding mechanisms between atoms are *ionic, covalent,* and *metallic*. Depending on the availability of states when atoms are bonded, molecules with only a few atoms may form or macromolecules with a huge number of atoms are formed. We shall now discuss the bonding mechanisms and the properties of the materials with each bond in the order given.

The *ionic bond* is between an electropositive and an electronegative atom. The electropositive atom gives up its electrons and the electronegative atom accepts electrons. As a result of this process, positive (with valence $+n_1$) and negative (with valence $-n_2$) ions are formed with closed-shell electron configurations. The ions with charges $+n_1$e and $-n_2$e are then attracted to each other. The repulsive force sets in when the ionic-closed shell electron configurations begin to overlap.

In the context of quantum mechanics, the valence electrons originally belonging to the neutral atoms are shared by both atoms comprising the molecule. The sharing is such that the probability density of the electrons, for example, the probability of finding where the electrons are in a small volume at a point in space at a particular time, is greatest around the electronegative atom and thus this atom acquires the negative charge as shown in Fig. 1-12. This is the basis of a qualitative property chemists call *electronegativity*— "the power of an atom in a molecule to attract electrons to itself". In order to maximize the attractive energy while minimizing the repulsive, the nearest neighbors of a positive ion (cation) must be negative ions (anions), and vice versa, as shown in Fig. 1-13 for NaCl. There is some freedom in packing the ions

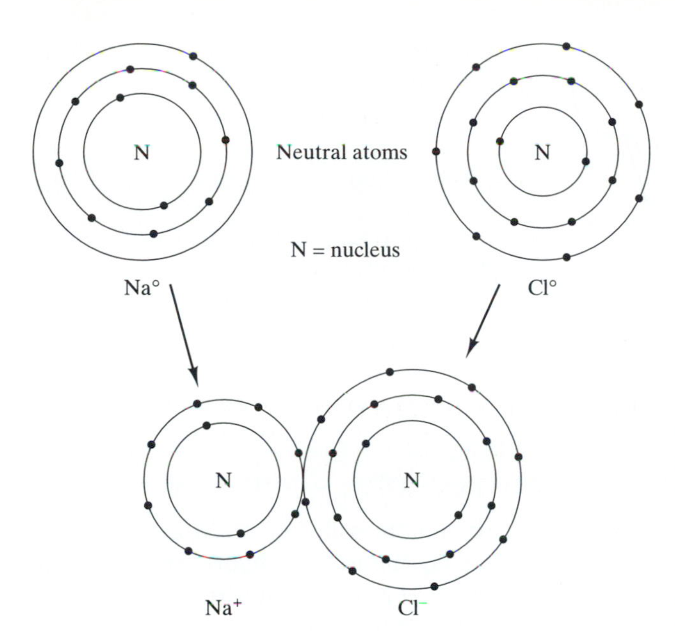

Fig. 1-12 Example of ionic bonding between Na and Cl atoms. Valence electrons are shared between the two atoms, but the probability density of electrons is greatest around the Cl; thus, Cl becomes negative and Na positive.

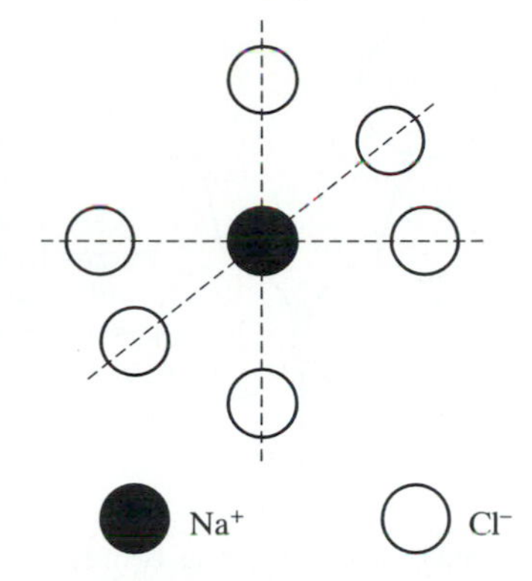

Fig. 1-13 In ionic bonding, positive ions are surrounded by negative ions, and vice versa, to minimize repulsion and maximize attractive energy.

around each other; therefore, the ionic bond has no directionality. Nonetheless, it is necessary to balance positive and negative charges, both locally and macroscopically, and consequently, the electrons are primarily localized with the electronegative atom. Because the electrons are localized, the electrical conductivity of ionic materials are relatively poor. For conductivity to occur, the ions have to move with the electric field and because they are heavy, they move very slowly and are, therefore, poor thermal and electrical conductors at low temperatures. The bonding energy as a result of ionic bonding is relatively high and thus we might expect high melting materials. Ceramics are examples of materials with ionic bonding and are used effectively to sustain high temperature environments and as insulators and dielectrics, but are very brittle because of the localized nature of the bond.

When electrons are equally shared by the atoms in a molecule, the resulting interatomic bonds are either *covalent* or *metallic*. An important feature of these bonds is that they can form between atoms of the same type, between which there can be little or no ionic bonding. For example, the atoms in H_2, Cl_2, diamond, and silicon are held by covalent bonds, while the atoms in copper are held by metallic bonds. In *pure covalent bonds,* the electrons from the atoms share available quantum states and they become shared between the nuclei to form a closed shell configuration. The classic example of covalent bonding is, of course, the hydrogen molecule. We recall that each atom of hydrogen has an empty energy state available in the 1s quantum state. The two electrons (with opposite spins) of the two atoms of hydrogen in the H_2 molecule occupy the $1s$ quantum state to give the closed shell configuration $1s^2$, the helium atom configuration. Each of the hydrogen nuclei share this configuration, and *saturation bonding* between the two hydrogen atoms is said to have formed. As such, the bond energy between the hydrogen atoms is very high but no strong bonds are available for other H_2 molecules as illustrated in Fig. 1-14. In other words, the *intramolecular bond* is very strong indeed, but the *intermolecular attraction* is very weak, which is why H_2 is a gas; this weak intermolecular attraction is the *van der Waals' secondary bond,* which will be discussed in more detail shortly. A similar behavior can be said for any molecule with saturated bonding.

In terms of the probability density of the electrons in H_2, the highest probability is located between the two nuclei. This is shown in Fig. 1-14 and as such, the electrons in a covalent bond are primarily localized between the two nuclei, and the covalent bond is very directional. The covalent bond is thus commonly represented as two dots between the atoms such as in H:H, or as a — as in H—H.

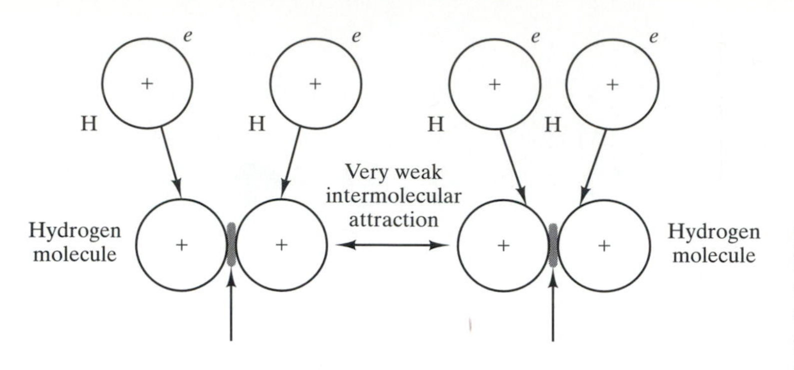

Fig. 1-14 Covalent bonding between two H atoms exhausts the available quantum state to produce a saturated bonding and a very strong intramolecular bond but very intermolecular attraction between two molecules.

The — signifies one single bond, in this case, a covalent bond.

In diamond, the sp^3 hybridization of each carbon atom leads to tetrahedral bonding as shown in Fig. 1-15 and requires four atoms as neighbors to form a closed shell configuration. However, the four neighbors in turn will each require four other neighbors, and so on. We note that there is always a bonding state available for carbon atoms to attach to. This is referred to as *unsaturated bonding* because of the availability of quantum states, illustrated as single bonds in the four atoms in Fig. 1-16a. Because of this, a very large aggregate (a *macromolecule*) of carbon atoms form and diamond is a solid. The situation is entirely different if we bond each carbon atom with four hydrogen atoms to form methane, CH_4; the four sp^3 hybridized quantum states of carbon will each be occupied by an electron from the hydrogen atom indicated in Fig. 1-16b. The carbon atom now shares eight electrons, two electrons with each of the hydrogen atoms. Both the carbon and hydrogen atoms form saturated bonds; again a case of very strong intramolecular (C—H) bonds but very weak intermolecular attraction. Because of this, CH_4,

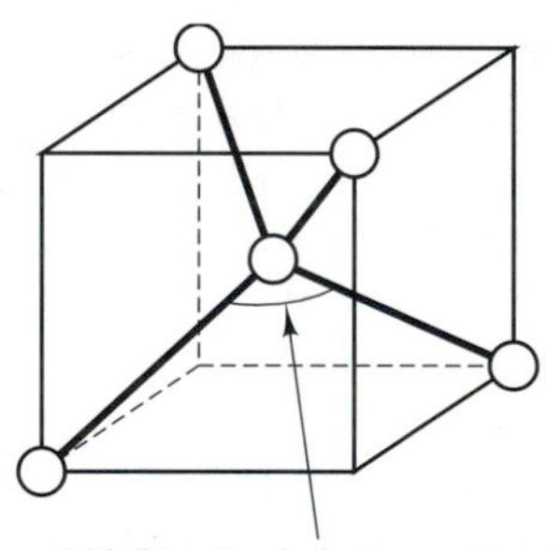

Fig. 1-15 Tetrahedral covalent bonding in C, Si, and Ge as a result of hybridization.

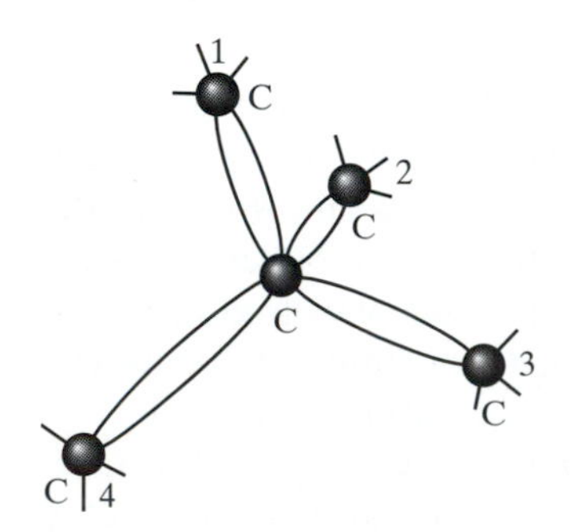

(a) "Unsaturated bond"– central atom has its quantum states filled with 8e but atoms 1, 2, 3, and 4 have unfilled states requiring more atoms to produce a diamond-macromolecule

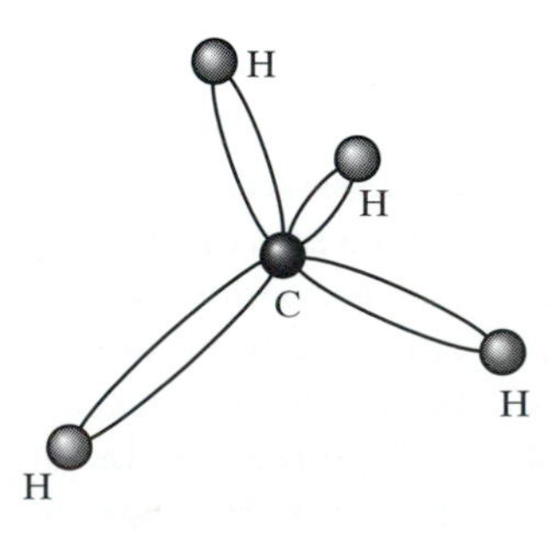

(b) "Saturated bond"–both the C and H atoms have their quantum states filled with electrons to produce methane, CH_4 – a small gas molecule

Fig. 1-16 (**a**) Unsaturated bonds in a diamond produce a solid, very large macromolecule; (**b**) saturated bonds in methane do not require more atoms, thus a single molecule is produced—a gas.

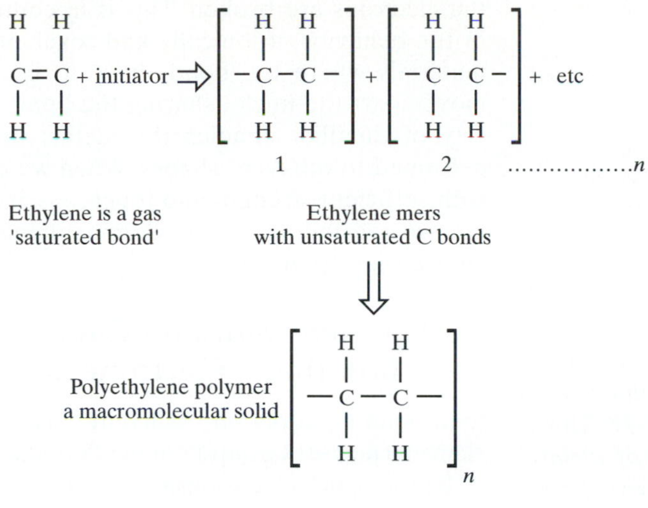

$$\begin{matrix} H & H \\ | & | \\ C & = & C \\ | & | \\ H & H \end{matrix} + \text{initiator} \Rightarrow \left[\begin{matrix} H & H \\ | & | \\ -C- & C- \\ | & | \\ H & H \end{matrix}\right]_1 + \left[\begin{matrix} H & H \\ | & | \\ -C- & C- \\ | & | \\ H & H \end{matrix}\right]_2 + \text{etc} \;\ldots\ldots n$$

Ethylene is a gas 'saturated bond'

Ethylene mers with unsaturated C bonds

$\Downarrow$

Polyethylene polymer a macromolecular solid

$$\left[\begin{matrix} H & H \\ | & | \\ -C- & C- \\ | & | \\ H & H \end{matrix}\right]_n$$

Fig. 1-17 The polymerization of ethylene (a gaseous molecule) to polyethylene, a polymer with a large number of molecules, a macromolecule.

methane, is a gas. Silicon, which is the main material in the semiconductor (electronics) industry is also solid, for the same reasons as for the diamond. When we bond silicon with four hydrogen atoms, SiH_4, silane, forms that is also a gas.

Another example of the effects of "saturated" and "unsaturated bonding" is shown in Fig. 1-17 for the polymerization process. Ethylene, the starting material for polyethylene and with the formula $H_2C{=}CH_2$, is a gas because all the quantum states between C and H are filled. The bonding in terms of the occupancy of the quantum states is saturated. If the double bond between the carbon atoms is broken, however, we get a molecule $-[H_2C-CH_2]-$ that has one unsaturated bond state from each of the carbon atoms. If there is a large number of this bracketed molecule, they can get bonded to each other to form a very large macromolecule and a solid at room temperature. The bracketed molecule is called the *mer* and the product $[-H_2C-CH_2-]_n$ containing many mers (n) is the *polymer*. The mer in this example is ethylene (a gas) and the polymer is polyethylene (a macromolecule and a solid).

We can explain some of the properties of covalently bonded materials, such as diamond and silicon, by the localized electrons between atoms. They can be very strong but are brittle because the atom nuclei cannot be moved apart much before the bond breaks. Diamond reflects the covalent bond strength in being the hardest natural occurring material. In addition, the localized electrons make covalently bonded materials very poor electrical conductors. Silicon exhibits some conductivity only at high temperatures and it is thus called a semiconductor. However, the electrical conductivity of silicon in integrated circuits comes from "impurities or dopants" purposely added to it and not from being pure, as we shall discuss in Chap. 3. Because of the strength in the covalent bond in diamond, it is very difficult to make it conductive as in silicon. Thus, it is an electrical insulator but curiously, it is an excellent thermal conductor. For this purpose, diamond is being investigated to ease the thermal load in integrated circuit boards. Polymeric materials are also covalently bonded and are, therefore, brittle as solids. Again, they are normally nonconductive but there are now so-called conductive polymers that contain groups in their chains that can free electrons. Polymeric materials can be mixed mechanically with metallic particles to make them conductive when the metallic particles form continuous contact within the polymer. Ceramic materials based on the silicates and the advanced ceramics SiN, SiC, and BN are all covalently bonded. The latter are being used for their hardness, wear, and high temperature resistance properties.

The *metallic bond* arises from the easy dissociation of metals into positive ions and free electrons. There are a lot of quantum states available in metals, all of which cannot be filled by electrons when the atoms are brought close to each other. Thus, they have unsaturated bonding and they are composed of a large number (a macromolecule) of atoms. The freed electrons easily move from the orbitals of one

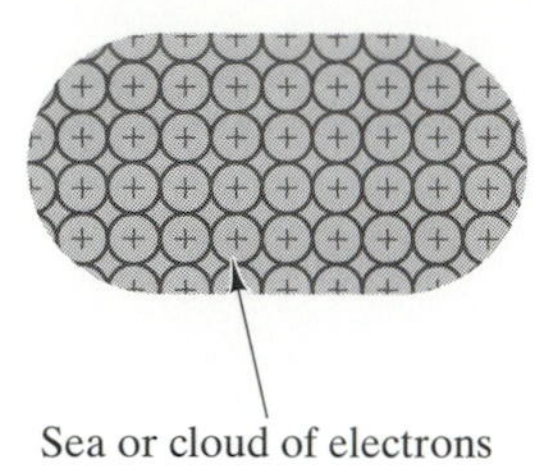

Fig. 1-18 The metallic bond consists of positive ion cores embedded in a sea or cloud of electrons.

atom to another and are no longer confined or localized to a pair of atoms; they are essentially "free flowing". Thus, metals are usually depicted as a large aggregate of positive ion cores embedded in a "cloud" or "sea" of electrons as in Fig. 1-18. This is the classic simple *"free electron theory of metals"* proposed by Drude and Lorentz in the period 1900 to 1916 to differentiate metals from nonmetals. Although this simple theory is now superseded by the quantum mechanical models in the 1930s and later years, the "free flowing electrons" explain the extremely good electrical and thermal conductivities of metals. The *optical opacity* and *reflectivity* are also explainable. The oscillation of the free electrons absorbs the energy of the incident light at all wavelengths and thus makes the metal opaque. In addition, the oscillating electrons also emit light waves (photons) and gives the metal reflectivity.

The unsaturated bonding in metals also explains the close-packing of atoms, their alloying properties, and the superior mechanical properties of toughness and ductility with sufficient strength. In metallic bonding, the number of nearest neighbors (which is called the coordination number) is only limited by the geometry. Metals, in general, are more densely packed than ionically and covalently bonded materials—one reason why metals have higher densities. This will be apparent in Chap. 2. Another characteristic of free electrons in metallic bonding are their insensitivity to the type of the positive ion core they are bonded to. It is thus possible to randomly substitute atoms of one metal with those of another and to form a so-called substitutional solid solution or alloy. A good example of this is the complete solid solution of copper and nickel in each other. Because the electrons are not confined to particular sites, the positive ion cores can be moved apart for some distance from each other, giving metals good ductility, before the bonds with the electrons are broken. This is in contrast to the brittle behavior of ionically and covalently bonded materials where the bonds have to be broken to move apart the nuclei sharing the bond. The property of ductility enables the metals to be easily deformed to different shapes. When we couple this with sufficient strength and toughness, it is no wonder we see metals as the dominant materials for engineering applications.

1.6.3 Mixed Ionic and Covalent Bonding—Electronegativity

Ionic bonding works very well only between atoms of elements nearest Group 18 in the Periodic Table, such as between the electropositive elements (Groups 1 and 2) and the electronegative ones (Groups 16 and 17). In contrast, pure covalent bonding exists only between atoms of the same element. Covalent bonding may also form between atoms of different elements, but in this case, the covalent bond is not symmetrical between them. The shared electron(s) tends to have a greater probability density in one atom than in the other atom. The power of an atom to attract electrons to itself is what chemists call *electronegativity*. Thus, in general, the bond between two different atoms will have a mixture of covalent and ionic bonding, which depends on the electronegativities of the elements; the presence of some ionic character generally strengthens the covalent bond. Linus Pauling proposed the following formula to approximate the fraction of the ionic character of a single bond between atoms A and B, with electronegativities x_A and x_B, respectively:

$$\text{Fraction of Ionic Character} = 1 - e^{-\frac{1}{4}(x_A - x_B)^2} \tag{1-16}$$

Figure 1-19 shows that the electronegativities of elements increase as their locations in the Periodic Table approach fluorine, from left to right and from bottom to top; fluorine has the highest electronegativity.

EXAMPLE 1-7

Compare the fraction of ionic character in the single bonds in H—C, in H—N, and H—O.

SOLUTION: We use Eq. 1-13 and the electronegativities of elements in Fig. 1-16. We tabulate

H																
2.1																
Li	Be	B											C	N	O	F
1.0	1.5	2.0											2.5	3.0	3.5	4.0
Na	Mg	Al											Si	P	S	Cl
0.9	1.2	1.5											1.8	2.1	2.5	3.0
K	Ca	Sc	Ti	V	Cr	Mn	Fe	Co	Ni	Cu	Zs	Ga	Ge	As	Se	Br
0.8	1.0	1.3	1.5	1.6	1.6	1.5	1.8	1.8	1.8	1.9	1.6	1.6	1.8	2.0	2.4	2.8
Rb	Sr	Y	Zr	Nb	Mo	Te	Ru	Rb	Pd	Ag	Cd	In	Sn	Sb	Te	I
0.8	1.0	1.2	1.4	1.6	1.8	1.9	2.2	2.2	2.2	1.9	1.7	1.7	1.5	1.8	2.1	2.5
Cs	Ba	La-Lu	Hf	Ta	W	Re	Os	Ir	Pt	Au	Hg	Tl	Pb	Bi	Po	At
0.7	0.9	1.1-1.2	1.3	1.5	1.7	1.9	2.2	2.2	2.2	2.4	1.9	1.8	1.8	1.9	2.0	2.2
Fr	Ra	Ae	Th	Pa	U	Np-No										
0.7	0.9	1.1	1.3	1.5	1.7	1.3										

Fig. 1-19 Electronegativities of elements [L. Pauling, *Nature of Chemical Bond,* 3d Ed., Used by permission of Cornell Univ. Press].

Single Bond	$x_A(H)$	x_B	$(x_A - x_B)^2$	Fraction ionic character
H—C	2.1	2.5	0.16	0.04
H—N	2.1	3.0	0.81	0.18
H—O	2.1	3.5	1.96	0.39

The fraction ionic character is obtained by:

for the H-C single bond:

$$\text{Fraction} = 1 - e^{-(0.16/4)} = 0.04\ (4\%)$$

for the H-N single bond:

$$\text{Fraction} = 1 - e^{-(0.81/4)} = 0.18\ (18\%)$$

For the H-O single bond:

$$\text{Fraction} = 1 - e^{-(1.96/4)} = 0.39\ (39\%)$$

We note that the H—C single bond, for instance, in CH_4, is essentially all covalent, only 4 percent ionic. On the other hand, the H—O single bond (as in the water molecule) has a very substantial 39 percent ionic character that explains the permanent dipole in water and water being a liquid.

1.6.4 Secondary Bonding—the van der Waals Bond

The van der Waals bond is the weak attractive force that may exist between atoms and molecules. It is responsible for the condensation of the noble gases and chemically saturated bonded molecules to liquids and solids at low temperatures. The attraction arises primarily from the emergence of weak dipoles and the electrostatic attraction between the oppositely charged ends of these dipoles. With the noble gases, such as He, and the chemically saturated molecules, such as H_2 and N_2, fluctuations and surges in the electronic charges will induce asymmetry in the normally spherical electron "cloud" distribution; this creates the dipoles, as illustrated in Fig. 1-20. When this dipole formation occurs in unison for all atoms or molecules, dipolar electrostatic attraction occurs over many atoms or molecules; the gases condense to liquids or solids at low temperatures. Examples are liquid He at 4 K, liquid hydrogen at about 20 K, and liquid nitrogen at 77 K. The dipolar attractive energy varies as $1/r^6$ and the van der Waals bond energy has the form

$$E_{vdW} = \underbrace{-\frac{C_A}{r^6}}_{\text{attractive}} + \underbrace{\frac{C_R}{r^n}}_{\text{repulsive}} \quad (n \approx 12) \qquad (1\text{-}17)$$

The thermal agitation of the molecules when temperature rises above their liquefaction temperatures is sufficient to "boil-off" the molecules.

With molecules consisting of two or more different atoms, permanent dipoles may be created between the atoms because of the differences in their electronegativities (see Example 1-7). The higher electronegative atom draws the electron to itself from the lower electronegative element; a polar

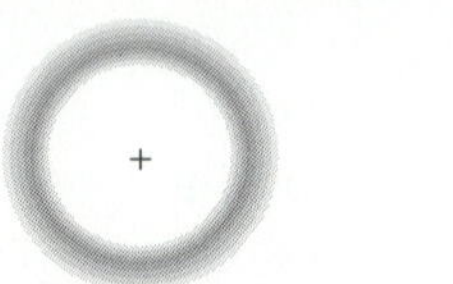

Normal spherical electron cloud of noble gases make centers of positive and negative charges coincide - no dipoles, no attraction

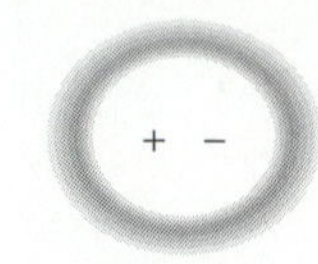

Asymmetrical electron cloud caused by fluctuations; center of negative charges is displaced from positive center, creating a dipole attraction that proceeds to condense gas

Fig. 1-20 van der Waals attraction of noble gas atoms that leads to condensation to liquid, such as helium.

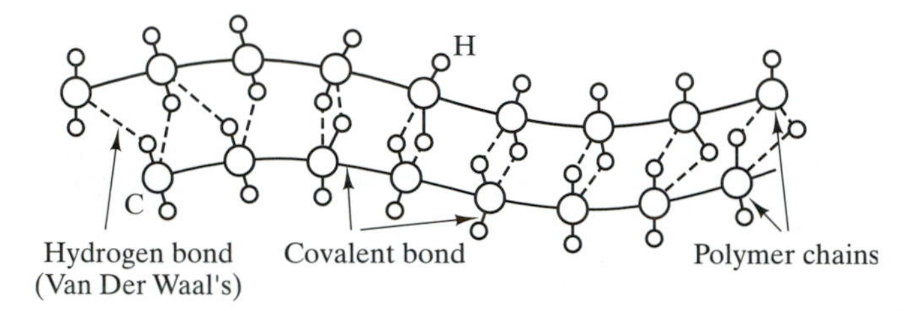

(a)

H
O — H
H
Hydrogen bond
O
Covalent bond
H
O — H
H

(b)

Fig. 1-21 van der Waals intermolecular bonding between (a) polymeric chains and (b) water molecules.

molecule with a permanent dipole is formed with the latter positively charged and the former negatively charged. A good example of this is the *hydrogen bonding* that binds water-polar molecules and keeps water liquid at room temperature. The dipoles consist of the negatively charged O atoms and the positively charged H atoms. The hydrogen bond is the bond between the hydrogen atoms of one water molecule with the negative O atoms of other water molecules, as shown in Fig. 1-21.

Hydrogen bonding is also the mechanism by which the macromolecular chains of polymers are held together. The low yield strength and low moduli of polymers (plastics) are explained by the weak van der Waals hydrogen bonds that are easily broken when the polymers are stressed.

1.7 The Sizes of Atoms and Ions

We consider now how atomic and ionic sizes depend on atomic number, Z, and the type of bonding. As the nuclear charge (Z, the atomic number) increases, electrons fill the quantum states in the outer shells with increasing *n* values; this tends to increase the atom size. However, because of the increase in positive charge, the occupied inner electron shells are pulled closer to the nucleus. These two effects approximately compensate each other such that the electron orbital radii (atomic size) of heavy atoms are not much larger than the light atoms. However, it is difficult to define an atomic size precisely because the electron cloud has no sharp boundary. The radius of the maximum of the probability density in Fig. 1-3 gives one indication. Experimentally, the measured distances between the nuclei of neighboring atoms in molecules and crystals (the r_o in Fig. 1-11) can be used as an indication of the atomic size. The measured internuclear distances involving a particular atom varies considerably, however, with the nature and the strength of the interatomic forces involved. The effects of these factors are indicated in Table 1-4, which is a summary of the atomic and ionic radii of the different elements, in Ångström units (1 Å = 10^{-8} cm = 10^{-10} m). We see that the atomic sizes of an element depend on the type of bonding, shown as vertical columns. For example, a sodium atom in covalent bonding has an atomic size of 1.57 Å, in metallic bonding 1.85 Å, and in ionic bonding 0.98 Å. The number in parenthesis next to the atomic size in the "metallic" is the coordination number (CN), the nearest neighbors of an atom in metallic bonding. For similar types of metals, such as the transition metals, the atomic size of an atom with CN = 12 is larger than that of an atom with CN = 8. The numbers in parenthesis in the "ionic" column signify the oxidation states or valences of the ions. A negative ion is much larger than a positive ion and a higher positive valence further reduces the ion size. The atomic and ionic sizes will be used in Chap. 2 to estimate the lattice parameters of crystalline materials.

Table 1-4 Atomic and ionic radii, in Ångström units [**Ref. 6**].

Notes: 1. Numbers in parenthesis in metalic bond column are coordination numbers of atoms.
2. Numbers in parenthesis under ionic bond are valences of ions.

	Type of Bond			
Element	**van der Waals**	**Covalent**	**Metallic**	**Ionic**
H	1.2	0.37		2.08(−1)
He	0.8			
Li		1.22	1.52(8)	0.7(+1)
Be		0.89	1.13(12)	0.34(+2), 0.2(+3)
B		0.88		0.2(+3)
C		0.77		2.6(−4), 0.20(+4)
N	1.5	0.74		1.7(−3), 0.16(+3), 0.15(+5)
O	1.4	0.74		1.35(−2), 0.09(+6)
F	1.35	0.72		1.33(−1), 0.07(+7)
Ne	1.59			
Na		1.57	1.85(8)	0.98(+1)
Mg		1.37	1.6(12)	0.75(+2)
Al		1.25	1.43(12)	0.55(+3)
Si		1.17		1.98(−4), 0.4(+4)
P	1.9	1.10		1.86(−3), 0.44(+3), 0.35(+5)
S	1.85	1.04		1.82(−2), 0.37(+4), 0.30(+6)
Cl	1.80	0.99		1.81(−1), 0.34(+5), 0.26(+7)
A	1.91			
K		2.02	2.25(8)	1.33(+1)
Ca		1.74	1.96(12)	1.05(+2)

(*Continued*)

Table 1-4 (*Continued*)

Element	Type of Bond			
	van der Waals	Covalent	Metallic	Ionic
Sc		1.44	1.63(12)	0.83(+3)
Ti		1.32	1.45(12)	0.76(+2), 0.70(+3), 0.64(+4)
V		1.22	1.31(8)	0.88(+2), 0.75(+3), 0.61(+4) 0.59(+5)
Cr		1.17	1.25(8)	0.89(+2), 0.65(+3), 0.36(+6)
Mn		1.17	1.3(12)	0.91(−2), 0.62(+3), 0.52(+4) 0.46(+7)
Fe		1.16	1.24(8)	0.83(+2), 0.67(+3)
Co		1.16	1.25(12)	0.82(+2), 0.65(+3)
Ni		1.15	1.25(12)	0.78(+2)
Cu		1.17	1.27(12)	0.96(+1), 0.72(+2)
Zn		1.25	1.37(12)	0.83(+2)
Ga		1.25	1.35(12)	0.62(+3)
Ge		1.22		2.72(−4), 0.65(+2), 0.55(+4)
As	2.0	1.21		1.91(−3), 0.69(+3), 0.47(+5)
Se	2.0	1.17		1.93(−2), 0.5(+4), 0.35(+6)
Br	1.95	1.14		1.96(−1), 0.47(+5), 0.39(+7)
Kr	2.01			
Rb		2.16	2.49(8)	1.49(+1)
Sr		1.91	2.13(12)	1.18(+2)
Y		1.61	1.81(12)	0.95(+3)
Zr		1.45	1.60(12)	0.80(+4)
Nb		1.34	1.42(8)	0.74(+4), 0.7(+5)
Mo		1.29	1.36(8)	0.68(+4), 0.65(+6)
Tc			1.35(12)	0.56(+7)
Ru		1.24	1.34(12)	0.65(+4)
Rh		1.25	1.34(12)	0.75(+3), 0.65(+4)
Pd		1.28	1.37(12)	0.80(+2), 0.65(+4)
Ag		1.34	1.44(12)	1.13(+1), 0.89(+2)
Cd		1.41	1.52(12)	0.99(+2)
In		1.50	1.57(12)	0.92(+3)
Sn		1.41	1.58(12)	2.94(−4), 1.02(+2), 0.74(+4)
Sb	2.2	1.41	1.61(12)	2.08(−3), 0.90(+3), 0.62(+5)
Te	2.20	1.37		2.12(−2), 0.89(+4), 0.56(+6)
I	2.15	1.33		2.20(−1), 0.94(+5), 0.50(+7)
Xe	2.20			Cs 2.35
2.63(8)	1.70(+1)			
Ba		1.98	2.17(8)	1.38(+2)
La		1.69	1.87(12)	1.15(+3)
Ce		1.65	1.82(12)	1.18(+3), 1.01(+4)
Pr		1.65	1.82(12)	1.16(+3), 1.00(+4)

(*Continued*)

Table 1-4 (*Continued*)

Element	Type of Bond van der Waals	Covalent	Metallic	Ionic
Nd		1.64	1.82(12)	1.15(+3)
Pm			1.8(12)	1.09(+3)
Sm		1.66	1.8(12)	1.13(+3)
Eu		1.85	1.98(8)	1.29(+2), 1.13(+3)
Gd		1.61	1.8(12)	1.11(+3)
Tb		1.59	1.77(12)	1.09(+3), 0.81(+4)
Dy		1.59	1.77(12)	1.07(+3)
Ho		1.58	1.76(12)	1.05(+3)
Er		1.57	1.75(12)	1.04(+3)
Tm		1.56	1.74(12)	1.04(+3)
Yb		1.70	1.93(12)	1.00(+3)
Lu		1.56	1.74(12)	0.99(+3)
Hf		1.44	1.59(12)	0.86(+4)
Ta		1.34	1.43(8)	0.73(+5)
W		1.30	1.36(8)	0.68(+4), 0.65(+6)
Re		1.28	1.38(12)	0.72(+4), 0.56(+7)
Os		1.25	1.35(12)	0.88(+4), 0.69(+6)
Ir		1.26	1.35(12)	1.66(+4)
Pt		1.29	1.38(12)	1.06(+2), 0.92(+4)
Au		1.34	1.44(12)	1.37(+1), 0.85(+3)
Hg		1.44	1.55(12)	1.27(+1), 1.12(+2)
Tl		1.55	1.71(12)	1.49(+1), 1.05(+3)
Pb		1.54	1.74(12)	2.15(−4), 1.32(+2), 0.84(+4)
Bi		1.52	1.82(12)	2.13(−3), 1.20(+3), 0.74(+5)
Po		1.53	1.7(12)	0.67(+6)
At				0.62(+7)
Rn				
Fr				1.80(+1)
Ra				1.42(+2)
Ac				1.18(+3)
Th		1.65	1.80(12)	1.02(+4)
Pa				1.13(+3), 0.98(+4), 0.89(+5)
U		1.42	1.5(8)	0.97(+4), 0.80(+6)
Np				1.10(+3), 0.95(+4)
Pu			1.63(12)	1.08(+3), 0.93(+4)
Am				1.07(+3), 0.92(+4)

Summary

We have learned in this chapter that the structure of an unbonded atom consists of negatively charged electrons moving around a relatively stationary nucleus that contains positively charged protons and zero charged neutrons; these particles are bound by the electrostatic attraction of positive and negative charges. The energy of the moving electron is not continuous. Rather, as explained by wave mechanics, the energy of the electron can only be at certain discrete energy levels dictated by the quantum states. The electrons of an atom (element) will occupy the available lowest-energy quantum states by adhering to the Pauli Exclusion Principle. The arrangement of electrons in this manner yields the electron configuration of an atom (element) in the ground state. Each atom (element) has a unique electron configuration and as the atomic number (number of protons = number of electrons) increases, the occupancy of the quantum states results in the Periodic Table of the Elements. The unique electron configuration of an atom reflects unique energy states. This explains the characteristic X-ray emissions that are used in chemical analysis of elements. The position of an element in the Periodic Table reveals some of its characteristics, such as being a metal or non-metal, its valence, its electronegativity, and possible chemical affinity (bonding) with other elements.

Gases, liquids, or solids may form when two or more atoms bond together. The principal bonding mechanisms of free atoms to form solids may be either metallic, ionic, or covalent. In addition, a secondary bond called the van der Waals's bond may occur. Gases are formed when the combination of a small number of atoms yield a molecule with saturated bonding. On the hand, kiquids and solids are formed when the combination of atoms yield unsaturated bonding requiring a huge number of atoms to form macromolecules. Liquids may also form as a result of van der Waals bonding such as that which is formed in water and in the condensation of helium, hydrogen, nitrogen, methane, and others. The strength (binding energy) of the principal bond gives an indication of the melting point, the coefficient of expansion, and the Young's Modulus of metals and ceramics. In polymers, however, the effect of the van der Waals's attraction shows up first as low strength and low modulus when polymers are stressed. The localization or delocalization of electrons between atoms during bonding determines the electrical conductivity and the ductility of the material. Because electrons are localized in ionic and covalent bonding, materials with these bonds are poor conductors and are brittle. In contrast, metallic bonded materials, with delocalized (free) electrons, are good conductors and are ductile and tough. During bonding, the interatomic distances, which are measures of the size of an atom, do not vary very much from the lightest to the heaviest atom.

References

1. *Handbook of Chemistry and Physics*, 76th ed., CRC Press, 1995–1996.
2. McGraw-Hill *Encyclopaedia of Science and Technology*, 7th ed., 1992.
3. A. Cottrell, *Introduction to the Modern Theory of Metals*, Institute of Metals, London, 1988.
4. J. I. Goldstein, et al, *Scanning Electron Microscopy and X-ray Microanalysis*, 2d ed., Plenum Press, New York, 1992.
5. M. F. Ashby and D. R. H. Jones, *Engineering Materials I*, Pergamon Press, 1980.
6. A. H. Cottrell, *An Introduction to Metallurgy*, 2d ed., E. Arnold Publisher, London, 1975.
7. L. Pauling, *The Nature of the Chemical Bond*, 3d ed., Cornell University Press, 1960.

Terms and Concepts

Atomic diameter
Atomic mass or weight
Atomic mass unit
Atomic number
Avogadro's number
Characteristic X rays

Covalent bonding
Dual behavior of electrons
Electron configuration
Electronegative
Electronegativity
Electron energy levels
Electropositive
Free electron theory
Hybridization
Ionic Bonding
Isotope
Macromolecules
Metallic bonding
Mixed ionic-covalent bonding
Pauli exclusion principle
Periodic Table
Probability density
Quantum numbers
Quantum or wave mechanics
Quantum states
Saturated bonding
Unsaturated bonding
Valence
van der Waals's bonding

Questions And Exercise Problems

1.1 What is the significance of the probability density in quantum mechanics? of the atomic number?

1.2 Indicate the difference between the atomic mass unit, atomic mass number, and the atomic weight?

1.3 The isotopes of copper are ^{63}Cu and ^{65}Cu. The atomic mass of ^{63}Cu is 62.9296 u and its natural abundance is 69.17 percent, while the atomic mass of ^{65}Cu is 64.9278 u and its natural abundance is 30.83 percent. What are the number of neutrons in each of the isotopes? Determine the atomic weight of copper and compare your answer to the value given in the Appendix.

1.4 The isotopes of gallium are ^{69}Ga and ^{71}Ga. What are the number of neutrons in each of the isotopes? The atomic mass of ^{69}Ga is 68.925580 and its abundance is 60.108 percent; the atomic mass of ^{71}Ga is 70.924700 and its natural abundance is 39.892 percent. Determine its atomic weight and compare it with that given in the Appendix.

1.5 The following are the only isotopes for each element with 100 percent natural abundance: ^{9}Be with atomic mass of 9.012182, ^{19}F with atomic mass of 18.99840, ^{23}Na with atomic mass of 22.9897, ^{89}Y with atomic mass of 88.90585, and ^{93}Nb with atomic mass of 92.906377. Determine the number of neutrons in each of these elements. What can you say about the number of neutrons relative to the number of protons in each atom? What are the atomic weights of each of the elements?

1.6 From the following table, determine the atomic weight of iron.

Isotope	Atomic mass, u	Abundance in %
^{54}Fe	53.9396	5.8
^{56}Fe	55.9349	91.72
^{57}Fe	56.9354	2.1
^{58}Fe	57.9333	0.28

1.7 For each of the isotopes of iron in Problem 1.6, determine the number of protons, the number of electrons, and the number of neutrons.

1.8 Yellow brass, which is an alloy of 70 weight percent copper and 30 weight percent zinc, is used to make cartridges. What are the mol percent and atomic percent of copper and zinc in the alloy?

1.9 Nitinol is a shape memory alloy with an equiatomic nickel and titanium composition, that is, 50 atomic percent nickel and 50 atomic percent titanium. If we have to melt this alloy, what are the minimum amounts (weights) of nickel and titanium that we have to melt to get one ton (1,000 kg) of the alloy?

1.10 What are the four quantum numbers an electron may have? Indicate the significance of each.

1.11 What are the quantum states of an electron? What is the Pauli exclusion principle? Is there a relationship of the quantum states with the Pauli exclusion principle? Which quantum number indicates the number of quantum states?

1.12 What is the electron configuration of a neutral atom in the ground state? What is the relationship, if any, of the electron configuration with the Periodic Table of the elements?

1.13 What are the elements from Groups 3 to 12 called? Which quantum states are being filled in the fourth period from Groups 3 to 12? In the fifth period from Groups 3 to 12?

1.14 Which states are being filled in the lanthanide series? In the actinide series? Explain the number of elements in each of these series?

1.15 What is meant by the "closed shell configuration" and its significance? What are the valence electrons?

1.16 Each electron has a distinct set of four quantum numbers. Indicate the set of quantum numbers of each of the valence electrons in carbon (C), aluminum (Al), magnesium (Mg), iron (Fe), titanium (Ti), chromium (Cr), germanium (Ge), bismuth (Bi), and tellurium(Te).

1.17 Indicate the electron configurations of the following ions: P^{+3}, P^{+5}, P^{-3}; Mn^{+2}, Mn^{+3}, Mn^{+7}, I^{+7}, I^{-1}; and Pt^{+2}, Pt^{+4}.

1.18 The characteristic K_α lines of copper and molybdenum are used as monochromatic radiation for X-ray diffraction. Calculate the wavelength of each of the lines.

1.19 What are electropositive elements? electronegative elements? What is electronegativity?

1.20 Differentiate the three principal bonding mechanisms in solids? What is the van der Waals's bonding? What are the relative binding energies of the different mechanisms?

1.21 Give examples of materials that exhibit each of the principal bonding mechanisms and the van der Waals's bond. Compare the relative ductilities and moduli of these materials, their electrical conductivities, and their melting points.

1.22 Determine the fraction of the ionic bond character in the compounds GaAs, $AsCl_3$, SiC, SiN, SiO_2, MgO, and NaCl. From Table 1-4, what are the appropriate atomic/ionic sizes for the different elements in the seven compounds.

2

Crystalline Structures —Ideal and Actual

2.1 *Introduction*

In Chap. 1, we learned how large aggregates of atoms can bond to form solids. Engineering materials, being solids, exhibit properties that reflect the packing or arrangement of these large aggregates of atoms. In particular, we shall learn in this chapter how the density and the deformability of materials depend significantly on the atomic arrangement.

2.2 *Crystalline and Amorphous (Noncrystalline) Materials*

In addition to the different types of bonding, materials are classified as crystalline or amorphous according to how the atoms or molecules are arranged in the solid. If there is a regular arrangement of atoms which results in a pattern that repeats itself in three dimensions, the material is said to be crystalline and to have a long-range order. If the arrangement of atoms is just localized and does not repeat itself in three dimensions, the material is said to be amorphous (noncrystalline) and to have a short-range order. The schematics of long-range and short-range orders in two dimensions are shown in Fig. 2-1. The points may represent a single atom or groups of atoms or molecules. In long-range order, the shaded pattern repeats itself as it is moved horizontally and vertically with the translation vectors indicated. Examples of two-dimensional patterns are found in wallpapers, floor tiles, textile weaves, and the like.

In short-range order, the atom arrangement in Fig. 2-1a is just localized within the envelopes shown in Fig. 2-1b. This short-range order is similar to the structure of liquids and may be inherited from the liquid state when the materials solidify. Thus, while most metals and alloys are ordinarily crystalline, some become amorphous when they are quenched very drastically from the liquid state. These are called amorphous metals and alloys and the conditions of their formation are discussed in Chap. 7. Therefore,

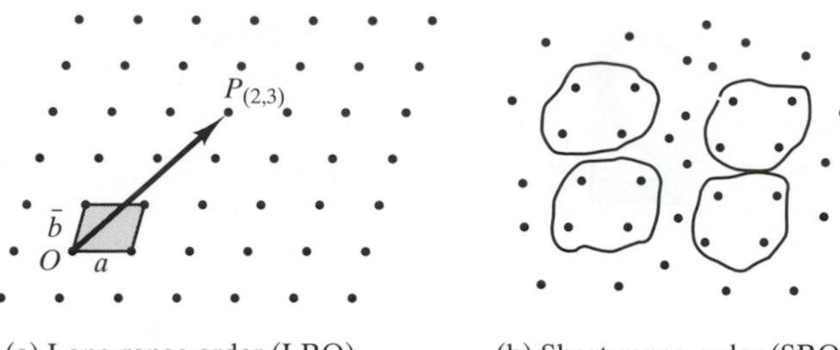

Fig. 2-1 Two-dimensional (a) LRO, and (b) SRO.

solids are not necessarily crystalline.* For the same reason, while most liquids have short-range order, some exhibit long-range order and are crystalline as in liquid crystalline polymers that are discussed briefly in this chapter and in Chap. 15.

2.3 The Unit Cell, Crystal Systems, and Point or Space Lattices

The study of the structures or arrangement of atoms in crystalline materials or crystals is the field of crystallography. We shall learn about the different crystal systems and structures into which three-dimensional patterns of atoms can be arranged. These crystal systems have specific dimensions and characteristics that determine the packing of the atoms and the density of the material. In addition, we shall learn about the specific planes and directions in each crystal that determine the deformability of materials.

We start by defining the pattern that repeats itself in the crystal structure as a unit cell. A two-dimensional unit cell is shown in Fig. 2-1a and *a three-dimensional unit cell is the parallelipiped* illustrated in Fig. 2-2. It is characterized by the three unit translation vectors **a**, **b**, and **c** by which it is repeated in three dimensions. Being vectors they have both magnitude and direction and are called the crystallographic axes because they define the size and shape of the unit cell. The magnitudes of the axes or vectors **a**, **b**, and **c** and the angles α, β, and γ between them are called the *lattice constants* or *lattice parameters* of the cell.

*The distinction between a crystalline and an amorphous material is made using diffraction techniques. A crystalline material will diffract and exhibit diffraction patterns, while an amorphous material does not.

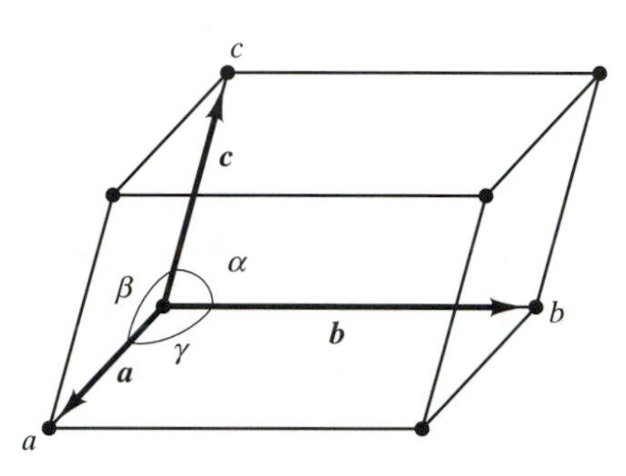

Fig. 2-2 A Parallelipiped Unit Cell. Three translation vectors, a, b, and c, and their angular relationships α, β, and γ define the shape and crystal system.

If we start from a point and then translate this point in all directions with the set of the three unit vectors in Fig. 2-2, we obtain what is called a *point space lattice*, Fig. 2-3. A point space lattice is defined as an array of points that are arranged in space so that each point has identical surroundings. If we choose a point in this lattice as the origin and apply each of the unit translation vectors any number of times to come to a second point, we obtain a vector from the origin to that second point. The vector is

$$\mathbf{P} = u\,\mathbf{a} + v\,\mathbf{b} + w\,\mathbf{c} \qquad (2\text{-}1)$$

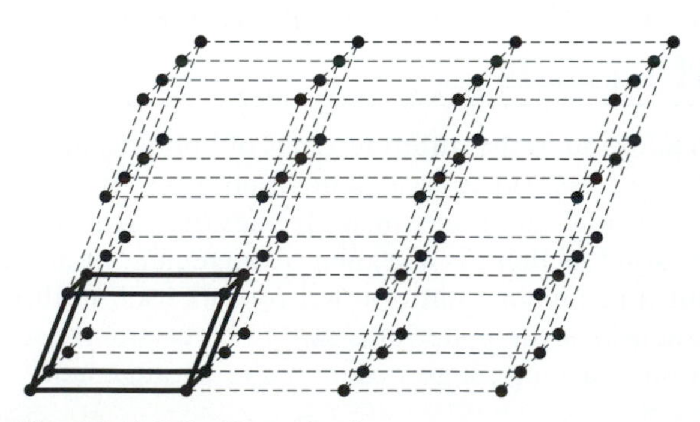

Fig. 2-3 A Point Space Lattice.

where u, v, and w are integers and indicate the number of times the respective axes were translated. From the chosen origin, they are referred to as the vector coordinates of the point P and indicated as (u, v, w). These are illustrated in the two-dimensional lattice in Fig. 2-1a where the vector is $\mathbf{P} = 2\mathbf{a} + 3\mathbf{b}$ from the chosen origin, and the vector coordinates of point P are (2,3) as indicated. In general, a vector in the two-dimensional lattice of Fig. 2-1a is

$$\mathbf{P} = u\,\mathbf{a} + v\,\mathbf{b} \tag{2-2}$$

The more general equation is the three-dimensional form, Eq. 2-1. We see, therefore, that each point in the space lattice can be represented by a vector. We should note, however, that while the unit cell has shape and orientation, its origin in the point space lattice is not necessarily fixed, except when defining or illustrating a certain problem, as in the example just illustrated. This means that we can choose and change the origin of the unit cell at our convenience to solve a particular problem. Furthermore, we should note that while u, v, and w are integers to represent points in the space lattice, locations of atoms or translations within a unit cell are only fractional translations of the unit axes; therefore, fractional vector coordinates are specified. For example, one-half translation of each of the axes is designated by the vector, $\mathbf{P} = {}^1\!/_2\,\mathbf{a} + + {}^1\!/_2\,\mathbf{b} + {}^1\!/_2\,\mathbf{c}$, and the vector coordinates are $({}^1\!/_2, {}^1\!/_2, {}^1\!/_2)$.

By specifying values for the axial lengths and the angles between them, unit cells of various sizes and shapes are produced. It turns out that there are only seven unique and independent ways we can specify the axial lengths and the different angles between them. In other words, we can only have seven types of unit cells or what are called crystal systems. These are

Table 2-1 Crystal Systems and Bravais Lattices.

System	Axial Lengths and Angles	Bravais Lattice	Lattice Symbol
Cubic	Three equal axes at right angles $a = b = c$; $\alpha = \beta = \gamma = 90°$	Simple Body-Centered Face-Centered	P I F
Tetragonal	Three axes at right angles, two equal $a = b \neq c$; $\alpha = \beta = \gamma = 90°$	Simple Body-Centered	P I
Orthorbombic	Three unequal axes at right angles $a \neq b \neq c$; $\alpha = \beta = \gamma = 90°$	Simple Body-Centered Base-Centered Face-Centered	P I C F
Rhombohedral	Three equal axes, equally inclined $a = b = c$; $\alpha = \beta = \gamma \neq 90°$	Simple	P
Hexagonal	Two equal coplanar axes at 120° Third axis at right angles $a = b \neq c$; $\alpha = \beta = 90°$, $\gamma = 120°$	Simple	P
Monoclinic	Three unequal angles, one pair not at right angles; $a \neq b \neq c$; $\alpha = \gamma = 90° \neq \beta$	Simple Base-Centered	P C
Triclinic	Three unequal axes, unequally inclined; none at right angles $a \neq b \neq c$; $\alpha \neq \beta \neq \gamma \neq 90°$	Simple	P

shown in Table 2-1 together with the crystallographic axes and the angular relationships to each other.

If we translate each of the seven crystal systems in the directions of their respective unit vectors, we generate seven different point lattices similar to that shown in Fig. 2-3. The points are at the corners of the translated unit cell. In terms of the definition of a point lattice given earlier whereby each point in the lattice is to have identical surroundings, points can be added to the interior and faces in some of the seven unit cells and still satisfy the definition of a point lattice. Thus we have additional point lattices aside from the original seven point lattices with points only at the corners of the unit cells. To differentiate these lattices from others, the original seven lattices are designated simple or primitive lattices and given a symbol P or R. If a point is added inside a unit cell, the point lattice generated is called body-centered and is given a symbol I. If a point is added at the center of a plane in the unit cell, the point lattice generated is either base- or face-centered, with symbols C or F, respectively; F is designated when the plane is a square. These point lattice designations and their symbols are also indicated in Table 2-1. Bravais, a French crystallographer, demonstrated that there are only 14 possible point lattices and no more. To recognize this contribution, the *point lattice* is also called the *Bravais lattice* and both names are used interchangeably. The 14 Bravais lattices are shown in Fig. 2-4.

In general, the points in the Bravais lattices may represent one single atom, two atoms, a molecule, or a cluster of atoms or molecules. For this introductory course, we shall take each lattice point to be represented by a single atom or ion, depending on the type of bonding. We shall also assume the atoms or ions to be "hard spheres" with their sizes given in Table 1-4 in order to illustrate the characteristics of the different crystal structures. This does not suggest, however, that the atoms or ions resemble hard spheres. In the following section, we will need to know the number of lattice points or atoms belonging to a unit cell (UC). This is given by

$$\mathbf{N}_T = \mathbf{N}_I + \frac{\mathbf{N}_F}{2} + \frac{\mathbf{N}_C}{8} \tag{2-3}$$

where $\mathbf{N}_T$ is the total atoms per UC, $\mathbf{N}_I$ is the number of atoms in the interior, $\mathbf{N}_F$ is the number of atoms on the faces, and $\mathbf{N}_C$ is the number of atoms at the corners of the UC. We shall have many opportunities to apply Eq. 2-3 when we consider the different crystal structures of metallic, ionic, and covalently bonded materials in the following sections.

2.4 The Common Metallic Structures

As indicated in Chap. 1, most of the elements in the Periodic Table are metals and the majority of these metals crystallize into one of three crystal structures: body-centered cubic (BCC), face-centered cubic (FCC), and hexagonal close packed (HCP).

2.4.1 The Body Centered Cubic (BCC) Structure

This structure is one of the cubic crystal systems with the Bravais lattice designation, I, in Table 2-1 and in Fig. 2-4; the axes are all equal ($a = b = c$) and the angles are all 90°. Thus, the lattice parameter is a. When an atom is placed at each point of the body-centered cubic (I) Bravais lattice in Fig. 2-4, we see that there are eight atoms at the corners and one atom at the center of the cube, as illustrated in Fig. 2-5a. The center (body-centered) atom touches all the atoms at the corners of the cube, as shown in Fig. 2-5b, and we say that the corner atoms are the nearest neighbors of the central atom. The number of nearest neighbors is called the *coordination number* (CN) of an atom in the structure, and in the BCC structure, the CN is 8. The number of lattice points or atoms per unit cell according to Eq. 2-3, is

$$\mathrm{N_{BCC}} = 1 + \frac{0}{2} + \frac{8}{8} = 2 \text{ atoms/unit cell.}$$

Assuming the atoms to be "hard spheres" with atomic radii equal to r, we can estimate the lattice parameter of the unit cell. *To estimate the lattice parameter of a unit cell, we look for a direction along which the atoms are in contact and based on the geometry of the unit cell, we establish the relationship between the atomic sizes and the lattice parameter along this contact direction.* This concept is also applicable to all types of unit cells. In the BCC structure, the atoms are in contact along the body diago-

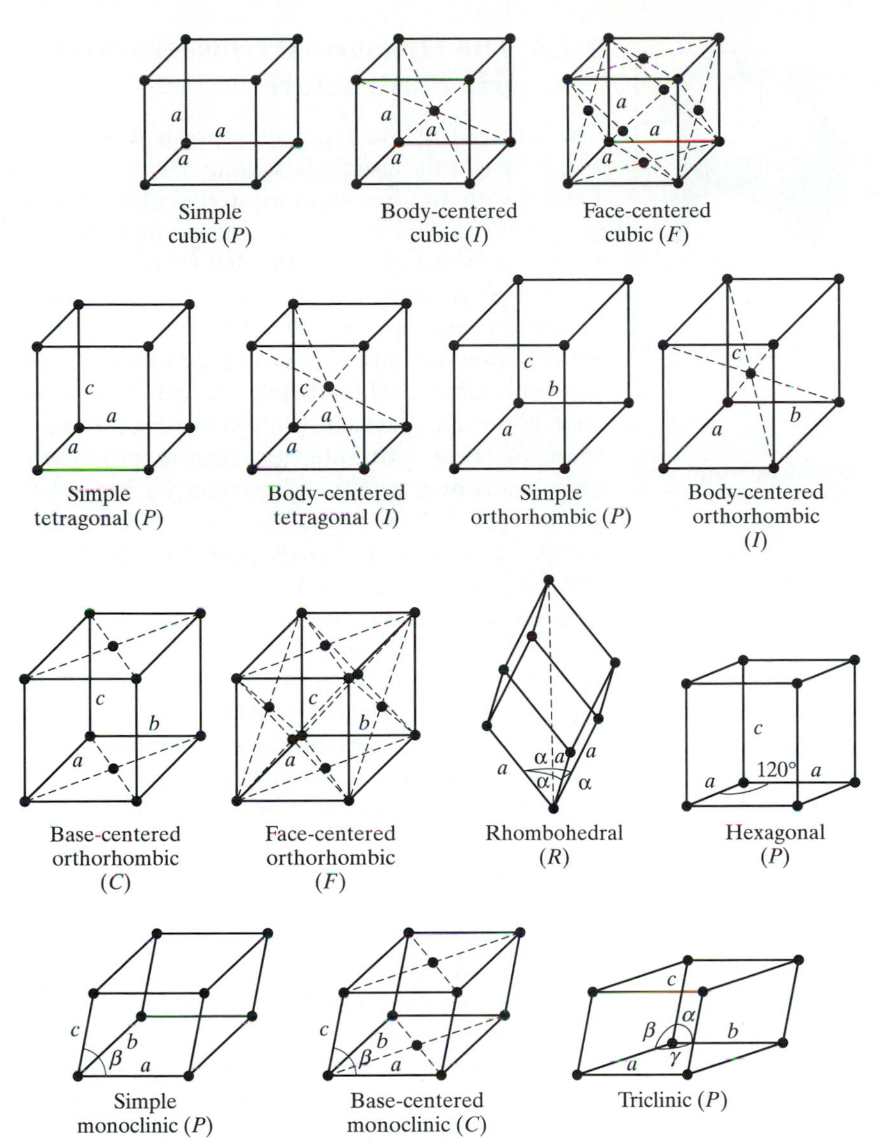

Fig. 2-4 The 14 Bravais Lattices.

nal, from point 1 to 2 in Fig. 2-5a. The relationship between the atomic size and the lattice parameter is thus,

$$\text{Body Diagonal Length} = \sqrt{3}a = 4r$$

$$a = \frac{4r}{\sqrt{3}} \qquad (2\text{-}4)$$

where a is the lattice parameter and r is the atomic radius.

2.4.2 The Face Centered Cubic (FCC) Structure

This structure is again one of the cubic crystal systems with the Bravais lattice designation, F; again all the unit axes are equal, all angles equal 90°, and the lattice parameter is also a. If we place one atom at each of the points in the face-centered cubic (F) Bravais lattice in Fig. 2-4, we have six atoms at the centers of the six faces, in addition to the eight atoms at the corners (C) of the cube, as shown in Fig. 2-6.

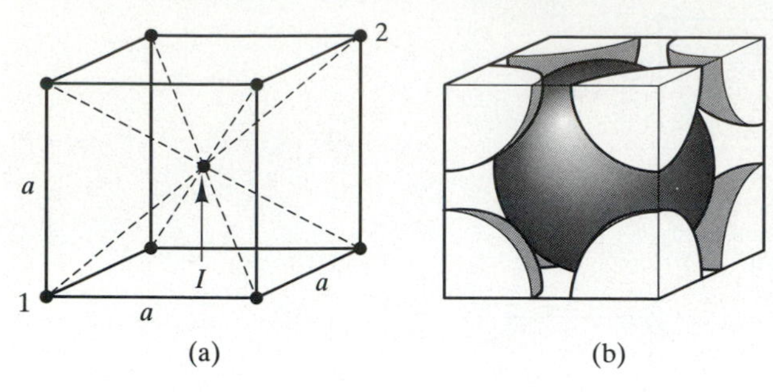

Fig. 2-5 The BCC structure: (a) Bravais lattice, (b) hard-sphere model.

The number of lattice points or atoms per unit cell according to Eq. 2-3 is

$$\mathbf{N}_{\mathrm{FCC}} = 0 + \frac{6}{2} + \frac{8}{8} = 4 \text{ atoms/unit cell.}$$

The atoms are in contact along the face diagonals and the CN in FCC is 12. We are able to see this by considering a face-atom such as that marked X in Fig. 2-6b. Eight of the 12 atoms in contact with X are numbered 1 through 8 in Fig. 2-6b. The other four are the mirror atoms of 5, 6, 7, and 8 in the next unit cell attached to the face containing X.

Since the atoms are in contact along the face diagonals, the relationship between the lattice parameter a and the atomic size, r, is

$$\text{Face diagonal length} = \sqrt{2}a = 4r$$

$$a = \frac{4r}{\sqrt{2}} \qquad (2\text{-}5)$$

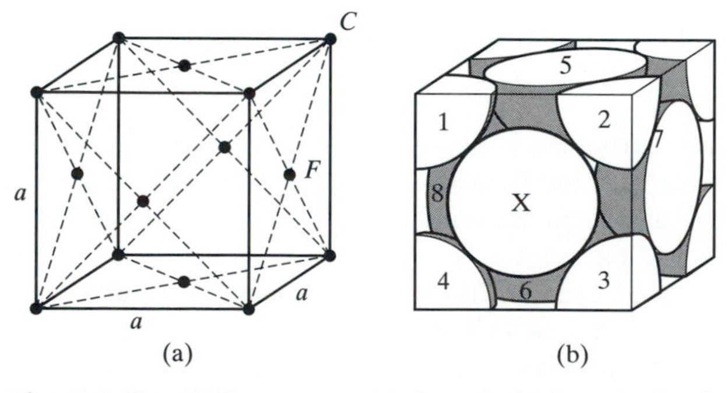

Fig. 2-6 The FCC structure: (a) Bravais lattice, (b) hard-sphere model.

2.4.3 The Hexagonal Close-Packed (HCP) Structure

The unit cell of the HCP structure is illustrated in Fig. 2-7a. It is the simple hexagonal Bravais lattice in Fig. 2-4 with another atom located at either the vector point $(\frac{2}{3},\frac{1}{3},\frac{1}{2})$ or $(\frac{1}{3},\frac{2}{3},\frac{1}{2})$. Recall that these are fractional translations of the lattice parameters of the unit cell $\mathbf{a}_1$, $\mathbf{a}_2$, and $\mathbf{c}$, where $\mathbf{a}_1 = \mathbf{a}_2 \neq \mathbf{c}$. The angular relationships are: $\alpha = \beta = 90°$, and $\gamma = 120°$, the angle between $\mathbf{a}_1$ and $\mathbf{a}_2$. It is difficult to see the characteristics of the HCP with this unit cell. We can visualize better the characteristics of the HCP if we put three of these unit cells together to produce the hexagon, shown in Fig. 2-7b. When we put spheres (atoms) at each point in Fig. 2-7b, we see how the atoms are packed and stacked in Fig.2-7c. The top and bottom planes of the hexagon are called the *basal planes* and the six sides or faces are the *prismatic planes.* The atoms in the lower basal plane (first layer) are all in contact and arranged in the same manner as freshly racked billiard balls in a pool game. Then the atoms inside the hexagon (second layer) fit in the depressions of the first layer and are arranged in the same manner as the first layer. The atoms in the top basal plane are positioned exactly as in the lower basal plane. Thus, if we call the basal plane A and the second layer B, the stacking of atoms in the HCP structure is ABABAB.... It is easy to see that the arrangement of atoms in this manner gives the closest packing of the atoms—thus the name close-packed structure.

Again, using Eq. 2-3, the simple HCP cell shown in Fig. 2-7a yields 2 lattice points or atoms per unit cell. The hexagon in Fig. 2-7b with three of the simple HCP cell will therefore have six atoms. The coordination number of an atom in the HCP is also 12 that is easily seen by considering the center atom in the top basal plane in Fig. 2-7c. In this plane, we see six atoms in contact. The three atoms at the plane beneath it are also in contact, as well as three more, if we put a similar plane on top.

It is also easy to see that the atoms are in contact along the unit axes $\mathbf{a}_1$ and $\mathbf{a}_2$. Therefore, one of the lattice parameters of the HCP cell is

$$\mathbf{a}_1 = \mathbf{a}_2 = \mathbf{a} = 2r. \qquad (2\text{-}6a)$$

We can determine the other lattice parameter, c, by noting (1) the vector coordinate point $(\frac{2}{3},\frac{1}{3},\frac{1}{2})$ of the

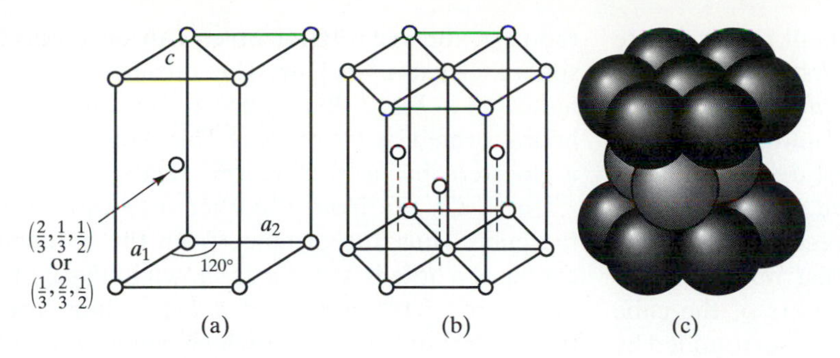

Fig. 2-7 The HCP structure: (a) the unit cell, (b) three unit cells, and (c) hard-sphere model.

atom inside the simple cell in Fig. 2-7a, and (2) that this atom is in contact with atoms in the basal plane and therefore the interatomic distance is also $2r = a$. We refer to Fig. 2-8 to determine the relationship of c to a. We note in Fig. 2-8a a right triangle, where

$$x^2 = \left(\frac{2}{3}a\right)^2 - \left(\frac{1}{3}a\right)^2 = \frac{1}{3}a^2$$

and in Fig. 2-8b, we get

$$a^2 = \frac{1}{3}a^2 + \left(\frac{1}{2}c\right)^2, \qquad \text{from which} \qquad \frac{c^2}{4} = \frac{2}{3}a^2$$

From this we obtain,

$$\left(\frac{c}{a}\right)^2 = \frac{8}{3} \qquad \text{or} \qquad \frac{c}{a} = 1.633 \qquad (2\text{-}6b)$$

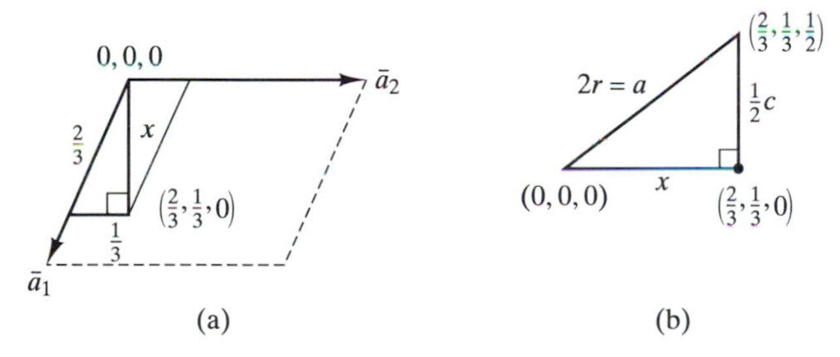

Fig. 2-8 Determination of c/a ratio in HCP, (a) right triangle in basal plane, and (b) right triangle perpendicular to basal plane.

EXAMPLE 2-1

Estimate the lattice parameter of (a) BCC molybdenum, (b) FCC gold, and (c) HCP cobalt.

SOLUTION: To reiterate the concept in estimating lattice parameters, we look for the direction where atoms are in contact and equate lengths along this direction in terms of atomic sizes and of the lattice parameter.

(a) For BCC, atoms are in contact along the body diagonal, and

$$\sqrt{3}a = 4r \qquad \text{or} \qquad a = \frac{4r}{\sqrt{3}}$$

We need r for Mo and we get this from Table 1-4. The correct value to use is 1.36 Å (1 Å = 10^{-10} m) because BCC has a coordination number of eight. Thus,

$$a = \frac{4 \times 1.136}{\sqrt{3}} = 3.1408 \text{ Å} = 0.31408 \text{ nm}$$

$$(10 \text{ Å} = 1 \text{ nm})$$

(b) For FCC, the atoms are in contact along the face diagonal and

$$\sqrt{2}a = 4r \qquad \text{thus,} \qquad a = \frac{4r}{\sqrt{2}}$$

The appropriate atomic radius of gold from Table 1-4 is 1.44 Å because FCC has a coordination number of 12. Thus,

$$a = \frac{4 \times 1.44}{\sqrt{2}} = 4.0729 \text{ Å} = 0.40729 \text{ nm}$$

(c) For HCP, $a = 2r$, $c = 1.633a$, and the atomic radius for cobalt is 1.25 Å. Thus,

$$a = 2 \times 1.25 = 2.5 \text{ Å} = 0.25 \text{ nm}$$

and

$$c = 1.633 \times 2.5 = 4.0825 \text{ Å} = 0.40835 \text{ nm}$$

2.4.4 Interstices or Holes in Crystal Structures

In the foregoing structures, there are empty spaces called interstices or holes where small-sized atoms can fit. The sizes of these interstices vary with the different structures and determine the extent that each structure can accommodate the small atoms. There are basically two types of interstices in these structures—*octahedral* and *tetrahedral interstices*—which are shown in Figs. 2-9, 2-10, and 2-11, for the FCC, the HCP, and the BCC structures. The names arise because they are each located at the center of an octahedron and a tetrahedron, respectively.

In the FCC structure, the largest holes are the octahedral holes located at the centers of the cube edges and at the cube center. They are surrounded by six atoms that form the corners of an octahedron, an example of which is drawn in Fig. 2-9a around the interstice (small open circle) at the cube center. In addition to the octahedral hole at the cube center, Fig. 2-9a indicates also 12 other similar holes located between two atoms at the centers of the cube edges. Assuming the atoms to be close-packed spheres of radius r, the octahedral holes can accommodate a sphere of radius 0.414r. The tetrahedral holes are indicated in Fig. 2-9b and are at the centers of tetrahedra formed by four atoms. The center of each hole is located by a translation vector of the form $<^1/_4\mathbf{a}, {}^1/_4\mathbf{a}, {}^1/_4\mathbf{a}>$ from a corner into the interior of the cube. Thus, the distance from the corner to the center of a hole is $\sqrt{3}/4\ \mathbf{a}$ as shown in Fig. 2-9b. Since there are eight corners in a cube, there are eight tetrahedral holes, each of which can accommodate a sphere of radius 0.225r, just about half the sphere that an octahedral hole accommodates.

The HCP structure has basically the same atomic arrangement as the FCC; therefore, the octahedral and interstitial holes in HCP that are shown in Fig. 2-10a and 2-10b will accommodate spheres of the same size relative to spherical atoms as in the FCC structure.

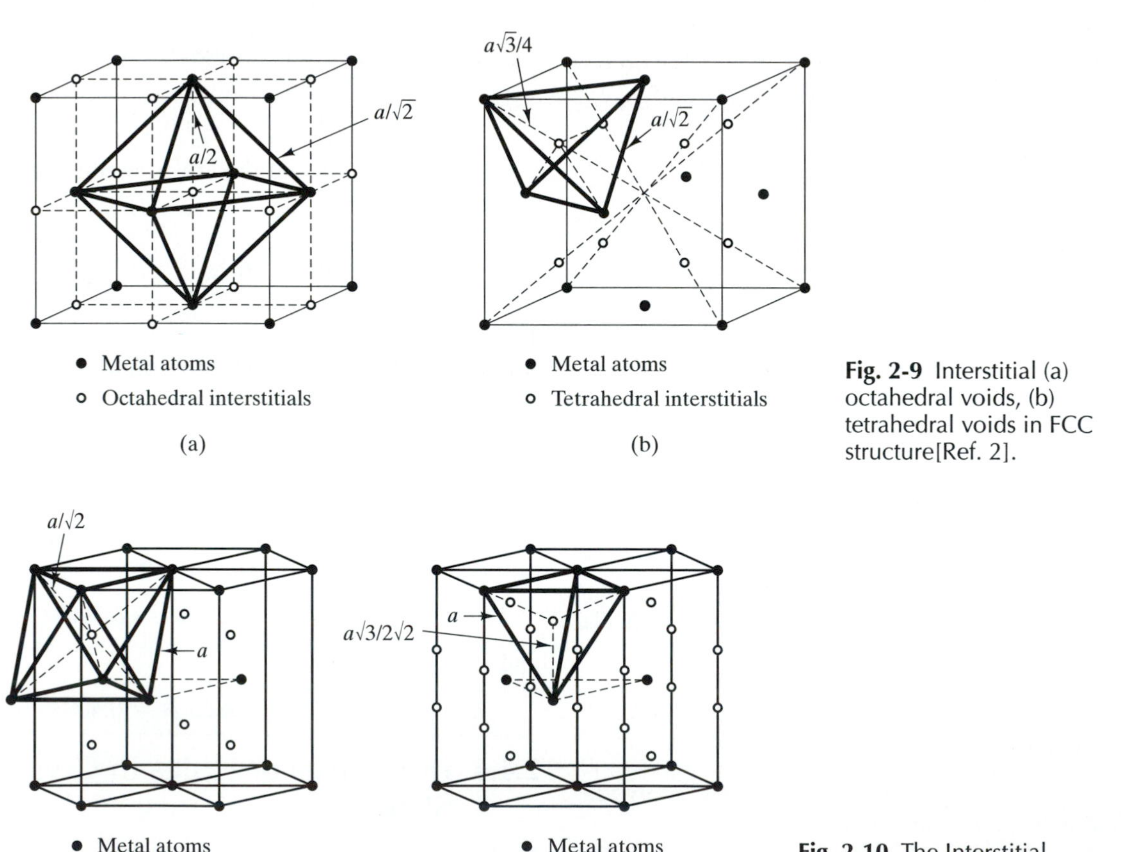

Fig. 2-9 Interstitial (a) octahedral voids, (b) tetrahedral voids in FCC structure[Ref. 2].

Fig. 2-10 The Interstitial (a) octahedral, and (b) tetrahedral voids in the HCP structure with an ideal c/a = 1.633[Ref. 2].

The octahedral and the interstitial holes in the BCC structure are shown in Figs. 2-11a and 2-11b. The larger of the two is now the tetrahedral hole located at vector coordinates of the form <$\frac{1}{2}$, $\frac{1}{4}$, 0>; a total of 24 tetrahedral holes, four on each face of the cube. These holes can accommodate a sphere of radius about 0.291r, where r is the radius of the atom in the BCC lattice sites. The smaller octahedral holes are located at the midpoints of the edges (0,0,$\frac{1}{2}$) . . ., and so forth, and at the equivalent positions ($\frac{1}{2}$, $\frac{1}{2}$, 0) . . ., et cetera, for a total of 18 sites. The octahedron enclosing each hole is slightly compressed, with two atoms nearer than the other four, as shown in Fig. 2-11a, and for this reason, the holes can only accommodate a sphere of radius about 0.155r.

Thus, we see that the BCC structure accommodates a smaller-sized spherical atom than can be accommodated by either the FCC or the HCP structure. The FCC can accommodate a sphere about 0.414r in an octahedral hole, which is 42.3 percent larger than the 0.291r sphere that can be accommodated in the larger tetrahedral hole in the BCC. This is reflected in the austenite phase (FCC) of steels having greater solubility for carbon atoms than the ferrite (BCC) phase and is the fundamental reason why steels can be hardened when they are heat-treated. This will be covered in Chap. 9.

EXAMPLE 2-2

Estimate the largest spherical atoms or ions that the tetrahedral and octahedral interstices in the BCC and the FCC can accommodate.

SOLUTION:

For the BCC structure:

Tetrahedral interstices are at <$\frac{1}{2}$, $\frac{1}{4}$, 0> and these are the centers of the atoms or ions that fit the interstices. The host atom with radius r will be in contact with the interstitial atom or ion with r_i. Taking the host atom at (0,0,0) and the interstitial atom at (0, $\frac{1}{2}$, $\frac{1}{4}$), the magnitude of the vector between these coordinates must equal ($r + r_i$),

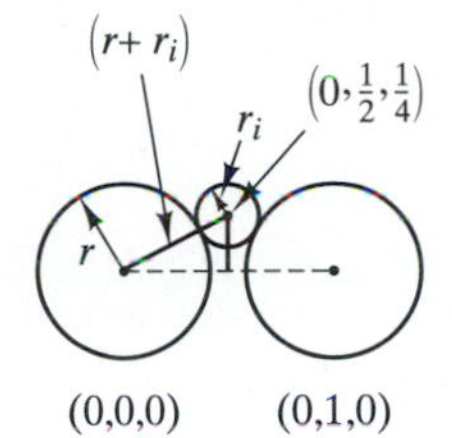

Fig. EP 2-2a

$$r + r_i = \sqrt{\left(\frac{1}{2}a\right)^2 + \left(\frac{1}{4}a\right)^2}$$

$$= a\sqrt{\frac{1}{4} + \frac{1}{16}} = 0.559a$$

$$= 1.291r, \quad \text{after using} \quad a = \frac{4r}{\sqrt{3}}$$

thus, $r_i = 0.291r$.

As an example of *octahedral interstices*, we take the location indicated in Fig. 2-11a that is at ($\frac{1}{2}$, $\frac{1}{2}$, 1). The host atoms in contact with the interstitial atom

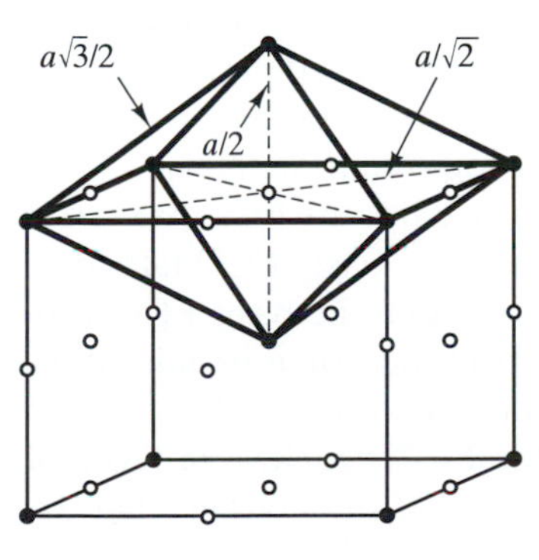

● Metal atoms

○ Octahedral interstitials

(a)

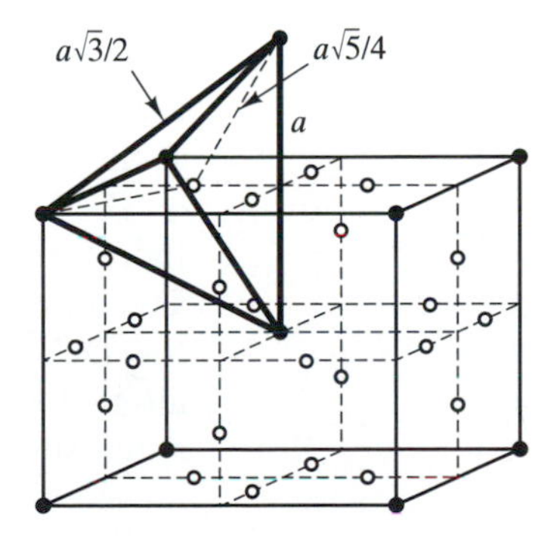

● Metal atoms

○ Tetrahedral interstitials

(b)

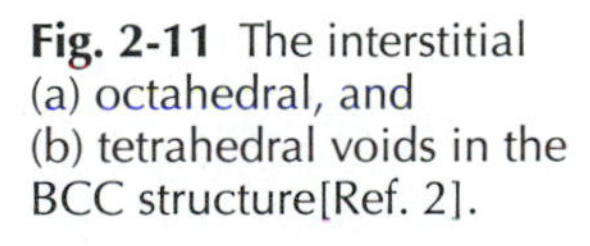

Fig. 2-11 The interstitial (a) octahedral, and (b) tetrahedral voids in the BCC structure[Ref. 2].

at this location are at ($\frac{1}{2}$, $\frac{1}{2}$, $\frac{1}{2}$) and ($\frac{1}{2}$, $\frac{1}{2}$, $\frac{3}{2}$). The distance between the centers of the host atoms is the lattice parameter, a. Thus,

$$2r + 2r_i = a = \frac{4r}{\sqrt{3}}$$

$$r_i = \frac{2r}{\sqrt{3}} - r$$

$$= 0.155r$$

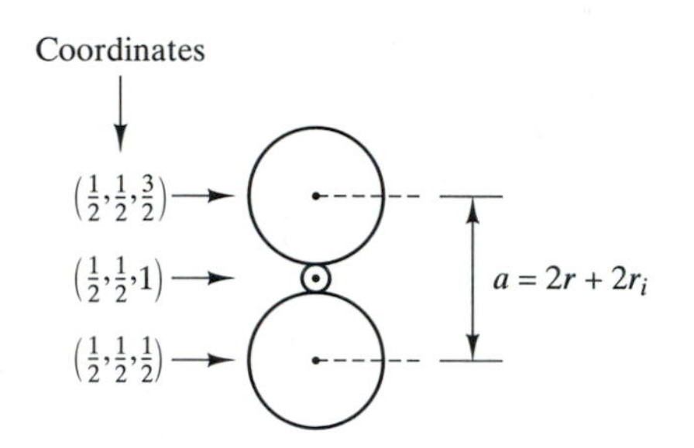

Fig. EP 2-2b

For the FCC structure:

The *tetrahedral interstices* are at <$\frac{1}{4}$, $\frac{1}{4}$, $\frac{1}{4}$>. The magnitude of the vector from (0,0,0) to ($\frac{1}{4}$, $\frac{1}{4}$, $\frac{1}{4}$) equals ($r + r_i$). Thus,

$$r_i + r = \sqrt{\left(\frac{1}{4}a\right)^2 + \left(\frac{1}{4}a\right)^2 + \left(\frac{1}{4}a\right)^2} = \frac{\sqrt{3}}{4}a$$

$$r_i = \frac{\sqrt{3}}{4}\left(\frac{4r}{\sqrt{2}}\right) - r, \quad \text{where} \quad a = \frac{4r}{\sqrt{2}}$$

$$= 0.225r$$

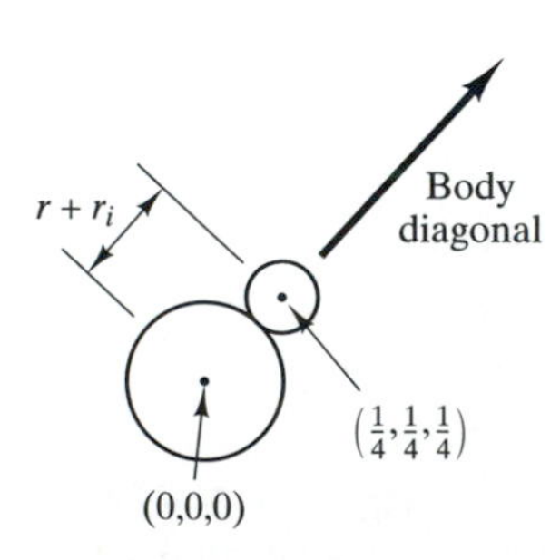

Fig. EP 2-2c

For the *octahedral interstices,* we take the location indicated in Fig. 2-9a at the center of the cube at ($\frac{1}{2}$, $\frac{1}{2}$, $\frac{1}{2}$). An interstitial atom located at this location will be in contact with all the host atoms at the centers of the faces. The distance between the centers of the host atoms is a, the lattice parameter. Thus, taking any two atoms on the same axis in contact, the relationship is

$$2r_i + 2r = a = \frac{4r}{\sqrt{2}}$$

$$r_i = r\left(\frac{2}{\sqrt{2}} - 1\right) = 0.414r$$

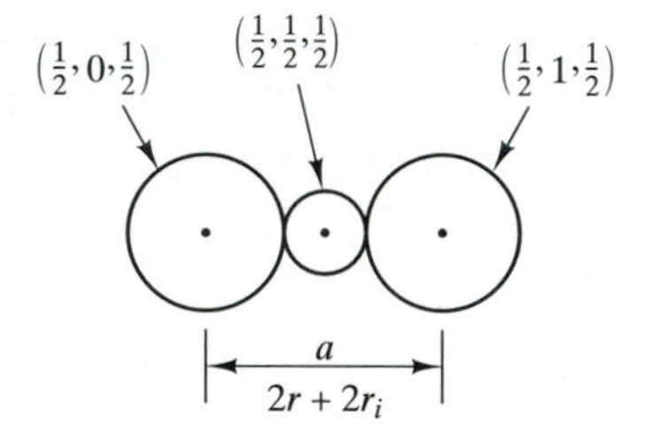

Fig. EP 2-2d

2.5 Covalent Structure — The Diamond Cubic Structure

The diamond cubic (DC) structure is displayed by the covalently bonded elements C, Si, and Ge and by crystals formed between atoms of similar electronegativities whose electron configurations are not closed to the closed shell configuration (inert gas). Examples are the advanced ceramics SiC and SiN, and the semiconductor compound, GaAs.

The DC structure has the FCC Bravais lattice and therefore displays the basic FCC lattice in Fig. 2-4. The difference between the basic FCC structure and the DC is the presence of four additional interior atoms located at four of the eight tetrahedral interstitial sites of the FCC structure in Fig. 2-9b and the tetrahedral bonding of these atoms that were shown in Fig. 1-12. From an origin, the vector coordinates of these occupied sites are ($\frac{1}{4}$, $\frac{1}{4}$, $\frac{1}{4}$), ($\frac{3}{4}$, $\frac{3}{4}$, $\frac{1}{4}$), ($\frac{3}{4}$, $\frac{1}{4}$, $\frac{3}{4}$), and ($\frac{1}{4}$, $\frac{3}{4}$, $\frac{3}{4}$) and are indicated by I for interior

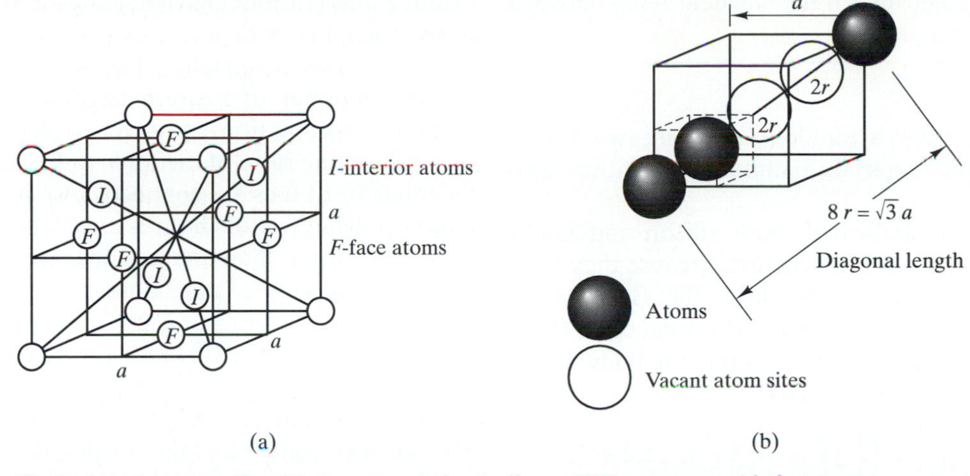

Fig. 2-12 (a) Diamond cubic structure is basically an FCC structure with four interior atoms in addition to the four regular FCC atoms/unit cell; (b) along the diagonal where atoms are in contact, two void atom sites exist.

atoms in Fig. 2-12a. Thus, the number of lattice points or atoms in the DC unit cell according to Eq. 2-3 is

$$N_{DC} = 4 + \frac{6}{2} + \frac{8}{8} = 8, \text{ atoms per unit cell.}$$

Technically, the DC structure is an FCC Bravais lattice with two atoms associated per lattice point. The coordination number in this structure is four, however, instead of 12 for the FCC with one atom per lattice point.

To estimate the lattice parameter of the DC unit cell from the atom size, we note that atoms touch along the body-diagonal between a corner atom and an interior atom at a vector location $<\frac{1}{4}\mathbf{a}, \frac{1}{4}\mathbf{a}, \frac{1}{4}\mathbf{a},>$ from the corner atom. The length or magnitude of this vector is $(\sqrt{3}/4)\,\mathbf{a}$ and the same length in terms of the atomic radius is 2r. Therefore,

$$\frac{\sqrt{3}}{4}\mathbf{a} = 2r,$$

from which we obtain,

$$\sqrt{3}\mathbf{a} = \text{body diagonal} = 8r. \qquad (2\text{-}7)$$

The atomic radius r in Eq. 2-7 is the covalent radius in Table 1-4. The 8r on the right side of Eq. 2-7 suggests that there are four atom diameters along the body diagonal of the DC structure. The body diagonal in the DC structure, shown in Fig. 2-12b with the locations of the atoms, contains only the equivalent of two atomic diameters and indicates that two atomic diameter spaces are vacant. This signifies that the DC structure is packed looser than the simple FCC structure and that the density of DC is much lower than the metallic structures.

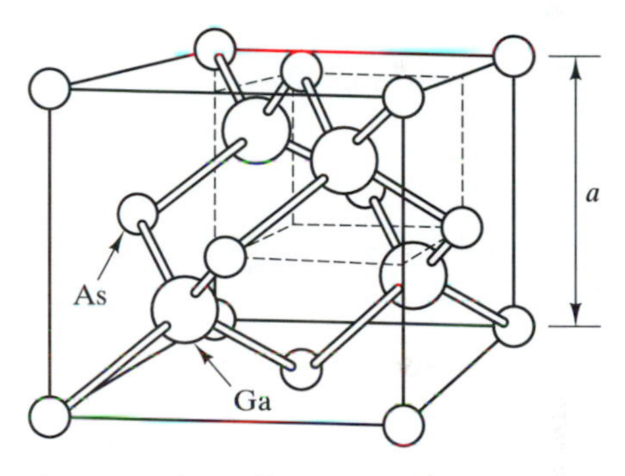

Fig. 2-13 The gallium arsenide (GaAs) structure

An example of a covalently bonded compound with a structure very similar to the diamond cubic structure is gallium arsenide (GaAs) shown in Fig. 2-13. Technically, GaAs has a zinc blende structure described in the ionic structures. However, it is included here because of the covalent bond between gallium and arsenic.

EXAMPLE 2-3

Silicon and gallium arsenide (GaAs) are two prominent electronic materials. Estimate their lattice constants.

SOLUTION: As indicated, both silicon and GaAs have the diamond cubic structure because they have predominantly covalent bonding. The GaAs structure is shown in Fig. 2-13. In the diamond cubic (DC) structure, atoms are in contact along the body diagonal and their centers are apart

$$\sqrt{\left(\frac{1}{4}a\right)^2 + \left(\frac{1}{4}a\right)^2 + \left(\frac{1}{4}a\right)^2} = \frac{\sqrt{3}}{4}a$$

For silicon,

$$\frac{\sqrt{3}}{4}\mathbf{a} = 2r$$

$$\mathbf{a} = \frac{8r}{\sqrt{3}}$$

For GaAs,

$$\frac{\sqrt{3}}{4}\mathbf{a} = \mathrm{r}_{\mathrm{Ga}} + \mathrm{r}_{\mathrm{As}}$$

$$\mathbf{a} = \frac{4}{\sqrt{3}}(\mathrm{r}_{\mathrm{Ga}} + \mathrm{r}_{\mathrm{As}})$$

All of the atomic radii are covalent radii from Table 1-4, that is,

$$r_{Si} = 1.17\ \text{Å},\ r_{Ga} = 1.25\ \text{Å},\ r_{As} = 1.21\ \text{Å}$$

from which

$$\mathbf{a}_{\mathrm{Si}} = 5.4040\ \text{Å} \qquad \mathbf{a}_{\mathrm{GaAs}} = 5.6811\ \text{Å}$$

2.6 *Ionic Crystal Structures — Ceramic Materials*

We now turn to structures of ceramic materials that are predominantly ionic compounds of metals and nonmetals (oxides, carbides, nitrides, borides, silicides, and halides). These structures are determined in large part on how positive and negative ions can be packed to maximize electrostatic attraction and minimize electrostatic repulsion. This means that positive ions (cations) have negative ions (anions) as nearest neighbors or are in contact with them, and vice versa. The most stable structure is that with the greatest number of nearest-neighbors or with the highest coordination number (CN); the latter depends on the ratio of the ionic radii. If the nearest-neighbor (n-n) ions are connected, we have a coordination polyhedron of the n-n ions. Since anions are generally larger than the cations, the ratio of the cation (+): anion (−) radii almost always determines the structure; and the first criterion for stability of ionic (ceramic) structures is for this ratio to be greater than or equal to a critical value for a given coordination number or polyhedron to be acquired. The critical ratios for the different coordination numbers and coordination polyhedra are given in Fig. 2-14, and although structures with lesser CN value can form, the most stable is that with the highest CN. Since the polyhedra enclose the cations, they are also referred to as cation coordination polyhedra (CCP). The ionic radii given in Table 1-4 may be used to determine the ratio. These values may vary, however, with the coordination number.

In addition to the critical radius ratio criterion, a stable ionic structure must also be electrically neutral, both on the macroscopic scale and on the atomic scale. The macroscopic scale simply means that the total positive charges must equal the total negative charges. The atomic scale involves the strength of local ionic bonds that are beyond the scope of this book. In addition, there are three more criteria, aside from charge neutrality and radii ratio, which are also beyond the scope of this text. These criteria deal with the linking of the polyhedra and the number of different constituents in the structure (see Kingery, et al, in the Reference section).

The Bravais lattice of an ionic structure is determined by the appropriate translation vector. The common translation vectors are: (a) the $(1, 0, 0)$ type (along the edge) for the simple cubic lattice, (b) the $(\frac{1}{2}, \frac{1}{2}, 0)$ type (face-diagonal) for the FCC lattice, and (c) the $(\frac{1}{2}, \frac{1}{2}, \frac{1}{2})$ type (body-diagonal) for the BCC lattice. To determine the lattice type, the appropriate translation vector must start and end at the same kind of ion. This will be illustrated when we discuss the specific ionic structures.

We shall now look at three of the simplest oxide ionic structures that are built up on the basis of nearly

Coordination number	Location of ions about central ion	Range of ratio cation radius to anion radius
8	Corners of cube	≥0.732
6	Corners of octahedron	≥0.414
4	Corners of tetrahedron	≥0.225
3	Corners of triangle	≥0.156
2	Linear	≥0

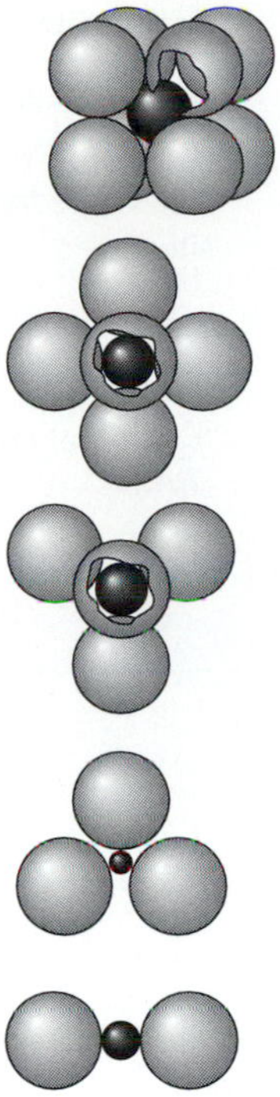

Fig. 2-14 Coordination number and cation coordination polyhedron as function of cation-to-anion ratio[Ref. 8].

close-packed oxygen ions with the cations placed in available interstices. The structure of silica is also discussed here because it is the basis of silicates—the most widely used ceramic, from window panes to chemical wares to refractories. The characteristics and examples of ionic oxide structures are summarized in Table 2-2. For the rock salt and zinc-blende structures, the cation, being the smaller ion, occupies the interstitial sites discussed in relation to Fig. 2-9.

2.6.1 The CsCl Structure

This is an example of an ionic structure with CN = 8 because the $\frac{r_{Cs^{+1}}}{r_{Cl^{-1}}} > 0.73$. This structure is shown in Fig. 2-15 in which the Cl^- ions are in a simple cubic array and the Cs^+ ion at the interstitial site (center of the cube). Since the Cs^+ ion is in contact with all the eight Cl^- ions, the coordination number is eight for both the cation and anion. At first glance, the structure can be mistaken as a BCC, but the structure is simple cubic. When we apply the BCC translation vector $(\frac{1}{2},\frac{1}{2},\frac{1}{2})\mathbf{a}$ to this apparent BCC looking structure, we start at a Cl^- ion and end at a Cs^+ ion. Thus, it is not a BCC lattice. When we apply the simple cubic translation vector $(1,0,0)\mathbf{a}$, we start and end at the same type of ion for either positive or negative ion. Therefore, the CsCl structure is *simple cubic*. The characteristics of the CsCl structure are:

- the coordination number is eight, the number of positive charge ions nearest a negative ion, or vice versa
- there is one positive ion and one negative ion per unit cell
- the ions are in contact along the body diagonal and the lattice parameter of the structure may be calculated from the body-diagonal relationship,

$$\sqrt{3}a = 2(r_+ + r_-) \tag{2-8}$$

2.6.2 The NaCl Structure

The sodium chloride (NaCl) or rock salt structure is shown in Fig. 2-16 where the large anions (Cl^-) are at the normal FCC lattice points and the cations (Na^+) are located at all the octahedral interstitial sites shown in Fig. 2-9a. From Fig. 2-9a, we see that the

Table 2-2 Simple Ionic Oxide Structures [8]

Structure Name	Anion Packing	Cation Sites	CN of M and O	Examples
Cesium Chloride	Simple cubic	All cubic sites	8 and 8	CsCl, CsBr, CsI
Rock Salt (NaCl)	Cubic close-packed (FCC)	All octahedral sites	6 and 6	NaCl, KCl, LiF, MgO, CaO, VO, BaO, FeO, NiO, MnO, CdO.
Zinc Blende	Cubic close-packed (FCC)	$\frac{1}{2}$ of tetrahedral sites	4 and 4	ZnS, BeO, SiC
Silica Types	Connected tetrahedra	———	4 and 2	SiO_2, GeO_2

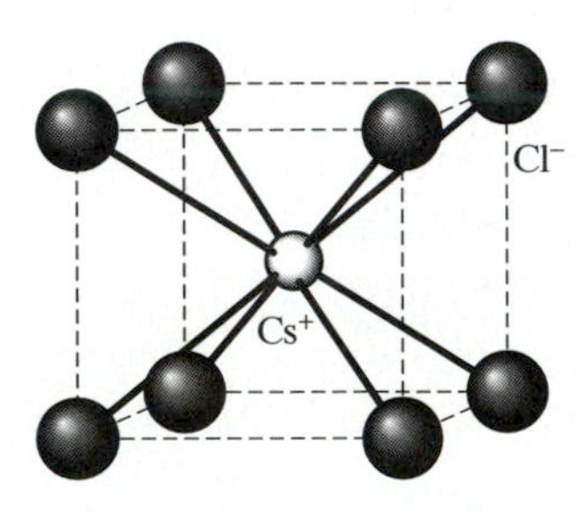

Fig. 2-15 The Cesium Chloride structure.

cation coordination polyhedron is an octahedron and the *coordination number* is 6 for both the cation and the anion. For stability, the radius ratio should be in the range 0.414 and 0.732. Some oxides shown in Table 2-2 having this structure are MgO (magnesium oxide), CaO (calcium oxide), SrO (strontium oxide), BaO (barium oxide), CdO (cadmium oxide), MnO (manganese oxide), FeO (ferrous oxide), CoO (cobalt oxide), and NiO (nickel oxide). The *structure* is FCC because the face-diagonal type translation vectors $(\frac{1}{2},\frac{1}{2},0)\mathbf{a}$ can be applied to both the Na^+ and Cl^- ions. In the normal FCC positions, there are 4 Cl^- ions and therefore the structure must also have 4 Na^+ ions (for *neutrality*). The Na^+ ion at the center of the cube accounts for one. The other 3 atoms come from the 12 atoms at the centers of the edges. Each of the edges is shared by four lattices. Thus the contribution of the atom at the edge to a unit cell is only $\frac{1}{4}$ and the number of atoms from the edges per unit cell is 12 x $\frac{1}{4}$ = 3. The ions are in contact along the cube edge directions and the lattice parameter is therefore

$$a = 2(r_+ + r_-) \tag{2-9}$$

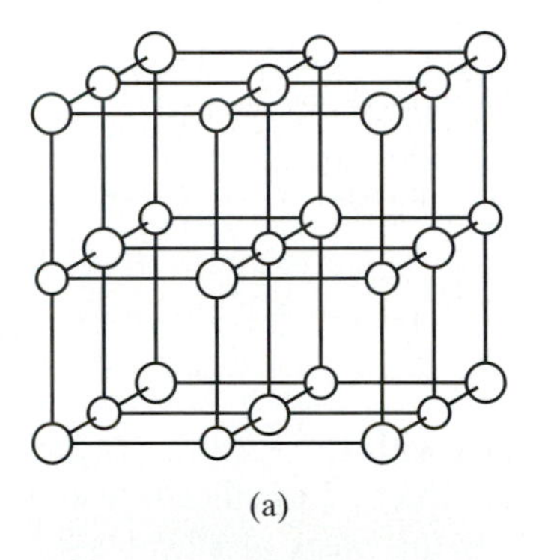

(a)

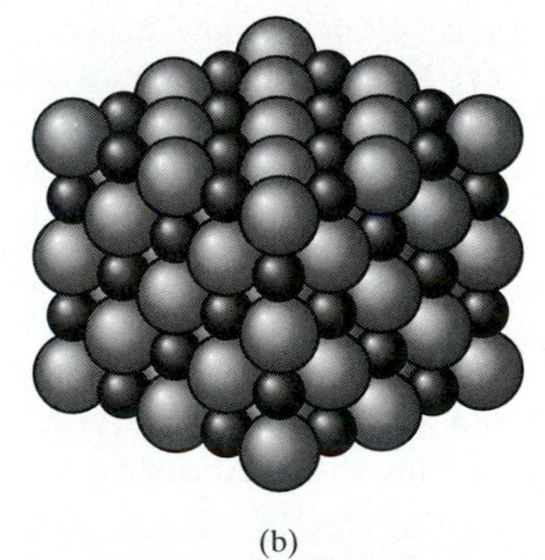

(b)

Fig. 2-16 The rock salt or sodium chloride structure.

2.6.3 The Zinc Blende Structure

The zinc blende structure is illustrated in Fig. 2-17 where the anions are in the normal FCC lattice points and the cations occupy four of the eight tetrahedral sites in Fig. 9-b. This is indicated in Table 2-2 as $\frac{1}{2}$ the tetrahedral sites in the "cation site" column.

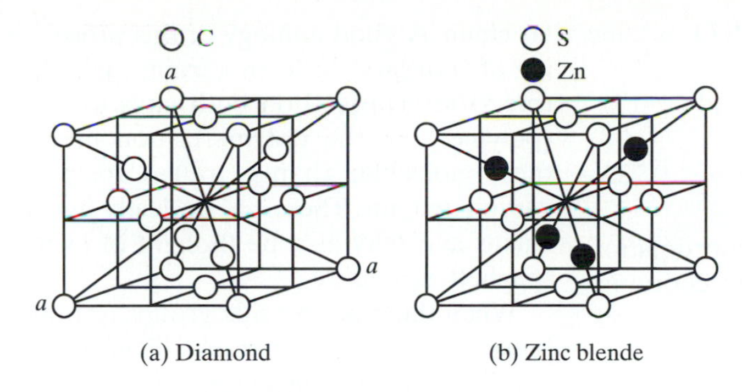

Fig. 2-17 The zinc blende structure in (b) compared with the diamond cubic structure in (a). Both structures are FCC.

The structure is in fact very similar to the DC structure, the exception being that the interior atoms in the DC are replaced by cations. Both the cations and the anions have tetrahedral coordination numbers (that is CN = 4) and the number of anions, as well as that of cations, is four for electrical neutrality. As shown in Fig. 9-b, the distance batween the centers of the two ions is $(\sqrt{3}/4)a$ along the body diagonal and this equates to the sum of the two ionic radii. Thus,

$$\frac{\sqrt{3}}{4}a = (r_+ + r_-) \tag{2-10}$$

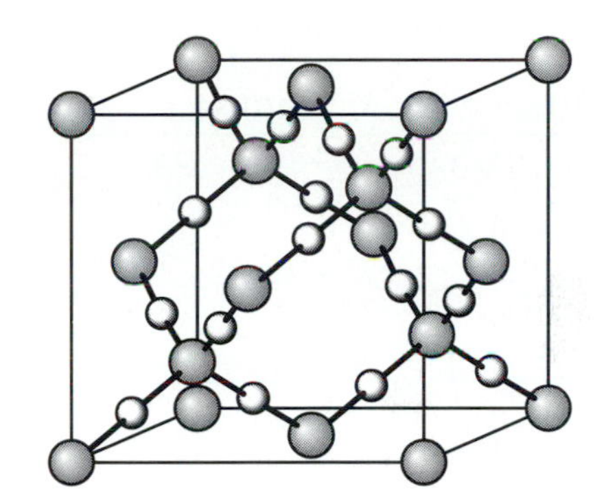

Fig. 2-18 The High (β) Cristobalite Structure.

2.6.4 The Silica Structure

The radius ratio of Si^{4+} to O^{2-} is 0.29 and indicates that the cation (Si^{4+}) requires a tetrahedral coordination, with four oxygen ions surrounding a central Si^{4+} cation. The SiO_4^{4-} tetrahedra can be linked such that the corners are shared in several ways to form various silicate glasses. A more detailed discussion of these silicates is deferred however to Chap. 14 on Ceramics and Glasses. We shall only discuss one form of the crystalline silica structure here.

The three basic structures of silica are quartz, tridymite, and cristobalite, each of which exists in two or three modifications. These modifications are called the allotropes or polymorphs of silica. The most stable forms are low (α) quartz, below 573°C; high (β) quartz, 573 to 867°C; high (β) tridymite, 867 to 1470°C; high (β) cristobalite, 1470 to 1710°C; and liquid, above 1710°C; the structure of high (β) cristobalite is shown in Fig. 2-18. This structure is obtained by starting with the diamond cubic structure of silicon and then inserting the oxygen ion between each pair of silicon. In general, the structure formed by the linking of the silicate tetrahedra leads to a more open structure or one with relatively more void space compared to the NaCl structure. Thus, the various forms of silica have relatively lower densities than MgO that has a density of 3.59 gm/cm^3; quartz has 2.65, tridymite 2.26, and cristobalite 2.32. This is in spite of the fact that the atomic mass of Si (28.08) is higher than that of Mg (24.30) and the molecular weight of SiO_2 is 60 compared to 40 for MgO.

EXAMPLE 2-4

Determine (1) the probable structure of BeO and MgO, and (2) their lattice parameters.

SOLUTION:

(1) The probable structure of ionically bonded solids is determined by the ratio of r_+/r_-. For BeO, $r_{Be^{2+}} = 0.34$ Å, and $r_{O^{2-}} = 1.35$Å from Table 1-4,

$$\frac{r_{Be^{2+}}}{r_{O^{2-}}} = \frac{0.34}{1.35} = 0.252, \qquad 0.225 < 0.252 < 0.414$$

that requires a coordination number of four from Fig. 2-14 and a tetrahedron cation coordination polyhedron. Therefore, the structure of BeO is zinc blende.

For MgO, $r_{Mg^{2+}} = 0.75$ Å, $r_{O^{2-}} = 1.35$ Å, and

$$\frac{r_{Mg^{2+}}}{r_{O^{2-}}} = \frac{0.75}{1.35} = 0.556, \qquad 0.414 < 0.556 < 0.732$$

which according to Fig. 2-14 requires a coordination number of six and an octahedron cation coordination polyhedron. Thus, the structure of MgO is rock salt or NaCl.

(2) The lattice parameters or constants are

For BeO, from Eq. 2-10,

$$a = \frac{4(r_{Be^{2+}} + r_{O^{2-}})}{\sqrt{3}} = \frac{4(0.34 + 1.35)}{\sqrt{3}}$$

$$= 3.9029 \text{ Å}$$

For MgO, from Eq. 2-9.

$$a = 2(r_{Mg^{2+}} + r_{O^{2-}}) = 2(0.75 + 1.35)$$

$$= 4.20 \text{ Å}$$

2.7 Crystallinity in Polymers

Figure 2-19 is the same as Fig. 1-17 and illustrates the conversion (polymerization) of a gaseous molecule ethylene into a macromolecular (polymer) solid by having a huge number of the starting molecules (mers) attached to each other to form a long molecular chain. A good analogy of this process is the linking of "boxcars" to form a train; each of the boxcars (mers) are coupled on both ends to form the train (polymer). In the polymerization process, a lot of these molecular chains (trains) are produced with varying lengths. These are basically linear molecular chains and may also be thought of as independent, spaghetti strings.

When there are no side groups or branches to the linear chains, these can stack and group together into parallel bundles to produce "crystalline" areas where the mers are perfectly aligned. In some instances, the molecular chains fold onto themselves as depicted in Fig. 2-20a, much like the folded electrical extension cords sold in hardware stores. The mers can again align themselves, although we can easily appreciate that the alignment of the mers cannot be perfect; thus, there is no perfect (100 percent) crystallinity in polymers as in metals. At best, a polymer is semicrystalline but the plastics industry calls it crystalline, nonetheless. As indicated earlier, the unique characteristic of a crystalline material is the ability to diffract. In polyethylene (PE), there is sufficient crystallinity to diffract X rays as in a metal crystal, and a unit cell can be defined as in Fig. 2-20b. Because the molecules are packed neatly in crystalline areas, the density of a polymer will be directly related to the amount of crystallinity. For example, low-density polyethylene (LDPE) with a density from 0.910 to 0.925 Mg/m^3 exhibits about 60 percent crystallinity while the high density variety (HDPE)

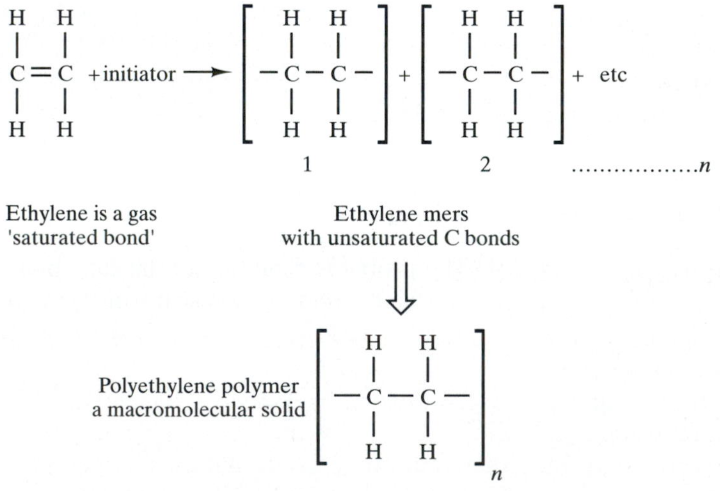

Fig. 2-19 The polymerization of ethylene to polyethylene (polymer).

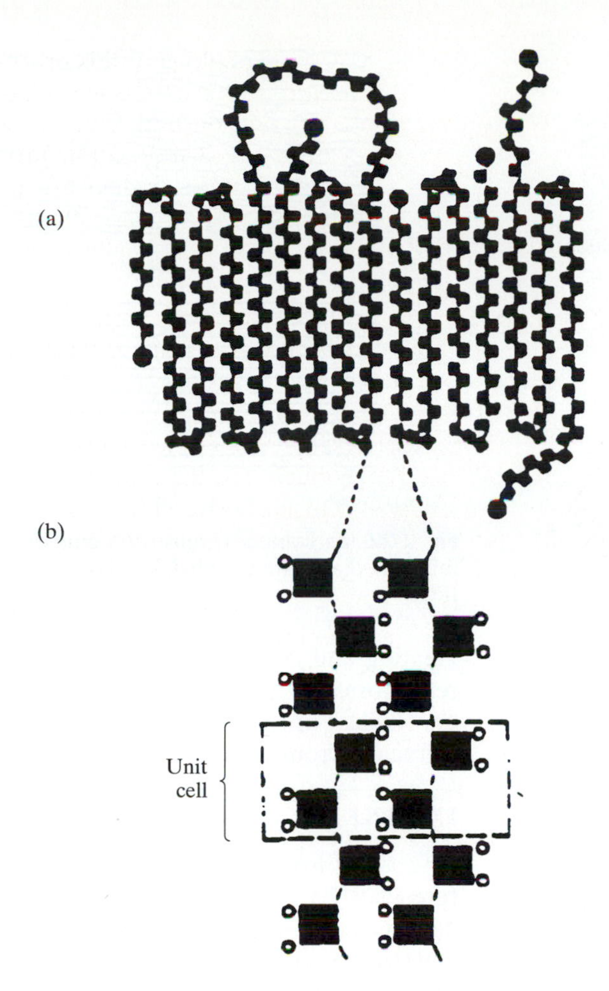

Fig. 2-20 Alignment of polymeric molecules to form (a) crystalline areas, and (b) unit cells [Ref. 1].

with a density from 0.941 to 0.965 Mg/m^3 exhibits a crystallinity of about 80 percent.

With the description of crystallinity in PE, what do you think is the characteristic of liquid crystalline polymers (LCPs)? If you say the polymeric chains are aligned in the liquid state, you are, of course, absolutely right! LCPs are a unique class of polymers composed of very stiff, rodlike molecular structures that are organized in large, ordered parallel arrays or domains in the liquid and solid states. The ordered domains diffract X rays and are, therefore, crystalline. Figure 2-21 shows the difference between the structure of flexible polymer molecules, such as PE, and that of a liquid crystalline polymer (LCP) in the liquid state. The stiff, rodlike LCP molecules do not bend and flex and can be aligned even in the liquid state. Figure 2-22 shows the LCP aligned fibers in (b) compared to the conventional polymers in (a) after extrusion through a spinneret. The straight, aligned fibers, with no folds in the structure, exhibit very high tensile strength as well as high tensile modulus. An example of a LCP is Kevlar that is used significantly in composite material fabrication.

2.8 *Allotropy or Polymorphism*

Allotropy or polymorphism is the ability of crystalline materials to assume two or more crystalline structures. Many metals and ceramic materials exhibit this phenomenon. Iron is a good example for metals and as discussed in Sec. 2.6.4, silica typifies ceramic materials; both materials are used extensively in

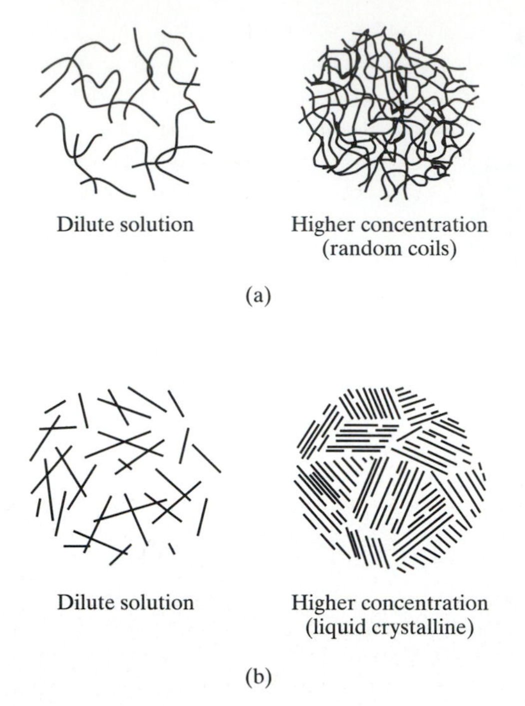

Fig. 2-21 Behavior of (a) common flexible polymeric chains compared to (b) stiff rod-like liquid crystalline polymers in liquid [Ref. 7].

engineering applications. Iron is the principal constituent in steels and cast irons and it assumes the BCC structure at low temperatures, the FCC structure at about 912°C, and then reverts back to the BCC at 1394°C. The alloys of iron (steels and cast irons) also exhibit these changes in structures, but at different temperatures depending on composition. Silica is used as a refractory in furnaces and it assumes the different forms stated earlier, which are: α-quartz to 573°C, β-quartz from 573 to 867°C, β-tridymite from 867 to 1470°C, and β-cristobalite from 1470 to 1710°C.

The significance of allotropy in engineering applications arises from the change in volume from one structure to another. This change in volume induces strains and stresses in materials while they are subjected to changes in temperature. When these strains and stresses are large, they may induce cracking of the materials. For example, in the welding of steels, the composition of the steel is carefully controlled so that the areas around the weld will not crack.

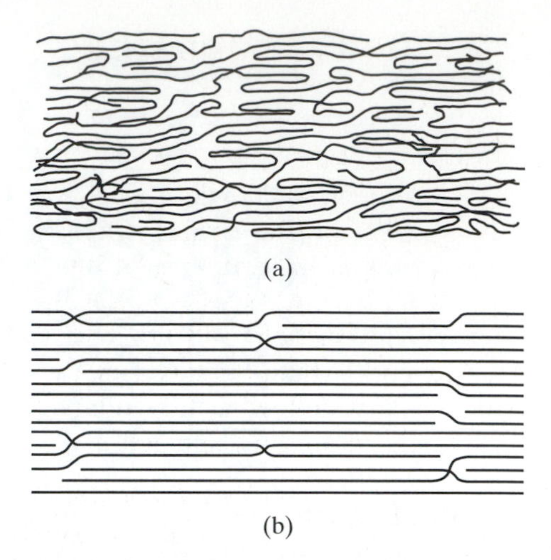

Fig. 2-22 (a) Folded chains in common polymers, and (b) aligned and straight chains in liquid crystalline polymer [Ref. 7].

Heating and cooling of the silica linings and other refractories in furnaces and in liquid metal containers (ladles) are done ever so slowly to prevent the refractory from fracturing (spalling).

EXAMPLE 2-5

On cooling, iron changes from an FCC to a BCC structure at 912°C. Estimate the change in volume and determine whether it is expansion or contraction.

SOLUTION: To estimate the change in volume at 912°C, we need lattice parameters of both structures at 912°C. In the absence of this high temperature data, we shall use room temperature data for both, i.e.,

$$a_{\text{FCC}} = 3.589\ \text{Å} \qquad \text{with 4 atoms per unit cell}$$

$$a_{\text{BCC}} = 2.866\ \text{Å} \qquad \text{with 2 atoms per unit cell}$$

and

$$\%\Delta V = \frac{V_{\text{BCC}} - V_{\text{FCC}}}{V_{\text{FCC}}} \times 100$$

The volumes must be based on the same number of atoms. Thus,

$$\%\Delta V = \frac{2 \times (2.866)^3 - (3.589)^3}{(3.589)^3} \times 100 = +1.845\%$$

Since $\%\Delta V$ is positive, it is an expansion from FCC to BCC.

2.9 *Packing Fraction and Density*

We have just learned the characteristics of the various metallic, covalent, and ionic crystal structures. We want now to see how the various structures affect one of the more important properties of engineering materials, that is, density. Density is dependent on the volume packing fraction (VPF) that reflects how the atoms are packed and arranged in the crystal structure. The VPF is defined as the volume fraction occupied by atoms or ions in a unit cell,

$$\text{Volume Packing Fraction} = \frac{\text{Volume of atoms (ions)}}{\text{Volume of unit cell}}$$

$$= \sum_{i=1}^{n} \frac{N_i \cdot \frac{4}{3}\pi r_i^3}{V_{uc}} \qquad \text{(2-11)}$$

where N_i is the total number of atoms or ions in the unit cell, and r_i may either be atomic or ionic radius. For the metallic and covalent structures with only one kind of atom, there is only one N. For the ionic structures, we need to add the total volume of each ion in the numerator. The volume of the unit cell, V_{uc}, is calculated from the lattice parameter(s) that is (are) estimated for each of the structures in terms of the atomic or ionic radii. Thus, we see that the density is also dependent on the atomic and ionic sizes. The calculated VPFs for the different structures with just one kind of atom are shown in Table 2-3, while Table 2-4 summarizes the formulas for the volume packing fractions and densities for the CsCl, the NaCl, and the zinc-blende structures. For the same atomic mass, a higher VPF means the crystal structure has a higher density. We see that both the FCC and the HCP yield the highest VPF; that is the reason why they are called the closest packed structures. It also says that the maximum packing hard spheres of the same size can attain is 74 percent of the volume. The BCC has a slightly lower VPF and thus will exhibit a lower density. The DC structure has a more open structure with a VPF equal to 0.34 and thus will be expected to have an even much lower density.

The other factor that determines the density is the mass of a material that in turn depends on the atomic masses of its constituents. In terms of the crystal structure, we estimate density, ρ, as follows:

$$\rho = \frac{\text{Mass of atoms/ions per unit cell}}{\text{Volume of unit cell}}$$

$$= \sum_{i=1}^{n} \frac{N_i \cdot \frac{M_i}{N_{Av}}}{V_{uc}} \qquad \text{(2-12)}$$

where N_i is number of atoms or ions per unit cell, M_i is the atomic mass, and N_{Av} is the Avogadro's number (6.0223×10^{23} atoms per mole).

Summarizing, we see that density depends on the atomic or ionic sizes, the volume packing fraction, and the atomic mass. Let us now assess the relative contributions of these factors to explain the range of densities of materials, which is from one to about 20 Mg/m^3 or g/cm^3. We noted in Chap. 1 that the atomic sizes do not change much from the light elements to the heavy elements. Thus, the atomic volumes are about the same. We note also in Table 2-3 that the VPF can vary by a factor of two. We see that this factor is not sufficient to explain the density range of materials. The factor that contributes most to the density is, therefore, the atomic mass. Metals are inherently heavy because of their high

Table 2-3 Volume Packing Fractions of Different Structures.

Structure	N, number of atoms/unit cell	Packing Fraction
BCC	2	0.68
FCC	4	0.74
HCP (unit cell)	2	0.74
(hexagon)	6 (three unit cells)	0.74
DC	8	0.34

Table 2-4 Characteristics of Simple Ionic Structures

Structure	Lattice parameter, a	Number of cations & anions	Volume packing fraction	Density
CsCl	$2(r_+ + r_-)/\sqrt{3}$	1 and 1	$4\pi(r_+^3 + r_-^3)/3a^3$	$(M_+ + M_-)/a^3$
NaCl	$2(r_+ + r_-)$	4 and 4	$16\pi(r_+^3 + r_-^3)/3a^3$	$4(M_+ + M_-)/a^3$
Zinc blende	$4(r_+ + r_-)/\sqrt{3}$	4 and 4	$16\pi(r_+^3 + r_-^3)/3a^3$	$4(M_+ + M_-)/a^3$

atomic masses and close-packing. Polymers and ceramics are relatively lighter because they are composed of light elements (C, H, O, Si, N, B, and others), in addition, of course to their more open structures.

EXAMPLE 2-6

Determine the volume-packing fraction (VPF) and estimate the density of (a) molybdenum, (b) gold, (c) cobalt, and (d) silicon.

SOLUTION:

(a) Molybdenum is BCC that has two atoms per unit cell, and from Example 2-1 the estimated a = 3.1408 Å, and the atomic mass (weight) is 95.94 g/mol. By definition,

$$\text{VPF} = \frac{\text{Volume of atoms per unit cell}}{\text{Volume of unit cell}}$$

$$= \frac{2 \times \frac{4}{3}\pi r^3}{\left(\frac{4r}{\sqrt{3}}\right)^3} \qquad \text{where } a = \frac{4r}{\sqrt{3}}$$

$$= \frac{\sqrt{3}\pi}{8} = 0.68$$

The estimated density of molybdenum is

$$\text{Density} = \frac{\text{mass of atoms per unit cell}}{\text{volume of unit cell}}$$

$$= \frac{2 \times \frac{95.94}{6.022 \times 10^{23}}}{(3.1408 \times 10^{-8})^3} = \frac{2 \times 95.94}{6.022 \times 3.098}$$

$$= 10.285 \text{ g/cm}^3$$

(b) Gold is FCC with 4 atoms per unit cell, estimated a = 4.0729 Å from Example 2-1, and atomic mass (weight) of 196.97 g/mol.

$$\text{VPF} = \frac{4 \times \frac{4}{3}\pi r^3}{\left(\frac{4r}{\sqrt{2}}\right)^3} \qquad \text{where} \qquad a = \frac{4r}{\sqrt{2}}$$

$$= \frac{\sqrt{2}\pi}{6} = 0.740$$

$$\text{Density} = \frac{4 \times \frac{196.97}{6.022 \times 10^{23}}}{(4.0729 \times 10^{-8})^3} = \frac{4 \times 196.97}{6.022 \times 6.756}$$

$$= 19.365 \text{ g/cm}^3$$

(c) Cobalt is HCP with two atoms per unit cell, estimated lattice parameters of a = 2.50 Å, and c = 4.0825 Å from Example 2-1, and atomic mass (weight) is 58.93 g/mol. The volume of the hexagonal unit cell is (a^2 c cos 30°).

$$\text{VPF} = \frac{2 \times \frac{4}{3}\pi r^3}{a^2 c \cos 30°} = \frac{\frac{8}{3}\pi r^3}{4r^2\, 1.633 \times 2r \times \frac{\sqrt{3}}{2}}$$

$$= \frac{2\pi}{1.633 \times 3 \times \sqrt{3}} = 0.740$$

$$\text{Density} = \frac{2 \times \frac{58.93}{6.022 \times 10^{23}}}{(2.50)^2 \times 4.0825 \times 10^{-24} \times \frac{\sqrt{3}}{2}}$$

$$= \frac{2 \times 58.93}{6.022 \times 2.210} = 8.856 \text{ g/cm}^3$$

(d) Silicon is diamond cubic with eight atoms per unit cell, the estimated lattice constant a = 5.404 Å, and atomic mass of 28.085.

$$\text{VPF} = \frac{8 \times \frac{4}{3}\pi r^3}{\left(\frac{8r}{\sqrt{3}}\right)^3} \quad \text{where} \quad a = \frac{8r}{\sqrt{3}}$$

$$= \frac{\sqrt{3}\pi}{16} = 0.34$$

$$\text{Density} = \frac{8 \times \frac{28.085}{6.022 \times 10^{23}}}{(5.404 \times 10^{-8})^3} = \frac{2 \times 28.085}{6.022 \times 15.781}$$

$$= 2.364 \text{ g/cm}^3$$

EXAMPLE 2-7

Determine the volume-packing fraction and estimate the densities of BeO and MgO. Use the lattice constants from Example 2-4, such as, a = 3.9029 Å for BeO, and a = 4.20 Å for MgO.

SOLUTION: For BeO, structure is zinc blende, with 4 Be^{2+} ions and 4 O^{2-} ions.

$$r_{Be^{2+}} = 0.34 \text{ Å}, \qquad r_{O^{2-}} = 1.35 \text{ Å}$$

$$\text{VPF} = \frac{4 \times \frac{4}{3}\pi[(0.34)^3 + (1.35)^3] \times (10^{-8})^3}{(3.9029 \times 10^{-8})^3}$$

$$= \frac{16\pi(0.0393 + 2.460)}{3 \times 59.451} = 0.704$$

$$\text{Density} = \frac{4\frac{(9.012 + 16)}{6.022 \times 10^{23}}}{(3.9029 \times 10^{-8})^3}$$

$$= \frac{4 \times 25.012}{6.022 \times 5.945} = 2.795 \text{ g/cm}^3$$

For MgO, structure is rock salt, $r_{Mg^{2+}} = 0.75$ Å, $r_{O^{2-}} = 1.35$ Å; 4 Mg^{2+} and 4 O^{2-} ions.

$$\text{VPF} = \frac{4 \times \frac{4}{3}\pi[(0.75)^3 + (1.35)^3](10^{-8})^3}{(4.20 \times 10^{-8})^3} = 0.652$$

$$\text{Density} = \frac{4 \times \frac{(24.305 + 16)}{6.022 \times 10^{23}}}{(4.20 \times 10^{-8})^3} = 3.613 \text{ g/cm}^3$$

2.10 Directions and Planes in Crystals—Miller Indices

The objective of this chapter, as stated in the Introduction, was to determine how the crystal structure influences the density and deformability of materials. In the last section, we have accomplished the first of the two objectives, that is, the density was related to the volumetric characteristics of the crystal structures (unit cells) and the volumetric atomic packing fraction. The deformability of materials, on the other hand, is dictated by their planar and linear atomic fractions. The deformation of crystalline materials occurs by shear (sliding of planes) only on specific crystal planes and only along specific directions, both depending on the crystal structure. There are also other properties, such as magnetization, that depend on crystal planes and directions. As a result, it is customary to indicate properties of materials on certain planes and along certain directions that are identified by numbers called *Miller indices*. Thus, it is important to learn about Miller index notations of directions and planes.

2.10.1 Directions

We start with directions because these are related to the unit vector translations of the unit cell discussed at the start of this chapter. We recall that when the unit vectors, **a**, **b**, and **c**, of the parallelipiped unit cell in Fig. 2-2 were translated, the point space lattice of Fig. 2-3 was established. Each point in this lattice can be represented by a vector and its coordinates (u, v, w) signify the vector

$$\mathbf{P} = u\mathbf{a} + v\mathbf{b} + w\mathbf{c}$$

from a chosen origin, as shown in Fig. 2-23. Since a vector has both direction and magnitude, a direction in the space lattice is a vector. The direction is identified with the vector coordinates (u,v,w) of P from a chosen origin and they are taken as the Miller indices of the direction from the origin to P and are enclosed

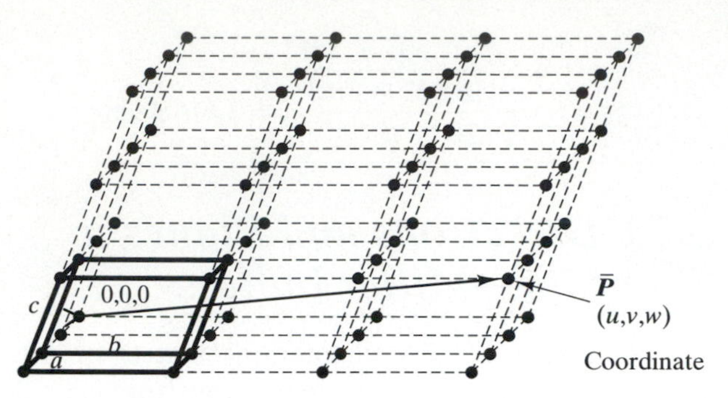

Fig. 2-23 A direction in a point space lattice is a vector represented by the vector P with coordinates (u,v,w).

in a square bracket as [uvw]. The numbers u, v, and w must be the smallest primed integers. Thus, [$\frac{1}{2}$ $\frac{1}{2}$ 1], [112], and [224] all represent the same direction, but [112] is the preferred form. Fractions are removed by multiplying with the least common denominator. Positive numbers signify translations of the unit axes in the positive direction and negative numbers signify translations of the axes in the negative direction. A negative number in the Miller index notation is indicated by a bar over the number. For example, (1, −2, 3) is indicated as $[1\bar{2}3]$.

Since a direction is represented by a vector, the direction between two crystal lattice points or vectors, P_1, and P_2, is obtained by the vector difference of the two vectors, taking into consideration the start and end points indicated by the tail and head, respectively, of the arrow in Fig. 2-24. Thus, the direction **P** between $\mathbf{P}_1$ and $\mathbf{P}_2$ is

$$\mathbf{P} = \mathbf{P}_1 - \mathbf{P}_2 \text{ (direction from } \mathbf{P}_2 \text{ to } \mathbf{P}_1\text{), or} \tag{2-13a}$$

$$= \mathbf{P}_2 - \mathbf{P}_1 \text{ (direction from } \mathbf{P}_1 \text{ to } \mathbf{P}_2\text{)} \tag{2-13b}$$

$$= [uvw]_{Head} - [uvw]_{Tail} \tag{2-13c}$$

We illustrate Eq. 2-13 with the two vectors $\mathbf{P}_1$ and $\mathbf{P}_2$ and their respective coordinates, (5, 1) and (1, 4)

$\bar{P}_1$ $\vec{P}$ $\bar{P}_2$

$[uvw]_{Tail}$ $[uvw]_{Head}$

$\vec{P} = [uvw]_{Head} - [uvw]_{Tail}$

Fig. 2-24 The direction (arrow) between any two points in a space lattice is the difference between the two vectors represented by those points.

in the two-dimensional lattice shown in Fig. 2-25, with unit axes **a** in x direction and **b** in the y direction. The direction **P** as indicated is

$$\mathbf{P} = (1-5)\mathbf{a} + (4-1)\mathbf{b} = -4\mathbf{a} + 3\mathbf{b}, \text{ or in short, } [\bar{4}3],$$

which signifies the vector coordinates from an origin. If we take advantage of an earlier statement relating to a free choice of origin to solve a particular problem, these vector coordinates are in fact those of point $\mathbf{P}_2$ if we choose $\mathbf{P}_1$ as the origin. The path taken by **P** from $\mathbf{P}_1$ is four negative translations of **a** and three positive translations of **b**, also indicated in Fig. 2-25. The reverse direction −**P** has, of course, the coordinates or direction $[4\bar{3}]$ and this is identified as taking four positive steps of **a** and three negative steps of **b** in Fig. 2-25 with $\mathbf{P}_2$ as the origin. The **P** direction would, of course, be the same had we translated them to the initial origin chosen, O. This signifies that parallel directions have the same Miller indices.

The above example can be universally applied and can be carried over to three-dimensional lattice points or vectors (u, v, w). Therefore, it is not necessary to move the direction to the initial common origin, O, to get the Miller indices of the direction. It must only be remembered that the indices of the resultant direction must be the least primed integers and cleared of fractions, if there are any.

Let us now consider the specific case of a cubic unit cell, Fig. 2-26, whose corners are identified with their *vector coordinates from a common origin.* We see that the direction from the origin along the +x axis is [100], along the +y [010], and along the +z [001]. The directions along the negative axes from the origin are, respectively, $[\bar{1}00]$, $[0\bar{1}0]$, and $[00\bar{1}]$. There are six specific directions; *the Miller indices of each of the specific direction, by convention, is enclosed with square brackets.* We note that all the six directions are along the edges of the cube and all have the same length. Their Miller indices arise from the specific choice of origin and axes. Using the same cube as in Fig. 2-26, we can rename the z axis as the x axis and without changing the cube, the direction would change from [001] to [100]. But that edge of the cube has not changed! The point is that all the edges in the cube are the same. They are examples of directions in the unit cell that are related by symmetry and are called *directions of a form or family of*

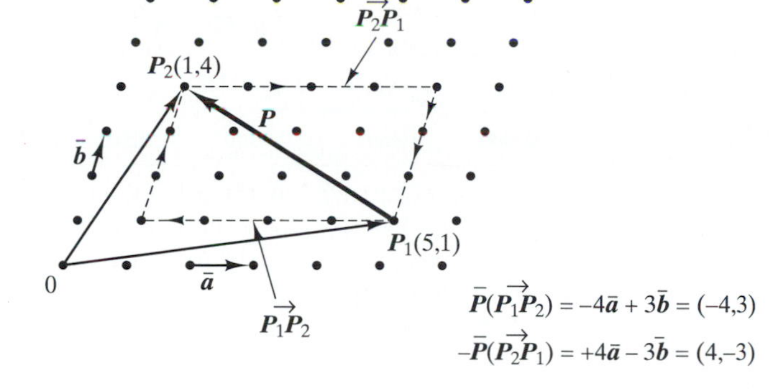

Fig. 2-25 Graphical illustration of Eq. 2-13 in two-dimensional point space lattice.

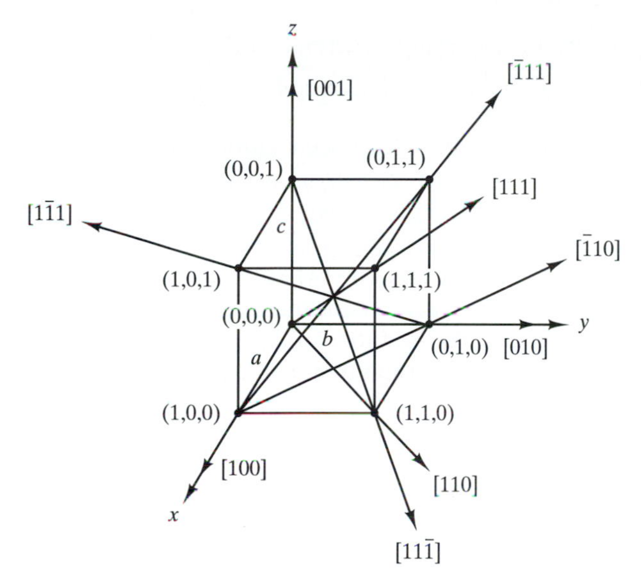

Fig. 2-26 The <100> (edges) and the <111> (body diagonal) directions in a cubic unit cell and two <110> directions.

directions. The six directions along the edges are a set of these directions whose Miller indices all have the same numbers. A set of directions is represented by taking any one of the Miller indices in the set and enclosing it within *angular brackets*. It is customary to choose one of the positive indices and in this case the directions along the edges of the cube are of the form <100>.

Figure 2-26 also shows the directions along the body diagonals as well as some directions along the face diagonals. The direction [111] is the direction from the origin (0,0,0) to the vector coordinates (1,1,1). The identification of the other body diagonal illustrate the application of Eq. 2-13. Thus, the directions of the other body diagonals, as shown in the figure, are obtained by

$$[1\bar{1}1] = [101] - [010]$$

$$[11\bar{1}] = [110] - [001], \text{ and}$$

$$[\bar{1}11] = [011] - [100].$$

(We recall in vector addition or subtraction that we add or subtract only the respective components.) Thus, we see that there are four body diagonal directions, (eight if we consider their reverse directions) which are [111], $[\bar{1}11]$, $[1\bar{1}1]$, and $[11\bar{1}]$. These directions constitute the set of body diagonal directions and are of the form <111>.

The other directions in Fig. 2-26 are the two directions along the face diagonal, $[\bar{1}10]$ and [110], shown on the bottom cube face. These directions have the same Miller indices as those face diagonal directions on the top face of the cube in Fig. 2-27; this is an example of the principle stated above that parallel directions have the same Miller indices. Thus, the top and bottom faces have two face diagonal directions (four if the reverse directions are considered). There are also two face diagonal directions on the left and right faces, as well as two from the front and back cube faces. There are, therefore, six face diagonal directions (12 if the reverse directions are included), which are all shown in Fig. 2-27. The specific Miller indices of each direction are obtained in the same manner as we obtained the body diagonal directions above. For example, $[\bar{1}01] = [011] - [110]$. The face diagonal directions constitute a set of directions that are of the form <110>.

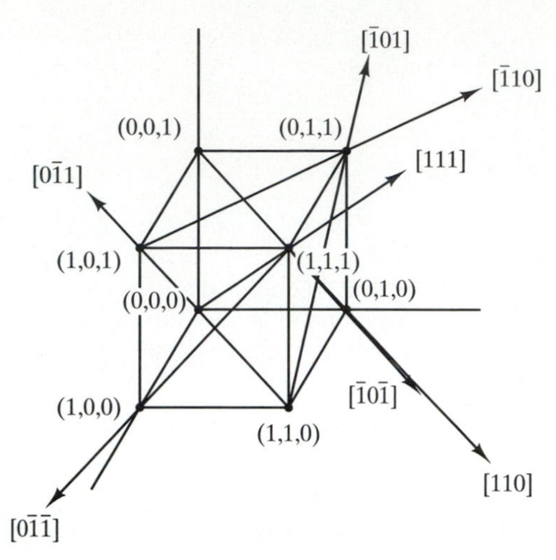

Fig. 2-27 The <110> family of directions in a cubic unit cell.

EXAMPLE 2-8

Determine the Miller indices of the directions shown within the cubic unit cell in Fig. EP 2-8 below.

SOLUTION: To determine the directions, we need to know the vector coordinates of the head and tail of each of the directions (arrows) to apply Eq. 2-13. We should remember that these coordinates must have the same origin. If we choose O as the origin, the vector coordinates are shown below.

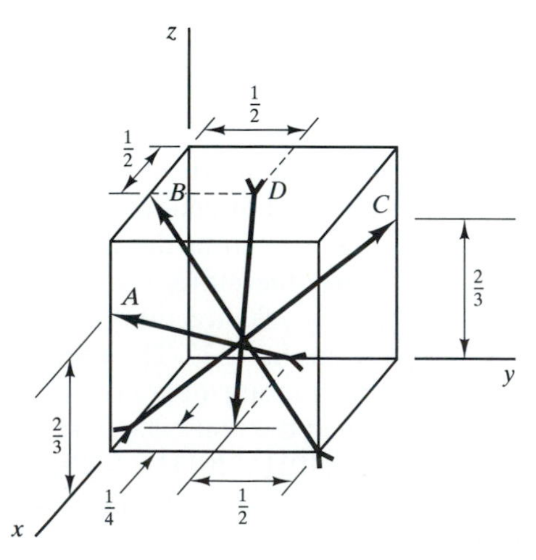

Fig. EP 2-8

Table EP 2-8

Direction	Vector Coordinates: Head	Vector Coordinates: Tail	[Head]−[Tail]	Miller Indices
A	(1, 0, 2/3)	(0, 1/2, 0)	[1, −1/2, 2/3]	[6 $\bar{3}$ 4]
B	(1/2, 0, 1)	(1, 1, 0)	[−1/2, −1, 1]	[$\bar{1}$ $\bar{2}$ 2]
C	(0, 1, 2/3)	(3/4, 0, 0)	[−3/4, 1, 2/3]	[$\bar{9}$ 12 8]
D	(3/4, 1/2, 0)	(1/2, (1/2, 1)	[(1/4, 0, 21]	[1 0 $\bar{4}$]

EXAMPLE 2-9

Sketch the following directions within a cubic unit cell.

(a) [1 2 $\bar{2}$], (b) [2 1 $\bar{3}$], (c) [$\bar{3}$ 0 1]

SOLUTION: We recall that directions are vector coordinates of a vector from a chosen origin in the point space lattice. Since we need directions within a unit cell, the translations of the unit of crystallographic axes must be one (1) or less. Thus, we need to transform the vector coordinates to one or less by dividing the coordinates of each direction by the largest number of the three. Thus,

Point Lattice Coordinates		Unit Cell Coordinates
(1, 2, $\bar{2}$)	Divide by 2 to get	($\frac{1}{2}$, 1, $\bar{1}$)
(2, 1, $\bar{3}$)	Divide by 3 to get	($\frac{2}{3}$, $\frac{1}{3}$, $\bar{1}$)
($\bar{3}$, 0, 1)	Divide by 3 to get	($\bar{1}$, 0, $\frac{1}{3}$)

We can sketch these directions either by using a different origin in the unit cell for each direction and then make the fractional translations of the unit axes, or by using the same origin and applying Eq. 2-13, Fig. EP 2-9. In the latter, the negative coordinate is made the tail coordinate. For example, in the first direction the head coordinate is ($\frac{1}{2}$, 1, 0) while the tail coordinate is (0, 0, 1).

2.10.2 Planes

Planes in the crystal lattice are also identified by *Miller indices*; this time as h, k, and l enclosed in parenthesis, as in (hkl), for a specific plane. Just as in crystal directions, there are *planes of a form*, which are equivalent and are related by symmetry. This set (family) of planes (of a form) is identified by enclosing one of the Miller indices in the set in braces, as in {hkl}.

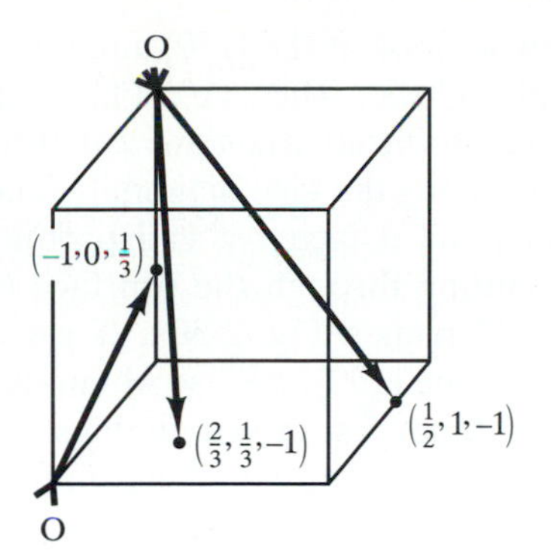

Method A. Change origin for direction

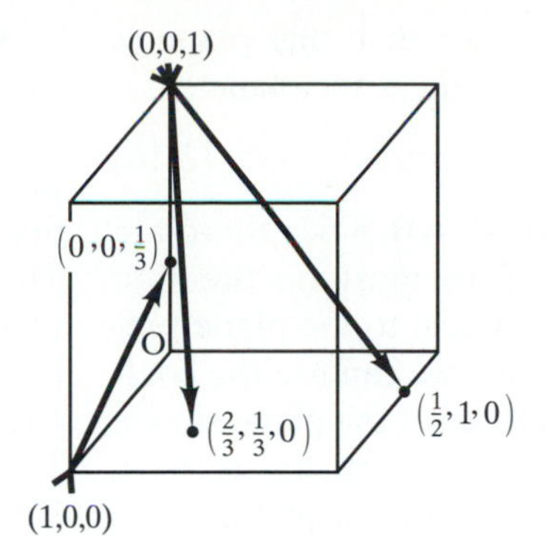

Method B. Common origin and using Eq. 2-13

Fig. EP 2-9

We learned that the Miller indices of a direction are the vector coordinates of a point or vector in the space lattice. Now, we ask, "What do the Miller indices of a plane signify?" To answer this question, we recall that planes generally have intercepts with the axes. For example, the rational plane, shown in Fig. 2-28a with intercepts A (3a), B (4b), and C (2c) on the three noncollinear axes, has the following equation in intercept form,

$$\frac{x}{A} + \frac{y}{B} + \frac{z}{C} = 1 \qquad (2\text{-}14)$$

The coefficients of x, y, and z in Eq. 2-14 identifies a particular plane; therefore, they can be used to identify the plane. The problem is they are fractions. We need to transform Eq. 2-14 such that the coefficients are integers and then the coefficients can be used as the indices of a plane. To do this we need to use two theorems from geometry.

- **Theorem 1.** For a rational plane with intercepts A, B, and C, there are ABC equally spaced, identical planes from the origin to this rational plane (provided that a common factor does not occur in any of the pairs AB, AC, or BC).
- **Theorem 2.** The equation $(ax + by + cz = m)$ represents a plane m times as far from the origin as $(ax + by + cz = 1)$.

With the above, we can multiply Eq. 2-14 by ABC, and obtain

$$\text{BC}\,x + \text{AC}\,y + \text{AB}\,z = \text{ABC} \qquad (2\text{-}15)$$

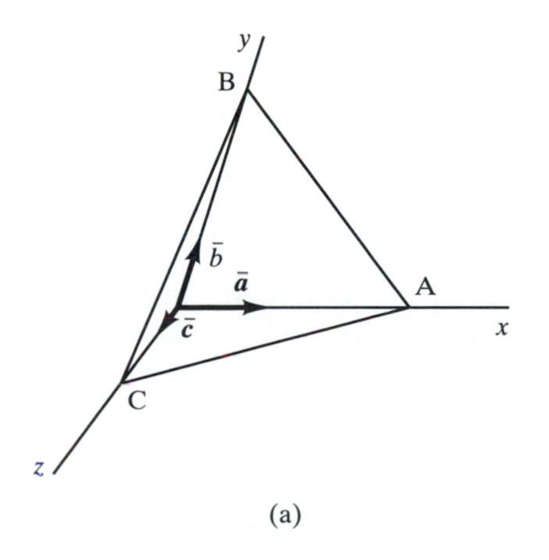

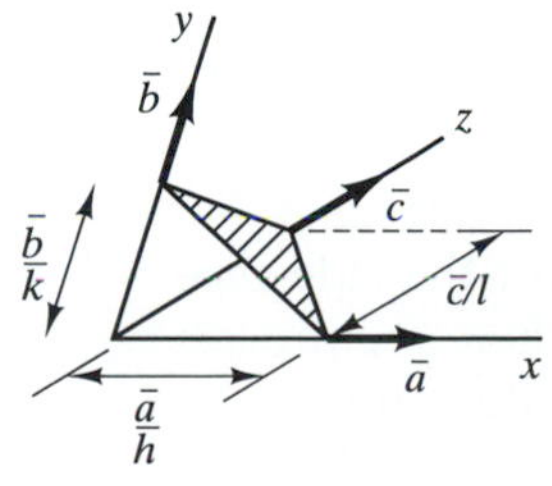

Fig. 2-28 (a) Rational plane passing through lattice points A (3a), B(4b), and C(2c), and (b) the first plane from the origin of the set of planes produced by plane in (a); intercepts of plane are, respectively, the unit axes divided by the Miller indices.

Then, using Theorem 2, we see that this plane is (ABC) times as far from the origin as the plane

$$hx + ky + lz = 1 \quad (2\text{-}16)$$

where $h = BC$, $k = AC$, and $l = AB$. Since there are ABC identical planes, Eq. 2-16 must be the first plane from the origin and identical to the plane represented by Eq. 2-14. Therefore, we can use the coefficients h, k, l as the Miller indices of the plane shown in Fig. 2-28a and enclose them in parenthesis as (hkl).

When we express Eq. 2-16 in intercept form, we obtain

$$\frac{x}{\left(\frac{1}{h}\right)} + \frac{y}{\left(\frac{1}{k}\right)} + \frac{z}{\left(\frac{1}{l}\right)} = 1 \quad (2\text{-}17)$$

Thus, we see that the Miller indices are the reciprocals of the fractional intercepts, $1/h$, $1/k$, and $1/l$, on the respective crystallographic axes of the first plane from the origin. If the axial lengths are a, b, and c, the actual intercepts of the plane are a/h, b/k, and c/l, as shown in Fig. 2-28b.

Thus, *to get the Miller indices (hkl) of a plane*, we

(1) identify the intercepts of the plane on the crystallographic axes,
(2) take the reciprocals of the intercepts,
(3) clear the fractions, if there are any, and
(4) represent the Miller indices as the smallest set of hkl enclosed in parenthesis, that is, (hkl).

If the plane is parallel to any of the axes, the intercept on that axis is assumed to be at infinity and the reciprocal is 0. A plane with Miller indices that are negative to those of a second plane is parallel to and is the same as the second plane.

We now look at the low index planes of the cubic structure. These are shown in Figs. 2-29, 2-30, and 2-31 for the {100}, {110}, and the {111}, respectively. In Fig. 2-29, the intercepts of the front face are $x = 1$, $y = \infty$, and $z = \infty$ and their reciprocals give the Miller indices (100). The intercepts of the side face are $x = \infty$, $y = 1$, and $z = \infty$, and their reciprocals yield the Miller indices (010). The intercepts of the top face are $x = \infty$, $y = \infty$, and $z = 1$, and thus the Miller indices (001). Since these planes represent the faces of the cube, they constitute a set or family of planes, that is, the {100}.

Now we look at the {110} and imagine solid plastic toy cube blocks. The {110} planes are the surfaces that are produced when we cut through the blocks into two along the face diagonals. Since there are two diagonals on a face, we can produce two surfaces. Thus, cutting through the top face (001), we obtain two {110} planes, Fig. 2-30a. If we cut through the (010) and the (100), we also obtain two surfaces from each of them, as shown in Figs. 2-30b and 2-30c. There are, therefore, six planes in the set or family of {110} in the cubic structure.

In Fig. 2-30, the letters O, x, or y in parenthesis next to the Miller indices of specific planes signify the origin chosen in the unit cell to be able to identify the intercepts of the planes. This is in reference to a statement made earlier that the unit cell does not have a specific origin, which means we can move the origin to solve the problem. In planes, we need to move the origin to get nonzero intercepts. In Fig. 2-30a, we use point O as the origin for the (110) and with this we see the intercepts on the $x = 1$, $y = 1$ and $z = \infty$, whose reciprocals yield the (110) indices. For the ($\bar{1}$ 1 0), we move the origin to the point marked x and see the intercepts at $x = -1$, $y = 1$, and $z = \infty$ from which we obtain the indicated Miller indices. With this example, the indicated Miller indices of the other four planes in Figs. 2-30b and 2-30c can be easily verified.

The four planes (111), ($\bar{1}$11), ($\bar{1}\bar{1}$1), and (1$\bar{1}$1) in a cubic structure, drawn in their quadrants in Figs. 2-31a,

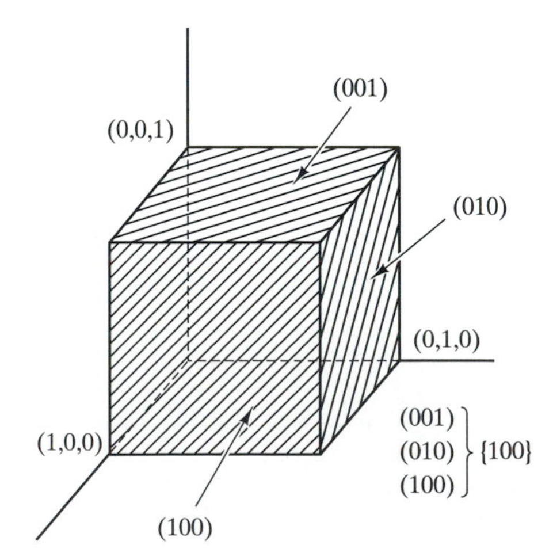

Fig. 2-29 The faces of a cube constitute the family of {100} planes.

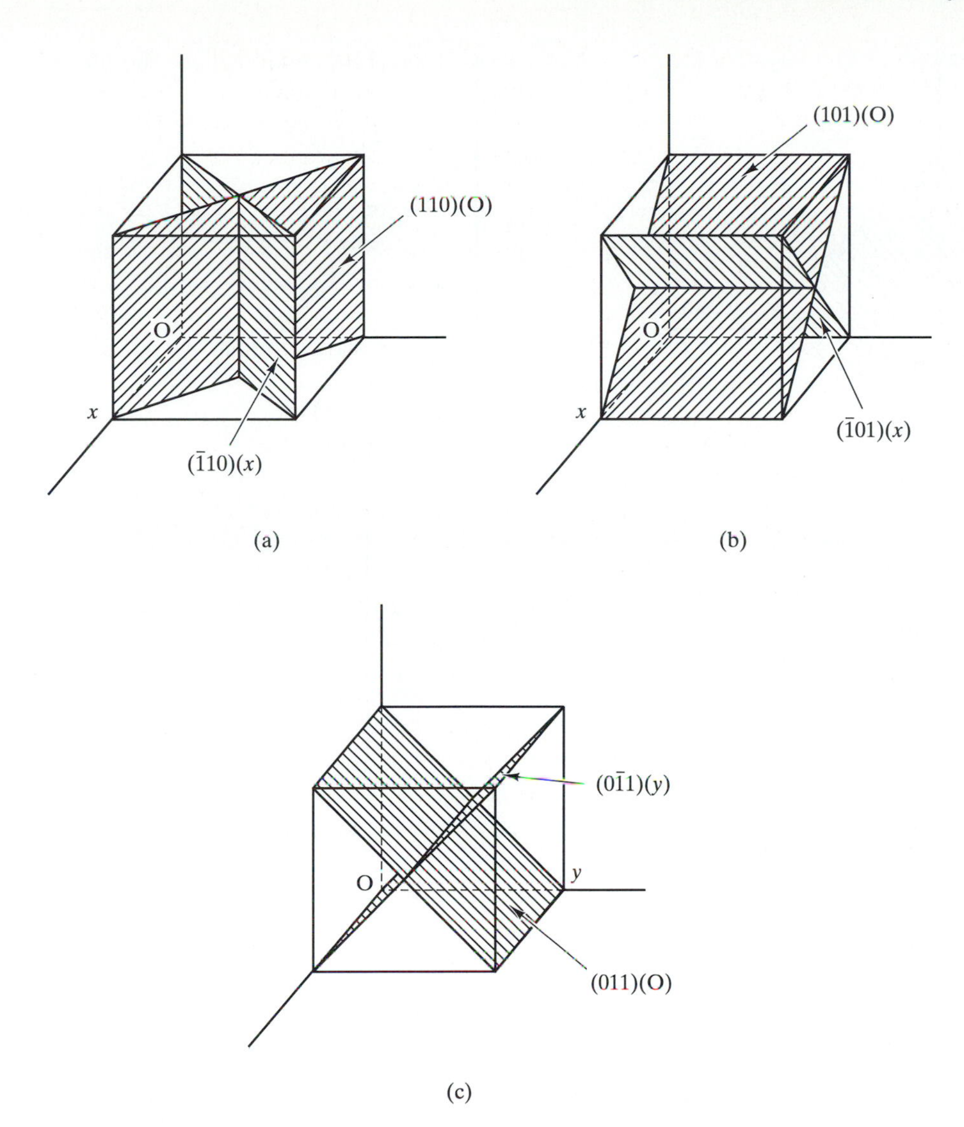

Fig. 2-30 The six {110} planes in a cubic unit cell.

2-31b, 2-31c, and 2-31d, constitute the set of {111} planes. Figure 2-31e is the top view of all the planes put together. The view is that of four equilateral triangular faces of the top portion of a polyhedron. The bottom portion of the polyhedron also has the same four equilateral triangular faces. When joined, the top and bottom portions yield an octahedron—a polyhedron with eight sides. For this reason, the {111} are also referred to as octahedral planes.

EXAMPLE 2-10

Determine the Miller indices of the planes A, B, and C shown in the cubic unit cells in Fig. EP-10.

SOLUTION: To determine the Miller indices, we must know the intercepts of the plane on the axes because the reciprocals of these intercepts yield the Miller indices. In order to get the intercepts, we must choose an appropriate origin for each plane. These are marked O_A, O_B, and O_C, respectively, for planes A, B, and C, in Fig. EP 2-10.

For Plane A with origin at O_A,

Intercepts	$x = 3/4$	$y = -1/2$	$z = -3/4$
Reciprocal	$4/3$	-2	$-4/3$
Clear fractions → $(h\,k\,l)$	4	-6	-4
least $(h\,k\,l)$		$(2\,\bar{3}\,\bar{2})$	

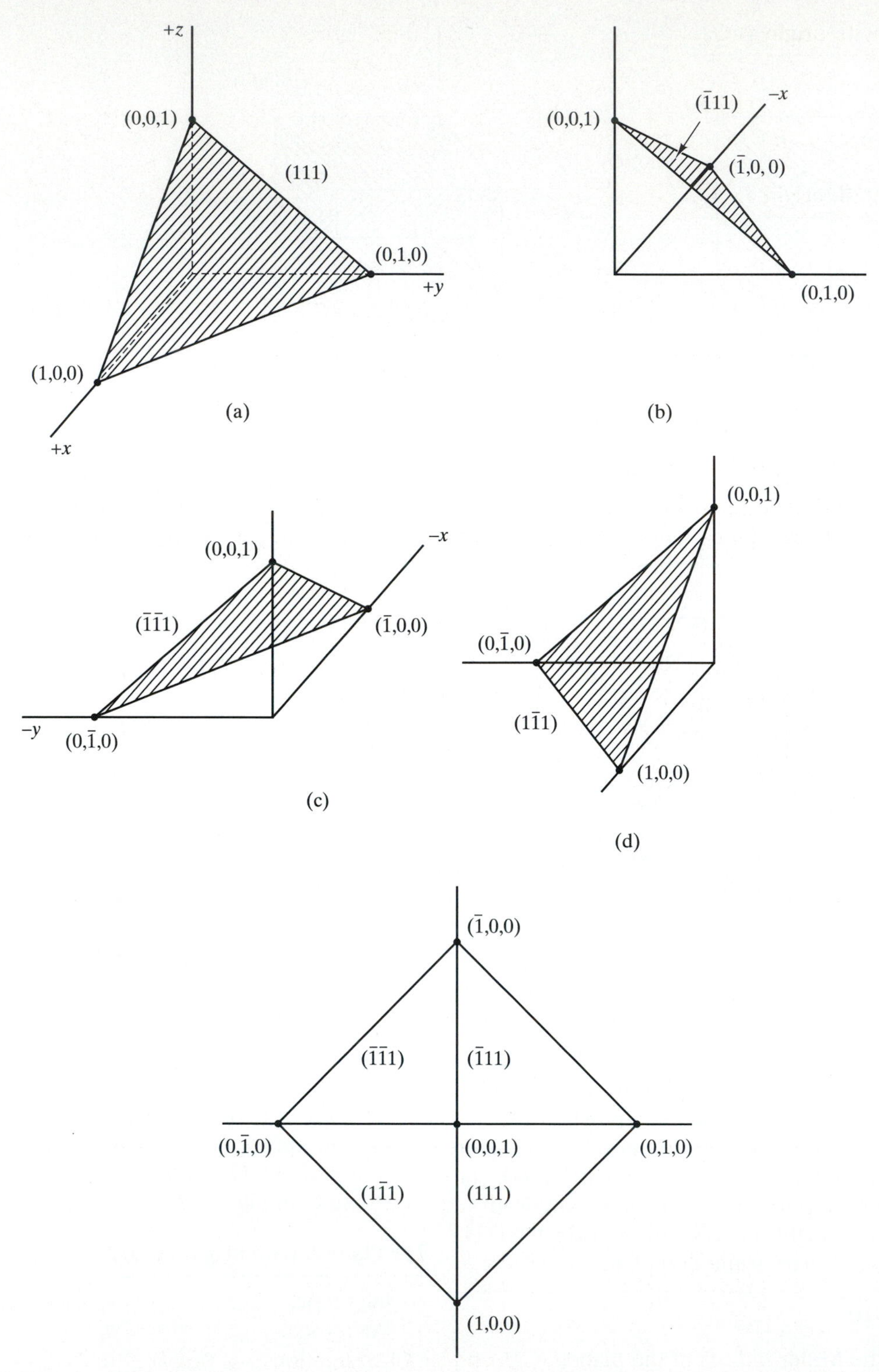

Fig. 2-31 The four {111} planes (a), (b), (c), (d), in a cubic unit cell, and (e) top view of planes put together, looking down directly on (0,0,1), and showing the top half of a octahedron formed by planes.

For Plane B with origin at O_B,

Intercepts	$x = -2/3$	$y = 3/4$	$z = \infty$
Reciprocals	$-3/2$	$4/3$	0
Clear fractions → $(h\,k\,l)$	$(\bar{9}\,8\,0)$		

For Plane C with origin at O_C,

Intercepts	$x = -2/3$	$y = \infty$	$z = -1/2$
Reciprocals	$-3/2$	0	-2
Clear fractions → $(h\,k\,l)$	$(\bar{3}\,0\,\bar{4})$		

EXAMPLE 2-11

Sketch the following planes within a cubic unit cell:

(*a*) $(1\,\bar{2}\,2)$ (*b*) $(\bar{2}\,1\,3)$ (*c*) $(\bar{3}\,0\,\bar{1})$

SOLUTION: To sketch the planes, we need to know the intercepts of the planes on the axes. For each plane, the reciprocals of the Miller indices are the fractional intercepts on the respective axes. Once we get the reciprocals we need to choose the appropriate origin in the unit cell to sketch the plane, see Fig. EP 2-11.

For plane $(1\,\bar{2}\,2)$, the intercepts are $x = 1, y = -1/2, z = 1/2$.

For plane $(\bar{2}\,1\,3)$, the intercepts are $x = -1/2, y = 1, z = 1/3$.

For plane $(\bar{3}\,0\,\bar{1})$, the intercepts are $x = -1/3, y = \infty, z = -1$.

2.10.3 HCP Planes and Directions

The procedures that were described above to identify the Miller indices of planes and directions apply to every crystal system. The same procedures are therefore used to get the Miller indices of planes

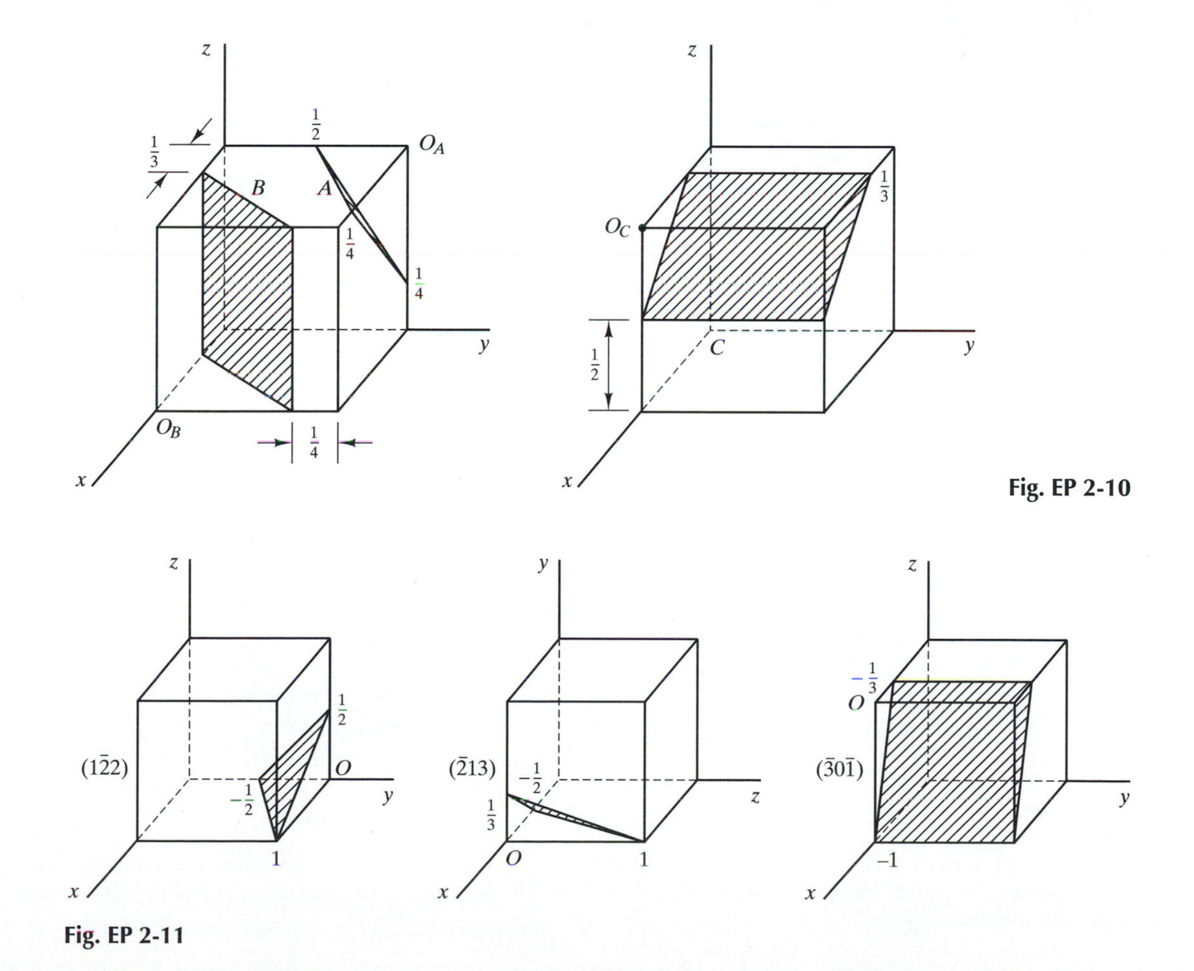

Fig. EP 2-10

Fig. EP 2-11

and directions in the HCP structure. However, in order to reveal the hexagonal symmetry, the three-digit (Miller indices) representation is converted to a four-digit representation called the Miller-Bravais indices. The modification arises from the presence of a third axis $\mathbf{a_3}$ lying in the basal plane of the hexagonal prism shown in Fig. 2-32a. $\mathbf{a_3}$ is symmetrically related to $\mathbf{a_1}$ and $\mathbf{a_2}$ and is used in conjunction with the basic unit axes. Thus, *the Miller–Bravais indices of a plane in the HCP refer to the four axes, a_1, a_2, a_3, and c, and are represented as (hkil). The index i is also the reciprocal of the fractional intercept on the a_3 axis.* The intercepts of a plane on the a_1 and a_2 axes determine the intercept on the a_3 *axis* and thus *the value of i depends on the values of h and k.* The relationship is

$$i = -(h + k) \tag{2-18}$$

The Miller-Bravais representation gives similar indices to similar planes; therefore, they may be indicated as planes of a form. For example, the sides or prismatic planes of the hexagon in Fig. 2-32b are all similar and

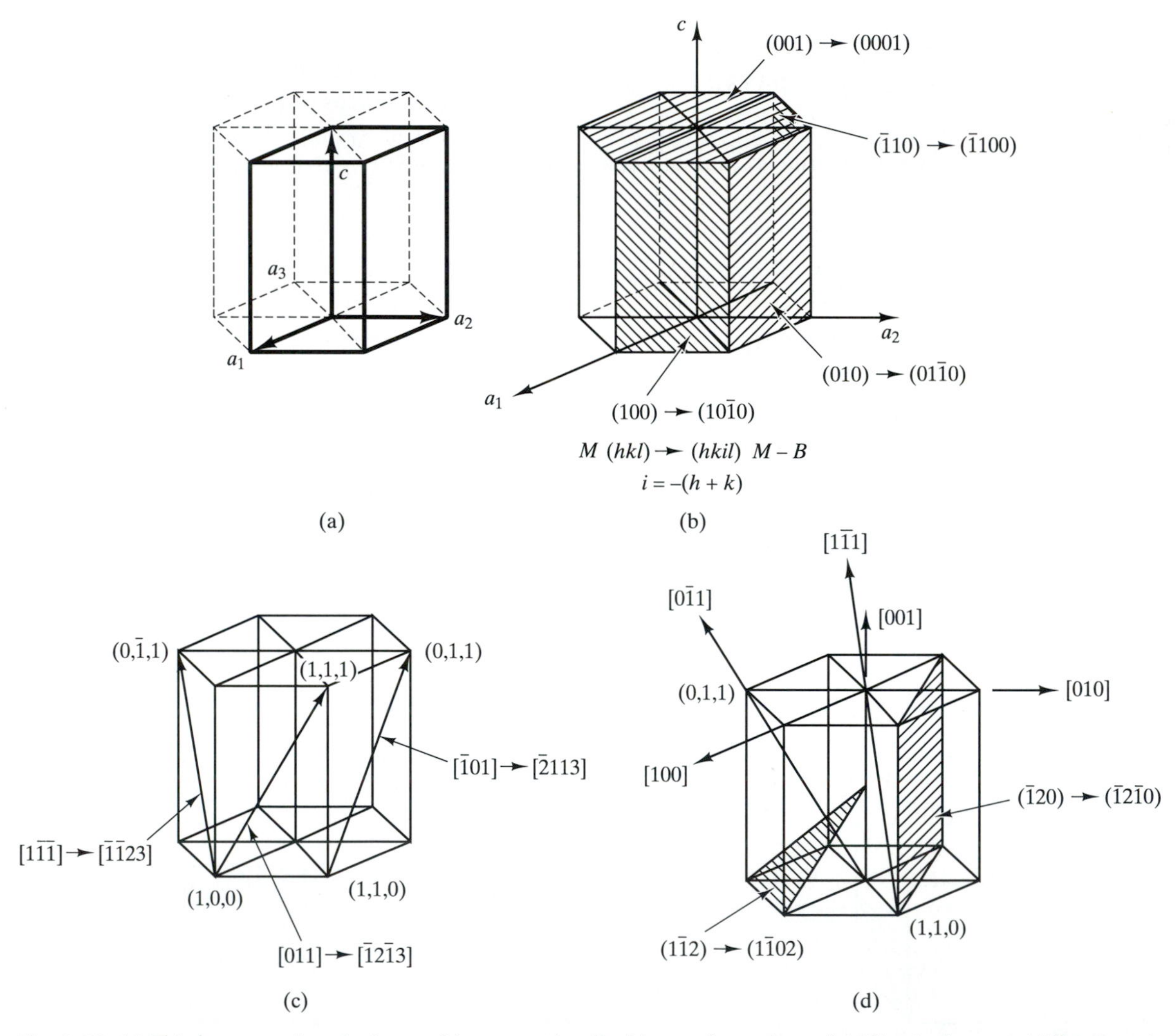

Fig. 2-32 (a) Third axes a_3 basal plane of hexagonal cell, (b) transformation of Miller indices to Miller-Bravais indices for planes, (c) Miller indices to Miller-Bravais indices for directions, (d) directions in Miller indices and planes in Miller-Bravais indices.

symmetrical. In the basic three-axes system, the Miller indices are (100), (010), $(\bar{1}10)$, $(\bar{1}00)$, $(0\bar{1}0)$, and $(1\bar{1}0)$. The similarity of these planes, in terms of having the same set of numbers, is not revealed because of the different indices; two indices have two 1s and one 0, and four indices with one 1 and two 0s. In contrast, in the Miller-Bravais four-axes indexing, the corresponding indices are $(10\bar{1}0)$, $(01\bar{1}0)$, $(\bar{1}100)$, $(\bar{1}010)$, $(0\bar{1}10)$, and $(1\bar{1}00)$. All the indices now have the same integers and the similarity of the planes is brought out. The planes can be represented as a set of {1100}.

In the same manner, the four-index system brings out the similarity in directions. This is illustrated in Fig. 2-32c where three diagonal directions in three prismatic planes are indicated. In the three-digit system, these directions are $[\bar{1}\,\bar{1}1]$, [011], and $[\bar{1}01]$. We use the following set of equations to convert the Miller indices to the four-digit representation.

$$\left.\begin{aligned} u &= \frac{1}{3}(2u' - v') \\ v &= \frac{1}{3}(2v' - u') \\ i &= \frac{1}{3}(u' - v') \\ w &= w' \end{aligned}\right\} \qquad (2\text{-}19)$$

where the primed letters represent the 3-digit system, $[u'v'w']$ and the unprimed are for the 4-digit system, $[uviw]$. The four-digit indices of the same directions are now $[\bar{1}123]$, $[\bar{1}2\bar{1}3]$, and $[\bar{2}113]$. Thus, the diagonals are in a set of <1123> directions.

It must be reemphasized that in order to obtain the four-digit representation, we must first obtain the Miller indices (the three-digit representation) of both planes and directions. As indicated above, the fourth digit i in the planar representation $(hkil)$ is actually the fractional intercept on the fourth axes a_3 and has a physical meaning. A similar physical meaning for the fourth digit in the Miller-Bravais indexing is not readily apparent, and because of the more complicated procedure to obtain it, *it is better to retain the Miller indices for directions* (at least for the purpose of this book) *and use only the Miller-Bravais indices for planes for the hexagonal structure.* Figure 2-32d shows Miller and Miller-Bravais indices for two planes and only Miller indices for directions.

EXAMPLE 2-12

Determine the Miller indices of the directions in the HCP structure in Fig. EP 2-12.

SOLUTION: We need to apply Eq. 2-13 and must identify the vector coordinates of the head and tail of each direction (arrow).

Direction	Head	Tail	Coordinates [Head−Tail]	Miller Indices
A	(−1,0,1)	(1,1,1)	[−2, −1, 0]	$[\bar{2}\,\bar{1}0]$
B	(1,0,1)	(0,1,0)	[1, −1, 1]	$[1\bar{1}1]$
C	(1, 0, 0)	(0, 0, 1)	[1, 0, −1]	$[10\bar{1}]$

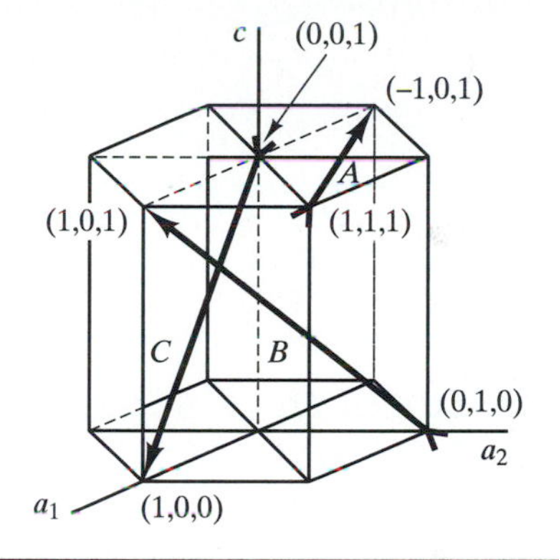

Fig. EP 2-12

EXAMPLE 2-13

Determine the Miller indices and then the Miller-Bravais indices of the planes D, E, and F in the hexagonal structure shown in Fig EP 2-13.

SOLUTION: As in Example 2-10, we need to identify the intercepts of each of the planes on the axes by the appropriate choice of an origin. The axes of the hexagonal structure are a_1, a_2, and c.

For Plane D,

Intercepts	$a_1 = 1$	$a_2 = \infty$	$c = \frac{1}{2}$
Reciprocals	1	0	2
Miller Indices (hkl)	(1 0 2)		
Miller-Bravais (hkil)	$(10\bar{1}2)$,	where $i = -(h + k)$	

For Plane E, we move the origin to O_E,

Intercepts	$a_1 = -1$	$a_2 = \infty$	$c = \infty$
Reciprocals	−1	0	0

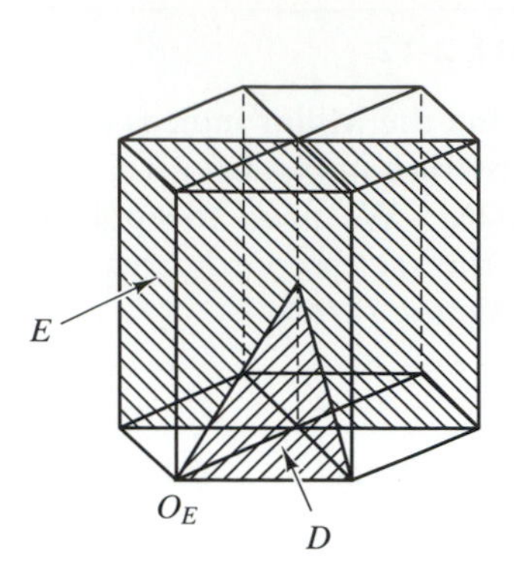

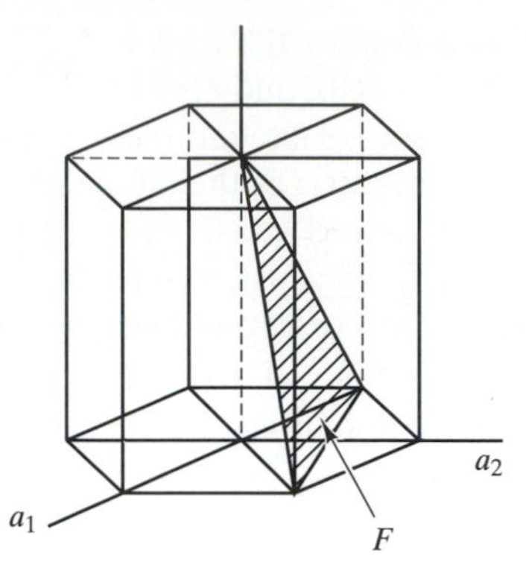

Fig. EP 2-13

Miller Indices (h k l)	$(\bar{1}00)$
Miller Bravais (h k i l)	$(\bar{1}010)$

For Plane F,

Intercepts	$a_1 = -1$	$a_2 = +\frac{1}{2}$	$c = 1$
Reciprocals	-1	$+2$	0
Miller Indices (h k l)	$(\bar{1}21)$		
Miller Bravais (h k i l)	$(\bar{1}2\bar{1}1)$		

2.11 *Slip Systems in Metals and Deformability*

Having learned the nomenclatures of crystal planes and directions, we are in a position to see how the crystal structure of a metal influences its deformability. The deformation of metals occurs by sliding or slip on preferred planes along certain preferred directions. These are called *slip planes* and *slip directions; a combination of a slip plane and a slip direction is called a slip system.* The preferred *slip planes* are *those with the highest planar atomic fraction* while the preferred *slip directions* are *those with the highest linear atomic fractions*. Therefore, we need to determine the slip planes and slip directions for the most common metallic structures to determine their slip systems. The greater the number of independent slip systems the more deformable is the metal.

2.11.1 Slip Planes

To determine the slip planes, we define planar atomic fraction (PAF) as

$$\text{PAF} \equiv \frac{\text{Area of atoms on the plane}}{\text{Area of the plane}} \tag{2-20}$$

The areas of the atoms are the sections the plane cuts passing through the center of the atom or the lattice point. We consider the {100}, the {110}, and the {111} planes in both the BCC and the FCC structures to determine which plane has the greatest PAF. The determination of the slip plane in each of these structures is summarized in Table 2-5. The dimensions of each plane and the atoms intersected by it for each structure are given under the planar section column. Just as in the volumetric packing fraction, we determine the fraction of the atomic area contributed by each atom at the corners, in the interior, and on the edge of the plane, and add them up. The total atomic area is given by

$$N_A = N_i + \frac{N_e}{2} + \frac{\theta}{360}N_c \tag{2-21}$$

where N_A is the total number of atoms in the plane, N_i is the number of atoms within the plane, N_e is number of atoms on the edge of the plane, N_c is the number of atoms at the corners of the plane, and θ is the inside angle at the corners of the planar section. For a square or rectangle, θ is 90° and each atom at the corners contribute $\frac{1}{4}$ the atomic area, πr^2. For an equilateral triangle, θ is 60° and each atom at the corners contributes $\frac{1}{6}$ the atomic area. The N_A is multiplied by (πr^2) to get the total area of atoms on the plane and then this is divided by the area of the plane to get the PAF. Since we are just interested in the highest PAF, we do not necessarily need to have absolute values. We look for a common factor in the fraction column for each structure and compare the coefficients for each plane section. In the BCC structure, the common

Table 2-5 Planar Atomic Fractions of the {100}, {110}, and {111} in the BCC and FCC.

Planes	BCC			FCC		
	Planar section/area	Area of atoms	Planar fraction	Planar section/area	Area of atoms	Planar fraction
{100}	a, a; $A = a^2$	$N_A = 1$; $A = \pi r^2$	$(1)\ \dfrac{\pi r^2}{a^2}$	a, a; $A = a^2$	$N_A = 2$; $A = 2\pi r^2$	$\dfrac{2\pi r^2}{a^2}$
{110}	$\sqrt{2}a$, a; $A = \sqrt{2}a^2$	$N_A = 2$; $A = 2\pi r^2$	$(2/\sqrt{2})\ \dfrac{\pi r^2}{a^2}$	a, $\sqrt{2}a$; $A = \sqrt{2}a^2$	$N_A = 2$; $A = 2\pi r^2$	$(1/\sqrt{2})\ \dfrac{2\pi r^2}{a^2}$
{111}	$\sqrt{2}a$, $\sqrt{2}a$, $\sqrt{2}a$; $A = \sqrt{3}a^2/2$	$N_A = \frac{1}{2}$; $A = \frac{1}{2}\pi r^2$	$(1/\sqrt{3})\ \dfrac{\pi r^2}{a^2}$	$\sqrt{2}a$, $\sqrt{2}a$; $A = \sqrt{3}a^2/2$	$N_A = 2$; $A = 2\pi r^2$	$(2/\sqrt{3})\ \dfrac{2\pi r^2}{a^2}$

factor is $(\pi r^2/a^2)$ and the coefficients are $1, 2/\sqrt{2}$, and $1/\sqrt{3}$, for the {100}, the {110}, and the {111} planes. We note that the highest planar atomic fraction is on the **{110}** and these are the slip planes for BCC structure. For the FCC structure, the common factor in the fraction column is $(2\pi r^2/a^2)$ and the coefficients of this factor are 1, $1/\sqrt{2}$, and $2/\sqrt{3}$, for the {100}, the {110}, and the {111} planes. The highest planar atomic fraction is now on **{111}** and these are the slip planes for the FCC structure.

2.11.2 Slip Directions

To determine the slip direction, we define the linear atomic fraction (LAF) as

$$\text{LAF} \equiv \frac{\text{Length of atoms along a direction}}{\text{Length of the direction}} \qquad \text{(2-22)}$$

The length of the atoms is given in terms of atomic radius when the direction passes through the center of the atom or the lattice point. The determination of the slip direction in each of the BCC and the FCC structures is summarized in Table 2-6 and the low index directions <100>, <110>, and <111> in both these structures are again examined. Their line sections that include the length of the direction and the atoms on them are illustrated for each direction of each structure. We note the number of atomic radii present on the direction and then divide it by the length of the direction to get the LAF. We see that the slip directions are **<111>** for the BCC, and **<110>** for the FCC because these are the directions with the highest LAF. We should also note that the distance between atoms in the slip direction is the shortest.

2.11.3 Number of Slip Systems and Deformability

Figure 2-33a illustrates the (110) plane of the BCC and Fig. 2-33b the (111) plane of the FCC. The number of BCC slip systems on the (110) is two; the two <111> directions on the (110). Specifically, the slip systems are (110)[$\bar{1}11$] and (110)[$\bar{1}1\bar{1}$]. We must note two things. First, the manner in which a slip system is always written, (plane)[direction]. Secondly, the reverse directions are not considered and there are only two systems per {110} slip plane. Since there are six of the {110} as shown in Fig. 2-30, there are 12 total slip systems in the BCC with the {110}. For the (111) FCC plane in Fig. 2-33b, there are three <110> directions and therefore, there are three slip systems per {111} plane. Since there are four of these {111} cubic planes, shown in Fig. 2-31, there are also 12 total slip systems in the FCC.

The above results are summarized in Table 2-7. The 48 total slip systems given for BCC include the 12 discussed here from the {110}, 12 from the {112}, and

Table 2-6 Linear atomic fractions of the <100>, <110>, and the <111> in the BCC and FCC structures.

Directions	BCC Section	BCC Atom length	BCC Lin. fraction	FCC Section	FCC Atom length	FCC Lin. fraction
<100>	a; Length = a	$2r$	$(1)\left(\frac{2r}{a}\right)$	a; Length = a	$2r$	$(1)\left(\frac{2r}{a}\right)$
<110>	$\sqrt{2}a$; Length = $\sqrt{2}a$	$2r$	$\frac{1}{\sqrt{2}}\left(\frac{2r}{a}\right)$	$\sqrt{2}a$; Length = $\sqrt{2}a$	$4r$	$\frac{2}{\sqrt{2}}\left(\frac{2r}{a}\right)$
<111>	$\sqrt{3}a$; Length = $\sqrt{3}a$	$4r$	$\frac{2}{\sqrt{3}}\left(\frac{2r}{a}\right)$	$\sqrt{3}a$; Length = $\sqrt{3}a$	$2r$	$\frac{1}{\sqrt{3}}\left(\frac{2r}{a}\right)$

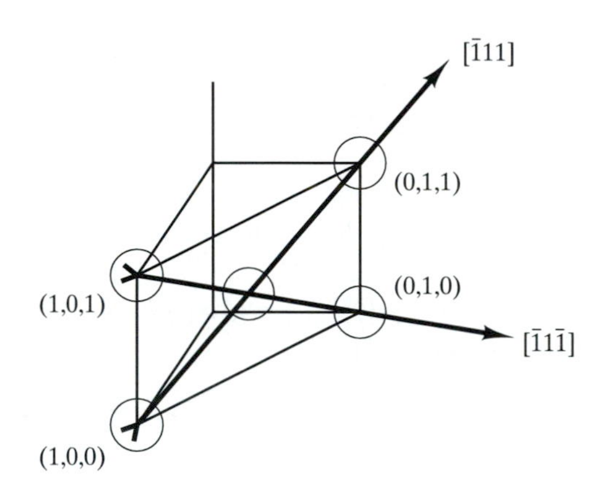

(a) (110) BCC plane

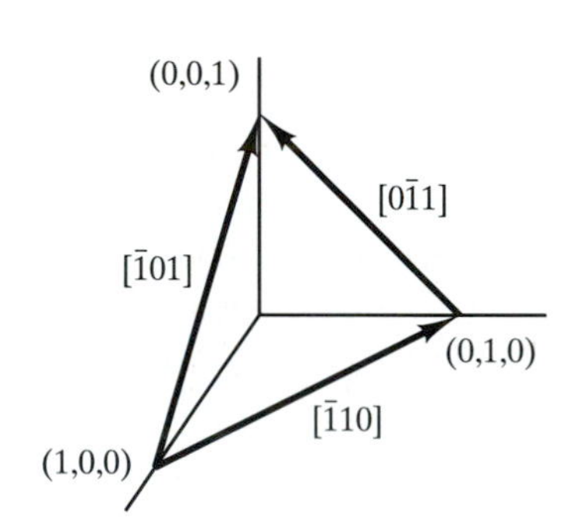

(b) (111) FCC plane

Fig. 2-33 Slip directions on the (a) (110) BCC plane, and (b) (111) FCC plane.

Table 2-7 Slip systems in metals.

Crystal Structure	Slip System Planes	Directions	Number
BCC (Fe, Mo, W)	{110}		12
	{112}	<111>	12
	{123}		24
FCC (Cu, Ni, Al)	{111}	<110>	12
HCP (Cd, Zn, Ti)	{0001}	<100>	3

24 from the {123}. The {112} and the {123} of the BCC are also possible slip planes. Included in Table 2-7 are also the slip systems in HCP that consist of the basal plane (0001) and the directions along the three unit axes a_1, a_2, and a_3 directions; a total of only 3 systems. The effect of this limited number of slip systems is the limited deformability of HCP metals (e.g., magnesium, zinc) compared to the cubic metals, BCC (e.g., steel) and FCC (e.g., copper, aluminum, and nickel). The HCP can only be shaped into products that require small deformation at room temperature. The BCC and FCC metals can withstand severe deformation and can be formed into deep drawn products, for

instance, cartridges produced from copper alloys and deep drawn oil pans from sheet steel.

It might be mentioned in passing that crystalline ceramics also exhibit slip systems. Magnesium oxide (MgO) that has the NaCl structure exhibits a {110}<1$\bar{1}$0> slip system; however, there are only two independent systems. Again, this makes ceramics very brittle and very hard to deform.

In plastics, deformation as indicated in Chap. 1 is through the sliding of the macromolecular chains in thermoplastics. This is accomplished through the breaking of the van der Waals bond. There are no slip planes or slip directions as in metals. The mechanical behavior of plastics is discussed in greater detail in Chap. 15.

EXAMPLE 2-14

Sketch the (a)(0$\bar{1}$1) BCC plane, and (b)(1$\bar{1}\bar{1}$) FCC plane and the atoms on them. Identify and write the slip systems in both these planes.

SOLUTION: See Fig. EP 2-14.

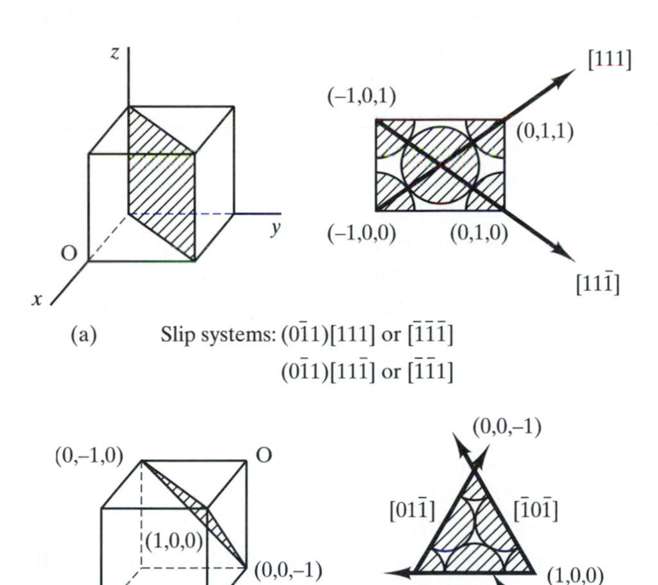

Fig. EP 2-14

2.12 Imperfections in Crystalline Structures—Actual Structures of Materials

Up to this juncture, we have discussed crystal lattices assuming that every point in the space lattice is occupied by an atom of the same kind. In addition, since we have not discussed them elsewhere, we were discussing only one crystal with no indication of size and orientation. We have just discussed the structure of an ideal, perfect infinite-sized single crystal. We shall now consider the actual structures of materials.

2.12.1 Surface Structures

The first thing we note is that engineering materials are produced to certain sizes and shapes. This means that they have surfaces and since the surface atoms have no neighbors on one side, they may take up equilibrium positions quite different from those beneath the surface. There are two effects that may arise—*surface relaxation or surface reconstruction.* Surface relaxation, which is schematically shown in Fig. 2-34, is the change in the interatomic spacing perpendicular to the surface and it is such that the surface atoms or ions usually move inwards beneath the surface to make the interatomic spacing smaller. Surface reconstruction is a more drastic change in which the atomic arrangement or pattern at the surface is altered from that beneath the surface. Examples of surface reconstructed structures are shown in Fig. 2-35 that shows the open circles as the inside and original square net of atoms, for example, a (100) BCC cubic face, and the solid circles as the atom positions of the reconstructed surface. Although the reconstructed surface structure is

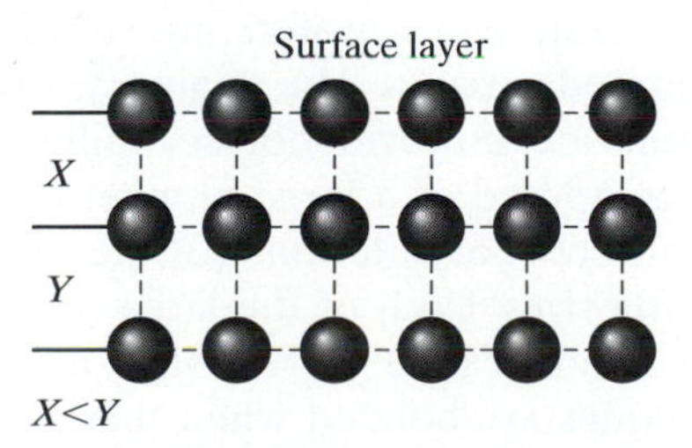

Fig. 2-34 Surface relaxation—change in interatomic spacing at the surface [Ref. 4].

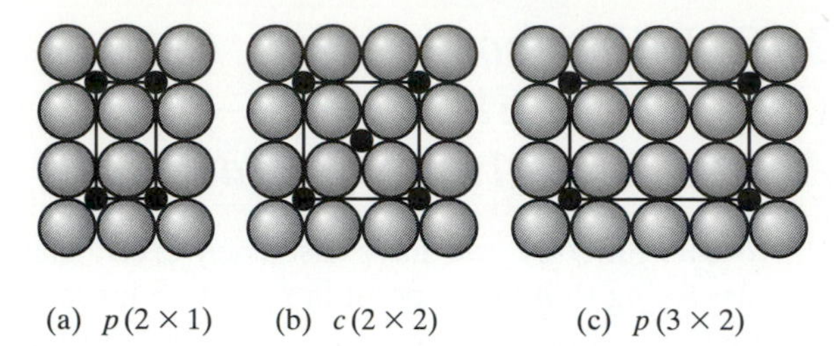

Fig. 2-35 Surface reconstruction—Solid circles represent surface structure; open circles represent bulk structure. [Ref. 4]

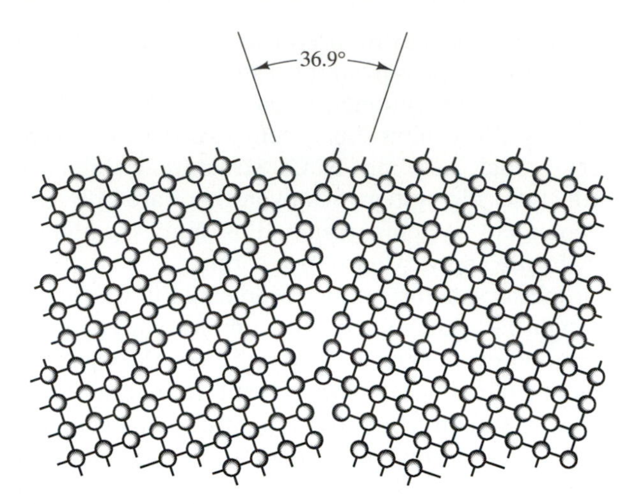

Fig. 2-36 Disorder in atomic arrangement in high-angle grain boundary where grains (crystals) of different orientations meet.

entirely different, it still bears a relationship with the atomic arrangement beneath the surface, the open circles. In the simple two-dimensional network in Fig. 2-35, let us say that the original lattice vectors are **a** and **b**. In terms of these unit vectors, the reconstructed structure may have unit axes that are multiples of **a** and **b**, such as $m\mathbf{a}$ and $n\mathbf{b}$. The reconstructed structure is said to have an $m \times n$ type of structure relative to the initial structure. An additional letter p or c is also used to indicate whether the reconstructed structure is *primitive* or has *an atom at the center of the structure*. Thus, the notation is $p(m \times n)$ or $c(m \times n)$ as the case may be. The notation $c(2 \times 2)$ in Fig. 2-35b shows an atom at the center of the square of solid circles. The surface structure is, of course, very important in the chemical interaction of crystalline materials with the environment, especially with regards to catalytic reactions. Also, since the bonding of the atoms is somewhat weaker, atoms are easier to break away and to get vaporized from the surface. It is also believed that surface "melting" occurs below the melting point of the bulk material.

A crystalline material with only one orientation of the unit cell throughout its size is called a *single crystal*. Examples of single crystal engineering materials are jet engine nickel superalloy blades and silicon semiconductors from which wafers are sliced off to produce integrated circuits. The majority of engineering materials are, however, polycrystalline materials that are composed of a large number of single crystals with different orientations (of the unit cell) and with varying sizes. Each of the large number of single crystals is now called a grain (like a grain of salt) and the grains are bonded when their surfaces or boundaries, called *grain boundaries*, butt together. Because of the random orientations of the grains, the atoms do not align and there is a great deal of misalignment, whereby imperfect surface structures are found at the grain boundaries. The angle of misorientation is usually large and an example is shown in Fig. 2-36. Just as in surface "melting," it is a well-known phenomenon that incipient melting of metals occurs first at the grain boundaries. Metallurgists call this phenomenon "burning".

Unless they are deformed, the grains are thought to be multifaceted polyhedra, which show almost equal size when plane sections are made through them; they are called equi-axed grains. The average sizes of grains are determined by standards (in the United States, this is ASTM E 112; other countries may have their own standards) and they are reported as a grain size number. Some examples of grains and their sizes are shown in Fig. 2-37. The curved or jagged lines in this figure are the grain boundaries and the clear enclosed areas are the grains. The original definition of the ASTM grain size number, G, is related to the number of grains, N, in the structure per square inch when the structure is magnified 100x. The relationship is

$$\mathbf{N} = 2^{(G-1)} \tag{2-23}$$

If another magnification, M, is used other than 100x, and the number of grains per inch is obtained at that

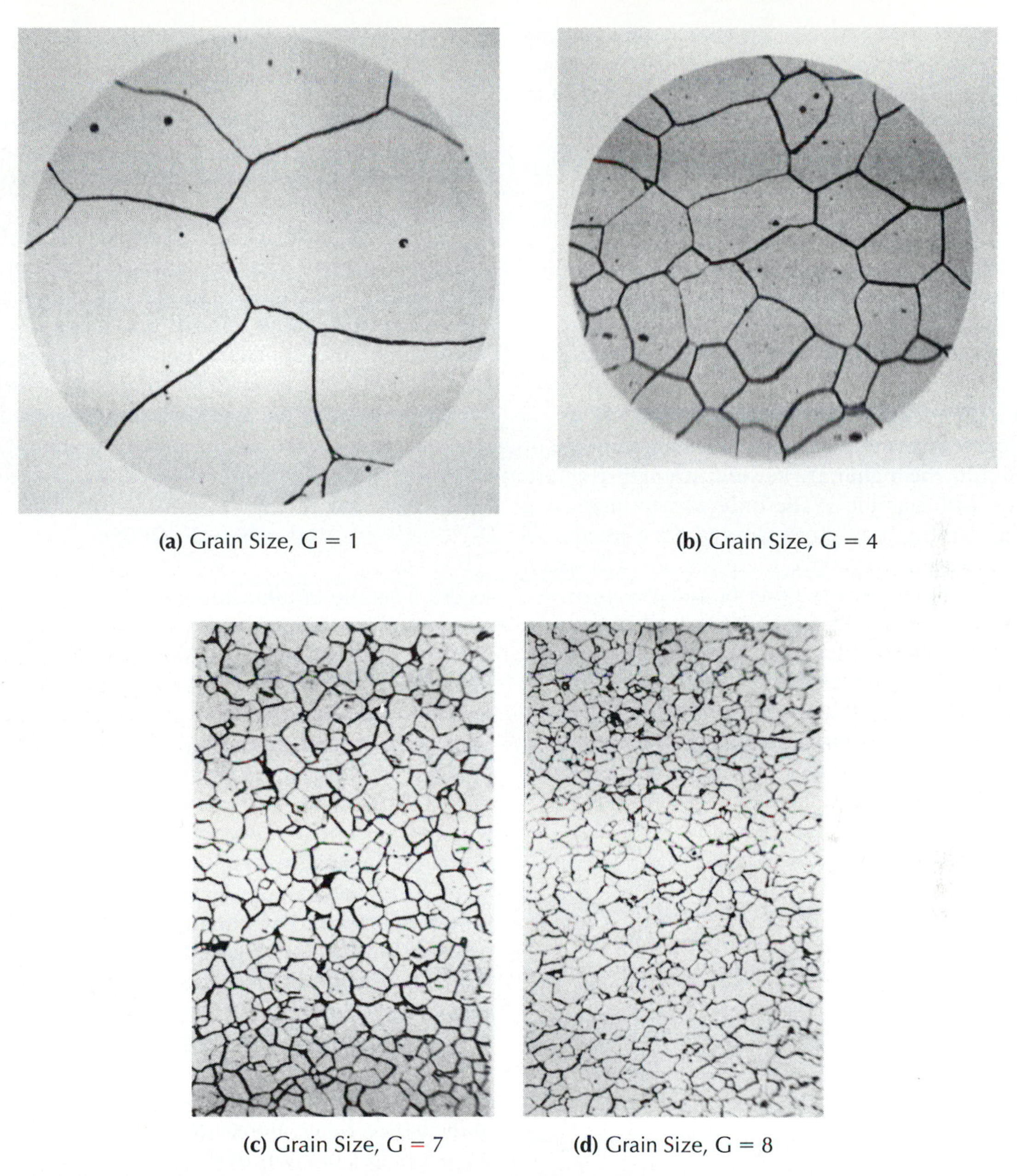

(a) Grain Size, G = 1

(b) Grain Size, G = 4

(c) Grain Size, G = 7

(d) Grain Size, G = 8

Fig. 2-37 Examples of grain sizes in low carbon steels. Mag. 100×.

magnification, the grain size number obtained using Eq. 2-23 is G_a, an apparent number. The real grain size number, G, is obtained by adding a factor Q to G_a, such as

$$G = G_a + Q \tag{2-23a}$$

where

$$Q = 2\log_2\left(\frac{M}{100}\right)$$

$$= 6.644\log_{10}\left(\frac{M}{100}\right) \tag{2-23b}$$

and M is the magnification used.

At low temperatures, the smaller the grain size of the material, the higher are its yield strength, fracture strength, and toughness. The only way to increase a material's strength, fracture strength, and toughness simultaneously is to reduce the grain size. However, the open structure at the grain boundaries provides sites for other atoms to segregate as well as for easy movement of atoms, especially at high temperatures. If a material is under load at high temperatures, the movement of atoms along grain boundaries enhances creep deformation (slow change in dimension of components). This is the reason for using single crystals for jet engine blades.

EXAMPLE 2-15

A metallographer determined the grain size number G to be eight. What is the number of grains per square inch at 100x magnification?

SOLUTION:

$$N = 2^{(G-1)} = 2^{(8-1)} = 128 \text{ grains/sq. inch}$$

EXAMPLE 2-16

The metallographer in Example 2-15 realized after he handed his results to his supervisor that he used 200x, magnification instead of the standard 100x. What is the correct grain size?

SOLUTION: Using Eq. 2-23a and 2-23b and noting that $G_a = 8$, and

$$Q = 6.644\log_{10}\left(\frac{200}{100}\right) = 2$$

$$G = 8 + 2 = 10$$

2.12.2 Imperfections within a Grain or Single Crystal

Within a grain or a single crystal there are also imperfections or defects in the arrangement of atoms. The kinds of defects that may occur are point defects, line defects, and planar defects.

Figure 2-38 illustrates the common point defects in metallic structures. In pure metals, the only probable point defect that can thermodynamically exist is the vacancy. As the name implies, a *vacancy* is a lattice point that is unoccupied by an atom. At a particular temperature, there is an equilibrium concentration of vacancies. At low temperatures, this concentration is low but it increases with increasing temperature, according to the formula

$$n_v = Ne^{-}(E_F/RT) \tag{2-24}$$

where n_v is the equilibrium concentration of vacancies per unit volume, N is the total number of lattice sites per unit volume; E_F is the energy of formation of vacancies per mole, R is the gas constant, and T is absolute temperature. As discussed in the previous section, an atom from the surface can easily vaporize to leave initially, a surface vacancy. Then the vacancy moves to the inside of the crystal when an atom from within the crystal reoccupies the surface vacancy, leaving a vacancy within the crystal where it came from. This process is repeated because atoms are always in motion. The energy of formation of vacancies is estimated to be about one-fourth to one-half the heat of vaporization and has been theoretically calculated to be of the order of 1 eV. Near the melting point of a metal, studies indicate that the vacancy concentration, $n_v \approx 10^{-4}N$.

The next two point defects are due to foreign atoms in the lattice. Some atoms are very small and can fit in the interstices or holes of the lattice; the foreign atom (point defect) is called an *interstitial defect.* We calculated earlier the sizes of spherical atoms that can fit in the interstices of the BCC and the FCC structures and have seen that the FCC can accommodate a larger size atom than the BCC. Expressed in another way, the FCC is capable of taking in more interstitial atoms than the BCC. For this reason, the maximum solubility of carbon (a small atom) in the iron FCC structure is much higher than in the BCC. The other way a foreign atom can be present in the lattice is to replace a host atom at a lattice site. The foreign atom is now called a *substitutional defect;* the substitutional atom in this case must have an atomic size within 15 percent of the size of the atom it replaces.

In ionic or ceramic structures, the Frenkel and the Schottky defects, shown in Fig. 2-39, may form. The Frenkel defect involves the motion of a cation from a normal site to an interstitial site, leaving a vacancy. The Frenkel defect consists of the pair of cation vacancy and a cation interstitial. The Schottky defect is a pair of cation and anion vacancies. This arises from the condition of electrical neutrality. Thus, if a cation vacancy is formed, an anion vacancy must also form; otherwise, there is an excess of negative charge.

Line defects in crystalline structures are called *dislocations.* They allow crystalline materials to be deformed into shapes. These are the defects that move or glide on the slip planes and along the slip direction and render the material ductile. Materials researchers control the strength of materials by blocking the motion of dislocations. When dislocations are hard to move or when there are no dislocations, such as in an amorphous material, the material becomes brittle.

Figure 2-40 shows schematically the formation of an *edge dislocation* in the crystal lattice. In Fig. 2-40a, a block of material is sliced halfway through to the line marked $\perp - \perp$, then the material above the cut is displaced or pushed to the right a distance b and perpendicular to the marked line relative to the material below, then glue the cut-and-displaced surfaces together. On a microscopic scale, the structure at the $\perp - \perp$ line is shown in Fig. 2-40b where the defect is seen as the line of atoms at the edge of an extra half-plane of atoms above the cut-plane; thus the dislocation is called edge dislocation and is given the symbol $\perp$. A dislocation line can be observed only by transmission electron microscopy (TEM). An example is shown in Figure 8-6.

If the cut surfaces were displaced a distance b relative to each other and parallel to the marked line, a screw dislocation is formed where spiral "screwlike" steps are formed by the atoms above and below the cut-plane as shown in Fig. 2-41.

The manner in which plastic deformation (or strain) of the crystal is produced when the dislocations move is schematically shown in Fig. 2-42. The

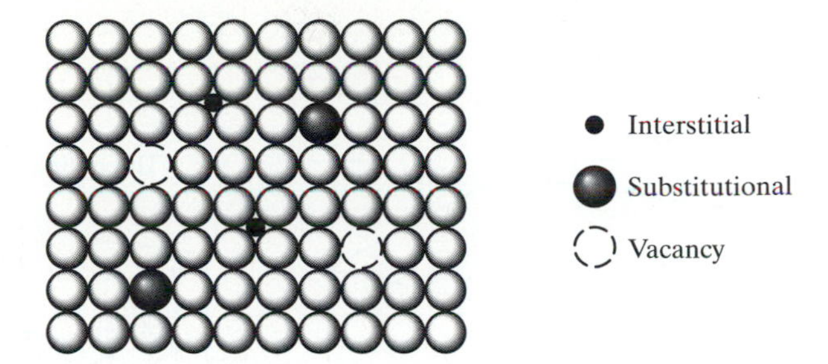

Fig. 2-38 Point defects in metallic crystalline structures.

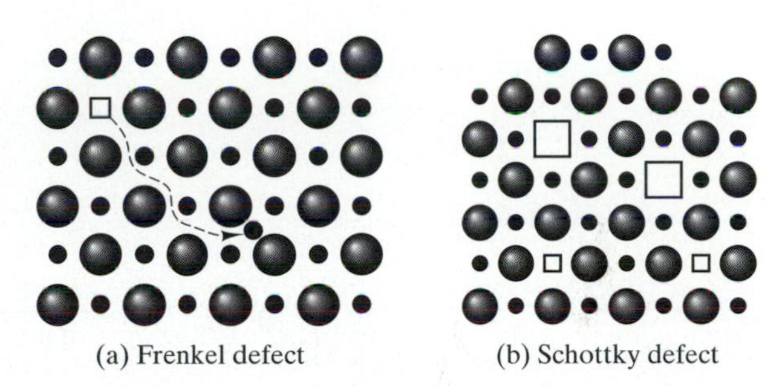

Fig. 2-39 Point defects in ionic solids.

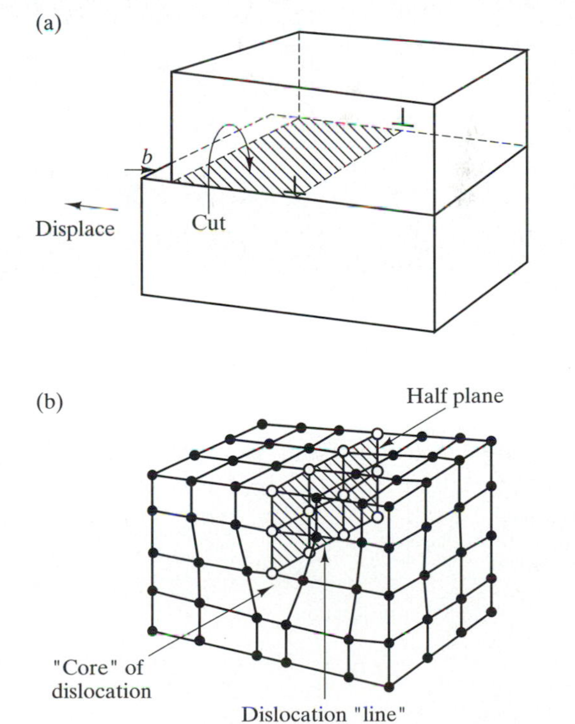

Fig. 2-40 Schematic of formation of edge dislocation.

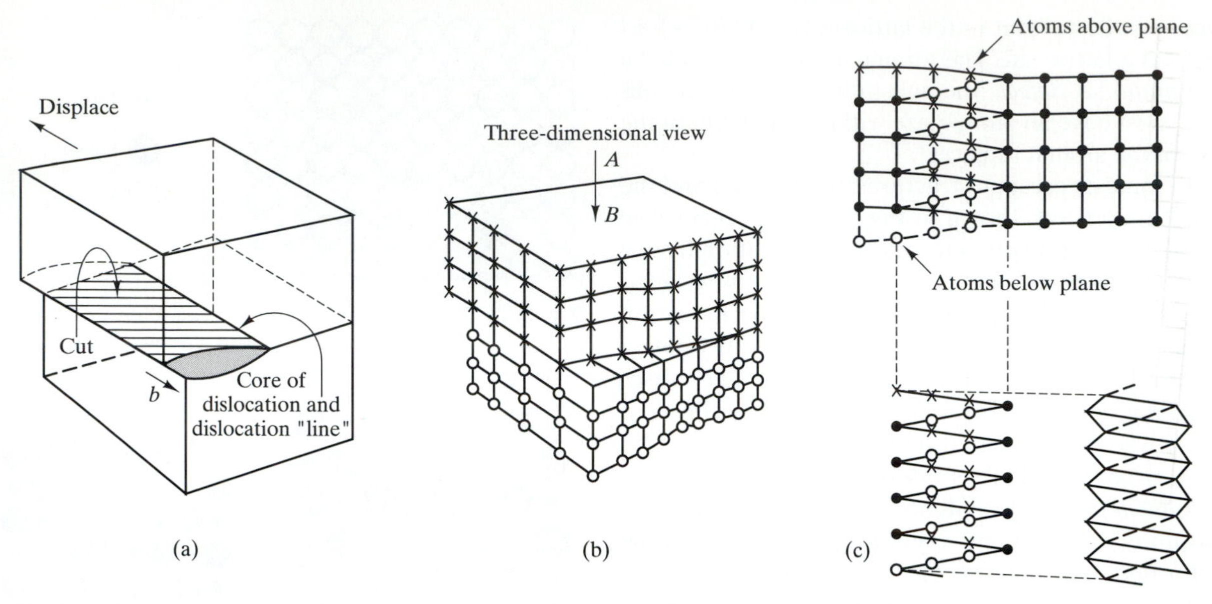

Fig. 2-41 Screw dislocation (a) macroscopic displacement, (b) microscopic displacement, and (c) screwlike arrangement of atoms above and below the plane of the cut.

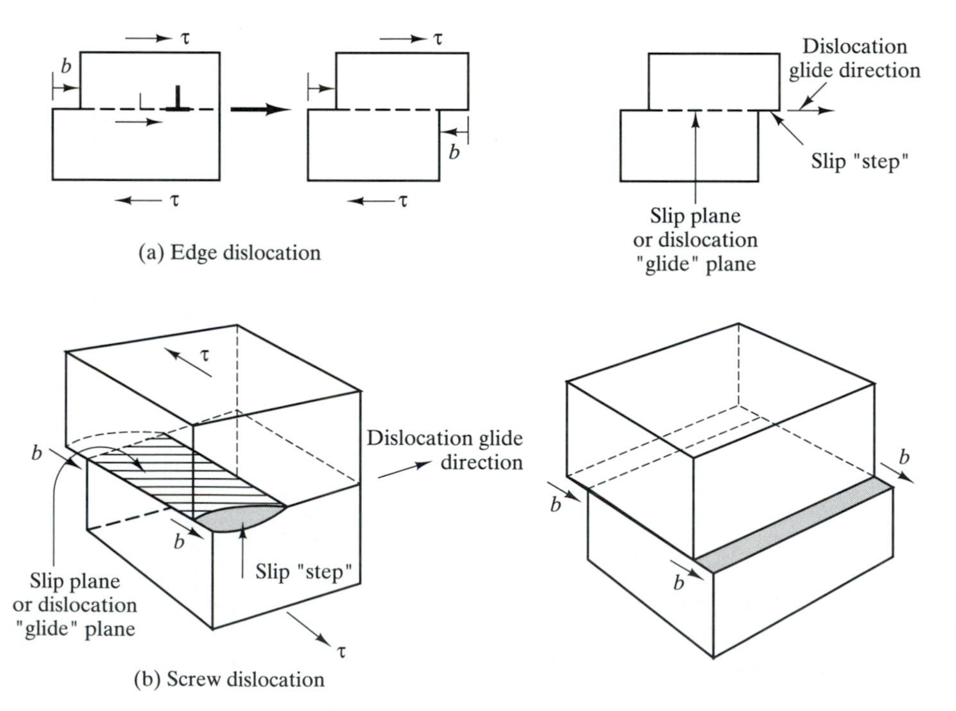

Fig. 2-42 Strain is manifested by "slip steps" produced by motion of (a) edge and (b) screw dislocation.

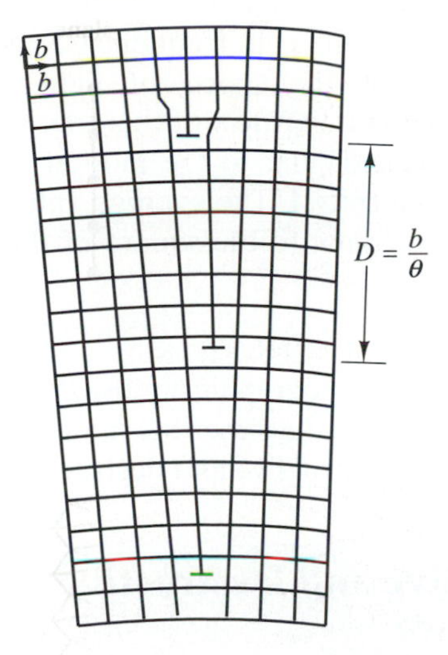

Fig. 2-43 Dislocations tilt or small angle grain boundary.

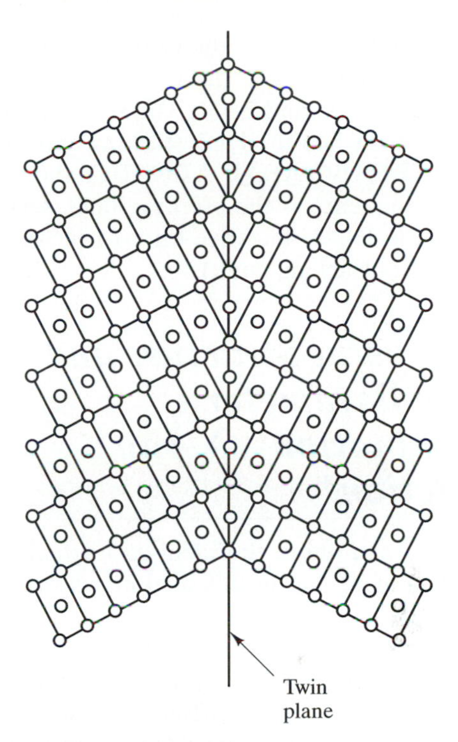

Fig. 2-44 Twin crystals have mirror images of atoms across plane.

strain is shown as a "slip step" after the dislocation passes through the block of material.

When edge dislocations are arranged on top of one another in the crystal as shown in Fig. 2-43, a small angle grain boundary is formed. If such a boundary encloses an area, the area enclosed is called a subgrain because it is a small "grain" within a larger grain (large angle of misorientation) and the *small angle grain boundary is also called a subgrain boundary.* The subgrain boundary is a type of planar defect within a grain. Other planar defects that may arise are *twins* and *stacking faults,* shown in Figs. 2-44 and 2-45. Twins are areas within the grain that show the atoms as mirror images of each other; the mirror plane is called the twin plane. These are predominantly observed in brasses as strips across the grain and are called annealing twins (See Photo 2 in Appendix). Stacking faults refer to the disorder in the stacking of atom layers in the FCC structure. As discussed earlier in this chapter, both the FCC and the HCP are close-packed structures, but they differ in the stacking of the close-packed planes, {111} for the FCC and {0001} for the HCP. In the FCC, the stacking delineates three distinct layers, A, B, and C, and the stacking sequence goes ABCABCABC ... and so on; the fourth layer is directly on top of and, therefore, identical to the first layer. In the HCP structure, the stacking delineates only two distinct layers, A and B, and the stacking

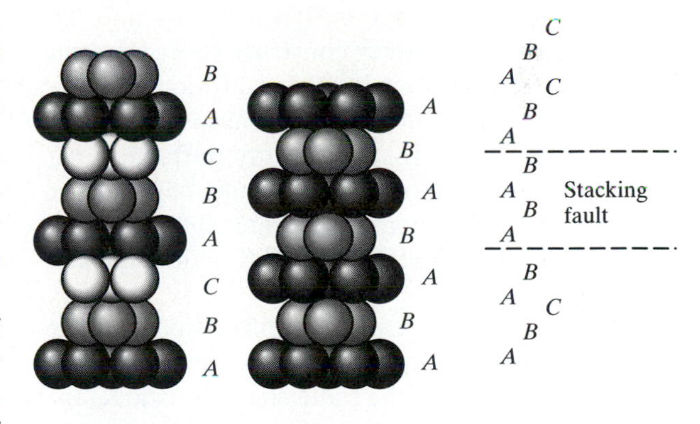

Fig. 2-45 Stacking of atoms in (a) FCC, (b) HCP, and (c) stacking fault (HCP stacking) in FCC structure.

goes ABABABABAB ... ; the third layer is directly on top of and identical to the first layer. These are shown in Figs. 2-45a and 2-45b. The stacking fault in the FCC occurs when the HCP stacking appears. For example, ABCABC-ABAB-ABCABC; the underlined HCP sequence is the stacking fault, schematically shown in Fig. 2- 45c. (See Photo 1 in Appendix for an image of stacking fault.)

2.13 Crystal Structure and Lattice Parameter Determination—X-ray Diffraction*

We have learned about the ideal and actual crystal structures of materials. In this section, we shall learn about how the cubic crystal structures and their lattice parameters are determined. The method commonly used to determine them is X-ray diffraction.

As indicated in Sec. 2.2, diffraction is used to differentiate crystalline materials from noncrystalline or amorphous materials. It is a phenemenon exhibited only by scattered waves from periodic scattering centers. The atoms or molecules in a crystal lattice constitute a set of periodically arranged scattering centers from which scattered waves emanate. The scattered waves will either constructively or destructively interfere with each other; diffraction occurs when constructive interference ensues. When the waves are X rays, it is called X-ray diffraction. The wavelike characters of electrons and neutrons are also used for diffraction studies and the techniques are, respectively, called electron diffraction and neutron diffraction.

The X ray that is used for diffraction is plane-polarized, monochromatic (single wavelength), and as indicated in Chap. 1, is the same characteristic K_α that is used to identify materials. As indicated in Sec. 1.5, the monochromatic K_α is obtained by filtering out the continuous spectrum and other emissions. It is also mentioned there that X ray is an electromagnetic wave that carries the energy expressed in Eq. 1-12. The latter is reproduced here as Eq. 2-25

$$\text{Energy} = h\nu = \frac{hc}{\lambda} \qquad (2\text{-}25)$$

where h is Planck's constant, ν is frequency, λ is the wavelength, and c is the speed of light = 3×10^8 meter per sec. The rate of flow of this energy through a unit area perpendicular to the motion of the wave is called the intensity, I. The average value of the intensity is proportional to the square of the amplitude, that is,

$$\text{Intensity, I} \propto A^2 \qquad (2\text{-}26)$$

This intensity shows up in photographic emulsions as the amount of darkening. We shall use this intensity to differentiate between constructive and destructive interference.

2.13.1 Constructive and Destructive Interference

When X rays hit an electron (charged particle), the electron oscillates and its oscillation also emits an electromagnetic wave of the same frequency as the X rays. But, while the X ray is plane polarized, the emitted (scattered) wave from the electron (scattering center) is a spherical wave front and spreads out symmetrically from it, as depicted in Fig. 2-46.

Thus, when an area of the crystal is hit by an X-ray beam, all the electrons of the atoms in the area emit spherical waves that expand and overlap one another. Where the wave crests of the scattered waves coincide, the amplitude of the total scattered wave is increased; the waves are said to be in phase and the amplitudes add up. This is the phenomenon of *constructive interference,* shown in Fig. 2-47a, for two waves each having an amplitude A. The total amplitude of the resultant wave after constructive interference is 2A and the intensity is $\propto 4A^2$. This

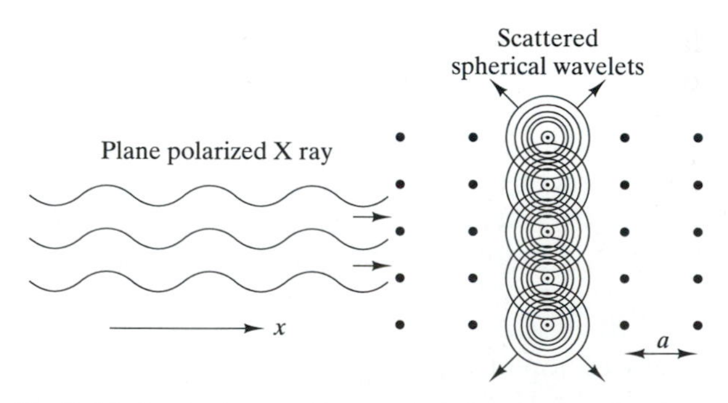

Fig. 2-46 Atoms in crystals, as scattering centers for X ray, produced spherical wavelets.

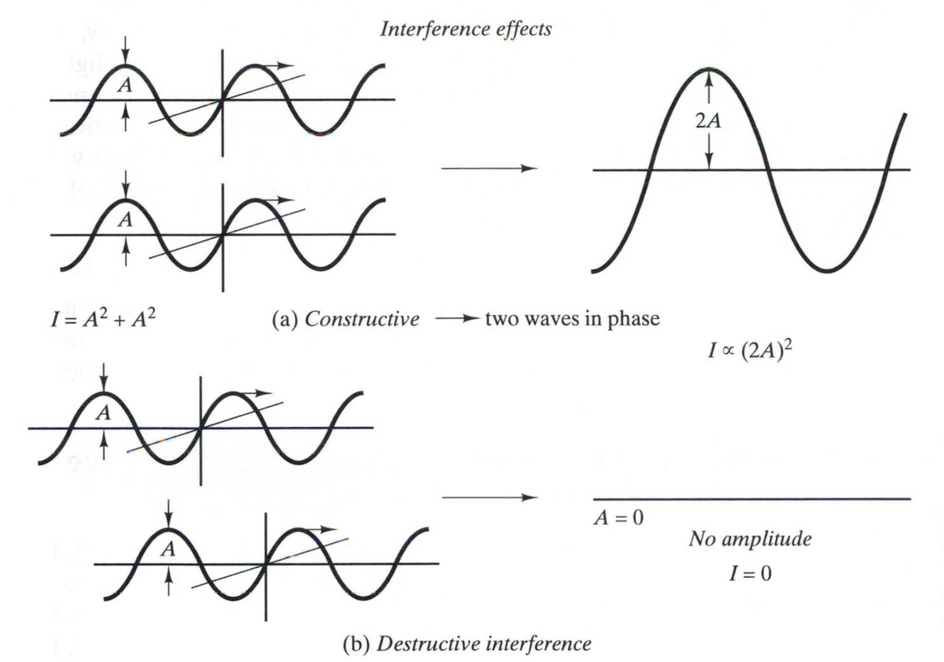

Fig. 2-47 Wave interference affects (a) constructive interference of in-phase waves, (b) destructive interference of out-of-phase waves (no intensity).

intensity is twice the sum of the intensities of the two waves, that is, $A^2 + A^2 = 2A^2$. Where crests and troughs do not coincide, the waves are out of phase, and in the specific illustration in Fig. 2-47b, they are 90° out of phase. In this case, there is mutual cancellation of the amplitudes and the result is there is no wave, that is, the amplitude is zero and the intensity is zero. This is the phenomenon of *destructive interference.* Diffraction is the result of the constructive interference of the scattered wavelets from the periodically arranged atoms in the crystal.

2.13.2 Bragg's Law

Having gone over the preliminaries, we are now in a position to learn the condition when constructive interference (diffraction) occurs in a crystal and then later, how to determine the crystal structure and lattice parameters of cubic structures. The condition for diffraction is Bragg's Law after a young English student physicist who simplified an earlier analysis by the German physicist von Laue. Von Laue had announced earlier in 1912 the proof that in fact crystals diffract X rays and had proposed a diffraction theory by considering the scattered waves from each atom and then adding the contributions from all atoms, allowing for phase differences in the superposed waves. Later in that same year, Bragg proposed a more simplified model which considered whole planes of atoms and added their contributions.

The essence of Bragg's analysis is shown in Fig. 2-48 where a section of a crystal is illustrated consisting of a set of parallel planes (where atoms are arranged) A, B, C, D, ..., normal to the plane of the drawing and spaced a distance d apart (interplanar spacing). A perfect plane monochromatic X-ray beam hits the surface of the crystal at an angle θ, where θ is measured between the surface and the X-ray beam. The atoms produce scattered waves that will travel in all directions. *The condition of constructive interference of the scattered waves is when the distances they travel are multiples of the wavelength of the X-ray beam* and happens only when the diffracted beam is also at an angle of θ with the surface. As an example, the incident rays **1** and **1a** strike atoms K and P, respectively, in plane A (surface) and are scattered in all directions. The scattered waves will be in phase only when they also make an angle θ with the surface in the directions **1′** and **1a′**, because the distance traveled by the two waves are exactly equal—QK for ray **1-1′** = PR for **1a-1a′**. This is true for all the rays scattered by the

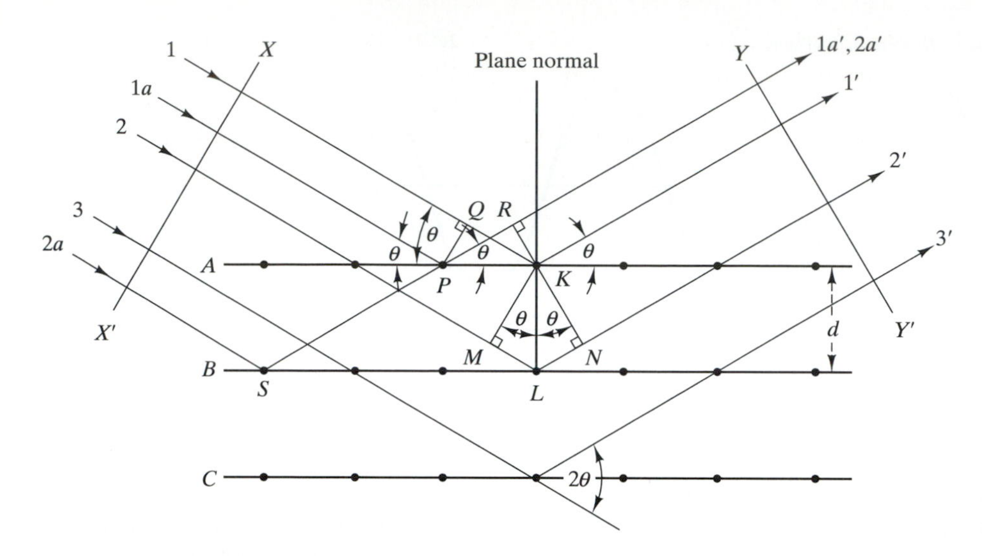

Fig. 2-48 Geometry of X-ray diffraction from layers of crystal planes A, B, C.

atoms in plane A and also prevails for each of the planes considered separately.

Bragg's law relates to the constructive interference of scattered waves from two adjoining parallel planes. Rays 1 and 2, for example, are scattered by atoms K and L, and the path difference for the rays 1K1′ and 2L2′ is

$$\mathrm{ML} + \mathrm{LN} = d\sin\theta + d\sin\theta = 2d\sin\theta \qquad (2\text{-}27)$$

The condition for constructive interference of 1′ and 2′ is for the path difference to equal multiples of the wavelength, that is,

$$n\lambda = 2d\sin\theta \qquad (2\text{-}28a)$$

This is Bragg's law, where λ is the wavelength of the incident wave (either X rays, electrons, or neutrons), d is the interplanar spacing of the parallel planes, θ is the angle of incidence, and n is the order of diffraction. It is customary to take the first order, $n = 1$, thus

$$\lambda = 2d\sin\theta \qquad (2\text{-}28b)$$

2.13.3 Interplanar Spacing, d

The interplanar spacing for cubic structures is given by

$$\mathrm{d} = \frac{a}{\sqrt{h^2 + k^2 + l^2}} \qquad (2\text{-}29)$$

where a is the lattice constant, and h, k, and l are the Miller indices. The interplanar spacings for other structures may be found in more advanced treatments of X-ray diffraction.

2.13.4 Diffraction Angle or Direction

Combining the interplanar spacing with Bragg's law, we find

$$\sin^2\theta = \left(\frac{\lambda^2}{4a^2}\right)(h^2 + k^2 + l^2) \qquad (2\text{-}30)$$

This equation predicts the *Bragg angle*, θ, from planes with Miller indices (hkl) for a cubic crystal of lattice parameter, a, when the incident wavelength is λ. What is experimentally measured is not θ, however, but rather 2θ, which is called the *diffraction angle.* The diffraction angle is the angle between the diffracted beam and the transmitted beam, as indicated by ray **3-3′** in Fig. 2-48.

2.13.5 Intensity of Diffracted Beam

The derivation of Bragg's law says nothing about the intensity of the diffracted beam and predicts that every plane (hkl) is capable of producing a diffracted beam. Even if the conditions of Bragg's law are satisfied, no diffraction may occur because of the particular arrangement of the atoms in the unit cell. Since the arrangement of atoms in the unit cell is specific to a crystal lattice, this forms the basis for determining the crystal structure of a material because there will only be certain planes that will yield diffraction intensities.

We cannot go much into the details and can only give a very brief summary of the theory. Basically, each atom independently has an atomic scattering factor, f, defined by

$$f = \frac{\text{amplitude of wave scattered by atom}}{\text{amplitude of wave scattered by one electron}} \quad (2\text{-}31)$$

The atomic scattering factor is a measure of the scattering efficiency of all the electrons in the atom. It is equal to Z if the scattering is in the same direction as the X ray, because all the Z electrons in the atom with atomic number Z will have scattered waves with the same phase. In other directions, however, f is a function of the direction of scattering.

When atoms are packed close to each other, the resultant scattered wave from one atom affects those emanating from nearby atoms. The resultant wave from all the atoms in the region will be the sum of all the scattered waves from all the atoms. Since the unit cell is a repeat pattern within a crystal, the interference of scattered waves from the atoms of the unit cell reflects the entire crystal. In the unit cell, each atom will have its coordinates (u, v, w) (vector coordinates in terms of the unit axes) and the scattered wave from this atom is expressed in exponential form as

$$Ae^{i\phi} = fe^{2\pi i(hu + kv + lw)} \quad (2\text{-}32)$$

where A is the amplitude and $e^{i\phi}$ is the complex exponential

$$e^{i\phi} = \cos\phi + i\sin\phi \quad (2\text{-}33)$$

with ϕ being the phase of the wave, (hkl) are indices of the diffracting plane, and (u, v, w) are the coordinates of the atoms in the unit cell. Each of the atoms in the unit cell will have a scattered wave expressed by Eq. 2-32 by virtue of its unique (u, v, w) in the unit cell. The resultant wave from the unit cell for the hkl diffracting plane is the sum of all the waves from the atoms of the unit cell, thus

$$F = f_1 e^{2\pi i(hu_1 + kv_1 + lw_1)} + f_2 e^{2\pi i(hu_2 + kv_2 + lw_2)} + f_3 e^{2\pi i(hu_3 + kv_3 + lw_3)} + \ldots$$

$$F_{hkl} = \sum_{1}^{n} f_n e^{2\pi i(hu_n + kv_n + lw_n)} \quad (2\text{-}34)$$

F is called the structure factor and, in general, is a complex number but its absolute value $|F|$ gives the amplitude of the resultant wave in terms of the amplitude of the wave scattered by a single electron, and defined as

$$F = \frac{\text{amplitude of resultant wave (all atoms of unit cell)}}{\text{amplitude of wave scattered by one electron}} \quad (2\text{-}35)$$

The intensity of the resultant diffracted coherent beams (waves with the same phase) from all atoms of the unit cell in a direction that obeys Bragg's law is proportional to $|F|^2$, the square of the amplitude of the resultant wave. Thus, if $F \neq 0$, the resultant wave will have intensity and, therefore, the diffraction from (hkl) will appear. On the other hand when $F = 0$, the resultant wave has no intensity and the diffraction from (hkl) will not appear. We shall now illustrate these with the cubic structures.

2.13.6 Structure Factors for Cubic Structures

In the following examples, the structures consist only of atoms of the same kind; therefore, there is only one f. But before we start, it is helpful to introduce some relationships of the complex exponential function. These are

$$\begin{aligned} e^{n\pi i} &= -1, \text{ when } n \text{ is odd} \\ &= +1, \text{ when } n \text{ is even} \\ e^{n\pi i} &= e^{-n\pi i}, \text{ where } n \text{ is any integer.} \end{aligned} \quad (2\text{-}36)$$

(a) We start with the *simple cubic structure* that contains only one atom per unit cell. The coordinates (u, v, w) of this atom in the unit cell can be taken as (0,0,0). Therefore,

$$F = fe^{2\pi i(0)} = f \quad \text{thus} \quad F^2 = f^2 \quad (2\text{-}37)$$

F is seen to be independent of h, k, and l and is the same for all diffracted beams. Thus, all planes will have a resultant diffracted beam (wave) and all (hkl) will appear.

(b) For the BCC structure, the unit cell has two atoms at (0,0,0) and at ($\frac{1}{2}$,$\frac{1}{2}$,$\frac{1}{2}$). For this case,

$$\begin{aligned} F &= fe^{2\pi i(0)} + fe^{2\pi i(h/2 + k/2 + l/2)} \\ &= f[1 + e^{\pi i(h + k + l)}] \end{aligned} \quad (2\text{-}38)$$

and

$$F = 2f \text{ when}(h + k + l) \text{ is even, and } F^2 = 4f^2. \quad \text{(2-38a)}$$

$$F = 0 \text{ when}(h + k + l) \text{ is odd, and } F^2 = 0. \quad \text{(2-38b)}$$

(c) For the FCC structure, there are four atoms located at (0,0,0), ($\frac{1}{2}$,$\frac{1}{2}$,0), ($\frac{1}{2}$,0,$\frac{1}{2}$), and (0,$\frac{1}{2}$,$\frac{1}{2}$). The resultant structure factor, F, is

$$F = f[1 + e^{\pi i(h+k)} + e^{\pi i(h+l)} + e^{\pi i(k+l)}] \quad \text{(2-39)}$$

When the Miller indices hkl are either all odd or all even, each of the exponential terms will equal one. Thus,

$F = 4f$ for unmixed indices, that is, either all odd or all even, and $F^2 = 16f^2$. (2-39a)

When the hkl are mixed, the sum of the exponentials is -1 and thus $F = 0$.

We note that the structure factors were derived for cubic lattices. However, the above structure factors are also applicable to any simple lattice with one atom, any body-centered lattice with two atoms, and any face-centered lattice with four atoms.

2.13.7 Determining the Crystal Structure and the Lattice Parameter

This task can be very demanding and is usually left to a specialist in crystallography. The process is illustrated here only for the cubic structures. The combined equations of Bragg's law and the interplanar spacing for cubic crystals can be expressed as

$$\frac{\sin^2\theta}{(h^2 + k^2 + l^2)} = \left[\frac{\sin^2\theta}{s}\right] = \frac{\lambda^2}{4a^2} \quad \text{(2-40)}$$

For diffraction studies, we usually start knowing the X-ray wavelength and a being the lattice constant will not vary. Therefore, the right side of the above equation is a constant. s is the sum of the squares of h, k, and l, and θ is the Bragg angle.

To determine the crystal structure and the lattice parameter of an unknown sample, we start by obtaining a diffraction pattern from powders of the unknown material. The powder diffraction pattern may either be recorded on films or on charts using X-ray diffractometers, a schematic of which is shown in Fig. 2-49. The powders are held on a thin rod and placed at the center of a carriage table. The rod is rotated as the X rays strike the powders and the diffracted beams are detected by a movable X-ray detector, moving from the exit port of the X rays on the circumference of the carriage table calibrated in angles. The X-ray detector detects the diffraction angle, 2θ, where diffraction (constructive interference) occurs as well as the intensity of the X rays; both are recorded on a chart. These are the raw data used by crystallographers to determine the crystal structure and the lattice parameter. We shall illustrate only the use of the diffraction angle, 2θ.

From the diffraction angles, our tasks are: (1) to identify the Miller indices of the diffracting planes for each of the diffraction angles, (2) to determine the crystal structure of the unknown sample, and (3) to determine the lattice parameter. To do these, we use Eq. 2-40 and note that the right side of the equation is constant and applies to all diffracting planes—each (hkl) corresponding to each of the diffraction angles, 2θ. We note that

$$\frac{\left[\dfrac{\sin^2\theta_n}{s_n}\right]}{\left[\dfrac{\sin^2\theta_1}{s_1}\right]} = 1 \quad \text{or} \quad \frac{\sin^2\theta_n}{\sin^2\theta_1} = \frac{s_n}{s_1} \quad \text{(2-41)}$$

where the subscript 1 refers to the first diffracting plane detected and n to the nth diffracting plane. We see that the ratio of the squares of the sine of the Bragg angles, θ's, equals the ratio of the sum of the squares of the h, k, and l of the Miller indices (hkl). The experimental values (left side) are used to determine the Miller indices (right side). Therefore, we determine the ratios of the sine squared values (experimental data); these are also the ratios of the sum of the squares of the Miller indices. These ratios are converted to the nearest whole numbers from which the Miller indices are obtained. From these Miller indices, we note the presence or absence of certain peaks and infer the crystal structure, see Fig. 2-50. After this, the value of the right side of the Eq. 2-40 is obtained for each of the (hkl); theoreti-

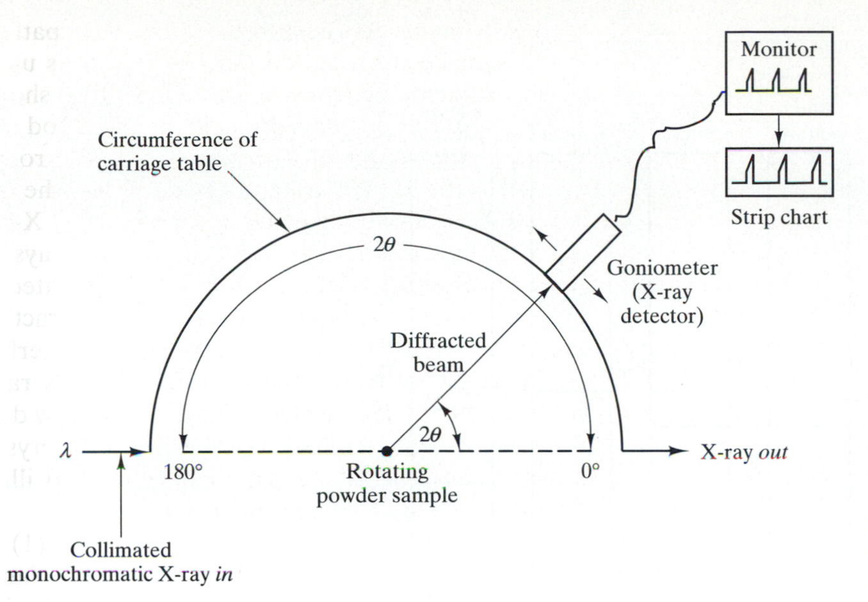

Fig. 2-49 Schematic of method to obtain diffraction patterns from crystal powders.

cally, the right side should be a constant. But since the left side values are experimental, there is some error. Thus, we average the experimental values on the left and use it to obtain the lattice parameter a on the right of Eq. 2-40.

The vertical strips in Fig. 2-50 represent the charts or films for the cubic lattices and the hexagonal lattice. The horizontal solid lines on these strips represent the positions of diffracting planes when the structure factor $F \neq 0$. Along the first strip (simple cubic), we see the values of s on the left and the corresponding (hkl) values on the right. We note that all the (hkl) have solid lines in the simple cubic strip, predicted in Eq. 2-36. In the other cubic structures, only certain (hkl) as indicated by Eqs. 2-38 and 2-39 will appear. It should be noted that the simple cubic structure is not exhibited by pure metallic materials. The experimentally determined (hkl)'s are compared with the BCC, FCC, and diamond cubic strips to determine the unknown's crystal structure.

EXAMPLE 2-17

A powder pattern from a metal containing a cubic structure was obtained with the following 2θ values. Determine the probable structure and the lattice constant of the metal. The wavelength of the X ray used is 1.542 Å (Cu K_α).

Line	2θ, deg.
1	40.6
2	58.4
3	73.4
4	87.2

SOLUTION: The solution is based on Eq. 2-43 where we take ratios of $\sin^2\theta$. The steps used in the solution are the following:

1. Obtain $\sin^2\theta$ values.
2. Use Eq. 2-41 to obtain the ratio $R = s_n/s_1$.
3. Multiply R by the smallest integer to make all the R values close to an integer, s.
4. From $s^2 = (h^2 + k^2 + l^2)$, we obtain the Miller indices of each plane (h k l).
5. Then we identify the probable structure of the metal with the help of Fig. 2-50 and the structure factors.
6. We use Eq. 2-40 to get $(\lambda^2/4a^2)$for each of the diffraction lines or peaks.
7. We use the average of the values in Step 6 to obtain the lattice constant.
8. We identify the metal by comparing the structure and the lattice constants with those given in tables.

The results of the steps are tabulated in Table EP 2-17.

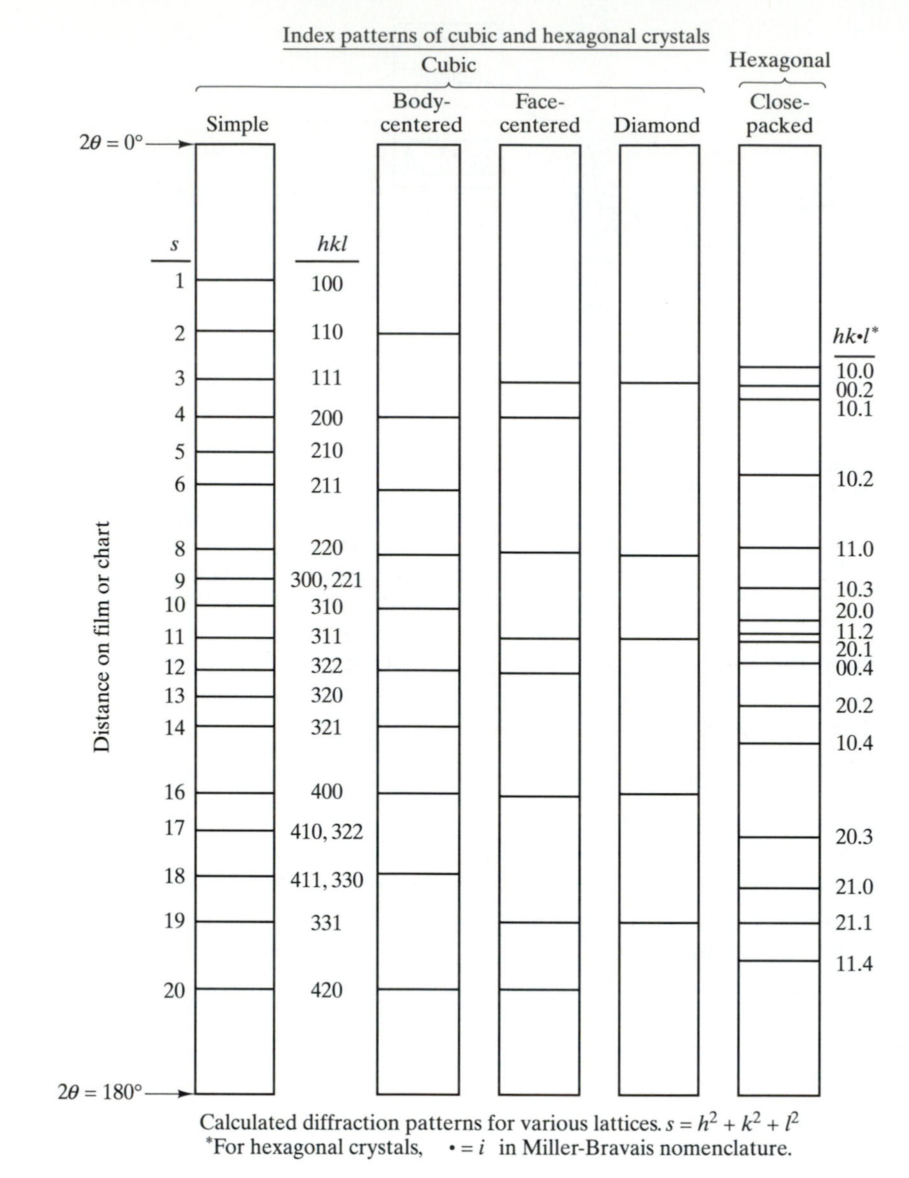

Fig. 2-50 Miller indices (hkl) of planes from simple cubic, BCC, FCC, diamond cubic, and HCP structures that will diffract.

Table EP 2-17

Line	2θ	θ	$\sin^2\theta$	$R = s_n/s_1$	s	s	$(h\,k\,l)$	$\lambda^2/4a^2$
1	40.6	20.3	0.1204	1	1	2	(110)	0.0602
2	58.4	29.2	0.2380	1.9767	2	4	(200)	0.0595
3	73.4	36.7	0.3572	2.9668	3	6	(211)	0.0595
4	87.2	43.6	0.4756	3.950	4	8	(220)	0.0594

Average value = 0.05965

Notes:

1. The first s values correspond to a simple cubic structure that does not exist for metals. These numbers are then multiplied by 2 to get the second s values from which the (hkl) are obtained.
2. The (hkl) values obtained are for a BCC structure , see Eq. 2-38 and Fig. 2-50.
3. The lattice constant calculated from the average value of the last column of 0.05965 and using the Cu K_α wavelength = 1.542 Å is 3.1568 Å.
4. The metal with BCC structure and a lattice constant closest to the experimentally observed value is tungsten that has a reported lattice constant of 3.1648 Å.

EXAMPLE 2-18

The diffraction peaks or lines from a powder pattern of a metal with a cubic structure, using Cu K_α radiation with $\lambda = 1.5418$ Å, are the following:

Line	2θ, deg.
1	38
2	44
3	64
4	78
5	82
6	98

Determine (1) the probable structure of the metal and the Miller indices of each of the planes corresponding to each diffraction peak, (2) the lattice constant, and (3) the identity of the metal.

SOLUTION: We follow the same steps as in Example 2-17. The tabulated results are:

Table EP 2-18

Line	2θ	θ	$\sin^2\theta$	$R = s_n/s_1$	$R \times 3$	s	$(h\,k\,l)$	$\lambda^2/4a^2$
1	38	19	0.1060	1	3	3	(111)	0.0353
2	44	22	0.1403	1.323	3.969	4	(200)	0.0351
3	64	32	0.2808	2.649	7.947	8	(220)	0.0351
4	78	39	0.3960	3.736	11.208	11	(311)	0.0360
5	82	41	0.4304	4.030	12.18	12	(222)	0.0359
6	98	49	0.5695	5.372	16.116	16	(400)	0.0356

Average value = 0.0355

Notes:

1. The (hkl) values are all odd or all even and correspond to an FCC structure. See Eq. 2-39 and Fig. 2-50.
2. The calculated lattice constant from the average value of the last column is 4.0918 Å.
3. The closest metal with an FCC structure and lattice constant of 4.0862 Å is silver.

Summary

The arrangement of atoms in the solid state may take one of the 14 Bravais space-point lattices. The characteristics of the most common metallic, covalent, and ionic structures were discussed. We have learned to determine the lattice parameters, the number of atoms (or ions) per unit cell, the volume-packing fraction, the directions, the planes, and the planar atomic and linear atomic fractions in cubic structures. We learned of the different imperfections, such as grain boundaries, dislocations, vacancies, interstitial, and substitutional atoms that may be present in actual crystals. We also learned to determine the structure and lattice parameters of cubic crystalline materials.

More importantly, we learned how the density and the deformability of a material are influenced by the crystal structure. We found out that while density is influenced by the volume packing fraction, the most significant factor that explains the range in densities of materials is the mass(es) of the atom(s) that comprise the unit cell. Metallic materials have high densities because they are packed denser and have higher atomic masses. Ceramic materials are lighter resulting from looser packing than metals and also the atoms in ceramics come partly from oxygen, carbon, nitrogen, boron, and silicon that have relatively low atomic masses. Polymers are light because they consist predominantly of C and H and are very loosely packed.

The planar and linear atomic fractions revealed the slip systems for the most common metallic structures. The total number of slip systems indicate the deformability of metals. HCP has only three; this reflects the difficulty in deforming HCP materials at room temperature.

References

1. M. F. Ashby and D. R. H. Jones, *Engineering Materials 2,* Pergamon Press, UK, 1986.
2. C. S. Barrett and T. B. Massalski, *Structure of Metals,* 3d ed., McGraw-Hill, 1966.
3. M. J. Burger, *Elementary Crystallography,* John Wiley, 1956.
4. A. Cottrell, *Introduction to the Modern Theory of Metals,* Institute of Metals, London, 1988.
5. A. Cottrell, *An Introduction to Metallurgy,* 2d ed., Edward Arnold, 1975.
6. B. D. Cullity, *Elements of X-ray Diffraction,* Addison-Wesley, 1956.
7. Composites, vol. 1, *ASM Engineered Materials Handbook,* 1987.
8. W. D. Kingery, H. K. Bowen, and D. R. Ullmann, *Introduction to Ceramics,* 2d ed., John Wiley, New York, 1976.
9. J. F. Shackelford, *Introduction to Materials Science for Engineers,* Prentice Hall, 1996.
10. J. Weertman and J. R. Weertman, *Elementary Dislocation Theory,* Macmillan, New York, 1964.
11. *ASTM Standard E* 112.
12. P.B. Hirsch, et al, *Electron Microscopy of Thin Crystals,* Plenum Press, New York, 1965.

Terms and Concepts

Allotropy or polymorphism
Amorphous materials
Anion
Basal plane
Body centered cubic
Body diagonal
Bragg's law
Bravais lattice (point lattice)
Cation
Cesium chloride structure
Constructive interference
Coordination number
Covalent structure
Crystallinity in polymers
Crystalline materials
Crystal structure
Crystal system
Deformability
Density
Destructive interference

Diamond cubic structure
Diffraction angle
Directions
Dislocation
Edge dislocation
Face centered cubic
Face diagonal
Frenkel defect
Grain boundary
Grain size
Hexagonal close packed
Ideal crystal structure
Imperfections
Intensity of diffracted beam
Intercepts of planes
Interplanar spacing
Interstitial
Interstitices
Ionic crystal structure
Lattice parameters
Linear atomic fraction
Liquid crystalline polymers
Long-range order
Miller-Bravais indices
Miller indices of directions
Miller indices of planes
Number of atoms per unit cell
Octahedral interstitice
Planar atomic fraction
Planes
Prismatic plane
Schottky defect
Screw dislocation
Short-range order
Single crystal
Slip direction
Slip plane
Slip system
Sodium chloride structure
Stacking fault
Structure factor
Subgrain boundary
Substitutional
Surface reconstruction
Surface relaxation
Tetrahedral interstitice
Tilt boundary
Translation vectors
Twin
Unit cell
Vacancy
Vector coordinates
Volume-packing fraction
Wave scattering
X-ray diffraction
X-ray wavelength
Zinc blende structure

Questions and Exercise Problems

2.1 If you were given a material and were asked to determine whether it is crystalline or amorphous, how will you determine it?

2.2 What is a unit cell? What do the crystal systems signify? What is a point space lattice and its relationship to the crystal systems?

2.3 Estimate the lattice parameter of each of the following elements from their atomic sizes and compare it with that given in the Appendix. (i) Cesium (Cs)—BCC; (ii) Aluminum (Al)—FCC; (iii) cadmium—HCP; and (iv) Diamond (carbon)—Diamond cubic.

2.4. Estimate the lattice parameter of each of the following elements from their atomic sizes and compare it with that given in the Appendix. (i) Iron—BCC; (ii) copper—FCC; (iii) magnesium—HCP; and (iv) germanium—diamond cubic.

2.5 Calculate the largest atom relative to the host atoms that can locate in the octahedral site of the HCP (see Fig. 2-10a).

2.6 Calculate the largest atom relative to the host atoms that can locate in the tetrahedral site of the HCP (see Fig. 2-10b).

2.7 Estimate the lattice parameter cubic silicon carbide with zinc blende structure as in GaAs.

2.8 Estimate the lattice parameter of InAs, a compound semiconductor, similar to GaAs, with a "zinc blende" type structure.

2.9 Determine the most probable structure and the lattice parameter of nickel oxide (NiO).

2.10 Determine the most probable structure and the lattice parameter of lithium fluoride (LiF).

2.11 Determine the most probable structure and the lattice parameter of zinc sulfide (ZnS).

2.12 Determine the most probable structure and the lattice parameter of cesium chloride (CsBr).

2.13 From the lattice parameters of the two allotropes of cobalt given in the Appendix, determine the volume change during cooling when the structure changes from FCC to HCP. Is it contraction or expansion?

2.14 From the lattice parameters of the two allotropes of titanium given in the Appendix, determine the volume change during cooling when the structure changes from BCC to HCP. Is it contraction or expansion?

2.15 Indicate and discuss the relative importance of the factors that determine the density of a material. Give examples of metallic, ceramic, and polymeric materials and explain the differences in their densities.

2.16 Determine the volume-packing fraction and estimate the density of each of the elements in Problem 2.3.

2.17 Determine the volume-packing fraction and estimate the density of each of the elements in Problem 2.4.

2.18 Estimate the theoretical density of the cubic silicon carbide in Problem 2.7.

2.19 Estimate the theoretical density of GaAs. Use parameter in Table 3-5.

2.20 Estimate the theoretical density of InAs. Use parameter in table 3-5.

2.21 Estimate the density of nickel oxide, NiO.

2.22 Estimate the density of lithium fluoride, LiF.

2.23 Estimate the density of zinc sulfide, ZnS. Use parameter in table 3-5.

2.24 Estimate the density of cesium bromide, CsBr.

2.25 Draw a two-dimensional point lattice with unit axes, $a = b$, and the angle between them is 90°. Indicate in this drawing, the direction from the vector point (2,5) to the vector point (−2,2). What is the direction, its angle with the horizontal, and its magnitude?

2.26 If the unit axes of a two-dimensional point lattice are such that $a = 2b$ and the angle between them is 90°, determine the direction from the vector point (2,5) to the vector point (−2,2), its angle with the horizontal, and its magnitude.

2.27 Determine the Miller indices of the direction: (a) from the point (2,0,3) to the point (4,5,6); (b) from (−1,3,5) to (3,2,1); (c) from (3, 5, 7) to (1, −3,5); (d) from (1,4,5) to (−2, 1, −1); and (e) from (2,3,6) to (5,0,2).

2.28 Determine the Miller indices of the direction from: (a) (0,1/2, −1/3) to (1/2,0,2/3); (b) (1/2, −2/3,1/4) to (2/3,1/2,1/4); (c) (2/3,1/3,1/4) to (−1/3,2/3,1/2).

2.29 Indicate within a unit cubic cell the directions in Problem 2.27.

2.30 Indicate within a unit orthorhombic cell the directions in Problem 2.28.

2.31 Determine the Miller indices of the directions within the unit cubic cell in Fig. P2.31.

2.32 Determine the Miller indices of the directions within the unit hexagonal cell in Fig. P2.32.

2.33 Determine the Miller and the Miller-Bravais indices of the planes in Fig. P2.33.

2.34 Sketch within a unit hexagonal cell the following directions and planes: (a) [021], (b) $[\bar{2}12]$, (c) $[2\bar{1}1]$, (d) $(11\bar{2}2)$, (e) $(1\bar{2}11)$, and (f) $(20\bar{2}3)$.

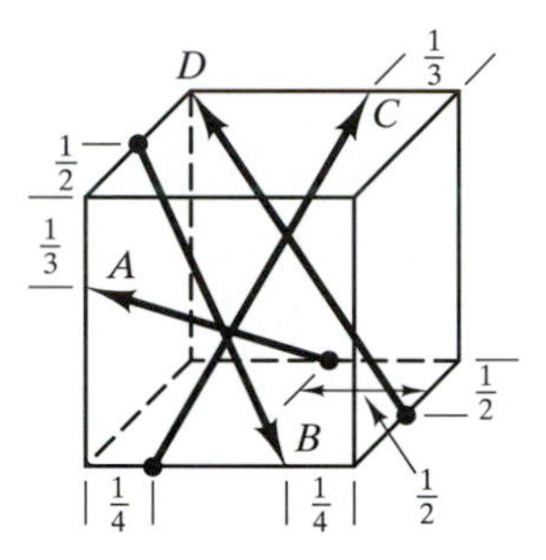

Fig. P2.31 Directions in a cubic unit cell.

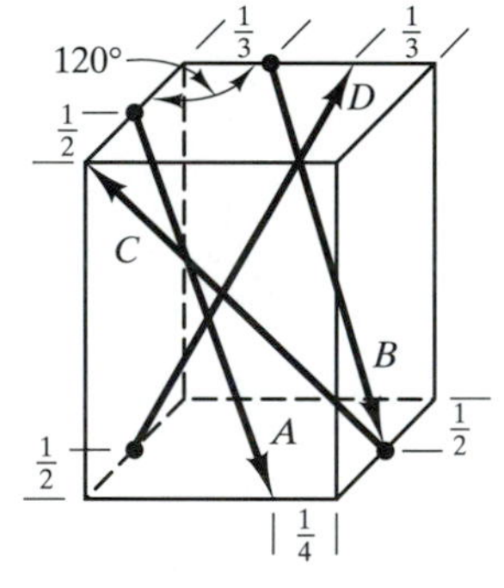

Fig. P2.32 Directions in a unit hexagonal cell.

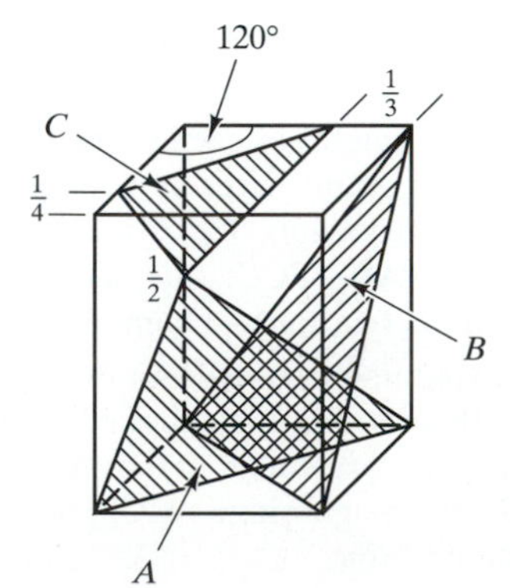

Fig. P2.33 Planes in a unit hexagonal cell

2.35 Sketch the {100}, the {110}, and the {111} planes of the diamond cubic structure and determine the planar atomic fractions in each of these families of planes.

2.36 Sketch the <100>, the <110>, and the <111> directions of the diamond cubic structure and determine the linear atomic fractions along these directions.

2.37 Sketch all the {111} slip planes and slip directions in a single crystal FCC metal.

2.38 Sketch all the {110} slip planes and slip directions in a single crystal BCC metal.

2.39 Sketch the {0001} plane and the slip directions in a single crystal HCP metal.

2.40 Determine the average number of grains per sq. in. at 100 ×, when the ASTM grain size number is 10? What is the average size (diameter) in millimeters of the grains?

2.41 The average number of grains per sq. in at 200 × was counted to be about 128 grains per sq. in. Determine the ASTM grain size number and the average size (diameter) of the grains in millimeters.

2.42 What is the interplanar spacing of the {211} in iron?

2.43 What is the interplanar spacing of the {311} in nickel?

2.44 What is the interplanar spacing of the {311} in silicon?

2.45 Using a monochromatic Cu K_α (with $\lambda = 0.1542$ nm), determine the expected diffraction angles (of a powder pattern) of the first three diffracting planes from iron.

2.46 Using a monochromatic Cu K_α (with $\lambda = 0.1542$ nm), determine the expected diffraction angles (of a powder pattern) of the first three diffracting planes from nickel.

2.47 Using a monochromatic Cu K_α (with $\lambda = 0.1542$ nm), determine the expected diffraction angles (of a powder pattern) of the first three diffracting planes from silicon.

3

Electrical Properties of Materials

3.1 Introduction

The electrical properties of materials relate to how much current is produced when the materials are subjected to an applied electric field. The resulting current is due to the motion of electrons or other charge carriers and depends on the type of bonding of the atoms in the material as discussed in Chap. 1. Materials producing current under an applied electric field are referred to as *electrically conductive or metals* while those that do not produce current are *electrically non-conductive, nonmetals, or insulators.* This traditional differentiation is done at absolute zero temperature. At temperatures above 0 K, some of the nonmetals or insulators are able to conduct current, and these are called *semiconductors.* The latter have influenced our daily lives in so many ways so as to justify calling our current historical period the "semiconductor age." Therefore, it is important to learn about the basic characteristics of these materials. We shall also learn in this chapter about the new and upcoming materials called *superconductors.*

We shall concentrate in this chapter on electronic electrical conductivity, that is conductivity due to "free electrons." We shall first relate how free electrons are generated in the different bonded materials, both crystalline and amorphous, before we discuss conductivity and semiconductivity of materials.

3.2 Metals and Nonmetals (Insulators)

At 0 K, materials are either metals or nonmetals. This classification is based on whether electrons are free or not free to move under an applied electric field. Whether the electrons are free or not depends on the bonding of the large aggregate of atoms and the availability of quantum states that electrons can occupy.

3.2.1 Relationship to Bonding and Energy Bands

As described in Chap. 1, metals are predominantly the elements that easily give away their valence electrons. We should recall that the valence electrons are those occupying the sp or d-states beyond the closed shell configuration (s^2p^6). When large aggregates of atoms are bonded metallically together, the atoms are packed very closely so that their electron orbitals overlap each other. This condition allows the free electrons to move from orbitals of one atom to orbitals of other atoms and constitutes the basic characteristic of metallic materials—the electrons are shared by all the atoms in the aggregate. However, the free electrons cannot all have the same energy. They must comply with the Pauli exclusion principle, which states that each quantum state can only be occupied by two electrons that have opposite spin numbers. Because of this, a "band of energy levels" is produced.

We illustrate the formation of the "energy band" with the Group 1 (or I) elements. To start, we recall from Chap. 1 that the electron configuration in the ground state of the free unbonded atom of these elements is s^1p^0 beyond the closed shell. For N unbonded atoms, there are N s-states and N free electrons that will be shared in metallic bonding. When the atoms metallically bond, all the N s-states cannot have the same energy level because of the Pauli Exclusion Principle and as a result, an energy band of discrete energy levels of s-states is formed; the difference between energy levels is very small. As in the free unbonded atoms, the shared valence electrons must distribute themselves in a manner whereby the lowest s-states are filled first. Each of the s-states can accommodate two electrons. Since there are N s-states and N free electrons, the "ground state" of the bonded atoms will have the shared free electrons occupying only $N/2$ s-states, thus $N/2$ s-states will be left unfilled. This is schematically shown in Fig. 3-1a and the highest energy level occupied by the electrons is called the *Fermi energy,* E_F. The vacant unfilled s-states are the quantum states the free electrons can easily occupy when the material is subjected to an electric field; therefore, current flows rather easily.

For elements in Group 2 (or II), the electron configuration of each unbonded atom is s^2p^0. Thus for N unbonded atoms, there are $2N$ free electrons and N s-states. When these N atoms are bonded, all the N s-states are all occupied by the $2N$ free electrons, as shown in Fig. 3-1b. However, the p-states are available and under an applied electric field, the free electrons will move to the higher energy vacant p-states and can assume the electron configuration, s^1p^1. This occurs easily because the s and p are subshells in the electron configuration and the energy difference between subshells is small. The promotion of an electron from the s-state to the p-state requires energy but this energy is somewhat compensated by the fact that the electron is being separated from the repulsive field of two negative charges in the s-state. This is very similar to the hybridization of the electrons in the s^2p^2 to s^1p^3 which was discussed in Chap. 1. Thus, Group 2 elements are also good electric conductors. We can extend this line of thought to other metals with partially filled sp-states and also to the transition metals (with electrons occupying d-states), as shown in Fig. 3-1c. We conclude that metals have empty quantum states that the free electrons can occupy when an electric field is applied to them.

We now contrast metallic bonded materials to covalently bonded materials. The predominant covalently bonded materials involve elements in Group 14 (formerly IV) in Fig. 1-5. These elements have the hybridized s^1p^3 condition. Each of these four states are directed to the corners of a tetrahedron, discussed and illustrated in Fig. 1-12 and shown again as Fig. 3-2. In covalent bonding, the atoms at the corners of the tetrahedron share locally one of their electrons with the central atom. The central atom shares a total of eight electrons and acquires the electron configuration s^2p^6. The four s and p states are fully occupied and there are no empty states unfilled. In terms of the energy band for N atoms, all the $4N$ states are filled with electrons and thus the energy band is full as shown in Fig. 3-3. Since there are no empty states left, electrons cannot move when an applied electric field is applied to the materials. Thus, covalently bonded materials are generally nonmetals at 0 K—examples are carbon and silicon. Other covalently bonded materials such as organic polymers, and silica or quartz are also predominantly nonconductive.

In order to make the covalently bonded materials conductive, it would be necessary to extricate and free electrons from the covalent bonded state, as shown schematically in Fig. 3-4b, in order that they can move when an electric field is imposed. In Fig. 3-4a, each of the lines between silicon atoms represents an electron shared between the two atoms. The extrication of an electron leaves a *vacant electron site which is called a hole* in Fig. 3-4b and which has a positive charge. The freed electrons are in a

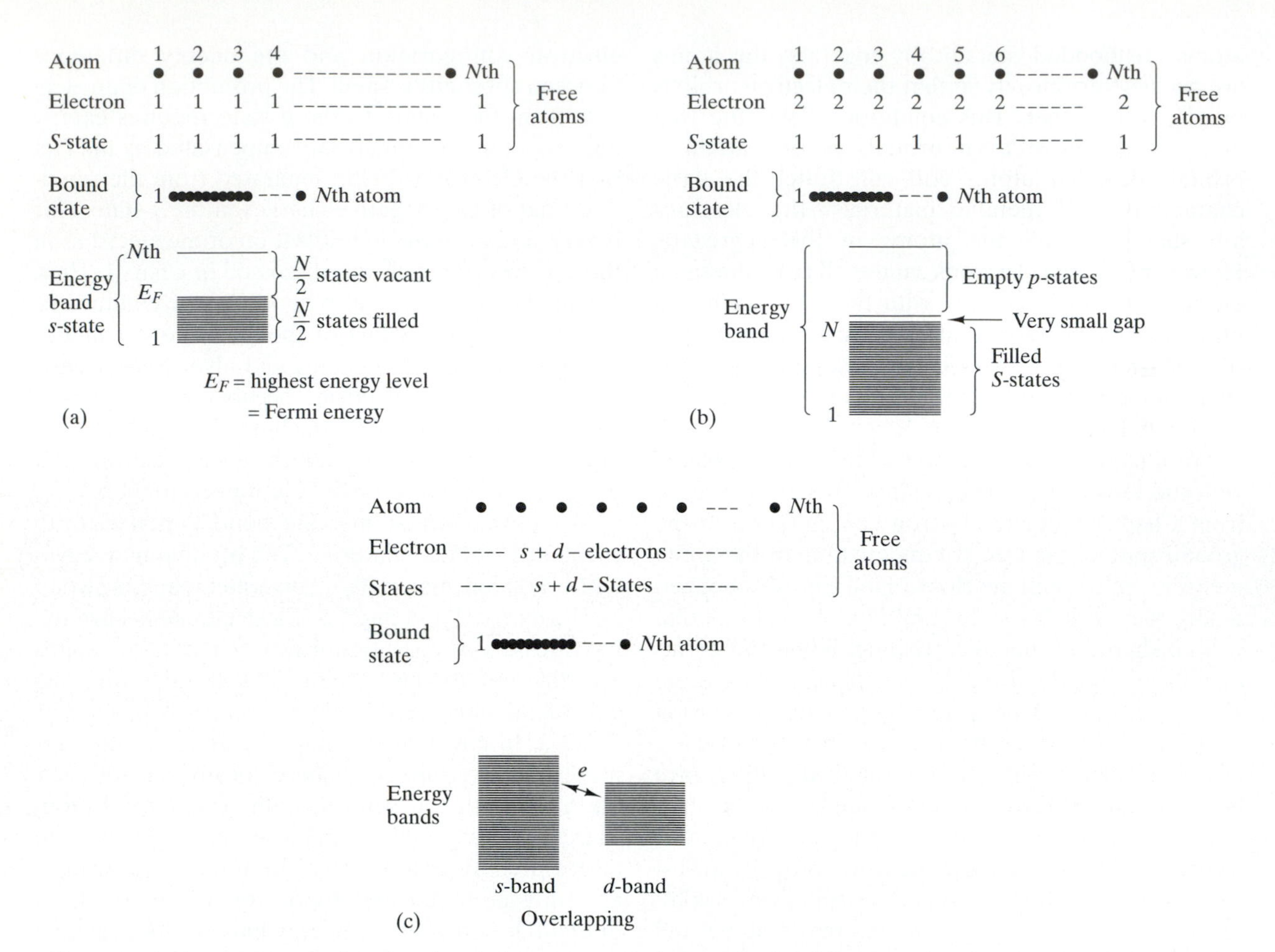

Fig. 3-1 (a) Electron energy band in bonded group 1 (alkali) metals. (b) Electron energy band in bonded group 2 (alkaline earth) metals. (c) Overlapping electron energy *s*- and *d*-bands in transition metals.

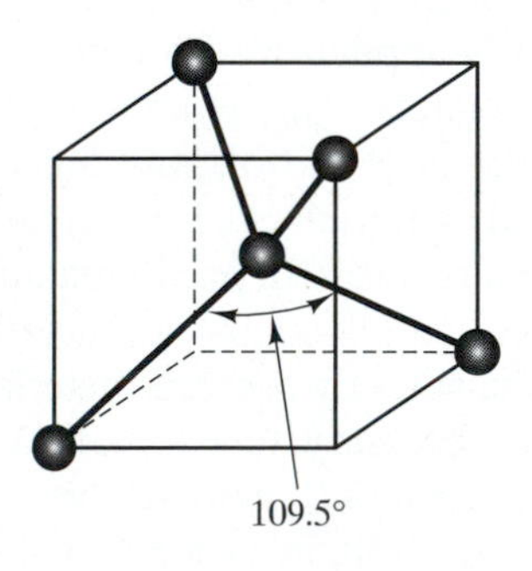

Fig. 3-2 Tetrahedral bonding in group 14 elements - C, Si, Ge, and Sn.

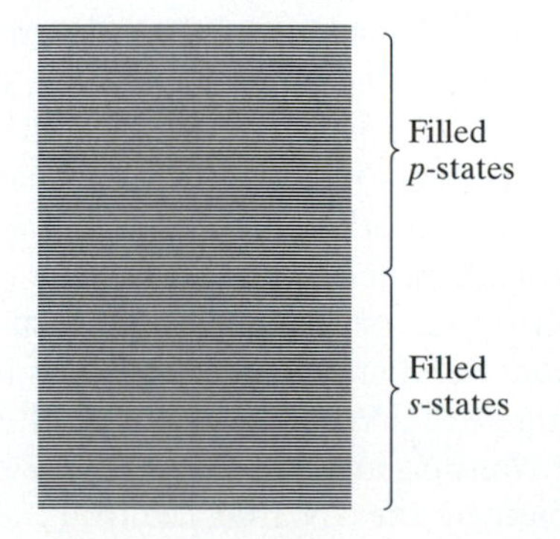

Fig. 3-3 Electron energy band in bonded group 14 elements.

conducting state and are at a much higher energy state. To differentiate these freed, mobile, and electrically conducting electrons from the covalently bonded, immobile electrons, the extricated electrons are said to be in a *conduction band* and the covalently bonded electrons are said to be in a *valence band,* as illustrated in Fig. 3-4c. These two bands are separated by an energy gap, E_g, which is equivalent to the energy required to extricate the electrons from the covalent bonded state. If the E_g is small, thermal

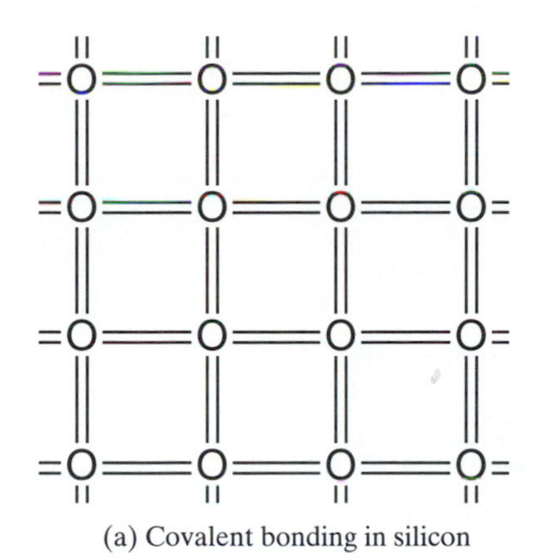

(a) Covalent bonding in silicon

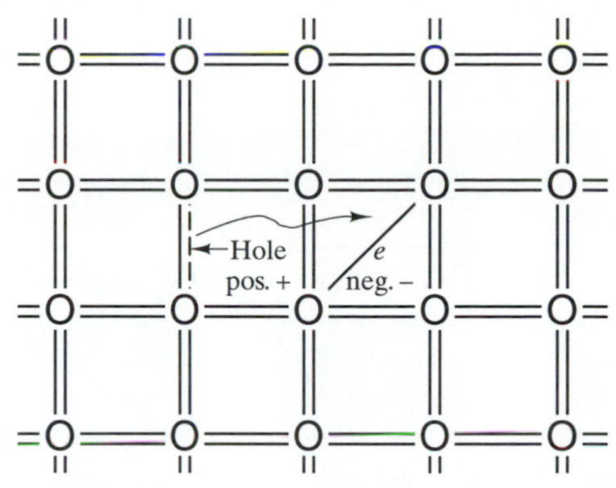

(b) Broken covlent bond creates negative free electrons and positive holes

(c)

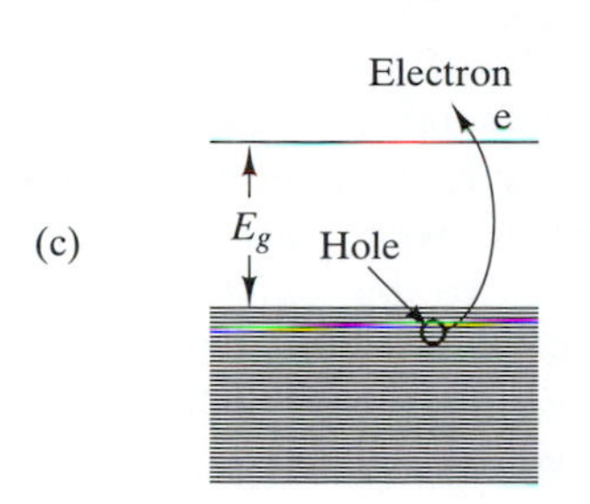

Fig. 3-4 (a) Two-dimensional covalent bond representation in silicon, (b) creation of positive holes and free electrons when covalent bonds are broken, (c) representation of a broken covalent bond relative to energy band.

Table 3-1 Group 14 Diamond Structure Elements.

Element	Symbol	Atomic Mass, g/mol	Lattice Parameter, Å	Density, g/cm^3	Energy Gap, eV	Mobility @ RT, cm^2/v.s Electron	Hole
Carbon	C	12.01	3.5668	3.51	5.4	1,800	1,400
Silicon	Si	28.09	5.4307	2.328	1.107	1,900	500
Germanium	Ge	72.61	5.6575	5.323	0.67	3,800	1,820
α-Tin	α-Sn	118.71	6.4912	5.765	0.0	2,500	2,400

energy or fluctuation of the covalent bonded electron is sufficient to bring electrons to the conduction band and these covalent bonded materials become metallic. A good example of this, of course, is tin that is in Group 14 with C and Si. Tin has essentially zero E_g and is ordinarily considered a metal. Germanium and silicon have E_g values that makes them conductive at high temperatures. As indicated earlier, these are semiconductors and of the two materials, silicon is primarily the commercial material because of its abundant supply. The strength of the covalent bond in diamond is so high that it is difficult to break the bond and, therefore, diamond remains nonmetallic and electrically nonconductive even at high temperatures. Table 3-1 shows the E_g values of the C, Si, Ge, and Sn as schematically indicated in Fig. 3-5.

3.2.2 Nonmetal to Metal Transition and Vice Versa*

There is now enough evidence to show that nonmetals can be made to be electrically conductive or have metallic character even at absolute zero. Likewise, metallic materials can be made to be non-

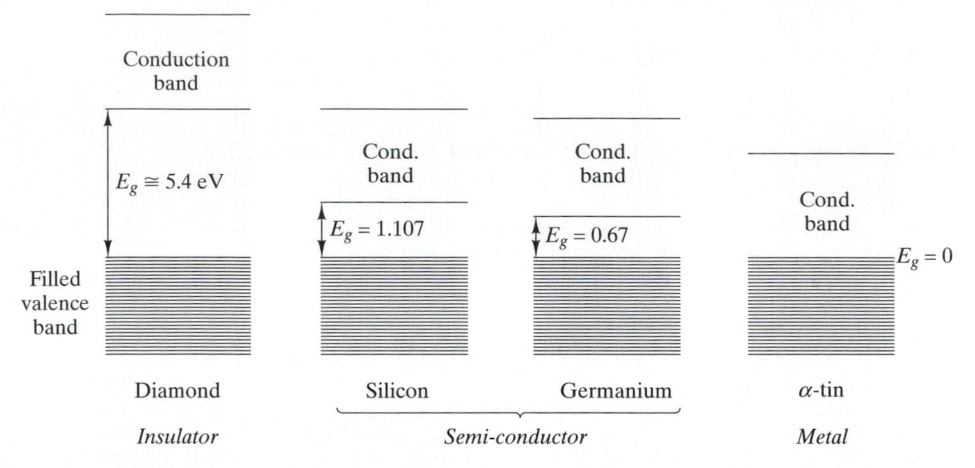

Fig. 3-5 Energy gaps in group 14 elements determine electrical behavior of the elements.

metals. In both instances, the condition for metallic behavior is a critical density of atoms per unit volume that forces the atomic electron orbitals to overlap each other and allows the electrons to be delocalized, free, and itinerant. Evidence of this concept is provided in the doping of silicon, in polymer composite, in doping of "conducting polymers," in "oxide bronzes," in the inert gases of hydrogen and xenon, and in the cesium metal.

As a first example let us take silicon. As indicated above, silicon is a semiconductor at high temperatures but at 0 K, it is an insulator or nonmetal. If we substitute phosphorus (P) atoms for silicon atoms, there is one excess electron for every substitutional (see Sec. 2.11.2) P atom because P is from Group 15 (formerly Group V) that has five valence electrons. Four of these five electrons participate in the covalent bonding, as the four electrons from the replaced silicon atom participated in covalent bonding. The excess electron is uncombined and is free to move. However, if the P concentration is low, the randomly distributed substitutional P atoms are too far apart and each excess electron hovers and is localized near the positively charged P atom, as shown in Fig. 3-6. As the concentration of P atoms increases, the distance between P atoms decreases and a critical concentration is reached when the electron clouds from each P atom will overlap. At this concentration, the electrons can move from one electron cloud (orbit) to neighboring electron clouds and they become itinerant. This situation is similar to metallic bonding where the positive ion cores (P atoms) are embedded in a sea or cloud of free electrons. Thus, *we have transformed a nonmetallic silicon to a metallic silicon at 0 K by adding a critical concentration of P atoms.*

The addition of P or any impurity atoms in silicon is called *"doping"* by the semiconductor industry and Figs. 3-7 and 3-8 show the experimental evidence of the nonmetal to metal transition. In Fig. 3-7, the resistivities of various phosphorus-doped silicon are plotted as a function of temperature as they are cooled from 100 K to 0 K. We note that the resistivity at ~0 K decreases dramatically by about four to six orders of magnitude when the concentration of

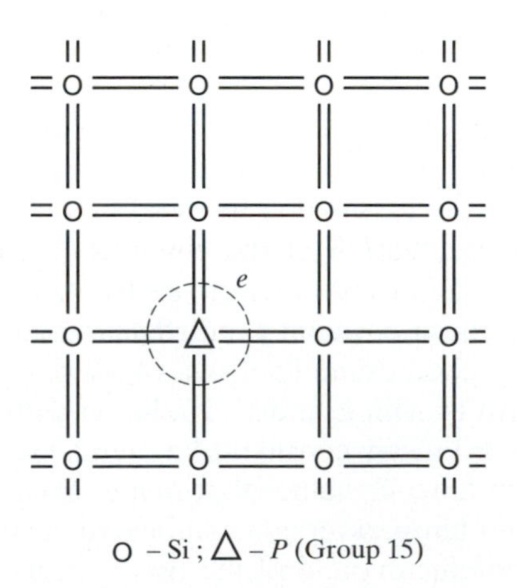

Fig. 3-6 Substitution of group 15 element (e.g., *P*) for Si provides a localized (almost free) electron.

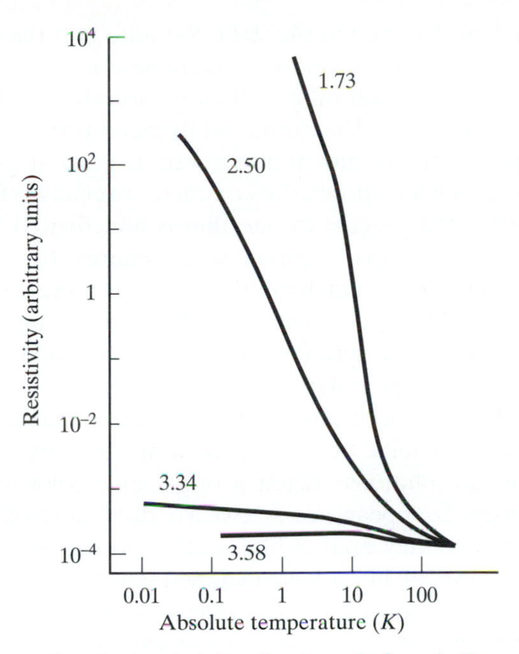

Fig. 3-7 Electrical resistivity of various P-doped silicon as a function of temperature; concentration given on each curve in 10^{24} P atoms/m^3 (10^{18} atoms/cm^3). [Ref. 1]

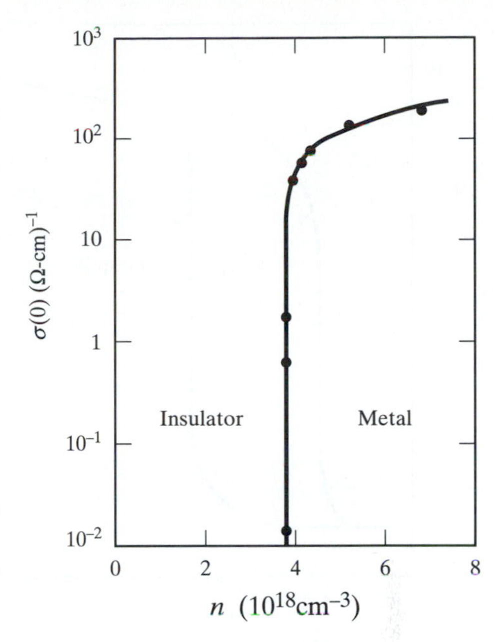

Fig. 3-8 Conductivity of P-doped silicon at 0 K, σ(0 K), as a function of concentration of P atoms; nonmetal to metal transition occurs at abrupt change in conductivity. [Ref. 2]

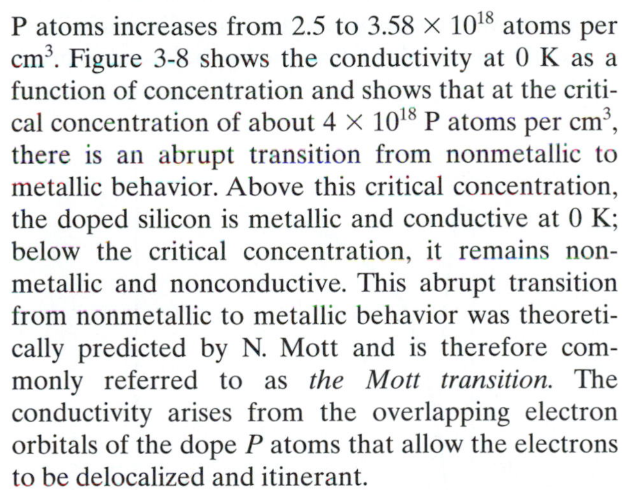

P atoms increases from 2.5 to 3.58 $\times$ 10^{18} atoms per cm^3. Figure 3-8 shows the conductivity at 0 K as a function of concentration and shows that at the critical concentration of about 4 $\times$ 10^{18} P atoms per cm^3, there is an abrupt transition from nonmetallic to metallic behavior. Above this critical concentration, the doped silicon is metallic and conductive at 0 K; below the critical concentration, it remains nonmetallic and nonconductive. This abrupt transition from nonmetallic to metallic behavior was theoretically predicted by N. Mott and is therefore commonly referred to as *the Mott transition.* The conductivity arises from the overlapping electron orbitals of the dope *P* atoms that allow the electrons to be delocalized and itinerant.

Another example is a composite of polymers and metals to make the polymers conductive. This is simply done by uniformly dispersing metallic fillers in the ordinarily nonconductive polymer. At a critical threshold concentration of the metallic filler, the polymer becomes conductive. This conductivity arises either from continuous contact of the metallic fillers or from electon tunneling through a thin layer of the polymer. In either mechanism, the conductivity arises from close proximity of the metallic fillers. The form of the filler has a significant influence on the critical concentration, as shown schematically in Fig. 3-9; this demonstrates that fibrous fillers have lower critical concentration than granular fillers. The critical concentration of granular fillers is about 40 to 50 percent by volume to form a continuous metallic phase. For the fibers, as low as 5 percent by volume with fiber diameter of 0.01 mm and length 1 mm, will make the composite conducting in Fig. 3-9.

Certain polymers can also be made conducting by doping and thus they are called *conducting polymers.* These are polymers with high molecular weights and have in their backbone structure alternating double and single bonds. A structure with alternating double and single bonds is called a *conjugated structure;* these polymers are thus called *conjugated polymers.* The first of these to exhibit conductivity is polyacetylene, $(CH)_n$, doped with either strong electron donors or strong electron acceptors, for example, I, AsF_5, IBr, SbF_6, ClO_4, Na, et cetera. There are two forms of $(CH)_n$, the cis and the trans forms, shown in Fig. 3-10; the cis form has the H atoms on the same side of the carbon atoms with the double bond, while

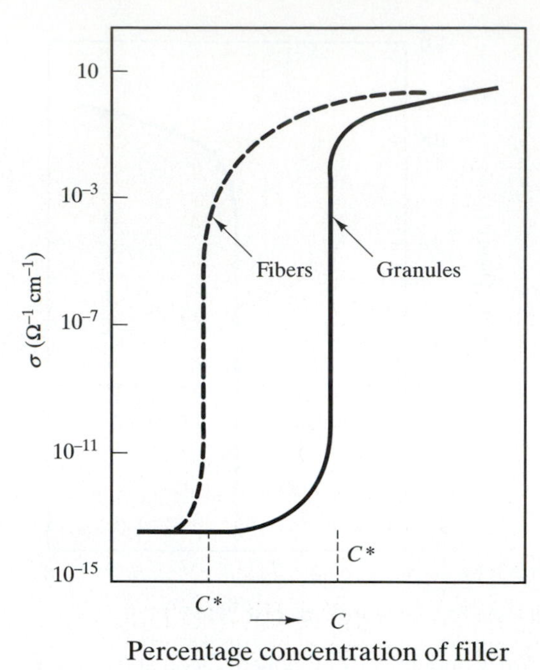

Fig. 3-9 Schematic of conductivity of polymer filled with metallic fibers or granules; sudden increase in conductivity is nonmetal to metal transition. [Ref. 3]

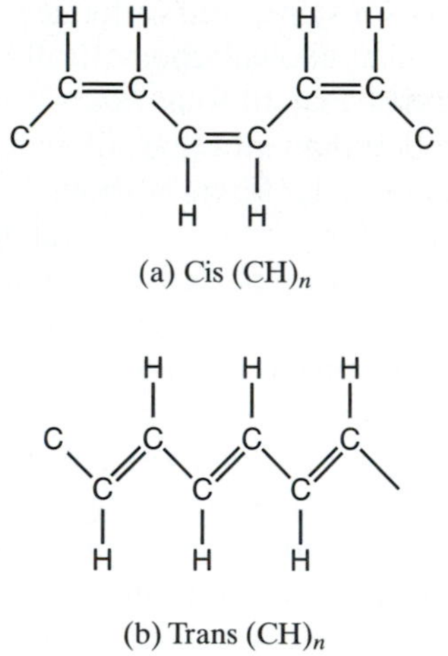

(a) Cis $(CH)_n$

(b) Trans $(CH)_n$

Fig. 3-10 Cis (a) and trans (b) forms of polyacetylene - a conjugated polymer. [Ref. 3]

the trans has H atoms on opposite sides of the carbon atoms with the double bond. The more stable form is the trans form. The conductivity of the undoped $(CH)_n$ varies from 10^{-9} (ohm-cm)$^{-1}$ for the cis form to 10^{-5} (ohm-cm)$^{-1}$ for the trans form. With doping, the conductivity can be raised to 10^3 (ohm-cm)$^{-1}$, as depicted in Fig. 3-11. We note that the conductivity of polyacetylene increases rapidly with dopant concentration and then it saturates to about 10^3 (ohm-cm)$^{-1}$. The nonmetal to metal transition in polyacetylene occurs at about 4 mole percent of the dopant at which point, the conductivity changes from an activated process to one that is unactivated. (An activated process requires some energy to get it started while an unactivated does not require energy or time.) After polyacetylene, other conducting polymers have been synthesized such as polyphenylene, polypyrrole, and polyphenylene sulfide.

It has also been observed that liquid ammonia can be made metallic by adding alkali metals. Very dilute ammonia solutions exhibit a bright blue color and a nonmetallic appearance. Above a certain range of concentration, they appear like molten bronze and conduct electricity like a heavily doped semiconductor.

Addition of alkali or electropositive metals to *"oxide bronzes"(ceramic materials)* also exhibits nonmetal to metal transition. The oxide bronzes are defined as having the formula M_xTO_n, where M represents an electropositive metal and T is a transition element. They are called oxide bronzes after the discovery by Wöhler in 1823 of the "metallic lustre" and the gold-yellow color exhibited by the reaction product between Na_2O_4 and tungsten oxide, WO_3, under a

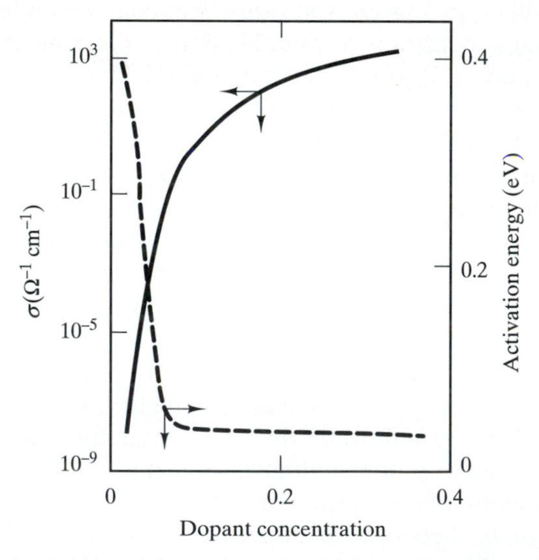

Fig. 3-11 Electrical conductivity and activation energy of polyacetylene $(CH)_x$ as a function of dopant concentration. [Ref. 3]

reducing atmosphere of hydrogen. It turns out that the product is Na_xWO_3, and is equivalent to alkali metal (Na) doping of WO_3 that is ordinarily a semiconductor and thus nonmetallic at 0 K. There are conflicting results but the nonmetal to metal transition occurs with a critical concentration, x_c, of about 0.2 to 0.25. This transition is again due to delocalization of electrons—in this case, the d-electrons of the transition element. Again, the delocalization occurs by the overlapping of the d-electron orbitals. But, without going into the details of the mechanism, the overlapping is not the direct contact of d-electron orbitals, but rather is achieved via the indirect overlapping with the oxygen 2p-orbitals as schematically illustrated in Fig. 3-12. This mechanism appears to hold as well for other types of oxides or ceramics.

The test of the concept that materials can be made conductive if we increase the atomic density is the case of the inert gases H_2 and xenon. These gases were observed to be electrically conductive at very high pressures: for hydrogen at 200 GPa and xenon at 150 GPa. Iodine was also observed to be metallic at 160 kbars. At high pressures, the atoms or molecules are packed closely together and their electron orbitals can overlap to allow the electrons to be delocalized.

The corollary of the above concept is that if we separate the atoms of a metallic material, we should be able to transform the metallic material to a nonmetallic material. This is, in fact, the case for the metal, cesium, shown in Fig. 3-13. The concentration of cesium is given as the reciprocal of the cube root of n, the electron density. Being a cube root, the abscissa is the average distance between atoms, and we see that if we separate the atoms, the metallic cesium becomes a nonmetal.

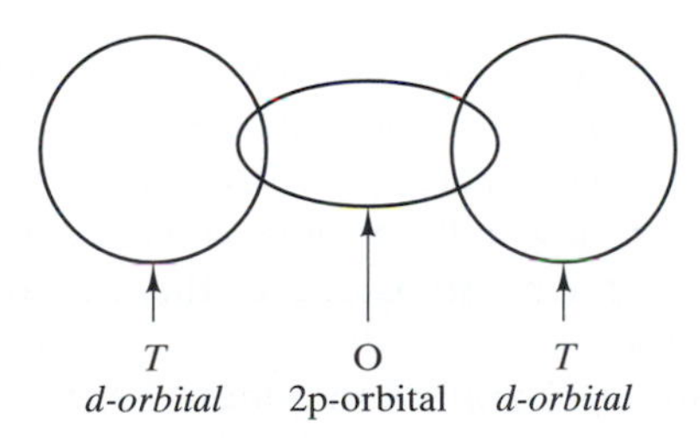

Fig. 3-12 Schematic of indirect overlapping of transition metal, T, d-orbitals via oxygen 2p-orbitals in oxide bronzes, M_xTO_n.

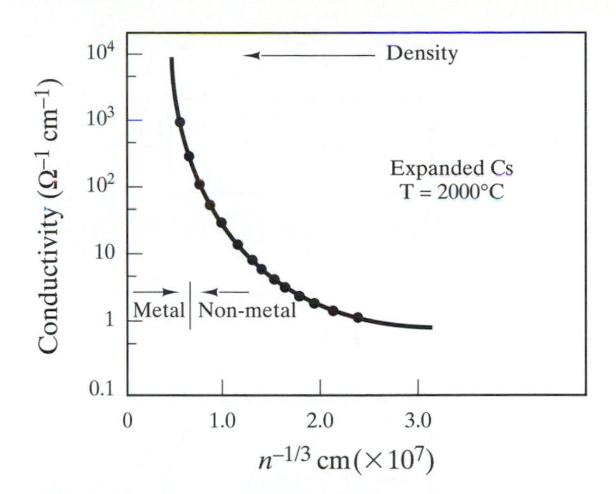

Fig. 3-13 Metal to nonmetal transition in cesium as distance between cesium atoms increases at 2000°C. [Ref. 2]

3.3 Electrical Conductivity

Electrical conductivity reflects the motion of a charge carrier in an electric field. The most prominent charge carrier is the electron because it is light (weight) and as indicated above, electrical conductivity occurs when electrons can be delocalized. These "free" electrons are in constant motion within the metallic material and in the absence of an electric field, their random motion produces no net flow of electrons.

The following is a general discussion of electrical conductivity or flow of charge when an electric field is imposed on any material. Consider a wire connected to a battery and a potential difference, V, in volts, is imposed on that wire, Fig. 3-14a. If the length of the wire is L, then an electric field, E, is established at every point in the wire. The strength of the field is

$$E = \frac{V}{L} \tag{3-1}$$

This electric field will exert a force on the electron and will set it in motion according to

$$F = e\,E = m\,a \tag{3-2}$$

where e is the electronic charge, m is the mass of the electron, and a is acceleration. The direction of electron motion is opposite that of E and we say that an electric current, I, is established by the net flow of electrons. If the potential difference, V, is increased, I increases, and if the current increases linearly with

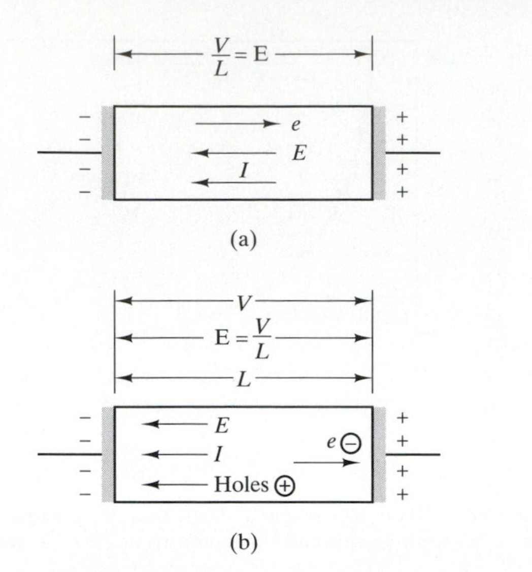

Fig. 3-14 Motion of charge carriers (a) electrons (metals), and (b) electrons and holes (semiconductors) under an applied electric field, E.

the potential difference, V, then the material is said to obey Ohm's Law, that is,

$$V = I R \tag{3-3}$$

where V is in volts, I is in amperes, and R is the resistance of the material in ohms. The resistance, R, is

$$R = \frac{\rho L}{A} = \frac{L}{\sigma A} \tag{3-4}$$

where ρ is the resistivity, σ is the conductivity, A is the cross-sectional area, and $\sigma = 1/\rho$.

It is convenient to express the current as current density, J, which is defined as

$$J = \frac{I}{A} \tag{3-5}$$

Combining Eqs. 3-3 and 3-4, and then using Eq. 3-1, we obtain

$$J = \frac{I}{A} = \sigma \cdot \frac{V}{L} = \sigma E \tag{3-6}$$

We now introduce the basic definition of current, that is,

$$\begin{aligned} I &\equiv \text{Net flow of charge per unit time} \\ &= \text{Coulombs/time} = \text{Amp. sec/sec} = n \cdot e \end{aligned} \tag{3-7}$$

where n is the total number of charge carriers flowing per unit time and e is the charge per carrier. Thus, the current density, J, is a charge flux, that is, the flow of charge per unit area per unit time.

Being a charge flux, the flow of charge through a conductor can be compared with the flow of fluids through a pipe. In fluid flow, the fluid mass flux (mass per unit area per unit time) is obtained by multiplying the mass density of the fluid with the average fluid velocity. Thus, by analogy,

$$J = (e\,N)\,v_d \tag{3-8}$$

where (eN) is the charge density, e is the charge of the carrier, N is the number of charge carriers per cm^3, and v_d is the drift velocity of the charge carrier. The drift velocity is

$$v_d = a\,\tau \tag{3-9}$$

where a is the acceleration as in Eq. 3-2 and τ is the mean free time of the charge carrier. The latter is the mean time between collisions or scattering of the charge carrier. Substituting the acceleration from Eq. 3-2, the drift velocity is

$$v_d = \frac{e\tau}{m} E \tag{3-10}$$

Thus,

$$J = \frac{e\tau}{m}(eN)E \tag{3-11}$$

and equating Eqs. 3-8 and 3-11,

$$\sigma = \left(\frac{e\tau}{m}\right)(eN) = \mu(eN) \tag{3-12}$$

where μ is the mobility of the charge carrier. *Thus, we see that conductivity may be regarded as the product of two factors, charge density (eN) and mobility (μ) of the carrier.* We may have high conductivities because there are lots of carriers or because they acquire high drift velocities (by having high mobilities). In metals, the charge carriers are the electrons and their mobilities are quite low (about two orders of magnitude lower than those of semiconductors); their high conductivity is due to the high density of free electrons.

Equation 3-12 applies to a material that has only one type of charge carrier—the electrons in metals. If there is more than one type of charge carrier, the total conductivity of the material will be the sum of the individual conductivities of the different charge carriers; that is, Eq. 3-12 is used for each type of carrier and the total conductivity is

$$\sigma_T = \sigma_1 + \sigma_2 + \sigma_3 + \cdots = \sum_{i=1}^{n} \sigma_i$$
$$= \sum_{i=1}^{n} \mu_i (eN)_i \quad (3\text{-}13)$$

Semiconductors and ceramic oxides have more than one type of charge carrier. As seen in Fig. 3-4b, when we extricate an electron from the covalent bond state, there is a vacant electron site in the bonding that acquires a positive charge since a negative charge was taken away. This is called a "hole" and this positive charge will also move when an electric field is imposed. The motions of holes and electrons under an applied electric field are schematically shown in Fig. 3-14b. Thus, in semiconductors, we add the conductivities of the electrons and the holes. In ceramic oxides, it is also possible to have free electrons and holes in the structure in addition to the charged cations (positive) and anions (negative). Thus, in ceramic oxides we have four types of carriers. We can get the relative contribution of each carrier type to the total conductivity by defining the transference number, t_i, for each carrier, as

$$t_i = \frac{\sigma_i}{\sigma_T} \quad (3\text{-}14)$$

and, obviously

$$\sum_{i=1}^{n} t_i = 1 \quad (3\text{-}15)$$

The relative contributions of the different charge carriers in some ceramic materials and compounds are shown in Table 3-2 where t_+, t_-, and $t_{e,h}$ signify the contributions of the cation, anion, and the electrons and holes. We see from this table that, except for CuCl, we obtain conductivities of these materials at high temperatures and that depending on the compound, the majority carrier (the one with the highest transference number) could be one of the four carriers. The conduction from the cations and anions is called ionic conduction and that due to the electrons and holes is called electronic conduction. We shall limit our discussion from here on to electronic conduction.

3.4 Electronic Conduction in Metals and Alloys

Again from Eq. 3-12, the two factors that contribute to conductivity is the mobility and the charge density. For pure metals, the charge density is constant and the only factor that affects conductivity will be the mobility of the electrons. Since e and m are constants for the electron, the factor that determines the mobility is τ, the mean free time between collisions of the electron with scattering centers in the lattice. The longer τ is, the higher are the mobility and the conductivity. τ may be expressed as

$$\tau = \frac{l}{v_d} \quad (3\text{-}16)$$

Table 3-2 Transference numbers of cations t_+, anions t_-, and electrons and holes $t_{e,h}$ in some ceramic compounds [Ref. 4].

Compound	Temperature (°C)	t_+	t_-	$t_{e,h}$
NaCl	600	0.95	0.05	
KCl	600	0.88	0.12	
AgCl	20–350	1.00		
BaF_2	500	···	1.00	
CuCl	20	0.00	···	1.00
	366	1.00	···	0.00
ZrO_2 + 18% CeO_2	1500	···	0.52	0.48
+ 50% CeO_2	1500	···	0.15	0.85
$NaO \cdot CaO \cdot SiO_2$ glass	···	1.00 (Na^+)		

where l = mean free path of the electron, that is, the average distance traveled between successive collisions. Since resistivity, $\rho = 1/\sigma$, we see that ρ is inversely proportional to l, the mean free path, that is $\rho \propto 1/l$. Thus, the shorter l is, the higher is the resistivity or resistance.

Using copper as an example, the variation of the resistance of metals with temperature is shown in Fig. 3-15. At temperatures above 100 K, the resistance increases linearly with T and this is what engineers are quite familiar with because most engineering applications are above this temperature. Below 100 K, the resistance is proportional to T^5 and at absolute zero, would still show some finite resistance as shown in the inset of Fig. 3-15. This metallic behavior is compared to superconductivity where the resistance goes to zero below the critical temperature. The increase resistance (or resistivity) with increasing temperature means that the mean free path, l, between collisions gets shorter and there are more collisions of the electron. These collisions with the lattice ions cannot be pictured, however, as the simple classical bumping of particles against each other. If these were the case, the mean free path will be of the order of the atomic spacing and would hardly change with temperature. The collisions are considered as the scattering of the "electron waves, ψ." In this way, we can understand why an ideally perfect lattice structure will have no resistance because the scattered waves from this structure are in phase with each other. This was called *coherent scattering* when we discussed X-ray diffraction. Electrical resistance or resistivity is due to the incoherent scattering of the electron waves at places where the lattice periodicity is disturbed. *Irregularities on an atomic scale are particularly effective* in scattering electron waves, while larger features such as strain and stress fields have little influence. *Thermal vibrations* throw the atoms out of alignment and they cause the scattering responsible for the increase in resistivity for pure metals shown in Fig. 3-15. *Other atomic irregularities are interstitial or substitutional solid solution atoms,* and an example of the effect of substitutional solute atoms on the resistivity of copper is shown in Fig. 3-16a. The addition of other elements to a pure metal is called alloying and, in general, alloys exhibit higher resistivity than pure metals as seen in Fig. 3-16b, where we see that copper nickel alloys have much higher resistivity than either of the pure metals,

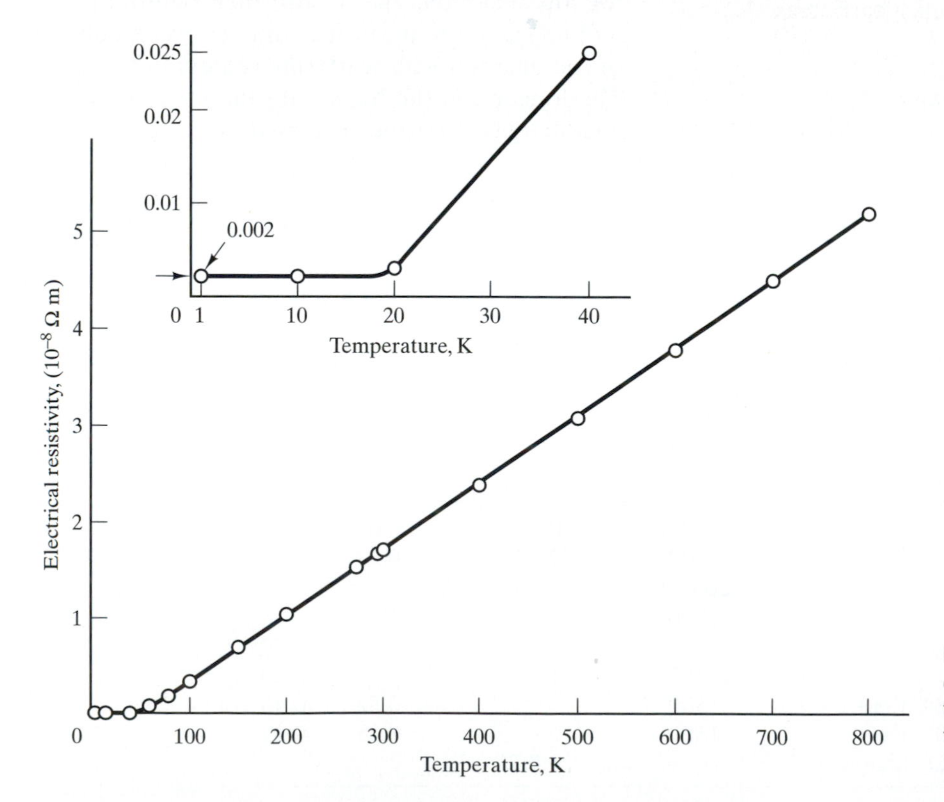

Fig. 3-15 Resistivity of pure copper as a function of temperature K (Inset from 0 K to 40 K). [Based on data from [Ref. 8]

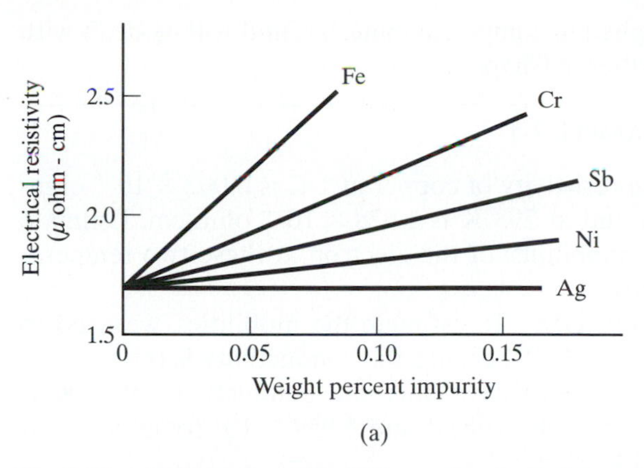

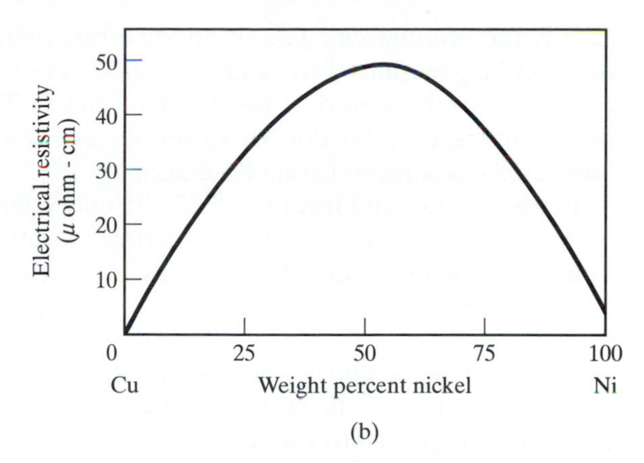

Fig. 3-16 (a) Effect of small additions of various elements on RT electrical resistivity of pure copper; (b) electrical resistivity of Cu-Ni Alloys at 0°C. [Ref. 9]

Cu or Ni. As discussed in Sec. 2.12.2 and illustrated in Fig. 2-38, atoms in solid solution are part of the lattice of the host material. In Fig. 3-16a, the host material is copper and the various elements shown in the figure are the atoms substituting for the copper in its lattice. In some instances, the addition of an element to another results in the formation of a compound. When this happens, the compound separates from the host lattice and while the compound is still within the material, it is no longer part of the regular arrangement of the host lattice. The effect of this compound on the electrical resistivity of a metal is far less significant than that when the alloying element is in solid solution. This phenomenon is shown in Table 3-3 for the resistivity of aluminum.

Table 3-3 Effect of elements in and out of solution on the electrical resistivity of aluminum.

Element	Maximum solubility in Al, %	Average increase in resistivity per wt%, μΩ · cm In solution	Out of solution(a)
Chromium	0.77	4.00	0.18
Copper	5.65	0.344	0.030
Iron	0.052	2.56	0.058
Lithium	4.0	3.31	0.68
Magnesium	14.9	0.54(b)	0.22(b)
Manganese	1.82	2.94	0.34
Nickel	0.05	0.81	0.061
Silicon	1.65	1.02	0.088
Titanium	1.0	2.88	0.12
Vanadium	0.5	3.58	0.28
Zinc	82.8	0.094(c)	0.023(c)
Zirconium	0.28	1.74	0.044

Note: Add above increase in the base resistivity for high-purity aluminum. 2.65 μΩ · cm at 20°C (68°F) or 2.71 μΩ · cm at 25°C (77°F).

(a) Limited to about twice the concentration given for the maximum solid solubility, except as noted.

(b) Limited to approximately 10%.

(c) Limited to approximately 20%.

Source: L.A. Willey, Alcoa Research Laboratories. [Ref. 10]

Let us now examine the quantitative effects of thermal vibration and the impurity solid solution elements on the resistivity of metals. Let us suppose that the mean free path, l, is due to thermal lattice vibrations and impurity atoms. Let l_{TH} and l_I be the mean free paths from thermal lattice vibrations and impurity atoms. If the electron has traveled a distance X, it will have experienced X/l_{TH} and X/l_I collisions with lattice vibrations and with impurity atoms. The mean free path, l, reflects the total number of collisions experienced over the distance traveled. Thus,

$$\frac{1}{l} = \frac{\frac{X}{l_{TH}} + \frac{X}{l_I}}{X}$$

$$\frac{1}{l} = \frac{1}{l_{TH}} + \frac{1}{l_I} \tag{3-17}$$

Since the resistivity ρ is inversely proportional to l, we see that the resistivity is the sum of the resistivities from thermal lattice vibrations and from impurity atoms,

$$\rho = \rho_{TH} + \rho_i \tag{3-18}$$

This is the *Mathiessen's rule* of additive resistivities that can be generalized to include other factors that contribute to the electrical resistivity of metals. The other factors may be due to vacancies, interstitial atoms, dislocations, and grain boundaries.

In pure metals and from Fig. 3-15, the only contribution to the resistivity is thermal lattice vibration. Thus, at any temperature above 0 K, the resistivity is

$$\rho = \rho_{0\,K} + \rho_{TH} \tag{3-19}$$

where $\rho_{0\,K}$ is the residual resistivity or the resistivity at the flat portion of the curve in Fig. 3-15. The thermal resistivity from 0 to 100 K is,

$$\rho_{TH} = aT^5 \tag{3-20}$$

and from 100 K upwards,

$$\rho_{TH} = bT \tag{3-21}$$

where a and b are constants. The latter linear relationship enables us to interpolate or extrapolate values of resistivity, if two resistivities are known at two temperatures above 100 K. The resistivity at any temperature can be found from the following equation, if a resistivity at a known or reference temperature is known

$$\rho_T = \rho_R + \beta(T - T_R) \tag{3-22}$$

where ρ_T and ρ_R are the resistivities at temperature T and reference temperature, T_R and β is the linear temperature resistivity coefficient.

The resistivity contribution from impurity atoms depends on their concentration. For dilute alloys, the resistivity increases linearly with concentration. This is known as the *Nordheim's rule* and is expressed as

$$\rho_I = I \cdot x_I \tag{3-23}$$

where I is a constant and x_I is the atomic fraction of the impurity atoms. For more concentrated single-phase solid solutions,

$$\rho_I = I \cdot x_I(1 - x_I) \tag{3-24}$$

This equation predicts the Nordheim's rule when $x_I \ll 1.0$.

For multiphase alloys, the resistivity is obtained by the *rule of mixtures,* which states that

$$\rho = \sum f_i \rho_i \tag{3-25}$$

where f_i is the volume fraction, and ρ_i is the resistivity of the individual phases. The rule of mixtures is also applied in composite materials and will be dealt with further in Chap. 15.

EXAMPLE 3-1

The resistivity of copper at 1 K is 0.002×10^{-6} ohm-cm and at 293 K is 1.678×10^{-6} ohm-cm. Estimate the mobilities of the electron at these two temperatures.

SOLUTION: To estimate the mobilities, we need to use Eq. 3-12, because the conductivity is the reciprocal of resistivity. Thus, the conductivity at 1 K is 5×10^8 (ohm-cm)$^{-1}$ and 5.959×10^5 (ohm-cm)$^{-1}$ at 293 K. Since,

$$\sigma = \mu\,(e\,N)$$

we need to know the density of charge carriers, N, per cubic cm. We recall that in metallic bonding free electrons are given off by the atoms such that the positive ion core has a closed shell configuration. From Table 1-2, we see that copper has an atomic number of 29 and the electronic configuration is $1s^2 2s^2 2p^6 3s^2 3p^6 3d^{10} 4s^1$. From this we see that each copper atom will give off one electron. Thus, we need to calculate the number of atoms per cu. cm, which will also give us the number of electrons (carriers per cu. cm). We may calculate the atoms per cm^3 in two ways: (1) using the density, and (2) using the characteristics of the unit cell.

(1) Using the density,

$$N = \text{Density, g/cm}^3 \times \frac{1}{\text{Atomic Mass, g/mol}}$$

$$\times \text{ Avogadro's Number, atoms per mol}$$

$$= \frac{8.96 \times 6.022 \times 10^{23}}{63.546}$$

$$= 8.491 \times 10^{22}$$

(2) Using the characteristics of the unit cell. Copper has an FCC structure, with four atoms per unit cell, and a lattice parameter of 3.6146 Å.

$$N = \frac{4 \text{ atoms per unit cell}}{(3.6146 \times 10^{-8})^3}$$

$$= 8.47 \times 10^{22}, \text{ atoms per cm}^3$$

The charge density is $eN = 1.6022 \times 10^{-19} \times 8.491 \times 10^{22} = 13{,}604$ C/cm^3. Thus, the mobilities are

(a) at 1 K,

$$\mu = \frac{5 \times 10^8}{\text{ohm-cm}} \times \frac{\text{cm}^3}{13{,}604\ \text{C (Amp-sec)}}$$

$$= 36{,}753\ \frac{\text{cm}^2}{\text{volt-sec}}$$

(b) at 293 K,

$$\mu = \frac{5.959 \times 10^5}{\text{ohm-cm}} \times \frac{\text{cm}^3}{13{,}604\ \text{C (Amp-sec)}}$$

$$= 43.80\ \frac{\text{cm}^2}{\text{volt-sec}}$$

Note the dramatic drop in the mobility of the electron as temperature increases.

EXAMPLE 3-2

The resistivity of pure copper at 100 K is 0.348×10^{-6} (ohm-cm) and the coefficient of resistivity is 0.0068 (ohm-cm) $\times 10^{-6}$/°C or K. Estimate the resistivity of copper at 300 K and at 700 K.

SOLUTION: The resistivity at a temperature is estimated by using Eq. 3-22,

$$\rho_T = \rho_R + \beta(T - T_R)$$

(a) at 300 K,

$$\rho_{300} = 0.348 \times 10^{-6} + 0.0068 \times 10^{-6}(300 - 100)$$
$$= 1.708 \times 10^{-6}\ \text{ohm-cm}$$

The observed value of resistivity at 300 K is 1.725 micro-ohm-cm.

(b) at 700 K,

$$\rho_{700} = 0.348 \times 10^{-6} + 0.0068 \times 10^{-6}(700 - 100)$$
$$= 4.428 \times 10^{-6}\ \text{ohm-cm}$$

The observed value of resistivity at 700 K is 4.514 micro-ohm-cm.

EXAMPLE 3-3

From the resistivities of copper at 1 K and at 293 K given in Example 3-1, determine the resistivity contribution from the thermal vibration of the atoms.

SOLUTION: We use Eq. 3-19,

$$\rho_{TH} = \rho_{293\,K} - \rho_{0\,K}$$
$$= (1.676 - 0.002) \times 10^{-6} = 1.674 \times 10^{-6}$$

EXAMPLE 3-4

Values of the resistivities at 293 K of various compositions of Cu-Ni solid solution alloys are given in Table EP 3-4. The accuracy of the values is within ±5 percent. Determine contributions of impurity atoms to the resistivity of the alloy compositions and the ranges in composition of the applicability of Eqs. 3-23 and 3-24. The resistivities of pure copper at 1 K and 293 K are given in Example 3-1 while the corresponding values for pure nickel are 0.003 and 6.93×10^{-6} ohm-cm.

SOLUTION: In solving this problem, we have (1) to separate the contributions of thermal vibrations from the impurity or solute atoms and (2) determine the solute in the range of compositions and express the concentration in atomic fractions or percents. The solute is considered the material with the lesser concentration.

Table EP 3-4 Resistivities of copper-nickel alloys at 293 K.

Composition, wt % Cu	Resistivity, Ω-cm $\times 10^{-6}$	Composition, wt % Cu	Resistivity, Ω-cm $\times 10^{-6}$
99 (1 wt% Ni)	2.85	1	8.08
95 (5 wt% Ni)	7.71	5	12.50
90 (10 wt% Ni)	13.89	10	17.82
85 (15 wt% Ni)	19.83	15	23.35
80 (20 wt% Ni)	25.66	25	35.11
70 (30 wt% Ni)	36.72	30	41.79
60 (40 wt% Ni)	45.38	40	47.73
		50	50.05

Thus, when we convert the composition to atomic percent or fraction, we find that the two left columns are for nickel being the solute and the two right columns are for copper being the solute. Thus, for nickel as the solute, the thermal contribution that has to be substracted from the resistivity is 1.674×10^{-6} ohm-cm (Example 3-3) for copper. The conversion to atomic percent or fraction, x_1, is given by

$$x_1 = \frac{\dfrac{(\text{wt. }\%)_1}{(\text{atomic mass})_1}}{\dfrac{(\text{wt. }\%)_1}{(\text{atomic mass})_1} + \dfrac{(\text{wt. }\%)_2}{(\text{atomic mass})_2}}$$

Thus, we need the atomic masses of copper and nickel; these are 63.546 and 58.693 g/mol. A conversion example for 1 wt% Ni shows the corresponding atomic fraction to be

$$x_{\text{Ni}} = \frac{1/58.693}{1/58.693 + 99/63.546} = .0108$$

The summary of the calculations are tabulated below with nickel as a solute (Example 3-4, Table EP 3-4 (S1)) and with copper as solute (Example 3-4, Table EP 3-4 (S2)).

The applicability of either Eq. 3-23 or Eq. 3-24 is determined by the value of I calculated from the data. Allowing for experimental error, I should be relatively constant in the range of composition. We note from the two tables that Eq. 3-23 is applicable up to 30 wt% of either Cu or Ni as the solute. Eq. 3-24 is applicable in concentrations greater than 30 wt% solute as clearly indicated in Table EP3-4 (S_2).

3.5 *Superconductivity and Its Characteristics**

Superconductivity was alluded to in the introduction to this chapter and we are now going to learn what it is, its characteristics, and some of its current and potential applications.

Superconductivity is electronic conduction with zero resistance below a certain critical temperature,

Table EP3-4 (S1)—Nickel as solute, $\rho_{TH} = 1.674 \times 10^{-6}$, ohm-cm.

Wt. % Ni	x_l Atom Fraction Ni	ρ, Resistivity, Ω-cm, 10^{-6}	Solute Resistivity $\rho_l = (\rho - \rho_{TH})$	$I = \rho_l/x_l$ Eq. 3-23	$I = \rho_l/[x_l(1 - x_l)]$ Eq. 3-24
1	0.0108	2.85	1.174	108.7	109.9
5	0.0539	7.71	6.034	111.9	118.3
10	0.1074	13.89	12.214	113.7	127.4
15	0.1604	19.83	18.154	113.2	134.8
20	0.2103	25.66	23.984	114.0	144.4
30	0.3169	36.72	35.044	110.6	161.9
40	0.4192	45.38	43.704	104.3	179.6

Table EP3-4 (S2)—Copper as solute, $\rho_{TH} = 6.93 \times 10^{-6}$, ohm-cm.

Wt. % Cu	x_l Atom Fraction Cu	ρ, Resistivity, Ω-cm, 10^{-6}	Solute Resistivity $\rho_l = (\rho - \rho_{TH})$	$I = \rho_l/x_i$ Eq. 3-23	$I = \rho_l/[x_l(1 - x_l)]$ Eq. 3-24
1	0.009	8.08	1.15	127.8	128.9
5	0.046	12.50	5.57	121.1	126.9
10	0.093	17.72	10.89	117.1	129.1
15	0.140	23.35	16.42	117.1	136.4
25	0.235	35.11	28.18	119.9	156.8
30	0.284	41.79	34.86	122.7	171.4
40	0.381	47.73	40.80	107.1	173.0
50	0.480	50.05	43.12	89.8	172.8

magnetic field, and current density. It is one of those serendipity discoveries that was not purposely sought after. The discovery came when Onnes and his collaborators at the University of Leiden (Netherlands) in 1911 were testing a hypothesis that at low temperatures the electons will freeze and thus will lead to an abrupt rise in resistivity. Realizing that impurities will play a vital role in resistivity measurements, mercury was chosen because it was very easy to purify. Instead of getting an abrupt rise in resistivity, there was an abrupt drop in resistivity to zero at about 4.2 K (this was later corrected to 4 K), as seen in Fig. 3-17.

Because its resistance is zero, a superconductor can conduct electric current with no loss in power. With no loss in power, superconducting systems do not require a continuous supply of power to replenish the losses and there is no heating effect. Can you imagine the far reaching consequences of this effect? It will enable higher currents to be carried through wires, higher fields in magnets, and further miniaturization as well as faster computers to be manufactured. The currents in interconnects of computer chips are low but the interconnects are very small in size, about 1 μm in width. Therefore, the current densities are very high and heating of the chips is currently a big problem. With no resistance, the size of the interconnects can be further reduced and more interconnects can be packed in the "silicon real estate". The closer these are, the faster will be the response of the computers.

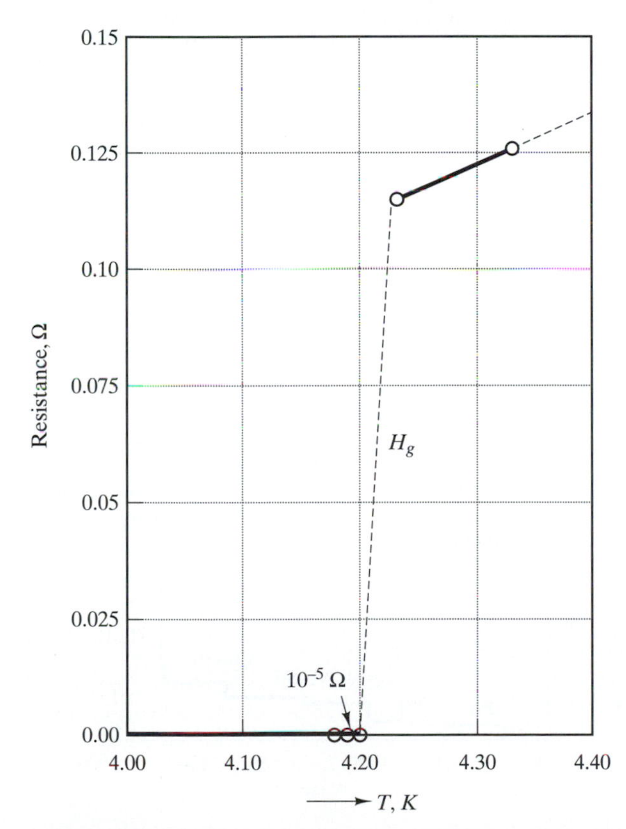

Fig. 3-17 Discovery of superconductivity in 1911 by Onnes. Plot is resistance in ohms (ordinate) as function of temperature, K, for mercury [Ref. 7].

Twenty years after Onnes's discovery, another important *unique property of superconductors* was discovered. This is the *expulsion of the magnetic field from the interior of the material* and is called the *Meissner effect,* see Fig. 3-18. Ordinary magnets have magnetic fields inside the material and they attach to magnetic objects; however, the Meissner effect will push away a magnet and the latter will get suspended or levitated in air. This effect is being explored in the Maglev (*mag*netic *lev*itated) trains. Another application is their usage as magnetic shields to screen electronics against electromagnetic interference from other nearby equipment.

3.5.1 Critical Properties

The first property that was observed to transform a normal conductor into a superconductor is the critical temperature, T_c. In applying the materials, it also became apparent that the current density and the strength of a magnetic field (which a material can

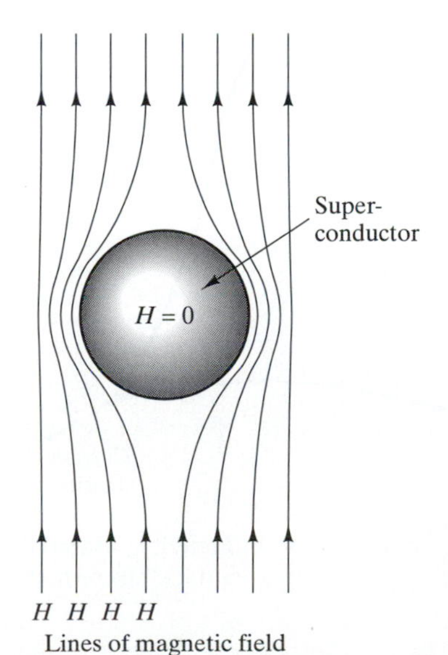

Fig. 3-18 Expulsion of magnetic field by superconductor—the Meissner effect-H = O inside superconductor.

sustain) are also important factors. Thus, there is a critical current density, J_c, and also a critical magnetic field strength, H_c. Putting all the three factors together, a schematic of the superconducting boundaries is shown in Fig. 3-19. For a given material, its superconducting properties are achievable when the operating temperature, current density, and magnetic field are within the volume limited by the critical surface.

Figure 3-20 shows the T_c's of superconducting materials discovered since 1911. Before 1986, most of the materials were metals and alloys (a combination of two or more pure metals) and they exhibited T_c's less than 30 K. In 1986, a ceramic material containing copper, barium, lanthanum, and oxygen was shown to have a critical temperature of about 35 K. This discovery spurred frenzied activity and in less than a year, in 1987, an oxide material, $YBa_2Cu_3O_7$, known as YBCO, was shown to have a 93 K T_c with a layered structure shown in Fig. 3-21. The highest listed T_c in the Handbook of Chemistry and Physics is 133 K for $HgBa_2Ca_2Cu_3O_8$. The metallic materials are now called the "older" low temperature superconducting (LTS) materials and the ceramic materials are called the high temperature superconducting (HTS) materials. The LTS materials have been well characterized and current theory and most of the current technology are based on them. The HTS materials are not completely characterized and the theory is still evolving.

Superconductors are also classified into two types according to their behavior in an applied magnetic field, see Fig. 3-22a. Type I superconductors, which include most pure metal superconductors, exclude magnetic flux (Meissner effect) until a maximum field, H_c, is exceeded at which point the material loses its superconductivity. These are not technologically important because H_c is very low—100 to 1,000 gauss (0.01 to 0.1 tesla). For example, one of the current applications of superconductors is in the nuclear Magnetic Resonance Imaging (MRI) equipment. This equipment requires a magnetic field of about 15,000 gauss (1.5 tesla) to be effective in imaging living tissues. None of the type I superconductors remains superconducting in such a high field.

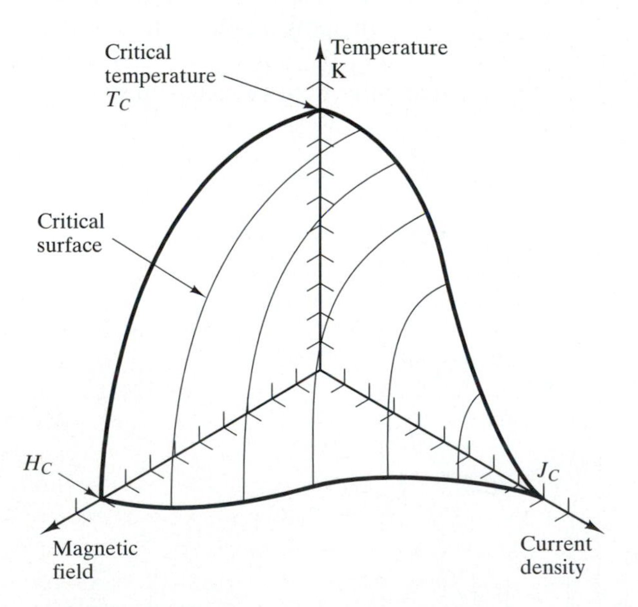

Fig. 3-19 Critical surface of superconductivity. Material must be maintained below the "critical surface" to remain superconducting.

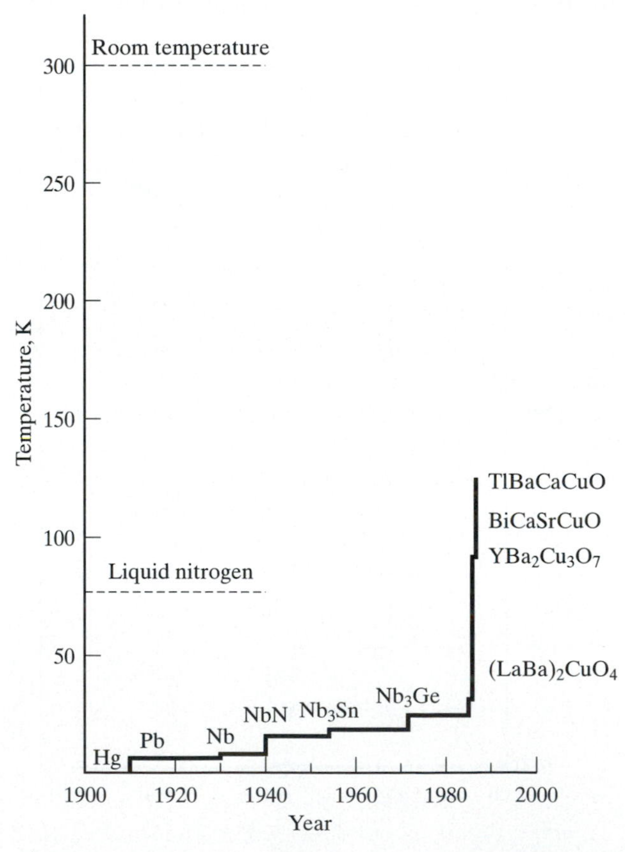

Fig. 3-20 Discovery of superconducting critical transition temperatures in the 20th century.

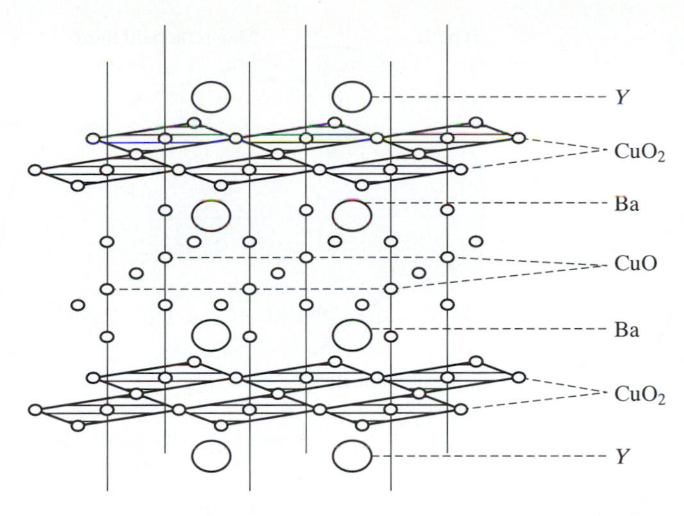

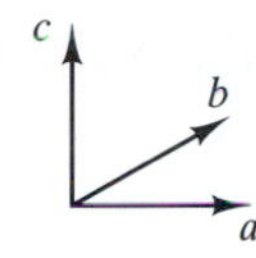

Fig. 3-21 Structure of $YBa_2Cu_3O_7$ showing stacking of planes along the C-axis. Supercurrent is conducted primarily in the CuO_2 plane.

Virtually all superconductors of technological importance are type II, including the new HTS materials. Type II superconductors exhibit two critical fields, H_{c_1} and H_{c_2}, Fig. 3-22b. At magnetic fields less than, H_{c_1}, they behave like the type I materials—at fields in excess of H_{c_2}, their superconducting properties are lost. In between the critical fields, there is partial penetration of the superconductor forming bundles of magnetic field lines "wrapped" and shielded from each other by vortices or "whirlpools" of supercurrents. These supercurrents repel one another and are arranged in a regular array to be as far from one another as possible. As the magnetic field increases toward H_{c_2}, more vortices are formed, the lattice spacing is decreased, until at H_{c_2}, superconductivity disappears. The HTS materials have much higher values of H_{c_2} values than the LTS materials. For example, the H_{c_2} of Nb_3Sn at 4 K is about 20 tesla, while the extrapolated value for the HTS YBCO (ytrium, barium, copper oxide) ceramic is about 200 tesla.

In many applications, superconductors also sustain high current densities, typically 10^5 to 10^6 Amps/cm^2. The maximum current density, J_c, that a superconductor can sustain depends not only on the composition of the material, but also on the impurities and defects in the crystal lattice. In type II materials, the flow of electric current exerts a force (called the Lorentz force) on the vortex lattice, see Fig. 3-23, and this force gets stronger as the electric current increases. Resistance appears when this force is strong enough to dislodge the vortex lattice from the pinning sites to which it is attached. Pinning sites of the vortices are defects, impurities, grain boundaries, and other points of weakened superconductivity. Unpinned vortices are locked in place at "lattice sites" due to mutual repulsion. Collectively, the defect pinning and the lattice locking are known as the pinning force in the superconductor. At the critical current, the Lorentz force overcomes the pinning force and the vortices begin to move. This movement constitutes resistance and eventually destroys the superconductivity.

Thermal lattice vibration can also affect the electrical behavior of superconductors. In the presence of a magnetic field, the HTS materials exhibit a small resistivity at temperatures considerably below T_c. This phenomenon is not observed in LTS materials and may mean that HTS materials need to be operated substantially below T_c in applications with a magnetic field. The residual resistivity is considered to be due to poor coupling between grains of HTS and to weak vortex pinning relative to thermal lattice

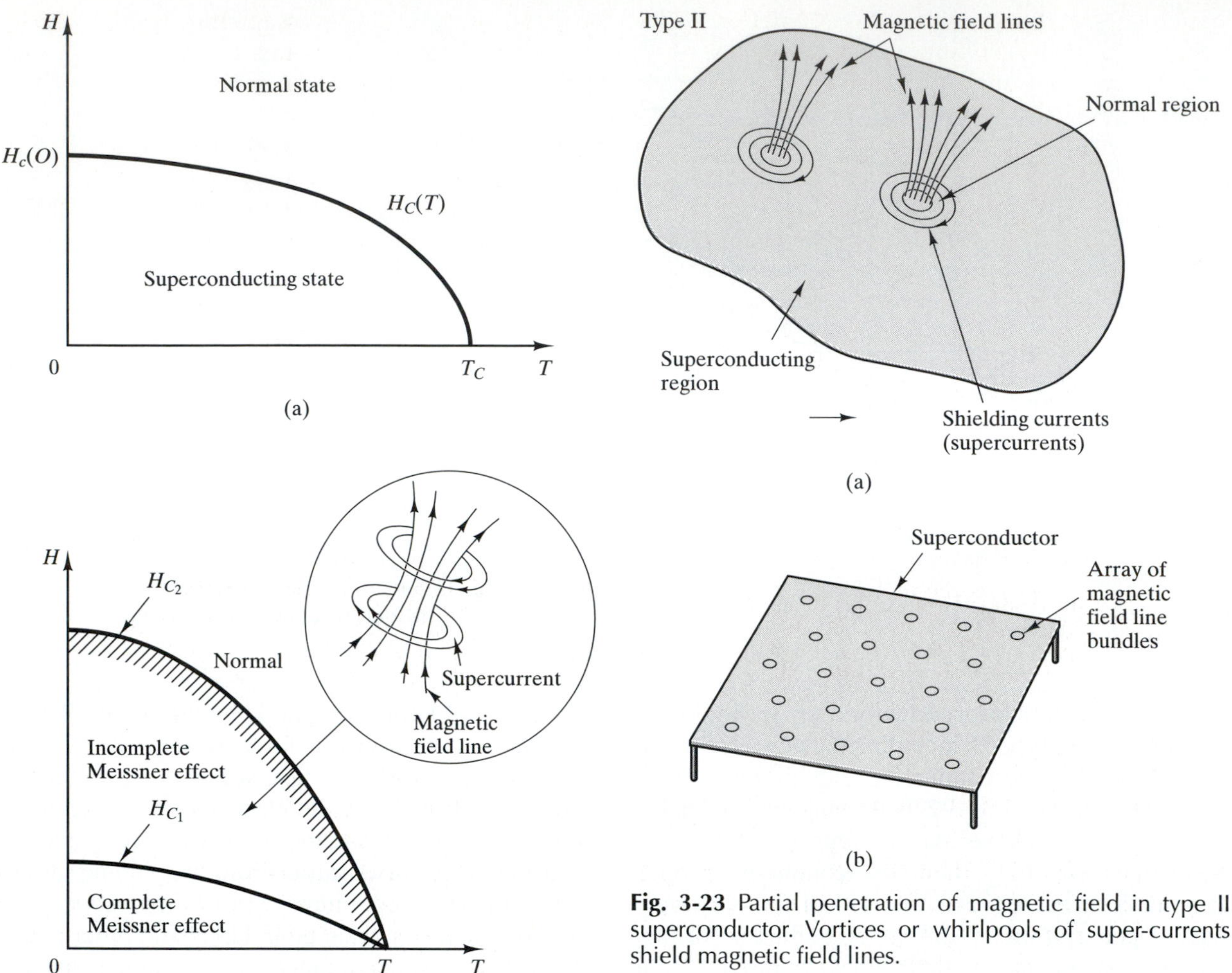

Fig. 3-22 (a) Type I superconductor—material must be below the H_C (T) curve to remain superconducting. (b) Type II superconductor—partial penetration of magnetic flux occurs between H_{c_1} and H_{c_2}.

Fig. 3-23 Partial penetration of magnetic field in type II superconductor. Vortices or whirlpools of super-currents shield magnetic field lines.

vibration. The lattice vibration causes vortices to jump from one site to another (flux creep).

3.5.2 Motion and Interaction of Electrons in LTS Materials

The motion and interaction of electrons in LTS materials explain their superconducting behavior. The development of the current theory of LTS materials relied on experimental facts to provide the necessary clues. The *two clues* that provided the insights are: the *Meissner effect* and the *isotope effect.* The Meissner effect suggests that the motion of the electrons must be coordinated over large distances and that the electrons as a group behave as a quantum superfluid, that is, a frictionless gas. The isotope effect suggests that the critical temperature is proportional to the inverse square root of the isotopic mass. This inverse square root relationship is precisely the same relationship between the vibrational frequency and an oscillating mass on an elastic spring. The clue provided by this experimental fact is the existence of an interaction between the electrons and the phonons (the quanta of lattice vibrations) because the energy of phonons is directly proportional to the vibration frequency and to the inverse square root of the vibrating mass.

In 1957, Bardeen, Cooper, and Schrieffer formulated their theory on superconductivity for which they received a Nobel prize. The *BCS theory* combined an earlier suggestion by Cooper that in order to attain the quantum superfluid state, the electrons must move in pairs. The manner in which they move together was not earlier explained and was the contribution of the BCS theory. The *superconductivity of LTS materials is due to electron-phonon-electron interaction.* This interaction is thought to occur in the following manner: *As one electron moves through the lattice, it passes between two positive ion cores; the positive cores are drawn slightly together by the passage of the electron and an excess positive charge density develops, which is represented as a packet of phonons; the excess positive charge in turn attracts another electron and so in an indirect way, the electron is attracted to another electron via a phonon interaction forming the Cooper electron pairs.*

3.5.3 Theory for HTS Materials

At this point, it is not clear how the BCS theory will be applied, if at all, to the new HTS materials. Judging from the development of the BCS theory for LTS materials, it is anticipated that more experimental facts need to be gathered. Nonetheless, there are theories being proposed that are based on the more recent metal-nonmetal transition theories discussed earlier. Most of the theories are all tentative and it is best to leave them to theorists to take out all the "cobwebs" before they are discussed in an elementary text.

3.6 Work Function and Contact Potential

In a lot of applications, different conductors need to be joined or in contact. The contact of two different conductors influences the flow of charge carriers at the junction and produces a potential that depends on the work functions of the conductors. While the following discussion is on metals, the principles are applicable to any pair of dissimilar conductors. The concepts are applied to junctions or contact points in semiconductors that are the bases for different semiconductor devices.

Although the valency electrons are free to move inside a metal, they cannot escape through the surface. For the electrons to escape through the surface, an energy has to be spent and this is called the *work function.* The work function is just like a latent heat of vaporization of the electrons from a metal. In a simple "potential box" model of the electron in a metal, shown in Fig. 3-24, the work function, Φ, is the difference in energy between the Fermi level at 0 K and the potential energy of an electron in space outside the metal.

The work function is related to the *photoelectric threshold* and to the *thermionic emission* of metals. Photoelectric threshold directly measures the work function because, by definition, it is the lowest frequency ν of an incident photon that can cause an electron to be emitted from the metal. Thus, the energy transfer is expressed by

$$h\nu = \Phi \tag{3-26}$$

Thermionic emission is the emission of electrons at high temperatures when the electrons acquire sufficient energy from thermal fluctuations to escape the surface. The probability of an electron acquiring the required energy is proportional to $\exp(-\Phi/kT)$, where k is the Boltzmann constant and T is the absolute temperature. The kinetics of the thermal electron evaporation (emission) process per unit area of the surface is given by the *Richardson–Dushman* equation,

$$j = C(1 - r)T^2 \exp(-\Phi/kT) \tag{3-27}$$

where j is the thermionic emission current density, $C = 4\pi mk^2e/h^3 = 1.2 \times 10^6$ A $m^{-2}K^{-2}$, and r is the reflection coefficient (typically ≈ 0.5) which measures the chance that an electron approaching the surface

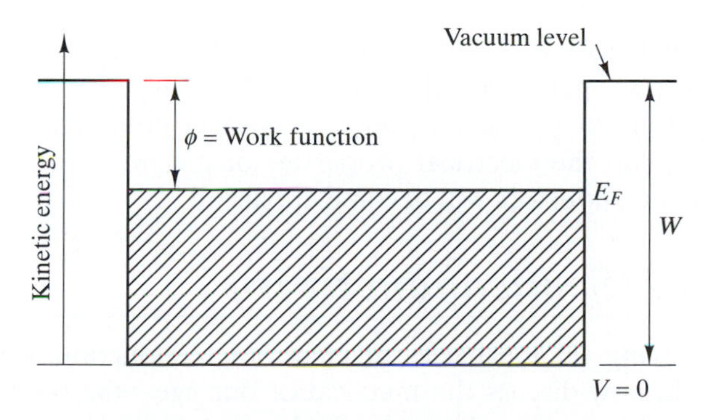

Fig. 3-24 Relation of the fermi energy, E_F, and work function ϕ relative to energy of free electron in vacuum.

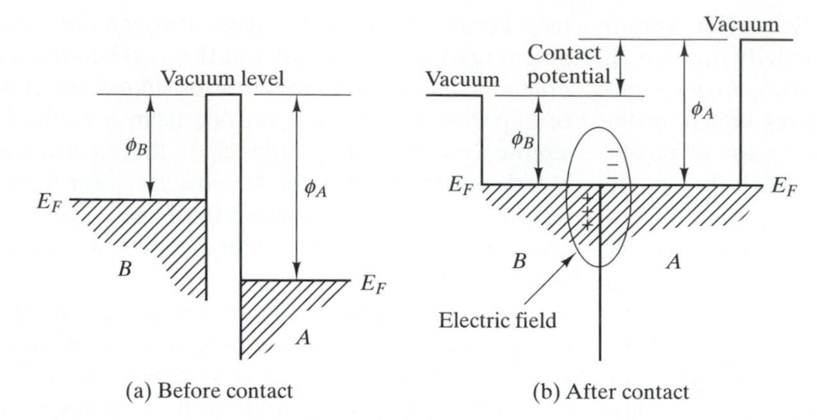

Fig. 3-25 Formation of contact potential when conductors of different Fermi levels are joined.

from the interior and with sufficient thermal energy is turned back rather than evaporate.

To describe the contact potential, the electrons in metals may be considered to be an "electron sea," with the Fermi level as the sea level. Just as in any fluid, the basic principle is that the Fermi levels (sea levels) of two different metals will equalize when the metals are brought into electrical contact and electrons can flow freely from one to the other. The flow of electrons and the equilibration process is schematically illustrated in Fig. 3-25. Figure 3-25a shows metal A has a larger Φ than metal B, relative to the potential in space. Metal B has a higher Fermi "sea" level. When contact is made between A and B, the electrons from B will flow towards A. As this happens, an electrostatic potential (voltage) is set up tending to reverse the flow of electrons. The potential energy of the electrons in A is raised relative to the potential energy of the electrons in B, Fig. 3-25b. Net flow of electrons from B to A stops when the electrostatic potential energy between A and B is equal to the difference in the work functions. The contact potential difference is $(\Phi_A - \Phi_B)$. As indicated above, the relevance of contact potentials to explain the electrical properties of p-n junctions in semiconductors will become apparent.

3.7 Semiconductivity

Having set the basics of electronic conduction, we can now discuss the material of our age—the semiconductor. No other material has revolutionized our way of life than the semiconductor. It is important, therefore, to know how these materials attain their properties and how the different semiconductor devices are built.

To begin, we should know that there are two general types of semiconductors—*intrinsic* and *extrinsic*. The intrinsic semiconductors are pure materials, while the extrinsic semiconductors contain controlled and purposely-added impurities called *dopants*. Depending upon the dopant, the extrinsic semiconductor may be either n-type or p-type. Most of the technologically commercial semiconductors are the extrinsic variety. We shall see why this is, but before we do, we need to see how intrinsic conductivity is obtained.

3.7.1 Intrinsic Semiconductors

As discussed earlier, the two semiconductors from the Group 14 elements are silicon and germanium; however, silicon is the most used due to its abundance and cost. Thus, we shall concentrate on silicon; although everything we say about Si is equally applicable to Ge.

We start by indicating that everything that hinders the motion of electrons in metals also have the same effect on semiconductors. Thus, defects and impurities must be kept to the minimum. For this reason, grain boundaries are avoided and all silicon integrated chips are single crystals. In addition, the silicon undergoes purification by what is called the zone-refining process to limit unwanted impurities. Typically, impurity concentration is kept below 1 in

10^9, or 1 part per billion. The single crystal produced can be as large as 200 mm in diameter from which slices called wafers are obtained to serve as the starting material for the controlled addition of impurities or dopants to produce the extrinsic semiconductors and the integrated circuits.

Now, we refer back to Fig. 3-4c where the valence band and the conduction band of pure silicon is separated by the energy gap, E_g. Intrinsic semiconductivity arises from the extricated electrons (from the covalent bonded state) that are free to move when an applied electric field is imposed. As seen also from Fig. 3-4b, for every electron extricated, there is also a positive vacant electron site, called a "hole," that is left in the covalent bond. Thus,

$$N_e = N_h \tag{3-28}$$

where N_e is the number of electrons per unit volume, and N_h is the number of holes (subscript h) per unit volume. As indicated earlier in Fig. 3-14b, both of these charge entities will move when an electric field is imposed, and, therefore, according to Eq. 3-13, the conductivity of the intrinsic semiconductor is the sum of the conductivities due to the electrons and the holes,

$$\sigma_i = \sigma_{i_e} + \sigma_{i_h} = N_{i_e} |e| \mu_e + N_{i_h} |e| \mu_h \tag{3-29}$$

where the subscripts i stands for intrinsic, e for electrons, and h for holes; e is the charge of an electron and μ stands for mobility. As in the electrical conductivity of metals, the semiconductor conductivity is controlled by the charge densities of electrons, $N_{i_e} |e|$ in the conduction band and the charge densities of the holes, $N_{i_h} |e|$, in the valence band as well as by their respective mobilities, μ_e and μ_h. For a particular element, the mobilities and the $|e|$ are constants and it only remains to determine N_{i_e} and N_{i_h}, the number of intrinsic electrons per unit volume and the number of intrinsic holes per unit volume in order to calculate the intrinsic conductivity.

In order to calculate N_{i_e}, we need to know the number of electrons that can be allowed in the conduction band and then multiply it by the probability of the electrons being in the conduction band. For N_{i_h}, we need to know the number of covalent electron sites that can exist near the top of the valence band and then multiply it by the probability that they will not be occupied. We obtain the number of allowable electrons in the conduction band by determining the allowable number or density of quantum states within a certain energy range—each quantum state when multiplied by two (Pauli Exclusion Principle) gives the number of electrons. The number of quantum states is related to the energy level of the electrons in the conduction band. In terms of energy levels, we shall also be able to determine the number of holes or unoccupied electron sites within a certain energy range in the valence band. The probability function used is the Fermi-Dirac distribution.

3.7.2 Density of States and the Fermi-Dirac Function

We start by recalling that the energy of a free electron in one dimension, Eq. 1-8 (now written in the x-direction) is

$$E(x) = \frac{h^2 n_x^2}{8\,mL^2} \tag{3-30}$$

In three dimensions, Eq. 3-30 becomes

$$E = \frac{h^2}{8\pi^2 mL^2}(n_x^2 + n_y^2 + n_z^2) \tag{3-31}$$

where h is the Planck's constant, L is the side of a cube, and n_x, n_y, and n_z are positive integers. It is convenient at this time to think of a vector, **n**, from the origin to any coordinate (n_x, n_y, n_z). In this way, we see that the quantity in parenthesis is the square of the magnitude of the vector **n**, that is

$$n^2 = n_x^2 + n_y^2 + n_z^2 \tag{3-32}$$

We immediately note that each coordinate in the n-space, that is, a point (n_x, n_y, n_z), corresponds to an allowable E and is, therefore, a quantum state. The number of states in n-space within a spherical volume of radius of n is equal to,

$$\text{number of states} = \frac{4}{3}\pi n^3 \tag{3-33}$$

From Eq. 3-31, we see that

$$n^2 = KE \qquad \text{where} \qquad K = \frac{8mL^2}{h^2}$$
$$n = (KE)^{1/2} \tag{3-34}$$

and from which we obtain the number of quantum states with energies less than E to be

$$\frac{4}{3}\pi n^3 = \frac{4}{3}\pi K^{3/2}E^{3/2} \quad (3\text{-}35)$$

Similarly, the number of states having energies less than (E + dE) is

$$\frac{4}{3}\pi K^{3/2}(E + dE)^{3/2} \quad (3\text{-}36)$$

The number of states with energies between E and E + dE is thus equal to

$$Z(E)dE = \frac{4}{3}\pi K^{3/2}[(E + dE)^{3/2} - E^{3/2}]$$

$$= 2\pi K^{3/2}E^{1/2}dE \quad (3\text{-}37)$$

The number of states is for the entire spherical volume. Since the allowable integers are only at the positive octant of the spherical volume, the number of states has to be divided by eight. Furthermore, because there are two electrons with opposite spins per energy state (Pauli principle), it is multiplied by two to get the number of allowable electrons. The total number or density of allowable electrons between E and E + dE is

$$Z(E)dE = CE^{1/2}dE \quad (3\text{-}38)$$

where $C = 4\pi L^3(2m)^{3/2}/h^3$. Equation 3-38 is applicable for an energy range in either the conduction or the valence band.

Now we want to know how many electrons are actually in the energy range E and E + dE. In quantum mechanics, this is expressed as the number of states occupied by electrons. To obtain the occupied states, we multiply the total number of states (number of electrons when multiplied by two) by an occupation probability function, F(E). This probability function is the *Fermi-Dirac distribution function*

$$F(E) = \frac{1}{\exp\left(\frac{E - E_F}{kT}\right) + 1} \quad (3\text{-}39)$$

where E_F is the Fermi energy, the highest energy level occupied by the electron in the bonded ground state. The properties of this distribution function are:

$$(1) \text{ At } E = E_F \qquad F(E) = 1/2, \quad (3\text{-}40)$$

and (2) At absolute zero, T = 0 K

$$F(E) = 1 \qquad \text{when } E < E_F, \quad (3\text{-}41a)$$

and

$$F(E) = 0 \qquad \text{when } E > E_F \quad (3\text{-}41b)$$

which means that at the Fermi level, the probability of occupation is 1/2, and the probability occupancy is one when $E < E_F$ and zero when $E > E_F$. This distribution is shown in Fig. 3-26 and is quite different from the classical Maxwell-Boltzmann exponential function, $\exp(-E/kT)$.

For electron energies much above the Fermi level, such that,

$$(E - E_F) \gg kT$$

the exponential term in Eq. 3-39 is much greater than one, therefore,

$$F(E) \cong \exp\left(-\frac{E - E_F}{kT}\right) \quad (3\text{-}42)$$

which is the classical Maxwell-Boltzmann distribution. Thus, for sufficiently large energies, the Fermi-Dirac distribution converts to the Maxwell-Boltzmann distribution, commonly referred to as the "Boltzmann tail." Equation 3-42 is the probability function we will use to obtain N_{i_e} in the conduction band.

For electron energies below the Fermi level such that $(E_F - E) \gg kT$, the Fermi-Dirac function may be approximated by

$$F(E) \cong 1 - \exp\left(-\frac{E_F - E}{kT}\right) \quad (3\text{-}43)$$

which means that the probability of a state being occupied by an electron is very close to one because

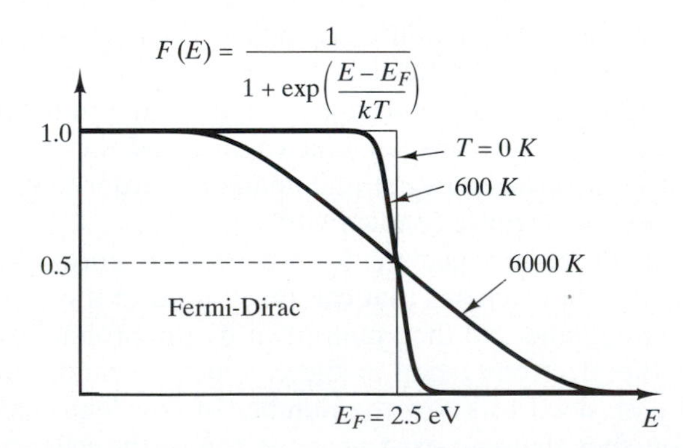

Fig. 3-26 Fermi-Dirac distribution function for a Fermi energy of 2.5 eV at temperatures 0 K, 600 K, and 6,000 K. [Ref. 6]

the exponential approaches zero. This is the probability function for the electron occupancy in the valence band and simply emphasizes the obvious that most of the electrons are in the covalent bonded state in the valence band. In terms of the vacant or unoccupied states, we need the probability that the state is unoccupied. Since F(E) is the probability of occupation, then 1 − F(E) is the probability of non-occupation by an electron, and we see that

$$1 - \mathrm{F(E)} = \exp\left(-\frac{\mathrm{E_F - E}}{\mathrm{kT}}\right) \qquad (3\text{-}44)$$

Equation 3-44 is the probability function we will use to obtain N_{i_h} in the valence band.

3.7.3 Intrinsic N_{i_e} and N_{i_h}

With the foregoing background, we are now in a position to calculate the density of electrons and holes in an intrinsic semiconductor. We start by defining the energy levels in the energy band structure as shown in Fig. 3-27. The top of the valence band is taken as $E = 0$ and the bottom of the conduction band is taken as $E = E_g$. The energy states close to the top of the valence band and at the bottom of the conduction band will be considered. It should also be mentioned that the electrons and holes in an intrinsic semiconductor are not the same as the free electrons discussed in the previous section. It can be shown that the energy states of the electrons near the bottom of the conduction band are essentially the same as Eq. 3-31, except that m, the mass of the free electron, is replaced by m_e^*, the effective mass of the electron in the conduction band,

$$\mathrm{E} = \frac{h^2}{8\,\pi^2\,\mathrm{m_e^*}\,\mathrm{L}^2}\,(n_x^2 + n_y^2 + n_z^2) \qquad (3\text{-}45)$$

The energy for the holes in the valence band is an identical expression if the effective mass of the holes, m_h^*, is used in Eq. 3-45. Except for these changes, the expressions for the density of states and the Fermi-Dirac distribution apply also to the electrons and holes in an intrinsic semiconductor.

Thus, using Eq. 3-38, the permissible density of electrons having energies between E_c and E_g in the conduction band is

$$\mathrm{Z(E)} = C_e(E_c - E_g)^{1/2} \qquad (3\text{-}46)$$

where $C_e = 4\pi(2m_e^*)^{3/2}/h^3$

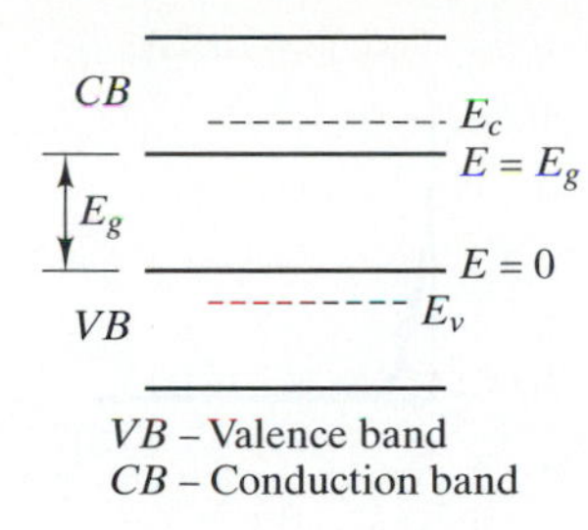

Fig. 3-27 Definition of energy levels in calculating density of states in valence and conduction bands.

Again, using Eq. 3-38, the permissible density of electrons in the valence band between E_v and 0 near the top of the valence band is

$$\mathrm{Z(E)} = C_h(0 - E_v)^{1/2} \qquad (3\text{-}47)$$

where $C_h = 4\pi(2m_h^*)^{3/2}/h^3$

These permissible densities of electrons (or states) are schematically illustrated in Fig. 3-28 and have meanings only in the allowed bands but not in the energy gap.

In order to obtain the actual density of electrons in the conduction band, we must multiply the permissible density of electrons, Eq. 3-46, by the Fermi-Dirac probability function, Eq. 3-42, and integrate over the energy range. For the density of electrons, N_{i_e}, we make the assumption that E_F, the Fermi level, lies somewhere between 0 and E_g, the forbidden gap,

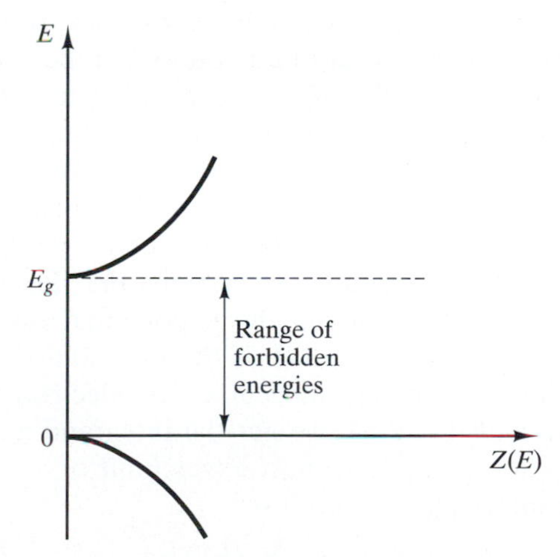

Fig. 3-28 Density of states' plots of Eqs. 3-46 (bottom of CB) and 3-47 (top of VB). [Ref. 6]

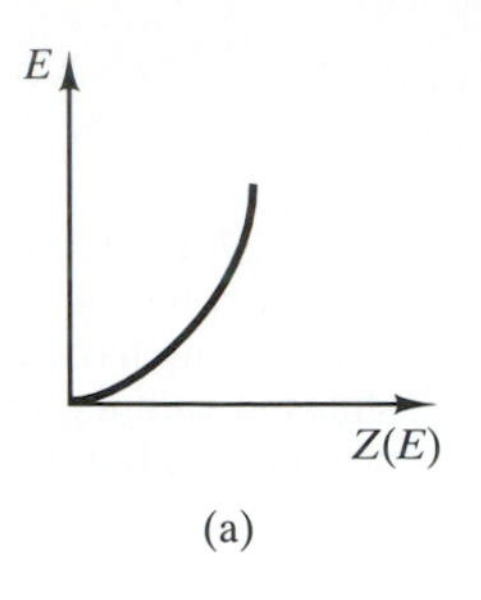

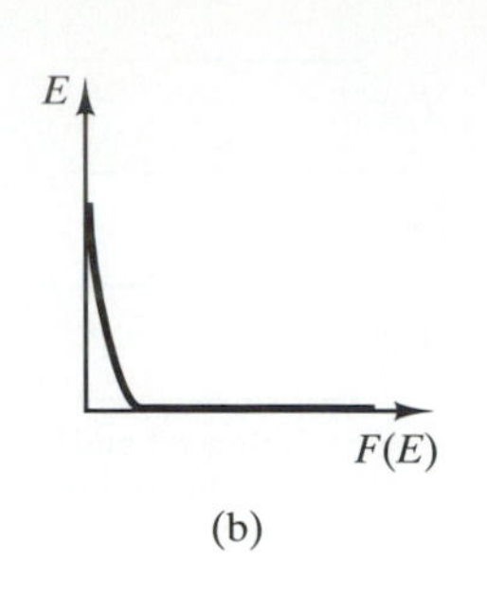

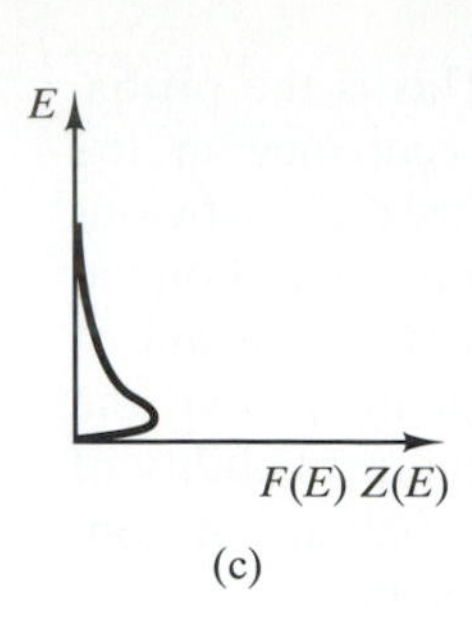

Fig. 3-29 (a) The density of states Z(E) as function of energy for bottom of CB, (b) the Fermi-Dirac distribution function F(E) for the same energy range, and (c) the plot of the product of F(E)Z(E). [Ref. 6]

and that $(E_c - E_F) \gg kT$. Combining Eqs. 3-42 and 3-46, we find

$$N_{i_e} = C_e \int_{E_g}^{\infty} (E - E_g)^{1/2} \exp\left(-\frac{E - E_F}{kT}\right) dE \quad (3\text{-}48)$$

The upper limit of integration, which is infinity ∞, is justified by Fig. 3-29. The sharply declining exponential function, F(E), Fig. 3-29b when multiplied by the density of states, Z(E) in Fig. 3-29a, results in a curve that peaks close to the bottom of the conduction band and declines sharply to practically zero above a certain energy, E_c, Fig. 3-29c. The integral, being the area under the curve in Fig. 3-29c, will essentially be the same whether the upper limit is made E_c or ∞. Using ∞ as the upper limit eases the difficulty of the integration because mathematical tables can be used. The details are left out with the result being,

$$N_{i_e} = N_c \exp\left(-\frac{E_g - E_F}{kT}\right) \quad (3\text{-}49)$$

where $N_c = 2(2\pi m_e^* kT/h^2)^{3/2}$

Thus, we obtain the density of intrinsic electrons (the number of electrons per unit volume) in the conduction band as a function of some fundamental constants, of temperature, of the effective mass of the electron at the bottom of the band, and of the amount of energy by which the bottom of the conduction band is above the Fermi level.

In a similar manner, we obtain the density of holes, N_{i_h}. By combining the exponential function, $[1 - F(E)]$, Eq. 3-44, which is the probability of the quantum state being unoccupied or electron being absent, with Eq. 3-47, we get the integrand and the integration is made from a lower limit of ∞ to an upper limit of 0. The result is

$$N_{i_h} = N_v \exp(-E_F/kT) \quad (3\text{-}50)$$

where $N_v = 2(2\pi m_h^* kT/h^2)^{3/2}$

We have obtained the densities of electrons and holes without specifying exactly where E_F is in the energy gap. We are now in a position to determine where the Fermi level is based on the fact that every time we excite an electron from the valence band to the conduction band, a hole is left in the valence band. Thus, $N_{i_e} = N_{i_h}$, and equating Eqs. 3-49 and 3-50 and subsequent algebraic manipulation yields

$$E_F = \frac{E_g}{2} + \frac{3}{4} kT \ln \frac{m_h^*}{m_e^*} \quad (3\text{-}51)$$

Because kT is small and the effective masses of the holes and the electrons are about equal, the second term in Eq. 3-51 is zero. We note that the Fermi level roughly lies midway between the top of the valence band and the bottom of the conduction band, about midway in the forbidden gap.

Therefore, the resulting N_{i_e} and N_{i_h} expressions are

$$N_{i_e} = N_c \exp(-E_g/2kT), \quad (3\text{-}52a)$$

and

$$N_{i_h} = N_v \exp(-E_g/2kT). \quad (3\text{-}52b)$$

For the density of electrons to be equal to the density of holes, the coefficients of the two expressions must be equal and that means that the effective masses of the holes and electrons must also be equal. From the above expressions, if we know E_g, we can calculate the densities of electrons and holes in an intrinsic semiconductor as a function of temperature. The values of E_g for silicon and germanium are given in Table 3-1. At a given temperature, smaller E_g values will yield higher densities of electrons and with Eq. 3-29, higher intrinsic conductivities. The E_g values are experimentally determined usually by plotting the $(\ln \sigma)$ versus $1/T$ as seen in Fig. 3-30. With the assumption that the coefficients N_c and N_v in Eq. 3-49 and 3-50 are constants, the E_g can be obtained from the slope of Fig. 3-30. Since these coef-

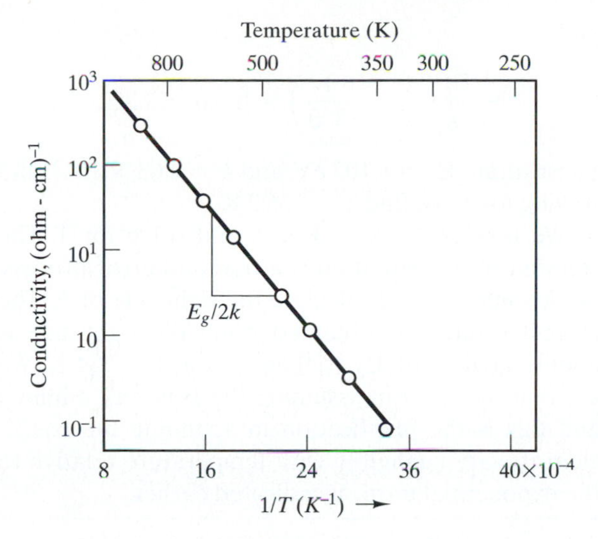

Fig. 3-30 Experimental determination of E_g from slope of plot of conductivity as a function of $1/T$, with T in K. [Ref. 9]

ficients have a $T^{3/2}$ dependency, the linearity of the ($ln\ \sigma$) versus $1/T$ plot suggests that the variability of the coefficients is small compared to the exponential function. This fact will also be demonstrated in Example 3-7.

As indicated earlier, we note that when $T = 0$ K, the electron and hole densities are zero and the material is an insulator. As T increases above 0 K, the exponential indicates that the densities increase; therefore, the conductivities increase. Herein lies the basic difference between a metal and a semiconductor—the conductivity of a metal decreases, while that of a semiconductor increases, as the temperature increases. Thus, temperature is the only variable that we can use to control the conductivity of intrinsic semiconductors.

EXAMPLE 3-5

Estimate the electrical conductivity of silicon at 300 K and compare this with the reported value of intrinsic resistivity of 2.3×10^5 ohm-cm at 300 K.

SOLUTION: The intrinsic conductivity is given by Eq. 3-29.

$$\sigma_i = N_{i_e} |e|\, \mu_e + N_{i_h} |e|\, \mu_h$$
$$= N_{i_e} |e|\, (\mu_e + \mu_h)$$

The values of $|e|$, μ_e, and μ_h can be obtained from tables; we need only to calculate N_{i_e}. We can do this by using Eq. 3-52,

$$N_{i_e} = N_c \exp(-E_g/2kT) \quad \text{where}$$
$$N_c = 2\left(\frac{2\pi m_e^* kT}{h^2}\right)^{3/2}$$

We assume the effective mass of an electron, $m_e^* = m$, the mass of free electron $= 9.11 \times 10^{-31}$ kg. Using values of constants from Tables,

$$N_c = 2\left[\frac{2\pi \times 9.11 \times 10^{-31} \times 1.38 \times 10^{-23}}{(6.63 \times 10^{-34})^2}\right]^{3/2}$$
$$\times T^{3/2} = 4.818 \times 10^{21} T^{3/2}$$

At 300 K,

$$N_c = 4.818 \times 10^{21}\,(300)^{3/2} = 2.503 \times 10^{25}\ e/m^3.$$

Thus,

$$N_{i_e} = 2.503 \times 10^{25} \exp\left(-\frac{1.107}{2 \times 8.63 \times 10^{-5} \times 300}\right)$$
$$= 1.299 \times 10^{16}\ e/m^3 = 1.299 \times 10^{10}\ e/\text{cm}^3$$

and

$$\sigma_i = 1.299 \times 10^{10}\ e/\text{cm}^3 \cdot 1.6 \times 10^{-19}\ \text{C}$$
$$(\text{amp-sec/e}) \cdot (1900 + 500)\ \text{cm}^2/\text{volt-sec}$$
$$= 4.988 \times 10^{-6}\ (\text{ohm-cm})^{-1}$$

This estimated value is compared to the observed intrinsic resistivity of 2.3×10^5 ohm-cm that gives the observed intrinsic conductivity of $1/(2.3 \times 10^5) = 4.348 \times 10^{-6}$ (ohm-cm)$^{-1}$.

The above solution gives a relation of N_c as a function of T that can be used with other problems.

EXAMPLE 3-6

At 300 K, what is the fraction of the electrons in silicon that is present in the conduction band?

SOLUTION: The fraction of electrons in the conduction band $= N_{i_e}/N_{total}$ The value of N_{i_e} at 300 K was calculated in Example 3-5 to be 1.299×10^{10} e/cm^3. We need to determine N_{total} to get the fraction. As in Example 3-1, we can determine N_{total} by using either the density or the unit cell characteristics.

Using the density,

$$N_{total} = 2.328 \times \frac{1}{28.085} \times 6.022 \times 10^{23} \times 4\, e/\text{Si atom} = 2.0 \times 10^{23}\, e/\text{cm}^3.$$

Using the unit cell characteristics, silicon is diamond cubic with 8 atoms per unit cell and has a lattice constant of 5.4307 Å.

$$N_{total} = \frac{8\,\dfrac{\text{Si atoms}}{\text{u.c.}} \times 4\,\dfrac{\text{e}}{\text{Si}}}{\text{Vol. u.c.}} = \frac{8 \times 4}{(5.4307 \times 10^{-8})^3} = 2.0 \times 10^{23}\, e/\text{cm}^3.$$

The fraction of electrons in the conduction i.e., Fraction band

$$= \frac{1.299 \times 10^{10}}{2 \times 10^{23}} = 6.5 \times 10^{-14}$$

EXAMPLE 3-7

Determine the temperature at which the intrinsic conductivity of silicon is 10 times that at 300 K.

SOLUTION: As in Example 3-5, the intrinsic conductivity is

$$\sigma_i = N_{i_e} |e|\, \mu_e + N_{i_h} |e|\, \mu_h = N_{i_e} |e|\, (\mu_e + \mu_h)$$

Since $|e|$, μ_e, and μ_h are constants, the intrinsic conductivity is directly proportional to N_{i_e}, thus

$$\sigma_i = CN_{i_e} = C\, 4.818 \times 10^{21} T^{3/2} \exp(-E_g/2kT)$$

where C is a constant.

The above equation is the intrinsic conductivity at any temperature. The ratio of the conductivity at T and at 300 K is

$$\frac{\sigma_T}{\sigma_{300}} = 10 = \left(\frac{T}{300}\right)^{3/2} \exp\left[-\frac{E_g}{2k}\left(\frac{1}{T} - \frac{1}{300}\right)\right] \quad \text{(A)}$$

We will first solve the problem by assuming that the coefficient of the exponential is approximately equal to one as we had done earlier in relation to Eq. 3-51. With this assumption,

$$\frac{\sigma_T}{\sigma_{300}} = 10 = \exp\left[-\frac{E_g}{2k}\left(\frac{1}{T} - \frac{1}{300}\right)\right] \quad \text{(B)}$$

and

$$-\frac{E_g}{2k}\left(\frac{1}{T} - \frac{1}{300}\right) = \ln 10 = 2.302$$

Substituting $E_g = 1.107\, e\text{V}$, and $k = 8.63 \times 10^{-5}$, and solving for T, we find T = 336.2 K.

We now go back to Eq. A and solve for T. The solution of this equation is a trial and error process. We assume a T and calculate the right side of A. The correct solution is obtained when the right side is about equal to 10. By trial and error, T ≈ 334 K. We see that the error in assuming Eq. B is very minimal and this is the justification in assuming the coefficient to vary negligibly with temperature relative to the exponential term, as indicated earlier.

3.7.4 Extrinsic Semiconductors

We obtain extrinsic semiconductors by *doping* or adding controlled amounts of impurities to zone-refined silicon or germanium. The common dopants come from either the Group 15 (or V) elements or from the Group 13 (or III) elements. Very minute (percentage-wise) amounts of added dopants from these two groups form a substitutional solid solution, with silicon or germanium as the host lattice. The common Group 15 (or V) dopants are phosphorus (P), arsenic (As), and antimony (Sb) and they produce *n-type semiconductors.* The common Group 13 (or III) dopants are boron (B), aluminum (Al), and gallium (Ga) and they produce *p-type semiconductors.* The prefixes *n* and *p* refer to negative (electrons) and positive (holes) majority charge carriers. These terms are explained below.

In Section 3.2.2 on nonmetal to metal transition, we saw in Fig. 3-6 how silicon, a semiconductor as well as an insulator at 0 K, can transform to a metal at 0 K when a critical concentration of phosphorus is added. Below this critical concentration, it is an *n* (negative)-*type extrinsic semiconductor* because each phosphorus atom *donates* an extra electron (negative charge), Fig. 3-31a, when it replaces the silicon in the covalent-bonded structure. Phosphorus (or any Group 15 atom) is said to be an *electron donor* and the extra donated electron is very close to being in an ionized state when it can move freely in the silicon lattice. While the extra electron is still attracted to the phosphorus atom, it requires only a small amount

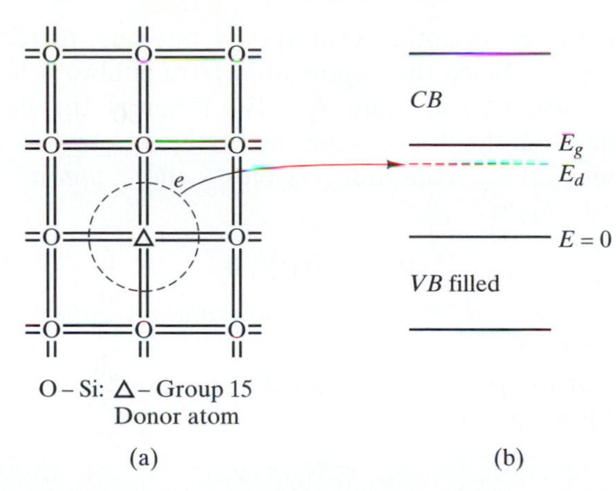

Fig. 3-31 (a) Group 15 element donates electron in n-type extrinsic semiconductor, (b) energy of donated electron relative to the intrinsic electron energy band diagram in silicon.

Table 3-4 Donor and acceptor impurity energy levels in silicon and in germanium.

(a) Silicon (E_g = 1.107 eV)

Donors		Acceptors	
Impurity	$(E_g - E_d)$	Impurity	$(E_a - 0)$
P	0.044 eV	B	0.045 eV
As	0.049	Al	0.057
Sb	0.039	Ga	0.067
Li	0.033	In	0.160

(b) Germanium (E_g = 0.670 eV)

Donors		Acceptors	
Impurity	$(E_g - E_d)$	Impurity	$(E_a - 0)$
Li	0.0093 eV	Zn	0.029 eV
P	0.012	B	0.0104
As	0.0127	Al	0.0102
Sb	0.0097	Ga	0.0108
Bi	0.012	In	0.0112
		Cu	0.040

of energy to get it ionized or brought to a conducting state as a free electron. Thus, in terms of the valence-conduction band energy picture, the electron is said to be at an energy donor state, E_D, very close to the bottom of the conduction band, as illustrated in Fig. 3-31b. If the bottom of the conduction band is at E_g as before, approximate quantum mechanical calculation shows the energy required or gap, $(E_g - E_D)$, to be about 0.025 eV. The experimentally observed value to make the electron from the phosphorus completely free to move in silicon is 0.045 eV. The different donor atoms (from Group 15 or V) will have different $(E_g - E_D)$ values in Si and Ge which are listed in Table 3-4.

If a Group 13 (or III) element replaces a silicon atom in the lattice structure, it brings with it only three electrons, one electron less than the silicon atom it replaces. In terms of the covalent bond picture, one of the electrons needed to make all four covalent bonds saturated (each bond with two electrons) is missing, Fig. 3-32a. Thus, a *vacant electron site—a hole* is created that can accept an electron and that acquires a positive charge—a *p-type extrinsic semiconductor.* The movement of the hole involves the filling of this vacant bond site by neighboring electrons. In actuality, the entities moving (diffusing in an applied electric field) are the electrons hopping to the empty sites. The motion of electrons from one site to the next takes more time (than a free electron in the conduction band) and thus the mobility of holes is considerably smaller than the mobility of the free electron. One can compare these processes to travels along the interstate highways (electron mobility) and along city streets through stop lights (hole mobility).

In terms of the valence-conduction band energy picture, the electrons of the silicon atoms will require only a small amount of energy to move to the vacant electron acceptor site created by the Group 13 (or III) acceptor dopant elements. Thus, the electron acceptor site or hole is said to be at an energy state, E_A, just above the top of the valence band, shown in Fig. 3-32b. With the energy at the top of the valence band taken as zero, the electrons near the top of the valence band need to acquire $(E_A - 0)$ to get to the acceptor levels. Values of $(E_A - 0)$ for the different Group 13 (or III) acceptor elements in Si and Ge are also given in Table 3-4.

The energy values in Table 3-4 are considered the activation energies of conduction in extrinsic semiconductors. Compared to the intrinsic values in Table 3-1 for silicon and germanium, these are very small and make conduction in the extrinsic types much easier than in the intrinsic. Thus, commercial semiconductors are invariably the extrinsic type.

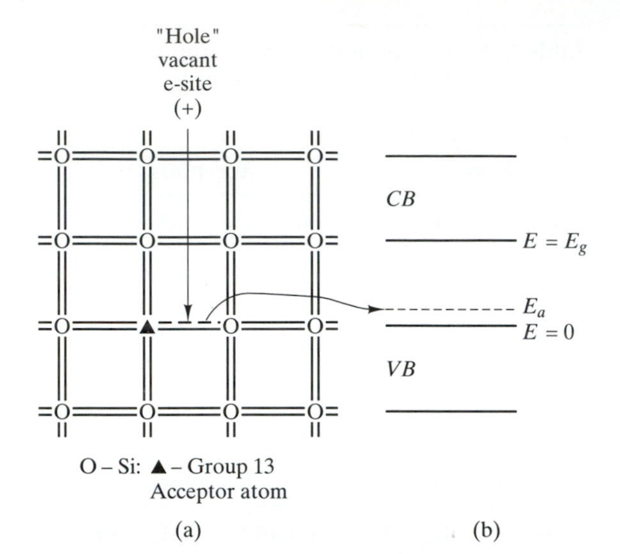

Fig. 3-32 (a) Group 13 element substituting for silicon creating a vacant electron site or hole with positive charge in p-type extrinsic semiconductor; (b) energy level of hole relative to the intrinsic electron energy band in silicon.

3.7.5 Conductivities of Extrinsic Semiconductors

The total conductivity, σ_T, of an extrinsic semiconductor is given by the general equation,

$$\sigma_T = \sigma_i + \sigma_x \tag{3-53}$$

which is the sum of the intrinsic, σ_i, and the extrinsic, σ_x, contributions. At low temperatures, the intrinsic contribution is relatively insignificant and is quite often neglected. For the *n-type semiconductors,* the extrinsic conductivity is predominantly,

$$\sigma_{x\mathbf{D}} = N_{x_e} |e| \mu_e \tag{3-54}$$

where σ_{xD} is the extrinsic conductivity from donor impurities, N_{x_e} is the density of electrons ionized from the donor atoms, $|e|$ is the charge of an electron, and μ_e is the mobility of electrons. If N_D is the total donor atoms per unit volume (or the density of donor atoms) then,

$$N_{x_e} = N_D \exp\left(-\frac{E_g - E_D}{kT}\right) \tag{3-55}$$

When the temperature is high, that is when $kT \gg (E_g - E_D)$, the exponent will approach zero and the exponential term approaches one, thereby $N_{x_e} = N_D$. Since the exponential term is always less than one, the maximum N_{x_e} is N_D. When all the electrons from the donor atoms are ionized to free electrons, this condition yields the *saturation conductivity,*

$$\sigma_{xSD} = N_D |e| \mu_e \tag{3-56}$$

where σ_{xSD} is *the extrinsic saturation conductivity from the donor atoms.*

For the *p-type semiconductor,* the conductivity is predominantly,

$$\sigma_{xA} = N_{x_h} |e| \mu_h \tag{3-57}$$

where σ_{xA} is the extrinsic conductivity due to acceptor impurities, N_{x_h} is the density of acceptor holes occupied by electrons, μ_h is the hole mobility. If N_A is the total density of acceptor atoms added, then

$$N_{x_h} = N_A \exp\left(-\frac{E_A}{kT}\right) \tag{3-58}$$

As with donor atoms, the saturation conductivity due to acceptor atoms is attained when $N_{x_h} = N_A$,

$$\sigma_{xSA} = N_A |e| \mu_h \tag{3-59}$$

where σ_{xSA} is the *extrinsic saturation conductivity from the acceptor atoms.* We should note, in both expressions for N_{x_e} and N_{x_h}, that these densities are zero when T is zero, which means that the extrinsic semiconductors are still insulators at absolute zero.

The typical behavior of extrinsic silicon semiconductors is illustrated schematically in Fig. 3-33 which is a plot of (ln σ) versus 1/T. The curves represent different concentrations of the dopant. At low temperatures (higher 1/T values), the slopes of all the curves for different dopant concentrations indicate essentially the same value, meaning that the activation energy, $(E_g - E_D)$ or E_A, is independent of dopant concentration. This is how the values in Table 3-4 of the extrinsic activation energies are obtained. The almost vertical line represents the intrinsic behavior of silicon and the very steep slope reflects the high "activation energy" for intrinsic conduction relative to extrinsic conduction. The low temperature region (higher 1/T) reflects the conductivity contribution from the dopants. The leveling off in the conductivity from the impurity indicates the exhaustion region. Fig. 3-33 indicates also that

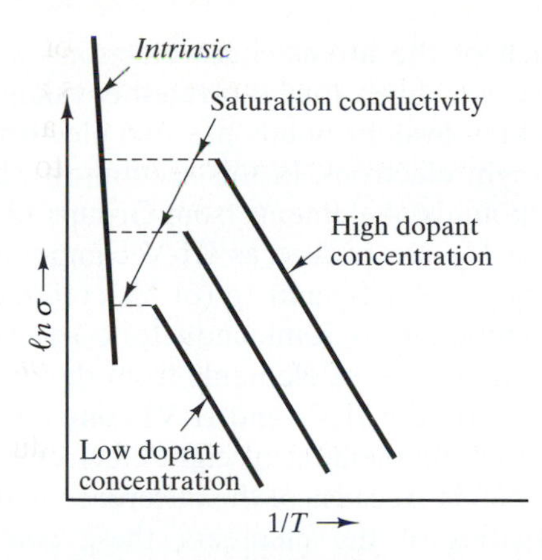

Fig. 3-33 Schematic of a semilog plots (1/T vs. log σ) of an extrinsic semiconductor for varying dopant concentration. Slope of plots at low T (high 1/T) gives the $(E_g - E_d)$ for n-type or $(E_a - 0)$ for p-type

the extrinsic conductivity can be controlled by varying the temperature and the dopant concentration. It is, of course, advantageous to use the semiconductor at the exhaustion region, and the higher the dopant concentration, the higher the exhaustion or saturation conductivity.

EXAMPLE 3-8

Determine the saturation conductivity of silicon that is doped with 0.0001 atomic percent phosphorus. What is the conductivity of this extrinsic semiconductor at 25°C?

SOLUTION: Since the dopant is phosphorus (Group 15 element), the semiconductor is an n-type extrinsic semiconductor. The saturation conductivity can be calculated using Eq. 3-56. Thus,

$$\sigma_{xSD} = N_D |e| \mu_e,$$

where N_D is the concentration of P atoms/cm^3 and is also the concentration of electrons/cm^3, because each atom of P donates one electron. Thus, we need to convert the concentration of P in atomic percent to atoms per cm^3. We can do this by calculating first the concentration of silicon in atoms per cm^3. We have done this in Example 3-6 when we calculated the number of total electrons per cm^3. This time we do not need the electron concentration, but only the atom concentration. Thus, if we remove the factor of 4 in the total electron concentration, the concentration of silicon atoms per cm^3 is 5×10^{22}. Since P is present in concentration of 1 atom per 10^6 atoms of Si,

$$N_D = (10^{-6} \text{ atom P/Si atom}) \times 5 \times 10^{22} \text{ atoms Si/cm}^3$$
$$= 5 \times 10^{16} \text{ atoms P/cm}^3$$

and

$$\sigma_{xSD} = 5 \times 10^{16} \times 1.6 \times 10^{-19} \times 1900$$
$$= 15.2 \text{ (ohm-cm)}^{-1}$$

At 25°C (298 K),

$$N_{x_e} = N_D \exp\left(-\frac{E_g - E_D}{kT}\right)$$

$$= N_D \exp\left(-\frac{0.044}{8.63 \times 10^{-5} \times 298}\right) = N_D \times 0.1805$$

We see that the conductivity at 25°C is 0.1805 that at saturation. Thus,

$$\sigma_{298\text{ K}} = 0.1805 \times 15.2 = 2.744 \text{ (ohm-cm)}^{-1}$$

EXAMPLE 3-9

The electrical conductivity of a germanium doped with antimony was found to be 1,000 (ohm-cm)$^{-1}$ at the saturation range. Determine the atomic percent of antimony in germanium.

SOLUTION: In this problem, the concentration of the dopant in atoms per cm^3 can be calculated from

$$\sigma_{xSD} = N_D |e| \mu_e$$

because σ_{xSD} is given. When we obtain N_D, we can convert it to atomic percent. Solving for N_D

$$N_D = \frac{1{,}000}{(1.6 \times 10^{-19}) \times 3800} = 1.64 \times 10^{18}\ e/\text{cm}^3$$

The charge density N_D is also equal to the concentration of dopant atoms per cm^3 because antimony is from Group 15 just like P, and each atom gives off one electron.

To get dopant concentration in atomic percent we need to know the concentration of germanium atoms per cm^3. We can obtain this in two ways, just as what was done with silicon in Example 3-6, using the density and the unit cell characteristics.

Using the density,

$$N_{ge} = \frac{5.323 \times 6.022 \times 10^{23}}{72.61}$$

$$= 4.42 \times 10^{22} \text{ atoms Ge/cm}^3$$

Using the unit cell characteristics,

$$N_{ge} = \frac{8 \text{ atoms/u.c.}}{(5.6575 \times 10^{-8})^3}$$

$$= 4.42 \times 10^{22} \text{ atoms Ge/cm}^3$$

$$\text{a/o Sb} = \frac{N_D}{N_D + N_G} \times 100$$

$$= \frac{1.64 \times 10^{18}}{4.42 \times 10^{22} + 1.64 \times 10^{18}} \times 100$$

$$\approx \frac{1.64}{4.42} \times 10^{-2}\% = 3.71 \times 10^{-3}\%$$

3.7.6 Compound Semiconductors

Up to this point, we have limited our discussion to the element semiconductors, silicon and germanium. Since they are from Group 14 (or IV) in the periodic table, each of the atoms share electrons with four other atoms to form four saturated covalent bonds. Each of the covalent bonds has two electrons for a total of eight electrons. In this section, we shall discuss compounds of elements from Groups 13 (or III) and 15 (or V), referred to as III-V compounds, and from Groups 2 (or II) and 16 (or VI), referred to as II-VI compounds, as semiconductors. The stoichiometric combination of elements from the respective groups to form the III-V and II-VI compounds provide the full complement of eight electrons to form covalent bonds. Because of the difference in the electronegativities of the elements, these compounds also show some ionic bonding characteristics. The partial ionic bond is reflected generally in higher energy band gaps. Some of these compounds are shown in Table 3-5 with their band-gap energies.

Of these compounds, gallium arsenide (GaAs) is the subject of much research and development work because GaAs integrated chips are known to have very fast response. The structure of GaAs was shown in Fig. 2-13. While it is classified as having a zinc-blende structure, the structure is not determined by the ratio of the cation or anion, as indicated in Fig. 2-14, because the bonding is principally covalent

Table 3-5 Some compound semiconductors with zinc blende structure.

Compound	Molecular mass, g/mol	Lattice Parameter, Å	Density, g/cm³	Energy Gap, eV	Mobility@ RT, cm²/v.s	
					Electron	Hole
II-VI						
ZnS	97.43	5.4093	4.079	3.54	180	5 (@400 C)
ZnSe	144.34	5.6676	5.42	2.58	540	28
ZnTe	192.99	6.101	6.34	2.26	340	100
HgS	232.65	5.8517	7.73	2.10	20,000	~ 1.5
HgSe	279.55	6.084	8.25	−0.06	25,000	350
III-V						
AlP	57.95	5.451	2.42	2.45	80	
AlAs	101.90	5.6622	3.81	2.16	1200	420
AlSb	148.73	6.1355	4.218	1.60	200–400	550
GaP	100.69	5.4505	4.13	2.24	300	150
GaAs	144.64	5.6532	5.316	1.35	8,800	400
GaSb	191.47	6.0954	5.619	0.67	4,000	1,400
InP	145.79	5.8688	4.787	1.27	4,600	150
InAs	189.74	6.0584	5.66	0.36	33,000	460
InSb	236.57	6.4788	5.775	0.163	78,000	750

between the atoms, as shown in two dimensions in Fig. 3-34. The three electrons from Ga and the five electrons from As are shared between the two atoms to form covalent bonds. Thus, it was discussed with the diamond-cubic structure because of the covalent bonding of the Ga and As atoms. A *perfect stoichiometric compound,* one with exact proportions of atoms of the elements in the compound, in this case, one Ga to one As, will be an *intrinsic semiconductor.* The conductivity of the stoichiometric compound depends on the E_g and the temperature. *Extrinsic semiconductors* may be obtained from either *nonstoichiometric compounds or by addition of dopants.* In nonstoichiometric compounds, excess As atoms will make n-type, while excess of Ga atoms will make p-type semiconductors. Substitution of elements from Group 16 (or VI), such as Se and Te, for As (Group 15 or V) will "donate" electrons to the structure. These elements are donor atoms and the semiconductor is an n-type. Substitution of elements from Group 2 or II, such as Zn, for Ga (Group 13 or III) will result in a lack of electrons for bonding and thereby creating "holes." These elements are therefore acceptor atoms and the semiconductor is a p-type. When silicon (Group 14 or IV) is substituted for Ga, the semiconductor will be n-type. On the other hand, when silicon is substituted for As, the semiconductor will be a p-type. The silicon is said to be an *amphoteric dopant.* The donated electrons (n-type) occupy energy levels just below the conduction band, while the holes created (p-type) occupy energy levels just above the valence band in the same manner as in silicon. Some of these energy gaps are shown in Table 3-6.

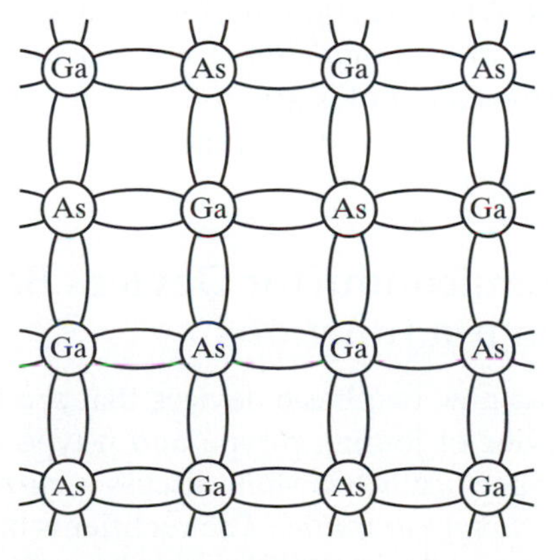

Fig. 3-34 Two dimensionsal representation of covalent bonding in pure GaAs.

EXAMPLE 3-10

An aluminum-antimony compound semiconductor was made by melting 0.825 kg of antimony and 0.175 kg of aluminum. Determine (a) whether this in an intrinsic or extrinsic semiconductor; (b) if extrinsic, which type?; and (c) the maximum majority carrier concentration per cm^3.

SOLUTION: We must first note that this is one of the III-V compounds. In these compounds, an intrinsic semiconductor is one with an equi-atomic composition of the Group 13 and Group 15 elements. If it is not equi-atomic, the semiconductor is extrinsic. If the Group 13 element is more than the Group 15 element, there is an excess of acceptor atoms and the semiconductor is a p-type. If the Group 15 element is more, there is an excess of donor atoms and the semiconductor is n-type. Thus, to answer the questions, we must convert the weight of each element into mols.

$$\text{Mols of Sb (antimony)} = \frac{825\text{ g}}{121.75\text{ g/mol}}$$

$$= 6.776\text{ mols}$$

$$\text{Mols of Al} = \frac{175\text{ g}}{26.981\text{ g/mol}} = 6.486\text{ mols}$$

When the number of mols is multiplied by the Avogadro number, we get the number of atoms of each element. Since mols of Sb is more than mols of Al, the semiconductor compound is extrinsic and it is n-type because Sb (Group 15) is more than Al. The majority carrier in this compound is therefore the electron whose concentration is equal to the excess

Table 3-6 Donor and acceptor levels in GaAs (min. E_g = 1.35 eV).

Donor		Acceptor	
Element	**($E_g - E_d$)**	**Element**	**($E_a - 0$)**
Si	0.006 eV	Si	0.035 eV
Se	0.006 eV	Zn	0.031 eV

Sb atom concentration. To get the excess atom concentration per cm^3, we need first to have the concentration in atomic percent.

$$\text{at. fraction, Sb} = \frac{6.776}{6.776 + 6.486} = 0.51093$$

$$\text{at. fraction, Al} = \frac{6.486}{6.776 + 6.486} = 0.48907$$

The excess Sb per one total (Al + Sb) = 0.51093 − 0.48907 = 0.02186 excess Sb/atom (Al + Sb).

The concentration of total (Al + Sb) per cm^3 is obtained from the characteristics of the unit cell,

$$N_{Al+Sb} = \frac{8 \text{ atoms/u.c.}}{(6.1355 \times 10^{-8})^3 \text{ cm}^3\text{/u.c.}}$$
$$= 3.464 \times 10^{22}$$

The concentration of excess Sb = concentration of electron charge carriers $= 0.02186 \times 3.464 \times 10^{22}$ $= 7.57 \times 10^{20}$ e/cm^3.

3.7.7 Interdependency of N_e and N_h

In either intrinsic or extrinsic semiconductors, the density of electrons depend on the density of the holes and vice versa. At any one temperature, these densities are in dynamic equilibrium, meaning that electrons will combine with holes to produce a covalent bond while at the same time a covalent bond will break to form electrons and holes. In chemistry, this relationship is described as the law of mass action and can be described by the following reaction.

$$A + B \rightleftarrows AB \downarrow \qquad (3\text{-}60)$$

Two species in solution react to form a precipitate AB. The dynamic equilibrium is described by

$$K = \frac{[AB]}{[A][B]} \qquad (3\text{-}61)$$

where K is the equilibrium constant and the quantities in the brackets refer to the activities of the products and reactants. The activities are related to the concentrations of the species and at low concentrations, they are equal to them. The activity of a precipitate is one and the constant K depends on temperature only. Therefore, we see that

$$[A][B] = 1/K = \text{constant} \qquad (3\text{-}62)$$

We cannot change the concentration of A, [A], without changing the concentration of B, [B].

The analogous reaction for electrons and holes is

$$\text{electrons} + \text{holes} \rightleftarrows \text{covalent bond} \qquad (3\text{-}63)$$

and the analogous equilibrium relationship is

$$N_e \cdot N_h = \text{constant} \qquad (3\text{-}64)$$

This relationship is valid for either intrinsic or extrinsic semiconductors. The concentrations or densities of electrons and holes are the total coming from both the intrinsic contribution and the impurity (dopant) contribution. In intrinsic semiconductors, where $N_e = N_h = N_i$, we have

$$N_e \cdot N_h = N_i^2 = \text{constant} \qquad (3\text{-}65)$$

and the electrons and holes are said to be co-carriers of current. In extrinsic semiconductors, the type determines the majority carrier and the other is the minority carrier. In n-type, the majority carrier is the electron and the minority carrier is the hole. In p-type, the hole is the majority carrier while the electron is the minority carrier. In both cases, Eq. 3-65 still holds. We note in passing that if we substitute Eqs. 3-49 and 3-50 in Eq. 3-65, and then solve for N_i which is equal to both N_e and N_h, we arrive at the same relationships as Eqs. 3-52a and 3-52b.

There are terms associated with the dynamic equilibrium, Eq. 3-63. The arrow to the right is the formation of a covalent bond. This is referred to as *recombination* in which both the electrons and holes are annihilated. On the other hand, the reaction to the left is the breaking of a covalent bond and this is referred to as the *generation* of electrons and holes. Where recombination occurs, the region is referred to as a *depletion zone,* because most of the electrons and holes are nonexistent. This depletion zone occurs at the p-n junction, which we shall now discuss.

3.7.8 Semiconductor Devices Based on p-n Junctions

We discuss now two basic devices that are based on the behavior of joining p-type and n-type semiconductors. Specifically, we shall discuss a *current rectifier* and a *voltage amplifier.* The rectifier is based on a p-n junction while an amplifier is based on a p-n-p or an n-p-n junction and their principles will now be discussed. These junctions are the foundations of integrated circuits in semiconductor devices.

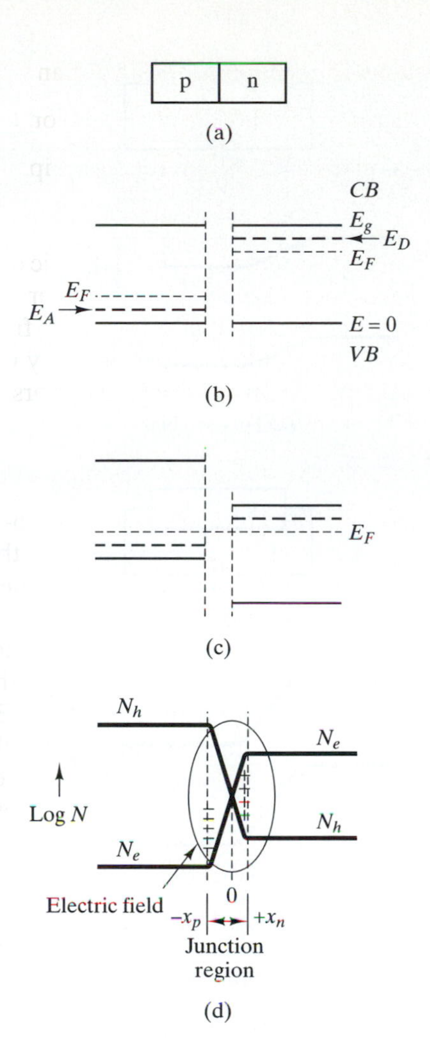

Fig. 3-35 Schematic of (a) a p-n junction; (b) Fermi levels and impurity energy levels (E_A for p-type; E_D for n-type) in the gap of valence (E = 0) and conduction (E = E_b) bands; (c) Equalization of Fermi levels; (d) Buildup of electric potential during diffusion of electron to p-side and holes to n-side.

We start by showing the p-n junction as a contact between a p-type and an n-type semiconductor in Fig. 3-35a. Note, however, that actual devices are not made in this manner. (The manufacture of actual p-n junctions is described in Sec. 6.2.5.) The simple picture given here is intended only to illustrate the principles. As discussed earlier with metals in Sec. 3-6 on Work Function and Contact Potential, the Fermi levels of the two conductor materials will equalize when they are joined together. Thus, the Fermi levels, indicated by E_F on the respective sides in Fig. 3-35b of the p- and n-type semiconductors, will equalize and thermal equilibrium will set in. In the process of getting to the equal Fermi-level state in Fig. 3-35c, electrons from the n-type material will flow to the p-type because there is an excess of electrons in the n-type, while the p-type is deficient of electrons. This is simply a process of diffusion (discussed in Chap. 6) to equalize the concentration of electrons. For the same reason, there is an excess of holes in the p-type and holes will diffuse to the n-type material. These processes are indicated in Fig. 3-35d. However, the electrons do not go very far into the p-type material and likewise, the holes do not go very far into the n-type material. As the electrons leave the n-type material, a net positive charge is created. In the same manner, a net negative charge is developed as holes leave the p-type material. The positive and negative charges create an electric field or potential that opposes the motion of electrons and holes. This potential increases until it is equal to the difference in the work functions of the two materials. When this is reached, the materials are said to be in thermal equilibrium, their Fermi levels are equalized, and there is no net flow of charges either way, see Fig. 3-35c. The distance of diffusion of electrons to the p-type is indicated by $-x_p$ and that of the holes to the n-type is $+x_n$. The sliver of material between $-x_p$ and $+x_n$ is called the junction region shown in Fig. 3-35d, and in this region, the recombination (forward) reaction of Eq. 3-63 ensues because the electrons and holes are coming together. Thus, the concentration of these mobile carriers will decrease abruptly and the junction is commonly called the depletion zone. This depletion zone is estimated to be on the order of 1 μm.

Also, in the process of equalizing the Fermi energy levels, the energy bands (valence and conduction) of the p-type material are raised and consequently, the potential energy of the p-type is higher relative to the n-type material. The resulting diagram is shown in Fig. 3-36. The electrons at the bottom of the conduction band at the p-side will roll down the slope because they decrease their potential energy this way. The flow of electrons from left to right, namely, the flow of current, is proportional to the density of electrons in the p-type material

$$I_{e\ (\text{left to right})} \propto N_{ep} \tag{3-66}$$

In order for electrons to move from right to left (from the n-type to the p-type), they will have to overcome the potential energy difference, eV_o. Since

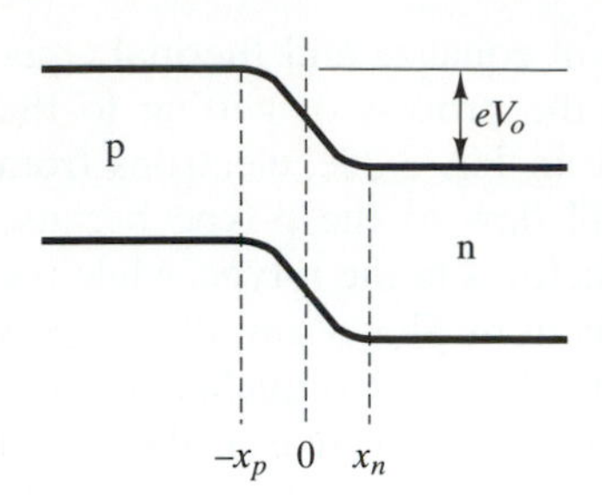

Fig. 3-36 Energy bands of p are raised relative to n energy bands at equilibrium. Electrons from n must overcome potential difference eV_o to move to p.

there are many more electrons in the n-type material, N_{en}, some of them can acquire enough energy to jump the potential hill. The number of electrons (flow of current) jumping the hill is proportional to

$$I_{\text{right ot left}} \propto N_{en} \exp(-eV_o/kT) \quad (3\text{-}67)$$

Substituting N_{en} from Eq. 3-49, we see that the flow of electrons (the flow of current from right to left) is proportional to

$$I_{e\,(\text{right to left})} \sim N_c \exp\left(-\frac{E_g + eV_o - E_F}{kT}\right) \quad (3\text{-}68)$$

At equilibrium, the current flowing from right to left equals the current flowing from left to right, that is, Eqs. 3-66 and 3-68 are equal. Thus, we get an expression of the density of electrons from the p-type, N_{ep}, which is

$$N_{ep} = N_o \exp\left(-\frac{E_g + eV_o - E_F}{kT}\right) \quad (3\text{-}69)$$

In this equation, N_o contains the N_c and, therefore, is temperature dependent. As before, we consider this dependency relatively small compared to the exponential term and consider it constant at low temperatures. eV_o equals the difference in the original Fermi levels and may be considered an "activation energy" for the flow of electrons from the n-type to the p-type material.

We can now illustrate the current rectifying characteristic of the p-n junction. For this, we look at Fig. 3-37 where a voltage, V_1, is applied to the junction to induce flow of electrons and holes. If the *p-type* is made *positive,* the flow of electrons is induced towards the p-type material and this is referred to as *forward bias.* The applied voltage V_1 is positive and the equilibrium activation energy is reduced by eV_1. The potential barrier is now $e(V_o - V_1)$ and for this reason more electrons (higher current) will flow towards the p-type

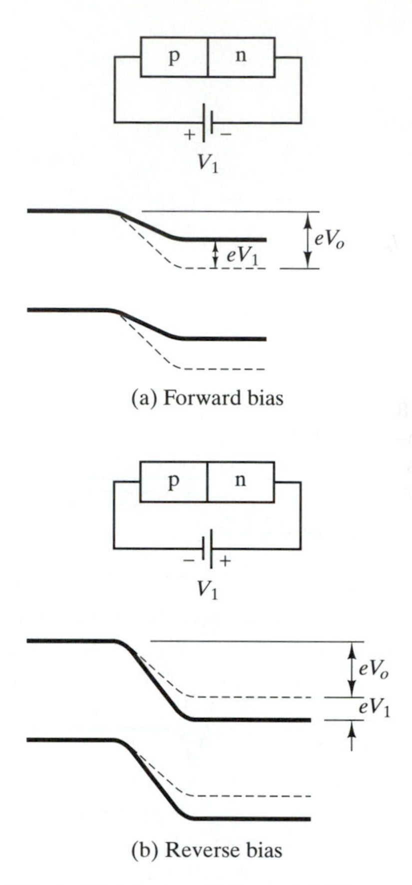

Fig. 3-37 Effect of (a) forward bias, and (b) reverse bias on the activation potential for flow of electrons from *n* to *p*. Forward bias lowers potential, while reverse bias increases potential.

material in forward bias. On the other hand, if the *p-type* is made *negative,* the negative electrode will repulse the flow of electrons to the p-type and this is called *reverse bias.* The potential barrier or activation energy is $e(V_o + V_1)$ for reverse bias and there is less electron flow (lower current) to the p-type.

The foregoing qualitative statements can be quantified by setting the dynamic equilibrium current, in Eq. 3-68, as I_o. Note that at equilibrium, I_o is the current flow from right to left as well as from left to right. Therefore, when $V_1 = 0$ (no voltage applied and in equilibrium), there is no net flow of electrons and $I = 0$. In forward bias, the flow of electrons (current) from right to left is,

$$I_{e\,(\text{right to left})} = I_o \exp\left(\frac{eV_1}{kT}\right) \quad (3\text{-}70)$$

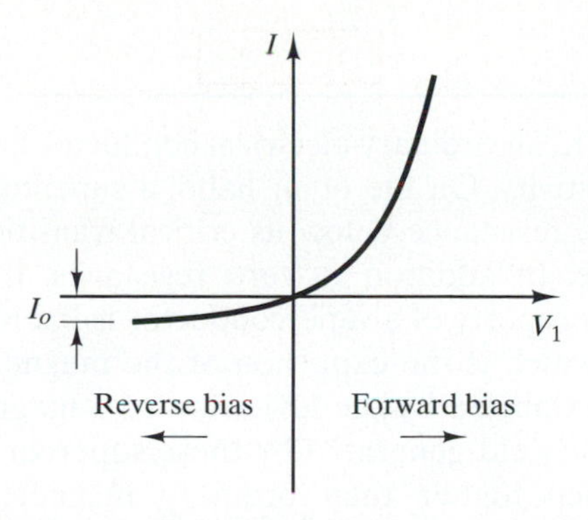

Fig. 3-38 Current in forward and reverse bias of p-n junction—very low current in reverse bias essentially converts an AC current to a DC current—p-n junction is a current rectifier.

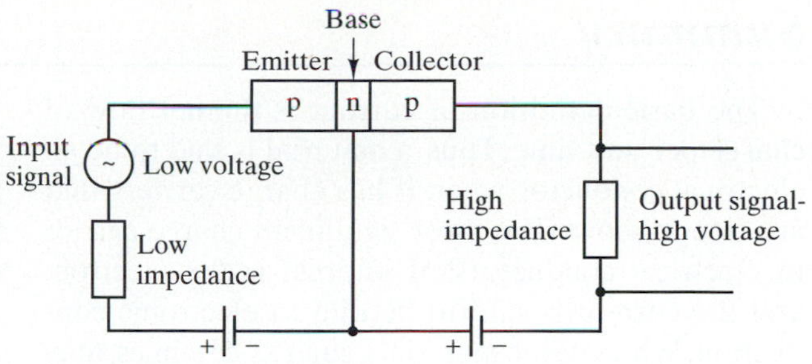

Fig. 3-39 Schematic of a p-n-p junction that produces a voltage amplifier transistor.

and we see that the current increases exponentially with V_1. Because the potential energy of the electrons in the p-type is higher than those in the n-type, the applied voltage does not affect the flow of electrons from left to right and the flow remains the same as when $V_1 = 0$,

$$I_{e\,(\text{left to right})} = I_o \tag{3-71}$$

and hence the net flow of current is,

$$I_{net} = I_o\left[\exp\left(\frac{eV_1}{kT}\right) - 1\right] \tag{3-72}$$

Equation 3-72 is known as the rectifier equation. For negative values of V_1, reverse bias, I_e will tend to $-I_o$, because the exponential term approaches zero as V_1 increases negatively. The resulting plot of the rectifier equation is shown in Fig. 3-38. Thus, when an AC voltage is applied to the junction (device), the output current is primarily unidirectional or direct current (DC). The *current rectifier* device that uses the principles of the p-n junction is called a *diode.* There are different kinds of diodes, but unfortunately we cannot cover all of them in this text.

We turn now to the *transistor (voltage amplifier)* that has two p-n junctions with one semiconductor common to both. The common semiconductor is called the *base* and the other two materials are the *emitter* and the *collector,* as shown in Fig. 3-39, for a p-n-p transistor. There are also n-p-n transistors and the ensuing discussion of the p-n-p transistor can be made to apply by a corresponding change of words.

First we discuss the emitter-base p-n junction and impose a forward bias on it (positive on p-side). As soon as the forward bias is applied, holes will move to the base and electrons will flow to the emitter. As the holes enter the base (which is n-type), the process of recombination starts. However, the recombination requires both time and space. For a very narrow (thin) base material ($\ll 1\,\text{mm}$), the holes can pass through it without much loss. Hence, the hole current leaving the base is essentially identical to the hole current entering the base from the emitter. Now, the holes leaving the base and entering the collector are exposed to a reverse bias, that is, the p-type collector is negatively charged, and therefore, they move freely towards the output to increase the signal.

Why is the transistor an amplifier? It is an amplifier since the voltage gets amplified by a large factor, although the current gain is less than unity. The amplification arises because the input circuit is a low impedance* circuit and thus a low voltage is sufficient to cause a certain current. The current reappears in a high impedance output circuit and is made to flow across a large load resistance, resulting in a high output voltage. Hence, the transistor in a common base circuit is a voltage amplifier.

*impedance is in AC circuitry as resistance is in DC circuitry; it is the ratio of the alternating voltage to the alternating current.

Summary

The basic definition of current is the net flow of charge per unit time. Thus, a material is said to be an electrical conductor when it has charge carriers that are free to move. The most significant charge carrier in electrical conductors of interest is the electron, and the current is said to be due to electronic conduction. While other materials, such as ceramics, may have ions as charge carriers, they are relatively poor conductors and are basically insulators at room temperature.

At absolute zero temperature (0 K), materials are classified as either metals or nonmetals. Metals are electrical conductors while nonmetals are not. Among the nonmetals are semiconductors and insulators—semiconductors become electrical conductors at higher temperatures, while insulators remain nonconductors. The conductivity of metals is due to the free electrons given up by each of the atoms in the large aggregate comprising the solid. The free electrons are called delocalized electrons (from their original atoms); the delocalization is due to the overlap of electronic orbitals of the tightly bonded atoms in the large aggregate. Electrons from one atom easily move from their own orbitals to available orbitals (quantum states) of the next atom. Nonmetals have electrons localized to their own atomic orbits because there are no available quantum states. Covalently bonded materials are basically all nonmetals or insulators at 0 K.

A metallic material is, therefore, one that has delocalized electrons. Thus, an ordinary nonmetal can be made metallic if there are delocalized electrons; silicon, a nonmetal at 0 K, was made metallic by containing a critical concentration of phosphorus atoms. The critical dopant concentration enabled the orbitals of the phosphorus atoms to overlap and hence to delocalize the electron from the confines of each atom. Organic and ceramic materials can also be made metallic by appropriate doping to produce linking of orbitals that allows electrons to be delocalized. Noble or molecular gases, such as xenon and hydrogen, which ordinarily have atoms or molecules very far apart, can be made metallic by forcing the molecules to come closer together under high pressure. The reverse transition from metallic to nonmetal behavior can also be made by separating the atoms.

At 0 K, an ordinary electrical conductor has residual resistivity. On the other hand, a superconductor has zero resistance below its critical transition temperature. In addition to zero resistance, the most unique property of a superconductor is the Meissner effect, which is the expulsion of the magnetic field and is exhibited by the levitation of a magnet. The magnetic field generated by these superconductors are much higher than ordinary magnets. These properties of superconductors are very exciting but for the most part unrealized because of their low critical temperatures. Up to the mid-1980s, most of the superconductors were metals or alloys and the maximum critical temperature was about 30 K. Consequently, the materials science world was excited when ceramic superconductors with critical temperatures up to 125 K were discovered. Based on this development, the metal superconductors are now called low temperature superconductors (LTS) and the ceramic superconductors are called high temperature superconductors (HTS). There are a lot of potential applications for superconductors in the semiconductor industry, in the transportation industry, and in electric power transmission. The LTSs are currently being applied in magnetic resonance imaging (MRI) due to their high magnetic field.

The electrical conductivity of materials is dependent on the charge density (number of charged particles per unit volume) and the mobility of the charge particle. In metals, there are a large number of charge particles (electrons) per unit volume but their mobility is low. This is due to the random collision of the electrons with point defects, such as vacancies and interstitial and substitutional atoms, and with the thermal vibrations of the atoms. Because thermal vibration of the atoms increases with temperature, the frequency of collision with electrons increases; moreover, for metals the conductivity decreases or the resistivity increases.

There are two types of semiconductors—intrinsic and extrinsic. Intrinsic semiconductors are pure materials (predominantly, silicon and some germanium) that depend on external energy, such as thermal, to free or delocalize some of its electrons from the covalent bonded state to be conductive. At room temperature, this intrinsic semiconductivity is relatively low, however. Commercial semiconductors are extrinsic and depend on impurity or dopant atoms to provide sufficient charge carrier density to have conductivities comparable to metals at room temperature. These impurities provide either positive holes (p-type) or negative electron (n-type) carriers that require minimal excitation energy (for instance, room temperature) to be conductive. Integrated circuits for many applications are based on the joining (p-n junction) of these two types of semiconductors. In addition to the silicon and germanium, there are also compound semiconductors that will achieve the same purpose.

References

1. A. Cottrell, *Introduction to the Modern Theory of Metals,* The Institute of Metals, London, 1988.
2. P. P. Edwards, "The Metallic State—Revisited," p. 93 in *Advances in Physical Metallurgy,* J. A. Charles and G. C. Smith, eds., The Institute of Metals, London, 1990.
3. P. P. Edwards and C. N. R. Rao, eds., *The Metallic and Nonmetallic States of Matter,* Taylor and Francis, 1985.
4. W. D. Kingery, et al, *Introduction to Ceramics,* 2d ed., John Wiley, N.Y., 1976.
5. J. W. Mayer and S. S. Lau, *Electronic Materials Science: For Integrated Circuits in Si and GaAs,* Macmillan, N.Y., 1990.
6. L. Solymar and D. Walsh, *Lectures on the Electrical Properties of Materials,* 5th ed., Oxford University Press, 1993.
7. G. Vidali, *Superconductivity. The Next Revolution?* Cambridge University Press, 1993.
8. *Handbook of Chemistry and Physics,* 76th ed., CRC Press, 1995–1996
9. C.R. Barrett, W. D. Nix, and A. S. Tetelman, The Principles of Engineering Materials, Prentice-Hall, 1973.
10. J.E. Hatch, Ed, Aluminum, Properties and Physical Metallurgy, ASM International, 1984.

Terms and Concepts

Acceptor atoms
Activation energy
BCS (Bardeen-Cooper-Schrieffer) theory
Charge carrier density
Charge density
Compound semiconductor
Conducting polymers
Conduction band
Conductivity
Conjugated polymers
Conjugated structure
Contact potential
Critical properties of superconductors
Current
Current density
Current rectifier
Density of electrons
Density of holes
Density of states
Depletion zone
Donor atoms
Dopant
Doping
Dynamic equilibrium
Electric field
Energy (electron) bands
Energy gap
Extrinsic semiconductivity
Extrinsic semiconductors
Fermi-Dirac function
Fermi energy
Generation
High temperature superconductors
Hole
Impedance
Insulators
Intrinsic semiconductivity
Low temperature superconductors
Matthiesen's rule
Meissner effect
Metallic
Metals
Mobility
Mott transition

Nonmetals
Nordheim's rule
n-type semiconductors
Oxide bronzes
p-n junction
p-n-p junction
p-type semiconductors
Recombination
Residual resistivity at 0 K
Resistivity
Rule of mixtures
Saturation conductivity
Semiconductivity
Semiconductors
Superconductivity
superconductors
superconductors
superconductors
Superconductors
Thermionic emission
Transference number
Valence band
Voltage amplifier
Work function

Questions and Exercise Problems

3.1 Differentiate (a) metals from nonmetals, and (b) insulators and semiconductors.

3.2 Explain the relationship of bonding in materials to the electron energy bands.

3.3 Differentiate a valence band from a conduction band? Can a valence band be a conduction band? Can a conduction band be a valence band?

3.4 On the basis of energy bands, differentiate metals, semiconductors, and insulators.

3.5 What is meant by the Mott transition? What happens when a nonmetallic material is transformed to a metallic material? What is the common method of converting a nonmetallic material to a metallic material at 0 K?

3.6 If we can make nonmetallic materials, conductive at 0 K, can we make metallic materials nonconductive? How?

3.7 Electrical conductivity is based on the motion of charge carriers when an electric field is applied to materials. What are the charge carriers in metals, semiconductors, and in ceramics?

3.8 From the results of Example 3-1, calculate the time interval between collisions of electrons at 1 K and at 293 K in copper.

3.9 Determine the mobility of electrons in silver that has an electrical resistivity of 1.617 micro-Ω-cm at 298 K. Determine also the time between collisions of the electrons.

3.10 Determine the mobility of electrons at 300 K in aluminum that has an electrical resistivity of 2.733 micro-Ω-cm at 300 K. Determine also the time between collisions of the electrons.

3.11 The resistivity of sodium (metal) is 4.2 micro-Ω-cm at 293 K. Calculate the charge carrier density and the mobility of the charge carrier (electrons) in sodium.

3.12 The resistivity of aluminum at 0 K is 0.0001 and at 300 K is 2.733 micro-Ω-cm. When cold-drawn, the resistivity at 300 K is 2.828 micro-Ω-cm. Determine the thermal and cold-work contribution to the resistivity of aluminum at 300 K.

3.13 The coefficient of resistivity of aluminum is 0.00429 micro-Ω-cm per °C. With an electrical resistivity of 2.828 micro-Ω-cm at 300 K, determine the resistance per meter of a No. 1 B & S gauge (0.7348 cm in diameter) cold-drawn aluminum wire at 350 K.

3.14 The coefficient of resistivity of copper is 0.0068 micro-Ω-cm per °C. With an electrical resistivity of 1.725 micro-Ω-cm at 300 K, determine the resistance per meter of a No. 1 B & S gauge (0.7348 cm in diameter) annealed copper wire at 350 K.

3.15 Using the resistivity data given in Example 3-1 for copper at 1 K and at 293 K, determine the resistivity of a copper-15 wt% zinc alloy at 293 K, when a copper-5 wt% zinc alloy is found to have a resistivity of 2.92 micro-Ω-cm at 293 K.

3.16 The resistivity of pure iron at 1 K is 0.0225 and at 273 K is 8.57 micro-Ω-cm. When alloyed with 5 wt% nickel, the resistivity of the alloy at 273 K is 18.7 micro-Ω-cm. Estimate the resistivity of an alloy with 15 wt% nickel at 273 K.

3.17 A No. 14 gauge (0.1628 cm in diameter) copper wire carries a steady current of 15 amp. What is the current density, the drift velocity, and the mean free path of the electrons in the wire? Assuming no heat loss to the surrounding, what would be the minimum increase in temperature of a one meter long, No. 14 copper wire after carrying the 15 amp current for 24 hours? The resistivity of copper at room temperature, 298 K, is 1.712 micro-Ω-cm and the specific heat is 0.092 cal/gm/°C.

3.18 A No. 14 gauge (0.1628 cm in diameter) aluminum wire carries a steady current of 15 amp. What is the current density, the drift velocity, and the mean free path of the electrons in the wire? Assuming no heat loss to the surrounding area, what would be the minimum increase in temperature of a one meter long, No. 14 cold-drawn aluminum wire carrying the 15 amp current for 24 hours? The resistivity of cold-drawn aluminum at room temperature is 2.828 micro-Ω-cm and the specific heat is 0.224 cal/gm/°C.

3.19 Differentiate between the following types of semiconductors and give examples of each type: (a) intrinsic and extrinsic semiconductors, (b) elemental and compound semiconductors, and (c) n-type and p-type semiconductors.

3.20 A 1 cm. cube of pure germanium was found to pass a current of 0.0064 amp when a 0.0010 volt was applied between two of its parallel faces. Determine the temperature of the germanium.

3.21 Determine the intrinsic resistivity of germanium at 300 K. What fraction of the total electrons in germanium account for the conductivity at 300 K?

3.22 The intrinsic resistivity of germanium at 455 K is 0.61 Ω-cm and at 715 K is 0.0274 Ω-cm. Estimate the energy gap, E_g, of germanium.

3.23 Estimate the intrinsic conductivity of GaAs and GaSb at 300 K.

3.24 The phosphorus concentration in silicon was determined to be 10 ppm (parts per million) by weight. Determine the saturation conductivity of the resultant semiconductor. What fraction of the saturation conductivity is the conductivity at room temperature (300 K)?

3.25 The boron concentration in silicon was determined to be 10 ppm (parts per million) by weight. Determine the saturation conductivity of the resultant semiconductor. What fraction of the saturation conductivity is the conductivity at room temperature (300 K)?

3.26 It is desired to have a conductivity of a silicon semiconductor of 5 $(\text{ohm-cm})^{-1}$ at room temperature (300 K). What is the minimum addition of lithium per kilogram of liquid silicon in order to have the desired conductivity?

3.27 Having the same conductivity as in Problem 3.26, what is the minimum addition of aluminum per kilogram of molten silicon?

3.28 Arsenic was purposely added in excess of the proportion needed for the equi-atomic gallium arsenide compound in order to produce a saturation conductivity of 5 $(\text{ohm-cm})^{-1}$. Because of the presence of an unknown acceptor impurity, the saturation conductivity was only 4 $(\text{ohm-cm})^{-1}$, the sample remained an n-type. What were the electron and hole densities in the compound semiconductor? What is the composition by weight of gallium and arsenic?

3.29 If an excess of gallium were added, what would be the composition by weight of gallium and arsenic in gallium arsenide to yield a saturation conductivity of 5 $(\text{ohm-cm})^{-1}$?

4

Materials Properties for Design of Structures and Components

4.1 *Introduction*

As the title indicates, we shall cover in this chapter most of the materials properties that are used for the design of structures (bridges, buildings, and ships) and components (pressure vessels such as chemical reactors and gears). The properties are listed in Table 4-1 and are classified as either microstructure insensitive or microstructure sensitive properties. Both types of properties are derived primarily from the strength of the atomic bonds (binding energy), and the atomic arrangement and the packing of the atoms in the solid. By microstructure, we mean the imperfections in actual crystals as discussed in Chap. 2, such as solutes, particles, grain size, inclusions, dislocations, surface imperfections, cracks, and so on. The microstructure insensitive properties are predominantly the physical and chemical properties that do not vary sizably with materials imperfections; on the other hand, the microstructure sensitive properties are predominantly the mechanical properties that change significantly with materials imperfections.

Table 4-1 Most materials properties for design of structures and components.

Microstructure Insensitive	Microstructure Sensitive
Density, ρ	Strength, σ (see Sec. 4.4.1 for different criteria)
Modulus of Elasticity, E	Ductility
Thermal Conductivity, κ	Fracture Toughness, K_{I_c}
Coefficient of Linear Thermal Expansion, α	Fatigue and Cyclic Properties
Melting Point, T_m	Creep
Glass Transition Temperature, T_g, for Polymers	Impact
Corrosion and Degradation	Hardness

Except for corrosion which we shall discuss in Chap. 10, all the properties will be discussed here and will be related to atomic bonding, atomic arrangement and packing, and materials imperfections.

4.2 Physical Properties—Microstructure Insensitive

4.2.1 Density

Density is one of the most important materials properties because it determines the weight of the structure or component. We shall see in Chap. 11 that it is a part of the structural performance factors for materials. It was related to the crystal structure of materials in Sec. 2.9. We saw that the *density of a solid basically depends on three factors: the mean atomic mass of its atoms or ions, their atomic or ionic sizes, and their crystal structure (the way they are packed).* The range in densities of materials arises primarily from the atomic mass (from one for hydrogen to 207 for lead) and the volume-packing fraction. Metals are dense because they consist of heavy atoms and are more or less closely packed. Polymers have low densities because they consist of light atoms, primarily carbon and hydrogen, in a linear, 2- or 3-dimensional loosely packed network. Ceramics have low densities because they contain C, O, or N atoms and also have lower packing fractions than metals. The minimum density of solids is about 1 Mg/m^3, which can be achieved with the lightest atoms packed in an open structure. Materials with lower densities than this are foams that contain substantial pore space[1].*

4.2.2 The Melting Point

As discussed in Sec. 1.5.1, the bond energy of the atoms is directly proportional to the melting point of the material. This is schematically illustrated in Fig. 4-1 where r_o is held constant for three different bonding curves to reflect the fact that the atomic sizes of different atoms do not vary much. The bond energy is essentially due to the long-range attractive (Coulombic) energy of the atoms between positive and negative charges. Figure 4-1 shows the bond

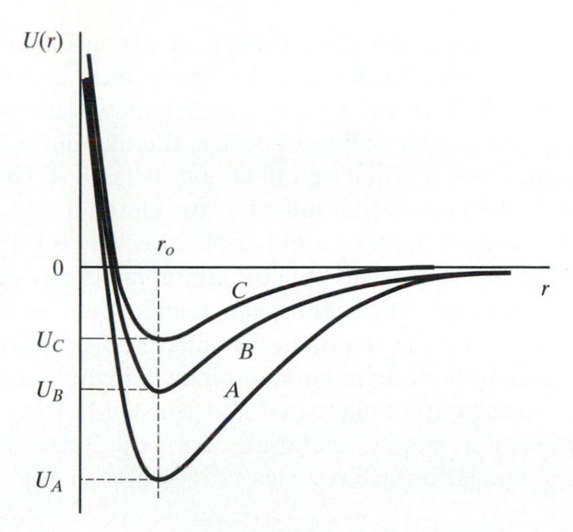

Fig. 4-1 Schematic binding energy curves for three different materials. Binding energy reflects amount of energy to separate atoms.

energy to be zero when the distance between the atoms is infinity. The attractive energy becomes significant only when the atoms are within a couple of atomic diameters. Because the liquid state does not have much resistance (strength) to flow, most of the bond energy in Fig. 4-1 has to be spent in the transition from the solid to the liquid state. The higher the bond energy, the higher will be the thermal energy or the melting point needed to separate the atoms.

The question is: How does the melting point of a material come into play in design? It becomes a factor when we are contemplating the use of materials at high temperature environments, such as that used in combustion engines. However, a high temperature for one material might be relatively low with respect to another material due to differences in melting points. To determine the effect of temperature on materials properties, we normalize the actual temperature, T, in absolute scale, by the absolute melting temperature, T_m. As a rule of thumb, the ratio of the environmental temperature, T, to the melting point (in absolute scale) of the material must be less than 0.3, that is,

$$\frac{T}{T_m} < 0.3 \tag{4-1}$$

must be true in order to neglect possible temperature effects on the material properties. In general, all the properties vary with temperature. If a material is under stress, the principal property of concern is

*Numbers in brackets refer to references at the end of the chapter.

creep, which is the slow change in dimension of a material under load when the above ratio exceeds 0.3. At 298 K (room temperature), lead will already creep but copper will not because the melting point of lead is 600 K while that of copper is 1,358 K. Thus, one of the factors that must be considered in design is the melting point of a material, especially for high temperature applications. To attain increased fuel efficiency and cleaner emissions, combustion engines should operate at the highest temperatures possible. Because of their high melting points, ceramic materials are currently being used and constantly tried as combustion engine materials, for both terrestrial (cars) and aerospace vehicles.

4.2.3 The Glass Transition Temperature

The glass transition temperature, T_g, is a property of noncrystalline or amorphous materials. The difference between T_g and T_m is schematically illustrated in Fig. 4-2. On cooling, if a crystalline structure of a material forms, it will occur just below the melting point of the material with an abrupt change in property. If a noncrystalline or amorphous structure forms, there is a gradual change in property below the melting point when the material is considered a supercooled liquid. At the glass transition temperature, T_g, the supercooled liquid transforms to a solid and this is exhibited by a slight change in property. The glass transition temperature originated from ceramic materials, specifically with silica where the amorphous silicate is called *glass.* This name has been carried over first to polymers and now to amorphous metals. In design, T_g is to amorphous solids as T_m is to crystalline materials. The consideration of T_g is especially crucial in polymers because T_gs (glass transition temperatures) of polymers are relatively low, such that even at room temperature, design with plastics or polymers involves long-term creep effects.

4.2.4 Coeffficient of Linear Thermal Expansion

The coefficient of linear thermal expansion (CLTE), like the melting point, can also be qualitatively related to the binding energy. In Fig. 4-1, the higher bonding energy material exhibits a much steeper curve than the lower bonding energy curves. The relation of the binding energy to the CLTE can be made by considering the equilibrium separation between the centers of atoms. The equilibrium separation between the centers of the atoms r_o in Fig. 4-3 represents the separation at 0 K. At higher temperatures, the atoms in a material vibrate or oscillate about their positions and their equilibrium separation distance changes, which is basically thermal expansion on the atomic scale. For a given temperature change, ΔT, this can be represented by a change in energy, ΔU. Because the separation distance fol-

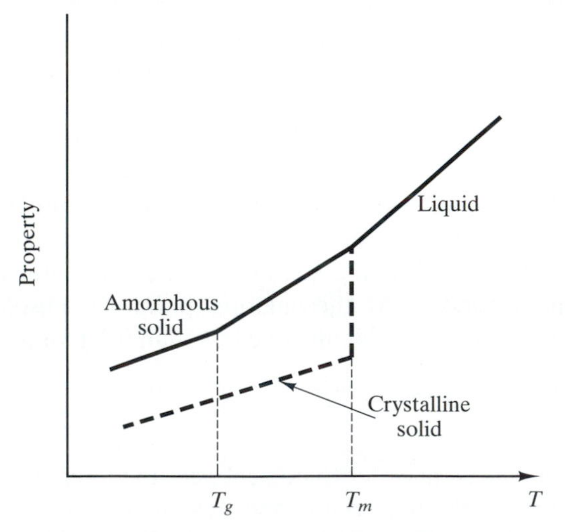

Fig. 4-2 Glass transition temperature, T_g, for amorphous solids relative to melting point, T_m, for crystalline solids.

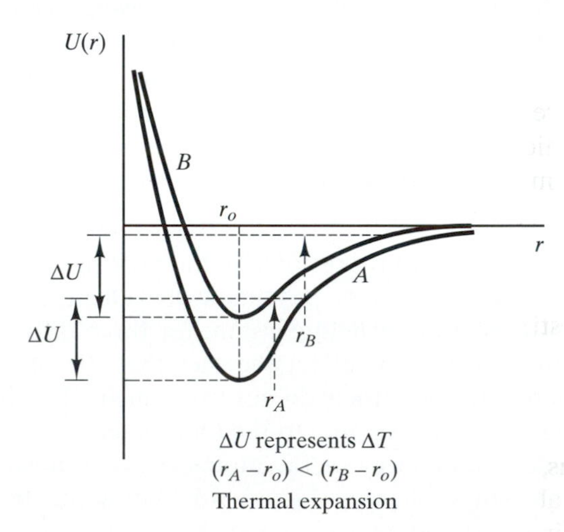

Fig. 4-3 Relation of binding energy curves to thermal expansion of solids. Higher binding energy material A has lower thermal expansion.

lows the bonding energy curve, the change in the mean vibrational position (per degree change in temperature) of the atoms for the same ΔT is less in the much steeper (higher bond energy) curve A than in the shallower (lesser bond energy) curve B, as schematically illustrated in Fig. 4-3.

Quantitatively, the coefficient of linear thermal expansion, α, is defined, macroscopically, as

$$\alpha = \frac{dL}{dT} \cdot \frac{1}{L} \tag{4-2}$$

where L is the linear dimension of the body. The coefficient of linear thermal expansion is also related to some physical properties and the modulus of elasticity, E, by,

$$\alpha = \frac{\gamma_G \rho C_v}{3E} \tag{4-3}$$

where γ_G is the Gruneisen's constant that has values ranging between 0.4 to 4, but for most solids it is about 1. C_v is the specific heat at constant volume and ρC_v is the volumetric specific heat at constant volume. The latter is a constant for most materials and, therefore, we see that α is proportional to 1/E. Diamond, with the highest modulus, has one of the lowest coefficients of thermal expansion—elastomers with the lowest moduli will expand the most[1].

The modulus E of a material may also be related to the melting point, T_m, according to the following[1],

$$E \approx \frac{100 k T_m}{\Omega} \tag{4-4}$$

where k is the Boltzmann's constant and Ω is the atomic volume in the structure of the material. The volumetric specific heat is also found to be

$$\rho C_v = \frac{3k}{\Omega} \tag{4-5}$$

Substituting Eqs. 4-4 and 4-5 into Eq. 4-3, we obtain

$$\alpha = \frac{\gamma_G}{100 T_m} \tag{4-6}$$

Thus, we see that α is also inversely proportional to the absolute melting point. For all solids, the thermal strain just before they melt is about the same.

In design, civil engineers take into account CLTE by incorporating expansion joints in structures such as bridges. In composite materials, the CLTEs of the constituent materials need to be matched very closely to reduce the risk of delamination at the constituents' interfaces due to thermal stresses induced by thermal cycling. This is especially of great concern in the semiconductor industry.

EXAMPLE 4-1

Determine the difference in length of a one metric-ton (1,000 kg) 10-mm diameter aluminum extrusion rod at room temperature (25°C) from the as-extruded length at 450°C.

SOLUTION: We need the coefficient of linear thermal expansion (CLTE) to use Eq. 4-2. This is found to be 27.5×10^{-6} per °C over the range of 20–500°C. Using Eq. 4-2,

$$\Delta L = \alpha \cdot L_{25°C} \cdot \Delta T$$

to get ΔL, we need $L_{25°C}$. We get this from the one-metric ton mass using the density of aluminum which is 2.70 g/cm^3.

$$\text{Mass} = (\text{Area} \times L_{25°C}) \times \text{density}$$
$$= (\pi/4)(1)^2 \times L_{25°C} \times 2.70$$
$$L_{25°C} = 1{,}000 \text{ kg} \times 1000 \text{ g/kg} \times 4/(2.70\pi) \text{ cm}$$

With $\Delta T = 450 - 25 = 425$,

$$\Delta L = 5{,}511 \text{ cm} = 55.11 \text{ m}$$

For the same weight and size of a rod, the ΔL for iron is only 979 cm or 9.79 m, because the mean CLTE for iron in the same temperature range is 14.25×10^{-6} per °C, and the density is 7.87 g/cm^3. We also compare the melting points of iron and aluminum, which are 1,536°C, and 660°C, and note that aluminum has a much higher CLTE than iron because it has a lower T_m.

4.2.5 Thermal Conductivity

In heat conduction, thermal conductivity, κ, is the analogous property to electrical conductivity. When a temperature gradient, $\Delta T/\Delta x$, is imposed on a material, Fig. 4-4, the flux of heat flow per unit area per unit time, Q, is

$$Q = -\kappa \frac{\Delta T}{\Delta x} \tag{4-7}$$

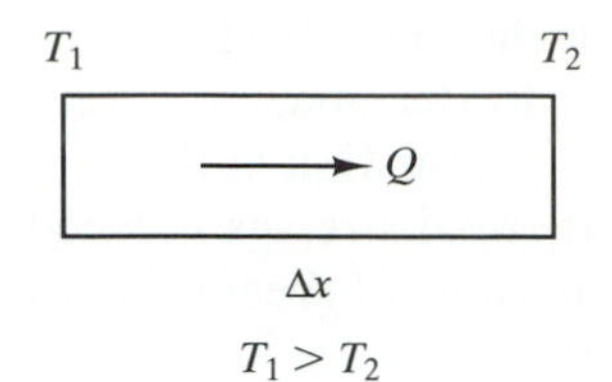

Fig. 4-4 Heat Flow, Q, with Temperature Gradient.

where κ is in watt per meter per Kelvin, that is, $W(mK)^{-1}$. This equation is to be compared with the flux of current (current density, J) in Chap. 3, which is

$$J = \sigma E = \sigma \frac{V}{x} \tag{4-8}$$

In metals such as copper, silver, and aluminum, electrons are also the carriers of heat. Because electrons conduct heat as well as electricity, there is a relationship between κ and σ, the electrical conductivity, in metals. This is

$$\frac{\kappa}{\sigma T} = L = 5.5 \times 10^{-9} \frac{\text{cal} \cdot \text{ohm}}{\text{s} \cdot \text{K}^2} \tag{4-9}$$

where L is called the *Lorenz constant*. (We must note the carryover of nomenclature of σ and E from Chap. 3 in Eqs. 4-8 and 4-9. These are the only two equations in this chapter that refer to these symbols as electrical conductivity and electric field. For the rest of the Chap. 4, the symbols signify stress and the Young's Modulus of Elasticity.)

Basically, the thermal conductivity can be described by[1]

$$\kappa = \frac{1}{3}(Cvl)_e \tag{4-10}$$

where C is the electron specific heat per unit volume, v is the electron velocity (2×10^5 m/s), and l is the electron mean free path, which is typically 10^{-7} m in pure metals. Just like electrical conductivity, substitutional solid solutes and other defects scatter electrons to reduce the mean free path and thereby reduce κ.

In ceramics and polymers, heat conduction is carried by phonons or lattice vibrations. The latter are scattered by each other (via an anharmonic interaction) and by impurities, lattice defects, and surfaces. These determine the phonon mean free path, l, and the thermal conductivity is still of the form in Eq. 4-10 and is now written as,

$$\kappa = \frac{1}{3}\rho C_p v l \tag{4-11}$$

with v now being the elastic wave speed (around 10^3 m/s), ρ is the density, and ρC_p is the volumetric specific heat. If the crystal is perfect and the temperature well below a characteristic temperature called the Debye temperature, as in diamond at room temperature, the phonon conductivity can be high. This phenomenon is exhibited by single crystals of diamond, silicon carbide, and even alumina to allow these materials to have thermal conductivities as high as copper. In recent years, there is a lot of interest in diamond deposition for application in the semiconductor industry because its thermal conductivity may be used to relieve the thermal load of integrated circuits. In contrast, the low thermal conductivity in glass is due to its irregular amorphous structure: the characteristic length of the molecular linkages (about 10^{-9} m) determines the mean free path. Likewise, polymers have low conductivities because the elastic wave speed v is low and the mean free path in the disordered structure is small[1].

The best thermal insulators are highly porous materials such as firebrick, cork, and foams. Their conductivity is limited by the gas phase in their cells and by heat transfer by radiation through the transparent cell walls. The thermal protection system (TPS) of the National Aeronautics and Space Administration's (NASA) shuttle is porous silica.

EXAMPLE 4-2

Estimate the thermal conductivity of aluminum at 20°C from the electrical resistivity of aluminum at 20°C which is $2.6548 \cdot 10^{-6}$ (ohm-cm). Compare this calculated value with the recommended value of thermal conductivity of 2.37 W $(\text{cm-K})^{-1}$.

SOLUTION: We use Eq. 4-9,

$$\frac{\kappa}{\sigma T} = 5.5 \times 10^{-9} \frac{\text{cal} \cdot \text{ohm}}{\text{s} \cdot \text{K}^2}$$

where σ is the electrical conductivity which is equal to $(1/2.6548) \times 10^6$ $(\text{ohm-cm})^{-1}$, and T = 293 K. Thus,

$$\kappa = 5.5 \times 10^9 \frac{\text{cal} \cdot \text{ohm}}{\text{sec} \cdot \text{K}^2} \times \frac{1}{2.6548} \times 10^6 \times 293$$

$$= 0.607\left(\frac{\text{cal}}{\text{s}}\right)(\text{cm} - \text{K})^{-1}$$

Now, the unit of W (watt) is Joule/sec, thus $\kappa = 0.607 \times 4.184 = 2.53\ \text{W}\ (\text{cm} - \text{K})^{-1}$. This is 6.75 percent more than the recommended value. For engineering purposes, this error is acceptable for rough estimates.

4.3 Concepts of Stress, Strain, and Young's Modulus

The heat flow and the current density equations discussed above are two examples of a more general equation relating stimulus to response, which is expressed as,

$$\textit{Response} = (\textit{Proportionality Constant}) \times \textit{Stimulus} \quad (4\text{-}12)$$

That is, if we impose a stimulus on a material, the material will respond depending on the proportionality constant—a property of the material. In the two examples above, the properties are thermal and electrical conductivities. In this section, we consider another example of stimulus and response. If we stretch (strain) a material, the material will respond by resisting the stretching stimulus. The resistance is manifested by the effort or the force (stress) we have to exert to stretch the material. This is expressed as follows,

$$\textit{Stress} = \textit{Young's Modulus} \times \textit{Strain}$$

or

$$\sigma = E\epsilon \quad (4\text{-}13)$$

As we shall see very shortly, strain is the normalized stretch stimulus and stress is the normalized force response. For very small strains, this relationship is known as Hooke's Law for simple tension or compression. The proportionality constant E in Eq. 4-13 is called the Young's modulus of elasticity and is a material property. Before we consider the relationship of E (Young's Modulus of elasticiy) to bonding, we need to introduce the concepts of stress and strain.

4.3.1 Concept of Stress

In Fig. 4-5, if a force, $F_\perp$, applied normal to the cross section of a bar, is divided by the cross section, A, of the bar, we obtain a normalized term we call stress, thus

$$\sigma \equiv \frac{F_\perp}{A} \quad (4\text{-}14)$$

This gives the intensity of the force experienced by the material as illustrated in Fig. 4-5a and 5b, where the same normal force is applied to two materials with cross sections A_1 and A_2, with $A_2 > A_1$. Although the force is the same, the intensity (stress) of the force in A_1 is greater than in A_2. When the stress stretches the filaments of a material, it is called a *tensile stress.* When the stress presses on the filaments to shorten them as in Fig. 4-5b, it is called a *compressive stress.* Tensile stress is conventionally taken as positive, while compressive stress is negative. Both stresses are referred to as normal stresses because the forces act perpendicular to the area.

There is another type of stress that results when the force acts parallel to the surface or area. This is referred to as *shear stress,* τ, shown in Fig. 4-6.

$$\tau \equiv \frac{F_\parallel}{A} \quad (4\text{-}15)$$

In these equations, forces in SI units are in newtons, N, and so the stresses are measured in newtons per meter squared, $N \cdot m^{-2}$. This unit is called Pascal (Pa) and is

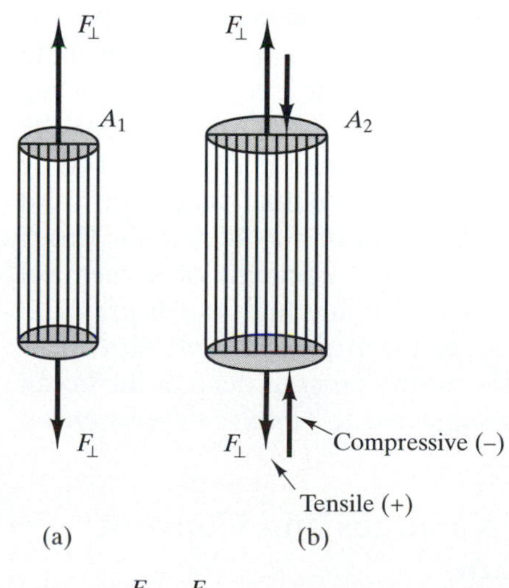

Fig. 4-5 Definition of normal stresses.

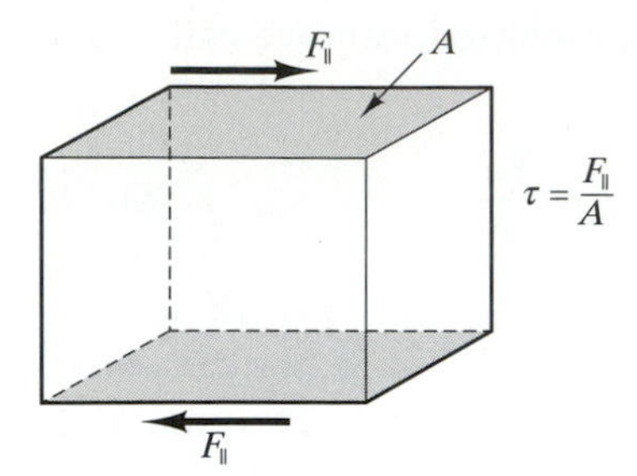

Fig. 4-6 Definition of shear stress.

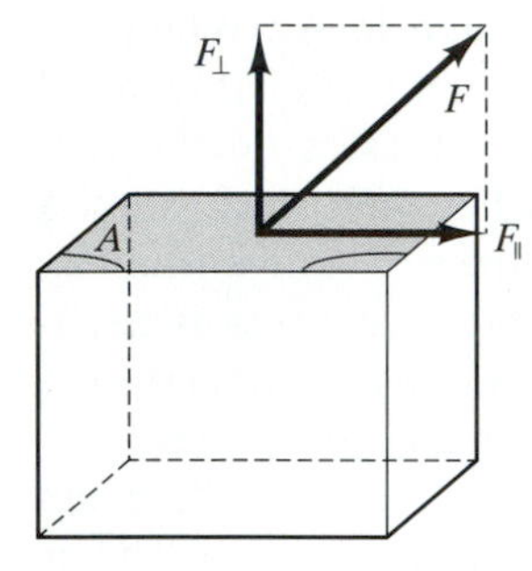

Fig. 4-7 Force applied to surface A is resolved to components normal and parallel to A to get stresses.

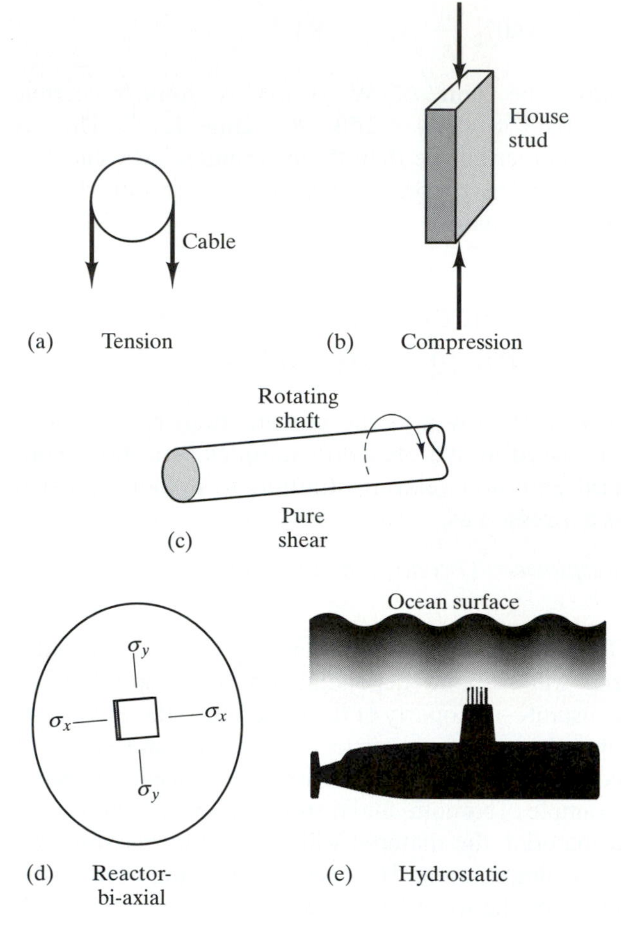

Fig. 4-8 Common stress modes in structures or components.

small. A stress is usually reported in meganewton per meter squared or megapascal ($MN \cdot m^{-2}$ or MPa).

In general, a stress on a body is neither normal nor parallel to the surface. In this case, the force on the body is resolved into normal and parallel forces to the surface to get the normal and the shear stresses, as illustrated in Fig. 4-7.

Figure 4-8 shows examples of the four commonly occurring stress states in a structure. The simplest is the *simple tension* or *compression.* The cable is in simple tension, while a stud or a column supporting the weight of a structure is under compression. The next stress state is *pure shear* that is experienced by the surface elements of a rotating shaft. The third common stress state is that of *biaxial tension.* This is exemplified by the surface elements of a chemical reactor or a line pipe carrying fluids under pressure. The fourth stress is the *hydrostatic pressure.* This occurs deep in the earth's crust or deep in the ocean when a solid is subjected to *equal compression* on all sides.

4.3.2 Shape Changes and Concept of Strain

Depending on the stress state, the material will deform or change its shape accordingly. Figure 4-9 illustrates the changes in shape (*deformations*) of a material when it is subjected to a tensile stress, a shear stress, or a hydrostatic stress. The deformations correspond to strains on the material.

A tensile stress induces the material to elongate parallel to the tensile force and to contract laterally along a direction perpendicular to the tensile force. The elongation and contraction are again normalized by dividing them by the original linear dimensions parallel (length) with and perpendicular (width) to the tensile force. The normalized quantities are called the *strains,* and the strain parallel to the force is called *longitudinal* or *tensile strain,* while the perpendicular strain is called the *lateral strain.* Thus,

$$\text{tensile strain, } \epsilon_L = \frac{\Delta L}{L} \qquad (4\text{-}16)$$

and

$$\text{lateral strain, } \epsilon_w = \frac{\Delta w}{w} \tag{4-17}$$

where L and w are the original dimensions parallel to and perpendicular to the tensile force. The longitudinal strain, ϵ_L, is positive while the lateral strain ϵ_w is negative because Δw is negative. The ratio of the lateral strain to the longitudinal strain is defined as the Poisson's ratio, ν, namely

$$\nu \equiv -\frac{\epsilon_w}{\epsilon_L} \tag{4-18}$$

A shear stress distorts a unit cube into a parallelipiped. As the shear stress acts on planes separated by a distance L, it is assumed that the bottom plane is

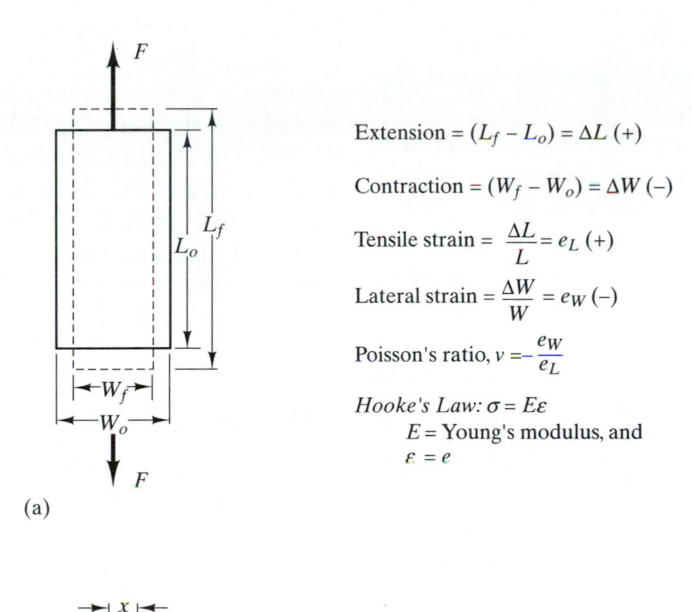

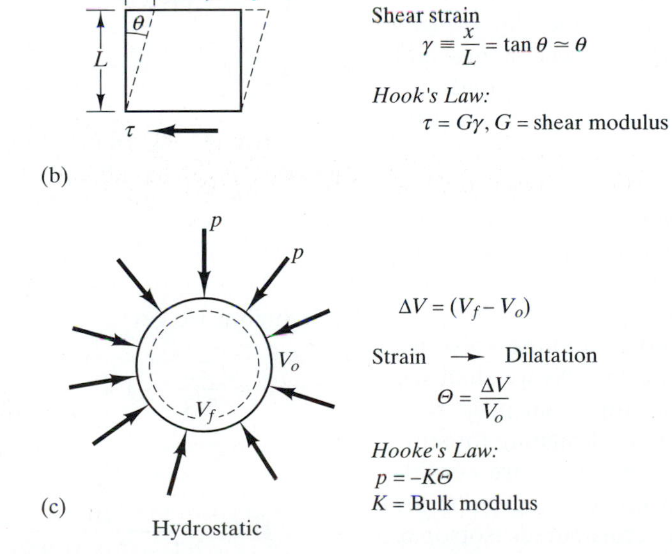

Fig. 4-9 Shape changes, strains, and Hooke's Law in (a) tension, (b) shear, and (c) hydrostatic compression.

fixed and the top plane moves relative to the bottom. The shear stress induces a displacement parallel to the shear direction that is proportional to the distance from the bottom plane. The engineering shear strain, γ, is defined as

$$\gamma = \frac{x}{L} = \tan\theta \approx \theta \tag{4-19}$$

for small strains. The corresponding Hooke's Law between shear stress and shear strain is

$$\tau = G\gamma \tag{4-20}$$

where G is another material property called the *shear modulus of elasticity.*

The hydrostatic stress or pressure, p, acting as a compressive stress equally in all directions causes a negative volume change called a dilatation. If the original volume is V and the negative change in volume induced by the hydrostatic stress is ΔV, then the *dilatation,* Θ, is defined as

$$\Theta \equiv \frac{\Delta V}{V} \tag{4-21}$$

The Hooke's Law for this stress state is

$$p = -K\Theta \tag{4-22}$$

where K is the bulk modulus.

The Hooke's Laws for the different stress states allow us to calculate the allowable stress for the elastic strain or deformation of materials with linear stress-strain relationships. Most engineering designs are based on this linear relation between stress and strain. The linear elastic strains for most solids are very small, usually about 0.001—beyond this, brittle materials will break and ductile materials will become plastic. (These terms are discussed in Sec. 4.4.4 in this chapter.) Rubbers or elastomers are elastic up to strains four or five, but they cease to be linearly elastic, that is, the stress is no longer proportional to strain after a strain of 0.01.

The three moduli, E, G, and K, and the Poisson's ratio, ν, are properties of materials. As we shall see in Chap. 11, E and G are the important properties we look for in materials if we are designing for stiffness. In engineering applications, these are considered constants because most engineering materials are polycrystalline and are considered isotropic. (Isotropic means the property does not vary with direction or is independent of direction.) The four constants are interrelated by the following equations

$$K = \frac{E}{2(1 - 2\nu)} \tag{4-23a}$$

$$G = \frac{E}{2(1 + \nu)} \tag{4-23b}$$

$$E = \frac{9KG}{(3K + G)} \tag{4-23c}$$

We see that we need to know only two of the four constants and the other two can be derived. Since the strains are dimensionless, the units of the above moduli are the same as the stresses, for instance, $N \cdot m^{-2}$. The English units are psi, pounds-force per square inch or ksi, kips per square inch, where 1 kip = 1,000 pounds. In SI units, the moduli are large quantities and are usually reported as giganewtons per meter squared (a giga being 10^9) or $GN \cdot m^{-2}$ or GPa. In addition, for most metals, ν is about 0.33.

EXAMPLE 4-3

A titanium alloy has a Young's Modulus of 110 GPa, and a shear modulus of 42 GPa. Calculate its bulk modulus.

SOLUTION: We use the relationships of elastic properties in Eq. 4-23. To get the bulk modulus, K, we use Eq. 4-23, which is

$$K = \frac{E}{2(1 - 2\nu)}$$

E, the Young's Modulus, is given, but we need to know the Poisson's ratio, ν. To get ν, we use

$$G = \frac{E}{2(1 + \nu)}$$

Rearranging, we get

$$\nu = \frac{E}{2G} - 1 = \frac{110}{2 \times 42} - 1 = 0.309$$

and

$$K = \frac{110}{2(1 - 2 \times 0.309)} = 144.0 \text{ GPa}$$

EXAMPLE 4-4

The Young's Moduli of steel and aluminum are 30×10^6 psi and 10×10^6 psi. If their yield strengths are 50,000 psi, and 25,000 psi, and a design tensile load of 5,000 lb-force is required, (a) determine the diameters of steel and aluminum bars required for elastic deformation, and (b) determine the maximum strain before plastic deformation sets in.

SOLUTION:

(a) For elastic deformation, the design stress $\leq \sigma_{y.s.}$. The minimum size of bars are obtained when the design stress $= \sigma_{y.s.}$. The minimum sizes for steel and aluminum needed are:

$$50{,}000 = \frac{5{,}000}{\frac{\pi}{4}(D)^2}$$

$$D = \sqrt{\frac{0.1 \times 4}{\pi}} = 0.357 \text{ in. for steel and}$$

$$25{,}000 = \frac{5{,}000}{\frac{\pi}{4}(D)^2}$$

$$D = \sqrt{\frac{0.2 \times 4}{\pi}} = 0.505 \text{ in. for aluminum}$$

(b) the maximum elastic strains are:

$$e_{st} = \frac{50{,}000}{30 \times 10^6} = 0.0017, \quad \text{and}$$

$$e_{Al} = \frac{25{,}000}{10 \times 10^6} = 0.0025$$

4.3.3 Young's Modulus and Interatomic Bonding [2]*

Let us now consider the relationship of E, the Young's Modulus, with interatomic bonding. We start with the binding energy curve, U = U(r), in Fig. 4-10a. If we take the derivative or slope at every point in the U(r) curve, we obtain the force curve, Fig. 4-10b, necessary to separate the atoms apart because

$$F = \frac{dU}{dr} \tag{4-24}$$

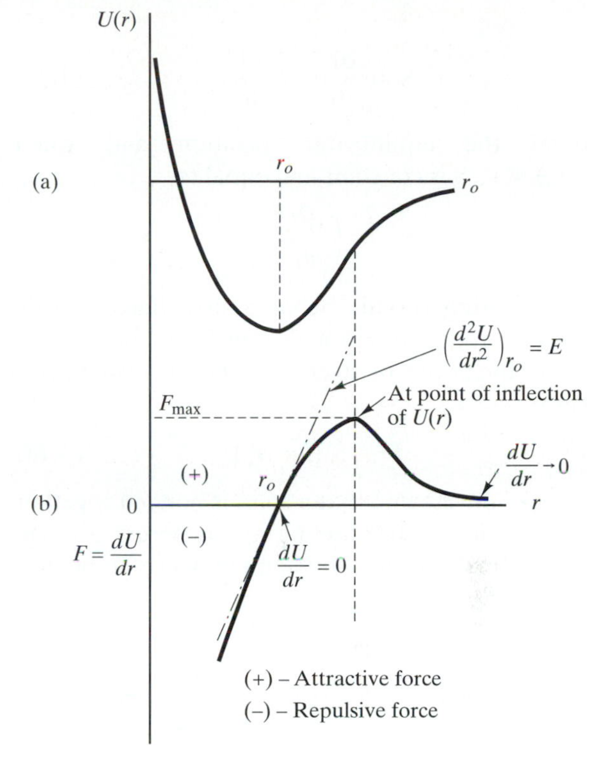

Fig. 4-10 Force curve (b) is the derivative of binding energy curve (a). Derivative of F or second derivative of U at ro relates to modulus, E.

In order to help understand the interatomic force curve in Fig. 4-10b, we shall assume that the bonds between the atoms can be represented by little springs, as in Fig. 4-11a. The equilibrium position, r_o, represents the natural length of the springs and at this length, the force on the spring is zero. Mathematically, the derivative (or slope of line) of the curve U(r) at r_o is zero. In order to move the atoms closer to each other, that is, make their interatomic distance $r < r_o$, a negative (compressive) force is needed that increases monotonically to $-\infty$ as we push the atoms closer together. To move the atoms apart, making $r > r_o$, a positive (tensile) force is needed; this force attains a maximum value at the point of inflection of the binding energy curve, U(r), and then decreases and approaches zero as $r \to \infty$. The stiffness of the springs, S, dictates the magnitude of the force exerted for a given change in length of the springs. This is expressed as

$$F = S(r - r_o) \tag{4-25}$$

or

$$S = \frac{dF}{dr} = \frac{d^2U}{dr^2} \tag{4-26}$$

Around the equilibrium position and when $(r - r_o) \ll r_o$, S is constant and equal to

$$S_o = \left(\frac{d^2U}{dr^2}\right)_{r=r_o} \tag{4-27}$$

and the spring (bond) behaves in a linear elastic manner accroding to Hooke's Law. Thus, the force to separate the atoms or stretch the bond between two atoms is

$$F = S_o(r - r_o) \tag{4-28}$$

Let us now assume that a solid is bonded together by such little springs, linking atoms across a plane within the material, as shown in Fig. 4-11b. To be precise, the atoms and their positions dictated by the crystal structure of the material have to be indicated in the plane. However, for simplicity, let us assume a simple cubic structure with atoms located only at the corners of the cube. In practice, very few materials have the simple cubic structure, but the results of this simple model [2] will not differ very much from the actual results. If we take a unit area in the plane, the total force needed to pull the springs (bonds) is the stress on that unit area, that is,

$$\sigma = \frac{F}{A} = NS_o(r - r_o) \tag{4-29}$$

where N is the number of bonds per unit area. In this simple model, the number of bonds per unit area is also the number of atoms per unit area. As indicated in Sec. 1.5, the equilibrium interatomic distance, r_o, is also taken as the average atomic size and thus, the area occupied by an atom is proportional to r_o^2. This being the case, the number of atoms per unit area (also the number of bonds, N, per unit area) is

$$N = \frac{1}{r_o^2} \tag{4-30}$$

Substituting the value of N, and noting that $(r - r_o)/r_o$ equals the tensile strain, ϵ, we get

$$\sigma = \frac{S_o}{r_o}\epsilon \tag{4-31}$$

Comparing this to Hooke's Law in tensile stress state, we recognize that

$$E = \frac{S_o}{r_o} = \frac{1}{r_o}\left(\frac{dF}{dr}\right)_{r_o} = \frac{1}{r_o}\left(\frac{d^2U}{dr^2}\right)_{r_o} \tag{4-32}$$

Thus, the modulus is seen to be the slope of the interatomic force-distance curve at the equilibrium distance divided by the equilibrium distance (atomic size).

Thus, in order to estimate E, we need to obtain S_o from theoretically determined binding energy curves, U(r). There are different U(r)s for the different types of bonding. An example U(r) for the ionic bonding in NaCl is

$$U = U_i - \frac{q^2}{4\pi\epsilon_o r} + \frac{B}{r^n} \tag{4-33}$$

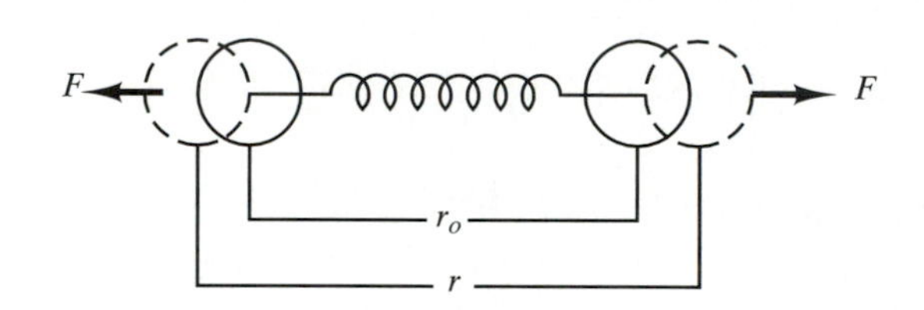

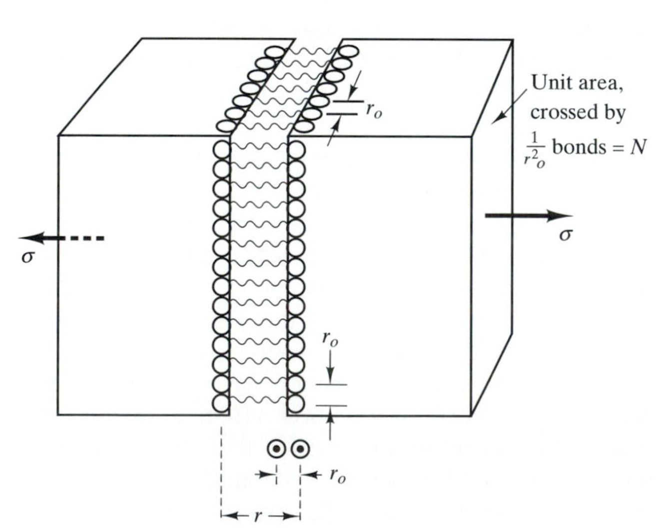

Fig. 4-11 Spring model of atomic bonding to estimate the Young's Modulus of Elasticity, E. [2]

where U_i is a constant, the second term on the right is the attractive part, and the last term is the repulsive part. Taking dU/dr and equating it to zero at r_o, we obtain

$$B = \frac{q^2 r_o^{(n-1)}}{4\pi n \epsilon_o} \tag{4-34}$$

Taking the second order derivative at r_o, we get

$$S_o = \frac{\beta q^2}{4\pi \epsilon_o r_o^3} \tag{4-35}$$

where β is a constant. In order to calculate S_o properly, the sum over all the nearest, the next nearest, the next to the next nearest neighbor bonds, of all attractive and repulsive energies are taken. In this way, β is found to be about 0.58. Substituting all the physical constants and taking $r_o = 2.5 \times 10^{-10}$ m (2.5Å), S_o is found to be 9.5 N/m for NaCl. If we divide S_o by r_o, we obtain an estimate of E.

Estimates of S_o and E are given in Table 4-2 for the different bonding mechanisms. Because r_o does not vary much, the wide range in moduli of materials is due to the range in S_o. We see in Table 4-2 that the highest modulus is that of the covalent bond in diamond. This is due to the small-sized carbon atoms that result in a greater number of bonds, N, per unit area, and a high S for the covalent bond that ranges from 20–200 N/m. The S values for the metallic and the ionic bonds are a little less, ranging from 15–100 N/m. Metals have high moduli because the close-packing of atoms in their structures gives a high bond density and the bonds are also strong. Ceramics have high moduli due to the fact that their bonding is a mixture of covalent and ionic and the presence of the smaller sized atoms of C, O, and N in their structures. Polymers contain both the strong covalent bonds and the weak hydrogen or van der Waals bonds. It is the weak bonds that stretch first when polymers deform and thus, most polymers exhibit low moduli.

The lower limit of E for true solids is estimated to be 1 GPa. This assumes that the atom size, r_o, is 3 Å and S_o is 0.5 N/m, a very weak bond. However, there are materials that have moduli less than this limit. These are either elastomers or foams. Foams are materials made up of cells containing a large pore (empty space) fraction. Elastomers have very low moduli because at room temperature they are above their glass transition temperature, T_g. At this temperature, the elastomers are in the supercooled liquid state and a good number of van der Waals bonds have broken or are melted. The rubber remains solid because of the cross-link bonds between the macromolecules. At temperatures much lower than the T_g, the elastomers become true solids and the calculated modulus is about 2 GPa. We shall cover more of the structures of polymers/plastics in Chap. 15. Foams have low moduli because the cell walls bend when the material is loaded.

4.4 Mechanical Properties of Materials and Test Methods

In this section, we shall discuss the structure-sensitive mechanical properties that relate primarily to the response of materials to stresses or loadings imposed on them. We shall also discuss some of the standard laboratory tests used to determine these properties to guarantee the reliability and dependability of the results for use in design. The results of these tests are usually compiled in handbooks. Users of the data should be cautioned, however, to realize and to keep in mind that the data are the results of

Table 4-2 Estimated moduli for different bond types [2].

Bond Type	S_o, N-m^{-1}	Estimated E from (S_o/r_o), GN/m^{-2}
Covalent, C—C bond	180	1,000
Pure ionic, e.g., Na—Cl bond	9–21	30–70
Pure metallic, e.g., Cu—Cu bond	15–40	30–150
Hydrogen bond, e.g., in water	2	8
van der Waals (waxes, many polymers)	1	2

using small laboratory-sized test samples. The actual properties of the materials used in actual large structures and components may differ from those given in handbooks.

4.4.1 Strength of Materials

The strength of a material determines the amount of force or load it can withstand before it fails. The failure criterion used in design can be different for different materials and, therefore, there are multiple criteria of strength. For *metals and thermoplastic polymers,* the design criterion is usually based on the *yield strength.* However, because the materials may have undergone previous cold work or plastic deformation, the range of yield strengths may range from the initial yield strength for the annealed material to the tensile strength after the material is strain-hardened (see Sec. 4.4.4a). For most design purposes, the yield strength is assumed to be the same in tension as in compression. For *brittle materials,* such as ceramics and concrete, the failure criterion is the *crushing strength in compression* and not that in tension, which is about 15 times smaller than that in compression. Because brittle materials behave elastically up to fracture, the tensile strength takes the place of the yield strength. In these materials, the tensile strength is quite variable because of the presence of variable defect sizes. Thus, in ceramics and brittle materials, we have to incorporate a probability function. For *elastomers*, the failure criterion is the *tear-strength* and for *composites*, it is the *tensile failure strength.*

The strength of metals and alloys, ceramics, and polymers/plastics depends on different factors. The factors that determine the strength of ceramics and polymers/plastics will be discussed in Chaps. 14 and 15. We shall now relate the strength of metals and alloys to microstructures.

4.4.2 Strengthening of Metallic Materials*

The yield strength of metals and alloys reflects the amount of force or stress to move dislocations in them. Thus, to increase their yield strength, obstacles to the motion of dislocations are "engineered or created" in them. The four most basic ways to strengthen metals and alloys are through: (1) solid solution, (2) decrease in grain size, (3) increase dislocation density, and (4) dispersed phases or particles. A short description of each mechanism is given in the following subsections.

4.4.2a Solid Solution Strengthening. Solute atoms in solid solution may either be interstitial or substitutional. The interstitial atoms are relatively small and fit in open spaces (interstices) of the host structure, while substitutional atoms locate at the host lattice sites. The effects of the solute atoms are due to dilatational, distortional, and/or stiffness misfits (with the host atoms) that interact with and retard the motion of dislocations. The interactions with dislocations may either be to anchor dislocations at rest (interstitial solutes) or to retard moving dislocations (substitutional solutes). *The anchoring or locking of the static dislocations is manifested during tensile test by a sharp yield point* that gives rise to an upper- and a lower-yield stress in the stress-strain curve. This phenomenon (an experimental observation) is manifested by low-carbon annealed steels and by annealed or heat-treated aluminum-magnesium solid solution alloys. If we remove the carbon in steel, the stress-strain curve becomes smooth. Accompanying the sharp yield point in mild steel is an initial *inhomogeneous plastic deformation* called the *lower yield elongation* at the lower yield stress. These effects are shown schematically in Fig. 4-12.

The effect of the substitutional solutes is analogous to frictional effects experienced by a body in motion. Extra energy or stress is exerted to overcome the fric-

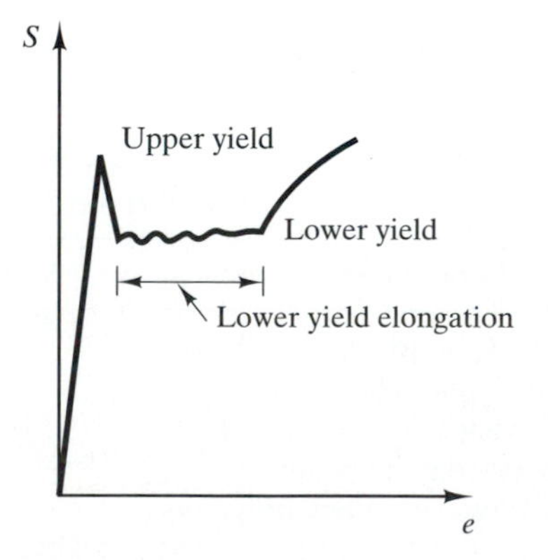

Fig. 4-12 Sharp yield point in stress-strain curve reflects the dislocation anchoring by interstitials, such as carbon in steels.

tion. The frictional stress to overcome is proportional to the solute concentration, that is, the strength will increase when solute concentration increases. The strengthening depends also on the differences in atomic sizes and the crystal structures of the solute and host atoms. These differences lead to elastic misfit strains that strengthen the host lattice—the same effect as in particle strengthening shown in Eq. 4-40.

Figures 4-13 and 4-14 show the strengthening effects of interstitial (carbon) and substitutional solutes, in pure iron. In Fig. 4-13, we see that the yield strength of pure iron (0% carbon) is less than 5,000 psi and this dramatically increases to about 28,000 psi with just 0.005%C. Commercially, iron with less than 2%C is steel. Of the substitutional solutes in Fig. 4-14, Cr, W, V, and Mo have the same body-centered cubic (BCC) structure as iron and their hardening or strengthening effects correspond to their differences in atomic sizes with iron; chromium exhibits the least difference in atomic size and molybdenum the largest. P, Si, Mn, and Ni reflect both differences in atomic sizes and crystal structure.

4.4.2b Grain Size Strengthening. This strengthening is primarily due to the large misorientation at the grain boundaries where grains (single crystals) meet in polycrystalline engineering materials; the tilt boundaries that form subgrains may also contribute to strengthening. (The description of the grain boundaries and the determination of the average grain size in polycrystalline materials were discussed in Sec. 2.11.) The effect of grain boundaries is schematically illustrated in Fig. 4-15. The deformation in a grain cannot continue to adjoining grains

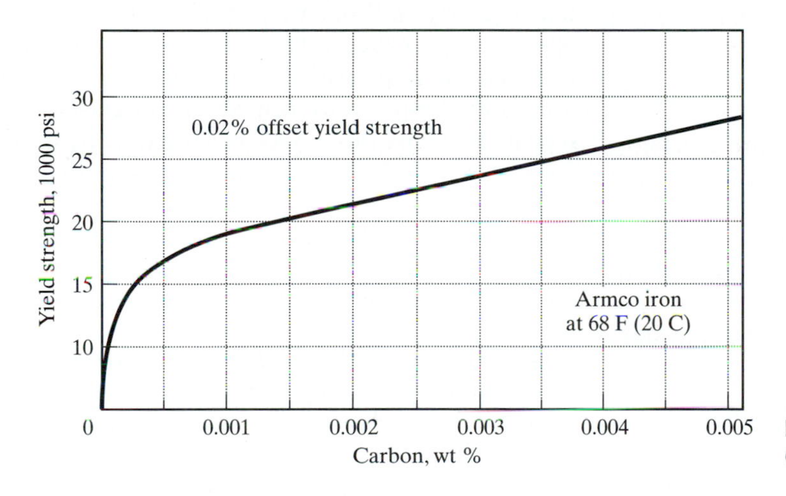

Fig. 4-13 Strengthening of ferrite (α-Fe) by interstitial carbon[3].

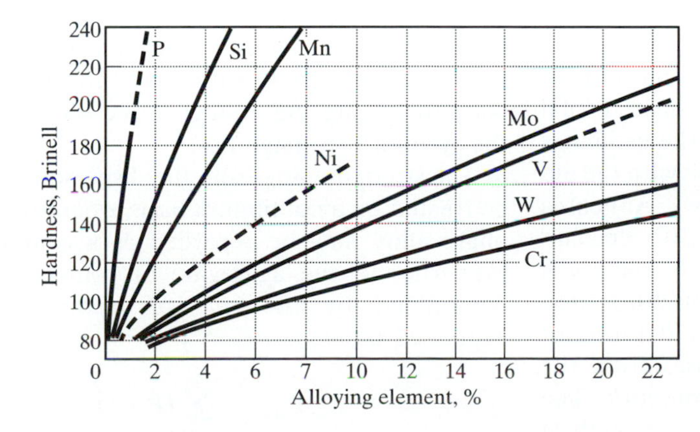

Fig. 4-14 Strengthening or hardening of ferrite by substitutional solutes in solution[3].

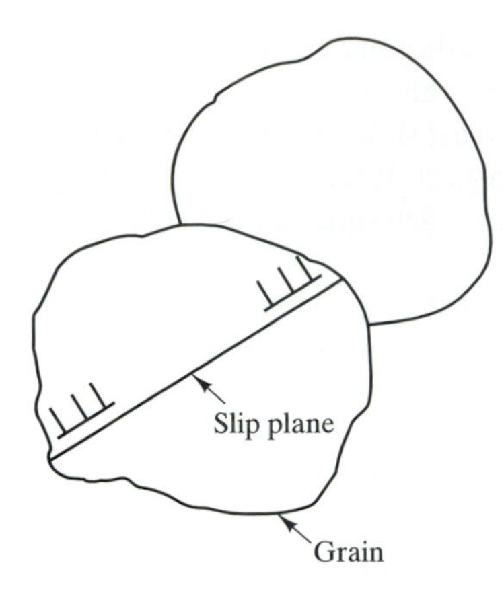

Fig. 4-15 Crystals are strengthened by grain boundaries blocking dislocations creating a pile-up.

because the motion of dislocations is blocked. The more grain boundaries (smaller grain size) there are, the more obstacles there are for dislocation motion.

The grain size strengthening has been verified experimentally and the empirical relation is called the Hall-Petch relation

$$\sigma_{ys} = \sigma_o + kd^{-1/2} \quad (4\text{-}36)$$

where σ_{ys} is the yield strength, σ_o and k are constants, and d is the average grain size. We note that the smaller d is, the higher will be σ_{ys}. Examples of optical microstructures of steels with their respective yield strengths are shown in Fig. 4-16. As indicated in Chap. 2, the average grain sizes are reported in standard ASTM numbers from which the absolute sizes in mm or μm can be obtained.

4.4.2c Dislocation Strengthening. The presence of other dislocations in the structure will interfere with the motion of a dislocation and thus increase the resistance or strength of the material. This is the *basis for strain-hardening* or *cold-work hardening*. The dislocation content is usually reported as *density of dislocations,* which is the estimated total length of dislocations in cm per cm^3. (The estimates of the densities of dislocations are made with the aid of transmission electron microscopy.) Annealed materials are quoted to contain 10^6 cm/cm^3 dislocation density. A fully cold-worked structure is said to contain about 10^{12} to 10^{13} cm/cm^3 dislocation density. Thus, as the crystal plastically deforms or strains, the dislocation density increases from the soft annealed state to the full hard condition. The increase in dislocation density raises the strength of the crystal. Thus, strain-hardening or cold-work hardening is achieved via plastic strain or cold-work. The yield strength increase due to dislocation density is

$$\Delta\sigma \propto \sqrt{\rho} \quad (4\text{-}37)$$

where ρ is the dislocation density.

The manner in which the dislocation density increases from 10^6 to 10^{13} is accomplished by a multiplication mechanism, shown in Fig. 4-17. In this figure, the dislocation is depicted as a taut string anchored at two points, Fig. 4-17a. During deformation, the dislocation is bowed by the shear force that acts perpendicular to the line at every segment. The shear force is τ b, where τ is the shear stress and b is the Burger's vector of the dislocation. As the line is bowed past the semicircular stage, segments of the line rotate around the anchor points. These dislocation segments will meet at a point where they have opposite signs, cancel out, and reform the original dislocation line. In reforming the original line, a circular dislocation loop is also formed. The original line can repeat this many times and thus multiply by producing a dislocation loop every time it reforms to its original state.

What are the anchor points? These may be particles on the path of the dislocation or intersection points with other immobile network dislocations.

4.4.2d Dispersed Phase or Particle Strengthening. In addition to being polycrystalline, all engineering materials are either multiphase or have dispersed particles in their microstructures. In reality, all engineering materials are composite materials, but they are not referred to as such because the term composite material is specifically reserved for man-made materials with macroconstituents—the constituents are big enough to be seen simply with the eye or a 10× magnifier.

The strengthening from the dispersed phases may be treated on a continuum basis, whereby, the multiphase alloy may be regarded as a composite. The properties of the aggregate or composite may be determined by the *rule of mixtures,* which is expressed as

$$P_{agg} = \sum_{1}^{n} (P_i \cdot f_i) \quad (4\text{-}38)$$

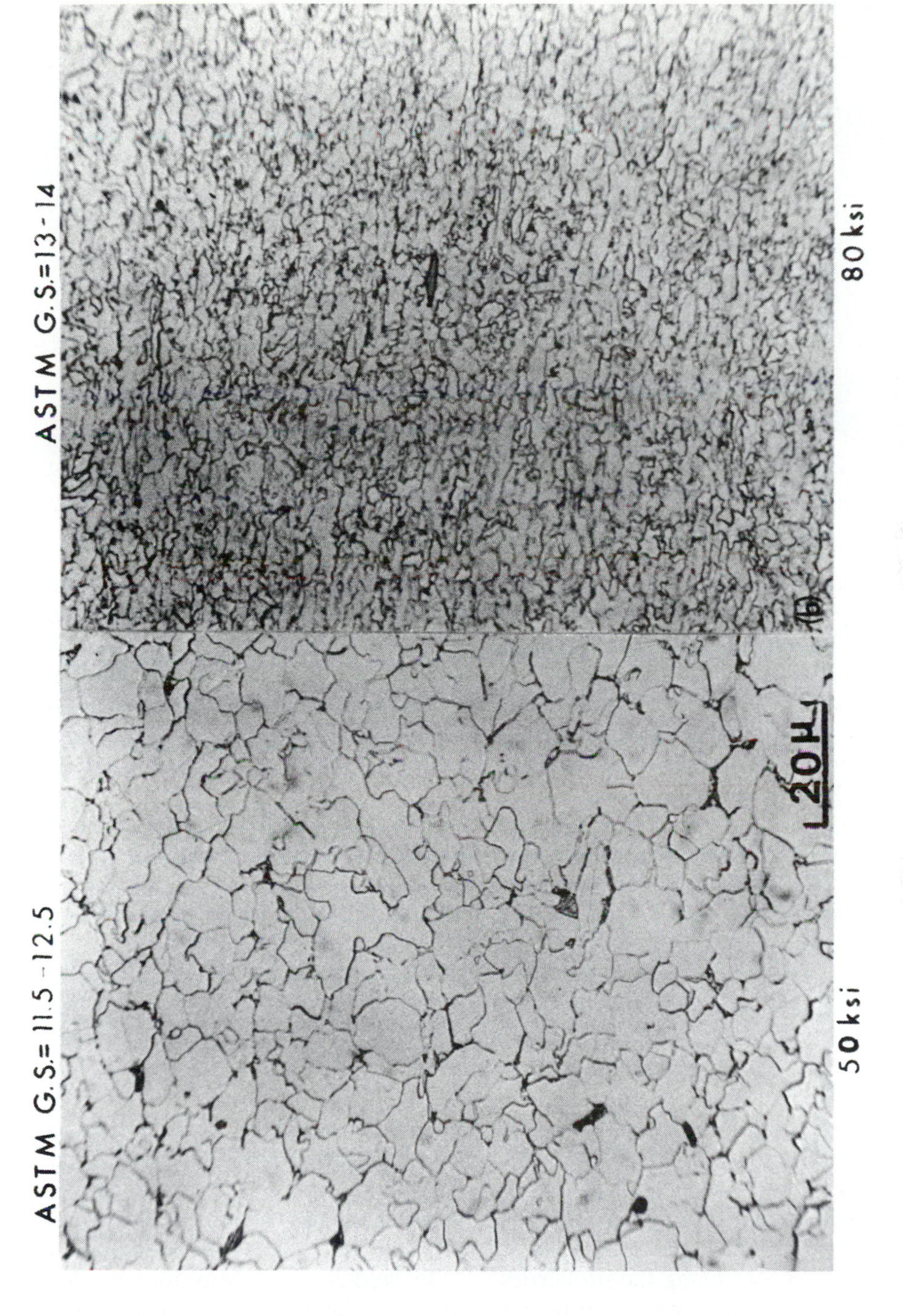

Fig. 4-16 Optical microstructures of (a) 345 Mpa (50 ksi), and (b) 550 Mpa (80 ksi) yield strength steels.

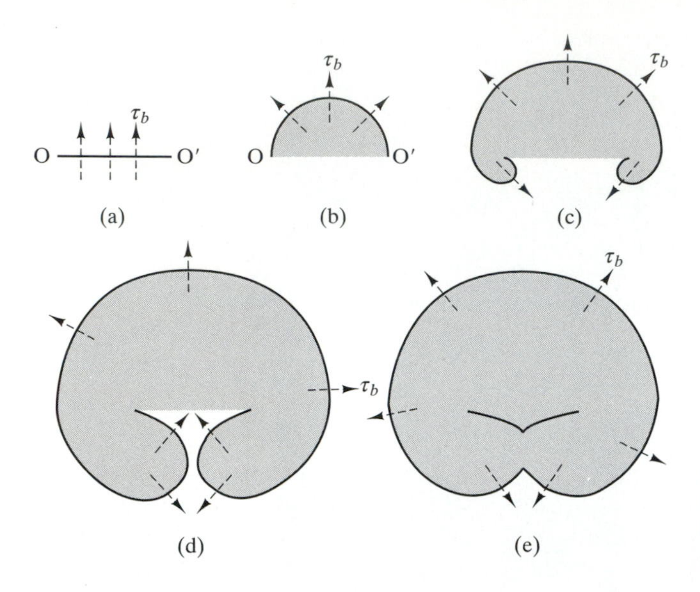

Fig. 4-17 Dislocation multiplication mechanism. Dislocation segment DD′ forms a circular dislocation loop in (e) as stress is applied; the process repeats itself with continued straining.

where P_{agg} is the aggregate or composite property, P_i is the property of the *i*th constituent and f_i is the volume fraction of the *i*th constituent in an aggregate containing *n* phases. The application of Eq. 4-38 to the structure-sensitive property of yield strength requires one of two hypotheses. One hypothesis assumes that each of the phases experiences equal strain. For this case, the average stress in the alloy for a given strain will increase linearly with the volume fraction of the strong phase. This is illustrated by an empirical result shown in Fig. 4-18, which is a plot of the 0.1% offset yield strength of an aggregate of austenite and martensite phases in a 304 stainless steel [4]. The amount (volume percent) of the hard-phase martensite in the 304 austenitic stainless steel was changed by a variety of thermal and deformation processes. The regression of the data shows a linear dependence of the yield strength of the austenite-martensite aggregate on the volume fraction of the martensite (hard phase).

The alternative hypothesis for an aggregate structure is that the phases are subjected to equal stresses. The average strain of the alloy can be calculated from Eq. 4-38 with the strains being the properties, P_i. The rule of mixtures will be discussed further in Chap. 15 in relation to composite materials.

The predominant strengthening from dispersed particles, which may be coherent or non-coherent, is treated as an interaction with dislocations. (The descriptions of coherent and non-coherent particles are covered in Chap. 9.) The influence of these particles is to block or retard the motion of dislocations. For coherent particles, the dislocation moves or cuts through the particle, as if it is part of the host lattice. However, this is acccomplished at a much higher force than when it moves through the host lattice. Thus, we observe a very high increase in strength of the material. For noncoherent particles, the particles block the dislocation motion by acting as anchor points, as depicted in Fig. 4-17. Figure 4-19 depicts schematically the shear stress pushing the dislocation anchored by hard particles with interparticle spacing of λ. The strengthening arising from these particles is inversely proportional to λ, namely

$$\Delta\sigma \propto \frac{1}{\lambda} \tag{4-39}$$

Thus, the closer the particles, the stronger is the material. The number of particles per unit volume controls the property. For a fixed weight percent of particles, smaller-sized particles will exhibit greater strengthening than larger-sized particles.

In addition to dislocation-particle interactions, Mott and Nabarro suggested that a source of strengthening is the elastic strain mismatch of the particle and the matrix. The increase in yield stress is given by

$$\Delta\sigma \propto \epsilon f \tag{4-40}$$

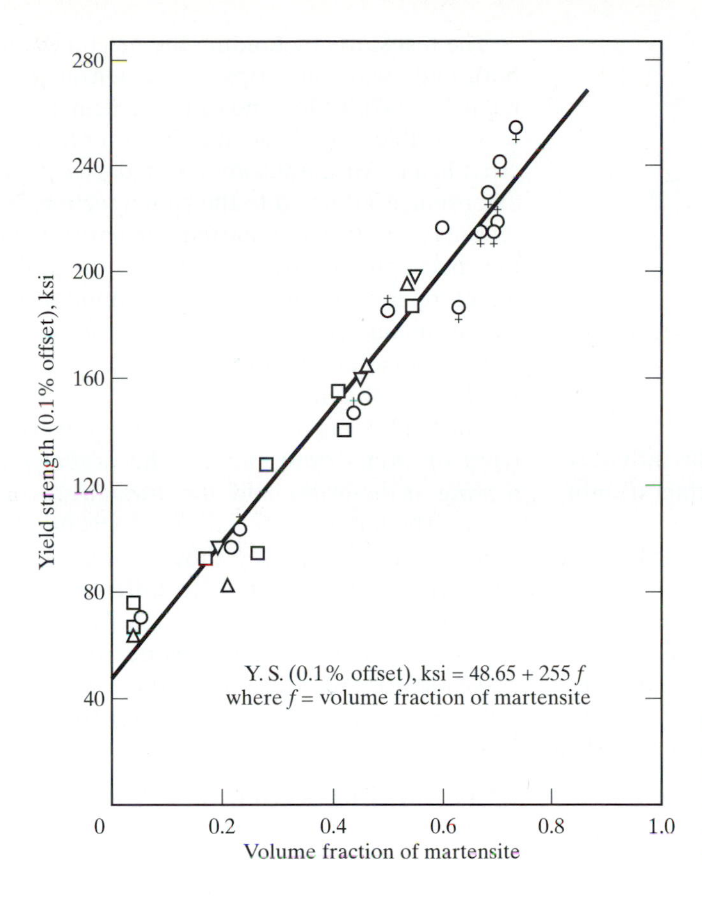

Fig. 4-18 Strengthening of austenite by martensite (dispersed phase) in strained and heat-treated 304 stainless steel that conforms to rule of mixtures [4].

where ϵ is the elastic strain field and f is the volume fraction of the dispersed particles. Equation 4-40 is especially true for coherent particles. The effect is similar to that discussed in substitutional solute strengthening in solid solutions, only this time the particles are coming out of solution. This is the strengthening observed in precipitation hardening that will be discussed further in Chap. 9.

The overall yield strength of metals and alloys is the result of the four factors just described. Neglecting the interactions between them, the yield strength of a crystalline material based on dislocation mechanisms may be taken as the additive effects of the four factors. Thus,

$$\sigma_{ys} = \sigma_o + \Delta\sigma_{ss} + \Delta\sigma_{gs} + \Delta\sigma_{sh} + \Delta\sigma_p \qquad (4\text{-}41)$$

where the *subscripts* of the terms on the right side of the equation stand for the following: o is due to the inherent resistance of the lattice to dislocation motion, ss is for solid solution, gs is for grain size, sh is for strain-hardening (dislocations), and p is for dispersed phases and/or particles.

As a final note, the foregoing principles are very basic. Materials engineers constantly use these concepts in materials processing to control materials microstructures and properties. The finer or the smaller the sizes of the features in the microstructures, the stronger the materials are. This is the direction in which materials scientists and engineers are actively doing research. There is a lot of research and development activity to produce *nanomaterials*, materials with crystal sizes in nanometers, 10^{-9} meters.

4.4.3 Ductility

The *ductility* is the property of a material that allows it to be formed into different shapes before it breaks. A material that can do this is said to be *ductile,* while a material that exhibits no change or difficulty in forming a shape before it breaks is called *brittle*. For

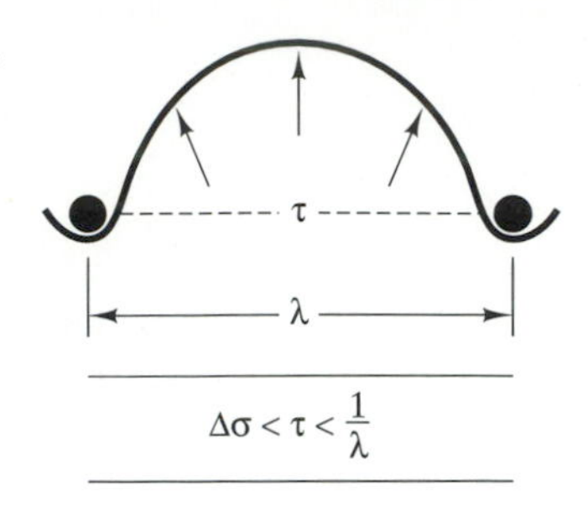

Fig. 4-19 Dislocation motion blocked by particles increases strength.

design considerations, ductility directly influences the fracture toughness and the manufacturability of a component—a ductile material has a higher fracture toughness than a brittle material. In addition, the selection of a material for engineering applications depends significantly on its ability to be formed to the desired component or product shapes. This is the reason for the engineering importance and usage of metals compared to ceramics because metals are more conducive to mass production than ceramics.

4.4.4 The Tensile Test—ASTM E-8 and Properties Derived From It

We shall now discuss how the *yield and tensile strengths and ductility of metals are determined with the tensile test.* The ASTM (American Society for Testing and Materials) E-8 standard test begins with the preparation and machining of the standard specimen, Fig. 4-20a. The example shown is for flat specimens. As much as possible, standard-sized specimens are prepared. However, if the original size of the product or sample is small or if only a limited amount of the material is available, equivalent subsize standard specimens are available and can be machined and tested. The machined specimens have larger sections at the end and a reduced section in the middle. The material in the reduced section will be the part of the specimen that is subjected to the loading and the stretching. Before testing, a certain length of the reduced section is marked from which to determine the ductility after the test. This length is called the *gauge length;* the standard length is 50 mm (2 inches).

The test starts by holding the machined sample at both ends with the grips of the tensile test equipment, Fig. 4-20b. One end of the specimen is attached to a movable cross-head and the other to a stationary cross-head. An extensometer, a device to measure extension, is attached to the gauge section. When the tensile test machine is started, the movable grip pulls the specimen and the resistance to the pull is determined by a calibrated load cell. Simultaneously, the extensometer senses the elongation of the gauge section. A load-elongation plot, P-ΔL, as shown in Fig 4-21a is obtained.

The load-elongation plot may be converted to *two types of stress-strain curves*—the *nominal* or *engineering stress-strain* and the *true stress-true strain* curves. The latter is also referred to as the *flow stress curve* because it represents the true plastic-flow characteristics of the material. To differentiate between these two curves, different symbols for stress and strain are used. For the engineering stress-strain curve, we use s for stress and e for strain. For the true stress-true strain curve, we use σ for true stress and ϵ for true strain.

4.4.4a The Engineering Stress-Strain Curve. The *engineering stress-strain curve* is obtained from the load-elongation, P-ΔL, data in the following manner. The engineering stress is obtained by dividing the load, P, by the original area of the specimen, A_o. The engineering strain is obtained by dividing the elongation, ΔL, by the original gauge length, L_o. Thus, each point in the P-ΔL curve is converted to s and e by the following

$$s = \frac{P}{A_o} \tag{4-42a}$$

and

$$e = \frac{\Delta L}{L_o} \tag{4-42b}$$

The converted points are used to trace the *s-e curve*, Fig. 4-21b, which *has essentially the same shape as the P-ΔL* curve.

The s-e curves of ductile materials exhibit three characteristic parts: (1) an initial linear portion referred to as the elastic region; (2) a portion with continually decreasing slope as the stress increases until it (the slope) reaches zero (referred to as the

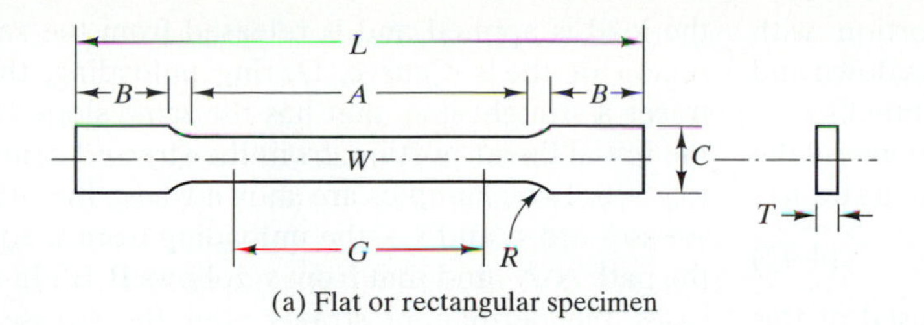

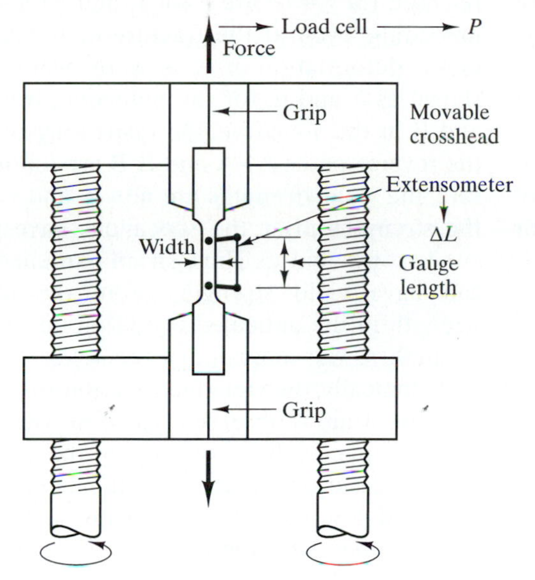

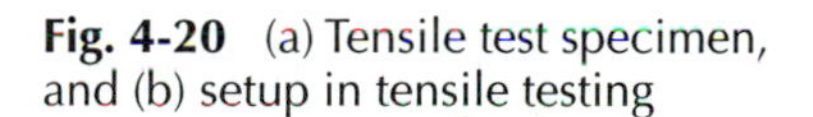

Fig. 4-20 (a) Tensile test specimen, and (b) setup in tensile testing

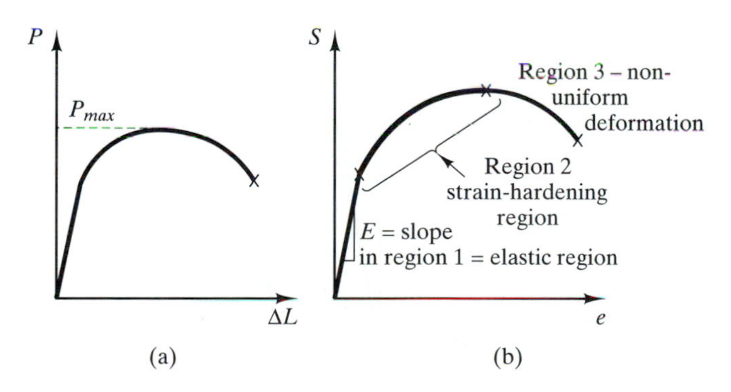

$(P, \Delta L)$ transforms to (s, e) by: $s = \dfrac{P}{A_o}$; $e = \dfrac{\Delta L}{L_o}$

Fig. 4-21 Conversion of (a) load-elongation curve, to (b) engineering stress-strain curve. Three regions of (b) are marked. Refer to text.

strain-hardening region); and (3) a portion with decreasing stress when the specimen necks down and exhibits nonuniform deformation until it breaks.

The *initial linear portion* reflects the response of the material according to *Hooke's Law* for small strains,

$$s = \mathrm{E}e \tag{4-43}$$

Thus, if the sensitivity of strain is increased in this region, as shown in Fig. 4-22, the *Young's Modulus* of the material is the *slope of this initial linear portion.* The *material* when stretched in this region *behaves elastically,* which means that the material will revert back to its original shape when the stress is removed. In other words, a material will not change shape if the applied stress is in this linear portion. In terms of the *s-e* curve, *e* will return to zero when the load or stress is released. The deformation in this region is called *elastic deformation.*

When the material passes the linear portion into the second region, the material undergoes *plastic deformation,* meaning that a *permanent shape change will occur after the load is released.* The shape change in this case is reflected in a permanent extension in length, ΔL or *e*. In terms of the *s-e* diagram, the *e* does not return to zero when the load or stress is released. The value of *e* when $s = 0$ is referred to as *permanent strain.* Figure 4-23 shows the *s-e* traces as the load is applied and is released from the second region of the *s-e* curve. During unloading, the *s-e* traces a straight line, that has the same slope (E) as the initial linear portion, from the stressed condition to $s = 0$. Two examples are shown where the original stresses are s_1 and s_2—the unloading from s_1 follows the path A-A′ and that from s_2 follows B-B′. In these cases, the permanent strains after the stresses are released (at $s = 0$) are *a* for s_1 and *c* for s_2. In every unloading, even at the fracture or break point, the elastic deformation or strain is recovered. These are shown as *b* and *d*, for the unloading from points A and B in the *s-e* curve. On restressing or reloading, the reverse paths A′-A and B′B are followed, therefore, the yield strengths are now s_1 and s_2. These are the strengths after the specimens have permanent strains of *a* and *c*. Thus, strain-hardened materials can have yield strengths from the initial yield strength in the annealed condition up to the tensile strength, as was indicated in Sec. 4.4.2.

Technically, the yield failure criterion occurs when a plastic strain is observed. The observance of plastic strain depends, however, upon the sensitivity of strain-measuring devices. Some materials, such as annealed low carbon steels, show a distinct point between the transition from the first part to the second part of the *s-e* curve. This was earlier depicted in

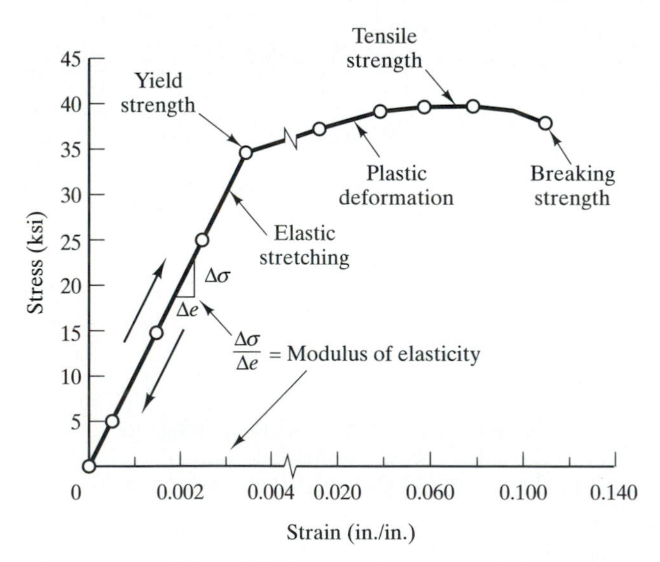

Fig. 4-22 Modulus of Elasticity is determined from stress-strain by expanding the scale for the initial strains. Note change in scale of strain in abscissa.

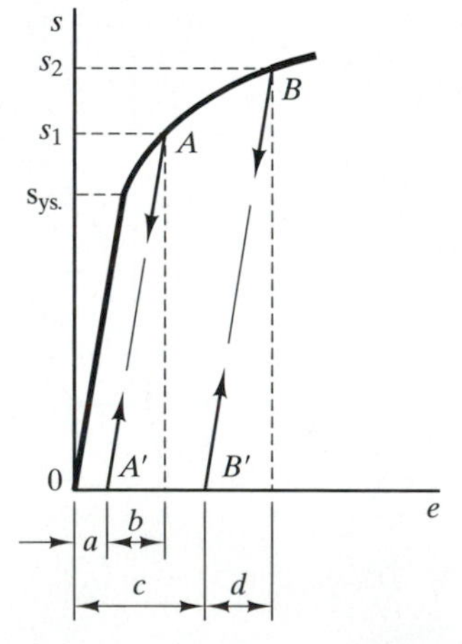

Fig. 4-23 Stressing material in region 2 to points A and B leaves permanent strains a and c. *b* and *d* are elastic recovered strains from A and B.

Fig. 4-12 and again shown in Fig. 4-24a where we can delineate clearly an upper-yield stress and a lower-yield stress. The *lower-yield stress* is reported as the conservative value. For most materials, there is no sharp delineation of where the linear portion ends, Fig. 4-24b. In this case, an offset yield strength is obtained by offsetting or moving the origin of the s-e curve by a specified amount of strain that is usually 0.001 (0.1%) or 0.002 (0.2%). A straight line is then drawn parallel to the linear portion of the *s-e* curve. The offset yield strength is the value of stress at the intersection of the offset line with the *s-e* curve. The yield strength is reported as *either 0.1% offset yield strength or 0.2% offset yield strength.* The yield strength (Y.S.) obtained by the offset method is commonly used for design and specification purposes because it avoids the practical difficulties of measuring the elastic limit or the proportional limit (the limit where Hooke's Law applies). Where there is essentially no linear portion in the *s-e* curve, the offset method cannot be applied. In this case, the yield strength is defined as that stress corresponding to some total strain, for example, $e = 0.005$. Gray cast iron exhibits this behavior.

The *tensile strength* (T.S.) or *ultimate tensile strength* (UTS) is the maximum load divided by the original cross section of the specimen.

$$\text{T.S or UTS} = \frac{P_{max}}{A_o} \qquad (4\text{-}44)$$

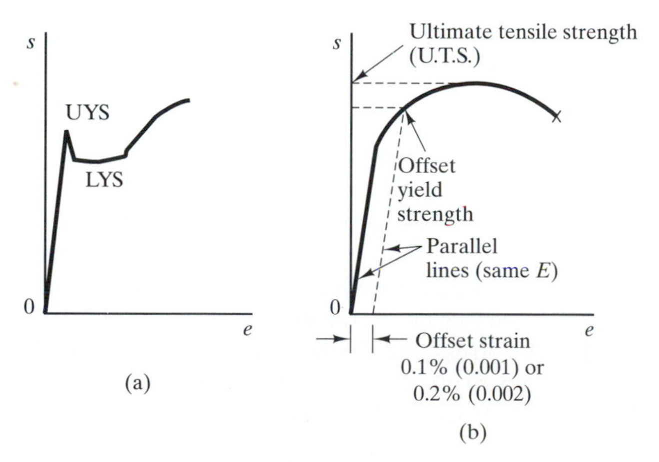

Fig. 4-24 Two types of stress-strain curves and how yield strengths are determined. (a) Upper- and lower-yield strengths, (b) Offset yield strength.

For many years, designers have used the tensile strength, reduced by a factor of safety, as a design failure criterion for the static-load bearing capacity of a material. Current design procedures use the yield strength for static-load bearing design failure criterion for ductile materials. For brittle materials, tensile strength is still a valid criterion. Most gray cast irons are specified by their tensile strengths. In addition, because of their previous extensive use, empirical correlations were established to relate other properties to tensile strength (for example, see the sections on hardness and fatigue in this chapter).

4.4.4b Uniform and Nonuniform Deformation of Test Specimen. At this point, let us examine what happens to the reduced section of the specimen, in particular, the material between the gauge marks during the tensile test. As shown in Fig. 4-25a, the material contracts laterally as it elongates along the tensile axis. Initially, it contracts uniformly between the gauge marks. Uniform contraction (deformation) means that the specimen exhibits the same lateral dimension (or the same cross-section) at every point in the reduced section, as the length between the marks L_o elongates. If we take the length of the gauge marks at any time and multiply it by the instantaneous area, the volume of the material within the gauge marks is obtained. Multiplying this by the density, we obtain the mass of the material between the marks. Because we conserve mass and because the density is essentially constant, the total volume of the material within the marks must also remain constant when deformation is uniform. That is,

$$A_oL_o = AL = \text{constant} \qquad (4\text{-}45)$$

Uniform deformation of the gauge section occurs up to the maximum tensile load, Fig. 4-25a. The uniform elongation is a measure of the stretch formability of materials and will be discussed in Chap. 8.

Shortly after reaching the maximum load, the gauge section will exhibit a preferential contraction in one area and *necking* of the specimen is observed, Fig. 4-25b. Because the necked area is smaller, the ensuing deformation will be concentrated in it requiring a lesser load. This is region 3 in the s-e curve; now, the deformation is *nonuniform* along the length of the gauge marks, and the load continually decreases

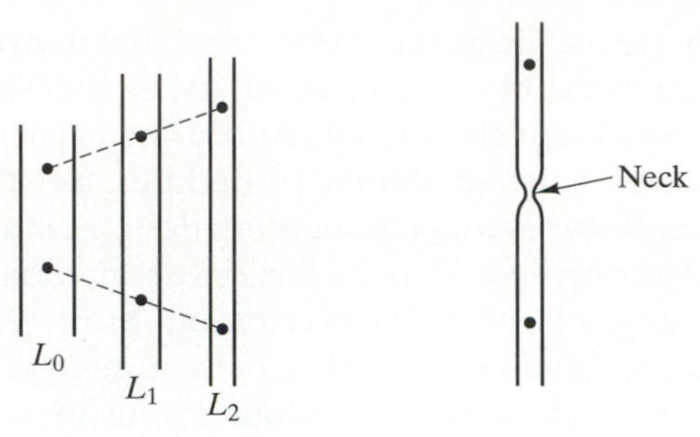

(a) Uniform deformation up to maximum load $A_0L_0 = A_1L_1 = A_2L_2$ Regions 1 and 2 of *s–e*

(b) Non-uniform deformation – Onset of necking at maximum load Region 3 of *s–e*

Fig. 4-25 (a) Uniform deformation of tensile specimen in regions 1 and 2 of *s-e* curve; (b) Necking and non-uniform deformation in region 3 of *s-e*.

until the specimen fractures or breaks. Thus, we see why the stress decreases in the third section of the *s-e* diagram in Fig. 4-21b. *In nonuniform deformation, Eq. 4-45 does not apply.*

4.4.4c Measures of Ductility. After the specimen fractures, we determine the ductility of the material using the final gauge length and area from the broken sample. The *measures of ductility are the percent elongation, % e, and the percent reduction in area, % RA.* These are defined as

$$\% \text{ e} = \frac{L_f - L_o}{L_o} \times 100 \qquad (4\text{-}46)$$

and

$$\% \text{ RA} = \frac{A_o - A_f}{A_o} \times 100 \qquad (4\text{-}47)$$

The subscripts *o* and *f* refer to the original and final dimensions. For % e, the final gauge length is measured after carefully fitting the fractured surfaces of the specimen, to reconstruct the specimen before it broke. For this data to be valid, the fracture surface must be within the original gauge length, L_o. The % e is always reported with the original gauge length because its value varies with the gauge length used—the smaller the gauge length used, the larger is the % e. For the %RA, the final area is obtained by measuring the dimensions of the specimen at the fracture area. The %RA is less dependent on the gauge length used and would thus be a better consistent measure of the ductility.

4.4.4d The True Stress—True Strain Curve. The engineering stress-strain curve does not portray an accurate indication of the *actual true stresses and true strains of the material* because it is based on the original dimensions, area, and gauge length of the specimen. These dimensions continually change in the course of the test and the actual or true stress at any one time is much higher than that depicted by the *s-e* curve. While the *s-e* curve is sufficient for design because it underestimates the actual capacity of the material, the true stress-true strain (σ-ϵ) curve is the more important property in forming or manufacturing operations because it portrays the real load requirement of the material for further deformation or flow of the solid elements. For this reason, the σ-ϵ curve is often called the *flow stress curve.*

The definition of true stress is the load divided by the actual or instantaneous area,

$$\sigma = \frac{P}{A_{inst}} \qquad (4\text{-}48)$$

and that of true strain is

$$d\epsilon = \frac{dL}{L_{inst}} \qquad (4\text{-}49)$$

and

$$\epsilon = \int_{L_o}^{L} \frac{dL}{L} = \ln \frac{L}{L_o} \qquad (4\text{-}50)$$

Thus, *to convert the P-ΔL curve to a true stress-true strain curve*, we obtain for each point (P, ΔL),

$$L = L_o + \Delta L \qquad (4\text{-}51)$$

and obtain ϵ with Eq. 4-50. From Eq. 4-45, $A_oL_o = AL =$ constant, we obtain the instantaneous area, with which we obtain σ from Eq. 4-47. The schematics of both the σ-ϵ and the *s-e* curves are shown in Fig. 4-26 for comparison. The σ-ϵ curve continually increases up to fracture.

The true stress and the true strain can be related to the engineering stress and the engineering strain. From Eq. 4-50 and 4-51, we see that

$$\epsilon = \ln\frac{L}{L_o} = \ln\frac{L_o + \Delta L}{L_o} = \ln(1 + e) \qquad (4\text{-}52)$$

For the σ-s relationship, we multiply and divide Eq. 4-48 by A_o,

$$\sigma = \frac{P}{A} \cdot \frac{A_o}{A_o} = \frac{P}{A_o} \cdot \frac{A_o}{A} = s\frac{L}{L_o} = s(1 + e) \quad (4\text{-}53)$$

Equations 4-52 and 4-53 are only good up to the maximum load. After the maximum load, the true stress should be determined from actual measurements of load and cross-sectional area. The true strain should be based on actual area or diameter measurements.

For small strains,

$$\epsilon = \ln(1 + e) \approx e, \qquad \text{and} \qquad \sigma \approx s \quad (4\text{-}54)$$

Thus, as seen in Fig. 4-26, the initial linear portion of the σ-ϵ curve is essentially the same as the s-e curve. Thus, in engineering design, we can use the yield strength in engineering stress or true stress. After the elastic deformation, the σ-ϵ curve starts to deviate from the s-e curve and always exhibits a higher stress than the engineering stress because the instantaneous area continually decreases. The true stress-true strain monotonically increases up to the fracture or break point.

Because of the importance of the flow or true stress in metalworking operations, many attempts were made to fit a mathematical equation to the σ-ϵ curve. The best-fit equation for the plastic deformation part (region 2 in Fig. 4-21b) up to the maximum load is

$$\sigma = K\epsilon^n \quad (4\text{-}55)$$

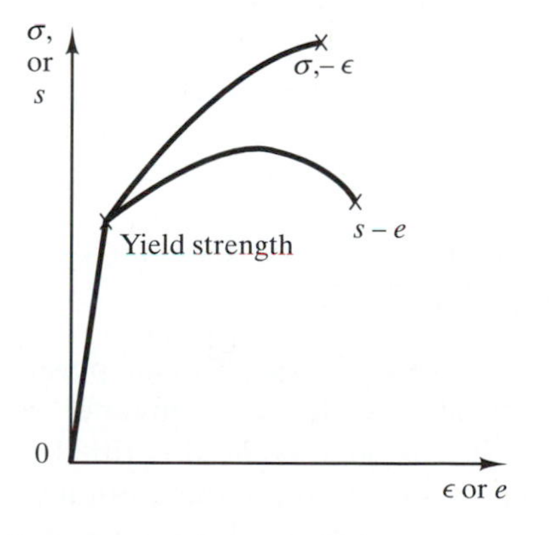

Fig. 4-26 Comparison of true stress-true strain curve (σ-ϵ) and engineering stress-strain curve (s-e).

where K is called the *strength coefficient* and n is the *strain-hardening exponent* or *index*. Equation 4-55 is empirical and is obtained when data points between the yield and tensile strengths from the σ-ϵ are plotted as $\log \sigma$ versus $\log \epsilon$. The plot is linear with n being the slope of the line and K the value of σ when $\epsilon = 1.0$, as shown in Fig. 4-27.

To gain a perspective on the importance of the strain-hardening exponent and strain-hardening, we consider a perfectly plastic material with $n = 0$ and a perfectly elastic (brittle) material with $n = 1$. The stress-strain curves for these materials are illustrated in Fig. 4-28. A perfectly plastic material exhibits no strain hardening and will deform continually when the load or stress reaches the yield strength of the material. Obviously, any material cannot deform indefinitely; eventually, it fractures and breaks. The material cannot sustain any load beyond the yield stress. In contrast, a perfectly elastic material cannot be shaped because it will revert back to its original shape if the load is released. Thus, in order to form a material it must have an n value between 0 and 1; most metals have n values between 0.10 to 0.50. Materials with higher n values are more stretch formable as will be discussed in Sec. 8.4.4.

This brings us to the importance of the strain-hardening of a material. As indicated in Fig. 4-23, strain-hardening is the increase in strength of a material after it is plastically deformed (formed into shape). The yield strength of a plastically-deformed or cold-worked material is the flow stress needed to continue deforming the material; it is higher than the annealed or initial condition. The formed material (product) is therefore able to sustain higher stresses than the as-received materials. Thus, if designers use the higher strength of a product, there is opportunity to save weight and cost of the material.

At this point, we have alluded to two measures of stretch formability: (1) the uniform elongation in Sec. 8.4.4b, and (2) the strain-hardening exponent, n as illustrated here. We shall now show that these two terms are identical. We should recall that uniform elongation occurs only up to the maximum load. At this point, the P-ΔL curve flattens out or mathematically we say $dP = 0$ shown in Fig. 4-29. Recall also that at this point necking starts forming, which is a

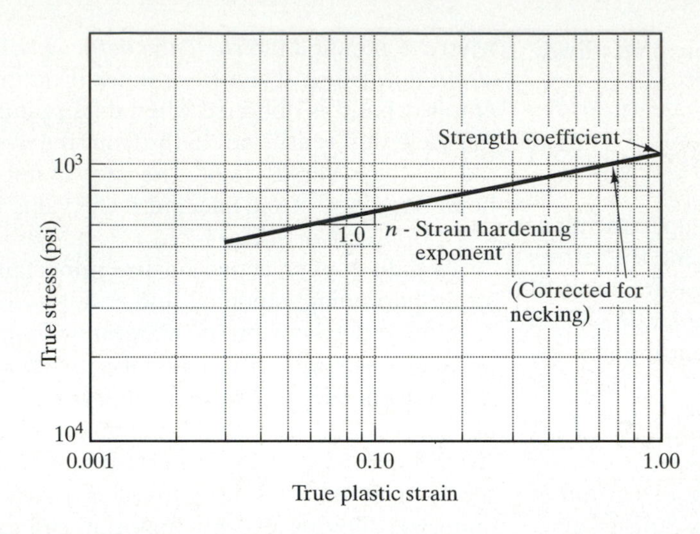

Fig. 4-27 Empirical determination of strength coefficient, K, and strain-hardening exponent, n, from a $\log \epsilon - \log \sigma$ plot.

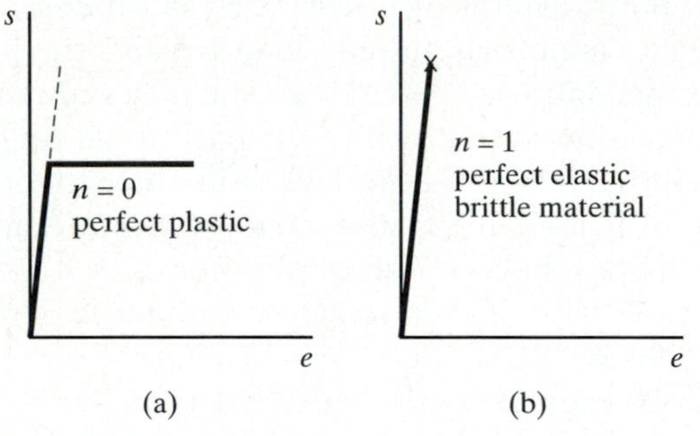

Fig. 4-28 Stress-strain curve for (a) perfect plastic material, (b) perfect elastic britlle material.

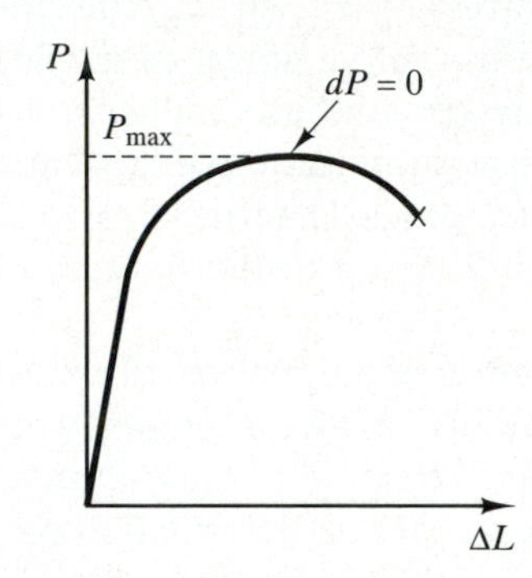

Fig. 4-29 Load-elongation curve showing dP = 0 at maximum load.

failure; therefore, it is also referred to as the *instability point.* Now, from the definition of true stress, the load at any time is

$$P = \sigma A \tag{4-56}$$

Differentiating and equating to zero at the instability point, we find

$$dP = \sigma\, dA + A\, d\sigma = 0 \tag{4-57}$$

and,

$$\frac{d\sigma}{\sigma} = -\frac{dA}{A} = \frac{dL}{L} = d\epsilon \tag{4-58}$$

Rearranging, we get

$$\frac{d\sigma}{d\epsilon} = \sigma \tag{4-59}$$

Using Eq. 4-55,

$$nK\epsilon^{n-1} = K\epsilon^{n} \tag{4-60}$$

From which we obtain,

$$n = \epsilon_u \tag{4-61}$$

We have therefore proven that *the strain-hardening exponent is equal to the uniform true strain up to the maximum load.* Because a higher uniform true strain is desired in forming operations, a material with a higher n value is therefore desirable.

4.4.5 Tensile Test for Plastics—ASTM D 68

Obtaining the tensile properties of plastics is conducted in essentially the same manner as that for metals, which has just been described. The yield strength, the tensile strength, the modulus, and elongation may also be obtained. Compared to metals, the stresses are much lower. Figure 4-30a compares the stress-strain behavior of a polycarbonate plastic with those of steel and Fig. 4-30b shows the polycar-

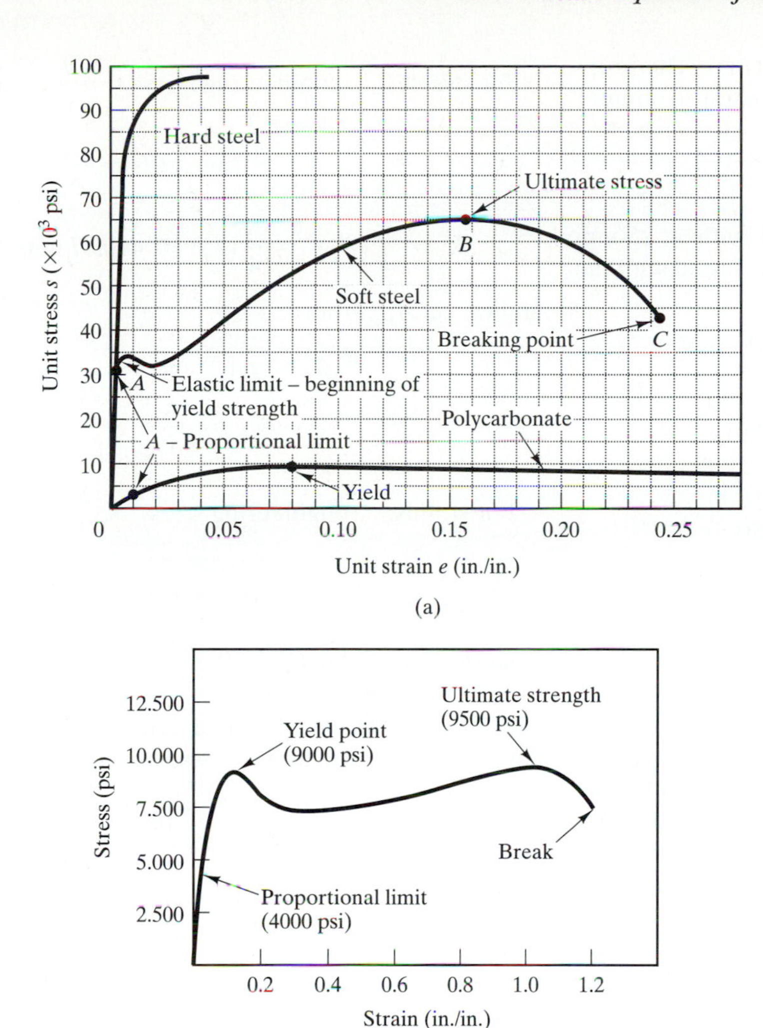

Fig. 4-30 Stress-strain curves for polycarbonates (plastics); (a) on a scale that compares with steels, and (b) s-e scale expanded.

bonate stress-strain behavior in a more expanded scale. Figure 4-30b exhibits the essential tensile properties that are obtained in a tensile test. The proportionality limit shown in Fig. 4-30b is the maximum stress that follows the Hooke's Law, the proportionality between the stress and strain. If we expand the strain scale in this linear portion, we can also obtain the modulus of elasticity. In many plastics and in gray cast iron, the initial linear portion does not exist and we cannot obtain a modulus of elasticity. The alternative is to obtain the tangent modulus or the secant modulus, as illustrated in Fig. 4-31. The tangent modulus is the slope of the tangent to the stress-strain curve passing through the origin. The secant modulus is the slope of the line connecting the origin and a

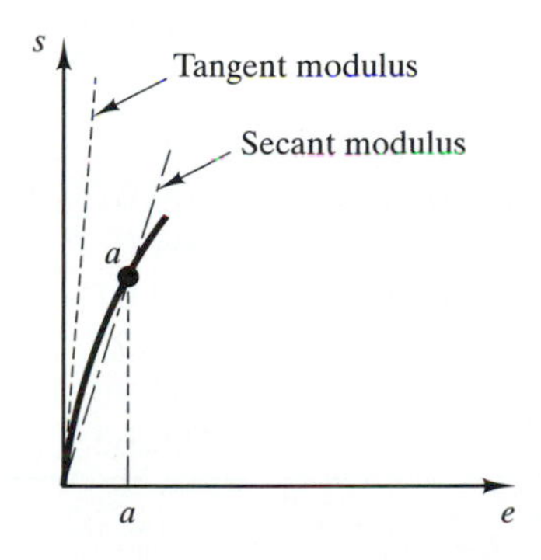

Fig. 4-31 Tangent and Secant Moduli when material does not exhibit initial linear portion.

prescribed strain in the stress-strain curve. The properties of polymers/plastics are discussed in greater detail in Chap. 15.

EXAMPLE 4-5

A tensile specimen was machined with a 25-mm diameter in the gauge section and was tested with the following load-elongation results:

Load, (N)	Gauge Length, (mm)
0	50.0000
50,000	50.0613
100,000	50.1227
150,000	50.1848
175,000	50.50
200,000	51.35
225,000	52.90
231,000 (break load)	53.40

After fracture, the gauge length was measured to be 53.1 mm, and the diameter was 23.25 mm. Convert the data to engineering stress-strain, plot it, and determine (a) the modulus of elasticity, (b) the 0.2 percent offset yield strength, (c) the tensile strength, (d) the ductility in % e, and in % RA, (e) the strain-hardening exponent, (f) the strength coefficient, and (g) the true stress and true strain at fracture.

SOLUTION: Because we are asked to plot the data, we need to convert the data to stresses and strains. We need both (1) the engineering stress-engineering strain, and (2) the true stress-true strain. The conversion is summarized in Table EP 4-5.

For stresses: $\text{engineering stress} = \dfrac{P}{A_o}$ and

$$\text{true stress} = \frac{P}{A_{inst}}$$

$$A_o = \frac{\pi}{4}(D_o)^2 = \frac{\pi}{4}(25)^2 = 490.87\text{ mm}^2$$

and we get A_{inst} from

$$A_{inst} \times L_{inst} = A_o L_o$$

$$A_{inst} = \frac{A_o L_o}{L_{inst}}$$

For strains: $\text{engineering stress} = \dfrac{\Delta L}{L}$ and

$$\text{true strain} = \ln \frac{L}{L_o}$$

from the data given, $L_o = 50$ mm.

(a) From the graph, the Young's Modulus = 83.5 GPa. When calculated directly from the given data, we can use the first two stresses and strains.

$$E = \frac{101.86}{0.001226} = 83.08\text{ GPa}$$

$$E = \frac{203.72}{0.002454} = 83.01\text{ GPa}$$

(b) the 0.2% offset yield strength = 332 N/mm^2, (MPa)

(c) Tensile strength = fracture strength = 470.59 MPa

Table EP 4-5

				Strain		Stress	
Load, N	**L_{gauge}, mm**	**ΔL, mm**	**A_{inst}, mm^2**	**Eng., e**	**True, ϵ**	**s, N/mm^2**	**σ, N/mm^2**
0	50.0000	0	490.87	0	0	0	0
50,000	50.0613	0.0613	490.27	0.001226	0.001225	101.86	101.98
100,000	50.1227	0.1227	489.67	0.002454	0.002451	203.72	204.22
150,000	50.1848	0.1848	489.07	0.003696	0.003689	305.58	306.70
175,000	50.50	0.50	486.01	0.010000	0.009950	356.51	360.07
200,000	51.35	1.35	477.97	0.027000	0.02664	407.44	418.44
225,000	52.90	2.95	463.96	0.0580	0.05638	458.37	484.96
231,000	53.40	3.40	459.62	0.0680	0.06579	470.59	502.59

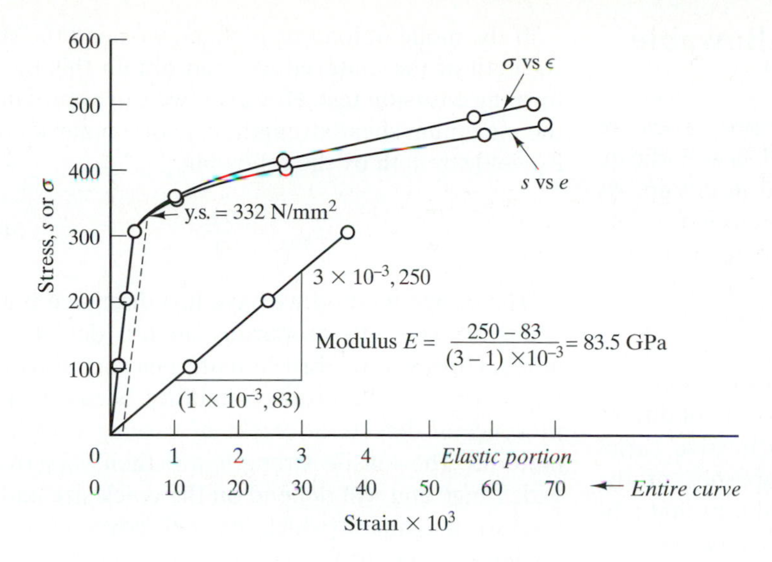

Fig. EP 4-5a Engineering stress-strain and true stress-true strain curves. Initial portion has expanded strain scale to get E.

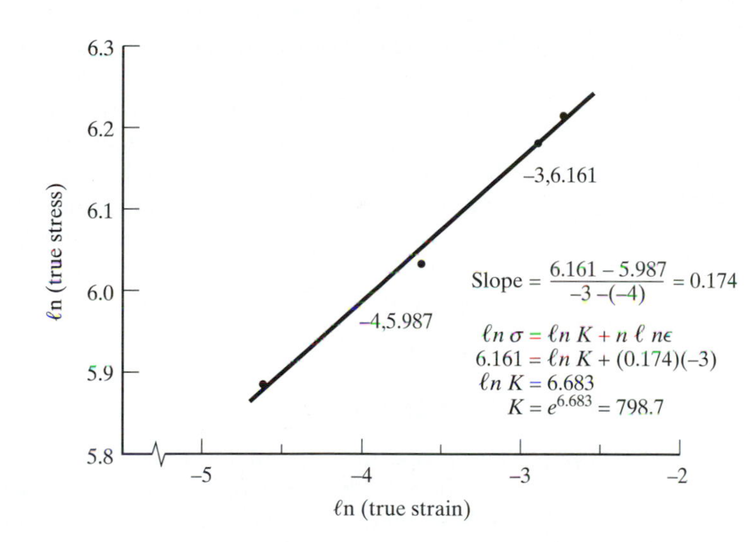

Fig. EP 4-5b Calculation of n and K from a log-log plot of true strain versuss true stress.

(d) Ductiliy:

$$\%e = \frac{53.1 - 50.0}{50} \times 100 = 6.2\%$$

$$\%RA = \frac{\frac{\pi}{4}(25)^2 - \frac{\pi}{4}(23.25)^2}{\frac{\pi}{4}(25)^2} \times 100 = 13.51$$

(e) from ln(true stress) versus ln(true strain) plot, $n = 0.174$;

(f) from $\sigma = K\epsilon^n$, K = 798.7 MPa

(g) From data true stress = 502.59; true strain = 0.06579;

From the broken specimen:

$$\text{true strain} = \ln \frac{L_f}{L_o} = \ln \frac{53.1}{50} = 0.060$$

The true strain from the broken specimen is lower than that taken directly from the data because of the elastic strain recovery.

4.4.6 Factor of Safety and Allowable Design Stress

We have just learned how the tensile properties are obtained in the laboratory. We shall now indicate how the strength properties are used in design. As stated in Sec. 4.4.1, the failure criterion used for ductile metals and plastics under static or steady tensile loading is the yield strength. However, the yield strength of the material is not taken as the design or working stress for the material. It is common for engineers to employ a so-called "factor of safety" to ensure against uncertainties or unknown conditions that may arise. Such conditions may involve variations and nonuniformity in the properties of the material. In designs using either yield strength or tensile strength, the usual factor of safety used for yield strength is two while that for tensile strength is four. Yield strength is used for ductile materials such as structural steels and tensile strength is used for brittle materials such as gray cast iron. In using a safety factor, we are designating the stress level (the applied or design stress) to be sustained by the material to be much lower than the failure criterion, either the yield or the tensile strength. The allowable design stress, σ_{des}, is

$$\sigma_{des} = \frac{\sigma_{ys}}{2} \tag{4-62}$$

or

$$= \frac{\sigma_{uts}}{4} \tag{4-63}$$

With the design stress, we can obtain the size (area) of the material from the maximum load, P_{max}, the structure or component is going to sustain.

$$A_{des} = \frac{P_{max}}{\sigma_{des}} \tag{4-64}$$

This method oversizes the material; this translates to higher costs and weights of material. The weight is not a particular problem for static structures such as buildings, bridges, or storage tanks, but it is a factor in transport structures such as airplanes and freight trucks where energy costs and net weights are important.

If the mode of loading is shear, we need the shear strength of the material. We can obtain this by performing a torsion test. However, we can approximate the maximum shear strength, τ_{max}, of a material from its yield strength by the following,

$$\tau_{max} = \frac{\sigma_{ys}}{2} \tag{4-65}$$

The design method we have just discussed is good when the strength properties do not deviate very much. However, for brittle materials such as ceramics, there is an inherent variability because of the presence of defects or cracks of varying sizes in the material. The tensile strength will then vary over a wide range and will depend on the crack size and the fracture toughness (which we shall cover in the following section) of the material. Design with brittle materials uses probability design methods that incorporate the size of the material. The most common of these is Weibull statistics, covered in greater detail in Chap. 14 on Ceramics and Glasses. Why, you might ask, do we need to design for tensile strength when the recommended mode of loading for ceramics is compression? The reason is that, while the loading is compression, there might arise tensile thermal shock stresses since ceramics are destined for high-temperature applications. Cool-down and heating of the ceramics will induce thermal stresses.

EXAMPLE 4-6

The properties of a material are: Young's Modulus, $E = 20 \times 10^6$ psi; Poisson's ratio, $\nu = 0.3$; and tensile yield strength, $\sigma_{y.s.} = 60{,}000$ psi. If the material is subjected to a shear stress, determine the maximum shear strain before a permanent shape change occurs.

SOLUTION: Under shear stress, we need to use the relationship,

$$\tau = G\gamma$$

and we are asked to determine γ, the shear strain. However, we note that neither G nor τ are given. We can derive these two from the given information. First,

$$G = \frac{E}{2(1+\nu)} = \frac{20 \times 10^6}{2(1+0.3)} = 7.69 \times 10^6 \text{ psi.}$$

Just as in tension, the maximum shear strength is the shear stress needed to make a shape change. The maximum shear strength of a material is

$$\tau_{max} = \frac{\sigma_{y.s.}}{2} = \frac{60{,}000}{2} = 30{,}000 \text{ psi}$$

Thus, maximum shear strain before shape change is

$$\gamma = \frac{30{,}000}{7.69 \times 10^6} = 0.0039$$

4.4.7 Fracture Mechanics and Fracture Toughness [5]

The allowable stress design method just described, which uses a fraction of the failure strength (yield strength) as the design stress, is sufficient for most designs. However, it does not guarantee that catastrophic failure will not occur. Catastrophic failures such as the collapse of an entire bridge structure, the burst of a pressure vessel or storage tank (molasses tank), the break up of a ship, and the break up of an airplane have occurred and cannot be explained by this method of design. Most of the explanations of these disasters (see the example case study in Chap. 16) are based on *fracture mechanics* and the *fracture toughness of the material.* Current design methods for safety and performance incorporate fracture toughness, and it is therefore important that we are familiar with this method.

Fracture mechanics is the field of mechanics that deals primarily with the study of crack propagation in a material. It is a relatively new field compared to statics and dynamics, having been developed only during and after the second World War to explain the break up of many merchant ships and tankers. The application of fracture mechanics has now expanded to many areas and is most crucial in aerospace structures. *The basic premise in using fracture mechanics in design is to assume that materials have defects or cracks in them. The material property that resists the propagation of these cracks is the fracture toughness.*

The fracture toughness property is analogous to the yield strength property. In the tensile test, the material sustains a stress and will remain elastic until the stress level applied exceeds the yield strength. If we use the yield strength as a failure criterion, the material fails after the stress level surpasses the yield strength of the material. In the presence of a crack in a material, the applied global stress away from the crack is intensified or increased at the tip of the crack. If σ_{app} is the applied stress on a material containing a crack size, 2a, within an infinitely wide plate material, stress analysis reveals that the local stresses at the tip of the crack are of the form,

$$\sigma_{local} = \frac{(\sigma_{app}\sqrt{\pi a})}{\sqrt{2\pi r}} \cdot f(\theta) \tag{4-66}$$

where the local stress is at the point (r,θ), r is the radial distance from the tip of the crack and θ is the angle from the plane of the crack, see Fig. 4-32, and a is a measure of the crack size. For a through-thickness crack in an infinitely wide plate material, a is one-half the actual crack size inside the material; at the edge of the plate, a is the actual crack size. The quantity in parentheses with the applied stress and the crack size is called the *stress intensity factor, K*, that is,

$$K \equiv \sigma_{app}\sqrt{\pi a} \tag{4-67}$$

To allow for the finite dimensions of a structure, a geometric factor Y is introduced, and the stress intensity factor becomes,

$$K \equiv Y\sigma_{app}\sqrt{\pi a} \tag{4-68}$$

$Y = 1$ for a through-thickness crack within an infinitely wide material, such as that in Fig. 4-32, and $Y = 1.12$ for a through-thickness edge-crack in an infinitely wide plate material, as indicated in Fig. 4-33. Width correction factors (Y) need to be multiplied, if the width of the material is not much bigger than the crack size (see Example 4-7). Different crack geometries in materials have different stress-intensity expressions. These expressions are compiled in handbooks [6] and the proper stress-intensity equation must be used for the particular crack geometry. Details of these relationships are covered in a fracture mechanics text[5].

The local stress at the tip of the crack have components that will enlarge the crack, as shown in Fig 4-34. The modes of enlargements are called Mode I for the opening mode, Mode II for the sliding or shear mode,

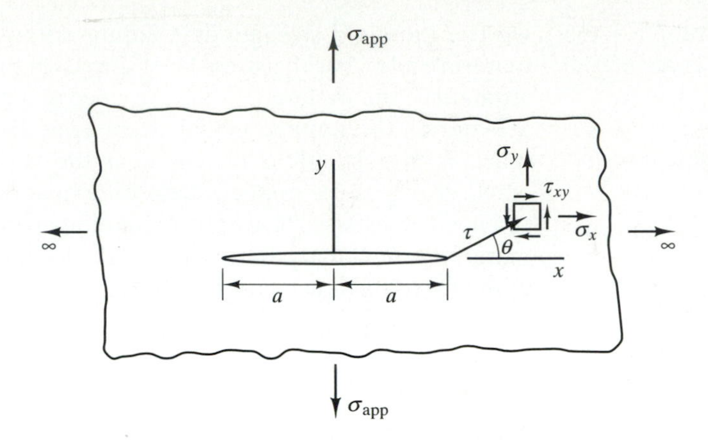

Infinitely wide plate containing a through thickness crack $2a$.

$$\sigma_{local} = \begin{Bmatrix} \sigma_y \\ \sigma_x \\ \tau_{xy} \end{Bmatrix} = \frac{\sigma_{app}\sqrt{\pi a}}{\sqrt{2\pi r}} f(\theta)$$

Fig. 4-32 Local stresses near the tip of a through-thickness crack in an infinitely wide plate with a gross-applied stress away from the crack.

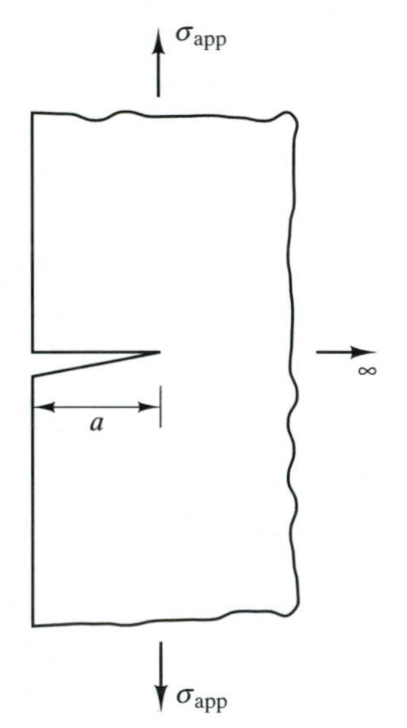

Fig. 4-33 Edge crack in an infinitely wide plate; $Y = 1.12$.

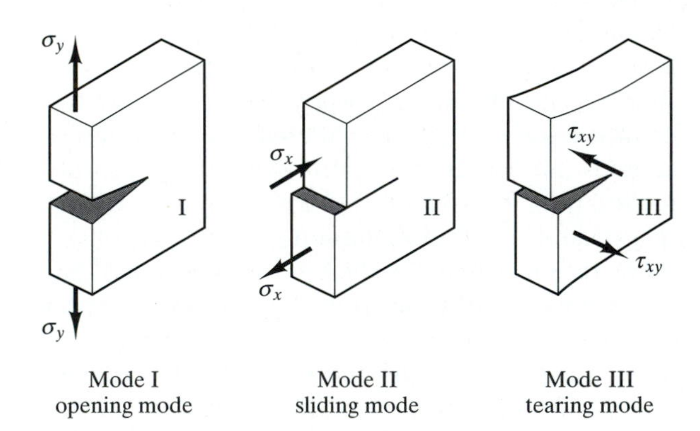

Fig. 4-34 Mode I, II, and III crack propagation modes.

and Mode III for the tearing mode. The stress intensity factors for these modes are K_I, K_{II}, and K_{III}—the most important of these is K_I with the tensile stress component perpendicular to the plane of the crack. From here on, we shall limit the discussion to this mode since it initiates the catastrophic failure.

Similar to the tensile stress in the elastic range, the stress intensity K_I at the tip of the crack increases linearly with increasing applied stress for a constant crack size. At the stress level when the crack propagates, the stress intensity at the tip of the crack is called the *critical stress-intensity factor,* K_{I_c}, and this is what is called the *fracture toughness of the material*, see Fig. 4-35. Similar to yield strength, when the stress intensity at the tip of a crack reaches the material's fracture toughness, K_{I_c}, the material fractures or fails. The fracture toughness is a material property because once it is determined, it *applies for all crack sizes.*

In any material, there will be cracks of different sizes. The largest crack in the proper orientation dic-

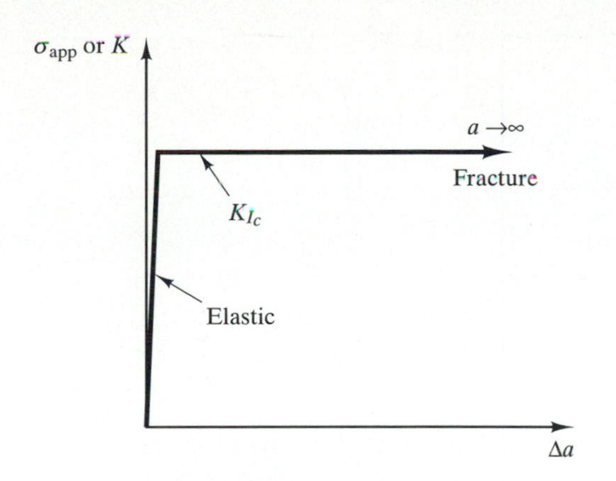

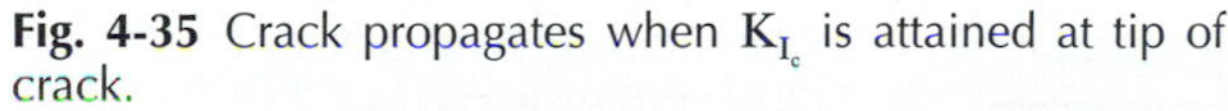
Fig. 4-35 Crack propagates when K_{I_c} is attained at tip of crack.

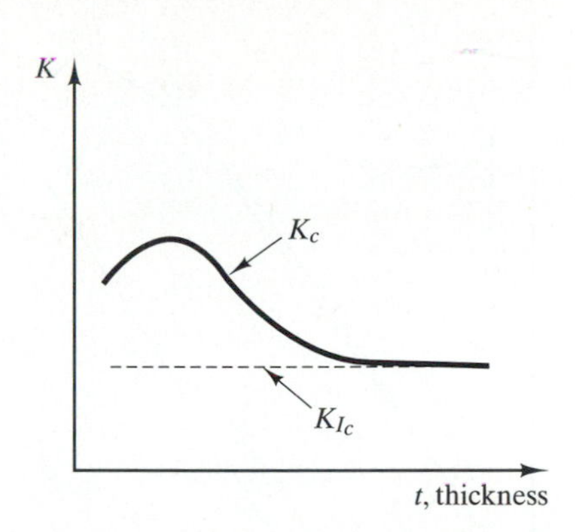

Fig. 4-36 Schematic of variation of *K* with thickness of material.

tates the fracture behavior because it has the largest stress intensity. A corollary to this is that the tensile strength (applied stress) of a brittle material depends on the largest crack size in the specimen tested. Since the sizes of cracks vary in each specimen tested, we obtain what is called the probable tensile strength for brittle materials using the Weibull statistics (see Chap. 14).

The discussion so far assumes that the crack does not increase in size before the critical stress is attained. In other words, *the material at the tip of the crack is* just *elastically deforming* and has been assumed to have an infinite yield stress. This treatment is called *linear elastic fracture mechanics,* which means that the *material is a perfect linear elastic solid* and exhibits only the first portion of the stress-strain curve. In addition, the thickness of the material was assumed to be very large compared to the crack size in order to ensure what is called *plane strain* deformation, which means that there is *no deformation in the thickness direction of the material.* The fracture toughness, K_{I_c} is called the *plane strain fracture toughness*.

The assumption that the material has infinite yield strength is obviously unrealistic. Therefore, there is always a small amount of plastically deformed material ahead of the tip of the crack. In order to essentially maintain plane strain deformation, this plastically deformed region must be small compared to the size of the crack and the thickness of the material. The thickness and the crack size must be at least 50 times the plastically deformed region to assure plane strain conditions. When the crack propagates at a critical stress level and the plane strain conditions are not met, the stress intensity is simply referred to as the critical stress intensity, K_c, not as the fracture toughness of the material. The K_c varies with the thickness of the material, as shown in Fig. 4-36. As the thickness increases, the plane strain condition will eventually prevail and at that point, the stress intensity becomes the material fracture toughness and subsequently independent of thickness.

4.4.8 Fracture Toughness Testing

Fracture toughness testing involves, therefore, the determination of the K_{I_c}, the plane strain fracture toughness, of the material. This test is standard ASTM E-399. The details of the test cannot be covered here, however. Suffice it to say that standard specimens for the test are precisely machined and ground to produce a very smooth surface, a very sharp crack induced by fatigue, and then tested with the same machines used for tensile testing. The raw data obtained is a load-crack opening or displacement P-ν, curve. The three types of the P-ν curve are shown in Fig. 4-37. From the P-ν curve, a load P_Q is obtained and substituted in the stress-intensity equation for the particular test specimen geometry to obtain a provisional stress-intensity factor, K_Q. If the test results meet all the required standard validity conditions, then K_Q is the desired K_{I_c}. If the test results do not meet the validity conditions, K_Q is just a critical stress intensity for the particular thickness

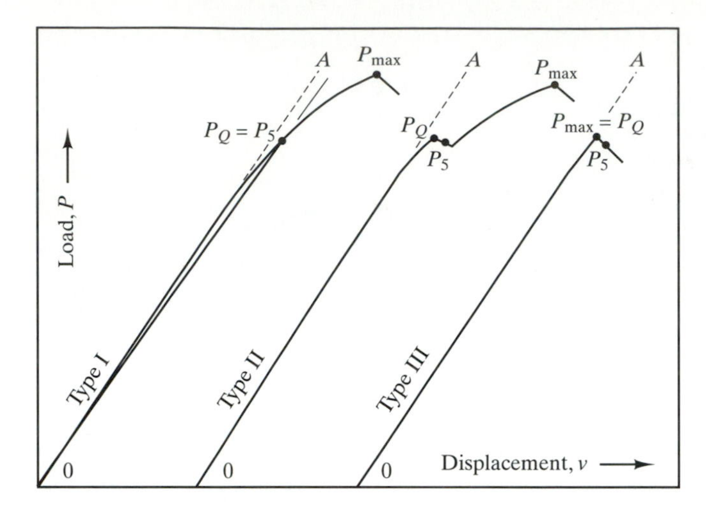

Fig. 4-37 Three types of load-crack opening displacement curves for fracture toughness testing according to ASTM E-399.

of the material. Thus, after a long arduous and expensive test, we cannot necessarily be assured that we can obtain a K_{I_c} value.

The K_{I_c} test can be successfully accomplished only with brittle and very high-strength materials. Most structural metals, in particular, steels, have low strengths and exhibit a lot of plastic deformation at the tip of the crack. In this case, the fracture involves both elastic and plastic deformation at the tip of the crack and this field is called *elastic-plastic fracture mechanics*. There are three tests to determine the resistance to crack propagation in ductile materials. These are: (1) ASTM E-561, the resistance curve (R-curve) determination, (2) ASTM E-813, a J-integral standard test for J_{I_c}, a measure of fracture toughness, and (3) ASTM E-1290, standard test method for crack-tip opening displacement (CTOD) fracture toughness measurement. Results of these tests are correlated to K_{I_c}. The results of the CTOD test have been successfully applied to design of pressure vessels.

4.4.9 Design Using Fracture Toughness

A good design of a structure or a component must consider the fracture toughness, in addition to the strength of the material. The key in using fracture toughness in design is to select a material with a high K_{I_c} because it can tolerate higher stress and/or larger crack size, as shown in Fig. 4-38. Figure 4-38 is a plot of applied stress versus crack size for materials A and B with different

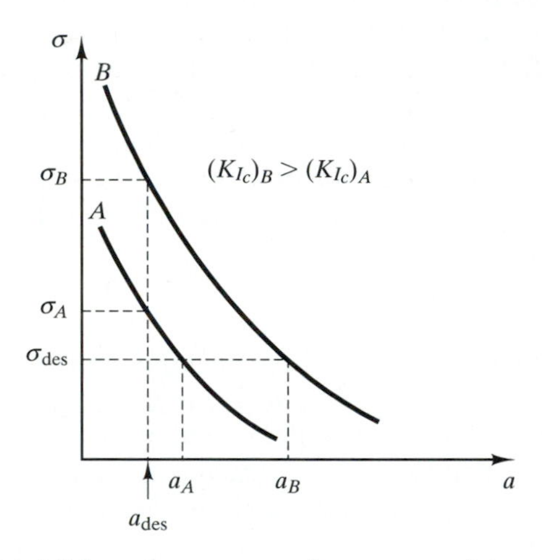

Fig. 4-38 Higher fracture toughness materials tolerate higher stress and larger cracks.

K_{I_c} values. Higher K_{I_c} materials are shown to tolerate higher stress at a certain design crack size or a larger crack size at a certain design stress level. The design stress will depend on the critical crack size we incorporate in the design. The critical crack size to be incorporated in the design is usually dictated by the non-destructive test (NDT) available to detect the crack. If the NDT can detect very small cracks, then we can use small cracks and the design stress can be high. On the other hand, if the NDT can only detect large cracks, then the design stress will be low. Quite often the design stress is kept lower to allow for routine inspection and maintenance before catastrophic failure of the structure. This is done in airplanes to allow for a routine inspection and maintenance. In pressure vessels, line pipes carrying fluids under pressure, or in storage tanks, the critical crack size is usually chosen to be larger than the thickness of the material used. When the crack exceeds the thickness, a leak can be detected before a catastrophe will occur. This is referred to as the leak-before-break criterion.

EXAMPLE 4-7

A structural material has a fracture toughness, $K_{I_c} = 120 \text{ ksi} \cdot \sqrt{\text{inch}}$, and a yield strength of 100 ksi. The design stress is $\sigma_{y.s.}/2$. Calculate the critical crack size in an infinitely wide plate for (a) a through-thickness crack within the material, and (b) a through-thickness edge crack.

SOLUTION: We use Eq. 4-68, $K = Y \cdot \sigma_{app} \cdot \sqrt{\pi a}$ and note that

$$\sigma_{app} = \sigma_{des} = \frac{\sigma_{y.s.}}{2} = \frac{100}{2} = 50 \text{ ksi}$$

(a) For an infinitely wide plate with a through-thickness crack, Y = 1. Thus,

$$a_{cr} = \left(\frac{120}{50}\right)^2 \cdot \frac{1}{\pi} = 1.833 \text{ inch}$$

The actual crack size within the material is $2a_{cr} = 3.666$ inch

(b) For an edge-crack in an infinitely wide plate, Y = 1.12. Thus, for this case,

$$a_{cr} = \left(\frac{120}{1.12 \times 50}\right)^2 \cdot \frac{1}{\pi} = 1.46 \text{ inch}$$

This example shows that edge or surface defects are much more deleterious than defects that are within the material.

EXAMPLE 4-8

For the same structural material as in Example 4-7 and the same design stress, calculate the critical crack size in a 10-inch wide plate for (a) a through-thickness crack, and (b) an edge crack.

SOLUTION: In using $K = Y \cdot \sigma_{app} \cdot \sqrt{\pi a_{cr}}$, we note that there are two unknowns, a_{cr} and Y. We, therefore, need another relationship. This relationship is given in Tables EP 4-8a and EP 4-8b. These may be expressed as $Y = f(a/b)$ but because this relationship is in tabular form, we cannot make a direct substitution in K. To solve for a_{cr} in both cases, we do the following iteration steps.

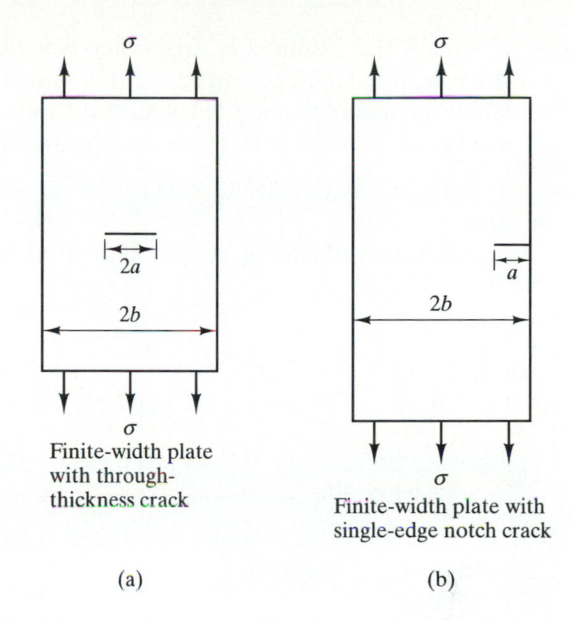

Fig. EP 4-8 (a) Finite-width plate with through-thickness crack. (b) Finite-width plate with single-edge notch crack.

Table EP 4-8a Correction factor ***Y*** for finite width plate with through-thickness crack for Fig. EP 4-8a.

a/b	factor *Y*
0.074	1.00
0.207	1.02
0.275	1.03
0.337	1.05
0.410	1.08
0.466	1.11
0.535	1.15
0.592	1.20

Step 1. We assume a a_{cr}.

Step 2. We get a_{cr}/b, and from which we obtain Y from the appropriate table.

Step 3. We substitute Y in the K equation and calculate a_{cr}.

Step 4. Compare the assumed a_{cr} from Step 1 with the calculated a_{cr}. The correct a_{cr} is obtained when the assumed and the calculated values are equal. If not equal, we repeat (iterate)

Steps 1-3, until the assumed and calculated values are equal.

(a) for a through-thickness crack in a 10-inch wide plate, $Y = f(a_{cr}/5)$.

First iteration:

1. Assume a_{cr} is 1.5.
2. $a_{cr}/\text{b} = 0.30, \text{Y} \approx 1.04$.

$$a_{cr} = \left(\frac{120}{1.04 \times 50}\right)^2 \times \frac{1}{\pi} = 1.695 \text{ inch}$$

We note $1.5 \neq 1.695$, so we must repeat. The correct value lies between the two values.

Final iteration:

Assume $a_{cr} = 1.67$ then $a_{cr}/b = 0.334$, and $Y = 1.05$

$$a_{cr} = \left(\frac{120}{1.05 \times 50}\right)^2 \times \frac{1}{\pi} = 1.663 \text{ inch}$$

Now we note that the assumed and calculated values are about equal. Therefore, $a_{cr} = 1.67$ inch. The through-thickness crack size is $2\,a_{cr} = 3.34$ inch.

Table EP 4-8b Correction factors, Y for Fig. EP 4-8(b)

a/b	factor *Y*
0.10	1.154
0.20	1.198
0.30	1.288
0.40	1.366
0.50	1.512
0.60	1.680
0.70	1.893
0.80	2.139
0.90	2.464
1.00	2.856

(b) for an edge crack in a 10-inch wide plate, $Y = f(a_{cr}/5)$.

First iteration:

Assume $a_{cr} = 1.5$, $1.5/5 = 0.30$, $\text{Y} = 1.288$

$$a_{cr} = \left(\frac{120}{1.288 \times 50}\right)^2 \times \frac{1}{\pi} = 1.105 \text{ inch}$$

We note that the assumed value (1.5) > the calculated value (1.105), therefore we need to repeat the procedure.

Final iteration:

Assume $a_{cr} = 1.20$, $1.20/5 = 0.24$, $\text{Y} = 1.234$

$$a_{cr} = \left(\frac{120}{1.234 \times 50}\right)^2 \times \frac{1}{\pi} = 1.204 \text{ inch}$$

The assumed value equals the calculated value, therefore the critical edge crack size is 1.20 inch.

4.4.10 Fatigue Properties and Fatigue Testing

Components of machines, vehicles, and structures are often subjected to repeated cyclic loads and the resulting *cyclic stresses induce damage* to the materials of construction. This damage, which occurs predominantly at stresses well below the yield strength, *accumulates with continued application until cracks initiate, propagate, and finally fracture the material.* This failure of materials under repeated cyclic stresses and strains is called *fatigue*. Because the cyclic stresses and strains are applied continuously and at high rates, *fatigue properties* are usually referred to as *dynamic properties*. In contrast, the *tensile properties* are referred to as *static properties* because they are applied very slowly and at a steady rate. The application of fatigue properties in the design process involves the prediction of the life (number of cycles) the material can sustain before cracks initiate and the life before one of these cracks propagates to the critical crack size. Since a critical crack size is involved, the design process involving fatigue is done in conjunction with the fracture toughness of the material.

The *fatigue properties* of materials can be obtained under either a *constant stress amplitude* or a *constant strain amplitude* test. As we shall see, the latter closely represents the behavior of materials with stress concentrations. For this reason, use of the older constant stress amplitude data in design is decreasing in favor of the more recent constant strain amplitude data. Nevertheless, we still have to discuss the characteristics of each test and learn how to use the data.

4.4.10a The Constant Stress Amplitude Test. We start by defining the terms used in the test as shown in Fig. 4-39. The cyclic stress is depicted as a

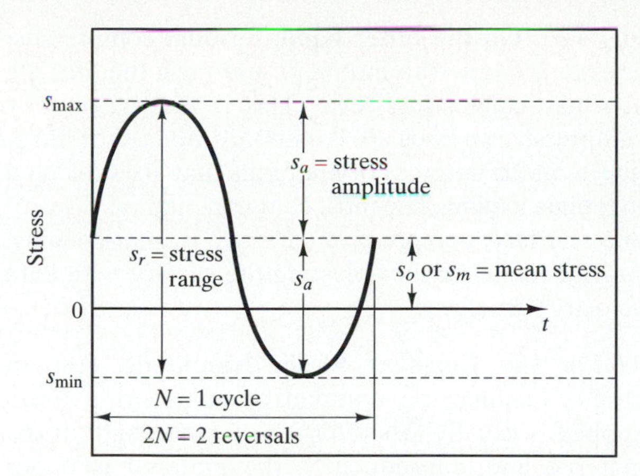

Fig. 4-39 Stress wave represents cyclic loading and the nomenclature use in fatigue [7a].

sinusoidal curve and one stress cycle is shown. In one cycle, the material is subjected to two reversals in stress, that is, from tension to compression, and then from compression to tension. Thus, there are $2N$ reversals with N cycles of stress. The *maximum stress*, s_{max}, is the highest algebraic value in the stress cycle, a positive value is tensile stress, and a negative value is compressive stress. The *minimum stress*, s_{min}, is the lowest algebraic value, again tensile stress is positive and compressive stress is negative. In this test, the stress is engineering or nominal stress. The stress range, s_r, is defined as

$$s_r = s_{max} - s_{min} \tag{4-69}$$

The stress amplitude, s_a, is defined as one-half the stress range, that is,

$$s_a = \frac{s_r}{2} = \frac{(s_{max} - s_{min})}{2} \tag{4-70}$$

The mean stress, s_m, is the algebraic average of the maximum and minimum stresses, that is

$$s_m = \frac{(s_{max} + s_{min})}{2} \tag{4-71}$$

The stress ratio, R, is the ratio of the minimum stress to the maximum stress, that is,

$$R = \frac{s_{min}}{s_{max}} \tag{4-72}$$

In fatigue testing, the maximum stress is almost always a tensile stress. R = −1 means that the minimum stress is compressive and is the same as the maximum tensile stress. This is referred to as completely reversed tension-compression testing. When the minimum stress is zero, we have a zero-to-tension testing and R = 0.

Schematic of typical constant stress amplitude test results for mild steel and aluminum in flexural beam bending fatigue tests are shown in Fig. 4-40. The test results, plotted in a semi-log plot are called *stress-life* (s-*N*) curves, and the characteristic of a typical s-*N* curve for mild steel is the appearance of a stress plateau at long life, that is, high number of cycles, *N*. The *stress plateau is defined as the fatigue limit* (*endurance limit* is an older term used and may still be used). The material with an applied stress at the fatigue limit has a 50 percent probability of failure. Thus, if the applied stress is below the fatigue limit, the probability of failure is much less. This value is used by designers and because of its importance, an empirical relationship between the fatigue limit and the ultimate tensile strength of quenched and tempered steels is found to be

$$\text{Fatigue Limit} \approx 0.5 s_{uts} \tag{4-73}$$

The stresses on the curve other than the fatigue limit are called the *fatigue strengths* of the material for the corresponding life, (the number of cycles), *N*. *In corrosive environments, the fatigue limit disappears for steels*. It has also been found that *large periodic overloads*, common to ground vehicles, *can also eliminate the fatigue limit* that appears in constant amplitude testing. In this case, the s-*N* curve continues downward and we refer only to a fatigue strength for a certain number of cycles, *N*. If all the carbon in steel is taken out, the fatigue limit will also not exist [8]. *For aluminum, there is no fatigue limit and the stresses are the fatigue strengths*.

There are three common modes for fatigue testing. These are (1) the push-pull axial test, (2) the rotating-beam bending mode, and (3) the flexural-beam bending mode. Of these three modes, the push-pull axial test yields the most conservative results, namely, the lowest stresses, fatigue limit, and fatigue strengths for a given life, *N*. Thus, we should be careful to use fatigue data from the bending modes for design. It is safe to use the axial push-pull test results for any design application.

If the semilog s-*N* curve is converted to a log-log plot of *log true stress*, σ_a, versus *log number of reversals*,

$2N$, the curve becomes linear, Fig. 4-41. The equation of the line is

$$\sigma_a = \sigma_f'(2N)^b \tag{4-74}$$

where b is the slope of the line and σ_f' is the intercept when $2N = 1$, that is, for one reversal. If there is any mean stress, σ_m, present in the material such as residual stresses, Eq. 4-73 becomes

$$\sigma_a = (\sigma_f' - \sigma_m)(2N)^b \tag{4-75}$$

The algebraic value of the mean stress is substituted in the above equation, with tensile being positive and compressive being negative. The effect of a tensile residual stress is to decrease the intercept and lower the life, $2N$, for a particular stress level as shown in Fig. 4-42. On the other hand, residual compressive stresses increase the intercept and raise the life, $2N$, at a particular stress level. This is the reason why compressive stresses are commonly induced or introduced on the surfaces of materials that are subjected to fatigue loading. The most common method to introduce surface compressive stresses is shot-peening, which is done by shot-blasting the surface with high velocity particles.

4.4.10b The Constant Strain Amplitude Test. In fatigue loading, the nominal or engineering stress applied is usually less than the yield strength of the material and, consequently, the material is under elastic loading. However, in areas where defects are present, stress concentrations are present and local

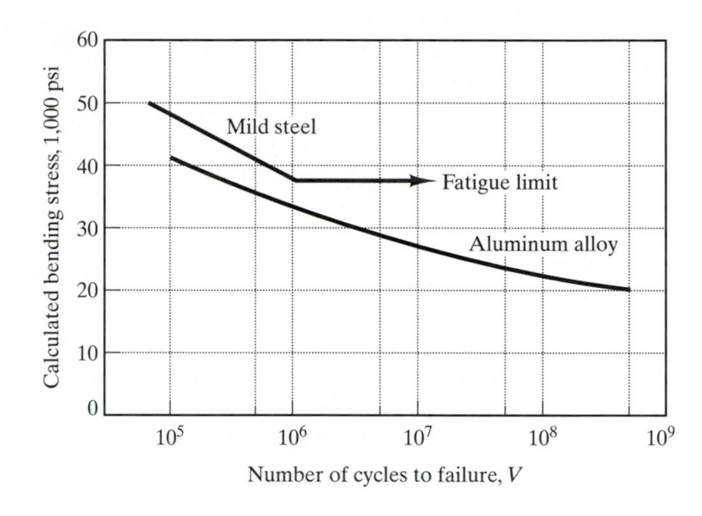

Fig. 4-40 Constant stress amptitiude S-N curves for mild steel and an aluminum alloy.

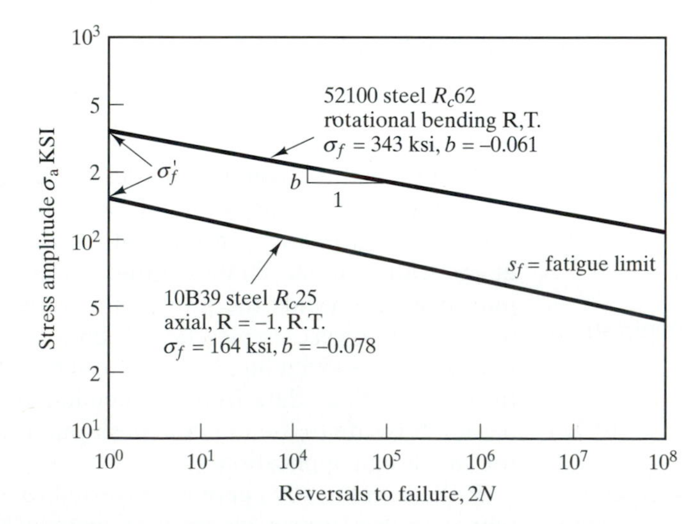

Fig. 4-41 Log stress-log reversal (log s-log $2N$) plot for constant stress amplitude [7a].

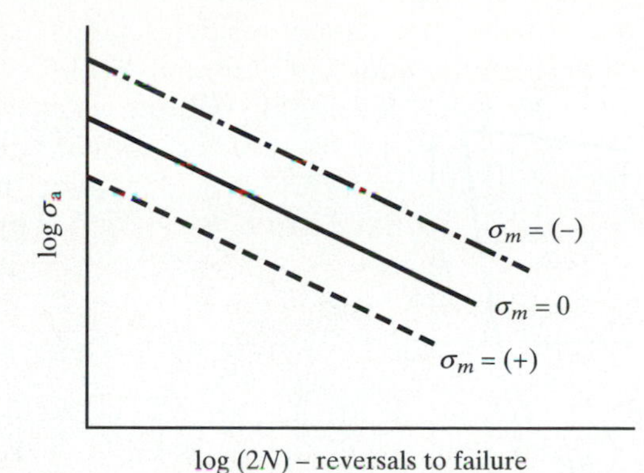

Fig. 4-42 Schematic showing effect of residual or mean stresses during fatigue; tensile mean stress is positive, and compressive mean stress is negative.

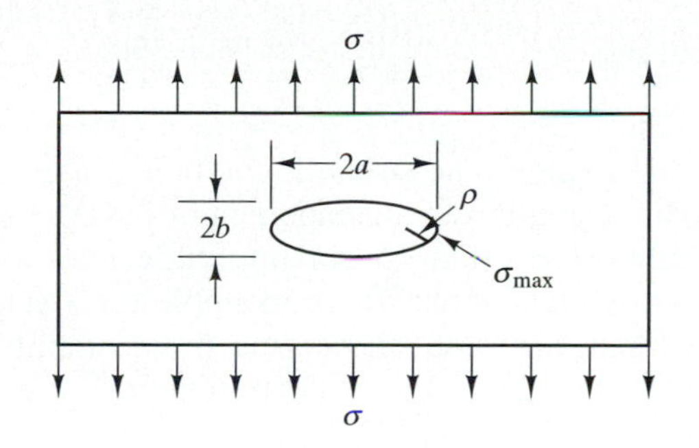

Fig. 4-43 Stress concentration at the tip of an elliptical crack.

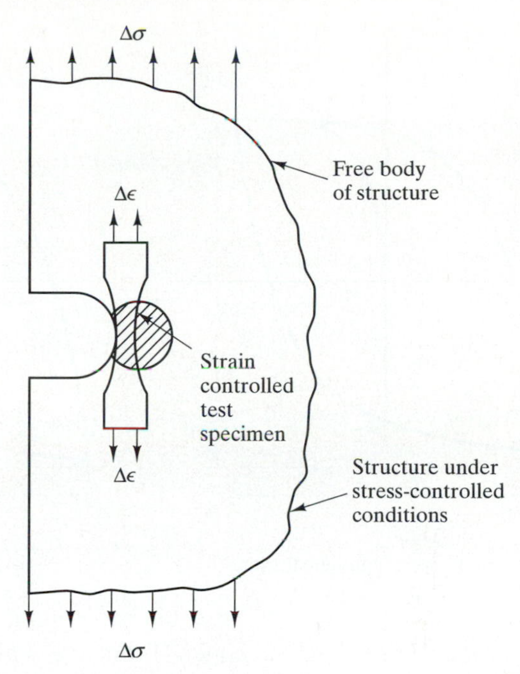

Fig. 4-44 Area around stress concentrations are under strain-controlled even when fatigue test is constant stress amplitude. [5]

areas can be plastically deformed. For the elliptical crack shown in Fig. 4-43, the stress-concentration factor, SCF, at the edge of the ellipse is defined as

$$\text{SCF} = \frac{\sigma_{max}}{\sigma_{app}} = 1 + \frac{2a}{b} \qquad (4\text{-}76)$$

where σ_{max} is the stress at the tip of the crack, σ_{app} is the nominal applied stress on the material, 2a is the major axis, and 2b is the minor axis of the elliptical crack. The SCF for very sharp cracks can be obtained by making ($a \gg b$) and for this,

$$\sigma_{max} \simeq \sigma_{app} \cdot 2\sqrt{\frac{a}{\rho}} \qquad (4\text{-}77)$$

where ρ is the radius at the tip of the elliptical crack shown in Fig. 4-43. As $\rho \rightarrow 0$, we see that σ_{max} approaches infinity. Thus, we see that σ_{max} can exceed the yield strength to plastically strain the material locally even though the applied stress is much below the yield strength. The true stress and the true strain in these plastically deformed areas with stress concentrations are constrained or limited by the overall surrounding elastically deforming material. Therefore, *even when the structure is stress-controlled below the yield strength, the localized plastic zones around the defects are approximately under strain-controlled conditions,* see Fig. 4-44. Because defects are present *in all engineering components*, the latter's *fatigue behavior is best simulated by testing smooth specimens under strain-controlled conditions.*

The *strain-life curves, ϵ-N,* are established by imposing a strain amplitude and noting the number of cycles, *N*, or reversals, 2*N*, when the material fails. The standard test is ASTM E-606. The initial response and the hysteresis loop of the first cycle of the material is shown in Fig. 4-45a. Subsequently, the material usually exhibits an initial transient behavior in which the hysteresis loop keeps changing. A stable stress-strain behavior is reached when a constant hysteresis loop is obtained, see Fig. 4-45b. This condition is usually attained in less than 50 percent of the

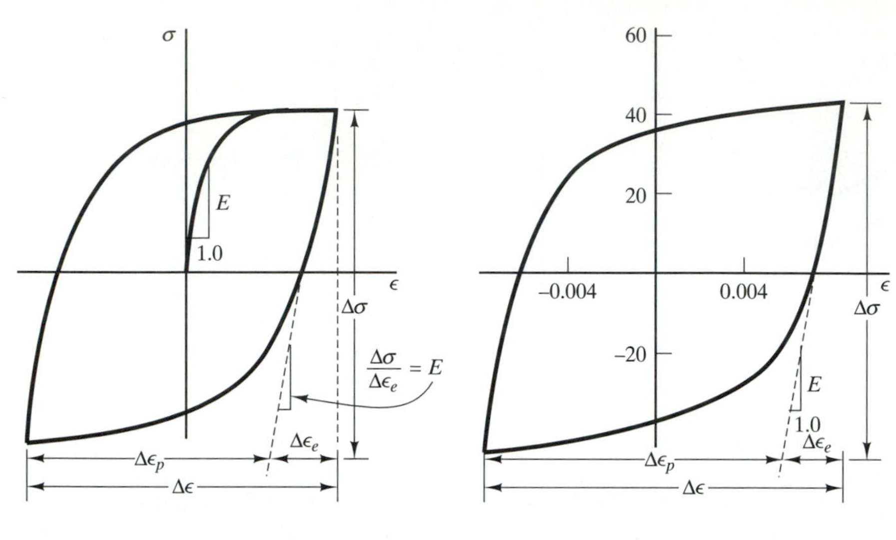

Fig. 4-45 Strain life testing (a) initial cycle, and (b) stable hysteresis loop. [5]

total fatigue life of the specimen. From the stable loop, the total true strain range, $\Delta\epsilon$, of the cycle is twice the strain amplitude, ϵ_a, and the total true stress range, $\Delta\sigma$, is also twice the stress amplitude, σ_a. In addition, the total strain amplitude can be represented by the sum of its elastic and plastic components, that is

$$\epsilon_a = \frac{\Delta\epsilon}{2} = \frac{(\Delta\epsilon)_e}{2} + \frac{(\Delta\epsilon)_p}{2}$$

$$= \frac{\Delta\sigma}{2E} + \frac{(\Delta\epsilon)_p}{2} \qquad (4\text{-}78)$$

where E is the Young's Modulus, $\Delta\epsilon/2$ is the total strain amplitude, $(\Delta\epsilon)_e/2$ the elastic strain amplitude, $(\Delta\epsilon)_p/2$ the plastic strain implitude, and $(\Delta\sigma)/2$ the stress amplitude.

A plot of the true stresses and the true strains at the tips of the stable hysteresis loops at various strain amplitudes yields the cyclic true stress-true strain curve, Fig. 4-46. A 0.2% offset cyclic yield strength, σ_{ys}, may also be obtained from this curve, as indicated.

As in the tensile test, a log-log plot of the true stress and the true strain in the plastic deformation region yields a straight line, Fig. 4-47, and the equation of this line is

$$\frac{\Delta\sigma}{2} = K'\left(\frac{\Delta\epsilon_p}{2}\right)^{n'} \qquad (4\text{-}79)$$

where K′ is the cyclic strength coefficient and n′ is the cyclic strain-hardening exponent or index.

Results of constant strain-amplitude tests show that log-log plots of plastic strain amplitude, $(\Delta\epsilon)_p/2$, versus $2N$ in Fig. 4-48, and elastic strain amplitude, $(\Delta\epsilon)_e/2$, versus $2N$ in Fig. 4-49, are linear. The equations of these lines are,

$$\frac{\Delta\epsilon_p}{2} = \epsilon_f'(2N)^c \qquad (4\text{-}80)$$

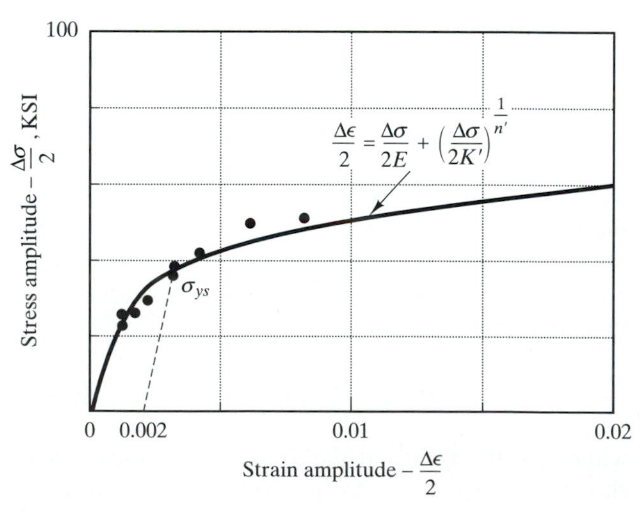

Fig. 4-46 Points at the tips of stable hysteresis loops produce the cyclic stress-strain curve. [7a]

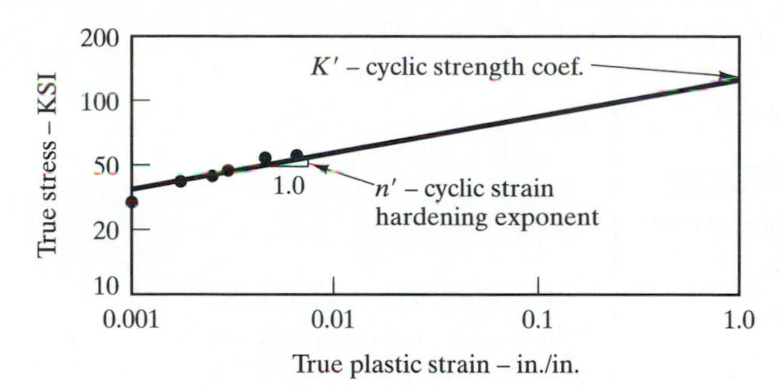

Fig. 4-47 Log cyclic strain versus log cyclic stress plot determines K′ and n′. [7a]

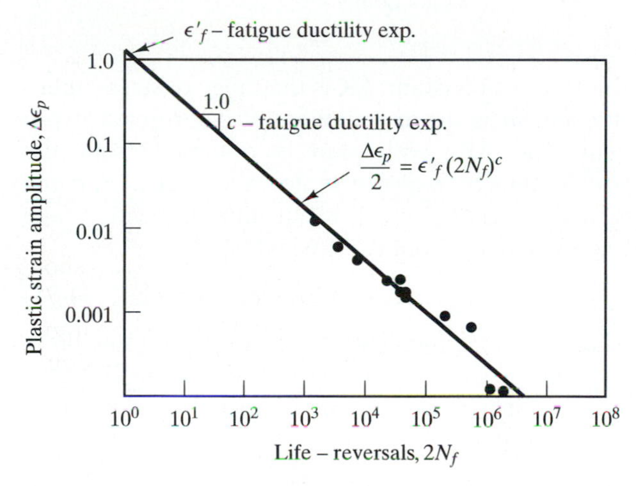

Fig. 4-48 Plastic strain amplitude versus reversals to failure. [7a]

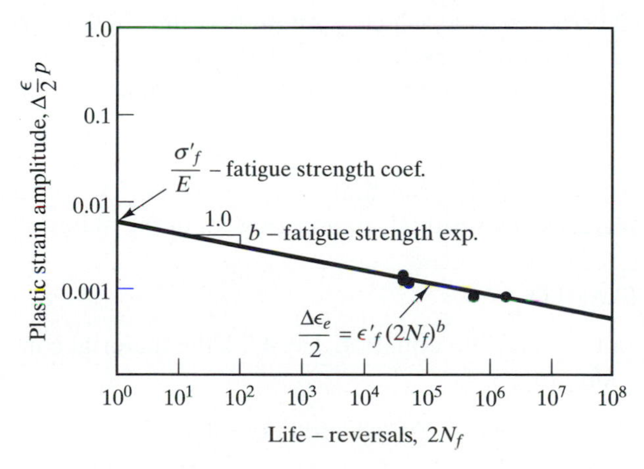

Fig. 4-49 Elastic strain amplitude versus reversals to failure. [7a]

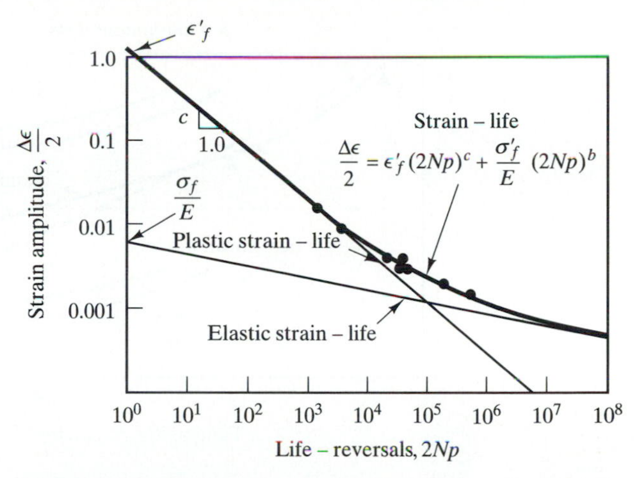

Fig. 4-50 Total strain amplitude versus reversals to failure. [7a]

$$\frac{\Delta\epsilon_e}{2} = \frac{\sigma_f'}{E}(2N)^b \qquad (4\text{-}81)$$

The total strain-life curve is obtained by adding the strains at each $2N$ value, Fig. 4-50. Thus,

$$\frac{\Delta\epsilon}{2} = \epsilon_f'(2N)^c + \frac{\sigma_f'}{E}(2N)^b \qquad (4\text{-}82)$$

The four constants in Eq. 4-82 are:

1. ϵ_f' is the fatigue ductility coefficient and is the true plastic strain required to cause failure in one reversal. It is the intercept of the log-log plot of Eq. 4-80 in Fig. 4-48 at $2N = 1$.
2. c is the fatigue ductility exponent and is the slope of the log-log plot of Eq. 4-80 in Fig. 4-48.
3. σ_f' is the fatigue strength coefficient and is the true stress required to cause failure in one reversal. It is the intercept of the log-log plot of Eq. 4-81 in Fig. 4-49 at $2N = 1$.
4. b is the fatigue strength exponent and is the slope of the log-log plot of Eq. 4-81 in Fig. 4-49.

The above four constants plus K' and n' in Eq. 4-79 are the fatigue properties under strain-controlled conditions. Knowing these cyclic properties of a material, we can predict the fatigue behavior of a component. An illustration of this is shown in Fig. 4-51, which shows that the strain-life model (local strain model) very closely predicts the actual fatigue behavior of a component. The constant stress-life model (s-N), when corrected for the stress

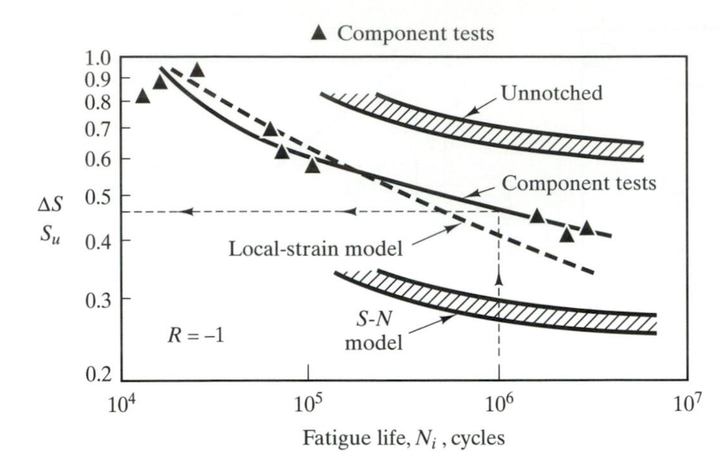

Fig. 4-51 Data points from actual fatigue component tests are closely predicted by the local strain model (strain-life); constant stress-life (s-*n*) with correction for stress-concentration underestimates data points [7b].

concentration factor, underestimates the actual component behavior. Figure 4-51 shows the fatigue life as N_i, the number of cycles to initiate a crack of a certain size, as compared to N_f, the total number of cycles to failure, used in the previous figures. Thus,the failure criterion adopted in Fig. 4-51 is the crack initiation stage. This stage is usually taken as the life to observe a 0.25 mm (0.01-inch) visible crack. We see that the *constant strain-amplitude test from a smooth specimen predicts very closely the crack initiation life, N_i, of a component or structure.*

The strain-life curve can be estimated from the six fatigue properties using Eqs. 4-80, 4-81, and 4-82. Take a value of N_i and substitute it in the equations to obtain $(\Delta\epsilon)_e/2$, $(\Delta\epsilon)_p/2$, and $\Delta\epsilon/2$. Repeat this for other N_i to generate other points along with the curve.

The total life of the component or structure is the sum of the crack initiation life, N_i, and the number of cycles or life, N_p, to propagate the crack until it fractures. That is,

$$N_T = N_i + N_p \qquad (4\text{-}83)$$

The initiation life, N_i, as we have just seen is estimated by the strain-life curve. Thus, in order to make an estimate of the total life of the component, we need now to estimate the propagation life, N_p, of the initial 0.25-mm (0.01-inch) crack size. To accomplish this, it is necessary to know the crack growth rate, da/dN, which is the change in crack length per cycle. This is obtained experimentally for the material and is of the form

$$\frac{da}{dN} = C(\Delta K)^m \qquad (4\text{-}84)$$

where C is a constant, ΔK is the range of stress intensity, and m is an experimentally determined exponent. The ΔK expression is obtained from the stress-intensity expression for the crack. For the through-thickness crack in an infinitely wide plate, the ΔK derived from Eq. 4-68 is

$$\Delta K = Y\Delta\sigma\sqrt{\pi a} \qquad (4\text{-}85)$$

where $\Delta\sigma$ is the live-load stress range (s_r in Eq. 4-69). By using the ΔK, rearranging and intergrating Eq. 4-84, we can calculate or estimate N_p with the following

$$N_p = \frac{1}{C}\int_{a_i}^{a_c} \frac{da}{[\Delta\sigma\sqrt{\pi a}]^m} \qquad (4\text{-}86)$$

The integration may be accomplished analytically or numerically. The limits of the integration are the initial crack size, a_i, which is about 0.25mm (0.01 inch) to the critical crack size, a_c. The latter is obtained from the fracture toughness of the material,

$$K_{I_c} = Y \cdot \sigma_{max}\sqrt{\pi a_c} \qquad (4\text{-}87)$$

where σ_{max} is the maximum load (s_{max}) which includes any residual stresses present in the material.

EXAMPLE 4-9

In the stress-life equation, Eq. 4-74, the material constants are

$$\sigma_f' = 255 \text{ ksi and } b = -0.0977$$

and the yield and tensile strengths are 160 and 170 ksi, respectively. Estimate

(a) the maximum stress amplitude that the material must be subjected to for it to have a minimum one million cycles service life.

(b) the fatigue life when the stress amplitude is 85 ksi.

(c) the maximum stress amplitude for a minimum one million cycles fatigue life when a residual tensile stress of 20 ksi is present in the material.

SOLUTION: We use Eq. 4-74 $\sigma_a = \sigma_f'(2N)^b$ and for $N = 1 \times 10^6$

(a) maximum stress amplitude,

$$\sigma_a = 255(2 \times 10^6)^{-0.0977} = 61.79 \text{ ksi}$$

(b) With $\sigma_a = 85$ ksi, $85 = 255(2N)^{-0.0977}$

$$2N = \left(\frac{85}{255}\right)^{-\frac{1}{0.0977}} = 76{,}442 \text{ reversals}$$

$$N = 38{,}221 \text{ cycles}$$

(c) with a mean tensile stress of 20 ksi, σ_a for one million cycles is

$$\sigma_a = (255 - 20)(2 \times 10^6)^{-0.0977} = 56.94 \text{ ksi}$$

EXAMPLE 4-10

With the value of K′ = 945 MPa and n′ = 0.22, estimate the 0.2 percent cyclic yield strength of the material and compare it to the tensile yield strength, which is equal to 228 MPa.

SOLUTION: We use Eq. 4-78, $\frac{\Delta\sigma}{2} = K'\left(\frac{\Delta\epsilon_p}{2}\right)^{n'}$ by substituting

$$\frac{\Delta\epsilon_p}{2} = 0.002 \text{ (0.2\% strain)}.$$

Thus,

Cyclic 0.2% offset yield strength

$$= \frac{\Delta\sigma}{2} = 945(0.002)^{0.22} = 240.8 \text{ MPa}.$$

We see that *the cyclic yield strength is greater than the tensile yield strength*. This phenemenon is called **cyclic hardening.** It is also possible to have **cyclic softening** when the cyclic yield strength is less than the tensile yield strength.

EXAMPLE 4-11

The strain-life fatigue properties of a steel are

Strain–Life Fatigue Properties

	Eq. 4-79	Eq. 4-80	Eq. 4-81
Coefficient	K′ = 945 MPa	$\epsilon_f' = 0.95$	$\sigma_f' = 828$ MPa
Exponent	n′ = 0.22	c = −0.64	b = −0.11

The tensile yield strength is 228 MPa and the modulus of elasticity of the steel is 2.10×10^5 MPa. Estimate (a) the constant strain range and (b) the corresponding cyclic stress range for an initiation life of one million cycles.

SOLUTION:

(a) We use Eqs. 4-80 and 4-81, to calculate the plastic strain and the elastic strain amplitudes. Then, if we add the results, we shall obtain the total strain amplitude.

the plastic strain amplitude

$$= \frac{\Delta\epsilon_p}{2} = 0.95(2 \times 10^6)^{-0.64} = 0.000088$$

the elastic strain amplitude

$$= \frac{\Delta\epsilon_e}{2} = \frac{828}{2.10 \times 10^5} \times (2 \times 10^6)^{-0.11} = 0.0008$$

the total strain amplitude

$$= 0.0008 + 0.000088 = 0.000888$$

the strain range

$$= 2 \times 0.000888 = 0.001776$$

(b) For the cyclic stress range, we use Eq. 4-78 to get the cyclic stress amplitude and then multiply it by two to get the desired result.

the cyclic stress amplitude

$$= \frac{\Delta\sigma}{2} = 945(0.000088)^{0.22} = 121 \text{ MPa}$$

the stress range

$$= 2 \times 121 = 242 \text{ MPa}$$

COMMENTS:

1. The above is an example of the calculation for a point (2×10^6 reversals, 0.000888 strain amplitude), in the estimation of the strain-life curve of a smooth specimen tested under constant-strain amplitude. The procedure can be repeated for other life values (number of cycles) to generate the strain-life curve.
2. In the long life, from approximately one million cycles and up, the elastic strain dominates. We see this above where the elastic strain amplitude is 10 times more than the plastic strain amplitude.
3. We see that the cyclic stress amplitude is about one-half that of the tensile yield strength.

4.4.11 Creep and Creep Testing

Up to this point, we have not introduced the influence of temperature. The stressing or loading of materials has been tacitly assumed to be at room or relatively low temperatures so that its influence can be neglected. In some applications, such as in steam turbines in power plants, jet and rocket engines, melting furnaces, and nuclear reactors, the materials are exposed to extremely high temperatures. For metals, the influence of temperature is based on T_m, the absolute melting temperature. When the temperature is in the range of 0.3–0.6T_m or above, movements of atoms (diffusion) in the material will induce a very slow deformation (strain), called *creep,* if the material is under stress. This strain will eventually cause dimensional problems in the structure or component to cause malfunction or failure. If left under stress at high temperatures, the material will ultimately rupture or fracture. T_m is also used for pure crystalline ceramics. However, in practice, most ceramics are multicomponent and there is no distinct melting temperature. The measure of refractoriness is the pyrometric cone equivalent or the softening point—the temperature at which a standard test cone of the ceramic softens or bends (see Sec. 14.5.3). For glasses, the softening point is the temperature at which glass will deform under its own weight (see Sec. 14.3.3). For polymers, the T_g (the glass transition temperature) or the deflection temperature under load (DTUL) is used (see Sec. 15.7.3). The latter temperature is relatively low such that at room temperature polymers/plastics exhibit substantial creep.

Creep testing to determine the creep behavior of a material is done by applying a constant axial load, usually in tension, to a bar or cylindrical sample of the material. The setup is schematically shown in Fig. 4-52. The creep strain is measured with time, and the time at rupture or fracture is recorded if this occurs during the test. Tests on a given material are usually done at various temperatures and stresses, and test durations can range from less than a minute to more than a year or a few years.

An idealized creep deformation or strain behavior response is schematically shown in Fig. 4-53. Upon loading (stressing) the material, the immediate response is an instantaneous strain, ϵ_o, which may be elastic or plastic (including an elastic part) at time t = 0. The creep deformation with time, the ϵ versus t curve, consists of three regions that are based on the strain rate, $\dot{\epsilon} = d\epsilon/dt$, which is the slope of the curve. The first region from ϵ_o to ϵ_1, is referred to as the primary or transient creep, and is characterized initially by a very high and then gradually decreasing slope or strain rate up to ϵ_1. The second region is from ϵ_1 to ϵ_2, referred to as secondary or steady-state creep, and is characterized by a constant slope or steady-strain rate, ϵ_s. The third region, from ϵ_2 to ϵ_r, referred to as tertiary or unstable creep, and is characterized by an increasing slope or strain rate until the material ruptures or separates at ϵ_r. The time, t_r, at ϵ_r is called the time to rupture. In the last region, the deformation becomes localized and the

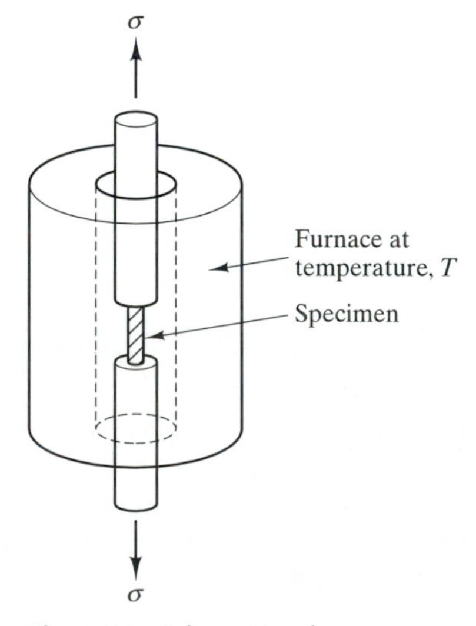

Fig. 4-52 Schematic of creep test.

specimen necks as in the tensile test, or voids may form inside; both may occur.

The results of a creep test are shown in a graph that indicates at least three of the four following quantities: the true stress, σ, the temperature, T, the steady-state creep rate, $\dot{\epsilon}_s$, and the time to rupture, t_r. Because of the long times involved and the range of the strains, the scales of the graphs are usually logarithmic. Two examples are shown: (1) in Fig. 4-54 are plots of true stress versus the steady-strain rate, σ versus $\dot{\epsilon}_s$, at constant temperature, T; and (2) in Fig. 4-55 are plots of the true stress versus time to rupture, σ versus t_r, at constant temperature, T. The latter plot is analogous to an s-*N* curve for fatigue, except that the life is in rupture time rather than in the number of cycles. In addition to rupture life (time), the life to reach a particular value of strain may be used. Stress-life plots for three values of strain and also for rupture, all for a single temperature, are shown in Fig. 4-56 for polyvinyl chloride at 20°C for the three strains.

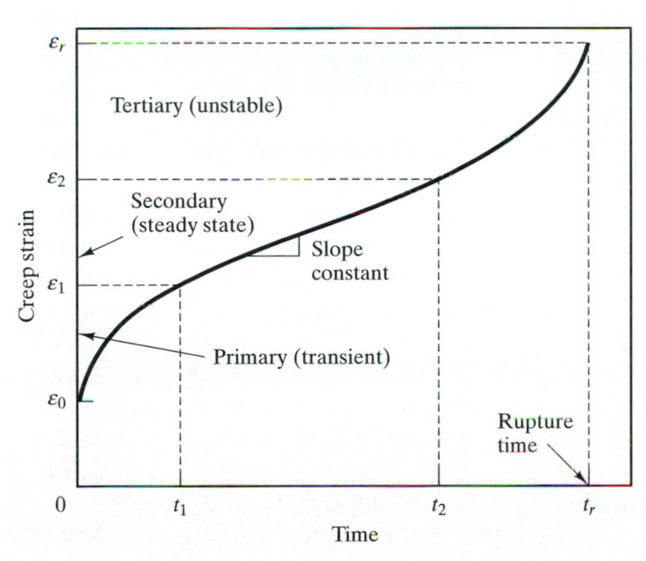

Fig. 4-53 Schematic of creep-rupture curve showing three stages.

4.4.11a The Mechanisms of Creep. We need to differentiate the creep behavior of amorphous solids from that of crystalline solids.

In Chap. 2 we characterized amorphous solids as having only short-range order structure very similar to liquids. Thus, we shall consider these solids as liquids with very high viscosity. This is generally the case for silica glass and thermoplastic polymers. In response to a stress applied to amorphous solids, molecules or groups of molecules slide relative to one another in a time-dependent manner, resulting in creep strain, Fig. 4-57a. This mechanism is called *viscous creep*. Because the materials also have either elastic or plastic strain, depending on the stress applied, creep is usually referred to as visco-elastic or visco-plastic deformation.

In liquid flow or fluid mechanics, the shear stress and the strain rate are related to the viscosity of the liquid. That is

$$\eta_\tau = \frac{\tau}{\dot{\gamma}} \tag{4-88}$$

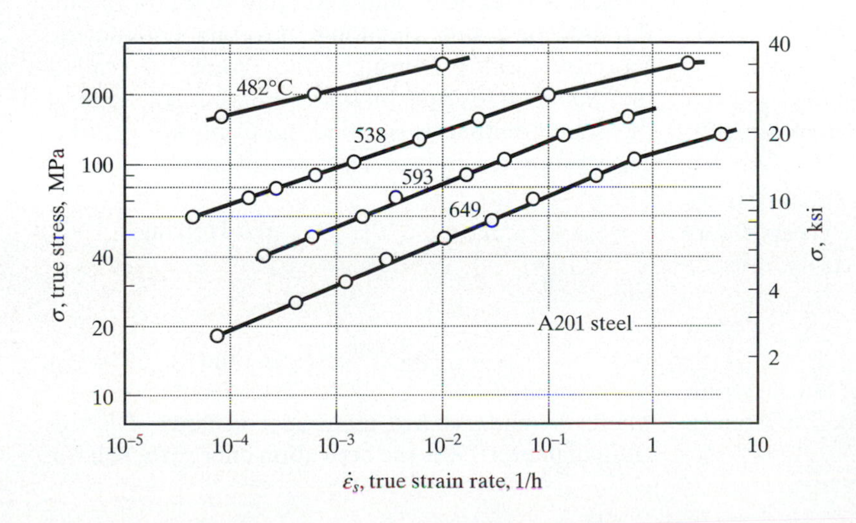

Fig. 4-54 Stress-steady state creep strain rate at various temperatures for carbon steel used for pressure vessels [11].

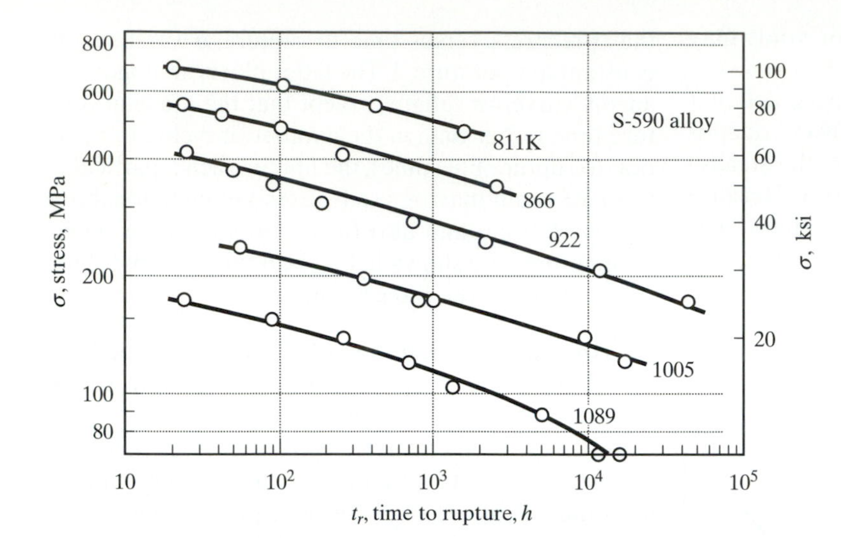

Fig. 4-55 Stress-time to rupture (life) curves for a heat-resistant alloy S-590 [9].

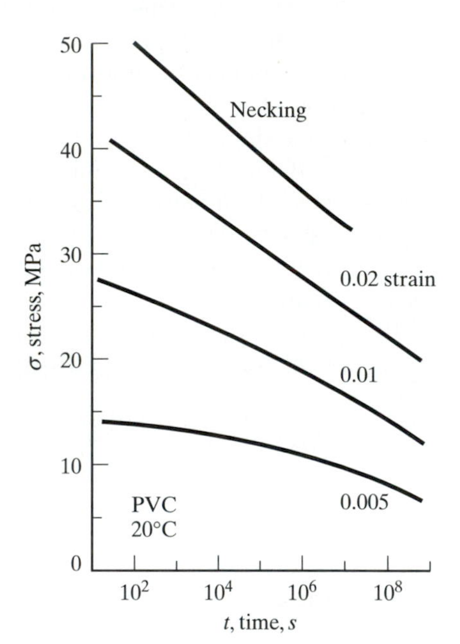

Fig. 4-56 Stress-time-to-rupture curves at three strains and onset of necking for an unplasticized polyvinyl chloride [12].

where η_τ is the shear viscosity, τ is the shear stress, and $\dot{\gamma}$ is the shear-strain rate. In tensile creep deformation, we can also define a tensile viscosity, η, as

$$\eta = \frac{\sigma}{\dot{\epsilon}} \tag{4-89}$$

where σ is the tensile stress and $\dot{\epsilon}$ is the tensile strain rate. For an incompressible ideal viscous substance, $\eta = 3\eta_\tau$.

The motion of molecules in amorphous solids (viscous creep) constitutes a diffusion process that is enhanced by increasing temperature. Thus, it is called a *thermally activated process* and as such is expected *to follow an Arrhenius type equation of the form*

$$\dot{\epsilon} = Ae^{-Q/RT} \tag{4-90}$$

where Q is called the activation energy in cal/mole, R is the gas constant ≈ 2 cal/mole/K, and T is the absolute temperature in K. Ideally, in this Arrhenius type equation, both Q and A are constants. However, in creep, both may be dependent on stress.

In *crystalline materials*, for instance, metals and their alloys and crystalline ceramics, creep occurs by either *diffusional creep* or *disclocation creep*, shown in Figs. 4-57b and c. Diffusional creep involves the motion of vacancies and this may occur primarily through the grains or along the grain boundaries. Vacancy motion through the grains is called the *Nabarro-Herring mechanism*, while that along the grain boundaries is called the *Coble mechanism*. The strain rates in these cases are given by

$$\dot{\epsilon} = \frac{A_2\sigma}{d^2T} e^{-(Q_v/RT)} \quad \text{Nabarro-Herring} \tag{4-91}$$

and

$$\dot{\epsilon} = \frac{A_2'\sigma}{d^3T} e^{-(Q_b/RT)} \quad \text{Coble} \tag{4-92}$$

where A_2 and A_2' are material constants, d is the grain diameter, Q_v is the activation energy for self-or

(a) Viscous creep

Nabarro-Herring

Coble

(b) Diffusional creep

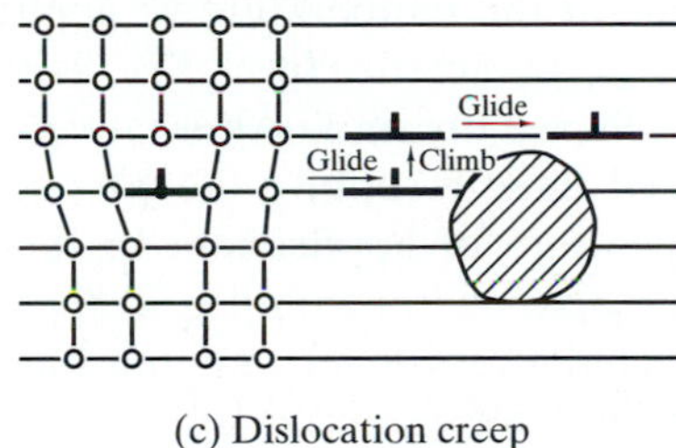

(c) Dislocation creep

Fig. 4-57 Suggested creep mechanisms. (a) Viscous creep for amorphous solids; diffusional and dislocation creeps (b and c) in crystalline solids [9].

volume-diffusion, and Q_b is the activation energy for grain boundary diffusion. We must note in the above two equations that the strain rate is inversely proportional to either d^2 or d^3. Thus, the smaller the grain diameter, the faster is the strain rate. To decrease the strain rate, large grain sizes are therefore favorable. The ultimate is to eliminate the grain boundaries and use single crystal materials at high temperatures for the least creep (strain) rate. Single crystal nickel superalloys are used for jet engine turbine blades.

Dislocation creep involves the motion of dislocations. The exact mechanism is not fully understood but it is believed that dislocation climb is an important factor. Dislocation climb means the edge of the extra plane moves up to another plane parallel to the previous plane that it was in before. This also involves diffusion of vacancies and thus the strain rate is also thermally activated and has the form

$$\dot{\epsilon} = A_3 \frac{\sigma^m}{T} e^{-(Q/RT)} \tag{4-92}$$

where the exponent m varies from one material to another, but is typically on the order of five.

4.4.12 Design Involving Creep—Time-Temperature Parameters

In engineering design involving creep, there must be neither an excessive deformation nor rupture within the desired life of the component or structure—a life could be 20 years or more. However, test data is obtained for a much shorter time than this expected life. For estimating the stress for very low strain rates at the service temperature, we may extrapolate the σ versus $\dot{\epsilon}$ curves in Fig. 4-54 to very low strain rates. Alternatively, we may extrapolate the σ versus t_r curves to very long times in Figs. 4-55 and 4-56. These extrapolations are, however, not advisable because the creep mechanisms may change.

A more successful approach is to use *time-temperature parameters* that are based on short time tests at higher temperatures than the expected service temperature. These parameters are then used to estimate the behavior for longer times at the lower service temperature. Two of these parameters, the Sherby-Dorn (S-D) and the Larson-Miller (L-M) parameters, are based on the viscous creep strain rate, Eq. 4-90, and not on the more recent accepted theories of diffusional creep, namely, the Nabarro-Herring or Coble creep. Nevertheless, both give reasonable estimates but neither one is clearly superior.

We start by rewriting Eq. 4-90 as,

$$d\epsilon = A(\sigma)e^{-(Q/RT)}dt \tag{4-94}$$

Integrating both sides of the equation and discarding the constant of integration, the steady-state creep strain is found to be

$$\epsilon_s = A(\sigma)\left[t\exp\left(-\frac{Q}{RT}\right)\right] \tag{4-95}$$

In Eq. 4-94, $A(\sigma)$ will be a constant for a particular stress. Therefore, the steady-state creep strain is a function of the quantity within the bracket. This term is designated θ, that is ,

$$\theta = t\exp\left(-\frac{Q}{RT}\right) \tag{4-96}$$

and is called the *temperature-compensated time (TCT)* parameter. Experimental data indicates that the creep strain at rupture or fracture is fairly constant for a given value of θ_r. Consequently, θ_r may be used as a design parameter.

Taking the natural logarithms (base e) of both sides of Eq. 4-96, then converting the natural logarithms to common logarithms (base 10), and using θ_r and t_r, we obtain

$$\log \theta_r = \log t_r - 0.217Q\left(\frac{1}{T}\right) \tag{4-97}$$

4.4.12a The Sherby-Dorn Parameter. This approach assumes θ_r as a function of stress only and that the activation energy, Q, is constant. The S-D parameter is

$$P_{S-D} = \log \theta_r \tag{4-98}$$

For a constant stress, $P_{S\text{-}D}$ is constant, and rearranging Eq. 4-97

$$\log t_r = P_{S-D} + 0.217\,Q\left(\frac{1}{T}\right) \tag{4-99}$$

Equation 4-99 suggests that if we plot 1/T versus log t_r for a constant stress, the plot should be linear from which we find Q from the slope and the $P_{S\text{-}D}$ parameter when $1/T = 0$, as shown for a particular stress in Fig. 4-58. Since Q is assumed constant, linear plots for different stresses exhibit the same slope but the $P_{S\text{-}D}$ varies. For various structural steels and high temperature nickel alloys, $Q \approx 90{,}000$ cal/mole. Once Q is known, the stress-time to rupture (life) data (Fig. 4-55) is used to construct a plot of $P_{S\text{-}D}$ versus stress. An example of this plot is shown in Fig. 4-59. The data from all stresses and temperatures fall along a single curve. The degree of fit or the correlation coefficient of the single curve with all the data determines the degree of success of the parameter in predicting the time-to-rupture life. Using a plot such as Fig. 4-59 and Eq. 4-99, the time to rupture (life) can be predicted for particular values of stress and temperature. For the particular stress that the material is expected to sustain at the temperature of application, we obtain the $P_{S\text{-}D}$ parameter from Fig. 4-59 and then using Eq. 4-99, we obtain t_r. Alternatively, we may select a desired time-to-rupture life for the structure at the temperature of application. In this case, we obtain the $P_{S\text{-}D}$ parameter from Eq. 4-99 and then obtain the predicted stress level that the material can sustain from Fig. 4-59.

Service lives in creep may also be determined for particular levels of creep strain or deformation. The

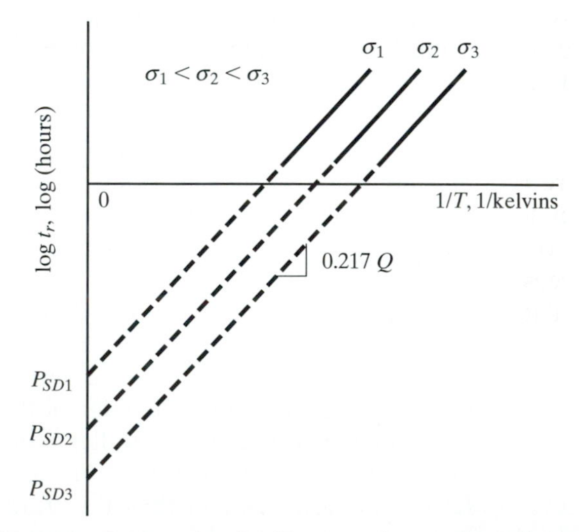

Fig. 4-58 Schematic of 1/T versus log t_r plot indicating how the Sherby-Dorn parameters for creep are obtained at various stresses. Slopes are constant because activation energy Q is assumed constant [9].

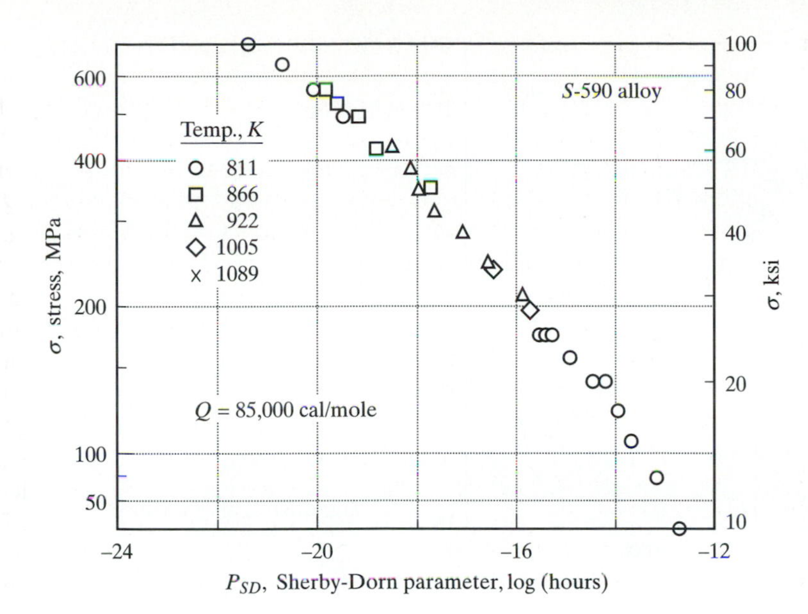

Fig. 4-59 Sherby-Dorn parameters for various creep stresses and temperature for an S-590 alloy [9].

creep strain specified (for example 1 or 2 percent) is considered the failure criterion and the service life is now called t_f, time-to-failure, rather than t_r. The $P_{S\text{-}D}$ parameter and Eq. 4-99 can also be used in this situation if t_r is replaced by t_f and if $P_{S\text{-}D}$ is determined from data of t_f rather than from t_r.

4.4.12b The Larson-Miller Parameter. In this approach, θ_r is assumed constant and does not vary with stress, but the activation energy is considered a function of stress. Rearranging Eq. 4-97 and defining the $P_{L\text{-}M}$ parameter = 0.217Q,

$$P_{L-M} = 0.217Q = T(\log t_r + C) \quad (4\text{-}100)$$

where $C = -\log \theta_r$. The unit of T in Eq. 4-100 is absolute K and t_r is in hours. However, the prevailing unit of T used in the literature for the $P_{L\text{-}M}$ is in degrees Fahrenheit, denoted by T_F. The parameter calculated using the latter is P'_{L-M} and is given as

$$P'_{L-M} = (T_F + 460)(\log t_r + C) \quad (4\text{-}101)$$

and

$$P'_{L-M} = 1.8P_{L-M} \quad (4\text{-}102)$$

For a certain stress, the activation energy is constant and therefore,

$$\log t_r = P_{L-M}\left(\frac{1}{T}\right) - C \quad (4\text{-}103)$$

and a plot of 1/T versus log t_r will yield the $P_{L\text{-}M}$ as the slope of the linear plot (which equals 0.217Q), Fig. 4-60. At different stresses, different $P_{L\text{-}M}$s are obtained but the intercept at 1/T = 0 will be a constant at −C. Values of the constant C for various steels and other structural engineering metals is about 20. Knowing C, the $P_{L\text{-}M}$ parameters, as abscissa in Fig. 4-61, are plotted against stress for the same material used in the

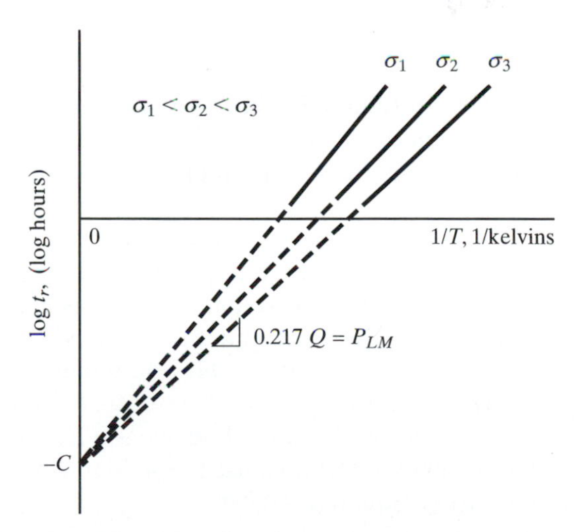

Fig. 4-60 Schematic plot of 1/T versus log t_r to determine the Larson-Miller parmeter at various creep stresses. C is held constant [9].

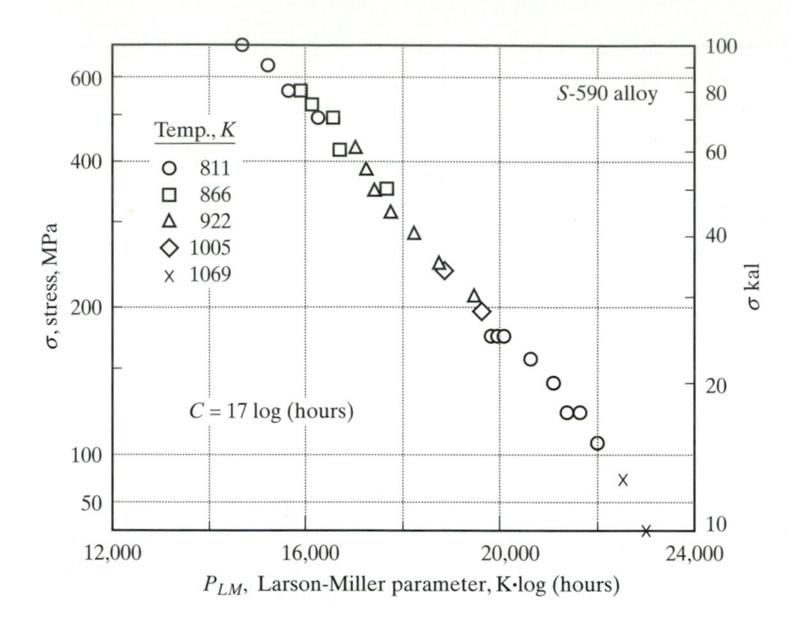

Fig. 4-61 Larson-Miller parameter for various creep stresses and temperatures for an S-590 alloy [9].

P_{SD} parameter in Fig. 4-59. Figure 4-61 is used in the design process in the same manner as Fig. 4-59 in conjunction with Eq. 4-103. In the absence of more specific information, a value of C =20 can be used to estimate creep service life. In the same manner as in the use of $P_{S\text{-}D}$, the $P_{L\text{-}M}$ may also be used to estimate service life, t_f, to a specific creep strain prior to rupture, provided of course that the t_f data is available.

EXAMPLE 4-12

It is desired to use the heat-resisting alloy S-590 for a component that is required to have a minimum service life of one-year under static stress at 500°C. Determine the maximum static stress the component must sustain.

SOLUTION: The true stress-time-to-rupture data for the S-590 is given in Fig. 4-55. It is very hard to get the true stress from this by interpolation of curves. The best estimate of the stress that must be sustained is to use the Sherby-Dorn and the Larson-Miller parameters, which are shown in Figs. 4-59 and 4-61, respectively.

(a) For the Sherby-Dorn method, we use Eq. 4-99 to get the parameter and then use Fig. 4-59 to get the maximum stress. Equation 4-99 is

$$\log t_r = P_{S-D} + 0.217Q(1/T)$$

We know the required time of one year = 365 × 24 hours, and the temperature T = 500 + 273 = 773 K. We need to know Q for the material to solve for $P_{S\text{-}D}$. We get this from Fig. 4-59 which says that Q = 85,000 cal/mole. When this is not known, we can use the average empirical data of 90,000 cal/mole given in the text. Thus,

$$P_{S-D} = \log(365 \times 24) - \frac{0.217 \times 85{,}000}{773}$$

$$= -19.92$$

From Fig. 4-59, the stress is about 500 MPa or 70 ksi.

(b) For the Larson-Miller method, we use Eq. 4-100, which is

$$P_{L-M} = T(\log t_r + C)$$

In this equation, we need to know C to get the parameter. We note again that this is given in Fig. 4-61 as C = 17 log hours. In the absence of this information, 20 may be used for C, as stated in the text. Thus,

$$P_{L-M} = 773\,(\log[365 \times 24] + 17) = 16{,}189$$

From Fig. 4-61, the stress is also found to be about 500 MPa or 70 ksi.

COMMENTS:

1. We see that there is no significant difference between the two parameters.
2. In design, we, of course, do not use the maximum stress. We need to use a safety factor with the

stress. Alternatively, we can also use a safety factor in the minimum time required. In this case, the time used in the equations will be longer and the parameters are solved to get the design stress.

4.4.13 Impact Properties and Tests

The impact property of a material is its resistance to fracture when a sudden and dynamic load is applied. The test to obtain impact properties is very simple, quick, and inexpensive. To simulate a defect in the material, a standard notch is machined in a standard specimen to be tested. For this reason, the test is also called *notch-impact testing* and the results from the test are called *notch-impact toughness properties* of the material.

The most common test is the Charpy V-notch test. A standard specimen is machined with a V-notch shown in Fig. 4-62, according to ASTM E-23. The specimen temperature is first stabilized and then the specimen is quickly placed and centered in an anvil and is broken by a swinging pendulum striker. A portion of the potential energy of the pendulum is used to break the specimen and is recorded as absorbed energy of the specimen in a dial. There are many factors controlling the accuracy of the results and because of these, testing at one temperature requires at least duplicate or triplicate results. Testing over a range of temperatures is usually done and results are schematically shown in Fig. 4-63. Most structural steels exhibit an upper-absorbed energy plateau and a lower energy plateau. These are commonly called

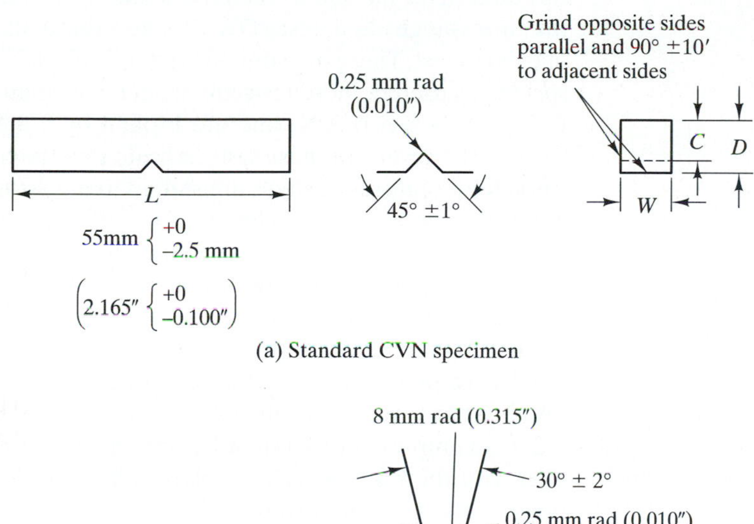

(a) Standard CVN specimen

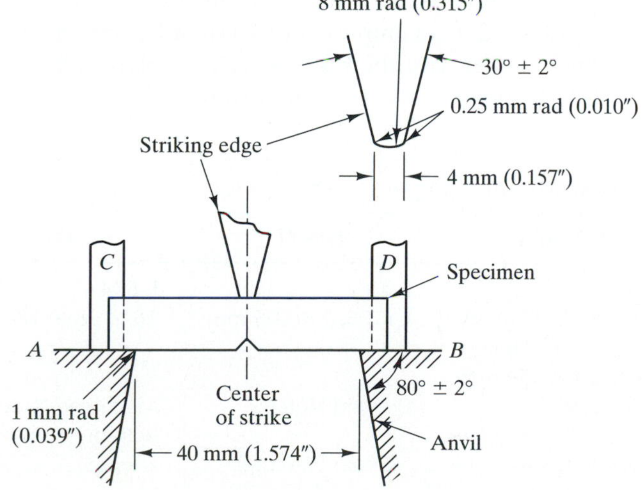

(b) CVN specimen in anvil when hit by striker

Fig. 4-62 Standard CVN impact test (a) specimen dimensions, and (b) breaking of specimen by pendulum striker.

upper-shelf energy and lower-shelf energy, respectively. The transition from the upper shelf to lower shelf for most structural (low-strength) steels occurs abruptly within a very narrow range of temperatures. Higher-strength steels (quenched and tempered) exhibit a more gradual transition. Other materials, notably FCC metals, do not exhibit this abrupt transition behavior.

At the transition temperature where the abrupt drop in absorbed energy is observed, the specimen exhibits also a change from ductile to brittle behavior. For this reason, the transition is also called the ductile to brittle transition temperature (DBTT). This temperature serves as a benchmark in the selection of materials, making sure that the lowest temperature to which the material will be exposed is well above the DBTT. For materials with gradual transition curve, the temperature at which a minimum 15 ft-lb or 20 ft-lb absorbed energy is specified. This is called the 15 ft-lb or 20 ft-lb transition temperature and was developed after the World War II ship fractures were analyzed. While the criterion was specifically for steels used in ships, it is likewise applied to many plate steel applications.

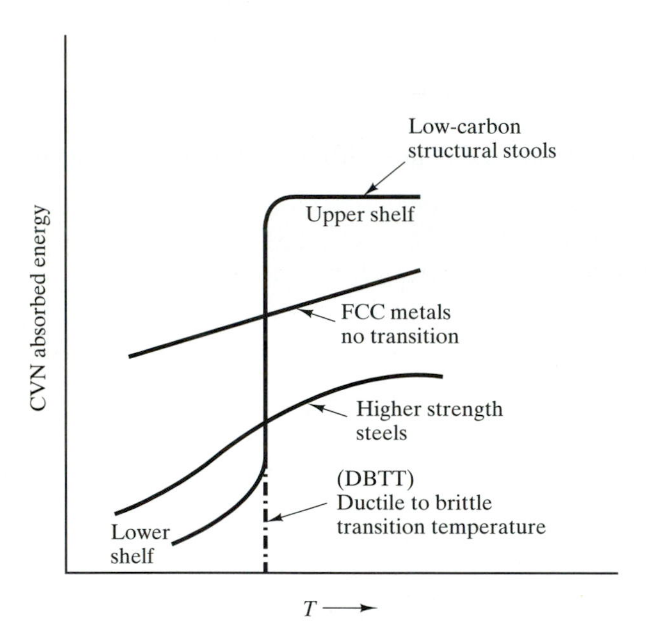

Fig. 4-63 Schematic of CVN impact test results over a range of temperatures. Low-carbon structural steels exhibit an abrupt change in absorbed energy at the ductile-to-brittle transition temperature(DBTT).

After the ship failures in World War II, the Naval Research Laboratory investigated other methods in an attempt to have a better criterion of materials performance. This has led to the nil ductility test (NDT), the drop-weight tear test (DWTT), and the dynamic tear (DT) test. These are summarized in Table 4-3. The specimens used in these tests are much larger than the Charpy V-Notch (CVN) and are broken by a falling weight. The results of these tests indicate that the transition temperatures of steels are shifted to higher temperatures for thicker specimens. Thus, the results of impact tests are used at best as a guide to selection of materials and for quality control purposes.

The initial portion of the transition behavior from the lower-shelf enegy to the upper shelf is also exhibited by the plane strain fracture toughness, K_{I_c}, values obtained at low temperatures. This observation has led to an empirical relation of K_{I_c} and the upper-shelf energy of the CVN tests. The relationship was developed at the U.S. Steel (now USX, U.S. Steel Group)

Table 4-3 ASTM standard dynamic impact toughness tests.

ACRONYM*	CVN	DWT-NDT	DWTT	DT
ASTM Standard	E-23	E-208	E-436	E-604
Speciment Size	10 × 10 mm	15.9 × 51 × 127 mm 19 × 51 × 127 mm 25.4 × 89 × 356 mm	t × 76.2 × 305 mm	16 × 38 × 181 mm
Notch Design	Machined	Brittle Weld	Pressed Notch	Machined Notch with press tip
Performance Criteria	Energy; %Shear Lateral Expansion	Break/No break	%Shear Fracture	Energy

*CVN—Charpy V-Notch; DWT-NDT—Drop-Weight-Nil Ductility Transition; DWTT—Drop Weight Tear; DT—Dynamic Tear

Applied Research Laboratory by Barsom and Rolfe [5] from data of steels with yield strengths from 110 to 246 ksi, CVN upper-shelf energies from 16 to 89 ft-lbs at room temperature, and K_{I_c} values from 87 to 246 ksi$\sqrt{in}$.. The relationship is shown in Fig. 4-64. The regression of the data leads to the following linear relationship

$$\left(\frac{K_{I_c}}{\sigma_{ys}}\right)^2 = 5\left[\frac{CVN}{\sigma_{ys}} - 0.05\right] \tag{4-104}$$

where CVN is the upper-shelf energy of the materials in ft-lbs, σ_{ys} is the 0.2 % offset yield strength in ksi, and K_{i_c} is in ksi$\sqrt{in}$. The importance of this correlation is that it provides approximate fracture toughness values to use in design from the CVN impact test results. As in any empirical correlation, it is valid only within the ranges of the values used in the correlation. Extrapolation outside the initial domain of the data must be done with caution.

EXAMPLE 4-13

A steel exhibits a 0.2% offset yield strength of 120 ksi, and a CVN absorbed upper-shelf energy of 50 ft-lbs. Estimate the K_{I_c}, fracture toughness, for the steel.

SOLUTION: We use the empirical relation Eq. 4-104,

$$\left(\frac{K_{I_c}}{\sigma_{ys}}\right)^2 = 5\left(\frac{CVN}{\sigma_{ys}} - 0.05\right)$$

$$= 5\left(\frac{50}{120} - 0.0.5\right) = 1.833$$

$$K_{I_c} = 120\sqrt{1.833} = 162 \text{ ksi} \cdot \sqrt{\text{inch}}$$

4.4.14 Hardness and Hardness Tests

In engineering applications, *hardness* is commonly defined as the *resistance to indentation*. Indentation is the pressing of a hard object with a definite shape and

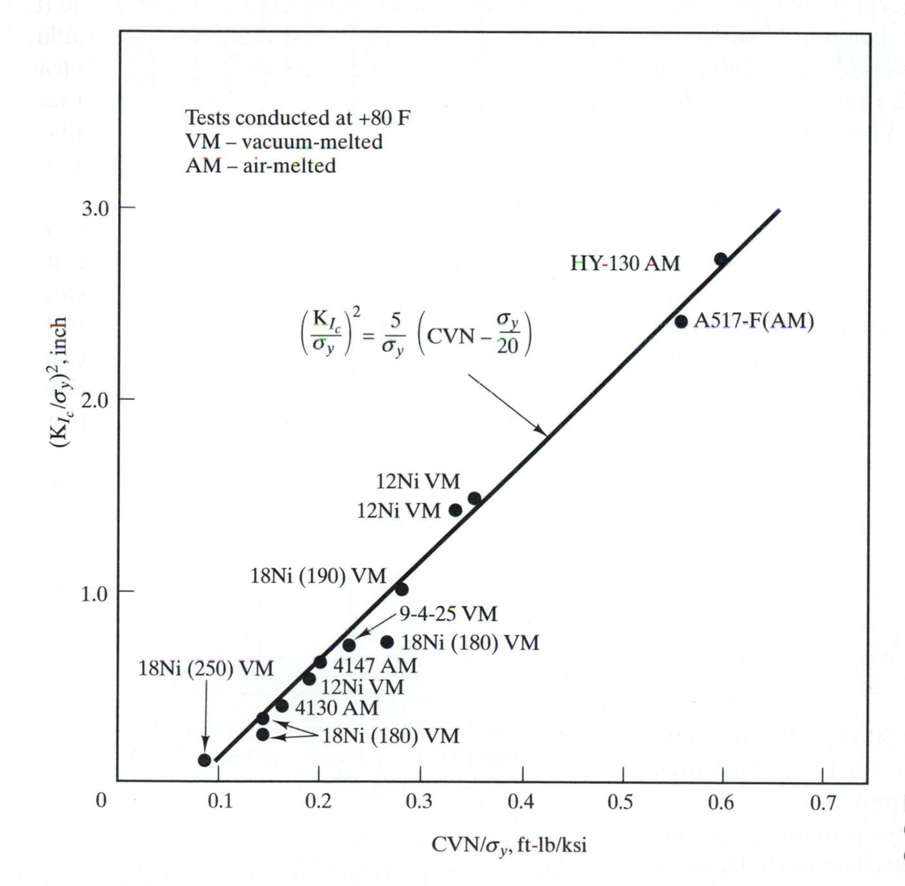

Fig. 4-64 Empirical Correlation of K_{I_c} and impact CVN upper-shelf values [5].

with a definite load against a material. The indentation plastically deforms the material and leaves a depression. A measure of hardness of the material is obtained from either the size or the depth of the depression.

The rebound of an object from the material surface is also used as a measure of hardness. This is the basis of the portable *Scleroscope hardness tester* that employs a hammer with a rounded diamond tip. The height of the rebound of the hammer is a measure of the hardness, the higher the rebound, the harder the material. The height of the rebound is calibrated so that the scale is 100 for a fully hardened tool steel. A modified version of the test is also used for polymers.

Mineralogists use the *scratch test* to determine the relative hardnesses of minerals. This test uses the *Mohs hardness scale* in which the diamond, the hardest material, is assigned a value of 10. Decreasing values are assigned to other minerals, down to one for the soft mineral, talc. In this scale, a mineral with a higher number will scratch minerals with lower numbers, but not vice versa. Decimal values between the standard minerals may be assigned as a matter of judgment. Various materials are ranked according to the Mohs scale in Fig. 4-65 as well as according to the Vickers or Brinell hardness scale, two indentation tests that will be described now.

The Brinell and the Vickers tests are indentation tests that measure the size of the depression or plastic deformation left by the indenter. The Brinell uses two loads with a spherical 10 mm hardened steel as the indenter. A 3,000 kg load is used for ferrous materials (steels and cast irons) while a 500 kg load is used for nonferrous materials, such as copper and aluminum alloys. For very hard materials, a tungsten carbide ball may be used. The Brinell hardness number, BHN, is obtained from the measured diameter of the indentation, according to the formula

$$\mathrm{BHN} = \frac{2\mathrm{P}}{\pi \mathrm{D}\left[\mathrm{D} - \sqrt{\mathrm{D}^2 - \mathrm{d}^2}\right]} \qquad \text{(4-105a)}$$

where P is the load, 3,000 kg or 500 kg, D is the diameter of the indenter, 10mm, and d is the measured diameter of the indentation, in mm.

The Vickers hardness indenter is diamond shaped as a pyramid with a square base. The angle between

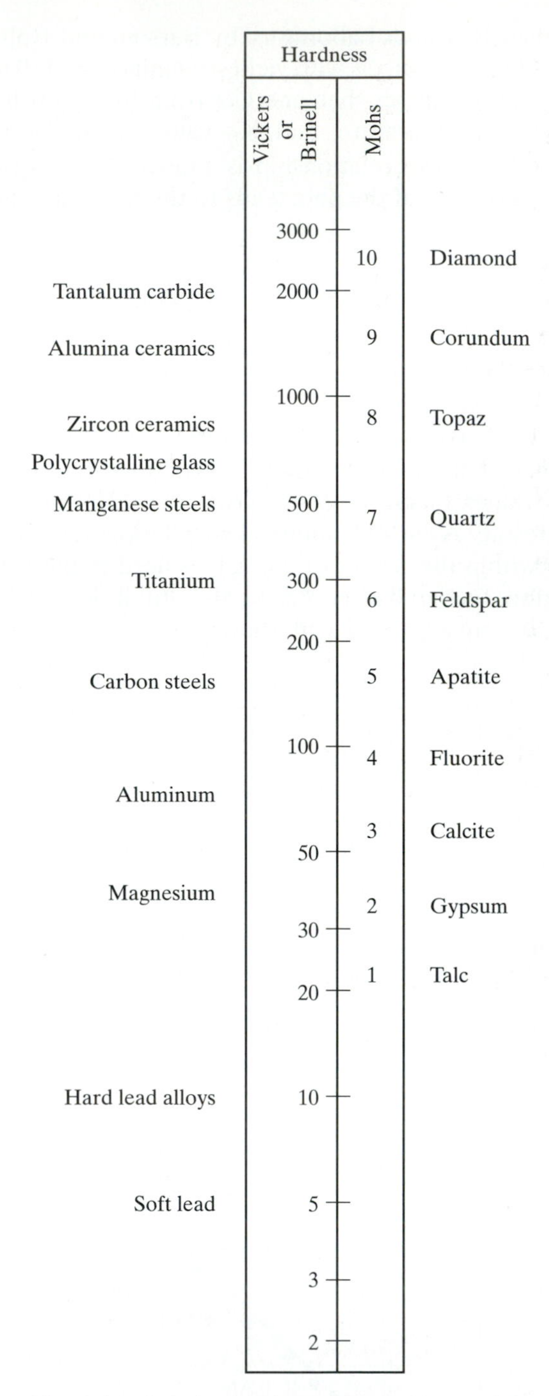

Fig. 4-65 Relative hardness of various materials and relationship of Vickers and Brinell hardness with Mohs scale [13].

the faces of the pyramid is $\alpha = 136°$. The diagonal length, *d*, of the depression left by the indenter is used to calculate the Vickers hardness number, VHN, according to

$$\text{VHN} = \frac{2P}{d^2}\sin\left(\frac{\alpha}{2}\right) \qquad (4\text{-}106)$$

where *d* is in mm, and *P* is the load used in kilogram(s). The standard pyramidal shape of diamond leaves depressions that are geometrically similar regardless of size. This geometric similarity results in a VHN that is independent of the magnitude of the load used. With a standard range of loads from 1 to 120 kg, all solid materials can be tested and given a VHN in wide-ranging scale. Within the more limited range of BHN, the VHN and the BHN are about equal as indicated in Fig. 4-65.

Another indentation test is the Rockwell hardness test that measures the depth of penetration of the indenter. For hard materials, a conical-shaped diamond "Brale" indenter is used. Steel balls ranging in size between 1.6 mm and 12.7 mm are used for softer materials. There are a number of scales used for Rockwell hardness tests that involves the combinations of the appropriate indenter and the required load. For very soft materials, a superficial scale is used. The hardness values from hardness testing are reported with the scale used. For example, 65 HRC means the material 65 hardness value in Rockwell C scale.

The hardness values from the different tests can be converted to other scales. The conversion charts for steels and other metals are given in ASTM standard E-140 and in various handbooks. In addition, an empirical relationship between the ultimate tensile strength and the BHN have been found for low- and medium-strength carbon and alloy steels. This relationship is,

$$s_{uts} = 3.45(\text{BHN}), \quad \text{in MPa} \qquad (4\text{-}106a)$$

or

$$= 0.5(\text{BHN}), \quad \text{in ksi} \qquad (4\text{-}106b)$$

EXAMPLE 4-14

A steel exhibits a 500 BHN hardness number. Using empirical relations, estimate the fatigue limit of the steel.

SOLUTION: The empirical relation of the fatigue limit of steels is Eq. 4-72,

$$\text{Fatigue limit} \approx 0.5\ s_{UTS},$$

and the empirical relation between s_{UTS} and hardness number (Eq. 4-106b) is

$$s_{UTS} = 0.5\text{BHN} = 0.5 \times 500 = 250\ \text{ksi}.$$

Thus, the rough estimate of fatigue limit $\approx$ 125 ksi. From design standpoint, we may use this as a rough starting point for the design stress. The fatigue limit indicates that the component has a 50 percent probability for failure. A design must incorporate also a safety factor on the fatigue limit. A safety factor of two yields 62.5 ksi for a first estimate of stress amplitude for fatigue.

Summary

We have two types of properties—the microstructure insensitive and the microstructure sensitive properties. The microstructure insensitive properties arise from the binding energy, the arrangement, and the packing of the atoms. In addition to the latter, the microstructure sensitive properties are also influenced by the presence of the imperfections in the crystal structure, such as solute atoms, dislocations, grain size, particles, phases, and the presence of cracks in materials.

The importance of the materials properties in design were discussed. The density and the Young's and shear moduli are important properties that determine the efficiencies and performance of materials in structures as will be discussed in Chap. 11. The melting point and the glass transition temperatures determine the application temperature of materials under load because of creep. The coefficient of linear thermal expansion determines the thermal stresses that the material experiences and is an important factor in composite materials. Thermal conductivity is important to dissipate or to absorb thermal energy.

We have learned how to determine all the microstructure sensitive properties in the laboratory and how these are incorporated into the design of structures or components. The strength, fracture toughness, fatigue, and creep properties are used in design to avoid failure, fracture, or rupture of the material in service. The strength is used for static loading while fatigue is used for dynamic cyclic loading. Design for both static and dynamic loading must incorporate the fracture toughness of materials. Creep properties involve the application of a material under load or stress at temperatures that are high relative to the material's melting, glass transition, or softening temperature.

The failure strength criterion for each type of material is different. For ductile metals and plastics, it is the yield strength. For brittle metals such as cast iron, it is the tensile strength. For ceramics, it is the compressive strength; although tensile strength is also important. Tensile strength of ceramics is variable, however and we need to use Weibull's probability statistics to determine the most probable tensile strength, as will be discussed in Chap. 14. Allowable stress design involves using only a fraction of the strength failure criterion in order to determine the size of the material to sustain the required load. A good design for failure prevention must incorporate the principles of fracture mechanics and the fracture toughness property of materials, especially for thick, high strength, and/or brittle materials.

Design for fatigue involves data from constant stress-life or constant strain-life curves. The strain-life curves approximate the fatigue performance of actual components better than the stress-life curves. There are six fatigue properties of materials from strain-life curves that can be used to predict the fatigue life to initiate a crack in a component. The predicted fatigue life is the life or number of cycles to initiate a surface crack of about 0.010-in. The entire fatigue life, which includes the number of cycles to propagate this initial crack size to the critical size, is determined in conjunction with the fracture toughness of the material.

Design with creep involves using parameters (Sherby-Dorn or Larson-Miller) that can predict the stress required for a certain rupture life or vice versa at a particular temperature. These parameters are obtained with short-time creep tests at higher temperatures than the expected service temperature of the material.

Impact and hardness test data are the easiest to obtain and they are predominantly good for quality control and for qualitative selection of materials. They are very useful because they have been empirically correlated to the fracture toughness and tensile strength of steels. The tensile strength of steels is also empirically related to the fatigue limit. These empirical relationships are very convenient because fracture toughness and fatigue tests are very expensive and time-consuming.

References

1. M. F. Ashby, "On the Engineering Properties of Materials," *Overview No. 80*, Acta Met. no. 5, p. 1273–1293, 1989.
2. M. F. Ashby and D. R. H. Jones, *Engineering Materials I*, 1st ed., Pergamon Press, 1980.
3. E. C. Bain and H. W. Paxton, *Alloying Elements in Steel*, 3d revised printing, ASM, 1966.
4. P. L. Mangonon and G. T. Thomas, *The Structure and Properties of Thermal-Mechanically Treated 304 Stainless Steel*, Met. Trans. A, 1, p. 1587, 1970.
5. J. M. Barsom and S.T. Rolfe, *Fracture & Fatigue Control in Structures*, 2d ed., Prentice Hall, 1987.
6. G. C. Shih, *Handbook of Stress Intensity Factors for Researchers and Engineers*, Institute of Fracture and Solid Mechanics, Lehigh University, Bethlehem, PA, 1973.
7. Proceedings of the SAE Fatigue Conference, P–109, SAE, Warrendale, PA, 1982.
   a. N. R. LaPointe, "Monotonic and Fatigue Characterization of Metals," p. 23.
   b. S. Reemsnyder, "Constant Amplitude Fatigue Life Assessment Methods," p.119.
8. R. E. Smallman, *Modern Physical Metallurgy*, p. 467, 4th ed., Butterworths, 1985.
9. N. E. Dowling, *Mechanical Behavior of Materials*, Prentice Hall, 1993.
10. G. E. Dieter, *Mechanical Metallurgy*, 3d ed., McGraw-Hill, 1986.

11. P. N. Randall, "Constant-Stress Creep Rupture Tests of a Killed Carbon Steel," Proc. of the ASTM, Vol. 57, pp. 854–876, 1957.
12. European vinyls Corp (UK), Ltd. (EVC), Cheshire, UK, "Engineering Design with Unplasticized PVC," 1989.
13. C. W. Richards, Engineering Materials Science, p. 402, Wadsworth, Belmont, Cal., 1961–(now PWS-Kent Publishing Co., Boston)

Terms and Concepts

Allowable design stress
Biaxial stress
Brinell test
Brittle materials
Bulk modulus
Charpy V-notch impact test
Coble creep
Coefficient of linear thermal expansion
Cold-work hardening
Compressive stress
Constant strain amplitude test
Constant stress amplitude test
Crack initiation life
Crack propagation life
Crack propagation modes
Creep
Critical crack size
Critical stress intensity factor
Crushing strength
CTOD (crack tip opening displacement) test
Cyclic strain hardening exponent
Cyclic strength coefficient
Cyclic stress-cyclic strain curve
Density
Dilatation
Dislocation locking
Dislocation multiplication
Dislocation strengthening
Dispersed phase strengthening
Ductile-brittle transition temperature
Ductile materials
Ductility
Dynamic tear test
Elastic constants
Elastic strain
Elastic deformation
Elastic limit
Elastic-plastic fracture mechanics
Elastic strain amplitude
Elastic strain recovery
Elongation, percent
Engineering strain
Engineering stress
Extensometer
Factor of safety
Fatigue constants
Fatigue limit
Fatigue limit to tensile strength relationship
Fatigue strength
Flow stress
Fracture mechanics
Fracture toughness
Fracture toughness—CVN shelf energy relationship Glass transition temperature
Grain size strengthening
Hardness test
Hooke's law
Hydrostatic stress
Instability point
J-integral test
Larson-Miller parameter
Lateral strain
Linear elastic fracture mechanics
Load cell
Lorentz constant
Mean stress
Mechanical properties
Melting point
Microstructure
Microstructure insensitive
Microstructure sensitive
Mohs scale
Nabarro-Herring creep
Necking
Nil Ductility test
Nonuniform deformation

Number of cycles
Number of reversals
Offset cyclic yield strength
Offset strain
Offset yield strength
Particle strengthening
Perfect elastic material
Perfect plastic material
Permanent strain
Phonon
Physical properties
Plane strain fracture toughness
Plastic deformation
Plastic strain
Plastic strain amplitude
Poisson's ratio
Primary creep
Proportional limit
R-curve test
Reduction in area, percent
Rockwell hardness test
Secant modulus
Secondary creep
Sharp yield point
Shear modulus
Shear-strain
Shear stress
Sherby-Dorn parameter
Solid solution strengthening
Stiffness
Strain
Strain hardening
Strain hardening exponent
Strain life curves
Strength
Strength coefficient
Stress
Stress amplitude
Stress concentration factor
Stress intensity factor
Stress-life curve
Stress range
Stress ratio
Stress reversal
Stress-rupture life
Tangent modulus
Tear strength of elastomers
Tensile strain
Tensile strength-hardness relationship
Tensile stress
Tensile test
Tertiary creep
Thermal conductivity
Total strain amplitude
True strain
True stress
Uniform deformation
Uniform true strain
Viscous creep
Yield strength
Young's Modulus (same as Modulus of Elasticity)

Questions and Exercise Problems

4.1 The cohesive or binding energies (in kcal/mole) of alkali metals are the following: Lithium (Li) = 36.5; Sodium (Na) = 26.0; Potassium (K) = 22.6; Rubidium (Rb) = 18.9; and Cesium (Cs) = 18.8. Look up the corresponding melting points of these elements and indicate a relationship of the binding energy and the melting point of a material. Give a reason why cesium has the lowest binding energy and lowest melting point.

4.2 The cohesive or binding energies of some common metals are listed in Table P4.2. Fill up the information on the crystal structure; the melting point, T_m; the modulus of elasticity, E; and the coefficient of linear thermal expansion, α.

Arrange the elements in succession according to their binding energies and their physical properties, and indicate the relationship of the binding energies and the physical properties. Now, group the elements with the same crystal structures and note the relationship of the binding energies and the physical properties. Next arrange the metals belonging to the same group in the periodic table and note their relationship.

4.3 If a design requires the material to be at 100°C, determine which of the metals in Problem 4.2 will likely creep under stress at this temperature.

4.4 A material 8 mm in diameter is pulled with a 1,000 *N* force. What is the stress? If the length of an original 50 mm mark is 50.05 mm under load, what is the strain?

Table P4.2 Binding energies and physical properties of some common metals.

Metal and Its Symbol	Binding Energy, in kcal/mole	Crystal Structure	Melting Point	Modulus of Elasticity	Coeff. linear thermal exp.
Aluminum, Al	74.4				
Beryllium, Be	76.6				
Chromium, Cr	80				
Copper, Cu	81				
Iron, Fe	97				
Magnesium, Mg	36				
Molybdenum, Mo	155				
Nickel, N_i	101				
Niobium, Nb or (Columbium, Cb)	184				
Tantalum, Ta	185				
Titanium, Ti	112				
Tungsten, W	201				
Vanadium, V	120				
Zirconium, Zr	125				

4.5 If the compressive strength of concrete is 3,000 psi, what will be the maximum load that an 8-inch diameter concrete column can sustain?

4.6 A bar, 10 mm in diameter and 100 mm long, was pulled with a 2,000 N force. The length of the bar under load is 100.01 mm. Assuming the material is loaded elastically, determine the modulus of elasticity.

4.7 A No. 0000 Browne & Sharpe (B&S) gauge, also American Wire Gauge (AWG), copper wire (1.000 cm. in diameter) has a yield strength of 210 MPa. Determine the maximum tensile force that can be imposed on the wire before permanent strain is observed. What would the length be of a 50 mm original length of wire under this load? when the load is released? The modulus of elasticity of copper is 110 GPa.

4.8 A 2024 aluminum alloy tie rod is required to carry a tensile load of 10,000 N. If the yield strength of the alloy is 207 MPa, determine the minimum cross section of the rod that is needed. Assuming the minimum cross section is used, what will be the length of an original 300 mm long tie-rod under load?

4.9 The modulus of elasticity of steel is 207 GPa and its Poisson's ratio is 0.33. Calculate its shear modulus as well as its bulk modulus.

4.10 A 10-mm diameter aluminum alloy rod is subjected to a 15,000 N tensile force. If the modulus of elasticity is 72 GPa, and the Poisson's ratio is 0.31, determine the change in diameter of the rod under load.

4.11 The coefficient of linear thermal expansion for aluminum is 23.6×10^{-6} (strain units) per °C. If a 200 mm long, 10 mm diameter aluminum rod is heated from 20°C to 100°C, what are the dimensions of the diameter and length of the aluminum rod at 100°C. Is there a stress on the rod?

4.12 If the aluminum rod in Problem 4.11 is snugly fitted, not fixed or joined, against two immovable nonconducting walls and then uniformly heated to 100°C, what would be the dimensions of the rod at 100°C? What is the stress acting on the rod? If the modulus of elasticity of aluminum at this temperature is 8×10^6 psi and the yield strength is 4,000 psi, determine if the rod will plastically deform? What will be the dimensions of the aluminum rod after it is cooled from 100°C to 20°C? Will it still be snugly fitted?

4.13 Consider a steel tie rod uniformly heated and with dimensions of 10 mm in diameter and 200 mm long at 200°C. The rod at 200°C is joined at both ends to immovable frames. (a) What is the stress on the rod when it has cooled to 20°C? (b) Determine the dimensions of the rod at 20°C. The modulus of elasticity at 20°C is 207 GPa and the Poisson's ratio is 0.33.

4.14 A copper single crystal is under a hydrostatic pressure of 103 MPa. Determine the lattice parameter of copper under this hydrostatic pressure. The modulus of elasticity of copper is 110 GPa and its Poisson's ratio is 0.33.

4.15 A shear force of 10,000 N is applied between two parallel faces of a 10-mm cuboid of copper. Determine the change in the angle of the edges (between the parallel faces) of the cube when the shear force is applied. The modulus of elasticity of copper is 110 GPa and its Poisson's ratio is 0.33.

4.16 A precisely ground 10-mm cuboid of an aluminum alloy is compressed along one axis with a 5,000 N force. Determine the new dimensions of the alloy under load. The modulus of elasticity is 72 GPa and the Poisson's ratio is 0.31.

4.17 A block of a material with initial dimensions L_o, w_o, t_o, is deformed to new dimensions L, w, and t. (a) Express the volume strain, $\ln(V/V_o)$ in terms of true strains ϵ_L, ϵ_w, and ϵ_t. (b) Assuming constant volume during deformation, what is the sum of the true strains? (c) If the constant volume deformation is done in the elastic range, what is the Poisson's ratio?

4.18 (a) A bar of initial length, L_o, is uniformly extended until its length $L = 2L_o$. What are the engineering strain and the true strain for this extension. (b) To what length, L, must a bar of initial length, L_o, be compressed if the strains are to be the same (except for sign) as those in part (a)?

4.19 A round tensile steel specimen with a reduced section diameter of 12.827 mm and a gauge length of 50.8 mm yielded the test results in Table P4.19.

From the broken test specimen, the measured gauge length = 72.898 mm and the diameter = 6.7564 mm.

(a) Plot the engineering strain-engineering stress curve.

(b) (i) Determine the modulus of elasticity.

(ii) In the elastic region, determine the Poisson's ratio and compare it with the value of 0.33 given in the literature. Rationalize your result.

(iii) Determine the 0.2% offset yield strength.

(iv) Determine the percent elongation and the percent reduction in the area.

(v) What is the recovered elastic strain after the specimen fractured?

(c) Plot the ln (true plastic) strain and the ln (true stress) curve on a log-log paper up to the maximum load.

(d) Determine the strength coefficient K and the strain-hardening exponent, n.

(e) Determine the true uniform strain and the true nonuniform strain up to fracture.

(f) Determine the recovered true elastic strain.

4.20 The tensile data obtained from a tensile specimen with original dimensions in the reduced section of 12.827 mm diameter and 50.8 mm gauge are given in Table P4.20.

After fracture the gauge length was 69.088 mm, and the final diameter was 9.195 mm.

Table P4.20 Tensile Data

Load, Newtons (N)	Gauge Length, mm
0	50.80000
4,448	50.81143
8,896	50.82286
13,345	50.83429
17,793	50.85080
26,689	51.30800
35,586	52.12080
44,482	53.64480
48,930	55.88000
40,034 (fracture)	69.850

Table P4.19 Tensile Test Data

Load, Newtons (N)	Gauge Length, mm	Load, Newtons (N)	Gauge Length, mm
0	50.800000	28,024	51.308
2,224	50.804064	31,138	52.324
4,448	50.808128	33,362	52.832
6,672	50.813208	37,810	53.848
8,896	50.818288	42,703	55.372
11,121	50.822606	44,482	57.404
13,345	50.826670	44,927	58.420
15,569	50.830988	45,372	63.500
17,793	50.835052	44,704	65.532
20,017	50.839116	42,925	66.548
22,241	50.844450	40,479	68.580
24,465	50.848514	36,031	70.104
26,689	50.851816	30,248	Break

(a) Plot the engineering strain-engineering stress curve.

(b) Determine:

(i) the modulus of elasticity.

(ii) the Poisson's ratio; rationalize your result.

(iii) the 0.2% offset yield strength.

(iv) the tensile strength.

(v) the ductility in percent elongation and percent reduction in the area.

(vi) the fracture stress.

(vii) the recovered elastic strain after fracture.

(c) Plot the ln (true plastic) strain and the ln (true stress) curve in a log-log paper up to the maximum load.

(d) Determine the strength coefficient, K, and the strain-hardening exponent, n.

(e) What is the true stress at fracture?

4.21 If the true strain-true stress curve of a material follows the relation $\sigma = 46{,}400\,\epsilon^{0.54}$, estimate the ultimate tensile strength.

4.22 In a tensile test, the material fractures at the maximum stress or load. The initial dimensions are $D_o = 20$ mm and $L_o = 50$ mm—the final dimensions are $D = 10$ mm and $L = 60$ mm. Determine the true strain ductility of the material using both the length and the diameter dimensions?

4.23 During a tensile test, the material sustains the maximum load and breaks thereafter. If the initial dimensions are $D_o = 20$ mm and $L_o = 50$ mm and the final dimensions are $D = 8$mm and $L = 75$ mm, calculate the true ductility, using both the diameter and length dimensions, separately.

4.24 The allowable design stress for a structural steel is one-half its yield strength. What is the factor of safety?

4.25 A one-meter long tie-rod is required to sustain a tensile load of 50,000 *N*. If a factor of safety of two is used, determine the design diameter of a structural steel with a 276 MPa yield strength. What would be the maximum strain anticipated for the structural steel in service? If uniform corrosion is anticipated at the rate of 0.075 mm per year, what would be the diameter if the tie-rod is designed for a 10-year service life.

4.26 If the cost per pound of a heat-treated aluminum alloy is 3.5 times the cost per pound of structural steel, determine the viability of using the aluminum alloy with a yield strength of 414 MPa with the same factor of safety in the application described in Problem 4.25. You can assume that the aluminum will corrode negligibly over the 10-year service period.

4.27 For a 2-inch thick material with $K_{I_c} = 50$ ksi $\sqrt{\text{inch}}$, $\sigma_{ys} = 100$ ksi, and a $\sigma_{des} = 40$ ksi, determine the critical-crack length in an infinitely wide and long plate for (a) an edge crack, and (b) a through-thickness crack.

4.28 For an edge crack in a 20-inch wide, 2-inch thick plate with $\sigma_{ys} = 50$ ksi and $K_{I_c} = 100$ ksi$\sqrt{\text{inch}}$, determine the critical crack size for $\sigma_{des} = \sigma_{ys}$, and $\sigma_{des} = \sigma_{ys}/2$.

4.29 A long 1-inch thick, 10-inch wide steel plate has a 2-inch deep edge crack and is loaded in tension. If the steel has a yield strength of 50 ksi and a $K_c = 200$ ksi$\sqrt{\text{inch}}$, what load can the plate sustain before failure? What is the mode of failure?

4.30 A structural steel with yield strength of 120 ksi has a CVN upper-shelf energy of 40 ft-lbs. Estimate the fracture toughness of the steel.

4.31 In constant stress amplitude fatigue testing, differentiate the methods and the test-results between the three tests: (a) push-pull, (b) rotating beam test, and (c) the bending test.

4.32 In a push-pull fatigue test, the maximum stress applied in tension is 80,000 psi and the minimum stress in compression is 20,000 psi. Determine the stress-range, the stress amplitude, the mean stress, and the stress ratio, R.

4.33 A steel has an ultimate tensile strength of 160 ksi. When $R = 0$, what is the maximum cyclic stress of a steel in a corrosion free environment to have a 50 percent probability of enduring the stress for the life of the component? With a safety factor of two, determine the diameter of the steel that has to be used to carry a fluctuating load of 20,000 lb-force.

4.34 A steel was found to have values of $\sigma_f' = 1{,}758$ MPa, and $b = -0.0977$. Estimate the fatigue limit of the steel at 10^7 cycles when the mean stress is (a) zero, (b) 175 MPa in tension, and (c) 175 MPa in compression.

4.35 Why is the strain-life test successful in predicting the initiation life of actual components?

4.36 Where is the regime of applicability of the (a) elastic strain-life curve, and (b) plastic strain-life curve?

4.37 In Fig. 4-50, determine the expression for the number of cycles, N_T, at the intersection of the plastic strain life with the elastic strain life.

4.38 For the same steel used in Example 4-11, determine (a) the total strain range and (b) the corresponding cyclic stress range for 1,000 and 10,000 cycles to initiate a fatigue crack on a smooth specimen.

4.39 Using Fig. 4-55 and stress levels of (a) 400 MPa and (b) 200 MPa, determine the activation energy, Q, in the Sherby-Dorn parameter equation.

4.40 Using the same figure and the stress levels in Problem 4.39, determine C in the Larson-Miller parameter equation.

4.41 A component that is made of S-590 alloy is to be subjected to a static stress of 400 MPa at a high temperature. Using a safety factor of two based on the stress, determine (a) the maximum temperature the component must be exposed to for it to sustain the stress for at least 1,000 days, and (b) the expected service time of the component at 600°C. Compare the answers obtained with the Sherby-Dorn and the Larson-Miller parameters.

4.42 The heat-resisting alloy S-590 is used for a component that will be exposed to 700°C. With a safety factor of two on the time, determine the maximum static stress the component must sustain in order to have a service life of 10,000 hours.

Part II

Processing

5

Equilibrium Phase Diagrams

5.1 *Introduction*

In Chapter 4, we discussed the dependence of the mechanical properties of materials on their microstructures. These microstructures are primarily the solutes in solid solution, the grain size, the dislocations, the dispersed phases, and precipitates or particles.

The question is: How do the microstructures arise and how can we obtain the optimum microstructures for the properties of the materials we desire? The response is: We obtain the desired microstructures by controlling the processing of the materials. These processings start with the melting, refining, and alloying (addition of desired composition), then the shaping (deformation) that may involve heating and cooling, then heat-treating, and finally the fabrication of the structure or component that may include forming, machining and welding. The key phrase is "controlling the processing" and for us to do this, we need to know when the solid will melt, when the liquid will solidify, whether the material is single phase or multiphase, whether it can be heat treated, and so on. We can get all this information from an *equilibrium phase diagram,* which for materials is *a plot of composition and temperature.* As its name implies, the phase diagram is a "map" of phases at particular compositions and temperatures. It serves as a guide in the selection and/or processing of materials. Using the phase diagram, we can anticipate what alloy compositions are likely to have good properties; what processes or heat treatments we must impose and control to attain the desired properties; and what treatments are likely to be harmful and to be avoided. Thus, it is important to learn the basic concepts and to know how to interpret a phase diagram.

In an introductory course on materials, the basic concepts pertaining to phase diagrams can be illustrated with a two-component binary system—three or more component systems will be left out for more advanced courses on materials. We shall look at different cases to illustrate what happens when two elements of varying characteristics are alloyed together.

5.2 Basic Concepts and Terms

We start by defining some terms. A *phase* is a distinguishable homogeneous state (either gas, liquid, or solid) of matter that manifests uniform composition and properties. For this course, we shall be concerned only with the liquid and solid states; each state may contain one or more phases. We shall consider a single homogeneous liquid phase, however, the solid state may contain one or more phases.

For a pure material, the solid state will transform to the liquid state at the melting point. If we want to melt a pure material, we need to raise its temperature past its melting point. We can represent this transformation by a vertical line with only the temperature as the variable because the composition is constant for pure materials. Figure 5-1 shows two vertical lines representing two pure materials, A and B. In each of these lines, the melting point of the pure material is indicated. *Each vertical line is a "phase diagram" of the pure material* because it shows the material is in the solid state (S) below its melting point and in the liquid state (L) above its melting point.

When we mix A and B to produce an alloy, another variable arises. This variable is the composition and is represented by the horizontal line connecting the two vertical lines in Fig. 5-1 to form the skeletal frame, shown in Fig. 5-2, of a binary (two-component) diagram. The two vertical lines retain their identities as those of pure (100 percent) A and pure (100 percent) B. Composition in solids and liquids is usually expressed in weight percent (abbreviated as wt%) and the horizontal line is scaled from 0 to 100%B from left to right or 0 to 100%A, from right to left. The space above the horizontal and between the two verticals will show areas or fields indicating the different phases for varied compositions of the mixtures (alloys) of A and B and at different temperatures.

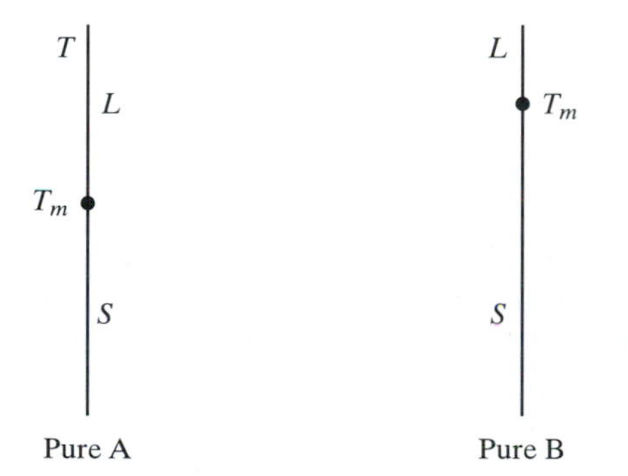

Fig. 5-1 "Phase diagrams" of pure components consist of vertical lines with temperature as the only variable.

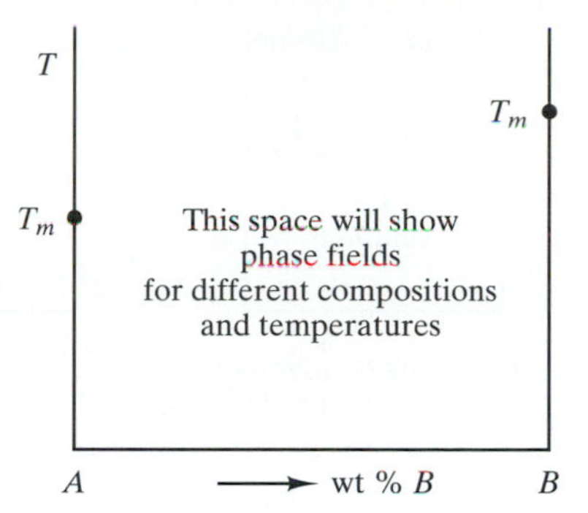

Fig. 5-2 Frame of binary phase diagram when two elements are alloyed.

To appreciate the different situations that may arise when two solid elements A and B are mixed (alloyed), we look at the similar situations in the liquid state. In the liquid state, the liquid with the larger concentration is called the *solvent* and the added liquid or solid is called the *solute*. For example, water is the solvent and salt is the solute in the ocean. In solids, the solute retains its meaning but the solvent is called the *matrix*. Thus, in yellow brass, which has a composition of 70%Cu and 30%Zn, copper is the matrix and zinc is the solute. The characteristics of the matrix and the solute determine whether: (1) they form a single continuous solid solution throughout the entire range of composition; (2) the solute is only partially soluble in the matrix and the matrix is saturated with the solute; (3) they may form a compound; (4) the solute is insoluble in the matrix. The examples of these situations in the liquid state are shown in Fig. 5-3 as (a) the complete solubility of alcohol and water; (b) the partial solubility of salt at room temperature; the water is saturated with salt and the excess salt forms a second phase; (c) the precipitation of a compound; and (d) the insolubility of oil and water forming two liquid phases.

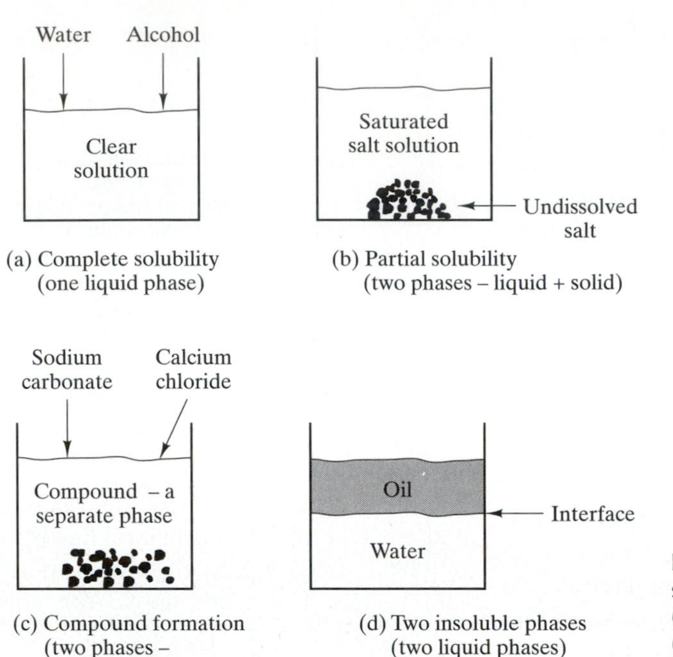

Fig. 5-3 Cases in the liquid state: (a) complete solubility, (b) partial solubility, (c) compound formation, (d) insolubility of phases.

The binary phase diagram of two pure materials is determined experimentally by studying the structures formed at various compositions and temperatures. These studies reveal areas (phase fields) in the phase diagram that may contain a single solid phase (a solid solution), a mixture of two or more phases, stoichiometric or nonstoichiometric compounds, the transformation temperatures, and the compositions of stable equilibrium phases. Thus, a *phase diagram* is a portrayal of *stable equilibrium phases* that are found in alloys of the two elements. By stable phases, we mean thermodynamic stability, and because thermodynamics refers to equilibrium conditions, the phase diagram is an *equilibrium diagram.* It is also sometimes referred to as a *constitutional diagram* because it shows the obvious—the phases and their compositions.

We consider now examples of binary phase diagrams exhibiting the equivalent situations of the different cases shown in Fig. 5-3 in the liquid state. With these diagrams, we introduce more basic terms and concepts. We shall also learn how the initial microstructures are developed when the homogeneous liquid phase transforms to the solid state.

5.3 Complete Solubility in the Solid State

The simplest binary (two-component) system is where the characteristics of the two elements are so similar that they form *solid solutions at all compositions.* Just as in an alcohol-water liquid solution in Fig. 5-3a, a solid solution is one in which the solute is completely compatible with the matrix and therefore no other solid phase appears. The conditions for the formation of complete solid solutions between two materials or elements were experimentally determined by Hume-Rothery and consequently are called the *Hume-Rothery rules.* These are:

1. The atomic radii of the two elements must differ by less than 15 percent. The closer the sizes of the atoms, the more favorable is the formation of solid solutions.
2. The electrochemical nature of the elements must be similar. If one is electropositive and the other is electronegative, compounds will form rather than solid solutions. Compounds form a second

solid phase, as in the equivalent liquid state illustrated in Fig. 5-3c.

3. Other things being equal, a metal of lower valency is more likely to dissolve a higher valency metal than vice versa. This rule was specifically found to be true for alloys of copper, silver, and gold with metals of higher valency.
4. The crystal structures are the same.

Going back to Fig. 5-2, the complete miscibility (100 percent solubility) in the solid state as well as in the liquid state is depicted by connecting the melting points of A and B with two curves, one lower and the other upper, as shown in Fig. 5-4. These curves reflect the behavior of any composition of the alloys of A and B during the transformation from liquid to solid or vice versa. Unlike the pure materials, the alloys do not transform completely from a solid to liquid at one temperature. Rather, during heating, as exemplified by the composition C_o in Fig. 5-4, the solid will start melting at T_S and will complete melting at T_L—A different composition will have a different T_S and T_L. For the different compositions, the lower curve is obtained by connecting every T_S and the upper curve by connecting every T_L. Note that these curves separate the phase diagram in Fig. 5-4 into three *phase fields* (areas): the area above the upper curve is one homogeneous liquid phase; the area below the lower curve is one homogeneous solid phase; and the area between the two curves contains solid and liquid phases. Thus, the curves are boundaries for 100 percent liquid or 100 percent solid. For these reasons, the upper curve is called the *liquidus curve* and the lower curve is called the *solidus curve.*

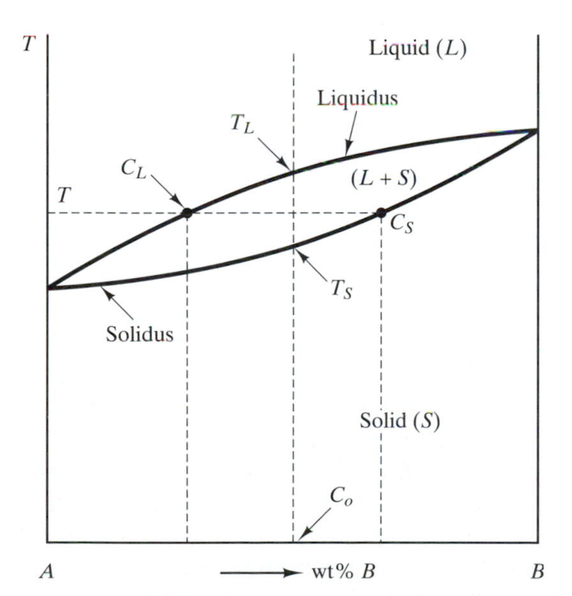

Fig. 5-4 Schematic binary phase diagram showing complete solubility of the two elements in both the solid and liquid states.

Citing the composition, C_o, in Fig. 5-4 as an example, we see that the alloy is 100 percent solid with composition C_o below T_S, and is 100 percent liquid with composition C_o above T_L. In the range of temperature, T_L–T_S, the solid and the liquid phases co-exist and are in equilibrium. The coexistence is brought about by the partial conversion of either the solid to liquid during heating or the liquid to solid during cooling. The composition of B in each of the phases changes as the temperature changes within that range. This change is necessitated by the thermodynamic equilibrium condition that the "chemical potential" of B in the solid phase must equal its "chemical potential" in the liquid phase. There is thus a relationship or a tie between the solid and the liquid phases in this two-phase field; the tie-line is the temperature. The composition of the liquid phase at any temperature, T, in the range, T_L–T_S, is the intersection of the tie-line with the liquidus curve at C_L. The corresponding composition of the solid at that same temperature in equilibrium with the liquid is the intersection of the same tie-line with the solidus curve, at C_S. In other words, if we take the tie-line as the horizontal temperature line within the two-phase region, the ends of this tie-line represent the compositions of the respective phases touching it.

We shall now look at the changes in compositions of the liquid and solid phases in the melting range, T_L–T_S, under equilibrium conditions. The alloy with composition, C_o, will start melting at T_S, see Fig. 5-5. The composition of the first liquid to form will be C_{LS}, while the composition of the solid is still essentially at C_o. As the temperature increases from T_S to T_L, the compositions of the liquid and solid phases change along the liquidus and solidus curves, respectively. At temperature T_L, the last trace of the solid will have a composition, C_{SL}, and the liquid will possess, essentially, the C_o composition. When the alloy is cooled from the liquid phase, solidification starts at T_L and the first solid nucleus has a composition of C_{SL}. The last trace of liquid will have the composition, C_{LS}.

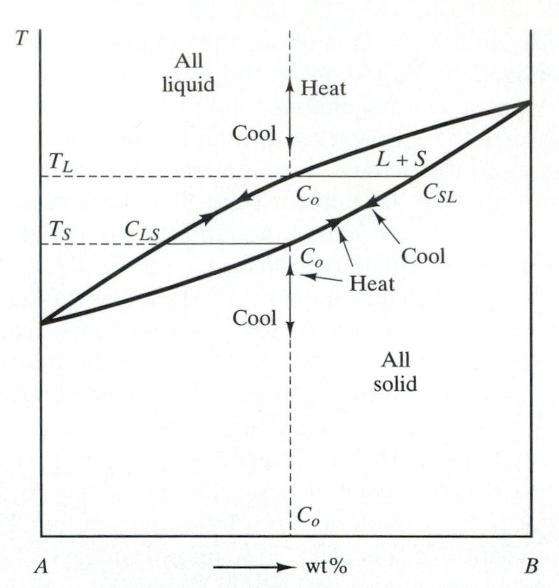

Fig. 5-5 During heating and cooling in the two-phase region, the liquid and the solid phases are in equilibrium and their compositions follow the liquidus and solidus.

Examples of binary systems exhibiting complete solubility in the solid state can be found in a compilation—*Binary Alloy Phase Diagrams,* published by ASM International. Some of these systems are Ag-Au, Bi-Sb, Cr-V, Cu-Ni, Ti-Hf, Hf-Zr, Mo-Ta, and Nb-V. Examination of these systems reveal that the two elements or components in each of the systems have the same crystal structure. Therefore, these systems are sometimes referred to as isomorphous. The Bi-Sb binary diagram is shown in Fig. 5-6. There are also phase diagrams for ceramic materials and these are found in another compilation—*Phase Diagrams for Ceramists.* Some ceramic phase diagrams are included and discussed in Chap. 14.

5.4 The Gibbs Phase Rule

Before we proceed further, it is important to consider two concepts that are universally applicable to any phase diagram. These are the *Gibbs Phase Rule* and the *Lever Rule.* This section describes the phase rule and the lever rule will follow.

The phase rule was enunciated by Willard Gibbs and deals with the number of variables that need to be specified to fix the thermodynamic state of a system. The common variables are pressure, temperature, and composition. Additional variables influencing the system may be due to electrical, magnetic, gravitational, elastic forces, and surface effects. In the study of phase diagrams, however, we shall consider only pressure, temperature, and composition. For a system in equilibrium, the number of independent variables, V, or degrees of freedom, to fix the thermodynamic state is given by the following *Gibbs Phase Rule,*

$$V = C - P + 2 \tag{5-1}$$

where C is the number of components, P is the number of phases, and 2 stands for the variables pressure and temperature. Liquids and solids, which are the phases we are concerned with, are incompressible at ordinary pressures and, therefore, pressure is an insignificant factor in their systems. Thus, the phase rule reduces to

$$V = C - P + 1 \tag{5-2}$$

for liquids and solids, where 1 stands for the temperature variable.

Let us apply Eq. 5-2 to specific cases. *For a pure component, $C = 1$, and $V = 2 - P$.* When this component is in a single-phase state, either a solid or a liquid, $P = 1$, and we find $V = 1$. There is one variable that has to be specified to define the thermodynamic state of the system. Since we know the composition, namely, pure component, the only other variable is temperature that we have to specify for either the solid or liquid phase. In relation to Fig. 5-1, the solid phase exists in the range of temperature on the vertical line, below the melting point of either A or B. We can vary the temperature in this range and the material will still be solid. Thus, we must specify the temperature to fix the thermodynamic state. If there are two phases in the system, such as during partial melting, then $P = 2$ and $V = 0$. This state is what is referred to as an *invariant (zero variable) state.* The only temperature where a pure component has solid and liquid phases in equilibrium simultaneously is at the melting point—a constant for all crystalline materials!

For binary alloys, $C = 2$ and $V = 3 - P$. In a single-phase state, below the solidus curve or above the liquidus curve in Fig. 5-4, $V = 2$, and means that we must specify both the composition and temperature to indicate the thermodynamic state in either the single-phase solid solution or liquid-phase field. If two phases exist as in the temperature range T_L–T_S in Fig. 5-4, then $V = 1$. This means that we need only to specify

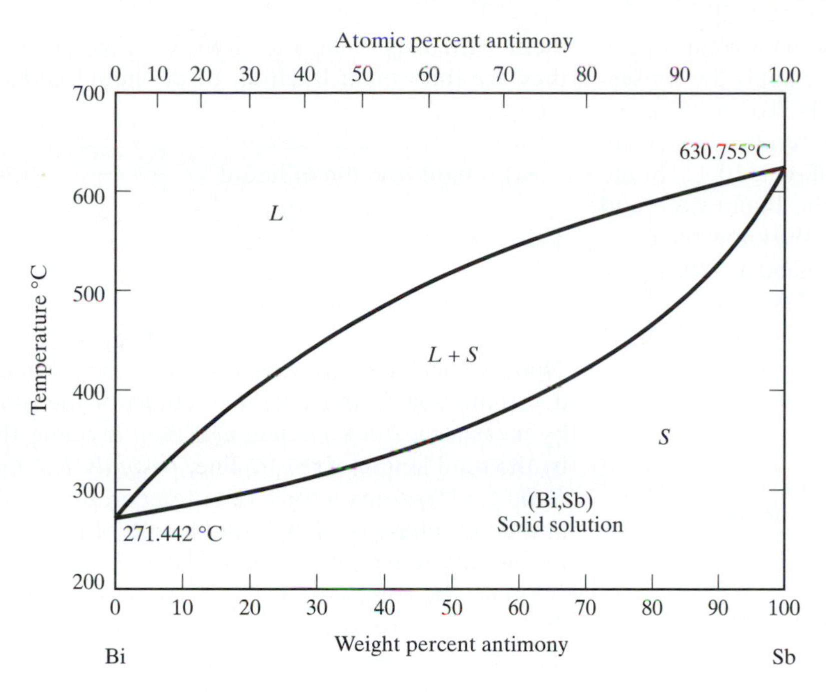

Fig. 5-6 Bismuth (Bi) and antimony (Sb) are completely soluble in both the liquid and solid states[Ref. 1].

one variable—either the composition or the temperature to indicate the thermodynamic state. The change of one variable will affect the other. For example, if we change the temperature in the two-phase state, the compositions and proportions of the two coexisting phases change in a fixed way. This was explained earlier as due to the need to have equal "chemical potential" in both phases. Thus, this two-phase state can exist over a range of temperatures and the variations of the compositions of the coexisting phases were described earlier, in relation to Fig. 5-4. The proportions of the coexisting phases will be determined by the Lever Rule, which we shall cover next. *If three phases coexist in a binary alloy, then $V = 0$.* This happens only at a fix composition and a fix temperature, that is, *a fix point in the phase field.* Such a point represents an *invariant three-phase reaction* and examples of this reaction are discussed in Sec. 5.6.

5.5 Lever Rule

As indicated in the previous section, there is only one independent variable in the two-phase field—temperature and composition are interdependent. If we change the temperature, the compositions of the phases in equilibrium change as well as the proportions or the amounts of each phase. The proportions or the amounts of each phase can be obtained by the *Lever Rule,* which is the result of a material balance of the components.

The material balance is an expression of the conservation of mass. As indicated in Fig. 5-7, an alloy of composition C_o when brought to the temperature T, between T_S and T_L, will produce a liquid phase with

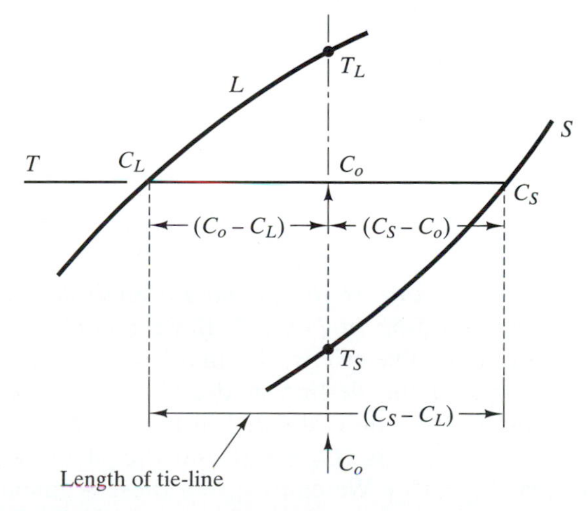

Fig. 5-7 Schematic illustration of the Lever Rule—the geometric representation of mass balance.

composition C_L and a solid phase with composition C_S. The Lever Rule will result if we apply the conservation of mass to component B. To do this, we assume 100 mass units (grams, pounds, etc.) of the alloy. The mass units of B in the alloy is $100C_o$. Some of these B units will end up in the liquid state and the rest will end in the solid state. We know the compositions of the phases and we need to know the amounts or relative proportions of the solid and the liquid phases. This can be calculated in the following manner:

Total mass balance: We assume a total combined 100 mass units of both A and B.
Let

$$x = \text{the mass units of the liquid}$$

then

$$(100 - x) = \text{the mass units of solid}$$

Component B Balance:

$$100C_o = xC_L \quad + (100 - x)C_s \qquad (5\text{-}3)$$

$$\begin{matrix}\text{total B units} \\ \text{in alloy}\end{matrix} = \begin{matrix}\text{B units} \\ \text{in liquid}\end{matrix} + \begin{matrix}\text{B units} \\ \text{in solid}\end{matrix}$$

Expanding the equation, and solving for x, we get

$$x = \frac{C_S - C_o}{C_S - C_L} \cdot 100, \quad \text{mass units of liquid} \qquad (5\text{-}4)$$

Since 100 mass units were originally assumed, $x/100$ = fraction of original mass that is liquid. Thus,

mass or weight fraction of liquid

$$= \frac{x}{100} = \frac{(C_S - C_o)}{(C_S - C_L)} \qquad (5\text{-}4a)$$

and the

weight fraction of solid

$$= 1 - \frac{(C_S - C_o)}{(C_S - C_L)} = \frac{(C_o - C_L)}{(C_S - C_L)} \qquad (5\text{-}4b)$$

The Lever Rule is the geometric equivalent of Eqs. 5-4a and 5-5b in Fig. 5-7. If we consider the temperature tie-line at T, we see that $(C_S - C_L)$ is the total length of the tie-line in the two-phase field. The length of the tie is divided in two segments at C_o. One segment is $(C_S - C_o)$ and the other segment is $(C_o - C_L)$. We can express these segments as fractions of the total tie length and by comparing them with Eqs. 5-4a and 5-4b, we recognize that they are the weight fractions of the liquid and solid phases,

$$\text{(a) weight fraction of liquid} = \frac{C_S - C_o}{C_S - C_L} \qquad (5\text{-}5a)$$

and

$$\text{(b) weight fraction of solid} = \frac{C_o - C_L}{C_S - C_L} \qquad (5\text{-}5b)$$

Thus, we see that the Lever Rule is the geometric determination of the weight fractions of the phases by measuring the segments and then dividing them by the total length of the tie-line. *To apply The Lever Rule:* (1) Draw the appropriate temperature tie-line in the two-phase field; the total length of the tie-line is the difference of the compositions at its ends; (2) mark the composition, C_o, of the alloy on the tie-line, so as to produce two segments and then determine the lengths (absolute value) of these segments by subtracting C_o from each of the compositions at the end of the tie-line; (3) the weight fraction of a phase in a two-phase field is the length of the segment of the tie-line farthest from the phase boundary divided by the total length of the tie-line.

EXAMPLE 5-1

Consider a Bi-50 wt% Sb alloy and determine the following:

(a) the liquidus and solidus temperatures of the alloy
(b) the composition of the last liquid to solidify
(c) the composition of the last solid to melt
(d) the compositions of the solid and liquid phases at 450°C
(e) the amounts of the solid and liquid phases at 450°C.

SOLUTION: The answers are explained in conjunction with Fig. 5-8. We first draw a vertical line at the original composition, $C_o = 50\%$.

(a) The liquidus temperature, T_L, is the intersection of C_o-vertical with the liquidus curve, and the solidus temperature, T_S, is the intersection with the solidus curve. These are 520°C and 360°C.

(b) The composition of the last liquid to freeze is the intersection of the 360°C tie-line and the liquidus curve that is 12%Sb.

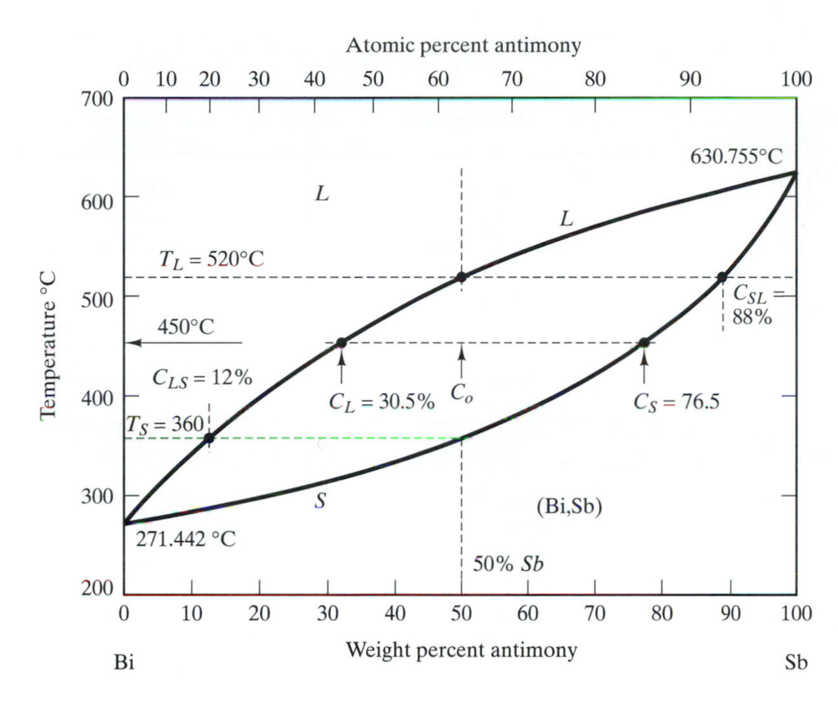

Fig. 5-8 Bi-Sb diagram accompanying Example 5-1.

(c) the composition of the last solid to melt is the intersection of the 520°C with the solidus curve that is 88%Sb.

(d) The compositions of the solid and liquid phases are at the intersections of the 450°C tie-line with the solidus and liquidus curves, which are 76.5% Sb and 30.5% Sb.

(e) the amounts of the phases are obtained using the Lever Rule. These are

$$\text{solid} = \frac{50 - 30.5}{76.5 - 30.5} \cdot 100 = 42.4\%$$

$$\text{liquid} = \frac{76.5 - 50}{76.5 - 30.5} \cdot 100 = 57.6\%$$

5.6 Three-Phase Reactions

When we applied the Gibbs Phase Rule for solids and liquids, Eq. 5-2, to a binary system (C = 2) with three phases (P = 3), we saw V = 0. This means we are left with no variables to specify and we can vary neither the temperature nor the composition. The coexistence of three phases arises when either a single phase transforms to two phases, or two phases combine to transform to a single phase. The process of transformation is from a high temperature to a lower temperature, such as, during cooling. When the transformation is not complete, all three phases coexist and all are in equilibrium. Because they are in equilibrium, the transformation is also called a three-phase reaction.

The distinguishing feature of the presence of a three-phase reaction in a phase diagram is a horizontal line. On cooling, *the three-phase reaction occurs at a fixed invariant point on this horizontal line that is also the temperature tie-line.* The point also gives the composition of the single phase, while the ends of the tie-line give the corresponding compositions of the other two phases. These are schematically illustrated in Fig. 5-9a for a single phase transforming to two phases—$P_1 \rightleftarrows P_2 + P_3$, and in Fig. 5-9b for two phases transforming to a single phase—$P_2 + P_3 \rightleftarrows P_1$. In both cases, the temperature tie-line is the horizontal line, and C_R is the composition of the single phase, P_1, involved in the reaction and is also where the three-phase reaction occurs. C_2 and C_3 at the ends of the tie-line are the compositions of P_2 and P_3.

Table 5-1 indicates the names and the reactions for the most common three-phase reactions in a binary phase diagram. In one column are the reactions involving the liquid phase, and in the other column are reactions involving all solids. The left column

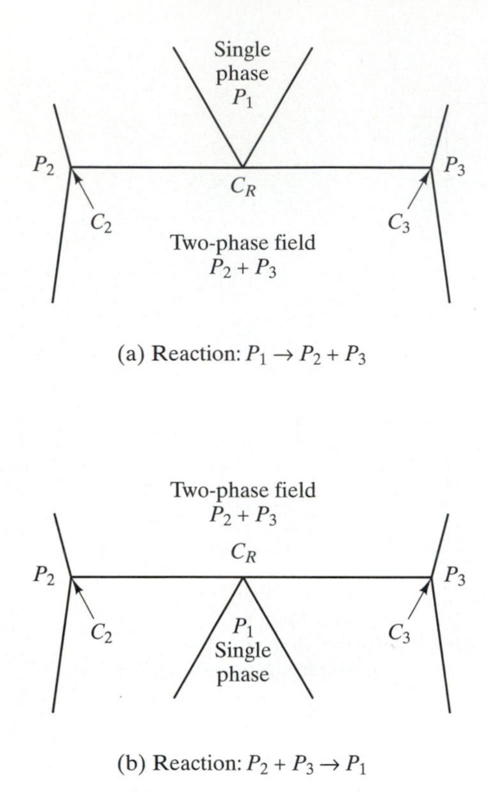

(a) Reaction: $P_1 \rightarrow P_2 + P_3$

(b) Reaction: $P_2 + P_3 \rightarrow P_1$

Fig. 5-9 Types of three-phase reactions.

Table 5-1 Three-Phase Reactions.

With liquid	No liquid-All solids
Eutectic (S_1, L, S_2) $L \rightarrow S_1 + S_2$	*Eutectoid* (S_1, S_1, S_3) $S_1 \rightarrow S_2 + S_3$
Peritectic (S_1, L, S_2) $L + S_1 \rightarrow S_2$	*Peritectoid* (S_1, S_2, S_3) $S_1 + S_2 \rightarrow S_3$
Monotectic (S_1, L_1, L_2) $L_1 \rightarrow S_1 + L_2$	*Monotectoid* (S_2, S_1, S_3) $S_1 \rightarrow S_2 + S_3$

reactions with liquid(s) have names ending with *"ic"* and those on the right with all solids have names ending with *"oid"*. The two most significant reactions in Table 5-1 are the eutectic and eutectoid reactions discussed below. Both reactions involve a single phase transforming to two phases. In the eutect*ic,* the single phase is a liquid transforming to two solid phases, while in the eutect*oid,* the single phase is a solid transforming to two other solid phases.

5.7 The Eutectic Systems

In addition to the characteristics of the three-phase reaction, the eutectic transformation is characterized by an alloy composition that has a freezing or melting temperature lower than either the melting point of the two pure components forming the alloy. This allows an alloy to be made that will have a much lower melting point and improved castability. A solder, an alloy of lead and tin, is a practical material that exhibits the characteristics of a eutectic system. Aluminum with silicon forms a eutectic and is used for casting automotive parts. Using eutectic systems, we shall illustrate complete insolubility, partial solubility, and learn more terms as well regarding phase diagrams.

5.7.1 Complete Insolubility in the Solid State

An example of a system with eutectic transformation is the Bi (bismuth)-Cd (cadmium) system shown in Fig. 5-10. The horizontal tie-line is 146°C and the reaction (three phases coexist) occurs at 60 wt.%Bi. At this point, (composition, 60 wt%Bi, and temperature, 146°C), the liquid with 60% Bi transforms to two solid phases, L $\Rightarrow$ (Cd) + (Bi). The invariance of the point makes the 60 wt%Bi liquid behave like a pure component because the temperature remains constant as long as all three phases are present. Because all three phases are in equilibrium, (Cd) and (Bi) are tied by the 146°C temperature tie-line (as in Fig.5-4). We see that the temperature tie-line is extended across the diagram to the vertical lines, which are the compositions of pure Cd and Bi. Thus, this system indicates that neither Bi nor Cd has a capacity to dissolve the other in the solid, although

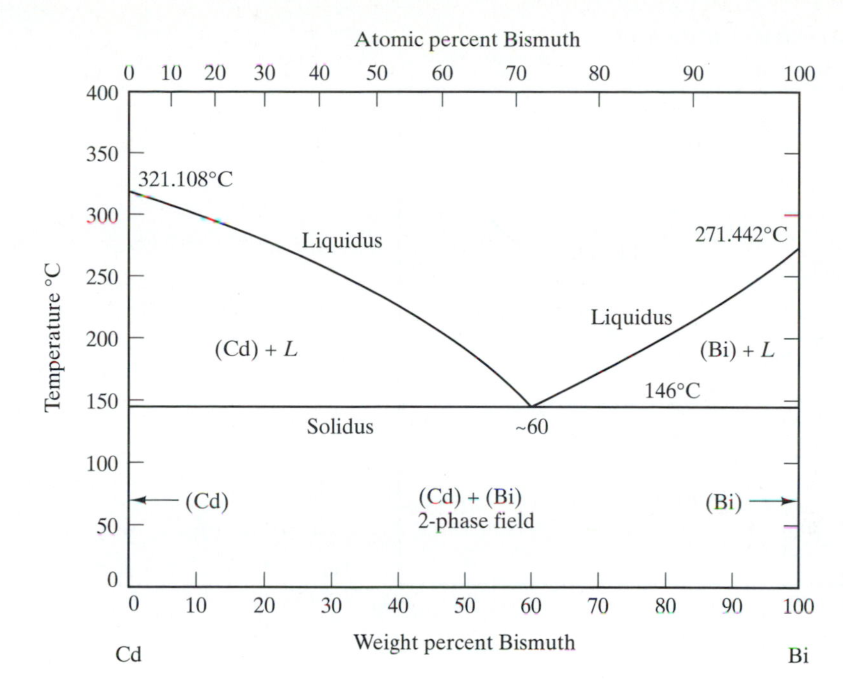

Fig. 5-10 Bismuth (Bi)-Cadmium (Cd) Alloys illustrate complete insolubility in the solid state[Ref. 1].

they are completely soluble in the liquid. *The Bi-Cd system is an example of complete insolubility in the solid state* of two elements, in the same way as oil and water are completely immiscible or insoluble in each other in the liquid state.

5.7.2 Partial Solubility in One Component

The Bi-Sn (tin) eutectic system, shown in Fig. 5-11, indicates that the eutectic temperature tie-line is 139°C and the eutectic reaction occurs at 57% Bi. The compositions at the end of the tie-line are 21% Bi and 99.9% Bi. This means that the solid phase (Sn) at 139°C contains 21% Bi and the other solid is essentially Bi (99.9%). *The Bi-Sn is an example showing one element with partial solubility in the other.* This case illustrates that Bi is partially soluble in Sn, while Sn is negligibly soluble in Bi at 139°C.

5.7.3 Partial Solubilities in Both Components

The last example of a eutectic system is the Pb (lead)-Sn shown in Fig. 5-12. *The Pb-Sn system is an example showing two elements that are partially soluble in each other.* In this case, (Pb) contains 19% Sn, while (Sn) contains 2.5% Pb at the eutectic temperature that is at 183°C. The composition for the eutectic reaction is indicated to be 61.9% Sn.

For the partially soluble systems, Figs. 5-11 and 5-12, the solubility of one element in another is generally maximum at the eutectic temperature. As temperature decreases, the solubility of an element decreases. The limit of solubility of one element (solute) in another (matrix) is depicted in the phase diagram as a *solvus curve.* The curves below the eutectic temperature tie-line, indicated in Figs. 5-10 and 5-11, are the solvus curves. *The decrease in solubility as the temperature decreases is the basis of precipitation hardening systems,* which will be covered in Chap. 9.

In Fig. 5-12, the solubility of Sn in Pb decreases from 19% to less than 1%, while the solubility of Pb in Sn decreases from 2.5% to virtually 0%, when the temperature decreases from 183°C to 0°C. Thus, at any temperature below the eutectic, the temperature tie-line ends at these solvus curves indicating that the compositions of the two solid phases, (Pb) and (Sn), are in equilibrium at that temperature. The symbols (Pb) and (Sn) indicate solid solutions with lead and tin as the matrices. The relative amounts of these phases for a given composition at a temperature below the eutectic are also calculated by the lever rule.

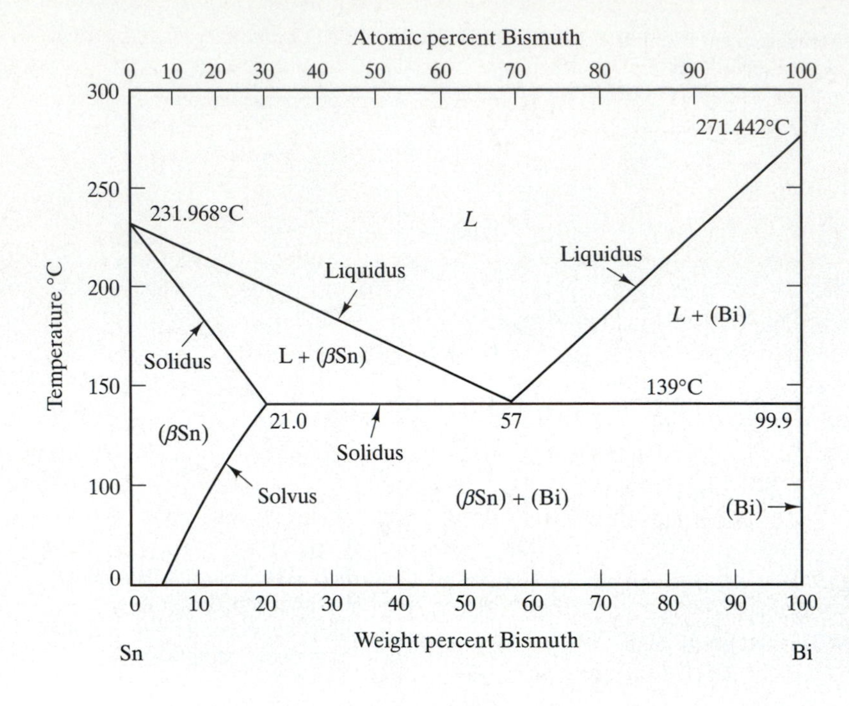

Fig. 5-11 Bismuth is partially soluble in tin (Sn) but tin is essentially insoluble in bismuth [Ref. 1].

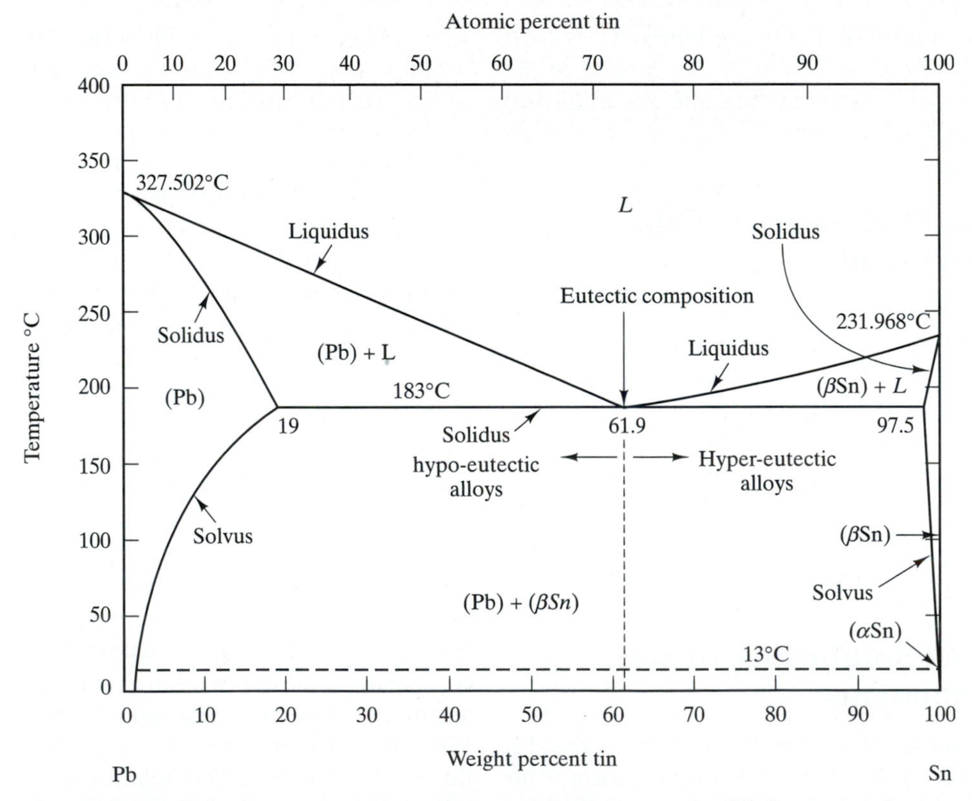

Fig. 5-12 The lead (Pb)-tin (Sn) binary diagram exhibiting partial solubilities one element in the other element.

The *eutectic tie-line is also a part of the solidus.* The other parts of the solidus are the lines connecting the end points of the tie-line with the melting points of the components. The phase fields bounded by the solvus, the solidus (from the eutectic to the melting temperature), and the vertical lines are single-phase solid solution fields, marked (Pb) and (Sn) in Fig. 5-11. These are called *primary solid solutions* because they are with the pure components. The field below the eutectic tie-line and bounded by the solvus curves is a two-phase field of (Pb) and (Sn). The fields above the eutectic temperature and bounded by the solidus and liquidus curves are also two-phase fields of (Pb) + L (for *hypoeutectic compositions*—less than the eutectic composition) and (Sn) + L (for *hypereutectic compositions*—greater than the eutectic composition).

As indicated earlier, the eutectic composition behaves as if it is a pure component with respect to melting or freezing; it all melts or freezes at the eutectic temperature. Any other composition will have a range of temperatures to completely melt or freeze the alloy (the same principles as those described with Fig. 5-4). In going through this range, the compositions (ends of tie-lines) and the proportions (using the Lever Rule) of the liquid and solid phases change.

5.7.4 Development and Control of Microstructures in Eutectic Systems

We shall now discuss how the microstructures will arise, and how we can control them when alloys in the eutectic system are cooled and transformed from the liquid state to the solid state.

We shall first start with an alloy with the eutectic composition. When a liquid with the eutectic composition solidifies, it does so by forming a nucleus of one of the solid phases and the second solid immediately nucleates beside it. This alternate nucleation and subsequent growth form 100 percent lamellae of the two phases, as sketched in Fig. 5-13. The thickness of the lamellae may be controlled by the cooling rate—the faster the cooling rate, the finer are the lamellae. Because the lamellae are platelike, some researchers are attempting to produce a composite of continuously aligned lamellae through directional eutectic solidification. Directional solidification has been successfully used to produce single crystals of superalloys for jet engine turbine blades.

For either hypoeutectic or hypereutectic compositions, namely, alloy compositions either less or greater than the eutectic composition, varying amounts of the eutectic lamellae will form depending on how close the alloy composition is to the eutectic composition. We shall first consider the solidification of a hypoeutectic lead-tin alloy with 40 wt.%Sn, which is also shown in Fig. 5-13. When this alloy is cooled from above its liquidus temperature, solid nuclei of (Pb) with composition of 14% Sn will form at the liquidus temperature. As discussed earlier in relation to Fig. 5-4, the temperature range of solidification in this case is from the liquidus, T_L, to the eutectic temperature, 183°C. As the temperature is decreased from T_L to the eutectic, the composition of the liquid changes and moves along the liquidus towards the eutectic composition, while the solid phase, (Pb) nuclei, grows in size and in amount while its equilibrium composition moves along the solidus. These are indicated in Fig. 5-13 by the arrows in the liquidus and solidus curves. At the eutectic temperature and because the eutectic temperature tie-line is also the solidus, the last liquid will have the eutectic composition and will solidify as lamellae. The lamellae will surround the (Pb) nuclei that grew and increased in amount as the temperature decreased from T_L to the eutectic temperature. The (Pb) that formed and grew before the eutectic lamellae occured, is called the *primary or the proeutectic phase.* The process of solidification in alloys with hypereutectic composition is similar to the hypoeutectic case. For a hypereutectic alloy, the primary or proeutectic phase is (Sn). The proportions (amounts) of liquid and proeutectic phases at a particular temperature are again determined by the Lever Rule.

In the solid state and for either hypo- or hyper-eutectic composition, the *phases or microconstituents in the structure are the proeutectic or primary phase* [(Pb) for a hypoeutectic alloy or (Sn) for a hyper-eutectic alloy] *and the eutectic lamellae;* the lamellae consist of alternate layers of (Pb) and (Sn). *To determine the relative proportions of the proeutectic phase and the eutectic lamellae*

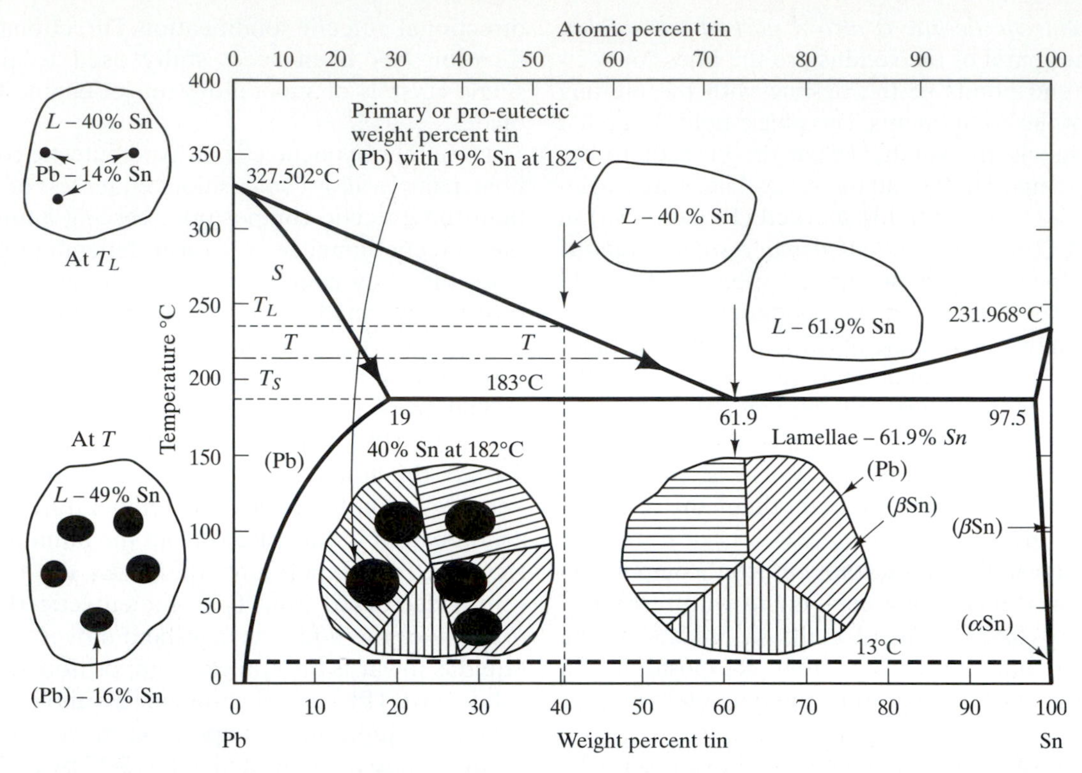

Fig. 5-13 Development of equilibrium microstructures in Pb-Sn alloys. An alloy with the eutectic composition will develop a 100 percent lamellar structure. A hypoeutectic alloy (e.g. 40%Sn) will develop a microstructure with proeutectic or primary phase (Pb) surrounded by lamellar structure.

immediately formed below the eutectic temperature, we use the eutectic composition at one end of the tie-line, and at the other end, the composition of the proeutectic phase. We let the alloy composition, C_o, cut the tie-line, and then we apply the Lever Rule. This is illustrated in Figs. 5-14a and 5-14b for hypoeutectic and hypereutectic alloys. The relative proportion (in wt%) of the eutectic lamellae in both cases is proportional to the segment $|C_o - C_p|$, where C_p is the composition of the primary phase. We see that the amount of eutectic lamellae increases as the alloy composition moves closer to C_E, the eutectic composition.

To determine the relative proportion of the phases constituting the eutectic lamellae, we take C_E, as the C_o, and then take the ends of the eutectic temperature tie-line as compositions of the two phases in the lamellae, as schematically illustrated in Fig. 5-15. To get the total amount of each of the two phases in the microstructures of either hypo- or hypereutectic alloys, we use the ends of the eutectic temperature tie-line as the compositions of the two phases, then the alloy composition, C_o, cuts the tie-line, as illustrated in Figs. 5-16a and 5-16b.

For compositions with less than 19% Sn and greater than 97.5% Sn, the equilibrium structures will not show any lamellae. This is shown in Fig. 5-17 for a Pb-10 wt%Sn alloy. The range of solidification for this alloy is from T_L to T_S and when the temperature reaches T_L, nuclei of (Pb) will form. On further cooling, the equilibrium compositions of the (Pb) and the liquid follow the solidus and liquidus curves. At T_S, the last liquid will have 24% Sn before it solidifies to (Pb). The equilibrium structure will not contain the eutectic lamellae. Between T_S and the solvus temperature (T_{solvus}), the microstructure con-

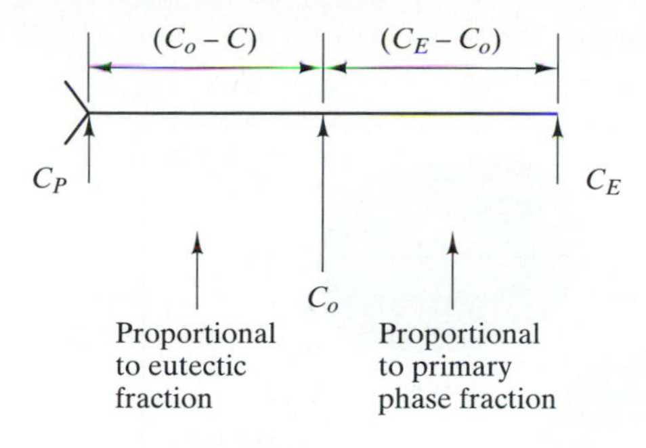

(a) Hypo-eutectic alloy

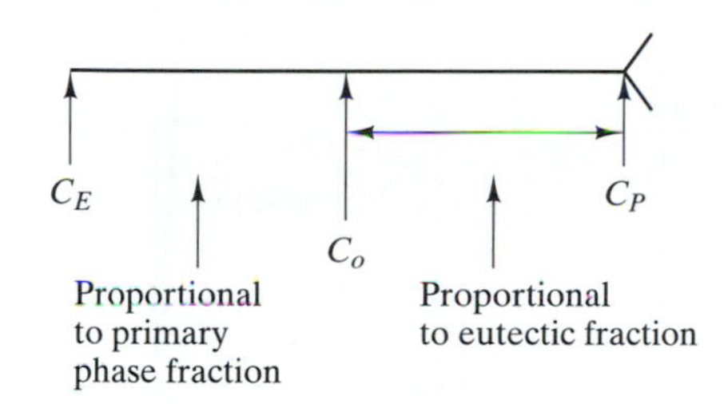

(b) Hyper-eutectic alloy

Fig. 5-14 Schematics in determining relative weight fractions of primary phase and eutectic lamellae in (a) hypoeutectic, and (b) hypereutectic alloys.

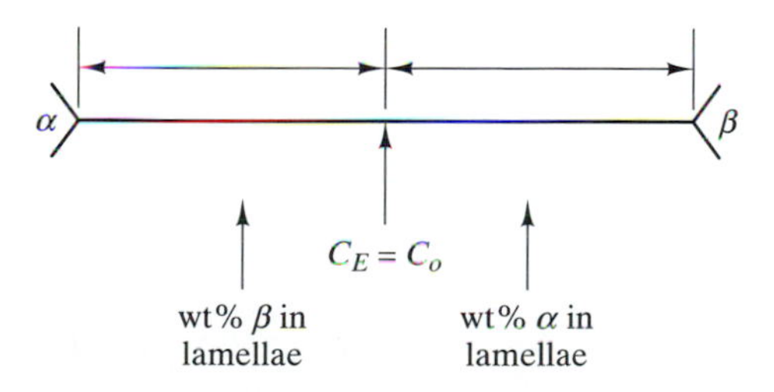

Fig. 5-15 Schematics to determine the relative weight fractions of the alternate α and β phases in the eutectic lamellae.

sists only of the primary solid solution of (Pb). Below the solvus temperature, (β-Sn) will separate as a second phase, which is shown as an angular phase.

The description of the development of the different phases or microconstituents in the structures of the eutectic Pb-Sn system is very similar to that of

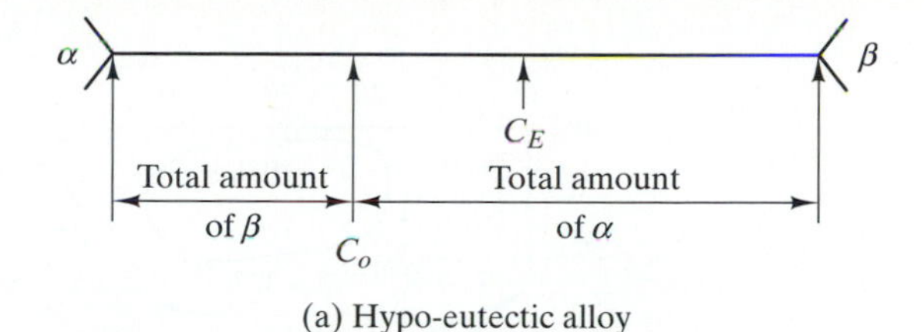

(a) Hypo-eutectic alloy

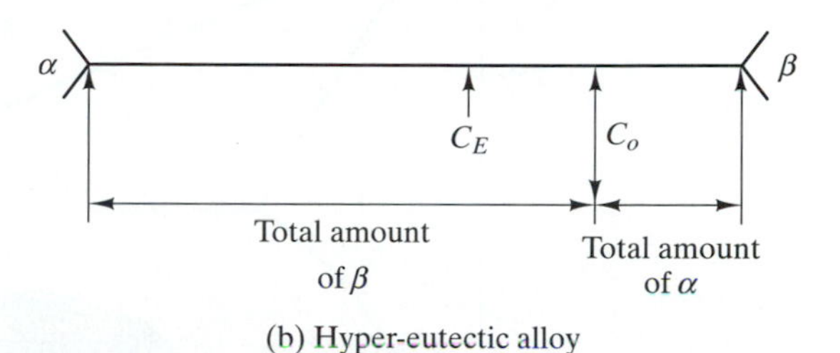

(b) Hyper-eutectic alloy

Fig. 5-16 Schematics to determine the total weight fractions of α and β phases in (a) hypoeutectic, and (b) hypereutectic alloys.

the Fe-Fe_3C eutectoid system, which we shall discuss as one of the iron-carbon systems following Examples. 5-2 and 5-3.

EXAMPLE 5-2

For a Pb-Sn alloy with the eutectic composition, calculate the equilibrium weight fractions of (Pb) and (Sn) in the microstructure at (a) 182°C, and (b) 0°C.

SOLUTION: As Fig. 5-13 indicates, the microstructure consists of 100 percent eutectic lamellae and we need to use Fig. 5-15 to calculate the relative weight fractions of α (Pb) and β (Sn).

(a) At 182°C, the tie-line is essentially the eutectic temperature and applying the lever rule,

$$\text{weight fraction of (Pb)} = \frac{97.5 - 61.9}{97.5 - 19.0} = 0.453$$

$$\text{weight fraction of (Sn)} = 0.547$$

(b) At 0°C, the ends of the tie-line are essentially 0.9%Sn in (Pb) and 100% Sn. Thus,

$$\text{weight fraction of (Pb)} = \frac{100 - 61.9}{100 - 0.9} = 0.392$$

$$\text{weight fraction of (Sn)} = 0.608$$

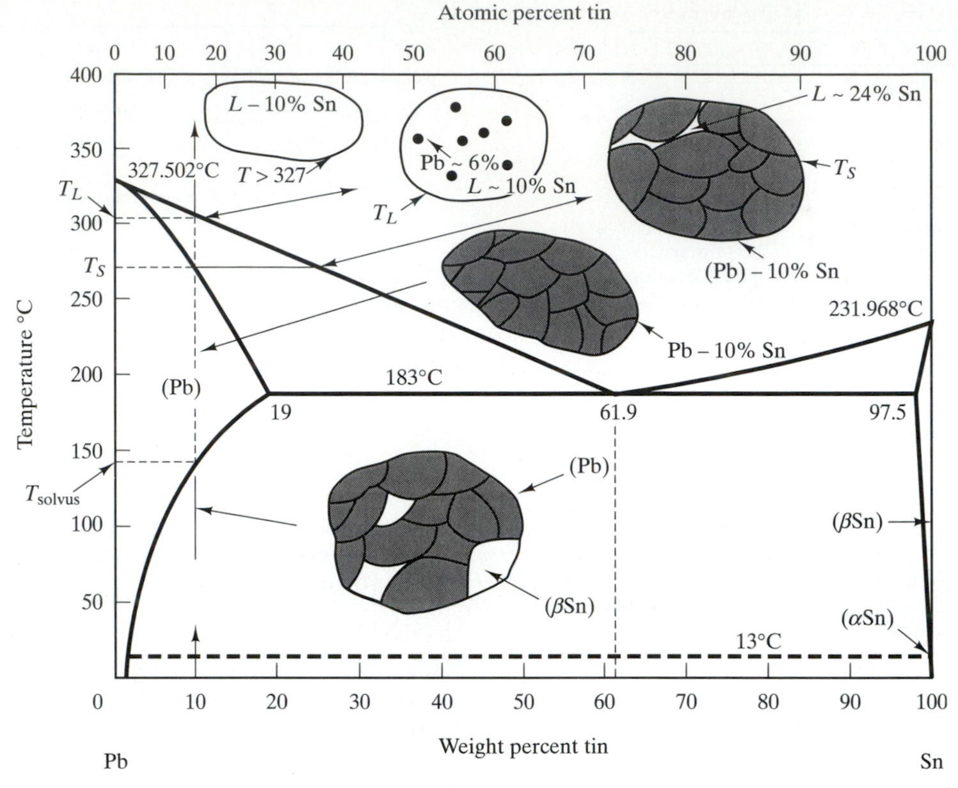

Fig. 5-17 Development of microstructure in alloys with less than the maximum solubility at the eutectic temperature—10%Sn., in Pb-Sn Binary system.

EXAMPLE 5-3

For the 40% Sn hypo-eutectic alloy indicated in Fig. 5-13, determine

(a) the equilibrium weight fractions of the proeutectic phase (Pb) and the liquid at indicated temperature, T = 215°C.

(b) the equilibrium weight fractions of the proeutectic (Pb) and liquid at 184°C.

(c) the weight fractions of the proeutectic (Pb) and the eutectic lamellae at 182°C.

(d) the total weight fractions of (Pb) and (Sn) at 182°C, and

(e) question (d) by using the results of (c) and Example 5-2.

SOLUTION: We apply the lever rule in cases (a) to (d) and the mass balance in (e). Thus,

(a) at T = 215°C shown in Fig. 5-13, the compositions at the ends of the tie-line are (Pb) = 16%Sn, and L = 49%Sn.

$$\text{weight fraction (Pb)} = \frac{49 - 40}{49 - 16} = 0.273$$

$$\text{weight fraction L} = \frac{40 - 16}{49 - 16} = 0.723$$

(b) at 184°C, the ends of the tie-line are essentially: (Pb) = 19%Sn and L = 61.9%Sn.

$$\text{weight fraction (Pb)} = \frac{61.9 - 40}{61.9 - 19} = 0.510$$

$$\text{weight fraction L} = \frac{40 - 19}{61.9 - 19} = 0.490$$

(c) at 182°C, ends of the tie-line are (Pb) = 19%Sn, and the eutectic lamellae = 61.9%Sn.

We use Fig. 5-14a and the answers are the same as in (b) except that the weight fraction of L is now the weight fraction of the eutectic lamellae. Thus, weight fraction (Pb) = 0.510 and that of the eutectic = 0.490.

(d) We use the schematic in Fig. 5-16a, and the ends of the tie-line are at (Pb) + 19%Sn and (Sn) = 97.5%. Thus,

$$\text{weight fraction (Pb)} = \frac{97.5 - 40}{97.5 - 19} = 0.732$$

$$\text{weight fraction (Sn)} = \frac{40 - 19}{97.5 - 19} = 0.268$$

(e) We use mass balance and assume 1 mass unit of alloy. From (c), mass of eutectic = 0.49 and that of (Pb) = 0.51. From Example 5-2, the eutectic consists of (Pb) = 0.453 and (Sn) = 0.547. Since total mass is assumed to be one, mass equals weight fraction. Thus,

$$\text{total weight fraction (Pb)} = 0.51 + 0.453 \times 0.49 = 0.732$$

$$\text{total weight fraction (Sn)} = 0.546 \times 0.49 = 0.268$$

5.8 The Iron–Carbon System

There are two important diagrams in the Fe-C system and both of these are indicated in Fig. 5-18. One is the iron-cementite (Fe-Fe_3C) or iron-carbide metastable equilibrium and the other is the iron-graphite (Fe-$C_{graphite}$) equilibrium diagram, which are indicated in Fig. 5-18 as the solid curve and the dashed curve. They represent the most technologically used metals and alloys called the ferrous alloys. The iron-cementite diagram is predominantly used for steels and the iron-graphite diagram is used for cast irons. Steels are ferrous alloys with as much as 2 wt% carbon, but the majority of steels produced have carbon contents less than 0.20 wt%C. Cast irons, on the other hand, are ferrous alloys with carbon contents from 2 to 4.3%C. Because the phase diagrams contained in Fig. 5-18 represent the most often used metal alloys, it is important to understand them, especially if we are going to do any heat-treating.

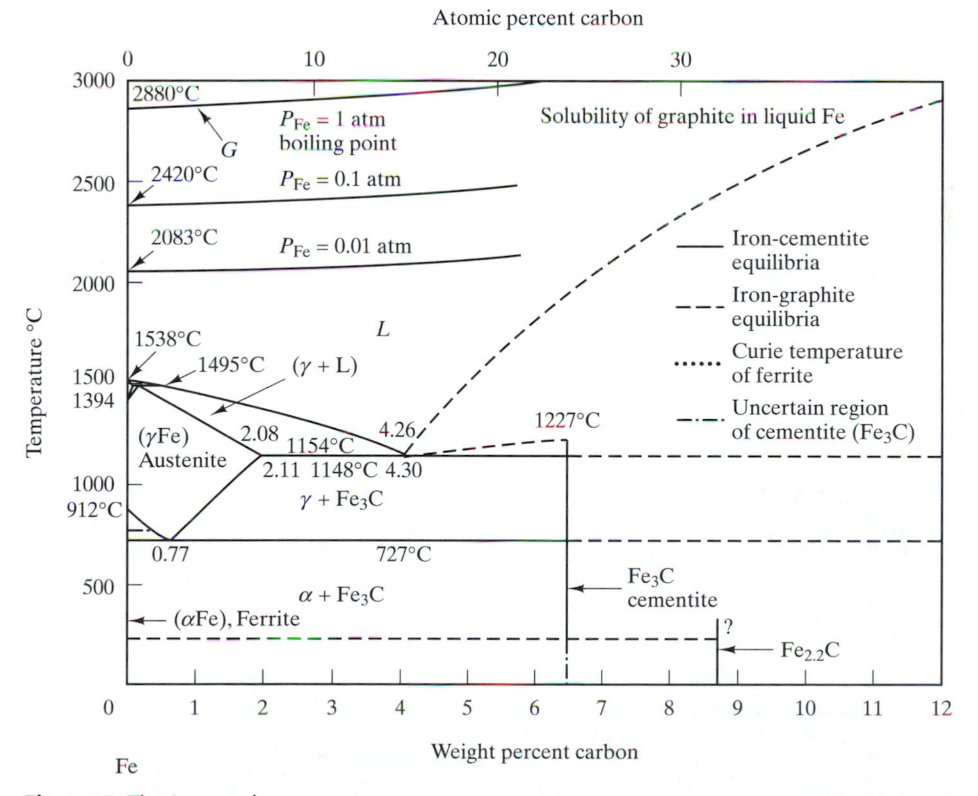

Fig. 5-18 The iron-carbon systems.

5.8.1 The Fe-Fe_3C Metastable System

The boundaries for this system are better shown in Fig. 5-19. The two verticals are the pure iron (100% Fe) on the left and the vertical marked, Fe_3C, cementite or iron carbide on the right. The Fe_3C vertical is however not marked 100% Fe_3C, but its composition is given as the weight percent of carbon in Fe_3C, which is 6.69 wt%C, shown in Fig. 5-18. The other vertical at temperatures below 230°C shown in Fig. 5-19 has a question mark. This means there is uncertainty in the presence of the χ-carbide ($Fe_{2.2}C$) phase. We shall not be concerned with this uncertainty and for the purpose of this text, we shall assume that the Fe_3C vertical line extends to 0°C.

We start by looking at pure iron, the left vertical boundary. As we go up in temperature, iron exhibits two allotropic changes (see Sec. 2.8) at 912°C and at 1394°C. These changes are

$$\alpha_{BCC}(\text{ferrite}) \rightarrow \gamma_{FCC}(\text{austenite}) \text{ at } 912°C \quad \text{and}$$

$$\gamma_{FCC} \rightarrow \delta_{BCC} \text{ (ferrite) at } 1394°C.$$

The melting point of pure iron is at 1538°C and the boiling point is 2880°C.

The allotropic forms of iron, (i.e., α, γ, and σ), form limited or partial interstitial solid solutions with carbon. Within these limits of solubility, the interstitial solid solutions are also called α-ferrite, γ-austenite, and δ-ferrite, the same names given to the allotropic forms of pure iron. In Chap. 2 on crystal structures, we have seen that the body-centered cubic (BCC) structure has a smaller spherical interstitial void than that in the face-centered cubic (FCC) structure. To accommodate an interstitial atom, the BCC has to be strained more than the FCC. This translates to a much lesser amount of solute that can be held in solution in BCC than in FCC. This is illustrated in Figs. 5-20, 5-21, and 5-22. The expanded α-ferrite solid solution phase field in Fig. 5-20 shows the maximum solubility of carbon in α-ferrite to be 0.0218% at 727°C. The δ-ferrite, which is also a BCC and at a much higher temperature, shows only a maximum carbon solubility of 0.09% at 1495°C in Fig. 5-21. These limits of solubilities in BCC are much lower than the 2.11% carbon solubility in γ-Fe (austenite), shown at the top right corner in Fig. 5-22 at 1148°C.

The *cementite* or *iron carbide, Fe_3C, is an example of a compound formation* when two elements are com-

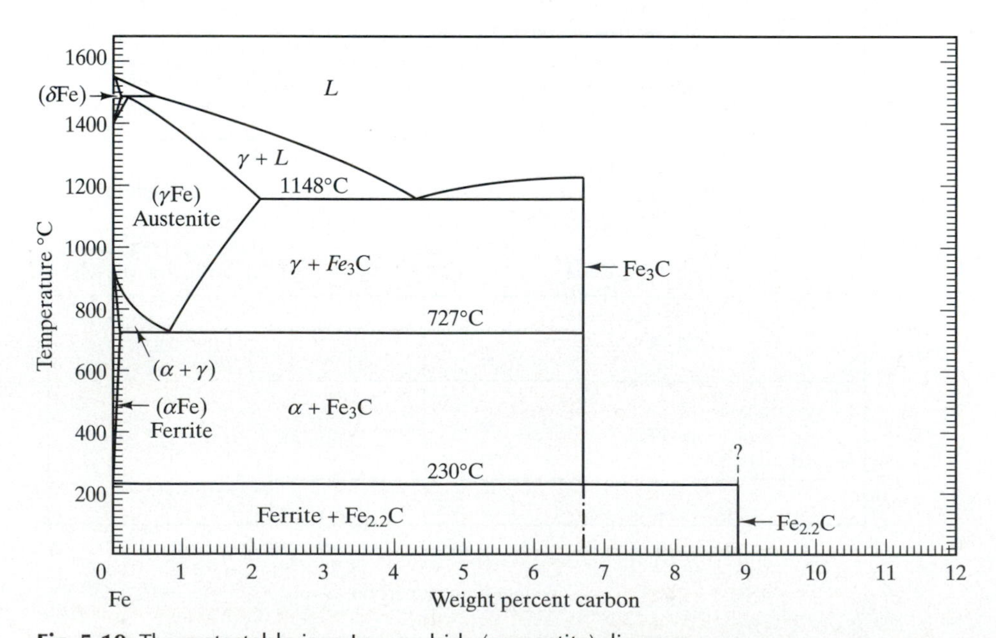

Fig. 5-19 The metastable iron-Iron carbide (cementite) diagram.

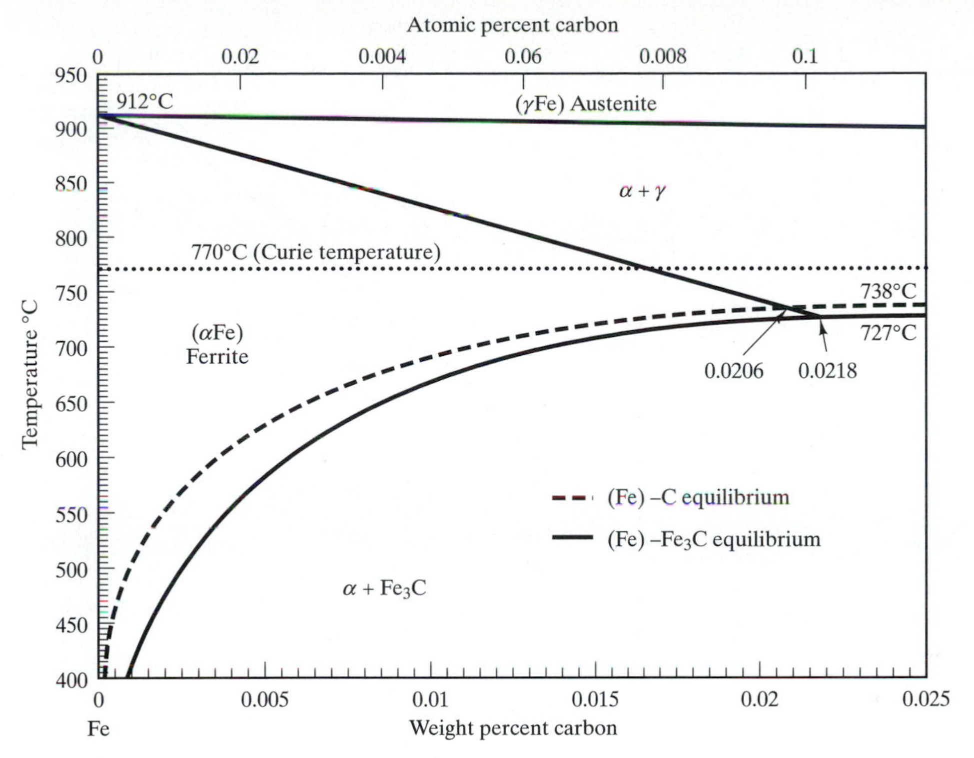

Fig. 5-20 Solvus of carbon in Ferrite.

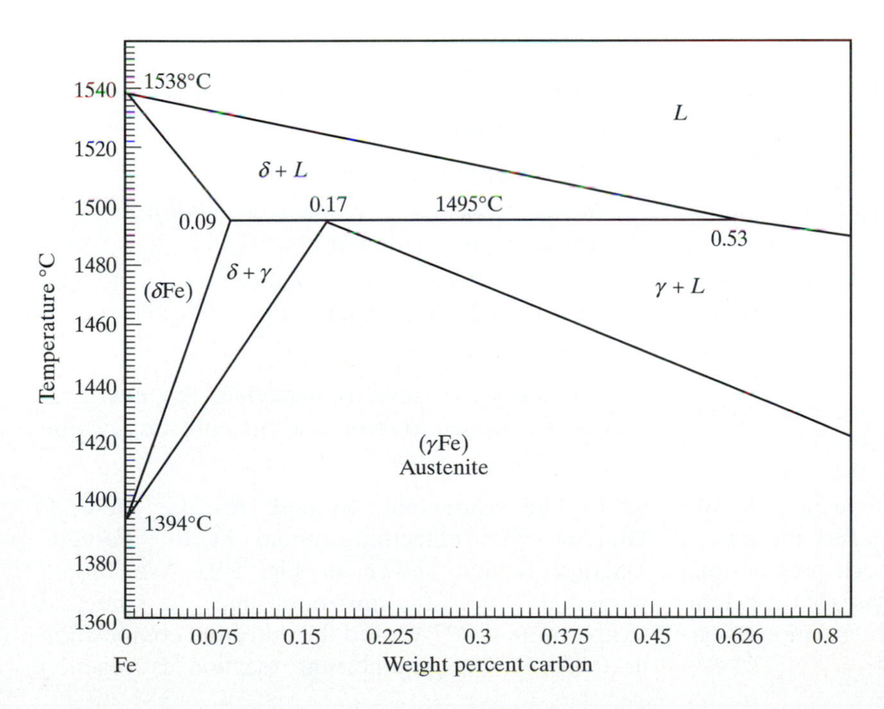

Fig. 5-21 Solvus of carbon in σ-ferrite and the peritectic reaction.

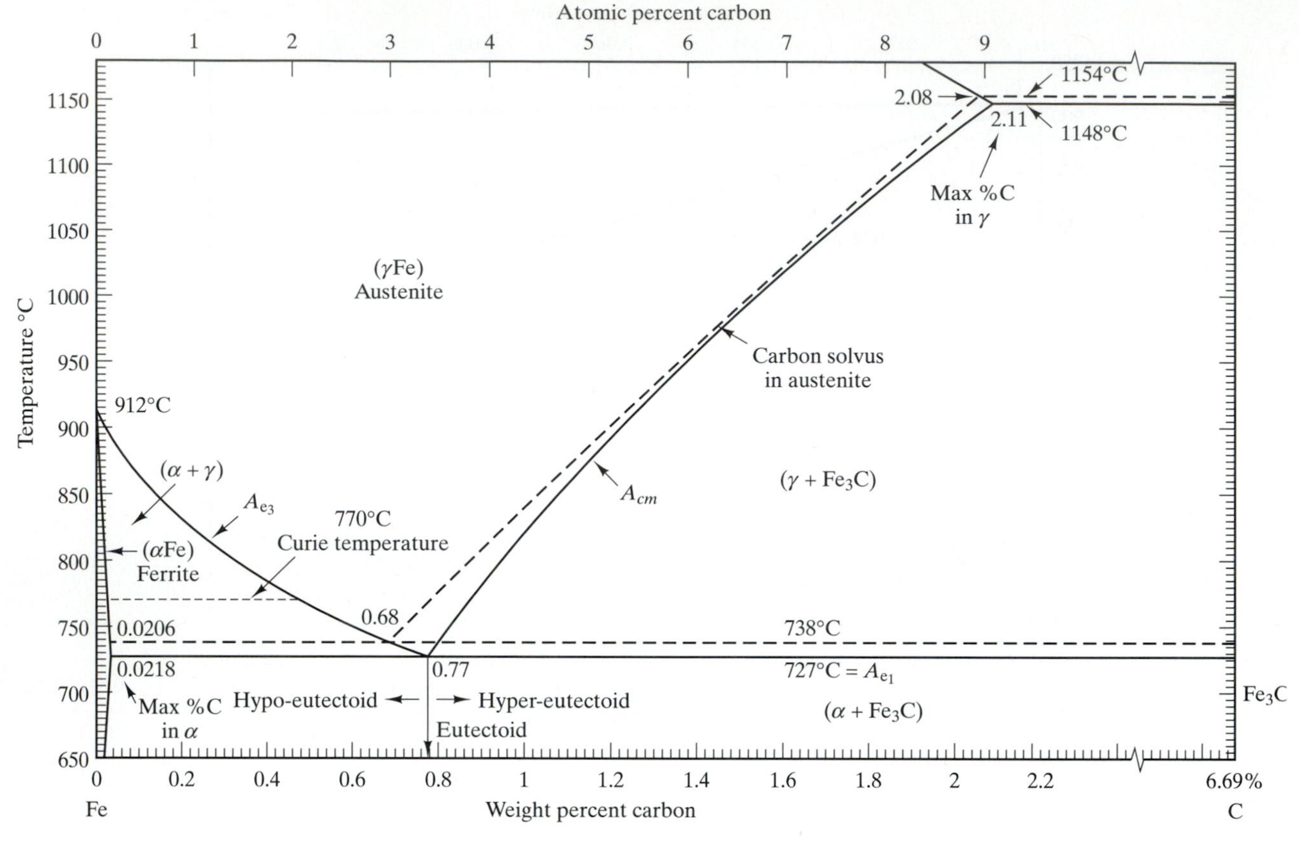

Fig. 5-22 Eutectoid portion of the iron-iron carbide (cementite) diagram.

bined. It is represented as a vertical line because it is a stoichiometric compound, one that has definite ratios of the elements in combination, and thus has a definite composition. The ratio of C:Fe in the compound is 1:3 or carbon is 25 atomic % (at%) in Fe_3C. The conversion of 25 at%C to wt%C, shown in Example 5-4, yields 6.69 wt%C as shown in Fig. 5-19. Any carbon in excess of the limits of solubilities (solvus curves) of carbon in ferrite from room temperature to 727°C and in austenite from 727°C to 1148°C (the A_{cm} line indicated in Fig. 5-22) will form the intermetallic compound Fe_3C. In the expanded diagrams of Figs. 5-20 and 5-22, we must not overlook that the Fe_3C vertical is the right boundary. This is indicated in Fig. 5-22 with a break in scale after 2.2 wt%C. Because the carbon content in commercial steels is much greater than 0.0218 wt%, steel has at least two phases at room temperature, α-ferrite (an interstitial solid solution of carbon in BCC iron) and Fe_3C.

Figure 5-18 shows three, three-phase invariant reactions. These are

(1) The eutectoid reaction at 727°C,
$\gamma_{0.77\%C} \rightarrow \alpha_{0.0218\%C} + (Fe_3C)_{6.69\%C}$ in Fig. 5-22.
(2) The eutectic reaction at 1148°C,
$L_{4.30\%C} \rightarrow \gamma_{2.11\%C} + (Fe_3C)_{6.69\%C}$.
(3) The peritectic reaction at 1495°C,
$\delta_{0.09\%C} + L_{0.53\%C} \rightarrow \gamma_{0.17\%C}$ in Fig. 5-21.

For processing of ferrous materials, the eutectoid reaction is applied to steels and the eutectic reaction is applied to cast irons.

5.8.1a The Eutectoid System in the Fe-Fe$_3$C Diagram. The eutectoid system in the Fe-Fe_3C diagram is that shown in Fig. 5-22. We should immediately note the horizontal line (the eutectoid temperature at 727°C) and the eutectoid composition at 0.77%C. The equilibrium reaction is reaction

(1) above and the horizontal eutectoid temperature tie-line indicates that Fe_3C is very minutely soluble in Fe but Fe is not soluble in Fe_3C. As indicated above, steels have at least the two phases of α-ferrite and Fe_3C at room temperature. Steels with less than 0.77%C are hypoeutectoid and those with more than 0.77%C are hypereutectoid. Since all the phases are in the solid, there are no solidus and liquidus curves. The transformation lines or curves in Fig. 5-22 are designated A_{e_1}, A_{e_3}, and A_{cm}. The subscript *e* signifies equilibrium conditions: 1 is for the eutectoid temperature, 3 is for the complete transformation of ferrite to austenite, and *cm* stands for the maximum solubility or solvus of carbon in austenite. The identification of the phase fields is done by noting that the intermetallic compound Fe_3C (@ 6.69%C) is a phase and is in equilibrium with ferrite below A_{e_1} and with austenite above A_{e_1} for hypereutectoid steels.

The equilibrium structures that develop during very slow cooling have exactly the same features as those in the eutectic system described in Sec. 5.7.4. These are schematically illustrated in Fig. 5-23. If a steel with the eutectoid composition (0.77%C) is cooled under equilibrium conditions from the γ region to room temperature, it will transform at 727°C according to reaction (1) above to $\alpha + Fe_3C$ in lamellar form, in the same manner as the eutectic system. In this case, the first to nucleate is the Fe_3C and then the α immediately nucleates beside it. While there is no name given to the lamellar product in the eutectic system, the lamellar microstructure of α and Fe_3C in steels is called *pearlite.*

If we cool a hypoeutectoid steel (%C < 0.77%) from the γ-region, the equivalent changes described in Fig. 5-13 are illustrated for a 0.40%C steel in Fig. 5-23. As soon as the temperature hits the A_{e_3} transformation curve, α-ferrite nuclei will form along the austenite grain boundaries. At temperatures less than the A_{e_3}, the α and the γ are in equilibrium. The compositions of these two phases are related by the temperature tie-line and are found at the ends of the tie-line intersecting the solvus and transformation curves. For example, at 750°C, austenite has 0.58%C and ferrite has about 0.02%C. The composition of the austenite follows the A_{e_3} curve as the temperature is decreased and the last austenite will attain the eutectoid composition that transforms to pearlite at A_{e_1}, the eutectoid temperature. The *α-ferrite* is also called the *primary* or *proeutectoid phase for the hypoeutectoid steel* and its composition follows the solvus line above the eutectoid temperature. The weight fractions of phases present in the equilibrium structure are also calculated using the Lever Rule in the same manner as those illustrated in Figs. 5-14, 5-15, and 5-16. We simply need to change the eutectic composition in those figures to eutectoid composition for the corresponding alloys to be hypoeutectoid and hypereutectoid. Also, we need to remember that the lamellae in steels are called pearlite.

If a steel with a hypereutectoid composition (%C > 0.77%) is cooled from the γ region, it will start to transform when the temperature reached the A_{cm} curve. The first phase to form this time is cementite, Fe_3C, which is the proeutectoid phase. As the steel cools further, the γ composition follows the A_{cm} and more cementite with constant composition at 6.69%C will form. At the A_{e_1} temperature, the remaining γ will transform to pearlite. Again, the Lever Rule is applied to calculate the equilibrium amounts of cementite and austenite at a particular temperature. The length of the temperature tie-line extends from the composition at the A_{cm} to 6.69%C, the composition of cementite and shown in Fig. 5-23 after a break in scale. The process of transformation is very similar to that of a material with a hypereutectic composition in the eutectic system.

EXAMPLE 5-4

Calculate the weight percent of carbon in the intermetallic compound, Fe_3C.

SOLUTION: In compounds, this is easily done by using one formula weight as the basis. In this case, the formula weight is

$$3\,\text{Fe} \times 55.85 = 167.55$$

$$1\text{C} \times 12.01 = \underline{12.01}$$

$$\text{Formula weight} = 179.56$$

$$\%\text{C by weight} = \frac{12.01}{179.56} \times 100$$

$$= 6.69\%$$

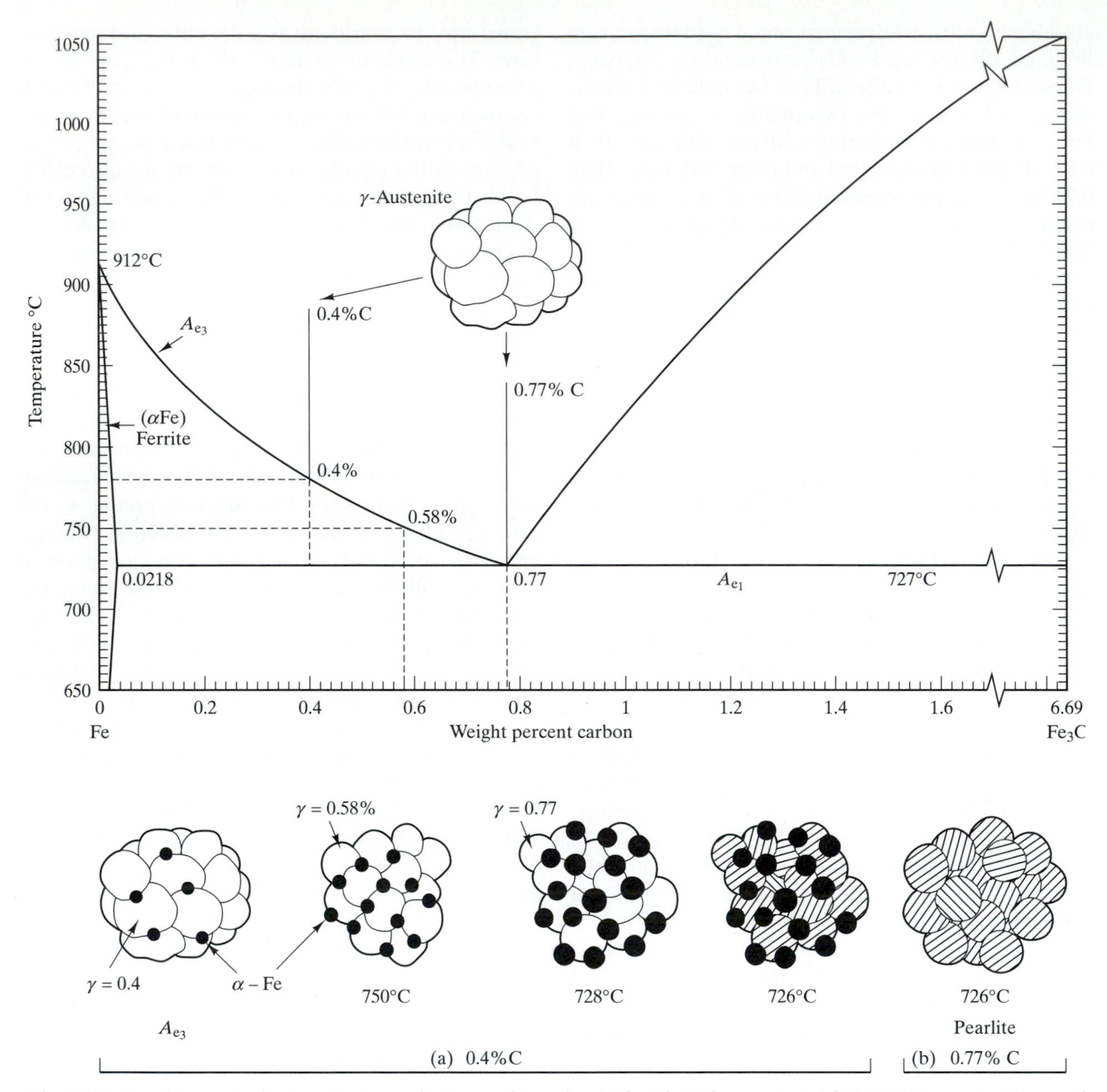

Fig. 5-23 Development of microstructures during cooling of a steel with (a) hypoeutectoid, 0.40%C, composition and (b) eutectoid, 0.77%C.

We shall use this example to illustrate the conversion from atomic percent to weight percent. Since the formula of Fe_3C contains 3 atoms of Fe and 1 atom of C, there are 4 atoms total and the atomic percent of C is 25% and that of iron is 75%. To convert to weight percent, we assume 100 atoms, and

$$\text{weight of carbon} = 25 \times 12.01 = 300.25$$

$$\text{weight of iron} = 75 \times 55.85 = \underline{4{,}188.75}$$

$$\text{Total weight} = 4{,}489.004$$

$$\text{wt\%C} = \frac{300.25}{4{,}489.00} \times 100$$

$$= 6.69\%$$

EXAMPLE 5-5

A steel with a eutectoid carbon composition was cooled under equilibrium conditions from the austenite phase. Determine: (a) the percent pearlite in the microstructure that is formed, (b) the percent of α-ferrite in the microstructure at 726°C, and (c) the percent α-ferrite in the equilibrium structure at 400°C.

SOLUTION: Since the steel composition is the eutectoid composition of 0.77%C, all the austenite phase will transform to the lamellar structure pearlite. Thus, the answer for

(a) is 100%. To get the percent α-ferrite in the next two questions, we apply the schematics in Fig. 5-15.

(b) The ends of the tie-line at 726°C are

$$\alpha\text{-ferrite} = 0.0218\%\text{C, and } Fe_3C = 6.69\%\text{C}$$

$$\text{wt.\% } \alpha\text{-ferrite} = \frac{6.69 - 0.77}{6.69 - 0.0218} \times 100 = 88.78\%$$

(c) At 400°C, we note the ends of the tie-line are α = 0.0009%C (see Fig. 5-20) and 6.69%C. Thus,

$$\text{wt.\% } \alpha\text{-ferrite} = \frac{6.69 - 0.77}{6.69 - 0.0009} \times 100 = 88.50\%$$

EXAMPLE 5-6

A hypoeutectoid steel with 0.4%C is cooled under equilibrium conditions from the austenite region to below the eutectoid temperature. Determine: (a) temperature at which the primary phase begins to form, (b) what is the primary phase? (c) the weight fraction of austenite and its composition at 750°C, (d) the weight fraction of pearlite at 726°C, and (e) the weight fraction of Fe_3C in the microstructure.

SOLUTION: We refer to Fig. 5-23 and follow the development of the microstructure in 0.4%C steel.

(a) the temperature at which the primary phase begins to form is at A_{e_3} = 780°C.

(b) the primary or proeutectoid phase that forms at the grain boundaries of austenite is the α-ferrite.

(c) As indicated in Fig. 5-23, the composition of austenite at 750°C is 0.58%C. The weight fraction of austenite at this temperature is obtained by the Lever Rule. The composition of the α-ferrite ≈ 0.02%C. Thus,

$$\text{wt. fraction } \gamma \text{ (austenite)} = \frac{0.40 - 0.02}{0.58 - 0.02} = 0.68$$

(d) The weight fraction of pearlite at 726°C is obtained in the same manner as illustrated in Fig. 5-14a. The ends of the tie-line are at α = 0.0218%C, and pearlite = 0.77%C.

$$\text{wt. fraction pearlite} = \frac{0.40 - 0.0218}{0.77 - 0.0218} = 0.51$$

(e) The weight fraction of Fe_3C is obtained as schematically shown in Fig. 5-16a. The ends of the tie-line are at α = 0.0218%C and at Fe_3C = 6.69%C.

$$\text{wt. fraction } Fe_3C = \frac{0.40 - 0.0218}{6.69 - 0.0218} = 0.06$$

EXAMPLE 5-7

Assuming that a perfect lamellar structure of pearlite can be obtained, determine the relative thicknesses of α and Fe_3C in pearlite.

SOLUTION: Up to this point, we have been dealing with weights of the phases. We are now being asked for relative dimensions. Thus, we must think of a correlation of dimensions and weight—achieved through the density and the volume. Thus, we need to know the densities of α and Fe_3C (found to be α = 7.87 g/cm^3 and Fe_3C = 7.66 g/cm^3). For the volume, we let

$$x = \text{thickness of } \alpha, \text{ and } y = \text{thickness of } Fe_3C.$$

If we take a unit length of the lamellar structure, the volume of each phase per unit depth is x and y for α

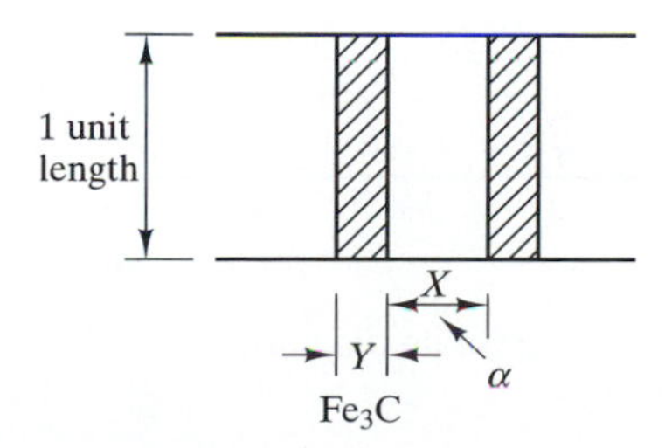

Fig. EP 5-7

and Fe_3C, respectively. In this volume of lamellar structure, the weight of $\alpha = 7.87x$, and the weight of $Fe_3C = 7.66y$. By taking the ratio of the weights, the

$$\text{relative weight of } \alpha : Fe_3C = \frac{7.87\,x}{7.66\,y}$$

We now have a relationship of weight to the thickness. We get the relative weight of α and Fe_3C from their weight fractions in the eutectoid structure that we found in Example 5-5—0.8878 for α and 0.1122 for Fe_3C. Thus,

$$\frac{7.87\,x}{7.66\,y} = \frac{0.8878}{0.1122} \quad \text{and} \quad \frac{x}{y} = \frac{7.70}{1}$$

5.8.2 The Stable Fe-C(Graphite) System

The stable iron-graphite system is shown in Fig. 5-24. The right-vertical boundary is now at 100% carbon or graphite, which is shown with a break in scale in Fig. 5-24. There are some minor differences in the temperatures and compositions of the eutectic and eutectoid reactions from those in the Fe-Fe_3C metastable diagram. These reactions are

(1) eutectic at 1154°C $\quad L_{4.26\%C} \rightarrow \gamma_{2.08\%C} + C_{graphite}$

(2) eutectoid at 738°C $\quad \gamma_{0.68\%C} \rightarrow \alpha_{0.02\%C} + C_{graphite}$

The peritectic reaction remains the same as in steels.

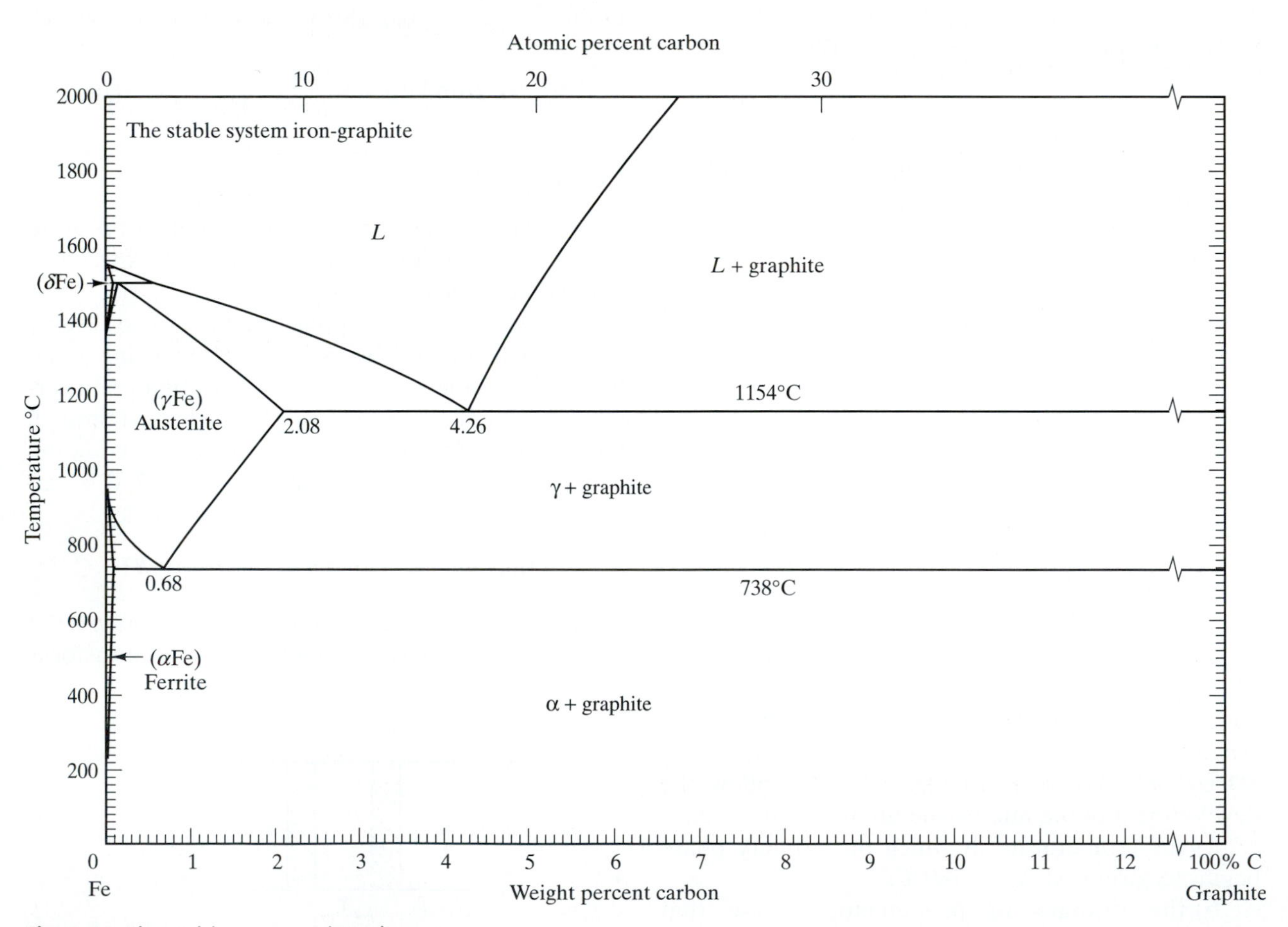

Fig. 5-24 The stable iron-graphite diagram.

In order to induce the eutectic reaction, cast iron compositions always contain a certain amount of silicon. Without silicon, the graphite component of the eutectic reaction is very difficult to induce completely because of the competing eutectic reaction at 1148°C that produces Fe_3C. The addition of silicon retards the Fe_3C formation and allows more time to form the graphite. The silicon content is converted to an equivalent carbon and the desired total carbon equivalent is about 4.30. This is expressed in the following equation,

$$\text{Carbon Equivalent (C.E.)} = \%C + \frac{\%Si}{3} = 4.3 \qquad (5\text{-}6)$$

Equation 5-6 is drawn in Fig. 5-25, which is a plot of silicon content versus carbon content in cast irons, and which indicates the ranges of silicon and carbon compositions for the different types of cast irons. The carbon and silicon contents in steel are also indicated for comparison. If phosphorus is added to the cast iron, the carbon equivalent is

$$\text{C.E.} = \%C + \frac{\%Si + \%P}{3} = 4.3 \qquad (5\text{-}7)$$

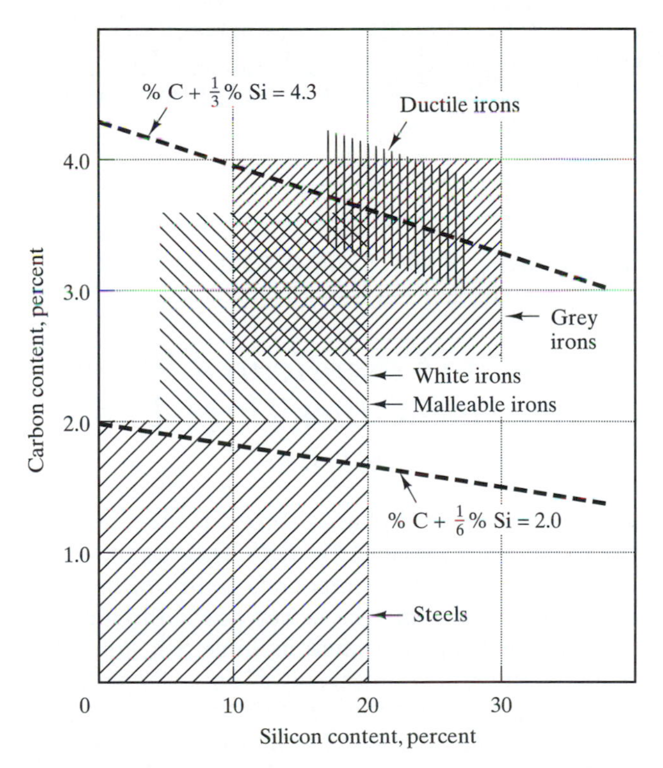

Fig. 5-25 Approximate ranges of carbon and silicon contents in ferrous alloys[Ref. 2].

As in the eutectic system, cast irons with C.E. less than 4.3 are hypoeutectic and those with C.E. more than 4.3 are hypereutectic. Carbon equivalents close to 4.3 induce the graphite reaction and tend to produce gray irons, while lesser C.E.'s will induce the formation of Fe_3C and form white cast iron. The terms gray and white arise from the color of the fracture surfaces. The gray iron produces a dull gray fracture surface, while the white cast iron produces a lighter color.

The eutectic reaction in cast iron does not yield the typical lamellar structures of other systems. The reason for this is the presence of some additives that are purposely added to alter the graphite shapes. Without any additives, the graphite forms as different flakes that may appear in the microstructure, as shown in Fig.5-26. The possible reason for this is that the graphite is very light and the induced convective stirring of the liquid during solidification might alter its form.

5.9 Intermetallic Compounds and Secondary Solid Solutions

As indicated by the Hume-Rothery rules, when the electrochemical nature of the components are dissimilar, intermetallic compounds can form. These *compounds* are usually designated as A_xB_y and may be *stoichiometric* or *nonstoichiometric*. They are stoichiometric when the subscripts x and y are definite numbers; otherwise, they are called nonstoichiometric. We have seen an example of an stoichiometric compound in cementite or iron carbide, Fe_3C, which is displayed in the phase field as a vertical line with a definite composition. The nonstoichiometric compound, on the other hand, does not have definite numbers for x and y and is usually depicted as a field, much like the field of a primary solid solution. For this reason, it is also called a *secondary solid solution*. An example of a phase diagram showing stoichiometric and nonstoichiometric compounds is the Ti-B system in Fig. 5-27. In this diagram, Ti_3B_4 is

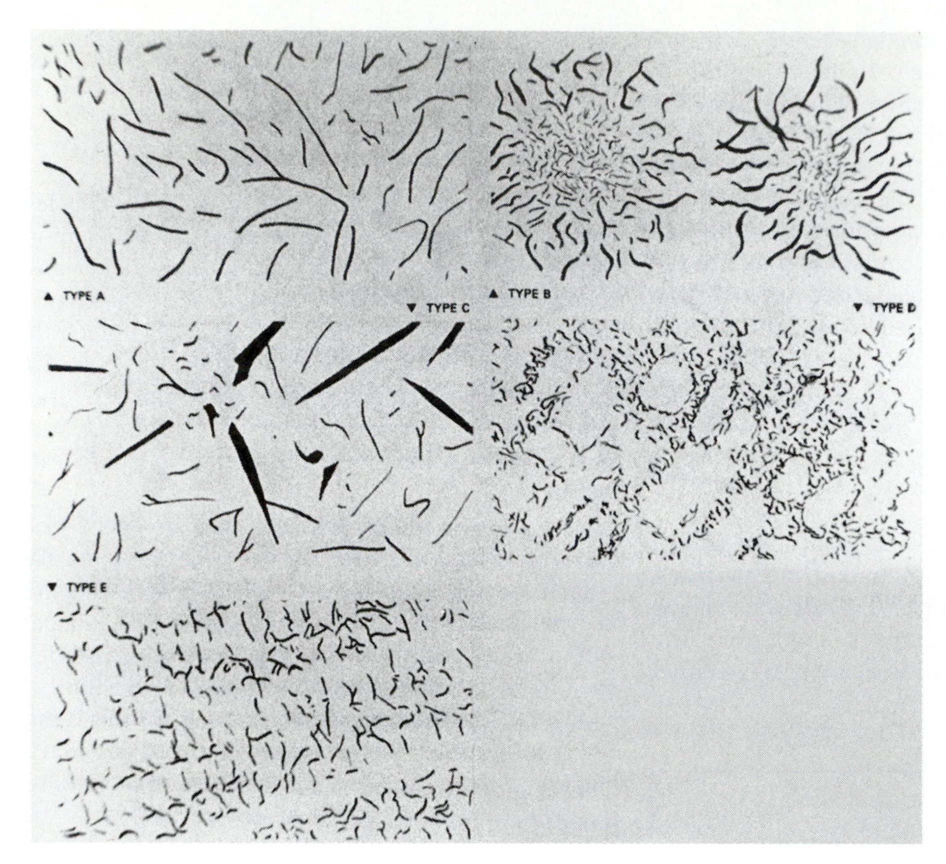

Fig. 5-26 Five types of graphite flakes in cast irons.

stoichiometric, while TiB and TiB_2 are nonstoichiometric and secondary solid solutions.

We note in Fig. 5-27 that the intermetallic compounds usually exhibit high melting points, higher than that of at least one component, and sometimes higher than that of either the A or B component. The reason for this arises from the bonding involved in intermetallic compounds that is very similar to ionic bonding. This characteristic makes them good candidate materials for very high temperature applications; however, because of the nature of the bonding, they are hard and brittle. An example where the use of intermetallic compounds is currently being explored is in both the National Aerospace Plane (X-30) and the High Speed Civil Transport (HSCT) plane. Both of these planes require high temperature, combustion jet engines in order to increase the thrust and the fuel economy. Composites of the intermetallic compounds help alleviate the brittleness problem.

5.10 More Complicated Phase Diagrams—Other Three-Phase Invariant Reactions

A more complicated phase diagram is illustrated by the Al-Mn binary system, Fig. 5-28. This system manifests the intermetallic compounds, both stoichiometric and nonstoichiometric, as well as the three-phase invariant reactions indicated in Table 5-1. Pure aluminum (left vertical) exists only as one solid phase up to its melting point of 660.452°C. Manganese (right vertical), on the other hand, exhibits changes in crystal structure (allotropy) from α-Mn to β-Mn to γ-Mn

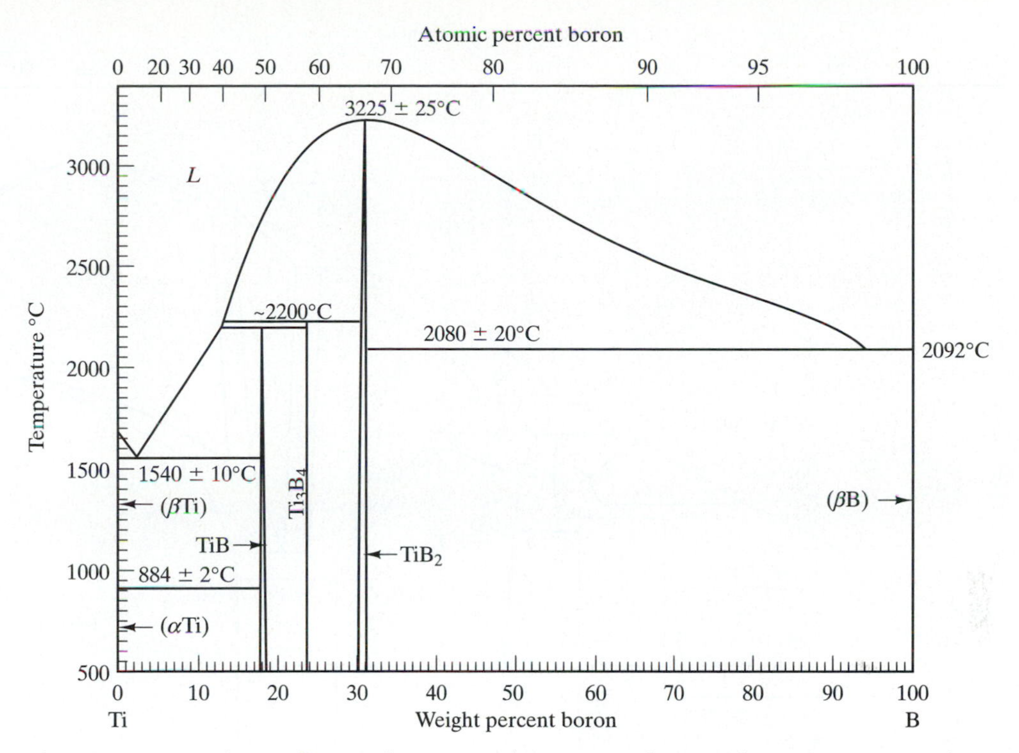

Fig. 5-27 Titanium-boron diagram showing stoichiometric and nonstoichiometric compounds [Ref. 1].

to δ-Mn as the temperature increases from low temperatures to its melting point at 1246°C. There are three stoichiometric compounds—Al_6Mn, λ, and $Al_{11}Mn_4$—and two secondary solid solution fields—μ and γ_2— at low temperatures, in addition to the primary solid solutions of (Al) and (Mn). At higher temperatures, the additional secondary solid solutions are γ_1, γ, and ϵ. As indicated earlier, the short horizontal lines in the diagram indicate three-phase reactions.

In order to consider the other three-phase reactions in Table 5-1, we need to name or identify the phase fields or areas in the phase diagram. We do this by drawing a horizontal temperature line across the phase diagram, and the phase fields that the horizontal line crosses are

$$1|2|1|2|1|.........|2|1$$

where 1 signifies a single-phase field and 2 signifies a two-phase field. The single-phase fields are usually identified in the phase diagram. In a two-phase field, the phases in equilibrium are at the end of the temperature tie-line. As an example, consider the 600°C temperature line. The first field is the primary solid solution, (Al), and is single phase. The second area, therefore, must contain two-phases, and the third must contain one phase. The latter is the stoichiometric compound, Al_6Mn. This vertical line is much like the vertical line in the Bi-Cd system and Fe_3C that signifies no solid solubility in the compound. The two phases in the second field are, therefore, (Al) and Al_6Mn. The fourth field must have two phases and the fifth field is a single phase. The latter is again the stoichiometric intermetallic compound, λ, and the fourth field must contain Al_6Mn and λ. If we continue, the successive phase fields at the 600°C are

$$(\mathrm{Al})|(\mathrm{Al}) + C_1|C_1|C_1 + \lambda|\lambda|\lambda + \mu|\mu|\mu + C_2|C_2|C_2 + \gamma_2|\gamma_2|\gamma_2 + \beta\mathrm{Mn}|\beta\mathrm{Mn}|\beta\mathrm{Mn} + \alpha\mathrm{Mn}|\alpha\mathrm{Mn}$$

where C_1 is Al_6Mn and C_2 is $Al_{11}Mn_4$.

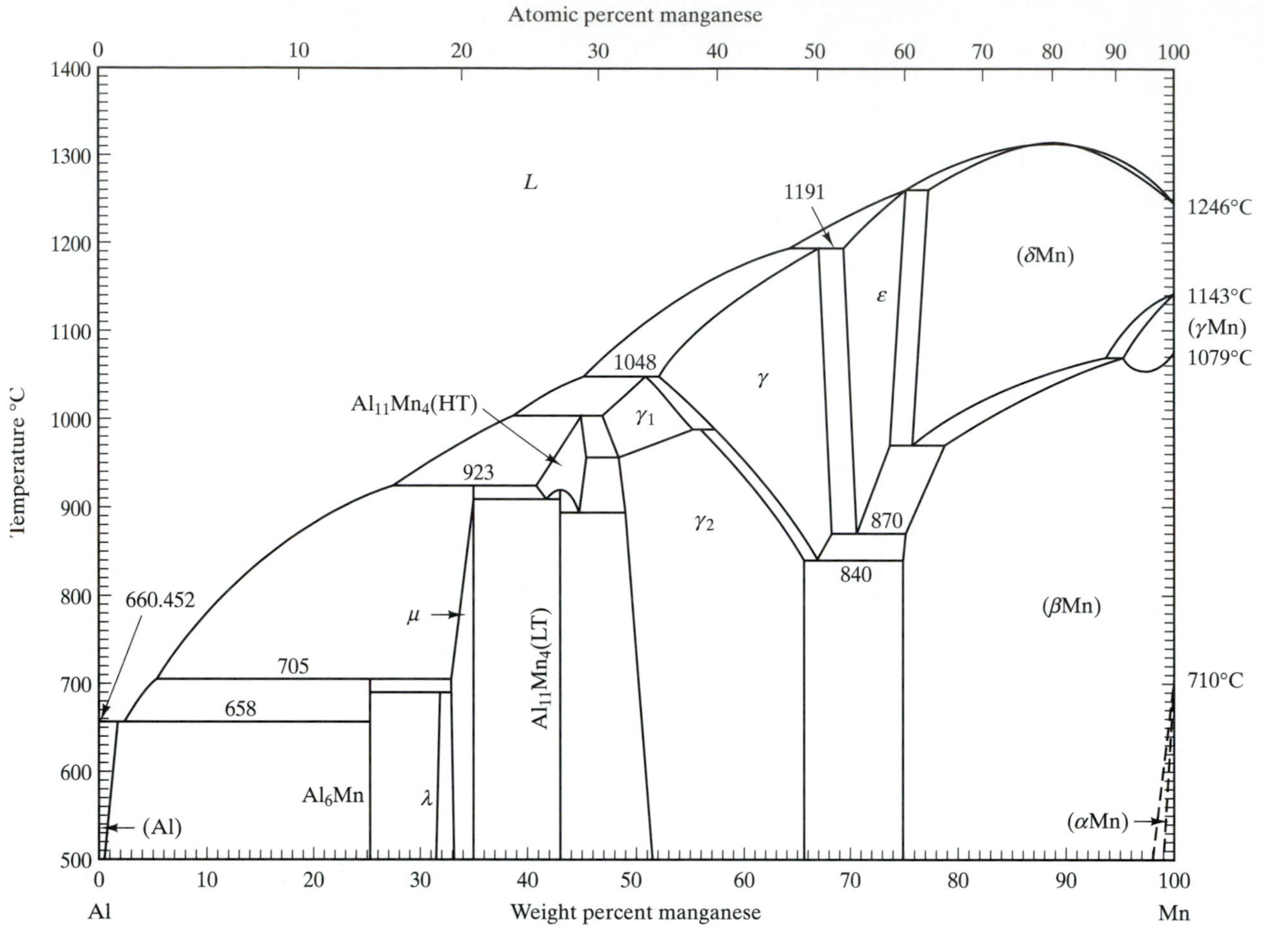

Fig. 5-28 Aluminum-manganese diagram[Ref. 1].

As specified earlier, *the short horizontal lines in phase diagrams indicate temperatures of three-phase reactions.* The 658 line is the eutectic temperature. The reactions at 705, 923, 1048, and 1191 are examples of the *peritectic reaction.* These reactions are

at 705°C $L + \mu \rightarrow C_1$

at 923°C $L + C_2(HT) \rightarrow \mu$

at 1048°C $L + \gamma \rightarrow \gamma_1$, and

at 1191°C $L + \epsilon \rightarrow \gamma$.

At 840°C and 870°C, the *reactions* are *eutectoid.* These reactions are

at 870°C $\epsilon \rightarrow \gamma + \beta\text{-Mn}$, and

at 840°C $\gamma \rightarrow \gamma_2 + \beta\text{-Mn}$.

There is a *peritectoid reaction* at the unmarked temperature between 658°C and 705°C. This reaction is

$$C_1 + \mu \rightarrow \lambda.$$

An example of the *monotectic reaction* is that shown in the Cu-Pb phase diagram, Fig. 5-29. This occurs at the point 37.4%Pb at 955°C. The reaction is

$$L_1 \rightarrow L_2 + Cu$$

In this system, copper and lead are insoluble with each other at temperatures below 326°C. The insolubility of lead in copper (and also in iron) was used to advantage to produce free-machining grades of brasses and steels. It enabled the high-speed machining of parts with very good surfaces. In addition, because the lead melts during machining, it acts as a lubricant between

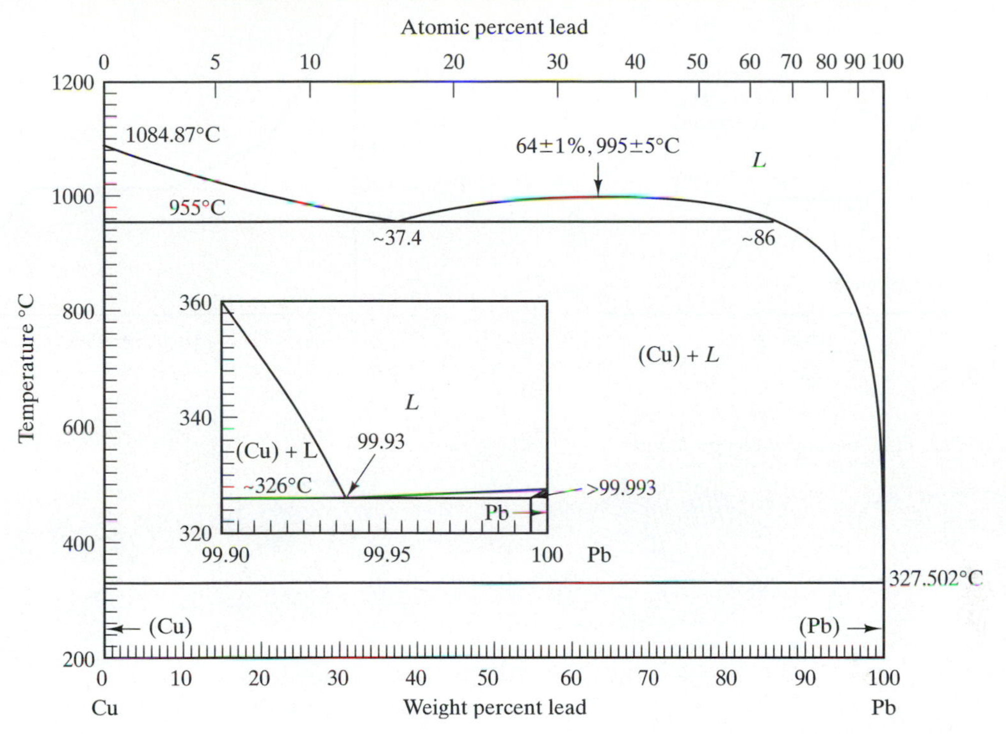

Fig. 5-29 Copper-lead diagram showing a monotectic reaction [Ref. 1].

the cutting tool and work piece. Because it melts, the temperature is high enough to vaporize some lead. The Occupational Safety and Health Act (OSHA) has now banned free-machining grades containing lead.

5.11 Determination of the Liquidus and Solidus Curves of the Phase Diagram

The phase diagram is established by determining the phase boundaries. Since the phase boundaries represent where changes in phases and/or compositions occur, they also imply changes in physical properties because of the different phases that form. Thus, changes in physical properties are the primary methods used in determining the phase diagram. These methods are usually done in conjunction with microscopic and X-ray techniques. Microscopic techniques involve the direct observation of the phases by either optical or electron microscopy. X-ray techniques identify the structures of the phases. Examples of physical properties are electrical resistivity, change in volume that translates to change in length studied by dilatometry, magnetic properties, and the heat of fusion or solidification, to name a few. Only the last property will be covered here and the technique is called *thermal analysis.*

Thermal analysis, in conjunction with microscopic and X-ray techniques, is applicable to a large number of alloy systems and is particularly suitable in determining the *liquidus* and *solidus phase boundaries.* The liquidus is determined from *cooling curves* while the solidus is determined from *heating curves.* A cooling curve is simply a record of the time (x-coordinate) versus the temperature (y-coordinate) of the liquid as it cools. A heating curve is a similar time-temperature plot during the heating of a solid. Changes in the slopes of these curves represent either solidification (cooling) or melting (heating) of the material.

5.11.1 Cooling and Liquidus Curves

When a pure metal is cooled from its liquid state and reaches its freezing point, latent heat is evolved.

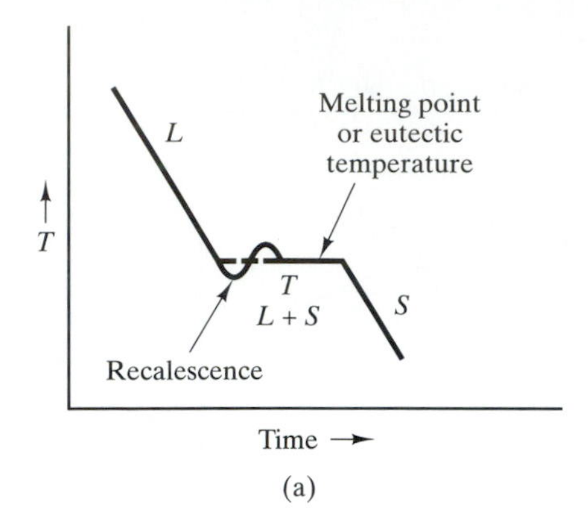

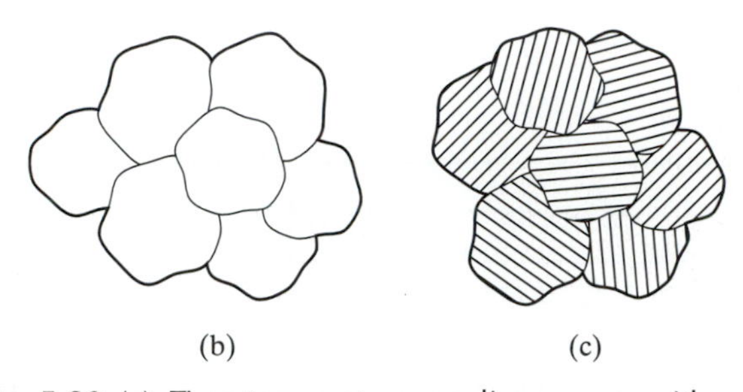

Fig. 5-30 (a) Time-temperature cooling curve with one arrest that may be a pure metal or a eutectic alloy. Microstructural examination will differentiate a pure metal structure in (b) versus a eutectic in (c).

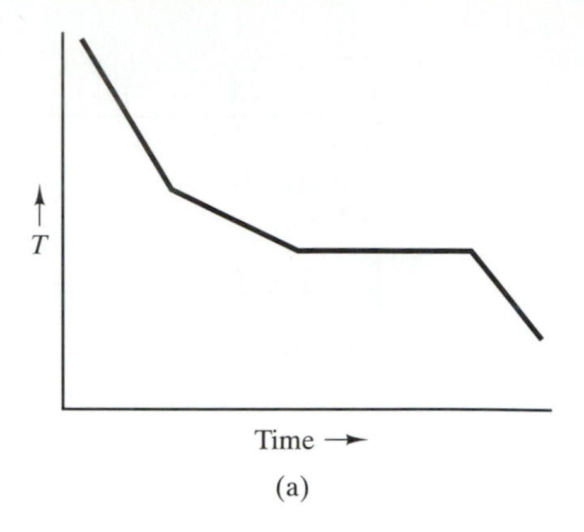

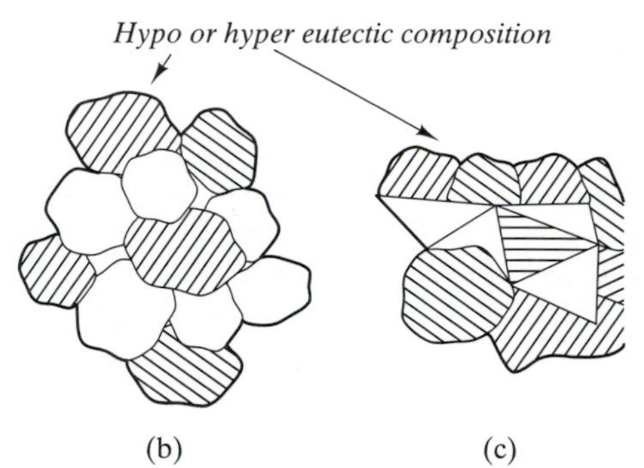

Fig. 5-31 (a) Time-temperature cooling curve showing a break and a plateau, which may be due to either a hypo- or a hypereutectic alloy, which may be differentiated by the shape of the primary phase in the microstructure.

A slow cooling rate of 1 to 1.5°C per minute will exhibit an arrest or a plateau of the time-temperature cooling curve when latent heat evolves, see Fig. 5-30a. The latent heat compensates for the heat loss of the system to the surroundings and the temperature remains constant until all the liquid has solidified. After all the liquid has solidified, the system starts cooling down again. A eutectic composition of an alloy will exhibit the same type of cooling curve as in Fig. 5-30a. Consequently, when we obtain a cooling curve as in Fig. 5-30a, we cannot say outright that it belongs to a pure metal or a eutectic composition. This can be done only after observing the optical microstructure of the sample. A pure metal will show a single-phase microstructure with grain boundaries, Fig.5-30b, while a eutectic composition will show a lamellar structure of two phases, Fig. 5-30c.

When either a hypoeutectic or a hypereutectic composition is cooled from its liquid state, latent heat is evolved when the liquidus is reached and this is depicted as a break (change in slope) in the cooling curve, see Fig. 5-31a. The reason for the break is the change in composition of the remaining liquid that changes the freezing temperature. Then the temperature decreases at a slower rate than the initial rate due to the latent heat evolution. This continues until the remaining liquid acquires the eutectic composition and, at that time, solidification will proceed at constant temperature until all the remaining liquid solidifies; after that, the temperature falls again. The alloys that will show this type of cooling curve have compositions between those at the ends of the eutectic temperature tie-line and the eutectic composition, (either hypo- or hypereutectic). Again, we cannot say outright, however, that Fig. 5-31a is from either a hypo- or a hypereutectic alloy. We can only say this after examining the microstructure of the sample. The characteristics of the primary phase or proeutectic phase of a hypoeutectic alloy will be different from those of the primary phase of a hypereutectic alloy. This difference might be in the form of shapes as schematically shown in Figs. 31b and 31c. See Photos 5, 6 and 7 for examples of a eutectic, a hypo-eutectic and a hypereutectic microstructure in a Cu–P binary system.

When the alloy composition is less than the maximum solubility shown by the solvus curve, the cooling curve again shows a break when the liquidus temperature is reached. It will continue to decrease but will not depict the plateau exhibited by either a hypo or hypereutectic alloy because the remaining liquid will not go through the eutectic composition. Theoretically, the next break in the cooling curve should be when the solidus curve is reached, as shown in Fig. 5-32a. However, because the amount of liquid left is very small when the last solid forms, this second break may not be as discernible as the first; therefore, its accuracy is not as satisfactory. An inverse rate-cooling curve such as that shown in Fig. 5-32b might help in delineating the breaks for the liquidus and solidus temperatures of an alloy. The inverse rate-1 cooling curve is a plot of the time interval required for a definite and arbitrary temperature change. It is tantamount to differentiating the direct time–temperature cooling curve with respect to the temperature. The type of cooling curve, shown in Fig. 5-32a is also exhibited by alloy systems with complete solubility in the solid state, similar to the bismuth-antimony system in Fig. 5-6.

Each of the cooling curves described above is for one composition only. As in pure elements, the changes for this composition can be represented by a vertical with only temperature as the variable. The arrest or break points in the cooling are indicated in the vertical. A number of alloys are investigated to yield the break or arrest points for each composition. A curve passing through the break- and arrest-points yield the liquidus and solidus (if this can be delineated) curves. This is illustrated for a series of Bi-Cd alloys in Fig. 5-33.

5.11.2 Heating Curves and Solidus Curves

Heating curves are used to more accurately determine the solidus curves, other than those shown by the eutectic tie-line. In order to attain the solidus, it is necessary to heat the alloy just below the suspected solidus to homogenize the composition and then heat the sample slowly, 0.5–1°C per minute from that point on. The solidus point will show as a break in the heating curve. Accurate determination of the solidus may be done by examining the microstructures of incrementally heated and quenched samples. The microstructure should show incipient melting, which is sometimes referred to as 'burning.' Figure 5-34 shows the determination of the solidus by heating and cooling and then examining the microstructure.

The solidus and liquidus curves are therefore determined in the following manner. Firstly, the constituents of the alloys are accurately weighed and are molten. The liquid is stirred and left for about 15 minutes at a temperature to insure homogeneity in composition. The liquid is cooled very slowly to obtain the cooling curve. If the break for the solidus curve is not easily discernible, then a heating curve

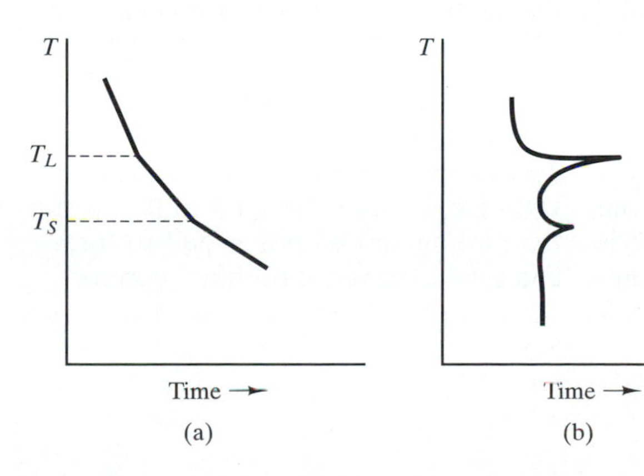

Fig. 5-32 (a) Time-temperature curve with two breaks, (b) an inverse rate curve that is more sensitive to changes in the slope of cooling curve.

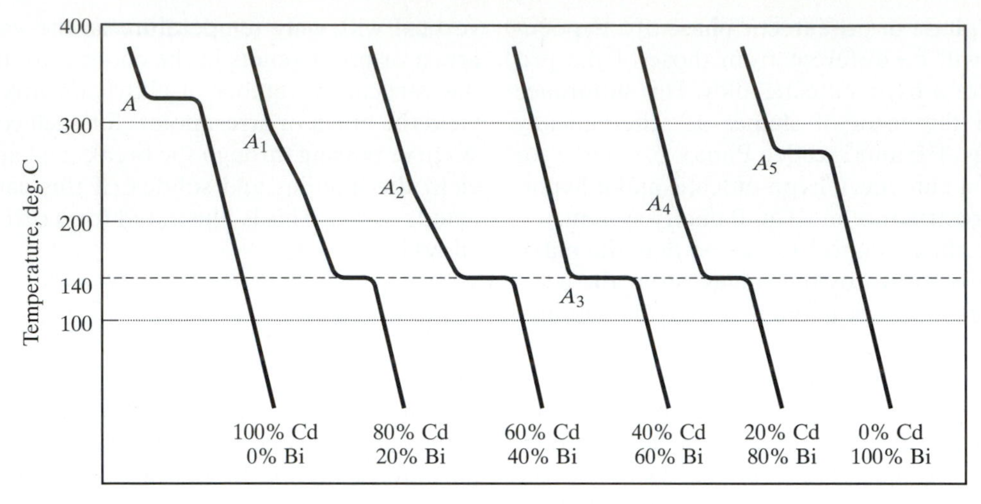

(a) Cooling curve of series of Bi-Cd alloys

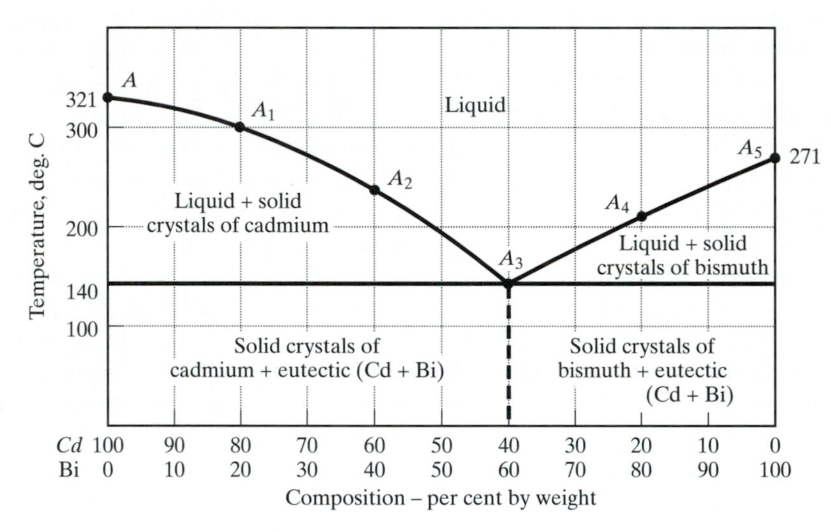

(b) Trace of break and arrest points

Fig. 5-33 (a) Cooling curves of series of Bi-Cd alloys (b) Trace of break and arrest points.

must be obtained. The breaks in the cooling and heating curves are the liquidus and solidus temperatures at the composition of the sample weighed. These temperatures are therefore indicated in the vertical line of the composition. This procedure is continued for other compositions. Each vertical composition line will have a liquidus and a solidus temperature. All the observed liquidus temperatures are connected with a smooth curve and all the observed solidus temperatures are also joined to form a curve. Both curves are connected to the melting points of the pure metals on either end to form the liquidus and solidus curves of the alloy system.

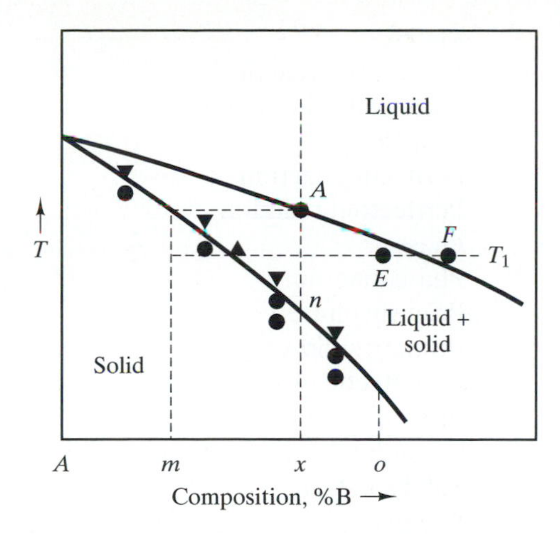

Fig. 5-34 Microstructural Examination after slow heating determines solidus; ▼–partial melting; ●–solid [Ref. 3].

Summary

We have presented the basic concepts and terminology associated with binary equilibrium phase diagrams. We have illustrated the principle of equilibrium between two phases with a tie-line from which we derived and then applied the Lever Rule. We have illustrated cases of complete solubility of metals in the solid, partial solubility, complete insolubility, and stoichiometric and nonstoichiometric compound formations. From the phase diagrams, we learned that a horizontal line signifies a three-phase reaction and a composition on this line locates where the reaction occurs. We have looked at two important systems—the eutectic and the eutectoid—and indicated how microstructures are developed during cooling from the liquid to the solid. The eutectic system is one that has an alloy composition that has a lower melting point than either of the two components. This system has practical application because it improves the castability of the alloy. The eutectoid system has practical significance in heat treating, especially for steels. Finally, we discussed how the liquidus and solidus curves of the phase diagrams are determined experimentally.

References

1. *Handbook of Binary Phase Diagrams,* vols. 1 and 2, ASM International.
2. *Iron Castings Handbook,* Iron Castings Society, Inc., 1981.
3. G. V. Raynor, "Phase Diagrams and Their Determination," p. 291 in *Physical Metallurgy,* R. W. Cahn, ed., North-Holland Publishing Company, Amsterdam, 1965.
4. G. L. Kehl, *The Principles of Metallographie Laboratory Practice,* 3rd Edition, McGraw Hill, 1949.

Terms and Concepts

Austenite
Carbon equivalent (cast iron)
Cementite
Chemical potential
Components
Composition
Constitutional diagram
Cooling curve
Degrees of freedom
Ductile cast iron

Equilibrium
Eutectic binary system
Eutectic composition
Eutectic reaction
Eutectoid composition
Eutectoid reaction
Ferrite
Gibbs phase rule
Graphite
Gray cast iron
Heating curve
Hume-Rothery rules
Hyper-(eutectic or eutectoid)
Hypo-(eutectic or eutectoid)
Intermetallic compound
Invariant reaction
Iron carbide
Lamellar structure
Lever Rule
Liquidus
Mass balance
Matrix
Monotectic reaction
Monotectoid reaction
Pearlite
Peritectic reaction
Peritectoid reaction
Phase
Phase diagram
Primary phase
Primary solid solution
Proeutectic phase
Proeutectoid phase
Secondary solid solution
Solid solution
Solidus
Solute
Solution
Solvent
Solvus curve
Tie-line
White cast iron

Questions and Exercise Problems

5.1 What are the different phases or states a pure material can be into? Construct a phase diagram for a pure material showing the transformations of the phases and describe the characteristics of these transformations.

5.2 What is a binary phase diagram and what information can we glean from it? In terms of thermodynamics, what do we mean by stable equilibrium, binary-phase diagram?

5.3 What do we mean by an alloy? In general, what is the difference in the melting behavior of an alloy from that of a pure component?

5.4 When we mix or alloy two elements together, cite the four cases that may result and indicate an example of each from the cases that were discussed in the chapter.

5.5 What is the significance of the Gibbs phase rule? Explain each of the terms in the rule and indicate the maximum number of phases that may be present in equilibrium in (a) a pure component system, and (b) a binary system. How many variables can we specify when the maximum number of phases occur?

5.6 What is the significance of the liquidus curve? the solidus curve? the solvus curve?

5.7 What is the significance of the Lever Rule and what does it really represent?

5.8 A one-metric ton of a brass alloy composed of 70 wt% copper and 30 wt% zinc (70Cu-30Zn) is desired. What are the minimum amounts of copper and zinc that need to be mixed and molten? What are the atomic percentages of copper and zinc in this alloy?

5.9 In the stoichiometric compound, Mg_5Y_2 (Mg = magnesium; Y = yttrium), what are the atomic percentages (at%) of Mg and Y? What are their weight percentages (wt%)?

5.10 Locate the following points in the Bi-Sb binary diagram (Fig. 5-8) and indicate the phases present and their relative amounts: (a) {40 wt%Sb, 250°C}, (b) {30 wt% Bi, 500°C}, and (c) {50 wt%Bi, 600°C}. Convert the compositions from wt% to at%.

5.11 An equiatomic alloy of Bi and Sb is heated under equilibrium conditions. Determine (a) the temperature at which incipient melting occurs and the composition of the liquid, (b) the compositions and relative amounts of the phases at 400°C, and (c) the temperature at which the last solid phase is observed and the composition of the solid phase.

5.12 A 40Bi-60Sb alloy is cooled under equilibrium conditions from the liquid. Determine (a) the temperature at which the first solid nucleus forms and its composition, (b) the compositions and relative amounts of the phases at 450°C, and (c) the temperature at which the last liquid phase is observed and the composition of the liquid phase.

5.13 When heated under equilibrium conditions, a Bi-Sb alloy produced a first liquid phase with 20 at%Sb. Determine (a) the composition of the alloy in wt%, (b) the melting range of temperatures, and (c) the compositions and relative amounts of the phases at a temperature half-way in the melting range.

5.14 When cooled under equilibrium conditions, a Bi-Sb alloy produced a first solid nucleus with 10 wt% Bi. Determine (a) the composition of the alloy in wt%, (b) the freezing range of temperatures, and (c) the compositions and relative amounts of the phases at a temperature half-way in the freezing range.

5.15 In the Cd-Bi binary system, Fig. 5-10, (i) describe with appropriate sketches the development of the microstructure during cooling from the liquid past the solidus temperature for the following alloy compositions: (a) a 40 wt% Bi, (b) a 40 wt% Cd, and (c) a 20 wt% Cd; (ii) for each of the latter alloys, indicate the primary phase, if any, and the relative amounts of the primary phase and the eutectic lamellar structure; and (iii) indicate an alloy composition when no eutectic lamella forms in the structure during cooling.

5.16 In the Bi-Sn binary system, Fig. 5-11, describe with appropriate sketches the expected cooling curves and the microstructures from the following alloys; (a) 15 wt% Bi, (b) 30 wt%Bi, (c) 57 wt% Bi, and (d) 80 wt% Bi.

5.17 For a 15Bi-85Sn alloy in Fig. 5-11, what is the relative amount of (β-Sn) at 139°C and at 50°C?

5.18 For a 30Bi-70Sn alloy, what are the relative amounts of the (i) primary phase, (ii) eutectic lamellae, and (iii) the (Bi) phase at (a) 138°C, and (b) 50°C?

5.19 What are the relative amounts of the (Bi) and the (β-Sn) in the eutectic lamellar structure at (a) 138°C, and (b) 50°C?

5.20 For an 80Bi-20Sn alloy, what are the relative amounts of the (i) primary phase, (ii) eutectic lamellar structure, and (iii) the total (Bi) phase at (a) 138°C, and (b) 50°C?

5.21 Sketch the hypothetical binary equilibrium phase diagram of A and B from the following data. Label all the phase fields and points in the diagram, the liquidus and the solidus curves, the solvus curves, and the primary solid solutions. You may connect points with straight lines.

Melting points: A = 700°C; B = 1,000°C

Eutectic Temperature = 500°C

Liquid composition in equilibrium at the eutectic temperature = 30 wt% A.

Solubilities at 500°C: B in A = 15 wt%; A in B = 20 wt%.

Solubilities at 25°C: B in A = 5 wt%; A in B = 7 wt%.

For this system, what are the relative weights of the phases present for a 70 wt% B at (a) 499°C, and (b) 25°C.

5.22 Sketch the hypothetical binary equilibrium phase diagram of A and B from the following information:

a) Melting points: A = 850°C, and B = 350°C
b) A and B form a stoichiometric compound, AB_2, which melts at 1,000°C; neither A nor B is soluble in the compound.
c) A eutectic invariant point is at (70 at%A, 700°C).
d) A peritectic invariant point is at (80 at%B, 500°C).
e) The liquid at 500°C has a composition of 95 at% B.
f) The solubilities of B in A are 10 at% at 700°C and 2 at% at 25°C.
g) The solubility of A in B is 25 at% at 25°C.

Connect points with straight lines. Label all the phase fields and write the three-phase reactions.

5.23 Indicate the difference and the applicability of the iron-iron carbide diagram from that of the iron-graphite diagram.

5.24 What is steel? Cast iron?

5.25 What are the allotropic forms of pure iron and the transformation points? What are the maximum solubilities of carbon in the various allotropic forms? Why does austenite have greater solubility for carbon? What type of solid solution is carbon in iron?

5.26 What do the lines A_{e_1}, A_{e_3}, and A_{cm} in the Fe-Fe_3C diagram signify?

5.27 Consider a steel (iron) with 0.2 wt% C. (i) What are its A_{e_1}, A_{e_3} temperatures? (ii) Describe the changes in phases and its structure as it is heated under equilibrium conditions to 850°C. (iii) At 750°C, determine the equilibrium compositions and the relative amounts of ferrite and austenite?

5.28 For a 0.30 wt% C in iron (steel), (i) what are the A_{e_1}, A_{e_3} temperatures? (ii) what is the primary or the proeutectoid phase and at what temperature does it initially form during equilibrium cooling from the austenite field? (iii) what are the relative amounts of the proeutectoid phase, the pearlite structure, and the total cementite (iron carbide) at 726°C?

5.29 For a 1.10 wt%C in iron (steel), (i) what are the phases present at 400°C? at 800°C? at 900°C? (ii) what are the relative amounts of the phases at these temperatures?

5.30 When the 1.10 wt% C steel is cooled under equilibrium conditions from 925°C, (i) what are the relative amounts of the primary or proeutectoid phase, the pearlite, and the total ferrite phase immediately below the eutectoid temperature at 726°C? and (ii) what is the weight fraction of pearlite at 400°C?

5.31 Given the density of austenite as 7.69 g/cm^3 and the densities of ferrite and cementite in Example 5-7, calculate the percent change in volume as a 0.77 wt%C-austenite transforms to pearlite at the eutectoid temperature?

5.32 A cast iron contains 3.5%C and 2.5%Si. What is its carbon equivalent?

5.33 A cast iron contains 3.2%C, 3.0%Si and 0.6%P. What is its carbon equivalent?

5.34 Where do the names gray, nodular, and white cast irons come from?

5.35 For the phase diagram in Fig. 5-28, indicate the phases present as we move from 100%Al to 100%Mn at (a) 800°C, and (b) 850°C.

5.36 What is metallography? thermal analysis? What is the significance of metallography in conjunction with thermal analysis in determining the liquidus and solidus curves?

6

Diffusion In Solids

6.1 *Introduction*

In Chap. 5, we discussed how the microstructures in the eutectic and eutectoid systems developed during the cooling from the liquid state. The initial development of the microstructures requires that the atoms from the liquid state (with amorphous or noncrystalline structure) move and rearrange to form small clusters of atoms with the crystalline structure of the solid. These clusters are called nuclei of the solid phase. The second step in the microstructure development is for the nuclei to grow and the growth requires that atoms move to the nuclei in order for the latter to increase its size. The whole process is referred to as a *nucleation and growth (N + G) process* and the movement of atoms is called *diffusion*. Thus, while the phase diagrams indicate the equilibrium phases that may form, the formation of these phases involves movement of atoms and requires time. The equilibrium phases are based on *thermodynamics,* but the actual formation of these phases depends on the rate of movement of atoms or *kinetics*. In the solid state, there are many instances when the thermodynamics (phase diagram) indicate an equilibrium phase but the kinetics does not allow its formation. This is in fact the whole basis of precipitation hardening (which we shall discuss in Chap. 9) in which nonequilibrium transition phases are formed to strengthen the material and these phases remain stable while the material is placed in service.

Diffusion is also involved in other thermal processing of materials, particularly if the temperature, $T > 0.4T_m$. It is involved in the homogenization process that evens up the composition of a casting of an ingot, in dissolution of particles, in stress-relieving, in annealing and recrystallization, in sintering and, as we saw in Sec. 4.4.11 in creep. Diffusion is also the process used in the doping of silicon, for example, with boron or phosphorus, to produce extrinsic semiconductors of either the p-type or the n-type. The doping of silicon is essentially the same process as the carburization or nitriding of steels by which carbon or nitrogen atoms are introduced on the surface of steel to create a hard surface for wear resistance. Thus, we note that all the thermal (heat treating) processes, that materials undergo to change their microstructures and properties involve diffusion. We need, therefore, to learn the basics of diffusion for us to at least understand how materials' microstructures and properties change during thermal processing and to intelligently process materials thermally, if we need to do it.

As indicated, diffusion is a kinetic or rate process and the distance the atoms move after a given time is dependent on the temperature. In order to be of practical and economic value, the process must be done in the shortest time possible—this means that high temperatures are used. The distance over which the effect of diffusion is realized is *macroscopic* but the manner in which this is achieved involves *microscopic atomic steps* through defects in the material. We start our discussion with the macroscopic aspects and then discuss the microscopic atomistic aspects.

6.2 Macroscopic Movements of Atoms—Fick's Laws of Diffusion

The macroscopic distance (over which matter moves) can be measured experimentally and is mathematically expressed by Fick's Laws.

6.2.1 Fick's First Law

Diffusion is one of the transport processes. The others are thermal conduction, electrical conduction, and fluid flow. All these processes follow Eq. 4-12 that involves a response to a stimulus. For electrical conduction, the stimulus is the electric field and the response is the current density, Eq. 3-6. For thermal conduction, the stimulus is a temperature gradient and the response is the heat flux, Eq. 4-7. In both of these equations, the coefficients, which are the electrical and thermal conductivity, are properties of the material.

In diffusion, the stimulus is a concentration gradient and the response is a flux of matter flowing through a unit area per unit time. Noting the similarity in the processes, the diffusion equations were developed by Adolf Fick in 1858 shortly after Fourier developed the heat conduction equations 10 years earlier. In one dimension, the flow of matter from a high concentration region to one of low concentration, is expressed by *Fick's First Law,* which is

$$J = -D\left(\frac{\partial c}{\partial x}\right)_t \tag{6-1}$$

where J is the flux of matter per unit area perpendicular to the x-direction at any instant, $(\partial c/\partial x)_t$ is the concentration gradient normal to the plane at the same instant, and D is called the *diffusion coefficient.* As in the other transport equations, D, being the proportionality constant in Eq. 6-1, is a property of the diffusing matter in the material. When the concentration at any point along x does not change with time, the process is said to be at steady state or stationary flow. The subscript t, which stands for time and the partial derivative in Eq. 6-1, can be removed and the expression becomes,

$$J = -D\frac{dc}{dx} = -D\frac{\Delta c}{\Delta x} \tag{6-2}$$

We note that when $(dc/dx) = 0$, $J = 0$. Thus, Fick's first law satisfies the requirement that there should not be a net flow of matter when there is no concentration gradient. The concentration gradient is a negative number because it is expressed as $(c_2 - c_1)$, where $c_2 < c_1$, in order for the flow of matter to occur—see Fig. 6-1. The negative sign on the right side always makes the flux a positive number. It should also be noted that since the concentrations do not change with time, (dc/dx) = constant, and since D is a constant, the flux, J, during a steady-state process is constant.

The units of J are (unit of matter)/(area × time) while those of concentration are (unit of matter)/volume. This makes the units of the *diffusion coeffi-*

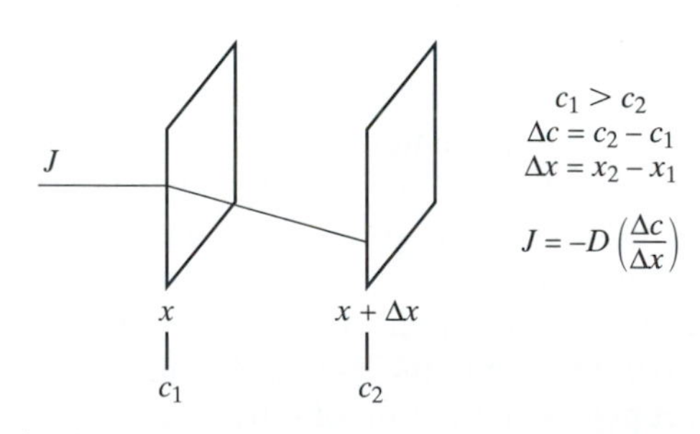

Fig. 6-1 Steady State Diffusion—Fick's First Law

cient to be in (length)^2/time, and is usually expressed as *cm^2/sec*. The unit of matter in both the concentration and the flux must be consistent, either as mass or mol or atom. As atoms, *concentration* will be in *atoms per cm^3*, while the *flux J* will be in (*atoms per cm^2 per sec*). As mass, the concentration is in mass per cm^3, while the flux J will be in (mass per cm^2 per sec). This will be illustrated in Example 6-1.

6.2.2 The Diffusion Coefficient

As indicated earlier, diffusion is a rate process that is temperature dependent. Following the lead from the electrical conductivity in Fig. 3-30 and because of the similarities, it was found experimentally that a plot of (1/T) versus (ln D) yields a straight line, as shown in Fig. 6-2. The equation for this line is

$$D = D_o \exp\left(-\frac{Q}{RT}\right) \tag{6-3}$$

where D_o is a constant with the same units as D, cm^2/sec, Q is the activation energy that may be in cal/mol or Joules/mol, R is the gas constant = 1.987 cal/mol/K or 8.314 J/mol/K, and T is absolute temperature, K. Q is obtained from the slope of the linear plot, as schematically indicated in Fig. 6-2.

Equation 6-3 expresses quantitatively the dependence of D on temperature. Because T is in the denominator of a negative exponent, we see that when we increase T, D also increases. Thus, any practical thermal process in which diffusion is of concern usually involves the highest temperature possible. On the other hand, if we do not wish or wish to limit diffusion, we need to keep the temperature as low as possible.

6.2.3 Fick's Second Law

Most practical diffusion processes involve changes of concentration with time (i.e., the concentration is a function of both time and position, c(x,t)). This is referred to as transient unsteady or nonstationary state of diffusion. The mathematical expression of this transient process is called Fick's second law and we can get an insight of the derivation of the equation with Fig. 6-3. Figure 6-3(top) shows the concentration profile in the x-direction. At any point on this curve, the partial derivative ($\partial c/\partial x$) is proportional to the flux at that point. In this case, the flux crossing the plane at x_1, $J(x_1)$, is larger than the flux passing the plane at $x_1 + dx$, $J(x_1 + dx)$. The difference in the fluxes (atoms or masses) crossing the two planes must be accumulated in the volume represented by

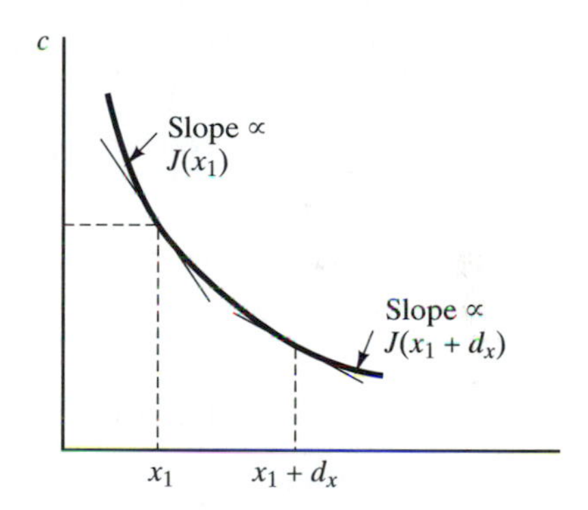

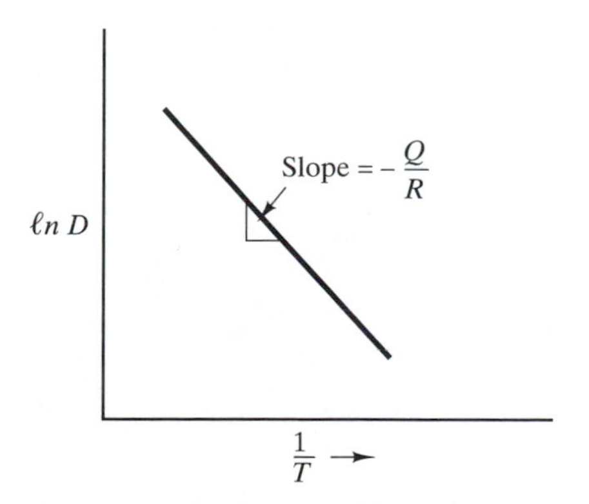

Fig. 6-2 Arrhenius plot of (1/T) versus in D is straight line whose slope yields Q.

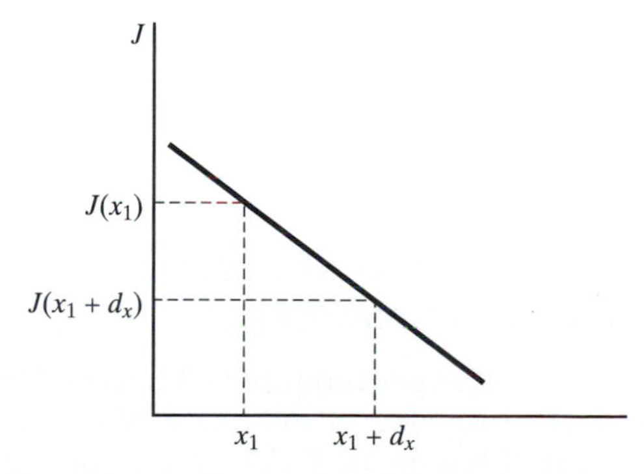

Fig. 6-3 Diffusion flux varies from point to point in unsteady state.

dx (i.e. thickness, dx, times a unit area). The mathematical expression for this is,

$$\left(\frac{\partial c}{\partial t}\right)_{x_1} dx = J(x_1) - J(x_1 + dx) \tag{6-4}$$

If dx is small, $J(x_1 + dx)$ can be related to $J(x_1)$ by the expression,

$$J(x_1 + dx) = J(x_1) + \left(\frac{\partial J}{\partial x}\right)_{x_1} dx \tag{6-5}$$

Substituting Eq. 6-5 into Eq. 6-4 and then using Eq. 6-1, we obtain

$$\frac{\partial c}{\partial t} = -\frac{\partial J}{\partial x} = \frac{\partial}{\partial x}\left(D\frac{\partial c}{\partial x}\right) \tag{6-6}$$

When D is constant,

$$\frac{\partial c}{\partial t} = D\frac{\partial^2 c}{\partial x^2} \tag{6-7}$$

where c is a function of position and time, that is, c(x,t).

The generalization of the last two equations in two or three dimensions is straightforward. In three dimensions, Eq. 6-7 becomes

$$\frac{\partial c}{\partial t} = \frac{\partial}{\partial x}\left(D_x\frac{\partial c}{\partial x}\right) + \left(D_y\frac{\partial c}{\partial y}\right) + \frac{\partial}{\partial z}\left(D_z\frac{\partial c}{\partial z}\right) \tag{6-8}$$

where D_x, D_y, and D_z, are the diffusion coefficients along the x, y, and z directions. In an isotropic medium,

$$D_x = D_y = D_z = D \tag{6-9}$$

and if D is constant, then

$$\frac{\partial c}{\partial t} = D \cdot \nabla^2 c \tag{6-10}$$

where

$$\nabla^2 = (\partial^2/\partial x^2 + \partial^2/\partial y^2 + \partial^2/\partial z^2).$$

6.2.4 Applications of Fick's Second Law in Processing

There are a lot of applications of Fick's second law in the thermal processing of materials. We shall now consider the solutions to Fick's second law in the homogenization of castings, the carburizing and decarburization of steels, and the doping of silicon to produce integrated circuits.

6.2.4(a) Homogenization [Ref. 1]. During solidification (to be discussed in Chap. 7), the preferential segregation of solutes to the liquid creates heterogeneities in composition in the ingot or the casting. The solid phase has a lower solute concentration than the remaining liquid because the solute atoms segregate to the liquid. When the latter solidifies, low and high solute concentration areas are produced; the true or average concentration is between these extremes. The homogenization process will level off those high and low concentrations resulting in an average concentration.

The variation in composition in the ingot or casting can be modeled as schematically shown in Fig. 6-4 and may be represented in one dimension as a periodic function

$$c(x) = c_o + c_m \cos\left(\frac{\pi x}{L}\right) \tag{6-11}$$

where c_o is the average composition, c_m is the amplitude of the fluctuating concentration function, and c(x) is the concentration of the solute at a point x in the x-direction. The cosine periodic function repeats every x = (2L)n, where n is an integer and 2L is the period. L is the distance along x between the maximum concentration, $c_o + c_m$, and the minimum concentration, $c_o - c_m$. Figure 6-4 shows L as half the period, but it is also the distance between maximum and the minimum. The solution to Fick's second law for this case, assuming D is constant, is

$$c(x,t) = c_o + c_m \cos\left(\frac{\pi x}{L}\right)\exp\left(-\frac{\pi^2 Dt}{L^2}\right) \tag{6-12}$$

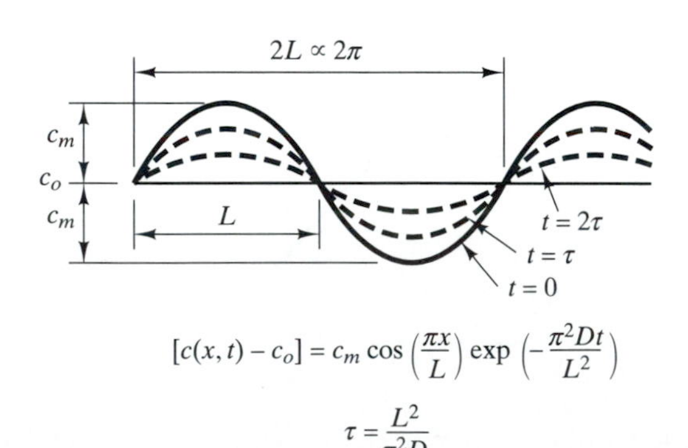

Fig. 6-4 Homogenization flattens composition fluctuation in material. [1]

From this solution, the difference in concentration, $(c - c_o)$, or the approach towards the homogenized state, can be expressed as,

$$c - c_o \approx \exp\left(-\frac{t}{\tau}\right) \tag{6-13}$$

where τ is a constant called the relaxation time and is given by

$$\tau = \frac{L^2}{\pi^2 D} \tag{6.14}$$

τ has a dimension of time. The composition profiles are schematically shown as dashed curves in Fig. 6-4, and we see that the fluctuations in composition decrease as homogenization proceeds.

The objective in homogenization is to accomplish it in as short a time, t, as possible. From Eq. 6-13, the goal is to make the left side of the equation equal zero. To achieve this situation, the ratio of (t/τ) needs to approach infinity. For this to happen, τ must be kept as small as possible and as Eq. 6-14 suggests, L must be kept as small as possible, while keeping D as large as possible. As will be discussed in solidification, L can be decreased by increasing the number of nuclei through an increased cooling rate. The increased number of nuclei per unit volume will decrease the average distance between the maximum and minimum concentrations. The way to increase D is to raise the temperature, as indicated in Sec. 6.2.2 and by Eq. 6-3.

In addition to the homogenization of ingots and castings, *homogenization is also the basis of the solution anneal for precipitation hardening and the austenitization during the heat treatment of steels.* Both of these thermal processes involve the dissolution of precipitates or second phases to homogenize them in a single phase. We must therefore remember that *these processes require time at temperature to homogenize the structure. We must allow homogenization to occur, otherwise the material will develop inconsistent microstructures and properties.*

In the electronic industry, homogenization is not a desired process. The integrated circuits depend on sharp boundaries of *n*- and *p*-type regions. When the chips are annealed and held at temperature for a long period, the boundaries will get diffused and not be sharp. This is the reason for the *rapid thermal annealing (RTA) processes in the semiconductor industry.* The chips are quickly brought to the annealing temperature and held there for only a short period and then are quenched rapidly to keep the p-n junction boundaries sharp.

6.2.4(b) Carburizing and Nitriding Processes[Ref. 2]. These are two very important steel heat-treating processes that produce a hard surface on a steel with a soft core. The hard surface is the result of the formation of a hard phase (martensite—see Chap. 9) and is desired for wear resistance, while the steel retains the soft core for toughness. All steel gears in machinery and equipment are either carburized or nitrided. In addition to its use for hardening, carburization is also used to restore or correct the carbon content on the surface of decarburized hot-worked steels. In the latter case, the decarburized layer will be a source of failure, particularly fatigue and wear, if it is not corrected.

The process consists of placing a steel with a low carbon content, C_o, in an atmosphere of high carbon or nitrogen. Modern heat-treating plants employ gaseous atmospheres of natural gas for carbon or ammonia for nitrogen. For the carburizing process, these atmospheres have what is called a carbon potential that is expressed in terms of carbon content of iron, C_a. The flux of carbon to the surface of the steel with a surface concentration, C_s, is governed by a first order surface reaction

$$J = -D\left(\frac{\partial C}{\partial x}\right)_s = k(C_a - C_s) \tag{6-15}$$

where $(\partial C/\partial x)_s$ is the carbon concentration gradient at the surface, k is the surface reaction rate constant, and D is the diffusion coefficient. Equation 6-15 indicates that the rate of carbon pickup at the surface is limited by the surface reaction for short times. For this reason, the surface carbon content changes with time, increasing rapidly at the beginning, and then more slowly at longer times.

The objective in solving Eq. 6-7 for the carburization process is to get a carbon concentration profile from the surface into the volume of the material after a certain carburization time. A carbon content at a certain depth from the surface is usually desired to give a certain hardness. This is referred to as the hardened case and the depth is the case depth. Because of the variation of the surface concentration with time, carbon concentration gradient changes with time. In addition, strictly speaking, D is also a function of carbon and it varies also with time because carbon varies with time. With these variations, solutions of Eq. 6-7 can only be achieved with numerical methods.

However, we can get an estimate of the desired case depth by assuming that the carbon surface concentration, C_s, and the diffusion coefficient, D, are constants. With these assumptions, the boundary conditions to solve Eq. 6-7 are the following:

$$C(0,t) = C_s = \text{constant} \qquad (6\text{-}16a)$$

$$C(\infty,t) = C_o = \text{constant} \qquad (6\text{-}16b)$$

and the solution for an infinite system is

$$\frac{C(x,t) - C_o}{C_s - C_o} = 1 - \text{erf}\left(\frac{x}{2\sqrt{Dt}}\right) \qquad (6\text{-}17)$$

where C_s is the surface concentration, C_o is the original carbon content of the steel, and C(x,t) is the carbon concentration at a distance x from the surface of the steel after a period of time t. *C_s, during the process, is the maximum solubility of carbon (the A_{cm} line) or nitrogen in iron at the carburizing or nitriding temperature.* Figure 6-5 shows the approximate limits of carbon solubility in austenite for eight common carburizing steels with their AISI-SAE designations. These designations are given in Table 12-3 and discussed there. For our purpose here, it is only sufficient to know that the last two digits in the codes, when divided by 100, gives the nominal wt%C. Thus, the steels 1020, 8620, and 8720 all contain nominal 0.20%C.

As time passes, the concentration profile of the carbon or nitrogen, shown in Fig. 6-6, changes from its initial step profile at time $t = 0$ to those shown by the curves mark t_1, t_2, and so forth, which are the carburizing or nitriding times. Each of these curves is a plot of C(x,t), that is, carbon or nitrogen concentration as a function of distance from the surface after a carburizing (nitriding) time. Any point on these curves satisfies Eq. 6-17.

Erf $(x/2\sqrt{Dt})$ is called the error function of $(x/2\sqrt{Dt})$ and a listing of its values is given in Table 6-1 with $z = (x/2\sqrt{Dt})$ being a dimensionless parameter. The value of the erf (z) goes from 0 to 1 as z goes from 0 to ∞. Note that Eq. 6-16b indicates $x = \infty$ and this is the reason for saying that the solution, Eq. 6-17, is for infinite systems. However, it is not the value of x, but rather the value of z, that determines when C(x,t) is close to C_o or not. Thus, x may just be 1 mm but if the value of $2\sqrt{Dt}$ is con-

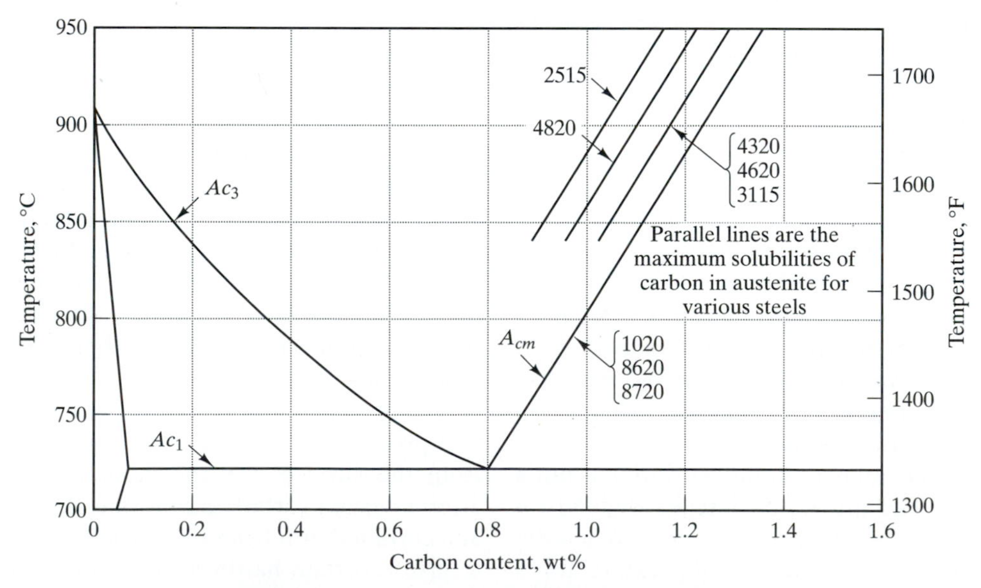

Fig. 6-5 Approximate maximum solubilities of carbon in austenite for various steels. The numbers indicated for the lines are the AISI-SAE designations for the steels [Ref. 2].

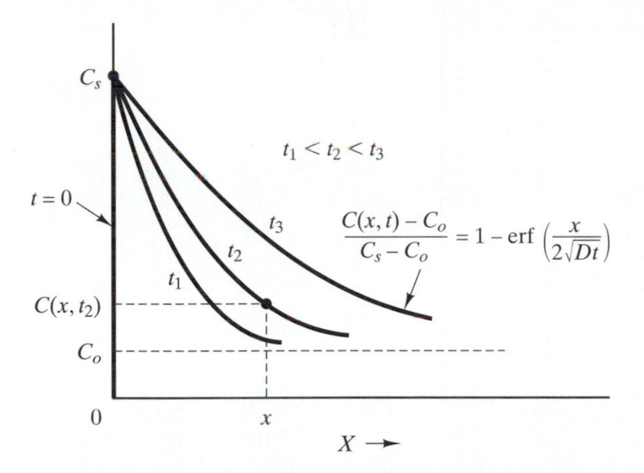

Fig. 6-6 Carbon concentration profiles with increasing time of carburization.

Table 6-1 Values of Erf z as function of z[6].

z	erf z	z	erf z
0.00	0.00000	1.20	0.91031
0.04	0.04511	1.24	0.92051
0.08	0.09008	1.28	0.92973
0.12	0.13476	1.32	0.93807
0.16	0.17901	1.36	0.94556
0.20	0.22270	1.40	0.95229
0.24	0.26570	1.44	0.95830
0.28	0.30788	1.48	0.96365
0.32	0.34913	1.52	0.96841
0.36	0.38933	1.56	0.97263
0.40	0.42839	1.60	0.97635
0.44	0.46623	1.64	0.97962
0.48	0.50275	1.68	0.98249
0.52	0.53790	1.72	0.98500
0.56	0.57162	1.76	0.98719
0.60	0.60386	1.80	0.98909
0.64	0.63450	1.84	0.99074
0.68	0.66378	1.88	0.99216
0.72	0.69143	1.92	0.99338
0.76	0.71754	1.96	0.99443
0.80	0.74210	2.00	0.99532
0.84	0.76514	2.20	0.99814
0.88	0.78669	2.40	0.9993115
0.92	0.80677	2.60	0.9997640
0.96	0.82542	2.80	0.9999250
		3.00	0.9999779
1.00	0.84270	3.20	0.9999940
1.04	0.85865	3.40	0.9999985
1.08	0.87333	3.60	0.99999964414
1.12	0.88679	3.80	0.99999992300
1.16	0.89910	4.00	0.99999998458
		∞	1.00000

siderably smaller, then z is very large, the system is infinite, and Eq. 6-17 applies. Since z is a dimensionless parameter, $\sqrt{Dt}$ must also have a dimension of length.

We can get an insight into how we can control the carburizing process by analyzing Eq. 6-17. The parameter that dictates the values of C(x,t) is z, which means that a particular value of C corresponds to a particular value of z. The parameter z has three process variables in it, x, D, and t, and we can have infinite combinations of these variables for a particular value of C. In industry, the parameter $x/\sqrt{Dt}$ is taken as the normalized carburized depth, and the left side of Eq. 6-17 as the normalized carbon content, C*. A plot of these two parameters, (i.e., C* as a function of $x/\sqrt{Dt}$) is shown in Fig. 6-7. This graph is the equivalent of Eq. 6-17 and we can also use it to estimate the carbon concentration at a particular case depth or vice versa. Carburizing practice designates the value of C* = 0.05 to indicate the depth to which carbon has diffused into the volume of the steel, (i.e., the total carburized case depth). Thus for

$$C^* \equiv \frac{C(x,t) - C}{C_s - C_o} \equiv 0.05 \qquad (6\text{-}18)$$

the value of the normalized carburized depth, $x/\sqrt{Dt}$, is found to be 2.75 from Fig. 6-7. From this, we obtain the *total case depth, x,* to be

$$\text{Total case depth, } x = 2.75\sqrt{Dt} \qquad (6\text{-}19)$$

While we can specify the total case depth, a more meaningful specification in carburized parts is to require a certain hardness at a specific depth, x, from the surface, which is called the *effective case depth.* In steels, the hardness specification is equivalent to carbon content. Thus, an effective case depth is defined as that depth at which a 0.40 wt%C concentration is attained. In both case-depth definitions, a particular value of x can be achieved by varying the product, Dt, to give the particular z value for the

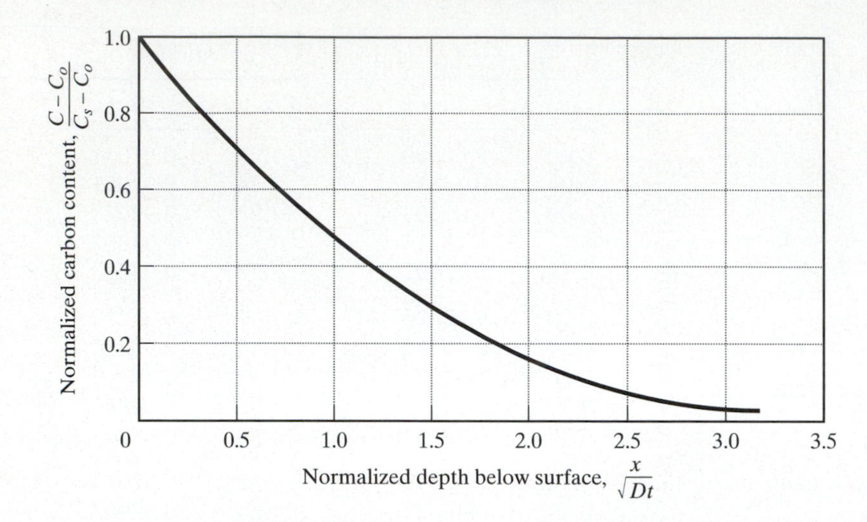

Fig. 6-7 Graphical equivalent of Eq. 6-17 for solving carbon concentration and carburized case depths [Ref. 2].

particular C. Thus, *for a desired case depth, x,* the parameters to be controlled are D and t. However, D depends on temperature, as seen in Eq. 6-3. Therefore, *the actual process parameters are the temperature and time of carburization.* While there is an infinite number of combinations of these two variables, the desired properties of the carburized case are obtained when the temperature of carburization is limited to about 900–950°C. Table 6-2 shows the diffusion coefficient expressions for carbon and nitrogen diffusion in austenite (γ-Fe) and in ferrite(α-Fe).

The case depth discussed up to this point is for a concentration gradient beneath a plane surface on a solid of infinite magnitude. For solid slabs with finite dimensions and diffusion from both surfaces, the equations are excellent approximations for case depth as long as

$$\frac{\sqrt{Dt}}{2l} < 0.20 \tag{6-20}$$

where $2l$ is the thickness of the slab. For Eq. 6-17 to be valid, the finite thickness of the slab must be greater than twice the total case depth.

The curvature of the surface being carburized also influences the case depth when the radius of curvature is comparable in magnitude with the case depth. For convex surfaces, the case depth obtained is greater than that expected on plane surfaces. The

Table 6-2 Diffusion Coefficients of Carbon and Nitrogen in Iron. [2]

Diffusing Specie	Iron Phase/Temp. Range	Diffusion Coefficient, in mm²/s
Carbon	in γ-Fe, 800–1000°C	$D = 16.2 \exp\left(-\frac{137{,}800}{RT}\right)^*$
	in γ-Fe, 500–800°C	$D = 33.5 \exp\left(-\frac{103{,}060}{RT}\right)^*$
Nitrogen	in γ-Fe,	$D = 0.5 \exp\left(-\frac{77{,}000}{RT}\right)^*$
	in α-Fe	$D = 91 \exp\left(-\frac{168{,}460}{RT}\right)^*$

*R = 8.314 J/mol. K; T = degrees K

case depth difference is greater for a doubly-curved surface, such as a sphere, than that obtained for a singly curved cylindrical surface of the same radius. These are illustrated in Fig. 6-8. For concave surfaces, the case depth is lesser than that expected from plane surfaces. The curves in Fig. 6-8 are for a constant Dt = 0.3906, a constant normalized concentration of $C^* = 0.25$, and a case depth of 1 mm (0.04 in.) for a slab of infinite thickness. This is approximately the effective case depth when the $C_s = 1.0\%$ and $C_o = 0.20\%$. Figure 6-8 indicates that the differences between the curved surfaces and the plane surface decrease as the radii of curvatures increase. It suggests that the error in using Eq. 6-17 for the carburization of 25.4 mm (1-in.) or greater diameter bars is minimal because the case depth approaches that of the plane surface.

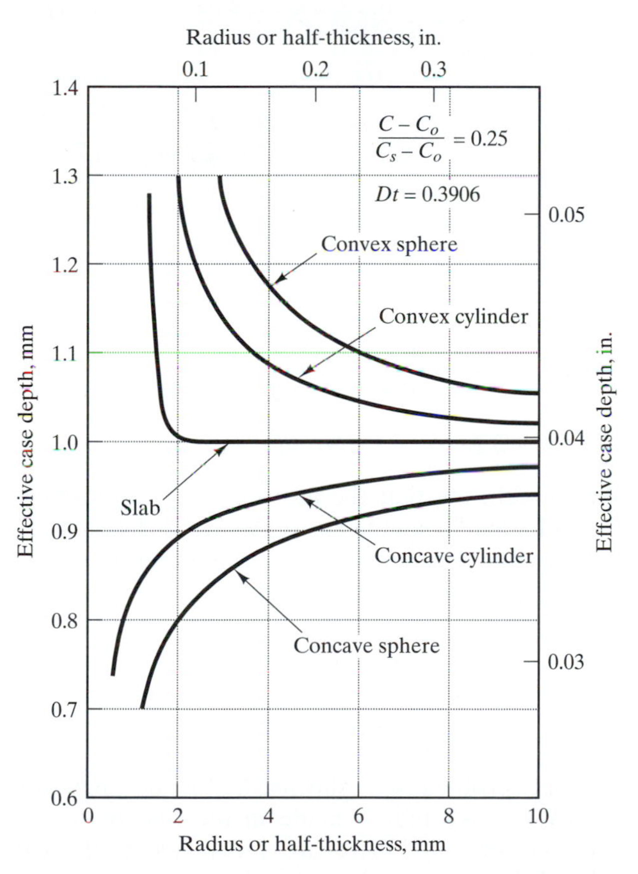

Fig. 6-8 Influence of surface curvature on the case depth [Ref. 2].

6.2.4(c) Decarburization [Ref. 1]. The reverse of carburization is decarburization, that is, instead of injecting carbon into the surface of the steel, carbon is taken out of the surface of steel. This happens usually when the heat treating atmosphere is not controlled. If the heat treating is done in an oxidizing atmosphere, decarburization of the surface will occur and the desired effects of carburization will not be realized. For heat-treating shops, the atmosphere for heat treating is carefully controlled to avoid decarburization. The problem arises in plants with no facilities to control the atmosphere when steels are normalized, annealed, or heated to high temperatures.

The solution to Fick's second law for decarburization at the heat-treating temperature is

$$\frac{C(x,t) - C_s}{C_o - C_s} = \text{erf}\left(\frac{x}{2\sqrt{Dt}}\right) \tag{6-21}$$

The concentration profiles, C(x,t), for different times are shown in Fig. 6-9.

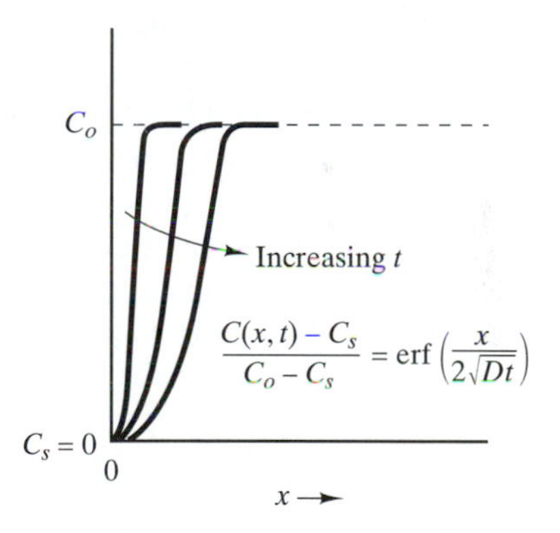

Fig. 6-9 Carbon concentration profiles during decarburization [Ref. 1].

EXAMPLE 6-1

A 3-mm thick iron sheet separates two ammonia chambers, both at 950°C. In one chamber, the ammonia imparts a constant 0.1 wt%N on the surface of the iron sheet, and in the other chamber, the ammonia imparts a constant 0.05 wt%N on the surface of the iron. Determine the flux of nitrogen through the iron sheet in (a) atoms per cm^2 per sec, and (b) gram per cm^2 per sec.

SOLUTION: Since the concentration at each surface is constant with time, this is an application of Fick's first law, which is

$$J = -D\frac{\Delta c}{\Delta x}$$

The first thing we should notice are the units of the required fluxes. This means that we must have the units of concentration in the units required. In this respect, we notice that the given concentration is in wt% and we need to convert this concentration to the appropriate concentrations of atoms N per cm^3 for (a) and grams N per cm^3 for (b).

(a) To convert the surface concentration from wt% to atoms/cm^3, we start by assuming a certain weight. We assume one gram and obtain

$$\text{(I) Atoms of } N = \frac{0.001}{14.007} \times 6.0223 \times 10^{23}, \quad \text{and}$$

$$\text{(II) Atoms of Fe} = \frac{0.999}{55.847} \times 6.0223 \times 10^{23}$$

We need the volume occupied by 1 gram of the iron sheet, and for all practical purposes, the volume of 0.999 g of Fe. We can get this by noting that a unit cell will contain a certain number of atoms. At 950°C, Fe is FCC, the lattice constant is 3.6467Å, and the unit cell contains 4 Fe atoms. Thus, the volume is

(III) Volume of 0.999 g Fe,

$$\text{cm}^3 = \frac{(3.6467 \times 10^{-8})^3}{4}\ \frac{\text{cm}^3}{\text{atoms}} \times \frac{0.999}{55.847} \times 6.0223 \times 10^{23} \text{ atoms}$$

Dividing atoms N by the volume, we convert the concentration 0.1 wt%N to

$$\text{Atoms N/cm}^3 = \frac{\text{I}}{\text{III}} = \frac{4 \times 0.001 \times 55.847}{14.007 \times (3.6467)^3 \times 0.999} \times 10^{24} = 3.29 \times 10^{20} \text{ atoms N/cm}^3$$

The concentration 0.05 wt%N is therefore $\frac{1}{2}(3.29 \times 10^{20})$ and which is also the $-\Delta c$. To get the flux, we need to know the D for N at 950°C. The expression for this is given in Table 6-2, which is

$$D = 0.5\exp\left(-\frac{77{,}000}{RT}\right) = 0.5\exp\left(-\frac{77{,}000}{8.314 \times 1223}\right)$$

$$= 2.571 \times 10^{-4}\ \text{mm}^2/\text{sec} = 2.571 \times 10^{-6}\ \text{cm}^2/\text{sec}$$

Thus, the flux

$$J = -2.571 \times 10^{-6} \times -\frac{3.29}{2 \times 0.3} \times 10^{20} = 1.41 \times 10^{15}, \text{ atoms N/cm}^2 \cdot \text{sec}$$

(b) For the concentration of gram per cm^3, we divide 0.001 g N by the volume of 0.999 g Fe, and get for 0.1 wt%N,

$$\text{gram N/cm}^3 = \frac{0.001}{\text{III}} = \frac{4 \times 55.847 \times 10^{21}}{48.5 \times 0.999 \times 6.0223 \times 10^{23}}$$

Then,

$$J = -2.571 \times 10^{-6}\frac{1}{2} \times -\frac{4 \times 55.847 \times 10^{21}}{48 \times 0.999 \times 6.0223 \times 10^{23}} \times \frac{1}{0.3} = 3.28 \times 10^{-8} \text{ g N/cm}^2 \cdot \text{sec.}$$

We should note that the answer to (b) could have been obtained from the answer in (a) by dividing the latter by the Avogadro's number and then multiplying by the atomic weight of N, thus

$$\frac{1.41 \times 10^{15} \text{ atoms}}{6.0223 \times 10^{23}} \times 14.007 = 3.28 \times 10^{-8} \text{ g N/cm}^2 \cdot \text{sec}$$

The only problem with this method is the carryover of any mistake that is made in the first part. The solutions provided here are independent of each other. It is also possible to use the density of iron at room temperature to get the concentration of N in atoms/cm^3, but we should be mindful that this density will be slightly higher than that at an elevated temperature. Thus,

Vol. of iron = 0.999/7.874 and

$$\text{atoms N/cm}^3 = \frac{0.001 \times 6.0223 \times 10^{23} \times 7.874}{14.007 \times 0.999}$$

and

$$J = 2.571 \times 10^{-6}$$

$$\times \frac{1}{2} \frac{0.001 \times 6.0223 \times 10^{23} \times 7.874}{14.007 \times 0.999} \frac{1}{0.3}$$

$$= 1.452 \times 10^{15}$$

We see that this answer is slightly higher than that in part (a) due to the higher density.

EXAMPLE 6-2

An 8620 and a 4820 steel are both carburized at 900°C for eight hours in an atmosphere whereby the maximum solubility of carbon in austenite is maintained on the surface from the start. For each steel, determine: (a) the total case depth and carbon concentration at this depth, and (b) the effective case depth for 0.40%C.

SOLUTION: We should first note that the nominal carbon composition C_o of these steels is 20/100 = 0.20%C. At T = 900 + 273 = 1173 K, the diffusion coefficient D is

$$D(\text{C in } \gamma\text{-Fe}) = 16.2 \exp\left(-\frac{137{,}800}{8.314 \times 1173}\right)$$

$$= 1.183 \times 10^{-5}, \text{ mm}^2/\text{sec}$$

and

$$t = 8 \times 3600 \text{ sec.}$$

(a) To calculate the total case depth, the definition requires the depth to have the normalized concentration $C^* = 0.05$ and the total case depth, x, in Eq. 6-19.

$$x = 2.75\sqrt{Dt} = \sqrt{1.183 \times 10^{-5} \times 8 \times 3600}$$

$$= 1.605 \text{ mm for both steels.}$$

The carbon concentration at this depth can be obtained from the normalized C*,

$$\frac{C(x) - C_o}{C_s - C_o} = 0.05$$

We need C_s for each steel. We use Fig. 6-5 for this and find $C_s = 1.24\%$ for 8620 at 900°C, and $C_s = 1.1\%$ for 4820 at 900°C. Substituting these values and $C_o = 0.20$, we get

$$C(x) = 0.252 \text{ for } 8620, \quad \text{and}$$

$$C(x) = 0.245 \text{ for } 4820$$

(b) For the effective case, we shall use both Fig. 6-7 and Eq. 6-17. For 8620, the normalized composition at the effective case depth is

$$\frac{0.4 - 0.20}{1.24 - 0.20} = 0.1923$$

From Fig. 6-7, we obtained the normalized case depth, $x/\sqrt{Dt} = 1.8$, thus

$$x = 1.8 \times \sqrt{1.183 \times 10^{-5} \times 8 \times 3{,}600} = 1.051 \text{ mm}$$

Using Eq. 6-17,

$$0.1923 = 1 - \text{erf}\left(\frac{x}{2\sqrt{Dt}}\right)$$

$$\text{erf}\left(\frac{x}{2\sqrt{Dt}}\right) = 1 - 0.1923 = 0.8077$$

From Table 6-1, we find that

$$\frac{x}{2\sqrt{Dt}} = 0.92 \quad \text{and} \quad \frac{x}{\sqrt{Dt}} = 1.84,$$

which is basically the same as that from Fig 6-7.

For 4820,

$$\frac{0.4 - 0.20}{1.1 - 0.20} = 0.2222$$

From Fig. 6-7, $x/\sqrt{Dt} = 1.7$, from which we obtained x = 0.992.

Using Eq. 6-17, erf $(x/2\sqrt{Dt}) = 0.7778$ and from Table 6-1, $x/2\sqrt{Dt} = 0.86$, and $x/\sqrt{Dt} = 1.72$, again basically the same as that from Fig. 6-7.

EXAMPLE 6-3

For the same effective case depth at 0.4%C in Example 6-2 for both steels at 900°C, determine how long it will take to achieve this at 850°C.

SOLUTION: From Example 6-2, effective case depth at 0.4%C,

$$x = 1.051\text{mm for } 8620 \quad \text{and}$$
$$x = 0.992 \text{ mm for } 4820,$$

and we are asked to determine the time of carburization, t, at T = 850 + 273 = 1123 K.

$$D = 16.2 \exp\left(-\frac{137{,}800}{8.314 \times 1123}\right)$$

$$= 6.305 \times 10^{-6}, \text{mm}^2/\text{sec}$$

and $C_s = 1.11$ for 8620 steel, and $C_s = 0.98$ for 4820 steel at 850°C. The approach is to determine the normalized concentration and then the normalized case depth, from which t can be calculated.

For 8620,

$$\frac{0.4 - 0.2}{1.11 - 0.20} = 0.22, \text{ and } \frac{x}{\sqrt{Dt}} = 1.7, \text{ and}$$

$$\sqrt{Dt} = \frac{1.051}{1.7}$$

$$t = \left(\frac{1.051}{1.7}\right)^2 \times \frac{1}{6.305 \times 10^{-6}}$$

$$= 60{,}621 \text{ sec} = 16.84 \text{ hours.}$$

For 4820,

$$\frac{0.4 - 0.2}{0.98 - 0.20} = 0.256, \text{ and } \frac{x}{\sqrt{Dt}} = 1.6, \text{ and}$$

$$\sqrt{Dt} = \frac{0.992}{1.6}$$

$$t = \left(\frac{0.992}{1.6}\right)^2 \times \frac{1}{6.305 \times 10^{-6}}$$

$$= 60{,}967 \text{ sec} = 16.94 \text{ hours.}$$

We see in this example that temperature is a very significant variable in carburization. By lowering the temperature 50°C, we doubled the time from 8 hours to more than 16 hours. We should also note that we cannot assume the surface concentration of carbon to remain constant when we change the temperature.

EXAMPLE 6-4

It is desired to carburize a 1010 steel to give an effective carbon composition of 0.30 wt% at a case depth of 0.15 cm beneath the surface. Assume that the maximum solubility of carbon is attained immediately on the surface of the steel as soon as carburization starts. Determine the carburizing temperature if the process must be done within an eight-hour shift.

SOLUTION: This is an example of designing a carburizing process. The steel in this case is unalloyed (plain) carbon steel because of the 10 prefix on the code, and the nominal carbon composition, C_o, from the last two digits is 10/100 = 0.10%C. We can vary both T and t with the constraint of maximum t = 8 hours to achieve a carbon composition, C(x,t) = 0.30% at x = 0.15 cm = 1.5 mm. We shall take as an example t = 8 hours, and for this the T is a minimum.

In examining Eq. 6-17,

$$\frac{C(x,t) - C_o}{C_s - C_o} = 1 - \text{erf}\left(\frac{x}{2\sqrt{Dt}}\right)$$

we note that we have two unknowns in C_s and D, and therefore need to have another relationship. This relationship is Fig. 6-5 where C_s depends on T (the line for 1020 steel will be used because 1010 is also unalloyed), from which we can obtain D from the relationship in Table 6-2. Because there is no explicit expression for C_s as a function of T, we need to do this by an iteration, trial-and-error method. The procedure is (1) assume T, (2) get C_s from Fig. 6-5 and solve for D, (3) determine the normalized concentration above (left side) and compare this with the value obtained for the right side. These two values must be equal to get the correct T. If they are not equal, we must assume another temperature and repeat steps (1) through (3), until the left and right values are equal or almost equal. Thus,

(a) we assume first, T = 900°C = 1173 K, and for 1010 steel, C_s = 1.23% (using 1020 curve) in Fig. 6-5. Also,

$$D = 16.2 \exp\left(-\frac{137{,}800}{8.314 \times 1173}\right)$$

$$= 1.183 \times 10^{-5}, \text{mm}^2/\text{sec}$$

left side: $$\frac{0.30 - 0.20}{1.23 - 0.10} = 0.2832;$$

right side: $$z = \frac{x}{2\sqrt{Dt}}$$

$$= \frac{1.5}{2\sqrt{1.183 \times 10^{-5} \times 8 \times 3{,}600}} = 1.285$$

From Table 6-1, for z = 1.285, erf z = 0.930, and (1 − erf z) = 1 − 0.930 = 0.07 and we note that the left-side value, 0.2832 > 0.07 the right side value. Thus we repeat.

(b) For the second iteration, we assume $T = 950°C = 1223$ K, for which $C_s = 1.37$, and

$$D = 16.2 \exp\left(-\frac{137{,}800}{8.314 \times 1223}\right)$$
$$= 2.108 \times 10^{-5}, \text{mm}^2/\text{sec}$$

Left side: $$\frac{0.30 - 0.10}{1.37 - 0.10} = 0.1575$$

Right side: $$z = \frac{x}{2\sqrt{Dt}}$$
$$= \frac{1.5}{2\sqrt{2.108 \times 10^{-5} \times 8 \times 3{,}600}} = 0.9626$$

from Table 6-1, for $z = 0.9626$, erf $z = 0.825$, and $(1 - \text{erf } z) = 1 - 0.825 = 0.175$.

We note that the left-side value, $0.1575 < 0.175$, the right-side value, and further note that the left value is now less than the right value. The left value swung its value from being greater to being lesser than the right value. This means that we have bracketed the correct temperature in the range 900° to 950°C.

(c) For the third iteration, we assume $T = 940°C = 1213$ K, for which $C_s = 1.35$, and

$$D = 16.2 \exp\left(-\frac{137{,}800}{8.314 \times 1213}\right)$$
$$= 1.885 \times 10^{-5}, \text{mm}^2/\text{sec}$$

Left side: $$\frac{0.30 - 0.10}{1.35 - 0.10} = 0.16$$

Right side: $$z = \frac{x}{2\sqrt{Dt}}$$
$$= \frac{1.5}{2\sqrt{1.885 \times 10^{-5} \times 8 \times 3{,}600}} = 1.0180$$

from Table 6-1, for $z = 1.018$, erf $z = 0.85$, and $(1 - \text{erf } z) = 1 - 0.85 = 0.15$

We note again that the left value, $0.16 > 0.15$, the right value. Since the change in temperature is only 10°C, we can rightly indicate that the minimum correct temperature, $T \approx 945°C$. We can increase the temperature to shorten the time of carburization. However, excessively high temperature might increase the austenite grain size to yield bad impact toughness properties.

6.2.4(d) Doping of Pure Silicon. Doping of pure silicon is the introduction of controlled amounts of impurities, called dopants, to produce either an n-type or a p-type semiconductor. The introduction of these impurities is most economically done by diffusion and is done in two steps. The first step is called *predeposition* whereby the dopants are introduced to a relatively shallow depth (a few tenths of a micrometer) into the substrate. After the predeposition, a *drive-in diffusion* step is used whereby the dopants are diffused deeper into the substrate to provide a suitable impurity distribution without additional impurity atoms.

Predeposition. The predeposition step is usually carried out by placing the single crystal silicon wafers in a furnace at a temperature between 800° and 1200°C. An inert carrier gas with the dopant is introduced into the furnace and in the same manner as in carburizing or nitriding, the surface concentration, C_s, and the diffusion coefficient, D, are assumed constant. Because the initial dopant concentration is zero, (i.e., $C_o = 0$), the solution to Fick's second law is simply,

$$\frac{C(x,t)}{C_s} = 1 - \text{erf}\left(\frac{x}{2\sqrt{Dt}}\right) = \text{erfc}\left(\frac{x}{2\sqrt{Dt}}\right) \tag{6-22}$$

where erfc is called the complementary error function. Here, the surface concentration, C_s, is also taken as the maximum solubility of the dopant in silicon at the predeposition temperature. The solubilities of some of the dopants used in silicon are shown in Fig. 6-10 as a function of temperature. The diffusion coefficient, D, of the dopants in silicon are also shown as a function of temperature in Fig. 6-11.

A schematic plot of x versus C(x,t) for a constant Dt is shown as an erfc curve in Fig. 6-12. The total amount, $N_x(t)$, of dopant atoms introduced during predeposition into the substrate after time, t, is the area under the curve C(x,t) after time t in Fig. 6-12. Thus,

$$N_x(t) = \int_0^\infty C(x,t)dx = \int_0^\infty C_s \text{erfc}\left(\frac{x}{2\sqrt{DT}}\right)dx = \frac{2(\sqrt{Dt})_{p-d}}{\sqrt{\pi}}C_s \tag{6-23}$$

The subscript p−d on ($\sqrt{Dt}$) indicates that this factor is for the predeposition step.

Drive-in. After the predeposition step, a drive-in diffusion is done to redistribute the dopant impurity atoms. Firstly, the surface concentration, C_s, which is the maximum solubility of the dopant in silicon obtained during predeposition, may be too high for device applications. The impurities can be redistributed and thereby lower the surface concentration. Secondly, as the name implies, the dopant atoms are driven further into the substrate to get a thicker diffusion layer.

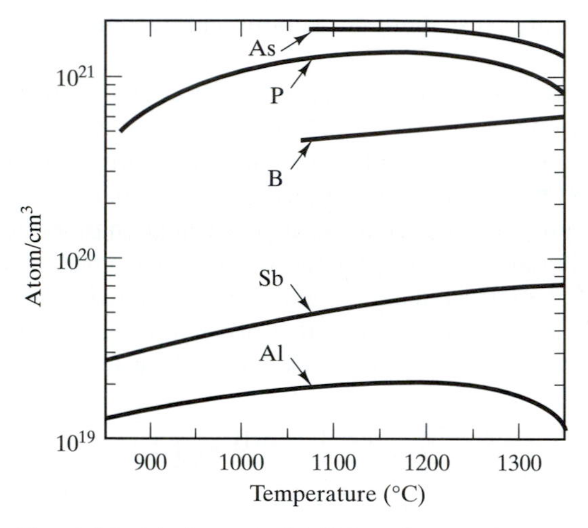

Fig. 6-10 Maximum solubilities of some dopants in silicon [Ref. 3].

In the drive-in step, the total amount of dopant atoms remains fixed and equals N_x from the predeposition step in Eq. 6-23. The solution to Fick's second law in this case assumes that all the N_x atoms are concentrated in a very narrow region and the drive-in step starts with a delta-function distribution of the dopant atoms, that is, the predeposition concentration profile is considered a "spike" as schematically shown in Fig. 6-13. The boundary conditions for this case are

$$N_x = \text{constant} \qquad \text{and} \qquad C(\infty,t) = 0 \qquad (6\text{-}24)$$

and the solution to Fick's second law is

$$C(x,t) = \frac{N_x}{(\sqrt{\pi Dt})_{d-i}} \exp\left(-\frac{x^2}{(4Dt)_{d-i}}\right) \qquad (6\text{-}25)$$

where the subscript d−i on (Dt) refers to the conditions of the drive-in diffusion. For x = 0, namely, at the surface,

$$[C_s(t)]_{d-i} = C(0,t) = \frac{N_x}{(\sqrt{\pi Dt})_{d-i}} \qquad (6\text{-}26)$$

which indicates that the surface concentration decreases as (Dt) increases. The concentration pro-

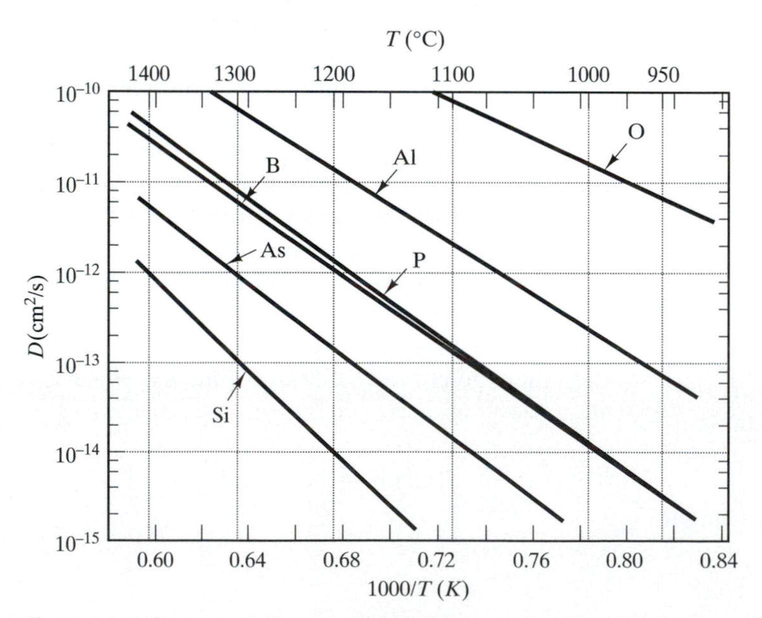

Fig. 6-11 Diffusion coefficients of some dopants in silicon [Ref. 3]

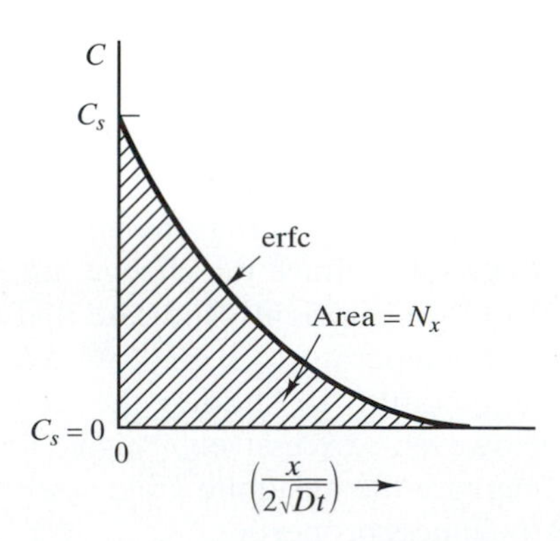

Fig. 6-12 Total number of dopant atoms, N_x, in predeposition step.

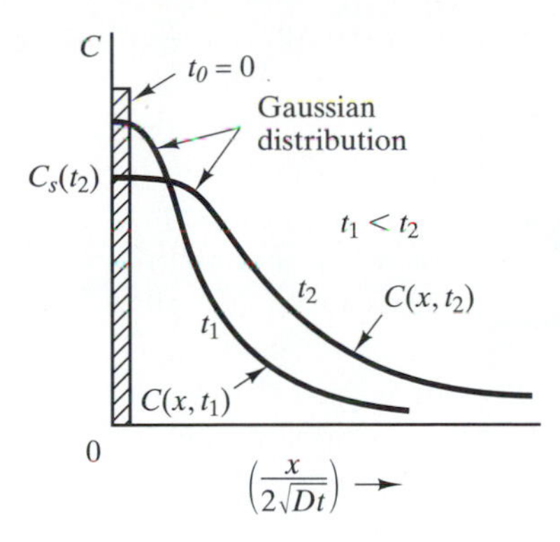

Fig. 6-13 Redistribution of dopant atoms during drive-in step.

files as a function of time of the drive-in process follow a Gaussion distribution profile and are schematically indicated in Fig. 6-13. Equation 6-26 indicates that the quantity (Dt) is the controlling factor in diffusion. We note that we can vary D (by changing the temperature) as well as the time, t. Thus, the drive-in step can be done either at high temperatures for a short time or at lower temperatures for longer times. As noted earlier, the $\sqrt{Dt}$ has a unit length and the *quantity* $2\sqrt{Dt}$ is referred to as the *characteristic diffusion length,* λ.

We should also note that the $\sqrt{Dt}$ for the predeposition step is different from that of the drive-in step. This distinction was made with a subscript p-d (i.e., $(\sqrt{Dt})_{p-d}$) when we calculate N_x in Eq. 6-23 and a subscript d−i, (i.e., $(\sqrt{Dt})_{d-i}$) in Eqs. 6-25 and 6-26. Because of the assumption that N_x is considered a "spike" in Fig. 6-13, the following condition must also be obeyed for drive-in diffusion.

$$(\sqrt{Dt})_{d-i} \gg (\sqrt{Dt})_{p-d} \tag{6-27}$$

6.2.5 Processing of Microelectronic Circuits

Since we now understand the diffusion process, we shall illustrate how actual p-n junctions are obtained in semiconductor circuits. As discussed in Sec. 3.7.8, the basis of a lot of electronic circuits in semiconductor devices is the p-n junction. In today's devices, the sizes of these circuits are literally on the order of micrometers ($1\ \mu m = 10^{-6}$ m) and thus the name, *microelectronic circuits.* The latter are interconnected to form what we refer to as *integrated circuits.* The terms—very large-scale integrated *(VLSI)* and ultra-large scale integrated *(ULSI)* circuits—refer to the number of circuits that are placed on a silicon "real estate." The area of silicon containing one unit of a manufactured device is called a *chip.*

The ability to produce a large number of circuits on such a small surface arises from the techniques of masking and then patterning by lithography. The sizes of the patterns determine the sizes of the interconnects and the areas where the dopants are predeposited and driven-in to the substrates. The procedure is illustrated in Fig. 6-14. The "mask" used is the oxide of silicon that is grown by thermal oxidation, referred to as thermox in the industry. The thickness of this oxide can be carefully controlled from previous experience. A layer of an organic material called photoresist is applied over the oxide layer on which lithography is done. Webster's Dictionary defines lithography as the process of printing from a plane surface (smooth stone or metal plate) on which the image to be printed is ink-receptive and the blank area ink-repellant. In microelectronics processing, a masking pattern is placed over the photoresist and ultraviolet light is passed through. Depending on whether the photoresist is positive (or negative), the area exposed (or unexposed) is washed away by a suitable developer to provide a window over the oxide. The oxide is etched away by hydrofluoric acid to expose the silicon surface onto which dopants are predeposited and driven-in or where metallic interconnect is deposited.

The process of making an n-p-n transistor is illustrated in Fig. 6-15. The starting substrate is a p-type silicon. An n-type epitaxial layer is deposited onto the p-type. (Epitaxy is the process of growing a thin crystalline layer on a crystalline surface.) The processes described in Fig. 6-14 are carried out from (b) to (e) when the photoresist is removed and the material is ready to be doped. Step (f) shows the introduction of a p-type element to produce p^+ regions below the windows (the oxide acts as a mask for dopant diffusion). The oxide is then removed and the whole surface is reoxidized, applied with photoresist and a pattern, exposed to UV light, the photoresist washed away, the oxide etched, and then a p-region is produced in step (g) by dopant diffusion. The steps are repeated and an n-type region is produced on top of the recently made p-region, step (h). Step (i) shows the metallization process to produce the interconnects

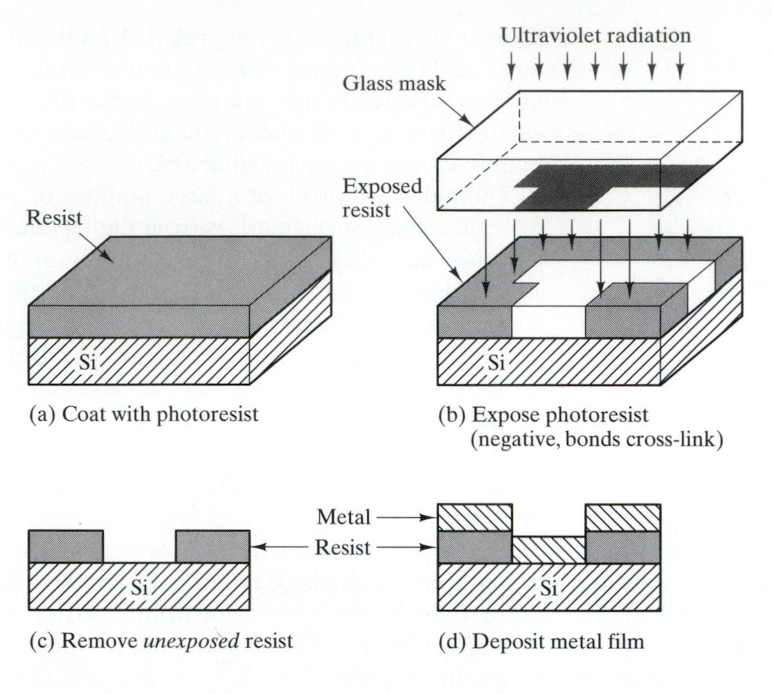

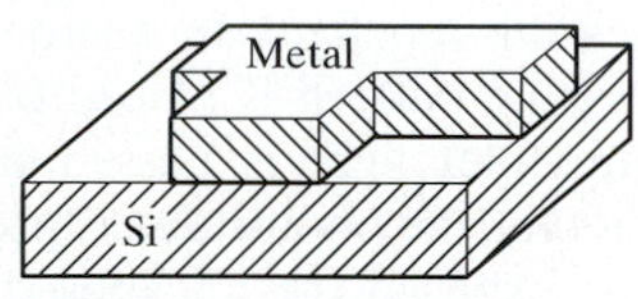

Fig. 6-14 Schematics of the lithographic methods to create metallic patterns and selected areas for infusion of dopant atoms [Ref. 3].

of the n^+-p-n regions, marked E, B, and C, which stands for the emitter, base, and collector regions of a transistor—an amplifier.

The location or depth of the junctions (i.e., n-p or p-n), after dopant diffusion is defined as that where the dopant concentration is the same as the bulk or substrate concentration. This indicates that the junction is a depleted region with no excess of either electrons or holes. The depletion region was previously discussed when p-n junctions of semiconductors were introduced in Sec. 3.7.8.

EXAMPLE 6-5

Using the temperatures 1200°C and 1300°C in Fig. 6-11, determine the activation energy for self-diffusion of silicon (i.e., diffusion of silicon in silicon).

SOLUTION: At 1200°C, $D = 1 \times 10^{-14}$, and at 1300°C, $D = 1 \times 10^{-13}$. From the equation,

$$D = D_o \exp\left(-\frac{Q}{RT}\right), \text{ we get}$$

$$\frac{D_{1300}}{D_{1200}} = \frac{\exp\left(-\dfrac{Q}{R.1573}\right)}{\exp\left(-\dfrac{Q}{R.1473}\right)}$$

$$= \exp\left[-\frac{Q}{R}\left(\frac{1}{1573} - \frac{1}{1473}\right)\right]$$

Substituting the D values and taking the natural log,

$$\ln 10 = -\frac{Q}{R}\left(\frac{1}{1573} - \frac{1}{1473}\right)$$

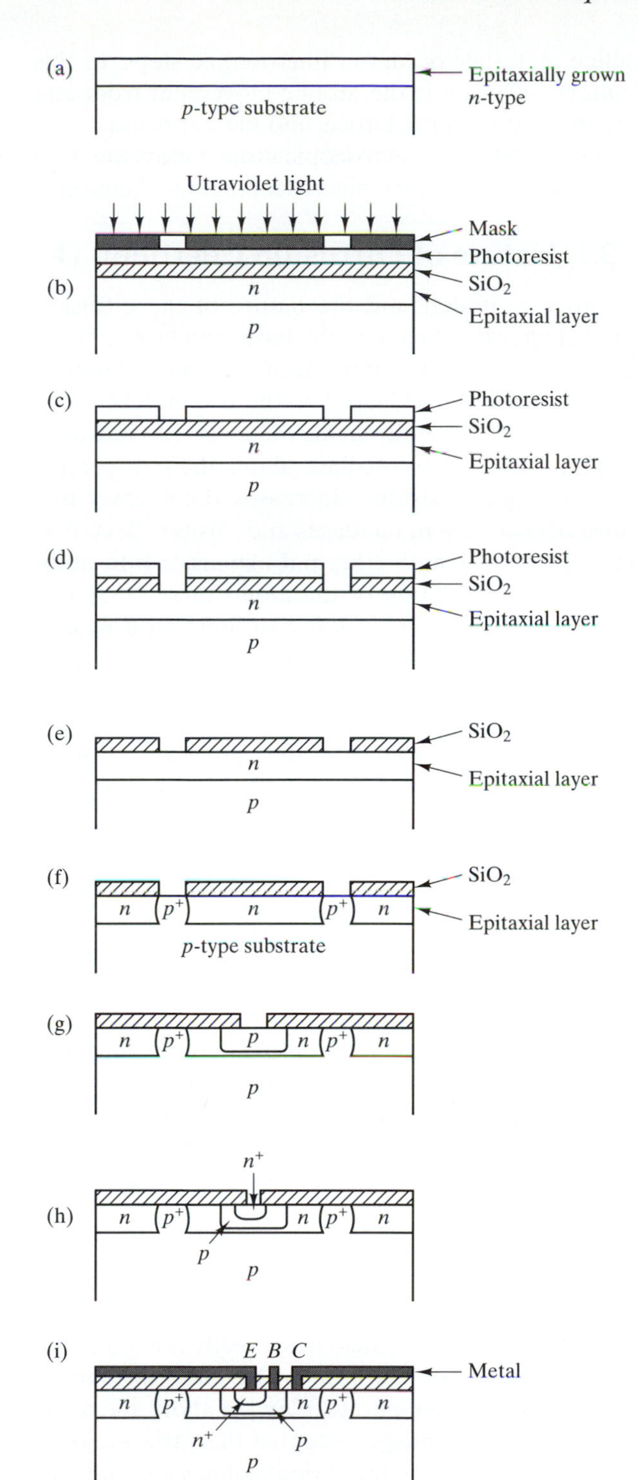

Fig. 6-15 Steps in the production of microelectronic circuits [Ref. 4].

$$Q = \frac{2.303 \times 1.987}{\left(\frac{1}{1473} - \frac{1}{1573}\right)} = 106{,}099 \text{ cal/mol}$$

EXAMPLE 6-6

A p-type silicon substrate with 5×10^{15} atoms boron per cm^3 was the starting material to manufacture integrated microelectronic circuits. To produce the n-p junctions, phosphorus atoms were diffused into the substrate. Predeposition was done at 1000°C for 60 minutes and drive-in diffusion was done at 1150°C for 5 hours. Determine the following:

(a) the junction depth after predeposition,
(b) the junction depth after the drive-in diffusion, and
(c) the surface concentration of phosphorus atoms after the drive-in.

SOLUTION: The very first item that we should understand is the meaning of n-p junction. In the discussion of p-n junction in Chap. 3, the junction was shown to be a depleted region, which means that there is an equal number of holes and electrons. Thus, because we are diffusing P atoms that donate electrons into a p-type substrate, the junction is where the concentration of P atoms is the same as the initial boron atoms in the p-type silicon substrate. Thus,

$$C(x_j) = 5 \times 10^{15} \text{ atoms P per cm}^3 \qquad \text{and}$$

$$C_o = 0 \text{ for P atoms.}$$

(a) At the predeposition temperature, T = 1000°C, and from Fig. 6-10, C_s of P in Si = 1×10^{21} atoms P/cm^3. Then using Eq. 6-22,

$$\frac{C(x_j)}{C_s} = \frac{5 \times 10^{15}}{1 \times 10^{21}} = 5 \times 10^{-6} = 1 - \text{erf}\left(\frac{x}{2\sqrt{Dt}}\right)$$

$$\text{erf}\left(\frac{x}{2\sqrt{Dt}}\right) = 1 - 0.000005 = 0.999995$$

$$\text{from Table 6-1, } \frac{x}{2\sqrt{Dt}} = 3.20$$

From Fig. 6-11, we obtain D of P in Si at 1000°C = 1.2×10^{-14} cm^2/sec, and with t = 60 × 60 sec,

$$x_j = 3.20 \times 2 \times 60\sqrt{1.2 \times 10^{-14}}$$
$$= 0.000042 \text{ cm} = 0.42 \ \mu\text{m}$$

(b) For the drive-in diffusion, we use Eq. 6-25,

$$C(x_j,t) = \frac{(N_x)_{p-d}}{(\sqrt{\pi Dt})_{d-i}} \exp\left(-\frac{x^2}{4\,Dt}\right)$$

and from Eq. 6-23,

$$(N_x)_{p-d} = \frac{2(\sqrt{Dt})_{p-d}}{\sqrt{\pi}} C_s$$

$$(N_x)_{p-d} = \frac{2\left(\sqrt{1.2 \times 10^{-14} \times 3600}\right)_{p-d}}{\sqrt{\pi}} \times 1 \times 10^{21}$$

$$= 7.416 \times 10^{15} \text{ atoms P/cm}^2$$

At the drive-in temperature of 1150°C, D of P in Si $= 4.5 \times 10^{-13}$ from Fig. 6-11 and for t = 5 hours = 5×3600 sec, solving for

$$\exp\left(-\frac{x^2}{4Dt}\right) = \left(\frac{5 \times 10^{15}}{7.416 \times 10^{15}}\right) \times \sqrt{\pi \times 4.5 \times 10^{-13} \times 5 \times 3600}$$

$$= 0.00010755.$$

Taking natural log on both sides, $-x_j^2/4Dt = -9.137$, from which

$$x_j = 2\sqrt{9.137} \times \sqrt{4.5 \times 10^{-13} \times 5 \times 3600}$$

$$= 0.000544 \text{ cm} = 5.44\ \mu\text{m},$$

We note that 5.44 μm $\gg$ 0.42 μm, as it should be.

(c) After the drive-in diffusion,

$$C_s(t) = C(0,t) = \frac{(N_x)_{p-d}}{(\sqrt{\pi Dt})_{d-i}}$$

$$= \frac{7.416 \times 10^{15}}{\sqrt{4.5 \times 10^{-13} \times 5 \times 3600}}$$

$$= 4.65 \times 10^{19} \text{ atoms P/cm}^3$$

*6.3 Microscopic Atomic Movements—Mechanisms Of Diffusion**

Up to this point, the treatment of diffusion has been confined to the macroscopic flow of matter, without regard for the crystalline structure and defects in materials. This is called the continuum approach. However, the actual movement of atoms in crystalline materials occurs in microscopic steps. In this context, diffusion is the atomic movement from one site to another in the lattice, and the net result of the infinite number of microscopic atomic movements is what we observe as the macroscopic flow of matter.

6.3.1 Nature of Diffusion Coeffient, D

In order to understand the nature of the diffusion coefficient, we fall back to the basic bonding in materials. The equilibrium interatomic distance between atoms described in Chap. 1 is the distance between the centers of the atoms at absolute zero. At any temperature, the atoms oscillate about their respective sites. As the temperature increases, the average distance between them increases and this is reflected as an increase in length (thermal expansion). In addition, as the oscillation of the atoms increases, some atoms acquire enough energy to jump to a nearby site. The problem is to relate the random jumps of an astronomical number of atoms from site to site to an overall net flow of atoms.

Leaving out the details of the derivation, in terms of atomic movements, Fick's first law is found to be

$$J = -\frac{1}{6}\Gamma\,\alpha^2\,\frac{\partial c}{\partial x} \tag{6-28}$$

where $\partial c/\partial x$ has the same meaning as in Fig. 6-1, α is the jump distance between two adjacent atomic planes 1 and 2 with concentrations c_1 and c_2, and Γ is the average jump frequency of any given solute atom in terms of the number of jumps per unit time. The factor 1/6 accounts for the assumption that the structure is cubic with six lattice sites available for every atom and that the site an atom jumps to is only one of the six sites. Comparing Eq. 6-28 with Eq.6-1, we see that

$$D = \frac{1}{6}\Gamma\,\alpha^2 \tag{6-29}$$

Thus, the diffusion coefficient is directly related to the jump frequency of the atoms. The jump distance α varies little with temperature and is about the same for most metals. Thus, we see that the variation of D with temperature and for various solutes and solvents must stem from the variation of the jump frequency, Γ.

The analysis leading to Eq. 6-28 did not assume any specific mechanism; therefore, it is quite general.

The only assumptions made were that the jumps were equally probable to all the six neighboring sites and that the jump distance α is the same for all the jumps. The assumptions are approximately true for cubic structures; hence D is isotropic in these structures. However, in non-cubic structures, the assumptions are not valid and as such, D is anisotropic.

6.3.2 Mechanisms of Diffusion, Activation Energy, and Γ

There are two basic mechanisms by which an atom moves in the structure—it can jump from one interstitial site to another or it can jump into an empty lattice site. The first mechanism is called *interstitial diffusion* and the second is called *vacancy diffusion.*

We shall discuss interstitial diffusion first and indicate what the activation energy of diffusion is and its relation to the jump frequency, Γ. Consider the interstitial solute atom (dark circle) shown in the two-dimensional lattice (light circles) shown in Fig. 6-16a. The stable positions of the interstitial atom are indicated by *A* and *B* and Fig. 6-16b indicates schematically their energy states. The movement of the solute atom from position *A* to *B* arises from its constant oscillation in all directions with a mean vibrational frequency of ν, while the solvent atoms oscillate

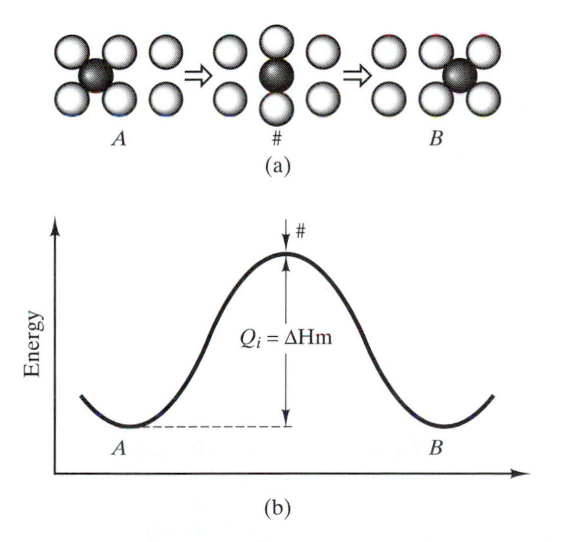

Fig. 6-16 (a) Schematic of movement of an interstitial solute from A to B through #, (b) activation energy, Q_i, required through #.

about their mean positions. When the chance to move arises, the solute will have to squeeze its way through the space between the solvent atoms. This requires work and the latter is called the activation energy, A_m, to move the interstitial atom. A_m is the energy required to bring the atom to the saddle point marked # in Fig. 6-16a and b. Not all the oscillations of the solute will bring it to this position. Only a certain number, n_ν, of the oscillations will acquire sufficient energy to get to the saddle point. The number n_ν that will acquire A_m to jump to one site is $\nu \exp(-A_m/RT)$. If N_s is the total number of neighboring sites, then the number of jumps per sec, Γ, is

$$\Gamma = N_s \nu \exp\left(\frac{A_m}{RT}\right) \tag{6-30}$$

Substituting Eq. 6-30 into Eq. 6-29, the expression for D becomes

$$D = \left[\frac{1}{6} N_s \alpha^2 \nu \exp\left(\frac{\Delta S_m}{R}\right)\right] \exp\left(-\frac{\Delta H_m}{RT}\right) \tag{6-31}$$

where A_m has been replaced by its thermodynamic equivalent, $(\Delta H_m - T\Delta S_m)$. ΔH_m is the enthalpy, ΔS_M the entropy, and T the temperature in absolute scale. Comparing Eq. 6-31 with the empirical Eq. 6-3, we note that D_o is the bracketed term in Eq.6-31, and Q_i is ΔH_m and is the energy to bring the interstitial to the saddle point. We should note that *interstitial diffusion is limited to small solute atoms, such as C, N, and H.*

Vacancy diffusion is the mechanism for movements of atoms in pure metals and of substitutional solute atoms in alloys. In pure metals, it is referred to as "self-diffusion" and in alloys, it is called substitutional diffusion. The vacancy diffusion in alloys that will be discussed here is limited to dilute alloys so that the solute atoms do not interfere with each other's movements.

In interstitial diffusion, the solute atoms jump to interstitial sites. These sites are open spaces in the crystalline structure and do not need to be created. In contrast, in substitutional or self-diffusion, the atoms move to lattice sites. Ordinarily, these sites are occupied by the solvent atoms. But because of the constant oscillatory motion of the solvent atoms, some of them will acquire enough energy to break their bonds and vacate their sites, thus we have vacancies or vacant sites. Now, in order for an atom to move to a lattice site, the lattice site must first be vacant, Fig. 6-17. Thus, for a substitutional solute or solvent atom to make a jump to a site, it must not only have the probability of having acquired the energy, A_m, to move as in the case of the interstitial solute, but it must also have the probability that the site it is jumping to is vacant. If A_v is the energy to create the vacant site, then the fraction of sites that is vacant and is in equilibrium at temperature T is,

$$N_v = \exp\left(-\frac{A_v}{RT}\right) \tag{6.32}$$

Now, the probability of two independent events (i.e., the creation and the motion of vacancy) occurring simultaneously is the product of their independent probabilities. Therefore, the probable number of jumps per second, Γ_v, in vacancy diffusion is the product of Eqs. 6-30 and 6-32, that is,

$$\Gamma_v = N_s \nu \exp\left(-\frac{A_m}{RT}\right) \exp\left(-\frac{A_v}{RT}\right) \tag{6-33}$$

The resulting equation for *D* is therefore

$$D = \left[\frac{1}{6} N_s \alpha^2 \nu \exp\left(\frac{\Delta S_m + \Delta S_v}{R}\right)\right] \exp\left(-\frac{\Delta H_m + \Delta H_v}{RT}\right) \tag{6-34}$$

Note that the observed activation energy of diffusion, Q_v, now consists of two terms: ΔH_m to move to a vacant site, and ΔH_v to create a vacancy. Thus, the activation energy for vacancy diffusion, Q_v, is always larger than that for interstitial diffusion, Q_i.

6.3.3 Diffusion along Defects and on Surfaces

The atomic movements just described occur in the lattice structure and the only defect considered is the vacancy. Because the atom movements require open areas or disorder in the lattice, we can infer that defects such as dislocations, grain boundaries, and surfaces will enhance atomic movements. This is exactly what happens. Dislocations, especially edge dislocation, have areas below the line defects that are expanded. Thus, a dislocation has an area resembling a pipe below the dislocation line where atoms can almost move freely without hindrance. In fact, this is referred to as "pipe diffusion." Grain boundaries have more atomic disorder while the surface of a material has the most defects. It is not too surprising therefore to find that, generally, atoms move easier along grain boundaries and easiest on surfaces. To differentiate these atomic movements, the diffusion within the lattice structure is called volume diffusion.

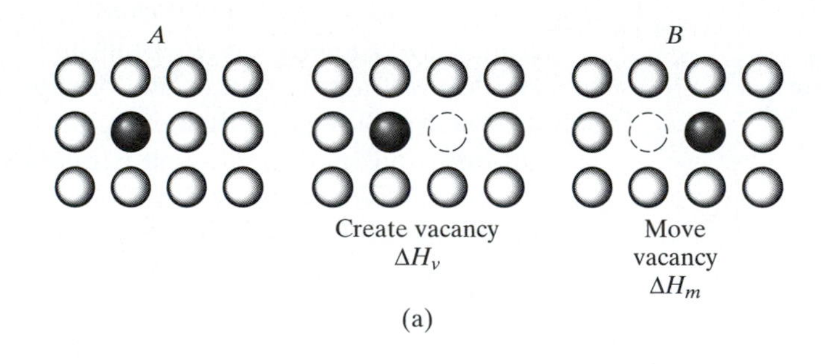

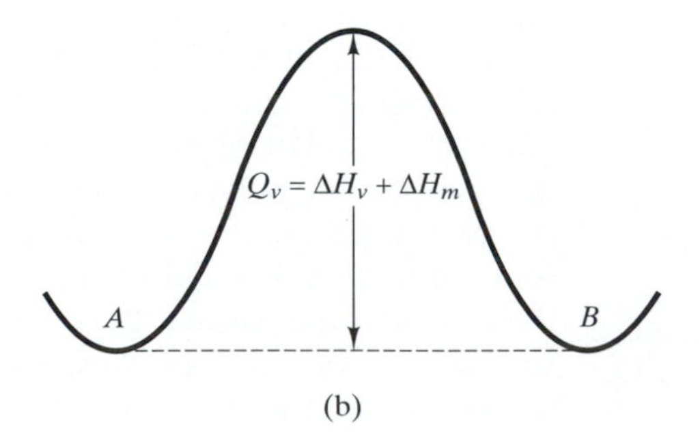

Fig. 6-17 Vacancy diffusion mechanism. (a) atom movement from A to B requires site B to be vacant first , (b) activation energy, Q_v, is the sum of energy to create and to move vacancy.

The effects of diffusion along dislocations, grain boundaries, and on surfaces are reflected by an increase in D_0 and/or a decrease in Q in Eq. 6-3 for D.

Sintering is a process that takes advantage of surface and grain-boundary diffusion and is described in greater detail in Sec. 14.3.2—Processing of Ceramics. It is the method used to produce parts from very hard and brittle materials or very expensive materials. It is commonly used for ceramic components and parts and is referred to as *a near net-shaped forming process* because it produces a product very close to its final dimension with only very little machining. The product is produced from very fine powders heated under pressure to force contact on the surfaces of the powders. The atoms on the surfaces of the powders rearrange and move to the contact areas. To facilitate the sintering process, more disorder is introduced on the surfaces by incipient melting, whereby the surfaces of the powders are selectively molten. The melting increases diffusion, helps the movement of the powders (grains), and shortens their consolidation time to the final product. The appplied pressure helps to close the spaces between the powders to decrease the porosity of the final product.

Sintering is an example of a process where atoms are taken from one part and deposited at another part, so that there is a change in shape (i.e., deformation), see Fig. 14-19. This slow deformation was referred to earlier in Chap. 4 as *vacancy creep* and is a special case of the general theory of diffusion in force fields. While it is a slow deformation, it is very important at high temperatures. It is enhanced by surfaces, such as grain boundaries, and this is the reason why single crystals of materials (not grain boundaries) are used in high temperature applications, for instance, jet engine materials

Summary

As a rate process, diffusion requires time; therefore, all processes that require atomic movements must be provided with time. Aside from time, the most significant parameter to influence diffusion is temperature. The macroscopic aspects of diffusion are expressed by Fick's laws. Homogenization or solution of precipitates and particles, carburizing and nitriding, and doping of silicon are examples of processes that can be described by Fick's second law. Aside from the soaking of ingots, solution anneal and austenitization are two processes exemplifying homogenization. Carburizing enables the production of a hard case with a soft core in steels. Diffusion enables the production of p-n junctions in semiconductors that are the bases for integrated circuits.

The microscopic aspects of diffusion indicate how the atoms move from one site to another and give physical significance to the diffusion coefficient and the activation energy of diffusion. Atomic movements occur via vacancy or interstitial mechanism within the lattice (volume diffusion) via defects such as dislocations (pipe diffusion) and via grain boundary and surface diffusion. Because of the disorder in grain boundaries and surfaces, diffusion is the fastest on them. Between interstitial and vacancy diffusion, interstitial diffusion has a lower activation energy and, therefore, occurs easier.

References

1. P. G. Shewmon, "Diffusion," p. 365 in *Physical Metallurgy,* R. W. Cahn, ed., North-Holland Publishing Company, Amsterdam, 1965.
2. "Case Hardening of Steel—Gas Carburizing," p. 135, in vol. 4, *Heat Treating, Metals Handbook,* 9th ed., ASM International, 1981.
3. J. W. Mayer and S. S. Lau, *Electronic Materials Science—for Integrated Circuits in Si and GaAs',* Macmillan Publishing Company, 1990.
4. L. Solymar and D. Walsh, *Lectures on the Electrical Properties of Materials,* 5th ed., Oxford University Press, 1993.
5. D. R. Poirier and G. H. Geiger, *Transport Phenomena in Materials Processing,* The TMS, Warrendale, PA 15086, 1994.
6. L. M. K. Boelter, et al, *Heat Transfer Notes*, McGraw-Hill, 1965.

Terms and Concepts

Activation energy, Q
Carburizing
Complementary error function
Concentration
Concentration gradient
Decarburization
Diffusion
Diffusion coefficient, D
Diffusion constant, D_o
Diffusion length
Doping of silicon
Drive-in diffusion
Effective case depth
Error function
Fick's First Law
Fick's Second Law
Flux
Grain-boundary diffusion
Homogenization
Interstitial diffusion
Jump distance
Jump frequency
Lithography
Macroscopic effect
Masking
Maximum solubility of carbon in austenite
Microelectronic circuits
Microscopic atomic steps
Nitriding
Normalized case depth
Normalized concentration
Photoresist
Pipe diffusion
Predeposition diffusion
Rapid thermal anneal
Sintering
Surface diffusion
Total case depth
Vacancy diffusion
Volume diffusion

Questions and Exercise Problems

6.1 A pure iron sheet, 1-mm thick, separates two chambers, both at 925°C. One chamber has a carburizing atmosphere and the other has a nitriding atmosphere. Assume that the surfaces attain the maximum solubilities of carbon and nitrogen, which are 1.25 wt%C and 2.2 wt% N in austenite. These values are from the iron-carbon and iron-nitrogen binary diagrams. Assuming no interaction between carbon and nitrogen atoms during diffusion, determine the flux of carbon and nitrogen through the iron sheet in (a) grams/m^2/sec, and (b) atoms/m^2/sec during steady state condition.

6.2 A 10-cm cubic aluminum container with a 5-mm thick wall contains hydrogen and is heated to 600°C. The solubility of hydrogen in aluminum at this temperature is 2.5×10^{-4} cm^3 (STP) of H_2 per gram of aluminum. The diffusion coefficient of hydrogen atoms through the aluminum is 1.13×10^{-5} cm^2 per sec. Assuming steady-state conditions, determine the loss in hydrogen from the container in (a) grams per day, and (b) liters per day. (STP stands for standard pressure and temperature, a term used in basic chemistry for gases. The volume of a mol of gas at STP is 22.4 liters.)

6.3 The solubility of gases in metals is given by the equation, $C = kp^{1/2}$, where C is the concentration of the gas in the metal, k is a constant, and p is the partial pressure of the gas in atmospheres over the metal. For hydrogen in copper at 1000°C, k = 1.4, C is in ppm (parts per million) by weight, and p is in atmospheres. Suppose one side of a 5-mm thick copper sheet forms the wall of two chambers. One chamber is at 9 atmospheres and the other has 4 atmospheres hydrogen pressure. Determine the flux of hydrogen in (a) atoms/sq. m./sec, and (b) liters/sq.m/day. The diffusion coefficient of hydrogen in copper at 1000°C is 10^{-6} cm^2 per sec.

6.4 From Fig. P6.4 below, calculate the D_o, and Q in Eq. 6-3 for each of the Group 13 and Group 15 elements in Ge, and for Ge in Ge.

6.5 Determine the activation energy of self-diffusion in silicon, that is silicon atoms diffusing in silicon from the data shown in Fig. 6-11. Use temperatures of 1,200°C and 1,300°C.

6.6 A 0.2 cm silicon wafer is exposed to phosphorus atom atmospheres at 1000°C so that it contains a constant P concentration of 10^{-2} atomic percent on one surface and 2×10^{-4} atomic percent on the other surface. Determine the flux of P in (a) atom-s/cm^2/sec, and (b) grams/cm^2/sec diffusing through the silicon wafer.

6.7 A 4320 steel is carburized in a gaseous atmosphere that immediately imparts saturated carbon content in austenite at the surface. If carburization is done for 8 hours at 950°C, determine the total case depth and the effective case depth.

6.8 If the effective case depth is desired at 2 mm below the surface with 0.40 wt%C in the 4320 steel in Problem 6.5, how long will the carburizing process be at 925°C?

6.9 It is desired to have an effective case depth of 1 mm from the surface with 0.40 wt%C in a carburized 8620 steel. If the carburizing must be achieved within 10 hours, what will be the carburizing temperature?

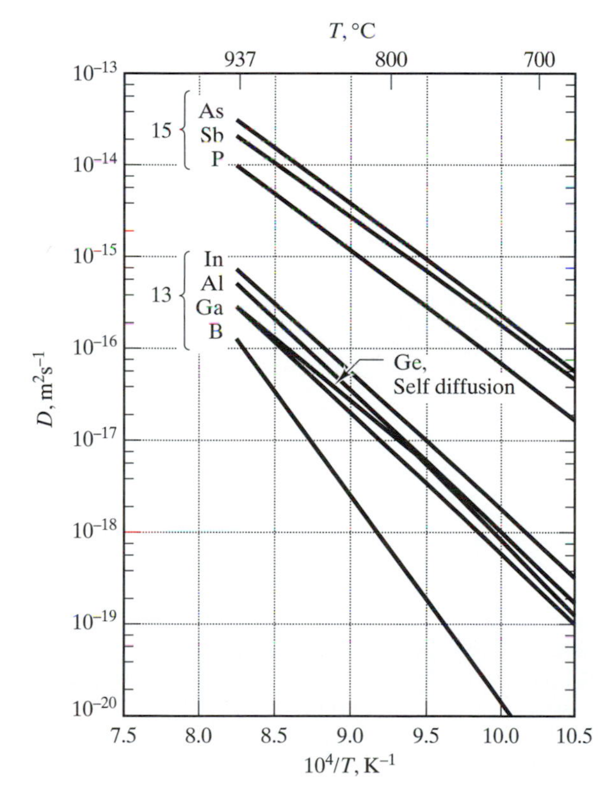

Fig. P6.4 Diffusion coefficients of Group 13 and Group 15 elements in Ge and self-diffusion of germanium.

6.10 A 1020 (0.20%C) steel was carburized at two temperatures to achieve the same effective carbon concentration of 0.40%C at 0.2 cm from the surface. It took 59.5 hours at 850°C and 14.4 hours at 950°C to obtain these results. From this data, calculate the activation energy of carbon diffusion in Joules per mol and the diffusivity constant, D_o, in mm^2 per sec.

6.11 During the manufacture of integrated circuits, arsenic (As) was predeposited and then driven-in to a silicon single crystal substrate containing 3×10^{16} boron atoms per cm^3. Predeposition of As was done at 1050°C for 2 hours, and the drive-in diffusion was done at 1,200°C for 6 hours. Determine (a) N_x, the total amount of As atoms predeposited, (b) the junction depth after the drive-in diffusion, and (c) the surface concentration of As after the drive-in.

6.12 Integrated circuits are manufactured using a p-type silicon semiconductor substrate with 5×10^{15} atoms boron per cm^3. Phosphorus atoms are predeposited at 1,000°C and then driven-in at 1,150°C. Determine (a) the time of the predeposition process to have a junction depth of 0.5 micrometer, (b) the time of the drive-in process to have a junction depth of 5 micrometers, and (c) the surface concentration of P-atoms after the drive-in process.

6.13 An analysis of a single crystal silicon wafer shows there is 1 part per million (ppm) by weight of phosphorus added to the silicon. In the manufacture of microelectronic circuits, boron was predeposited at 1,100°C for two hours. (a) If the final desired junction depth is 10 micrometer from the surface, determine how long it will take the drive-in diffusion at 1,200°C to achieve this depth. (b) What is the concentration of boron at the surface after the drive-in diffusion, expressed in ppm by weight? Atomic weight of phosphorus is 30.97.

6.14 The starting silicon material for the production of microelectronic circuits contains 0.5 ppm (parts per million) by weight of boron. Phosphorus is predeposited at 1,100°C for 1 hour. (a) If the final desired junction depth is 10 μm, determine the time for the drive-in diffusion at 1,200°C. (b) What is the surface concentration of phosphorus in ppm by weight after the drive-in? Atomic mass of phosphorus is 30.97.

6.15 In terms of microscopic atomic movements, what does the diffusion coefficient signify?

6.16 Describe and differentiate the two basic mechanisms of atomic diffusion in metals as to their applicabilities.

6.17 Based on the activation energies, why is it harder for substitutional atoms to diffuse than interstitial atoms?

6.18 Based on the structures, arrange in the order of difficulty the atomic movement (diffusion) along grain boundaries, surface, dislocations, vacancies, and interstitial sites.

7

Melting, Solidification, And Casting

7.1 *Introduction*

In this chapter, we shall discuss the very first steps in the processing of materials. These are very important steps and a judicious selection of materials must start with the knowledge of how the materials are produced. Most materials, except for the very high melting ceramics, are produced by melting them first because it is much easier to make alloying (additions) and to attain a homogeneous composition in the liquid state. The liquid has a more open structure to accommodate solutes and a lower viscosity for mixing. In addition, most of the undesired elements that can produce detrimental properties are also easier to remove or to reduce in the liquid state. These undesired elements or materials are called inclusions when they are entrapped in the solid material and are detrimental to the material properties such as fracture toughness, fatigue, ductility, and formability. A "cleanliness" rating is used to indicate how much of these inclusions are present in the materials. If we expect high-performance materials, such as excellent toughness, fatigue, and formability, the material must be "very clean". We shall discuss how these inclusions arise in the material and the processings to remove them.

In addition, we shall also discuss other defects that may arise during the solidification and casting of the liquid. A homogeneous liquid composition when solidified does not necessarily produce a homogeneous solid composition. Actual engineering materials are not cooled and solidified according to the equilibrium conditions required to produce the uniform equilibrium compositions indicated in the phase diagrams. Thus, castings, which are the forms of solids produced from the liquid, exhibit generally nonhomogeneity in composition and require the homogenization process described in the last chapter in order for the castings to exhibit good properties. We shall see also other possible defects that may arise. In order to produce good sound castings, we shall have to consider the fluidity of the liquid, the pouring practice, and the mold design.

7.2 Processing of Melts

The process of transforming a solid to a liquid is called *melting* and the starting liquid material is called the *melt*. Depending on the size (weight) of the melt, the starting materials are either recycled materials—such as scrap iron, recycled aluminum, or plastics, that need to be molten—or are in the molten state, "hot-metal" from previous processing steps. As indicated above, the various processings with melts are predominantly done to produce the desired composition and to remove or reduce the undesired impurities and gases. We shall indicate the processings of melt in relation to steel products—the most widely used technological metals and alloys.

7.2.1. Primary Processing

The primary processing is done in the melt furnace. For recycled scrap iron, the melting is usually done in an electric arc furnace. The scrap iron materials are usually sorted out as much as possible into different grades (compositions) of steel before they are placed in the furnace, but, inevitably, undesired impurities are included with the recycled materials. The facilities that primarily use scrap iron are called the nonintegrated steel plants or minimills. Integrated steel plants, on the other hand, use "hot metal" from the blast furnace and supplement this melt with scrap iron. We shall discuss the processings in these integrated plants.

The "hot metal" that is used for the primary processing of steel is the product after reducing the iron ore raw materials with carbon or coke in a blast furnace. Figure 7-1 is a schematic of the iron blast furnace showing the temperature zones and the chemical reactions that occur to reduce or extract the iron from iron ores. The products of these reactions consist of the slag and the hot metal that are taken from the bottom of the furnace and the gaseous products that are passed through various environmental protection equipment before they are discharged. The slag consists essentially of the molten oxides and is on top of the hot metal because it is lighter. The hot metal at this stage has a very high sulfur content and a carbon concentration of about 4.3 percent, both of which have to be reduced to desired levels during primary and secondary processing. The sulfur content may come from the iron oxides and the coke. Some of the sulfur may oxidize as part of the gaseous

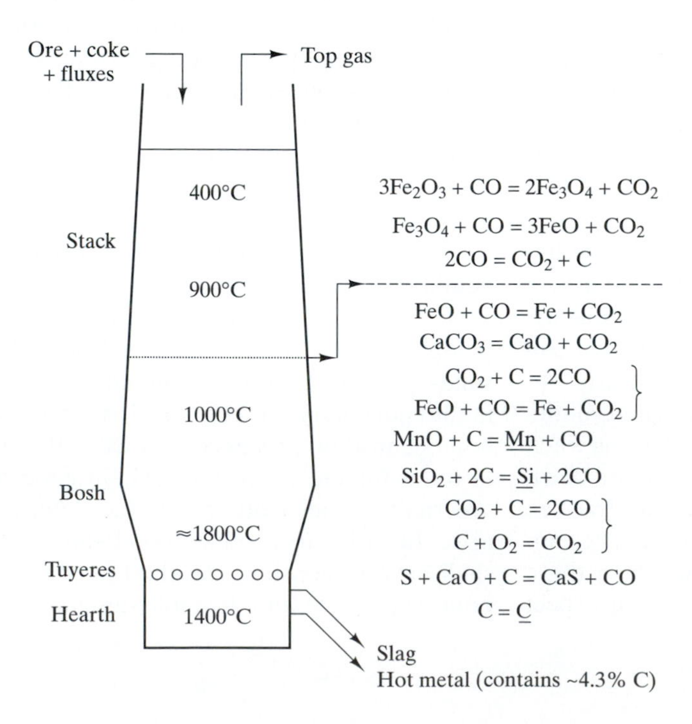

Fig. 7-1 Raw materials, reactions, temperature zones, and effluents, slag, and "hot metal" in iron blast furnace [Ref.1].

products but most of it is retained in the liquid with the hot metal. The carbon comes from the coke used to reduce the iron oxides and forms the eutectic composition with iron as indicated in Fig. 5-24.

The hot metal is collected in a transfer ladle called the "gondola" and transported to the steel making facility. During the transport stage, the metal is desulfurized in the gondola by the addition of elements such as magnesium that will react with sulfur (desulfurizing agents). As shown in Fig. 7-2 (the schematic of a basic oxygen process steelmaking shop with the various complementary operating units), the hot metal is transferred or charged to the basic oxygen furnace (BOF) and is mixed with some scrap that comprises about one-third of the total charge. The hot metal is blown or lanced with high purity oxygen primarily to reduce the carbon content, and the heat of combustion helps to melt the scrap and keep the temperature high. A 250-ton heat (batch of steel) can have its carbon content reduced to the desired level in about 20 minutes of oxygen lancing. When the desired carbon content (determined by chemical analysis) is attained, additional alloying additions can be added from the raw material storage bins to

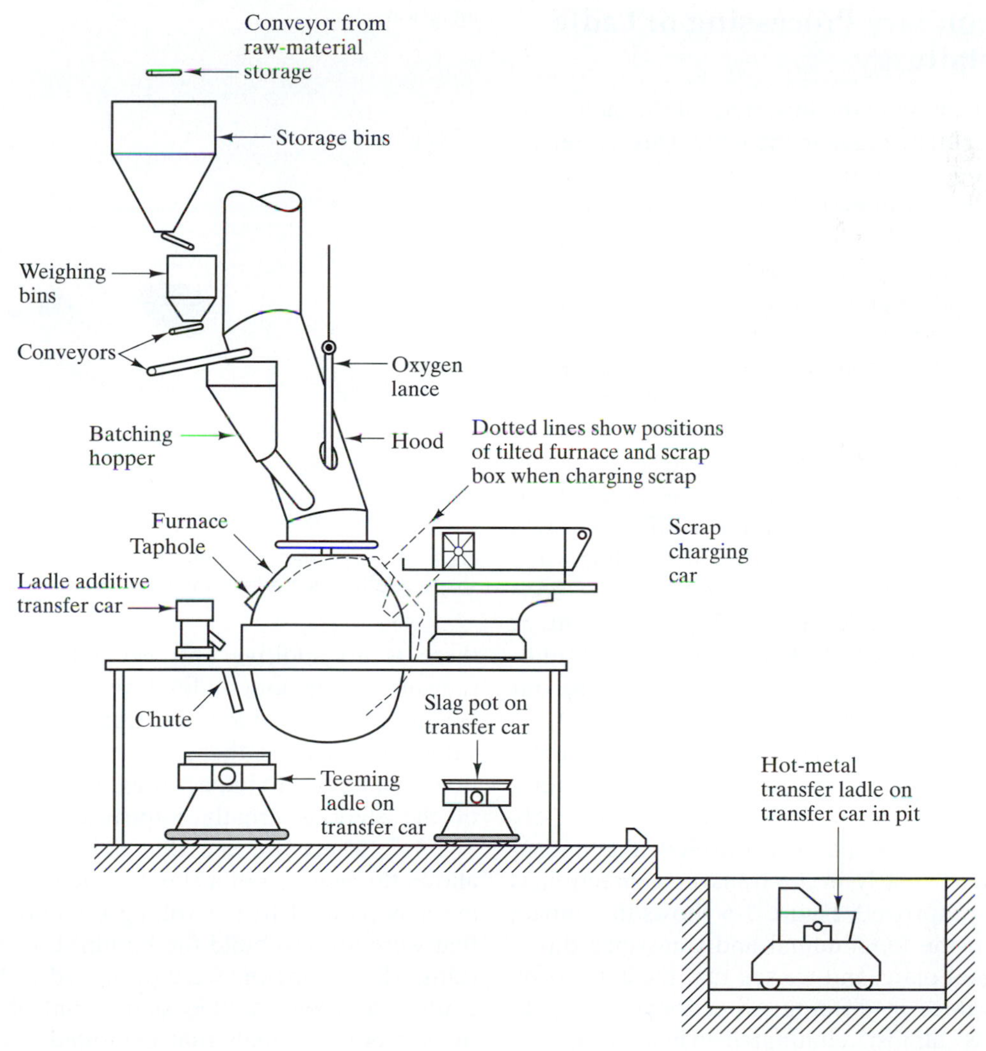

Fig. 7-2 Schematic of elevation of basic oxygen steelmaking process showing the locations of various operating units. [2]

the weighing bins, and through the batch hoppers just before the heat is tapped or transferred to the furnace. This is the extent of the *primary processing* and the melt is ready to be tapped and poured to a ladle. The taphole of the BOF is opened and before pouring the melt to a ladle, most of the slag is removed to a slag pot. Some slag is left to cover the melt in the ladle or a synthetic slag can be thrown to the ladle to cover the melt from oxidation. If the composition and extent of removal of impurities are satisfactory, the melt (in the ladle) is brought to the casting shop to be poured into ingot molds or into a continuous casting machine.

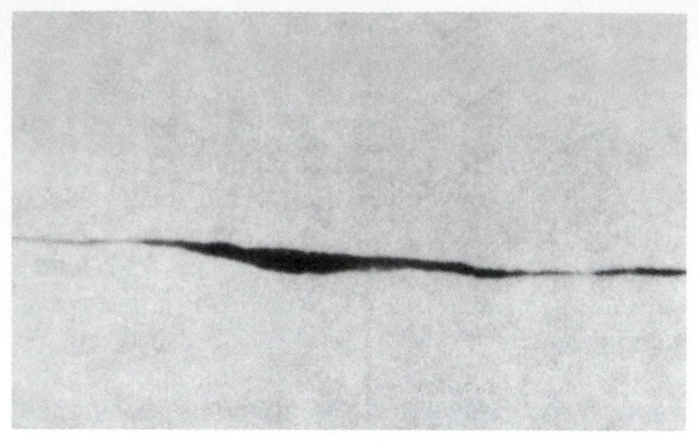

Fig. 7-3 A stringer of manganese sulfide (MnS) nonmetalic inclusion.

7.2.2 Secondary Processing or Ladle Metallurgy

Further refinement and adjustment of the melt composition can still be made in the ladle. This is referred to as *secondary refining process* or *ladle metallurgy*. Small adjustments in carbon can be done by adding coke to the melt. More expensive alloying additions are also added. In the last three decades, injection of magnesium and calcium through a lance is done to bring down the sulfur content to extremely low levels of less than 10 ppm and then a sulfide shape-controlling additive, such as cerium and rare-earth, is added. The latter addition changes the shape of the sulfur in the solid. Ordinarily, the sulfur in the composition shows up as a manganese sulfide nonmetallic inclusion. As manganese sulfide, it forms into large and long stringers after hot-rolling, such as that shown in Fig. 7-3, to produce very poor toughness materials. Reducing the sulfur content reduces the stringers and when a sulfide shape additive is made, the sulfide particles will not string out and will tend to be globular, as shown in Fig. 7-4.

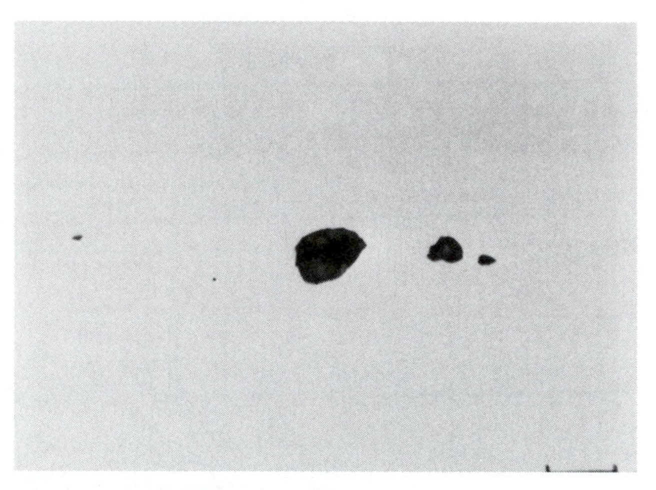

Fig. 7-4 Globular nonmetallic inclusion when a sulfide shape additive is made to melt.

The effect of inclusions in general is to reduce the Charpy upper-shelf toughness values and the total true strain ductility, as shown in Fig. 7-5a and 7-5b. When the sulfide-shape is controlled, the impact toughness, particularly, in the transverse direction, is significantly improved. Figure 7-6 shows the impact toughness in the longitudinal and transverse directions in a steel plate, and we see that the anisotropy (different values in different directions) of impact toughness is almost eliminated when cerium is added to control the shape of the sulfide. In this figure, the zero value in the cerium-sulfur ratio means there is no additive and we note the difference (anisotropy) in longitudinal and transverse values. When cerium is added, the transverse shelf energy values approached those in the longitudinal direction. In addition to impact toughness, the formability of the steel is greatly improved as indicated in Fig. 7-7, where we see that sulfide shape control allows the bending of a sheet of steel with the bending axis parallel to the rolling direction. The steels that were used to build the hundreds of miles of the trans-Alaska pipeline were produced with low sulfur content and with sulfide shape control. These features produced steels that exhibited extremely low impact transition temperatures appropriate for the Alaskan environment.

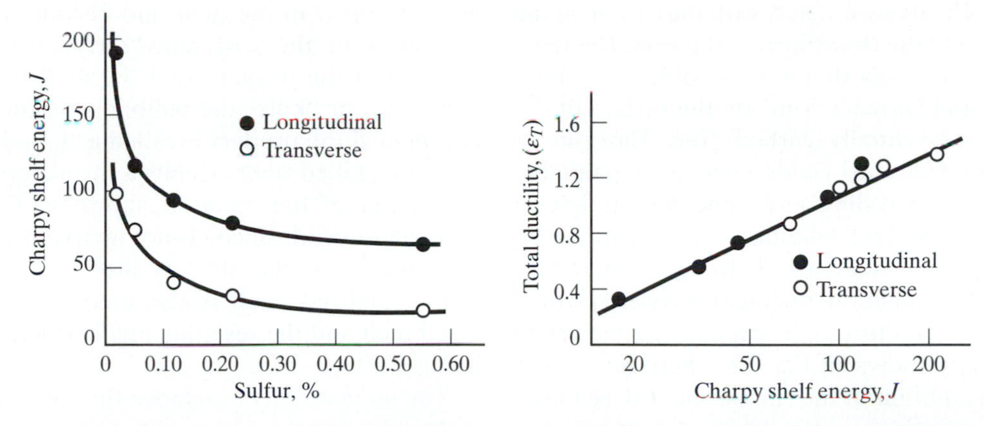

Fig. 7-5 Effect of nonmetallic inclusions on CVN upper-shelf energy and total ductility [Ref. 4].

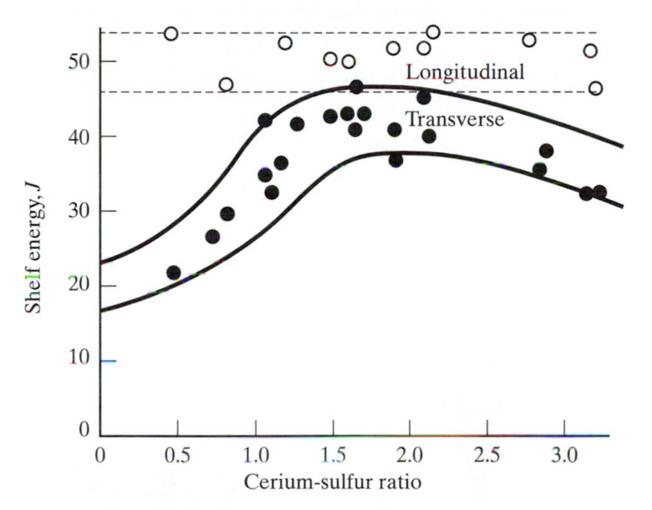

Fig. 7-6 Influence of rare-earth (cerium) additions on the transverse CVN upper-shelf energy; inclusions are globular.

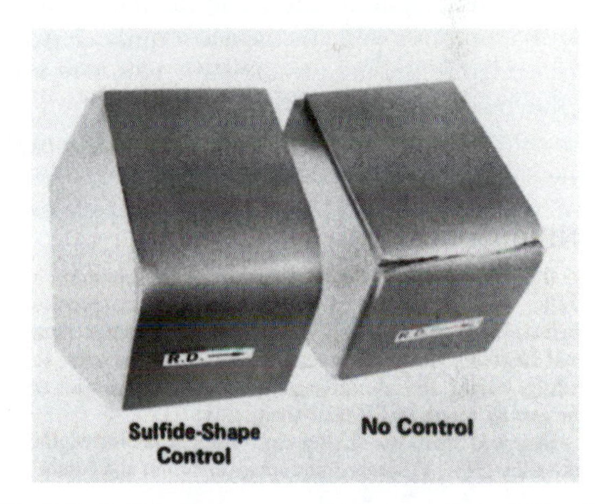

Fig. 7-7 Sulfide shape control improves formability. [5]

In addition to the alloy additions and impurities, gases in the melt can be controlled in the ladle. The most common gases in alloy melts are oxygen, nitrogen, and hydrogen that are usually dissolved in the melt in atomic form (i.e, as O, N, and H). When these atomic gases are left in the melt, they may combine to evolve as molecular gases.Thus pockets or porosities of gases may form in the casting because solids can accommodate less gas than in liquids. For wrought products, these porosities are not particularly detrimental within the volume of ingots because they can be welded together during hot-deformation, but they are obviously bad in castings sold as as-cast products. In thin wrought gauges, the gases may create blisters or flaking on the surfaces—hydrogen is known to cause blistering or thermal flaking and embrittlement. In addition to molecular gases, most of the atomic oxygen and nitrogen react with constituents of the steel to form nonmetallic inclusions. The oxygen forms oxides and the nitrogen forms cyanonitrides or nitrides that remain in the steel.

In low carbon drawing-quality steels that are finished by painting and enameling, the oxygen is left in the liquid to produce what is called a *rimmed steel.* When the liquid comes into contact with the mold, it forms solid nuclei and at the same time liberates the oxygen (O). The oxygen reacts with the carbon in the solid and essentially decarburizes the iron. The reaction produces CO gas that is noticeable by its bubbling sound and leaves a "rim" on the surface of the ingot that is essentially carbon free. This rim of almost carbon-free steel yields a very good paintable surface after cold-rolling and annealing. If left to proceed until complete solidification, the core of the ingot is chemically segregated. In order to reduce this chemical segregation, the ingot is covered with a metal cap. The structure of a "capped" rimmed steel is schematically shown in Fig. 7-8 where we see gas porosities in addition to the rim on the surface of the ingot. These porosities, within the ingot, weld together during the hot deformation.

The oxygen in the melt may be removed by addition of what are called deoxidizers. These deoxidizers are very strong oxide formers and the most common ones are aluminum and silicon. Most of the deoxidizers are added to the melt in the ladle before pouring, and the reaction products of the deoxidations are oxides, alumina or silica, that should float to the slag on the surface of the melt. However, because of the convective mixing of the liquid, some of these oxides get entrapped in the melt and become nonmetallic inclusions in the cast and "dirty" the steel. The removal of the oxygen from the melt by aluminum eliminates or "kills" the bubbling action during the casting and the product is called a "killed" steel. The ingot of a killed steel exhibits a shrinkage pipe at the top center of the ingot, as shown in Fig. 7-9. The deoxidation with silicon is not nearly as complete as aluminum and the steel is just "semikilled." The term "balanced steel" is also used for silicon deoxidized melt and the resulting ingot structure is shown in Fig. 7-10.

The ultimate way to remove the gases in the liquid melt is by vacuum degassing—this is the only way hydrogen can be taken out. As noted above, the deoxidation process leaves the possibility of entrapping nonmetallic inclusions that "dirty" the steel. A vacuum degassed melt produces a very clean steel. This principle is based on lowering the partial pressure of hydrogen (or any gas) in the atmosphere to reduce the amount of gas in the liquid, according to Sievert's Law,

$$[C_{gas}]_L = K[P_{gas}]^{1/2} \quad (7\text{-}1)$$

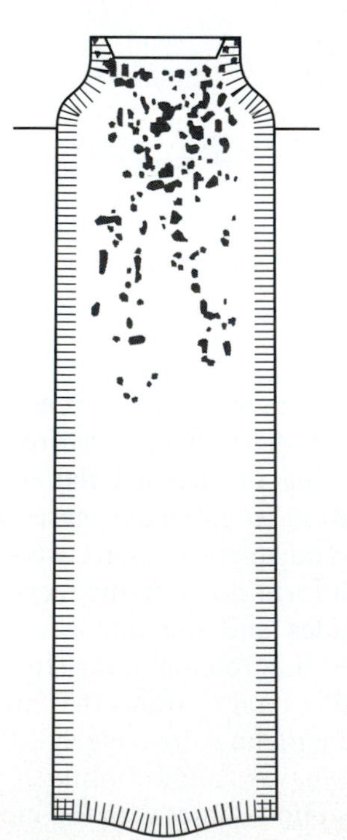

Fig. 7-8 Rimmed steel ingot.

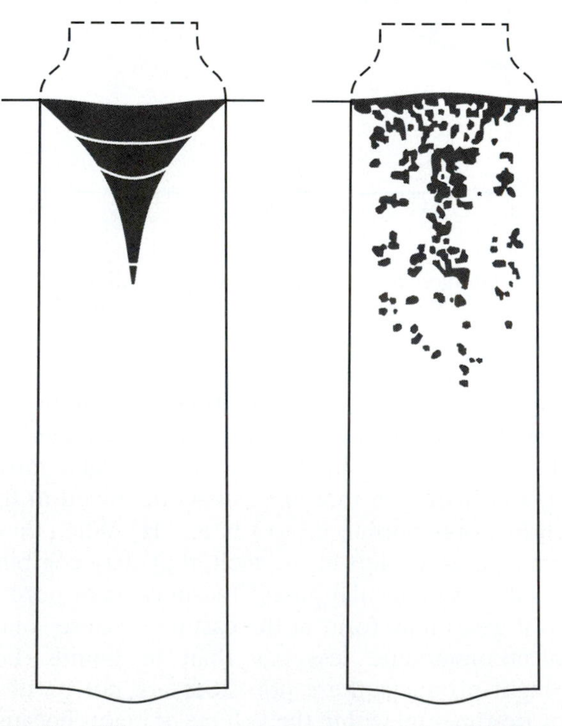

Fig. 7-9 Killed steel ingot.

Fig. 7-10 Semikilled steel ingot.

where $[C_{gas}]_L$ is the atomic concentration of the gas in the liquid, P_{gas} is the partial pressure of the gas in atmospheres over the liquid, and K is the proportional constant. Vacuum degassing involves reducing P_{gas} in order to reduce $[C_{gas}]_L$.

Another way of decreasing the hydrogen content in the liquid is to bubble inert gas. This was found to be a secondary benefit in the *argon-oxygen decarburization (AOD) process*. The AOD was specifically developed to decrease the carbon content in stainless steels with higher chromium contents. This process has allowed the production of what are called super-stainless steels with chromium contents as much as 26 percent or higher and with extra low carbon content of 0.02 percent. It consists of simply injecting the liquid in a special AOD vessel with the proper ratio of argon and oxygen to lower the carbon. The secondary benefits of this process are the decrease in hydrogen and nitrogen contents, as well as some desulfurization. Because of these, the process has found wide application also in the refining of low- and high-alloy nickel grades.

We note that the foregoing processings are directed towards producing a "clean" steel. Thus, every attempt is made to retain this cleanliness henceforth and to prevent reoxidation. During the casting operation, the "melt" is protected from the oxidizing environment by shrouding or by vacuum as indicated above. Figure 7-11 shows the shrouding of the melt in a continuous casting operation.

7.2.3 Remelting

During solidification and as indicated in Fig. 7-12, the liquid will contain more of the solute than the solid phase that forms from it. This fact is expressed by the following equation:

$$\kappa = \frac{C_S}{C_L} \tag{7-2}$$

where κ is the partitioning or distribution coefficient and is less than one, C_S is the concentration of the solute in the solid, and C_L is the concentration of the solute in the liquid phase. This is what leads to heterogeneity in composition of castings and necessitates the homogenization process to be done in all alloy castings.

The principle embodied in Eq. 7-2 is used in the remelting processes of zone refining used in the semi-conductor industry, vacuum arc remelting (VAR), and electroslag remelting (ESR) shown in Figs. 7-13, 7-14, and 7-15. All of these processes involve partially remelting the ingot, or material, and then resolidifying to a much cleaner or purer material. In zone refining, the ingot is locally heated and melted under vacuum by an induction coil. As the induction coil is slowly moved from one end to the other, the local or zone melt moves with it and the previous locally-melted material solidifies. The liquid retains most of the impurities, as indicated by Eq. 7-2, and thus a very high-purity solid can be

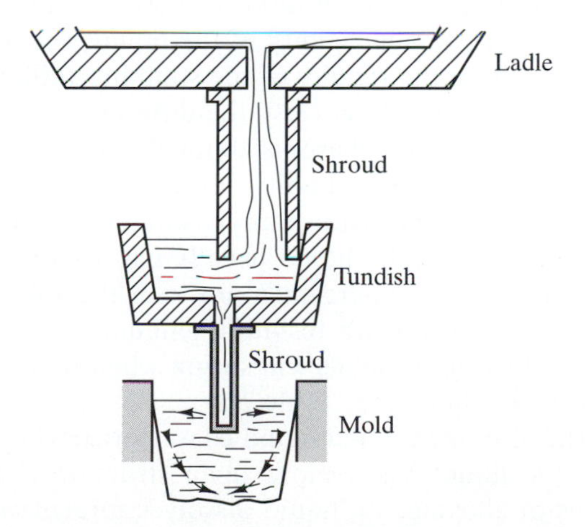

Fig. 7-11 Melt streams are shrouded from the atmosphere to prevent reoxidation and retain cleanliness. [6a]

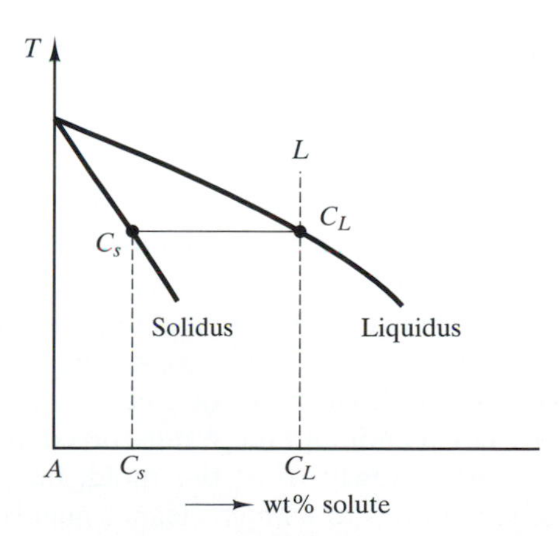

Fig. 7-12 Section of phase diagram showing solute content in liquid and solid during solidification.

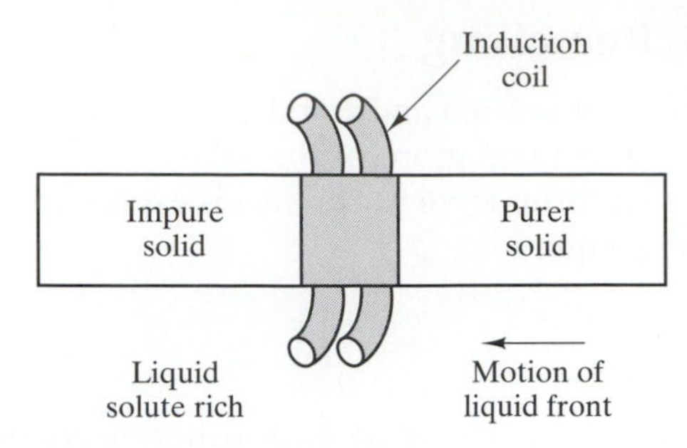

Fig. 7-13 Zone refining involves partial local or zone melting and resolidification of much cleaner or purer material.

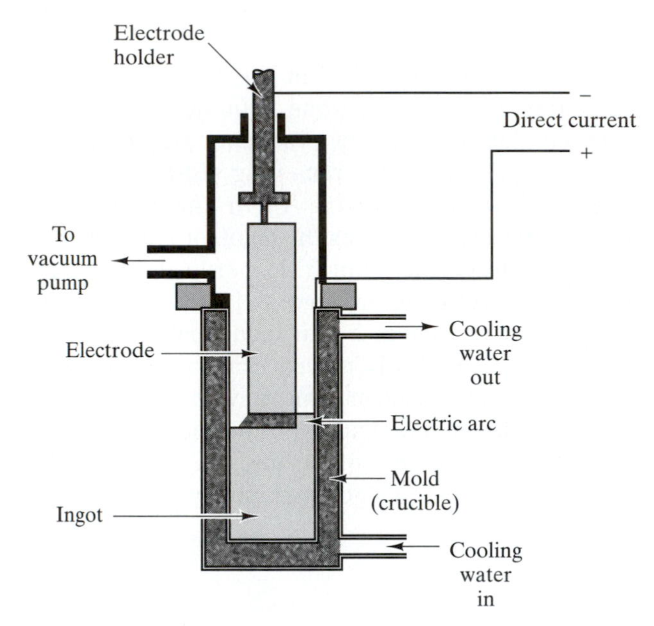

Fig. 7-14 Vacuum arc remelting (VAR) or consumable-electrode process.

obtained—a process used to produce very high-purity silicon for the electronics industry.

The VAR is also called the consumable electrode process under vacuum. The original ingot is made the electrode and is electrically connected with the water-cooled mold. The ingot or electrode is molten by the arc between it and the water-cooled mold. As it moves away slowly from the mold, the *molten* material resolidifies to a much cleaner material. The ESR process is essentially the same as the VAR except that it is not done under vacuum. Thus, to protect the liquid from reoxidizing and to help in the further refinement, a synthetic slag is used to cover the melt. The ingot is again made into one of the electrodes and melting is achieved by the arcing between it and the water-cooled mold. With the heat generated, the slag also becomes molten and covers the melt. As the ingot is slowly pulled up, the liquid retains most of the impurities and a much cleaner solid is formed.

Before we leave this section, we should note that metals are very reactive with oxygen and that the natural state of most metals is in the oxide form. Engineering metals and alloys are obtained from their oxides by a considerable cost of energy, and great effort is made during the processing to prevent them from reoxidation. The stability of the oxide also points to the natural tendency of metals and alloys to oxidize and the certainty of the oxidation problem called corrosion (which will be covered in Chap. 10). Thus, to continue or prolong the useful life of components and structures, we need to practice corrosion control.

7.3 Characteristics of the Liquid State

Before we discuss the solidification process, it is necessary to indicate some of the characteristics and structure of the liquid in relation to the solid phase. The most dramatic difference between the two phases is the loss in rigidity of the liquid. The liquid does not have the rigidity of the solid but rather has fluidity. In terms of viscosity, the difference between the solid and liquid phases is about 20 orders of magnitude (i.e., 10^{20}). The fluidity of the liquid allows the production of intricate-shape castings. In addition, for most metals, the liquid usually has a larger volume, about 2 to 8 percent more than the solid; the exceptions to this are bismuth, gallium, and plutonium. Thus, most metals will shrink when they solidify from the liquid, see Table 7-1.

The aforementioned liquid characteristics suggest that the liquid has a more open structure and, in terms of alloying, the liquid dissolves foreign atoms easier than the solid phase. The fluidity also allows

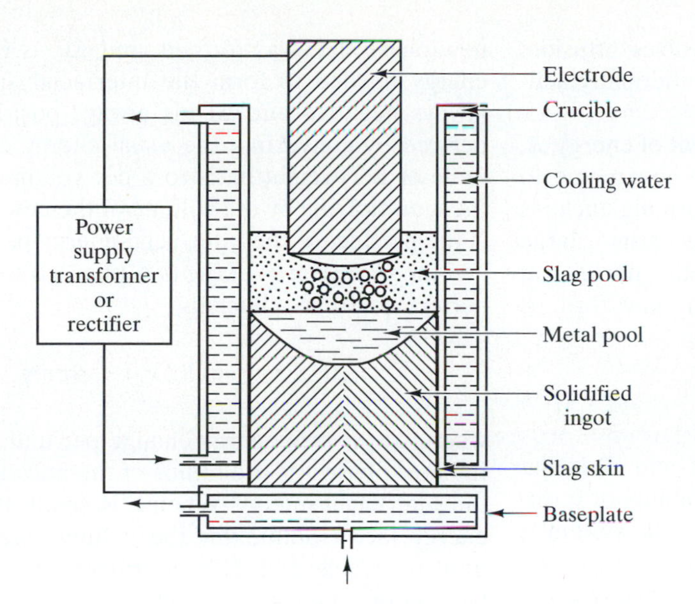

Fig. 7-15 Electroslag remelting (ESR)—remelted ingot is covered by slag to prevent reoxidation.

Table 7-1 Solidification shrinkage or expansion of some metals.

Element	% Shrinkage/Expansion(+)	Element	% Shrinkage/Expansion (+)
Aluminum—Al	7.0	Bismuth—Bi	+3.32
Copper—Cu	3.32	Gallium—Ga	+3.24
Indium—In	2.5	Iron—Fe	3.4
Magnesium—Mg	4.2	Manganese—Mn	1.7
Plotunium—Pu	+ 1 to + 2	Silver—Ag	5.58
Thallium—Tl	3.23	Tin—Sn	2.7

the stirring of the melt by convective flow of the liquid to easily obtain a homogeneous composition.

X-ray diffraction results from liquids exhibit diffused haloes that indicate that there is no order in the structure. However, the disorder is not completely random because the change in volume from the liquid to the solid is small. The diffraction results suggest a short-range order structure, whereby an atom has a correlation with the positions of neighboring atoms, and the interatomic distance with the nearest-neighbor atoms is very close to that in the crystalline structure. The positions of the atoms in the liquid do not correlate, however, with the positions of the atoms in the solid. This indicates that the order in the liquid is different and has no relationship with the crystallinity of the solid.

7.4 *Solidification—A Nucleation and Growth Process*

We now discuss the transformation of the liquid phase to the solid phase. The very first thing that this transformation (solidification) must have is a negative free energy change and when it does, it occurs by a nucleation and growth process, which we discussed earlier in the Introduction of Chap. 6. The nucleation involves the clustering of atoms arranged in the order of the new phase to form small nuclei (nucleation) in the old phase. The stable nuclei then grow in size (growth) by the transport or movement and rerrangement of atoms (diffusion) from the old phase to the new phase. Since the process involves diffusion, it is temperature-dependent and is a thermally activated rate process.

In any rate process, a certain amount of energy, *A*, called the *activation energy*, has to be overcome to start the process. The probability of forming nuclei is proportional to $exp(-A/kT)$. If there exists a large number *N* of similar places where the nuclei can form, the number of nuclei, *n*, that may form is approximately equal to

$$n = N \exp\left(-\frac{A}{kT}\right) \quad (7\text{-}3)$$

where *k* is the Boltzmann constant and *T* is the absolute temperature. When the probability of forming a nucleus is the same everywhere, the system is said to be in a state suitable for *homogeneous nucleation.* This mode of nucleation is hardly encountered, however. In most practical cases, there are always places in the system where nucleation would preferentially occur because the activation energy is less in these places. The latter is said to be in a state suitable for *heterogeneous nucleation*.

7.4.1 Homogeneous Nucleation

As indicated above, this mode of nucleation is hardly encountered in practice. Nonetheless, going through a simple theory of homogeneous nucleation exposes us to a couple of important principles in the process of solidification. We start by examining the necessity of a negative free energy change of the transformation. In thermodynamics, a change in property is always the final state minus the initial state. The negative free energy change means the initial state has more (excess) energy than the final state. This excess energy is called free energy because it is available to do work.

For the solidification process, the negative free energy change, ΔF_v , provides the energy (work) of forming the nuclei (nucleation). The needed work is the creation of an interfacial area between the new and the old phase. Whatever is left of the free energy after doing the work of nucleation is the net free energy change and this has to be negative for the process to continue.

$$\Delta F_{net} = \Delta F_v + \Delta F_s \quad (7\text{-}4)$$

where ΔF_v is the volume free energy change due to the change in state for the volume (amount) of the new phase that was formed, and ΔF_s is the surface energy needed to form the interfacial area. ΔF_v is always negative and ΔF_s is always positive. ΔF_v is derived by converting the usual energy change per mole or per weight basis to a per volume basis and then multiplying by the volume of the new phase. ΔF_s is the product of the interfacial energy per unit area and the surface area of the new phase. If we assume a perfect sphere as a nucleus,

$$\Delta F_{net} = \frac{4}{3}\pi r^3(\Delta F_v) + 4\pi r^2\gamma \quad (7\text{-}5)$$

where ΔF_v is free energy change per unit volume of the new phase, *r* is the radius of the nucleus, and γ is the interfacial energy. When *r* is small, the surface energy factor dominates the volume energy factor until the nucleus reaches a critical size, r_c. The net free energy at the r_c is the activation energy, *A*, of the process.

Figure 7-16 schematically shows the plots of the two terms on the right side of Eq. 7-5 as a function of the size of the nucleus, *r*. The ΔF_{net} curve is the algebraic sum of the two plots at each *r* and it peaks at the point (r_c, A). To obtain the coordinates of this maximum (peak), we take the derivative of ΔF_{net} and equate it to zero. Thus,

$$\frac{d(\Delta F_{net})}{dr} = 4\pi r^2(\Delta F_v) + 8\pi r\gamma = 0$$

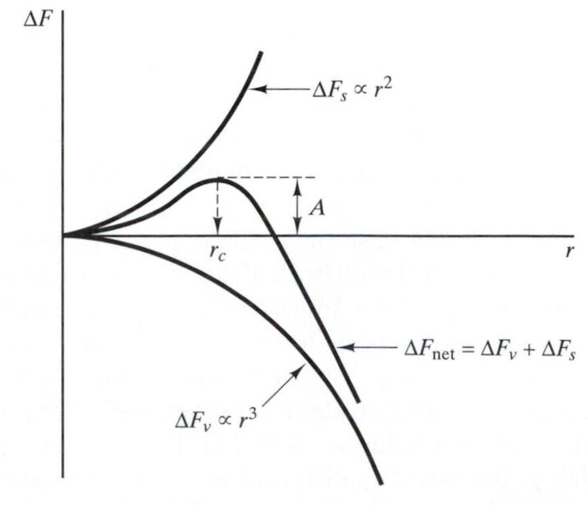

Fig. 7-16 Surface energy, volume free energy, and net free energy curves during nucleation of solid in liquid.

and find that

$$r_c = -\frac{2\gamma}{\Delta F_v} \quad (7\text{-}6)$$

where γ is in units of energy/area while ΔF_V is energy/volume of the new phase.

The value of ΔF_{net} at r_c is A. Thus, substituting r_c from Eq. 7-6 into Eq. 7-5,

$$A = \frac{16\pi\gamma^3}{3(\Delta F_v)^2} \quad (7\text{-}7)$$

Equations 7-6 and 7-7 indicate that the activation energy must be overcome before the solidification can proceed or stable nuclei can form. Thus, *we note as the first principle during solidification is that the nuclei must first have a certain critical size before they can grow.* The nuclei with sizes less than r_c will have a lower free energy if they decrease in size, therefore, they tend to dissolve back into the liquid. The nuclei with sizes greater than r_c will decrease their net free energy by increasing in size and consequently, will grow.

The second point that we should know about solidification can be also gleaned from Eq. 7-7. At the melting or freezing point, T_m, the solid and liquid phases are in equilibrium and have the same free energy. Thus, ΔF_v is equal to zero and we see from Eq. 7-7 that A will be infinity. Thus, it is impossible to overcome an infinite activation energy. *The second principle we learn from this simple treatment is that ΔF_v must be nonzero and that the temperature of the liquid must be lower than the freezing temperature to start the nucleation process.* The difference of the freezing temperature and the nucleation or the solidification temperature is called the *undercooling*. There must always be some undercooling and the greater the undercooling is, the lesser is the activation energy.

In more familiar terms,

$$\Delta F_v = \frac{\theta L_m}{MT_m} \quad (7\text{-}8)$$

$$A = \frac{16\pi\gamma^3}{3}\left(\frac{MT_m}{\theta L_m}\right)^2 \quad (7\text{-}9)$$

where $\theta = (T_m - T)$ is the undercooling, T_m is the melting or freezing temperature in K, T is the actual temperature of the liquid in K, L_m is the latent heat per mole, M is the molar volume (i.e., volume per mole), and γ is the interfacial energy. Because θ is the difference between two temperatures, its value in K or °C is the same.

When only one nucleus forms (i.e., n = 1), and using Eq. 7-3, and $N = 10^{23}$, we find also

$$A = kT \ln N \approx 50kT \quad (7\text{-}10)$$

Thus, equating the two expressions of A, Eqs.7-9 and 7-10, freezing by homogeneous nucleation should begin approximately at the temperature *T* in the expression,

$$\beta\gamma^3\left(\frac{MT_m}{\theta L_m}\right)^2 = 50kT \quad (7\text{-}11)$$

β in the above equation is assumed to be about 4.

Results of studies on homogeneous nucleation are shown in Table 7-2. From these results, the following approximate values are obtained: $\theta = 0.2T_m$. For metals where T_m = 1,000 to 2,000 K, $L_m/M = 1.5 \times 10^9\ \mathrm{J\,m^{-3}}$ and $kT_m = 2.5 \times 10^{-20}$ J, $\gamma = 0.1\ \mathrm{J\,m^{-2}} = 10^{-4}\ T_m$, (K). The average entropy of melting in metals is about 10 J per mol per K. Hence, L_m (J per mol) = $10\,T_m$ (K).

EXAMPLE 7-1

Calculate the critical radius of the nucleus and the number of atoms in the nucleus of lead when it homogeneously solidifies from the liquid. The latent heat of fusion is $2.37 \times 10^8\ \mathrm{J/m^3}$, and lead is FCC with a lattice parameter of 4.9489 Å.

SOLUTION:

(a) Since we are asked for the critical radius of the nucleus, we write down Eq. 7-6,

$$r_c = -\frac{2\gamma}{\Delta F_v}$$

and we need to know γ and ΔF_V. From Table 7-2, we find $\gamma = 33 \times 10^{-3}\ \mathrm{J/m^2}$. And from Eq. 7-8

$$\Delta F_v = \frac{\theta L_m}{MT_m}$$

$\theta = 80$ K, $T_m = 600$ K, and $\dfrac{L_m}{M} = -2.37 \times 10^8\ \mathrm{J/m^3}$.

Substituting all this into r_c,

$$r_c = \frac{2 \times 33 \times 10^{-3}\ \mathrm{J/m^2} \times 600}{2.37 \times 10^8\ \mathrm{J/m^3} \times 80}$$

$$= 2 \times 10^{-9}\ \mathrm{m} = 20\ \text{Å}$$

Table 7-2 Melting points, undercooling, and interfacial energies of pure metals during solidification.

Pure Metal	T_m, °C	T_m, °K	θ, °K	$\frac{\theta}{T_m}$	γ, 10^{-3} J/m^2
Mercury	−38.36	234.6	58	0.247	24.4
Gallium	29.8	302.8	76	0.251	56.0
Tin	231.9	504.9	105	0.208	54.5
Bismuth	271.3	544.3	90	0.165	54
Lead	327	600.0	80	0.133	33
Antimony	630.5	903.5	135	0.149	101
Germanium	937	1210	227	0.188	181
Silver	960.8	1233.8	227	0.184	126
Gold	1063	1336	230	0.172	132
Copper	1083	1356	236	0.174	177
Manganese	1245	1518	308	0.203	206
Nickel	1453	1726	319	0.185	255
Cobalt	1495	1768	330	0.187	234
Iron	1536	1809	295	0.163	204
Palladium	1552	1825	332	0.182	209
Platinum	1769	2042	370	0.182	240
				Ave = 0.186	

Source: D. Turnbull, *J. Chem. Phys.*, 18, p. 769, 1950.

(b) To get the number of atoms in the critical size nucleus, we get the volume of the nucleus and then divide by the volume of a unit cell to determine the number of unit cells. Since FCC contains four atoms per unit cell, we multiply the number of unit cells by 4. Thus,

$$\text{Number of atoms in nucleus} = \frac{4\pi/3(20)^3}{(4.9489)^3} \times 4 = 1{,}106 \text{ atoms}$$

7.4.2 Heterogeneous Nucleation

The undercooling of the entire melt (liquid) observed in homogeneous nucleation, Table 7-2, is not normally seen in practice because the liquid is usually poured into a mold that is always at a much lower temperature than the liquid. The portion of the liquid that gets in contact with the mold surface acquires very quickly the necessary undercooling. Thus, there is preferential nucleation at the mold surface. In addition to the mold surface, the liquid phase almost always contains suspended particle impurities or nonmetallic inclusions. They provide surfaces on which nucleation or crystallization will start. This preferential nucleation on mold and impurity surfaces is called *heterogeneous nucleation.*

In order for heterogeneous nucleation to proceed, the nucleus must stick to the surface. The sticking tendency can be described by the contact angle, Ψ, shown in Fig. 7-17c. The adhesion of the nucleus is provided by the surface tension on the solid impurity. The nucleus replaces the liquid on the solid surface before nucleation and at the same time creates a liquid-nucleus interfacial area, see Figs. 7-17a and 7-17b. To be stable, the surface tensions on the three surfaces must be in equilibrium. This means that

$$\gamma_{SL} = \gamma_{SN} + \gamma \cos\Psi \tag{7-12}$$

where γ_{SL} is the interfacial energy between the solid and the liquid, γ_{SN} is the interfacial energy between the solid and the nucleus, and γ is between the liquid and the nucleus. The surface tensions are the same as the interfacial energies.

In order to determine the critical size of the nucleus, we need an expression of ΔF_{net} as a function of the nucleus size, r, similar to Eq.7-5. To do this, we assume the nucleus to be a spherical cap of radius r, see Fig. 7-17b. For $0° \leq \Psi \leq 90°$, and using standard formulae, the following are found:

liquid-nucleus interfacial area

$$= 2\pi r^2(1 - \cos\Psi) \tag{7-13}$$

solid-nucleus interfacial area

$$= \pi r^2(1 - \cos^2\Psi) \tag{7-14}$$

and,

volume of the nucleus

$$= \frac{2\pi r^3}{3}\left(1 - \frac{3}{2}\cos\Psi - \frac{1}{2}\cos^3\Psi\right) \tag{7-15}$$

Knowing the interfacial areas and the volume of the nucleus, the ΔF_{net} expression becomes

$$\Delta F_{net} = \frac{2\pi r^3}{3}\left(1 - \frac{3}{2}\cos\Psi + \frac{1}{2}\cos^3\Psi\right)\Delta F_v + 2\pi r^2(1 - \cos\Psi)\gamma + \pi r^2(1 - \cos^2\Psi)[\gamma_{SN} - \gamma_{SL}] \tag{7-16}$$

The last two terms in brackets in Eq. 7-16 were not present in Eq. 7-5. The first of these, $\pi r^2(1 - \cos^2\Psi)\gamma_{SN}$ is the energy needed to create the new interface between the nucleus and the solid. The second, $\pi r^2(1 - \cos^2\Psi)\gamma_{SL}$, is the energy released by the solid-liquid surface that was replaced by the nucleus. Note that the solid-liquid interfacial area replaced by the nucleus is assumed to be the same as the nucleus-solid area. In reality, the latter area is smaller or larger depending on whether contraction or expansion occurs during nucleation.

Substituting Eq.7-12 into Eq. 7-16, and then simplifying, ΔF_{net} becomes

$$\Delta F_{net} = \left(1 - \frac{3}{2}\cos\Psi + \frac{1}{2}\cos^3\Psi\right)\cdot\left[2\pi r^2\gamma - \frac{2\pi r^3}{3}(\Delta F_v)^2\right] \tag{7-17}$$

Taking the derivative $d(\Delta F_{net})/dr$ and equating to zero, the critical nucleus size, r_c, for heterogeneous nucleation is found to be

$$r_c = \frac{-2\gamma}{\Delta F_v} \tag{7-18}$$

Comparing Eq. 7-18 with that of the homogeneous nucleation, Eq. 7-6, we note that both the expressions for homogeneous and heterogeneous nucleation are identical. We should note, however, that for the same r_c value, the volume of the nucleus for homogeneous nucleation is much larger than that for heterogeneous nucleation. The volume of the nucleus for homogeneous and heterogeneous nucleation, $(V_c)_{hom}$ and $(V_c)_{het}$ are

$$(V_c)_{hom} = \frac{4}{3}\pi r_c^3 \tag{7-19}$$

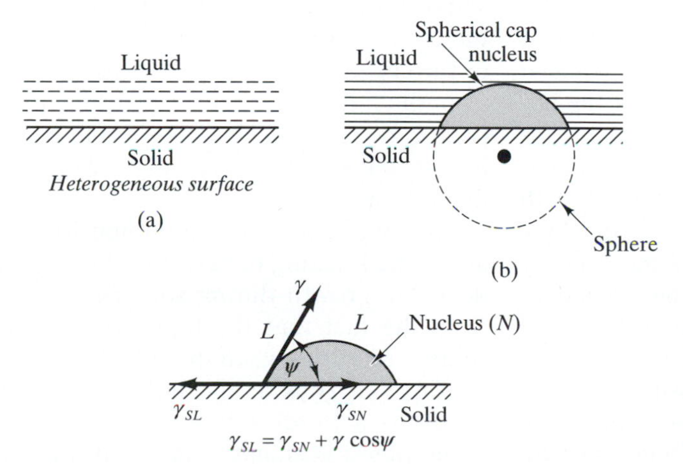

Fig. 7-17 Spherical cap nucleus sticks on solid heterogeneous surface replacing the liquid formerly on the surface. For the cap to be stable, Eq. 7-12 must be maintained.

$$(V_c)_{het} = \frac{2\pi r_c^3}{3}\left(1 - \frac{2}{3}\cos\Psi + \frac{1}{2}\cos^3\Psi\right) \quad (7\text{-}20)$$

This means that the volume-free energy needed to initiate a stable nucleus is much lower in the heterogeneous case. This translates further to a smaller undercooling and thus, easier nucleation.

7.4.3 Growth of Nuclei

The nucleation process is accompanied by the evolution of the latent heat of solidification that will increase the liquid-nucleus interfacial temperature. If there was a favorable undercooling for nucleation, the liberated heat would undoubtedly change that to a less favorable condition. In order for the nucleus to grow in size, favorable conditions for solidification must be restored at the liquid-nucleus interface. This means that the heat must be extracted from the interface. In addition, atoms from the liquid must be transported to the interface and must rearrange in the solid. At temperatures near the melting temperature, the transport of atoms and the rearrangement of atoms occur easily (i.e., they need very little activation energy). This is the case in practical conditions, such as in foundries, when heterogeneous nucleation dominates and requires only a small amount of undercooling. Thus, the interfacial temperature is very close to the melting temperature. Subsequently, the growth of the nucleus or the advance of the liquid-nucleus interface is determined primarily by the rate at which the latent heat of solidification is taken away.

Of the many nuclei that initially form at the mold walls, only a few grow in size from the mold walls to the interior of the liquid. Transformation of the liquid to a solid typically exhibits a treelike structure as shown in Fig. 7-18. Such a structure is called a dendrite and is characterized by a long "stem" with short stems or branches extending from the long stem. The long stem is referred to as the primary dendrite—the branches are called the secondary dendrites. The primary dendrite is one of the nuclei from the mold wall that grew favorably in size perpendicular to the mold wall and opposite the direction of the heat extraction from the mold wall. When the cast structure is sectioned and polished,

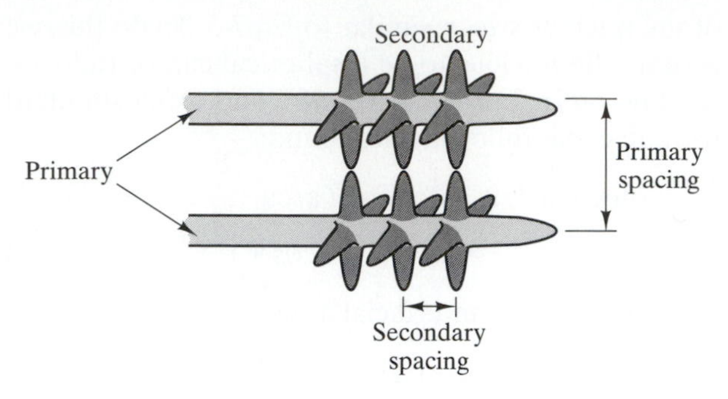

Fig. 7-18 Schematics of primary and secondary arms of "dendrites"—treelike crystals and their spacings.

the primary dendrite will show as a column of crystal (columnar grain). The sizes and spacings between the primary dendrites, as well as the secondary dendrites, can be controlled by the cooling rate. In slow-cooled castings, the sizes of these columnar grains may easily be seen macroscopically, moreover, spacings between them are also large. When the cooling rate is increased, more nuclei grow and subsequently the sizes and spacings of the primary and secondary dendrites decrease as shown in Fig. 7-19. The refinement of the structures leads to an increase in strength as shown in Table 7-3.

The formation of dendrites in pure materials is enhanced by the anisotropic properties of crystals. The growth of dendrites are found to occur in preferred crystallographic directions, indicating that crystal growth occurs preferentially on certain planes and directions. In cubic crystals, (i.e., FCC and BCC), the crystals grow in the directions of the cube edges of the crystal, namely <100>. This means that atoms from the liquid have a better sticking probability on {100} cube planes. This is also a common observaton in epitaxial growth of crystals during chemical vapor deposition.

Alloying elements and impurities profoundly affect the freezing behavior of liquids and promote dendritic growth during solidification. This comes from the fact that the liquid retains more of the solute impurities when the solid forms, as expressed by Eq. 7-2. Thus at the liquid-solid interface, the solid has a lower solute concentration, while the liquid phase is higher than the average. Because of

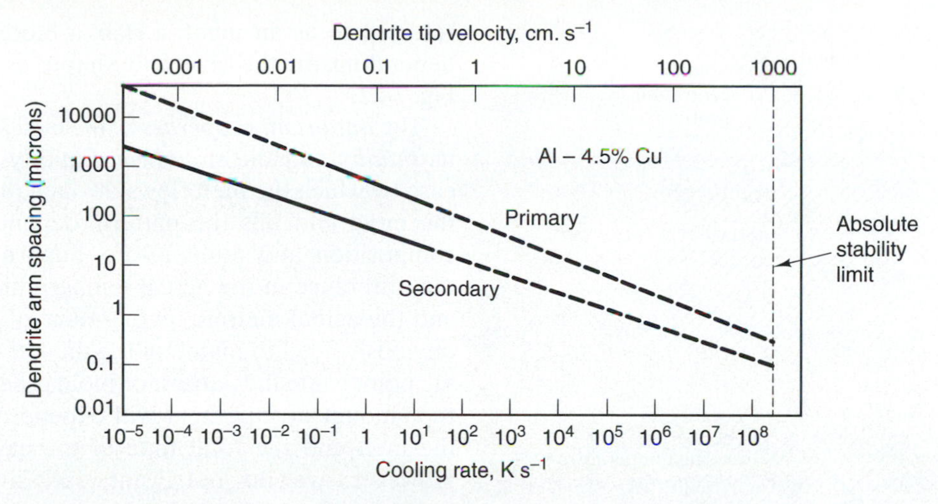

Fig. 7-19 Primary and secondary dendrite arm spacing as a function of the cooling rate in an Al-4.5%Cu alloy. (After M. C. Fleming) [Ref. 6b]

Table 7-3 Mechanical properties of cast Al-4.5%Cu alloy. (M. C. Fleming) [Ref. 6b]

	Yield Strength (psi)	Tensile Strength (psi)	Elongation (%)
Typical properties from commercial sand castings	22,000	26,000	2
Rapidly solidified	35,000	55,000	20
Rapidly solidified modified compostiton*	55,000	65,000	10
Guaranteed minimum properties in premium quality castings, modified composition*	50,000	60,000	5

*Compositions modified by adding small amounts of Si and Mn as strengthener

the solute enrichment in the liquid at the interface, diffusion will ensue and a concentration profile as that shown in Fig. 7-20a develops in the liquid. With higher solute concentration, the freezing point of the liquid decreases and the corresponding freezing temperature profile to the concentration profile in Fig. 7-20a is presented in Fig. 7-20b. In Fig. 20b, T_1 is the temperature at which the original average liquid composition C_B froze and resulted in the concentration profile shown in Fig. 20a. T_{L_1} was the temperature gradient in the liquid phase when the solid froze at T_1 , while T_{S_1} was the temperature gradient in the solid phase after freezing at T_1. In order for the solid to grow, the temperature at the solid-liquid interface must be brought down to at least T_2, the new freezing point of the liquid at the interface. When this is done, the temperature gradients of the liquid and the solid phases are those shown as T_{L_2} and T_{S_2}. We see that the temperature gradient T_{L_2} is lower than the freezing temperature of the liquid immediately ahead of the interface up to indicated distance, x. This portion of the liquid is undercooled and will therefore freeze, so that the solid can grow as far as where x is indicated. This growth spurt is the origin of the primary dendrite and because the undercooling in x is the result of a compositional gradient, it is referred to as *constitutional undercooling*—and is the result of the segregation of solute to the liquid expressed by Eq. 7-2 in alloys. Thus, we expect dendrites to form in alloys

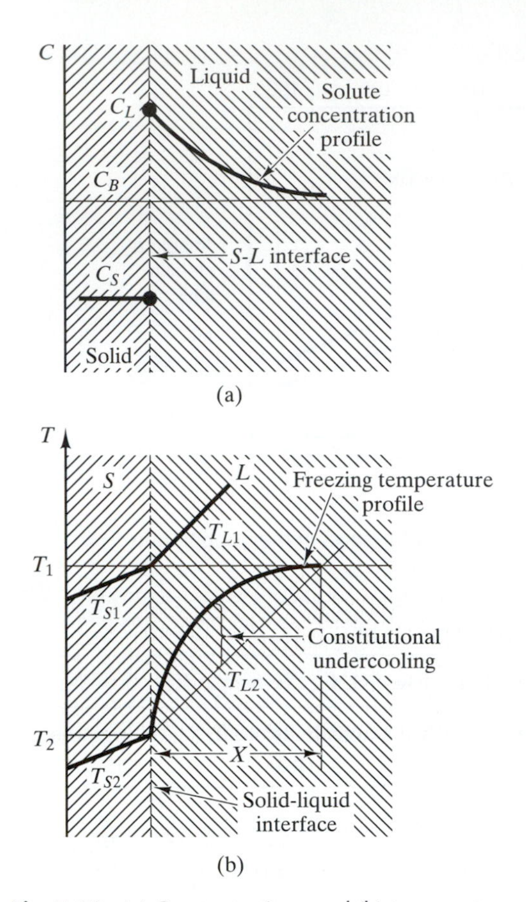

Fig. 7-20 (a) Concentration, and (b) temperature profiles in liquid that leads to "constitutional undercooling" as indicated in (b).

and their formation results in a periodic composition fluctuation that was modeled in the quantitative treatment of homogenization diffusion in Chap. 6.

7.5 Casting and Control of Cast Structure

The word *casting* is used to describe both (1) the process of pouring or forcing the melt to conform to a pattern or mold and the subsequent solidification of the melt, and (2) the as-cast product of the process that is commercially used. Some of the prominent castings for the automotive engine are shown in Figs. 7-21a, b, and c. If the solid will undergo further deformation and forming, the casting product is referred to as an ingot, a slab, a bloom, or a billet depending on the size and shape, as indicated in Fig. 7-22.

The *important properties of the melt for casting* are its *fluidity* and the *superheat*. Fluidity refers to the ease by which the melt flows through the passages of the mold and fills the pattern, dependent on both composition and temperature. Superheat refers to the difference in the actual temperature of the melt and the actual melting temperature of the alloy. We obviously need to maintain the melt above the melting point while the pattern or mold is being filled and the amount of superheat will depend on the size of the melt and the total time of the casting process. However, we do not want excessive superheat because the melt will likely have more dissolved gases and will take longer to solidify.

7.5.1 Possible Macroscopic Defects in Castings and Ingots

To obtain a good sound casting or ingot (i.e., one that is free from *macroscopic shrinkage cavities and gas porosities*), we must start with a good design of the mold or pattern. A good design allows for the solidification shrinkage (or expansion) and for the elimination of gases, either from the melt or from binders used with the mold, such as organics in sand casting. In castings, allowance for shrinkage of the liquid is accomplished with a *riser*; gases are allowed to escape with a *vent* in the mold or pattern. These are shown schematically in Figs. 7-23 (a) and (b) as being attached to the pattern or mold. The same figures also illustrate the two types of risers used in molding—the open and the closed or blind type. The riser acts as a reservoir for molten metal to supply the extra metal that feeds the shrinkage of the liquid metal as it solidifies in the mold. *Cavities* due to shrinkage effects and *porosities* due to gases cannot be tolerated in castings.

In ingots, shrinkage of the melt shows up usually as a *shrinkage pipe* and when exposed to the environment, its surfaces will oxidize and cannot be welded back together when the ingots are deformed hot. Thus, before deformation, this section of the ingots is cut and becomes a yield loss. To prevent much loss, the mold is provided with a "hot top" that also acts as the reservoir for molten metal and functions in the

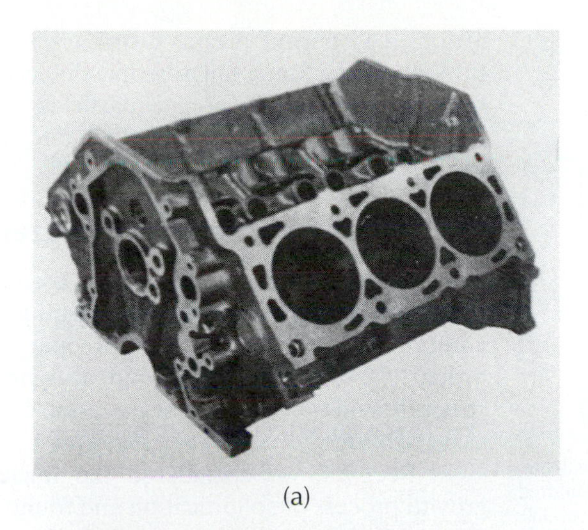

(a)

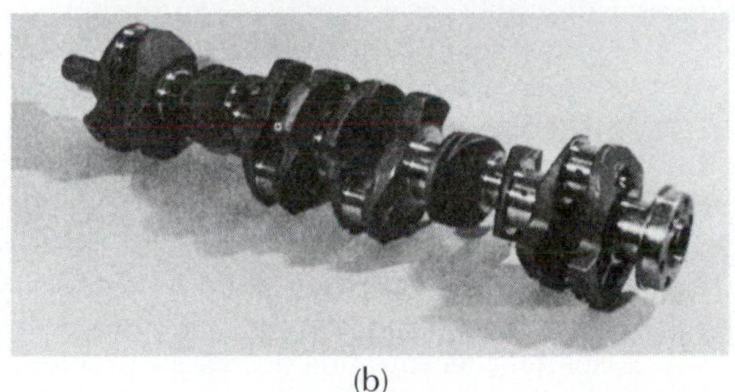

(b)

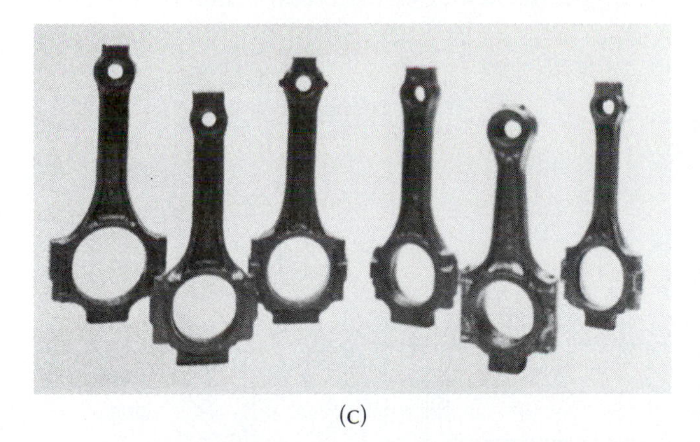

(c)

Fig. 7-21 (a) A gray iron casting of a 3.8 liter V-6 auto engine. (b) A ductile iron casting of an auto crankshaft. (c) Malleable iron one-piece castings of connecting rods and bearing caps that are later separted by machining. [9]

same manner as the riser in castings. Figure 7-24 demonstrates the macroscopic ingot structures with and without the use of a hot-top. Porosities due to gases when located inside the ingots, will close up during hot deformation.

The other consideration in producing a good sound casting or ingot is the *gate system,* that is, the passageway or channel through which the metal flows in filling the mold. Two examples of gating systems are shown in Fig. 7-25 (a) and (b). In order to function properly, the gate system must (1) result in the least turbulence of the liquid metal flowing into the mold, (2) have sufficient volume and pressure of the liquid metal to fill the mold quickly, and (3) be

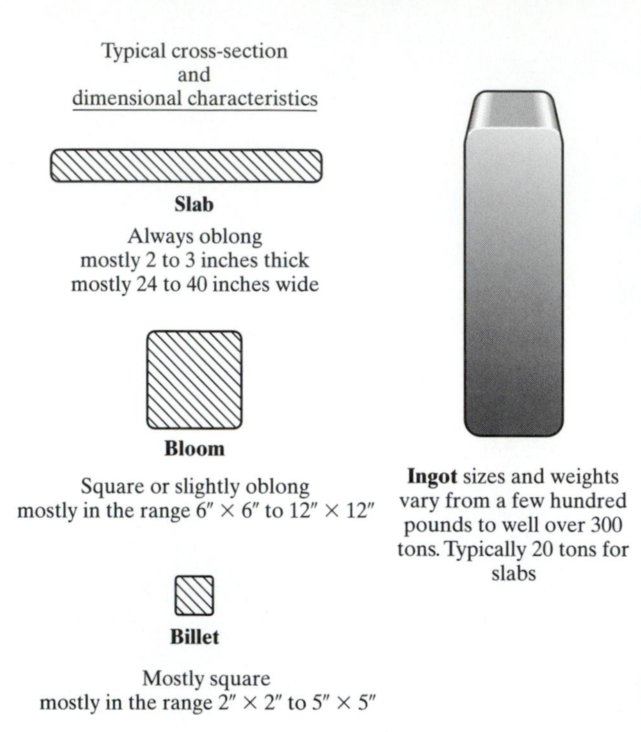

Fig. 7-22 Typical cross sections and dimensions of slab, bloom, billet, and ingot weights.

arranged to permit proper distribution of temperature gradients. Many macroscopic defects in castings such as imbedded foreign particles, cracks, and shrinkage cavities have been attributed to poor gating practice.

7.5.2 Cast Structure and Segregation

The preceding discussion illustrates the *macroscopic structure* that may develop in the casting or ingot. In addition to the possible macroscopic defects, the properties of a casting depend also on the grain structure and on the solute segregation existing in as-cast structures.

In Sec. 7.4, we considered the nucleation and growth process of solidification and found that alloys invariably form dendrites and the solute segregates to the liquid close to the dendrites. We saw that the primary dendrites are in fact the columnar grains that are sometimes observable by the naked eye in the grain structure revealed by etching. The secondary dendrites are seldom seen with the eye because they are relatively small. The segregation of solutes around the dendritic grains is called *microsegregation,* and *coring* is observed by which the solute concentration increases progressively from the center of the cross section of a dendritic arm to

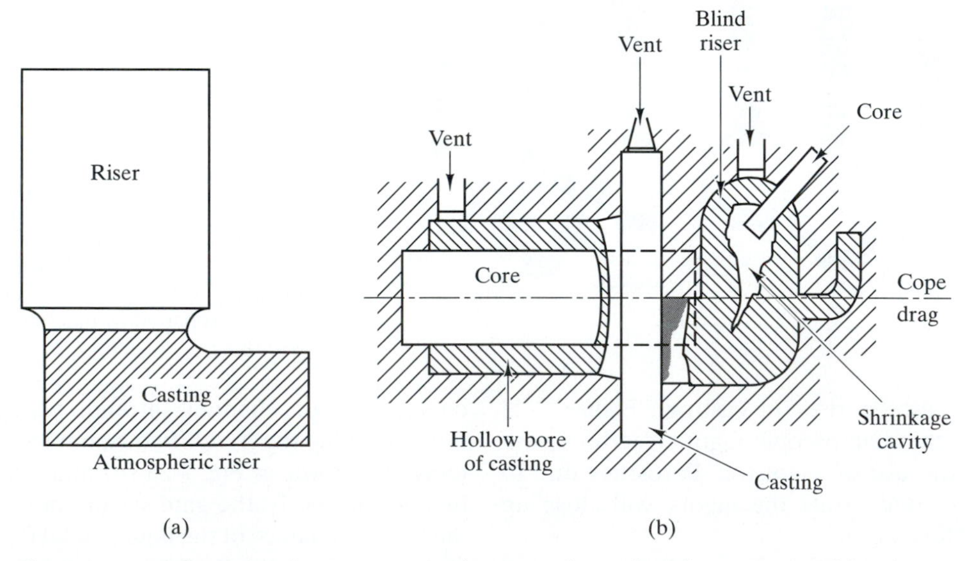

Fig. 7-23 (a) Schematic of a riser open to the atmosphere. (b) Schematic of mold with vent and blind riser. [2]

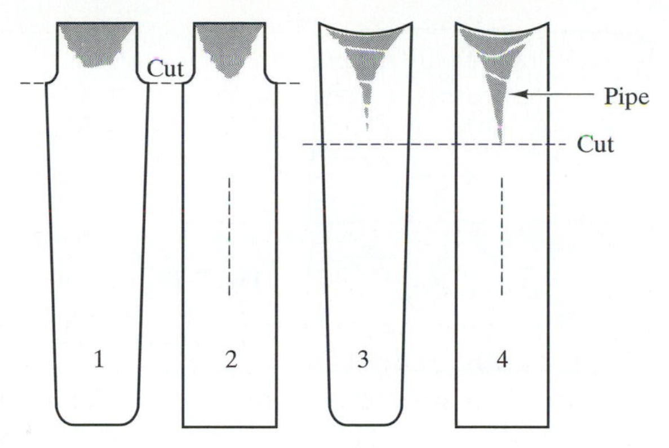

Fig. 7-24 Ingots 1 and 2 are "hot-topped" with no shrinkage pipe; ingots 3 and 4 are not hot-topped and exhibit the shrinkage pipe.

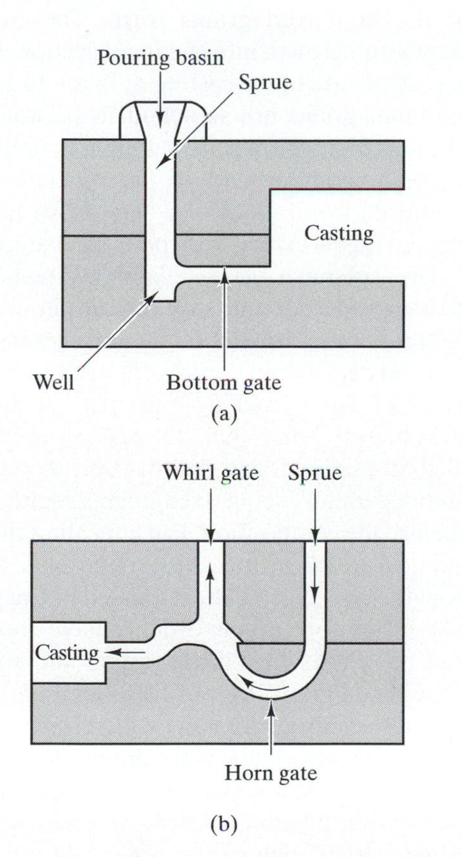

Fig. 7-25 Two examples of gate systems. [2]

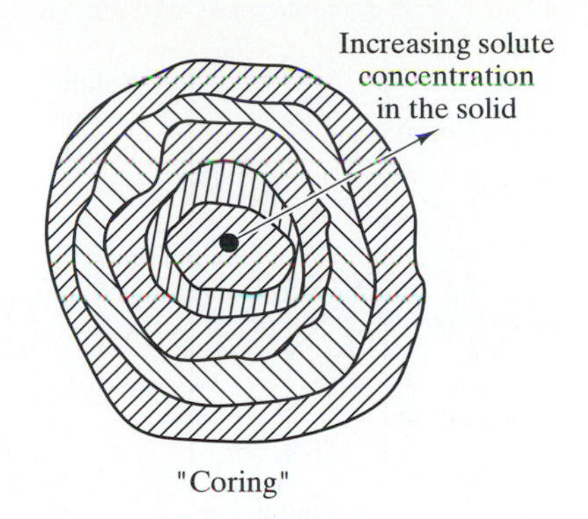

Fig. 7-26 Schematic of "coring" due to solute enrichment in the liquid before solidification.

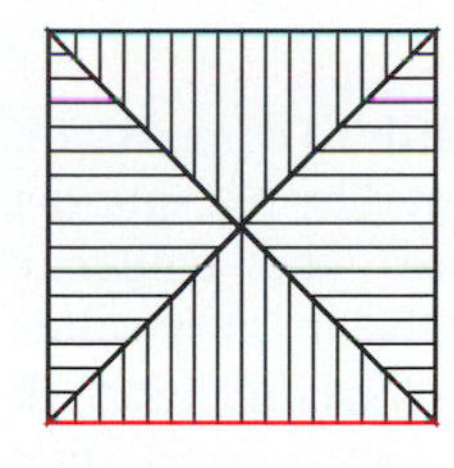

Fig. 7-27 A schematic of an ingot section with fully columnar grains.

the surface, illustrated in Fig. 7-26. Considering no other effects other than the partitioning of the solute to the liquid, primary dendritic grain growth may extend to the last liquid and fully columnar grains across the section of the ingot or casting may be attainable, as shown in Fig. 7-27.

However, in addition to solute partitioning, solidification in ingots and castings involves convection and stirring of the melt within the mold. The convection and stirring may arise from thermal gradients, gravity, and the difference in densities due to solute segregation, liquid shrinkage, and gas evolution. The result is the presence of *macrosegregation* in the casting or the ingot and the *formation of equi-axed grains*. *Macrosegration* is defined as the segregation of alloy elements that occur over distances that are large compared with the dendrite arm spacings. Examples of macrosegration are the "centerline" or positive cone ($\wedge$) segregation, and the inverted or negative cone ($\vee$) segregation, shown schematically across the section of an ingot, in Fig. 7-28. The $\wedge$-segregation shows that the solute concentration at the centerline of the ingot is greater than the outside part of the section. The $\vee$-segregation, on the other hand, shows a more solute-deficient center than the outside.

The extent of segregation in a steel ingot depends on (1) the chemical composition, (2) the type of ingot (i.e., whether killed, semikilled, or rimmed), and

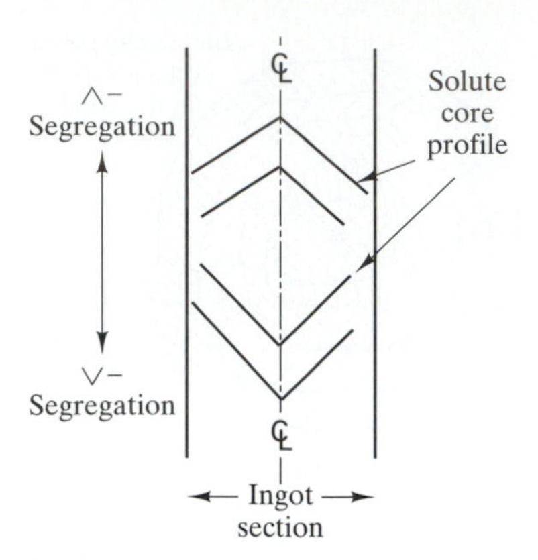

Fig. 7-28 Λ- and V-macrosegregation across section of ingot or casting.

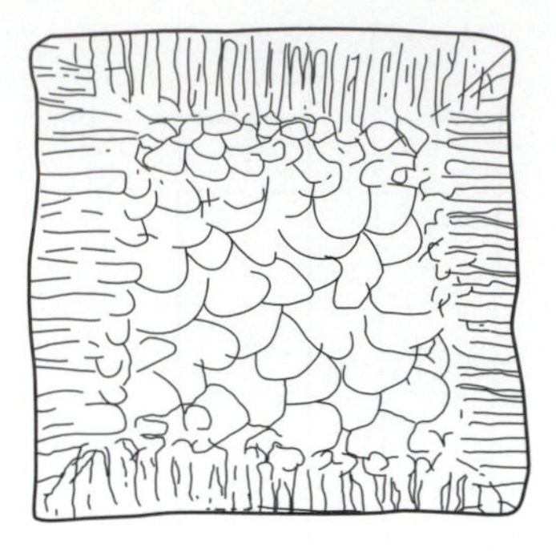

Fig. 7-29 Formation of equi-axed grains aided by convective liquid flow.

(3) size of the ingot. With regards to composition, sulfur segregates the most, followed by phosphorus, carbon, silicon, and manganese. The tendency for elements to segregate while the ingot is solidifying increases with increased time of solidification—larger ingots exhibit more severe segregation than small ingots. For the same size ingot, turbulence or convective flow current increases the tendency for elements to segregate. Therefore, killed steels are the least segregated, followed by semikilled, and rimmed steels are the most segregated. In the rimmed ingot, the rim zone is an illustration of negative segregation while the core zone is an illustration of positive segregation. The terms positive and negative segregation refer to the positive or negative deviation from the average composition. The rim and the core zones are so different that they resemble different steels; therefore, the boundary between them can be sharply delineated.

The formation of equi-axed grains with stirring of the melt in the ingot may arise from heterogeneous nucleation on non-metallic inclusions and on broken secondary dendritic arms. The stirring may entrap and bring some inclusions, that will act as nucleating agents, from the surface to the body of the melt. The stirring of the melt has also been shown to break secondary dendritic arms that may be brought to the body of the melt to start nucleation and growth there. These heterogeneous nucleation sites are tantamount to the inoculation process (introduction of particles in the melt) in foundry melts to induce equi-axed grains in the microstructures of castings. With these occurrences, the as-cast structure of the ingot may exhibit equi-axed grains as schematically shown in Fig. 7-29; the columnar grains are formed at the walls and extend to the inside of the ingot or casting before the equi-axed grains form. The extent of columnar grain growth into the ingot section depends on the cooling rate of the casting or ingot. In Fig. 7-29, the columnar grains are sketched to be much finer than those in Fig. 7-27 to reflect a higher cooling rate.

The columnar grains and the heterogeneity (segregation effects) in the as-cast or "green" structures of castings and ingots exhibit very poor mechanical properties. The columnar grains exhibit directional or anisotropic properties and easy fracture in one direction. When there is excessive columnar grains, ingots are given very small initial drafts (percent reduction in thickness) to avoid cracking and splitting the ingot. Castings before being put to service are usually annealed, normalized, and heat-treated to produce a more refined uniform equi-axed grain structure and to homogenize the composition. The annealing, normalizing, and heat-treating processes are the same as those for wrought steels that will be discussed in Chap. 9.

We see, therefore, that in order to have good properties in the as-cast condition, we should strive for more equi-axed grain structure and less segregation. Both of these can be achieved by increasing the cooling rate of the melt while it is solidifying. With regards to segregation, we saw above that smaller-sized ingots have lesser segregation because it takes a shorter time (increased cooling rate) to solidify a smaller ingot than a larger ingot. We should expect similar beneficial results as those illustrated in Fig. 7-19 and Table 7-3 when the cooling rate in an aluminum-4.5% copper alloy casting was increased. In the consideration of equi-axed grains, increased cooling rate will increase nucleation sites and with stirring of the melt, a more equi-axed structure will be induced.

In addition to cooling rate, a common practice in foundries is *inoculation*, which is the introduction of foreign particles to the solidifying melt. The most common inoculants are the refractory metals, such as tungsten, with high-melting points.

7.5.3 Continuous Casting

Small ingots and increased cooling rates are both achieved in continuous-casting and, therefore, this process will yield better products. In addition to this, continuous-casting also increases the product yield from melt to finish because of less in-house scrap and at the same time offers savings of time, thermal energy, and cost in the processing. Because of less heterogeneity, long homogenization treatment is avoided. The time and cost to reheat ingots for slabbing operation are nonexistent. In fact, it is possible to go from the continuous-casting of the slab (about 9 inches thick) to rolling the slab to the strips (about 0.100-in.) with a reheat furnace on the line. For billets or blooms, it is possible to directly roll to finish bar and wire products without the need for intermediate heating.

Fig. 7-30 shows the schematic vertical cross section of a modern slab caster. The process is accomplished in one station. (In ingot shops, the ladle is moved from one ingot to another.) The ladle is brought from the melt shop to a nearby casting shop. The melt is tapped by opening the plug of the tap hole at the bottom of the ladle and is poured into a tundish and from there into an oscillating water-cooled mold, as seen in Fig. 7-11. The melt streams are protected from the atmosphere by shrouds, again seen in Fig. 7-11, and the melt in the ladle, the tundish, and the mold are covered with synthetic slag to prevent reoxidation of the liquid. The tundish quiets the turbulence of the melt being poured from the ladle before the melt flows to the mold. The mold is relatively small and thus contains only a small amount of liquid that can be cooled very quickly to form a solid outer shell. The oscillation of the mold provides some stirring but primarily prevents the solidified shell from sticking to the mold. To start the casting, a plug is placed at the bottom of the mold. The plug is slowly pulled down when a solidified shell is formed. The thin solidified shell is cooled further with water sprays and guided through pinch rolls (see Fig. 7-30) as the cross section of the slab completely solidifies. Electromagnetic stirring of the melt in the mold or just below the mold may be provided,

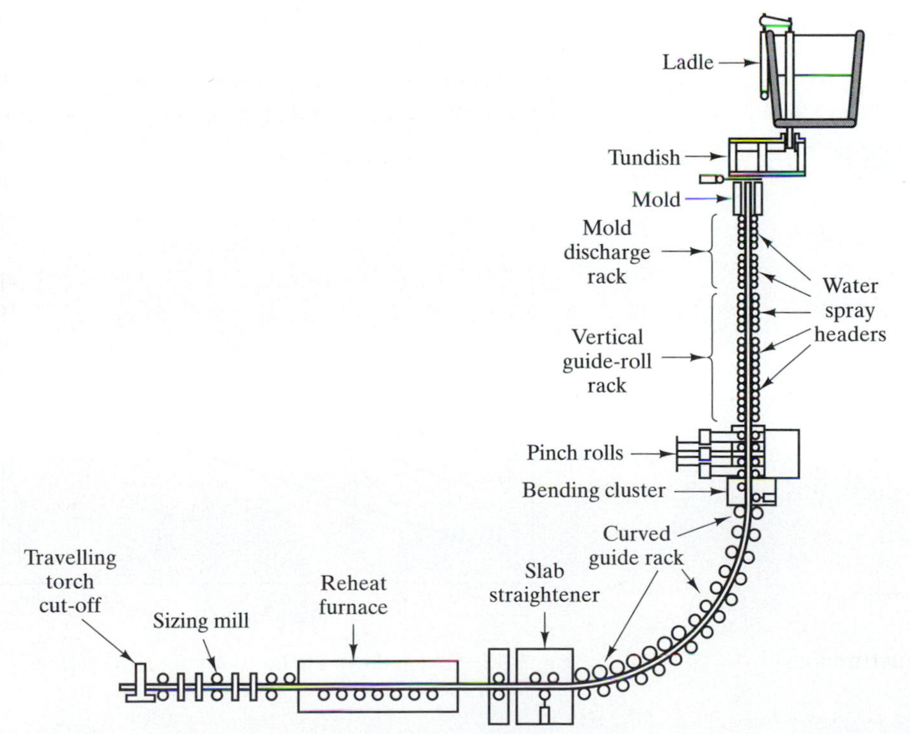

Fig. 7-30 Schematic of a modern slab caster.

see Fig. 7-31. It will take time to empty the contents of the ladle (~ 200–300t) and this is taken into account by the superheat. Once a ladle is empty, another ladle comes around to take its place.

7.6 Metallic Glasses or Amorphous Metals [6c]

We have seen how increasing the cooling rate can minimize the defects during solidification. If we can freeze the liquid entirely, then the problem of heterogeneity in composition will be entirely eliminated.

Freezing the liquid, however, means that the liquid structure is also frozen in the solid. This phenomenon is a fairly common occurrence in polymers and ceramics, in particular, glass. Glass is essentially SiO_2 mixed (alloyed) with other oxides with no long-range order. The absence of long range order makes glass transparent, while the crystalline state of SiO_2 (quartz sand) is opaque. The transformations of liquid SiO_2 to crystalline quartz and to glass are shown schematically in Fig. 7-32 and are essentially those described in Chap. 4 in relation to T_g, the glass transition temperature. When the liquid transforms to the crystalline state, the transformation exhibits a dramatic change in volume and enthalpy (latent heat of condensation) at the freezing or melting point, T_m. When the liquid transforms to glass, it cools below T_m as a supercooled liquid with a gradual change in volume or enthalpy as its temperature decreases. At T_g, there is a break in the gradual change and this corresponds to the liquid becoming glass (solid) (i.e., as a frozen liquid). Polymers also exhibit the glassy state and glass transition temperatures in the same manner as glass. Thus, it has become accepted to refer to a solid material with no three-dimensional long-range order as *glass* or *amorphous*.

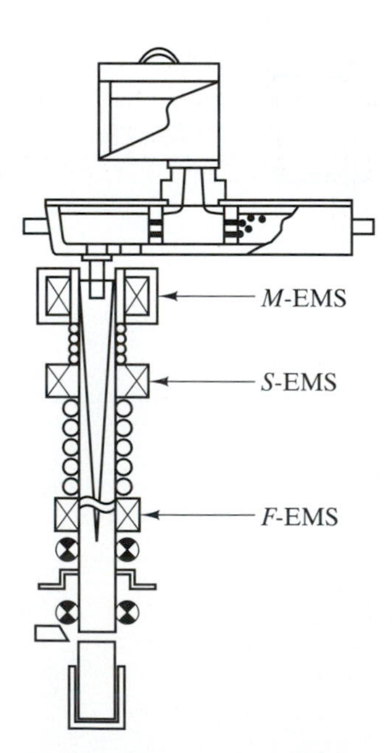

Fig. 7-31 Electromagnetic stirring system in continuous-casting. [11]

In the early 1960s, Pol Duwez and his colleagues demonstrated for the first time the ability to produce amorphous Au-25 a/oSi from the melt by splat cooling. This consisted of pouring the liquid on to a metallic sink surface. Following this lead, numerous studies have confirmed the *production of amorphous metals by rapid solidification technique* with cooling rates on the order of 10^6 K per sec or more. Fig. 7-32 describes essentially the same phenomenon in metals, for instance, the by-passing of the crystalline state and the existence of a glass transition temperature. Research is still ongoing on this subject, but it appears that the kinetic trapping of the melt into the glassy state is confined only to certain types of alloys and it is practically impossible to obtain the glassy state for pure metals. Alloy compositions at and around deep eutectic com-

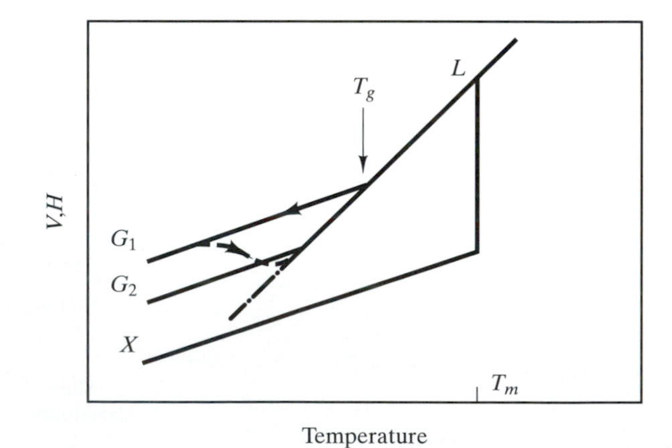

Fig. 7-32 Schematic of the transformation to crystalline and amorphous structure from a liquid.

positions have the easiest tendency to transform to the glassy state. The amorphous state will revert to the crystalline state when the solid is heated.

Alloys based on the ferromagnetic materials Fe, Ni, and Co have been extensively studied and an example of the range of compositions in the Fe-B binary is shown in Fig. 7-33. Metallic glasses retain the metallic bonding characteristics and the associated electronic behavior in metals, with slight modification to account for the spatial rearrangement of the atoms. Thus, they retain good electrical conductivity and for the alloys with Fe, Ni, and Co, magnetic properties. Currently, Metglas (a tradename for metallic glass by Allied Chemical Corporation) is being commercially developed as a soft magnet for power and electronic applications.

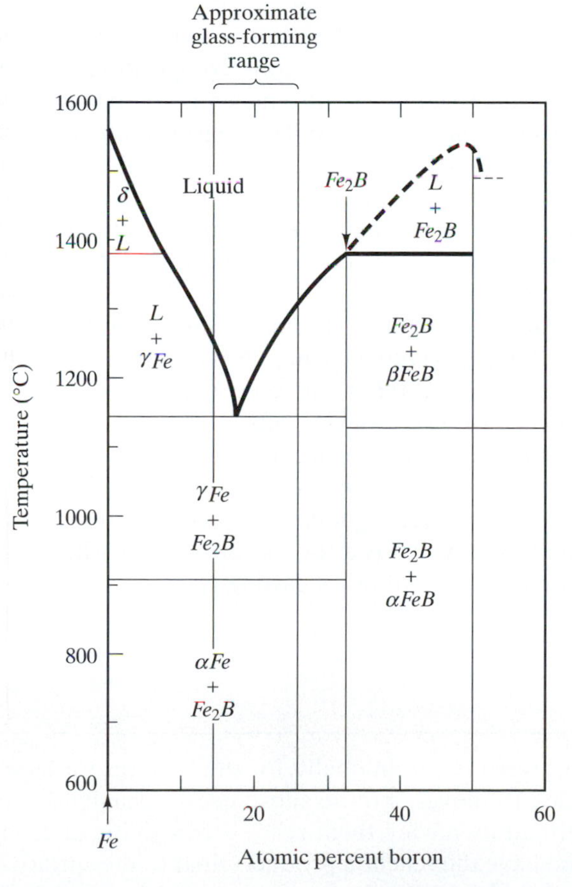

Fig. 7-33 Part of the iron-boron binary diagram showing the range of compositions that forms glasses.

7.7 *Single Crystals*

For structural and low temperature applications, grain boundaries in materials are desired—in fact, the smaller the grain size the better. Yield strength and fracture strength as well as fracture toughness are increased. The development of microalloyed high strength low-alloy (HSLA) steels is based primarily on the refinement of the ferrite grain size. As discussed in Sec. 4.4.2, the grain boundaries are obstacles to the motion of dislocations (for strength) and provide tortuous paths for fracture to proceed (fracture toughness).

For electronic properties and high-temperature applications, grain boundaries are unwanted defects. Thus, silicon single crystals are produced for semiconductors and nickel superalloy single crystals are made for jet engine turbine blades and vanes. Grain boundaries decrease the electronic properties of materials, and at high temperatures, they are an easy path for diffusional creep and promote grain boundary sliding (see discussion on creep properties). There are a number of techniques to produce single crystals but the common method for producing single crystal round sections of silicon is by the Czochralski technique or crystal pulling technique, Fig. 7-34. Single crystal is produced by dipping a seed crystal with the proper orientation, usually the (100), on the surface of the melt and then slowly and vertically pulled out of the melt. Diameters of the single crystal bar can be as large as 178 mm (7") or larger. Nickel superalloy single crystals for turbine blades

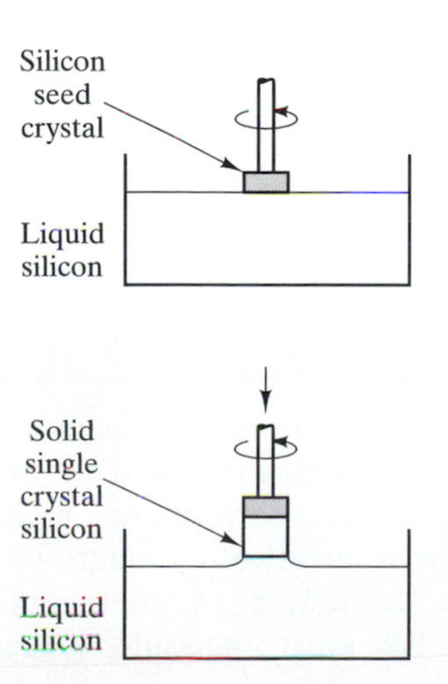

Fig. 7-34 Czochralski technique for producing single crystal, for example, silicon.

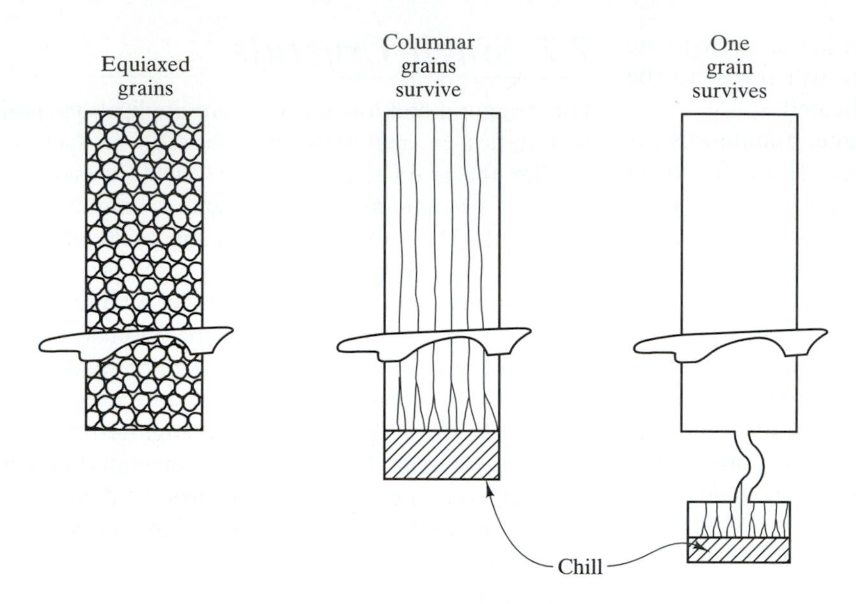

Fig. 7-35 Directional solidification developed by Pratt and Whitney to produce superalloy single crystal jet engine turbine blades.

and vanes are produced by directional solidification, a method developed by Pratt & Whitney, a division of United Technologies, Fig. 7-35. The single crystal is achieved by the selective growth of a grain from a chill zone. In simple terms, the technique induces the growth of a columnar grain. The sizes produced are varied depending on the size of the blades and vanes.

The silicon single crystals are further refined by *zone refining,* as described earlier in relation to Fig. 7-13.

7.8 Homogenization

As a result of solute segregation to the liquid during solidification, there is uneven composition from location to location in the ingot or casting. Because composition is an important factor in the resultant properties of the material, it is imperative that the ingot or casting has the same composition at every location. This process of equalizing the composition is called *homogenization*. The degree of homogenization dictates the variation in the properties, the quality, and the number of eventual rejects of the product. For a small foundry company, it may mean the difference between having a profit or a loss in the bottom line.

The quantitative treatment of homogenization has been covered in Chap. 6. An example of this process is the reheating of cooled steel ingots in homogenizing or soaking pits at 1200° to 1350°C for about 24 hours to prepare them for hot-rolling and at the same time to homogenize the composition. It is obvious that the ingot-casting process requires a lot more energy and is a lot less productive when compared to the continuous-casting process.

Summary

The quality and performance of material we select depend on the refining, solidification, and casting controls we require. Clean, high-performance materials can be obtained by removal of elements that may contribute to the formation of nonmetallic inclusions in the solid. For most materials, oxygen forms oxides; additionally, for steels, sulfur forms sulfides. Removal of these substances is usually accomplished by having them react with specific elements and have the reaction products float to the surface of the melt. Gases, which include oxygen, hydrogen, and nitrogen may also be removed by vacuum techniques. Shrouding techniques are used to protect the liquid from reoxidation during pouring, unless the latter is done in vacuum.

During solidification and casting, the liquid must be undercooled to start solidification. In all casting operations, the mold walls readily provide the undercooling. On solidification, solute atoms in the alloy segregate to the liquid. Because the process is nonequilibrium, there is an inherent chemical (solute) segregation. Such segregation can be minimized by increasing the cooling rate of the liquid in the mold. Increasing the cooling rate also increases the formation of nuclei that minimize the formation of columnar grains and induce the formation of small equi-axed grains. Stirring of the melt also helps. These factors are utilized in continuous casting. If we signifigantly increase the cooling rate, the liquid (amorphous) structure may be frozen that gives a very homogeneous product. For castings (products used in the as-cast condition), mold design that includes the riser and the gating system is important to obtain a sound product. We can also produce single crystals. Castings are homogenized to alleviate the chemical segregation.

For extra-clean, high-performance materials, the solidified materials can be remelted to take advantage of the principle that more solutes will go to the liquid during solidification and to remove entrapped inclusions by allowing them to float. This process is called zone-refining for silicon in the electronics industry. For other materials, there are the vacuum arc remelting (VAR) and electroslag remelting (ESR) processes.

References

1. T. Rosenquist, *Principles of Extractive Metallurgy*, 2d ed., McGraw-Hill, 1983.
2. H. E. McGannon, *The Making, Shaping and Treating of Steel*, 9th ed., U.S. Steel, 1971.
3. *Sulfide Inclusions in Steel, Proceedings of an International Symposium*, ASM, 1974.
4. F. B. Pickering, "High-Strength, Low-Alloy Steels—A Decade of Progress," p. 9, in *Micro-Alloying '75*, Proceedings of an International Symposium on High-Strength, Low-Alloy Steels, Union Carbide Corporation, 1977.
5. J. D. Grozier, "Production of Microalloyed Strip and Plate by Controlled Cooling," p. 241, in *MicroAlloying '75*, ibid.
6. *Metallurgical Treatises*, J. K. Tien and J. F. Elliot, eds., TMS-AIME, 1981.
   (a.) R. D. Pehlke, "Steelmaking Processes," p. 229.
   (b.) M. C. Flemings, "Segregation and Structure in Rapidly Solidified Cast Metals," p. 291.
   (c.) L. A. Davis and R. Hasegawa, "Metallic Glasses: Formation, Structure, and Properties," p. 301.
7. M. F. Ashby & D. R. H. Jones, *Engineering Materials 2*, 1st ed., Pergamon Press, 1986.
8. A. Cottrell, *An Introduction to Metallurgy*, 2d ed., Edward Arnold, 1975.
9. *Iron Castings Handbook*, C. F. Walton and T. J. Opar, ed/co-ed., Iron Castings Society, Inc., 1981.
10. M. V. Balakrishnan and R. O'Neil, "Metallurgical Factors which Influence the Formability of Sheet Steels, p. 100, in *Source Book on Forming of Steel Sheet*," ASM, 1976.
11. H. Takagi and T. Ohnishi, *Operational Techniques of Bloom Caster for High Quality Wire Rod and Bar*, p. 281, Second Process Technology Conference, Chicago, IL, ISS-AIME, 1981.

Terms and Concepts

Activation energy
Amorphous metal
Argon-oxygen-decarburization
Basic oxygen furnace
Blast furnace
Casting
Cleanliness rating
Columnar grains
Constitutional undercooling
Continuous casting
Control of cast structure
Coring
Critical size of nucleus
Czochralski technique

Dendrite
Deoxidizers
Desulfurization
Directional solidification
Electromagnetic stirring
Electroslag remelting (ESR)
Fluidity
Gas porosities
Gate system
Heterogeneous nucleation
Homogeneous nucleation
Homogenization
Hot-metal
Inclusion shape control
Inclusions
Inoculation
Killed steel
Ladle
Latent heat of fusion
Macroscopic segregation
Melt
Melting
Metallic glass
Microscopic segregation
Nucleation and growth process
Oxygen lancing
Primary dendrite
Primary refining
Rapid solidification
Reduction of oxides
Rimmed steel
Riser
Secondary dendrite
Secondary refining
Segregation
Shrinkage cavities
Shrinkage pipe
Shrouding
Sievert's law
Single crystals
Slag
Solidification
Solute partitioning
Superheat
Surface free energy
Tundish
Undercooling
Vacuum arc remelting (VAR)
Vacuum degassing
Volume free energy
Zone refining

Questions and Exercise Problems

7.1 Differentiate the blast furnace from the basic oxygen furnace. Indicate the processes in these furnaces.

7.2. Explain why the hot-metal from the blast furnace contains about 4 wt%C.

7.3 What is meant by ladle metallurgy and indicate what is done?

7.4 Why is it important that steels contain manganese in their compositions?

7.5 What are nonmetallic inclusions? Indicate ways how these inclusions can be present in steels.

7.6 Indicate the influence of nonmetallic inclusions and explain why it is important to have a clean steel for high performance?

7.7 What is deoxidation and explain why a melt has to be deoxidized? What are the common deoxidizers for steel?

7.8 Explain the differences in the ingot structure of a "killed" steel and a "rimmed steel."

7.9 What is Sievert's law? Explain how it is used during the vacuum degassing of the melt.

7.10 What is the argon-oxygen decarburization (AOD) process?

7.11 What is the purpose of remelting and what is the principle involved in attaining a cleaner or purer material? Indicate three processes that are used in applying the principle?

7.12 For most metals, why is it necessary to have a "hot top" for an ingot and a riser for a casting?

7.13 A hollow cylindrical aluminum ingot mold measures 300 mm inside diameter and 2 m high. If the mold is filled with liquid aluminum at 935 K, what would be the largest spherical cavity that may form in the ingot? If solidification produces no cavity at all, what would be the dimensions of the ingot? Use data in Table 7-1.

7.14 The mean coefficient of linear thermal expansion, α, for aluminum is given below for the respective ranges in temperature. From this data and the data in Table 7-1, estimate the density of liquid aluminum. The latticie constant for face-centered cubic aluminum is 4.0496 nm at 298 K and its melting point is 933 K.

Table P7.14 Mean coefficient of linear thermal expansion, α, in micro-strain/K.

Temperature range, K	α, in 10^{-6} per K
293–400	23.95
400–500	25.7
500–600	27.35
600–700	29.30
700–800	32.95
800–900	35.40
900–933	~38.5

7.15 What is the difference between homogeneous nucleation and heterogeneous nucleation?

7.16 From a thermodynamic point of view, what two things must occur in order for solidification from a liquid to proceed?

7.17 If the nuclei from the liquid were cuboids, determine the critical radius and the activation energy of solidification, as in Eqs. 7-6 and 7-7 for the spheroidal nuclei.

7.18 Calculate the critical radius and the number of atoms in the nucleus for the homogeneous nucleation in pure iron. The latent heat of fusion for iron is 3,658 cal per g-mole (atom). Note the allotrope of iron that forms at the temperature.

7.19 Calculate the critical radius and the number of atoms in the nucleus for the homogeneous nucleation in germanium. The latent heat of fusion for germanium is 8,100 cal per g-mole (atom) and the coefficient of linear thermal expansion is 5.75×10^{-6} per K.

7.20 Calculate the critical radius and the number of atoms in the nucleus for the homogeneous nucleation in gold. The latent heat of fusion for gold is 16.1 cal per gram and the coefficient of linear thermal expansion is 14.2×10^{-6} per K.

7.21 What is a dendrite and why do dendrites form during solidification?

7.22 When dendrites form, describe the resultant macrostructure of castings or ingots.

7.23 Explain what is meant by constitutional undercooling.

7.24 What is coring and why does it occur?

7.25 Explain the inverted $\wedge$ and the $\vee$ macrosegregation in an ingot.

7.26 What is meant by inoculation and what is its purpose?

7.27 What is continuous-casting and its advantages?

7.28 What does metallic glass or amorphous metal mean? How does it form? What system(s) is (are) most likely to form in metallic glass during cooling?

8

Plastic Deformation and Annealing

8.1 Introduction

When we discussed the tensile test in Sec. 4.4.4, we defined *plastic deformation* as that which exceeds the yield strength of the material and which *effects a change in the shape of the material and leaves a permanent strain.* We shall now consider plastic deformation as the process of shaping the material and as such, it is one of the most important technological processes involved in the production of many consumer goods such as the automobile and household appliances. The shaping process involves the flow and redistribution of materials through a "die or form" under the action of stresses. Thus, the process is also called *mechanical deformation* or *plastic forming* and the material undergoes changes in microstructure and properties.

In industry, the common term used is *working* and the material is either hot-worked or cold-worked depending on whether the plastic deformation is done at a temperature above or below what is called the *recrystallization temperature,* which is $\sim T_m/2$, where T_m is the melting point in K. During hot-working, the material remains very soft and the shaping process can be done at relatively low stresses. In cold-working, the material strain hardens as described in Sec. 4.4.4 and eventually loses all its ductility. In order to continue the forming process, the material needs to be softened and this is done by *annealing,* which is the process of heating the material at a temperature to change its structure. To completely soften the material, it must undergo a process called *recrystallization.*

In this chap., we shall learn about some of the plastic forming processes and their complex relationships with the structures and properties of materials. We shall learn about: (1) materials properties that are good for certain forming processes; (2) how the structures and properties change during deformation, annealing, and recrystallization; (3) how we can take advantage of the change in properties; and (4) how we can use this knowledge to control the microstructures and properties of materials. We shall also learn how to specify the cold-work condition of materials when we select and procure them.

8.2 Deformation Processes

Deformation processes are broadly classified into two categories: primary and secondary processes. Primary processes are generally done at metal producers' and suppliers' facilities and generally involve the hot-working operations, such as the breakdown of the ingots to slabs, blooms, or billets, and the subsequent hot-forming to various structural shapes, bars, and flat products, as those shown in Fig. 8-1. The various structural shapes shown and some of the bar products, notably the reinforcing bars, are predominantly used as produced. Otherwise, most of the bars and the flat products will undergo secondary processing and/or fabrication processes, such as drawing, forging, swaging, welding, machining, punching, and shearing. These are the secondary deformation processes, some of which are shown in Fig. 8-2, that are performed at metal users' or manufacturers' plants to form the final shape of the material, and may include both hot- and cold-working processes. In both primary and secondary deformation, the shape of the product is achieved by passing the material (workpiece) through a shaped die, roll, or a punch in the shape of the final product. The shape change is accompanied by internal as well as surface deformation due to frictional constraints.

The actual sequence of deformations in industry is what has just been described (i.e., hot-working, and then cold-working, and then possibly annealing and final shaping). Hot-working is usually carried out at temperatures ~$0.6T_m$ or higher, while cold-working

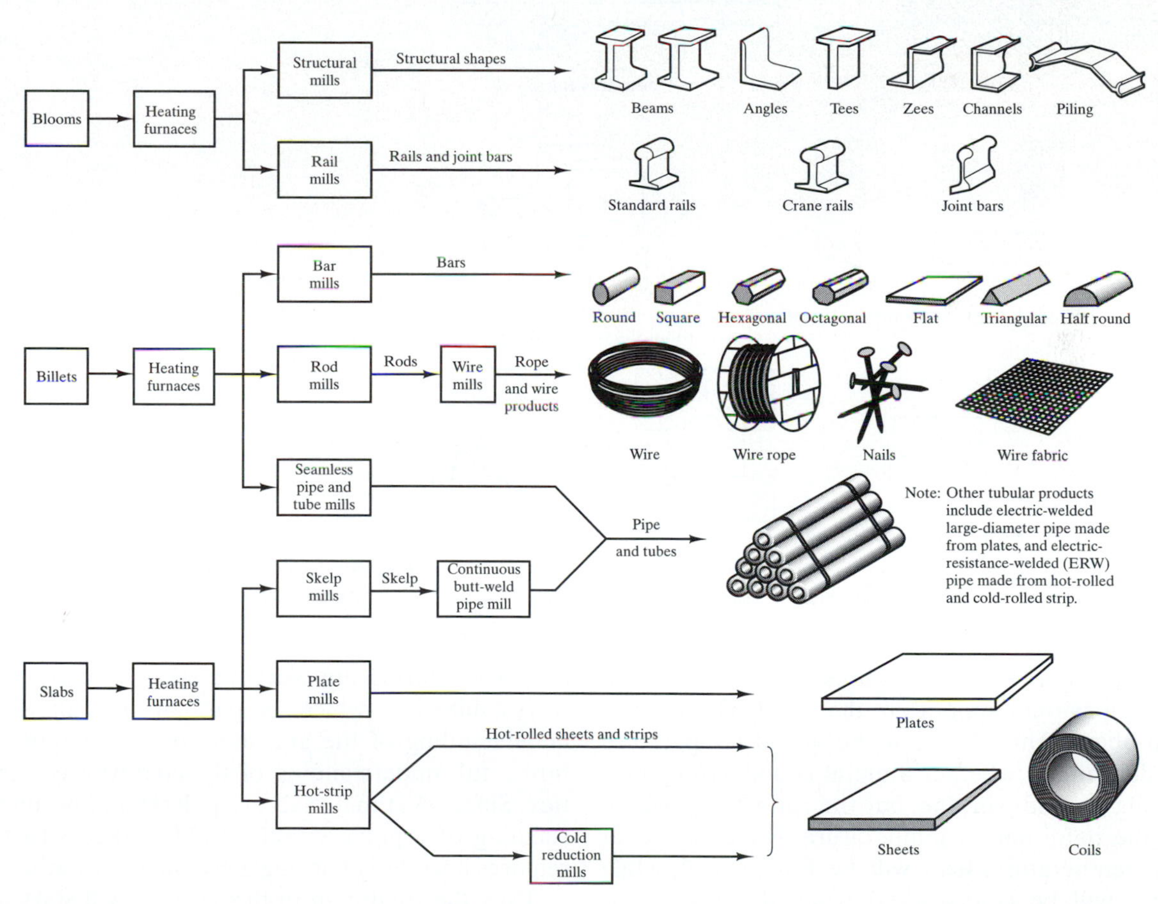

Fig. 8-1 Major product forms of primary deformation of blooms, billets, and slabs in an integrated steel plant.

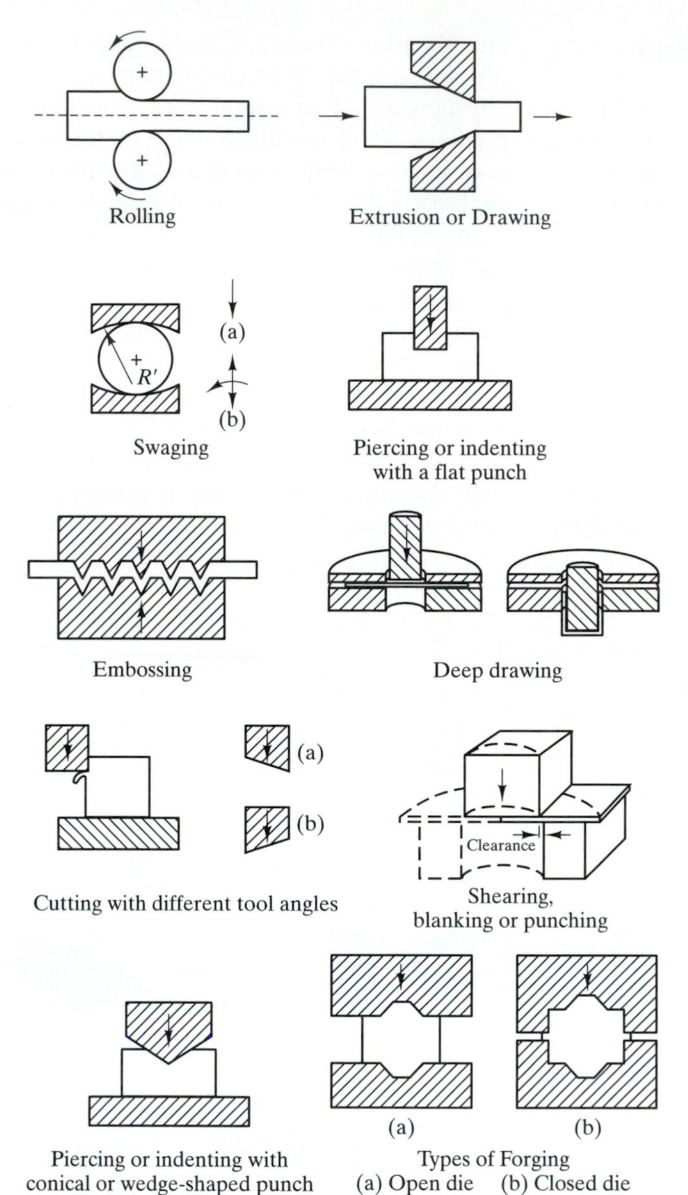

Fig. 8-2 Examples of secondary deformation processes.

is carried out at temperatures much below the recrystallization temperature that is $\sim 0.5T_m$, as indicated above. Thus, it is not the actual temperature that determines whether a metal is undergoing hot-working or cold-working, but the ratio T/T_m, where T is the deformation temperature. For example, at room temperature, lead will be hot-worked, while copper will be cold-worked when deformation is applied. Hot-working involves recrystallization, and in order to understand the changes in structure and properties during hot-working, we must understand recrystallization which is part of annealing. An understanding of the annealing process requires in turn a full understanding of the cold-worked structure. Since most materials are polycrystalline, understanding of a polycrystalline cold-worked structure requires knowledge of single crystal deformation.

Thus, the treatment in this chapter will start with concepts on the deformation of single crystals, then cold-working, annealing, and then hot-working. An example of a process design in the development of controlled thermal-mechanical processing will be described. This process incorporates the principles of cold-work and annealing during hot-rolling, and is instrumental in the development of high-strength materials, in both the ferrous and nonferrous industries. Finally, the concepts of superplasticity and superplastic forming will be covered. We shall confine our discussion on the control of and changes in structures and properties. Complete treatment of mechanical deformation that includes the interactions of the material and the dies or rolls is a subject beyond the scope of this book.

8.3 Plastic Deformation of Single Crystals

In Sec. 2.11, we learned that crystal deformation occurs predominantly by slip on the densest plane and along the densest direction, both of which comprise the *slip system*. In Sec. 2.12.2, we learned that *dislocations* are the defects that move on slip systems to allow the material to be shaped, and the number of these slip systems directly relate to the deformability of the crystal. In Sec. 4.4.2, we learned that if we block the motion of dislocations, we can strengthen the crystal. As indicated there, the stress to move dislocations in polycrystalline materials is the yield strength of the material. In single crystals, this stress is referred to as the critical resolved shear stress (CRSS) and is a property of single crystals, as the yield strength is to polycrytalline materials.

8.3.1 Existence and Evidence of Dislocations*

Before we discuss the CRSS of single crystals, we need first to answer the question: Why do dislocations exist in crystals?

The existence of dislocations as an imperfection in the lattice was postulated by Taylor, Orowan, and others in the 1930s to explain the low-yield strength of pure materials. This may be shown by calculating the shear strength of a perfect crystal. Consider the two rows of atoms, in Fig. 8-3, representing the edges of two planes of atoms in a perfect lattice. The interplanar spacing is a, while the interatomic spacing is b.

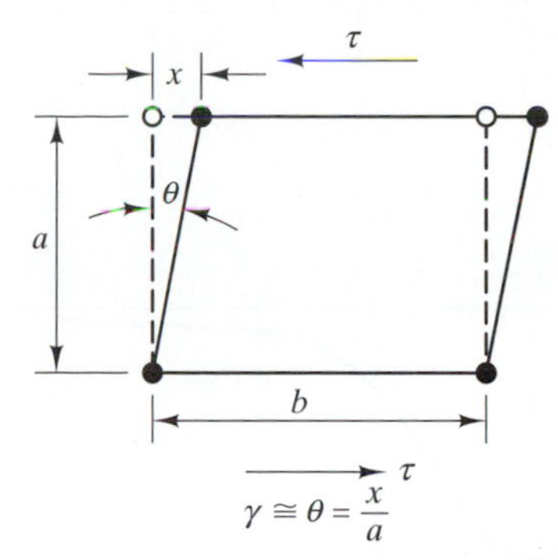

Fig. 8-3 Shear of a perfect lattice.

When the upper plane is moved (sheared) an amount x relative to the lower plane, the shear strain, γ, as we discussed in Sec. 4.3.2, is

$$\gamma \approx \frac{x}{a} \tag{8-1}$$

and the shear stress, τ, is

$$\tau = G\gamma \tag{8-2}$$

where G is the shear modulus. The shear strength of the crystal is the maximum, τ_m, when the bonds break. This may be derived by using the bonding curve and its derivative, the force curve, shown in Figs. 4-10a and b, respectively, and that are reproduced in Figs. 8-4a and b. Per unit area, the force curve is also the stress curve. We can approximate the force or stress curve by a sinusoidal function, and because the stress we want is shear, the function is

$$\tau = \tau_m \sin\left(\frac{2\pi x}{b}\right) \tag{8-3}$$

where τ_m, is the amplitude of the function and is the shear stress needed to create a "slip step" in a perfect lattice, see Sec. 2.11.2. Again, for small angles, $\sin y \simeq y$, and by combining Eqs. 8-1, 8-2, and 8-3, we obtain

$$\tau_m = \frac{Gb}{2\pi a} \tag{8-4}$$

which can be further approximated as

$$\tau_m = \frac{G}{2\pi} \tag{8-5}$$

because $b \approx a$ in most lattices.

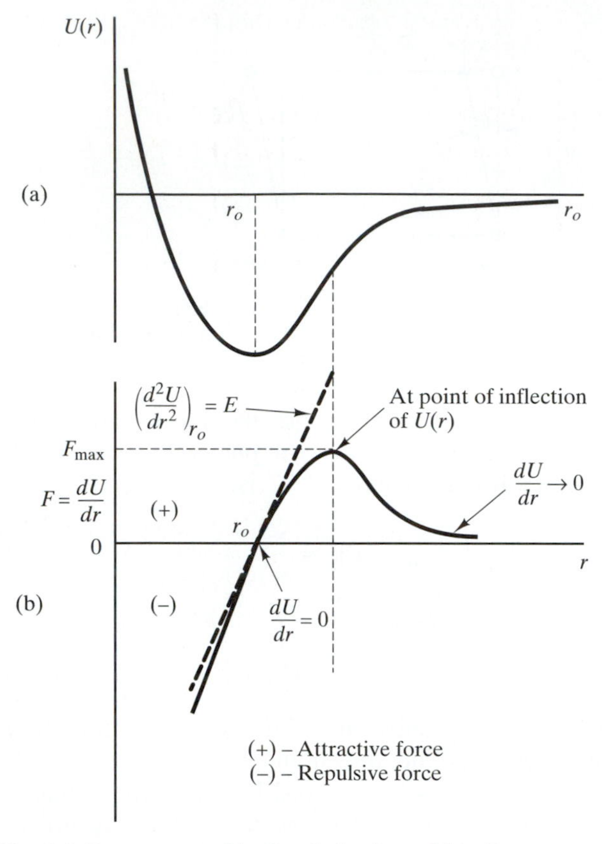

Fig. 8-4 Force curve (b), the derivative of binding energy curve (a), also represents stress, and the positive portion of the curve may be represented as a sine curve.

The shear modulus for metals is on the order of 10^7 psi. Therefore, the theoretical shear strength of materials from Eq. 8-5 is on the order of 10^6 psi. The actual values of the shear strength of single crystals are, however, only on the order of 1,000 to 10,000 psi, very much less than the theoretical strength predicted. An example of this is shown in Fig. 8-5 where the estimated yield strength of pure iron single crystal is only 4,000 psi. Since the yield strength is about twice the shear strength, the latter (shear strength) is only 2,000 psi. More refined calculations cannot reconcile the difference between the predicted and the observed values. Because of this discrepancy, the postulate was made in the 1930s that defects (dislocations) are present in the lattice and that they offer lower resistance to move. Years later, and with transmission electron microscopy, the image of a dislocation line was obtained. The image appears as a dark line, as illustrated in Fig. 8-6, because of diffraction contrast.

A dislocation requires energy to form and thus it is thermodynamically unstable. Nonetheless, it still forms in the crystal. It forms during solidification because, even for a single crystal, it is much easier for atoms to attach to ledges on a surface. This is shown in Fig. 8-7 where a screw dislocation provides the faster spiral growth of the crystal than when the defect is not present. Another source of a dislocation

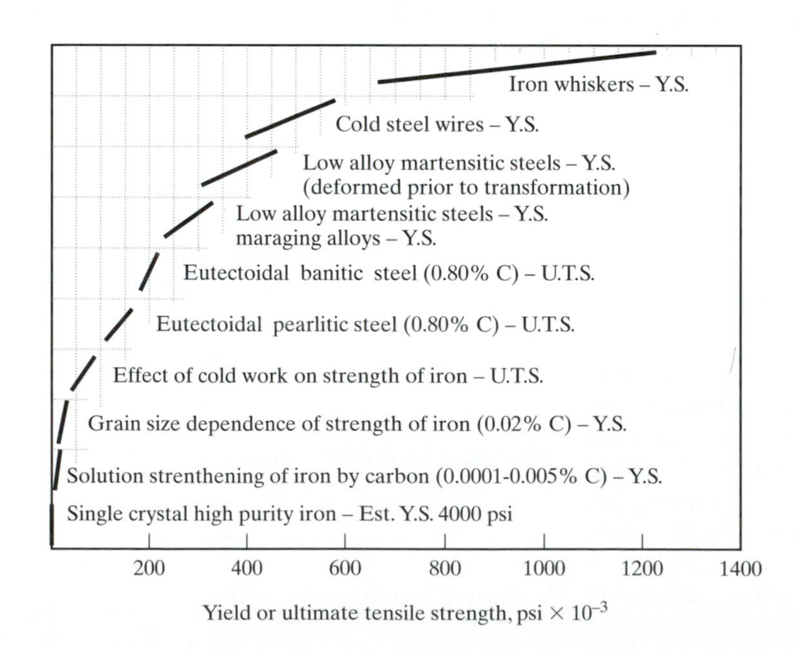

Fig. 8-5 Yield strengths of iron single crystals and iron whiskers show the range of obtainable properties.

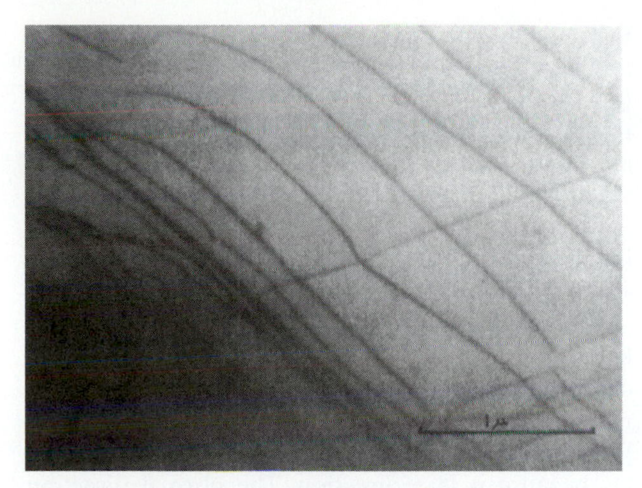

Fig. 8-6 Dislocation line images in aluminum using transmission electron microscopy. [13]

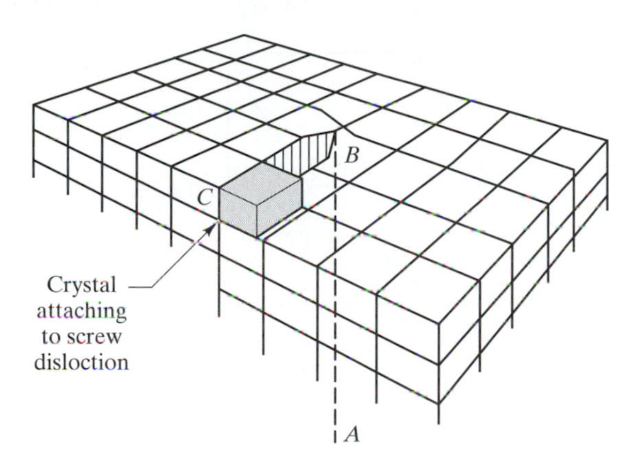

Fig. 8-7 Screw dislocation A-B provides a ledge on a surface where crystals can attach to easily.

is the coalescence of vacancies formed at high temperatures to form voids at room temperature. The boundaries of these voids are dislocation loops. In polycrystalline materials, the grain boundaries are defects from where dislocations can be generated.

Figure 8-5 also indicates the influence of the size of a material. It is a common observance in materials that larger size objects have more defects than smaller sizes. More dislocations mean more chances for mobile dislocations to exist and, therefore, low strength. Thus, we see in Fig. 8-5 that the iron single crystal has the lowest strength while the iron whisker has the highest strength, approaching the theoretical strength of the material. This is the reason why fiber composite materials use very thin fibers—in order to produce very high-strength materials.

8.3.2 Schmid's Critical Resolved Shear Stress (CRSS) for Slip

The CRSS depends on (1) the crystal structure (i.e., whether BCC, FCC, or HCP), (2) the orientation of the active slip plane with the external load, and (3) on the size of the single crystal. The critical stress must be a shear stress acting on the active slip plane. Therefore, for single crystals with different orientations, the resolved shear force or stress parallel to the plane must exceed the critical shear stress. It has been found that single crystals of the same size and material but with different orientations required different tensile loads to start the slip (shear) process. This is *Schmid's CRSS Law.*

Schmid's Law can be derived by considering a cylindrical single crystal with the cross-hatched, cross-sectional area, A_n, in Fig. 8-8. The angle between the tensile axis, which is normal to A_n, and the slip direction on the slip plane is λ, and the angle between the tensile axis and the normal to the slip plane, N, is Φ. If P is the applied tensile load, the resolved force parallel to the slip plane is $P\cos\lambda$. The area of the slip plane, A_s, (with respect to A_n) is $A_n/\cos\phi$. Therefore, the CRSS is

$$\text{CRSS} = \frac{P\cos\lambda}{\dfrac{A_n}{\cos\phi}} = \frac{P}{A_n}\cos\lambda\cos\phi = \sigma\cos\lambda\cos\phi \tag{8-6}$$

where σ is the applied normal stress, either tensile or compressive. It follows that for a constant CRSS,

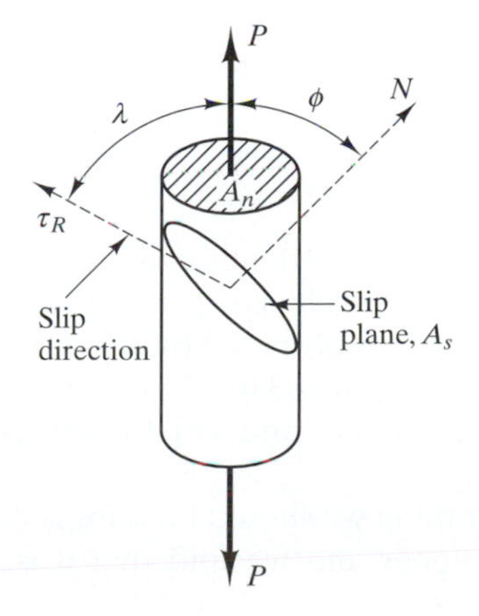

Fig. 8-8 Determination of critical resolved shear stress of a single crystal.

different tensile loads or stresses, σ, may be observed because

$$\sigma = \frac{\text{CRSS}}{\cos\lambda\cos\phi} \quad (8\text{-}7)$$

and σ depends on the orientation (angles, λ and ϕ) of the slip plane.

Example 8-1 also illustrates that the maximum resolved shear stress occurs when $\lambda = \phi = 45°$. Thus, in tensile fractures of thin sheets of materials, the fracture surface is about 50° from the tensile axis. This is, in fact, referred to as *shear fracture* that indicates ductile behavior of a material as opposed to brittle behavior when the fracture surface is normal to the tensile axis.

Equation 8-6 further indicates when slip is not favored. When either λ or ϕ equals 90°, CRSS is zero. The crystal is not going to deform by slip because the shear stress is zero. We can also deduce from Eq. 8-7 that the applied tensile stress will be infinite to start slip because cos 90° is zero. Thus, if we pull a single crystal in this orientation, it fractures (brittle condition) before it slips (ductile condition) because it cannot have an infinite shear strength.

EXAMPLE 8-1

In Fig. 8-8, determine the orientation of a slip plane for the CRSS to be maximum.

SOLUTION: In Fig. 8-8, we see that $\lambda = \pi/2 - \phi$, thus Eq. 8-6 becomes,

$$\text{CRSS} = \sigma\cos\phi\cos\left(\frac{\pi}{2} - \phi\right) = \frac{\sigma}{2}\sin 2\phi$$

For the CRSS to be maximum, $\sin 2\phi$ = maximum. The maximum value of $\sin 2\phi = 1$, and $2\phi = 90°$, so that the orientation of the slip plane for maximum shear is $\phi = 45°$.

EXAMPLE 8-2

A single crystal of copper rod, 10 mm in diameter is pulled in tension with the tensile axis parallel to the $[\bar{1}\,0\,1]$ of the crystal. Slip was observed on the $(\bar{1}\,\bar{1}\,1)$ plane when the applied load was 3,000 N. Determine (a) the slip systems activated, and (b) the critical resolved shear stress.

SOLUTION: The first thing we should look for is the crystal structure of copper and we find that it is a FCC metal. With this information, we should go back to Chap. 2 and find that the slip sytems for FCC are {1 1 1}<1 1 0>. Thus in this problem we need to identify the <1 1 0> in the $(\bar{1}\,\bar{1}\,1)$ and in question (a) we are being asked which of these directions are activated or will operate.

We can identify the <1 1 0> directions by sketching the $(\bar{1}\,\bar{1}\,1)$ and then from the sketch, we find the <1 1 0> directions as shown in Fig. EP 8-2.

We can also identify the directions by using the following concepts. In cubic crystals, and only in cubic crystals, the normal to a plane has the same indices as the plane. Thus, the normal to $(\bar{1}\,\bar{1}\,1)$ is $[\bar{1}\,\bar{1}\,1]$, noting the change from parenthesis to brackets because it is a direction. The normal direction being a vector means that it is perpendicular to any

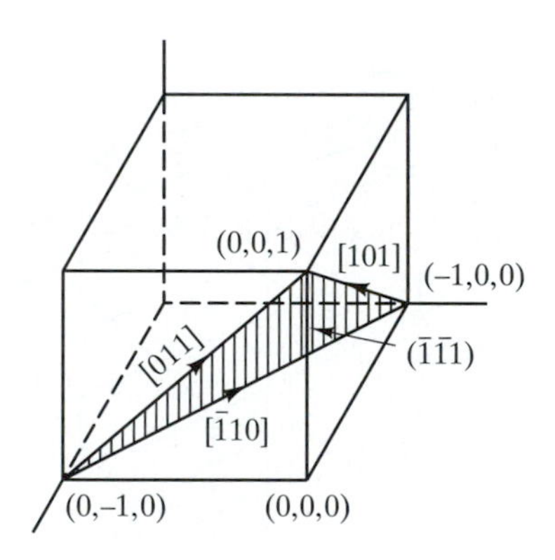

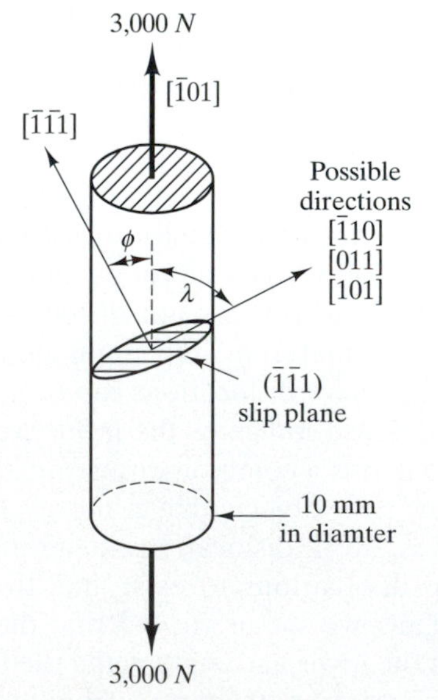

Fig. EP 8-2

vector on the plane. Thus, we can get the <1 1 0> by their dot products with $[\bar{1}\,\bar{1}\,1]$, with the correct desired directions resulting in zero dot products. Thus,

$$[\bar{1}\,\bar{1}\,1] \cdot [0\,1\,1] = 0$$

$$[\bar{1}\,\bar{1}\,1] \cdot [\bar{1}\,1\,0] = 0$$

$$[\bar{1}\,\bar{1}\,1] \cdot [1\,0\,1] = 0$$

Having identified the possible slip systems, we proceed now to determine which of the three directions on the slip plane will be activated. From Eq. 8-6,

$$\text{CRSS} = \sigma \cos \lambda \cos \phi$$

we now proceed to determine the quantities on the right. First,

$$\sigma = \frac{3{,}000 \text{ N}}{\frac{\pi}{4}(10)^2 \text{ mm}^2} = 38.2 \text{ MPa}$$

Cos λ and $\cos \phi$ can again be obtained by the dot products with the tensile axis, $[\bar{1}\,0\,1]$ of the possible slip directions and the normal to the slip plane, respectively. Thus,

$$\cos \phi = \frac{[\bar{1}\,0\,1] \cdot [\bar{1}\,\bar{1}\,1]}{\sqrt{2}\sqrt{3}} = \frac{2}{\sqrt{6}}$$

$$\cos \lambda_1 = \frac{[\bar{1}\,0\,1] \cdot [\bar{1}\,1\,0]}{\sqrt{2}\sqrt{2}} \frac{1}{2}$$

$$\cos \lambda_2 = \frac{[\bar{1}\,0\,1] \cdot [1\,0\,1]}{\sqrt{2}\sqrt{2}} = 0$$

$$\cos \lambda_3 = \frac{[\bar{1}\,0\,1] \cdot [0\,1\,1]}{\sqrt{2}\sqrt{2}} = \frac{1}{2}$$

From these we identify and answer question (a) that the two slip systems are $(\bar{1}\,\bar{1}\,1)[\bar{1}\,1\,0]$ and $(\bar{1}\,\bar{1}\,1)[0\,1\,1]$ because they yielded nonzero cosines. And for question (b),

$$\text{CRSS} = 38.2 \times \frac{1}{2} \times \frac{2}{\sqrt{6}} = 15.6 \text{ MPa}$$

EXAMPLE 8-3

The CRSS of a single crystal of a BCC metal is 35 MPa. Determine the force to initiate slip on the (1 1 0) of a 10-mm diameter rod with the tensile axis parallel to the [1 1 1] of the crystal.

SOLUTION: Again we should go back to Chap. 2 and find out that the slip systems in BCC are {1 1 0}<1 1 1>. Proceeding in the same manner as in Example 8-2, the normal to (1 1 0) is [1 1 0] and the specific <1 1 1> on the (1 1 0) are those that yield dot products with [1 1 0] equal to zero. Thus, the directions $[\bar{1}\,1\,1]$ and $[1\,\bar{1}\,1]$ are found to be the two directions on the (1 1 0) plane because

$$[1\,1\,0] \cdot [\bar{1}\,1\,1] = 0 \qquad \text{and} \qquad [1\,1\,0] \cdot [1\,\bar{1}\,1] = 0$$

and we find that

$$\cos \phi = \frac{[1\,1\,1] \cdot [1\,1\,0]}{\sqrt{2}\sqrt{3}} = \frac{2}{\sqrt{6}}$$

and

$$\cos \lambda_1 = \frac{[1\,1\,1] \cdot [\bar{1}\,1\,1]}{\sqrt{3}\sqrt{3}} = \frac{1}{3} \qquad \text{and}$$

$$\cos \lambda_2 = \frac{[1\,1\,1] \cdot [1\,\bar{1}\,1]}{\sqrt{3}\sqrt{3}} = \frac{1}{3}$$

The two slip directions are equally probable. Thus,

$$\sigma = \frac{35 \text{ MPa}}{\frac{2}{\sqrt{6}} \times \frac{1}{3}} = 128.6 \text{ MPa}$$

and the force, P, is

$$P = 128.6 \times \frac{\pi}{4}(10)^2 = 10{,}100 \text{ N}$$

8.3.3 Rotation of Crystal Planes and Effects of Constraints During Deformation

When allowed to move freely, the crystal planes will rotate during deformation. In tensile deformation, the rotation is such that the slip direction will tend to align or be parallel with the tensile axis, Fig. 8-9a. In compression, the slip direction will tend to be normal to the compression axis. Deformation proceeds without the slip planes getting distorted, that is, if one scraches a line on the surface, the line stays straight during deformation.

In the uniaxial tensile test described in Sec. 4.4.4, we saw that the ends of the specimen were gripped before

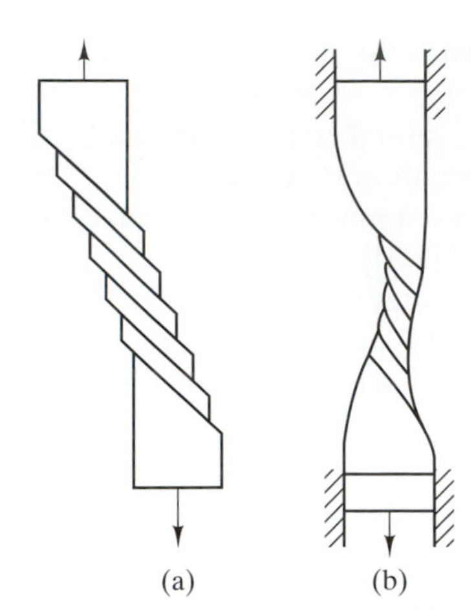

Fig. 8-9 Rotation of slip planes in single crystals (a) without, and (b) with constraints (grips) during tensile test.

conducting the test. If we tensile test a single crystal, the grip section will not deform and will constrain or restrict the rotation of the slip planes. This constraint distorts the planes trying to rotate and the slip planes get distorted or bent, as depicted in Fig. 8-9b. This phenomenon is also observed in deforming a polycrystalline material. The slip planes in the grain (crystal) would like to extend but are constrained by the adjoining grains. As observed in cold-worked microstructures of polycrystalline materials, the slip planes get distorted during deformation. An example of a cold-worked structure is shown in Fig. 8-10b; compare this with the annealed structure in Fig. 8-10a. The deformation is the result of rolling and we see that the grains in the cold-worked structure are flattened or pancaked relative to the annealed structure. The distortion is exhibited by the curved or distorted lines within each cold-worked grain.

8.4 Cold and Warm Working

As indicated earlier, cold-working is done below the recrystallization temperature of the material that is roughly one-half its melting temperature in K. In the sequence of manufacturing operations, this is usually the last operation and is called *cold-finishing* the manufactured parts and consumer products *to impart strength, close dimensional tolerance, and good surface finish.* Originally, cold-finishing operations were used only for dimensional tolerance and good surface finish. As more cold-finished products were produced, the substantial strengthening from cold-working was hard to overlook. In other words, designers or engineers should be using the properties of the as-fabricated product because of its increased strength, rather than the as-received properties of the material. It is now estimated that, in about 40 percent of applications of cold-worked materials, the imparted increase in strength is an important consideration.

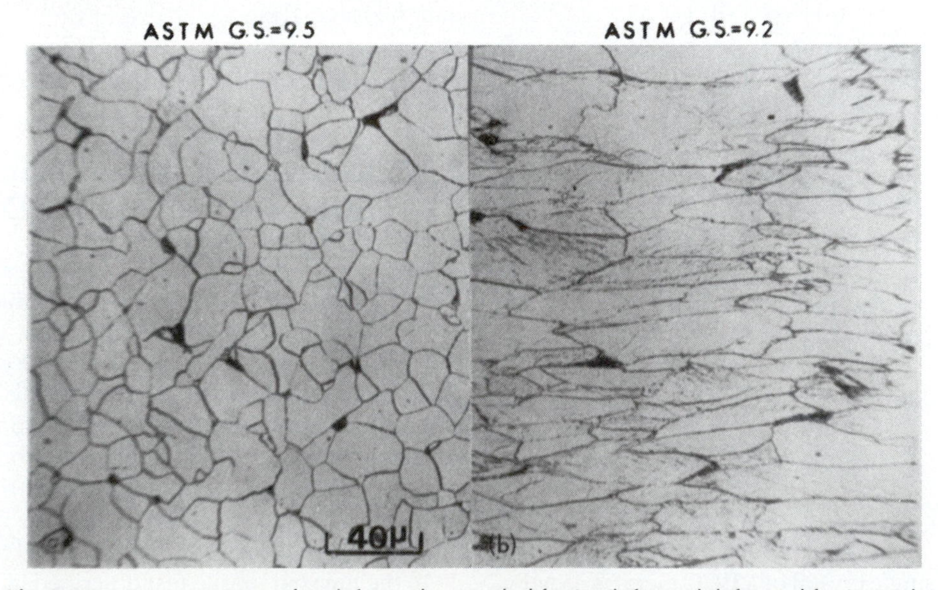

Fig. 8-10 Microstructures of undeformed, annealed ferrite (left), and deformed ferrite (right).

Most of the secondary deformation processes shown in Fig. 8-2 are done by cold-working. In general, these are done at room temperature, but warm temperatures are also used to enhance deformability and to increase tool life. The warm temperatures are still below the recrystallization temperature and to differentiate from the room temperature deformation, the process is called *warm-working*. Compared to cold-working, warm-working offers the potential for reduced loads, fewer steps in the forming of a part, and the possibility of energy savings. Compared with hot-working, warm-working offers better dimensional control, higher quality surfaces, and lower energy costs.

8.4.1 Strengthening with Cold-Work

The concept of cold-working or strain-hardening was introduced with the tensile stress-strain curve in Chap. 4. Figure 8-11 is a reproduction of Fig. 4-23 where we see that the yield strengths of materials are increased if the initial straining was done past their annealed yield strength. Accompanying this strength increase is, however, a loss in ductility. Because of cold-work, it is possible to have yield strengths of one material up to its initial tensile strength, but with no ductility left. This variable yield strength of a material due to cold-work was mentioned regarding the strength criterion to use for design in Sec. 4.4.1.

In terms of the secondary deformation processes, the phenomenon of cold-working in the tensile stress-strain curve means that there is a limit of cold-working before the material fractures because of the loss of ductility. It means also that a strain-hardened material will require higher stresses to form the material into parts and means furthermore, that higher capacity load machines will be required. For this reason, a low-initial yield strength material in the annealed state is preferred as a starting material. In addition, as indicated in Chap. 4, what we need in the deformation processes is the true stress-true strain characteristics, such as that given in Eq. 4-53, which is rewritten below.

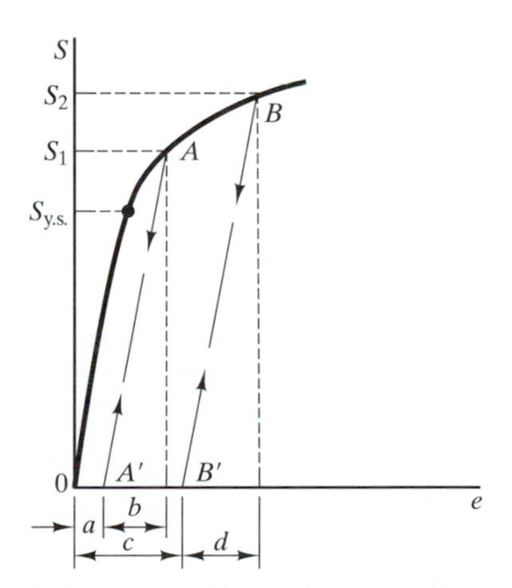

Fig. 8-11 Stressing past its yield strength cold-works or strain-hardens the material, leaves a permanent strain, or changes the shape of the material.

$$\sigma = K\,\epsilon^{n} \tag{8-8}$$

where σ is the true stress, ϵ is the true strain, K is the strength coefficient, and n is the strain-hardening exponent. Because we need good ductility, a high value of n is desired because

$$n = \epsilon_u \tag{8-9}$$

where ϵ_u is the uniform true strain up to the maximum load.

Equations 8-8 and 8-9 are strictly true only for low carbon steels (i.e., that there is only one value of n for the entire plastic strain range). Most materials exhibit a three-stage n behavior, whereby a high n value is bounded by lower values at both the low and high strain regions. The first n value at low plastic strains is used for springback calculations and the last n value is used for fracture and necking problems.

The other consideration in determining the load required to deform a material is the strain rate, $\dot{\epsilon}$ (i.e., how fast we shape the material). Most materials are very sensitive to the strain rate as reflected in the higher flow stresses required to deform the material for high strain rates. There are many different relationships—called constitutive relationships—between the flow (true) stress and the combined effect of strain hardening and the strain-rate hardening. These are classified as product or additive types and an example of each of these are given below.

$$\sigma = K'\,\epsilon^{n}\,(\dot{\epsilon})^{m} \tag{8-10}$$

$$\sigma = K''\epsilon^{n}(1 + \alpha\,(\dot{\epsilon})^{m})^{p} \tag{8-11}$$

where the exponent m is the strain-rate sensitivity index or exponent. The K′ and K″ in Eqs. 8-10 and 8-11 may be different from the K in Eq. 8-8 and the α and p in Eq. 8-11 are constants. The applicabilities

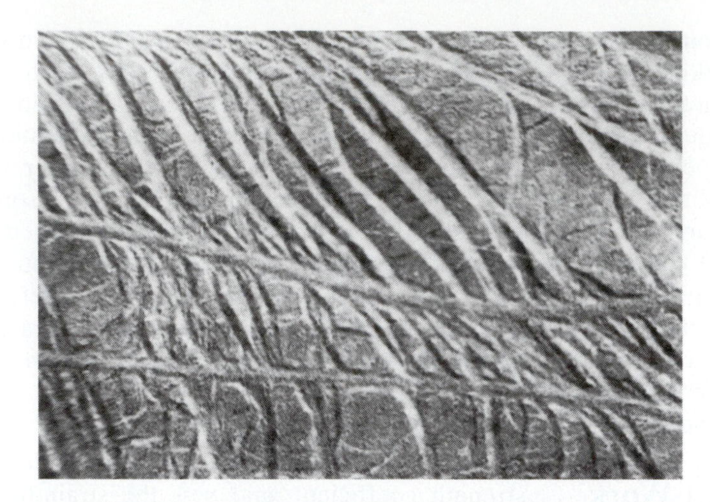

Fig. 8-12 Lüder bands or stretcher strain formation when annealed mild steel is deformed before temper rolling. [14]

of Eqs. 8-10 and 8-11 are for elevated temperatures and ambient processings, respectively.

While we want to start forming processes with the softest condition (i.e., annealed) of the material (in order to avoid high loads), the forming of low-carbon sheet steels in the annealed state results in strain markings on the surface called *Lüder bands* or *stretcher strains,* demonstrated in Fig. 8-12. Such markings are unacceptable aesthetically in fabricated consumer goods, for example, automobile fenders, even when painted, and they should be avoided. The markings are due to the presence of the yield point elongation (YPE) in the stress-strain curve of the annealed steel. Thus, the annealed sheets are plastically strained only slightly by what is called *temper rolling* at the steel suppliers' plants to remove the YPE before they are shipped to the customers for the press-forming operations.

The *real cause of the YPE,* as indicated in Chap. 4, is the interstitial carbon atom in solution in steels. *The carbon atoms diffuse to the core of the dislocations and prevent the motion of the dislocations.* At the upper yield strength, the dislocations are able to break away from the carbon atoms and they require less stress (lower yield stress) to propagate. The deformation proceeds non-uniformly and produces the YPE. Thus, by temper rolling, the dislocations are moved away from the carbon solute atmospheres and are immediately made mobile when the press-forming operations are imposed on the steel.

However, if we allow the carbon solutes to diffuse back to the dislocation cores by waiting a long time before processing the temper-rolled steel, the YPE will reappear, and the steel shows also a slight increase in strength. This strengthening is called *strain-aging* and is again due to carbon atoms locking or preventing the motion of the dislocations. This strengthening is not particularly bad once the forming processes are over. In fact, this phenomenon is advantageous during the paint-bake step of the car bodies by providing greater dent resistance to the fenders. This time the increase in strength is referred to as *paint bake hardening,* but it is still the same phenomenon as strain-aging. The disappearance and reappearance of the YPE (strain-aging) are depicted in the tensile stress-strain curves shown in Fig. 8-13.

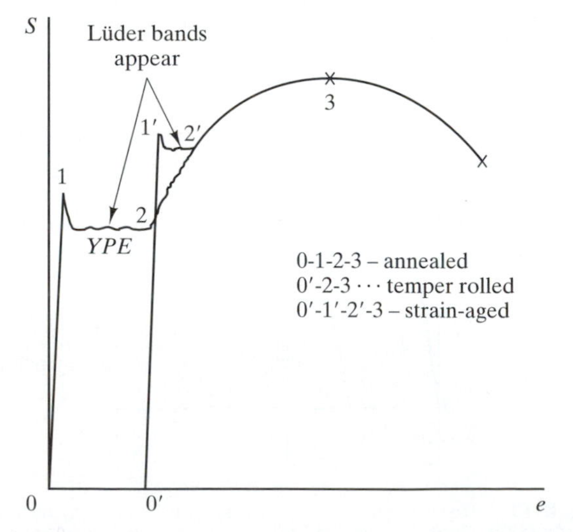

Fig. 8-13 Tensile stress-strain diagrams after annealing, temper-rolling, and strain-aging of mild steel. Appearance of YPE is responsible for Lüder bands.

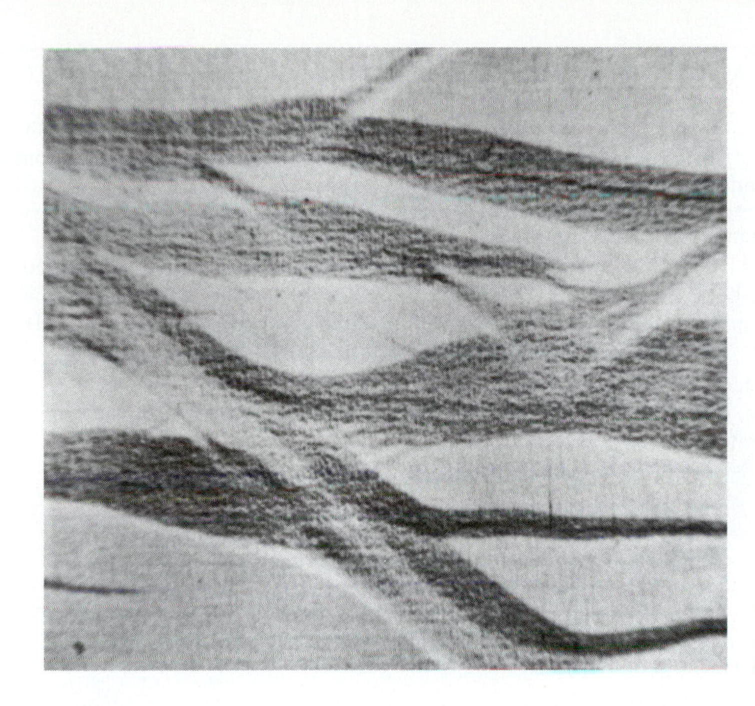

Fig. 8-14 Lüder bands in aluminum-magnesium alloy sheet. [6]

Some solid solution aluminum alloys, such as aluminum-magnesium, also exhibit the Lüder band or stretcher-strain formation on the surface. In fact, there are two types of Lüder bands, A and B, in aluminum alloys. Similar to annealed sheet steels, the type A markings are associated as well with yielding and the YPE of annealed or heat-treated solid solution alloys and an example is shown in Fig. 8-14. To prevent the occurrence of the type A bands, the alloys are also strained before fabrication or are formed at about 150°C (300°F). The type B Lüder bands may form when the alloys are stretched beyond the yield point. In contrast to type A, the type B bands occur in strain-hardened as well as annealed conditions of some alloys. The bands appear as shear bands oriented approximately 50° to the tension axis, and they increase in number as stretching progresses. In the stress-strain curve, this phenomenon, which is called the Portevin–Le Chatelier effect, shows up as serrations, shown schematically in Fig. 8-15, that continue until necking of the specimen.

8.4.2 Specifying the Cold-Work Condition

In the selection and purchase of cold-worked materials, the condition of their cold-work is specified rather than the strength or mechanical properties. The condition is called the *temper* of the material and is specified in the "degree of hardness." For each temper condition, the expected ranges in properties of the material, including strength, are included in the specification. An example of such a specification is ASTM A109, Table 8-1, which is a specification for cold-rolled, low-carbon steel strips. Each temper condition is designated by a number from No. 1 to No. 5 indicating the condition from "hard" to "dead

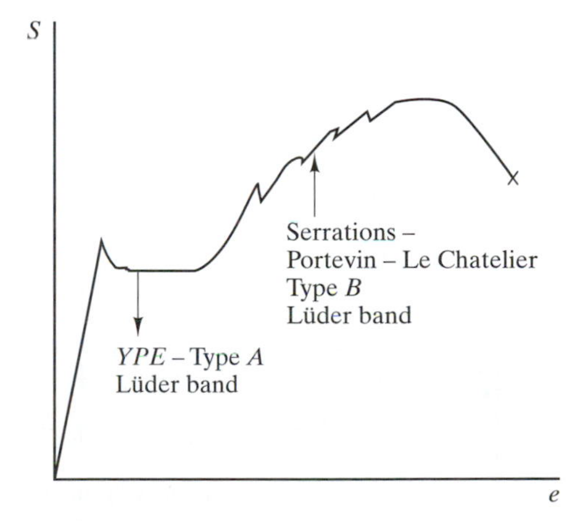

Fig. 8-15 Schematic of stress-strain curve of an aluminum alloy showing YPE and serrations that are responsible for the Type A and B Lüder bands on surfaces of deformed sheets.

Table 8-1 ASTM A 109—Mechanical properties of cold-rolled low-carbon steel strips. [7]

Temper	Rockwell B Hardness Required	Bend Test Requirements	Approx. Tensile Str. mPa	Approx. Tensile Str. Ksi	Elongation in 50 mm or 2 in, %(d)
No. 1 (hard)	90 min (a), 84 min (b)	No bending in either direction	550–690	80–100	Not specified
No. 2 (half-hard)	70–85	90° bend across rolling direction around a 1t radius (c)	380–520	55–75	4–16
No. 3 (quarter-hard)	60–75	180° bend across rolling direction, and 90° bend along rolling direction, both around a 1t radius (c)	310–450	45–65	13–27
No. 4 (skin-rolled)	65 max	Bend flat on itself in any direction	290–370	42–54	24–40
No. 5 (dead-soft)	55 max	Bend flat on itself in any direction	260–340	38–50	31–47

Notes: (a) For strip less than 1.78 mm (0.070 in.) thick; (b) for strip 1.78 mm (0.070 in.) thick and greater; (c) t = thickness of strip; (d) for strip 1.27 mm (0.050 in.).

soft" annealed. It is one of these numbers that is specified when we order. Each condition has its ranges of mechanical properties specified, as well as a bend test requirement, which must obviously be met by the supplier to be accepted by the customer.

The origin of such a specification arises from the observed mechanical properties of cold-worked materials as indicated schematically in Fig. 8-16, a plot of the degree of cold-work versus increase in hardness or strength. In metalworking operations, the degree of cold-work is usually specified as the percent reduction in area. Thus,

$$\%CW = \%RA = \frac{A_o - A_f}{A_o} \tag{8-12}$$

where CW is cold work, RA is reduction in area, and the subscripts o and f refer to the original and final areas. For sheets, strips, and plates, Eq. 8-12 is a reduction in thickness because in plastic deformation of these shapes, the width does not change much, thus

$$\%CW = \%RA = \frac{(wt_o - wt_f)}{wt_o} = \frac{t_o - t_f}{t_o} \tag{8-13}$$

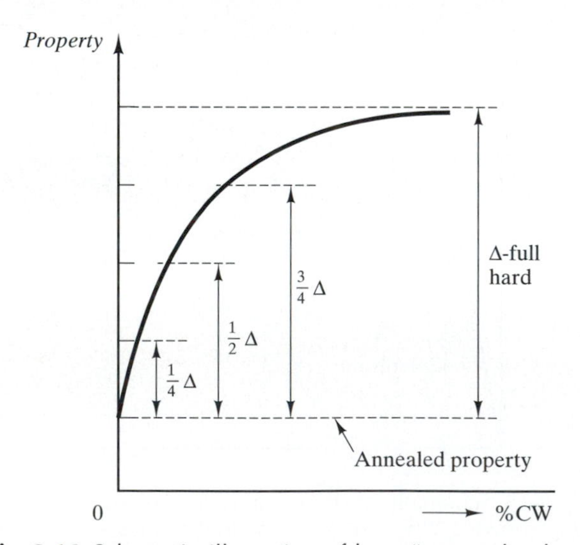

Fig. 8-16 Schematic illustration of how "temper hardness condition" is specified.

We see from Fig. 8-16 that the change in hardness or strength during the first few %RA from the annealed dead-soft condition is large and then it levels off. The point where the change levels off is designated the hard condition. If we take the total change, Δ, from the dead-soft to the hard condition and divide it by 2, the condition for that change (i.e., $^1/_2\Delta$), is designated half-hard. The condition for $^1/_4\Delta$ is quarter-hard and for $^3/_4\Delta$ is three-quarters hard.

The response of properties to cold-drawing of steel bars, 25.4 mm (1 in.) or less in cross-section with

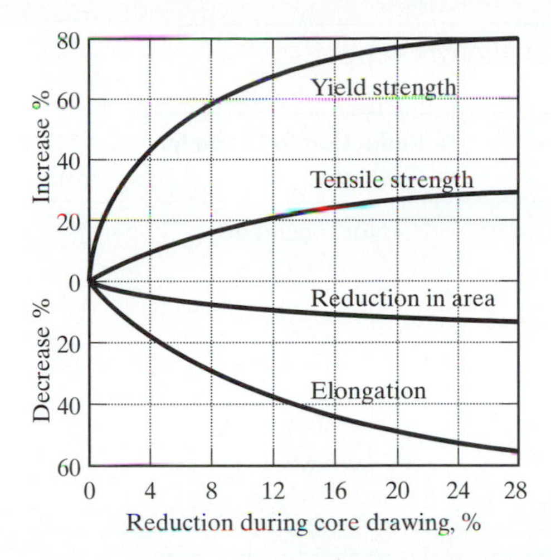

Fig. 8-17 Changes in tensile properties of cold-drawn steel bars, less than 25 mm (1 in.) cross-section with tensile strength of less than 690 MPa before drawing. [7]

a tensile strength of 690 MPa (100 ksi) or less before cold-drawing, is shown in Fig. 8-17. Based on the schematic of Fig. 8-16, the hard condition is taken at 80 percent increase in yield strength over the annealed condition and the 20, 40, and 60 percent increases are quarter-hard, half-hard, and three quarters-hard.

EXAMPLE 8-4

Using the tensile strength data in Table 8-1 for the No. 5 (dead-soft) and the No. 1 (hard) tempers for low-carbon sheet steels, verify the tensile strength specifications for the No. 2 (half-hard) and the No. 3 (quarter-hard) tempers.

SOLUTION: The tensile strength specifications are ranges of values. We can take the averages of the range and use them to verify the concept illustrated in Fig. 8-16. Thus,

for the No. 1 (hard) temper

$$\text{the average T.S.} = (550 + 690)/2 = 620 \text{ MPa}$$

for the No. 5 (dead-soft) temper

$$\text{the average T.S.} = (260 + 340)/2 = 300 \text{ MPa}$$

$$\text{the } \Delta = 620 - 300 = 320 \text{ MPa}$$

Based on this, the No. 2 (half-hard) temper average T.S. $= 300 + 320/2 = 460$ MPa, and the No. 3 (quarter-hard) temper average T.S. $= 300 + 320/4 = 380$ MPa. We now look at the specifications given for these two tempers, and verify that the average T.S. for the No. 2 and the No. 3 tempers are $(380 + 520)/2 = 450$ MPa, and $(310 + 450)/2 = 380$ MPa. We see that the estimates show a difference of 10 MPa for the No. 2 temper and 0 for the No. 3 temper.

For aluminum alloys, the cold-worked or strain-hardened condition is designated H1x, the x indicating the degree of strain-hardening or cold-work. The full-hard condition is determined when the %RA $\cong$ 75% and x = 8, thereby having a temper designation of H18. When we take the difference, Δ, in the tensile strengths of the full-hard and annealed condition, which is given a temper of O, and then use the concept in Fig. 8-16, Table 8-2 lists the temper designations for aluminum alloys.

For copper and copper alloys, the temper designations are based on the increase from the annealed condition in the Browne & Sharpe (B&S) gauge number for wires and sheets for non-ferrrous metals and alloys. These are indicated in Table 8-3.

Figures 8-18a and b are plots of the mechanical properties of 70Cu-30Zn cartridge brass strips as a function of the H numbers in Table 8-3, given as an increase of B&S gauge numbers. These figures may also be used for wires as long as we use the numbers in the abscissa as the percent reduction in diameter as indicated in the last column of Table 8-3. These figures show that the hard condition is not related to the leveling of tensile or yield strengths but appears to be related to the leveling of the Rockwell F numbers—significantly different from what Fig. 8-17 shows for steel bars. Thus, we should not be interrelating temper designations from one material to another. We should also note from Fig. 8-18 that a finer grain size material (0.015 mm) has greater strength and hardness values than a coarser grain (0.070 mm) material, but has

Table 8-2 Cold-work tempers of aluminum alloys.

Temper	Increase in Tensile Strength from Annealed State O	Degree of Hardening
H12	$^1/_4\Delta$	Quarter-hard
H14	$^1/_2\Delta$	Half-hard
H16	$^3/_4\Delta$	Three-quarter hard
H18	Δ	Full-hard
H19	$>\Delta$	Extra-hard

Table 8-3 Cold-worked tempers of wrought coppers and brasses.

Tempers-H	Increase in B&S Gauge Number	Temper Names	% Reduction in Diameter or Thickness
H00	Not specified	1/8-Hard	Not specified
H01	1	1/4-Hard	10.9
H02	2	1/2-Hard	20.7
H03	3	3/4-Hard	29.4
H04	4	Hard	37.1
H06	6	Extra hard	50.1
H08	8	Spring	60.5
H10	10	Extra Spring	68.6
H12	12	Special Spring	75.1
H14	14	Super Spring	80.3

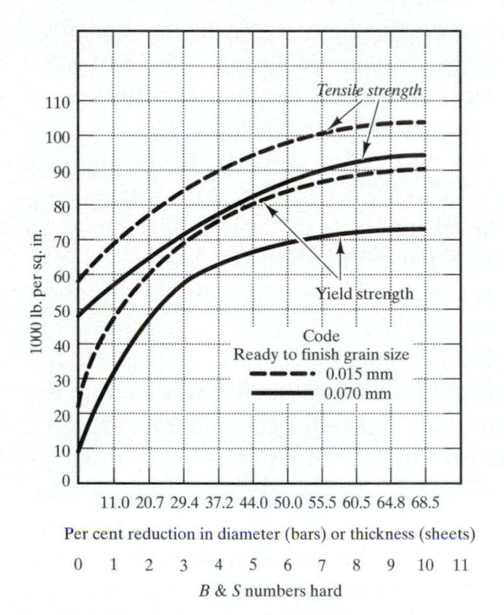

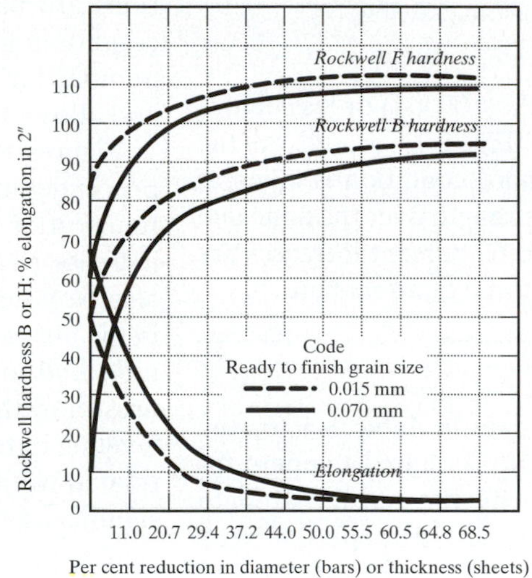

Fig. 8-18 (Left) Yield and tensile strengths of 70Cu-30Zn yellow (cartridge) brass at different percent reduction in diameters for bars, or thicknesses for sheets or strips, and B&S hardness numbers for two different annealed grain sizes of 0.015 mm and 0.070 mm; (Right) Rockwell B and F, and elongation properties.

lesser ductility. Photos 2, 8, and 9 in the Appendix illustrate the microstructure of 70Cu-30Zn cartridge brass in the annealed, half-hard, and full hard conditions, respectively.

EXAMPLE 8-5

A hot-rolled 25-mm diameter 1025 steel bar has a minimum 230 MPa yield strength. If the steel is cold-drawn 30 percent, determine (a) the final diameter and (b) the expected yield strength of the as-cold drawn bar.

SOLUTION:

(a)

$$0.30 = \frac{\frac{\pi}{4}(25)^2 - \frac{\pi}{4}D_f^2}{\frac{\pi}{4}(25)^2} \quad \text{and}$$

$$D_f^2 = 0.70(25)^2$$

$$D_f = 25\sqrt{0.70} = 20.92 \text{ mm}$$

(b) Using Fig. 8-17, we can estimate the yield strength of the as-cold drawn bar. The expected increase in yield strength is 80 percent. Thus, the yield strength will be

$$\text{Expected yield strength} = 1.80 \times 230 = 414 \text{ MPa}$$

EXAMPLE 8-6

A portion of the Browne & Sharpe table of gauge numbers and wire diameters or sheet thicknesses, in inches, is shown in Table EP 8-6. Two wires, A and B, of yellow brass with 0.070 mm average grain size, had original gauge numbers of 0 and 2 in the annealed condition. Each was cold-drawn, such that its gauge number increased by 3. Using Table EP 8-6, determine (a) the initial and final diameters of each of the wires, (b) the percent reduction in diameter, (c) the percent reduction in area, (d) the temper designation of the as-cold drawn wires, and (e) the yield strength of the as-cold drawn wires.

Table EP 8-6

B&S Gauge Number	Wire dia. or sheet thickness, in.
0	0.324861
1	0.289297
2	0.257627
3	0.229423
4	0.204307
5	0.181940

SOLUTION:

(a) for wire A

$$\text{initial diameter} = 0.324861$$

$$\text{final diameter} = 0.229423$$

for wire B

$$\text{initial diameter} = 0.257627$$

$$\text{final diameter} = 0.181940$$

(b)

$$\%\ \text{red. dia.} = \frac{0.324861 - 0.229423}{0.324861} \times 100$$

$$= 29.4\%$$

for wire A, and

$$\%\ \text{red. dia.} = \frac{0.257627 - 0.181940}{0.257627} \times 100$$

$$= 29.4\%$$

for wire B.

(c) for wire A,

$$\%\ \text{R.A.} = \frac{\frac{\pi}{4}(0.324861)^2 - \frac{\pi}{4}(0.229423)^2}{\frac{\pi}{4}(0.324861)^2} \times 100$$

$$= 50.1\%$$

for wire B,

$$\%\ \text{R.A.} = \frac{\frac{\pi}{4}(0.257627)^2 - \frac{\pi}{4}(0.181940)^2}{\frac{\pi}{4}(0.257627)^2} \times 100$$

$$= 50.1\%$$

(d) The temper designation of both as-cold drawn wires is H03 (three-quarter hard) because each had an increase in B&S number of 3.

(e) The yield strength of each wire is about 58,000 psi from Fig. 8-18a.

8.4.3 Deformation (Crystallographic) Texture

The deformation of single crystals demonstrated that the slip plane rotates as deformation proceeds. The deformation of polycrystals also results in the rotation of planes and axes. However, because of the interactions with adjoining grains, the rotations are very complex and the alignments of the grains with respect to the axis and the plane of deformation are determined empirically. These alignments are called crystallographic textures and since the alignments are due to deformation, they are specifically called *deformation textures*. This is to differentiate from another texture that arises from recrystallization that is called the *annealing* or *recrystallization texture*.

Textures arise when the material deformation is very high. For bars, this deformation results in wire products while for plates or strips, the products are sheets. The deformation textures in metals are shown in Table 8-4 together with the annealing or

Table 8-4 Deformation and annealing textures of different metallic crystal structures.

Crystal Structures	Deformation/Rolling Textures	Annealing/Recrystallization Textures
BCC—steel sheet	{100} <110> - main component {112} <110> and {111} <110> also present.	{111} <110> dominant component {332} and {211} also present, less {100}
FCC—aluminum sheet	{110} <112>, {112} <111>, and {123} <112>.	{100} <100> and {123} <112>
HCP—titanium sheet	{0001} <1 0 $\bar{1}$ 0>	{0001} <1 0 $\bar{1}$ 0>

recrystallization textures for the most common metallic crystal structures of BCC, FCC, and HCP. The texture in wires, called the *fiber texture,* is the average crystallographic direction of the grains with respect to the length of the wire. Thus, it is sometimes called the *fiber axis.* The texture in cold-worked sheets, called the *rolling texture,* is the alignment of the crystal planes to the rolling plane and of the crystal axes to the rolling direction. In order to produce this alignment of planes and axes, the sheets have to be cold-worked at least 50%RA or thickness, and the common range of deformation is from 50 to 70 percent. Because these sheets have to be formed subsequently into parts, they have to be softened by annealing and recrystallization. The alignment of planes in the cold-worked condition is completely obliterated and a new alignment called the *annealing* or *recrystallization texture* appears that dictates the formability of the sheet. The deformation (rolling) and annealing textures are related to each other, but we need not dwell on it here. It is sufficient only to indicate that it is important for sheet steels to have a 50 to 70 percent cold reduction in order for them to exhibit the optimized annealed properties needed for the forming process such as drawing and stretching. We shall learn what these optimized properties are by considering the material flow and redistribution during sheet forming.

8.4.4 Sheet Properties for Good Formability

Sheet forming is a very vital industry in an industrialized economy. It is essential in the manufacture of automobiles, airplanes, appliances, office and business machines, beverage cans, cartridges, and many more. It accounts for the large usage of primary sheet products of steel, aluminum, brass, titanium, and other metals. The most common sheet forming process is *press forming,* whereby a flat blank is pressed into shape between two matched dies. There are other methods of forming, but, in all these processes, it has been found that the deformation may be resolved into *two types: drawing and stretching.* The contribution of drawing or stretching in a forming process depends on the shape of the part being formed, and on other factors such as die design and lubrication. Now, the properties of the sheet for good drawability are different from those for good stretchability. Thus, the formability of a sheet cannot be expressed by one property but depends rather on a combination of several properties that may be different from part to part or from process to process. Nonetheless, by considering the drawing and stretching processes separately, it gives us the insight into what sheet properties are pertinent in sheet forming. We shall consider drawing first and then stretching.

8.4.4a. Drawing and Properties for Good Drawability. The drawing process is schematically illustrated in Fig. 8-19 and consists of pressing a flat blank into a die opening with a punch to produce a cup, cone, box, or shelllike product. The blank is held down by a blank-holder or a hold-down die with just enough force to allow the material to flow radially into the die cavity without wrinkling.

The progressive stages in the metal during drawing are schematically shown in Fig. 8-20. The punch area indicated in Fig. 8-20a is that area of the blank

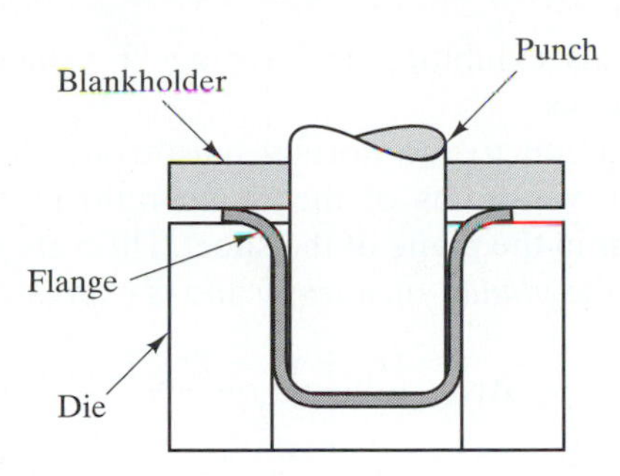

Fig. 8-19 The draw process showing the essential components.

that initially contacts the punch. As the punch presses on the blank, section 1 of the blank bends and wraps around the punch nose (Fig. 8-20b). Simultaneously as well as sequentially, the outer annular sections 2 and 3 move radially towards the center of the blank, bend around the punch nose, and form the straight wall of a cup (Figs. 8-20c and d).

The system of forces acting on the blank at the intermediate stage, such as in Fig. 8-20b or c, is shown in Fig. 8-21. The flange of the cup is under a biaxial stress state, circumferential compression, and radial tension in the plane of the sheet into the straight wall of the cup. The biaxial stress state in the flange is equivalent to pure shear in the plane of the flange. This process is analogous to the wire drawing process (Sec. 8.4.6b) in that a larger cross section from the flange area is drawn into a smaller and longer cross section in the wall of the cup. This is the reason we call this forming process drawing to differentiate it from stretching. To differentiate further from stretching, the punch area indicated in Fig. 8-20a undergoes no deformation during the drawing process.

The success of the drawing operation depends on the properties of the material to flow easily in the flange region in the plane of the sheet under pure shear conditions and to resist thinning in the sidewall. The first property requires that the material exhibits low-flow strength in all directions in the plane of the sheet, and the second property requires that it exhibits high flow strength in the thickness direction in order to resist thinning. The material must have the optimized combination of both properties for drawing. Low flow strength in the plane of the sheet is not helpful if the material also has low flow strength in the thickness direction.

Obviously, it is difficult to directly determine the flow strength of the sheet in the thickness direction because there is not enough material to prepare a tensile specimen. However, the flow strength in the thickness direction or the resistance to thinning of the sheet is obtained indirectly through the ratio of the true strains in the width and thickness directions when the sheet is tensile tested. This ratio is called the *plastic anisotropy strain ratio* r, because its value varies with different directions in the plane of the sheet. It is expressed as

$$r = \frac{\epsilon_w}{\epsilon_t} \tag{8-14}$$

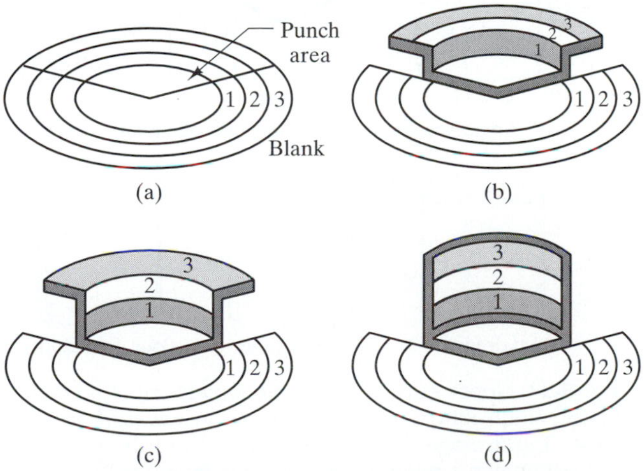

Fig. 8-20 Progressive stages in drawing a blank into a cup. [8]

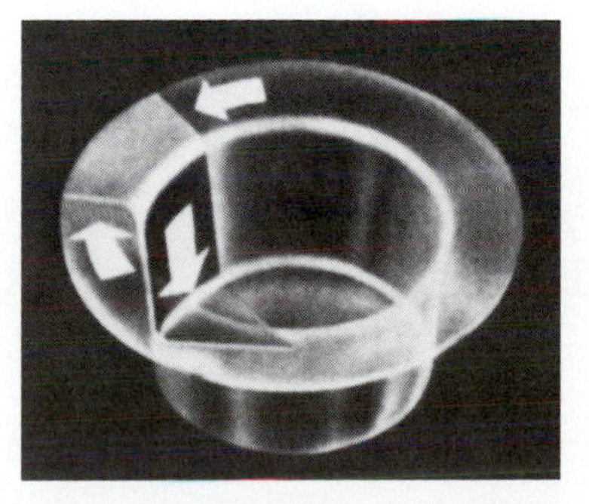

Fig. 8-21 Stress state in cup drawing; flange is in pure shear; wall in tension.

where ϵ_w and ϵ_t are the true strains in the width and thickness dimensions of the sheet. Because of the anisotropy, the measured values are different for specimens taken from parallel with, transverse to, and 45° to the rolling direction. The average of these values is called the *average strain ratio,* $\bar{r}$, and is equal to

$$\bar{r} = \frac{(r_L + 2r_{45°} + r_T)}{4} \tag{8-15}$$

where the subscripts L, T, and 45° of *r* indicate parallel with, transverse to, and 45° to the rolling direction, as indicated schematically in Fig. 8-22. Equation (8-15) indicates the ratio of the average flow strength in the plane of the sheet to the normal of the sheet and is called the *normal anisotropy.* A value of $\bar{r} = 1$ indicates equal flow strengths in the plane and thickness directions of the sheet. When $\bar{r} > 1$, the average strength in the thickness direction is greater than the average strength in the different directions in the plane of the sheet. In this case, the sheet is resistant to thinning. The relationship of this parameter to drawability is indicated in Fig. 8-23 which illustrates the relative sizes of cups produced with sheets exhibiting the average $\bar{r}$ values in the upper curves.

In addition to the normal anisotropy, there may exist also variations of the strain ratio in different directions in the plane of the sheet. This variation, Δr, is called the *planar anisotropy* and is expressed as

$$\Delta r = \frac{(r_L + r_T - 2r_{45°})}{2} \tag{8-16}$$

A nonzero value of Δr indicates nonuniformity in extensions of the material in some directions and leads to a defect in the formed part called "earing." Δr and $\bar{r}$ constitute the properties that indicate the anisotropy of a sheet material. A completely isotropic material has $\bar{r}$ equals 1 and Δr equals zero.

The properties for good drawability are $\bar{r} > 1$ and $\Delta r = 0$, both of which depend on the crystallographic annealing texture, which in turn depends on the composition and the processing operations, the last of which is the amount of cold-work prior to annealing. The annealing textures that induce good drawability in different materials are those shown in Table 8-4. For steels, a dominant {111} annealed texture is required. This means that the {111} are parallel to the plane of the sheet and thus the normal to these, the <111>, are parallel to the thickness direction. This texture is commonly called "cube-

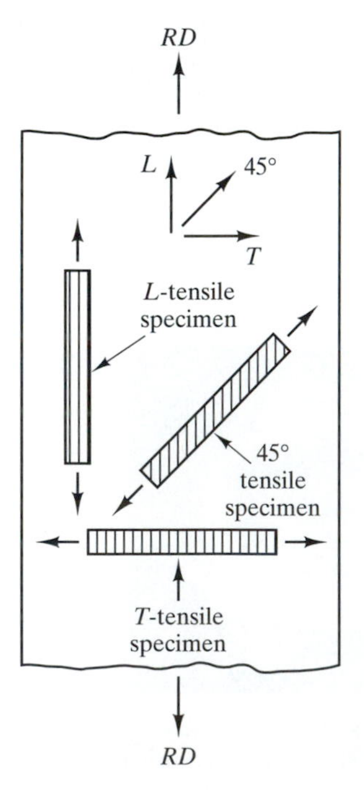

Fig. 8-22 Test specimens in the L, T, and 45° to the rolling direction.

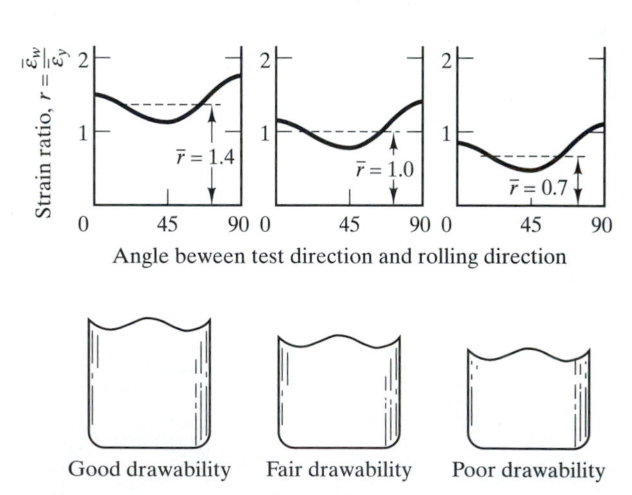

Fig. 8-23 Upper curves depict the typical variation of the plastic strain ratio with test direction in low carbon steel sheet. The average strain ratio, $\bar{r}$, in each of the plots and this parameter is related to drawability by the relative sizes of the cups below. [Ref. 8]

on-corner" because if we stand a cube on one of its corners on a surface, the <111> direction is perpendicular to the surface. This direction is also the densest pack direction in the BCC and thus would provide the greatest resistance in compression and in thinning.

The effect of the amount of prior cold-work of sheet steel on $\bar{r}$ and Δr are depicted in Figs. 8-24 and 8-25. The bottom plot in Fig. 8-24 shows the increase in the annealed {111} texture as the amount of cold-work (prior to annealing) increased up to about 90 percent reduction. With the increase in {111}, we see (from the top curve in Fig. 8-24) a corresponding increase in the r of the sheet. We also see that the highest $\bar{r}$ is at about 75 percent cold reduction; therefore, we see that we need to cold-work from 50 to 70 percent prior to annealing to produce drawing quality sheet steels. In Fig. 8-25, we see $\Delta r = 0$ at approximately 25 percent and 90 percent cold reduction in thickness. Accordingly, we should expect no "earing" from materials subjected to these cold reductions. Figure 8-26 shows cups for different reductions in areas or thickness from 0 to 95, and indeed, we see that the cups with 25 and 90 percent reduction in thickness (marked 25 and 90) exhibit no earing.

We should note in both Figs. 8-24 and 8-25 that both the requirements for $\bar{r} > 1$ and $\Delta r = 0$ are generally not satisfied at the same time for a particular cold-reduction. Thus, we have to optimize and choose a cold-reduction that will yield the best result for our purpose. The suggestion that we cold-work sheet steels between 50 to 70% means that we are maximizing the resistance to thinning with allowance for some earing, see the cups marked 50, 60, and 70 in Fig. 8-26. These "ears" obviously need to be sheared or cut if we have to use the cups for components.

8.4.4b Stretching and Properties for Stretchability. A simple stretch-forming process is that shown on the right of Fig. 8-27. Here a blank of sheet material is clamped firmly around the flange area to prevent the material in the flange from moving into the cavity as the punch descends. In this case and different from the drawing operation, the material over the punch experiences all the deformation by elongating and thinning. If deformation is carried out far enough, it will eventually be localized and then fracture. The behavior is very similar to that observed in a tensile test and the figures on the left show schematically the equivalent material response during a tensile test.

The actual stress system acting on the blank during stretching is biaxial. Nevertheless, we note that

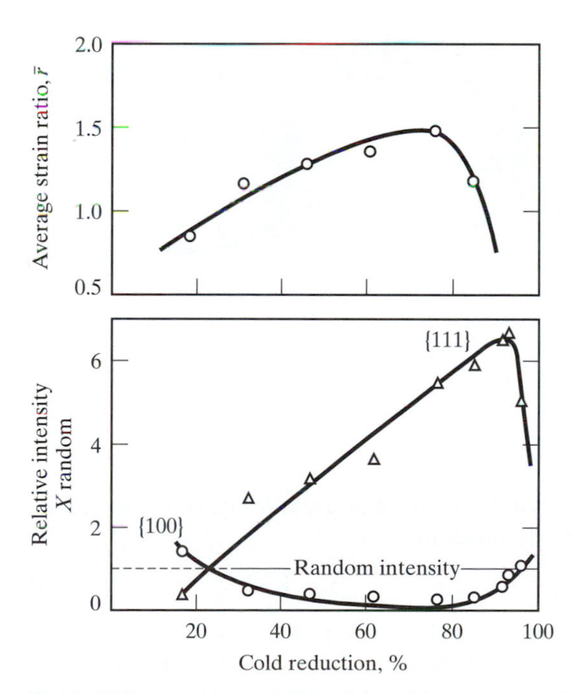

Fig. 8-24 Dependence of $\bar{r}$ and the {111} annealed texture on amount of cold work. [Ref. 8]

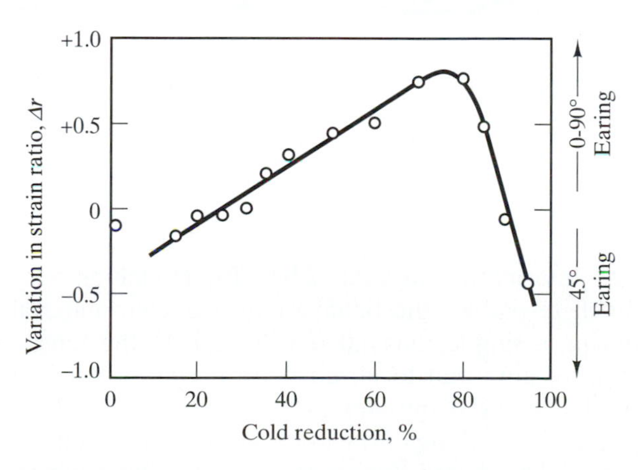

Fig. 8-25 Dependence of planar anisotropy, Δr, on cold reduction [Ref. 8]

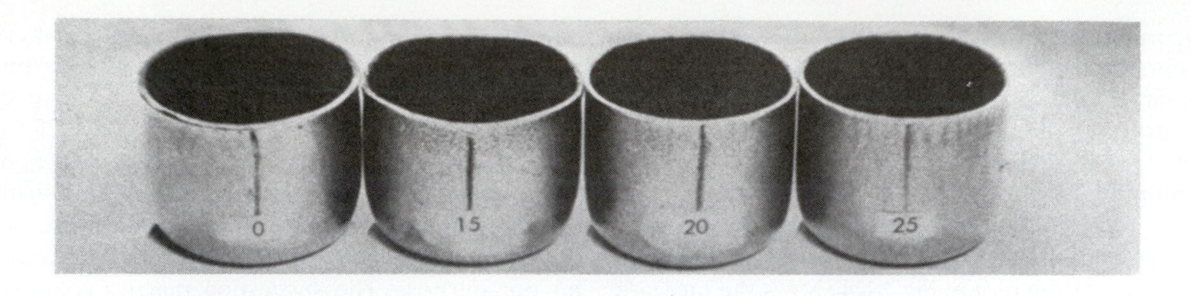

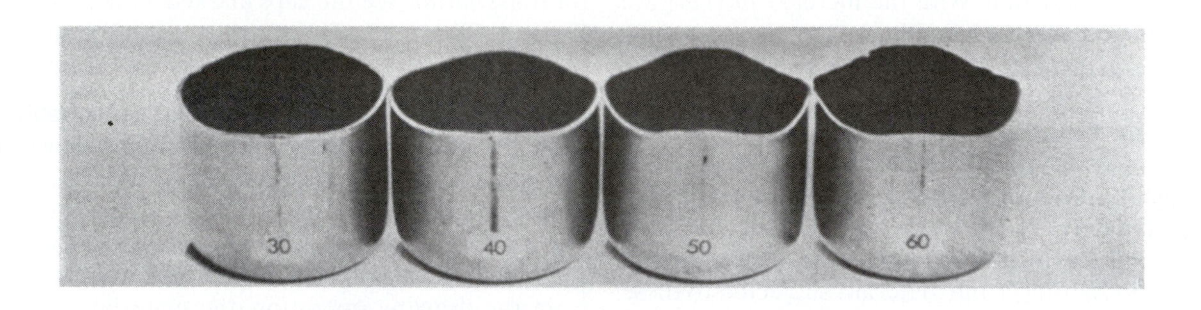

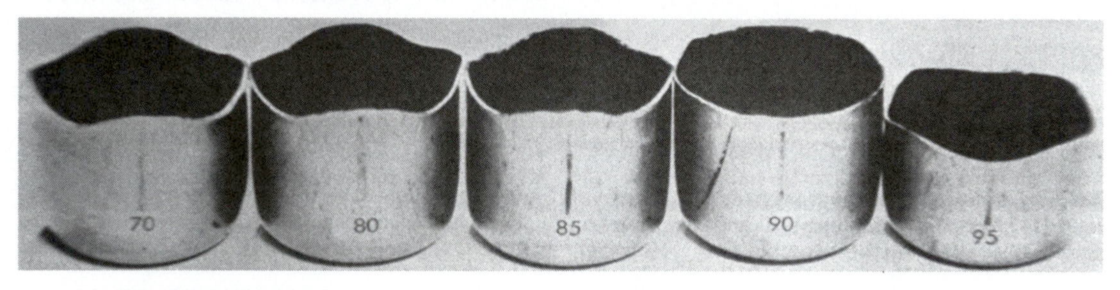

Fig. 8-26 The appearance or nonappearance of "earing" depends on the plastic anisotropy ratio, Δr, which in turn depends on cold reduction prior to annealing as shown in Fig. 8-25. The numbers on the cups correspond to cold-reduction, and the Δr values corresponding to these reductions are obtained from Fig. 8-25. The line on each cup indicates the rolling direction. [From D. J. Blickwede] [8]

the deformation just described for stretching very much resembles the behavior of a ductile material during a simple uniaxial tensile test. In the tensile test, we note that localized deformation or necking of the material initiates at the maximum load. In stretching, necking of the materials indicates that the material has failed. We can thus say that the true uniform strain, ϵ_u, up to the maximum load in the tensile test is a measure of stretchability. Because steels follow the relation,

$$\sigma = K\,\epsilon^n \tag{8-8}$$

and we have shown from Chap. 4 that $n = \epsilon_u$, as indicated in Eq. 8-9, it follows that the strain hardening exponent n is also a measure of the stretchability of materials. This conclusion from the simple tensile test

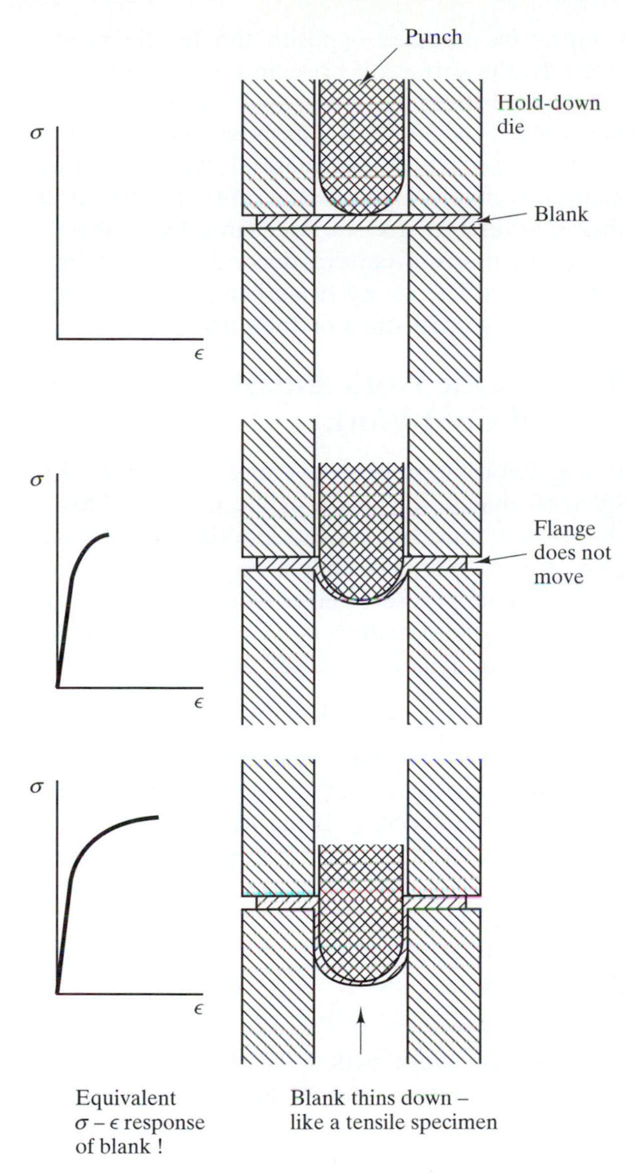

Fig. 8-27 (Right) Schematic of the stretch process showing the blank being stretched such as in a simple tensile test; (left) the conceptual response of the blank in a tensile test.

is confirmed in actual stretch tests, namely, that a high strain-hardening exponent acts to distribute plastic strain and thereby increases the total stretchability of the material.

Studies on what influences the strain hardening of steel indicate that the effects of composition and structure on n arise from the ferrite grain size. The following empirical relation was found as the relation between n and the average annealed grain diameter, d, in millimeters.

$$n = \frac{5}{10 + d^{-1/2}} \tag{8-17}$$

This expression suggests that we can increase n by increasing the grain size. However, practical considerations limit the grain size because grain sizes larger than ASTM 6 result in poor surface appearance and adversly affect the local strain at fracture under biaxial stress condition.

8.4.4c Actual Forming Operations. The actual forming operations are neither simple draw nor simple stretch operation, as described above, but are rather a complex combination of both operations. The extent of stretch and draw involved in a forming operation cannot be easily quantified, however, because of the complex interaction between the die, the workpiece, and the press. Nevertheless, it has been shown that the combination of the parameters n and $\bar{r}$ is an important measure of the formability of sheet steel.

8.4.5 Residual Stresses

In cold working, the deformation of the material is nonhomogeneous and sets up residual stresses in the material. A nonhomogeneous deformation means that the deformation across the section of a workpiece is not uniform. This arises primarily from the friction between the workpiece and the forming tool or die. This is illustrated in Fig. 8-28 in the rolling

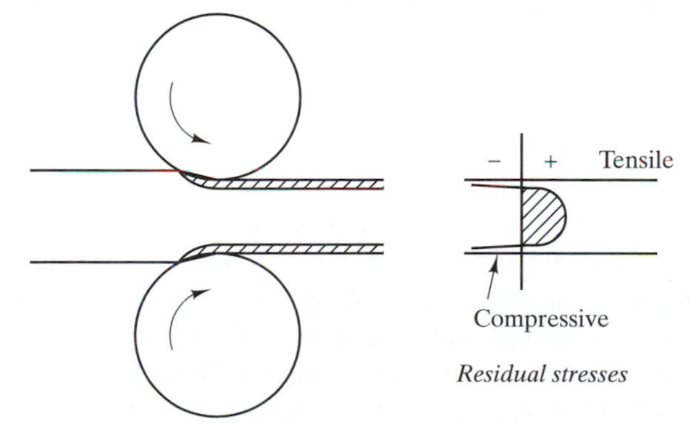

Fig. 8-28 Friction between rolls and workpiece sets up inhomogeneous deformation and residual stresses in workpiece, especially with low drafts (%RA).

process. At low percent reduction in area (%RA), the deformation is confined predominantly on the surface layers because of the friction effect between the rolls and the material, while the rest of the section is undeformed. The surface layers may want to extend but the inner layers prevent them. Because of this, they are under compressive stress and the inner layers are under tension. Figure 8-29 demonstrates the residual stress patterns for drawn 1050 steel bars with different percent reduction in area. We see that for 0.41%RA the residual stress in the longitudinal direction on the surface is compressive and in the inner layers, it is tensile. The residual stress pattern changes with different amounts of deformation and cannot readily be predicted because it depends on a lot of variables. It is sufficient to say only that the stresses are elastic and they can approach the yield strength of the material.

The residual stresses in the cold-work material are in equilibrium. If this equilibrium is not disturbed, the shape of the material will remain straight or flat as the case may be. However, if any of the surface material is removed by machining, this equilibrium will be disturbed, and the material will change shape by warping or changing its curvature to reach a new equilibrium. For example, if a round hole is drilled through a material with residual stress, the hole may become elliptical. Thus, residual stresses will impact the dimensional stability of manufactured parts and may affect the performance of the material. For example, tensile residual stress on the surface of a part is undesirable, particularly if the part is subjected to fatigue loading, as we saw in Chap. 4. Tensile residual (mean) stresses lower the fatigue life. In contrast, a compressive stress on the surface is known to improve the fatigue performance of materials. As an example, in the manufacture of gears, shot peening is used to introduce compressive stress. Figure 8-29 suggests also that light draft or surface rolling can introduce compressive stress on parts or components.

In general, there is a need to reduce or eliminate the residual stresses caused by cold-work. The stress-relief treatments may be done either thermally by annealing (to be covered shortly) or mechanically. The latter may be done either by introducing stresses of opposite signs or by further plastic deformation to give an opportunity for the relief of the nonuniform stresses. Shot peening is an example of introducing compressive stresses opposite the tensile residual stress on the surface, if present. Figure 8-30a shows the residual stress pattern in an as-drawn 1045 steel-bar, while Figs. 30b and 30c show examples of the results of mechanical and thermal stress relief treatments. Thermally, higher temperature treatment further relieves the stresses. Mechanically, changes in the residual stress pattern depend on the type of deformation. The rotary straigthening leaves a compressive tangential stress on the surface.

8.4.6 Plastic Work and Stored Energy of Cold-Work

In any deformation process, energy (plastic work) is spent to change the shape of the material. Most of this work is, however, converted to heat and to overcome frictional effects. Only a very small fraction of the spent energy is retained or stored in the material. For the deformation process, we may invoke the first law of thermodynamics and say that

$$\Delta E = Q + W \tag{8-18}$$

where ΔE is the change in internal energy of the material, and in this case, is the stored energy of cold-work of the material. Q is the heat associated with the process and in this case is negative because it is transferred from the material; W is the work or energy spent and is positive since it is done on the material. The stored energy of cold work, ΔE, is a positive quantity and as such, $Q < 0 < W$ and the $|Q| < |W|$.

8.4.6a Plastic Work with a Tensile Specimen. The plastic work done during deformation may be demonstrated with the simple elongation of the specimen in the tensile test. For this case, the tensile bar is stretched both plastically and uniformly by the tensile force. At any instant, the instantaneous length, area, and yield strength or flow stress of the material are l, A and σ. If the length increases from l to $l + dl$, the plastic work done is $dW = (\text{Force} \times dl) = \sigma A dl$. The total work needed to stretch the tensile bar from l_1 to l_2 is then

$$W = \int_{l_1}^{l_2} \sigma A dl = \int_{l_1}^{l_2} \sigma A \frac{l}{l} dl = V\int_{\epsilon_1}^{\epsilon_2} \sigma d\epsilon \tag{8-19}$$

where V is the volume of the material and is constant during the process, σ is the true stress, and ϵ is the

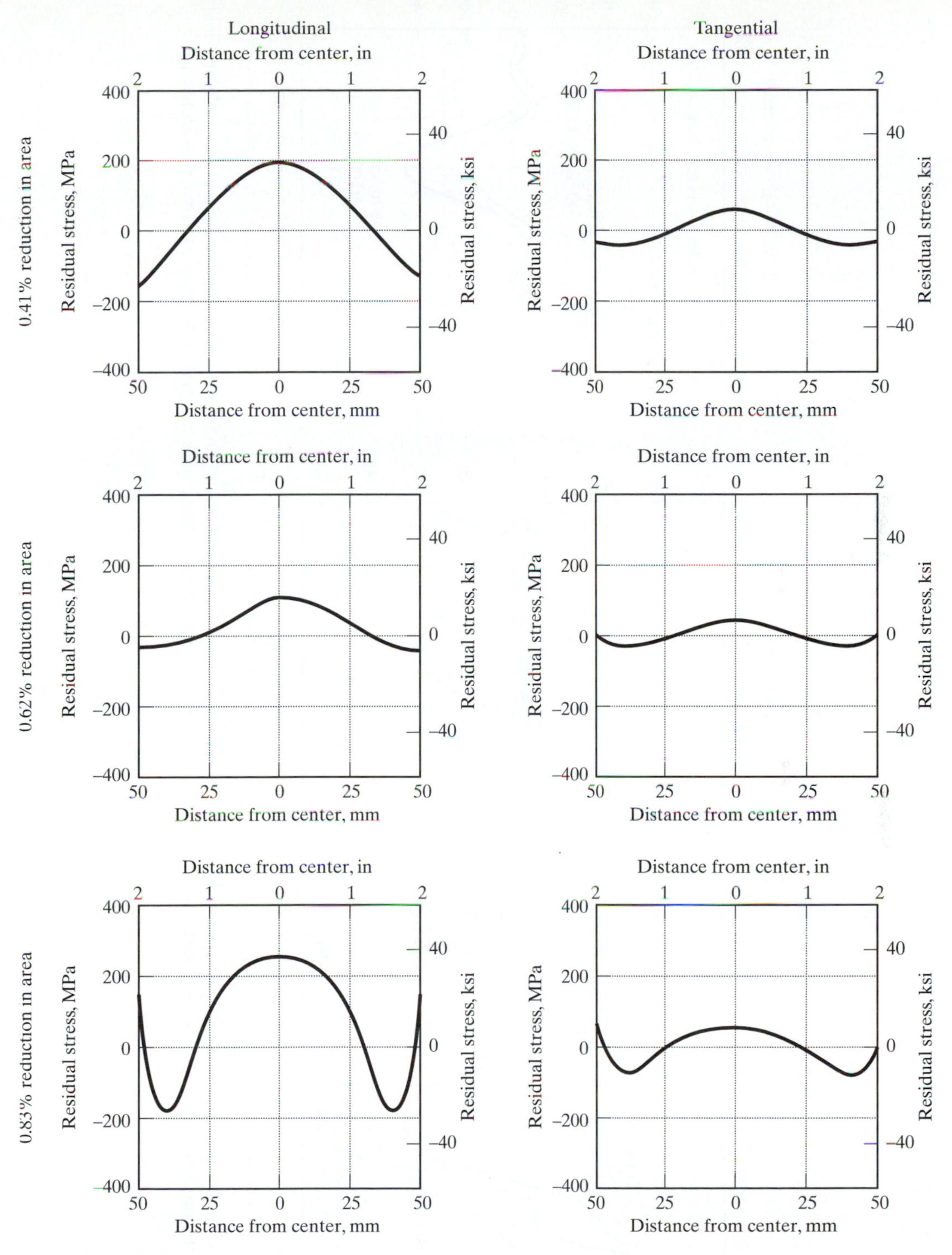

Fig. 8-29 Residual stress patterns in cold-drawn 1050 steel bars. Patterns change with amount of cold-reduction. [7]

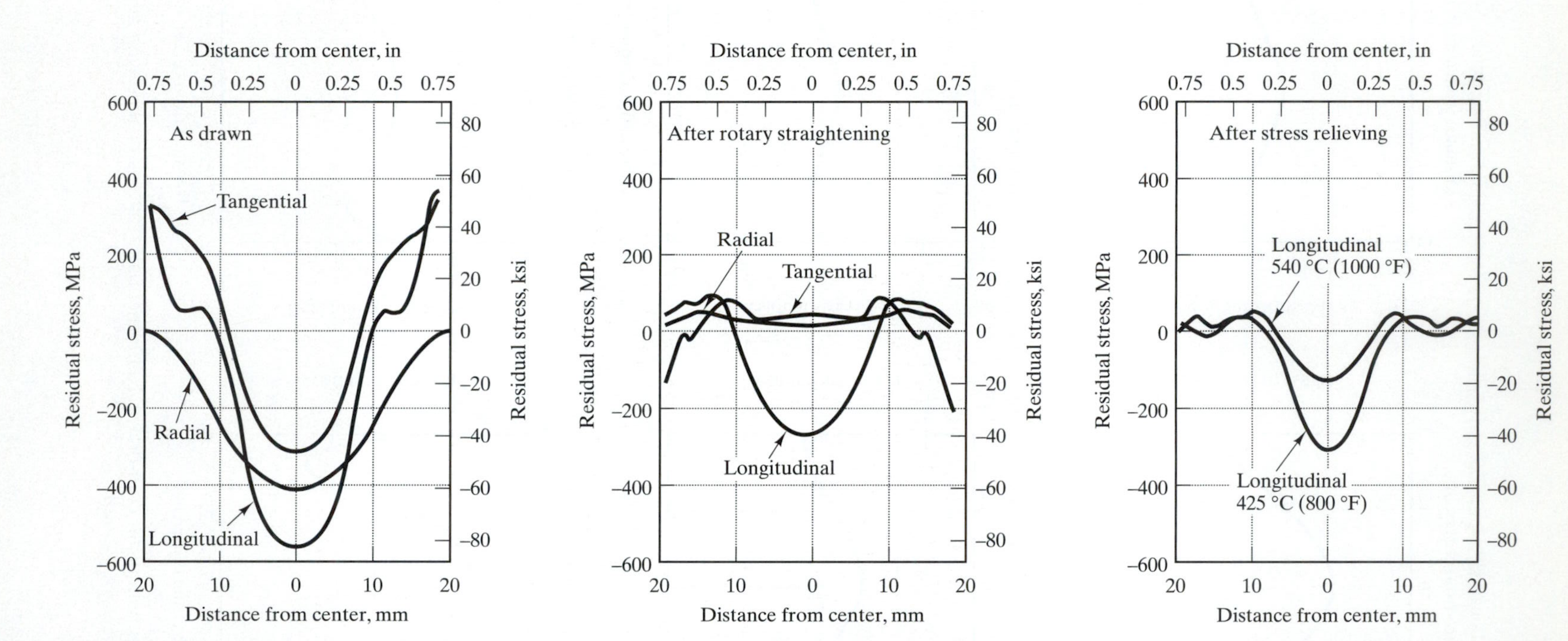

Fig. 8-30 (Left) Residual stress patterns set up in cold drawn 1045 steel bars with 20 percent reduction from 43 to 38 mm in diameter; (Middle) residual stress patterns after mechanical stress-relieving with rotary straightening; (Right) residual stress patterns after thermal stress-relief at two temperatures. [7]

true strain. The work done per unit volume of the material, $w = W/V$, and is the term under the integral sign. We should also recognize that this integral is the area under the true stress-true strain curve between the true strains ϵ_1 and ϵ_2. In order to be able to integrate Eq. 8-19, we need to have a relationship between σ and ϵ of the material, for example, Eq. 8-8.

In problems of plastic deformation (mechanical working), Eq. 8-19 is often simplified by taking the average value of σ and thus w, the work done per unit volume, becomes

$$w = \sigma_{av}(\epsilon_2 - \epsilon_1) = \sigma_{av} \ln\left(\frac{l_2}{l_1}\right) = \sigma_{av} \ln\left(\frac{A_1}{A_2}\right) \quad (8\text{-}20)$$

where σ_{av} is

$$\sigma_{av} = \frac{\sigma_1 + \sigma_2}{2} \quad (8\text{-}21)$$

and σ_1 and σ_2 are the tensile flow stresses at A_1 and A_2. Equation 8-20 is strictly valid for only nonstrain-hardening materials. It applies reasonably well in hot-working and also for heavily cold-worked material before the deformation process begins. The latter can be seen from the leveling of the yield strength in Fig. 8-16.

8.4.6b Extrusion and Wire Drawing. Extrusion and wire drawing are processes that resemble the tensile test (i.e., the workpiece elongates along one axis and reduces in the transverse direction). Thus we can use the above equations derived for a tensile bar. These processes are shown in Fig. 8-31 and the difference between them is in where the deformation force is applied. In extrusion, the force pushes on the larger initial workpiece, while with wire drawing, the force pulls on the final product or wire. In either case, F acts on a constant area, A, (A_1 for extrusion and A_2 for drawing). The applied stress, S, on the material is F/A_1 for extrusion, and F/A_2 for wire drawing. In both cases, the applied stress must exceed the yield strength of the initial workpiece before it goes through the die. The required work to push or pull the material a distance l is $W = (F \times l) = SAl = SV$. The work is done on a volume of material V. Hence, w, the work per unit volume is equal to S. Equation 8-20 becomes

$$w = S = \sigma_{av} \ln\left(\frac{A_1}{A_2}\right) = 2.303\, \sigma_{av} \log\left(\frac{A_1}{A_2}\right) \quad (8\text{-}22)$$

Equation 8-22 states that the applied stress, S, is always less than the average tensile yield or flow strength of the material for as long as $\ln(A_1/A_2) < 1.0$. This fact is important for it allows the wire drawing process to proceed with an applied tensile stress S that is greater than the initial yield strength, σ_1, of the material but less than the tensile yield strength, σ_2, of the wire product. It also reiterates the importance of the strain hardening of the material to acquire a higher flow stress (tensile yield strength), σ_2, after the deformation. This being the case, all the plastic deformation in wire drawing is accomplished within the die where the forces from the die are predominantly compressive, even though all the mechanical work is done by the tensile force pulling on the end of the exit wire.

The treatment so far has not included any frictional effects. Thus, Eqs. 8-20 and 8-22 are for ideal plastic working in which there is no waste in energy. In addition to frictional losses, there is also loss due to redundant plastic deformation. This is schematically sketched in Fig. 8-32 where an element is shown to be bent one way and then reverse-bent as it passes the die. Considering these losses and if ξ is the efficiency, then the actual stress, S_ξ, to be applied is

$$S_\xi = \frac{w}{\xi} \quad (8\text{-}23)$$

Most of the work energy spent in deforming the material is converted to heat. Only a small fraction is retained as stored energy of cold-work. The fractions or percents given in the literature vary from a few percent to 10 to 15 percent. According to current thinking, the stored energy exists in the material as point defects

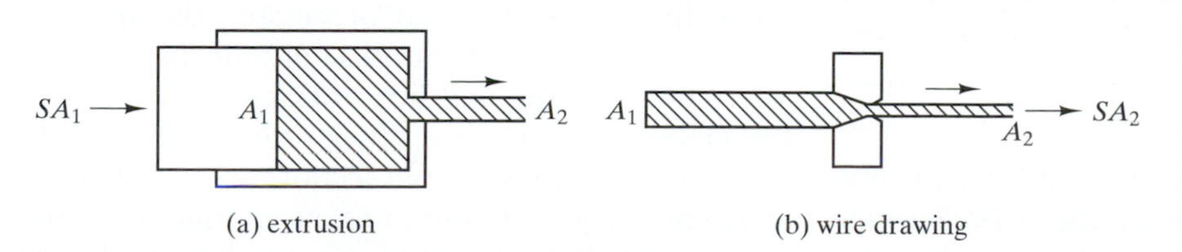

Fig. 8-31 The extrusion and wire drawing processes resemble the tensile test in that the workpiece elongates along one axis and reduces in the transverse direction.

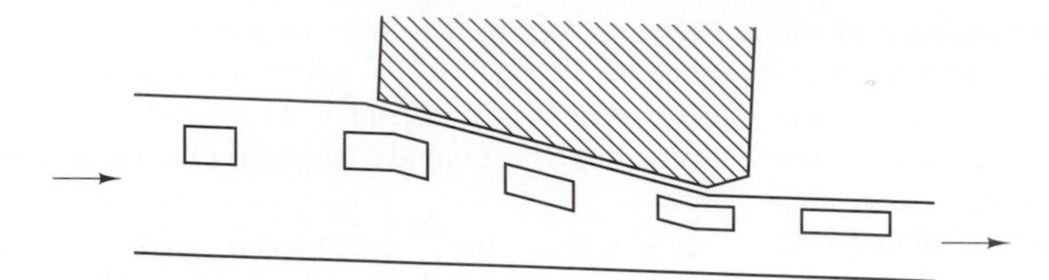

Fig. 8-32 Element in workpiece bent one way and reverse-bent as it passes the die is called redundant deformation.

such as vacancies, dislocations, and stacking faults. Thus, a cold-worked material is at a higher energy state and has become more thermodynamically unstable. With thermal activation, this state will transform to states of lower energy. This is the basis of annealing that will be covered following the example problems.

EXAMPLE 8-7

An annealed 70-30 yellow brass with a 0 B&S gauge number and a grain size of 0.015 mm is cold-drawn to B&S gauge number 1 in one step. Determine the (a) minimum force needed to draw the wire, and (b) the stress on the final wire. Compare it with the yield strength of the brass after deformation.

SOLUTION: From the data in Table EP 8-6, we find the diameter of the 0 B&S gauge to be 0.324861 in., and that for the 1 B&S gauge is 0.289297 in.

(a) From Fig. 8-18a, we obtain 22,000 psi as the annealed yield strength of a brass with a grain size of 0.015 mm. Thus,

Minimum force,

$$F = S \times A = 22{,}000 \times \frac{\pi}{4} \times 0.324861^2 = 1{,}823.51 \text{ lb}_f$$

(b) The stress on wire product

$$\frac{1{,}823.51}{\frac{\pi}{4}(0.289297)^2} = 27{,}741 \; psi$$

From Fig. 8-18, the yield strength of the material after deformation is 47,000 psi (H01 condition for 0.015 mm grain size). Because the yield strength is greater than the stress to draw the wire, the draw process is stable.

EXAMPLE 8-8

Using the same starting material as in Example 8-7, determine the smallest diameter the 0 B&S gauge wire can be cold-drawn in one pass through the die.

SOLUTION: To solve this problem, we have to assume different reductions and determine the reduction that will produce the wire that is still able to sustain the stress on it by the draw force that is used. From the previous example, the minimum force for drawing is 1,823.51 and the original wire diameter is 0.324861 in. We first assume a 50 percent reduction in diameter, thus the final diameter = 0.50 × 0.324861, and the stress on the wire is

$$\frac{1{,}823.51}{\frac{\pi}{4}(0.5 \times 0.324861)^2} = 88{,}000 \text{ psi}$$

with this reduction, the yield strength of the wire is about 82,000 psi after 50 percent reduction in diameter. Therefore, the draw process is unstable because the wire will still yield or deform past the die.

Thus, we have to lower the reduction and assume 44 percent reduction in diameter. For this, the final diameter = 0.56 × 0.324861 = 0.181922 in., and the stress on the wire is

$$\frac{1{,}823.51}{\frac{\pi}{4}(0.181922)^2} = 70{,}153 \text{ psi}$$

The yield strength of wire is 79,000 > 70,153. Thus, the smallest wire the 0 gauge can be drawn to is 0.181922 or B&S gauge number 5. The temper condition is H05.

8.5 Annealing—Heating to Soften Materials

Annealing is the thermal processing (heating) of materials with the sole purpose of relieving stresses after cold-working and hot-working, and to soften the material to improve other properties. As indicated in the Introduction section of this chapter, annealing may be required as an intermediate step in the forming of a part to restore the ductility of the material. This is called "process anneal" or "recrystallization anneal" which we shall cover in this chapter. Other annealing processes will be discussed in Chap. 9. The common characteristic of all these processes is that they all soften the material to improve their properties for further processing, such as cold forming and machining. This is in contrast to another thermal processing, commonly called heat-treating, which is predominantly used to strengthen or harden the material.

During recrystallization anneal, the cold-worked structure is transformed to states of lower energy. The process is customarily divided into *three stages*, namely, *recovery, recrystallization,* and *grain growth.* The driving force for the recovery and recrystallization stages is the decrease or the release of the stored energy of cold-work (Sec. 8.4.6); that for grain growth is the decrease in grain boundary surface energy. The changes in properties during these stages are schematically illustrated in Fig. 8-33. The properties are obtained by heating (annealing) the cold-work metal at a particular temperature and then holding it at that temperature for a certain time, usually 30 minutes or one hour, and then cooling the specimen to room temperature. Then the structures and properties of the cooled specimens are determined.

8.5.1 Recovery—The First Stage

At low temperatures, the annealing is called recovery. At this stage, there is essentially no change in the mechanical properties (i.e., hardness and strength) of the material, as depicted in Fig. 8-33; in fact, in carbon steels, there might even be an increase in strength due to strain aging. The term recovery comes from the observation that some of the physical properties of the material are recovered. For example, during cold-working, the electrical conductivity and density decreased very slightly. During the recovery stage, the electrical conductivity and the density increased to the values of the material in the annealed condition, Fig. 8-34. A fraction, which depends on the previous percent cold-worked (%CW), of the stored energy from cold-work is released at this stage. At low %CW, a large fraction of the relatively small stored energy is released, leaving only a small fraction for the recrystallization stage. At high %CW, the fraction of stored energy released during recovery is small compared to the fraction left for recrystallization.

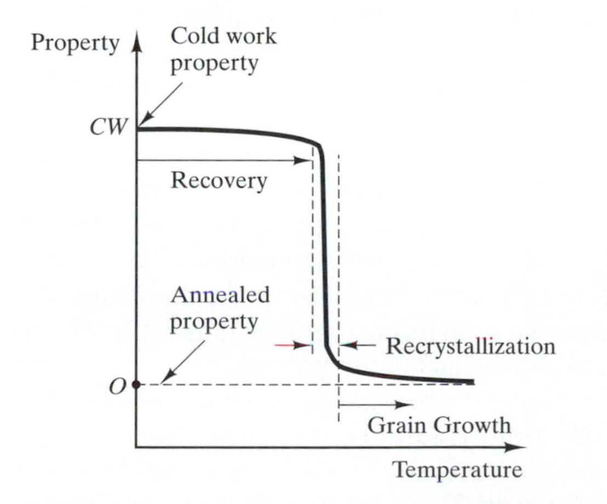

Fig. 8-33 Schematics of three stages of annealing a cold-worked structure; O is property of annealed soft material before cold-worked; CW is the as cold-worked property before annealing.

The microstructural changes in the material during recovery do not involve the movement of the high angle boundaries. For flat products, the original grains that were flattened (pancaked), because of the deformation or cold-work given to the material, remain flat and seemingly undisturbed during recovery. However, submicroscopic changes occur within the grains that are not usually observable with the optical microscope but are easily seen with the transmission electron microscope (TEM). These include the annealing-out of point defects and their clusters, the annihilation and rearrangement of dislocations, and the formation of subgrains and their growth. These microstructural changes relieve mostly the internal stresses and, consequently, this stage of the annealing process is used for *stress-relief treatment.* This is the process used in reducing or eliminating the residual stresses after cold-worked, see Fig. 8-30c. The loss of some dislocations means that the strength will decrease. However, this decrease is compensated by the formation of subgrains, which are grains with low-angle boundaries (~2-3° misorientation, described in Sec. 2.11.2) within the flattened cold-worked grains, illustrated in Fig. 8-35. The two effects balance out to retain predominantly the strength and hardness of the cold-worked structure. The recovery stage is applied to produce high-strength aluminum wires for electrical transmission lines; the recovery restores the lost electrical conductivity but retains the high strength obtained during cold-work.

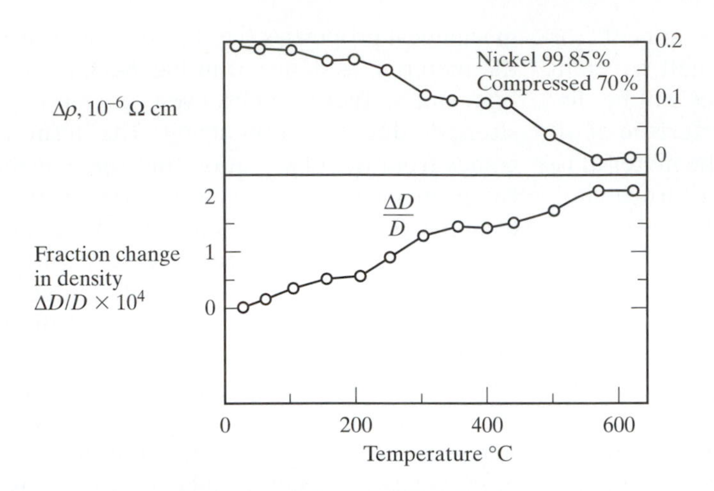

Fig. 8-34 Changes in electrical resistivity (top curve, read left) and in density (bottom curve, read right) during recovery.

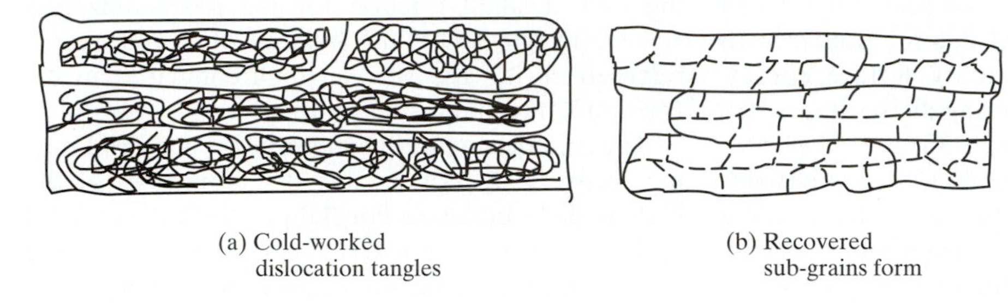

Fig. 8-35 Dislocation tangles in the cold-worked structure (a) rearrange to form polygonized structure or subgrains within the CW pancaked grains in (b) during recovery.

8.5.2 Recrystallization or Primary Recrystallization—The Second Stage

Towards the end of the recovery stage, nuclei formation of new grains at the old flattened cold-worked grain boundaries becomes observable with optical microscopy and indicates the initiation of the recrystallization stage, Fig. 8-36 (left). Similar to the solidification process, recrystallization is also a nucleation and growth transformation process and heterogeneous nucleation proceeds if surfaces are available. The nucleation of new grains (crystals) occurs by the coalescence of subgrains, and the surfaces for heterogeneous nucleation are microstructural defects, such as grain boundaries and coarse particle (such as inclusions) surfaces. The new grains form at the expense of the old, flattened, cold-worked grains. They are essentially strain-free and are bounded by high-angle grain boundaries with great mobilities that sweep (grow) and obliterate all traces of the old crystals. When the old crystals are all *transformed to the new, strain-free, equi-axed grains,* the cold-worked structure is said to have *fully recrystallized*, as depicted in Fig. 8-36 (right). This annealing stage is referred to as *static primary recrystallization*. Static because it occurs after the application of load (cold-work) and primary because it produces the first softened grains. (There is a *secondary recrystallization* that refers to the "abnormal or discontinuous grain growth" during the third stage.) As indicated, the driving force for this stage is the residual stored energy of cold-work after recovery. For low %CW, this is relatively small; consequently, more thermal energy (higher temperature) is required for recrystallization—for high %CW, this is comparatively large and thus lesser thermal energy (lower temperature) is required for recrystallization. Photo 10 in the Appendix shows a sequence of light micrographs illustrating the recrystallization in an aluminum alloy.

Because the new grains are strain-free, the strengthening from cold-work disappears. This explains the sudden drop in strength and hardness of the material depicted in Fig. 8-33. At this point, the material is essentially brought back to the annealed dead-soft condition prior to cold-working. Changes in strength

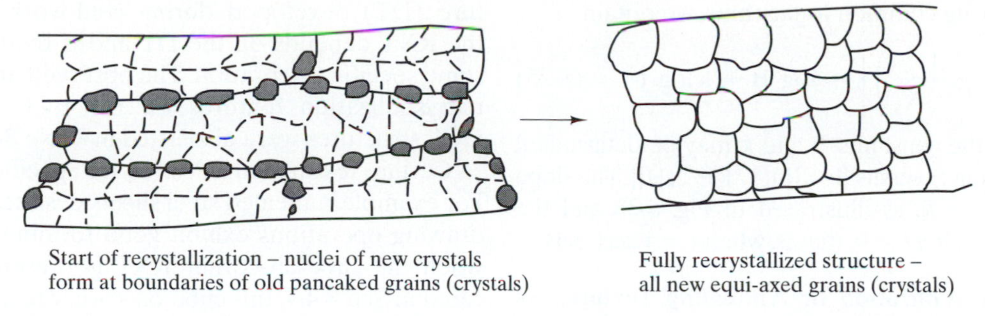

Fig. 8-36 Schematic of start of recrystallization (left), and fully recrystallized microstructure (right).

and hardness after recrystallization are small and gradual.

The recrystallization process allows the engineer to control the grain size and the mechanical property of the material. The grain size of the new recrystallized structure will depend on the amount of cold-work, the temperature of anneal, the time at temperature, and the composition of the material, according to the following "*laws of recrystallization*". [11]

1. A minimum, critical cold-work is necessary before recrystallization will occur.
2. The smaller the %CW, the higher is the temperature to cause recrystallization.
3. Increasing the annealing time decreases the temperature for recrystallization.
4. *The grain size when primary recrystallization has just been completed depends chiefly on the %CW, and to a lesser extent, on the annealing temperature. The grain size is smaller the greater the %CW and the lower the annealing temperature.*
5. The larger the original annealed grain size, the greater is the %CW required to give equivalent recrystallization temperature and time.
6. Solutes and fine dispersion of particles inhibit recrystallization.

8.5.2a Kinetics of Recrystallization. The rate (kinetics) of recrystallization depicts the characteristics of a typical nucleation and growth transformation process. When the annealing is followed isothermally, the fraction of the volume that recrystallizes as a function of time follows a typical sigmoidal type curve shown in Fig. 8-37. The kinetics shows a length of time before the transformation starts. This time is referred to as the *incubation period* that is followed by a slow transformation rate when the stable nuclei are being formed. Then a high rate of transformation ensues at approximately a constant rate and finally is followed by a slow rate towards the end of the transformation. The classical phenomenological theories of this type of transformation can be expressed by

$$X_v = 1 - \exp(-Bt^k) \qquad (8\text{-}24)$$

where X_v is the fraction of the volume recrystallized or transformed, t is the time in sec., and B and k are constants of the process. If we rearrange Eq. 8-24 and take the natural logarithm on both sides, we obtain

$$\ln(1 - X_v) = -Bt^k$$

or

$$\ln\left(\frac{1}{1 - X_v}\right) = Bt^k$$

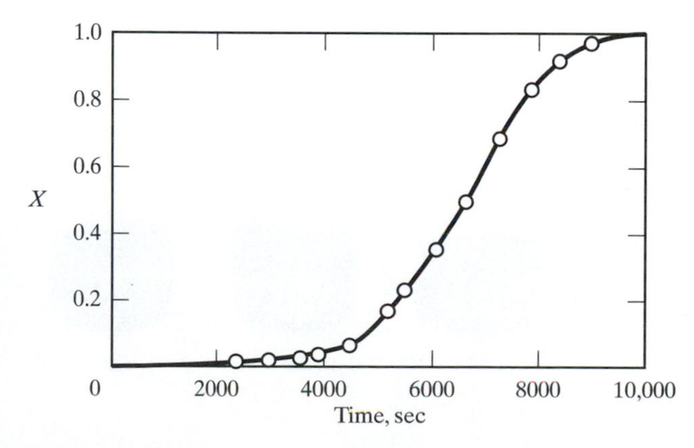

Fig. 8-37 Fraction, X, of volume that recrystallizes as a function of time during an isothermal process. [11]

and then taking common logarithms, we obtain

$$\log\left[\ln\left(\frac{1}{1 - X_v}\right)\right] = \log B + k \log t \qquad (8\text{-}25)$$

We see that the constants B and k may be determined by plotting ($\log t$) versus {log [ln(1/1 − X_v)]}. The slope of this curve is k, as illustrated in Fig. 8-38 and the intercept when log t = 0, that is, when t = 1 sec., is B.

8.5.2b. Recrystallization or Annealing Texture. As indicated in Sec. 8.4.4, a *recrystallization texture* (RXT) develops from heavily cold-worked structures. The texture is the preponderance of a specific plane parallel to the surface of the material and a specific direction parallel to the rolling direction. The RXT is entirely different from the deformation texture (DT) developed during cold-work. However, the RXT depends on the DT and is related to it by some specific orientation. The observed annealing or recrystallization textures for the BCC, FCC, and HCP structures were also listed in Table 8-4.

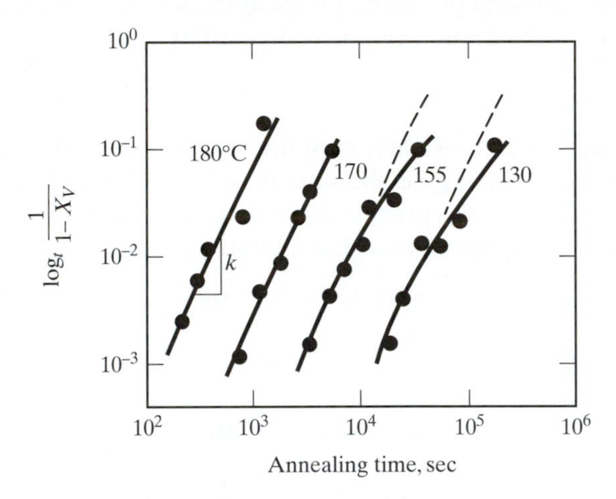

Fig. 8-38 Plots to determine B and k in Eq. 8-24.

Certain textures are desired for specific purposes. For example, annealed steel sheets destined for deep-drawing operations exhibit good forming properties when the cube-on-corner RXT is present. As indicated in Sec. 8.4.4, this cube-on-corner texture has the {111} parallel to the (rolling) plane of the sheet. With this texture, the body-diagonal <111> is normal to the surface and is obtained by letting the cube stand on its corner, thus the name, cube-on-corner. The <111> is the strongest direction in the BCC structure and therefore offers the most resistance to the thinning of the thickness dimension during cold-drawing—*a very desirable feature.* An ideal deep-drawing sheet material must have both $\bar{r}$ (normal anisotropy, Eq. 8-15) value > 1, and a Δr (planar anisotropy, Eq. 8-16) = 0. We saw from Figs. 8-24 and 8-25 that both $\bar{r}$ and Δr increased when the {111} texture increased, which in turn depends on the %CW. The optimum $\bar{r}$ value for sheet-steel drawability requires 50 to 70%CW, in spite of the fact that Δr ≠ 0 which may produce earing. Earing is a nonuniform extension in deep-drawing, as seen in Fig. 8-26, and is believed to be due to the {100} cube-on-face orientation for the sheet steel. For aluminum, which is face-center cubic, the resulting RXT, as seen in Table 8-4, is the {100} <100> and it also produces ears in deep-drawn cups, as seen in Fig. 8-39.

While the {100} <100> texture is undesirable for deep-drawing application, it is desired for Fe-Si steel

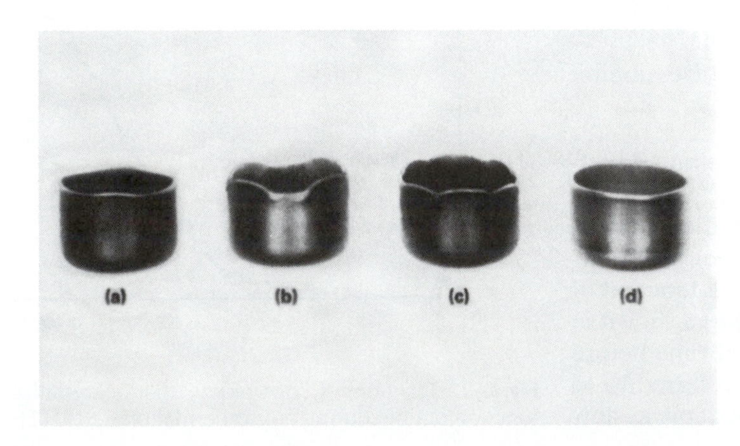

Fig. 8-39 Drawn aluminum cups showing "ears" in (a) 90°, (b) 45°, and (c) mixed, and no "ears" in (d). [6]

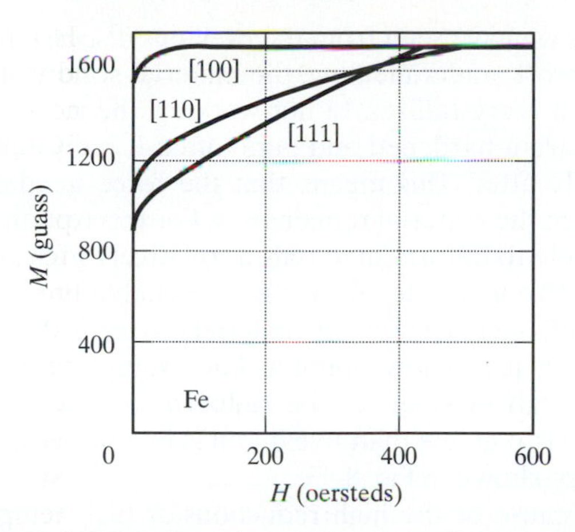

Fig. 8-40 [100] is the easy magnetization direction in iron.

sheets used for laminations in electric motors. The <100> are the easiest magnetization directions, as illustrated by Fig. 8-40.

8.5.3 Grain Growth—The Third Stage

Grain growth starts when the primary recrystallization is complete. It is characterized by a gradual decrease in strength of the material commensurate to the increase in grain size. While the driving force for recovery and recrystallization is the stored energy of cold work, the driving force for grain growth is the minimization of grain boundary interfacial energy. When the growth occurs uniformly among all the grains, (i.e., the average sizes of the grains are about the same), it is called *normal grain growth.* When a number of grains acquire grain sizes much larger than others, this is called *abnormal grain growth* or *secondary recrystallization.* These two growth mechanisms are illustrated schematically in Fig. 8-41. Secondary recrystallization is induced by dispersed second phases or particles because these same particles prevent the normal grain growth process. Like the primary recrystallization, the secondary recrystallization is also a nucleation and growth process that exhibits the typical sigmoidal transformation as shown in Fig. 8-42.

The most important commercial application of secondary recrystallization is the production of the {1 0 0} <1 0 0> texture in Fe-3%Si steel sheets used for laminations in electric motors. Traditionally, the

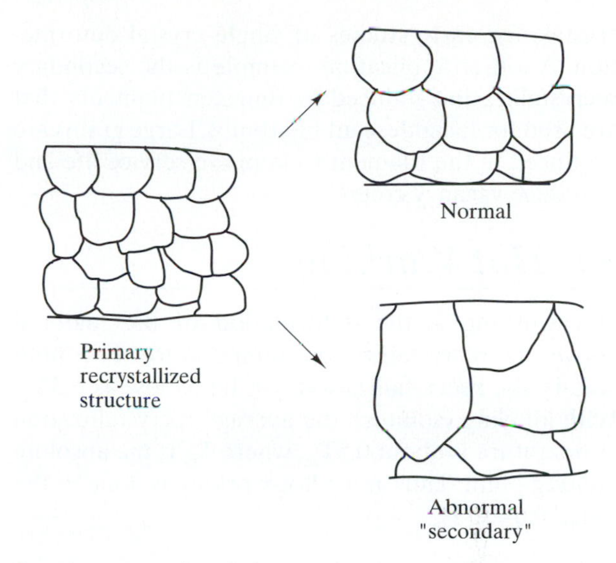

Fig. 8-41 Normal and abnormal grain growth or secondary recrystallization after primary recrystallization.

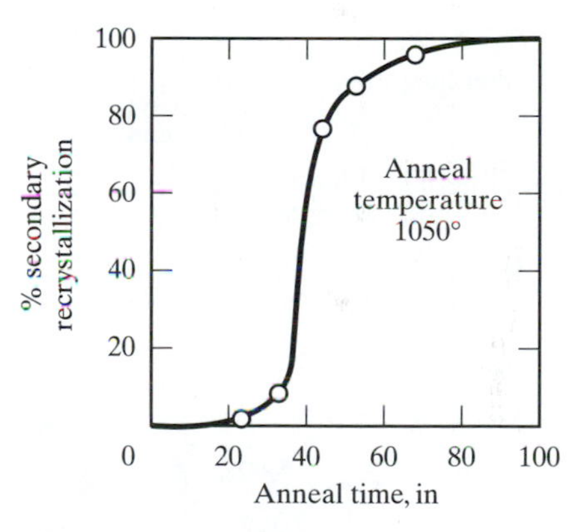

Fig. 8-42 Sigmoidal formation of cube texture in Fe-3%Si during secondary recrystallization.

particles thought to be responsible in the development of this texture are the manganese sulfide (MnS) nonmetallic inclusions. But, more recently, aluminum nitride (AlN) particles are also found to produce high permeability and low-core loss silicon steels. These particles are important in producing the intermediate orientation {1 1 0} <1 0 0> needed to produce the desired final texture. The final desired texture is produced by a final high temperature anneal by controlling the furnace atmosphere.

Before single crystals were produced, secondary recrystallization was used to grow sufficiently large crystals for early studies of single crystal deformation. Another application example is the secondary recrystallization induced in tungsten filaments that are used for incandescent lightbulbs. Large grains are produced in the filament to improve service life and to reduce vacancy creep.

8.6 Hot Working

Hot working is the deformation of the material above its recrystallization temperature. For pure metals, the recrystallization can be as low as $0.3T_m$. With alloying (solutes), the average recrystallization temperature is about $0.5T_m$, where T_m is the absolute melting point. Thus, most hot-working is done in the range $0.6–0.7T_m$.

As we have seen from the previous discussions on cold-work and annealing, the material softens again after it recrystallizes. In hot-working, the material is also strain-hardened and quite often it recrystallizes shortly after. This means that the force needed to deform the material remains low. For this reason, it is possible to deform materials at relatively high strain rates (from 10^{-1} to 10^3 per sec) to enable the reduction of thick sections of materials to very thin sections in just a few minutes. For example, a 9-inch thick slab of steel can be reduced to a 0.100-inch thick strip in less than five minutes in a hot-strip mill facility, shown in Fig. 8-43.

Because of the high reductions at high temperatures, elements of the material are forced together and the gas porosities inside an ingot or slab weld

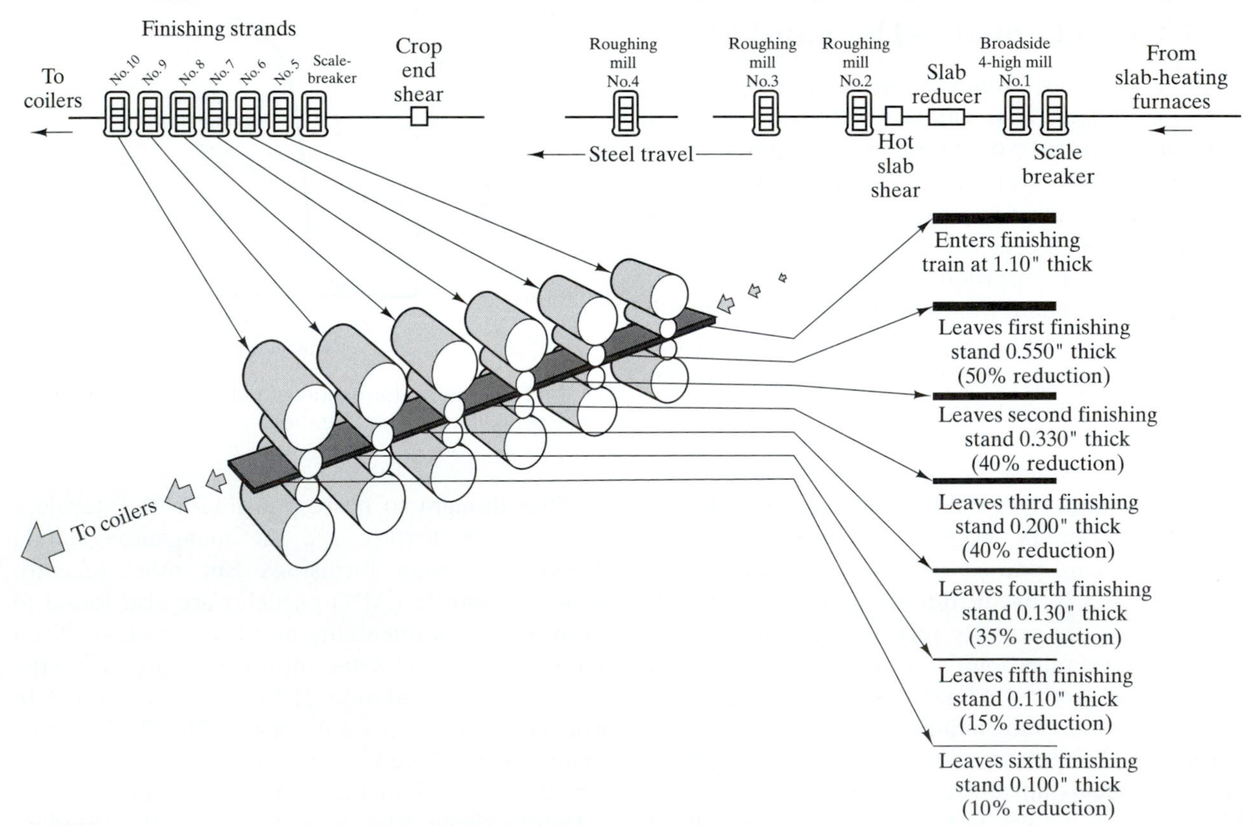

Fig. 8-43 Hot-strip mill produces wide strips suitable for stamping and press-forming of automotive and appliance parts.

back together. Any foreign materials, such as inclusions, will also deform if they are deformable. For example, manganese sulfide (MnS) is very deformable at the rolling temperatures used for steel. It elongates and in this shape, imparts poor toughness and fracture properties to steel (see Figs. 7-3 and 7-5). To avoid the elongated MnS inclusions in hot-rolled plates, rare earth metals, such as cerium, are added to the melt in the ladle to produce rare-earth sulfides (RES) (see Fig. 7-4). The latter are nondeformable at steel hot-rolling temperatures. They break, however, into small pieces and improve the toughness and fracture properties of the steel (see Fig. 7-6).

The high temperatures during hot-working induce oxidation and the loss of alloying elements in the layers of the material close to the surface if the atmosphere is not controlled. The oxide (scale) is a source of yield loss and needs to be removed before further processing. For steels, the reheating and hot-working in the air also decarburizes the surface. Because of this, a carbon restoration process is usually done to restore the carbon level at the surface for critical applications. This is especially important for thermal-hardening processes. The restoration process is essentially the carburizing process described in Chap. 6.

The current principles of hot-working arose from extensive torsion, tension, and compression testing of different kinds of materials at high temperatures. Essentially, the same processes, as those observed in cold-working and annealing are delineated. In addition to *static recrystallization,* there is also *dynamic recrystallization.* Dynamic recrystallization occurs while the material is under load and being deformed in the rolls, while static recrystallization occurs after the material passes the rolls and is not under load, illustrated in Fig. 8-44.

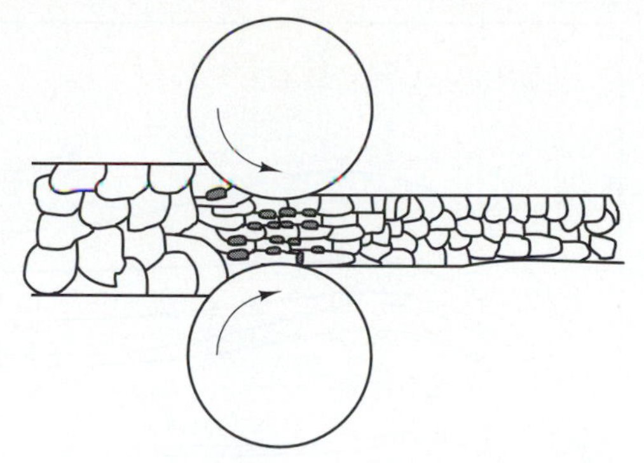

(a) Dynamic recrystallization

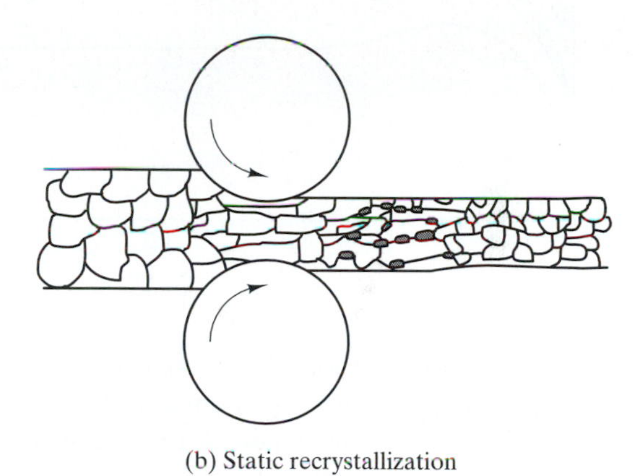

(b) Static recrystallization

Fig. 8-44 Schematic of (a) dynamic, and (b) static recrystallization during hot-rolling.

8.6.1 Dynamic Recovery and Recrystallization

The characteristic stress-strain curves for a low alloy steel are shown in Fig. 8-45 during torsion testing at hot-working temperatures and at a constant strain rate, $\dot{\epsilon}$. The stresses and strains are reported as equivalent true tensile stress and true tensile strains and strain rates, which are 1.155 times the strain values in rolling flat products. The curves are similar to those obtained for other steels in the austenitic condition tested in torsion, tension, or compression.

The initial rise in the flow-stress curve up to a peak level signifies strain-hardening. The observed microstructure during strain hardening consists of poorly developed subgrains as a result of the simultaneous action of strain-hardening (due to dislocation multiplication) and dynamic recovery. At a critical strain, ϵ_c, recrystallization nuclei form, but the flow stress does not decrease at this point. It still increases up to a peak strain value, ϵ_p, after some fraction of recrystallization has occurred, and then decreases and levels off when the effects of strain-hardening are balanced by the softening processes of dynamic recovery and recrystallization. The relationship between these two strains is

$$\epsilon_c = a\epsilon_p \tag{8-26}$$

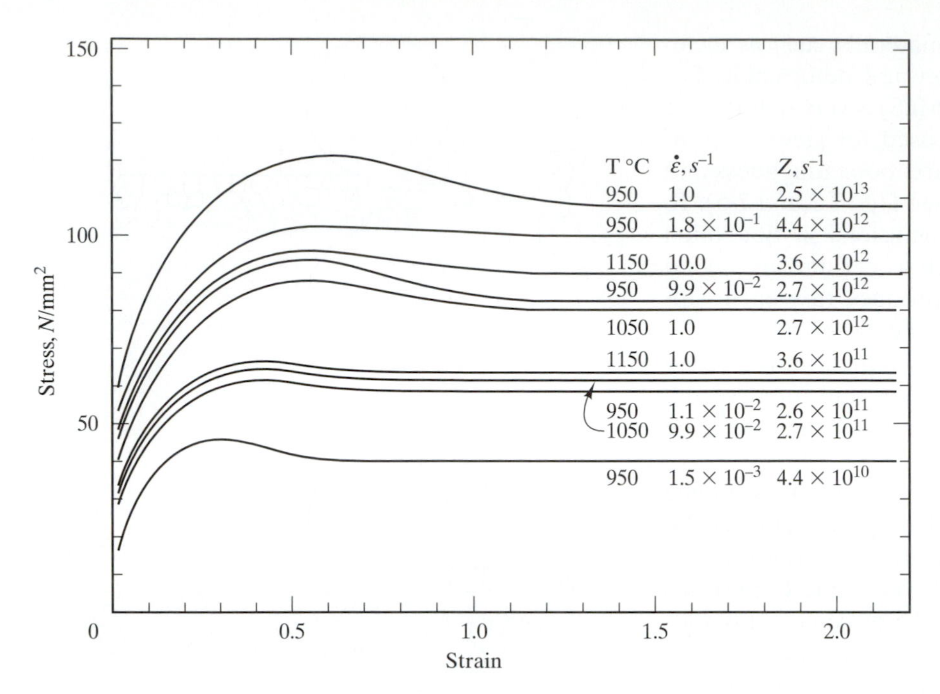

Fig. 8-45 Hot-torsion test results reported as equivalent true tensile stress-true strain curves as function of temperature, strain rate, $\dot{\epsilon}$, and Z, the Zener-Hollomon parameter. [12]

where the values of *a* are found to be between 0.67 to 0.86. The structural changes in the material are: the increase in dislocation density and stored energy until ϵ_c, the critical strain for dynamic recrystallization; and the dynamic recovery and recrystallization at ϵ_c. In dynamic recovery, the dislocations rearranged to subgrain boundaries and the coalescence of the subgrains initiate recrystallization. These processes, which are thermally activated, lead to a flow stress that depends on the strain, strain rate, and temperature. It is found that ϵ_p depends on a parameter, Z, which is called the Zener-Hollomon parameter and is expressed by

$$Z = \dot{\epsilon} \exp\left(\frac{Q_{def}}{RT}\right) \tag{8-27}$$

where $\dot{\epsilon}$ is the strain rate and Q_{def} is the activation energy. This parameter is very similar to the Sherby-Dorn and Larson-Miller parameters discussed with creep in Chap. 4, and when this is plotted versus ϵ_p, all the data points can be regressed to a relationship

$$\epsilon_p = 4.9 \times 10^{-4} d_o^{1/2} Z^{0.15} \tag{8-28}$$

where d_o is the initial grain size. Equation 8-28 is the mean curve drawn through the data in Fig. 8-46. The peak-strain value, ϵ_p, and thus ϵ_c, can be two to three times higher when particles or solutes are present to inhibit recrystallization—for example, niobium (Nb) carbides in microalloyed steels versus plain C-Mn steels.

8.6.2 Static Recrystallization

The dynamic structural changes leave the material in an unstable state and provide the driving force for static recovery and recrystallization to take place after the deformation pass. The typical sigmoidal curves for static recrystallization are shown in Fig. 8-47. These curves also demonstrate the effects of strain (deformation) and deformation temperature. Higher strains induce recrystallization much faster at a certain temperature; for example, at 950°C, a 0.5 strained material recrystallizes much faster than a 0.25 strained one. Higher temperatures induce recrystallization faster for a certain strain; compare the curves with 0.25 strain at 950°, 1050°, and 1150°C. The sigmoidal curves follow Eq. 8-24 and similar plots as those shown in Fig. 8-38 are illustrated above the sigmoidal curves in Fig. 8-47.

The sigmoidal static recrystallization curve in Fig. 8-47 may be followed by the decrease (softening) in yield or flow stress during a second deformation after holding for a certain time. This technique was

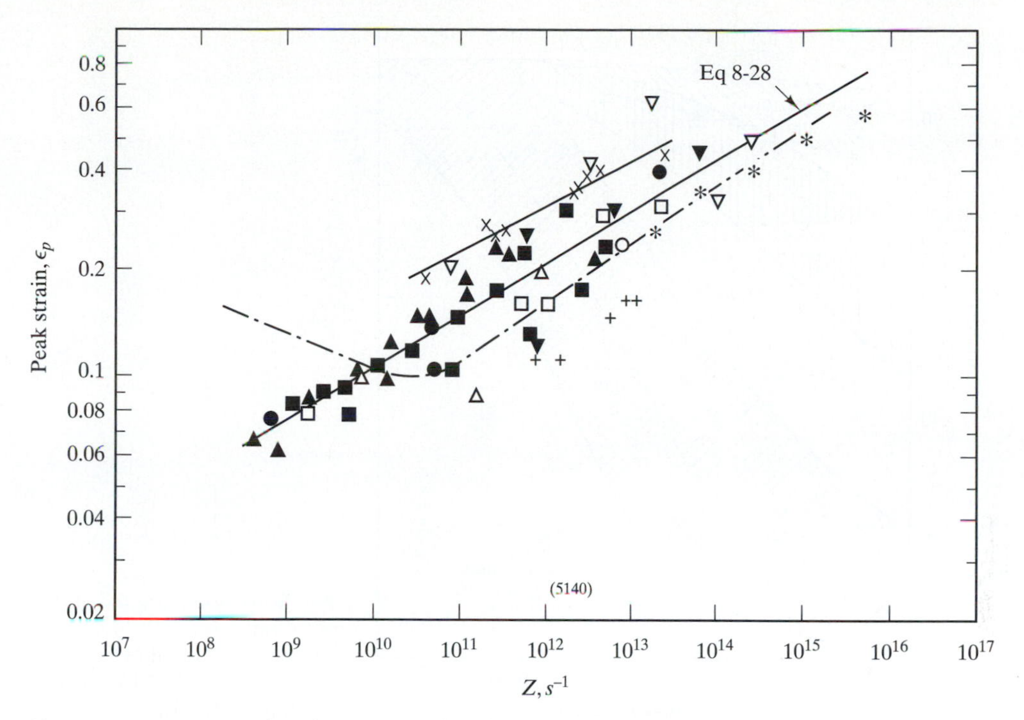

Fig. 8-46 Regression of peak strain as a function of Zener-Hollomon parameter, Z. [12]

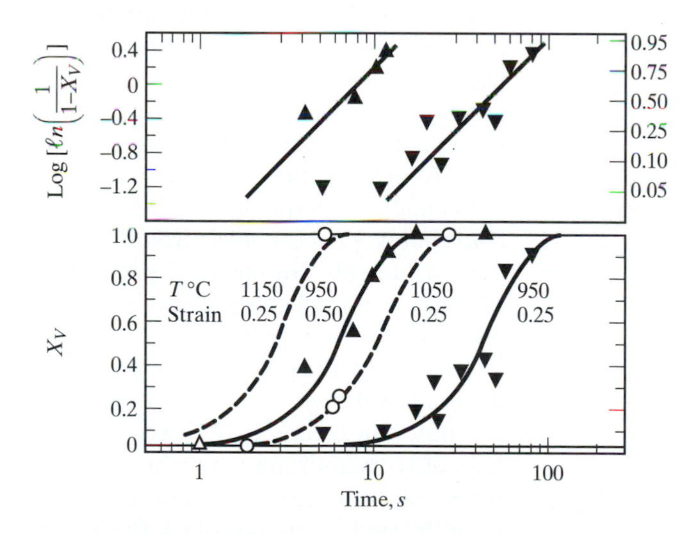

Fig. 8-47 Static recrystallization of AISI 5140 deformed at the temperatures indicated at an equivalent strain rate of 1 per sec and annealed at the same temperatures. [12]

employed to determine the degree of softening observed in microalloyed steels containing niobium (also called columbium) and vanadium. The results are compared in Fig. 8-48 with steels containing solutes that are in solid solution at the hot-working temperatures. The hold time at temperature after one deformation and before the second deformation is 3.5 seconds; this simulates the time it takes for the material to move from one stand to the next in the finishing stands in the hot-strip mill (Fig. 8-43). We note that the degree of softening is the least for the niobium-containing steels, while the degree of softening in steels containing elements that are in solid solution is at least 90 to 100 percent. In the microalloyed steels,

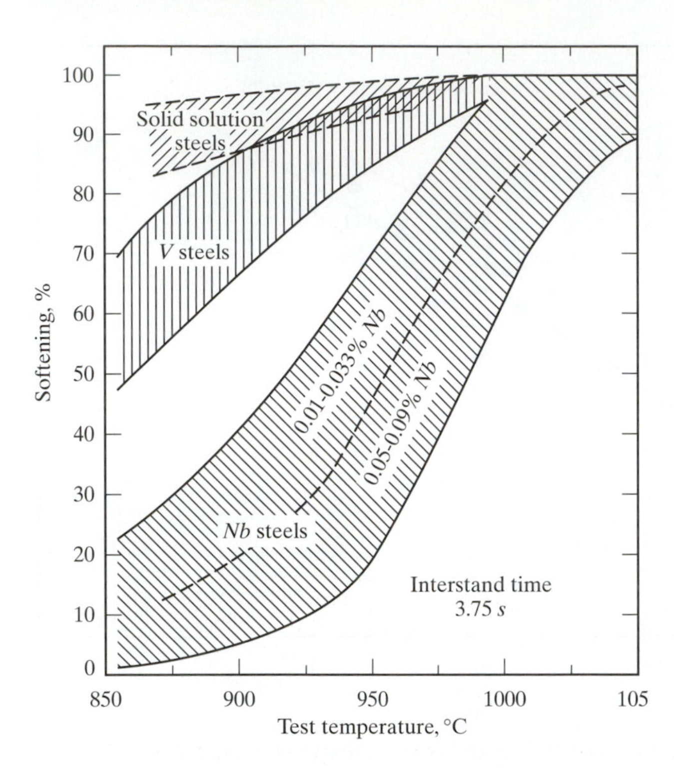

Fig. 8-48 Softening of solid solution steels in 3.75 s between deformation compared to Cb (Nb) and V containing steels. [15]

some of the niobium carbides and vanadium carbides precipitate out during the deformation; niobium carbides precipitate out more often because they have a lesser solubility limit (solvus) than vanadium carbide. Thus, because more particles are out of solution, recrystallization is greatly inhibited in Nb (Cb) steels; therefore, these steels show the least softening (see the No. 6 of laws of recrystallization in Sec. 8.5.2). Structurally, this means the austenite in Nb(Cb) steels is not completely recrystallized in the time between two consecutive passes (interstand time). This condition is reflected in the steels exhibiting higher flow stresses that are recorded as higher rolling loads when they are deformed the second time.

8.6.3 Controlled Thermal Mechanical Processing—A Process Design

Prior to the microalloying additions of Cb (or Nb), V, and Ti, the high-strength low-alloy (HSLA) steels were primarily alloyed with solid solution elements such as manganese, chromium, and nickel. In the hot-rolling of these solid solution steels, there was no particular attention given to the control of deformation and of the finishing temperature (i.e., the temperature after the last pass). Most of the rolling temperatures were done at the highest temperatures possible, ≥1000° C. The strengths of these steels were predominantly derived by increasing the carbon content. To achieve the strength levels of current *high-performance microalloyed HSLA steels,* the older *conventional HSLA steels would have to have high carbon contents.* For the same strength, the latter will exhibit lower impact toughness, as shown in Fig. 8-49, and poor weldability (see Chap. 12).

In contrast, the processing of the high performance microalloyed high-strength low-alloy (HSLA) steels illustrates an example of a process design using the laws of recrystallization to achieve the desired structures and properties. The term microalloying comes from the small additions of Cb (Nb), V, and/or Ti that seldom exceed 0.20 wt% singly or in combination. The carbon content is also very low, usually less than 0.10%. The strength of these steels predominantly arises from the very fine ferrite grains produced and a little increment from precipitation hardening. In order to achieve the desired structures and properties, specific hot-rolling conditions must be designed, controlled, and adhered to. These conditions involve

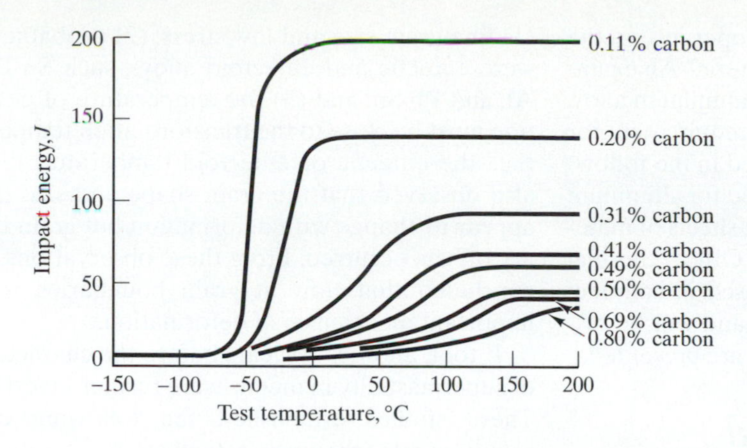

Fig. 8-49 Impact toughness of steels decreases with carbon content through the increase in pearlite content or carbide phase.

control of temperature (thermal) and deformation (mechanical) conditions—*controlled thermal mechanical processing.*

The controlled processing is illustrated in Fig. 8-50. First, the slabs, blooms, or billets are heated at a temperature high enough to dissolve all the carbide particles of vanadium, titanium, or niobium. Then they are hot-rolled at high temperatures to an intermediate thickness to produce recrystallized austenite structures. The temperatures for this initial hot-rolling retain the microalloy carbides in solution. The finished rolling is done at temperatures to induce the precipitation of very fine microalloy carbides to inhibit austenite recrystallization, and with a high amount of total reduction, to produce pancaked, unrecrystallized austenite grains prior to transformation into ferrite. (The cold-worked unrecrystallized austenite condition is reflected in the very high deformation loads during the hot-rolling process. This was a big problem for the older rolling mills because they did not have the load capacity.) Temperatures close to the A_{r_3}, sometimes in the intercritical region, A_{r_1} to A_{r_3} are used. (This is the equivalent of the A_1 to A_3 range in non-equilibrium cooling—see Sec. 9.2.2). After rolling, the austenite transforms to ferrite via a nucleation and growth process. As in recrystallization, grain boundaries, particles, and other defects are preferred ferrite nucleation sites for the transformation of austenite to ferrite. The unrecrystallized austenite provide a multitude of preferred nucleation sites for the new phase, ferrite, and the result is a very fine-grained, ferrite product. The strengthening comes predominantly from the very fine ferrite grain size with some incremental strengthening from the fine carbide particles. Because these steels have very low carbon and very fine grains, they exhibit superior toughness and excellent weldability properties. These steels are used in very severe climatic conditions, such as the TransAlaska pipeline.

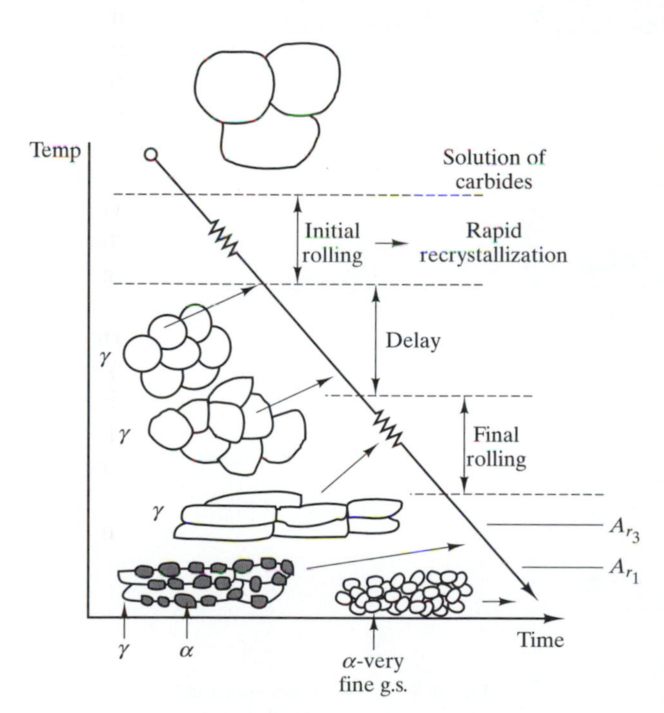

Fig. 8-50 Controlled thermal-mechanical processing of microalloyed steels.

The principles of controlled thermal mechanical processing are also applied to other materials, such as aluminum and titanium. The dominant purpose is to improve the mechanical properties of strength and toughness. The only method that simultaneously achieves both of these desired properties is the refinement of the grain size of the material. An example of this is shown in Fig. 8-51 for an aluminum alloy. An extremely fine grain size is also a requirement for *superplastic forming* which is discussed in the following section. This is a process that is used for aluminum and titanium in the production of thin sheets of materials for the aerospace industry. Other thermal mechanical treatments (TMT) are used in conjunction with other hardening processes and will be discussed when the hardening processes are presented.

8.7 Superplasticity and Superplastic Forming

In 1934, *superplasticity* was discovered when almost *2000 percent tensile elongation* was achieved with a tin-bismuth alloy without fracturing. The original requirements of superplasticity were thought to be: (1) fine grain size and low stress; (2) probable alloys were eutectic and eutectoid alloys, such Sn-Bi, Zn-Al, and Pb-Sn; and (3) the temperature of deformation must be close to the transformation temperature (i.e., the eutectic or eutectoid temperature). It was also observed that the grain shape and size did not appear to change with deformation but grain boundary offsets occurred. From these observations, it was concluded that flow at grain boundaries was the important mechanism of deformation.

It took another 30 years before the characteristics of superplasticity in metals were further investigated. These studies established the following current requirements for superplasticity:

(1) *Material must have fine stable uniform grain size.* It was also discovered that a two-phase alloy system is not a prerequisite.
(2) *Material must have a high strain-rate sensitivity*,
(3) *Strain (deformation) rate must be low.*

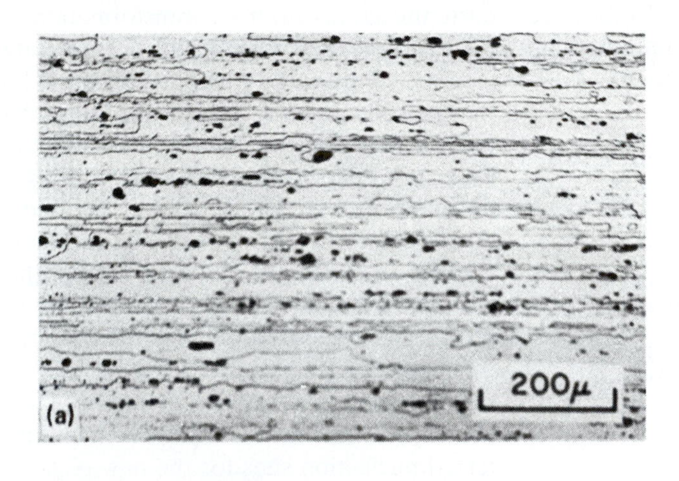

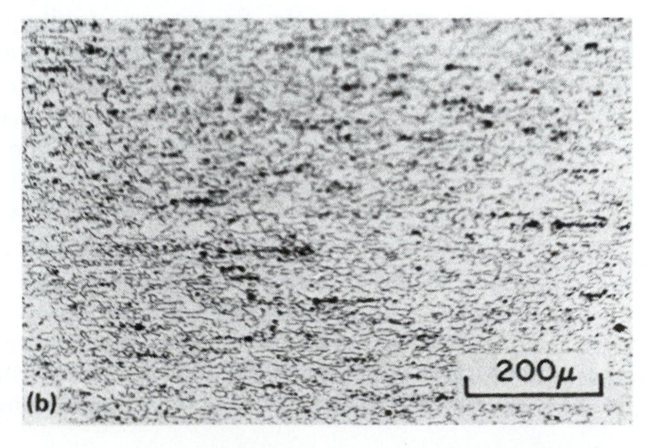

Fig. 8-51 Grain refinement in aluminum alloys by thermal mechanical processing, (a) conventional, (b) after TMT. [2]

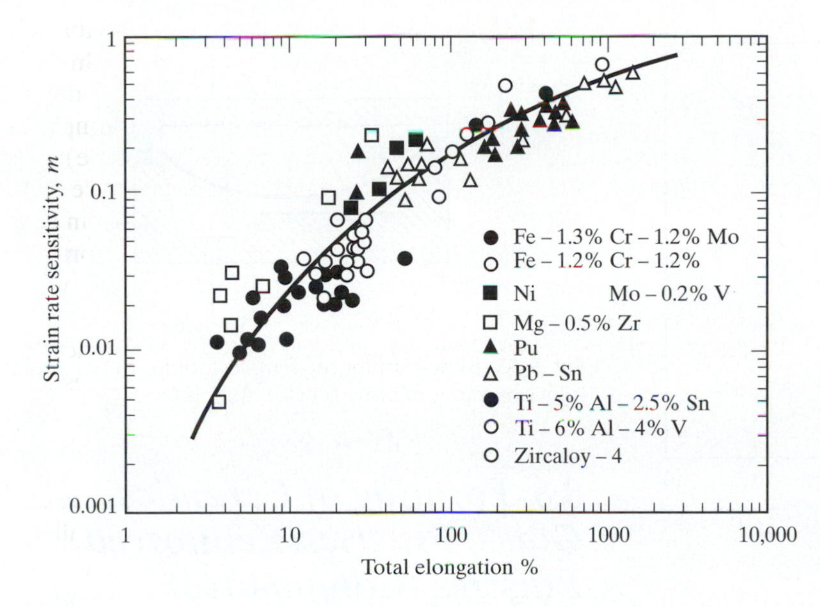

Fig. 8-52 Correlation of strain-rate sensitivity index, *m* and total elongation of some superplastic alloys. [2]

The requirement of a very fine, stable uniform grain size can only be achieved through controlled thermal mechanical processing. This can be achieved through appropriate hot-working, warm-working, and cold-working sequences with intermediate recrystallization anneal steps for wrought products. An example is shown in Fig. 8-51 for aluminum alloys. For hard to prepare alloys, it may be possible to achieve this through sintering of very fine powders, such as a nickel-based superalloy.

The strain-rate sensitivity index, m, of a material is related to its tensile elongation. Figure 8-52 shows that a high m results in a high total tensile elongation. The strain-rate sensitivity is defined as

$$m = \frac{d(\log \sigma)}{d(\log \dot{\epsilon})} \tag{8-29}$$

where σ is the true stress and $\dot{\epsilon}$ is the strain rate. Experimentally, m is obtained during the tensile test by a sudden change in strain rate (pulling speed) and by observing the change in load (stress), as shown in Fig. 8-53. m is not a constant but varies with strain rate and is descriptive of the strain-rate sensitivity of flow stress only over a small variation of strain rate. A high m value may be obtained either through a phase stability criterion where two phases are in equilibrium at the deformation temperature or through pinning or locking of the grain boundaries of the superplastic material with an insoluble particle or dispersoid phase. Example of the first technique is the eutectic Pb-Sn alloy and of the second technique, the superplastic aluminum alloy, Supral 100, or very fine grain 7000 series.

The low strain rate at high temperatures of deformation translates to low stress on the material, as shown in Fig. 8-54 for a Ti-6Al-4V alloy deformed at 927°C.

The name superplasticity arose from the similarity in tensile elongations of some metals to that of plastics. Because of this similarity, one of the superplastic forming techniques is blow-forming that is used in the plastics industry. An example is shown in Fig. 8-55, referred to as female forming, with vacuum created from below the sheet or pressurized from above. Figure 8-56 is a structural aircraft component that is

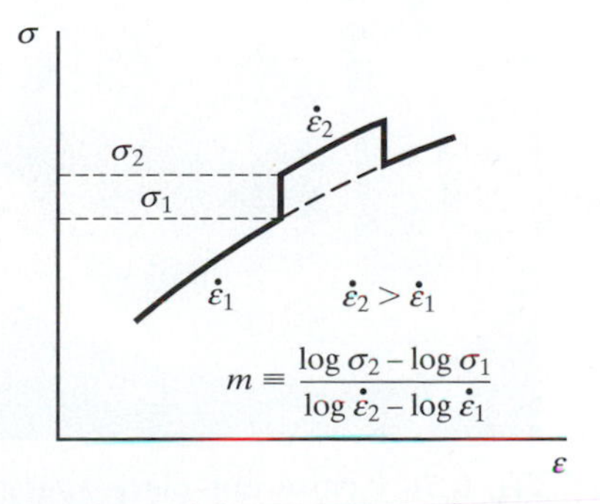

Fig. 8-53 Experimental determination of *m* in a tensile test.

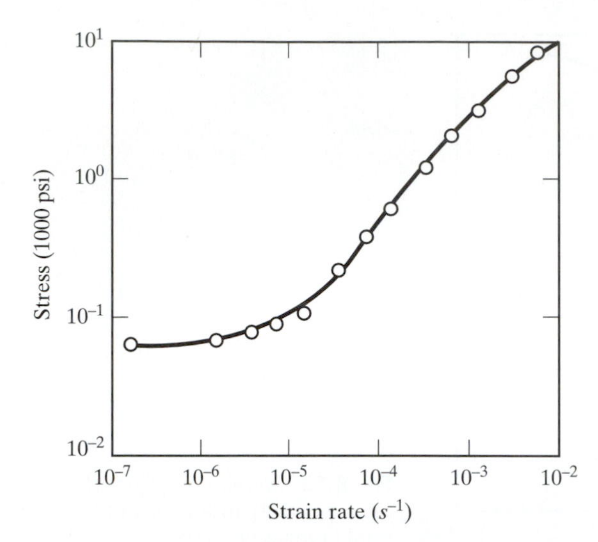

Fig. 8-54 Stress is very low at very low strain-rate superplastic forming of Ti-6Al-4V alloy. [2]

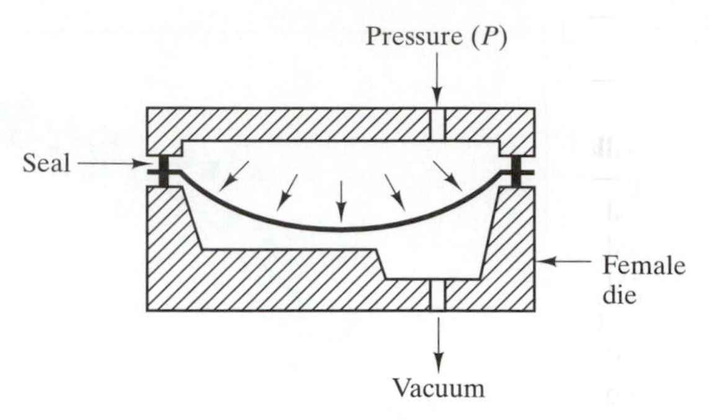

Fig. 8-55 Blow-forming or female forming of superplastic alloys by vacuum or by pressurizing. [2]

superplastically formed in one piece (left) from fine-grained 7475 aluminum alloy using this blow-forming technique to replace a former riveted component (right). From the standpoint of cost, superplastic forming is at a disadvantage because the process is slow. This limits the rate at which components can be formed and the production rate of a press to close the forming die. Nonetheless, superplastic forming is used effectively for a lot of aircraft components.

A partial listing of some of the more important superplastic alloys and their properties is shown in Table 8-5.

8.8 Forming of Ceramics, Glass, Plastics/Reinforced Plastics (Composites)

So far, we have discussed the forming of metallic products. The forming of ceramics and glass is discussed in Chap. 14, and that of plastics and reinforced plastics (composites) is described in Chap. 15. The forming of ceramics described in Chap. 14 refers to the traditional ceramics. The forming of advanced structural ceramics involves many other processes that could not be covered there. Plastics have hot-forming processes similar to those used for ceramics and glasses, such as extrusion, injection-molding, and blow-molding. In addition, some thermoplastics can be cold-formed and the forming may involve reaction such as reaction injection-molding.

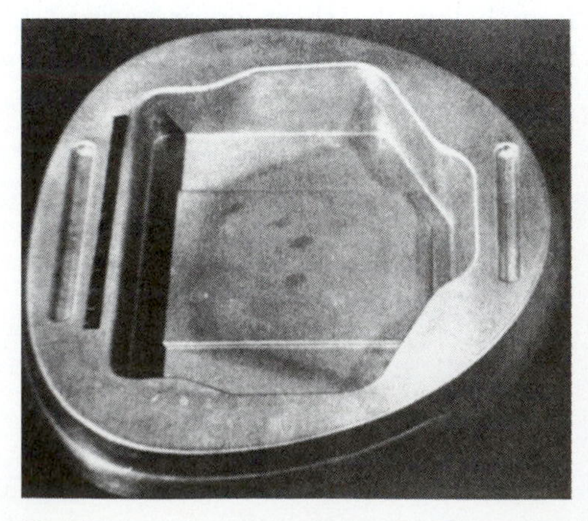

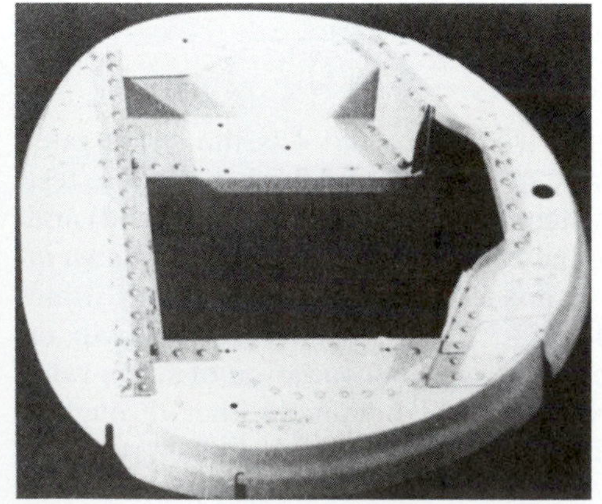

Fig. 8-56 (Left) A one-piece aircraft structural component superplastically formed from fine grained 7475 aluminum alloy; (Right) conventional riveted component it replaced. [2]

Table 8-5 Selected superplastic structural alloys and their properties. [2]

Alloy	Grain Size, μm	Deformation Temperature, °C	Elongation, %	Max. $\dot{\epsilon}$ Sensitivity Index, m
Al - 88Zn	2.5	230	2,850	0.45
Al - 6Ca - 0.4Zr	~10	400	800	0.50
Al - 6Zn - 2.5Mg -1Cu (7475 Al alloy)	10	515	600	0.85
Al - 5Ca - 5Zn	1–2	450	7,200	0.29
Fe - 6.5Ni - 25Cr	2–3	650	470	0.5
Mg - 5.8Zn - 0.5Zr (ZK60)	0.55	275	1,700	0.4
Ni - 12.5Cr - 3.2Mo - 4.9Al (IN 100)	2.5	1,038	850	0.66
Ti - 6Al - 4V	3.7	927	675	0.58

Summary

Plastic deformation is the process of changing the shape or producing the final shape of the product. It may be done either by cold-working or hot-working. Cold-working refers to plastic deformation below the recrystallization temperature of a metal or alloy, while hot-working refers to plastic deformation above the recrystallization temperature. Cold-working pancakes the original annealed equi-axed grains, increases the dislocation density, and increases the strength (strain-hardens) of the metal. Thus, when a product is manufactured via cold-working of the metal, a smart designer can take advantage of this strength increase at the design stage. In addition, cold-working produces very close dimensional control and very good surface finish. Plastic deformation with at least a 50 to 70 percent reduction in area also causes rotation of planes and alignment of direction, and it produces what is called the deformation texture in sheets or fiber texture in wires.

Annealing of a cold-work metal or alloy undergoes three stages: recovery, recrystallization, and grain growth. Recovery anneal exhibits no major change in the optical microstructure and no significant change in hardness or strength, and also is the thermal process used to relieve residual stresses. It involves some annihilation and rearrangement of dislocations to form subgrains. Recrystallization (primary) exhibits a major change in the cold-worked optical microstructure. It involves the nucleation and growth of new equi-axed grains (crystals) and obliterates the old pancaked grains (crystals). It produces an annealing or a recrystallization texture that has some dependence on the deformation texture. A texture with a normal anisotropy ratio of greater than one is desired for good drawability. There are laws of recrystallization and when these are followed, cold-working and recrystallization can be used to control the structures and properties of the materials. Grain growth can be either normal or abnormal; the latter is also called secondary recrystallization and may cause a change in the primary recrystallization texture. This is the method to produce the {100} texture in Fe-3%Si sheets that are used as laminae in electric motors for easy magnetization.

Hot-working also exhibits characteristics similar to those observed in cold-working. While a cold-worked material has to be heated to temperatures at which recrystallization can occur, the temperature used in hot-working is high enough to induce recrystallization and allows the plastic deformation (very quick reduction in thickness or shape) to be done at relatively low stresses. Nonetheless, during hot-working, a material also undergoes strain-hardening before it undergoes recovery and recrystallization. The latter may be either dynamic or static—dynamic occurs when the material is undergoing deformation, while static occurs shortly after the deformation. When recovery and recrystallization are inhibited and controlled, the strength of the material at the hot-working temperature also increases (strain-hardened) and is reflected in the high loads registered during deformation. When the strain-hardened material is allowed to recrystallize or to transform to another phase, for example, from face-centered cubic (FCC) austenite to body-centered cubic (BCC) ferrite at lower temperatures, the resultant microstructure consists of very fine grains of a recrystallized structure or a transformed phase that exhibits much higher strength and much improved toughness. This is the basis of the controlled thermal-mechanical processing of the high-performance microalloyed high-strength low-alloy (HSLA) steels that have very fine, grain size ferritic, bainitic, and/or martensitic structures. Controlled thermal-mechanical processing is also applied to produce extremely fine grain sizes that are needed for superplasticity in aluminum and titanium alloys.

References

1. *The Making, Shaping, and Treating of Steel,* Edited by Harold E. McGannon, 9th ed., United States Steel, 1971.
2. N. E. Paton and A. K. Ghosh, "Metalworking and Thermomechanical Processing," p. 361, in *Metallurgical Treatises,* Edited by J. K. Tien and J. F. Elliot, TMS-AIME, 1981.
3. S. Kalpakjian, *Mechanical Processing of Materials,* Litton Educational Publishing, Van Nostrand Company, 1967.
4. G. E. Dieter, *Mechanical Metallurgical,* 3d ed., McGraw-Hill, 1986.
5. D. J. Schmatz, F. W. Schaller, and V. F. Zackay, "Structural Aspects and Properties of Martensite of High Strength," p. 613, in *The Relation between the Structure and Mechanical Properties of Metals,* Proceedings of Conference Held at the National Physical Laboratory, Symposium No. 15, Teddington, Middlesex, UK, 7–9 January, 1963.
6. *Aluminum-Properties and Physical Metallurgy,* Edited by J. E. Hatch, ASM, 1984.
7. *Vol. 1 Properties and Selection: Irons and Steels, Metals Handbook,* 9th Ed., ASM, 1978.
8. D. J. Blickwede, Sheet Steel—"Micrometallurgy by the Millions," An ASM Campbell Memorial Lecture, p. 149, in *Source Book on Forming of Steel Sheet,* Edited by J. R. Newby, ASM, 1976. (This paper was originally published in Trans. ASM, vol. 61, 1968.)
9. "Deep Drawing," p. 52, in *Source Book on Forming of Steel Sheet,* see ref. 8.
10. A. Cottrell, *An Introduction to Metallurgy,* 2d ed., Edward Arnold, 1975.
11. H. Hu, "Recovery, Recrystallization, and Grain Growth," p. 385, in *Metallurgical Treatises,* Edited by J. K. Tien and J. F. Elliot, TMS-AIME, 1981.
12. C. M. Sellars, "The Physical Metallurgy of Hot Working," p. 3, in *Hot Working and Forming Processes,* Edited by C. M. Sellars and G. J. Davies, Proceedings of an International Conference held at the University of Sheffied, U.K., on 17–20 July, 1979. The Metals Society, London, 1980.
13. P. B. Hirsch, et al, *Electron Microscopy of Thin Crystals*, Plenum Press, 1965.
14. Vol. 7, Atlas of Microstructures of Industrial Alloys, *Metals Handbook*, 8th Ed., ASM, 1972.
15. J. R. Everett, et al, "The Effect of Microalloys in Hot Strip Mill Rolling", p. 16 in Proceedings cited in Ref. 12.

Terms and Concepts

Abnormal grain growth
Annealing
Annealing texture
Cold working
Controlled Thermal Mechanical Processing
Critical resolved shear stress
Cube-on-corner texture
Cube-on-face texture
Deformation texture
Drawability
Drawing
Dynamic recovery
Dynamic recrystallization
Earing
Existence of dislocations
Extrusion
Grain growth
High strength low alloy (HSLA) steels
Hot working
Kinetics of recrystallization
Laws of recrystallization
Lüder band
Mechanical deformation
Mechanical stress relief
Microalloyed HSLA steels
Normal anisotropy ratio, $\bar{r}$

Normal grain growth
Paint bake hardening
Percent cold work
Planar anisotropy ratio, Δr
Plastic anisotropy strain ratio, r
Plastic work
Press forming
Primary deformation
Primary recrystallization
Recovery
Recrystallization
Recrystallization texture
Residual stresses
Rolling texture
Rotation of crystal planes
Schmid's law
Secondary deformation
Secondary recrystallization
Shear fracture
Stages of annealing
Static recrystallization
Stored energy of cold work
Strain aging
Strain hardening exponent, n
Strain rate
Strain rate sensitivity index, m
Stress relief
Stretchability
Stretcher strains
Stretching
Subgrains
Superplastic
Superplasticity
Temper
Temper rolling
Texture
Theoretical shear strength
Thermal stress relief
Type A Lüder band in aluminum
Type B Lüder band in aluminum
Warm working
Wire Drawing
Working
Yield point elongation
Zener-Hollomon parameter

Questions and Exercise Problems

8.1 An FCC (face-centered cubic) single crystal is oriented such that the (100) is perpendicular to the applied stress. Determine the critical resolved shear stress on the $(1\bar{1}1)$ when slip was observed to start with an applied stress of 5,000 psi. Which of the following directions, [110], $[10\bar{1}]$, [011], are activated?

8.2 A single crystal rod of a BCC (body-centered cubic) metal has its axis parallel to the [001]. If the rod is pulled in tension, the following slip systems may be activated, (a) $(1\bar{1}0)[111]$, (b) $(0\bar{1}1)[111]$, and (c) $(\bar{1}01)[111]$. Determine which of the three slip systems are activated, and the shear stress(es) on the planes when the applied tensile stress is 100 MPa.

8.3 A 10-mm diameter aluminum single crystal rod was determined to have a critical resolved shear stress of 6.9 MPa. If the tensile axis was parallel to the [001] and the slip plane was (111), determine (a) the slip system(s) that were activated, and (b) the force needed to initiate slip.

8.4 A 10-mm. diameter single crystal rod of an FCC (face-centered cubic) metal was pulled in tension along the [001]. Slip was observed when the load was 5 kilogram-force. When the slip plane is the $(1\bar{1}1)$, determine the possible slip directions on this plane and the critical resolved shear stress.

8.5 An iron single crystal with a critical resolved shear stress of 35 MPa was pulled in tension along the [100]. Determine which of the following slip systems will be activated: (a) $(110)[\bar{1}11]$, (b) $(101)[\bar{1}11]$, and (c) $(01\bar{1})[\bar{1}11]$. What will be the applied tensile force if the diameter of the single crystal rod is 25 mm?

8.6 Show that the maximum tensile yield strength of single crystals is twice the critical resolved shear stress.

8.7 What is the theoretical shear strength of material? Why is this strength not exhibited by single crystals?

8.8 What is cold-working? What features does a material have after it has been cold-worked?

8.9 What is strain-hardening and what is it due to?

8.10 In metalworking, the stress that is of significance is the true or flow stress, according to Eq. 8-8. In this equation, the true strain is properly the true plastic strain, $\epsilon_P = \epsilon_{TOTAL} - \epsilon_E$, where the subscripts P and E, refer to plastic and elastic. For copper, n = 0.54, and K = 46.4 ksi in Eq. 8-8, and its modulus of elasticity is 16×10^6 psi. From this information, and

Eq. 8-9, determine (a) the 0.2% offset yield strength, (b) the ultimate tensile strength, (c) the total true strains at the yield strength and the tensile strength, and (d) the engineering strain at the maximum load.

8.11 If a 0.505-inch diameter of the copper in Prob. 8.10 were pulled in tension to a true plastic strain of 0.30, determine (a) the applied tensile force, (b) the extension of the 2-inch gauge length, (c) the engineering strain, and (d) the recovered elastic strain after the load is released.

8.12 The amount of cold-work is expressed in percent reduction in area in either Eq. 8-12 or 8-13. If a 25-mm diameter steel bar is cold-extruded first to 22-mm diameter and then from 22-mm to 20-mm diameter, determine (a) the amount of cold-work for each step, and (b) compare the sum of the amounts of cold-work for each of the steps to the cold-work from 25-mm to 20-mm, if this were done in one step.

8.13 If the amount of extrusion in Prob. 8.12 were expressed in true strains, determine (a) the true strain for each of the extrusion steps, and (b) compare the sum of the true strains of each extrusion step to the true strain from 25-mm to 20-mm diameter, if this were done in one step.

8.14 Compare the results of Probs. 8.12 and 8.13 and determine the additivity of the amounts of cold-work and the true strains. Can you generalize the results to multistep deformations, (i.e., more than two steps)?

8.15 The properties of 3105 aluminum are shown on Fig. P8.15 as a function of cold-work. What would be the required amount of cold-work to have a minimum half-hard temper, that is, H14 temper? For a quarter-hard, H12? For a H15 temper?

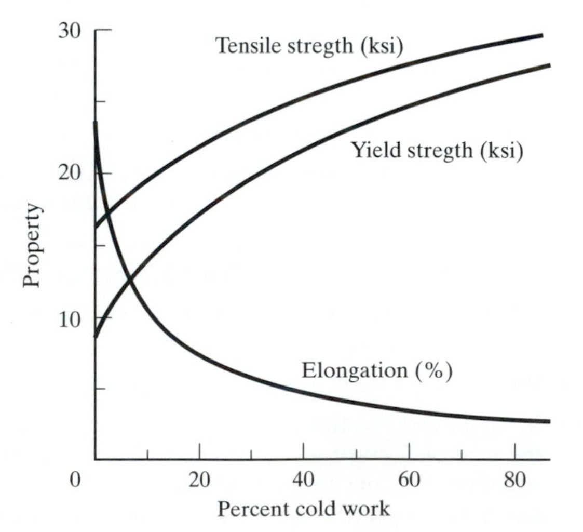

Fig. P8.15 —Properties of 3105 aluminum alloy as function of percent cold-work—75 percent is full hard.

8.16 If the final diameter of a 3105 aluminum bar is to be 20 mm, what would be the initial diameter of the annealed bar to give the tempers required in Prob. 8.15?

8.17 If a final strip thickness of 2 mm is required of the 3105 aluminum alloy, what would be the initial thickness of the annealed strip to give the tempers asked for in Prob. 8.15?

8.18 It is desired to have a 3105 aluminum alloy with a minimum yield strength of 15 ksi and a minimum elongation of 5 percent. What temper designation must be specified when procuring such a material?

8.19 The lowest Browne & Sharpe (B&S) gauge number for copper and copper alloys is 0000, followed by 000, 00, and 0, and then from 1 to 40. Based on the fact that an increase of one in the B&S represents a 10.9 percent reduction in diameter or sheet thickness, and on No. 0 gauge being 0.3249 in. (from Example 8-6), determine (a) the diameter of a No. 14 copper wire, which is specified for a 15-ampere service, and (b) the 0000, 000, and 00 diameter or sheet thickness.

8.20 Determine the true strains corresponding to (a) the percent reduction in diameter of a wire, and (b) the percent reduction in thickness in Table 8-3.

8.21 A 70-30 brass wire with a No. 1 B&S gauge and a grain size of 0.070 mm is needed to have a temper of H02. Because of the limitation in the draw machine, the B&S gauge number can only be increased by one for each draw. (a) determine the minimum draw force for each draw step, (b) determine the stress on the draw wire after each step and whether the draw process is stable, and (c) what is the yield strength of the brass wire in the H02 temper?

8.22 A 5-mm 70-30 brass strip is desired with a minimum yield strength of 50,000 psi yield strength and a minimum 10 percent elongation in two inches. Using Fig. 8-18, (a) determine the window of processing for the brass that has an initial annealed grain size of 0.070 mm, and (b) specify the temper condition of the brass strip and determine the initial thickness of the annealed strip.

8.23 Using Fig. P8.15, determine whether it is possible to draw a 25-mm 3105 annealed aluminum rod to 12.5 mm.

8.24 What is the difference between drawing and stretch forming? What material properties are important for drawing and for stretch forming?

8.25 What is the difference between cube-on-corner and cube-on-face annealing texture? What is cube-on-corner texture good for? cube-on-face texture good for?

8.26 What is plastic strain anisotropy ratio? What is the difference between normal and planar anisotropy?

8.27 What is earing and what causes it in steel sheets? in aluminum sheets?

8.28 What is the difference between extrusion and drawing?

8.29 What is the difference between thermal and mechanical stress relief?

8.30 What are the three stages of annealing? Describe the changes in structures and properties of a deformed material during the three stages?

8.31 What is the difference between primary and secondary recrystallization? between static and dynamic recrystallization?

8.32 Describe the kinetics of recrystallization?

8.33 What is the difference between normal and abnormal grain growth? How is abnormal grain growth used in practice?

8.34 What are the microalloyed high-strength low-alloy (HSLA) steels? What is the primary strengthening mechanism in them? How is it achieved?

8.35 What is superplasticity? What are the requirements for superplasticity? How is the strain rate sensitivity index determined?

9

Thermal Processes—Heat-Treating

9.1 Introduction

In the broadest sense, thermal processing is the heating and cooling for the purpose of achieving the desired mechanical properties and performance of a material by changing its microstructure and/or its residual stress pattern. Thermal processes are classified as *hardening* or *nonhardening,* depending on whether the processes harden or do not harden the material. In industry, the thermal process that results in the hardening or strengthening of a material is commonly called *heat treating.* Thus, when we say a material is heat-treated, it usually means that the material is strengthened when heated and cooled. In fact, aluminum alloys are customarily classified as either heat-treatable or nonheat-treatable depending on whether they can be or cannot be hardened when heated and cooled.

The two commonly used nonhardening thermal processes for all metals and alloys are recovery or stress-relief anneal and process or recrystallization anneal, both of which were briefly discussed in Secs. 8.5.1 and 8.5.2. We shall first broaden the discussion on these thermal processes and learn of other nonhardening processes that are specific to steels. Then, the general principles will be presented concerning the thermal hardening (heat treating) processes along with specific cases of hardening, such as precipitation hardening, spinodal decomposition, and the martensite transformation and thermal hardening in metals and alloys, especially ferrous alloys.

9.2 Nonhardening Thermal Processes

9.2.1 Thermal Stress-Relief of Residual Stresses

The occurrence of residual stress was specifically described in Sec. 8.4.5 as the result of the uneven or heterogeneous deformation of a material such as in rolling, forging, bending, and machining. In general, residual stresses may form or may be altered during the processing of materials from ingot to the final product, such as those listed in Table 9-1, that involve

Table 9-1 Processes that induce residual stresses. [1]

Thermal action	Mechanical action	Chemical action
Heat treatment	Machining, grinding, and polishing	Etching
Stress relieving	Mechanical surface treatments	Corrosion
Annealing	Shot peening	Chemical machining
Hardening	Surface rolling	Surface coating and plating
Tempering	Hammer peening	
Diffusion treatment	Ballizing	
Carburizing	Cold forming	
Carbonitriding	Stretching	
Nitrocarburizing	Drawing	
Cyaniding	Upsetting	
Nitriding	Bending and straightening	
Decarburizing	Twisting	
Fabrication with heat	Autofrettage	
Welding	Interference fitting	
Flame Cutting	Service overloads	
Hot forming	Explosive stressing	
Casting	Cyclic stressing	
Shrink fitting	Wear, chafing, bruising, gouging, and cracking	
Operation at elevated temperatures		
Electrical discharge machining		

thermal, mechanical, and/or chemical actions. For example, processing or fabrication involving heat (welding, flame cutting, and so forth) and quenching or cooling, with no transformation in crystal structure, usually result in the surface being under compressive stress. However, when an allotropic change in crystal structure occurs, such as that in iron and steels from FCC to BCC, the transformation of the surface ahead of the core or interior of a section results in tensile stresses on the surface. If the surface transforms after the core or the interior, a compressive stress develops on the surface. In welding, the very steep thermal gradients in the fusion zone and in the heat-affected zone (HAZ) produce residual stresses on a macroscale (see Sec. 12.4.1 for discussion of the terms on welding). The level of residual stresses in the HAZ approaches the yield strength of the material at room temperature.

Residual tensile stresses are particularly bad for the performance of a material and are specifically relieved or removed during the stress relief treatment. If these tensile stresses are sufficiently large, they may induce cracking immediately or just after the manufacturing process. A dramatic illustration of an effect of residual stress is the sudden longitudinal split of a 40-foot long I-beam shown in Fig. 9-1 that was lying on a shop floor after being torch-cut on both ends. Drastic quenches given during heat-treating may induce cracking. Cracks may also form around the heat-affected zone during the welding of plates. If not relieved, the residual stress may induce delayed cracking as in Fig. 9-1, especially if the material is exposed to a corrosive environment. With the latter, this phenomenon is called *stress-corrosion* that requires that the material is under tensile stress and exposed to a specific corrosive environment. A corrosive environment does not necessarily mean acids and bases but may just be the humidity in the atmosphere. (The phenomenon of stress-corrosion cracking was first discovered when brass cartridges stored in cases were observed to crack, particularly during the monsoon season. The cracking was first called

Fig. 9-1 Forty-foot long I-beam split spontaneously due to residual stresses set up after torch cutting both ends. [1]

season cracking, not because of the monsoon season, however, but because the cracking resembles that of seasoned wood.) As discussed in Sec. 4.4.7, Eq. 4-66 and Sec. 4.4.10, Eq. 4-73, for fracture toughness and fatigue properties, residual tensile stresses will lower the stress that can be applied to a component or structure; for fracture, the applied stress will be lower for a certain crack size to propagate, while for fatigue, the applied stress will be lower for a desired life, number of cycles.

The *thermal stress-relief* consists of uniformly heating a structure or component to a suitable temperature below the recrystallization temperature, holding it at this temperature for a predetermined period of time, and then followed by uniform cooling. Uniform cooling refers to every section of the structure or component, particularly when the component has variable sections. If the cooling is not constant and not the same for all sections of the component, then new residual stresses will arise. In addition to relieving the high stresses, stress-relief reduces the possible distortion of the component.

The process is designed to relieve residual stresses without major changes in the structures and properties of the material. In terms of the cold-worked structure discussed in Sec. 8.4.1, stress-relief was observed to be in the recovery anneal stage. We saw there that the structural change is the formation of subgrains within the pancaked grains but this change did not cause a significant change in the hardness or strength of the cold-worked structure. In the formation of subgrains, dislocations are annihilated, rearranged, and aligned. While the latter have been specifically related to cold-worked or deformed structures, it is generally believed that these are the same microstructural changes that occur in stress-relief.

The rearrangement of dislocations implies movements (diffusion) of atoms and the formation of subgrains involves the "climb" of dislocations, as discussed in Secs. 2.12.2 on imperfections and 4.4.11 on the mechanisms of creep. In general, we may think of these microstructural changes during stress-relief as microscopic creep, and thus expect an interdependence of time and temperature and the applicability of the parameters in creep. This was found to be the case and the thermal effect during stress-relief for steels was found to follow the Larson-Miller type relationship shown below.

$$\%\text{Relief of Stress} = T(\log t + 20) \times 10^{-3} \quad (9\text{-}1)$$

where T is temperature in Rankine (Rankine = °F + 460) and t is in hours. This effect is illustrated in Fig. 9-2 where the average percent removal or relief of residual stress after welding is shown as a function of both temperature and time at the stress-relief temperature. It is evident that similar relief of residual stresses can be achieved with various combinations of time and temperature.

Because the process involves dislocation motion, stress-relief for materials depends on the composition, particularly the presence or absence of particles or precipitates that interact with dislocation motion. Creep-resistant materials that contain particles or dispersed phases purposely added to retard dislocation motion will require higher stress-relief temperatures than low-temperature materials. For example, typical stress-relief temperatures for low-alloy steels are between 595° and 675°C and those of high-alloy steels may range from 900° to 1065°C.

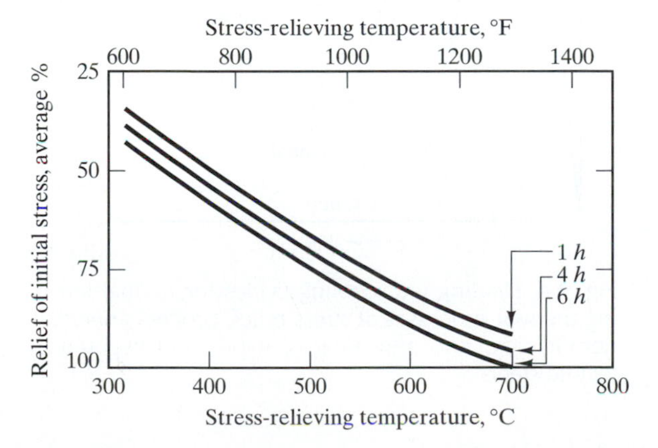

Fig. 9-2 Relief of residual stress is a function of both time and temperature according to the Larson-Miller equation.

9.2.2 Annealing

This refers to the process whereby major changes in the structure and properties occur after the material was heated to and held at a suitable temperature, and then cooled at a fairly slow rate. The objective of annealing is primarily to soften the material to prepare it for further processing, such as cold-forming and machining. In previously cold-worked structures, annealing induces the recrystallization of the hardened, pancaked, and deformed grains to new soft and equi-axed grains. In addition, there may also be coarsening of precipitate particles when these are present in the material. In previously hot-worked materials, the hot- worked grains may not recrystallize when annealed, but the properties will still be softer than the hot-worked condition.

For steels, there are three annealing processes that are used to achieve certain specific objectives. These are subcritical annealing, intercritical annealing, and full annealing. The names of the processes refer to the ranges of temperatures used relative to the critical transformation temperatures in the eutectoid portion of the iron-cementite diagram, as shown again in Fig. 9-3. The subcritical annealing is done below the first or lower critical temperature, A_{e_1}. The intercritical annealing is done in the range between A_{e_1} and A_{e_3} for hypoeutectoid steels or between A_{e_1} and $A_{e_{cm}}$ for hyper-eutectoid steels. The full annealing is done above the second or upper critical temperature, either A_{e_3} or $A_{e_{cm}}$ for hypo- or hypereutectoid steels, respectively.

The subscript "e" in the critical temperatures indicates equilibrium conditions of heating and cooling at infinitely slow rates. The heating and cooling rates during actual thermal processing are not infinitely slow; therefore, the equilibrium critical temperatures are not applicable. During heating, the critical temperatures are higher than the equilibrium values and "c" is used in place of "e" (i.e., A_{c_1}, A_{c_3}, and $A_{c_{cm}}$). During cooling, the critical temperatures are lower than the equilibrium values and "r" is used in place of "e" (i.e., A_{r_1}, A_{r_3}, and $A_{c_{cm}}$). The values of these critical temperatures depend on the heating and cooling rates, but the average of the two is approximately

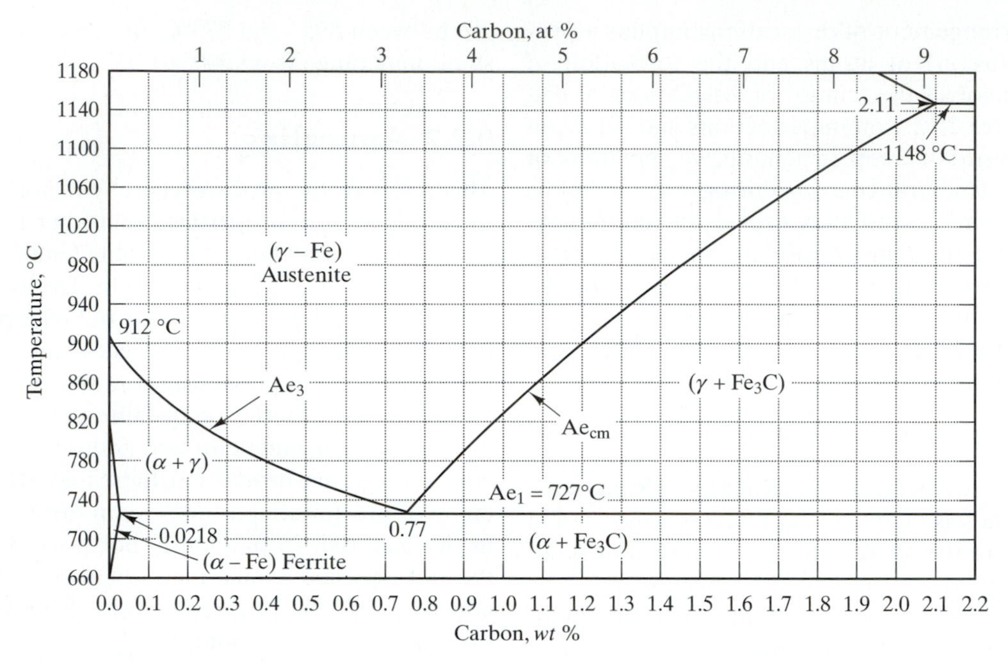

Fig. 9-3 Eutectoid portion of the iron-iron carbide (cementite) diagram showing the equilibrium critical temperatures.

equal to the equilibrium values. To simplify subsequent notations, the critical temperatures will be simply referred to as A_1 (lower critical), A_3, and A_{cm}, with the understanding that during heating they refer to A_c and during cooling to A_r.

The heating and cooling cycles involved in the three annealing processes are schematically drawn in Fig. 9-4; while the ranges in temperatures for these processes are indicated in the iron-cementite diagram in Fig. 9-5.

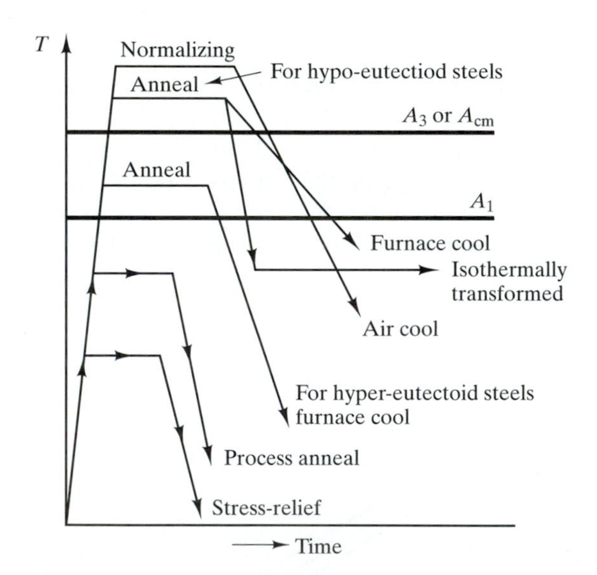

Fig. 9-4 Heating and cooling cycles for nonhardening thermal processes of stress-relief, process anneal, normalizing, and annealing of hypo- and hypereutectoid steels.

The *subcritical annealing processes* are the stress-relief and the process or recrystallization anneal that we have discussed previously. The stress-relief is accomplished at the lower temperatures of the range of subcritical annealing. The process anneal is specifically referred to as the heating of the steel just below the A_1, in most instances 11° to 22°C below, holding it for an appropriate time, and then followed by cooling, usually in air. The annealing of sheet and strip steels, which accounts for the most tonnage of annealed materials, is done by this method. In the wire industry, process annealing is used as the treatment before the last light drawing to the final size. The product is called "annealed in process".

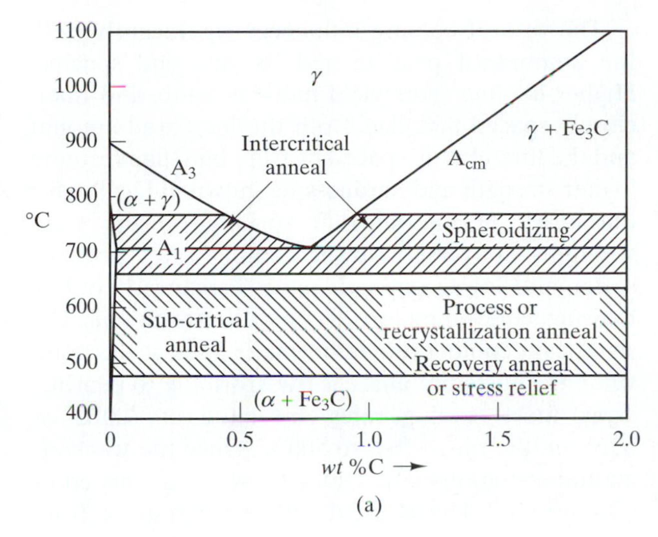

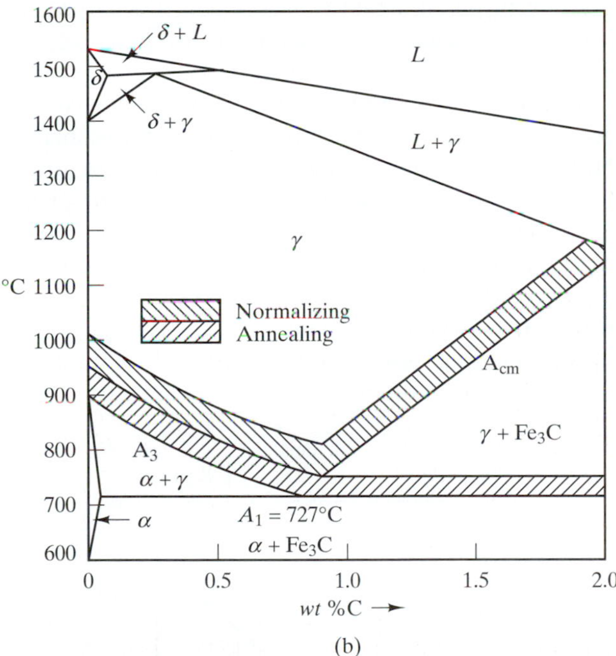

Fig. 9-5 (a) Ranges of temperatures for subcritical annealing (stress-relief and recrystallization anneal) and intercritical annealing (spheroidizing). (b) Ranges of temperatures for normalizing and annealing of steels.

Intercritical annealing followed by isothermal transformation produces the softest condition of the steel. Specifically, heating the steel to within 55°C (100°F) above the A_1 and then transforming it at a temperature less than 55°C below the A_1 produces the softest condition. The structure produced in this case consists of *spheroidized carbide particles* in a ferrite matrix. The *spheroidized structure* has the minimum hardness, the maximum ductility, as well as maximum machinability. Low-carbon steels are spheroidized for maximum ductility but not maximum machinability because they become excessively soft and gummy. Hypereutectoid steels are spheroidized for machinability and the *annealing conditions indicated above are the optimum for the spheroidization process.* For hypereutectoid steels, higher heating temperatures in the intercritical regime induces more carbide to dissolve in austenite thereby inducing the formation of lamellar carbide (pearlite) structures. Annealing of hypereutectoid steels is usually not performed above A_{cm} because subsequent slow cooling produces a carbide (Fe_3C) network around the prior austenite grain boundaries that is detrimental for toughness and ductility.

Full annealing is the heating to and the holding at a temperature above the upper critical temperature to obtain full austenitization, followed by either slow cooling or isothermal transformation below the lower critical temperature. Small forgings of carbon and alloy steels are given full annealing treatment to produce structures that are appropriate for subsequent processing. The austenitizing temperature is in the range of up to 50°C over the A_3 temperature. Temperatures towards the upper end of this range tend to homogenize the austenite better, and subsequent cooling will produce pearlitic structures. Fine pearlitic structures provide the good carbide distribution needed for optimum control for a subsequent selective hardening of the forging. In the lower end of this range, the austenite does not homogenize completely and transforms predominantly to a ferritic and spheroidized carbide structure that is suitable for machinability when cooled. For most of the latter applications, it is sufficient to simply specify that the material be cooled in the furnace after a minimum hold time of one hour at the proper temperature. To produce pearlitic structures, the material is quickly cooled to a desired temperature below the lower critical and then isothermally transformed.

9.2.3 Normalizing

Normalizing is a nonhardening process specifically for steels and consists of austenitizing the material or component, held for a period of time at a certain temperature, and then cooled in natural or slightly forced air-convection. The typical normalizing temperatures are shaded, shown in Fig. 9-5b. For hypoeutectoid steels, these are usually about 55°C (100°F) above A_3—for hypereutectoid steels about 27°C (50°F) above the A_{cm}. After the process, a normalized steel exhibits a uniform fine-grained microstructure with pearlitic areas, called the normal microstructure, and a characteristic of hypo- and hypereutectiod steels. The strength and hardness of a normalized steel may increase or decrease from the properties prior to the normalization process depending on the thermal and mechanical history of the material. Steels that do not produce the pearlitic microstructure are not usually normalized, for example, stainless steels, maraging steels, and air-hardening steels.

Among the benefits attained after normalizing are improved machinability, grain size refinement, homogenization of composition, and modification of residual stresses. Normalizing is used as a conditioning treatment for alloy steel forgings, rolled products, and castings before final heat treatment. Homogenization of castings by normalizing breaks up or refines the dendritic structure. For wrought steels, normalizing obliterates most of the banded structure produced during hot-rolling and the large grains or mixed grain sizes after forging. Steels to be thermally hardened are normalized first to homogenize the composition before they are austenitized, quenched, and tempered. Steels that are going to be carburized are usually normalized above the carburization temperature to minimize distortion. Hypereutectoid alloy steels are normalized for partial or complete elimination of carbide networks; the resultant structure is more susceptible to 100 percent spheroidization during subsequent spheroidizing anneal discussed previously.

The rate of heating to the normalizing temperature is ordinarily unimportant; however, for a component with varying sections, distortion of the component may result because of thermal stresses. The time at the normalizing temperature should be sufficient to produce a completely homogeneous austenite phase. Time must be allowed to completely dissolve the carbides and then to homogenize the composition via diffusion as discussed in Sec. 6.2.4. The common practice is to allow one hour at temperature for every inch of part thickness.

The rate of cooling influences significantly both the amount of pearlite and its size and spacing. Higher cooling rates yield more pearlite, and finer, closely spaced lamellae. Both the increased amount and the fine, closely spaced pearlite lamellae result in higher strength and hardness, as shown in Fig. 9-6 for the effect of interlamellar spacing on the yield strength. Conversely, lower cooling rates produce softer and lower strength parts. Enhanced cooling may be achieved by forced convection with fans. The significant cooling rate in normalizing is at the point when the austenite phase is transforming to pearlite. Thus, most cooling rates are cited at 700°C or between the range 700° to 500°C. Once the transformation is complete, they may be water-quenched or oil-quenched. This is usually done when all sections of the component turn to "black heat." For large section sizes, the center material must turn black before any drastic quenching may be done.

Carbon steels with 0.20% C or less usually receive no further treatment after normalizing. However, medium- or high-carbon steels are often tempered after normalizing to obtain specific properties, such as lower hardness for straightening, cold-working, or machining. The tempering process is just another heating process in the subcritical range to soften the structure in steels that have hardened during heating and cooling, especially for steels

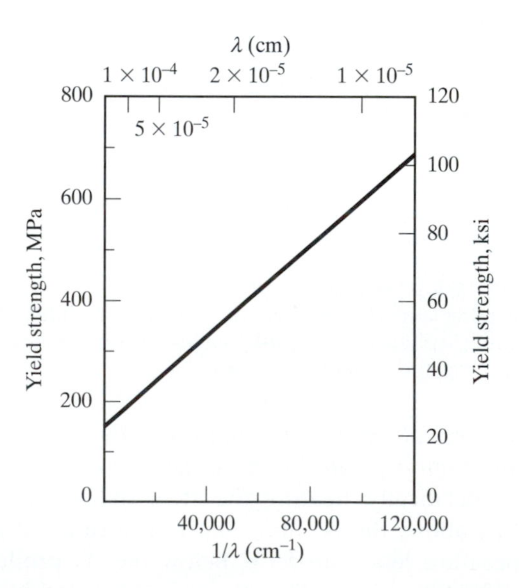

Fig. 9-6 Effect of interlamellar spacing, λ, of pearlite on the yield strength of pearlite. [17]

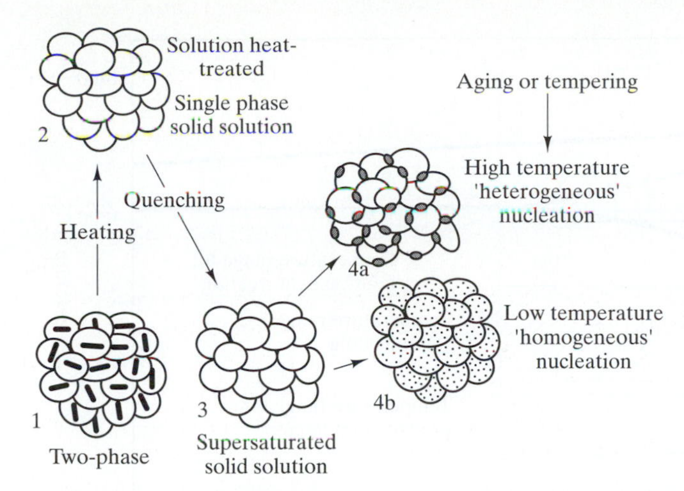

Fig. 9-7 Generic heat-treating processes for hardening—two-phase alloy (1) is solution-heat-treated to a single solid solution (2), quenched drastically to produce a supersaturated solid solution (3), and then aged or tempered at a high temperature (4a) or at a low temperature (4b) to induce either heterogeneous or homogeneous nucleation (distribution) of precipitates.

hardened with the martensite structure, as described in Sec. 9.7.2a.

9.3 Hardening (Heat Treating) Processes—General Principles

We shall now discuss the general principles of the thermal process to harden materials. Firstly, metals and alloys that exhibit hardening or strengthening when heated and cooled have the following common characteristics:

1. They are at least two-phase alloys at a low temperature and one of these phases has increasing solubility in the other as the material is heated to a higher temperature.
2. When cooled from the heated condition, the high-temperature state of the alloy can be frozen and retained at low temperatures by fast-cooling (quenching) techniques to produce a supersaturated solution of the second phase.
3. Hardening or strengthening may be achieved in the as-quenched condition or during subsequent heating to produce very fine nonequilibrium transitional coherent precipitates or phases that can block the motion of dislocations in the alloy. The coherent precipitates are clusters of solute atoms that are in the initial stages of forming the crystal structure and composition of the equilibrium precipitate particles. These transition clusters may have different crystal structures and composition from those of the equilibrium precipitates and do not detach from the host lattice (i.e, they are coherent with the host structure).

Because of the common characteristics above, the heat treating of a material consists of the following *generic heat-treating processes,* which are schematically illustrated in Fig. 9-7.

1. *Homogenization or solution treatment.* The two-phase or multiphase alloy at room temperature is heated to a temperature at which a single solid solution or more of the second phase is dissolved than at room temperature. This treatment may or may not involve a change in phase or crystal structure of the host or solvent material (state 1 to 2 in Fig. 9-7).
2. The alloy is *cooled at a rate greater than a critical cooling rate* to freeze the high-temperature state when brought to room temperature or lower. The cooling may or may not change the phase or crystal structure of the solvent or host material (state 2 to 3 in Fig. 9-7).
3. The alloy is *aged or tempered to get the desired properties.* The aging or tempering process may be done at room temperature (*natural aging*) or at elevated temperatures (*artificial aging*) to effect a phase change (i.e., to allow precipitates or phases to come out from the supersaturated solid solution—state 3 to 4a or 4b in Fig. 9-7).

To illustrate the above principles, we look at three cases: the eutectoid portion of the metastable iron-cementite in Fig. 9-3 for steels, the aluminum-rich section of the aluminum-4 wt%Cu in Fig. 9-8a, and

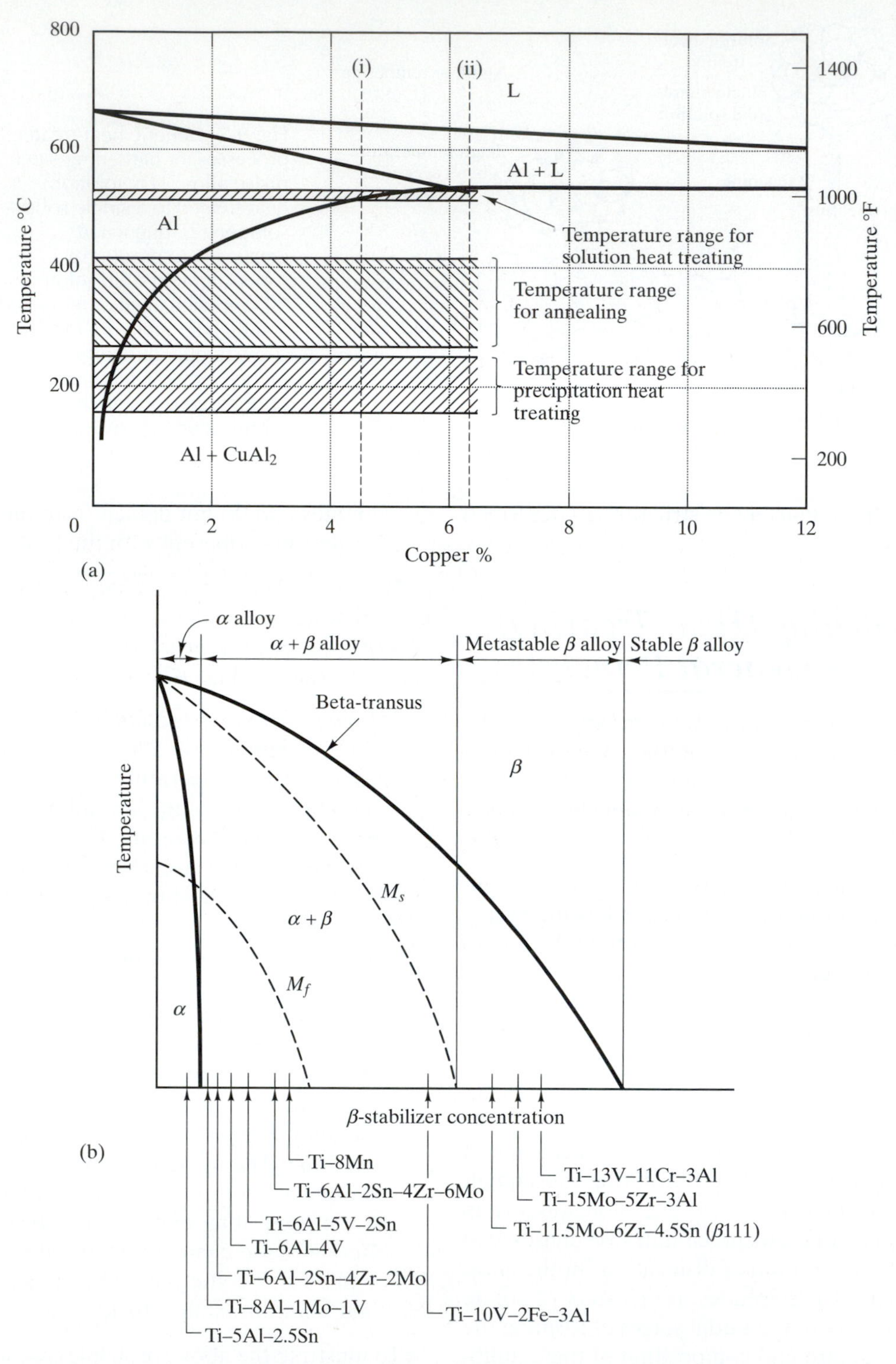

Fig. 9-8 (a) Aluminum rich section of Al-Cu binary alloy phase diagram indicating the solution heat-treating range for hardening. (b) Pseudo-binary β-isomorphous phase diagram of titanium alloys indicating location of U.S. titanium alloys. [17]

the pseudo-binary β-isomorphous titanium alloy equilibrium phase diagram in Fig. 9-8b. These are probably the three metal alloy systems that are most heat-treated and of these three, steels are heat-treated the most. In steels, the process first transforms the two-phase, $\alpha + Fe_3C$, structure to a single homogeneous austenite, γ, phase. When the γ is quenched below A_1, it becomes unstable and will transform to either the ($\alpha + Fe_3C$) structure or to a supersaturated ferrite solid solution with carbon in it. In the aluminum system, the two-phase structure consisting of the solid solution (Al) + intermetallic compound $CuAl_2$ is brought to a temperature above the solvus but below the eutectic temperature to dissolve all the second phase $CuAl_2$ in (Al). The upper temperature range in Fig. 9-8a is indicated as *"solution heat-treating"* because $CuAl_2$ is dissolved and put into solid solution. The high-temperature single phase (Al) is quenched to room temperature or lower and becomes a supersaturated (Al); then precipitation of particles from the supersaturated (Al) is done at room temperature or at the precipitation heat-treating range. In most practical heat-treatment of titanium alloys, the two-phase structure consisting of $\alpha + \beta$ is not brought up to the single phase β-region, but rather just below the β-transus temperature (the temperature above which 100% β is present or no α exists). Nonetheless, more of the second phase β is brought into solution, which, when quenched, produces a supersaturated condition with β that will decompose to α.

9.3.1 Transformation C-Curves and Their Implications

The critical cooling rate that is needed to retain the high temperature composition depends on the transformation characteristics of each alloy. The transformation that we shall discuss is that of the quenched supersaturated solid solution to the equilibrium phases or the precipitation of particles. This is a solid-solid transformation as opposed to the liquid-solid transformation during solidification. In general, this is also a nucleation and growth process in which clusters (nuclei) of the second phase or precipitate form and then grow. As in solidification, the formation of the nuclei depends on a favorable free energy change (thermodynamics) and their growth depends on the movement of atoms or diffusion—a rate process (kinetics) that in turn depends on both temperature and time. The effects of both these factors leads to a C-shaped transformation curve which we shall now discuss.

The C-shaped curve, which is characteristic of a nucleation and growth transformation process, arises from the interaction of the thermodynamics and the diffusion factors during the transformation. We start with thermodynamics and indicate that, as in solidification, the thermodynamics for the change must be favorable (i.e., the net free energy change, ΔF_{net}, must be negative). The net free energy change for solid-solid phase transformations is

$$\Delta F_{net} = \Delta F_v + \Delta F_s + \Delta F_\epsilon \qquad (9\text{-}2)$$

where the subscripts *net, v, s,* and ϵ signify the changes in total, volume, surface, and strain free energies, ΔFs. The first two terms on the right side of Eq. 9-2 have exactly the same meaning as in Eq. 7-2 for solidification. The strain energy (third) term comes from the distortion of atomic arrangement and bonding of the host lattice. During solidification, the liquid has a more open structure and when a cluster of atoms forms a solid nucleus, no resistance is encountered. In contrast, when a cluster of solute atoms forms in the solid, the host lattice must accommodate the cluster. The accommodation is usually achieved by stretching or contracting the host interatomic distance depending on the difference in atomic sizes and in the crystal structure. This is a resistance to the transformation as strain energy that must be supplied by the volume free energy.

Thus, in order for the transformation to proceed, there must be a sufficient ΔF_v (which is negative) to overcome the positive surface and strain energies. Using the concept expressed in Eq. 7-8 during solidification, the amount of ΔF_v is directly proportional to the undercooling expressed by the temperature difference, $(T_{eq} - T)$, where T_{eq} is the equilibrium transformation (critical) temperature and T is the actual transformation temperature. The farther the temperature is from the equilibrium temperature, the greater is the undercooling and the ΔF_v.

A negative ΔF_{net} in Eq. 9-2 signifies that the phase transformation can potentially occur. It does not indicate, however, when the phase transformation will occur. The time when transformation will start at a particular temperature depends both on the ΔF_{net} *and on the diffusion of atoms* to form the clusters of solute atoms. The *interaction of these two factors leads to the C-curve or the isothermal transformation (I-T) diagram,* illustrated in Fig. 9-9. At temperatures just below the equilibrium temperature, the ΔF_v is relatively small and as a result, the activation energy is large, according to Eq. 7-5. Thus, it takes time to form nuclei and relatively few will form. For the small ΔF_v , they will form preferentially at the grain boundaries or on surfaces of inclusions (heterogeneous nucleation). At these temperatures, the diffusion of atoms is relatively high and the time it takes for atoms to move is short. Thus, the controlling factor to start the transformation is the small ΔF_v to form a few stable nuclei. When the phase change is allowed to proceed to completion, only a few large phases or particles form, and are located mostly at grain boundaries or impurity particle interfaces, as indicated by 4a in Fig. 9-7. The latter are sites for heterogeneous nucleation because they lower both the surface and strain-energies required.

At low transformation temperatures, the ΔF_v will be large and the activation energy is low. Thus, a lot of nuclei can form and the large driving force induces homogeneous transformation within the crystal or grain, as indicated by 4b in Fig. 9-7. However, at low temperatures, the diffusion of atoms is relatively slow and the formation of stable nuclei is controlled by how fast the solute atoms can cluster. Thus, the time to start the transformation is also long. The resulting transformation structure will exhibit a much larger number of small particles or phases that are uniformly distributed within the matrix lattice, indicated by 4b in Fig. 9-7.

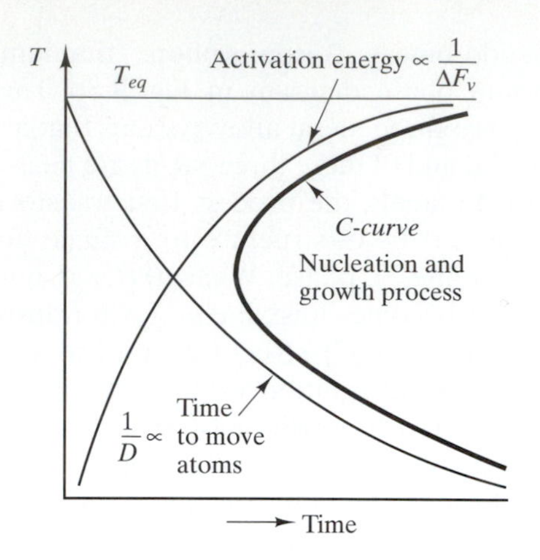

Fig. 9-9 Kinetics of a nucleation and growth process is typified by a C-curve that is the result of the interaction of free energy change and the diffusion of atoms.

Quite often the phases or precipitates that come out at these low transformation temperatures are coherent with the matrix (host lattice). This means that the cluster of atoms forming the nuclei of precipitates do not form an interface (surface) with the matrix, as schematically illustrated in Fig. 9-10b. The clusters are the transitional particles that strain and harden the matrix and the shape of these coherent clusters are disklike, as shown in Fig. 9-11, to minimize the strain energy.

At intermediate temperatures, there is a sufficient driving force, ΔF_v, and the mobility of the atoms is still high. These factors contribute to start the transformation at shorter times, and thus, the C-curve forms as indicated in Fig. 9-9. The C-curve is also called the isothermal-transformation (I-T) diagram

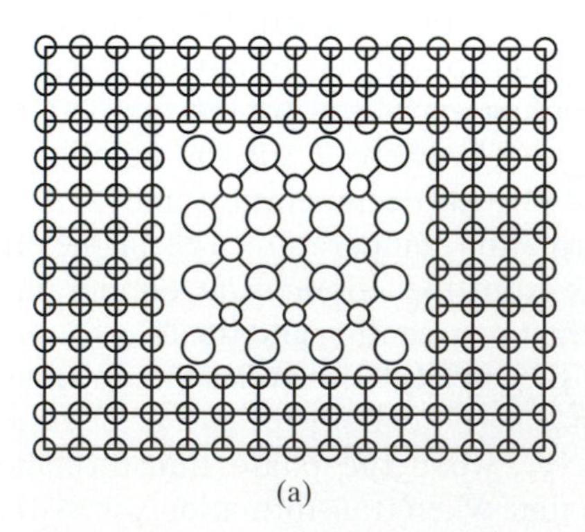

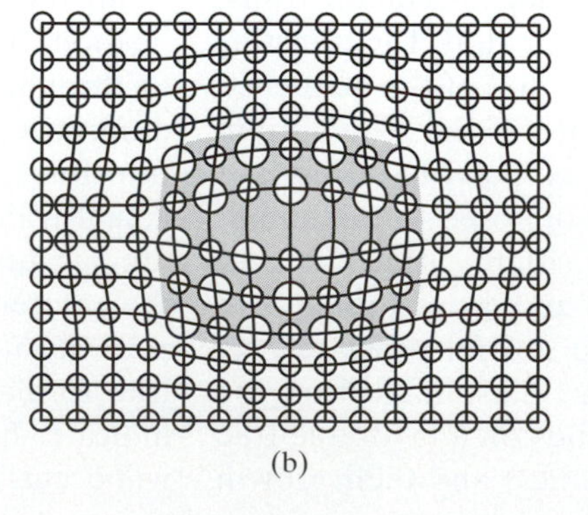

Fig. 9-10 Schematics of precipitate coming out from matrix (a) in incoherent forms with an interface to minimize strain energy; and (b) coherent forms having no interface to minimize surface energy, but increases strain energy. [17]

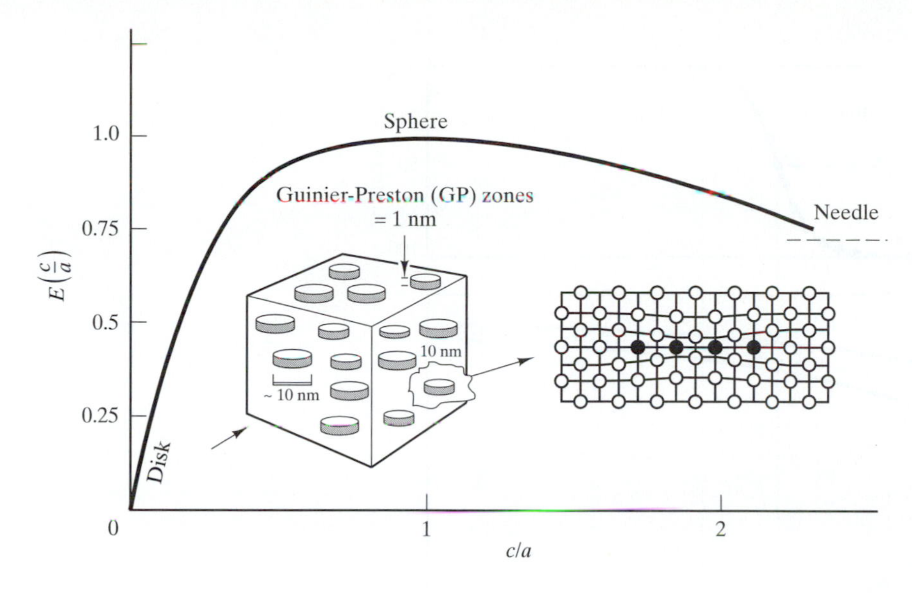

Fig. 9-11 The strain energy in the matrix due to coherent precipitation is minimized when the precipitate is disk-shaped (like coins in inset).

because the transformation is conducted at a constant temperature. The I-T diagram is established by bringing the single-phase solid solution from the solution treatment stage to a temperature in the two-phase field and noting the times to begin (start) and to end (finish) the transformation, as illustrated by a eutectoid steel at the bottom of Fig. 9-12. The complete I-T diagram is established by conducting isothermal transformations at different temperatures, noting the start and finish times of the transformation for each of these temperatures, and then connecting all of these times to produce the two C-curves in Fig. 9-12. Since this is a nucleation and growth process, the progress of the isothermal transformation follows a sigmoidal curve shown at the top of Fig. 9-12. This is the same type of curve as in the recrystallization process described in Sec. 8.5.2 and also follows the empirical relation Eq. 8-24.

Examples of I-T diagrams for aluminum and titanium alloys are shown in Figs. 9-13 and 9-14. The shortest start time of the transformation of these I-T diagrams is called the "nose" or "knee" of the C-curve. The *nose* of the diagram *indicates the minimum cooling rate the material has to be cooled from the homogenizing or solution treatment temperature* (Step 1 of the heat-treating process) in order to retain the solute in solution at the solution-treating temperature to obtain a supersaturated solution at room temperature or lower. The minimum cooling rate is the *critical cooling rate* referred to earlier and is illustrated as well in Figs. 9-12 and 9-13—a short time at the nose requires a fast cooling rate, a longer time means a relatively slower cooling rate. These effects are shown in Figs. 9-13a and b with Al-5.5wt%Mg and Al-4 wt%Cu. Noting the time scales in these diagrams, we see that the nose of the Al-5.5wt%Mg alloy is about an order of magnitude longer than that of the Al-4 wt%Cu alloy, thus, the diagram of the Al-5.5wt%Mg alloy indicates a more moderate quench or cooling rate.

The maximum hardness or strength attainable in materials depends on keeping all or most of the supersaturation before the aging or tempering step by avoiding any transformation at high temperatures. A high cooling rate means higher supersaturation and higher strength for the material; an example of this is illustrated by noting the tensile strengths of eight aluminum alloys in Fig. 9-15. (For a description of the alloy and temper designations for aluminum alloys, see Secs. 13.2.1 and 13.2.2.) The strength will also be higher if the particle sizes are small and are located within the matrix. These are obtained at low isothermal transformation temperatures. In contrast, the fewer, relatively large particles located at grain boundaries are obtained at high isothermal transformation temperatures. These lead to low strength, poor toughness, and low ductility.

There is another effect when precipitation or transformation from the supersaturated solution occurs as

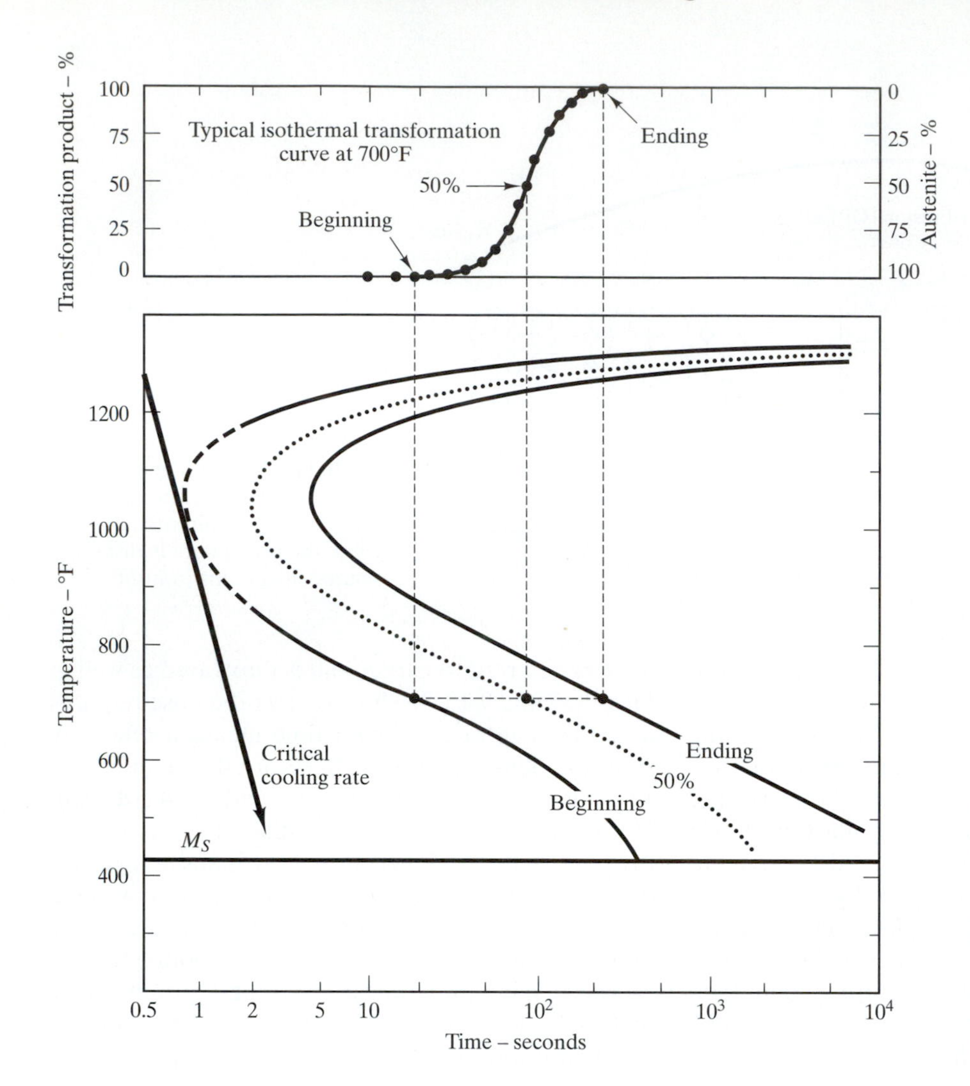

Fig. 9-12 An example of an isothermal transformation at 700°C for a eutectoid steel. Top—sigmoidal curve showing amount transformed; Bottom—shows start and finish times. C-curves are for start and finish of transformation; critical cooling rate and M temperature are also shown. [9]

the alloy is being quenched to room temperature. In aluminum alloys, this leads to intergranular corrosion as opposed to pitting corrosion when precipitation or transformation is prevented, as indicated in the C-curve for an Al-Cu alloy in Fig. 9-16.

9.3.2 The Martensite Transformation—A Displacive Transformation

We have been discussing the nucleation and growth (N-G) transformation. We shall now discuss another type of transformation called the *martensite transformation* that may occur in materials. This is a shear or displacive transformation that occurs when the transformation of the supersaturated solution involves a change in the crystal structure of the host lattice, such as from the face-centered cubic (FCC) to the body-centered cubic (BCC) structure in iron and steels. The difference between this and the N-G transformation is the possibility of transforming a larger volume of the material during a shear transformation. The latter occurs when a large undercooling provides a sufficient ΔF_v to supply the shear-strain energy that drives the transformation of a large volume of the original phase at one time and traps the solute in the supersaturated product phase. In steels, the retention of the carbon solute atoms strains the lattice to a body-centered tetragonal (BCT) rather than the BCC, and provides the primary hardening

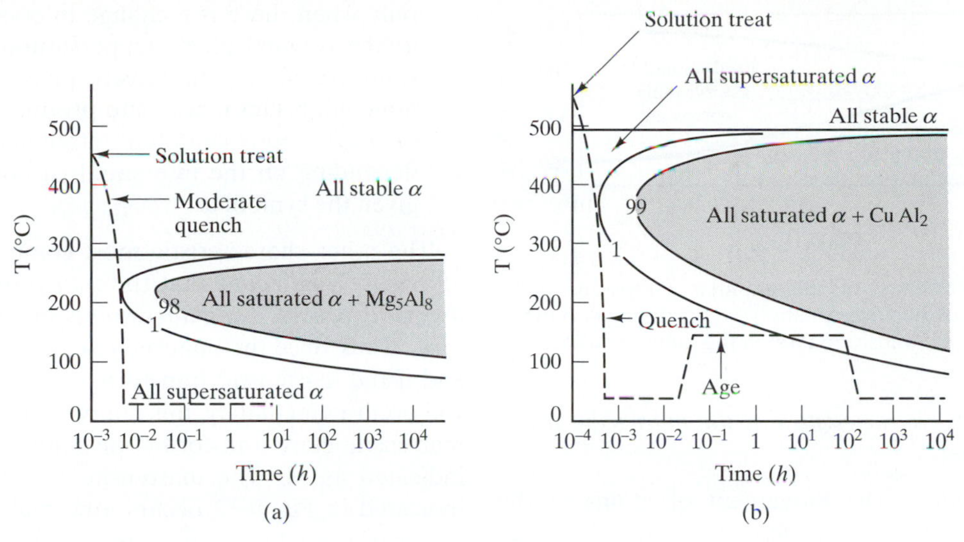

Fig. 9-13 Kinetics (C-curves) of precipitation in aluminum alloys, (a) Al-5.5wt%Mg, (b) Al-4wt%Cu. [8]

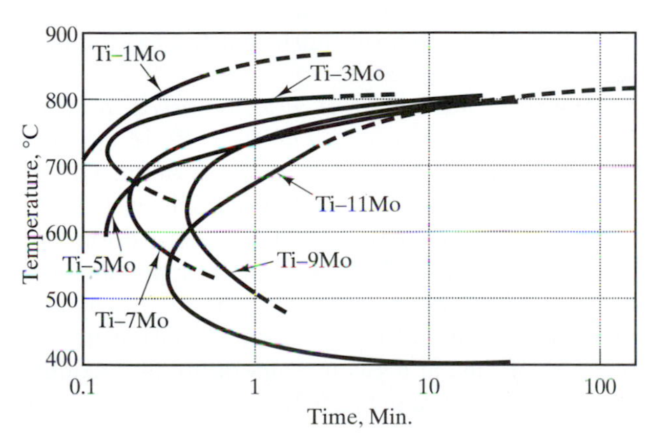

Fig. 9-14 C-curves (start times) for beta-to-alpha transformation of various titanium alloys with molybdenum. [18]

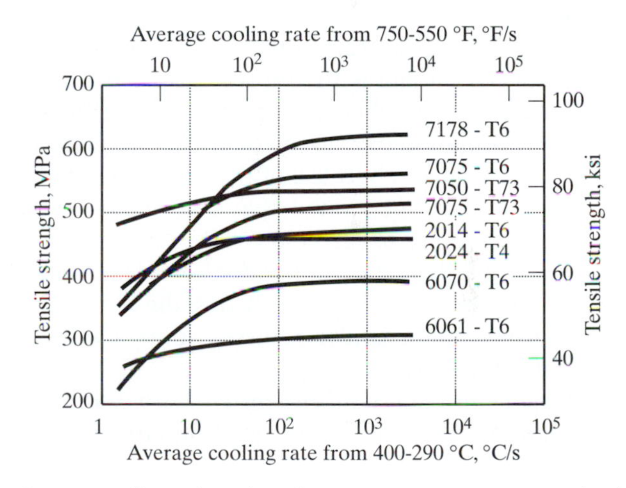

Fig. 9-15 Effect of cooling from 400–290°C on strength of heat-treated aluminum alloys. [4]

or strengthening. We shall discuss in greater detail in Sec. 9.7.2 the strengthening mechanism and other aspects of the martensite transformation. We shall only explain here the basic characteristics of the martensite transformation that apply to all systems including ceramics.

The term *martensite* was also used initially as the name of the *platelike microconstituent in hardened steel,* as the result of the *martensite transformation* of the austenite to the BCT structure. The technological importance of the hardening and the properties of martensite in steels stimulated studies to characterize this transformation. These investigations have shown that the martensite transformation occurs also in nonferrous alloy systems as well as in ceramics, and the product of this transformation in these systems is also called *martensite* because the transformation characteristics are the same as those in steels.

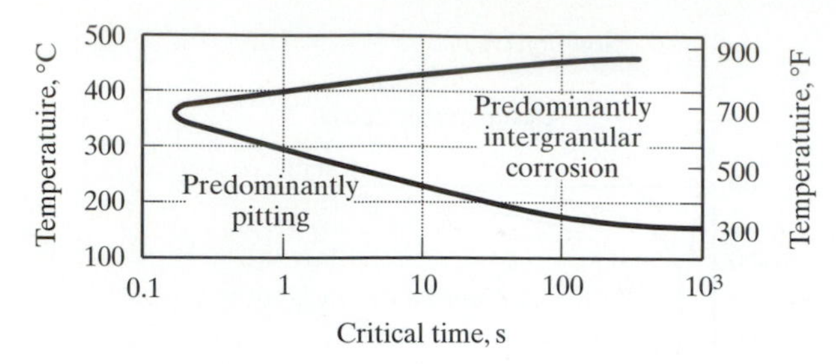

Fig. 9-16 Occurrence of intergranular corrosion in aluminum alloys when precipitation occurs during cooling or during isothermal transformation in the range indicated. [4]

The common *characteristics of the martensite transformation are:*

1. *Diffusionless.* The movement of atoms in the transformation is very short-range (i.e., less than one interatomic distance). This characteristic implies that the compositions of the original phase and the product of transformation are the same.
2. *Shear-like.* The transformation involves shear deformation and is therefore a *displacive* transformation. This characteristic makes possible the transformation of a large volume of the material at one time and because of the deformation, dislocations are generated that contribute to the strengthening, as in cold work. The shearlike mode involves the cooperative movement of a lot of atoms and because of this, the martensite transformation is also referred to as a *military transformation.*
3. *Change in the crystal structure of supersaturated solution.* The martensite transformation occurs only when there is a change in crystal structure as the original phase (supersaturated solution) transforms to the martensitic phase. In steels, the nonequilibrium martensite product with a BCC or BCT (body-centered tetragonal) structure, depending on the interstitial carbon content, is given the symbol α'.

The other characteristic most commonly cited is *athermal.* This means that the *transformation is not thermally activated and does not require time to transform.* Thus, once the material reaches a temperature and if the martensite transformation is favorable, it will occur immediately; this temperature is called the martensite start transformation temperature and is indicated as M_s. The martensite transformation as indicated in Fig. 9-12 occurs athermally (i.e., when the critical cooling rate is exceeded and the M_s is reached, the transformation will occur immediately). A certain amount of the product will form and will not increase with time at a given temperature. The amount of the martensite product will increase only if the temperature of the alloy is lowered further until the transformation finishes at the M_f, the martensite finish temperature.

However, *martensite is also observed to form isothermally, as shown in Fig. 9-17. Thus, athermal transformation is not a unique property* of the martensite transformation. The isothermal martensites also exhibit the three unique martensite transformation characteristics cited above and the C-curves shown in Fig. 9-17, which are indicative of a thermally activated process. The amount of transformed martensite as a function of time at a given temperature also exhibits the sigmoidal curve in Fig. 9-12.

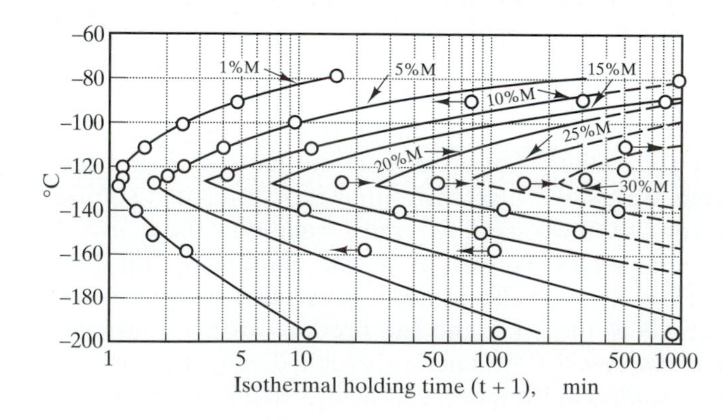

Fig. 9-17 C-curves for the isothermal transformation of martensite in an iron-nickel-manganese alloy. [19]

We shall now discuss the effect of composition on the properties of martensite. The strengthening derived during the martensitic transformation of most steels is due to the excess interstitial carbon retained in solution. For most steels, the strength of the as-quenched martensite is the highest and the aging or tempering indicated in Fig. 9-7 decreases the strength. For other materials containing no interstitial carbon in solution, the as-quenched martensite does not exhibit the strengthening attained with most steels. Included in this category are the nonferrous martensites, such as the titanium martensites described in Sec. 9.5, and the martensite produced with the maraging steels. For these, the strengthening comes during the subsequent aging or tempering, as is the case for precipitation hardening which we shall now discuss.

9.4 Precipitation Hardening

Having gone over the general principles of heat-treating, we shall now discuss the specific hardening processes and start with precipitation hardening. Historically, the accidental discovery of precipitation hardening occurred with aluminum alloys. The discovery was made in Germany when *Duralumin,* an aluminum-copper alloy, was retested for hardness weeks after it was laid aside in the laboratory. The retesting yielded much higher hardnesses. The first name given to the phenomenon was *age-hardening* (i.e., hardening as the alloy aged in time). Studies on the phenomenon revealed that it also occurred in other alloy systems and that the reason for the hardening was due to precipitates coming out of supersaturated solutions. Thus, the proper name for the phenomenon is *precipitation hardening,* although it is still referred to as age-hardening.

The basic requirements for an alloy to exhibit precipitation hardening are:

(1) That the alloy exhibits an increasing solubility of a solute or a second phase as the temperature increases,
(2) That the high-temperature condition, in which more solutes are in solution, can be quenched or frozen when the alloy is cooled to room temperature or lower. This entails that the C-curve for the alloy exhibits a reasonable, attainable critical cooling rate with available quench media. Since the quenched alloy contains more solute at room temperature than when it is in equilibrium, it is a supersaturated solution, is unstable, and will tend to precipitate out the excess solute or phase.
(3) That the precipitates can assume metastable (with respect to the equilibrium structure) transition structures that are coherent with the host lattice.

From the above requirements, the heat-treating process to achieve precipitation hardening consists of the following steps, schematically illustrated in Fig. 9-18, and follows the general principles indicated in Sec. 9.3:

(1) *Solution Anneal.* The term "solution" indicates that the alloy is heated to a temperature where more solute is put in solid solution. "Anneal" indicates that the heating also softens the alloy.
(2) *Quenching to form a supersaturated solution.* This is the most critical step in the sequence of

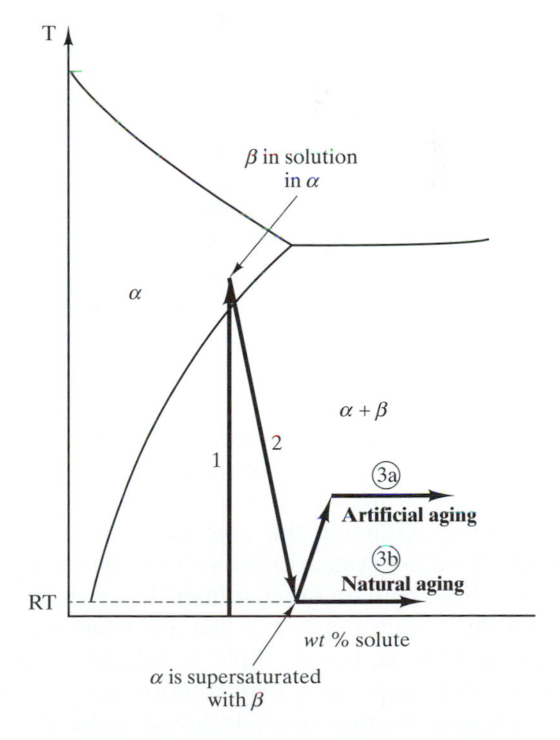

Fig. 9-18 Heat-treating steps in precipitation hardening.

heat-treating processes. The quenching rate must exceed the critical cooling rate to retain the composition at the solution anneal temperature and to form a supersaturated solution of the solute phase. This creates the driving force for the precipitation of the excess solute or phase.

(3) *Precipitation of excess solute or phase.* Hardening of the alloy is achieved by precipitating the excess solute or phase as a coherent metastable transient precipitate. *Hardening* arises from the lattice strain (*coherency strain*) induced by the coherent precipitate. When precipitation is done at room temperature, it is called *natural aging.* When it is done at higher temperatures, it is called *artificial aging.*

9.4.1 Aluminum Alloys

Since the discovery of the phenomenon was made with the aluminum-copper alloy, the results from early studies of this alloy have provided most of the basic understanding of precipitation hardening. The binary Al-Cu equilibrium phase diagram is shown in Fig. 9-19. Looking at the aluminum-rich end of the diagram, we see the equilibrium phases, α-(Al), the primary solid solution of aluminum, and θ, the secondary nonstoichiometric intermetallic phase, $CuAl_2$. The solvus indicates an increasing solubility of Cu in α-(Al) with increasing temperature from 0.20%Cu at 250°C to a maximum of 5.65%Cu at the eutectic melting temperature. (The solubility is considerably lower at temperatures below 250°C.) Accordingly, it should be possible to solution-anneal or heat-treat an Al-Cu alloy, for example, 3%Cu, at temperatures past the solvus, quench the alloy, and then precipitation harden it, according to the steps indicated above.

In solution-annealing, we guard against overheating and underheating. In overheating, the solution-anneal temperature used is over the eutectic melting temperature. Although the eutectic temperature of the Al-Cu binary system is shown to be 548°C, commercial alloys contain other solutes than copper and these form complex eutectics that are actually much lower than 548°C. For example, a 2024 Al-Cu alloy also contains magnesium as an addition and the complex eutectic melting point is found to be 502°C; therefore, the solution temperature for this alloy is 495°C. Without magnesium, the Al-Cu alloys may be solution treated up to 535°C. If overheated and appreciable eutectic melting were to occur, the tensile strength, ductility, and fracture toughness will be impaired. The phenomenon of eutectic melting is called burning and the liquid (which spreads along the grain boundaries) deteriorates the properties mentioned. We see therefore that the range of solution anneal temperatures for Al-Cu alloys shown in Fig. 9-8a is from 495° to 535°C. Because of the lower 495°C limit for solution anneal, the maximum heat-treatable composition of copper in aluminum is 4 percent.

In underheating, the solution anneal temperature is below the solvus of the alloy so that the solution of the second phase particle is incomplete. If underheating occurs, the expected strength of the alloy will not be achieved because not all the solute phases were put into solution. Underheating may also occur if insufficient time is given for the solute to dissolve.

Quenching of the solution annealed alloy is the most critical step of the hardening process. The cool-

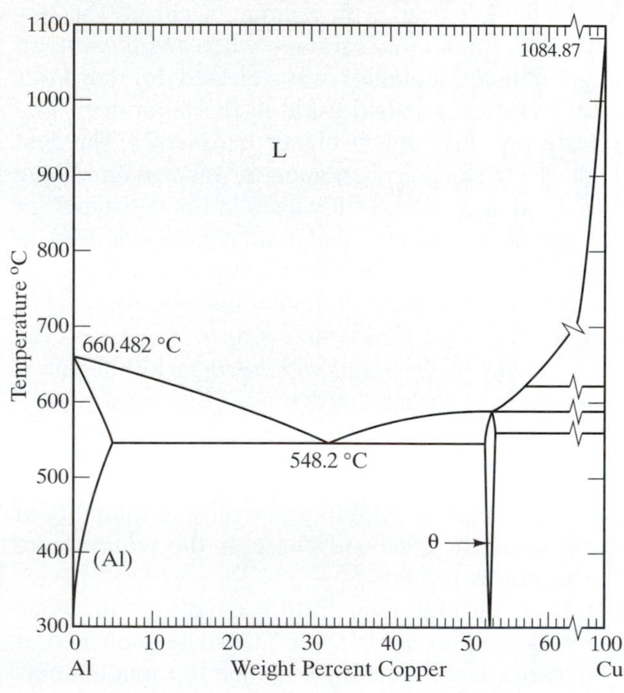

Fig. 9-19 Equilibrium binary aluminum-copper phase diagram. [10]

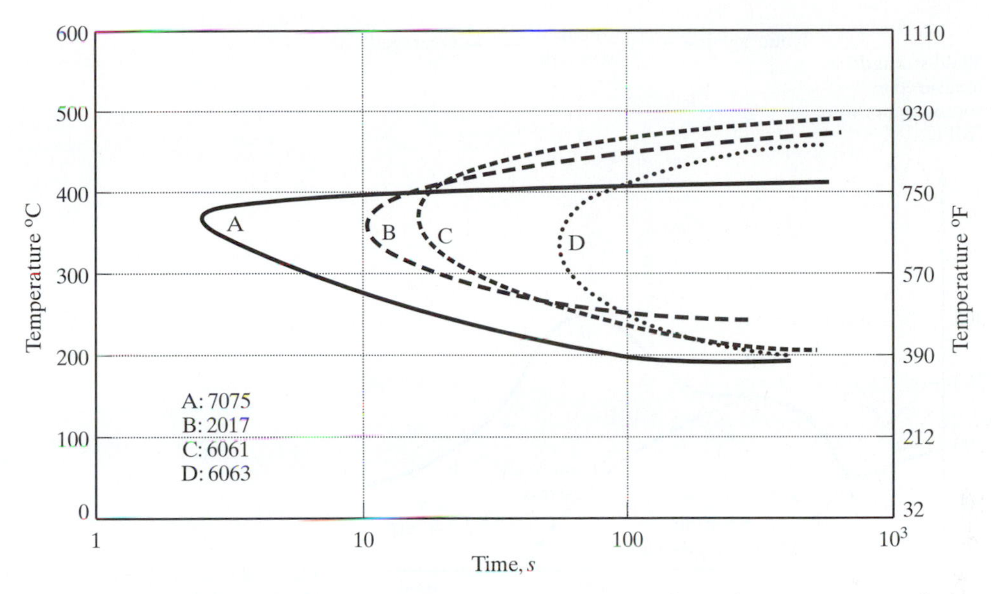

Fig. 9-20 C-curves of 7075, 2017, 6061 and 6063 aluminum alloys showing 7075 needs the most drastic quench and 6063 the least severe quench. [4]

ing rate must exceed the critical cooling rate past the nose of the C-curve in order *to retain all of the solute in solution.* The C-curves are also dependent on the type and amount of solute (alloying) present. Figure 9-20 demonstrates the C-curves of some aluminum alloys. From these curves, the 7075 (A) alloy requires the most drastic or severe quench in order to retain the solute atoms in solid solution and is considered the most sensitive alloy to cooling rate. Alloy D (6063) requires the slowest cooling to retain the solid solution and is considered the least sensitive alloy to cooling rate. The fastest cooling rate is needed in the range of 400° to 290°C ("nose" of the C-curve) where the fastest precipitation of solute occurs during cooling. If there is precipitation during cooling, the precipitates that form do not contribute to hardening and their potential contribution to strength is lost. This was shown in Fig.9-15 which shows the effect of the average cooling rate in the range of 400 °to 290°C for different alloys. In addition, intergranular corrosion is induced as shown in Fig. 9-16.

At the aging temperatures, natural or artificial, the stable equilibrium phases are α(Al) and θ($CuAl_2$); however, the stable θ phase does not precipitate immediately. Rather, the precipitation proceeds in stages that require time to accomplish. The first stage is the formation of solute(Cu)-rich clusters of atoms within the parent or host (Al) lattice. The clusters fit in the crystal structure of the parent (Al) lattice and are completely coherent with it. These clusters are called GP (Guinier-Preston) zones because they were first detected independently by Guinier and by Preston. At temperatures less than 180°C, two types of these clusters were detected and were thus named as GP1 and GP2 zones. The GP2 zone is also referred to as θ'' precipitate. Peak hardness or strength is achieved when θ'' is formed. Above 180°C, the θ'' (GP2) does not form in Al-4%Cu. Another metasble precipitate, θ', forms before the stable θ is observed. When the stable θ forms, the alloy is said to be *overaged* and the strength decreases. These precipitation stages are illustrated in Fig. 9-21 and are summarized by the precipitation sequence,

$$\alpha_{sss} \rightarrow \alpha + GP1 \rightarrow \alpha + \theta''(GP2) \rightarrow \alpha + \theta' \rightarrow \alpha + \theta$$

Because the clusters are embedded in the crystal structure of the parent metal and because of the atomic size difference between solute and solvent atoms, the coherent structure induces a lot of strain on the lattice. This is referred to as coherency strains and is reflected in an increase in hardness and strength. The strengthening comes from the particles acting as obstacles to dislocation motion, as shown in Fig. 9-22.

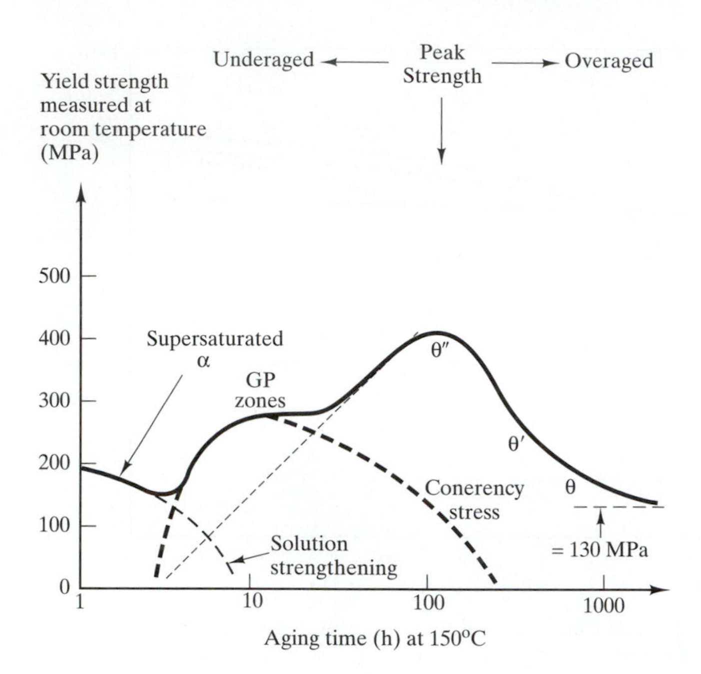

Fig. 9-21 Changes in the yield strength of Al-4 wt%Cu during aging at 150°C. [8]

Precipitation heat treatments of aluminum alloys are generally low-temperature long-term processes. Temperatures range from 115° to 190°C and times may vary from 5 to 48 hrs. Precipitation hardening (aging) curves show changes in hardness or strength with time at constant temperature. Figure 9-23 demonstrates the characteristics of these aging curves for the 2014 (aluminum-copper-magnesium-silicon) alloy. The aging curves indicate an initial decrease in strengths at the start of the isothermal aging. This is referred to as reversion and is due to the initial dissolution of small GP zones. The rehardening or increase in strength is ascribed to the formation of stable GP zones. The peak strengths are highest at low aging temperatures but it takes very long to attain them. Note that at the lowest aging temperature (105°C) indicated, the time to reach the peak strength approaches one year. Thus, the attainment of properties usually involves the compromise of attainable strength within a reasonable time at the aging temperature.

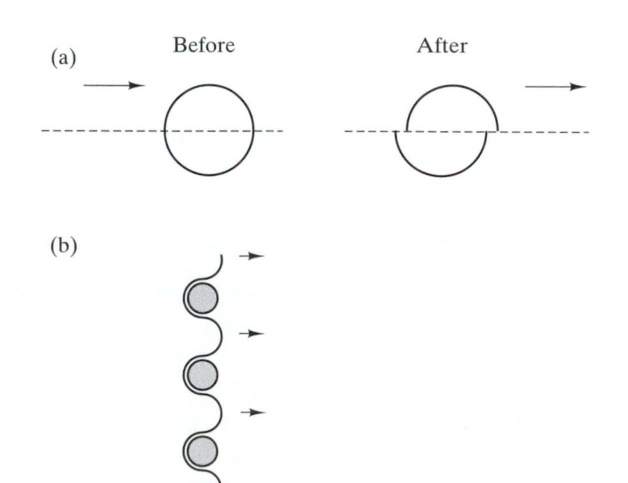

Fig. 9-22 Precipitate particles block or hinder dislocation motion. Dislocation (a) cuts through, and (b) bows around the particles. [8]

If the precipitation process does not produce the coherent transition stages, no amount of precipitation will exhibit hardening or strengthening. Alloys exhibiting precipitation but no hardening are called nonheat-treatable. Binary aluminum-silicon and aluminum-manganese alloys exhibit considerable precipitation when heat-treated, but the observed changes in mechanical properties are relatively insignificant and are thus non-heat-treatable alloys. The major heat-treatable aluminum alloy systems are:

- Aluminum-copper system with strengthening from $CuAl_2$, (2XXX)
- Aluminum-magnesium-silicon system with strengthening from Mg_2Si, (6XXX)

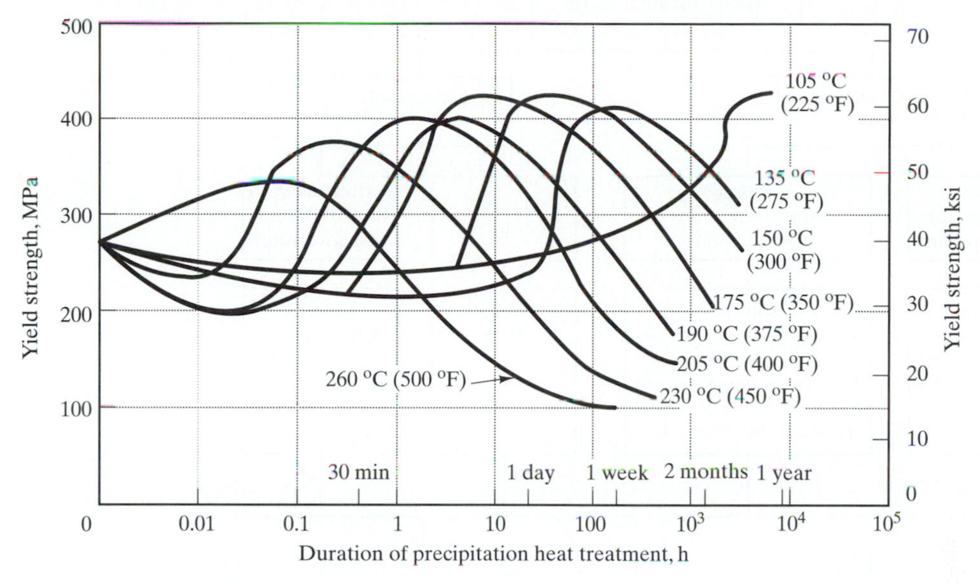

Fig. 9-23 Aging curves at various temperatures showing the changes in yield strength of 2014 aluminum alloy. [4]

- Aluminum-magnesium-zinc system with strengthening from $MgZn_2$, (7XXX)
- Aluminum-lithium system with strengthening from Al_3Li, a metastable, ordered σ' phase.

The aluminum-lithium system is an interesting system because it hardens as well as exhibits an increase in modulus of elasticity and a decrease in density (see Sec. 13.2.3).

After solution annealing, the alloy is in its dead-soft condition and the as-quenched condition provides the best opportunity to deform the alloy to any shape. During quenching, it is possible to induce warping and distortion. Therefore, in practice, cold-working, forming, shaping, and straightening follow the quenching process almost immediately and are completely finished before aging hardens the material. If scheduling makes this impossible, aging may be avoided by refrigerating to subzero temperatures. For example, it is conventional practice to refrigerate 2024 (Al-Cu) rivets before their use to maintain good driving characteristics. Full-size wing plates for jet aircrafts are solution-treated and quenched at the supplier's plant, packed in dry ice in specially designed shipping containers, and transported about 2,000 miles to aircraft manufacturers for forming.

9.5 Heat-Treating of Titanium Alloys [6]

Pure, unalloyed titanium is allotropic. At 885°C, the low temperature α-phase HCP crystal structure of pure titanium transforms to the high-temperature, β-phase, BCC structure. When the titanium is alloyed with another metal, the transformation of α to β occurs over a range of temperature and a two-phase ($\alpha + \beta$) field is formed. The alloying elements are classified as either α or β stabilizers depending on whether they raise the α to ($\alpha + \beta$) transus or lower the β to ($\alpha + \beta$) transus. (The transus is the transformation point. In terms of the binary alloys involving liquid and solid, the α to ($\alpha + \beta$) transus is the solidus and the β to ($\alpha + \beta$) transus is the liquidus. The plural of transus is transi.) Figure 9-24 shows this classification scheme of binary titanium alloys according to the alloying elements (solutes) raising the α to ($\alpha + \beta$) transus (α-stabilized) or lowering the β to ($\alpha + \beta$) transus (β-stabilized).

Most of the technical commercial titanium alloys are not simple binary, however, but rather multi-component alloys containing at least two solutes. They are classified primarily as *α and near α*, *$\alpha + \beta$*, *or β (metastable) and near β alloys* based on the

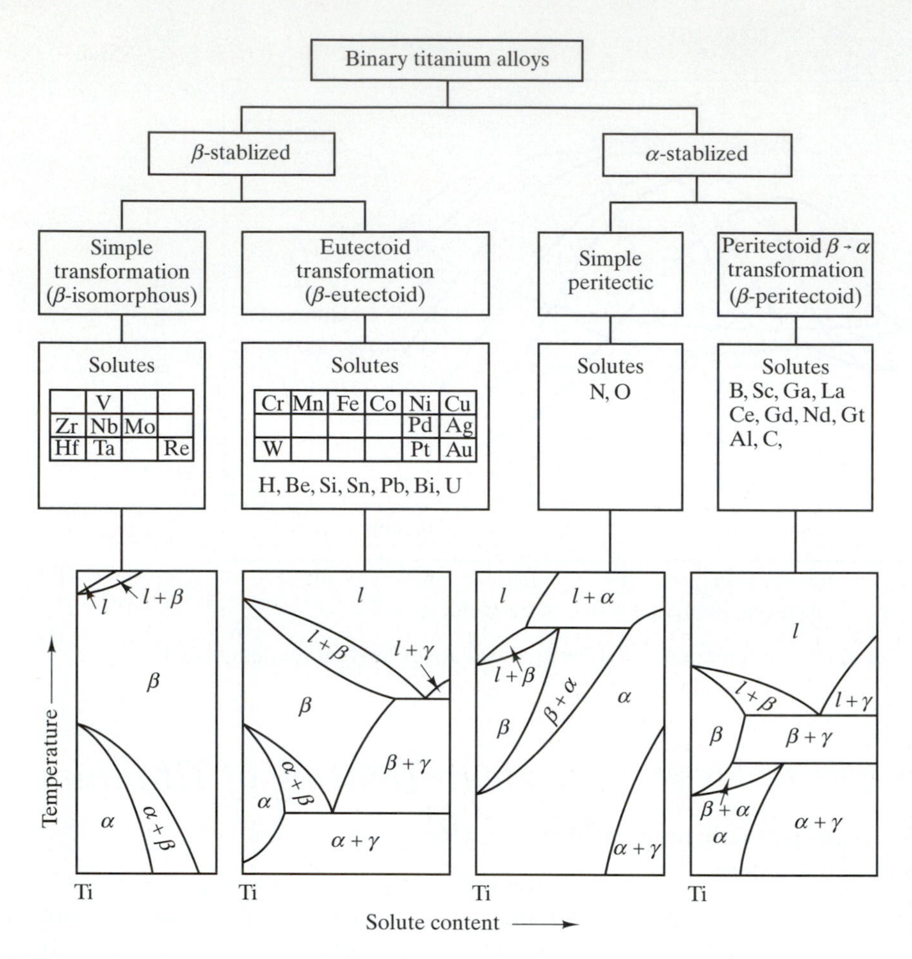

Fig. 9-24 A classification scheme for binary titanium alloy phase diagrams—α is HCP solid-solution alloys, β is BCC solid-solution alloys, and γ represents an intermetallic compound. [6]

predominance of the phases at room temperature. These classifications are shown in Table 9-2 and may be conveniently depicted in a pseudo-binary diagram showing the α, α + β, or β phases in Fig. 9-25 where the alloys are located in the respective phase fields. The number before each element's symbol indicates its weight percent. Thus, alpha and near-alpha alloys contain predominantly the α (HCP) phase at room temperature; they undergo only the nonhardening thermal processes of stress-relief and annealing and cannot be hardened by any type of heat treatment.

The commercial β alloys contain, in reality, a mostly metastable β phase and are strengthened when the metastable beta phase decomposes during aging after solution treatment. For these alloys, the aging and stress-relief treatments can be combined, and the annealing and solution treatment may be done at the same time. When these alloys are hot-worked and air-cooled, they are in a condition comparable to a solution-treated state and may not require solution treatment. When required, solution treatment at 790°C produces good uniform properties and after aging at 480°C for 8 to 60h, tensile strengths of 1.10 to 1.38 GPa (160 to 200 ksi) can be obtained. Aging for times longer than 60h may provide higher strengths, but at the expense of ductility and fracture toughness; if the alloy contains chromium, titanium-chromium compounds are formed. Slow-heating rate to the aging temperature may produce a significant increase in strength relative to placing the beta alloys directly in a hot furnace. This is due to the formation of a very fine alpha structure or a transitional "omega" phase. The omega phase is, however, undesirable because it reduces the ductility of the alloy.

Most of the significant heat treatable titanium alloys are the alpha-beta alloys and a summary of heat

Table 9-2 Classification of multicomponent titanium alloys. [6]

Composition, wt%	Classification
Ti-5Al-2.5Sn	α
Ti-8Al-1Mo-1V Ti-6Al-2Sn-4Zr-2Mo	Near-α
Ti-6Al-4V Ti-6Al-6V-2Sn Ti-3Al-2.5V	α + β
Ti-6Al-2Sn-4Zr-6Mo Ti-5Al-2Sn-2Zr-4Cr-4Mo Ti-10V-2Fe-3Al	Near-β
Ti-13V-11Cr-3Al Ti-15V-3Cr-3Al-3Sn Ti-3Al-8V-6Cr-4Mo-4Zr Ti-8Mo-8V-2Fe-3Al Ti-11.5Mo-6Zr-4.5Sn	β

treatments for α + β alloys are given in Table 9-3. The convenience of Fig. 9-25 comes to fore when it is used to illustrate the heat treating of α + β titanium alloys, which is based on the instability of the high temperature β phase at lower temperatures. These alloys can be solution treated to the single β phase if the temperature is increased above the β transus curve. However, solution annealing of these alloys are seldom brought to the single β phase field, as done in the Al-Cu alloys. Experience has indicated that when the temperature is in the single-phase field, the resulting tensile properties deteriorate, especially the ductility, and subsequent heat treatment cannot restore the properties.

Normally, solution annealing is done about 28° to 83°C below the transus temperature, as illustrated in Fig. 9-26. When we apply the lever rule we see that there is more of the β phase than α at the solution anneal temperature; this is now the reverse of the case at the low temperature where α is more than β. The α at the solution anneal temperature is called *primary* α. When the alloy is cooled to a temperature

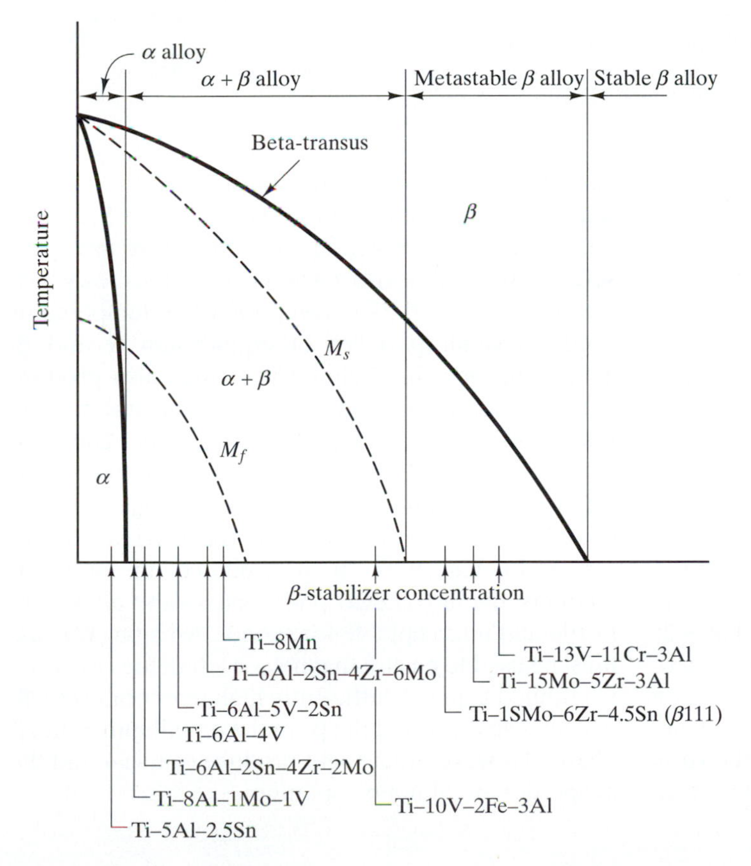

Fig. 9-25 Alloys in Table 9-2 are indicated in a pseudo-binary β-isomorphous phase diagram. [6]

Table 9-3 Heat treatments of α-β alloys. [6]

Heat-treatment designation	Heat-treatment cycle	Microstructure
Duplex anneal (or overage)	Solution treat at 50–75°C (90–135°F) below T_β(a)*, air cool and age for 2–8 h at 540–675°C (1000–1250°F)	Primary α, plus Widmanstätten α–β regions
Solution treat and age	Solution treat at 40°C (70°F) below T_β, water quench (b)* and age for 2–8 h at 535–675°C (996–1250°F)	Primary α, plus tempered α′ or α β–α mixture
Beta anneal	Solution treat at 15°C (30°F) above T_β, air cool and stabilize at 650–760°C (1200–1400°F) for 2h	Widmanstätten α–β colonies
Beta quench (Beta STOA —solution treated and overaged)	Solution treat at 15°C (30°F) above T_β, water quench and temper at 650–760°C (1200–1400°F) for 2h	Tempered α′
Recrystallization anneal	925°C (1700°F) for 4h, cool at 50°C/h (90°F/h) to 760°C (1400°F), air cool	Equi-axed α with β at grain-boundary triple points
Mill anneal	α–β hot work plus anneal at 705°C (1300°F) for 30 min to several hours and air cool	Incompletely recrystallized α with a small volume fraction of small β particles

*(a) T_β is the beta transus temperature for the particular alloy in question.
(b) In more heavily β-stabilized alloys such as Ti-6Al-2Sn-4Zr-6Mo or Ti-6Al-6V-2Sn, solution treatment may be followed by air cooling. Subsequent aging causes precipitation of α phase to form an α–β mixture.

lower than 650°C, there will be an excess β phase (over the equilibrium amount) that will precipitate α. The precipitation of α from the β is a nucleation and growth process and the precipitate is incoherent. In spite of the incoherency, the precipitation of α causes strengthening of the alloy. The strengthening arises from the mutual deformation of both the α and the β phases during the precipitation process. The α is deformed by the β matrix when it precipitates out from the β matrix; the latter is, in turn, deformed. The deformed phases act as obstacles for dislocation motion.

When the solution annealed material is quenched, the high-temperature β phase may transform to martensite as it crosses the M_s line shown in Fig. 9-25 and Fig. 9-26. There are two types of martensite, α′ (hexagonal) and α″ (orthorhombic), that can form from the high temperature β phase. Both of these are symbolized by α^m in Fig. 9-26. The M_s line shown in both figures represent the transformation start temperatures. The martensite transformation has the same characteristics as those mentioned in Sec. 9.3.2 but the α^m martensites in titanium alloys are quite soft. Strengthening arises when the martensites, which are supersaturated βs, decompose, when tempered or aged, to small particles of equilibrium α and β. Tempering or aging below 550°C may also produce the transition complex hexagonal ω (omega) precipitate and β′, which is a β phase with a leaner composition of the β-stabilizer element. The formation of β′ is referred to as phase separation and the regions where ω and β′ may form relative to the martensites are shown in Fig. 9-27. If precipitated in sufficient amounts, the ω (omega) phase makes the alloy very brittle and unacceptable for service. An aging process must be used to ensure that no ω is left in the structure. Continued aging at sufficiently high temperatures will decompose the ω and the β′ to the equilibium α and β phases. However, this might produce coarse α and the properties might not be optimum.

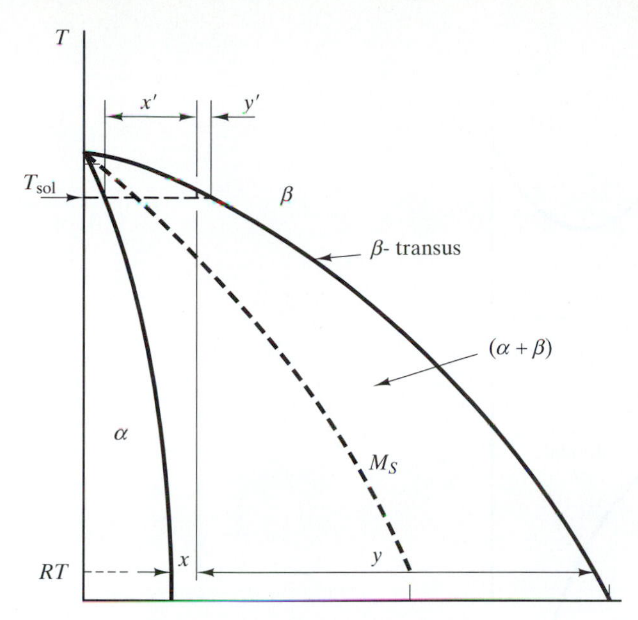

Fig. 9-26 Schematic of heat-treating titanium alloys below their transi temperatures in a pseudo-binary β-isomorphous diagram. x/y is relative amount of β at RT and x′/y′ is relative amount at the solution temperature.

9.6 Hardening Due to Spinodal and Order Transformations

One of the temper designations for copper alloys described in Sec. 13.5.1 refers to a spinodal heat-treated condition. We shall learn briefly about spinodal decomposition and ordering in alloys.

The characteristic feature of an alloy that will undergo the spinodal transformation is the presence of a solubility gap (a domelike boundary) in the phase diagram, as shown in Fig. 9-28b. At any temperature T_1 within the dome, two solid solutions α_1 and α_2 co-exist with compositions C_1 and C_2. The free energy curve of this alloy at T_1 is shown in Fig. 9-28a. The minima in this curve occur at the compositions of the stable phases at C_1 and C_2. The free energy curve also shows two inflection points and the compositions at these points are noted and are indicated inside the dome as shown. When the compositions at the inflection points are determined for different temperatures, a *boundary within the dome is established that is called the chemical spinodal.* A supersaturated alloy that will undergo spinodal transformation must have a composition within the chemical spinodal boundary at the temperature of transformation.

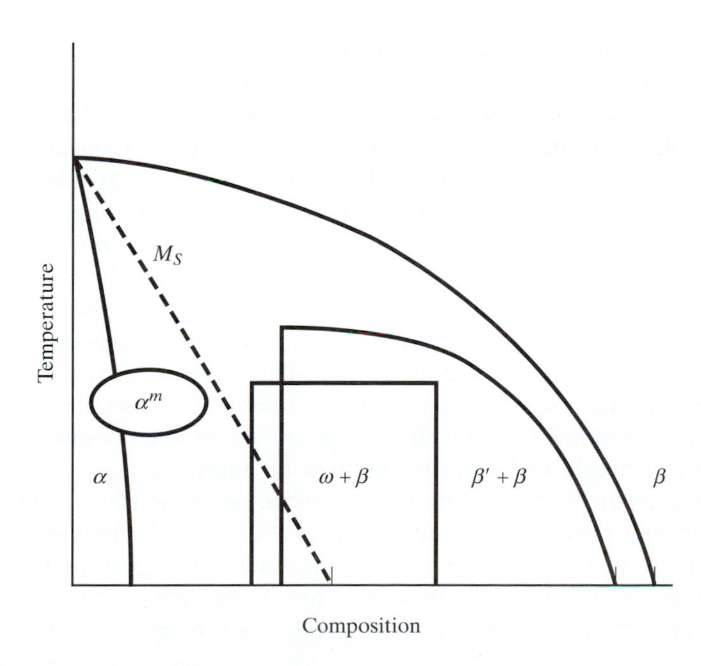

Fig. 9-27 Schematic of location of metastable phases (ω + β) and (β′ + β) phases in β-isomorphous diagram.

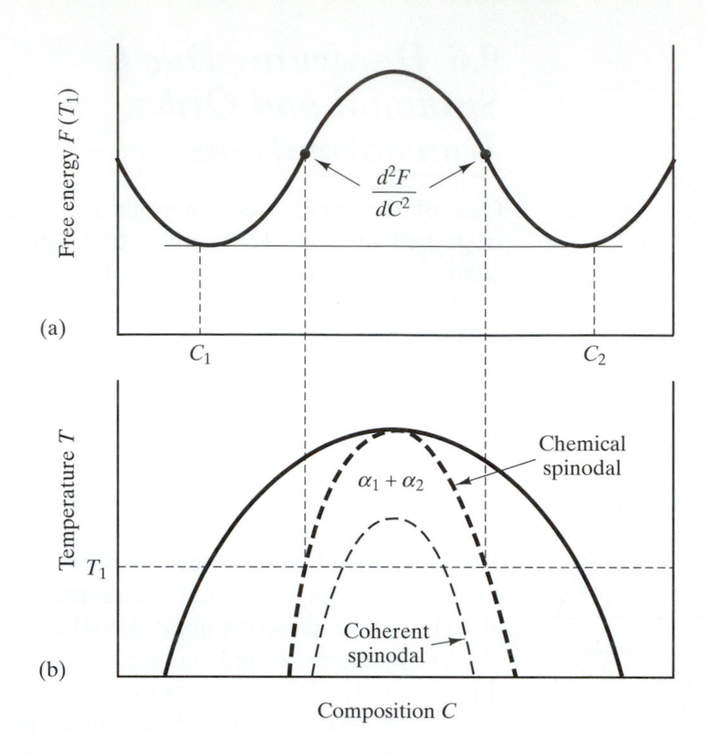

Fig. 9-28 Features of an alloy system undergoing spinodal decomposition. [11]

The heat treatment for hardening due to *spinodal transformation* is basically the same as in precipitation hardening (i.e., solution treatment above the dome, quenching, and aging within the chemical spinodal boundary). The hardening arises from the very fine structure (observable only by transmission electron microscopy) that develops when an alloy of *supersaturated composition,* C_o, *decomposes or transforms to a modulated structure of two solid solutions of compositions,* C_1 *and* C_2. The composition of the solute is lower in C_1 and higher in C_2 than the original composition, C_o. The stages of this transformation are indicated in Fig. 9-29. The supersaturated solution is indicated in (a); the progress of the transformation proceeds to (b), then to (c), and finally (d) which is also a representation of modulated composition and structure.

Thus, the heat treating for spinodal hardening consists of solution annealing an alloy of composition C_o within the chemical spinodal boundary (ChSB) to a temperature above the dome, quenching to a temperature within the ChSB, and then aging. The spinodal decomposition will proceed as illustrated in Fig. 9-29 to produce a modulated structure in which the *wavelength of the composition modulation is* λ. As the process proceeds, an *uphill diffusion* of solute atoms (indicated by arrows in Fig. 9-29) occurs in which the regions richer in solute than the original composition become richer, and the leaner regions become leaner, until the equilibrium compositions, C_2 and C_1, are attained at the respective regions. When λ is short, the concentration gradient is too sharp; when λ becomes longer, the diffusion distance becomes longer. *In the limit* $\lambda = \infty$, the chemical volume free energy balances the strain energy that results from the transformation. This condition *is depicted by another boundary within the ChSB which is called coherent spinodal boundary (CoSB).* The λ decreases as the decomposition (aging) temperature is lowered from the CoSB. A λ value in the range of 5–10nm is considered favorable and may be considered a *nanostructured material,* in using current terminology. Examples of spinodal hardening alloys are Cu-Ni alloys (about 10-30%Ni) with chromium or tin additions.

An *ordered structure* is one in which atoms of A and B occupy specific locations in the lattice, now called a *superlattice,* as opposed to being in random locations in a solid solution. Examples of an ordered and a disordered crystal structure are shown in Fig. 9-30.

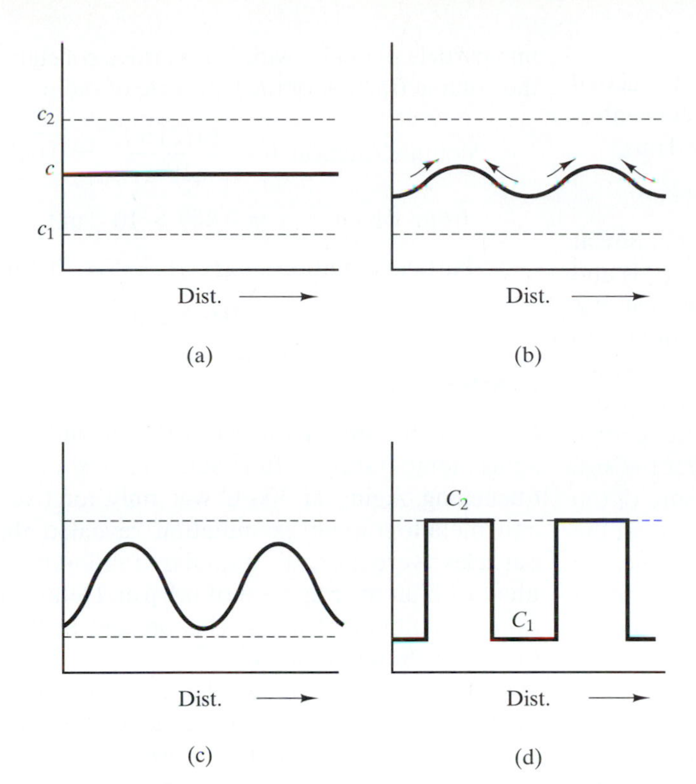

Fig. 9-29 Spinodal decomposition of an alloy with composition c, into two phases of compositions c_1 and c_2. [11]

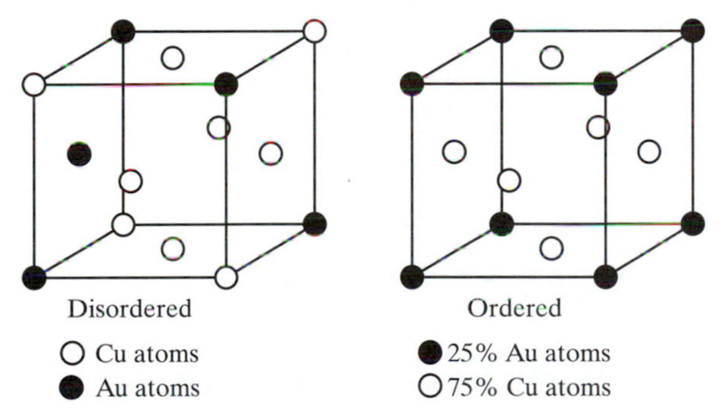

Fig. 9-30 Crystal structures—disordered (left) and ordered (also called superlattice)—Cu-25a/oAu. [12]

The dislocation in the ordered structure is no longer a single line, but rather a pair of dislocations called *superdislocation* and it is this pair that moves in the superlattice. Thus, when localized regions have ordered structure, they block the dislocation motion and make the material stronger. An example of localized ordered structures is the ordered γ'-Ni_3Al particles in nickel-based superalloys used in high temperature applications, such as in jet engines. The γ' is an equilibrium second phase in both the binary Ni-Al and the ternary Ni-Cr-Al systems and a metastable phase in the Ni-Ti and Ni-Cr-Ti systems. During the heat-treating of the $\alpha + \beta$ titanium alloy, Ti-6Al-4V, an ordered Ti_3Al may also form. It is thus not surprising that strengthening is achieved when Ni, Al, and Ti are simultaneously present in a composition because either Ni_3Al or Ti_3Al is produced. This is the basis of strengthening for both the maraging and the PH stainless steels. Copper alloys containing Al with Ni, Fe, Zn, and/or Co are also quoted as being hardened by ordering of solute atoms.

EXAMPLE 9-1

An Al—4wt%Cu was solution heat-treated, quenched to room temperature, and was artificially overaged at 300°C for 24 hours. Metallographic examination revealed that the θ precipitate particles were spheroids and were on the average 1 μm apart. The density of the θ particle is 4.26 g/cm^3, and that of aluminum is 2.7 g/cm^3. (a) What was the initial supersaturation of the θ phase when aging started? (b) Determine the average size of the θ particles.

SOLUTION:

(a) The initial supersaturation of the θ is the actual concentration of 4 wt% minus the solvus concentration at 300°C that is 0.5 wt% from Fig. 9-19. Thus,

$$\text{Supersaturation} = 4.0 - 0.5 = 3.5 \text{ wt\% } \theta$$

(b) After artificial aging of the aluminum alloy at 300°C for 24 hours, we can assume that the (Al) and θ phases have attained equilibrium. Thus, we can use the equilibrium diagram in Fig. 9-19. We should stop and think that we are being asked for the average size of the precipitates. From our knowledge of phase diagram and the lever rule, we can obtain the relative weight fraction of the phases. In order to get the size of the particles, we must obtain some measure of the volume of the precipitates. Thus, we must convert the weight fraction to volume fraction of the phases.
At 300°C, the tie-line with the over-all composition of the alloy is shown in Fig. EP9-1. Thus,

$$\text{Weight fraction, } \theta = \frac{4 - 0.5}{53.3 - 0.5} = 0.0663$$

$$\text{Volume fraction, } \theta = \frac{0.0663/4.26}{0.0663/4.26 + 0.9337/2.17} = 0.0431$$

From the average distance or interparticle spacing, we can represent the distribution as that of a cubic lattice, in Fig. EP 9-1. For this lattice, there is an average of one particle per cube with 1 μm lattice constant. Thus, the volume fraction of the θ particle of radius, *r*, is

$$\text{Volume fraction, } \theta = \frac{(4/3)\pi r^3}{(10^{-6})^3} = 0.0431$$

$$\text{from which} \quad r = 0.469 \times 10^{-6} \text{ m3}$$

$$\text{Particle diameter} = 2r = 0.938 \times 10^{-6} \text{ m} = 0.938 \ \mu\text{m}.$$

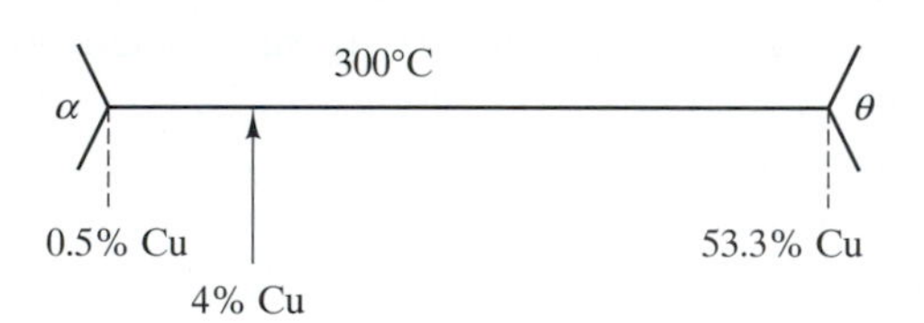

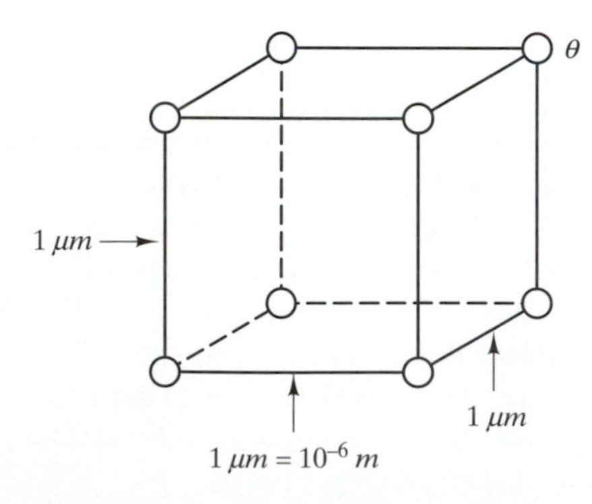

Fig. EP 9-1.

EXAMPLE 9-2

Consider the same alloy, Al-4wt%Cu, and the same aging temperature after solution treatment and quenching. Aging at 300°C was only for five hours and metallographic examination revealed that the particles were already globular, randomly distributed, with an average size of 0.1 μm. Determine the average number of θ particles present in the structure per cubic centimeter?

SOLUTION: Even though the aging was done only for five hours at 300°C, we can assume equilibrium conditions for calculation purposes. Thus, from Example 9-1, the volume fraction of θ is 0.0431. Since we are being asked, the number of particles per cubic centimeter, we can assume a cubic centimeter volume of the alloy. Then the volume of all the θ particles in 1 cm^3 equals 0.0431 cm^3. If n = number of particles,

$$n \cdot \frac{4}{3}\pi \left(\frac{10^{-5}}{2} \text{ cm}\right)^3 = 0.0431$$

$$n = 8.23 \times 10^{13} \text{ particles/cm}^3$$

9.7 Heat-Treating of Steels

There are two thermal hardening processes that are used for steels and these are:

(1) *Through Hardening Processes* that produce hardening through the entire cross section of the material.
(2) *Surface Hardening Processes* that harden only the surface.

In the heat-treating of steels, the hardening depends on having the carbides at room temperature (both cementite or iron carbide, Fe_3C, and alloy carbides) dissolved in austenite at the austenitizing temperature (the solution heat-treating temperature).

As indicated in Secs. 2.4.4 and 5.8.1, the solution of the carbides in austenite is due to its structure (FCC) being able to accommodate more carbon atoms because it has larger interstitial holes than the ferrite (BCC) phase. When the austenite is quickly cooled fast below the critical temperatures (Sec. 9.2.2 and Fig. 9-3), it becomes an unstable supersaturated solid solution that would want to transform to the equilibrim phases, ferrite and cementite (carbide). We can control the strength of the ferrite-carbide aggregate by controlling the isothermal temperature at which we transform the austenite. When the austenite is cooled below the M_s, the martensite start temperature, a supersaturated body-centered tetragonal (BCT) structure (called martensite) forms. The martensite structure is the hardest that can be produced with steels and in order to get the maximum hardening, the critical cooling rate for the steel must be exceeded and all the austenite must transform to martensite. We can roughly estimate this critical cooling rate from the I-T diagram (Sec. 9.3.1) of the steel. Actually, since the steel is continuously cooled from the austenitizing temperature, the critical cooling rate is better indicated by the continuous-cooling transformation (CCT) diagram. Both the I-T and the CCT diagrams are the nonequilibrium diagrams that are actually consulted when we heat treat steels. Thus, they are the actual processing diagrams, that we refer to during heat-treating not the equilibrium phase diagram. Therefore, we start with a discusssion on the development of these diagrams to understand them and to enable us to use them effectively during heat-treating.

9.7.1 Nonequilibrium Processing Diagrams

The equilibrium iron-carbon phase diagrams discussed in Sec. 5.8 are meant only to be rough guides. Actual heat-treating conditions cannot be done in equilibrium (i.e., very slow heating and cooling). Thus, it is important to have nonequilibrium diagrams that show the phases present during nonequilibrium cooling conditions.

9.7.1a Isothermal Transformation (I-T) Diagrams (C-Curves) of Steels. The simplest of the steel I-T diagrams is that of the eutectoid composition, shown as the bottom figure in Fig. 9-12 and with labels in Fig. 9-31. The ordinate on the left is the temperature and the abscissa is in [log (time), in seconds] to accommodate very short and very long times of transformation. Times of one minute, one hour, one day, and one week are indicated for comparison. The horizontal line A_s is the eutectoid temperature at 727°C shown in Fig. 9-3. The *A* indicates austenite and the subscript *s* signifies "start" and A_s means "start of austenite formation." Above this line, the steel is in the stable austenite region indicated by *A*. The development of a typical I-T diagram, as in Fig. 9-31, consists of heating several small specimens to the single phase austenite field *A* and then quenching them to an intermediate temperature below the equilibrium line(s), in this case, A_s for the eutectoid steel. Below 727°C, the austenite (A) becomes unstable, and if held at a constant (isothermal) temperature, will eventually transform to the stable ferrite (F) and iron carbide (C) phases. After a certain time at the intermediate temperature, one specimen is quenched to room temperature and then examined metallographically for the presence of the F and C phases. After another time, another specimen is quenched and examined. The time when the first F and C phases appear is noted as the start of transformation at this temperature and the time when all the austenite has transformed to the F and C phases is noted as the "finish of transformation." The first (left) C-curve is the trace of all the times at different temperatures when the start of the transformation of A $\Rightarrow$ F + C occurs. Likewise, the right C-curve is the trace of all the times at different temperatures when all the austenite has transformed to F + C. In between these two curves, the three phases, A + F + C, co-exist because the transformation is not complete. The field to the right of the second C-curve is marked only F + C because all the austenite has completely transformed to those stable phases. The area to the left of the first C-curve is marked A because the unstable austenite has yet to transform. Photos 11 through 16 in the Appendix form a series that illustrates the isothermal transformation of 1080 steel from austenite to pearlite. [16]

Figure 9-31 (the I-T diagram) indicates the different forms of the transformation products from austenite at various times and temperatures. For this reason, it is also referred to as the time-temperature-transformation (T-T-T) diagram. The F + C (stable phases) aggregate takes different forms or morphology, depending on whether the transformation is

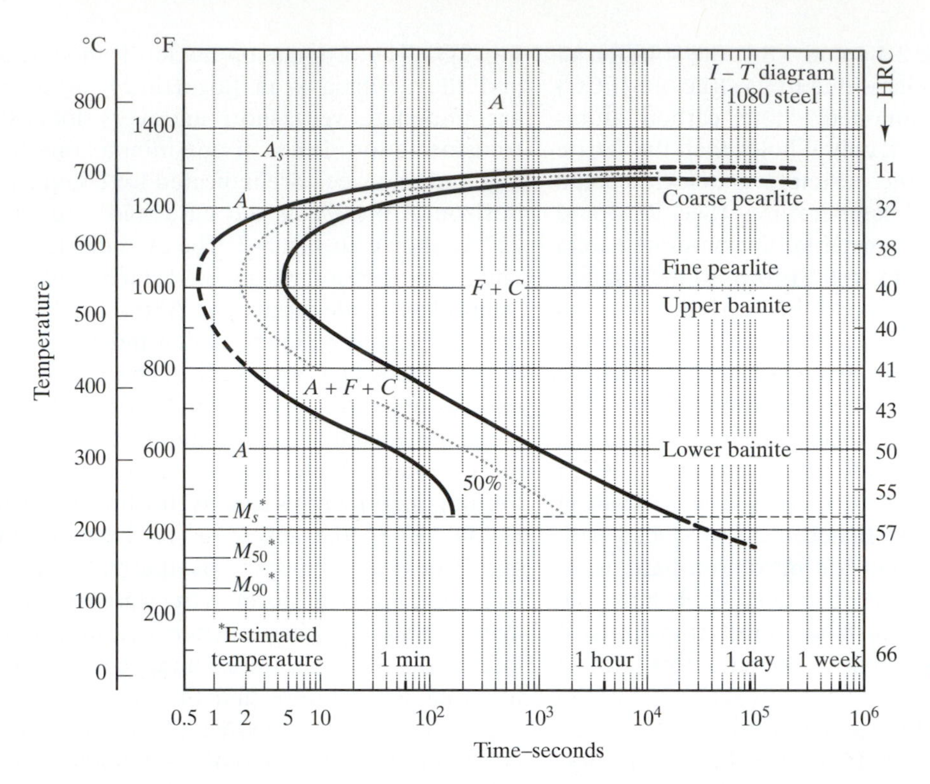

Fig. 9-31 Isothermal transformation diagram of a nominal eutectoid (1080) steel. [9]

done at temperatures above the nose or below the nose. *Above the nose,* the aggregate is in the *lamellar form of* F + C called *pearlite* in Sec. 5.8—*below the nose,* the aggregate is in the *nonlamellar form* of F + C called *bainite.* The pearlite that is formed very close to the A_s is *coarse pearlite* and closely resembles the equilibrium structure. The pearlite that is formed near the nose is *fine pearlite.* The difference between the two morphologies is the spacing between the two microconstituents. A pearlite is coarse when the ferrite (F) can be clearly delineated from the carbide (C) at 500x magnification or less. If they are not resolved, the morphology is fine pearlite. The bainite that is formed just below the nose is called *upper bainite* and that formed at the lower end of the C curves is called *lower bainite.* The morphology of upper bainite consists of very fine elongated carbide particles (unresolvable in optical microscopy) between laths of ferrite. The cluster of ferrite laths appears feathery and is the identifying feature of upper bainite in optical micrographs (photographs of structures using optical microscopes). The morphology of lower bainite resembles very closely a tempered martensite structure with the carbides being located within and oriented about 55° with the long axis of the acicular or platelike ferrites. Photos 17 and 18 in the Appendix illustrate coarse pearlite, Photo 19 fine pearlite, Photo 20 upper bainite, Photo 21 lower bainite and Photo 22 the martensite in 1080 steel. [16]

The isothermal transformation is a nucleation and growth process. Thus, the sizes of the ferrite and carbide microconstituents decrease as the isothermal transformation temperature decreases following the discussion we had in Sec. 9.3.1. This is reflected in *the hardnesses of the 100 percent isothermally formed aggregates shown at the right ordinate of Fig. 9-31*—the finer the structure, the harder the material. The strengthening may be thought of as the refinement (fine grain) of the ferrite grain size. The finest structure produced isothermally is the lower bainite and it exhibits the highest hardness of about 56 HRC (Rockwell Hardness C).

Used as a processing diagram, we can therefore isothermally transform austenitized materials at temperatures that will produce the desired hardness or strength. The full annealing process described in Sec. 9.4 makes use of isothermal transformation to produce pearlitic structures, as indicated schematically in Fig. 9-4. In addition, the relative times at the noses of the I-T diagrams indicate qualitatively the relative critical cooling rates that materials must undergo when martensite forms to retain all the solutes from the high-temperature treatment in order to get the maximum hardness or strength attainable with the composition. The longer the time is, the lower the cooling rate to retain the supersaturation and to form the martensite—in other words, it is easier to form martensite. The relative times at the noses depend on the carbon content, the amount of alloying, and the austenite grain size of the steel. In general, the higher or larger each of the three factors, the longer is the time at the nose. This is schematically illustrated in Fig. 9-32.

When the critical cooling rate is exceeded, the austenite can transform to the hardest microstructure in steels, the martensite structure, as indicated in Sec. 9.3.2. The martensite formation in Fig. 9-31 is athermal and as soon as the M_s temperature is reached, the transformation starts (subscript s) and will proceed further if we lower the temperature down to the M_f, when all of the austenite transforms to martensite. It is easy to see that if the nose of the I-T diagram indicates a very short time, it will take a very high cooling rate or a very severe quench to produce martensite. A very high cooling rate or severe quench can be obtained with a small size or thin material. However, small sized or thin materials are prone to have shape or distortion problems and possibly even cracking of the material due to thermal stresses induced by the severe quench. Thus, to avoid these practical problems, less severe quench must be used as much as possible. This is the reason for using alloy steels. We pay extra for alloying elements added to steel, but this extra cost obviates all the shape, distortion, or cracking problems that may be encountered during heat-treating by being able to use a lesser quench severity.

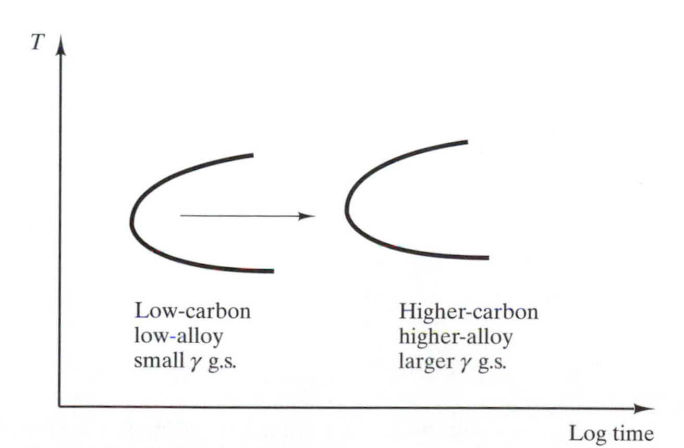

Fig. 9-32 Schematic of the influence of carbon and alloy contents and the austenite grain size on the position of the nose of the C-curve; the nose moves to longer times.

9.7.1b Continuous Cooling Transformation (CCT) Diagrams. These diagrams show the microstructures that will form after the steel was austenitized and then cooled continuously to room temperature at a certain rate. The cooling rate given in the diagrams is the rate at 700°C (1300°F) or the average cooling rate between 700° and 500°C. The diagrams are developed using dilatometry in conjunction with metallography (microstructural examination). Dilatometry follows the dilation (thermal expansion or contraction) of small steel specimens. Transformation points are accentuated because there is a volume expansion when austenite (FCC) transforms to ferrite (BCC) during cooling.

Examples of CCT diagrams are shown in Figs. 9-33a, b, and c for AISI 1040, 1541, and 15B41. The AISI-SAE steel codes or designations are listed in Table 12-3 and explained in Sec. 12.2.1. In these designations, with the exception of 9XX SAE steels, the XX or XXX (after the first two digits in the codes) when divided by 100 is the nominal weight percent of carbon. When no alloy is added to the steel, except Mn $\leq$ 1.0%, the steel is called plain carbon (i.e., only carbon is added to iron) and is designated with the first two digits, 10. Thus, 10XX is a plain carbon steel with XX/100 wt%C. Other two-digit prefixes in the designations represent different alloying additions to steels. A letter between the two-digit prefixes and the XX indicates a special addition to a grade of steel. Thus, the steels as shown in Figs. 9-33a, b, and c are (a) plain carbon steel with nominal 0.40%C, (b) a plain carbon steel with a nominal 0.41%C and a maximum manganese content of 1.65%, and (c) the same as (b) with boron added to it.

The CCT examples in Fig. 9-33 are the darker heavier curves—superimposed on them are cooling curves with their cooling rates indicated. The areas on the diagrams marked F, P, B, and M indicate where

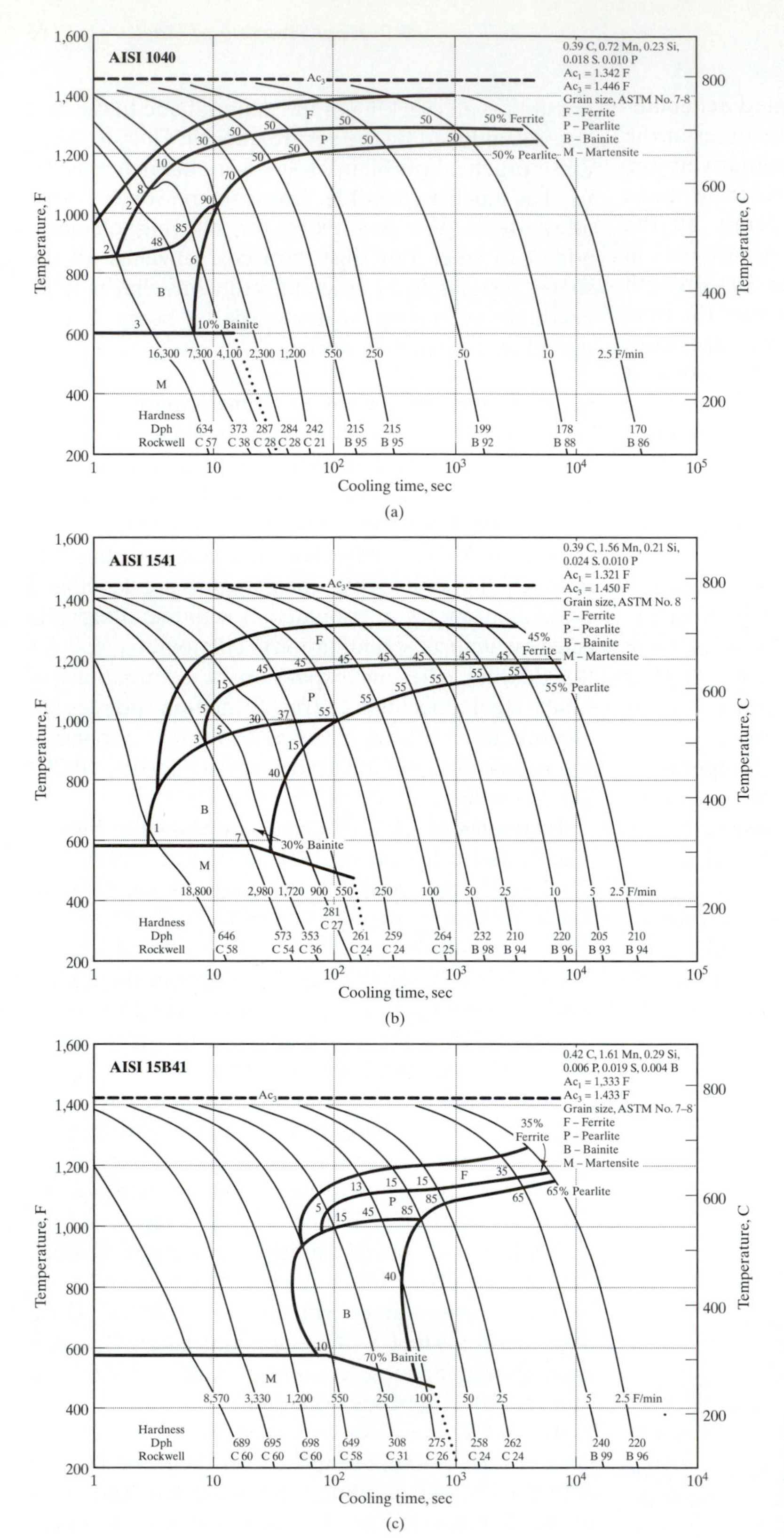

Fig. 9-33 (a) Continuous-cooling transformation diagram of AISI 1040 steel, (b) CCT of AISI 1541with nominal 0.41%C and 1.5%Mn, (c) CCT of AISI 15B41 steel with nominal 0.41%C, 1.60%Mn, plus boron. [9]

ferrite, pearlite, bainite, and martensite phases are formed. When a cooling curve passes these areas, it means that these phases will be present in the microstructure of the continuously cooled steel. The hardnesses of the aggregate structure are indicated at the end of the cooling rate curve as Rockwell B (HRB) or C (HRC) numbers or DPH (diamond pyramid hardness). Corresponding to the structures in the I-T diagrams, the coarse structures (low cooling rate) have low hardnesses while the bainite and martensite structures (high cooling rates) exhibit the highest hardnesses. These figures illustrate also the influence of alloying elements on the CCT diagram and how much easier martensite forms in AISI 15B41 than in either 1541 or 1040 (no alloy added). Full martensitic structure already forms in 15B41 with a cooling rate as low as 1,200°F/min but does not form in 1040 with a cooling rate as high as 16,300°F/min. Small boron additions dramatically lower the critical cooling rate to produce martensite. As a processing diagram, the CCT indicates what cooling rate we must give the steel from the austenitized condition to obtain the desired microstructures and properties. A corollary is the ability to indicate what hardness and structure prevail if we are given the cooling rate. We can use the CCT diagram of a steel, if available, in conjunction with the Jominy hardenability test (see Sec. 9.7.2c) to get the microstructures at different locations of the Jominy bar.

A modified way of showing the CCT is demonstrated in Figs. 9-34a, b, and c. In these diagrams, the cooling rate is represented by the approximate bar diameter that yields the microstructure at the center of the bar when it is quenched in oil or water, or air-cooled. We can easily note that a large diameter bar will cool slower than a smaller diameter bar when quenched in the same medium. The microstructures at the center of any bar diameter are determined by looking for the bar size in the scale for the appropriate quenching medium and then drawing a vertical line from that bar size through the CCT diagram. The microstructures at the center of the bar are those indicated in the areas crossed by the vertical line. For example, for a 10-mm air cooled bar, the microstructures at the center are ferrite and pearlite for a 0.18%C mild steel in Fig. 9-34a; ferrite, pearlite and a small amount of bainite for a 0.40%C plain carbon steel in Fig. 9-34b; and bainite and martensite for a 0.40%C with chromium, moybdenum, and nickel additions in Fig. 9-34c. The same size bar when oil-quenched will have bainite and ferrite seen in Fig. 9-34a, martensite and a small amount of bainite, in Fig. 9-34b, and only martensite, in Fig. 9-34c. As in the CCT diagrams, these diagrams indicate what quenching media we need to use to obtain structures and properties of austenitized bars.

The steel for Fig. 9-34a has 0.18%C with no alloy, for Fig. 9-34b has 0.41%C with no alloy, and for Fig. 9-34c has 0.41%C with alloy additions of Cr, Mo, and Ni. Thus, these diagrams show that it is much easier to form martensite with higher carbon content and with alloy additions. This type of diagram shows clearly that unalloyed plain carbon steels with less than 0.25%C can only be hardened in very small diameters or thin gauges, ≪6mm (1/4-in). Drastic quenching of these sizes usually result in distortion or shape problems for parts and components made of low-carbon plain carbon steels. Each of these diagrams is also accompanied by a plot of hardness versus the bar diameters in the as-cooled or sometimes tempered condition. The strength of the bar may be determined from the correlation of hardness and tensile strength, as shown in Fig. 9-35.

9.7.2 Through Hardening

Through hardening is the hardening of the component or part throughout its cross-section, as opposed to surface hardening that hardens only the surface of the component or part. As indicated earlier, before we do the heat-treating process, we must normalize the steel in order to obtain a more refined uniform grain size, with a homogeneous composition. The heat-treating consists of austenitizing the whole component or part at a prescribed-temperature, holding at the temperature for a length of time to insure homogeneity in the austenite, and then quenching the part or component in a medium with sufficient quench-severity (H) to produce the desired microstructure and properties. Quite often, the desired microstructure is martensite because this yields the maximum strength and hardness. Why is martensite very strong? The answer follows.

9.7.2a Hardness of As-Quenched Martensite. As indicated earlier, the hardest of the structures produced from austenite is the as-quenched martensite.

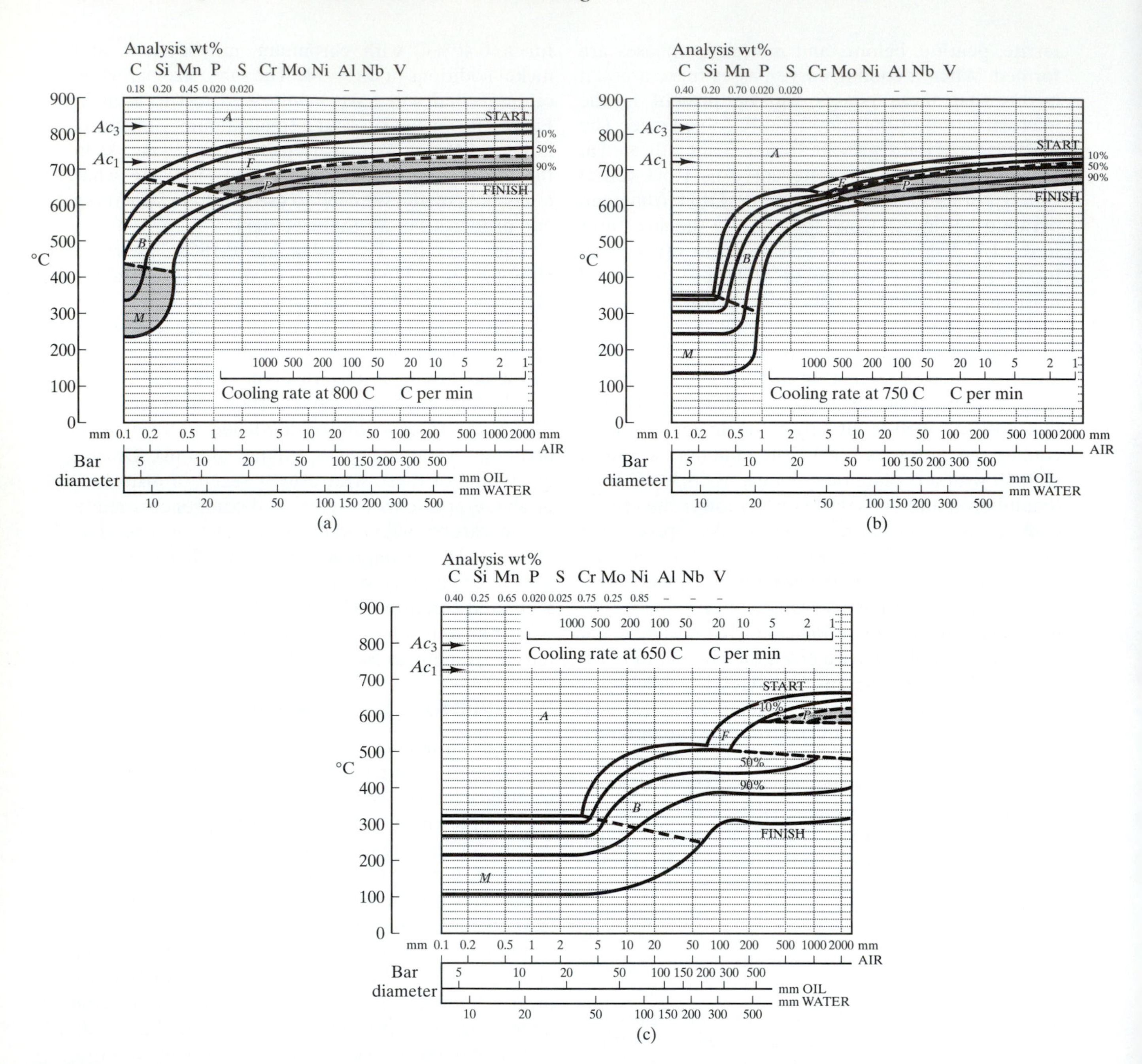

Fig. 9-34 (a) CCT diagram of a 0.18%C showing the microstructures produced when different bar diameters of the steel are quenched in air, oil, and water. (b) CCT of nominal 0.40%C plain carbon steel. (c) CCT of a 0.40%C steel alloyed with chromium, molybdenum, and nickel. [13]

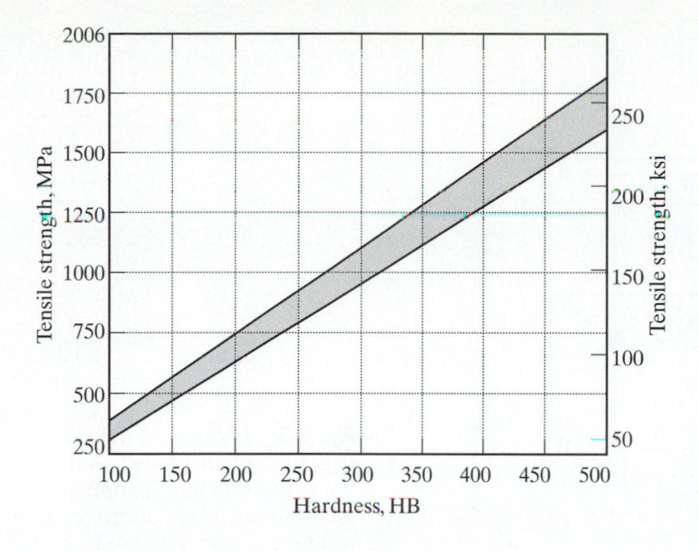

Fig. 9-35 Tensile strength as a function of Brinell hardness number for steels in the as-rolled, normalized, or quenched and tempered condition. The tensile strength in ksi is about one-half the BHN; in MPa is about 3.5 times the BHN.

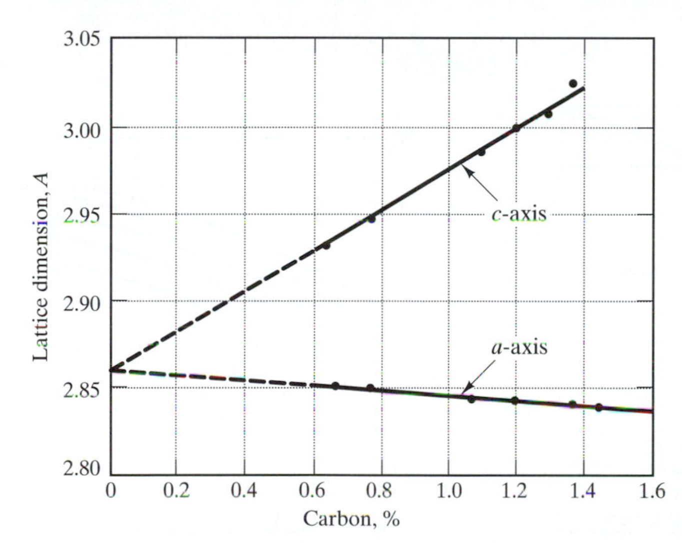

Fig. 9-36 Change in the c and a lattice parameters of BCT martensite with carbon content in iron-carbon alloys.

Structurally, *as-quenched martensite is a supersaturated solution of carbon in either a body-centered cubic (BCC) or body centered tetragonal (BCT) ferrite.* To differentiate it from the equilibrium form of ferrite, α, as-quenched martensite is designated as α′. We recall that the tetragonal structure has the crystallographic axes relationship, a = b ≠ c. A measure of tetragonality of the structure is the (c/a) ratio and the extent of tetragonality of α′ depends on the carbon content, as shown in Fig. 9-36. The tetragonality arises from the interstitial carbon atoms preferentially located along one of the edges of the cube. The result is the BCC lattice being elongated along one direction and contracted in the two directions perpendicular to the elongated axis to produce the tetragonal structure as illustrated in Fig. 9-37. It is easy to see that the *BCC structure is strained* and this is *the main reason or mechanism for the strength of as-quenched steel. Empirically, it is shown that carbon content is the sole determining factor for the hardness or strength of martensite.* This is illustrated in Figs. 9-38a and b. Figure 9-38a gives the as-quenched hardness of martensite in both Vickers and R_C hardnesses as well as comparing the hardnesses of pearlitic and spheroidized structures as a function of carbon content. Figure 9-38b also

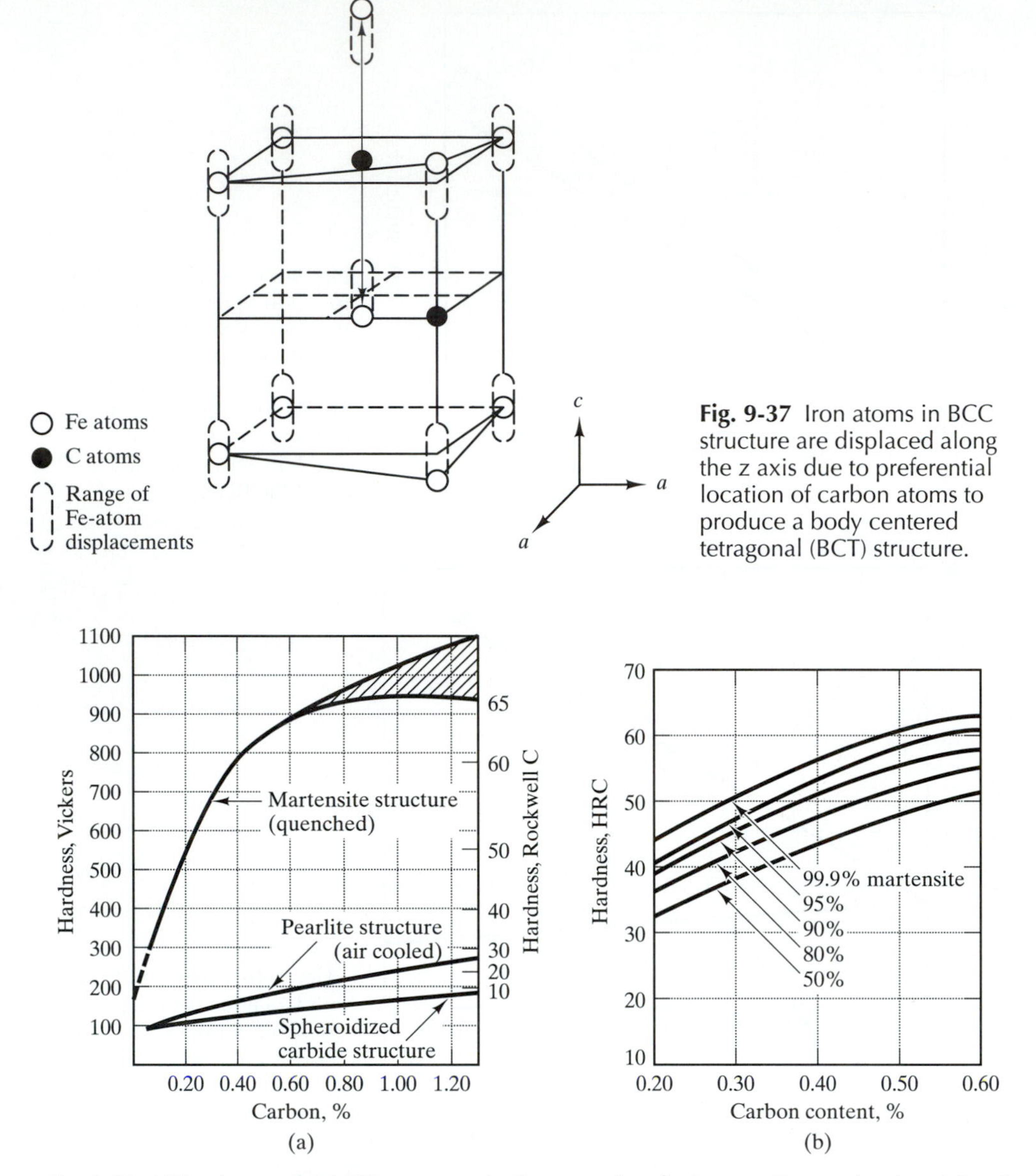

Fig. 9-37 Iron atoms in BCC structure are displaced along the z axis due to preferential location of carbon atoms to produce a body centered tetragonal (BCT) structure.

Fig. 9-38 (a)Hardness of 99.9% as-quenched martensite, ferrite-pearlite, and spheroidized microstructures as function of carbon content., (b) Hardness of as-quenched martensite as a function of carbon content and per cent of martensite in microstructure [2,3].

includes other as-quenched structures containing less than 100% (99.9%) martensite (i.e., 50, 80, 90, and 95 percent), as a function of carbon content. The information in Fig. 9-38b is replotted in Fig. 9-39 with the as-quenched hardness of steel as a function of percent martensite from 50 to 100 for different carbon contents.

The hardness dependence of as-quenched martensite on carbon content only is a very important concept. This indicates that other alloying elements do not influence the hardness of as-quenched martensite. This is the fundamental strengthening of quenched steels with carbon and is referred to as *the primary hardening.* With no carbon, as-quenched

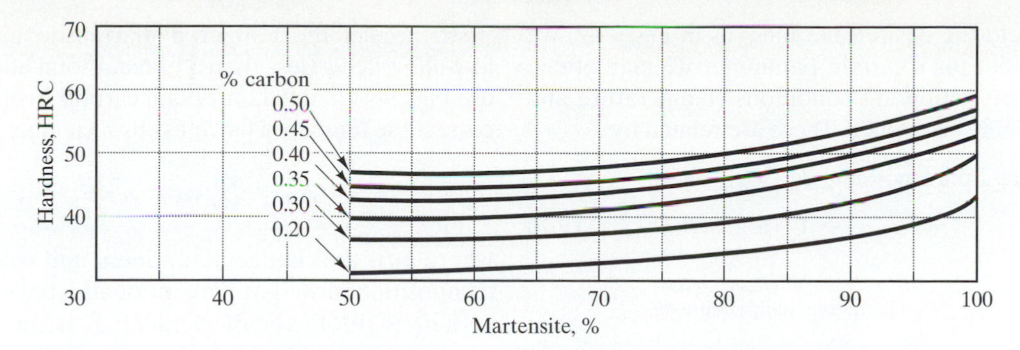

Fig. 9-39 As-quenched hardness of steel as function of percent martensite for different carbon contents. [2]

martensites of high-alloy steels or nonferrous alloys, such as titanium, will exhibit minimal strengthening, if any.

9.7.2b Tempering and Predicting Tempering Condition. The *as-quenched martensite* in its *hardest condition signifies that an as-quenched steel with a martensitic structure is very brittle.* Again, this is a significant difference from other alloy systems. With nonferrous alloys and high-alloy steels with no carbon, the as-quenched condition is the softest condition; strengthening is achieved after reheating the quenched supersaturated solution to an appropriate temperature. For these alloys, the best time to deform and shape the material is immediately after quenching when they can be easily straightened and formed into the desired shape. In contrast, as-quenched steels with martensitic structures cannot be easily formed because of their strength and brittleness. In order to give it some ductility, the as-quenched steel is heated to a temperature to soften it a little and to make it less brittle. *The process of heating the as-quenched steel (with α' martensitic structure) to give it some ductility* is called *tempering.* After tempering, the steel is in the *quenched and tempered* (Q+T) condition. Heat-treated carbon and low-alloy steels are always used and selected in the Q+T condition. The tempering process involves the decomposition of the α' to the stable phases of ferrite, α, and iron carbide, Fe_3C,

$$\alpha' \Rightarrow \alpha + Fe_3C \qquad (9\text{-}3)$$

At low temperatures, between 100° and 200° C, a transition carbide called ϵ-carbide, Fe_2C, may form, especially in high-carbon martensites. When this happens, there will be a slight hardening of the tempered martensite—the same principle as in precipitation hardening when nonequilibrium transition particles form. Once the stable Fe_3C forms, the hardness decreases steadily with increasing temperature as indicated in Fig. 9-40.

For low-carbon steels, the M_s and the M_f are quite high. During the cooling of α' from these temperatures, reaction 9-3 may occur and thus the martensite is essentially tempered. This phenomenon is called *auto-tempering* and the martensite is called *auto-tempered martensite.*

For fully hardened AISI-SAE carbon and low-alloy steels (see Table 12.3) containing 0.2 to 0.85%C and less than 5 percent total alloying elements, it is possible to estimate a tempering condition (temperature and time at the temperature) and to predict the hardness of the tempered martensite produced. This is done using the Hollomon-Jaffe (H-J) tempering parameter when the tempering temperature is in the range 345° to 650°C (~650° to 1200°F). Again, since tempering involves diffusion of atoms, this parameter is very similar to the Larson-Miller parameter in creep and takes the form of Eq. 9-1. Thus,

$$\text{H-J Parameter} = T(\log t + 18) \times 10^{-3} \qquad (9\text{-}4)$$

where T is in Rankine = deg F + 460, t is time in hours. Figure 9-41 illustrates the dependence of the H-J parameter Eq. 9-4 as a function of carbon content and the desired tempered hardness in DPH for *plain carbon steels.* From the carbon content, we go to the appropriate desired hardness curve and obtain an H-J parameter on the left. The parameter can now be used to obtain the temperature and time for tempering of the steel to yield the desired hardness as in Fig. 9-42. We see there that for a certain parameter we may obtain two or more tempering conditions (temperature and time) that may be applied. These are related by

$$\text{H-J parameter} = T_1 (\log t_1 + 18)$$
$$= T_2 (\log t_2 + 18) \quad (9\text{-}5)$$

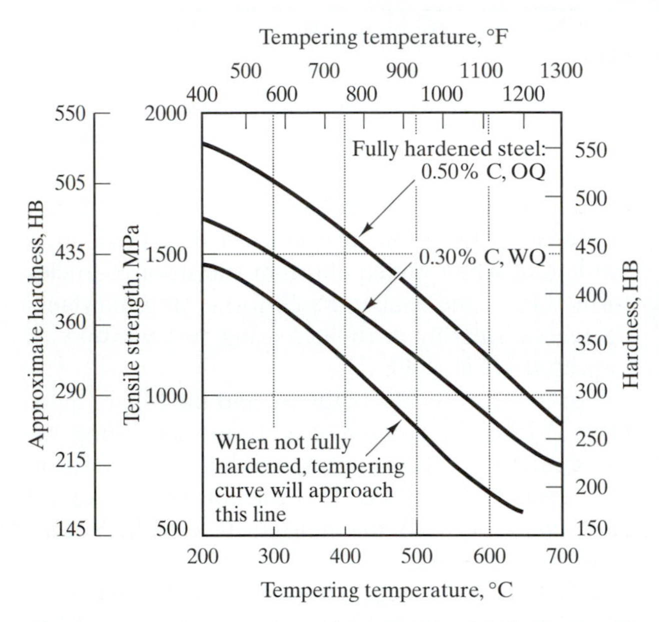

Fig. 9-40 Hardness and tensile strength of fully hardened (100% martensite) carbon and low alloy steels with 0.30% and 0.50% after tempering [2]

To predict the tempered martensite hardness of low-alloy steels (less than 5 percent total alloying), we use Figs. 9-41 and 9-42 for plain carbon steels and then correct the tempered hardness by a change, Δ, which is

$$\Delta_P = \sum_P \text{wt\% X} \cdot F_X \qquad (9\text{-}6)$$

where Δ_P is the change in hardness, and wt% X is the composition in weight percent of alloying element *X*, and F_X is the factor of element *X* at the particular parameter used. The correction factors, F_X, for the different alloying elements within certain composition ranges are given in Table 9-4 for varying parameters.

In high-alloy steels with carbon, such as in tool steels, tempering at about 550° to 600°C will precipitate out alloy carbides and this results in additional hardening of the steel, called *secondary hardening,* to differentiate it from the primary hardening of as-quenched steel martensites. The secondary hardening occurs at the approximate temperature of the tool bit during machining, and it keeps the high hardness of the tool to be able to cut the workpiece—thus producing a good machined surface. This phenomenon is illustrated in Figs. 9-43a and b. Figure 9-43a shows the effect of molybdenum (Mo) addition on tempering of a 0.35 wt%C plain carbon steel. Figure 9-43b is the tempering of a modified H11 tool steel. The secondary hardening is due to the precipitation of coherent alloy carbides and in the case of Fig. 9-43b, is due to chromium and molybdenum carbides.

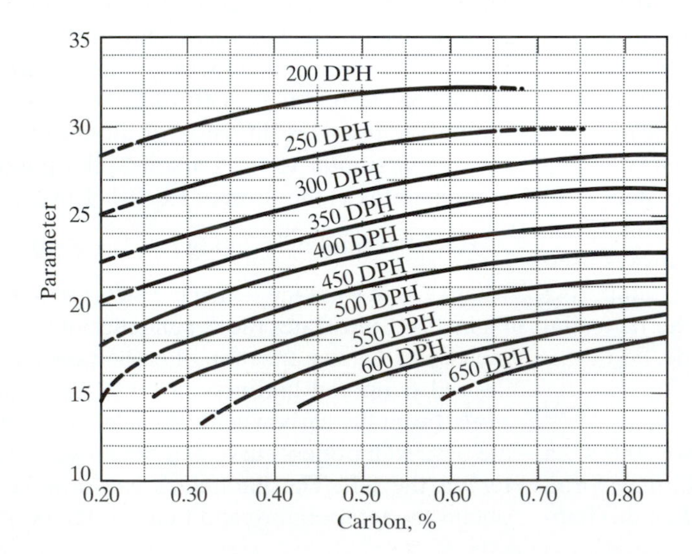

Fig. 9-41 Hollomon-Jaffe tempering parameter as a function of carbon content and desired hardness for plain carbon steels. [14]

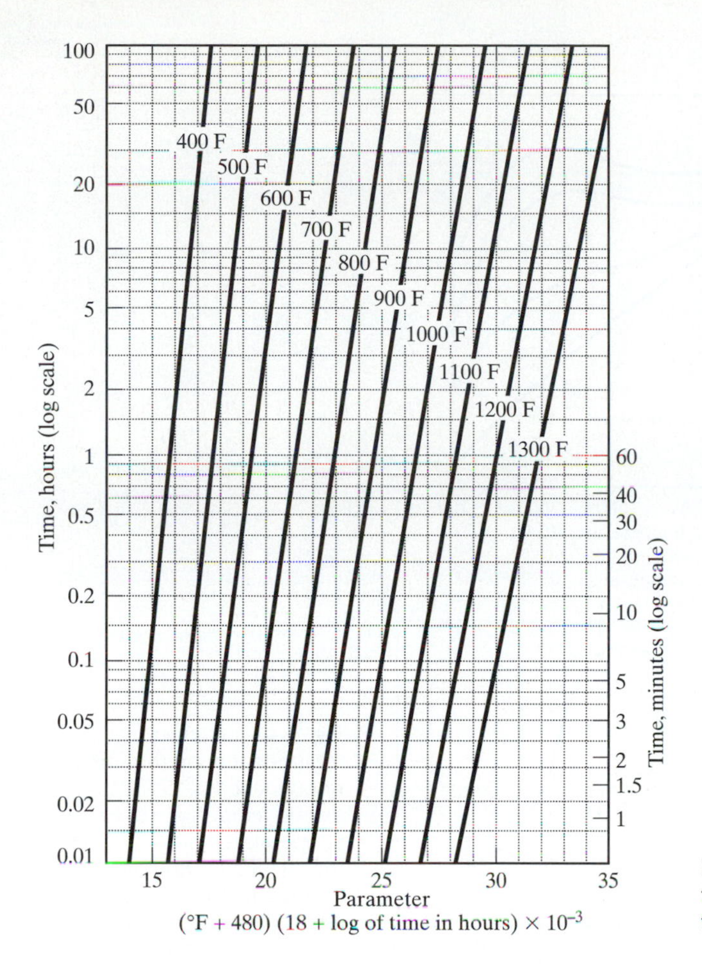

Fig. 9-42 Tempering time and temperature as function of tempering parameter. [14]

Table 9.4 Tempering hardness factors, F_X, of alloying elements added to tempered hardness of plain carbon steels. [14]

		Factor at Indicated Parameter Value					
Element	**Range**	**20**	**22**	**24**	**26**	**28**	**30**
Manganese	0.85–2.1%	36	25	30	30	30	25
Silicon	0.3–2.2%	65	60	30	30	30	30
Nickel	Up to 4%	5	3	6	8	8	6
Chromium	Up to 1.2%	50	55	55	55	55	55
Molybdenum	Up tp 0.35%	40	90	160	220	240	210
		(20)*	(45)*	(80)*	(110)*	(120)*	(105)*
Vanadium†	Up to 0.2%	0	30	85	150	210	150

*If 0.5–1.2% Cr is also present, use this factor.

†For AISI-SAE chromium-vanadium steels; may not apply when vanadium is the only carbide formerly present.

Note: Boron factor is 0.

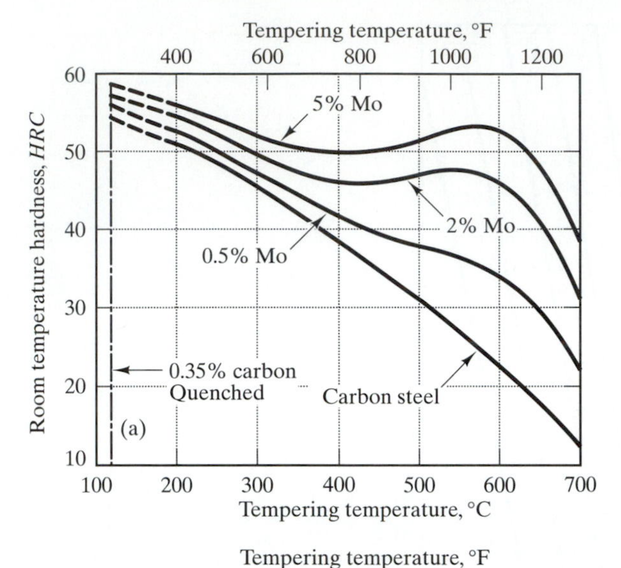

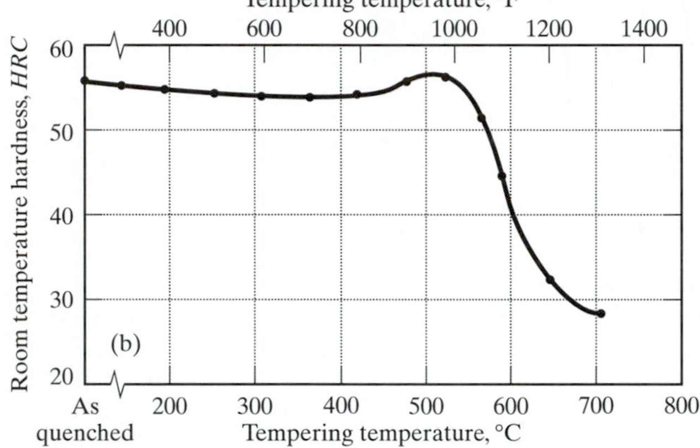

Fig. 9-43 (a) Effect of molybdenum on the softening of quenched 0.35%C steel during tempering. (b) Room temperature hardness of H11 tool steel after double tempering 2h + 2h at temperature. H11 contains 0.35%C, 5%Cr, 1.5%Mo, and 0.4%V. [2]

As indicated, the strengthening of quenched high-alloy steels with no carbon and nonferrous alloys occurs during the tempering process and thus tempering is actually an artificial aging process (precipitation hardening, Sec. 9.4.1) whereby coherent, ordered, or non-coherent precipitates form to harden the materials. This is the mechanism of hardening in the maraging and the PH (precipitation hardening) grades of stainless steel.

For the same steel, the Q + T fully martensitic structure offers the best combination of strength and toughness properties over other non-martensitic structures, such as pearlite and bainite. The strength or hardness achievable in the Q + T condition depends on the tempering temperature used. From the foregoing discussions on Figs. 9-40 through 9-42 and Table 9-4, we see that the tempered hardnesses and strengths of quenched carbon and low-alloy steels are controlled primarily by the carbon content, with only minor contributions from the alloy content. It also demonstrates the attractiveness of steels in that, from a fully hardened condition, we can achieve a variety of properties from one steel composition by varying the tempering temperature.

In the selection of steels for design in the Q + T condition, the desired tensile strength of the steel is converted to an equivalent Q + T hardness by using Fig. 9-35, and by using Fig. 9-44, we get the desired as-quenched martensite hardness. The latter will determine the carbon content of the steel, according to the desired percent of as-quenched martensite in the steel, as shown in Figs. 9-38b and 9-39. Since the

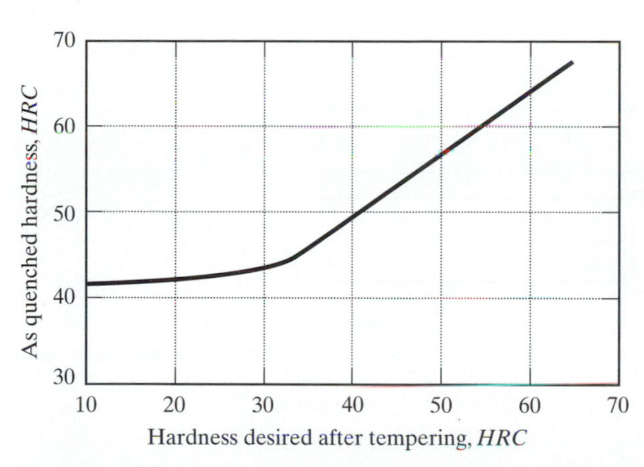

Fig. 9-44 As-quenched hardness of steel as a function of the desired hardness (HRC) after tempering. [2]

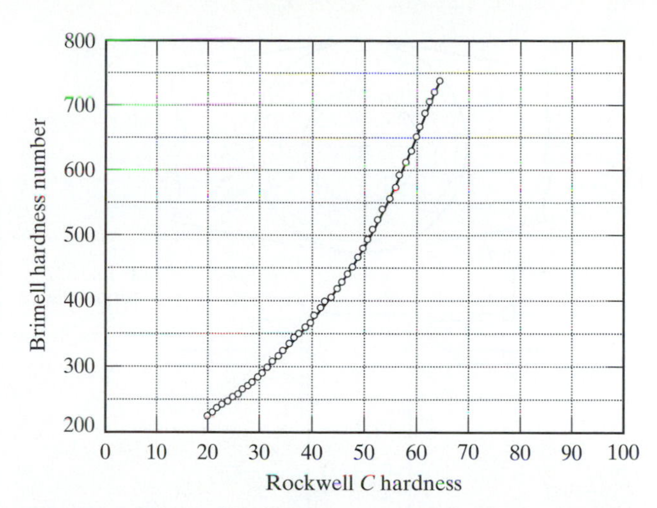

Fig. 9-45 Relationship between BHN and HRC for steels.

hardness scale used in Fig. 9-35 is BHN (Brinell Hardness number), we need to convert the hardness to HRC (Rockwell C) in order to use Fig. 9-44. The conversion can be done with Fig. 9-45 or any conversion table that may be available.

9.7.2c Hardenability and the Jominy Test. Through-hardening of a steel depends on: (1) its composition, (2) the section size of the component, and (3) the severity of the quench. The steel composition indicates how easy martensite will form. The ease of forming martensite is the property of the steel called *hardenability*. Qualitatively, we have discussed this property in connection with the I-T and the CCT diagrams. With I-T diagram, the time at the nose is a qualitative indication of the ease of forming martensite, which is the hardenability of the steel. With the CCT diagram, a steel that requires a lower cooling rate to form martensite is said to be more hardenable. For example, for the same section size, a steel that can be hardened by air-cooling is more hardenable than one that requires water-quenching to be hardened. Thus, we use the I-T and the CCT diagrams to indicate the relative hardenabilities of steels and the cooling rate necessary to avoid the nose of the curve to form martensite.

A more quantitative measure of hardenability is obtained with the *Jominy hardenability test.* This is a standard test adopted by the American Society for Testing and Materials (ASTM) as Method A255 and by the Society of Automotive Engineers (SAE) as Standard J406. Being a standard test, the specifics of the test must be adhered to. Figure 9-46 shows the details of the test specimen and the quenching fixture. The SAE specified normalizing and austenitizing temperatures for this test are indicated in Table 9-5 for the different carbon contents of the different AISI-SAE steel grades in Table 12-3. As indicated earlier, the normalizing treatment before austenitizing is imperative to ensure that the steel composition is homogenized. After normalizing, the specimen is austenitized in an atmosphere that controls the decarburization of the surface. After austenitization, the specimen is placed in the quenching fixture and quenched with a water jet at one end of the Jominy bar. Two flat surfaces, 180° or diametrically opposite to each other, along the length of the austenitized-and-quenched bar are made by very lightly grinding the surface (with grinding cuts less than about 0.013 mm {0.0005 in}) to a minimum depth of about 0.38 mm (0.015 in.) to remove any decarburized layer that may have formed. Hardness tests on the flat surfaces are made every 1/16 inch from the quench end for the first inch. The distance(s) between hardness tests for the remaining length of the bar is at the discretion of the tester.

The hardness data are plotted at each distance from the quench end on standard charts prepared for

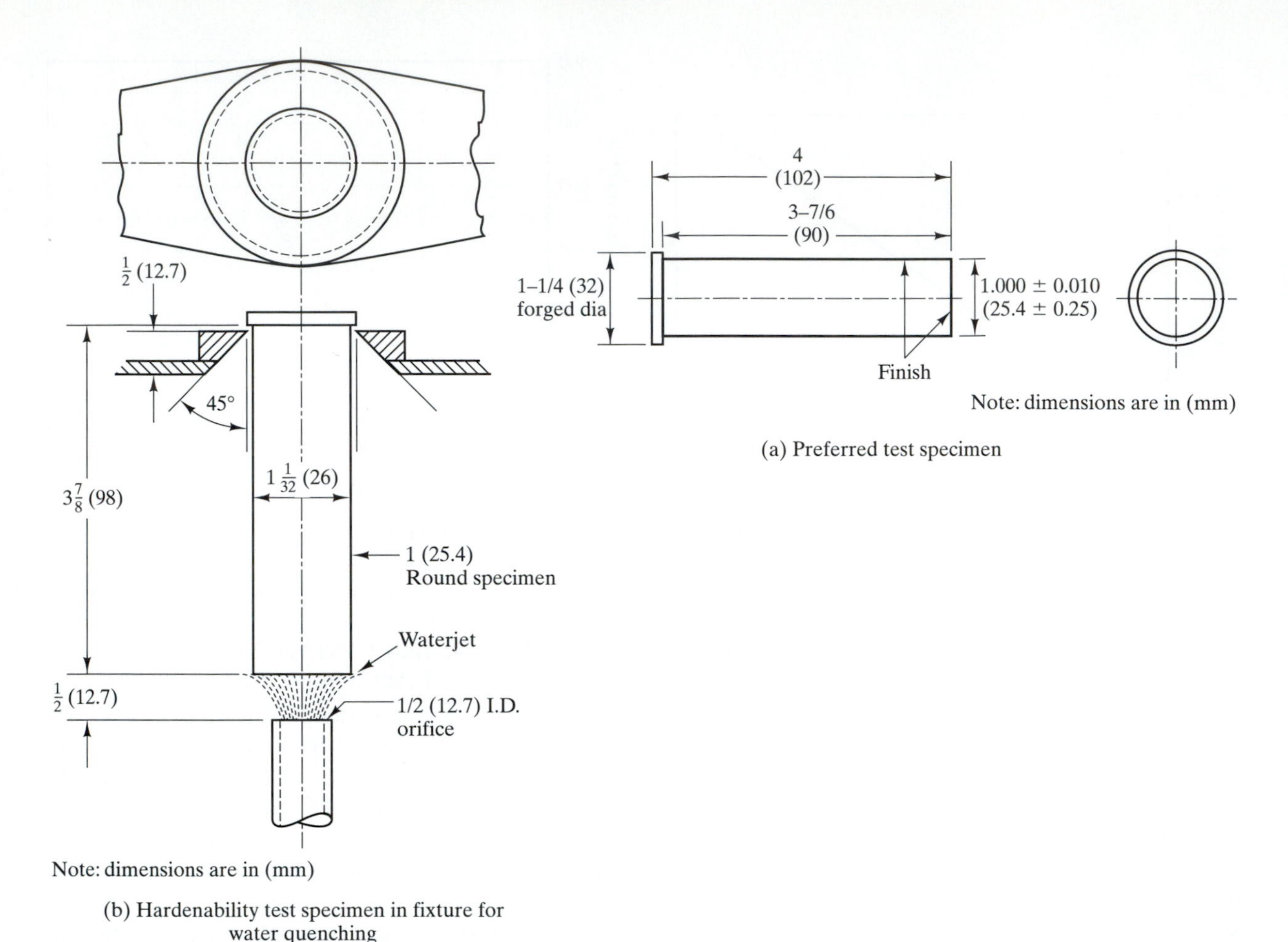

Fig. 9-46 (a) Standar Jominy Specimen, (b) Quenching Fixture. [20]

this purpose and the resulting curve is called the Jominy hardenability curve for the steel. An example is shown in Fig. 9-47. Included in this figure are the corresponding approximate cooling rates at 700°C (1300°F) at different locations along the length of the bar. The distance along the length of the bar is plotted in units of 1/16 in. and is referred to as the Jominy (J) distance from the quenched end. For example, 1J is 1/16 in. while 16J is 16/16 in from the quenched end. Implicit with the changes in hardness are the corresponding changes in microstructure. The CCT diagram of the steel is used to determine the microstructures at each J-position corresponding to its cooling rate.

The *relative hardenabilities of steels* are judged by how fast the hardness dips from the initial hardness at 1J. The hardness at 1J, assuming ~100% martensite, is a function of the carbon content only; higher carbon contents exhibit higher hardnesses at 1J. These effects are shown in Fig. 9-48 where steels 1, 2, and 3 have the same nominal carbon content but have different alloying additions, while steels 2 and 4 have the same alloying content but different carbon contents. The effects of the other two factors mentioned earlier (i.e., alloying addition and grain size for the same carbon content) are reflected in higher hardness readings at subsequent J distances from the quenched end. Between steels 1, 2, and 3 with the same carbon content (0.40%C) in Fig. 9-48, 1(4140) is the most hardenable, while 3(1340) is the least hardenable. The higher hardness readings (i.e., greater hardenabilities) after 1J for steels 1 and 2 are

Table 9-5 Normalizing and quenching temperatures[a,b] applicable to steel ordered to end-quench hardenability requirements. [20]

Max. ordered carbon content, %	Normalizing temperature		Austenizing temperature	
	°F	°C	°F	°C
Steel series 1000, 1300, 1500, 4000, 4100, 4300, 4600, 5000, 5100, 6100, 8600, 8766, 9400				
Up to 0.25 incl.	1700	925	1700	925
0.26 to 0.36 incl.	1650	900	1600	870
0.37 and over[c]	1600	870	1550	845
Steel series 4800, 9200				
Up to 0.25 incl.	1700	925	1550	845
0.26 to 0.36 incl.	1650	900	1500	815
0.37 and over	1600	870	1475	800
0.50 and over (9200)	1650	900	1600	870

[a]A variation of ±10°F (±5°C) from the above temperatures is permissible.

[b]When testing H steels, the normalizing and austenizing temperatures should be the same as for the equivalent standard steels. EXAMPLES: For 8622 H, the normalizing and austenitizing temperature should be the same as for SAE 8622; for 4032 H (carbon 0.30/0.37), the temperature should be the same as for SAE 4032 (carbon 0.30/0.35).

[c]Normalizing and austenizing temperatures should be 50°F (30°C) higher for the 6100 series.

due to the greater fraction of martensite formed, as seen in Figs. 9-38b and 9-39.

Specifications of steels based on hardenability (SAE J406c) require usually a minimum specified hardness at a certain distance from the quenched end or the minimum and the maximum hardness values at a certain J distance. For example, if a minimum HRC 36 is required at 8/16 in. from the quenched end (8J), it is written as J36 min = 8/16 in (13 mm). If, in addition to the minimum value, a maximum value is specified, then the specification is J36–50 = 8/16 in (13 mm), where HRC 50 is the maximum value specified for 8J. [Note: others specify the Jominy distance as a number following J. For example, J10 means 10/16 from the quenched-end or J32 means 32/16 from the quenched-end. Because of SAE J406C, the Jominy distance is indicated as the number before J in order not to confuse it (Jominy distance) with the specified hardness.]

The Jominy hardenability curve has a very important use in relation to the processing of actual components or parts. The actual locations in the component or part will acquire the same hardness as that in the Jominy curve with the same equivalent cooling rate. Thus, if we know the cooling rate at the location in the component, we can obtain the hardness at that location with the same cooling rate from the Jominy curve of the same steel used to construct the component. The Jominy bar used in obtaining the hardenability curve must have been processed the same way as the part (i.e., the same thermal history). This is called the Jominy equivalent condition, J_{ec}, and the procedure to obtain it is schematically illustrated in Fig. 9-49. The equivalent cooling rate is that corresponding to the Jominy distance that yields the hardness of the part. We apply this concept when we know the desired hardness of the part in the following manner. We go to the Jominy curve of the steel and determine the Jominy distance corresponding to the desired hardness of the part. The equivalent cooling rate at the Jominy distance must be achieved when quenching the part to attain the desired hardness.

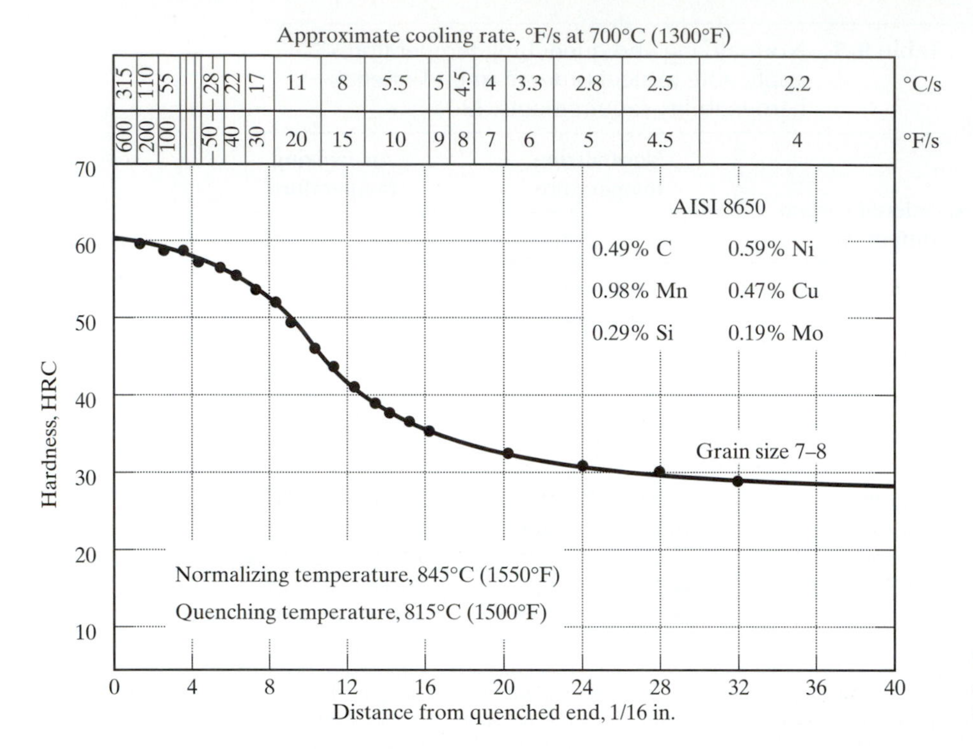

Fig. 9-47 Jominy end-quenched hardenability test results. [20]

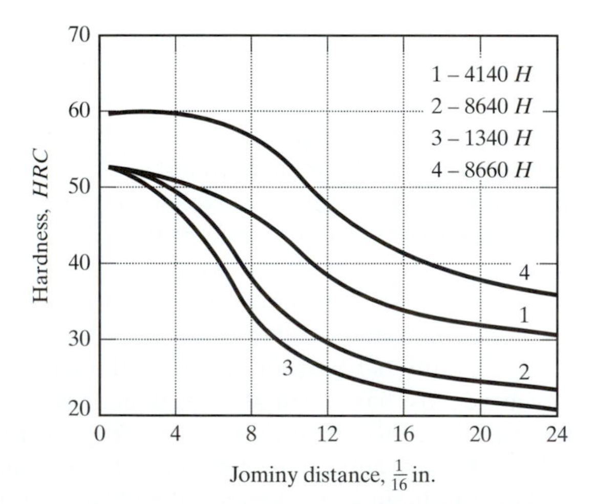

Fig. 9-48 Minimum hardenability curves for different steels showing effects of carbon content on the hardness at the 1J position and of alloy contents beyond 1J.

9.7.2d Hardenability Bands. There are certain standard steels that are specified for their hardenability performance. These are called the H-steels and their AISI-SAE designation ends with the letter H (see Sec. 12.2.1). The AISI-SAE steel designations specify a range of composition (for each element) that results in a hardenability band, as that shown in Fig. 9-50. To be on the conservative side, selection of steels is based on the minimum of the hardenability band and Fig. 9-51 shows the minima of the hardenability bands for six grades of H-steels with different carbon contents. The number on each of the curves for each grade indicates the percent of carbon in the steel and is the xx in the AISI-SAE designated steel. For example, the number 60 in Fig. 9-51f for the 86xx-H grade means it is the hardenability curve for 8660H with a nominal carbon content of 0.60%.

9.7.2e Severity of Quench. In very simple terms, the ability of a quenching medium to cool a heat-treated part is classified as its quenching power or severity of quench. This is given a symbol H and the faster it cools the part, the greater is the severity of quench. The severity of quench correlates directly to the thermal stresses set up in the heat-treated part.

Table 9-6 lists experimentally determined H values for a number of commonly used quenching media. This shows the range of H from 0.02 (still air) to 5 (violent circulation -brine).

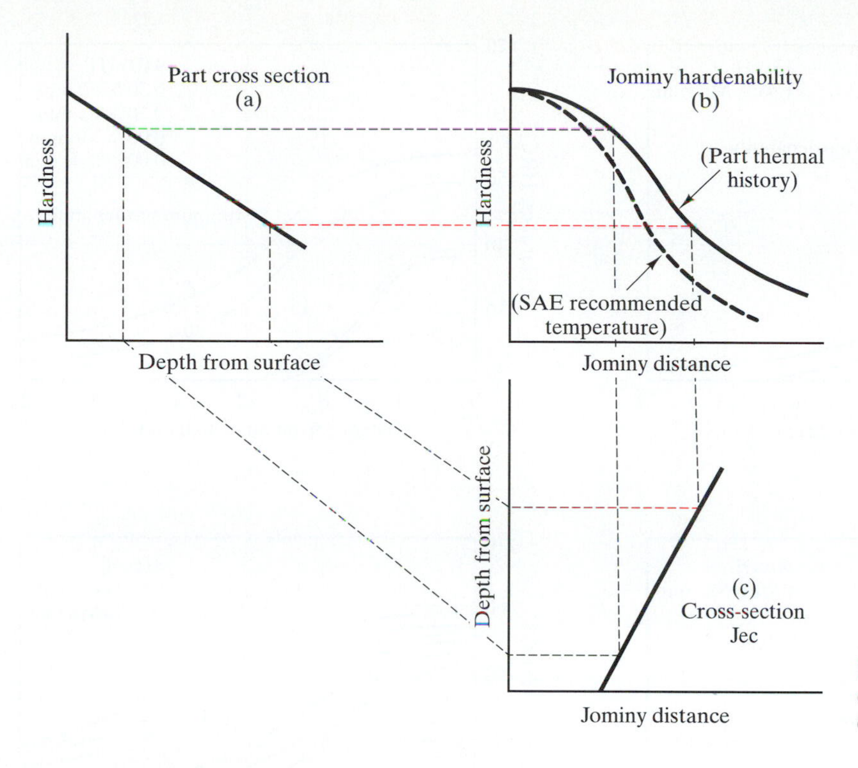

Fig. 9-49 Method of determining the Jominy equivalent condition (J_{ec}). [2]

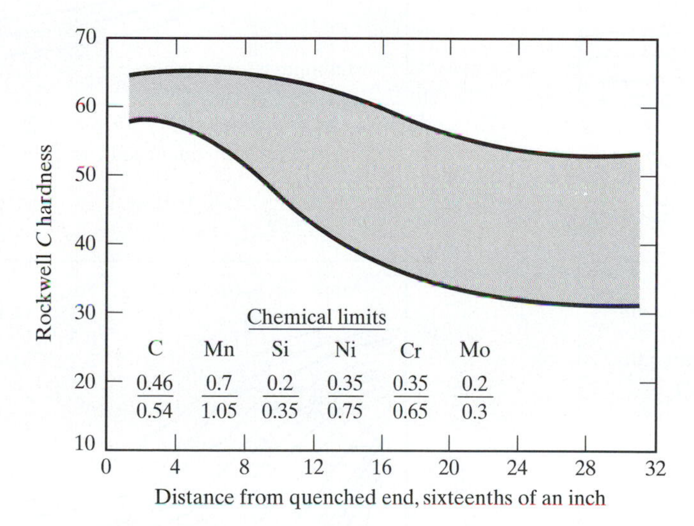

Fig. 9-50 Hardenability band arises because of the range in compositions of carbon and alloying elements. Example is for 8750H. [2]

The resultant cooling rate of the steel depends on both the quenching power, H, of the quenching medium and on the size of the steel. Figures 9-52 and 9-53 show the cooling rates at the surface and at other section locations of different steel bar diameters quenched in agitated water and in agitated oil. The section locations are illustrated in Fig. 9-54. The cooling rates are given as the Jominy equivalent cooling rate, J_{ec}, expressed as Jominy distance from the quenched end of the Jominy bar, bottom abscissa, and the cooling rates at 700°C, top abscissa in Figs. 9-52 and 9-53. We should note that a higher Jominy distance indicates a lower cooling rate. We note that for the same section size (diameter), the equivalent

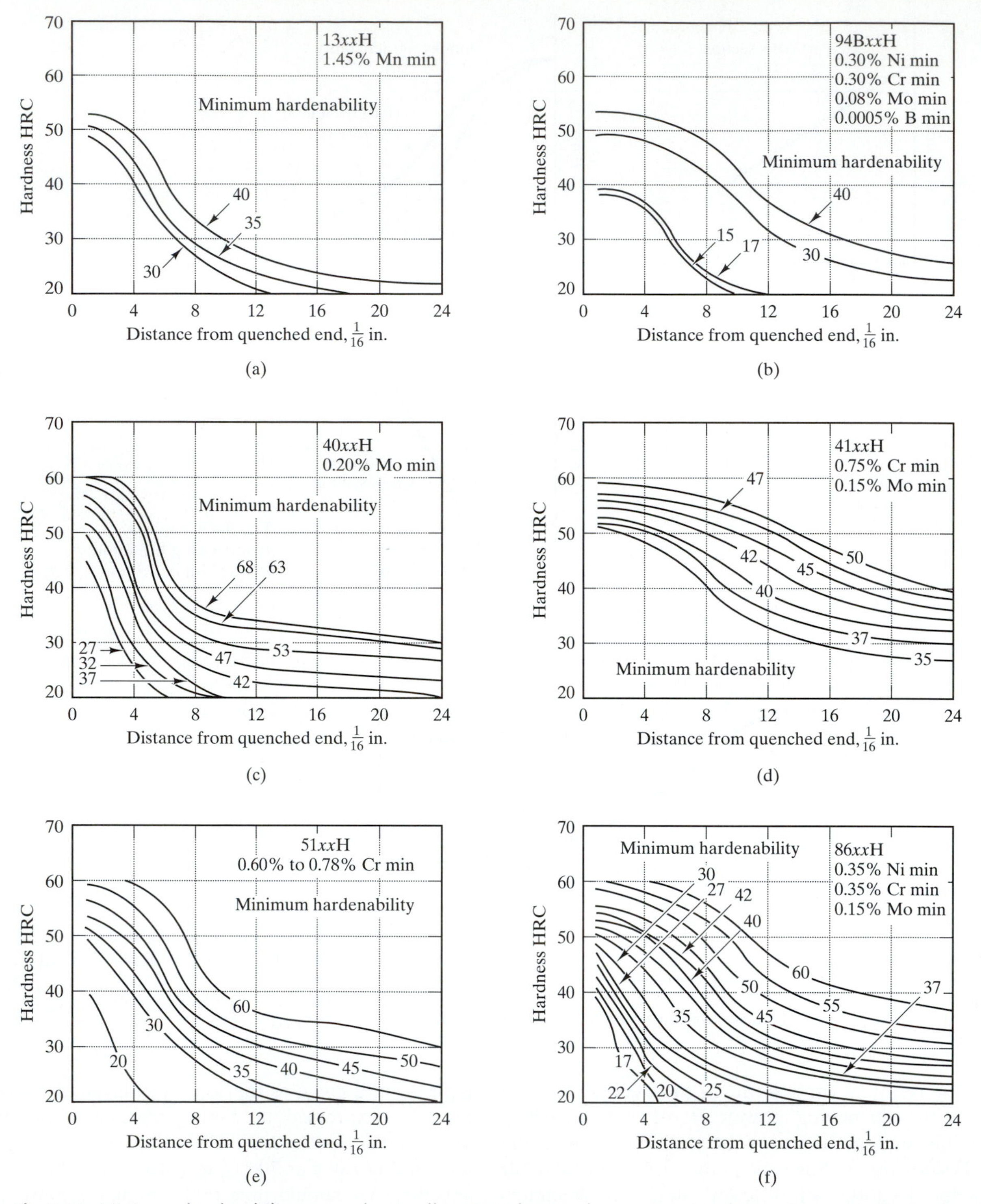

Fig. 9-51 Minimum hardenability curves for six alloy H steels—number on or pointed to each curve indicates the carbon content, to be inserted in place of xx in the alloy steel designation. [2]

Table 9-6 Quench severity (H) of various quenching media. [2]

	Air	Oil	Water	Brine
No circulation of fluid or agitation of piece	0.02	0.25 to 0.30	0.9 to 1.0	2
Mild circulation (or agitation)	...	0.30 to 0.35	1.0 to 1.1	2 to 2.2
Moderate circulation	...	0.35 to 0.40	1.2 to 1.3	...
Good circulation	...	0.4 to 0.5	1.4 to 1.5	...
Strong circulation	0.05	0.5 to 0.8	1.6 to 2.0	...
Violent circulation	...	0.8 to 1.1	4	5

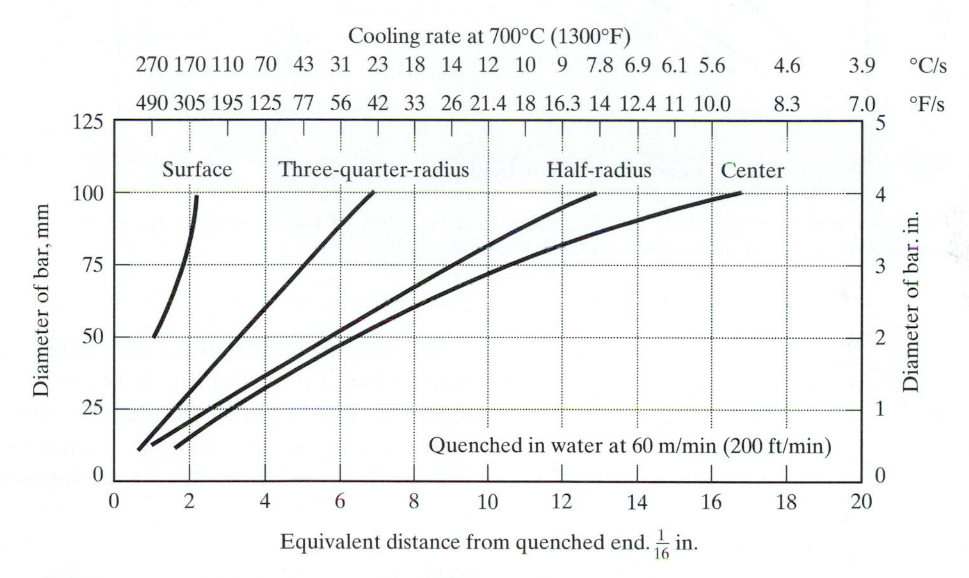

Fig. 9-52 Jominy equivalent cooling rates as a function of bar diameters quenched in agitated water at different locations of the bar sections. [2]

Jominy distance in a certain location of the bar, for example, half-radius, is higher in oil than in water; the quench severity or cooling power of oil is lower than water. For the same location, such as half-radius, the equivalent Jominy distance is higher for larger diameters; larger diameters cool slower because of their greater masses. Using the J_{ec} obtained from Fig. 9-52 or 9-53 for the section size and location, we obtain the hardness at that location from the hardenability curve of the steel, if the latter is available. Thus, we can use Figs. 9-52 and 9-53 in conjunction with the hardenability curve of the steel to determine if we can attain a desired hardness at a location, for example, the center of the bar. In this way, we get to determine the maximum size of a bar that we can quench in either agitated water or agitated oil.

EXAMPLE 9-3

A tensile strength of 150 ksi is required for an application. Determine the isothermal processing of a 1080 steel that will yield this property and the resultant microstructure.

SOLUTION: Since the property given in Fig. 9-31 for the 1080 steel is Rockwell C hardness, we have to figure out how we can translate the 150 ksi tensile strength requirement to a Rockwell C hardness number requirement. We first note the relationship of tensile strength to Brinell hardness number in

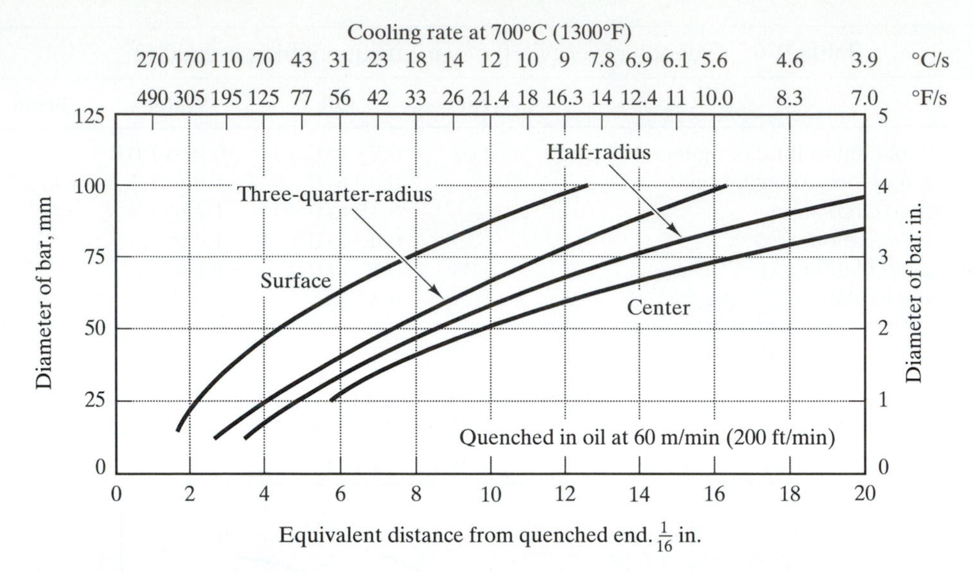

Fig. 9-53 Jominy equivalent cooling rates as a function of bar diameters quenched in agitated oil at different locations of the bar section. [2]

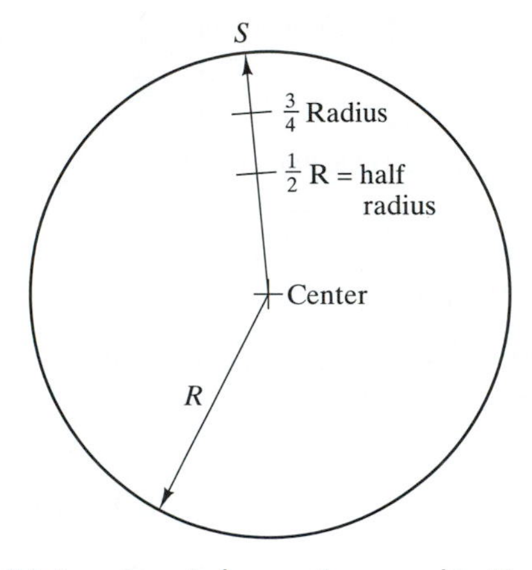

Fig. 9-54 Locations in bar sections used in Figs. 9-54 and 9-55.

Eq. 4-105b, which is also stated in the caption of Fig. 9-35, that is,

$$\text{Ultimate tensile strength} = 0.5 \times \text{BHN}$$

Therefore the BHN requirement for the steel is $150 \times 2 = 300$ BHN. Then using the relationship between BHN and HRC in Fig. 9-45, we obtain 32 HRC. With the latter, we go to Fig. 9-31 to indicate the processing in order to get this property with a 1080 steel.

We must austenitize the 1080 above the $A_s = 727°C$, homogenize the austenite (hold for a time), quench to 650°C (1200°F), and then let the austenite isothermally transform. The diagram indicates that about 100 seconds is sufficient for transformation and the microstructure is from coarse to medium pearlite. At this temperature, the quenching medium will most likely be a salt bath and depending on the size of the material, we should allow the material to attain the bath temperature. Because of the convection of the liquid bath, this will most likely take a couple of minutes. Thus, the material must be in the salt bath no more than 10 minutes and then be quenched to room temperature.

EXAMPLE 9-4

For the same tensile strength requirement of 150 ksi, determine the cooling rate needed from the austenitizing temperature and the resulting microstructure to meet the requirement for a (1) AISI 1040, (2) AISI 1541, and (3) AISI 15B40. For this, we use the CCT diagrams in Fig. 9-33.

SOLUTION: Again, we must translate the tensile strength requirement into a Rockwell C hardness number. We have found this in Example 9-3 to be 32 HRC.

(1) For the 1040 steel, we see from Fig. 9-33a a resulting 28 HRC when the cooling rate is about 2,260°C (4,100°F)/min and a 38 HRC when the cooling rate is about 4,020°C (7,300°F)/min. Thus, the cooling rate must be between these cooling rates. The microstructure will consist of ferrite, pearlite, and bainite.

(2) For the 1541 steel, we see from Fig. 9-33b a 24 HRC when the cooling rate is about 482°C (900°F)/min, and a 36 HRC when the cooling rate is about 940°C (1,720°F)/min. The cooling rate must be closer to the second cooling rate to meet the requirement. The microstructure based on Fig. 9-33b is a mixture of ferrite, pearlite, and bainite.

(3) For the 15B40 steel, we see from Fig. 9-33c a 31 HRC at a cooling rate of about 121°C (250°F)/min and 58 HRC at a cooling rate of 288°C (550°F)/min. The cooling rate is going to be closer to the first cooling rate. The microstructure is again ferrite, pearlite, and bainite.

We note that for each steel, the microstructure is a mixture, with varying proportions, of ferrite, pearlite, and bainite. The proportions cannot be established, however, with the CCT diagrams.

If we were to select a material among the three steels for ease of obtaining the properties and, most likely less distortion problems, the best choice would be the 15B40 steel. The cooling rate of about 150°C (300°F)/min or 2.5°C/sec should be sufficient. This cooling rate probably can be attained with air-cooling or certainly in forced-air convection with a fan. Thus, the treatment will be normalizing and with minimal distortion.

EXAMPLE 9-5

For the 1040 plain carbon steel, determine the maximum bar diameters that can produce (a) 100% martensite, and (b) 100% bainite through its section when (1) water-quenched, (2) oil-quenched, and (3) air-cooled from the austenitization temperature. Use Fig. 9-34b.

SOLUTION: As explained in the text, we simply draw a vertical line from the bar diameter to get the microstructures. For the questions, we locate the vertical line with the largest diameters where 100% martensite and 100% bainite are obtained. The answers are tabulated below

	Water-Quenched	Oil-Quenched	Air-Cooled
100% Martensite	15 mm	10 mm	0.3 mm
100% Bainite	50 mm	40 mm	3 mm

EXAMPLE 9-6

A plain carbon steel was held at a temperature for five hours and then quenched. The as-quenched specimen was found to contain equal volumes of ferrite and martensite, α', and the microhardness on the as-quenched martensite was 870 Vickers Hardness Number (VHN or DPH). Assuming the densities of ferrite and martensite are the same, determine

(i) the temperature at which the steel was held and the AISI-SAE designation of the steel
(ii) considering the microstructure to be a composite, determine its hardness
(iii) specify a processing sequence to convert the ferrite-martensite microstructure to 100 percent martensite. What is the hardness of the as-quenched martensite and the tensile strength in the quenched and tempered condition?

SOLUTION: The presence of ferrite and martensite in the microstructure indicates that the steel was held at a temperature in the $(\alpha + \gamma)$ field and on quenching, the γ transformed to α'. We shall recall a couple of things about the martensite transformation:

(1) That it is a diffusionless transformation, Sec. 9.3.2. This characteristic signifies that the composition of the martensite is the same as the austenite from which it transformed, and the amount of the martensite is the same as the amount of austenite.

(2) That the hardness of the martensite is a function of the carbon content only, Sec. 9.7.2a.

With (2), we translate the 870 VHN to a 0.60%C in the martensite by using Fig. 9-38a, and with (1) we see that the austenite also has 0.60%C. Because the densities of α' and ferrite are assumed equal, the volume percent is also the weight percent, and again with (1), this means that the weight fraction of austenite is also 0.50. Holding the steel for five hours at temperature means that the ferrite and austenite are pretty much in equilibrium; therefore, we can use the equilibrium phase diagram. With the austenite containing 0.60%C, we determine the holding temperature to be about 750°C and the composition of the ferrite is about 0.02%C (Fig. 9-3 or 5-22). The overall composition of the steel can be determined by using the lever rule at 750°C, thus

$$0.50 = \frac{C_o - 0.02}{0.60 - 0.02} \qquad \text{from which} \qquad C_o = 0.31\%$$

Thus, the AISI-SAE designation for the steel is 1030. The actual composition is within the range of the composition of a nominal 0.30%C steel.

To get the hardness of the composite, we need to know the hardness of the ferrite phase. We can get this by extrapolating the hardnesses of both the pearlitic and spheroidized structure to 0%C in Fig. 9-38a. The hardness is about 90 VHN. Thus,

$$\begin{aligned}&\text{Hardness of the composite} =\\&(50/50 \text{ ferrite/martensite}) = 0.50 \times 90 + 0.50 \times 870\\&\qquad = 480 \text{ VHN}\end{aligned}$$

We can convert the 50/50 ferrite/martensite structure to 100% martensite by austenitizing it and then quenching. For the 0.31%C, the A_{e_3} is about 800°C, therefore austenization should be from 830° to 850°C, holding there for about an hour and then quenching to water.

The as-quenched hardness of 100%-0.31%C martensite is about 650 VHN or 50–54 HRC. The tempered hardness from Fig. 9-44 is from 42–47 HRC. Using Fig. 9-45, the BHN is from 400–450 BHN. The tensile strength is roughly 200–225 ksi.

EXAMPLE 9-7

An alloy steel has a composition of 0.40%C, 0.75%Mn, 0.20%Si, 0.50%Cr, 0.60%Ni, and 0.25%Mo. The desired quenched and tempered hardness for this alloy steel is 48 HRC. Determine the tempering conditions for the steel.

SOLUTION: To determine the tempering condition, we need to use Figs. 9-41, 9-42 and Table 9-4. We see however that the hardness given in Fig. 9-41 is in DPH. Therefore, we must first convert 48 HRC to DPH or VHN and this is 484 DPH. The objective is to get a tempered hardness of 484 DPH for the alloy steel. To get this, we use Fig. 9-41 to get a tempered hardness for a plain carbon steel with 0.40%C. Then we add the contribution of the alloying elements using Eq. 9-6 and their factors in Table 9-4. To get the factors, we need to have a tempering parameter. But we don't know the latter, and in fact, the parameter is what we need to determine the temperature and time. Thus, the solution is one of trial and error. The method is as follows:

1. We assume a tempering parameter.
2. We use Fig. 9-41 to get the tempered hardness for a plain carbon steel with 0.40%C.
3. We use Eq. 9-6 and Table 9-4 to get the contributions of the alloying elements.
4. Add the results of Steps 2 and 3, and check to see if it equals or approximates closely 484 DPH. If yes, the assumed parameter is correct. If no, the parameter is wrong and we repeat Steps 1, 2, and 3.

For this problem, we take as our first assumption that the parameter = 22. From Fig. 9-41, at 0.40%C and 22, the DPH is 400. Using Eq. 9-6 and Table 9-4,

$$\Delta_p = 0.5(55) + 0.60(3) + 0.25(45) = 40.55$$

Therefore, with the Parameter 22, the predicted tempered hardness of the alloy is 400 + 40.55 = 440, which is much less than 484. Note in the Δ_p calculation that the contribution of Mn and Si were not included because the composition of the alloy is below the range given in Table 9-4.

We repeat the procedure, and now assume the parameter to be 20. From Fig. 9-41, at 0.40%C and 20, the DPH is 450 for the plain carbon. For the alloying contribution,

$$\Delta_p = 0.5(50) + 0.60(5) + 0.25(20) = 33$$

Therefore, the estimated tempered hardness of the alloy steel = 450 + 33 = 483 ≈ 484. Therefore the correct parameter is 20. With this parameter, we can use Fig. 9-42 or use Eq. 9-4. The usual tempering time used is about 30 min or 0.5 h. Using Eq-9-4,

$$20 = T(\log\{0.5\} + 18) \times 10^{-3}$$

from which T = 1130 R or 670°F. We shall see however in Sec. 9.7.3b that this temperature is within the range of tempered martensite embrittlement (TME) which is 260° to 370°C (500° to 700°F). It might not be bad, but to be sure, it is wise to get out of the range. Thus, we can use 710°F, and we solve for the time at 710°F (1170 R) and

$$1170(\log t + 18) = 20{,}000$$

$$t = 0.123 \text{ hour} = 7 \text{ min}$$

EXAMPLE 9-8

The end-quenched hardenability curve for a 4140 steel is shown in Fig. EP 9-8a. Draw the hardness profile across the diameter of a 75-mm (3-in.) diameter of the steel that was austenitzed and then quenched in (a) agitated water, and (b) agitated oil.

SOLUTION: In order to get the hardness profile for a 75 mm diameter 4140 steel, we need to get the Jominy equivalent cooling rate for the various locations across the diameter of the steel. We need to use Fig. 9-52 for the agitated water quench, and Fig. 9-53 for the agitated oil quench to get the Jominy equivalent cooling rate. Once we get the J_{ec}, we use the hardenability curve in Fig. EP 9-8a to get the hardnesses, and then we draw the hardness profile. This is summarized in Fig. EP 9-8b.

9.7.2f Ideal and Actual Critical Diameters. These two parameters are the indexes of hardenability and are the bases of estimating and comparing relative hardenabilities of steels of various compositions. We can use them to estimate the diameter that can attain a certain hardness at the center of the bar in the absence of the hardenability curve. The measure of the hardness is in terms of the 50 percent martensite structure. The *ideal critical diameter, D_I*, is the maximum diameter that produces 50 percent martensite at the center of the bar when it is quenched in a bath with $H = \infty$. The *actual critical diameter, D_I*, is the maximum diameter of a steel bar that produces 50 percent martensite at the center when quenched in a bath of known quench severity, H.

The 50 percent martensite criterion is based microstructurally on the transition of martensitic and nonmartensitic (pearlitic) structures or hardened and unhardened zones, as illustrated in Fig. 9-55. The light etching area is martensitic (hardened), while the dark etching area is nonmartensitic (unhardened). The transition between the two areas is the 50 percent martensite criterion. Hardness traverse measurements show that the 50 percent martensite is located close to the inflection point of the Jominy hardenability curve. This is the basis of the hardness data of the 50 percent martensite in Figs. 9-38b and 9-39. It is also the basis in indicating the inflection points in the Jominy hardenability curves as the 50 percent martensite location.

Figure 9-56 shows the relationship of the actual critical diameter, D_c, (the ordinate) as a function of the ideal critical diameter, D_I, for various H values. In this Figure, D_c, and D_I are equal when $H = \infty$. For a certain D_I, D_c decreases as H decreases. Because of the variation of D_c with different H values, the ideal critical diameter, D_I, is used as a better measure of hardenability. If the ideal critical diameter is known or estimated, the maximum diameter of a bar that can be through hardened (50 percent martensite at the center) in a given quench medium (known H value) is determined with this diagram.

Thus, if we can determine D_I, we can determine the actual diameter of the steel that can be through-hardened with 50 percent martensite at the center. One of the classical methods to determine D_I from the steel composition is the Grossman procedure and is based on the following:

$$D_I = F_C \cdot \prod_{1}^{n} F_i \qquad (9\text{-}7)$$

where F_C is the factor due to carbon and the austenite grain size, Π is the mathematical symbol indicating multiplication of the F_is, and F_is are the individual factors of each of the alloying elements Mn, Si, Ni, Cr, and Mo in the steel. The F_C is obtained from Fig. 9-57 that is shown as the ideal critical diameter as a function of carbon content, and the F_is are obtained from Fig. 9-58.

The procedure for the selection process of steels based on hardenability is therefore as follows: (1) Select a steel grade (composition); (2) determine its D_I; (3) determine H, the quench severity of the available quench medium; and (4) determine the actual diameter Dc that can be hardened through using Fig. 9-56. If the D_c is less than what is needed, then we repeat the procedure by selecting another steel grade or composition. Of course, we should realize that the results are estimated values and they need to be verified experimentally. Having said this, it must also be mentioned that the factors are the results of multitude of empirical data, with numerous modifications, and the predicted values are given a lot of confidence in industry. This confidence goes to the extent of predicting the end-quench hardenability curve, which we shall not get into here.

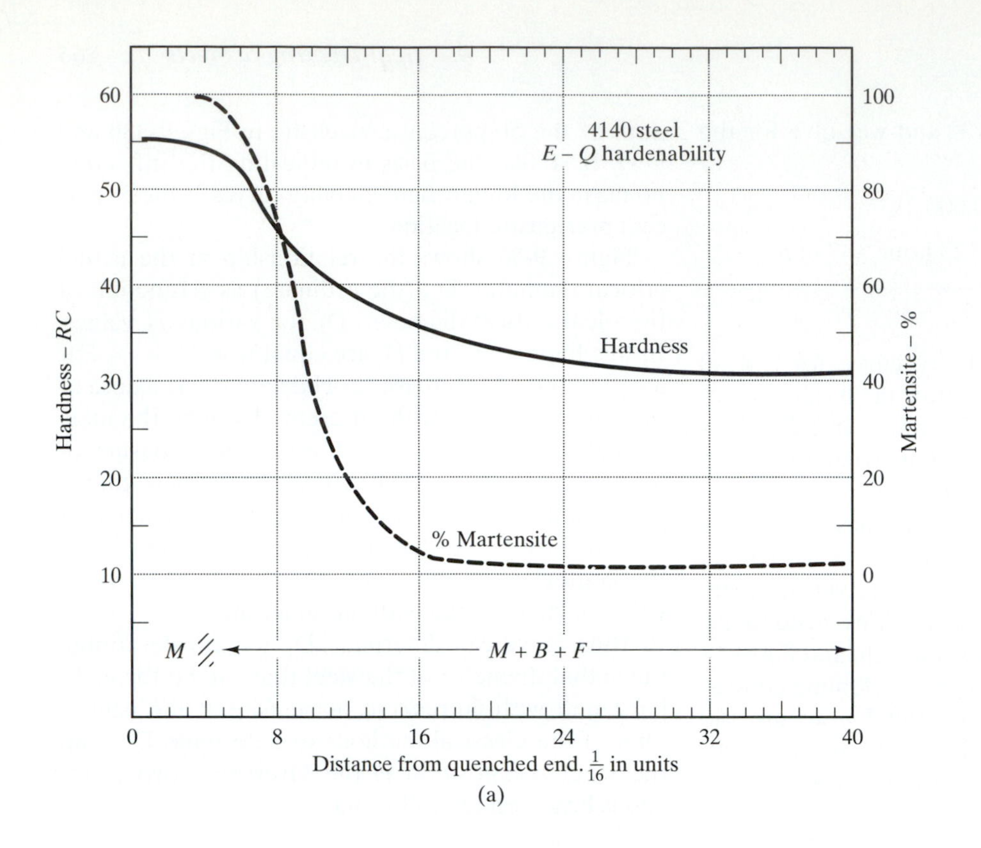

(a)

In agitated water

	J_{ec}	*HRC*
S	1.7	56
$\frac{3}{4}$ R	5	54
$\frac{1}{2}$ R	9	44
C	10.6	42

Hardness profile

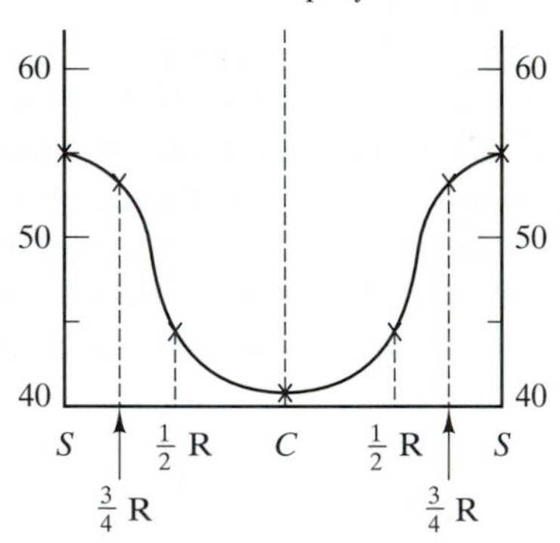

In agitated oil

	J_{ec}	*HRC*
S	8.3	46
$\frac{3}{4}$ R	11.7	40
$\frac{1}{2}$ R	14.5	37
C	17.5	35

50 50
40 40
30 30
S $\frac{1}{2}$ R *C* $\frac{1}{2}$ R *S*

(b)

Fig. EP 9-8 (a) Hardenability Curve and %Martensite for 4140 along E-Q bar. (b) Jominy Equivalent Cooling Rate, Hardnesses, Hardness Profile Across 75 mm diameter 4140 steel-(top) in agitated water, (bottom) in agitated oil.

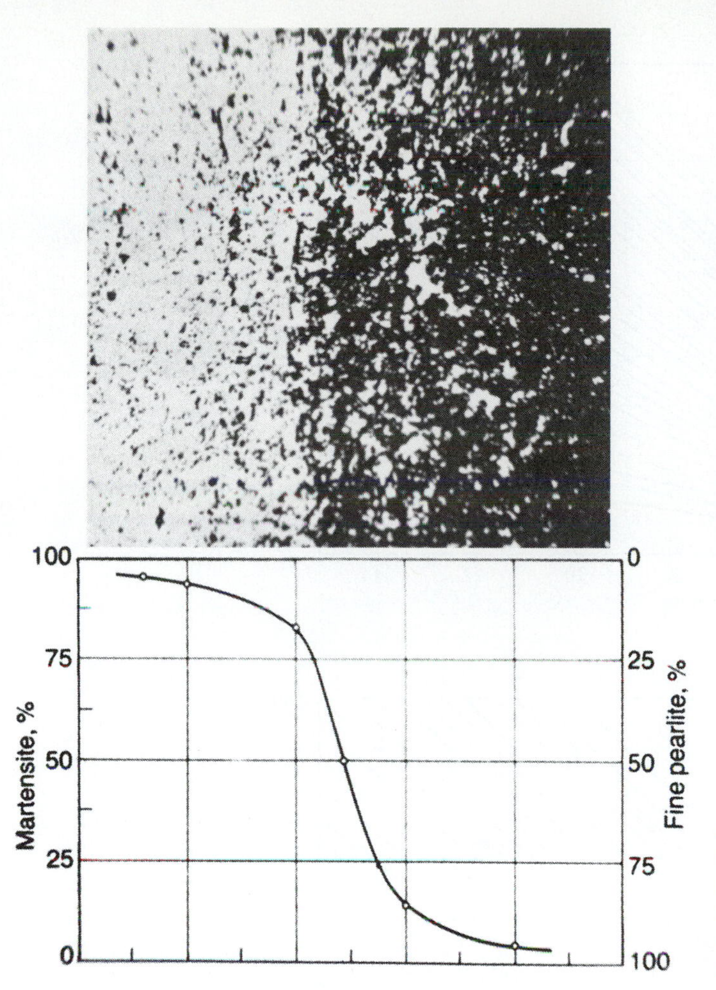

Fig. 9-55 Transition in microstructure (top-light to dark) and hardness (bottom-drop in hardness) from martensitic (hardened) to nonmartensitic areas in quenched steel. [3]

9.7.2g Special Processes Using the I-T Diagrams. So far we have been discussing direct quenching of a section size into the quenching medium to transform the austenite to martensite. Also, in using the I-T diagram as a processing diagram, we have only used the time to the nose of the C-curve. We have assumed implicitly that the cooling rate at the center of the section also exceeds the critical cooling rate. We have not considered the difference in the temperatures at the surface and at the center of the section and the thermal stress this difference might introduce. We shall now describe other processes that specifically use the other features of the I-T diagrams and shall take into consideration the difference in temperatures at the surface and at the center of the section.

The conventional direct quenching, as shown in Fig. 9-59a, may induce distortion, high residual stresses, and/or cracking in the material, especially for high-carbon steels. In order to reduce these problems, the material may be martempered. *Martempering,* which may be applied to both steels and cast irons, is not a simple tempering of martensite. Rather, it consists of a series of steps, with tempering of the martensite as the final step. The steps are schematically shown in Fig. 9-59b and consist of: (1) quenching from the austenitizing temperature into a hot fluid medium, for example, hot oil, molten salt or metal, or a fluidized particle bed, at a temperature above, but very close, to the M_s temperature; (2) holding the steel in the medium until the temperature of the steel is uniform substantially throughout its section; (3) cooling

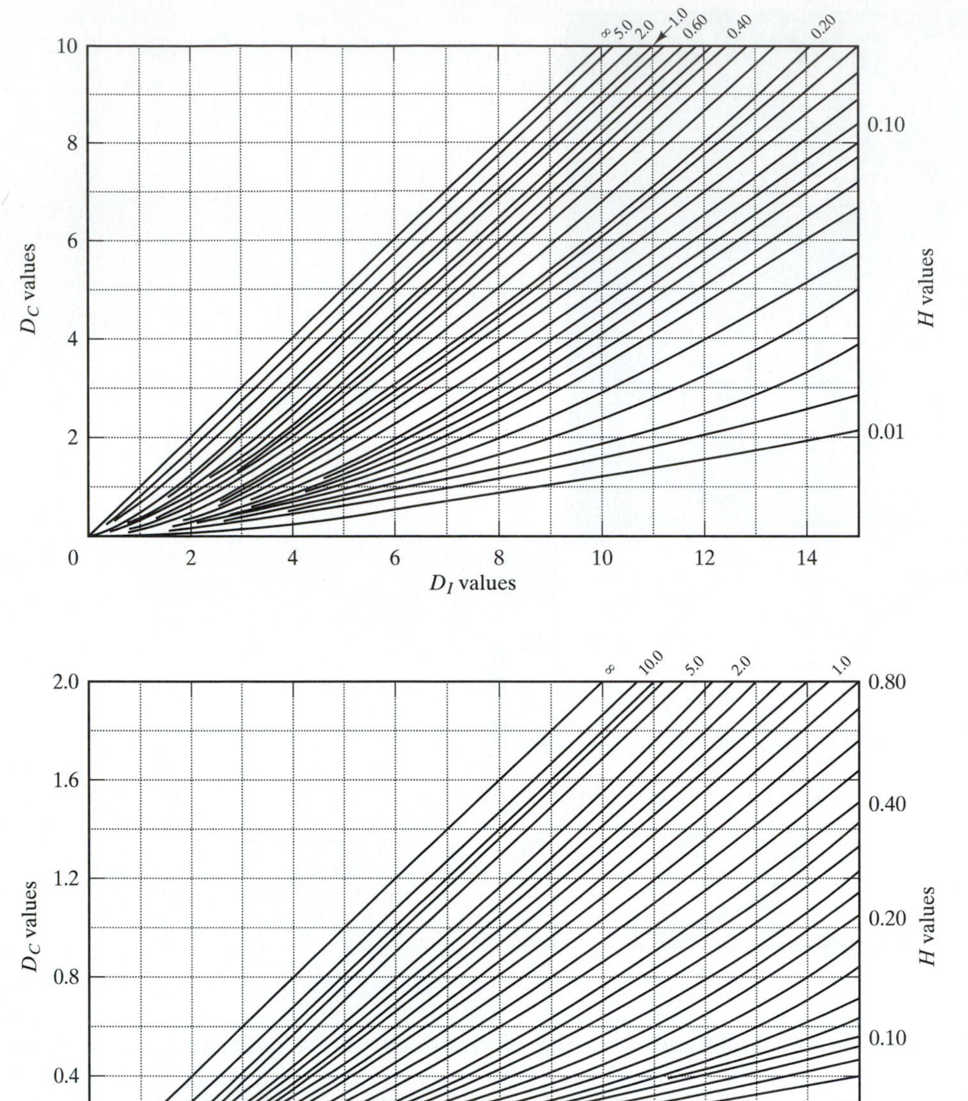

Fig. 9-56 (a) Relationship of actual critical diameter, D_C, and ideal critical diameter, D_I, with various quench severities, H. (b) Lower left corner of Fig. 9-58a shown in an expanded scale. [16]

of the steel through the martensite transformation range at a moderate rate (usually air cooling is sufficient) to prevent large temperature differences between the surface and the center of the section; and (4) finally, tempering the as-quenched martensite, as in direct quenching. The advantage of this process over the direct quenching method is the attainment of a uniform temperature through the section of the steel before it transforms to martensite. This induces uniform martensite formation through the section and hence reduces significantly the residual stresses developed, the susceptibility to cracking, and the distortion or shape problems.

A very similar treatment to martempering, and for essentially the same purpose to reduce residual stresses and distortion, is *austempering*. The process

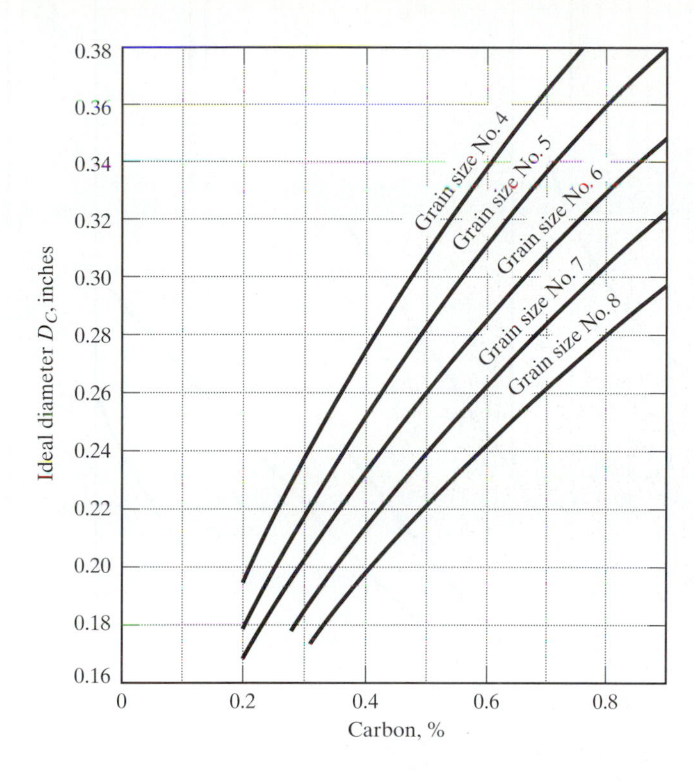

Fig. 9-57 Carbon factor, F_C, as ideal critical diameter and as a function of carbon content and the austenite grain size. [16]

is schematically shown in Fig. 9-60. It is also applicable to steels and cast irons and also consists of: (1) quenching the steel from the austenitizing temperature to a hot fluid medium (the same as in martempering) at a temperature above the M_s; (2) the steel is held at the temperature and is allowed to transform completely to a lower bainite microstructure; and (3) after transformation, the steel is air-cooled. There is no need to temper the structure because the microstructure is an aggregate of the stable microconstituents of ferrite and carbide, and not much thermal stress is introduced. The lower bainite microstructure exhibits high strength, with good ductility, and notch or impact toughness. Steels suitable for austempering must have short times to completion of the transformation of austenite to lower bainite.

In using both Figs. 9-59 and 9-60, the cooling rate at the center of the section must still exceed the critical cooling rate (i.e., miss the nose of the C-curve) in order to through-harden the section. Thus, the steels must have sufficient hardenability. In general, steels that are directly oil-quenched are suitable for both martempering and austempering, and alloy steels are better suited than carbon steels. Examples of suitable and unsuitable steel compositions for austempering are shown in Fig. 9-61.

Another process using the I-T diagram is *ausforming* to produce ultra-high strength alloy steels. Steels that can be ausformed exhibit two C-curves, as in Fig. 9-62—one for the pearlite transformation and the other for the bainite transformation. The resulting I-T diagram exhibits a "bay" where the untransformed austenite is stable for a long period of time when it is brought to this area after austenitizing. This allows us to do something with the austenite before it is quenched to produce martensite. Ausforming is derived from *aus*tenite de*forming* and is thus the deformation of the austenite in the bay before it is quenched to produce martensite. Additional strengthening is achieved with the increased dislocation density and with the very fine carbide particles produced after tempering.

9.7.3 Precautions in Processing and Use of Through Hardened Steels*

The structures, properties, and performance of through-hardened steels can only be maximized if

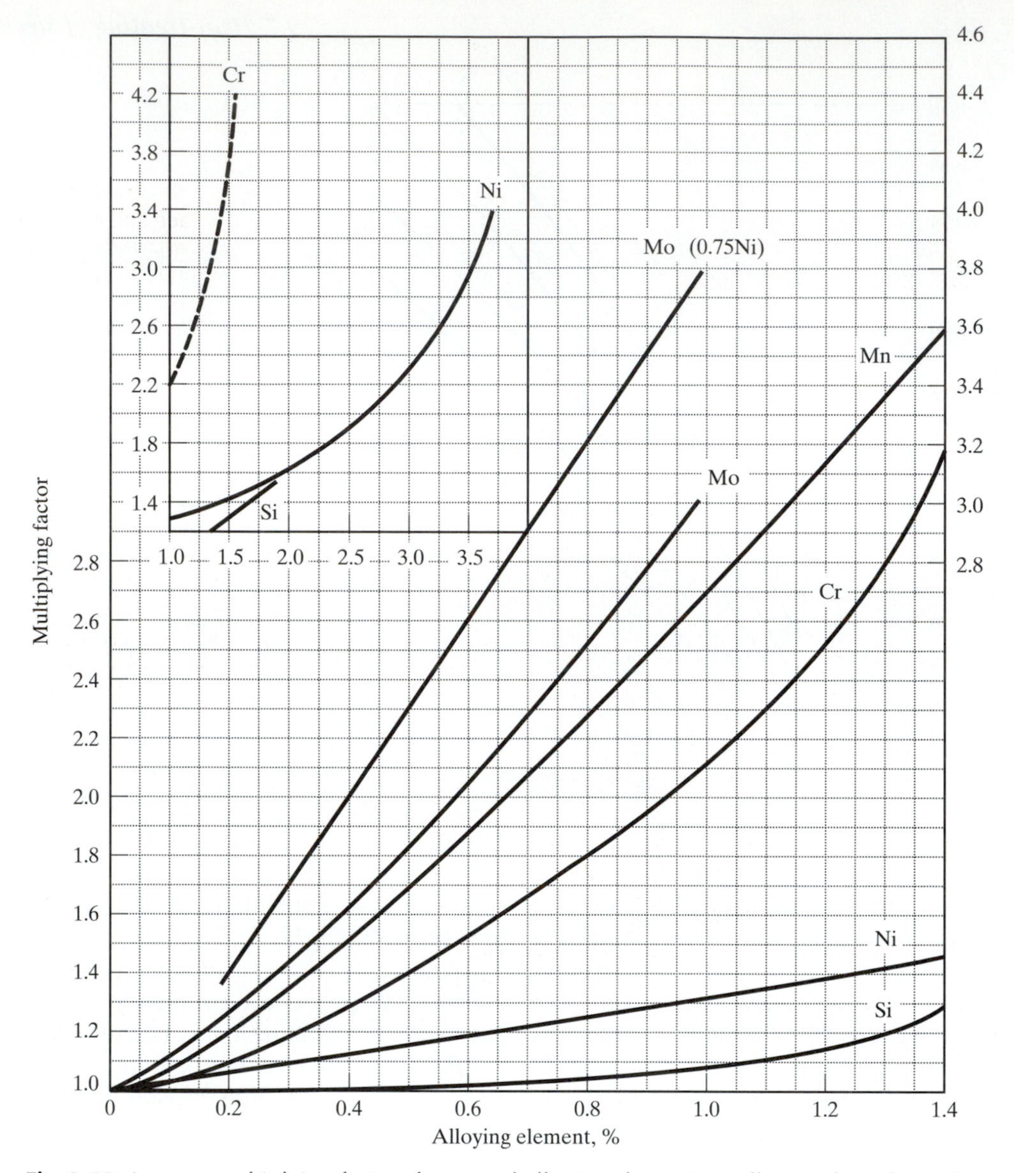

Fig. 9-58 Average multiplying factors for several alloying elements in alloy steels to determine the ideal critical diameter of steels. [15]

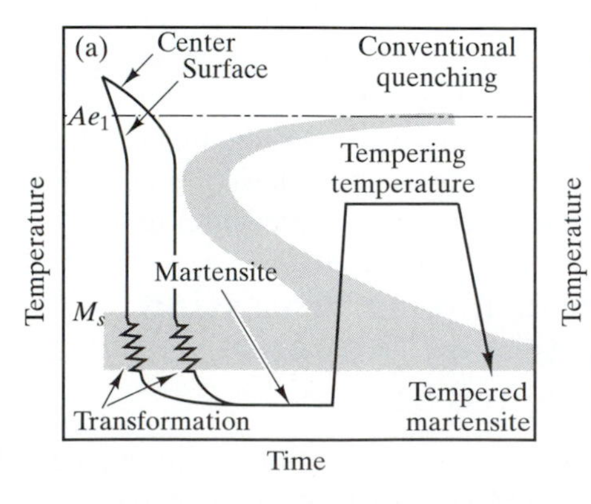

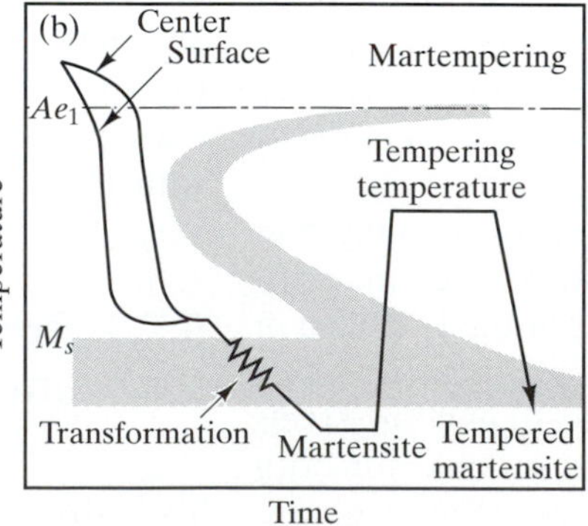

Fig. 9-59 Cooling rates of surface and center of a section superimposed on an I-T diagram showing (a) the conventional quenching and tempering process, and (b) the martempering process.

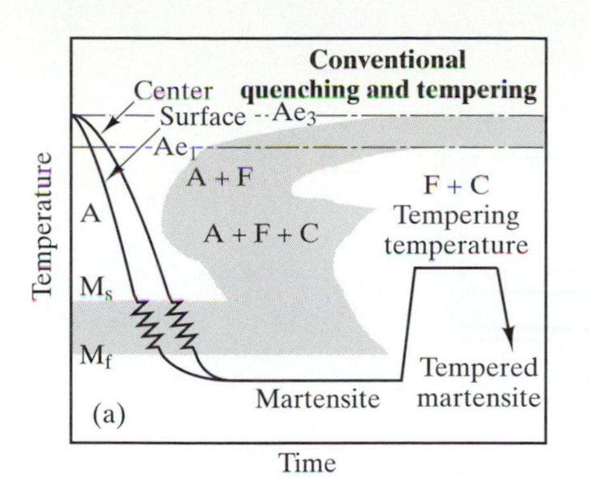

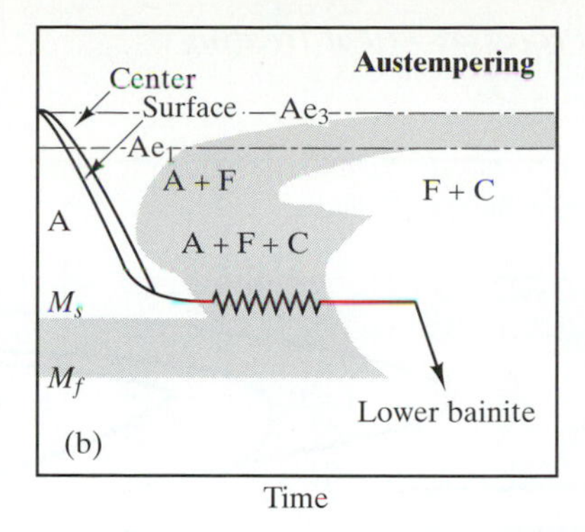

Fig. 9-60 Cooling rates of surface and center of sections superimposed on an I-T diagram showing (a) the conventional quench and tempering process, and (b) the austempering process.

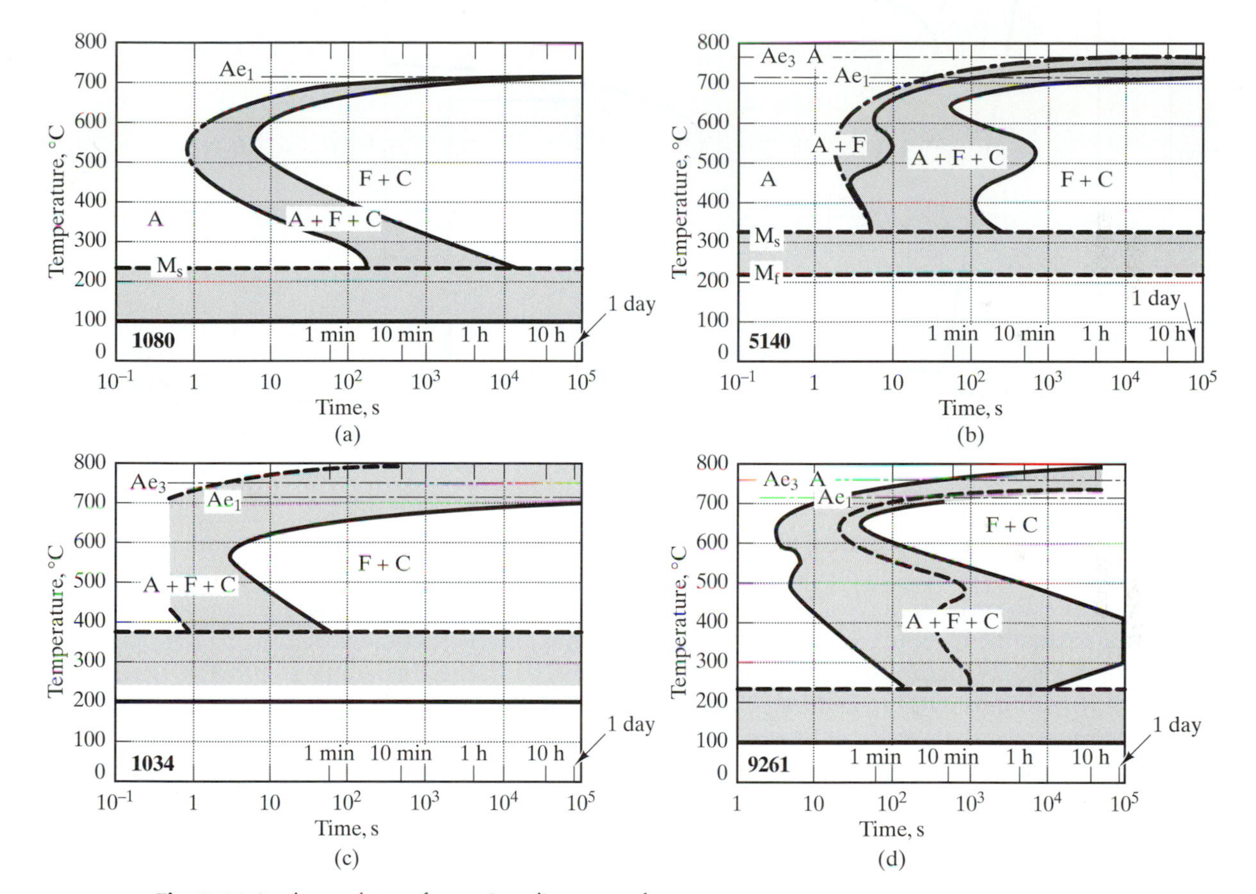

Fig. 9-61 Isothermal transformation diagrams of (a) 1080, (b) 5140, (c) 1034, and (d) 9261 steels illustrating limited suitability of 1080, very good suitability of 5140, unsuitability of 1034 due to the extremely fast pearlite reaction, and unsuitability of 9261 because of the extremely slow reaction to bainite.

they are processed and used correctly. We must, therefore, be aware of some problems that may arise during their processing and their use.

9.7.3a Homogeneous and Retained Austenite. Up to this point, we have just been talking about austenitizing temperatures without considering the condition of the austenite. When we talked about hardenability, the calculation of the ideal critical diameter, the hardness of martensite, the I-T, and the CCT, the composition we refer to is the composition of the austenite. When we use the composition of the

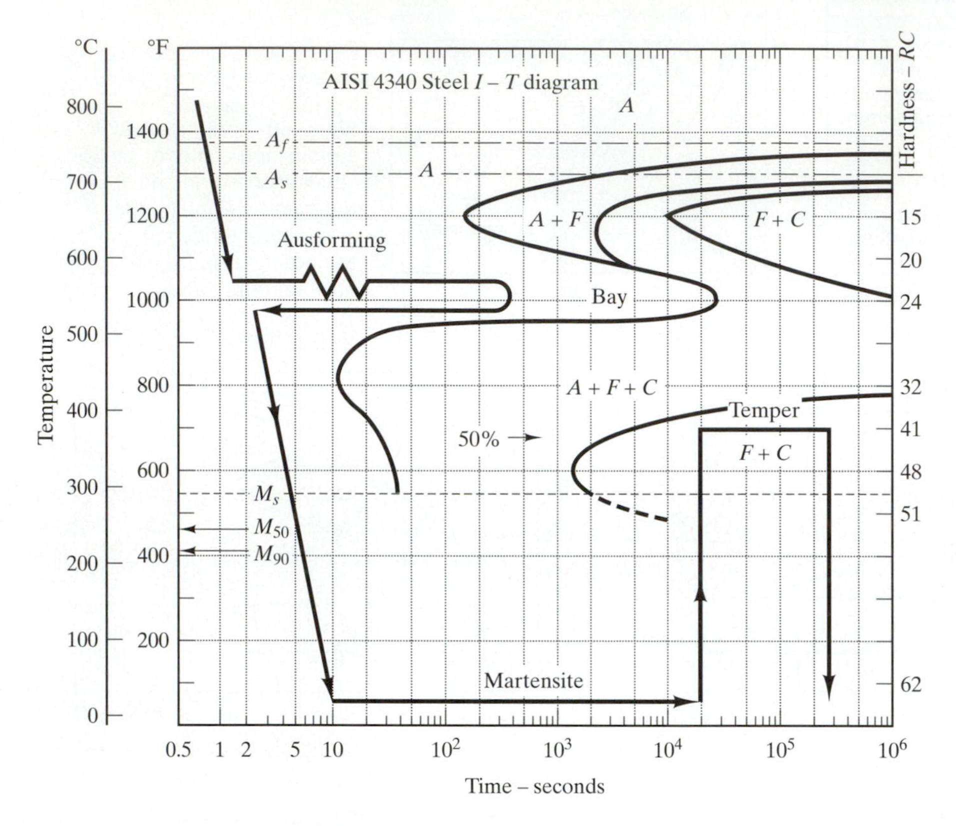

Fig. 9-62 Schematic of the Ausforming process (i.e., austenite deformation), at a temperature in the bay of an I-T diagram, such as that above for 4340 steel.

given steel, we assume implicitly that all the elements in the steel are present homogeneously in austenite prior to quenching. In order to have a homogeneous austenite, the austenitizing conditions must be such that the austenitizing temperature is high enough to put all the alloy additions and the carbon content in solution in austenite, and that sufficient time is allowed to dissolve all the alloy carbides and homogenize the composition. Austenitizing conditions include also atmosphere control to prevent decarburization. We must also be careful not to austenitize at very high temperatures because austenite grain-size coarsening will occur. Large austenite grain size results in poor toughness of the hardened steel. Another effect of exceedingly high austenitizing temperatures is that the excess heat that needs to be extracted during cooling makes it difficult to achieve critical cooling rates to obtain martensite and may induce more thermal stresses. The specified, standard austenitizing temperatures, such as those in Table 9-5, when followed, obviates this problem. With the austenitizing temperature selected, it is imperative that we allow all the carbides and alloy additions to dissolve completely and to homogenize the composition by holding the steel at the austenitizing temperature for a sufficient time. The common practice is to hold the steel at the austenitizing temperature for one hour for every inch of thickness. This provides sufficient time for the center section to reach temperature and for carbon and alloying elements to diffuse.

The control of austenitizing temperature and hold time at temperature is especially critical in tool steels containing high-carbon and high-alloying elements that form stable carbides. The austenitizing temperatures are usually much higher than for plain carbon steels and can go as high as 1250°C to 1300°C in high-speed tool steels. The high temperatures are used to induce dissolution of the stable alloy carbides. A compromise is made not to dissolve all the carbides by keeping the hold time at temperature for tool steels low. The undissolved carbides help to prevent excessive austenite grain growth. Complete dissolution of all carbides is not imperative because there is enough alloying and carbon in solution in austenite in these steels to produce the desired hardening.

The high alloy and high carbon in homogeneous solution in austenite will lead to lower martensite start (M_s) and finish (M_f) temperatures. The martensite start and finish temperatures decrease when the carbon and alloy contents are increased; consequently, they are pushed to lower temperatures. When the M_f temperature is lower than room temperature, retained austenite will be present in the microstructure. When a steel is used with retained austenite in its microstructure, it can exhibit distortion, shape or dimensional tolerance problems, or possibly even cracking. Distortion or dimensional tolerance problems arise if the retained austenite transforms during use of the steel (component). The austenite, being FCC, will expand when it transforms to BCC ferrite. If the austenite transforms to martensite, the martensite is as-quenched, untempered, and brittle. Thus, the component might crack. To avoid these problems, it is imperative to eliminate this retained austenite before the steel (component) is put to service. To eliminate the retained austenite, cryogenic cooling may be used and double or sometimes triple tempering of the component is accomplished. The cryogenic cooling is intended to complete the martensitic transformation by having the austenite pass through the M_f temperature. Double or triple tempering will also induce isothermal transformation of the austenite at the same time giving ductility and toughness to the hardened steel.

9.7.3b Embrittling Phenomena. Embrittling phenomena are those in which steels have been observed to deteriorate in toughness and ductility after quenching and tempering, or after having been placed in service; these may lead to brittle fracture. Certain heat treatments and elevated temperature service make steels susceptible to many forms of embrittling phenomena. The most common observation when a steel is embrittled is a shift in the impact transition to higher temperatures. The two common embrittling phenomena in Q+T low alloy engineering steels are *tempered martensite embrittlement (TME)* and *temper embrittlement (TE)*. Both exhibit a shift in impact transition temperature and the embrittlements lead to poor toughness and intergranular fracture along prior austenite grain boundaries.

The TME phenomenon occurs after tempering between 260°C to 370°C (500°F to 700°F) and is sometimes referred to as *350°C embrittlement* or *500°F embrittlement.* It occurs very quickly in the time normally used to temper the as-quenched martensite and because of this, it is also characterized as a "one-step embrittlement" and is independent of the size of the structure or steel and/or cooling rate through the range of temperatures. Thus, we must avoid tempering in this range of temperatures to achieve the desired tempered properties. The cause of the embrittlement appears to be the interaction of residual impurities, such as phosphorus, with the precipitation of cementite at the grain boundaries during tempering in this range of temperatures.

The TE phenomenon occurs after tempering in or cooling through the temperature range of 375°C to 575°C (707°F to 1070°F). Sometimes, it is also called a two-step embrittlement phenomenon because it requires two tempering treatments in the range of temperatures or a combination of a heating step and a cooling step to induce the embrittlement. It is a slow process that takes many hours or even days to develop. TE is therefore of major concern in heavy sections that are tempered or are used at high temperatures. In particular, large shafts and rotors for power generating equipment are susceptible to TE because after use at high temperatures, the heavy sections cool slowly through the critical range of embrittlement. Through the temperature range, TE exhibits a C-curve behavior in tempering temperature and time. The nose or minimum time for embrittlement occurs at about 550°C. One study shows that it took about one hour at 550°C to notice the first shift in transition temperature. In contrast, it took several hundred hours to notice the first signs of embrittlement at around 375°C, the lower limit of the range. TE is reversible and deembrittlement may be accomplished on heating to slightly above 575°C after holding for only a few minutes.

TE is observed only when specific impurities are present in the steel. The impurities most detrimental are antimony, phosphorus, tin, and arsenic. Relatively minute amounts, on the order of 100 ppm (parts per million) or 0.01 wt percent, of these elements are sufficient to trigger TE.

Silicon and manganese in large amounts also appear to induce TE. Plain carbon steels are not susceptible to TE when the manganese content is less than 0.5 percent; however, alloy steels, especially the Cr-Ni steels frequently used for heavy rotors, are very susceptible. Molybdenum, in the amounts of ≤0.5 percent, reduces and minimizes the susceptibility to TE.

Unhardened plain carbon and some alloy steels acquire an increase in strength but with a marked decrease in ductility and impact strength when they are heated between 230°C and 370°C. This range of temperatures is in the blue-heat range and the embrittlement is called "*blue brittleness*" caused by *strain aging.*

Embrittling Phenomena in Stainless Steels. In the heat treatment of stainless steels, *heating or slow cooling in the 425°C to 900°C range for the austenitic grades, and above 925°C and in the 370°C to 540°C range for the ferritic grades produce embrittlement.* The embrittlement in the austenitic grades is due to precipitation of chromium carbides along the grain boundaries that induces intergranular corrosion and embrittlement. This precipitation of the carbides is referred to as *sensitization.* The sensitizing range for ferritic stainless steels lies above 925°C. Sensitization may be minimized by using very low carbon contents or through the addition of carbide formers, such as titanium and niobium, to stabilize the carbides. The embrittlement of the ferritic grades in the 370° to 540°C range is caused by the formation of the brittle untempered chromium-rich martensite, α', phase, and the effects of this embrittlement increase rapidly with chromium content. To avoid this ferritic embrittlement, control of heat treatment is imperative. Both phenomena are reversible and de-embrittlement may be achieved by reheating the grades above 900°C for austenitic, and below 925°C for ferritic grades, followed by very rapid cooling through the respective ranges of temperature. For the ferritic grade, the range for deembrittlement or desensitization is 650° to 815°C to avoid both sensitization and α' embrittlement.

Ferritic and austenitic stainless steels may also be embrittled after they are cooled to room temperature after prolonged exposure in the temperature range of 560° to 980°C. The cause of this *embrittlement* is the formation of the *sigma-phase*, an approximately equiatomic iron-chromium, FeCr, intermetallic compound. The sigma-phase formation occurs after initial heating in the range of 1040° to 1150°C and either by slow cooling in the 560° to 980°C embrittling range or by quenching from the initial heating range, followed by heating in the embritting range; 850°C produces the greatest effect. Embrittlement is most detrimental when the steel has cooled to 260°C and below. Boiler tubes made of high-chromium ferritic stainless steels, such as type 446, have been known to shatter when dropped following removal after extended service.

As in tool steels, the martensitic stainless steel grades can be embrittled if *retained austenite* is present in the microstructure of the steel. In addition to its embrittling effect, the retained austenite leads to unacceptable dimensional and distortional changes, as discussed earlier. Cryogenic cooling to −75°C immediately after quenching to room temperature transforms some of the retained austenite to martensite. To eliminate most of the retained austenite, double tempering may be necessary.

9.7.4 Surface Hardening Processes

There are a lot of materials applications that require good, wear resistance, and resistance to high contact stresses at the surface, while at the same time providing good toughness at the core of the material. These requirements need high hardness and strength at the surface while keeping the core or inner section of the material soft and low strength. This gradient in properties through the section of the part or component can be achieved through *surface hardening* in which a hard martensitic structure is produced at the surface while nonmartensitic structures, pearlite or bainite, are at the inner section or core of the part. Surface hardening is usually specified to a certain depth from the surface resulting in a shell or case of hard material. Thus, *it is commonly referred to also as case hardening.* It may be *achieved with or without change of the surface composition* from the original steel composition.

9.7.4a Processes with No Change in Surface Composition. These processes are also called *localized heat treatment processes* because the surface is selectively heat-treated (i.e., the surface is austenitized and then quenched very quickly to produce martensite). The obvious requirement for these processes is that the original composition must have sufficient carbon and hardenability to achieve the required hardness at the surface. Medium carbon steels are usually suited for these processes. The processes are classified according to the heating source and these are *flame hardening, induction hardening, laser hardening, and electron-beam heat-treating.* In addition to increased wear resistance, the surface hardening (due to the martensite formation) also induces residual compressive stresses that result in improved bending and torsional strength as well as fatigue properties.

These processes are applied because of one or more of the following reasons: (1) parts to be heat-treated are so large as to make conventional furnace heating and quenching impractical and uneconomical—examples are large gears, large rolls, and dies; (2) only a small segment, section, or area of the part needs to be heat-treated—typical examples are ends of valve stems and push rods, and the wearing surfaces of cams and levers; (3) better dimensional accuracy of a heat-treated part; and (4) overall cost savings by using inexpensive steels to have the wear properties of alloyed steels.

Flame hardening employs direct impingement of a high-temperature flame or a high-velocity composition-product gas to rapidly austenitize a thin shell of the surface of the steel and then immediately quench to martensite. The high-temperature flame is obtained by a combustion of a mixture of fuel gas with oxygen or air. Examples of fuel gases used and the resulting flame temperature after combustion with oxygen or air are shown in Table 9-7. Depths of hardening from about 0.8 mm (1/32 in.) to 6.4 mm (1/4 in.) or more can be obtained. The process may be used for through-hardening of a steel 75 mm (3-in.) or less in cross section, depending on the hardenability of the steel. Medium carbon steels with 0.40–0.50 percent carbon are ideal for flame hardening, but steels up to 1.50%C can be flame-hardened with special care. The surface-hardened shell needs also to be tempered and flame tempering may also be used.

The heating for *induction hardening* comes from electromagnetic induction or simply induction of electrical eddy currents in the material to be heated—that is, the workpiece. The eddy currents dissipate energy and give rise to the heating. The basic components of an induction heating system are: (1) an induction coil, (2) an alternating-current (AC) power supply, and (3) the workpiece itself. The coil is commonly a copper tubing through which cooling water passes and takes a variety of shapes to suit the part to be heated. The AC current flows through the coil, generates an electromagnetic field that cuts through the workpiece, and which induces the eddy currents to heat the workpiece. The advantages of induction heating are rapid heating, less scale loss, fast start up, high production rates, and energy savings. Examples of inductor coils and their heating patterns are shown in Fig.9-63.

A common analogy used to explain the phenomenon of electromagnetic induction is the transformer effect. A transformer consists of two coils placed in close proximity to each other. When a voltage is impressed across one of the coils, known as the "primary," an AC voltage is induced across the other coil, the "secondary." In induction heating, the induction coil serves as the primary coil and when AC current is passed through it, eddy current is induced in the workpiece, the secondary.

Surface hardening is just one of the many applications of induction heating. Table 9-8 lists the other

Table 9-7 Characteristics of fuel gases used for flame hardening. [2]

	Heating Value		Flame Temperature			
			With Oxygen		With Air	
Gas	MJ/m^3	Btu/ft^3	°C	°F	°C	°F
Acetylene	53.4	1433	3105	5620	2325	4215
City Gas	11.2–33.5	300–900	2540	4600	1985	3605
Natural Gas (Methane)	37.3	1000	2705	4900	1875	3405
Propane	93.9	2520	2635	4775	1925	3495

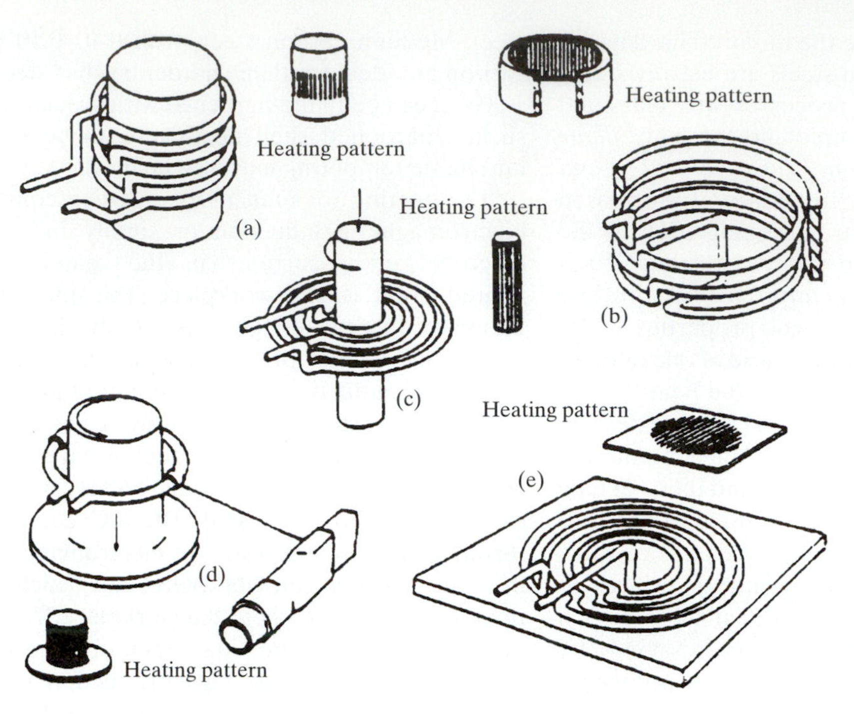

Fig. 9-63 Typical inductor coils and the heating pattern they produced on the surface of a material. [2]

Table 9-8 Applications of Induction Heating and Typical Products. [2]

Preheating prior to metalworking	Heat treating	Welding	Melting
Forging Gears Shafts Hand tools Ordnance	Surface Hardening, Tempering Gears Shafts Valves Machine tools Hand tools	Seam Welding Oil-country tubular products Refrigeration tubing Line pipe	Air Melting of Steels Ingots Billets Castings
Extrusion Structural members Shafts	Through Hardening, Tempering Structural members Spring steel Chain links		Vacuum Induction Melting Ingots Billets Castings "Clean" steels Nickel-base superalloys Titanium alloys
Heading Bolts Other fasteners			
Rolling Slab Sheet (can, appliance, and automotive industries)	Annealing Aluminum strip Steel strip		

applications of induction heating in the metals-processing industries. Since induction heating can be used for through hardening, the depth of case hardening can be varied, depending upon the needs or specifications. The depth is a function of the frequency of the AC current, with lower frequency producing a deeper case. A shallow-hardened case from 0.25 (0.01 in.) to 1.5 mm (0.06 in.) in depth, provides a part with good wear resistance in light to moderate loading. Rocker-arm shafts, sucker-rod couplings, and pump shafts are examples of parts requiring a shallow case depth. Case depths in the range from 1.5 mm (0.06 in.) to 6.4 mm (0.25 in.) are needed to sustain heavy or impact-type loading in addition to providing wear resistance. Examples of parts requiring these depths are gears, track pins, camshafts, heavy camshaft bearings, and bearing races. Induction-hardening bars and shafts to depths of 3.2 to 12.7 mm has greatly improved torsional and bending fatigue strength. Long bars and shafts are rotated as they pass through the inductor coil where the surface is austenitized and then immediately quenched. Truck, tractor, and automobile axle shafts and hydraulic piston rods are typical products processed in this manner. All surface-hardened parts need also to be tempered.

Laser is a coherent collimated light beam whose power density can be varied. With sufficiently high power density, usually given as watts per square centimeter, heat is generated as the laser impinges on the surface of the workpiece at a rate much faster than heat conduction into the interior can dissipate it. In a very short time, a thin surface layer will have reached the austenitizing temperature while the interior of the workpiece is still cool. In fact, in laser surface hardening, the heating period is frequently shorter than the cool-down time; hence the expression "up-quench" is often used. The quenching of the heated surface is by self-quenching (i.e., the mass of the cool workpiece is used as a heat sink in the same manner as in welding). The focused laser power density for surface hardening is much lower than the power density of the small, intensed focused spots used for welding and cutting.

The resulting case depth will depend on the hardenability of the material, but this is rarely more than 2.5 mm (0.1 in.). For steels with low hardenability, such as plain carbon steels, the case depth obtainable is much smaller, varying from 0.25 mm in low carbon steels (up to 0.2%C) to 1.3 mm in medium-carbon steels. Because of the very fast heating and self-quenching cooling rates obtainable in laser hardening, surface hardening may be achieved in steels not normally considered for through hardening, such as a 0.18 percent C plain carbon steel.

In *electron-beam heat treating,* a focused beam of high-velocity electrons is used to heat the surface. The power density of the electron beam may also be varied and can be used for purposes other than surface hardening such as welding and melting of alloys. As in laser hardening, the heating rate is very rapid and self-quenching of the heated surface is achievable. This means that the mass of the part must be large enough to be a good heat sink and to induce high cooling rates to produce martensite at the surface. The part to be heat-treated must be demagnetized to avoid interference with the electron beam. The case depth achievable in electron beam hardening depends on the time the surface temperature is maintained above the critical temperatures.

9.7.4b Processes with Change in Surface Composition. These are the processes of carburizing and nitriding that were discussed in Chap. 6 in regard to macroscopic diffusion (i.e., the application of Fick's second law). They involve the injection of the interstitial carbon and nitrogen atoms on the surface of low-carbon base steels (usually $\leq 0.20\%$C) to levels where the martensite produced will yield the hardness desired from the case. When the carbon atoms are being injected alone, the process is called *carburizing*; when nitrogen atoms are being injected, it is called *nitriding*. The carburizing process is usually conducted at stable austenite temperatures, whereas the nitriding process is done at subcritical temperatures (i.e., below the A_1), where the ferrite phase is stable. When nitrogen is added simultaneously with carbon at stable austenite temperatures, the process is called carbonitriding. When carbon is added simultaneously with nitrogen at subcritical temperatures, the process is called *nitrocarburizing*. Both carburizing and nitriding may be done when the sources of carbon and nitrogen, are either in the gaseous state, the liquid state, or the solid state. The corresponding process to reflect the states of carbon is either gas carburizing, liquid carburizing, or pack carburizing. The corresponding terms for nitriding are gas nitriding, liquid nitriding, and pack nitriding. In addition to these, nitrogen atoms may also be introduced to the surface directly from discharged plasma glow. This is done in vacuum and is the same process as the ion-implantation process used in the semiconductor industry in which the atomic elements are bombarded onto the surface—a process now commonly called ion nitriding. Gas carburizing may also be done under vacuum (i.e., pressures under atmospheric pressure) and it is designated *vacuum carburizing.*

Mechanism of Carburization. In Chap. 6, we were concerned only with the surface concentration of carbon and did not consider how this carbon concentration got to the surface. We shall now discuss the mechanism of gas carburizing in order to learn the principal processing variables that need to be controlled. The process of introducing carbon atoms into a steel involves a series of steps:

1. Transport of gas molecules containing the carbon atom from the carrier gas to the surface of the steel,
2. Reaction of the molecules at the surface to raise the carbon content of the steel to C_s, and
3. Diffusion of the carbon into the steel.

The primary gas molecule that is responsible for raising the carbon content of the steel at the surface is CO, carbon monoxide. This is shown by *either of the following two principal surface reactions:*

$$2CO(g) \rightleftarrows C_s(\text{in Fe}) + CO_2(g) \qquad (9\text{-}8)$$

$$CO(g) + H_2(g) \rightleftarrows C_s(\text{in Fe}) + H_2O(g) \qquad (9\text{-}9)$$

The two reactions are called heterogeneous reactions because they involve two phases—the solid phase (carbon in Fe at the surface, C_s) and the gaseous phase. All of the reactants and products of the reactions are gases, except C_s. These reactions are considered reversible and in equilibrium and the concentrations of the gas species will control the carbon content on the surface of the steel. Thus the carbon content may be *controlled by either a constant CO_2 content or a constant water vapor content determined by the dew point of the gas.*

The principal carrier gas in carburizing is natural gas that contains more than 80 percent methane, CH_4. Methane reacts with carbon dioxide and water in the following manner to generate CO and H_2.

$$CH_4(g) + CO_2(g) \rightleftarrows 2CO(g) + 2H_2(g) \qquad (9\text{-}10)$$

$$CH_4(g) + H_2O(g) \rightleftarrows CO(g) + 3H_2(g) \qquad (9\text{-}11)$$

Combining Eqs. 9-8 and 9-10, and Eqs. 9-9 and 9-11, we obtain for both combinations,

$$CH_4(g) \rightleftarrows C_s(\text{in Fe}) + 2H_2(g) \qquad (9\text{-}12)$$

The rate of transfer of carbon atoms from the gas phase to the surface may be represented by

$$J = -D(\delta C/\delta x)_s = k(C_a - C_s) \qquad (9\text{-}13)$$

where J is the flux, atoms/cm^2/sec, D is the diffusion coefficient, $(\delta C/\delta x)_s$ is the carbon concentration gradient from the gas phase to the surface, k is the surface reaction rate constant, *C_a is called the carbon potential of the gas carrier,* and C_s the surface carbon content. From the equation, we note that the carbon potential must always be larger than the surface carbon content in order for carbon to diffuse to the surface. The carbon potential is controlled by either the CO_2 or the dew point of the gas, as indicated by Eqs. 9-8 and 9-9 and shown in Figs. 9-64a and b. At the start of the process, the surface concentration is C_o and as time progresses it builds up to C_s. The actual build-up of C_s with time is illustrated in Figs. 9-65a and b for two different methane concentrations. We note that the surface carbon concentration is not constant and is a function of both time and the methane concentration. If we had considered this variability of C_s, the solution to Fick's second law in Chap. 6 could only be obtained numerically.

Thus, as indicated in Chap. 6, the surface carbon concentration, C_s, was assumed constant in order to solve Fick's second law. The solution allows us to estimate the carbon gradient from the surface to the inside of the steel,with the following equation,

$$\frac{C(x,t) - C_o}{C_s - C_o} = 1 - \text{erf}\left(\frac{x}{2\text{v}Dt}\right) \qquad (9\text{-}14)$$

where C(x,t) is the carbon concentration at a distance x from the surface after time t, in seconds, at the carburizing temperature T in K by which we can determine the diffusion coefficient D. The C_s is assumed to be the saturated carbon content of austenite in solution at the austenitizing temperature at the A_{cm} line, shown in Fig.6-5. The justification of maintaining C_s constant is that the time to build up the carbon concentration at the surface from C_o to C_s is considered much shorter than the total time of carburizing.

Summarizing, the variables that control the carburizing process are the carbon potential, the carburizing temperature, and the time at the temperature. The purpose of the process is to establish a carbon gradient profile through the section of the carburized structure in order to achieve the desired properties. An example of the carbon gradient in a carburized case and its corresponding hardness profile, is shown in Fig. 9-66 after carburizing a 1024 steel at 900°C for $2^1/_4$ hours. Both of these curves are used to specify the case depth for the required carbon content or hardness level of the hardened case.

Specifications are usually stated as "effective case depth to 50 HRC hardness" or "case depth to 0.40%C," as the case may be. The definitions for total case depth and effective case depth were given in Chap. 6.

The data points shown in Fig. 9-66a are the actual data from the carburizing process at 900°C. We note that the surface carbon concentration is slightly less than 1.2 percent and this compares favorably with the saturated carbon content of about 1.22 percent taken from the A_{cm} at 900°C in Fig. 6-5. Thus, the assumption of constant C_s at the surface is quite satisfactory. We should also note that the shape of the carbon gradient in Fig. 9-66a reflects the complementary error function solution of Fick's second law (see Chap. 6). This is called the normal carbon gradient during carburization.

The constant surface concentration of saturated carbon in austenite, for instance, 1.2 percent in Fig. 9-66a, is maintained constant because the carbon potential in the gaseous phase is maintained high enough in accordance with Eq. 9-13. Such a high carbon concentration at the surface when dissolved in austenite is, however, undesirable because it leads

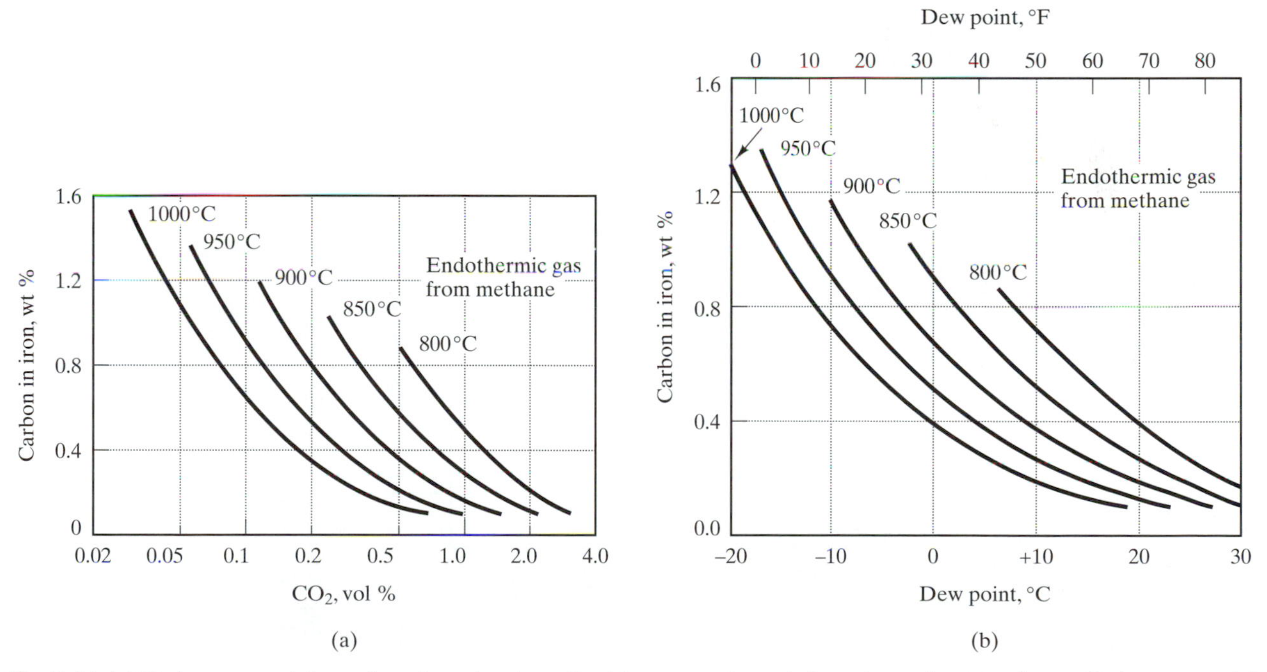

Fig. 9-64 (a) Carbon potential as a function of carbon dioxide content in endothermic gas from methane. Carbon potential given as weight percent carbon in iron, (b) Carbon potential as a function of the dew point in endothermic gas from methane.

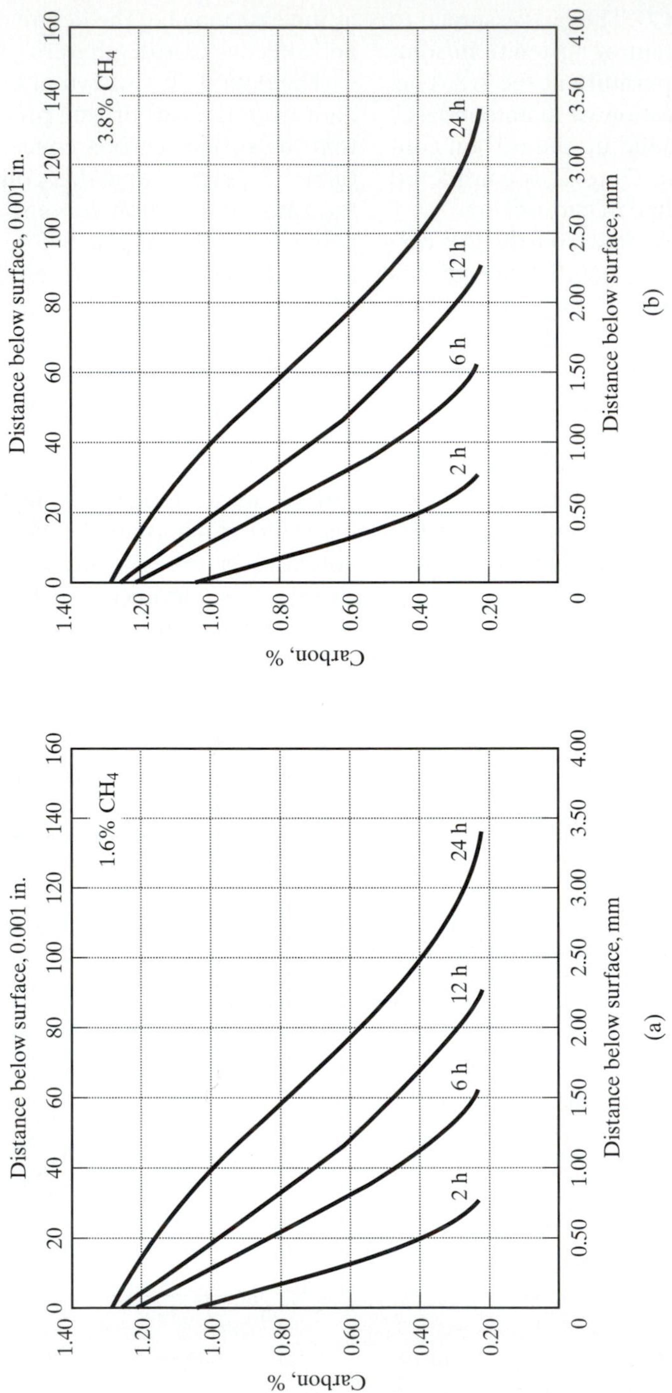

Fig. 9-65 Carbon gradients for carburized 1022 steel bars; carburized at 918°C in 20%CO–40%H_2 with (a) 1.6% methane, and (b) 3.8% methane.

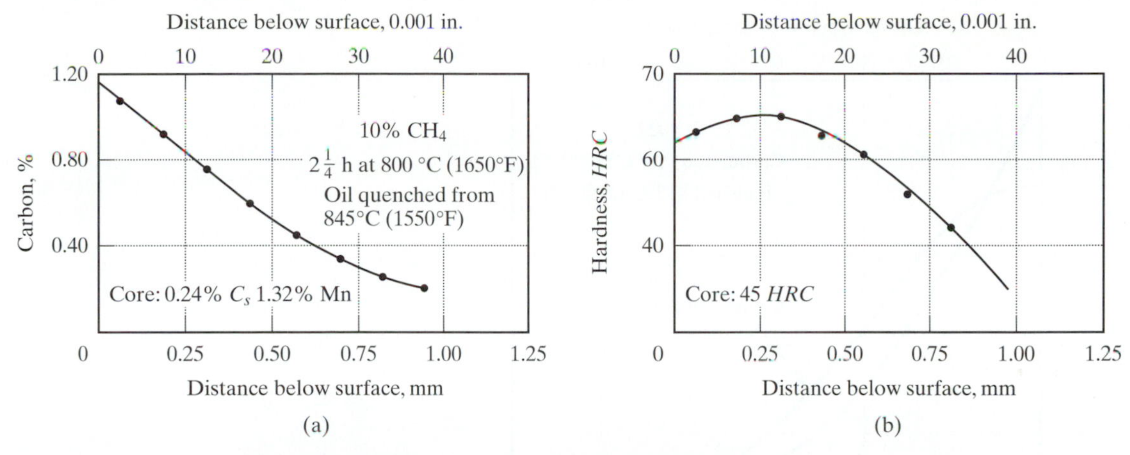

Fig. 9-66 Actual (a) normal carbon gradient and (b) hardness profile in a carburized case of a 1024 steel. Normal carbon gradient maintains saturated carbon percent in austenite at the surface. Low hardness at the surface in (b) is due to retained austenite.

to some retained austenite when the carburized case is quenched. If there is retained austenite on the surface of the carburized case, the hardness will be lower. This is shown in the case hardness profile from the surface to the interior of the section in Fig. 9-66b where the maximum hardness is obtained about 0.25 mm from the surface. The lower hardness at the surface is due to some retained austenite.

The influence of retained austenite on the fatigue properties of case-hardened components is to decrease the bending fatigue strength and to permit surfaces to deform plastically under heavy loads, resulting in ripples, or "orange peels." Thus, it is necessary to control the carburization procedure to produce what is called the modified carbon gradient by which the surface carbon concentration is lowered to below the saturated carbon content of the austenite at the carburizing temperature. This is similar to the drive-in diffusion in the doping of silicon. In fact, the modified carbon gradient is obtained in a continuous carburizing furnace by having three controlled zones: a carburizing zone, a diffusion zone, and a cooling zone. The carburizing zone is analogous to the pre-deposition step in the doping of silicon, and the diffusion zone is analogous to the drive-in step in the doping of silicon. In the diffusion and cooling zones, the carbon potential in the gaseous atmosphere is controlled at much below that needed to maintain a saturated carbon content in austenite. The result is a redistribution of the carbon within the carburized case and a lower surface carbon than the saturated carbon content in austenite on the surface. The result of this zone control, shown in Fig. 9-67, includes for comparison the normal gradient (without zone control). We should note that the shape of the modified carbon gradient is Gaussian and this is very similar to the profile of the drive-in diffusion step used in the semiconductor industry (see Chap. 6—Doping of silicon).

EXAMPLE 9-9

With the definition of critical diameter, D_c, as the diameter containing 50 percent martensite at the center of the round bar, determine the critical diameter of a 4140 when quenching is done in (a) agitated water, and (b) agitated oil. Use the percent martensite and the hardenability curves given in Fig. EP 9-8a with Example 9-8.

SOLUTION: From Fig. EP 9-8a, the 50 percent martensite content is found to be at 10J. In order to get 50 percent martensite, we need to use the equivalent cooling rate at 10J.

(a) For quenching in agitated water, we use 10J as the abscissa in Fig. 9-52 and its intersection with the curve marked "center" suggests that the critical diameter is ~70 mm.

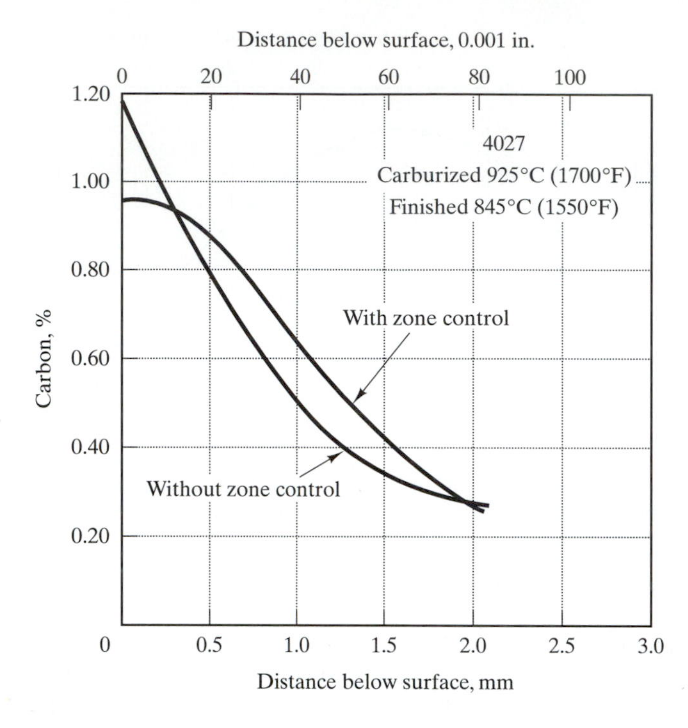

Fig. 9-67 Modified carbon gradient is achieved with zone control similar to drive-in diffusion is doping silicon.

(b) For quenching in agitated oil, the intersection of 10J in Fig. 9-53 with the "center" suggests that the critical diameter is ~50 mm.

EXAMPLE 9-10

The composition of the steel is 0.40 percent C, 1.00 percent Mn, 1.00 percent Cr, and 0.25 percent Mo. Estimate the maximum (critical) diameter that will produce 50 percent martensite at the center of this steel when it is quenched in a medium with H = 4.0. Assume the austenite grain size is 7.

SOLUTION: To estimate the critical diameter, D_c, we need to estimate first the ideal critical diameter, D_I, according to Eq. 9-7,

$$D_I = F_C \cdot \prod_1^n \cdot F_i$$

With 0.40%C and the austenite grain size of 7, we obtain $F_C = 0.215$. (The latter is called the ideal diameter in the ordinate because it will be the ideal diameter when the carbon is the only alloying element.) We obtain the F_i's from Fig. 9-58; $F_{Mn} = 2.72$, $F_{Cr} = 2.10$, and $F_{Mo} = 1.35$. Thus,

$$D_I = 0.215 \times 2.72 \times 2.10 \times 1.35 = 1.66 \text{ in.}$$

From this D_I, we obtain D_c using the curve in Fig. 9-56b with H = 4.0. We get $D_c = 1.35$.

EXAMPLE 9-11

Given the isothermal transformation diagram for 4140 in Fig. EP 9-11 indicate the processing steps for (1) martempering, and (2) austempering.

SOLUTION: For both martempering and austempering, we must first normalize and then austenitize the steel. The normalizing and austenitizing temperatures are given in Table 9-5 and are 870°C and 845°C. In both cases, the austenite must be allowed to homogenize. After austenitization, martempering is accomplished by quenching to 360°C (680°F), holding there for about 5 sec, then air-cool to room temperature, temper below or above the tempered martensite embrittlement range (which is 260° to 370°C (500–700°F) depending on the hardness desired.

After austenization, austempering is accomplished by quenching to 360°C (680°F) and letting the steel transform to bainite for about 10 minutes (600 sec) and then air-cooling. Bainite does not need to be tempered.

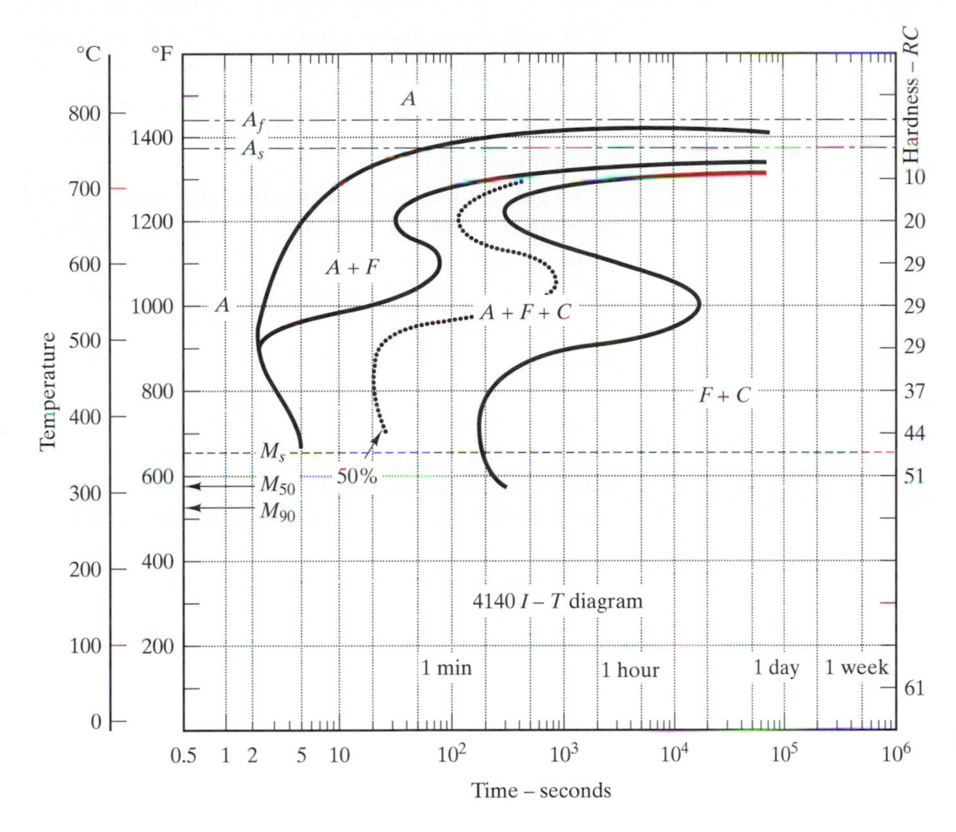

Fig. EP 9-11 Isothermal transformation diagram of 4140.

EXAMPLE 9-12

A 25-mm diameter 8620 steel bar was carburized at 900°C for eight hours. The diffusion coefficient of carbon in austenite, as given in Table 6-2, is

$$D = 16.2 \exp\left(-\frac{137{,}800}{8.314T}\right) \quad \text{mm}^2/\text{sec}$$

Determine (1) the location in the carburized case where a quenched and tempered hardness of 48 HRC may be obtained with a minimum of 95 percent martensite, and (2) whether the desired hardness can be obtained by quenching in agitated water or in agitated oil?

SOLUTION:

(1) The question that is really being asked is find the carbon content corresponding to the 48 HRC quenched and tempered (Q+T) hardness with 95 percent martensite and its location in the carburized case. To determine the carbon content, we first convert the Q+T hardness to as-quenched hardness by using Fig. 9-44. The as-quenched hardness is 54 HRC. Then, we use Fig. 9-38b or Fig. 9-39 to convert the as-quenched hardness to carbon content because the hardness of as-quenched martensite is only a function of carbon content. The carbon content is found to be 0.45 percent C = C(x,t) in the carburizing equation

$$\frac{C(x,t) - C_o}{C_s - C_o} = 1 - \text{erf}\left(\frac{x}{2\sqrt{Dt}}\right)$$

from which we can solve for x, the location of C(x,t) = 0.45 percent C. At 900°C (1173 K), C_s = 1.24 percent C for 8620 steel from Fig. 9-65 and the diffusion coefficient is

$$D = 16.2 \exp\left(-\frac{137{,}800}{8.314 \times 1173}\right)$$

$$= 1.183 \times 10^{-5}\ \text{mm}^2/\text{sec}$$

Then

$$\frac{0.45 - 0.20}{1.24 - 0.20} = 1 - \text{erf}\left(\frac{x}{2\sqrt{Dt}}\right)$$

and

$$\text{erf}\left(\frac{x}{2\sqrt{Dt}}\right) = 1 - 0.24 = 0.76$$

and from Table 6-1,

$$\frac{x}{2\sqrt{Dt}} = 0.84$$

and

$$x = 1.68\sqrt{1.183 \times 10^{-5} \times 83{,}600} = 0.9806 \text{ mm}$$

(2) We need to know now whether we can obtain the desired as-quenched hardness of 54 HRC at 0.98 mm from the surface by quenching in water or oil. We need to know the Jominy equivalent cooling rate at this location and then use a hardenability curve. Rounding the location to 1 mm, this location from the center of the 25-mm diameter bar is (11.5/12.5 = 0.92R).

For agitated water quenching, the Jominy equivalent rate at 0.92R (very close to surface) is 1J. When the S (surface) curve in Fig. 9-52 is extrapolated to 25 mm diameter from where it ends at 50 mm diameter, the J value will definitely be less than one. The hardenabilty curve starts at 1J, however, and therefore the smallest value of the Jominy equivalent cooling rate is 1J.

To get the hardness at this location with C = 0.45 percent, we look for the hardenability curve of 8645, which is shown in Fig. 9-53f. Note that we started with an 8620 steel. However at the location of interest, the composition is 0.45 percentC and therefore the alloy at that point is 8645. The minimum hardness at 1J for 8645 from Fig. 9-51f is ~57 HRC.

Doing the same for the agitation in oil and using Fig. 9-53, the Jominy equivalent cooling rate at the 0.92R location is found to be ~2.5J. Going to Fig. 9-51f and using 0.45 percentC, the minimum hardness at 2.5J is ~56 HRC.

We see that 57 HRC for water quenching and 56 HRC for oil quenching both exceed the required 54 HRC. Therefore, we can use either quench, but because oil quenching is less drastic, we should use oil quenching.

Summary

We have discussed nonhardening and hardening thermal processes, primarily for metals and alloys. The primary nonhardening processes for all metals and alloys are stress-relief and annealing. As the name implies, stress-relief relieves the residual stresses that form in the material after almost any processing. Annealing softens the material for the final fabrication or additional shaping steps. Normalizing and spheroidizing treatments are specific for steels and are applied for specific purposes. Normalizing homogenizes the composition and microstructure and is always a treatment given before austenitizing for the thermal hardening process. Spheroidization is applied for high-carbon steels to improve machinability and formability.

The general principles of thermal hardening processes were discussed. These consist of converting a two-phase alloy at room temperature into a single phase alloy at higher temperatures, quenching the single phase to room temperature or lower to form a supersaturated solution, and then aging or tempering. For most metals and alloys, the supersaturated solution is the softest condition and hardening is achieved with the precipitation or separation of a second phase or phases that may be coherent or noncoherent during aging. Precipitation hardening alloys depend on coherent transition precipitates to strengthen the alloys. Hardening of duplex alpha-beta titanium alloys depends on the mutual deformation of the precipitating noncoherent alpha phase and the supersaturated beta phase. Spinodal decomposition in some copper alloys is the separation of a single phase alloy at high temperatures into very fine modulated richer and leaner regions of solute than the average composition at lower temperatures. The strengthening or hardening arises from the fineness and the coherency of the modulating phases. In some instances, ordered phases separate and strengthen the alloys because deformation in these phases is due to superdislocations.

After austenization, the hardening of steels is due to the refinement of the microstructures, from coarse pearlite to lower bainite, when they are isothermally transformed. When quenched to produce martensite, the hardening is due to the straining of the ferrite equilibrium BCC to the BCT lattice when the ferrite is supersaturated with carbon. In contrast to most metal alloys, the as-quenched martensite in steels containing carbon is the hardest microconstituent, and therefore as-quenched steels are in their hardest condition. For this reason, quenching to form martensite is the primary strengthening mechanism for steels. When quenched steels are tempered, their hardness decreases steadily but their ductility and toughness increase. Steels are never used in the as-quenched condition and are always tempered. The presence of high concentration of alloying elements will induce precipitation of transition-coherent alloy carbides during tempering. As in precipitation hardening, this produces additional strengthening that is referred to as secondary hardening and is specifically beneficial for tool steels in order to be able to maintain their hardness when they are used to cut workpieces.

Empirically, the hardness of as-quenched martensite in steels is a function of carbon content only. Furthermore, the as-quenched hardness is related to the tempered hardness and the latter correlates with the tensile strength. Thus, it is possible to determine the required carbon content of the steel when we are required to have a certain tensile strength (and fatigue limit because of its relationship to tensile strength). Plain carbon steels cannot be through-hardened, however, in thick gauges or larger diameters than 10 mm. The ease of forming martensite is called the hardenability of the steel and this is improved by alloying elements. The hardenability of the steel is quantitatively measured by the Jominy test, and the point of inflection of the Jominy hardenability curve indicates where the 50 percent martensite structure exists. A Jominy distance in the hardenability curve not only indicates the hardness at that point, but it also corresponds to a cooling rate—in other words, each hardness on the hardenability curve is produced by a unique cooling rate. This concept is the basis of the Jominy equivalent cooling rate that says that a hardness of a part or component made of a steel will be achieved if it achieves the cooling rate corresponding to that hardness in the Jominy curve. Another measure of hardenability is called the ideal critical diameter that is defined as that which will contain 50 percent martensite at the center of the diameter when quenched ideally with the severity of quench, $H = \infty$. This can be estimated from the composition and the actual critical diameter can be obtained for the severity of quench H available. Hardenability of steels is one of the factors used in the selection of low alloy steels for components.

References

1. D. J. Wulpi, *Understanding How Components Fail,* ASM International, 1985.
2. Heat Treating, vol 4, *ASM Metals Handbook,* 9th ed., ASM International, 1981.
3. G. Krauss, *Principles of Heat Treatment,* ASM International, 1980.
4. "Aluminum and Aluminum Alloys," J. R. Davis, Editor, *ASM Specialty Handbook,* 1993.
5. *Aluminum—Properties and Physical Metallurgy,* J. E. Hatch, Editor, ASM, 1984.
6. *Titanium Alloys—a Materials Property Handbook,* ASM International, 1994.
7. A. Cottrell, *An Introduction to Metallurgy,* 2d ed., Edward Arnold, 1975.
8. M. F. Ashby & D. R. H . Jones, *Engineering Materials 2,* 1st ed., Pergamon Press, 1986.
9. *Atlas of Isothermal Transformation and Cooling Transformation Diagrams,* ASM, 1977.
10. *Binary Phase Diagrams,* ASM.
11. R. E. Smallman, *Modern Physical Metallurgy,* 4th Ed., Butterworths, 1985.
12. C. S. Barrett and T.B. Massalski, *Structure of Metals,* 3d ed., McGraw-Hill, 1966.
13. M. Atkins, *Atlas of Continuous Cooling Transformation Diagrams for Engineering Steels,* revised U.S. edition, ASM, 1980.
14. R. F. Kern and M. E. Suess, *Steel Selection,* 1st ed., John Wiley, 1979.
15. C. A. Siebert, D. V. Doane, D. H. Breen, *Hardenability of Steels,* ASM, 1977.
16. E. C. Bain and H. W. Paxton, *Alloying Elements in Steel*, 2nd ed., ASM, 1996.
17. D. R. Askeland, *The Science and Engineering of Materials*, 3rd Ed., PWS Publishing, 1994.
18. R. I. Jaffee, The Physical Metallurgy of Titanium Alloys, p. 65 in *Progress in Metal Physics*, Vol. 7, B. Chalmers and B. King, Eds., Pergamon Press, 1958.
19. L. Kaufman and M. Cohen, Thermodynamics and Kinetics of Martensitic Transformation, p. 165 in *Progress in Metal Physics*, Vol. 7, (See Ref. 18).
20. SAE Standard J406C.

Terms and Concepts

Actual critical diameter
Aging
AISI-SAE steel designation
Artificial aging
As-quenched martensite
Ausforming
Austempering
Austenitizing
Auto-tempering
Bainite
Blue brittleness
C-curve
Carbonitriding
Carbon potential
Carburizing
CCT (continuous cooling transformation) diagram
Coarse pearlite
Coherent precipitates (precipitation)
Critical cooling rate
Dew point
Equilibrium precipitates
Fine pearlite
Flame hardening
Full anneal
GP (Guinier-Preston) zones
H-steels
Hardenability
Hardenability band
Heat-treatable
Heat treating
Homogeneous austenite
I-T (Isothermal transformation) diagram
Ideal critical diameter
Induction hardening
Intercritical anneal
Jominy end-quench test
Jominy equivalent cooling rate
Laser hardening
Lower bainite
Martempering
Martensite
Martensite transformation
M_f—martensite finish temperature
M_s—martensite start temperature
Natural aging
Nitorcarburizing
Nitriding
Noncoherent precipitates (precipitation)
Nonheat-treatable
Normalizing
Nucleation and growth transformation
Order transformation
Patenting
Pearlite
Primary hardening
Process anneal
Q+T (Quenched and tempered)
Quenching
Quench severity
Retained austenite
Secondary hardening
Sensitization
Sigma-phase embrittlement
Solution anneal
Solution heat treat
Spheroidizing
Spinodal transformation
Strain aging
Subcritical anneal
Supersaturated solution
Surface hardening
Tempered martensite
Tempered martensite embrittlement
Temper embrittlement
Tempering
Tempering parameter
Thermal hardening (heat treating)
Thermal stress relief
Through hardening
Transition precipitates
TTT (Time temperature transformation) diagram
Upper bainite
Vacuum carburizing
Zone control in carburizing

Questions and Exercise Problems

9.1 What are the three types of annealing specific to steels? What is the purpose of doing each of the types of annealing?

9.2 What is meant by a "normal" microstructure in steels? What is the normalizing (non-hardening) process for steels? What type of steels can be normalized?

9.3 What is the process of spheroidizing and why do steels have to be spheroidized?

9.4 What do the C-curves indicate during solid state transformation? Explain the shape of this transformation curve.

9.5 What are the generic steps taken during thermal hardening (heat-treating) of metals and alloys?

9.6 The position of the "nose" of the C-curve is an important consideration during thermal hardening (heat-treating). What does it indicate and why is it important?

9.7 What are the specific steps followed during the heat-treating of (a) precipitation hardening aluminum alloys? (b) ($\alpha + \beta$) titanium alloys? and (c) steels? Why is the ($\alpha + \beta$) titanium alloy ordinarily not brought to the single-phase beta-field?

9.8 What are the unique characteristics of the martensite transformation? Explain the significance of each characteristic.

9.9 What is the primary source of hardening (strengthening) during heat-treating of (a) aluminum alloys, (b) titanium alloys, and (c) steels?

9.10 The yield strength of a precipitation hardened 2014 aluminum alloy (aluminum- 4.5 wt%Cu) after aging at 150°C is 450 MPa. If this total yield strength is primarily due to particle strengthening plus the inherent yield strength of pure aluminum (i.e., $\sigma_{ys} = \sigma_o + \Delta\sigma_{ppt}$), determine the average size of the precipitate particles (assumed spheroids). The yield strength, σ_o, of pure aluminum is 35 MPa and the solubility of copper at 150°C is essentially zero. The strengthening can be assumed to be due to bowing of dislocations as suggested in Eq. 4-38, and specifically by

$$\tau = \frac{Gb}{\lambda}$$

where τ is the shear stress needed to bow a dislocation with Burger's vector, $b = 2.86 \times 10^{-10}$ m, between two particles with average interparticle spacing, λ. The Young's modulus of aluminum is 70 GPa and the Poisson's ratio is 0.31.

9.11 If the average size of the spheroidal particles in the 2014 aluminum alloy in Prob. 9.10 after aging at 150°C is 5 nm, determine the yield strength after aging. Use the same assumptions and the data in Prob. 9.10, except the yield strength.

9.12 From the isothermal transformation diagram of 1080, Fig. 9-31, indicate the processing of the 1080 steel to obtain a structure with a HRC 50. What is the structure? If this structure is used for a fatigue application with constant stress amplitude, what would be the maximum cyclic stress it should be designed for?

9.13 A 1080 steel was austenitized at 775°C for one hour, quenched to 550°C, and then allowed to isothermally transform at 550°C. How long must the austenite be held at 550°C to complete the transformation? What is the expected microstructure and the expected tensile strength?

9.14 For a continuous average cooling rate of 11°C per sec (1200°F per min) from the austenitizing temperature, determine (a) the microstructure, and (b) the hardness expected from a (i) 1040, (ii) 1541, and (iii) 15B41 steel. Explain the differences in microstructure. (Use Figs. 9-33a, b, and c.)

9.15 It is desired to have a tensile strength of 875 MPa from the 1541 and 15B41 steels. Determine the cooling rate for (a) 1541, and (b) 15B41 from the austenitizing temperature to obtain this strength.

9.16 A design requires that a 100 percent martensite structure be present through a 40 mm diameter bar when quenched in oil from the austenitizing temperature. Which steel must you select from those in Fig. 9-34?

9.17 A 1020 plain carbon steel (0.20 wt%C with no other alloy addition) was austenitized and quenched from 770°C. (a) What phases are present in the quenched structure? (b) What are the weight percents of each of the phases? the hardnesses of each phase?

9.18 A 1030 steel (0.30 wt%C) was austenitized at 850°C and then quenched and transformed to a 100 percent martensite structure. What is the as-quenched hardness? If the quenching produced 80 percent martensite, what will be the resulting hardness?

9.19 An as-quenched microstructure shows a dual phase ferrite-martensite structure. Ferrite is 30 percent by weight and the martensite hardness is 830 VHN. Determine the austenitizing temperature and the overall composition of the steel.

9.20 A quenched and tempered steel with a minimum shear strength of 100 ksi and a yield to tensile strength ratio of 0.80 is required. The microstructure must be at least 90 percent martensite. What is the minimum carbon content of the steel?

9.21 A 4340H steel with a composition of 0.40%C, 0.75%Mn, 0.20%Si, 0.80%Cr, 1.80%Ni, and 0.25%Mo is desired to have at least 95 percent martensite structure with a quenched and tempered hardness of 45 HRC (~ 450 DPH). Determine the tempering temperature and time necessary to obtain the desired quenched and tempered hardness.

9.22 What is the difference between hardenability and hardness?

9.23 Describe the Jominy end-quenched test.

9.24 In relation to Figs. 9-33a, b, and c, what would be the expected microstructures and hardnesses of the three steels at the 6J and 36J positions in their Jominy (hardenability) curves?

9.25 In Fig. 9-48, (a) why do the Jominy curves of 4140H, 8640H, and 1340H merge at 1J? (b) What do the inflection points in the Jominy curves indicate? (c) based on the answer in b), which of the three steels is more hardenable?

9.26 Based on the minimum hardenability curve in Fig. 9-51f, plot the expected minimum hardness across the cross section (diameter) of a 75-mm diameter 8650H steel bar quenched in (a) agitated water, and (b) agitated oil.

9.27 Based on the minimum hardenability in Fig. 9-51d, plot the expected minimum hardness across the cross section (diameter) of a 50-mm diameter 4150H steel bar quenched in (a) agitated water, and (b) agitated oil.

9.28 A 25-mm diameter 8620 steel bar was carburized at 950°C for 30 hours and then was quenched in agitated oil. Determine the minimum hardness of the effective case depth (with 0.40%C)?

9.29 What is the difference between an ideal critical diameter and an actual critical diameter?

9.30 Estimate the ideal critical diameter of a steel when the austenite grain size is ASTM No. 7, and the composition is: 0.40%C, 0.75%Mn, 0.30%Si, 2.00%Ni, 1.00%Cr, and 0.30%Mo. What would be the actual critical diameter when quenching is done in violent agitation in brine?

9.31 Estimate the ideal and actual critical diameters of a 1040 steel with the composition shown in Fig. 9-34b when quenched in water. Compare the estimated result with the bar diameter found in Fig. 9-34b that will yield 100 percent martensite when quenched in water.

9.32 Differentiate a martempering process from an austempering process. What is the purpose of both of these processes?

9.33 What is an ausforming process? What is expected of an ausformed steel?

9.34 What is the difference between tempered martensite embrittlement and temper embrittlement? What is blue brittleness?

9.35 What is sensitization in stainless steels? sigma-phase embrittlement? How does the sensitization of the austenitic stainless steels differ from the ferritic grades?

9.36 What is the difference between carbonitriding and nitrocarburizing?

9.37 What is "zone control" in carburizing? What is the difference in carbon gradient between carburizing cycles with and without zone control?

III

Performance and Materials Selection for Engineering Design

10

Corrosion and Corrosion Control

10.1 Introduction

So far we have learned how to select materials based on: the composition and "cleanliness" (Chap. 7); the shape (bars, structural, and flats) and condition (hot-rolled, cold-worked, temper-rolled) (Chap. 8); and the thermal processing (stress-relief, normalized, annealed, precipitation-hardened, quenched and tempered, hardenability, carburized or nitrided) (Chap. 9). The next thing we need to know to select materials is their performance in the service environment they are intended for. This is the field of corrosion.

Corrosion has traditionally been referred to as the degradation of metals by chemical or electrochemical reaction with the environment, which may include the atmosphere, fluids, temperature, pressure, and stress. If we include the degradation of plastic by its chemical interaction with the ultraviolet rays of sunlight, we may say in general that corrosion is the degradation of materials. Glasses undergo also a degradation process called weathering (described in Sec. 14.3.3) that involves the leaching of ions under high humidity conditions or alternating cycles of moisture condensation and evaporation—the same conditions prevalent in aqueous corrosion of metals. Refractories (ceramics) degrade also under action of pressure, stress (thermal stress induces spalling of bricks), and the environment that may include gases, such as CO and SO_2, slags, metals, and dusts in gases (see Sec. 14.5.3). We shall, however, confine our discussion in this chapter to the traditional corrosion or degradation of metals.

Corrosion, as we shall see shortly, is the conversion of an element from a zero or lower valence state to a positive or higher valence state. This process is called oxidation. In the refining of metals, we have seen that most metals occur naturally in their oxidized state, except perhaps for gold and the nobler metals, like platinum. Thus, the stable state of a metal is in the oxidized state and corrosion being an oxidation process is, therefore, as inevitable as death and taxes. Without any protection, most metals will oxidize or corrode. Thus, we need to learn to control the corrosion (oxidation) of metals. When a metal corrodes, its thickness may decrease uniformly or locally; this compromises the integrity or reliability of the structure or component using the metal. Eventually, this thickness decrease will be a cause of failure of the structure or equipment. If it is an independent structure, it can be repaired. If the equipment is part of a manufacturing process, then this translates to an economic loss in productivity of the plant. The United States Department of Commerce estimated in 1982 that losses due to corrosion was upwards to $126 billion! Thus, it is imperative to control corrosion.

We can start to control corrosion by selecting the proper materials and by proper design. To do these, we must be cognizant of situations in which corrosion will occur so that we can avoid them. We shall learn about the basic principles of corrosion first and then learn about the methods to control corrosion.

10.2 Forms of Corrosion and Environments

Corrosion manifests itself in forms that have certain similarities and can therefore be classified into certain specific groups. Table 10-1 shows a classification of corrosion according to the mechanisms of attack and the types of corrosion under each mechanism.

General corrosion is the corrosive attack that leads to *the uniform thinning of the material. Atmospheric corrosion* implies the natural air, but the actual component of air causing corrosion is moisture or the relative humidity. In very dry environments such as in deserts, steel parts can remain bright and tarnish-free for a long period of time. It is found that corrosion occurs only when a certain minimum relative humidity of about 50 to 70 percent is present in the air. Corrosion is enhanced when the surface of the metal is wet with visible water pockets or water layers formed on the metal surface from sea spray, rain, or drops of dew. Therefore, we see that, at ambient temperatures, atmospheric corrosion is due to aqueous corrosion. In fact, aqueous solutions including natural waters and other man-made solutions are the environments frequently associated with corrosion problems. These are predominantly the environments we shall discuss in this chapter. In addition to relative humidity, the other factors that influence the corrosivity of the atmosphere are the dust content and the gases in the atmosphere. The major types of atmospheres may be classified into urban, rural, industrial, marine, arctic, and tropical.

All the other forms of corrosion in Table 10-1 will be discussed in conjunction with an aqueous environment. At this point, we need to introduce the basics of corrosion in an aqueous environment which is predominantly electrochemical in nature.

10.3 Electrochemical Aqueous Corrosion

The principles of corrosion are based on electrochemistry and may be illustrated by an electrochemical cell, shown in Fig. 10-1. The cell consists of two connected electrodes (solid electronic conductors—electrons are charge carriers), one anode and one cathode, which are in contact with an electrolyte (ionic conductor—ions carry charge). In this case, the electrolyte is the aqueous solution. When corrosion occurs, an *ox*idation process occurs simultaneously with a *red*uction process. Thus, it is called a *redox process.* The *oxidation process occurs at the anode* while the *reduction process occurs at the cathode. The overall cell reaction differs from an ordinary chemical reaction because it involves the transfer or giving up of electrons between the electronic and the ionic conductors. Thus, it is referred to as an electrochemical reaction and current flows through the circuit because charges (electrons and ions) move.*

Table 10-1 Forms and mechanisms of corrosion.

General Corrosion	Localized Corrosion	Metallurgically Influenced Corrosion	Mechanically Assisted Degradation	Environmentally Assisted Damage
Atmospheric	Filiform	Intergranular	Erosion	Stress-corrosion
Galvanic	Crevice	Dealloying	Fretting	Hydrogen damage
Stray-current	Pitting		Cavitation	Corrosion Fatigue

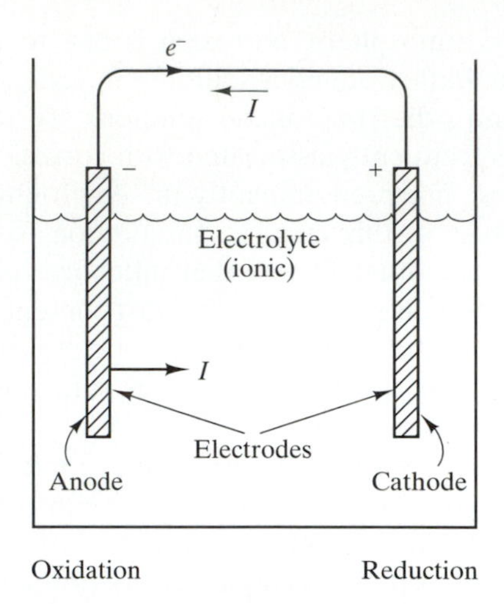

Fig. 10-1 Schematic of an electrochemical cell indicating the basic elements of corrosion.

Figure 10-1 illustrates the *four essential elements of aqueous electrochemical corrosion.* These are:

- the presence of an anode
- the presence of a cathode
- physical contact of the anode and the cathode
- the presence of an electrolyte.

All of the four elements must be present in order for corrosion to proceed. To prevent or control corrosion, it is only necessary to eliminate one of the four.

10.3.1 Electrochemical Reactions at the Anode and at the Cathode

In most instances, the initial state of the element is the zero valence state and *the oxidation or anodic reaction* is

(A)

$$M \rightarrow M^{+n} + n\,e, \qquad (10\text{-}1)$$

where *M* stands for the metallic element. Reaction A (for anodic) constitutes the corrosion or oxidation of metal *M* and *M* is called the anodic metal. The simultaneous reduction or cathodic reaction C (for cathodic) may be one of the following reactions:

(C1) Metal deposition

$$M^{+n} + n\,e \rightarrow M \qquad (10\text{-}2)$$

(C2) Metal ion reduction

$$M^{+n} + e \rightarrow M^{+(n-1)} \qquad (10\text{-}3)$$

(C3) In deaerated (no air or oxygen) solutions, hydrogen gas liberation

$$2H^+ + 2e \rightarrow H_2 \qquad (10\text{-}4)$$

(C4) Reduction of dissolved oxygen in aerated solutions,

$$O_2 + 2H_2O + 4e \rightarrow 4OH^- \quad \text{neutral or basic solution} \qquad (10\text{-}5a)$$

$$O_2 + 4H^+ + 4e \rightarrow 2H_2O \quad \text{in acidic solution} \qquad (10\text{-}5b)$$

(C5) In the absence of other reductions in deaerated solutions, water reduction

$$2H_2O + 2e \rightarrow H_2 + 2OH^- \qquad (10\text{-}6)$$

Reactions (C3) and (C5) are equivalent if we consider that water dissociates to H^+ and OH^-.

The overall electrochemical cell reaction is the combination of Reaction A with one of the cathodic reactions (C1 to C5) and because of this, the above reactions are called half-cell reactions. The number of electrons given up by the half-cell anodic Reaction A must be the same as the number of electrons involved with the corresponding half-cell cathodic reaction, because there must not be any accumulation of electrons. The flow of the electrons is from the anodic material (the negative electrode) through the external wire (electronic conductor) connection to the cathodic material (the positive electrode). At the same time, the current flows the opposite way—it leaves the anode, enters the electrolyte and then enters the cathode from the electrolyte. This is a very important concept in corrosion—*the anodic material is where the current leaves and enters the electrolyte.* Thus, in *stray current corrosion* in Table 10-1, which is due to the unintentional current passing through the metal structure, corrosion or damage to the structure occurs where the current leaves the structure, as schematically shown in Fig. 10-2.

We have to note that the electrochemical cell in Fig. 10-1 is used only to illustrate the half-cell reactions, to indicate which is the anode and the cathode, and to indicate the flow directions of electrons and

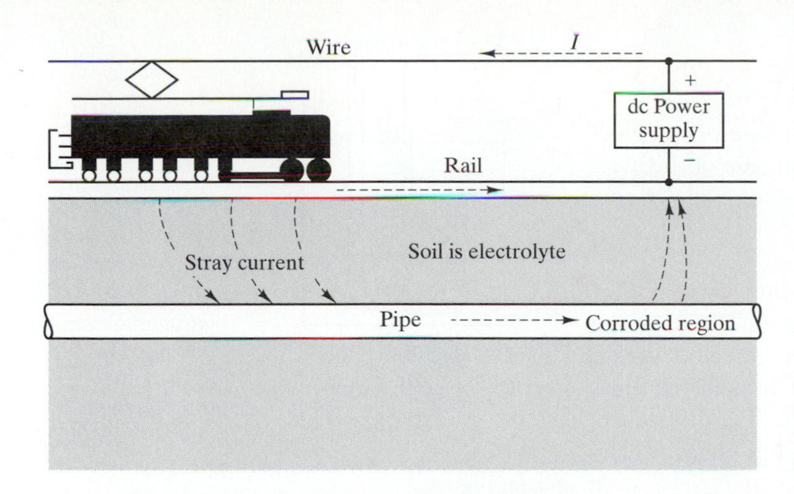

Fig. 10-2 Stray current from electric rail car corrodes buried pipe where the current leaves the pipe to enter soil.

current in the cell during a corrosion process. Actual corrosion of a material occurs because there are anodic and cathodic areas on the surface of the material and the moisture or water on the surface is the electrolyte needed to complete the cell. Thus, in atmospheric corrosion the time of wetness is an important parameter because that is when the electrolyte (water) is present. In order to control corrosion, we must try to keep surfaces of materials dry and the humidity below 50 percent.

10.3.2 Cell Voltage and Electrode Potential

In the electrochemical cell of Fig. 10-1, we note again that we have electronic conductors (the electrodes and lead wires) and an ionic conductor (the electrolyte). We have learned earlier in Sec. 3.6 that whenever two electronic conductors come into contact, there arises a potential difference across the contact interface which we called contact potential. If we have a circuit of metallic conductors such as that shown in Fig. 10-3, where the Ms stand for metals, the potential difference between M_3 and M_1, is

$$\psi^3 - \psi^1 = \phi^{2,1} + \phi^{3,2} \qquad (10\text{-}7)$$

M_1	M_2	M_3	$M_{1'}$

Fig. 10-3 Schematic of circuit consisting only of electronic conductors.

where ψ^3 and ψ^1 are the potentials of M_3 and M_1 and $\phi^{2,1}$ is the contact potential between M_2 and M_1, and $\phi^{3,2}$ is the contact potential between M_3 and M_2. If we take the potential difference, however, between M_1 and M_1', we obtain what is called an *open-circuit voltage,* ξ, and for this series of metallic conductors, ξ is zero,

$$\xi_m = \psi^1 - \psi^{1'} = \phi^{2,1}+\phi^{3,2} + \phi^{1',3} = 0 \qquad (10\text{-}8)$$

The M_1 and $M_{1'}$ in the open-circuit in Fig. 10-3 represent the lead wires and in effect, Eq. 10-8 is the measured potential or voltage between M_2 and M_3. Thus, Eq. 10-8 says, in effect, that at equilibrium the potential between two dissimilar metals (electronic conductors only) is equal to zero, while the contact potential is nonzero.

When we insert an electrolyte (ionic conductor) between M_2 and M_3, we obtain the equivalent open circuit in Fig. 10-4 of the electrochemical cell, where M_1 and $M_{1'}$ are still the lead wires and M_2 and M_3 are the electrodes. The open-circuit voltage for this circuit is

$$\xi_{ec} = \psi^1 - \psi^{1'} = \phi^{2,1} + \phi^{E,2} + \phi^{3,E} + \phi^{1',3} \neq 0 \qquad (10\text{-}9)$$

M_1	M_2	Electrolyte (E)	M_3	$M_{1'}$

Fig. 10-4 Schematic of a circuit consisting of electronic and ionic conductors as in the electrochemical cell.

It is convenient to group

$$\psi_L = \phi^{2,1} + \phi^{E,2} \quad \text{and} \quad \psi_R = \phi^{3,E} + \phi^{1',3}$$

to indicate the potential on the left side of the electrolyte and the potential on the right side of the electrolyte. Thus, Eq. 10-9 becomes

$$\xi_{ec} = \psi_L + \psi_R \neq 0 \qquad (10\text{-}10)$$

Not being equal to zero means that ψ_L is either more positive or more negative than ψ_R. Thus, we see that dissimilar metals electrically connected via an electrolyte (an ionic conductor) such as in the electrochemical cell will have a potential difference, quite different from the wholly metallic electrical circuit.

Leaving out the lead wires, the electrochemical cell is usually represented as

$$M_A\,|\text{Electrolyte}|\,M_C$$
$$\rightarrow I$$

with the subscripts *A* and *C* representing the anode and the cathode. With this representation, the current flows from the anode to the cathode (in this case, from left to right) and the open circuit voltage ξ is positive and equals

$$\xi = E_C - E_A \qquad (10\text{-}11)$$

where *E* is the electrode potential. While ξ can be measured, the Es which include the contact potentials in Eq. 10-10 are not directly measurable.

The electrode potentials represent the half-cell voltages of the half-cell reactions, and it would be very convenient if these can be measured or can have values. In order to do this, one of the electrode potentials in Eq. 10-11 must have a value and the other electrode potential can be determined from the measured ξ and this known value. Thus, the hydrogen reference electrode arose that is given a standard electrode potential of 0. The *standard hydrogen electrode* (SHE) consists of pure hydrogen at 25°C and one atmosphere pressure bubbling in an electrolyte over a platinized electrode, as shown in Fig. 10-5. If we use the hydrogen electrode with another electrode in an electrochemical cell, as shown in Fig. 10-6, we see from Eq. 10-11 that the measured ξ is the standard electrode potential of the metal. The values of the electrode potentials may be positive or negative with respect to the hydrogen electrode and Table 10-2 lists the standard potentials of some elements. *The potentials on the right column correspond to the half-cell reduction reactions given on the left column.*

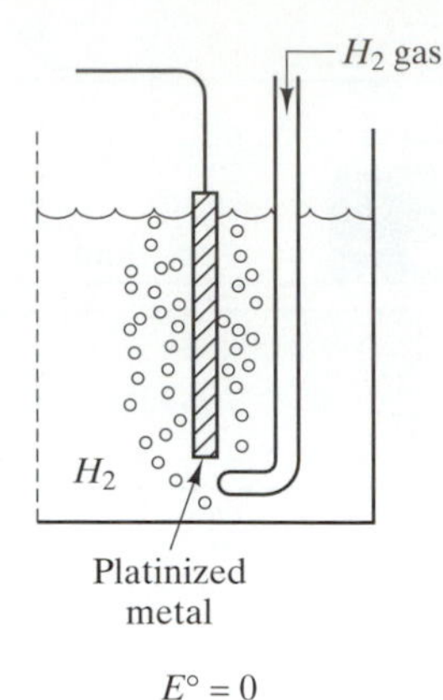

Fig. 10-5 Standard hydrogen electrode consists of H_2 gas at 1 atmosphere bubbling over a platinized electrode.

10.3.3 Thermodynamics and Cell Voltage

In the preceding section, we saw that the electrochemical reaction due to corrosion produced a potential—the open circuit voltage, ξ. While in general we want to avoid corrosion of a metal for structural applications, a useful application of corrosion is the conversion of the chemical energy of the electrochemical reaction into an electrical energy. This is, of course, the basis of batteries that we use to start our cars and a lot of toys, and in cordless tools as well as

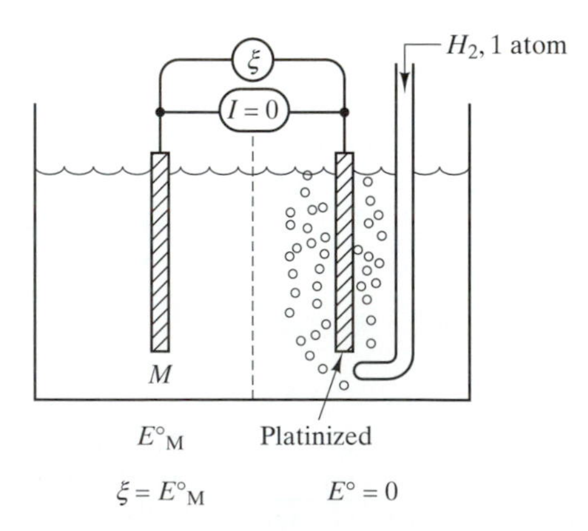

Fig. 10-6 Schematic of an electrochemical cell to measure electrode potentials.

Table 10-2 Standard electrode potentials (also called the EMF series) of metals for reduction half-cells relative to standard hydrogen electrode (SHE). [1]

Electrode reaction	Standard potential at 25°C (77°F), volts versus SHE
$Au^{3+} + 3e^- \rightarrow Au$	1.50
$Pd^{2+} + 2e^- \rightarrow Pd$	0.987
$Hg^{2+} + 2e^- \rightarrow Hg$	0.854
$Ag^+ + e^- \rightarrow Ag$	0.800
$Hg_2^{2+} + 2e^- \rightarrow 2Hg$	0.789
$Cu^+ + e^- \rightarrow Cu$	0.521
$Cu^{2+} + 2e^- \rightarrow Cu$	0.337
$2H^+ + 2e^- \rightarrow H_2$	0.000 (Reference)
$Pb^{2+} + 2e^- \rightarrow Pb$	−0.126
$Sn_2 + 2e^- \rightarrow Sn$	−0.136
$Ni^{2+} + 2e^- \rightarrow Ni$	−0.250
$Co^{2+} + 2e^- \rightarrow Co$	−0.277
$Tl^+ + e^- \rightarrow Tl$	−0.336
$In^{3+} + 3e^- \rightarrow In$	−0.342
$Cd^{2+} + 2e^- \rightarrow Cd$	−0.403
$Fe^{2+} + 2e^- \rightarrow Fe$	−0.440
$Ga^{3+} + 3e^- \rightarrow Ga$	−0.53
$Cr^{3+} + 3e^- \rightarrow Cr$	−0.74
$Cr^{2+} + 2e^- \rightarrow Cr$	−0.91
$Zn^{2+} + 2e^- \rightarrow Zn$	−0.763
$Mn^{2+} + 2e^- \rightarrow Mn$	−1.18
$Zr^{4+} + 4e^- \rightarrow Zr$	−1.53
$Ti^{2+} + 2e^- \rightarrow Ti$	−1.63
$Al^{3+} + 3e^- \rightarrow Al$	−1.66
$Hf^{4+} + 4e^- \rightarrow Hf$	−1.70
$U^{3+} + 3e^- \rightarrow U$	−1.80
$Be^{2+} + 2e^- \rightarrow Be$	−1.85
$Mg^{2+} + 2e^- \rightarrow Mg$	−2.37
$Na^+ + e^- \rightarrow Na$	−2.71
$Ca^{2+} + 2e^- \rightarrow Ca$	−2.87
$K^+ + e^- \rightarrow K$	−2.93
$Li^+ + e^- \rightarrow Li$	−3.05

many other gadgets. Batteries are also getting more important in the effort to curb pollution because they emit no pollutants and are currently being used in electric cars.

The relationship between the chemical and electrical energy is given by the following:

$$\Delta G = -nF\xi \qquad (10\text{-}12)$$

where ΔG is the Gibbs free energy of the overall electrochemical reaction, in joules, ξ is in volts, n is the number of electrons involved in the electrochemical reaction, and *F is the Faraday constant which is ~96,500 coulombs (C) per equivalent weight.* If ΔG is in the standard state (i.e., $\Delta G°$), then ξ is also in the standard state, $\xi°$, and,

$$\Delta G° = -nF\xi° \qquad (10\text{-}13)$$

In order for an electrochemical process to occur, the cell voltage ξ of the corrosion cell or redox reactions must be positive. We shall first start with $\xi°$ and then apply Eq. 10-11 using the values of E° from Table 10-2. To apply Eq. 10-11, we should know which metal in Table 10-2 is cathodic and which is anodic in order to get a positive $\xi°$. If it is positive, then ΔG is negative and the process may occur. To determine which metal is anodic or cathodic relative to another metal, we can start by assuming a half-cell reaction to be anodic and the other cathodic, and then determine the cell voltage with Eq. 10-11.

For an example, let us assume iron and zinc are the electrodes and let us further assume that zinc is the cathode and iron is the anode. Using Eq. 10-11, the cell voltage

$$\xi° = E^{\circ}_{Zn} - E^{\circ}_{Fe} = -0.763 - (-0.440)$$
$$= -0.323 \text{ volt}$$

Substituting this value in Eq. 10-13, we see that the $\Delta G°$ will be positive, that means that the overall reaction (as we have assumed) will not have any tendency to proceed. In other words, there is no tendency for iron to corrode (when it is assumed as the anode with respect to zinc) when it is connected with zinc in the presence of an electrolyte. If we now assume zinc to be the anode, and the iron as the cathode, the cell voltage is

$$\xi° = E^{\circ}_{Fe} - E^{\circ}_{Zn} = -0.440 - (-0.763)$$
$$= 0.323 \text{ volt}$$

Because $\xi°$ is positive, the $\Delta G°$ is negative and the overall reaction has a tendency to proceed. This means that zinc will be anodic and will tend to oxidize and corrode with respect to iron, when the two are made electrodes in an electrochemical cell.

The foregoing example illustrates the tendencies given in Table 10-2; the tendencies are only relative because the electrode potentials are directly related to the thermodynamics (overall chemical free energy change). *When we connect two metals from Table 10-2 as electrodes in an electrochemical cell, the anodic material will be the metal with the more negative (lower algebraic value) electrode potential.* However, when they are not connected, the values listed indicate only the relative tendencies of the metals to corrode, but do not indicate how fast they will corrode. For example, when we connect iron and aluminum in sea water, the aluminum will corrode because it is anodic. However, if we do not connect the two and exposed them in the atmosphere, aluminum will corrode slower than iron. Why? The aluminum forms a very tight adherent oxide on the surface that acts to prevent the electrolyte from further contact with the metal.

10.3.4 Dependence of Equilibrium Cell Voltages and Electrode Potentials on Concentration—Nernst Equation

The standard electrode potentials given in Table 10-2 are for the reactants and products in their standard states. We shall now see what we mean by standard states and how to determine the electrode potentials as a function of the activities or concentrations of gases or ions in the aqueous solution.

Consider the overall electrochemical reaction

$$bB + cC + \cdots \rightarrow dD + fF + \cdots$$

where *B* and *C* are the reactants with their respective coefficients, *b* and *c*, and *D* and *F* are the products with their respective coefficients, *d* and *f*. The reaction indicates that *b* moles of *B* and *c* moles of *C* react to form *d* moles of *D* and *f* moles of *F*. The free energy change, ΔG, of the above reaction is

$$\Delta G = (d \cdot G_D + f \cdot G_F + \cdots) - (b \cdot G_B + c \cdot G_C + \cdots) \quad (10\text{-}14)$$

where the *G*s are the free energies per mole of the reactants and products. In the standard state,

$$\Delta G^\circ = (d \cdot G_D{}^\circ + f \cdot G_F{}^\circ + \cdots) - (b \cdot G_B{}^\circ + c \cdot G_C{}^\circ + \cdots) \quad (10\text{-}15)$$

where the G°s are the standard free energies per mole of the reactants and products. The difference between *G* and G° of a substance is related to its activity, *a*, which can be regarded as a corrected concentration or pressure. For example,

$$d(G_D - G_D^\circ) = d\,RT \ln a_D = RT \ln a_D^d \quad (10\text{-}16)$$

where *R* is the gas constant = 8.314 J/deg-mole, and *T* is the absolute temperature in *K*. Subtracting Eq. 10-15 from Eq. 10-14 and based on Eq. 10-16, we obtain the expression,

$$\Delta G - \Delta G^\circ = RT \ln\left(\frac{a_D^d \cdot a_F^f \cdots}{a_B^b \cdot a_C^c \cdots}\right) \quad (10\text{-}17)$$

We see that the ΔG of the reaction equals that at the standard state ΔG° when the right side of Eq. 10-17 is equal to zero. This happens only when all the activities, the "*a*"s, in Eq. 10-17 are all equal to one. This is the *definition of the standard state of a substance, that is, when its activity a = 1.*

The quantity in parenthesis in Eq. 10-17 is defined as the equilibrium constant *K* of the overall reaction,

$$K \equiv \frac{a_D^d \cdot a_F^f \cdots}{a_B^b \cdot a_C^c \cdots} \quad (10\text{-}18)$$

and at equilibrium, $\Delta G = 0$, and

$$\Delta G^o = -RT \ln K \quad (10\text{-}19)$$

Applying the foregoing to an overall electrochemical reaction and substituting Eq. 10-12 and 10-13 into Eq. 10-17, we obtain

$$\xi = \xi^\circ - \frac{RT}{nF} \ln\left(\frac{a_D^d \cdot a_F^f \cdots}{a_B^b \cdot a_C^c \cdots}\right) \quad (10\text{-}20)$$

This is the Nernst equation that relates the cell voltage to the activities (concentrations) of the substances involved in the overall electrochemical reaction. We see again that the cell voltage is equal to the standard cell voltage if the activities of all the substances are all equal to 1.0—the standard state. The activity, *a*, of a substance dissolved in solution in the electrolyte is

$$a_i = \gamma_i c_i, \quad (10\text{-}21)$$

where γ is called the activity coefficient, and *c is its concentration in the electrolyte in molality. Molality is the number of gram-moles of the solute per 1,000 grams (≈ one liter of water) of the solvent.* For ideal and very dilute solutions, the activity coefficient γ is equal to 1 and the activity is equal to the concentration in molality. For ideal gases, the activity is the partial pressure of the gas in atmospheres. Pure reactants and products (solids and liquids) are assigned activities equal to 1. Assuming ideal solutions and ideal gases, the standard state in Table 10-2 means that the ionic concentration in the electrolyte is 1 gram-ion per liter or one atmosphere for gases at 25°C (298 K).

We can apply the above treatment also to the half-cell reactions and in this manner obtain the Nernst equation for the electrode potentials. We generalize the half-cell reactions listed in Table 10-2 with the following,

$$Ox + ne \rightleftarrows Red \qquad (10\text{-}22)$$

Ox in the reaction signifies the oxidized state of the metal and *Red* signifies the reduced state. *n* is the number of electrons involved in the reaction. Using Eq. 10-17 for the free energy, ΔG, of reaction (10-22), we obtain

$$\Delta G = \Delta G^\circ + RT \ln \frac{a_{Red}}{a_{Ox}} \qquad (10\text{-}23)$$

Substituting $\Delta G = -nFE$, and $\Delta G^\circ = -nFE^\circ$, we obtain the Nernst equation for the electrode potentials of the half-cell reactions.

$$E = E^\circ + \frac{RT}{nF} \ln \frac{a_{Ox}}{a_{Red}} \qquad (10\text{-}24)$$

10.3.5 Types of Electrochemical Corrosion Cells

10.3.5a Galvanic cells. In corrosion, a Galvanic cell arises when two dissimilar metals are connected and their surfaces are wetted. The electrochemical cell we have been discussing thus far is a Galvanic cell, and the corrosion or degradation of the anodic material is the *Galvanic corrosion* indicated in Table 10-1. We note from Sec. 10.3.3 that the anodic material has a more negative or lower electrode potential in Table 10-2. Because Table 10-2 is based on thermodynamics, it does not reflect the kinetics of corrosion and in practice, its use is quite limited. In its place, a Galvanic series for a particular environment is used and the environment that is widely quoted is sea water. The Galvanic series in sea water is shown graphically in Fig. 10-7; this may also be applied in natural waters and in uncontaminated atmospheres.

So far we have limited the discussion to two different materials (electrodes) in contact and the anodic material is the one with a lower or more negative electron potential. However, within one material, it is possible to have areas with different compositions that may have different electrode potentials. Galvanic corrosion will therefore also ensue, and localized corrosion may occur. Examples of these in materials are the sensitization of stainless steels discussed in Sec. 9.7.3b and the formation of precipitate-free-zones (PFZ) in aluminum alloys; both of these are shown schematically in Fig. 10-8. The areas close to the grain boundaries in both cases are depleted of solutes in the alloy and this depletion of solute results in *intergranular corrosion and fracture.* This is classified in Table 10-1 as a metallurgically influenced corrosion because the thermal processing in both cases changed the composition and electrode potentials of the localized areas.

10.3.5b Electrolyte Concentration Cells. These cells illustrate the effect of differences in concentration of the electrolyte and other reactants. Some examples are illustrated in Fig. 10-9 that include (1) a difference in concentration of the electrolyte in contact with the electrodes, and (2) differences in concentration of the gas reactants, which are called differential aeration cells.

1. The difference in concentration of the electrolyte is illustrated by an electrochemical cell with copper electrodes, one in contact with a dilute copper sulfate solution and the other in contact with a concentrated copper sulfate solution. The electrode in contact with the dilute solution is anodic and will be consumed, and plating (cathodic) will occur at the copper in contact with the concentrated solution. We can demonstrate this by using the Nernst equation, Eq. 10-24, for the electrode potentials. The half-cell reaction for both copper electrodes is

$$Cu^{+2}(ox) + 2\,e \rightarrow Cu(red)$$

The activity of Cu (red) being pure solid is 1 and thus, Eq. 10-24 becomes

$$E = E^\circ + \frac{RT}{nF} \ln a_{Cu^{+2}}$$

We see that the electrode potential of copper is proportional to the activity of copper ions, which in turn is proportional to the concentration of copper ions, according to Eq. 10-21. We see, therefore, that

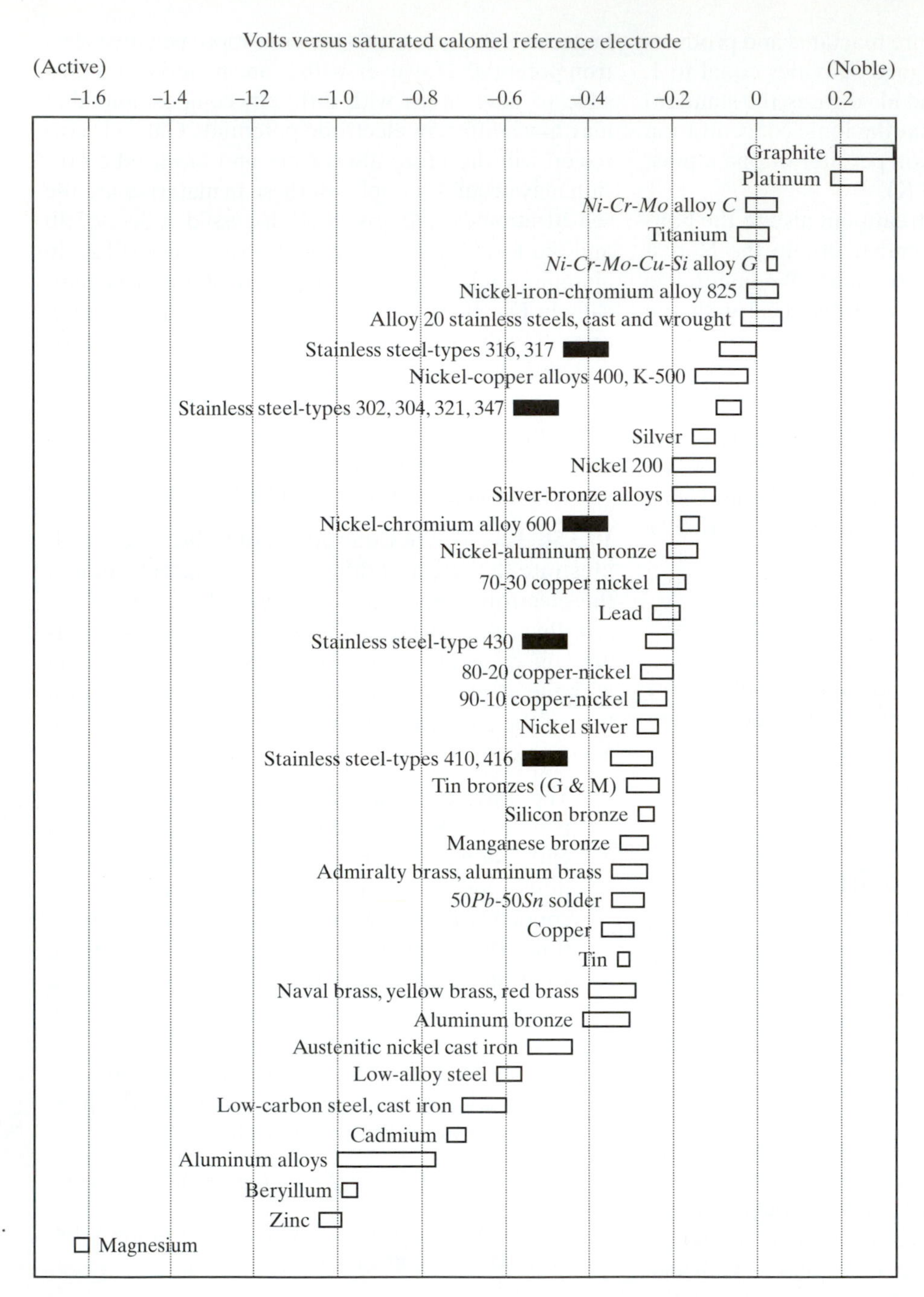

Fig. 10-7 Galvanic series in sea water; dark boxes indicate the active state of the passive state (on the right) of the alloys. [1]

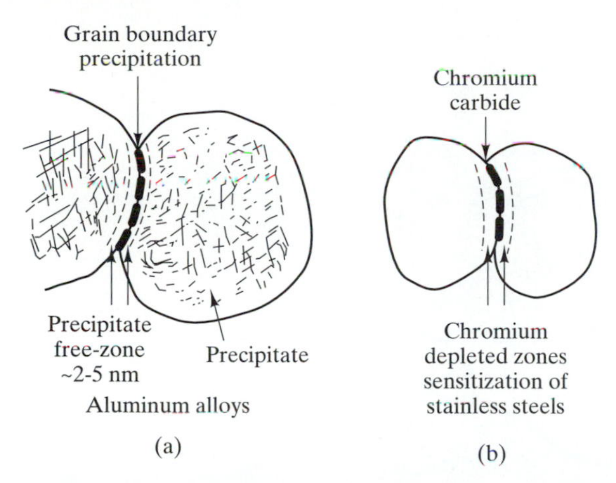

Fig. 10-8 Schematics of (a) precipitate-free-zone (PFZ) in aluminum alloys, and (b) chromium-depleted zones in sensitized stainless steels. Both are very close to the grain-boundary and induce intergranular corrosion and fracture.

the electrode potential of copper in contact with the concentrated copper sulfate electrolyte is higher than that in contact with the dilute copper sulfate solution. Thus, the copper in contact with the dilute solution that has a lower electrode potential becomes anodic when the two copper electrodes are connected to form an electrochemical cell.

2. The differential aeration cell is illustrated by the other cases in Fig. 10-9 which are: (a) iron electrodes dipped in the same sodium chloride solution with air, being bubbled around one electrode, and pure nitrogen bubbled around the other; (b) the atmospheric rusting of iron and steel, and (c) water-line corrosion in stagnant waters. In all cases the pertinent half-cell reaction is Eq. 10-5a,

$$O_2 + 2H_2O + 4e \rightarrow 4OH^- \quad \text{neutral or basic solution} \qquad (10\text{-}5a)$$

The electrode potential for this half-cell reaction is

$$E = E^\circ + \frac{RT}{4F} \ln \frac{a_{O_2} \cdot a_{H_2O}}{a^4_{OH}}$$

In this case, the activity of water is one (pure liquid reactant) and the activity of OH^- is kept constant; thus, E depends directly and solely on the activity of oxygen which is equal to the partial pressure of oxygen. Thus, E is higher when the partial pressure of oxygen in the electrode environment is higher.

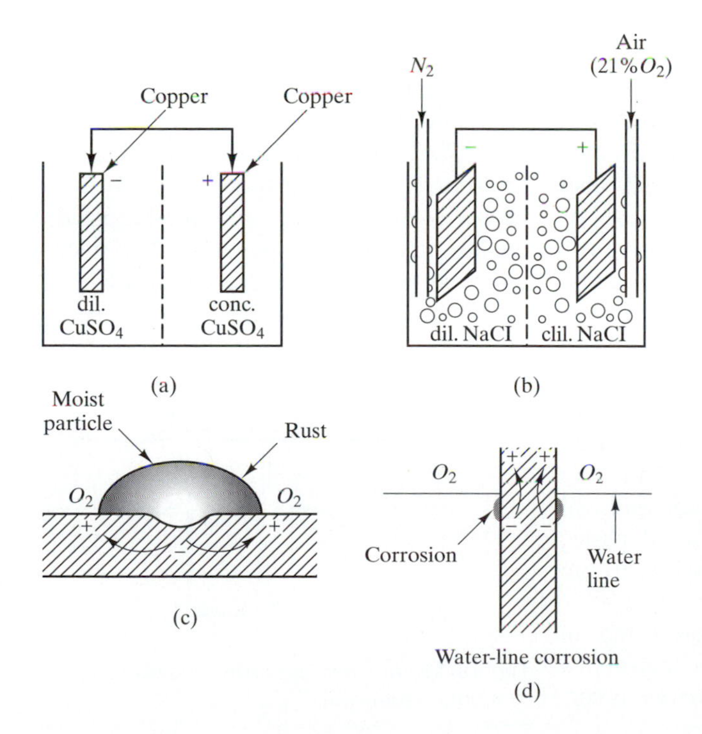

Fig. 10-9 Examples of corrosion cells showing effect of difference in concentration of electrolyte (a), differential aeration cells in (b), (c) , and (d) with different concentrations of ozygen; (c) is the rusting of iron. [5]

Fig. 10-10 Filiform corrosion underneath clear varnish on steel starting from the scratch on varnish (top right corner). [5]

In the first case, where air and nitrogen are bubbled through the same dilute sodium chloride solution, the electrode potential of iron with air (~0.21 atm. oxygen partial pressure) bubbling has a higher electrode potential than that with nitrogen (0 atm. oxygen partial pressure) bubbling. Thus, the iron with nitrogen bubbling will be anodic and will corrode. The rusting mechanism of iron proceeds also via the differential aeration cell. Dust particles or any covering on the surface of iron will deprive the area underneath with oxygen and will make the area anodic. If the dust is hygroscopic (attracts moisture easily), the area will be kept moist and rusting (corrosion) will begin there. In the waterline corrosion, the water below the waterline contains dissolved oxygen that is only a small fraction of the partial pressure of oxygen in air. Thus, the material just below the waterline is anodic and will corrode first because of what is called the distance effect: the anode will be as close as possible to the cathode.

All the localized forms of corrosion listed in Table 10-1 are essentially based on the mechanism of a differential aeration cell. The *filiform corrosion* is corrosion underneath coated surfaces when a break in the coating occurs and appears as meandering thread-like filaments that do not cross each other, shown in Fig. 10-10. *Crevice corrosion* occurs in narrow openings of joints between metal-to-metal or metal- to nonmetal components that are deprived of oxygen. This is schematically illustrated in Fig. 10-11a and in an actual crevice-corroded washer in Fig. 10-11b. *Pitting corrosion* is the extreme localized attack producing pits that can penetrate the thickness of the metal without appreciable uniform loss of the thickness. It is very hard to design for, except possibly to

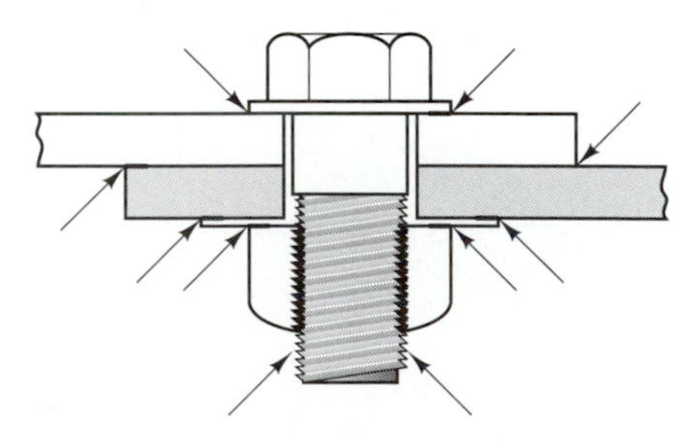

Fig. 10-11 (a) Crevice corrosion sites (pointed by arrows) in a bolted joint. [3]

Fig. 10-11 (b) Crevice corrosion at a metal-to-metal crevice site between components of 304 stainless steel in sea water. [1]

select the right material. It predominantly occurs in metals that produce passive surfaces. These are surfaces consisting of oxides of the metals, such as aluminum, stainless steels, and titanium that become impervious to the electrolyte. Pitting in these materials occurs when oxygen is deprived on the surface, especially in chloride environments because the chloride ion replaces the oxygen on the surface. A sketch of a pit is shown in Fig. 10-12 and the severity of pitting is measured by a pitting factor shown in the figure.

10.3.5c Differential Temperature Cells. Components of these cells are electrodes of the same metal, which are at different temperatures and are immersed in an electrolyte of the same initial composition. The fundamental theory of differential temperature cells is not quite defined yet because it involves many factors that we cannot get into in this text. Suffice it to say that they may be found in heat exchangers, boilers, immersion heaters, and similar equipment. For example, in heat exchanger tubes, the outside diameter surface is at a different temperature than the inside diameter surface. In addition, it might be in contact with a different fluid or electrolyte. The difficulty in arriving at a fundamental theory is due to conflicting results. For example, in copper sulfate solution, the copper electrode at a high temperature is cathodic; lead acts similarly. For silver, the polarity is reversed. For iron immersed in dilute aerated sodium chloride solution, the hot electrode is anodic to the colder metal of the same composition, but after a matter of hours the polarity may reverse. In aerated hot waters, reversal of polarity between zinc and iron occurs at about 60°C (140°F). Below 60°C (140°F), zinc is anodic to iron, but above 60°C (140°F), zinc becomes cathodic to steel and pitting of the steel occurs.

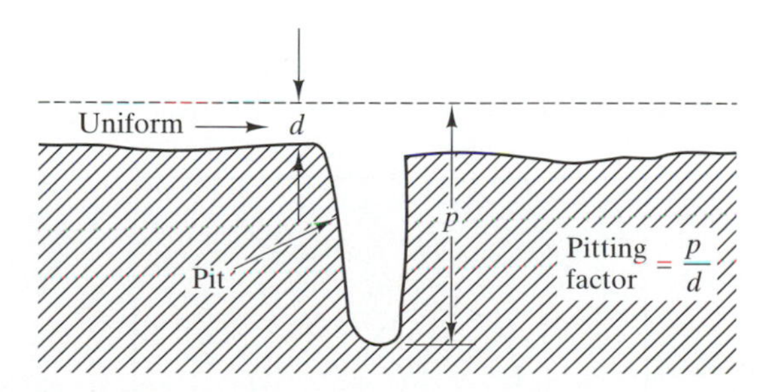

Fig. 10-12 Schematic of a pit. Severity of pit is measured by the pitting factor, p/d, where *d* is the uniform thickness loss and *p* is the depth of pit.

10.3.5d Differential Cold-Work or Stress Cells. The electrodes in this cell consist of areas with different degrees of cold-work. The area with more cold work will be anodic and will corrode faster. An example of this is the bent nail shown in Fig. 10-13 with the darkened areas being anodic. Another example is shown in Fig. 10-14 illustrating the variation in corrosion of a stainless steel spring; the middle section of the spring is thinner because it experiences more stress. Automobile thieves often grind the vehicle identification number (VIN) that is stamped at the factory in the engine block and then restamp another number in its place. Police laboratories reveal the original VIN by corroding the ground area with an acid and the original stress cell etches out the VIN. Cold-work may also introduce residual

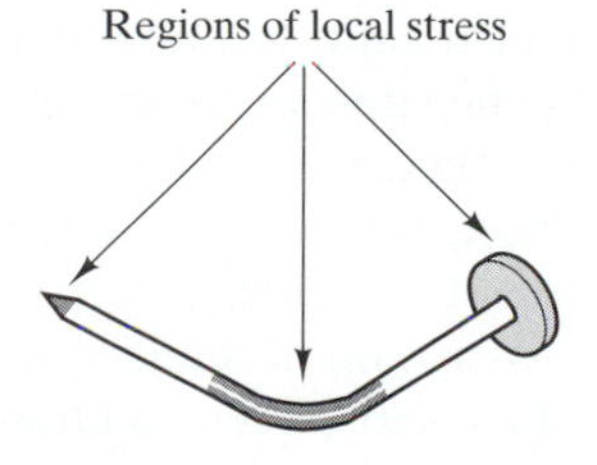

Fig. 10-13 Anodic areas in nail illustrating stress cells. [3]

Fig. 10-14 Variation in corrosion of stainless steel spring due to stress cell. [3]

stresses that may induce stress-corrosion cracking, as was mentioned for the season-cracking of brass cartridges in Chap. 8. *Stress-corrosion cracking (SCC),* under the environmentally induced cracking category in Table 10-1, occurs only in the presence of tensile stress, applied or residual, and only with specific environments for each metal. Some of the environment-alloy combinations that are known to result in SCC are shown in Table 10-3.

The different forms of corrosion listed in Table 10-1 may be associated with one type or a combination of the types of corrosion cells mentioned above. Thus, in Galvanic corrosion and in most cases, the contact of two different metals acting as electrodes is not required. Anodes and cathodes are present in a single material to set up local cells, as illustrated in Fig. 10-15. In these cells, the negative areas are anodic and the positive areas are cathodic. As indicated earlier, the metallurgically induced corrosion in Table 10-1 may be thought to be due to Galvanic corrosion. Examples of intergranular corrosion in stainless steels and aluminum alloys have heretofore been mentioned. Actual damage due to intergranular corrosion in sensitized stainless steel is shown Fig. 10-16 where the attack is clearly shown to be along the grain boundaries and the location of the damaged area in a welded structure is shown in Fig. 10-17. The explanation of the corroded location is shown in Fig. 10-18 where the temperature profile from the center of the weld is sketched in Fig. 10-18b and the sensitization temperature range of 425° to 900°C for austenitic steels is indicated. The corroded area is where the sensitization range existed in the base metal.

The two most prominent examples of dealloying, sometimes called parting, are dezincification of brasses

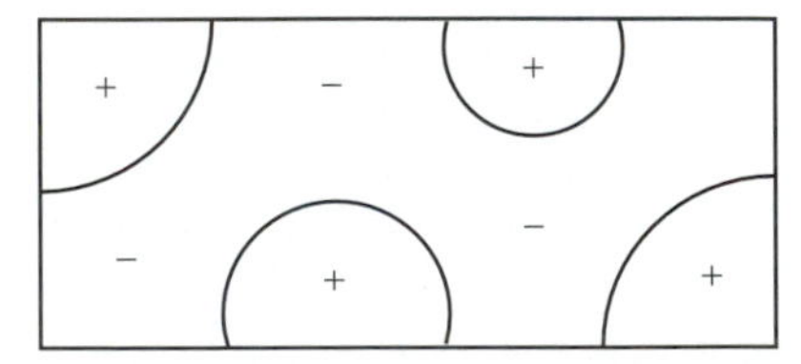

Fig. 10-15 Local anode and cathode areas on surface of metal that sets up electrochemical corrosion.

Table 10-3 Some environment-alloy combinations that are known to result in stress corrosion cracking (SCC). [1]

Environment	Alloy system								
	Aluminum alloys	Carbon steels	Copper alloys	Nickel alloys	Stainless steels: Austenitic	Stainless steels: duplex	Stainless steels: Martensitic	Titanium alloys	Zirconium alloys
Amines, aqueous	...	•	•	...	...	...	...	...	...
Ammonia, anhydrous	...	•	...	...	...	...	...	...	...
Ammonia, aqueous	...	...	•	...	...	...	...	...	...
Bromine	...	...	...	...	...	...	...	...	•
Carbonates, aqueous	...	•	...	...	...	...	...	...	...
Carbon monoxide, carbon dioxide, water mixture	...	•	...	...	...	...	...	...	...
Chlorides, aqueous	•	...	...	•	•	•	...	...	•
Chlorides, concentrated, boiling	...	...	...	•	•	•	...	...	...
Chlorides, dry, hot	...	...	...	•	...	...	...	•	...
Chlorinated solvents	...	...	...	...	...	...	...	•	•
Cyanides, aqueous, acidified	...	•	...	...	...	...	...	...	...
Fluorides, aqueous	...	...	...	•	...	...	...	...	...
Hydrochloric acid	...	...	...	...	...	...	...	•	...
Hydrofluoric acid	...	...	...	•	...	...	...	...	...
Hydroxides, aqueous	...	•	...	...	•	•	•	...	...
Hydroxides, concentrated, hot	...	...	...	•	•	•	•	...	...
Methanol plus halides	...	...	...	...	...	...	...	•	•
Nitrates, aqueous	...	•	•	...	...	...	•	...	...
Nitric acid, concentrated	...	...	...	...	...	...	...	...	•
Nitric acid, fuming	...	...	...	...	...	...	...	•	...
Nitrites, aqueous	...	...	•	...	...	...	...	...	...
Nitrogen tetroxide	...	...	...	...	...	...	...	•	...
Polythionic acids	...	...	...	•	•	...	...	...	...
Steam	...	...	•	...	...	...	...	...	...
Sulfides plus chlorides, aqueous	...	...	...	...	•	•	•	...	...
Sulfurous acid	...	...	...	...	•	...	...	...	...
Water, high-purity, hot	•	...	...	•	...	...	...	...	...

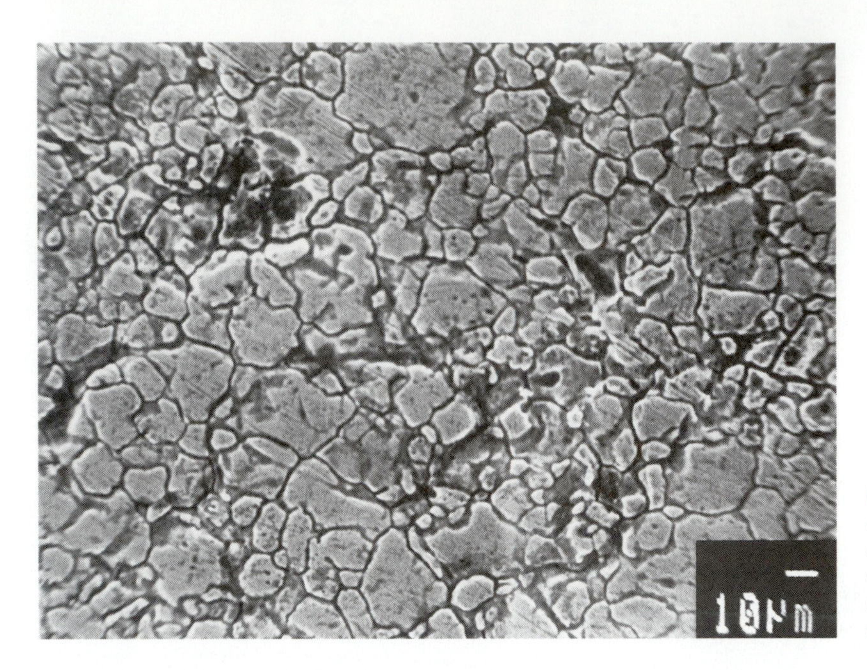

Fig. 10-16 Scanning electron micrograph of intergranular corrosion in sensitized stainless steel. [4]

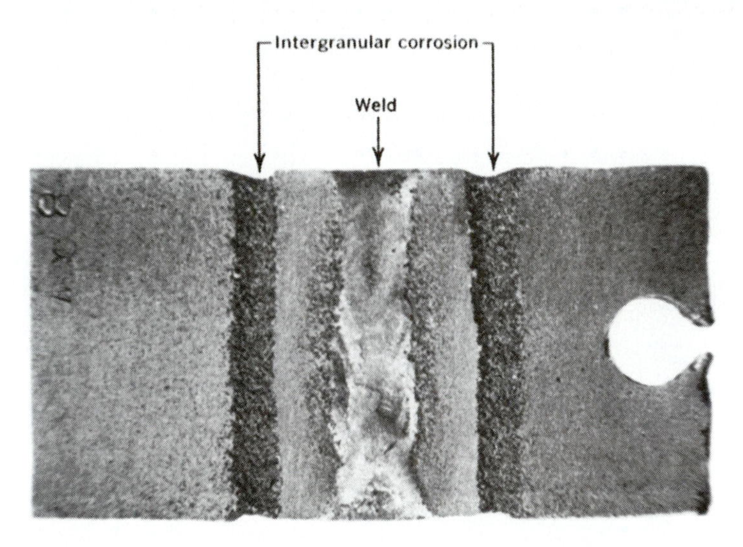

Fig. 10-17 Location of intergranular corrosion in a welded austenitic stainless steel. Area is where the sensitization range of temperature is located in the material (Fig. 10-18). [5]

(copper zinc alloys) with greater than 15%Zn and graphitic corrosion of gray iron. In both of these, shown in Figs. 10-19 and 10-20, the more cathodic material (based on the electrode potential or the Galvanic series) remains and retains the shape of a structure, for example, a pipe, while the anodic material is dissolved or leached out. Alloys of gold with copper and silver will undergo dealloying or parting below a certain gold content, depending on the electrolyte. Copper and silver, being anodic with respect to gold, dissolve.

10.3.6 The Pourbaix Diagram

So far we have been discussing the half-cell reaction

$$M^{n+} + ne \rightarrow M \tag{10-25}$$

as the only half-cell reaction in an aqueous environment. The Nernst equation for this reaction is usually written as

$$E_{M^{n+}/M} = E^\circ + \frac{RT}{nF} \ln \frac{a_{M^{n+}}}{a_M} = E^\circ + \frac{RT}{nF} \ln a_{M^{n+}} \tag{10-26}$$

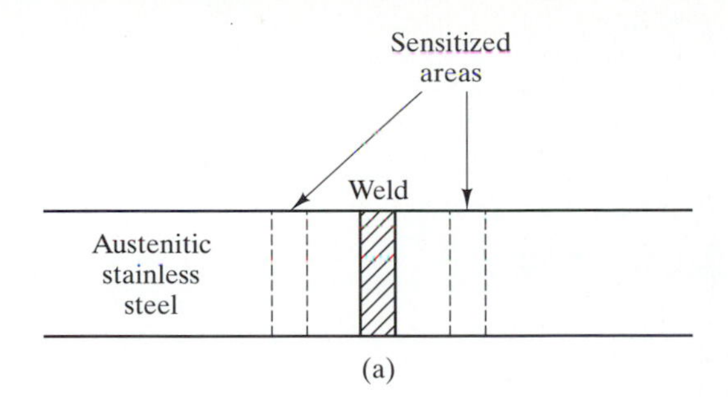

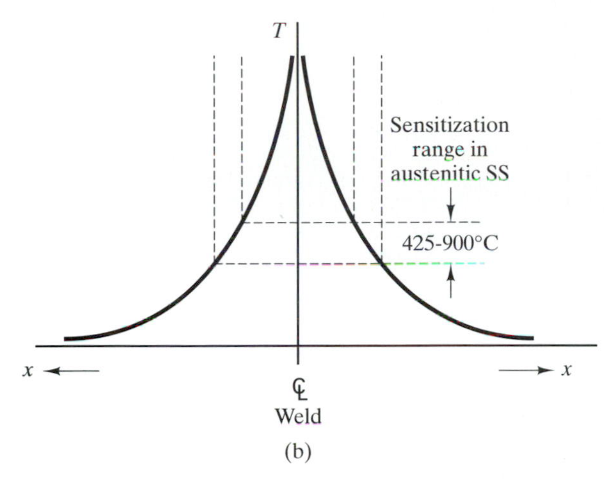

Fig. 10-18 Schematic of temperature profile in stainless steel after welding showing the sensitization range and the corrosion away from the weld.

Fig. 10-19 Plug type dezincification of brass pipe (inside diameter). [5]

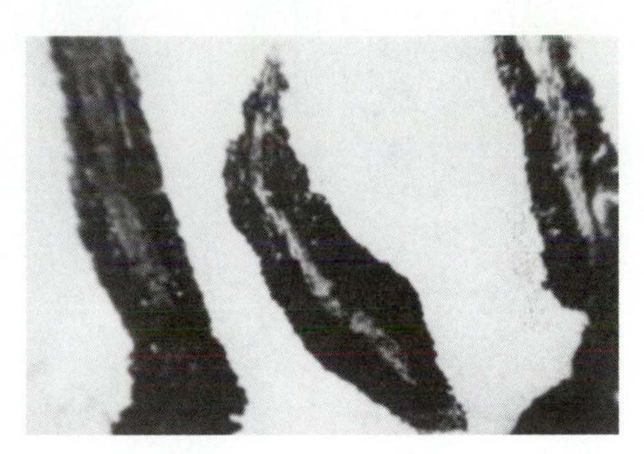

Fig. 10-20 Envelopes of corroded gray iron surrounding graphite flakes. [1]

where the activity of M is taken as 1 because it is pure metal and in the solid.

In an aqueous environment, metal hydroxides or oxides may also form with the following reaction,

$$M(OH)_n + nH^+ + ne \rightarrow M + nH_2O \qquad (10\text{-}27)$$

The Nernst equation for Eq. 10-27 is

$$E_{M(OH)_n} = E^\circ + \frac{RT}{nF} \ln \frac{a_{M(OH)_n} \cdot a^n_{H^+}}{a_M \cdot a^n_{H_2O}}$$

$$= E^\circ + \frac{2.303\,RT}{F} \log a_{H^+}$$

$$= E^\circ - \frac{2.303RT}{F}\, pH \qquad (10\text{-}28)$$

where $pH \equiv -\log a_{H^+}$. We see in Eq. 10-28 the relation between E and pH.

We can plot the Nernst Eqs. 10-26 and 10-28 as a pH-E (potential relative to the standard hydrogen electrode, SHE) Pourbaix diagram. There are now Pourbaix diagrams for most metals and two examples are shown in Fig. 10-21a and b for aluminum and iron (which is the basis for steels). These diagrams are very useful in that they show at a glance the specific conditions of potential and pH by which the metal either corrodes, is passive, or is immune. The areas in the diagrams are marked "corrosion," "passive," and "immune." In the corrosion area, the stable state of the metal is in the form of ions or complex ions, thus corrosion will occur. The passive areas assume that a stable oxide or hydroxide is formed to act as barrier for the electrolyte, and the immune area shows the metal un-ionized. The numbers along the lines are the base 10 exponents of the concentrations of the metallic ions. For example, −6 means a 10^{-6} g-ion per liter concentration and a 0 means 10^0 or activity equals 1. The horizontal lines are plots of Eq. 10-25 for different concentrations of ions in solution or dissolved. Each horizontal line represents the equilibrium electrode potential with the given ion concentration. For example, if the electrolyte concentration contains 10^{-6} g-ion per liter concentration,

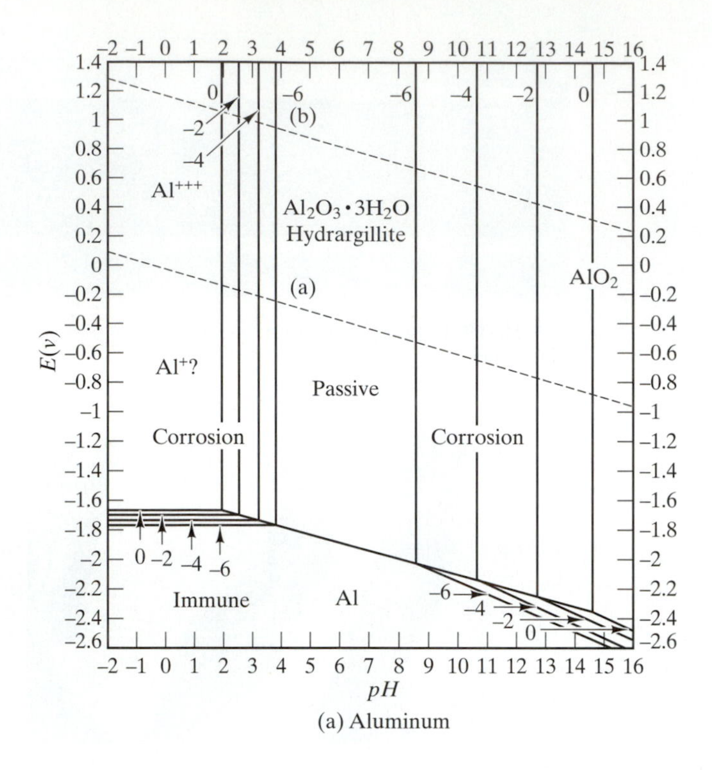

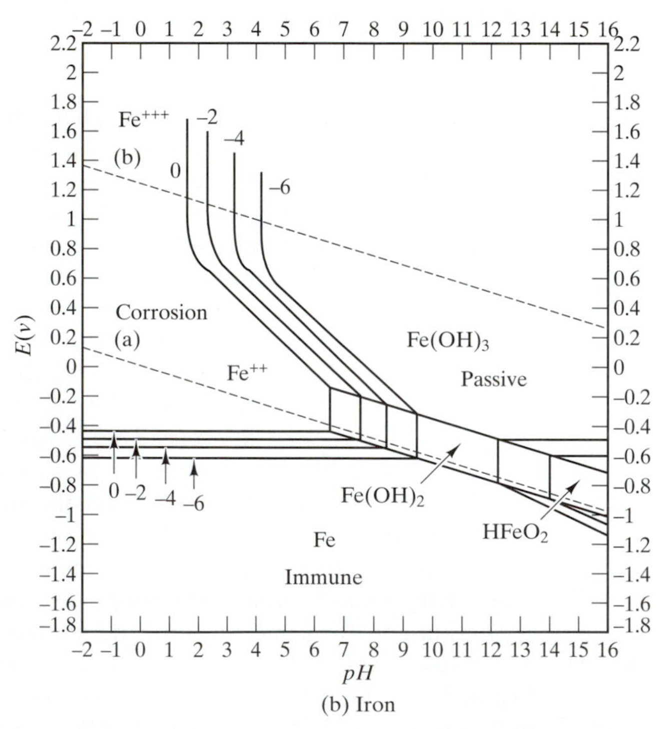

Fig. 10-21 (a) Pourbaix diagram (pH-E) of pure aluminum. (b) Pourbaix diagram (pH-potential) of pure iron. [9]

below the horizontal line (potential) marked −6, the metal is immuned and above it, the ion is stable and the metal will corrode. The solid, slanted lines are plots of Eq. 10-28 that indicate a reaction involving a transfer of the electron and the hydrogen ion. The vertical lines are reactions involving no transfer of electrons, strictly chemical reactions. The two slanting dashed lines represent water pH-potential and indicates the regions where hydrogen, water, and oxygen are stable.

We should realize and keep in mind that the Pourbaix diagrams are based on thermodynamics and they do not convey any information on the rates, either rapidly or slowly, of the possible reactions. Furthermore, they do not indicate the effectiveness of the passive or barrier oxide film that is actually produced. With these limitations in mind, they indicate, nonetheless, what might happen in equilibrium conditions when a metal is immersed in acids or alkalies, or when a given potential is impressed on the metal, as in cathodic protection to be discussed in Sec. 10.8.4. We should also note that the lines are for specific ions or species that were considered. When other ions such as sulfides and chlorides are present in the electrolyte, these lines must be established through their equilibrium reactions with the metals.

EXAMPLE 10-1

Calculate the equilibrium half-cell potential of zinc in 0.01M (molal) $ZnCl_2$ solution at 25°C. The activity coefficient of 0.01M $ZnCl_2$ is 0.71.

SOLUTION: We should first write the reduction half-cell reaction

$$Zn^{2+} + 2e \rightarrow Zn \qquad E^\circ = -0.763\text{ V}$$

and the Nernst equation,

$$E_{Zn^{2+}/Zn} = E^\circ + \frac{RT}{2F} \ln a_{Zn^{2+}},$$

and we see that we need to know $a_{Zn^{2+}}$, which we can obtain from 0.01M $ZnCl_2$, because

$$\begin{array}{ll} ZnCl_2 \rightarrow & Zn^{2+} + 2Cl^- \\ 0.01M & 0.01M \end{array}$$

and the $a_{Zn^{2+}} = 0.71 \times 0.01 = 0.0071$. Thus,

$$E_{Zn^{2+}/Zn} = -0.763 + \frac{8.314 \times 298}{2 \times 96{,}500} \ln 0.0071$$

$$= -0.8265\text{ V}$$

EXAMPLE 10-2

Calculate the theoretical tendency of nickel to corrode (in volts) in deaerated water at pH = 7. Assume that hydrogen is set free and Ni $(OH)_2$ forms with a solubility product of 1.6×10^{-16}.

SOLUTION: When we are asked for the theoretical tendency, we are being asked to determine the overall tendency of the electrochemical reaction for the possible corrosion of nickel. Thus, the electrochemical cell is

$$Ni \mid \text{water, pH} = 7 \mid Ni, H_2 \uparrow (1\text{ atm})$$

assume the anodic half-cell reaction is

$$Ni^{2+} + 2e \rightarrow Ni \qquad E^\circ = -0.250$$

and the equilibrium potential is

$$E_{Ni^{2+}/Ni} = E^\circ + \frac{RT}{2F} \ln a_{Ni^{2+}}$$

the cathodic half-cell reaction is

$$2H^+ + 2e \rightarrow H_2 \qquad E^\circ = 0$$

$$E_{H^+/H_2} = 0 + \frac{RT}{2F} \ln a_{H^+}^2 = 2.303 \frac{RT}{F} \log a_{H^+}$$

$$= -0.0591\text{ pH} = -0.0591 \times 7 = -0.4144$$

to get $a_{Ni^{2+}}$, we use the solubility product,

$$Ni(OH)_2 \rightleftarrows Ni^{2+} + 2\,OH^-$$

$$K_{sp} = 1.6 \times 10^{-16} = a_{Ni^{2+}} \cdot a_{OH^-}^2$$

and for a_{OH^-},

$$K_w = a_{OH^-} \cdot a_{H^+} = 1 \times 10^{-14}$$

and $a_{OH^-} = 1 \times 10^{-7}$ because pH = 7.

$$a_{Ni^{2+}} = \frac{1.6 \times 10^{-16}}{a_{OH^-}^2} = \frac{1.6 \times 10^{-16}}{10^{-14}} = 0.016,$$

and then

$$E_{Ni^{2+}/Ni} = -0.250 + \frac{0.0591}{2} \log 0.016 = -0.3032$$

and the open circuit voltage,

$$\xi = -0.4144 - (-0.3032) = -0.1112 \text{ V}.$$

Since ξ is negative, the ΔG will be positive and nickel will not tend to corrode.

EXAMPLE 10-3

Calculate the open-circuit cell voltage of the following cell at 40°C, and indicate which is the anode.

$$O_2 \text{ (1 atm)}, \text{Pt} \,|\, \text{water} \,|\, O_2 \text{ (0.1 atm)}, \text{Pt}$$

The appropriate half-cell reaction is the oxygen electrode, which is

$$O_2 + 4H^+ + 4e \rightarrow 2H_2O \qquad E° = 1.229 \text{ V}$$

SOLUTION: This is an example of the differential aeration cell and we start by assuming the left electrode is the anode. The equilibrium potential (Nernst equation) is

$$E = E° + \frac{RT}{4F} \ln \frac{a_{h^+}^4 \cdot a_{O_2}}{a_{H_2O}}$$

$$= E° + \frac{RT}{4F} \ln \frac{a_{H^+}}{a_{H_2O}} + \frac{RT}{4F} \ln a_{O_2}$$

$$= C + \frac{RT}{4F} \ln a_{O_2}$$

For the left electrode, $E_L = C$ since the partial pressure or activity of oxygen is 1. For the right electrode,

$$E_R = C + \frac{8.314 \times 313}{4 \times 96{,}500} \ln 0.1 = C - 0.0155 \text{ V}$$

Thus, the cell voltage

$$\xi = E_R - E_L = -0.0155 \text{ V}$$

Since the cell voltage is negative, the assumption we made (that the left electrode is the anode) is wrong. The right electrode should therefore be the anode and the current will flow from right to left in the representation of the electrochemical cell above. Again, this is an illustration of the differential aeration cell with the oxygen deficient electrode being the anode.

EXAMPLE 10-4

In Table 10-2, tin is cathodic with respect to iron. However, in tin plated food cans, the tin becomes anodic because tin forms complex ions. Determine the maximum concentration of Sn^{2+} in the food cans before it becomes cathodic to iron.

SOLUTION: This is an example of why Table 10-2 should be used with caution. When the concentration of the ions is very small, a cathodic material can become anodic because the equilibrium potential becomes less. The electrode potentials of the half-cell reduction cells for tin and iron are:

$$E_{Sn^{2+}/Sn} = -0.136 + \frac{0.0591}{2} \log a_{Sn^{2+}}$$

$$E_{Fe^{2+}/Fe} = -0.440 + \frac{0.0591}{2} \log a_{Fe^{2+}}$$

If the concentration of Sn^{2+} is very, very low, it is possible for the tin to be anodic with respect to iron. At the highest concentration of Sn^{2+} in the food before it turns cathodic, the equilibrium potentials above will be equal. Thus, equating the two potentials above,

$$-0.136 + \frac{0.0591}{2} \log a_{Sn^{2+}}$$

$$= -0.440 + \frac{0.0591}{2} \log a_{Fe^{2+}}$$

$$\frac{0.0591}{2} (\log a_{Sn^{2+}} - \log a_{Fe^{2+}})$$

$$= -0.440 + 0.136 = -0.304$$

$$\log \frac{a_{Sn^{2+}}}{a_{Fe^{2+}}} = \frac{-0.303 \times 2}{0.0591} = -10.288$$

$$\frac{a_{Sn^{2+}}}{a_{Fe^{2+}}} = 10^{-10.288} = 5.152 \times 10^{-11}$$

EXAMPLE 10-5

Calculate the equilibrium potential of iron versus the SHE when the Fe^{2+} ion concentration in the electrolyte is 10^{-6} gram-ion per liter of solution.

SOLUTION: The electrode potential of the iron half-cell reduction reaction is, as above,

$$E_{Fe^{2+}/Fe} = -0.440 + \frac{0.0591}{2} \log a_{Fe^{2+}}$$

At the given concentration, the activity is about equal to the concentration in molality. Thus,

$$E_{Fe^{2+}/Fe} = -0.440 + \frac{0.0591}{2} \log 10^{-6}$$

$$= -0.440 + \frac{0.0591}{2} \times (-6)$$

$$= -0.6173 \text{ volt}$$

This is the voltage corresponding to the horizontal line marked −6 in Fig. 10-21b, the Pourbaix diagram of iron.

EXAMPLE 10-6

A copper storage tank containing dilute H_2SO_4 (sulfuric acid) at pH = 0.1 is blanketed with hydrogen at 1 atm. Calculate the maximum Cu^{2+} contamination in moles per liter. What is the corresponding contamination if the hydrogen partial pressure is reduced to 10^{-4} atm.? Assume temperature is at 25°C.

SOLUTION: This is an example of the effect of corrosion on the quality of products. In this case, we are looking at the contamination of the sulfuric acid. Because we are interested in the corrosion of copper, the electrode potential of the half-cell reaction is

$$E_{Cu^{2+}/Cu} = 0.337 + \frac{0.0591}{2} \log a_{Cu^{2+}}$$

The electrolyte being an acid, the cathodic reaction is the hydrogen half-cell reaction and the potential is

$$E_{H^+/H_2} = 0 - \frac{RT}{2F} \ln \frac{a_{H_2}}{a^2_{H^+}}$$

$$= -\frac{0.0591}{2} \log p_{H_2} - 0.0591 \text{ pH}$$

For maximum concentration, the electrode potentials will be equal.
For $p_{H_2} = 1$ atm,

$$0.337 + \frac{0.0591}{2} \log a_{Cu^{2+}}$$

$$= -0.0591 \times 0.1 = -0.00591$$

$$\frac{0.0591}{2} \log a_{Cu^{2+}} = -0.00591 - 0.337$$

$$\log a_{Cu^{2+}} = \frac{2(-0.00591 - 0.337)}{0.0591} = -11.60$$

$$a_{Cu^{2+}} = 2.512 \times 10^{-12}$$

For $p_{H_2} = 10^{-4}$ atm,

$$0.337 + \frac{0.0591}{2} \log a_{Cu^{2+}}$$

$$= \frac{-0.0591}{2} \times (-4) - 0.0591 \times 0.1 = 0.11229$$

$$\log a_{Cu^{2+}} = \frac{2(0.11229 - 0.337)}{0.0591} = -7.604$$

$$a_{Cu^{2+}} = 2.5 \times 10^{-8}$$

10.4 Amount and Rate of Corrosion

The discussion of the electrochemical cell has so far been with the thermodynamics and cell voltages in an open circuit. An open circuit implies there is no current flowing and when there is no current, there is no flow of electrons or charges; therefore, corrosion or degradation of the anodic material does not occur. We shall now look into what happens when current flows through the electrochemical circuit and the anodic material gets to be consumed or dissolved into the electrolyte.

The amount of anodic material corroded or consumed when current passes through the circuit is governed by two Faraday's laws, which are:

(1) First Law: "In electrolysis, the quantities of substances involved in the chemical change are proportional to the quantity of electricity that passes through the electrolyte."
(2) Second Law: "The masses of different substances set free or dissolved by a given amount of electricity are proportional to their chemical equivalents."

In honor of Faraday, *the amount of charge that sets free or dissolves one chemical equivalent of a substance is called the Faraday constant and this is equal to 96, 485 C/equivalent (~96,500 C/eq.).* Thus, if I is the current in amperes passing through the circuit for a time, *t*, in seconds, the

$$\text{Amount of Electricity or Charge} = Q = I \times t \text{, in coulombs (C),} \quad (10\text{-}29)$$

and,

$$\frac{Q}{F} = \frac{I \cdot t}{F} \quad (10\text{-}30)$$

$$= \text{number of equivalents of corroded or dissolved material}$$

When we combine both the anodic and cathodic half-cell reactions, the overall electrochemical reaction can be generalized by the following balanced reaction, in which the number of electrons in the anodic half-cell reaction equals that in the cathodic half-cell reaction.

$$\sum_{1,ox} (\nu X)_j + \sum_{2,red} (\nu Y)_j \rightleftarrows \sum_{1,red} (\nu X)_j + \sum_{2,ox} (\nu Y)_j \quad (10\text{-}31)$$

where the *1* refers to the cathodic substances being reduced and 2 refers to the anodic substances being oxidized. The ν for each substance refers to the number of moles of the substances involved in the electrochemical reaction. For each of the substances involved,

$$\frac{\nu_j}{n} \text{ is referred to as the chemical equivalent,} \quad (10\text{-}32)$$

and

$$\frac{\nu_j}{n} M_j \text{ is the equivalent mass or weight} \quad (10\text{-}33)$$

where *n* is the total number of electrons involved in the overall reaction in Eq. 10-31, and M_j is the atomic mass or the molecular weight of the substance. Thus,

$$F = \frac{\nu_j}{n} M_j \qquad M_j = \frac{nF}{\nu_j} \quad (10\text{-}34)$$

If *m* is the mass of the substance dissolved (oxidized) or set free (reduced), and using Eqs. 10-33 and 10-30,

$$\text{number of equivalent weight} = \frac{m}{M_j} \times \frac{n}{\nu_j} = \frac{Q}{F} \quad (10\text{-}35)$$

$$\text{the mass of substance, } m = \frac{\nu_j Q}{nF} M_j = \frac{\nu_j I t}{nF} M_j \quad (10\text{-}36)$$

$$\text{The number of moles} = \frac{m}{M_j} = \frac{\nu_j Q}{nF} = \frac{\nu_j I t}{nF} \quad (10\text{-}37)$$

and the specific reaction rate,

$$r_j \equiv \frac{\text{mass of substance reacted}}{\text{Area, A,} \times \text{time}} = \frac{\nu_j i}{nF} \cdot M_j \quad (10\text{-}38)$$

where $i = I/A$ is called the current density and *A* is the total surface area of the anodic or cathodic material. We note that the rate of reaction is determined with the current density, *i*. We should note, however, that the current *I* in amperes is the quantity that is the same for each electrode, not the current density, *i*. The current density at the cathode is

$$i_c = \frac{I}{A_c} \quad (10\text{-}39a)$$

and that at the anode is

$$I_a = \frac{I}{A_a} \quad (10\text{-}39b)$$

When the current density, *i*, is in amperes per square meter, the rate in Eq. 10-38 is in grams per square meter per sec. The corrosion rate commonly used in the literature is grams per square meter per day (gmd). To obtain the rate in gmd, the rate in Eq. 10-38 must be multiplied by (24 h/day × 3600 sec/h). If the corrosion is uniform, we can also convert the gmd to a uniform depth of penetration in either mils (1 mil = 0.001 in) per year (mpy) or mm per year by dimensional analysis, and using the density of the material. The relative uniform corrosion resistance ranking in terms of mpy and its equivalents for ferrous and nickel alloys is shown in Table 10-4. The uniform depth of corrosion is considered in design when corrosion is important.

EXAMPLE 10-7

Aluminum is anodized (anode and oxidized) to form aluminum hydroxide in an aerated aqueous solution

Table 10-4 Rankings of uniform corrosion resistance of iron and nickel alloys in terms of mils per year (mpy) and its equivalents. [6]

Relative Corrosion Resistance[a]	mpy	mm/yr	μm/yr	nm/h	pm/s
Outstanding	<1	<0.02	<25	<2	<1
Excellent	1–5	0.02–0.1	25–100	2–10	1–5
Good	5–20	0.1–0.5	100–500	10–50	20–50
Fair	20–50	0.5–1	500–1000	50–150	20–50
Poor	50–200	1–5	1000–5000	150–500	50–200
Unacceptable	200+	5+	5000+	500+	200+

[a]Based on typical ferrous and nickel alloys. Rates greater than 5 to 20 mpy are usually excessive for more expensive alloys, while rates above 200 mpy are sometimes acceptable for cheaper materials (e.g., cast iron) of thicker cross section.

of pH = 7. Determine the equivalent weights of aluminum, oxygen, and aluminum hydroxide from the overall electrochemical reaction. Aluminum hydroxide when dehydrated forms alumina, Al_2O_3. What is the equivalent weight of Al_2O_3?

SOLUTION: We need to write the overall electrochemical reaction and use Eq. 10-33. The pertinent half-cell reactions are:

$$\text{Anodic:} \quad Al^{3+} + 3e \rightarrow Al \qquad (A)$$

$$\text{Cathodic:} \quad O_2 + 2H_2O + 4e \rightarrow 4OH^- \qquad (B)$$

To get the overall balanced reaction, we need to make the number of electrons in both equations equal. Thus, we multiply Reaction *A* by 4, and Reaction *B* by 3, and the total number of electrons for both reactions *A* and *B* will become 12. Then, we eliminate the electrons by subtracting Reaction *A* from Reaction *B* and obtain the overall balanced electrochemical reaction

$$4Al + 3O_2 + 6H_2O \rightarrow 4Al(OH)_3$$

or

$$4Al + 3O_2 + 6H_2O \rightarrow 2Al_2O_3 + 6H_2O$$

(The last reaction show the dehydration of aluminum hydroxide to form alumina.) We now use Eq. 10-33, remembering that $n = 12$ (total number of electrons),

The equivalent weight of aluminum

$$= \frac{4}{12} M_{Al} = \frac{M_{Al}}{3}$$

The equivalent weight of oxygen

$$= \frac{3}{12} M_{O_2} = \frac{M_{O_2}}{4}$$

The equivalent weight of $Al(OH)_3$

$$= \frac{4}{12} M_{Al(OH)_3} = \frac{M_{Al(OH)_3}}{3}$$

The equivalent weight of Al_2O_3

$$= \frac{2}{12} M_{Al_2O_3} = \frac{M_{Al_2O_3}}{6}$$

EXAMPLE 10-8

If the density of alumina, Al_2O_3, is 3.97 gm/cm^3, (a) determine how long it will take to produce a 10 μm thick anodized layer (Al_2O_3) if the current density is 0.1 A/cm^2, and (b) determine the number of moles of oxygen that was consumed?

SOLUTION:

Assuming 1 cm^2 surface area,
$I = 0.1$ A, and the

Volume of 10 μm thick
$Al_2O_3 = 10 \times 10^{-6}\,\text{m} \times 100\,\text{cm/m} \times 1 = 10^{-3}\,\text{cm}^3$

Mass of alumina
$= 3.97 \times 10^{-3}$ gram

Using Eq. 10-36,

$$m = \frac{v_j It}{nF} \times M_j = \frac{2}{12} \times \frac{0.1}{96{,}500} t \times M_{Al_2O_3}$$

$$M_{Al_2O_3} = 2(26.98) + 3(16) \simeq 102$$

$$t = \frac{96{,}500 \times 6 \times 3.97 \times 10^{-3}}{0.1 \times 102} = 225.36 \text{ sec}$$

For the number of moles of oxygen consumed, we use the concept that the number of equivalents of the substances involved in the reaction should be the same. Thus, the number of equivalent weights of alumina is the same as that of oxygen.

Number of equivalent weight of alumina

$$= \frac{m_{Al_2O_3}}{1/6\, M_{Al_2O_3}}$$

Number of equivalent weight of oxygen

$$= \frac{m_{O_2}}{1/4\, M_{O_2}}$$

Equating the two and solving for m_{O_2}/M_{O_2} = number of moles

$$\frac{m_{O_2}}{M_{O_2}} = \frac{3.97 \times 10^{-3} \times 6}{102 \times 4} = 5.838 \times 10^{-5} \text{ mol}$$

EXAMPLE 10-9

Derive the conversion factor from gmd to mm/y.

SOLUTION:

$$\text{mm/y} = \text{gmd} \times \text{factor}$$

$$= \frac{\text{gram}}{\text{m}^2 \cdot \text{day}} \cdot \frac{1000 \text{ mm}}{\text{m}} \cdot \frac{\text{cm}^3}{\rho, \text{gram}} \cdot \frac{\text{m}^3}{10^6 \text{ cm}^3} \cdot \frac{365 \text{ days}}{\text{year}}$$

$$\text{mm/y} = \text{gmd} \times \frac{0.365}{\rho}$$

where ρ is the density in gram/cm^3.

10.5 Corrosion Current and Change in Electrode Potentials → Polarization

We have just learned that corrosion can only proceed if there is current passing through the electrochemical cell. With current passing, it also means that the equilibrium conditions at the electrodes have changed and therefore the electrode potentials change during corrosion. The change in the equilibrium potentials is called *polarization* and quite often referred to as *overvoltage.* We shall determine what these changes are and how we can obtain the corrosion current that we shall use in Eq. 10-38. First we need to introduce the concept of the exchange current density that is related to equilibrium conditions.

We need therefore to go back to the equilibrium conditions and consider a metal electrode in equilibrium with its ions in an electrolyte. Equilibrium does not mean static conditions. If we write the possible half-cell reactions,

$$M^{n+} + ne \rightleftarrows M,$$

and

$$2H^+ + 2e \rightleftarrows H_2$$

equilibrium means that the forward direction (→) and the reverse direction (←) of each of the above reactions are occurring at the same rate. Since the rate is proportional to a current density in Eq. 10-38, at equilibrium, the current density of the forward direction equals that of the reverse direction. This current is called the *exchange current density* and is given the symbol, i_o. The exchange current density of the metal reaction is dependent on the concentration of electrolyte. The exchange current density of the hydrogen electrode depends on both the electrolyte and the cathode or metal surface.

A net current develops when the rate in one direction is greater than the other and if this happens, the equilibrium conditions are altered and corrosion starts. We start with the Nernst equation, the equilibrium condition for the electrode potential, in Eq. 10-24

$$E = E^\circ + \frac{RT}{nF} \ln \frac{a_{Ox}}{a_{Red}} \qquad (10\text{-}24)$$

and let us assume that the $a_{Red} = 1$ for both reactions. When there is current in the cell, the electrode potentials are determined by the activities or concentrations of the electrolyte in contact with the electrode. For anodic reactions, for example the reverse metal direction, there will be more oxidized ions at the surface because these are being transferred from the metal to the electrolyte, and diffusion takes time to even the composition of the electrolyte. This means that a_{ox} increases and Eq. 10-24 says that the electrode potential when corrosion occurs is greater than the equilibrium E. The change in potential (polarization) is positive and we say, the anodic material becomes more cathodic.

For the cathodic reaction, the ions in the electrolyte in contact with electrode will be reduced and their concentrations will therefore be depleted. Thus, according to Eq. 10-24, the electrode potential with current will be less than the electrode potential at equilibrium. The polarization (change in potential) is negative and thus the cathodic material becomes more anodic.

We can show the changes in potentials in a polarization diagram—a plot of the electrode potentials as a function of corrosion current. For the reactions indicated above, the metal is acting both as the anode and the cathodic surface for the hydrogen reaction. The metal is said to be a mixed electrode. Let us assume the metal is zinc and the activities of the ions and the hydrogen pressure is 1 atm. (i.e., all in their standard states). The starting points for the anodic and cathodic reactions in the polarization diagram are (E°_{A}, i_{oa}), and (E°_{C}, i_{oc}), shown in Fig. 10-22. When corrosion starts, the potential of the anode increases and the potential of the cathode decreases. This process continues until the potential of both the cathode and anode reactions are equal—called the mixed potential or *corrosion potential*, E_{corr}. The current at the corrosion potential is called the *corrosion current*, i_{corr}, and is the value we use in Eq. 10-38.

The exchange current densities can affect the corrosion current as shown in Fig. 10-23 for iron. The exchange current density of the hydrogen electrode is about four orders of magnitude on iron than that on zinc in acid solutions. The polarization diagram

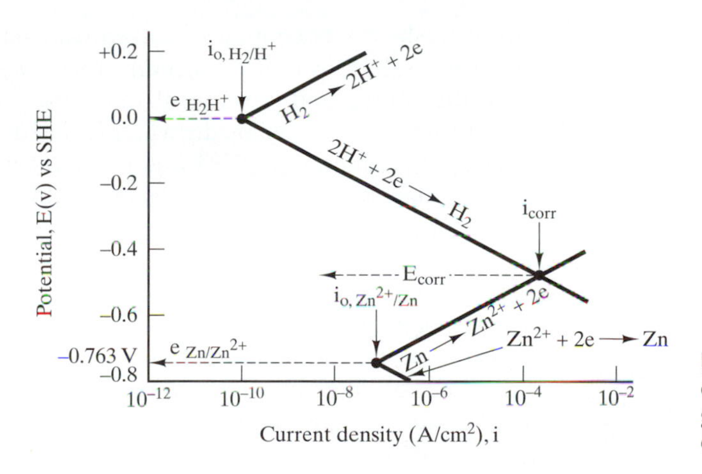

Fig. 10-22 Polarization of hydrogen cathode and zinc anode potentials to give corrosion potential, E_{corr}, and corrosion rate (current density) i_{corr}. [6]

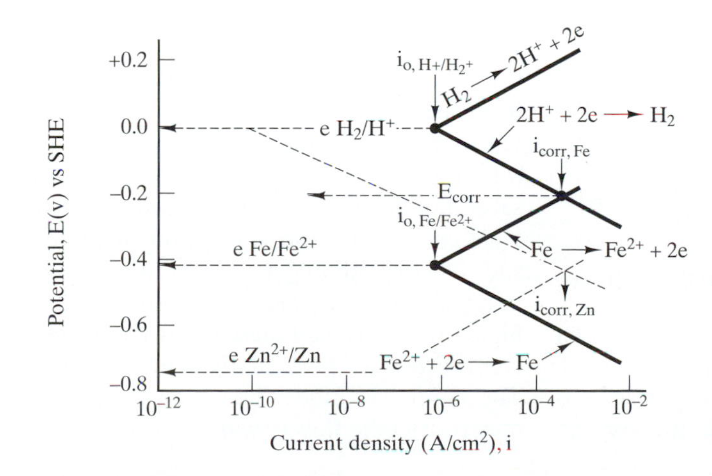

Fig. 10-23 (Solid lines) polarization of hydrogen cathode and iron anode potentials with Fig. 10-23 superimposed (dashed) for comparison to show effect of exchange current densities on E_{corr} and i_{corr}. [4]

for zinc is also dashed in Fig. 10-23 and we see that the corrosion current of iron is slightly higher than that of zinc. Thus, the corrosion rate of zinc is slightly lower in acids than that of iron, in spite of the fact that the standard potential of zinc is more anodic than iron. This behavior is largely due to the exchange current density of the hydrogen electrode on zinc being much lower than that on iron.

The effect of the exchange current density is but one of the many factors that can affect the polarization diagram and the corrosion rate of a metal. It is discussed here to indicate that we should be careful not to use Table 10-2 to form conclusions on the corrosion rate or on the kinetics of the reaction. The major factors that cause polarization are the slowness of reaction at the electrodes and the diffusion of ions to the electrodes—commonly referred to as activation and concentration polarization. Persons interested in learning more about these topics are referred to more advanced books on corrosion.

10.6 Some Factors Affecting Aqueous Corrosion Rates

10.6.1 Presence of Oxygen in the Electrolyte—Aerated Solutions

The presence of oxygen in the electrolyte is a significant contributor to corrosion rates. In deaerated (no air–no oxygen) solutions, the cathodic reaction will be C3 (Eq. 10-4)

$$2H^+ + 2e \rightarrow H_2$$

In acids this reaction proceeds rapidly, but it is a very slow reaction in alkaline or neutral media. In the presence of oxygen, the reaction is accelerated and reaction C4 (Eq. 10-5a) proceeds. In combination with the anodic reaction, the hydroxide of the metal forms, as illustrated for iron below.

$$2\,Fe + 2H_2O + O_2 \rightarrow 2\,Fe(OH)_2 \quad (10\text{-}40)$$

Thus, in order to reduce corrosion rates, the oxygen in the electrolyte needs to be reduced or removed. Removal of dissolved oxygen from water is accomplished by either chemically reacting the oxygen beforehand, called *deactivation,* or by distilling it off in a suitable equipment, called *deaeration.* Chemical deactivation can only be done with industrial waters because the chemicals used may be toxic. The water so treated is much less corrosive to a metal pipe distribution system. An example of a chemical used is sodium sulfite that results in the following reaction,

$$2\,Na_2SO_3 + O_2 \rightarrow 2\,Na_2SO_4 \quad (10\text{-}41)$$

The reaction is relatively fast at high temperatures, but slow at room temperature. It may be catalyzed by adding cupric (Cu^{2+}) salts; however, this addition might induce corrosion (Sec. 10.6.3). Deactivation of water for a steel heat-exchanger system reduced the corrosion rate from 0.2 mm/y before treatment to 0.004 mm/y after treatment.

10.6.2 Hard versus Soft Water

Natural fresh waters contain dissolved calcium and magnesium salts in varying concentrations, depending on the source and location of the water. If the concentration of salts is high, the water is called *hard;* otherwise it is called *soft.* Hard waters are much less corrosive than soft waters. The reason for this is the natural deposition on the metal surface of a thin diffusion-barrier film, composed largely of calcium carbonate, which acts to protect the metal surface from corrosive damage. The ability of calcium carbonate to precipitate on the metal surface depends on the total acidity or alkalinity, pH, and on the concentration of dissolved solids in water. This is indicated by the *saturation index,* which is

$$\text{Saturation index} = pH_{measured} - pH_{eq} \quad (10\text{-}42)$$

where pH_{eq} is the pH at which the water is in equilibrium with solid calcium carbonate and is calculated from the total solid content and the calcium salt contents. In order for the water to be protective, the saturation index must be positive. A water with negative saturation index does not provide the diffusion-barrier scale and will be corrosive. If this is the case, lime or soda ash or both may be added to raise the saturation index. A saturation index of about 0.5 is satisfactory—higher than this may cause excessive deposition of calcium carbonate, particularly at elevated temperatures. The latter will hinder heat transfer and if excessive, may restrict the flow of water.

10.6.3 Presence of More Noble Ions in Water

The presence of more noble metal ions in aqueous solutions than the container or pipe material will trigger corrosion of the container or pipe material. This is referred to as the replacement or substitution reaction, whereby the active metal will dissolve (corrode) to replace the nobler metal in the electrolyte. Again, the position of the metal in the Galvanic series determines whether it is anodic or cathodic with respect to another metal. For example, the copper ion in water flowing inside a carbon or low-alloy steel pipe will deposit on the inside surface of the steel pipe and ferrous ions will replace the copper ions in the electrolyte. Such a situation actually happens in a recirculating steam condensate in a boiler system. The steam from the boiler passes through predominantly copper condenser tubes. In the presence of some oxygen and carbon dioxide in the water, the condensate may leach out copper from the condenser tubes and will cause what is called the return line corrosion that pits or corrodes the boiler tubes. When the noble metal deposits on the active metal surface, it creates a microgalvanic cell that will continue to operate as long as there is an electrolyte in contact. The active metal surface acts both as the anode and as the cathode on which the metal can deposit. If there are sufficient noble metal ions in the electrolyte, the whole active surface will eventually be plated with the nobler metal and eventually the corrosion and plating may stop. This is the basis of the electroless plating process—the plating of a metal onto another metal substrate without the use of an external current supply. This specific process is also called by other names such as ion-exchange or charge exchange, cementation, or immersion plating. The half-cell reactions involved are A (Eq. 10-1) and C1 (Eq. 10-2) in Sec. 10.3.1.

10.7 Effect of Flowing Electrolyte

Flowing electrolytes can cause erosion and/or cavitation-erosion. Erosion or wear is the removal of surface material by the numerous impacts on the surface by solid or liquid particles in the electrolyte. In its mildest form, erosion polishes the surface of a material, as in metallographic polishing to create a mirrorlike or shiny surface. In severe erosion, protective scales on the surface are removed and the material gets eroded or worn away. Erosion damage may be localized or spread through an area and the factors that affect the resistance of materials to erosion are difficult to define. It appears, however, that the damage is purely mechanical in nature in that it is primarily due to the impacts of solid or liquid particles on the surface. This is the reason for categorizing this as mechanically assisted degradation in Table 10-1. An example of an erosion wear damage is shown in Fig. 10-24.

If the flow conditions are such that repetitive low (below atmospheric) and high-pressure areas are developed, gas bubbles form and collapse at the metal-liquid interface. This phenomenon is called cavitation and the damage to a metal caused by cavitation is called cavitation-erosion. The damage shows up as a deeply pitted surface and appears spongy as shown in Fig. 10-25. The damage can be purely mechanical in nature, as is the experience with glass or plastics or damage to a metal that occurs in organic liquids. The damage may involve chemical factors when protective films are destroyed to induce faster corrosion rate. But even in such a case, the mechanical damage factor is readily apparent from the distortion of the metal grains at the immediate surface, as if the metal were cold-worked. Cavitation-erosion occurs typically on rotors of pumps, on the

Fig. 10-24 Boiler tube inserts eroded by particle-laden flue gas, which was forced to turn to the tube inserts.

Fig. 10-25 Cavitation-erosion damage appears spongy.

trailing faces of propellers and of water-turbine blades, and on the water-cooled side of diesel-engine cylinders. Austenitic 18-8 stainless steel is relatively resistant to cavitation-erosion.

10.8 Effect of Corrosion on Mechanical Properties

There is no direct effect of corrosion on the tensile properties of materials, that is, the yield strength, the tensile strength, and the ductility are not directly affected. However, because the material loses thickness or develops localized pitting after corrosion, the load that the material can sustain when immediately tested after corrosion will be lower because of the lesser area and/or the stress concentration effects of localized corrosion. The uniform thickness reduction can be easily allowed for in design but the localized pitting is very difficult to allow for.

The fracture toughness and the fatigue performance of the material are, however, significantly affected by corrosion. With stress-corrosion cracking (SCC), the fracture toughness, K_{I_c}, in static loading is now referred to as the $K_{I_{SCC}}$, the stress corrosion cracking threshold. One of the test methods to determine $K_{I_{SCC}}$ is the cantilever-beam test that is schematically shown in Fig. 10-26. As in the K_{I_c} test, the machined specimen is fatigue precracked before it is placed in the corrosion cell. A dead weight is placed at the end of the cantilever loading arm and if the material is susceptible to the electrolyte in the corrosion cell, the fatigue crack will grow and the time when the crack starts to grow or extend is recorded. The dead weight is converted to a K_I, stress intensity value by using the formula (included in the standard test) for the specimen and the mode of loading. The lower the dead weight (lower K_I) the longer the time to failure (crack extension). The results are plotted as shown in Fig. 10-27. Specimens

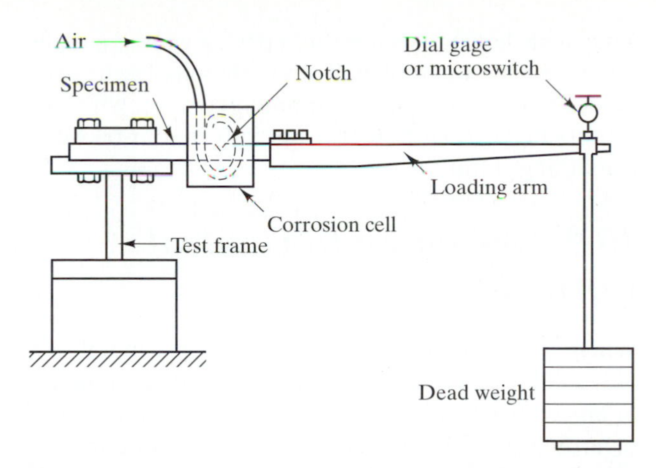

Fig. 10-26 Schematic cantilever beam setup for the testing of $K_{I_{scc}}$. Specimen is part of the beam and is immersed in corrosion cell. [7]

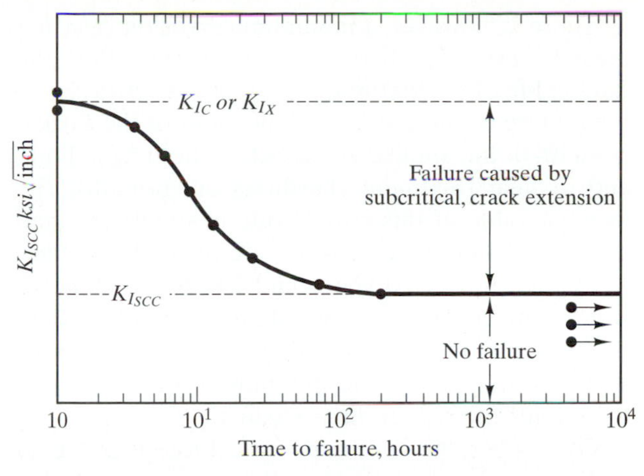

Fig. 10-27 Apparent $K_{I_{scc}}$ is the value that does not change with time. [7]

that did not fail after a long period of time, usually 1,000 hours for steels, are fractured and examined for the extension of the fatigue crack. The highest plane-strain K_I value at which there is no observed extension of the fatigue crack is designated as the $K_{I_{SCC}}$. The test result is, however, reported only as the apparent $K_{I_{SCC}}$ for a particular duration of test, because the value decreases as the test duration is prolonged as shown in Table 10-5. We see in Table 10-5 that the fracture toughness value is dramatically reduced by corrosion and there appears to be no true value of $K_{I_{SCC}}$.

The fatigue strength of materials are considerably decreased in a corrosive environment. In addition, for steels, the fatigue limit that is about half the ultimate tensile strength is nonexistent when corrosion is involved, as shown in Fig. 10-28. In terms of environment, *corrosion fatigue* is not specific to a particular environment and occurs in almost all corrosive environments. It occurs in natural water, sea water, general chemical environment and others and in general, the higher the uniform corrosion rate is, the shorter is the resultant fatigue life. Therefore, the resistance of a metal to corrosion fatigue is associated more nearly with its inherent corrosion resistance than with its high mechanical strength. Thus, copper, stainless steels, and nickel or nickel-base alloys will have better corrosion fatigue strength than carbon steels in fresh waters and brackish waters.

Table 10-5 Influence of cutoff time on apparent $K_{I_{scc}}$, ksi · $\sqrt{\text{inch}}$. [7]

Elapse Time, hrs.	Apparent $K_{I_{scc}}$
100	170
1,000	115
10,000	25

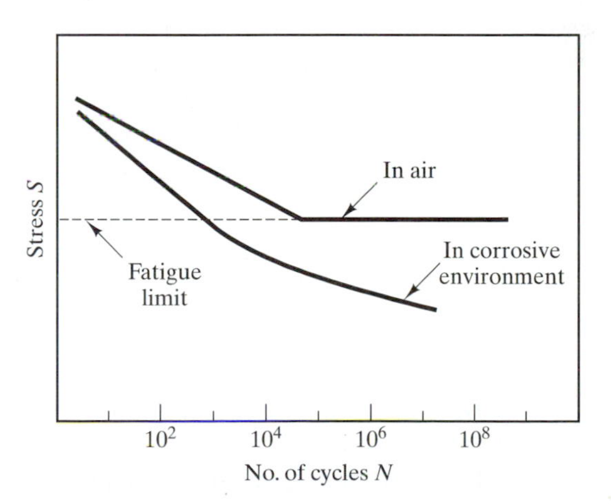

Fig. 10-28 S-N curves for steel in air and in corrosive environment.

There is, however, a minimum corrosion rate that must be exceeded before corrosion can affect the fatigue life of the material. For steels, the critical corrosion rate is found to be independent of the carbon content, of the applied stress below the fatigue limit, and of heat treatment (hardness or strength). The average value of this critical rate is about 0.58 gmd and is less than the corrosion rate of steels of about 1–10 gmd in aerated water and 3% NaCl. Thus, we would expect the corrosion fatigue to be lower than that in air. However at pH = 12, the uniform corrosion rate falls below the minimum critical rate and the fatigue life regains its value in air.

For copper, the minimum critical rate is 28.5 gmd and is much larger than the uniform corrosion rate in aerated water and 3% NaCl; hence, the fatigue life of copper is observed to be about the same in air as in fresh and saline waters.

Fretting corrosion is another form of corrosion that may lead to a shortened fatigue life or corrosion fatigue. This is the damage that occurs at the interface of two contacting surfaces, one or both being metal, subject to a slight relative slip that is oscillatory, such as that caused by vibration. The damage is characterized by a discoloration of the metal surface and, in case of oscillatory motion, by the formation of pits. These pits become fatigue crack nucleation sites.

During most corrosion processes, the anodic material also acts as the surface for one of the cathodic reactions. An example is the replacement or ion-exchange reaction discussed in Sec. 10.5. If the cathodic reaction is the liberation of hydrogen, Reaction C3 or C5 in Sec. 10.3.1, *hydrogen embrittlement* may occur. Reaction C3 is believed to proceed in the following steps:

Step 1. $H^+ + e \rightarrow H_{ads}$

Step 2. $H_{ads} + H_{ads} \rightarrow H_2$

The H_{ads} is atomic hydrogen adsorbed on the surface of the anodic material. The combination of two adsorbed hydrogen atoms is a very slow process, especially when there are "poison" elements such as sulfur that make the combination even harder. Thus, Step 2 above requires activation energy, but before the combination can occur, the hydrogen atoms have the opportunity to diffuse into the anodic material, just like carbon atoms diffusing into the lattice in the carburizing process. *The hydrogen in the lattice of high-strength steels induces hydrogen cracking and embrittlement.* This phenomenon is predominant with steels having hardness 40 HRC; a good example is the embrittlement of high-strength bolts. Hydrogen-induced cracking is also a common concern with oil-country high strength steels, especially in sour-oil wells containing sulfur.

10.9 Designing to Control Corrosion

When we need to control corrosion of a metal, we only need to eliminate one of the essential elements of an electrochemical reaction discussed in Sec. 10.3 regarding Fig. 10-1. We shall consider them one at a time.

10.9.1 Elimination of the Electrodes

Elimination of the electrodes means we should avoid situations where anodes and cathodes are formed that will lead to a Galvanic cell. The very obvious advice would be to avoid using different metals in a construction. However, this might not be very practical. If two different metals have to be used and joined, there are a few choices we can make in the design. The first thing we can do is to choose two metals that have the least open circuit voltage between them (i.e., that ξ in Eq. 10-11 is the least). Two dissimilar metals can be compatible when the difference in solution (anodic) potential is 0.25 V or less, as indicated in Table 10-6. The compatible couples are indicated by downward arrows on the right of Table 10-6 with potential or anodic index difference $\leq$ 0.25V. The open dot is the cathodic material and the solid dots are the anodic materials in one arrow. The second thing we can do is to make the anodic surface area much larger than that of the cathodic area. In this way, the current density of the anode is much smaller than that in the cathode and thus the rate of corrosion will be minimized. The third thing we can do is to try to keep the anode and the cathodic material as far apart as possible. This is what is called the distance effect and is illustrated in Fig. 10-29. The corrosion cell voltage will be diminished because of the voltage drop in the electrolyte and can be very effective if the resistance of the electrolyte is high.

What is not very obvious is the effect of fabrication and thermal processing in inducing localized galvanic cells in the selected materials. Differences in composition from one area to another will set up localized

Table 10-6 Compatible metals and alloys in dissimilar-metal couples. [8]

Group number	Metallurgical category	emf, V	Anodic Index (a), V	Compatible couples (b)
1.	Gold, solid and plated; gold-platinum alloys; wrought platinum	+0.15	0	
2.	Rhodium plated on silver-plated copper	+0.05	0.10	
3.	Silver, solid or plated; high-silver alloys	0	0.15	
4.	Nickel, solid or plated; monel metal, high-nickel-copper alloys	–0.15	0.30	
5.	Copper, solid or plated; low brasses or bronzes; silver solder; German silvery high copper-nickel alloys; nickel-chromium alloys; austenitic corrosion-resistant steels	–0.20	0.35	
6.	Commercial yellow brasses and bronzes	–0.25	0.40	
7.	High brasses and bronzes; naval brass; Muntz metal	–0.30	0.45	
8.	18% Cr type corrosion resistant steels	–0.35	0.50	
9.	Chromium plated; tin plated; 12% Cr type corrosion-resistant steels	–0.45	0.60	
10.	Tin-plate; terneplate; tin-lead solder	–0.50	0.65	
11.	Lead, solid or plated; high lead alloys	–0.55	0.70	
12.	Aluminum, wrought alloys ot the 2000 series	–0.60	0.75	
13.	Iron, wrought, gray or malleable; plain carbon and low-alloy steels; armco iron	–0.70	0.85	
14.	Aluminum, wrought alloys other than 2000 series aluminum, cast alloys of the silicon type	–0.75	0.90	
15.	Aluminum, cast alloys other than silicon type; cadmium, plated and chromated	–0.80	0.95	
16.	Hot-dip-zinc plate; galvanised steel	–1.05	1.20	
17.	Zinc, wrought; zinc based die-casting alloys; zinc based	–1.10	1.25	
18.	Magnesium and magnesium-base alloys, cast or wrought	–1.60	1.75	

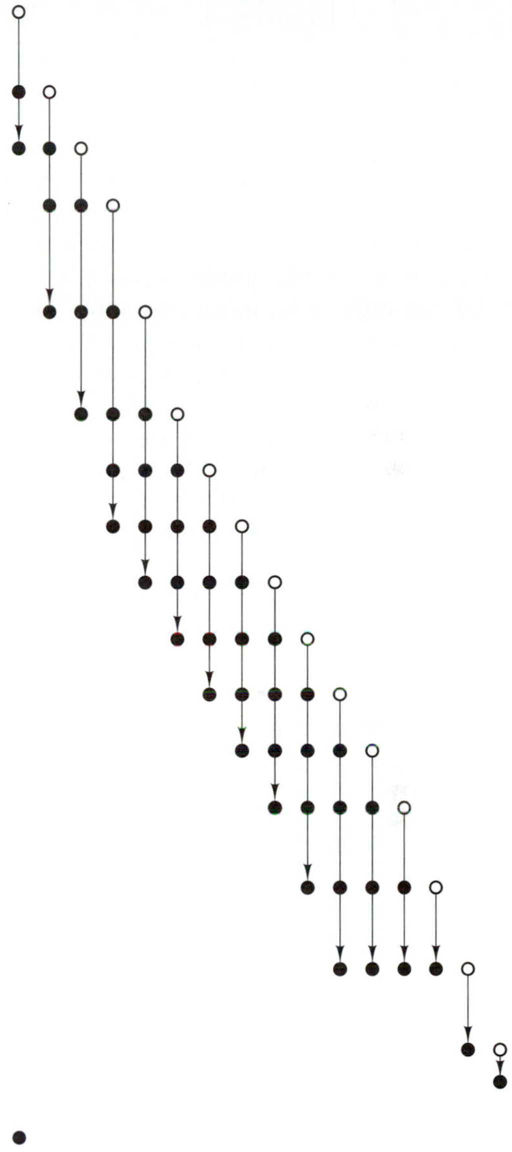

(a) Anodic index is the absolute value of the potential difference between the most noble (cathodic) metals listed and the metal or alloy in question. For example, the emf of gold (group 1) is +0.15 V, and the emf of wrought 2000-series aluminum alloys (group 12) is –0.60 V. Thus, the anodic index of wrought 2000-series aluminum alloys is 0.75 V. (b) "Compatible" means the potential difference of the metals in question, which are connected by lines, is not more than 0.25 V. An open circle indicates the most cathodic member of a series; a closed circle indicates an anodic member. Arrows indicate the anodic direction.

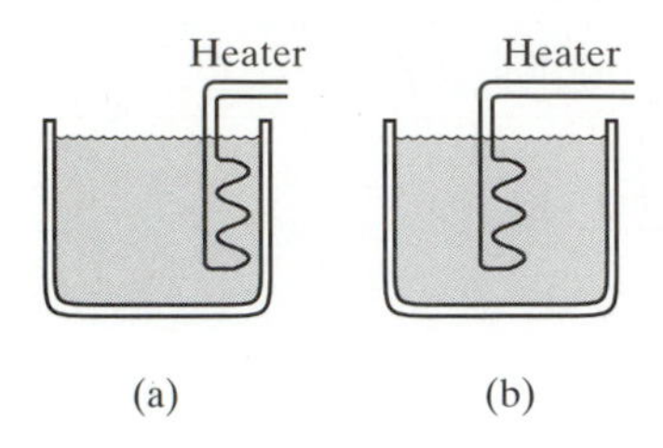

Fig. 10-29 (a) Galvanic corrosion of steel tank due to copper heater, (b) corrosion reduced due to distance effect.

cells, as in the sensitization of stainless steels, discussed in Sec. 9.7.3b—Embrittling Phenomena. In welding, we should specify filler wires with the same or slightly more noble composition than the base metal, not only to be more corrosion resistant but also to attain the mechanical properties of the base metal. As discussed earlier, the slow cooling of the weld can induce sensitization. Along the same line as the sensitization of stainless steel is the improper aging or heat-treating of aluminum alloys to precipitate particles along the grain boundaries that will form a precipitate-free zone (PFZ) and will induce intergranular corrosion and fracture. This was also discussed earlier. Finally, we should homogenize the composition of materials, especially as cast materials.

10.9.2 Eliminate or Avoid Contact of the Electrodes

If two dissimilar metals are to be joined, we can eliminate contact by using insulating washers between the surfaces and to use insulating nuts and bolts or bushings as illustrated in Fig. 10-30. Using an intermediate connector between the two dissimilar metals that can be easily replaced and maintained later may be a compromise situation when complete insulation cannot be achieved, illustrated in Fig. 10-31. The connector should have an electrode potential between the two dissimilar metals (see Fig. 10-7 and Table 10-6).

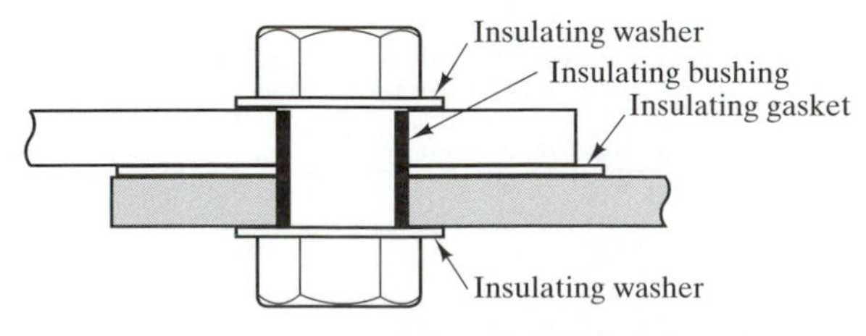

Fig. 10-30 Method of insulating bolted joints.

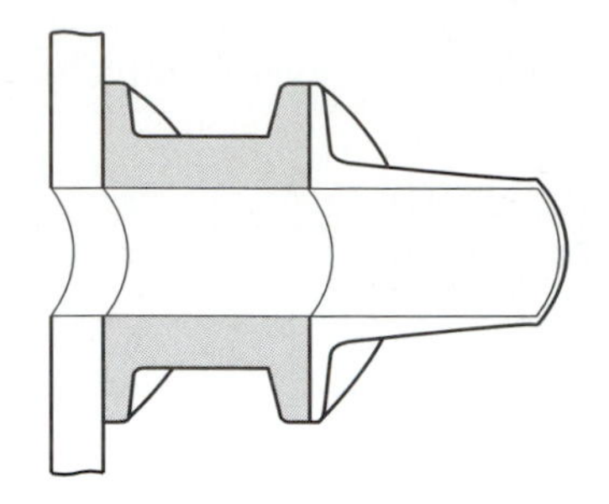

Fig. 10-31 Replaceable connector of two different metals.

10.9.3 Eliminate or Control the Electrolyte

This is probably the most applied technique in controlling corrosion. The electrolyte is eliminated by providing a *protective coating* on the surface of the metal or material. The coatings may be (1) *organic coatings,* (2) *inorganic coatings,* or (3) *metallic coatings.* The *organic coatings are paints* and there are a variety of paints available. The paints may have suspended sacrificial filler metals that will further protect the metal. An example of this is zincrometal paint that is used successfully for steels and irons. When using dissimilar metals, both the anodic and the cathodic materials should be painted. Painting of the anodic material can be disastrous because if a break occurs in the paint, the exposed area of the anodic material will be small and will result in a very high current density and very fast corrosion damage. The *inorganic coatings* are the *enamels, glass, and cement coatings.* Enamels are used on steel sheets for sinks, bathtubs, and basins. The glazed enamel is actually a layer of ceramic material on the steel surface. The inside of residential water heaters is glass (ceramic) lined. The *passivation of aluminum by anodizing* is a sort of an application of ceramic on aluminum because a *thick aluminum oxide layer* is induced by the anodizing process. Cement coatings are also ceramic materials and are applied generally to steel to improve the steel's corrosion resistance. They may be applied on both inside and outside surfaces, as in pipes. Some examples of *metallic coatings* are zinc on steel to produce galvanized steel, tin coating on tin cans, pure aluminum on precipitation-hardened aluminum alloys (ALCLAD), and stainless steel cladding on plain carbon or low-alloy steel. There are two types of coatings: *sacrificial coating or noble coating.* A sacrificial coating is a more active or anodic metal than the substrate, while the noble coating is more cathodic than the substrate. An example of a sacrificial coating is zinc on steel sheets to produce galvanized sheets, while nickel plating is an example of a noble coating on steel. The quality of the coating is very important in the noble coating because a break or scratch in it will lead to corrosion of the substrate. In the sacrificial coating, a break or scratch will not corrode the substrate.

If the structure cannot be coated, then we must *control the electrolyte.* This is done by "buffering" the electrolyte by adding inhibitors to the electrolyte that slows either the anodic or cathodic reaction or both. A formulation of an inhibitor is usually made to protect a variety of metal surfaces. This is especially critical for recirculating water in a heat transfer system, as the boiler steam condensate mentioned in Sec. 10.5 where copper ions can leach out of the copper condenser tubes. A good example of this formulation is that used as antifreeze coolants in cars. In addition to preventing freezing of the coolant, there are additives in the antifreeze that control the rusting of the engine (cast iron) and the radiator that may be copper or aluminum. The water going through cooling towers is also treated with additives to prevent corrosion of the structure.

The atmosphere surrounding the component may be controlled by what are called vapor phase inhibitors. They are used to protect critical components like semiconductor integrated circuits against corrosion during shipment.

10.9.4 Cathodic Protection

If we cannot put a coating on the structure or are unable to add inhibitors to control the electrolyte, then we can intentionally create a Galvanic electrochemical cell consisting of the metal to be protected as the cathode and another metal that is more anodic in the Galvanic series. This is called cathodic protection and this particular method is called the *sacrificial anode* because we are purposely sacrificing the damage (corrosion) of one metal to protect the cathode. A schematic of a setup to protect underground pipes is shown in Fig. 10-32. The most common sacrificial metals and alloys for buried pipe protection are magnesium-based and aluminum based alloys, and to a lesser extent, zinc, which is essentially pure, 99.99 percent. The reason for this is, of course, the high-cell voltage that magnesium produces with any cathode. Magnesium is also used as a sacrificial anode in domestic hot-water heaters. However, the efficiency

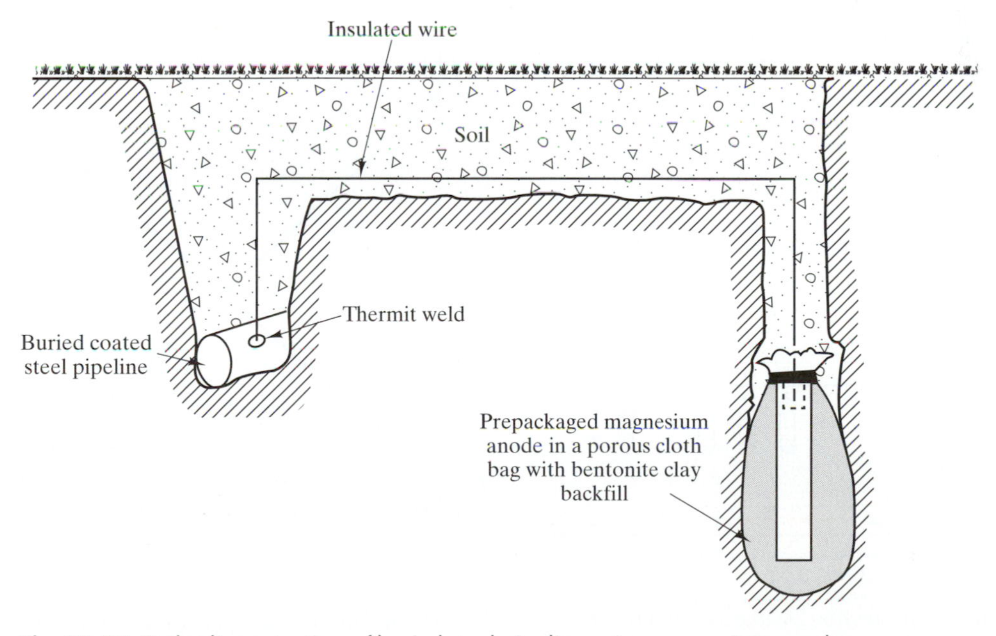

Fig. 10-32 Cathodic protection of buried steel pipeline using magnesium anode.

in using magnesium is lower than that for zinc and for this reason, zinc is used in marine (sea water) environments. (The efficiency is measured by the ratio of the theoretical amount needed to the actual amount needed.) Zinc is attached to the outside hull surface of ships and sufficient amount is attached to last inspection intervals. Aluminum is not used on ships because it will passivate when the ships are dry-docked and may remain passivated when the ships are put to service. Aluminum alloys are finding applications where their light weight is put to advantage such as in off-shore platform structures.

Another method of cathodic protection is the *impressed direct current method* that is schematically illustrated in Fig. 10-33. In this method, the alternating current (AC) is rectified to a direct current (DC) and is impressed on to the anode, then leaves the anode to enter the electrolyte (soil) and then leaves the electrolyte to enter the cathode or the structure to be protected. We note that in this method the anode is positive, whereas in the sacrificial anode method, the anode is negative. One thing is common in both cases, however. This is that the current leaves the anode to enter the electrolyte and then goes to the cathode from the electrolyte. The anode that is used is predominantly graphite and the required current to be impressed on it is of such a magnitude to make the resultant voltage lower than the equilibrium voltage of the cathode (structure to be protected) in the environment. This is schematically illustrated in the polarization diagram in Fig. 10-34.

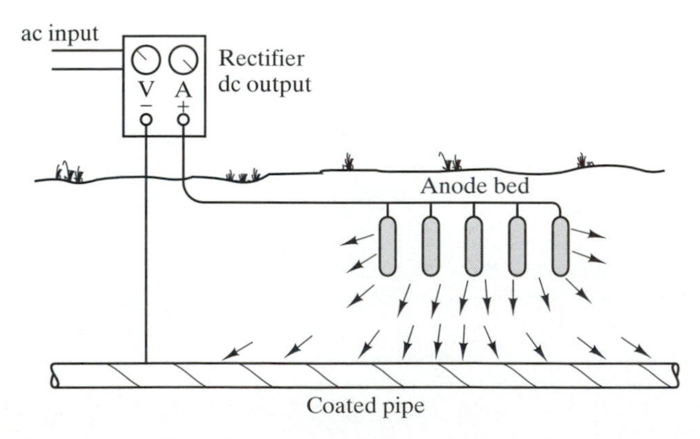

Fig. 10-33 Cathodic protection by impressed direct current method of buried steel pipeline.

10.9.5 Anodic Protection

Anodic protection depends on the same principle as the anodizing of aluminum (i.e., the formation of a passive layer on the surface of the material). Thus, the method is only applicable to metals that can be passivated. Among the structural metals, the ease of passivity decreases in the sequence:

Ti, Al, Stainless steel, Ni, carbon and
low alloy steels, copper

The passivation of titanium is so strong that titanium can be used in chloride environments when stainless steels and aluminum alloys cannot tolerate them.

The formation of the passive layer consists of actually making the material anodic (i.e., allowing it to corrode). The corrosion is in fact so fast that when the current passes a critical current density, the passive oxide barrier layer forms on the surface. Once the barrier oxide is formed, the corrosion current drops abruptly to a very small value. For steel in 1N H_2SO_4 (sulfuric acid), the critical current is 0.2 A/cm^2 and the passive corrosion current is 7 μA/cm^2, five orders of magnitude less than the critical current. Because of this small current, we see that the passivated material is still undergoing corrosion, although at an extremely

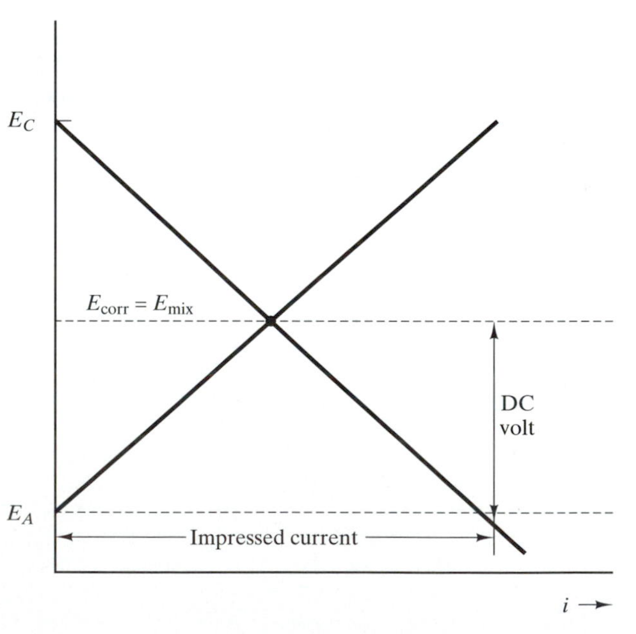

Fig. 10-34 Polarization diagram schematically showing voltage and current necessary for cathodic protection.

slow rate. This is in contrast to the cathodic protection method when the metal is completely protected from corrosion (i.e., zero corrosion current).

Although passivity has been known for some time to be the corrosion protection mechanism for titanium, aluminum, and stainless steels, the application of anodic protection to other systems has been slow in coming. The most successful application of anodic protection is in protecting the cheap, low carbon steel storage vessels and pipings in the handling and manufacture of sulfuric acid. It has lowered the dissolution of the steel, thus preserving the product (H_2SO_4) quality through reduction of metal pickup, extending the useful life of the equipment, and allowing the use of a lower cost alloy for the construction of the equipment. In addition, the method is also used to protect steel kraft digesters in the pulp and paper industries where the environment is NaOH + Na_2S at 175°C. The life of the kraft digesters is extended sevenfold and the economic value of this extension made it possible to

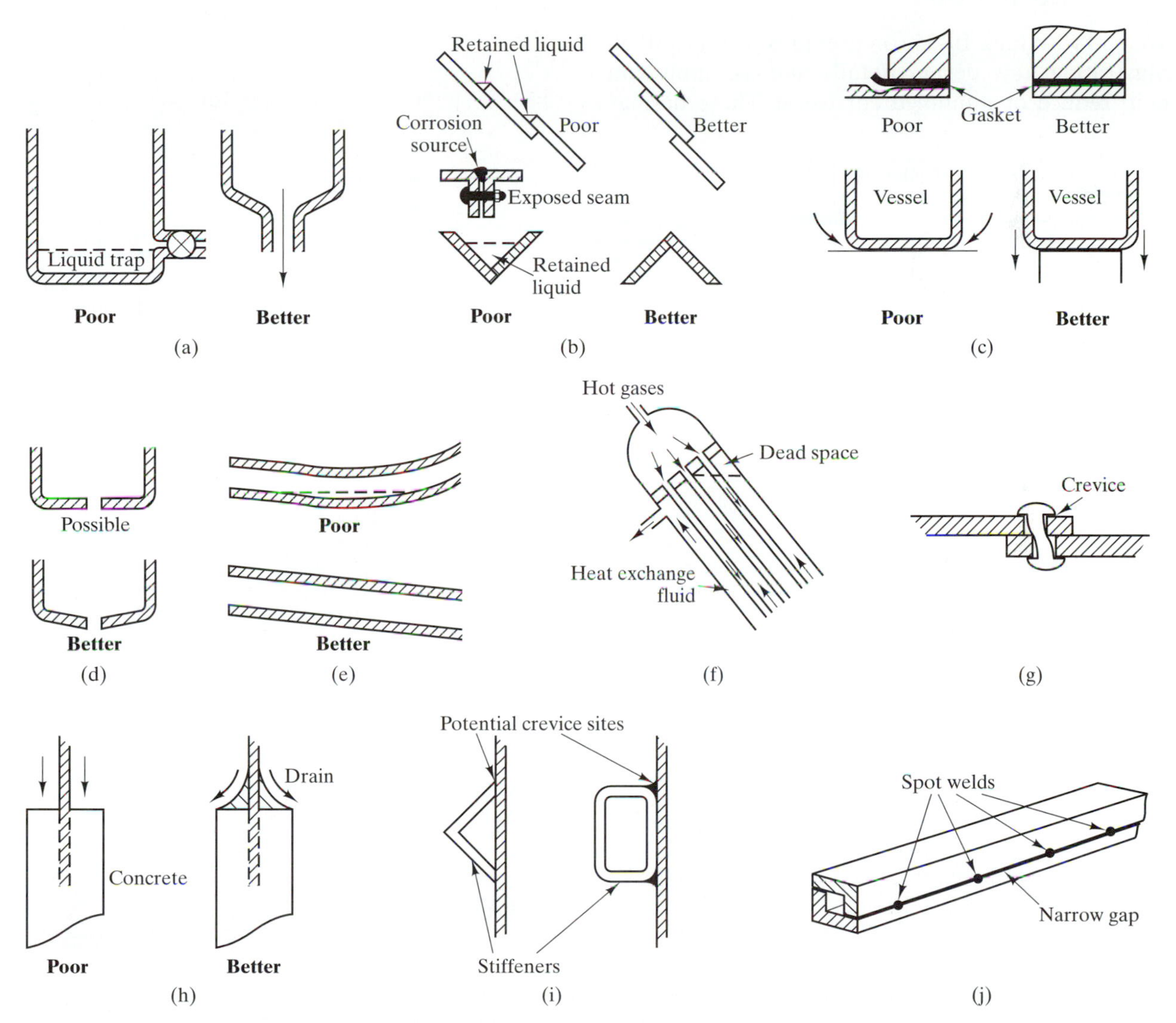

Fig. 10-35 Examples of design and assembly that can affect localized crevice corrosion. (a) and (d) liquids must be allowed to drain out of storage containers, (b), (e), and (h) structural members should be designed to avoid retention of liquids, (c), (g), (i), and (j) reduce chance of crevice corrosion, and (f) dead space can induce hot-spots. [1]

pay for the anodic protection system in about two years. Phosphoric acid plants used anodic protection of stainless steel tanks to prevent the danger of hydrogen gas accumulation due to corrosion. Replacing aluminum or stainless steels with anodically protected carbon steel in aqueous ammonia solutions cut the cost in half.

10.9.6 Design Details and Assembly of Components to Minimize Corrosion

We end with Figs. 10-35 (on previous page) to 10-38 illustrating how design details and assembly can help reduce or minimize corrosion. These designs incorporate the principles mentioned above about eliminating the electrolyte (keeping the surfaces dry) and avoiding crevice corrosion, minimizing the effects of the flow of liquids, avoiding contact of two dissimilar materials, and avoiding stress concentration areas. With moving fluids, designs for erosion and cavitation are illustrated. The figures are obtained from the *Corrosion Handbook,* Vol. 13, 9th Edition of Metals Handbook, published by the ASM International.

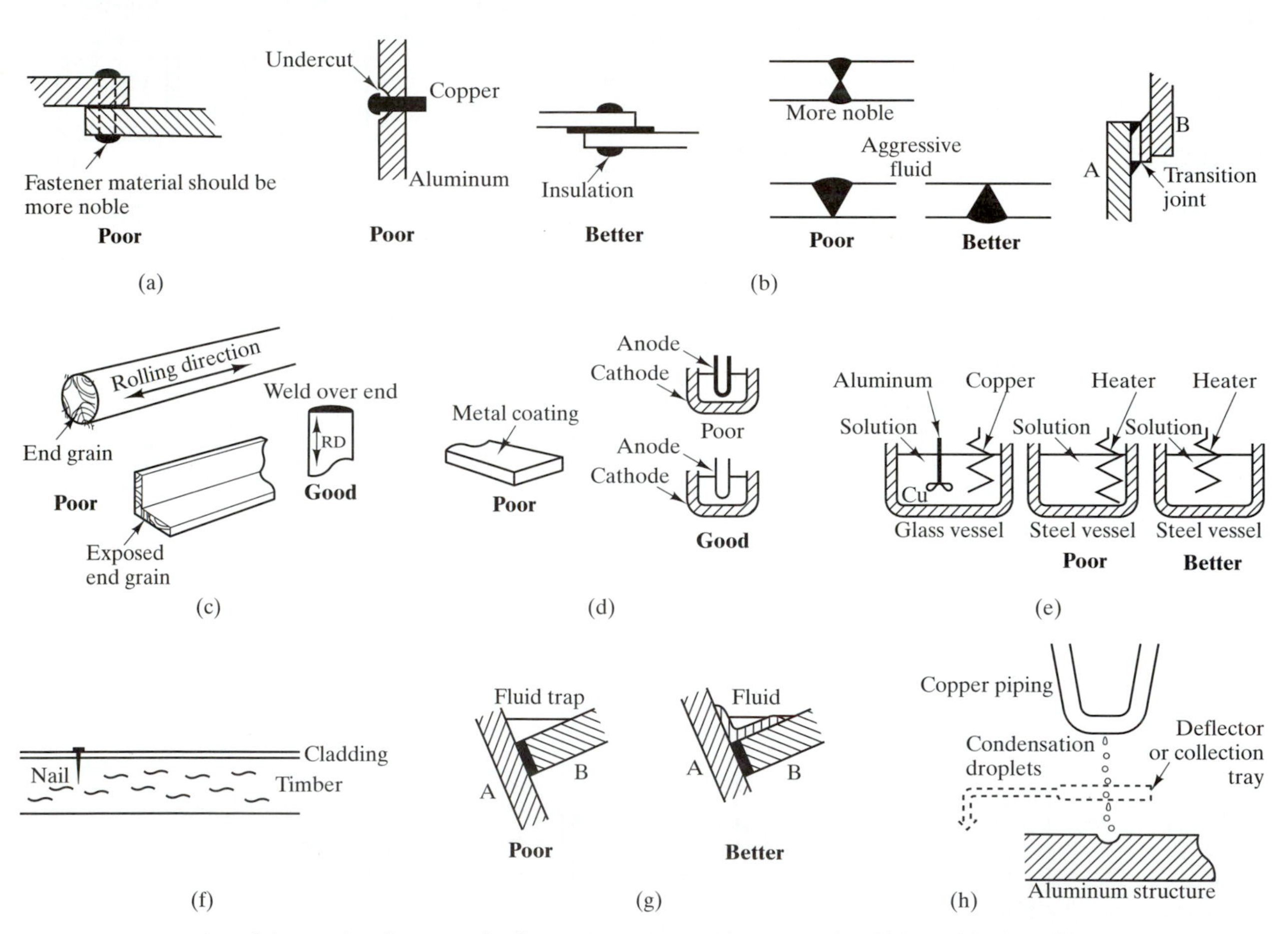

Fig. 10-36 Examples of design details to avoid galvanic corrosion. (a) Fasteners should be noble than the components (area effect), avoid undercutting, and use insulation; (b) weld filler should be more noble and smallest area exposed to corrosive environment and use transition joint; (c) avoid exposure of cut ends - weld over end; (d) apply coating on cathodic material, rather than the anodic; (e) distance effect; (f) Corrosion of nails in wood may be reduced by aluminum cladding; (g) Fluid trap can induce contact between dissimilar metal - use seal, mastic or a coating, (h) condensation droplets from a more noble structure induces impingement attack on active underlying structure - use deflector or collection trays. [1]

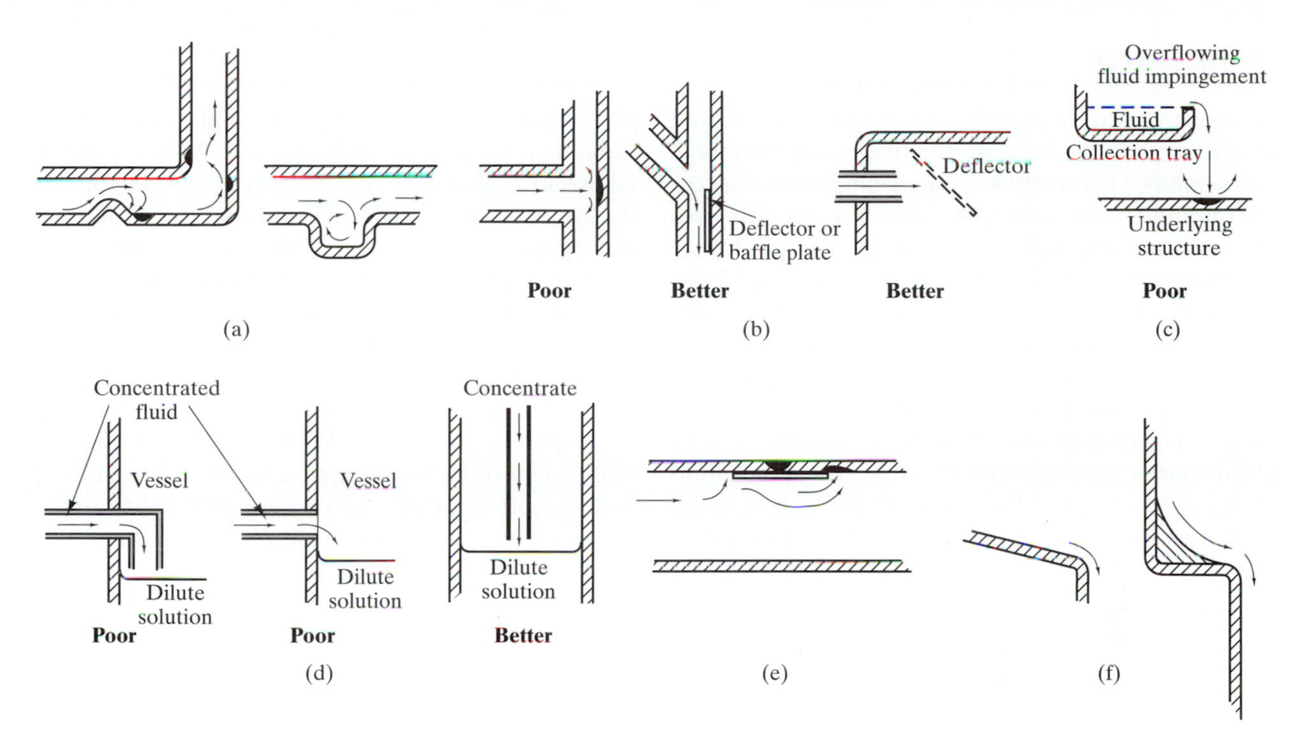

Fig. 10-37 Examples in design to avoid or reduce effects of flow of electrolyte. (a) Avoid disturbance of flow, and (b) avoid direct impingement of flow; (c) avoid overflow directly to underlying structure: (d) avoid splashes on container walls; (e) weld backing plates or rings can create local turbulence and crevices; (f) avoid fluid retention. [1]

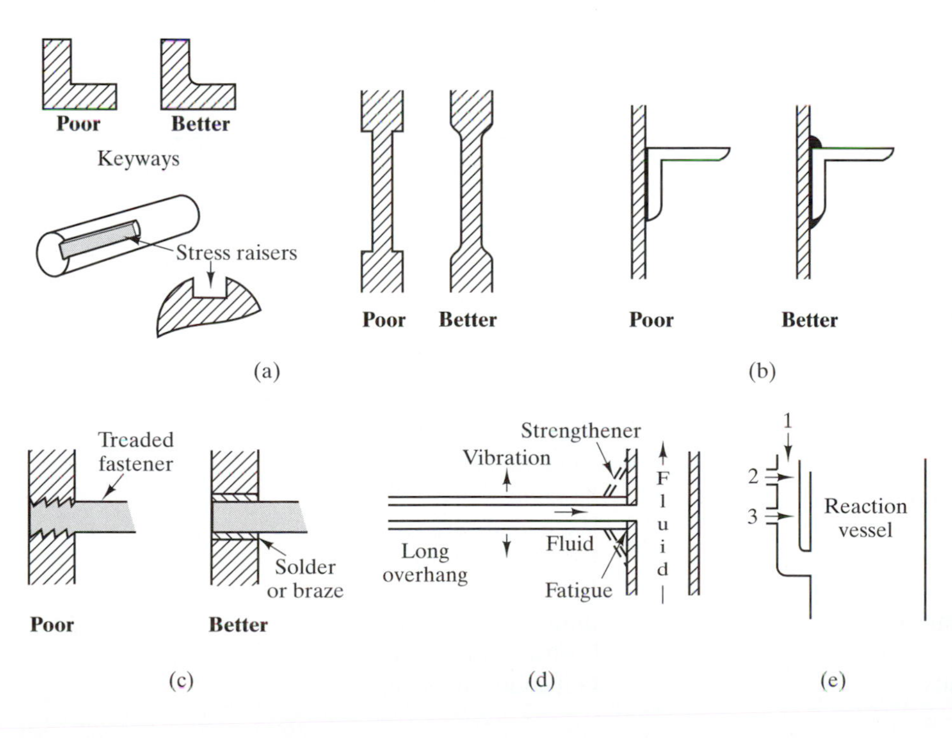

Fig. 10-38 Examples of designs showing stress concentrations (a) provide generous radius at corners, (b) avoid sharp corners in welds—make welds continuous, (c) avoid sharp profiles and use alternate fastening, (d) avoid long overhand that will induce vibration and fatigue at the junction, (e) side-supply pipework may not sustain thermal shock from air under pressure (1), steam (2), and cold water in sequence (3). [1]

Summary

Being an oxidation process, corrosion of metals and alloys is inevitable because the oxides are the stable compounds. Therefore, it is important to control corrosion and it is best to have these control measures incorporated during the design stage. We can do this because we now know the basics of corrosion and the different forms of corrosion to make us aware of situations when corrosion will be a significant factor. Basically, corrosion occurs because of the presence of all the following four items: an anode, a cathode, a contact between the anode and the cathode, and an electrolyte. Corrosion control depends upon eliminating at least one of these four. Composition of the electrodes (anode and cathode) and the electrolyte play significant roles. If the electrodes are not the same metal, we have a galvanic cell. Within a material, we may also have a difference in solute concentration from one area to another and this difference sets up the local anode and cathode areas for corrosion. This usually occurs because of improper thermal processing or insufficient homogenization of the alloy. Difference in electrolyte concentration induces crevice corrosion or differential concentration cells. Stress cells may also develop during assembly or use of the part or component. We have discussed the different measures to control corrosion, the design details, and assembly of components to minimize corrosion were presented.

References

1. "Corrosion," vol. 13, *ASM Handbook,* ASM International, 1987.
2. V. S. Bagotsky, *Fundamentals of Electrochemistry,* English Edition, Plenum Press, 1993.
3. S. A. Bradford, *Corrosion Control,* Van Nostrand Reinhold, 1993.
4. D. A. Jones, *Principles and Prevention of Corrosion,* Macmillan, 1992.
5. H. H. Uhlig and R. W. Revie, *Corrosion and Corrosion Control,* 3d ed., John Wiley, 1985.
6. M. G. Fontana, *Corrosion Engineering,* 3d ed., McGraw-Hill, N. Y., 1986.
7. J. M. Barsom and S. T. Rolfe, *Fracture & Fatigue Control in Structures,* 2d ed., Prentice Hall, Inc., Englewood Cliffs, N.J., 1987.
8. L. J. Korb, "Corrosion in Manned Spacecraft" p. 1058 of Reference 1.
9. M. Pourbaix, Atlas of Electrochemical Equilibria in Aqueous Solutions, NACE, 1974.

Terms and Concepts

Activity
Activity coefficient
Aerated solution
Anode
Anodic protection
Anodic reaction—oxidation
Aqueous corrosion
Atmospheric corrosion
Cathode
Cathodic protection
Cathodic reaction—reduction
Cavitation
Cell voltage
Corrosion fatigue
Corrosion potential (voltage)
Crevice corrosion
Current (corrosion) density
Dealloying
Dezincification
Electrochemical reaction
Electrode
Electrode potential
Electrode potential series
Electrolyte
Electrolyte concentration cells
Electronic conductor
Equivalent weight (mass)
Erosion
Exchange current density
Faraday constant
Faraday's laws
Filiform corrosion
Fretting corrosion
Galvanic corrosion

Galvanic series
Gmd—grams/m^2/day
Graphitic corrosion
Half-cell reactions
Hydrogen cracking
Impressed current
Inhibitors
Inorganic coatings
Intergranular corrosion
Ionic conductor
Molality
MPY—mils (0.001 in.) per year
Nernst equation
Noble metal coating
Open circuit voltage
Organic coatings
Overvoltage
Passivators
Pitting corrosion
Polarization
Pourbaix diagram
Redox (reduction oxidation) reactions
Sacrificial anode
Sacrificial metal coating
Saturation index
SCC—stress corrosion cracking
SHE—standard hydrogen electrode
Standard state
Stray current corrosion
$K_{I_{SCC}}$—critical stress intensity threshold for stress corrosion cracking
Stress cells
Stress-corrosion

Questions and Exercise Problems

10.1 Corrosion or degradation of metals is predominantly electrochemical in nature. Why is it called electrochemical?

10.2 What four elements must be simultaneously present in order for corrosion to occur? Describe the four elements and indicate how the flow of current and electrons proceed.

10.3 Why is the corrosion process called a redox process?

10.4 What is the difference in the electrical circuit consisting only of different electronic (metallic) conductors from that consisting of electronic and an ionic (electrolyte) conductors? What is the electrode potential?

10.5 What is the significance of the standard hydrogen electrode (SHE) and how are the electrode potentials of the different metals obtained?

10.6 What is the standard electrode potential series and what is its significance and limitation?

10.7 What is a Galvanic series and how does it differ from the standard electrode potential series?

10.8 What is the activity of a substance and what is meant by the standard state? What is the activity coefficient and its value in an ideal solution? What unit of concentration is used in electrochemical reactions? Differentiate it from other units of concentrations.

10.9 Calculate the standard free energy of the electrochemical reaction involving iron and zinc in ideal solutions containing one molal (1M) each of Fe^{2+} and Zn^{2+} ions.

10.10 Calculate the standard free energy of the electrochemical reaction involving copper and iron in ideal solutions containing one molal (1M) each of Fe^{2+} and Cu^{2+} ions. What is the open circuit voltage when copper and iron are used as electrodes in the ideal solutions?

10.11 Calculate the standard free energy of the electrochemical reaction involving aluminum and nickel in ideal solutions containing one molal (1M) each of the Ni^{2+} and Al^{3+} ions. What is the open circuit voltage?

10.12 Given

$$Fe^{2+} + 2e \rightarrow Fe \qquad E^\circ = -0.440\text{ V}$$

and

$$Fe^{3+} + e \rightarrow Fe^{2+} \qquad E^\circ = 0.771\text{ V}$$

Calculate the standard electrode potential, E°, of $Fe^{3+} + 3e \rightarrow Fe$.

10.13 Given

$$Cu^{+} + e \rightarrow Cu \qquad E^\circ = 0.521\text{ V}$$

and

$$Cu^{2+} + 2e \rightarrow Cu \qquad E^\circ = 0.337\text{ V}$$

Calculate the E° for $Cu^{2+} + e \rightarrow Cu^{+}$ and the equilibrium constant, K, for $Cu + Cu^{2+} \rightarrow 2\,Cu^{+}$
Which copper ion predominantly forms?

10.14 In a similar manner, estimate the equilibrium constant and determine which ion predominantly forms when chromium, Cr, dissolves anodically. Use the following information.

$$Cr^{2+} + 2e \rightarrow Cr \qquad E^\circ = -0.91\ V$$

and

$$Cr^{3+} + 3e \rightarrow Cr \qquad E^\circ = -0.74\ V.$$

10.15 Using the data given in Prob. 10.12 and similar to Probs. 10.13 and 10.14, estimate the equilibrium constant and determine which ion predominates when iron dissolves anodically.

10.16 Salt water such as in the ocean contains about 3.5 wt% salt. What is the concentration in molality? in molarity?

10.17 Calculate the equilibrium half-cell potential of iron in a 0.05 N (normal) ferrous chloride solution with an activity coefficient of 0.60.

10.18 The activity of the hydrogen ion, a_{H^+}, is commonly expressed as pH, which is defined as,

$$pH = -\log a_{H^+}$$

where log is logarithm to the base 10. Using the Nernst equation, show that the equilibrium hydrogen electrode potential varies with pH at room temperature and 1 atm hydrogen pressure according to

$$E_{H_2} = -0.0592\ pH.$$

10.19 Calculate the potential of an electrochemical cell consisting of a copper electrode in an electrolyte with copper activity equals 0.5 and pH = 2 (see Prob. 10.18 for definition of pH), and a hydrogen electrode with 2 atm hydrogen pressure. Will copper have a tendency to corrode?

10.20 Using the standard half-cell oxygen electrode in Example 10-3, calculate the equilibrim half cell voltage for oxygen using air at 1 atm. Air is 21% oxygen and 79% nitrogen by volume.

10.21 Using the same method as that used in Example 10-9 (dimensional analysis), show that the conversion of the uniform corrosion rate of a metal in gmd (grams/m^2/day) to corrosion current density in amps per sq. m. is,

$$i,\ amps/m^2 = gmd \times 1.117 \times \frac{M_j}{n}$$

where M_j is the atomic weight of the metal, and n is the number of electrons or valence of oxidized metal, as in the reaction,

$$M \rightarrow M^{n+} + n.$$

10.22 A 100-mm square steel sheet, with a thickness of 2 mm, is completely dipped in an acidic solution and the corrosion current density was determined to be 0.001 Amps per cm^2. (a) Determine the remaining thickness of the steel after 24 hours in the acid solution, and (b) the number of moles of hydrogen gas that was liberated. Neglect the thickness area. The density of steel is approximately that of iron which is 7.87 g/cm^3.

10.23 Steel A when immersed in a deaerated acidic solution produced a mixed potential of ξ_A and a corrosion current density of i_A. Steel B when immersed in the same acidic solution had a mixed ξ_B and a corrosion current density of i_B. The values for Steel B are higher than the corresponding values for Steel A. (a) Sketch the polarization diagrams for the two steels, assuming that the polarization of the anode is the same but that of the cathode polarizes differently but starts from the same exchange current density. (b) Which steel corrodes faster in the acidic solution? (c) When the steels are connected together in the acidic solution, which will corrode faster?

10.24 What is passivity and its general applicability to metals? How is a metal surface passivated?

10.25 What is the difference between cathodic protection and anodic protection?

10.26 What are the two methods used for cathodic protection?

10.27 In the cathodic protection of steel line pipes using magnesium as the sacrificial anode, calculate the standard free energy of the reaction at 25°C.

10.28 What is the saturation index of natural water and its effect on corrosion?

10.29 In terms of corrosion, what is meant by the replacement or substitution reaction? Give an example.

10.30 An aluminum container was used to store ferrous chloride ($FeCl_2$) salt. Because of the high relative humidity in the environment, the salt became wet and the aluminum was perforated. Indicate the possible anodic and cathodic reactions.

10.31 Describe how the different forms of corrosion in Table 10-1 occur.

11

Design, Selection, and Failure of Materials

11.1 Introduction

Engineering design of an engineering product, component, or structure is a difficult, complex, multidisciplinary, open-ended, problem-solving activity. The already difficult task of satisfying engineering and commercial requirements imposed on the design of a product or component becomes even more difficult with the very recent addition of legislated environmental requirements. With such varied, and often quite conflicting requirements, we should appreciate that design is rarely a one-person activity. A product's design is usually a combined effort of a team consisting of members from pertinent engineering disciplines who can critically evaluate the conformance of the product to the design requirements. Vital to this team is a materials engineer who can provide the expertise in the optimum selection of materials that will be used to construct or make the product. The optimum selection should lead to optimum properties and least overall cost of materials, ease of fabrication or manufacturability of component or structure, and environmentally friendly materials.

In this chapter, we shall discuss the general methodology of the design process and its relation to materials selection. For the most part, we shall discuss the factors involved in materials selection based on their performance, properties, processing, and their effect on the environment.

11.2 General Methodology of Design

The design process always starts with a need or some form of a problem statement that needs to be solved. This need may come from a client or a customer when an engineer works independently or for a consulting firm, or it may come from the company who employs the engineer. Since we are interested in the use of materials, the need is usually a tangible product that may be a component or a structure. The product may be completely new or may be a redesign of an older component because of size or performance limitations. To attain the desired objective, the design process typically goes through stages (or phases) such as those shown in Fig. 11-1. The phases

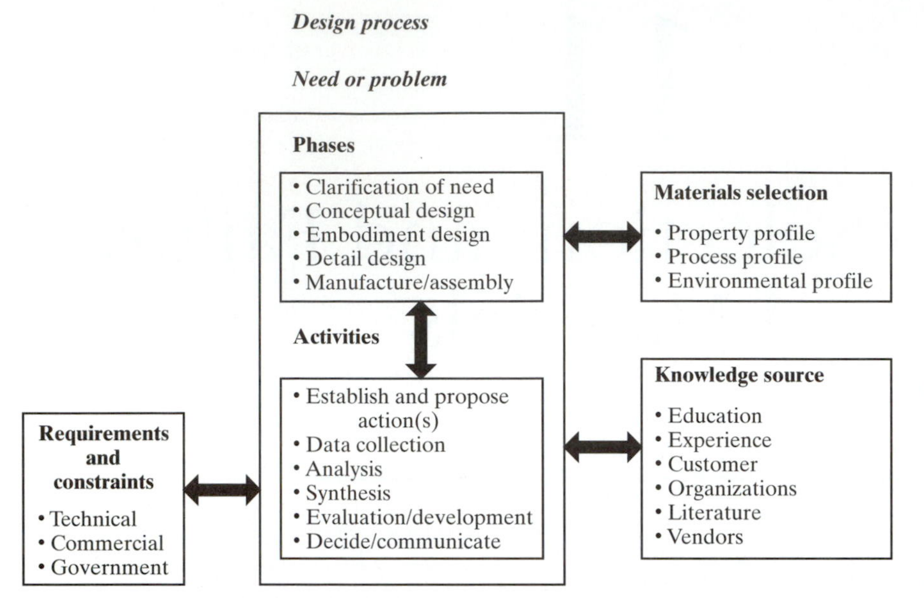

Fig. 11-1 Stages or Phases of Design Process with the associated activities (central portion); materials selection which includes the property, process, and environmental profiles is concurrently considered at every phase.

are: (1) clarification of the need, (2) conceptual design, (3) embodiment design, (4) detail design, and (5) manufacture/assembly. For each of these phases, the activities include establishing a forthcoming action, collecting information or data, analysis, synthesis, evaluation and development, and finally decision making and communicating. The design process is an iterative process; that means that we have to constantly weigh the results of each phase against the constraints and requirements.

11.2.1 The Phases of Design

The *clarification of the need* involves analyzing and clearly stating the problem. At this stage, the designer (team) collects and writes down all the requirements and constraints, the standards to be adhered to, statutory requirements, the projected date of completion of the design project, and the like. The target market of the product is delineated and identified because the type and quality of components that are put in a Chevrolet, for example, will be different from those going to a Cadillac. Safety considerations should also be paramount.

The objective of the *conceptual design* activity is to generate possible solutions, schemes, or methods to solve the problem. To achieve these possible solutions, the team should conduct unconstrained discussions to get the members of the design team to express all possible solutions. All "crazy" ideas are accepted and are evaluated according to their merits. If done properly, this activity imposes the greatest demands on the designer (team) and its output may result in many striking innovations or improvements. In this phase, all information and previous knowledge are put together in order that the proper decisions can be made. Examples of these are engineering science principles, laws, previous experiences, production methods, costs, and others. The output of the conceptual design stage should be two or three schemes that have the greatest chance of achieving the desired objective.

The *embodiment design* stage embodies and evaluates the two or three conceptual solutions or schemes selected in greater detail and makes a final choice of the scheme or method to be used. The selection is done with considerable feedback to the conceptual design activity. The output of this stage should be rough drawings, specifications, and broad compliance to the needs and specifications of the product.

The *detail design phase* considers all the large number of small but important details to make or fabricate the product or component. The quality of this work must be very good; otherwise delays, higher costs, and/or failure may ensue. The output of this activity is a set of very detailed drawings and final specifications including tolerance, precision, joining methods, finishing, and so on, to produce the component or product.

11.2.2 Design Activities

The design activities can be classified into analytical, creative, and execution types. The establishment of a proposed action and the collection of the data are analytical activities because they are the results generated from all the knowledge sources on the right of Fig. 11-1. The analysis, synthesis, and evaluation and development activities are the creative activities that engineers are expected to do. They involve deductive reasoning to analyze, synthesize, and evaluate the data from established basic principles and experiences in order to develop the product according to all applicable codes. The results of these activities are constantly matched with the requirements and constraints. The design product is seldom the ideal, perfect one that the engineer (designer) would like to have. Nonetheless, the designer, in conjunction with the client and/or company management, must make a decision and communicate it by means of detailed drawings, specifications, and other methods (execution activities). Figure 11-2 shows a flowsheet of a more detailed listing of some of the activities and the outputs that are typically involved during design. The phases during design in Fig. 11-2 are indicated on the right.

11.2.3 Materials Selection

Ultimately, the product has to be made of certain materials, and it is crucial that the materials selection should be a part in the decision process at each stage of the design. This is indicated on the right side of Fig. 11-1. The simultaneous consideration of all pertinent factors about a product in the "drawing board stage" (design) is known as holistic, simultaneous, or concurrent engineering, and it is motivated by productivity and global economic competitiveness. It allows the designer to find out early in the process if there is any problem in the availability, the cost, or processability of the material. For example, it makes a big difference if the material needed is a ceramic instead of a metal—the processing of a ceramic is entirely different from that of a metal. Designing with ceramics also takes a very different approach because of the inherent variability in their strength properties. Thus, it is important to make a decision on the materials to use early enough because it can affect the outcome of the design details. If we have to use two different materials, we need to know this in advance of the detail design stage in order to incorporate measures to control corrosion in the drawings. In addition, we also need to consider the difference in their thermal expansion coefficients.

Choice of a material should start at the conceptual stage when a very broad class(es) of material(s) is (are) identified as possible material(s). Constraints of temperature and corrosion may easily identify a class of materials and for illustration, let us say the material class might be the stainless steels. In the embodiment stage, we then look in greater detail at which of the stainless steels we need to use (i.e., whether it should be ferritic, austenitic, duplex or martensitic stainless steel. Preliminary designs can be made based on the published range of properties in handbooks. In addition to making the decision on the type of material, the designer at this stage should identify the materials vendor. In the final detail design stage, final design is made based on the data of the actual material to be used, such as that coming from the identified vendor; it also might be advisable to ask for statistical data indicating the range of properties of that material.

11.3 Factors in Materials Selection

From here on, we shall concentrate on materials selection in design. We shall first consider the various interrelated factors that are considered in the selection process, then discuss the performance or efficiency index of a material, and finally how selections are made and the considerations that are involved in this selection process.

11.3.1 Interrelated Constraints

There are many factors or constraints to be considered in selecting materials and most of these factors are interrelated. These may be grouped into the following factors:

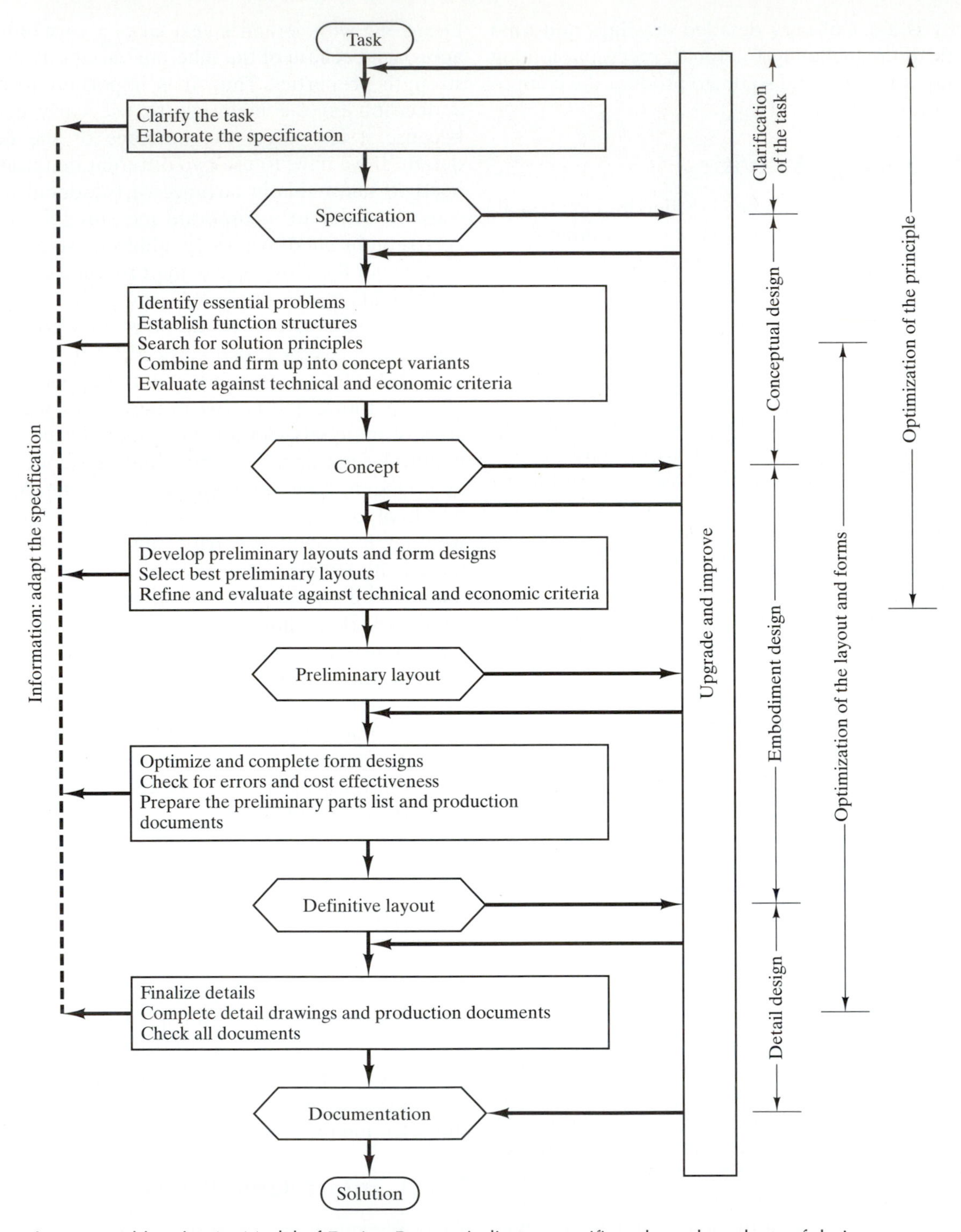

Fig. 11-2 Pahl and Beitz Model of Design Process indicates specific tasks and products of design process at each stage or phase, indicated on the right. [1]

11.3.1a Physical Factors. The factors in this group are the *size, shape, and weight of the material* needed and the space available for the component. All of these factors interrelate to the processing of the material. The size and shape might constrain the heat-treating of the material. A large size component may not be provided as a whole piece and consideration may have to be given to joining smaller subcomponents. If the product or component is large, we need to consider its transport as it might necessitate making subcomponents that are transportable. The shape of the material will dictate whether a casting or a wrought product will be required; this will influence the performance of the material when it is stressed in different modes. The weight of the material has implications not only in initial costs but also during applications, such as in the transport industry, where lightweight structures save energy costs and increase profitability. The space available for the component will dictate whether other materials can be considered. For example, plastics or aluminum will require a larger space to achieve the same structural performance as steel.

11.3.1b Mechanical Factors. The factors in this group relate to the ability of the material to withstand the types of stresses imposed on it. *These are the mechanical properties of the material that are used as the failure criteria in the design—the strength, the modulus, the fracture toughness, the fatigue strength, creep, and so on,* as discussed in Chap. 4. The *mode of loading*—(such as tensile, compression, bending, and cyclic) dictates which of the properties have the major influence. For example, high strength steels save a lot of weight when stressed in tension, but when they are loaded in bending and stiffness is required, these materials offer no weight savings over the lower strength grade steels because the property of importance is the modulus, as we shall see shortly in Sec. 11.3.3. The mechanical properties are also affected by the environment to which the materials are exposed, discussed in Sec. 10.8, corrosion.

11.3.1c Processing and Fabricability. These factors relate to the ability to form or shape the material. *Casting and deformation processing* are commonly used. Very intricate shapes are usually produced as castings and small objects may have to be investment casted. Ductile metals and thermoplastic materials are shaped normally by deformation processing because it is fast and is amenable to mass production. Ceramic materials that are brittle and have very high melting points are usually shaped by a much slower *sintering or powder metallurgy process. Spray-forming and lay-up techniques* are used in composite materials. *Fabricability embraces the joining processes,* (such welding, brazing, and soldering), *forming, and machining processes.* The weldability of steels has been empirically correlated to their carbon equivalents. Other materials have their own welding characteristics—the common characteristic in welding is the heat-treatment cycle that the material undergoes during the process. Aluminum alloys are often brazed, rather than welded. The high temperatures used in the joining process alter the "base properties" of the material (before it was welded or brazed). This brings us to *thermal* or *heat-treating processes* that are commonly implemented last or second to last. The *finishing processes* include coating and polishing and they are intended to give the material protection against corrosion, oxidation, and wear as well as improving its aesthetic appeal.

11.3.1d Life of Component factors. These factors relate to the length of time the materials perform their intended function in the environment to which they are exposed. The properties in this group are the *corrosion, oxidation, and wear resistance, creep, and the fatigue or corrosion fatigue life properties in dynamic loading.* The performance of a material based on these properties is the hardest to predict during the design stage.

11.3.1e Cost and Availability. In a market-driven economy, these two factors are inseparable. In addition, *quantity and standardization* are related to cost. Even if the materials are readily available, it matters whether orders are made in tonnages or in pounds or grams. The customer also pays a cost-penalty when orders are nonstandard items requiring special processing or are nonstocked items due to very little demand from other customers.

11.3.1f Codes, Statutory, and Other Factors. *Codes* are sets of technical requirements that are imposed on the material or the component. These are usually

Table 11-1 Materials properties which are most used for structural and component design.

Microstructure Insensitive	Microstructure Sensitive
Density, ρ	Strength, σ (see Sec. 4.4.1 for different criteria)
Modulus of elasticity, E	Ductility
Thermal conductivity, κ	Fracture toughness, K_{Ic}, or K_{Iscc}
Coefficient of linear thermal expansion, α	Fatigue and cyclic properties; corrosion fatigue
Melting Point, T_m	Creep
Glass Transition Temperature, T_g, for Polymers	Impact
Uniform Corrosion, mm/y or gmd	Hardness
Cost, $ per mass	

set by the customer, or are based from those of technical organizations such as the ASME, ASTM, SAE, and so on, which must be complied with. *Statutory* factors relate to local, state, and federal regulations about the materials and processes used or the disposal of the material. These relate to health, safety, and environmental requirements. The OSHA (Occupational Safety and Health Act) has banned the use of leaded brasses and steels used for bolts and screws. Toxic chemicals or solvents used in cleaning and processing cannot be disposed with the regular sewage. Because of dwindling fossil fuel resources, increased mileage per gallon (liter) for cars have been legislated. Pollution has led to legislation on emissions (pollutants) from cars. Solid waste disposal has also been legislated that requires materials to be recycled as often as possible. These are a few examples that have to be considered in materials selection.

11.3.2 Criteria and Tools for Materials Selection [2]

Current engineering designers choose materials to meet three general criteria. These are: property profile, processing profile, and environmental profile. In this context, factors 11.3.1(b) and (d) belong to the property profile, factors 11.3.1(a), (c), and (e) are in the processing profile, while 11.3.1(f) is in the environmental profile. The selection of the material is done with all three criteria considered simultaneously or concurrently with the performance and the cost of the material as paramount factors.

Selection based on property profile is the process of matching the numeric values of the properties of the material to the constraints and requirements. A combination of properties may have to be considered depending on the mode of loading and the service environment. The properties that are considered most are the properties in Table 4-1 and are shown once again as Table 11-1 with the cost factor included. The process of matching the property profiles of materials with the requirements can be very tedious and time-consuming if we have to search handbooks and data brochures of suppliers. Fortunately, databases that compile properties of different materials are now available to aid in the selection process. Databases are either provided by independent organizations interested in materials or by suppliers. Static design constraints such as fracture toughness, strength, stiffness, and weight can be matched easily with materials properties. Material charts (to be discussed in Sec. 11.3.4) are also available that enable us to select materials based on a performance or efficiency index, which is a combination of properties. We need only to be aware that the listed properties are the results from testing of small laboratory test specimens and may not be applicable in large-sized structures or components because of the presence of defects. In addition, vendors and suppliers data usually reflect the best properties of the material and do not indicate statistical variation. Furthermore, the dynamic properties of corrosion and its influence on the mechanical properties are very difficult to match because small changes in concentration of the electrolyte (or environment) cannot be predicted in the design stage.

Selection based on processing is aimed at identifying the process that will form the material to the desired shape, and then join and finish it at a minimum cost. It is a complex task because of the very large number of processing methods and sequence possibilities. This is made even more difficult due to new processes and materials that are continually being developed, whereby old processes are discarded. The complexity of integrating process selection into the design process is also manifested by the different approaches used in determining the cost of a process. Nonetheless, a decision has to be made on the processing in order to optimize the cost and performance of the material(s) selected.

One of the approaches and the easiest one to understand is an attempt to estimate the cost of the selection based on a manufacturing cost of a standard simple component. This standard cost is modified by a series of multipliers, each of which allows for an aspect of the component that is being designed. These aspects may include size, shape, complexity, precision, surface roughness, batch size, material, and so on. Processes are then ranked by the modified cost calculated in this way that allows the designer to make the final choice. The final choice is made with local factors taken into account.

Selection based on the environmental profile relates to the impact of the material—its manufacture, its use and reuse, and its disposal—on the environment. There is no question that this added constraint increases the cost of the product or structure. Designers and companies feel that if the costs of incorporating them in design are prohibitive, the environmental aspects are usually laid aside, unless it is mandated by law. In spite of the added cost, however, designing for the environmnent is a good philosophy because it can be a good marketing tool to environment-conscious consumers. Environmental awareness is so high in today's society that eventually the public is going to clamor for the protection of the environment; moreover, legislation will eventually have to be passed. So, why not get one step ahead of the crowd and be the leader at that time when it is forced on every company? A good example of this is the problem the U.S. automotive companies were confronted with when the oil crisis in the 1970s surfaced. These companies were milking good profits from large, heavy gas guzzlers. When mandated to cut the consumption of gasoline, they had difficulty producing small cars that could compete with foreign-made cars. For many years, Japanese and European car makers had addressed fuel consumption due to traditionally exorbitant gasoline prices in these countries relative to the United States. When the oil crisis surfaced, they were in a position to take advantage of the situation, thereby capturing a good portion of the U.S. auto market. While the U.S. companies have caught up, they have not recovered the market share that they held before the oil crisis.

Because of the many environmental aspects, the impact cannot be characterized in a straightforward way. Various approaches can be taken that view and assess environmental factors from different perspectives. The most popular of these approaches is the life-cycle analysis. The factor commonly used in the life-cycle analysis is the total energy associated with the product that includes:

1. the process used to extract the material from its ore and then to bring it to its usable form
2. the total energy spent during the use of the component (used in transport industry)
3. the recoverable energy after its use.

The energy factor is used because the energy to extract and to bring it to usable form comes predominantly from fossil fuels that are in danger of being depleted. In the transport industry, it is again fossil fuels that propel the vehicles; the consumption of energy is directly related to the inertia weight of the vehicle. The more gasoline and/or diesel fuel are burned, the greater is the pollution of the atmosphere due to the emissions from all vehicles. It is also important if the material can be recycled or some of the energy used recovered because this does not require solid waste disposal sites.

In fact, designing for ease of recycling of automotive components is gaining wide acceptance, especially in terms of plastic components. Up to the late 1980s, plastics have annually increased their usage in the automotive industry because of their light weight, ease of processing, and corrosion resistance. However, because thermoset plastics cannot be recycled, plastics usage has registered a decrease in usage in the early 1990s because of the application of design for recycling. The use of plastics in the automotive industry is now restricted to thermoplastics that are recyclable.

The manner in which design for recycling is incorporated during the design process is illustrated in Fig. 11-3, where recycling-oriented tasks are indicated at each of the steps of the design process. The objectives of design for recycling are the ease of disassembly of components, the use of unitary parts or

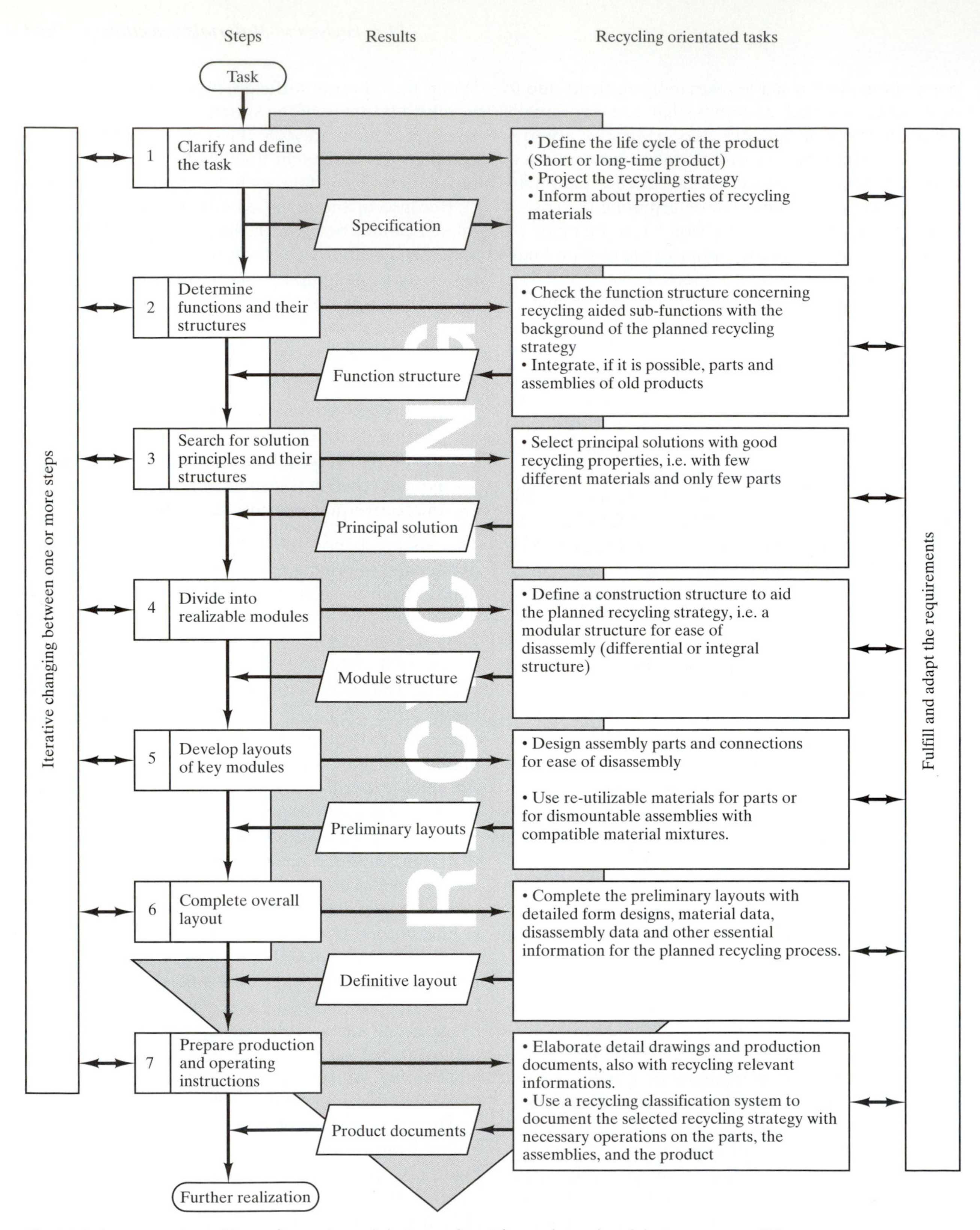

Fig. 11-3 Incorporation of Recycling-oriented design tasks with regular tasks of design process. [8]

products, the minimization of the number of materials used, and the use of compatible materials, in terms of being able to mix the recoverable materials to produce a new material that have very similar properties as the original component. An example of a compatibility matrix for plastics is shown in Fig. 11-4.

11.3.3 Performance and Efficiency of Materials

One of the factors in the selection of a material based on its property profile is its performance or efficiciency index. Materials are expected to perform based on their properties, such as modulus, strength, fatigue, fracture toughness, creep, corrosion, cost, and others in Table 11-1. Quite often, the best choice of a material for a component is based not on specific properties but rather on a combination of properties, in the effort to minimize the weight and/or cost.

A component is designed to do specific functions (*functional requirements*) with a specified geometry (*geometric requirements*) of size and shape due to constraint of space. Its performance can only be as good as the performance of the selected materials. The *performance* of a structural material may be defined by its *efficiency in performing the function* it is designed for. In mechanical loading, this *efficiency* may be defined as the *ratio of the load* a material can sustain *over the mass or weight of the material* for a particular structural geometry, that is,

$$\text{Efficiency} = \frac{\text{load on structural member}}{\text{weight of material}} = \frac{\text{P}}{\text{m}} \qquad (11\text{-}1)$$

The load P is the maximum the material can sustain based on the chosen failure criterion of the material and on the mode of loading. The general complex loading of a component or structure can usually be resolved into four basic modes—axial tension, compression and buckling, bending, and torsion or twisting, as shown in Fig. 11-5.

Efficiency in Axial Tension—Maximum Strength per Unit Mass. When a material is pulled in tension, it elongates while it sustains the load. The failure criterion in this case is the yield strength after which the material will undergo plastic deformation or permanent change in shape. There are two measures from which we can judge the materials performance: (1) it sustains the maximum load per unit mass, and (2) it

Matrix material	Add material											
Engineering materials	PE	PVC	PS	PC	PP	PA	POM	SAN	ABS	PBTP	PETP	PMMA
PE	●	○	○	○	●	○	○	○	○	○	○	○
PVC	○	●	○	○	○	○	○	●	◐	○	○	●
PS	○	○	●	○	○	○	○	○	○	○	○	○
PC	○	◔	○	●	○	○	○	●	●	●	●	●
PP	◔	○	○	○	●	○	○	○	○	○	○	○
PA	○	○	◔	○	○	●	○	○	○	◔	◔	○
POM	○	○	○	○	○	○	●	○	○	◔	○	○
SAN	○	●	○	●	○	○	○	●	●	○	○	●
ABS	○	◐	○	●	○	○	◔	○	●	◔	◔	●
PBTP	○	○	○	●	○	◔	○	○	◔	●	○	○
PETP	○	○	◔	●	○	◔	○	○	◔	○	●	○
PMMA	○	●	◔	●	○	○	◔	●	●	○	○	●

● Compatible
◐ Compatible within limits
◔ Compatible in small quantities
○ Non-compatible

Fig. 11-4 Compatibility Matrix for Plastics. [7]

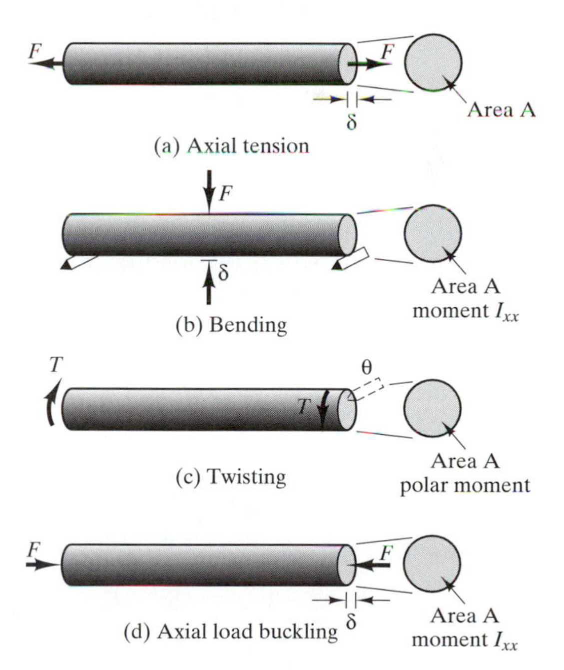

Fig. 11-5 Illustration of the basic modes of loading on a cylindrical bar. The mode of loading dictates the combination of properties of materials that enhance their performance. [3]

elongates the least per unit mass. For both we consider the bar, shown in Fig. 11-6, which has an original length L_o and an elongation ΔL when the load P is applied. The mass of the material, $m = AL_o\rho$, where A is the cross-sectional area of the bar, and ρ is the density of the material. For maximum load per unit mass, the efficiency or performance is thus,

$$\text{Efficiency} = \frac{P}{AL_o\rho} = \frac{\sigma_{des}A}{AL_o\rho} = \frac{1}{S_fL_o} \cdot \frac{\sigma_{ys}}{\rho} \quad (11\text{-}2)$$

σ_{des} is the design stress, σ_{ys} is the yield strength of the material, and S_f is the factor of safety chosen. Since S_f and L_o are constants chosen for the design, σ_{ys}/ρ is the combined material's property used to maximize the load that the material can sustain per unit mass. This ratio is called the *specific strength of the material.*

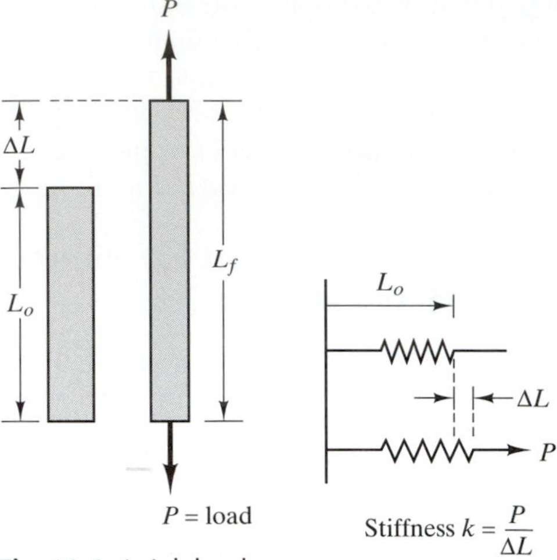

Fig. 11-6 Axial load, P, and extension ΔL in axial tension.

Fig. 11-7 Definition of stiffness of spring

Efficiency in Axial Tension—Maximum Stiffness per unit Mass. For the performance of least extension or strain, we start with,

$$\text{strain} \leq \frac{\Delta L}{L_o} \leq \frac{\sigma_{ys}/S_f}{E} \leq \frac{P}{AS_fE} \quad (11\text{-}3)$$

from which we obtain,

$$P = \frac{AS_f(\Delta L)E}{L_o} \quad (11\text{-}4)$$

and by definition, the efficiency is

$$\text{Efficiency} = \frac{P}{AL_o\rho} = \frac{S_f\Delta L}{L_o^2} \cdot \frac{E}{\rho} \quad (11\text{-}5)$$

The tensile bar is often compared to a spring, shown in Fig. 11-7, in which the load or force P on the spring is proportional to ΔL,

$$P = k \cdot \Delta L \quad (11\text{-}6)$$

where k is the stiffness of the spring. From this we see that the stiffness is

$$k = \frac{P}{\Delta L} \quad (11\text{-}7)$$

and rewriting Eq. 11-5,

$$\text{Efficiency} = \frac{k}{AL_o\rho} = \frac{S_f}{L_o^2} \cdot \frac{E}{\rho} \quad (11\text{-}8)$$

We see that the stiffness of the material is maximized when the performance of minimum elongation for a certain load is desired. The combined material property that yields the maximum stiffness is the ratio, E/ρ which is called the *specific modulus.*

Maximizing the specific strength yields a light, strong material while maximizing the specific modulus yields a light stiff material. Both of these factors are specifically sought for aerospace materials. For example, composite materials exhibit high specific moduli and specific strengths and are used significantly in aerospace structures.

Cost Performance. Another measure of performance of a material is its cost and we can define the cost-effectiveness of a material by,

$$\text{Cost-effectiveness} = \frac{\text{Load on structure}}{\text{Cost of structure}} = \frac{P}{C_m \cdot m} \quad (11\text{-}9)$$

where C_m is the cost of the material per unit mass and m is the mass. Comparing Eq. (11-9) to Eqs. 11-2 and 11-8, we see that in order to obtain a measure of the cost-effectiveness of the material, we need to divide the factors in Eqs 11-2 and 11-8 by C_m. Thus, to maximize the cost-effectiveness of a light strong material, the factor is $\sigma_{ys}/(C_m \cdot \rho)$ and that for a light stiff material is $E/(C_m \cdot \rho)$.

The efficiency and cost-effectiveness factors for the other modes of loading and for different shapes of structures or components may also be obtained.

The loading and deformation in these modes and shapes of structures or components are topics in mechanics of materials and are beyond the scope of this course. Results of some examples of the other modes of loading and shapes of components are given in Table 11-2 to illustrate the variety of the maximizing factors for maximum stiffness and maximum strength for the least weight of a material. The table is just a small part of many more tables for other properties in M. F. Ashby's articles and book listed in the reference section. [3–5]

11.3.4 Materials Charts [3, 5]

Perhaps the most time-consuming task in the selection of materials is the collection of the information on their properties to match the requirements and constraints of the design. This is the matching of the materials' property profile that was discussed in Sec. 11.3.2. As indicated earlier, most of the data is now in software form and has made the selection process much easier, at least in the conceptual design stage. Some of this information is compactly summarized in material charts that are plots of the properties that form the maximizing factors discussed in Sec. 11.3.3. *These charts were developed by Ashby and are meant only to be used at the conceptual stage of selection of materials.*

Figures 11-8 and 11-9 are simplified versions of two of Ashby's materials charts, which are plots of *density versus strength* for the selection of light, strong materials, and *density versus modulus* for the selection of light, stiff materials, respectively. To include the enormous range of properties (roughly over five orders of magnitude) of the different groups of materials, they are plotted on log-log scales. The failure strength criterion used in Fig. 11-8 was discussed earlier in Sec. 4.4.1 and is the yield strength for metals and polymers, the compressive strength for ceramics and glasses, the tensile tear for elastomers, and the tensile failure for composites.

The charts show that the properties of each of the different classes of materials are clustered together and are shown within envelopes or fields in the charts. The envelopes signify the different classes of materials, and within each of these envelopes are smaller ones that signify the variation of properties of specific materials within each class. These are not shown here. However, the details can be seen in Ashby's works referred to at the end of this chapter. Despite the wide range in the strengths and densities associated with metals, they (envelopes) occupy a field that is distinct for each class of materials, such as polymers, ceramics, or composites. This statement applies also for moduli, toughnesses, and other properties. There is sometimes an overlap in the envelopes or fields, but each class of materials has its own field in the chart.

The position of the envelopes or fields and the relationship of the properties (for example, modulus versus density) are all based on principles we have

Table 11-2 Combined properties to maximize efficiency or performance indices. [3, 5]

Component Shape and Mode of Loading	For Stiffness	For Strength
Bar-Tensile Axial Loading—load, stiffness, length specified, section area variable	$\frac{E}{\rho}$	$\frac{\sigma_{ys}}{\rho}$
Torsion Bar or Tube—torque, stiffness, length specified, section area variable	$\frac{G^{1/2}}{\rho}$	$\frac{\sigma_{ys}^{2/3}}{\rho}$
Beam—externally loaded or by self-weight in bending; stiffness, length specified, variable section area	$\frac{E^{1/2}}{\rho}$	$\frac{\sigma_{ys}^{2/3}}{\rho}$
Column–Axial Compression—elastic buckling or plastic compression; compression load and length specified; variable section area	$\frac{E^{1/2}}{\rho}$	$\frac{\sigma_{c}}{\rho}$
Plate—externally loaded or by self-weight in bending; stiffness, length, and width are specified; thickness is variable	$\frac{E^{1/3}}{\rho}$	$\frac{\sigma_{ys}^{1/2}}{\rho}$

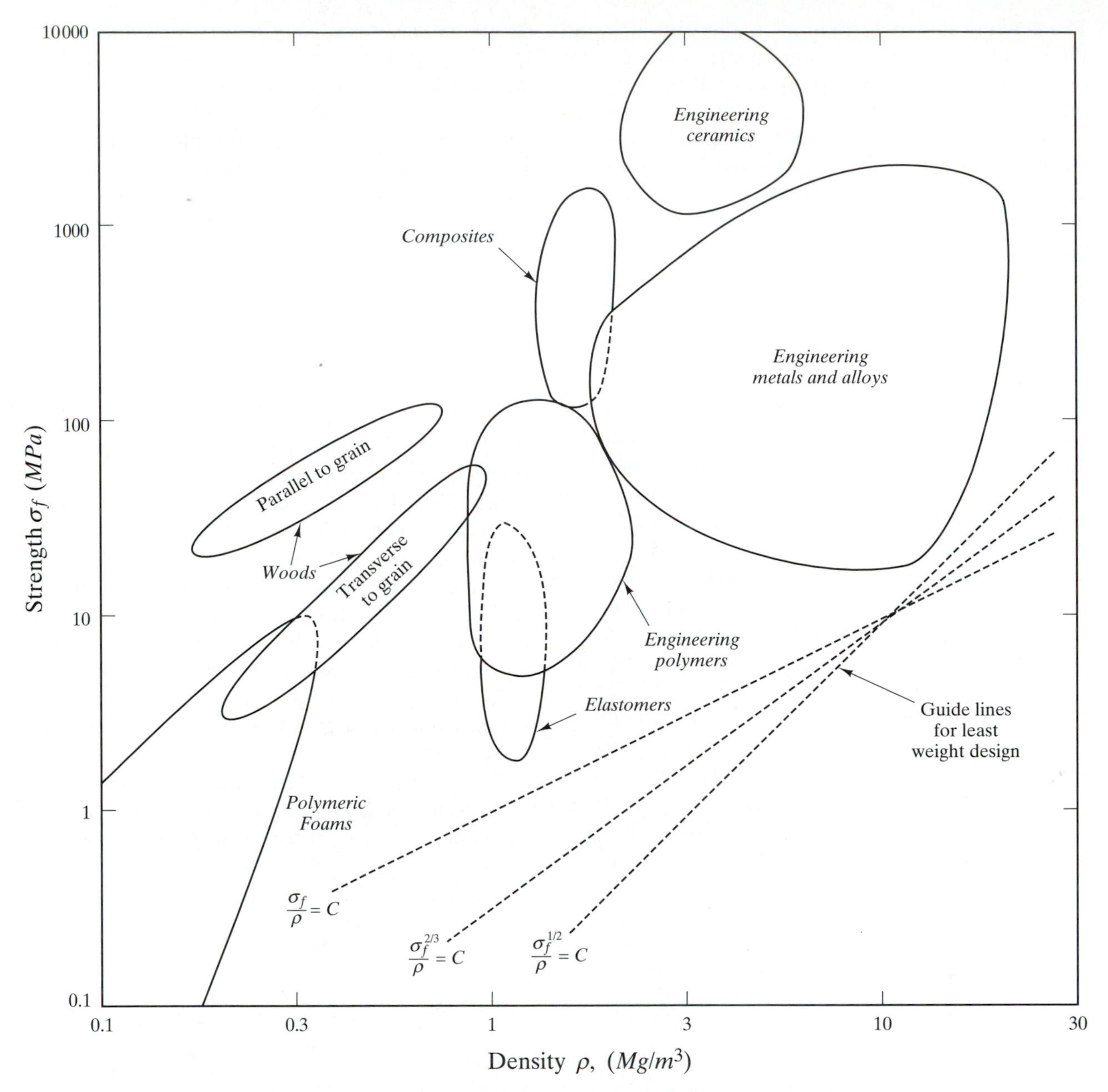

Fig. 11-8 Density versus strength materials chart. [3, 5]

discussed in earlier chapters on atomic bonding, on crystal structure, and on the processing of the materials. The large differences in the properties or fields of the different classes of materials arise primarily from differences in the masses of the atoms, the nature of their interatomic forces or bonding, and the geometry (crystal structure) of their packing. Alloying (composition), heat treatment, and mechanical working (processing) all influence the microstructure, and through this, the variation of the properties of each of the materials within the envelope, in particular those of the metals. However, the magnitude of their (composition and processing) effects is much lesser, by factors of 10, than that due to bonding and crystal structure.

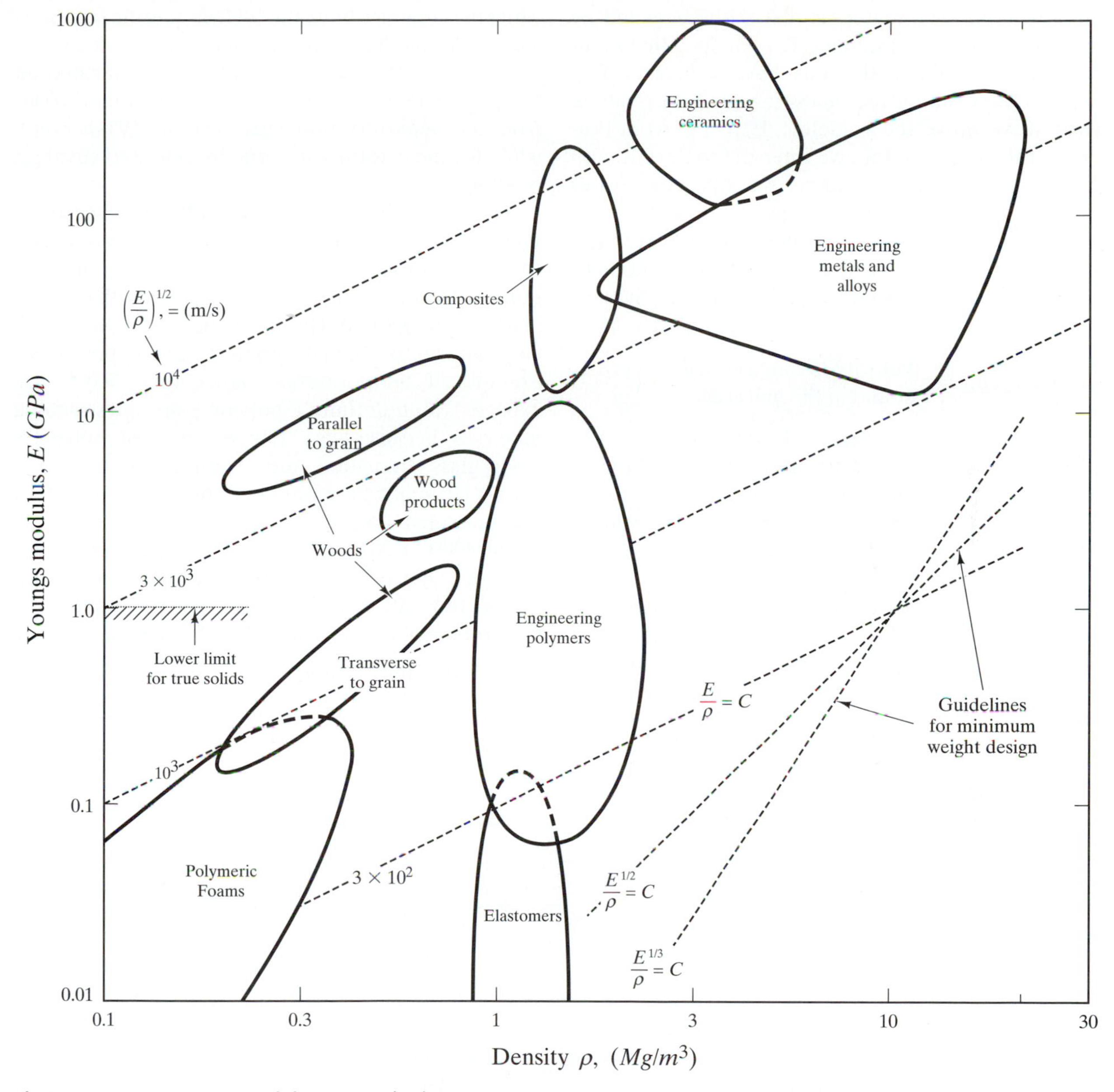

Fig. 11-9 Density versus modulus materials chart. [3, 5]

The application of these plots in materials selection will now be illustrated using Fig. 11-9. Let us consider first the maximizing factor for the maximum stiffness in axial tension and give it a value of C. Thus,

$$\frac{E}{\rho} = C \quad \text{or} \quad E = C\rho \qquad (11\text{-}10)$$

Taking log on both sides,

$$\log E = \log \rho + \log C \qquad (11\text{-}11)$$

which we should recognize as the equation of a family of straight lines, y = x + b, with a slope of 1. The value of *C* is obtained when ρ = 1 and equals E, in Fig. 11-9 and dictates the location of the line called the *guide line.* The materials that are intersected by this guide line have the same E/ρ, or specific modulus. Materials above this line have values of E/ρ higher than C and those below it have lower values. Thus, if we move the guideline, $E/\rho = C$ from bottom right to the top left, we intersect materials with higher specific moduli and we see that composites and ceramics have the highest specific moduli.

The values given on the lines are, however not E/ρ but rather as $\sqrt{E/\rho}$. The ratio of Young's modulus to density has dimensions of $(\text{velocity})^2$ (i.e., $m^2\ s^{-2}$), and thus

$$\sqrt{E/\rho}, \text{m/s}, = \begin{matrix}\text{velocity of longitudinal}\\ \text{sound in the material}\end{matrix} \qquad (11\text{-}12)$$

and the numbers indicated on the dashed lines parallel to $(E/p) = C$ are the square root values of C or the velocity of sound in the material. The sound propagates through solids by way of the elastic vibrations of the atoms and the higher the sound speed, the higher is its pitch (recall tuning forks!). In addition, the higher the strength of the material, the higher is the pitch of the sound. Thus, materials that are traversed by a certain value of the guide line have proportionately similar specific strength. Thus, Figs. 11-8 and 11-9 show that the elastomers exhibit the lowest specific strengths and specific moduli while the ceramics have the highest specific strengths and specific moduli, with diamond having the highest—a reflection of covalent bonding as indicated in Chap. 4. The resisting force in elastomers arises from the weak intermolecular van der Waals bond, while for the ceramics it is due to ionic and covalent bonding.

The Ashby charts are very useful in the *conceptual stage of design* because they show the broad classes of materials that can be used for a certain application. At the very start, the constraints of the design problem are defined. These may be in the form of the environment to which the component is exposed. For example, if temperatures higher than 300°C are expected, then definitely polymers are questionable materials because very few commercial polymers have glass transition temperatures greater than 300°C. (It might be mentioned, however, that there are polymers under development that exhibit T_g about 400°C.) Also, because of creep consideration and softening of precipitation-hardened aluminum alloys, the latter might have to be eliminated. Constraints of modulus and density such as,

$$E \geq 10\ \text{GPa} \qquad \text{and} \qquad \rho \leq 3\ \text{Mg/m}^3 \qquad (11\text{-}13)$$

are plotted as horizontal and vertical lines, respectively, as shown in Fig. 11-10. The viable materials at

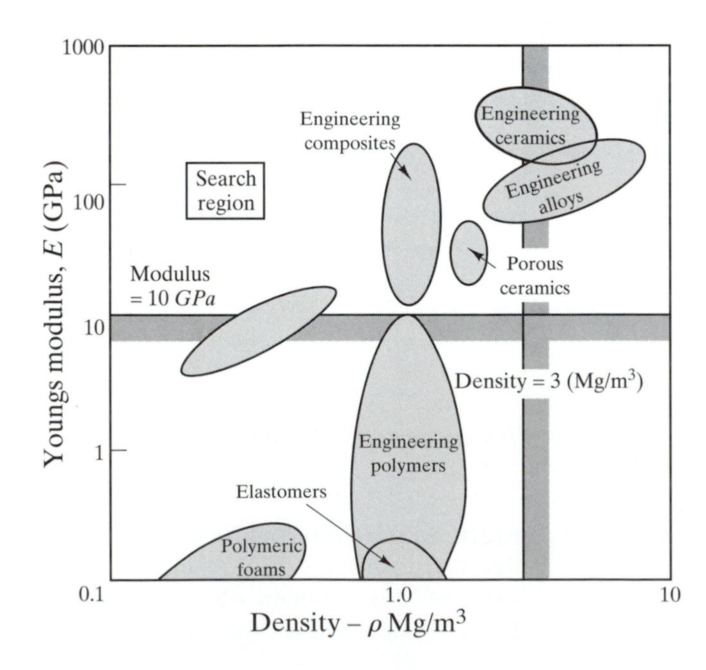

Fig. 11-10 Search for materials start by drawing the constraints of design; vertical and horizontal lines in the chart and candidate materials are in search region. [3, 5]

the conceptual stage are those shown in the search region in Fig. 11-10. The guide line, $E/p = C$, is moved to the search region and the materials in this region are evaluated on the basis of other constraints such as temperature, impact property, corrosion, and so on, for selection as probable materials.

Figure 11-9 also shows two other guide lines (efficiency factors),

$$\frac{E^{1/2}}{\rho} = C \quad \text{is the maximizing factor for design of stiff beams, shafts, and columns,} \tag{11-14}$$

and

$$\frac{E^{1/3}}{\rho} = C \quad \text{is the maximizing factor for design of stiff plates} \tag{11-15}$$

The factor

$$\frac{E^{1/2}}{\rho} = C \tag{11-16}$$

gives a family of straight lines with slope of 2 in a log ρ versus log E plot, and

$$\frac{E^{1/3}}{\rho} = C \tag{11-17}$$

gives another set with a slope of 3 in the same plot. One guide line of each of the families of lines is shown in the lower right of Fig. 11-9. The selection of materials based on these two efficiency factors is done in the same manner as the E/ρ factor.

Similarly, in addition to the efficiency factor, $(\sigma_f/\rho) = C$, in Fig. 11-8, the other two factors (guide lines) are

$$\frac{\sigma_f^{2/3}}{\rho} = C \quad \text{is maximizing factor for design of strong beams and shafts,} \tag{11-18}$$

and

$$\frac{\sigma_f^{1/2}}{\rho} = C \quad \text{is the maximizing factor for design of strong plates} \tag{11-19}$$

Again, a line representing each of the families of lines is shown at the lower right of Fig. 11-8. The *failure strength criteria for different materials were discussed in Sec. 4.4.1 and are:*

(a) the yield strength for metals and polymers
(b) the compressive strength for ceramics and glasses
(c) the tensile tear strength for elastomers (rubbers)
(d) the tensile failure for composite materials.

In both Figs. 11-8 and 11-9, the values of the constants C increase as the lines are displaced upwards and to the left.

We note that the efficiency factors to be maximized for the other component shapes and modes of loading given in Table 11-2 are all represented by one of the three guide lines in Figs. 11-8 and 11-9, except for $G^{1/2}/\rho$. We note, however, that G is related to E and in fact is approximately equal to $3E/8$ when ν, the Poisson's ratio is about 1/3 and which is a constant. Thus, $G^{1/2}/\rho$ is essentially $E^{1/2}/\rho$.

The cost-effectiveness factors, $E/(C_m \cdot \rho)$, and $\sigma_f/(C_m \cdot \rho)$, suggest that if we plot $C_m\rho$ versus either E or σ_f, we can evaluate the cost-effectiveness of materials to sustain a certain load. An example of this is shown in Figs. 11-11, where C_m, the cost per unit weight of the material is normalized to the cost of mild steel, that is

$$C_R = \frac{C_m}{C_{ms}} \quad \frac{\text{cost per unit weight of material}}{\text{cost per unit weight of mild steel}} \tag{11-20}$$

and $C_R \cdot \rho$, the relative cost per unit volume, is plotted versus modulus. As in Fig. 11-9, the most cost-effective materials in Figs. 11-11 are those towards the upper left-hand corner of the chart — low relative cost per unit volume and high modulus. This figure confirms that the most common construction materials of cement, concrete, and stones in the porous ceramic group, cast irons and mild steels (ms) in the engineering alloy group, and wood products are relatively cheap compared to the other materials. Their usage in houses, buildings, bridges, roads, and the like, brings down the costs and makes the structures, particularly houses, affordable to many.

Consider the structural materials used in single-family homes. A home is no doubt the most expensive item an average person or family will purchase in a lifetime. About half the cost of a home goes to the construction materials and a great many materials are used. The most significant function of materials in homes are their structural components that hold the buildings together. These are the studs, the floor and ceiling joists, the roof trusses, and the foundation. In these components, *the materials must be*

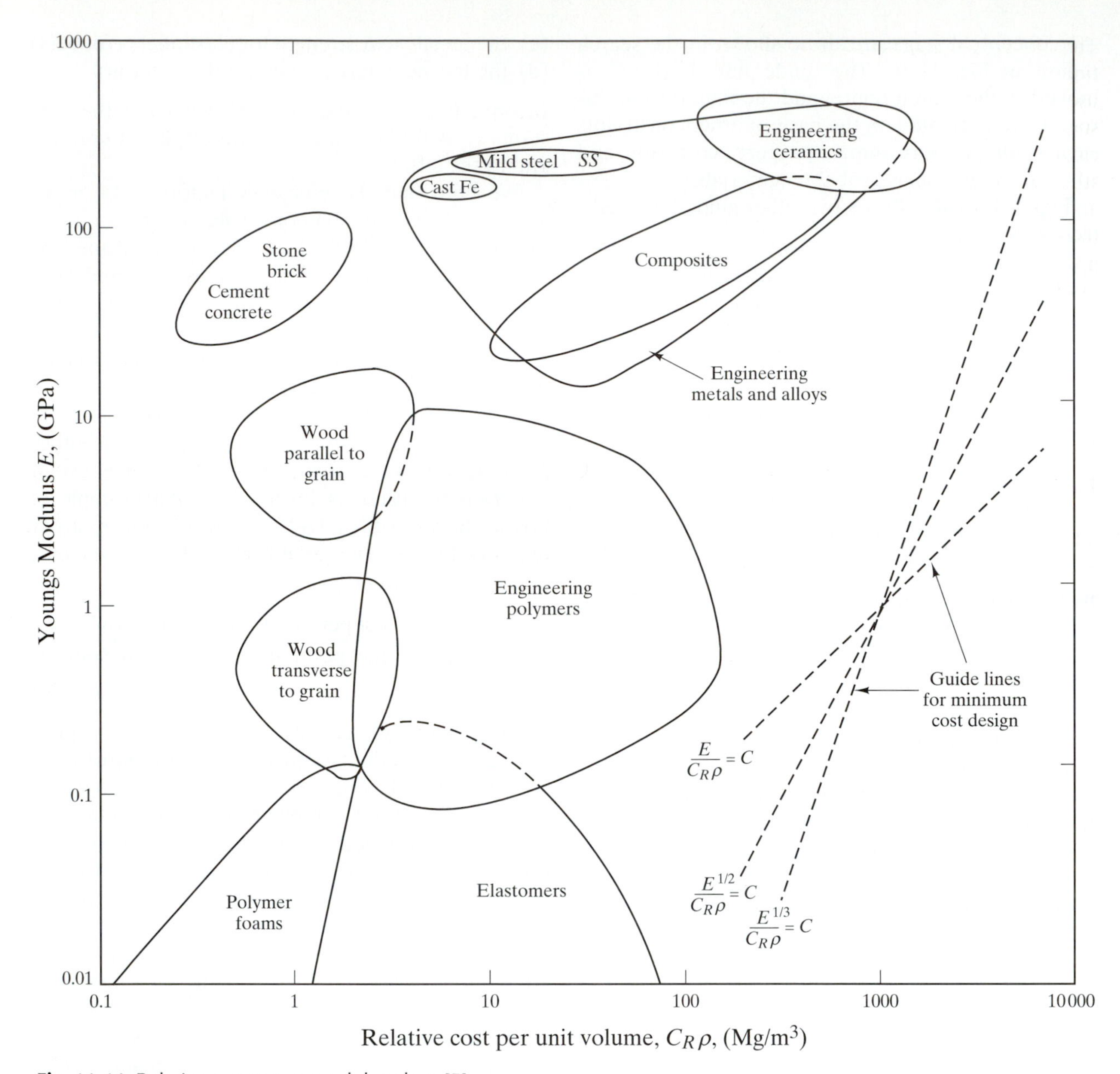

Fig. 11-11 Relative cost versus modulus chart [5]

stiff, strong, and inexpensive. *Stiff,* so that the building does not flex too much under wind loads or internal loading; *strong,* so that there is no risk of collapse; and *inexpensive,* to make the home affordable and salable. Thus, the selection of materials is based on the *efficiency factors of strength and stiffness at minimum cost.* The mode of loading for the studs is compression such as in columns; the joists are in bending as in beams; the trusses are used as ties; and for the foundation, predominantly in compression. Thus, we can pick the various *cost-effectiveness factors* for the components. For beams, these factors are

$$\frac{E^{1/2}}{\rho C_R} \quad \text{for stiffness} \qquad (11\text{-}21)$$

$$\frac{\sigma_f^{2/3}}{\rho C_R} \quad \text{for strength} \qquad (11\text{-}22)$$

The equivalent factors for the trusses and the foundation are $M_1 = E/\rho C_R$ and $M_2 = \sigma_f/\rho C_R$. For the studs, use the factors for columns in Table 11-2 with the substitution of ρC_R for ρ. The guide lines are indicated in the material chart shown in Fig. 11-11 for the stiffness failure criterion, and another chart is available for the strength failure criterion. The values of the constants C increase as the lines are displaced upwards and to the left. The least costly materials offering the greatest stiffness per unit cost are found towards the upper-left corner of Fig. 11-11. Note the locations of concrete, wood, brick, and mild steel (ms) —the common construction materials.

11.3.5 Shape Factors [4, 5]

The efficiency factors we have just discussed did not consider the effect of the section shape of the structure or component. We were just comparing materials of the same shape. The performance of a material is also a function of its section shape and on the mode of loading. Figure 11-12 shows the optimum section shapes that offer the best performance for the different indicated modes of loading. In axial tension, the area of the cross-section is important but not its shape; all sections with the same area will sustain the same load and thus, there is no shape factor. In bending, on the other hand, the shape of the section is important. Beams with hollow-box or I-sections are better than solid sections. This is also true in torsion; for example, circular tubes are better than either solid or I-sections. A hollow square section performs well with axial compressive loads.

To incorporate the influence of shape on the structural efficiency, a *shape factor,* ϕ, is introduced. The shape factor is a dimensionless number, which characterizes the mechanical efficiency of the section shape relative to a solid bar with a circular section and the same mass, regardless of size, in a given mode of loading. There is no shape factor in axial tension. We now look into the shape factors for the other modes of loading.

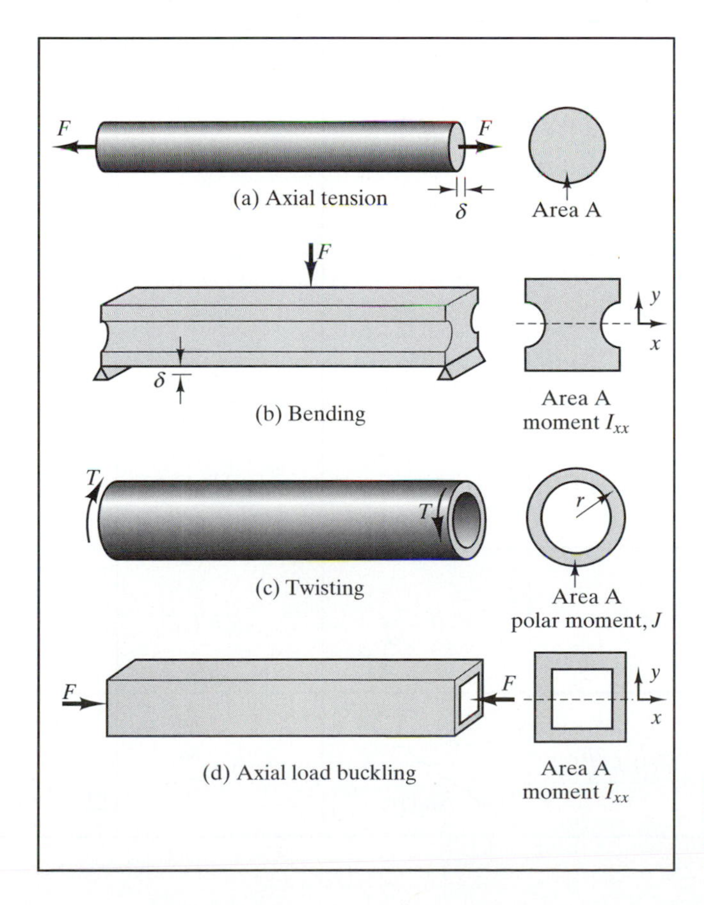

Fig. 11-12 Section shapes for the best performance of materials under the common modes of loading. [4, 5]

Bending Mode. As in previous discussions, we may have two criteria for performance—one for stiffness and one for strength; for each of these, there is an associated shape factor. The shape factor for the stiffness performance is the ratio of the stiffness of the shaped section to the stiffness of a solid circular section of the same mass and is designated ϕ_B^e. The subscript B is for bending and the superscript "e" refers to elastic; thus, the factor is for elastic bending. Since $\phi_B^e > 1$, it is a measure in the stiffness gained by the shaped section over the solid circular cross section. For the same reason, the shape factor for the strength performance is a measure of the gain in failure load of the shaped section over the failure load of a solid circular section of the same mass, provided local buckling does not occur first. This is designated ϕ_B^f; again B is for bending and the superscript "f" is for failure. The failure strength may either be the yield strength or the tensile strength. Since the factors are ratios of the same quantities (stiffness or loads), they are dimensionless.

The shape factors may be calculated from the geometry of the shape. They enter through the second moment of area, I, about the bending axis (assumed as the x axis):

$$I_{xx} = \int_{section} y^2 dA \tag{11-23}$$

where y is the normal distance to the bending axis and dA is the differential element of the area at y. The stiffness or elastic shape factor is obtained as,

$$\phi_B^e = \frac{4\pi I}{A^2} \tag{11-24}$$

and the strength or failure shape factor is

$$\phi_B^f = \frac{16\pi}{A^3}\left(\frac{I}{y_m}\right)^2 \tag{11-25}$$

Table 11-3 Areas, moments of selected sections, and I/y_m. [4, 5]

Section shape	$A\ (m^2)$	$I_{xx}\ (m^4)$	$K\ (m^4)$	$I_{xx}/y_m,\ (m^3)$	$Q,\ (m^3)$
$2r_0$	πr^2	$\frac{\pi}{4}r^4$	$\frac{\pi}{2}r^4$	$\frac{\pi}{4}r^3$	$\frac{\pi}{2}r^3$
h, b	bh	$\frac{bh^3}{12}$	$\frac{b^3h}{3}\left(1-0.58\frac{b}{h}\right)$ $h>b$	$\frac{bh^2}{6}$	$\frac{b^2h^2}{3h+1.8b}$ $(h>b)$
t, $2r_o$, $2r_i$	$\pi(r_o^2-r_i^2)$ $\approx 2\pi rt$	$\pi(r_o^4-r_i^4)$ $\approx \pi r^3t$	$\frac{\pi}{2}(r_o^4-r_i^4)$ $\approx 2\pi r^3t$	$\frac{\pi}{4r_o}(r_o^4-r_i^4)$ $\approx \pi r^2t$	$\frac{\pi}{2r_o}(r_o^4-r_i^4)$ $\approx 2\pi r^2t$
t, b, b	$4bt$	$\frac{2}{3}b^3t$	$b^3t\left(1-\frac{t}{b}\right)^4$	$\frac{4}{3}b^2t$	$2b^2t\left(1-\frac{t}{b}\right)^2$
t, h, t, $2t$, b, b	$2t(h+b)$	$\frac{1}{6}h^3t\left(1+\frac{3b}{h}\right)$	$\approx\frac{2tb^2h^2}{h+b}$ I / □ $\frac{2}{3}bt^3\left(1+\frac{4h}{b}\right)$	$\frac{h^2t}{3}\left(1+\frac{3b}{h}\right)$	$2tbh$ I / □ $\frac{2}{3}bt^2\left(1+\frac{4h}{b}\right)$

Values of A, I_{xx}, and I/y_m are given for selected sections in Table 11-3 and the shape factors are listed in Table 11-4. The values of ϕ_B^e and ϕ_B^f are 1 for solid circular sections. The values of other solid sections are not much greater than 1, but the values for shapes can be much greater and they do not depend on size. These are illustrated in Fig. 11-13. The three rectangular sections all have the same factor, $\phi_B^e = 2$, while the three I-sections all have $\phi_B^e = 10$. Since the shape factors represent the gains of a shaped section over that of a solid circular section of the same mass, the shape factors in Fig. 11-13 mean that the rectangular solid section is twice as stiff and the I-section 10 times stiffer than a solid circular section with the same mass.

Torsion Mode. As in the bending mode, we also have two shape factors: (1) for stiffness or elastic performance, and (2) for strength or failure performance. The shape factors again represent gains in torsional stiffness and torsional failure load of the shaped section over the torsional stiffness and torsional failure

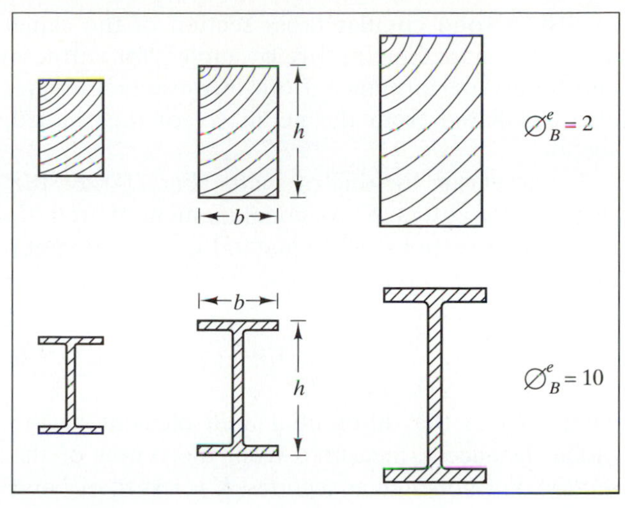

Fig. 11-13 Shape factors do not depend on size; three rectangular sections with the same $\phi_B^e = 2$, and three I sections with the same $\phi_B^e = 10$ [4, 5]

Table 11-4 Shape factors of selected sections. [4, 5]

	STIFFNESS		STRENGTH	
Section shape	ϕ_B^e	ϕ_T^e	ϕ_B^f	ϕ_T^f
Solid circle ($2r_0$)	1	1	1	1
Solid rectangle (h, b)	$\frac{\pi}{3}\frac{h}{b}$	$\frac{2\pi}{3}\frac{b}{h}\left(1-0.58\frac{h}{b}\right)$	$\frac{4\pi}{9}\frac{h}{b}$	$\frac{4}{9}\pi\frac{(b/h)}{\left(1+0.6\frac{b}{h}\right)^2}$ $h>b$
Tube ($2r_o$, $2r_i$, t)	$\frac{r}{t}$	$\frac{r}{t}$	$\frac{2r}{t}$	$\frac{2r}{t}$
Square box (b, t)	$\frac{\pi}{6}\frac{b}{t}$	$\frac{\pi}{8}\frac{b}{t}\left(1-\frac{t}{b}\right)^4$	$\frac{4\pi}{9}\frac{b}{t}$	$\frac{\pi}{4}\frac{b}{t}\left(1-\frac{t}{b}\right)^4$
Rectangular box / I-section (t, h, $2t$, b)	$\frac{\pi}{6}\frac{h}{t}\frac{\left(1+\frac{3b}{h}\right)}{\left(1+\frac{b}{h}\right)^2}$	□: $\frac{\pi b^2h^2}{t(h+b)^3}$ I: $\frac{\pi}{3}\frac{t}{b}\frac{\left(1+\frac{4h}{b}\right)}{\left(1+\frac{h}{b}\right)^2}$	$\frac{2\pi}{9}\frac{h}{t}\frac{\left(1+\frac{3b}{h}\right)^2}{\left(1+\frac{b}{h}\right)^3}$	□: $\frac{2\pi h^2}{bt\left(1+\frac{h}{b}\right)^3}$ I: $\frac{2\pi}{9}\frac{t}{b}\frac{\left(1+\frac{4h}{b}\right)^2}{\left(1+\frac{h}{b}\right)^3}$

load of a solid circular cross section of the same mass. The shaped factors are ϕ_T^e and ϕ_T^f, for stiffness and failure performance. Both of these factors are also calculated from the geometry of the shaped section.

In the elastic twisting of shafts (Fig. 11-12c), the shape factor enters the torsional moment of area, *K*. For circular sections, *K* is equal to the polar moment of area

$$J = \int_{section} r^2 dA \qquad (11\text{-}26)$$

where *dA* is the differential area element at the radial distance *r*, measured from the center of the section. For noncircular sections, *K* is less than *J* and is defined such that the twist θ is related to the torque *T* by

$$\theta = \frac{Tl}{KG} \qquad (11\text{-}27)$$

where *l* is the length of the shaft and G is the shear modulus of the material of which it is made. The shape factor in elastic twisting is defined by

$$\phi_T^e = \frac{2\pi K}{A^2} \qquad (11\text{-}28)$$

The shape factor for strength is a bit more complex. For circular sections (tubes and cylinders), the maximum shear stress occurs at the outer surface, at the radial distance r_m from the axis of twist,

$$\tau = \frac{Tr_m}{J} \qquad (11\text{-}29)$$

where *T* is the torque, and the quantity J/r_m in twisting has the same significance as I/y_m in bending. For noncircular sections with ends that are free to warp, the maximum surface shear stress is

$$\tau = \frac{T}{Q} \qquad (11\text{-}30)$$

where *Q*, with units of m^3, now plays the role of I/y_m in bending. With this, the definition of the shape factor for strength or failure in torsion is given by

$$\phi_T^f = \frac{4\pi Q^2}{A^3} \qquad (11\text{-}31)$$

The values of *A*, *K* and *Q* are given in Table 11-3 and those of the shape factors in Table 11-4. The values of ϕ_T^e and ϕ_T^f for circular solid sections are 1.

Axial Compressive Loading—Column Buckling. A compressively loaded column fails by buckling when the load exceeds the critical load (Euler load),

$$P_{Euler} = \frac{n^2\pi^2 EI}{l^2} \qquad (11\text{-}32)$$

where *n* is a constant that depends on the constraints, *E* is the Young's modulus, *I* is the second moment of area, and *l* is the length of the column. We see that the resistance to buckling is directly proportional to *I*; the same as in bending. Thus, the appropriate shape factor is the same as that in bending—ϕ_B^e in Eq. 11-24.

11.3.6 Efficiency of Shape Sections

We shall leave out the derivation for the efficiency factors (i.e., the combination of properties to be maximized), and just simply indicate the results. We shall also illustrate how the factors are used with the materials charts. For bending of beams, the efficiency based on stiffness design is

$$\text{Eff}_B^e = \frac{(E\phi_B^e)^{1/2}}{\rho} \qquad (11\text{-}33)$$

and that based on strength or failure is,

$$\text{Eff}_B^f = \frac{(\phi_B^f)^{1/3}\sigma_f^{2/3}}{\rho} \qquad (11\text{-}34)$$

Dividing the numerator and the denominator in Eq. 11-33 by ϕ_B^e, we obtain

$$\text{Eff}_B^e = \frac{(E\phi_B^e)^{1/2}/\phi_B^e}{\rho/\phi_B^e} = \frac{(E/\phi_B^e)^{1/2}}{\rho/\phi_B^e} = \frac{(E^*)^{1/2}}{\rho^*} \qquad (11\text{-}35)$$

We see that the efficiency of the shaped beam section has the same expression as that of an unshaped beam in Eq. 11-14 (or in Table 11-2), except that E* is substituted for E and ρ* for ρ in Eq. 11-35.

In a similar manner, if we divide the numerator and denominator of Eq. 11-34 by ϕ_B^f, we get

$$\text{Eff}_B^f = \frac{(\phi_B^f)^{1/3}\sigma_f^{2/3}/\phi_B^f}{\rho/\phi_B^f} = \frac{(\sigma_f/\phi_B^f)^{2/3}}{\rho/\phi_B^f} = \frac{(\sigma_f^*)^{2/3}}{\rho^*} \qquad (11\text{-}36)$$

We see again that the above equation is the same expression as in Eq. 11-18, except for σ_f^* substituting for σ_f and ρ^* substituting for ρ. Thus, we see that the efficiency factors are very similar to those without the shape factor, if the true values of E, σ_f, and ρ of the material are replaced by the effective values of E^*, σ_f^*, and ρ^*, respectively. In effect, the shape factors has essentially transformed the solid round material with real properties E, σ_f, and ρ to another "material" of the same weight with lower values of E^*, σ_f^*, and ρ^*.

The effect of the above transformation of a solid circular material to another material is shown in the schematic modulus-density chart shown in Fig. 11-14. In Fig. 11-14, a certain value of $E^{1/2}/\rho = C$ is imposed as a minimum constraint and is shown as the dashed guide line. The efficiency of the solid round material with properties E and ρ and $\phi_B^e = 1$ is shown to be below this line and is therefore unacceptable. The other material with effective properties E^* and ρ^* (after dividing E and ρ by the shape factor $\phi_B^e = 10$ for an I-beam) is above the guide line and is therefore acceptable.

For design based on the strength, the strength-density chart is used and the effect of the shape factor is schematically shown in Fig. 11-15 for a material with a shape factor equal to $\phi_B^f = 10$ compared to that with a shape factor equal to 1. Again the constraint guide line, $\sigma_f^{2/3}/\rho = C$, is established in the figure for a minimum performance. The shape has transformed the unacceptable circular solid material (based on the constraint imposed) to one that is acceptable .

For torsional loading, we use the shape factor ϕ_T^e for stiffness design, and ϕ_T^f for strength design. The efficiency factors are

$$\mathrm{Eff}_T^e = \frac{(G\phi_T^e)^{1/2}}{\rho} = \frac{(G^*)^{1/2}}{\rho^*} \quad \text{for stiffness} \quad (11\text{-}37)$$

and

$$\mathrm{Eff}_T^f = \frac{(\phi_T^f)^{1/3}\sigma_f^{2/3}}{\rho} = \frac{(\sigma_f^*)^{2/3}}{\rho^*} \quad \text{for strength} \quad (11\text{-}38)$$

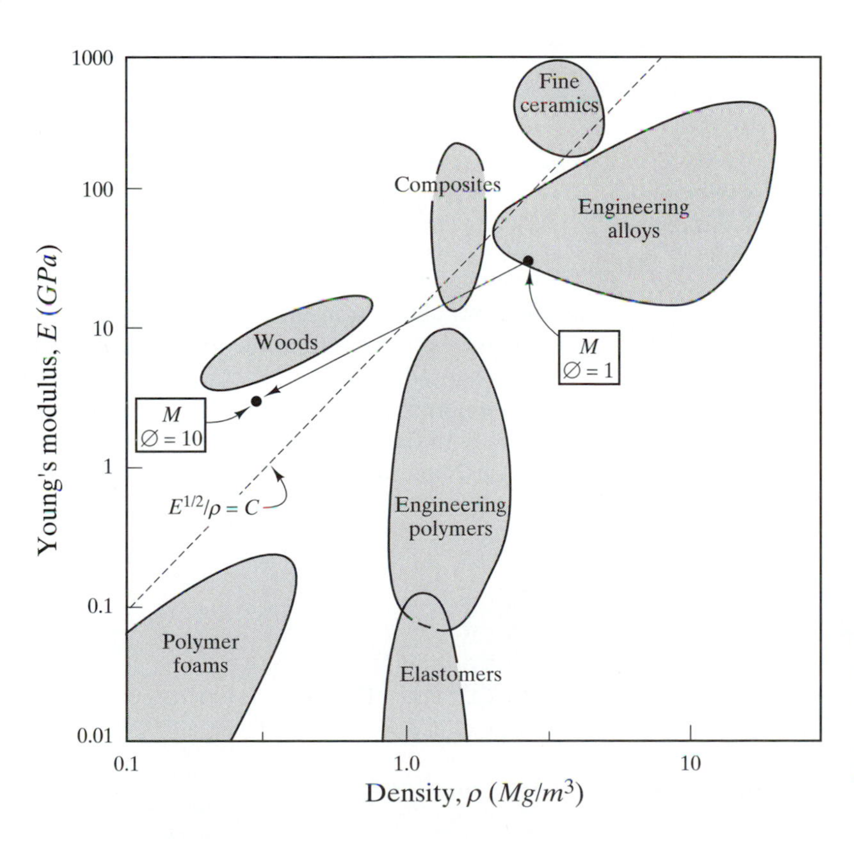

Fig. 11-14 Effect of the shape factor on the performance of a material. An unacceptable material (below the guideline) with a shape factor of 1 satisfies the constraints when its shape factor is 10. [4, 5]

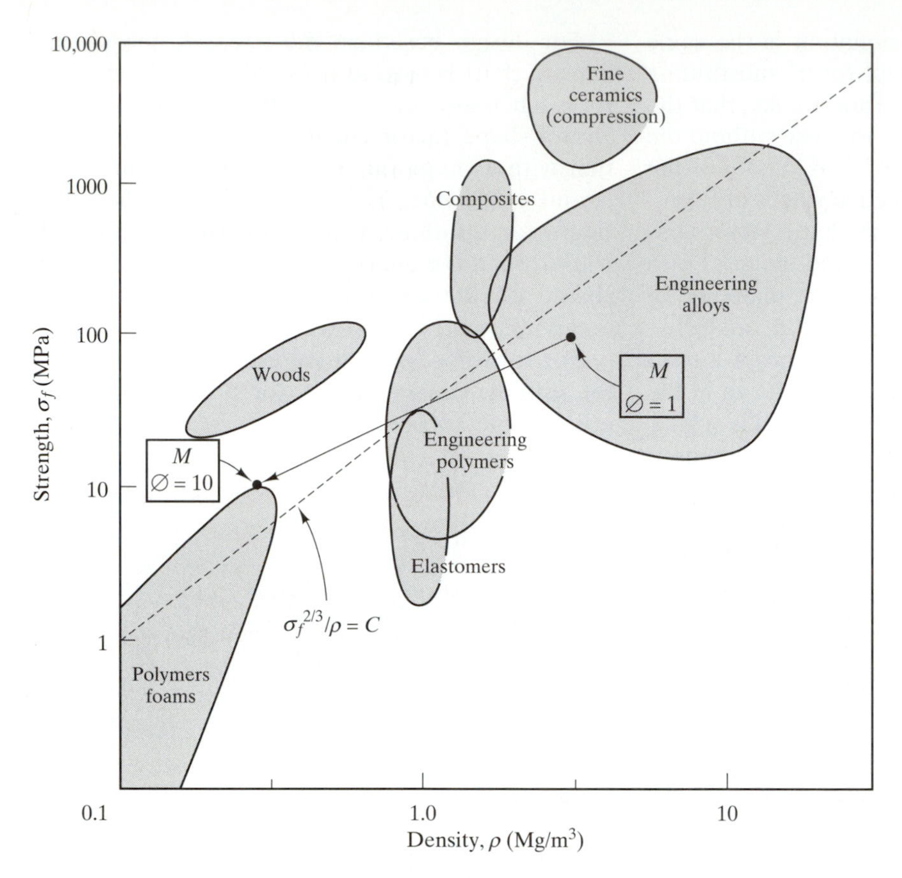

Fig. 11-15 Effect of shape factor—density versus strength chart—a solid circular performs below the guideline but an I-section is above the guideline. [4, 5]

where $G^* = G/\phi_T^e$, $\sigma_f^* = \sigma_f/\phi_T^f$, and $\rho^* = \rho/\phi_T^e$ in Eq. 11-37 and $\rho^* = \rho/\phi_T^f$ in Eq. 11-38. Because G is approximately equal to (3/8) E, Eq. 11-38 is approximately equal to

$$\mathrm{Eff}_T^e = \frac{(E\phi_T^e)^{1/2}}{\rho} = \frac{(E^*)^{1/2}}{\rho^*} \qquad (11\text{-}39)$$

Thus, we can use the modulus-density chart for stiffness for this. The interpretations of Eqs. 11-37, 11-38, and 11-39 are the same as in the bending of beams.

11.3.7 Influence of Internal Regular Shaped Features

The shape factors discussed so far deal with the external shape of the material. If the material contains internal regular arrangement of shaped features, these features contribute to the mechanical efficiency that is exhibited by a shape factor. Figure 11-16 illustrates the combination of the internal honeycombed structure with the external I-section shape. The resultant total shape factor for this case is the product of

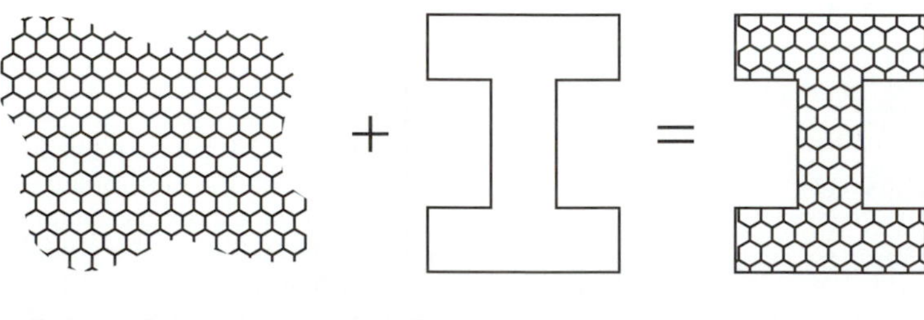

Fig. 11-16 Combining regular internal features with external shape increases the mechanical efficiency of structure or component. [4, 5]

the external and the internal shape factors. For example, for the elastic beam, the total shape factor is

$$(\Phi_B^e)_{total} = \phi_B^e \psi_B^e \tag{11-40}$$

where ψ_B^e is the internal shape factor in the bending of beams. When the prismatic cells as shown in Fig. 11-16 are long and have their axes perpendicular to the bending axis, the microscopic shape factor is

$$\psi_B^e = \frac{\rho_s}{\rho^*} \tag{11-41}$$

where ρ_s is the density of the solid material and ρ^* is the density of the honeycombed structure. In the limit of a solid (when $\rho^* = \rho_s$), $\psi_B^e = 1$ which is expected. Thus, the effect of the internal shape factor is to increase the total shape factor and in essence creates a new material when the total shape factor is used in Eqs. 11-35 and 11-36. Internal shaped features can also be used to improve stiffness in torsion (ψ_T^e) as well as to improve strength in both bending (ψ_B^f) and torsion (ψ_T^f).

11.4 Practical Issues in Engineering Design

The foregoing discussions on materials properties and shape factors suggest clearly that selections of materials must be done with concurrent considerations of the property, processing (whether it is possible to achieve the desired shape), and the environmental impact of materials. The materials charts provide an opportunity at the conceptual stage of design to adapt innovative and cost-effective selections of materials for increased profitability for the company. This is especially true for new starting companies with new products. However, for established companies, there are other issues that may dampen the use of new and innovative materials.

11.4.1 Risk Issues

An established company has an existing goodwill with its clients and customers through years of producing reliable products and providing reliable service. The stage at which the company gets repeat customers for their products and services takes years to develop and certainly the company is not going to risk its business for the sake of innovation only. The risk of innovation must be minimized or eliminated, especially if we think in terms of product liability cases. The innovation must be first proven reliable before it is incorporated into the system.

Engineering is a profession that relies heavily on experience. Designers rely on previous experiences on any aspect of engineering, including the selection of materials. Thus, before a new material can be used in the manufacture of a product, companies with research and development departments (R&D) try these materials in their actual service environments in order to generate engineering experience with them before they are incorporated in the design. The materials are tested and retested in order to eliminate or minimize the risk and it will take years before they can be incorporated. However, for small established companies with no R&D departments, designers will be very reluctant to use new materials and will just rely on the old and proven materials that have low risk factors and have data that can easily be found and accessed. Companies would also like to limit the number of materials in their inventory in order to minimize mixups in using a material for manufacture. Reducing the number of materials and increasing the usage of a material for diverse applications allows the company to buy in bulk quantities to reduce the material cost.

Reluctance of designers to specify new and untested materials is primarily due to the lack of reliable design data, especially in reference to creep, fatigue, and corrosion. Laboratory data on small specimens is very difficult to generalize to apply for design purposes of much larger structures. Also the data may vary from source to source and may have resulted from testing at different conditions. Suppliers also tend to publish the best data of their material and do not indicate statistical variations in databases and datasheets. This is especially true for plastics and ceramics where large variations in properties may exist. Therefore, designers rely frequently on empirical curve fits that are available.

From the standpoint of risk, the designer needs also to consider the consequences of the failure of the component. For example, the criteria for the design of a pressure vessel or a reactor containing life-endangering fluids should be entirely different from those used to construct other reactors. A nuclear reactor may have two or more shells to contain the fission materials in the event the first shell fails. This design procedure is called redundancy and allows for a backup system to take over in case the primary structure fails. This type of design is very common when safety is a paramount factor. The NASA Space Transportation System (commonly known as the Shuttle) contains numerous redundant systems, as do most commercial airlines..

11.4.2 Performance and Cost

In the concurrent considerations of the material's property, processing, and environmental impact, performance and cost play overriding roles. Unless legislated and when the cost for incorporating the environmental factor is expensive, it is usually excluded from the design. Thus, the major considerations are (1) the material's property that determines primarily its performance via the combination of properties, and (2) its processing that can determine the cost of the component or structure.

Industries that produce end-user products can be generally classified as performance-driven or cost-driven. In the performance-driven industries such as industries related to defense, the performance of the material is chosen first and the cost is secondary. For example, the stealth bombers and airplanes are made of the most advanced composite materials—the cost for the construction of the airplane or the processing of the materials is of secondary concern. In contrast to this, the cost-driven industries, such as the automotive and metal industries, have to choose processing simultaneously with material in order to cut costs. In these large volume industries, being profitable is literally accomplished by cutting the cost of materials and processing by as little as one cent per pound of material. In either the performance-driven or cost-driven case, the factor that is chosen first is the most cost-effective for that particular application.

11.5 Materials and Component Failures—As Sources of Engineering Experience [10]

A vital source of so-called engineering experience for design purposes is the information on how a component failed after it has been placed in service. The failure of a component is actually due to the failure of the material that the component is made of and the fundamental reasons why materials fail have already been compiled, after many failure analyses have been made of failed parts or components. It is found that the sources of failures of materials and components may be classified into one or a combination of the following:

- Deficiencies in design
- Deficiencies in material selection
- Imperfections in the material
- Deficiencies in fabrication and processing
- Errors in assembly
- Improper service conditions

11.5.1 Deficiencies in Design

Perhaps the most frequently observed design deficiency and one that can easily be avoided is the presence of mechanical notches that are sources of high-stress concentrations. An example of this design defect is the use of very sharp filet radius at a change in section of a shaft or a similar part that is subjected to bending or torsional loading. This is illustrated in the early failure of a tube-bending machine shaft made of A6 tool steel that failed by fatigue failure, Fig. 11-17. The original filet radius at the change in a section of the shaft was 0.010-inch maximum. The stress concentration created by this original design initiated the fracture of the shaft. The corrective measure taken was the change in the filet radius to 3/32-inch minimum. The larger filet radius minimized stress concentration and prevented recurrence of failure. It should also be mentioned that the windows of the early British Comet airplanes were also cutouts with very sharp filet radii, and these caused fatigue fractures. The solution was also to increase the filet radii.

Another design deficiency might be insufficient design criteria. This may arise from the impossibility of making reliable stress calculations and from insufficient information about the types and magnitude of the loads to which the part is to be exposed in service. An example of a complex part for which stress calculations were difficult to obtain is the hollow splined steel aircraft shaft shown in Fig. 11-18. The analysis of the failed shaft concluded that the fracture area was subjected to severe static radial, cyclic torsional, and cyclic bending loads. The failure is due to fatigue and in particular, corrosion-fatigue in which small fatigue

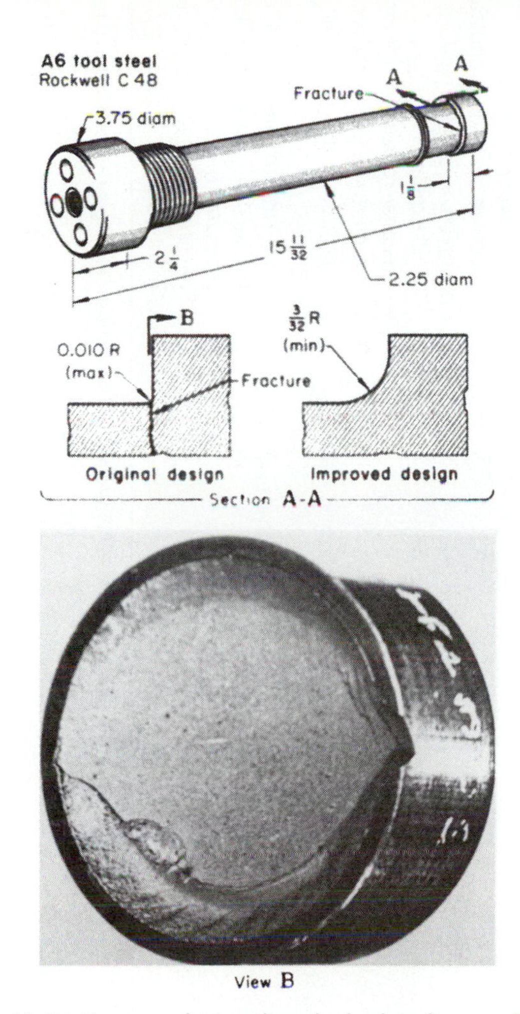

Fig. 11-17 Fracture of A6 tool steel tube-bending machine-shaft was due to very sharp filet radius (original design). Increased filet radius in improved design prevented recurrence. Sharp v-like notch (right) in View B is origin of fracture. [10]

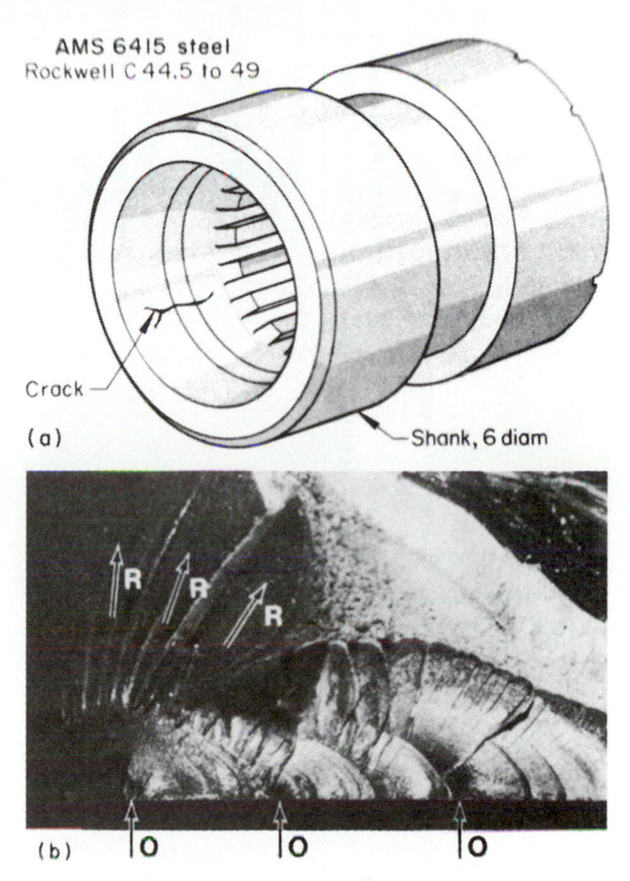

Fig. 11-18 (a) Shaft cracked on the tapered section due to unpredicted combined static radial, cyclic torsional, and cyclic bending loads; (b) crack orginated from multiple corrosion pit sites (arrows O), and propagated along R. [10]

cracks originated from corrosion pits. The shaft was lubricated originally by a noncorrosive oil—the oil was contaminated with water and became corrosive. The small fatigue cracks from the corrosion pits grew in time and then combined to form a large crack that propagated through the section. To avoid future failure, the recommendation was to take more precautions in the operation and maintenance to prevent the contamination of the oil with water and to polish the corrosion pits. Inspection and polishing of the pits were incorporated into the maintenance and overhaul schedule.

11.5.2 Deficiencies in Materials Selection

No generalizations can be made that will be applicable to all materials selection problems. Each problem must be considered individually or on the basis of closely related experiences. Materials must be selected on the basis of possible failure mechanisms, the types of loading and stress, and the environmental conditions, such as temperature and corrosion environment, to which the materials are likely to experience. Table 11-5 is a guide to the criteria or properties of materials needed for the selection of materials based on the failure mechanism and the environmental conditions. The most difficult problem to deal with is the material interaction with the environment. As indicated earlier, the most common of these are:

Table 11-5 Property criteria (a) used as a guide to materials selection according to types of loading and stress, and intended service temperatures. [10]

	Types of loading			Types of Stress			Operating temperatures			
Failure mechanisms	**Static**	**repeated**	**impact**	**Tension**	**Compression**	**Shear**	**Low**	**Room**	**High**	**Criteria generally useful for selection of material**
Brittle fracture	X	X	X	X	..	..	X	X	..	Charpy V-notch transition temperature. Notch toughness. K_{I_c} toughness measurements.
Ductile fracture (b)	X	..	..	X	..	X	..	X	X	Tensile strength. Shearing yield strength.
High-cycle fatigue (c)	..	X	..	X	..	X	X	X	X	Fatigue strength for expected life, with typical stress raisers present.
Low-cycle fatigue	..	X	..	X	..	X	X	X	X	Static ductility available and the peak cyclic plastic strain expected at stress raisers during prescribed life.
Corrosion fatigue	..	X	..	X	..	X	..	X	X	Corrosion resistance of material in environment.
Buckling	X	..	X	..	X	..	X	X	X	Modulus of elasticity and compressive yield strength.
Gross yielding (b)	X	..	..	X	X	X	X	X	X	Yield strength.
Creep	X	..	..	X	X	X	..	..	X	Creep rate or sustained stress-rupture strength for the temperature and expected life (d).
Caustic or hydrogen embrittlement	X	..	..	X	..	..	..	X	X	Keep HRC of material < 41.
Stress-corrosion cracking	X	..	..	X	..	X	..	X	X	Residual or imposed tensile stress and resistance to specific environement. $K_{I_{scc}}$ measurements (d).

(a) Adapted from T. J. Dolan, *Experimental Mechanics,* January 1970, pp. 1–14. (b) Applies to ductile metals only. (c) Millions of cycles. (d) Items strongly dependent on elapsed time.

fatigue, the effect of elevated temperatures on properties such as creep, and the resistance to wear, corrosion, stress-corrosion, corrosion-fatigue, and radiation.

The most common available property of a material is the yield strength or tensile strength because it is very easy to obtain. We should be careful, however, that we do not base the design on strength alone. Except for gross-yielding or ductile fracture, the strength property is often inadequate in fatigue and sometimes high strength can be a detriment, especially in a corrosive environment. Examples of this are the hydrogen-embrittlement and hydrogen-induced cracking that are triggered only when a certain threshold of hardness or strength is attained. High-strength materials also induce brittle fracture, and the appropriate property to consider then is the fracture toughness, K_{I_C} or $K_{I_{SCC}}$.

11.5.3 Imperfections in the Material

A very common problem in materials selection is to assume that materials of the same class or type have the same properties. An uninitiated designer is usually unaware that defects or imperfections in materials do exist. Imperfections may arise as a result of processing, fabrication, handling, as well as during the service of the material. Such imperfections may be solute or impurity segregation, inclusions, porosity, and voids resulting from the original melting (liquid), casting, and solidification conditions. These factors determine the original quality of the material. During shaping and forming operations, lamination, flow-line patterns, laps, and seams may arise. During fabrication, as in welding, the fusion zone is much like a small casting and thus the defects of casting and solidification may again arise. The heat-affected zone may crack after the welding. Improper handling of materials during assembly or fabrication may create surface defects such as scratches, nicks, and indentations. During the service of materials, a perfect material starting in service may develop fatigue cracks, corrosion pits, or induce segregation or coarsening of particles due to service temperature.

In view of the many defects that may arise, the design engineer must apply concurrent or holistic engineering with respect to materials. This means that materials selection must be done not only based on the loads and stresses they must carry but also based on the mode of loading and on the quality, processing, fabrication, handling, and the service conditions of the material. The quality of the material relates to the cleanliness from undesired inclusions (as discussed in Chap. 7) that induce poor performance, especially in fatigue. During processing, these inclusions can also affect the surface characteristics of the material. During fabrication, inclusions can induce cracking such as in delamination caused by sulfur after welding.

11.5.4 Failures Arising from Processing and Fabrication

In addition to the aforementioned defects that may arise in the materials during their processing and handling, susceptibility to failure may also result from unsuitable processing conditions, changes made in specifications without complete evaluation, failure to follow specified procedures, and operator error or accidental damage. Cold-forming and related operations—such as deep drawing, stretching, expanding, reducing, and bending—produce residual stresses and sometimes alter gross or local mechanical properties and cause localized depletion of ductility. Shearing, blanking, and piercing may produce rough or torn edges that constitute stress risers. In addition to notches that machining and grinding may produce, these operations may produce overheating and consequent local softening, and craze cracks in hardened steels.

Improper heat-treatment occurs in a variety of forms, such as overheating, undertempering, use of unacceptably low hardening temperatures, introduction of excessive temperature gradients, and use of quenching, tempering, annealing, and aging conditions unsuitable for the specific alloy or part. Decarburization during heat treatment induces failure by (1) fatigue because it greatly reduces the surface endurance limit, and (2) by distortion in small parts because it reduces the average strength of the cross section. It is particularly detrimental to the service life of springs and small shafts on which the surface stresses are ordinarily very high. Improper heat-treatment in age-hardening aluminum alloys usually induce coarsening of particles at the grain boundaries that produces a region called precipitate-free-zones (PFZs) that embrittle the material.

The fusion welding during fabrication produces a heat-affected-zone (HAZ) closed to the molten pool that undergoes a heating and cooling cycle (heat-treating). Thus, in addition to the casting and solidification problems that may arise in the weld pool, the heat-treating problems may arise in the HAZ. The most common of these in steels is the formation of the very brittle as-quenched martensite that may produce crack in the HAZ. The weldability of a steel is its ability to avoid martensite formation in the HAZ. In austenitic stainless steels, the ordinarily thin sections used and their low thermal conductivity contribute to the slow cooling in the HAZ and induce sensitization, as discussed in "embrittling phenomena" in heat-treating. Sensitization renders the stainless steel susceptible to stress-corrosion and intergranular corrosion cracking.

11.5.5 Failures Due to Errors in Handling and Assembly

Failures in service may result from errors in assembly that were undetected in inspection by the manufacturer or the purchaser and that did not prevent normal operation when the assembled components or equipment were first put into service. The errors usually result in short performance lives of moving parts of mechanical assemblies or with electrical assemblies or structural components. For example, small errors in the placement of rivet holes have caused fatigue failures in structural members of airplane wings. The errors are sometimes related to inaccurate, incomplete, or ambiguous assembly specifications, but they occur most frequently as a consequence of operator error or negligence.

Examples of errors are (1) using the wrong part for a particular application, (2) improper and nonuniform torque application, and (3) misalignment of shafts, gears, bearings, seals, and couplings.

11.5.6 Failures Due to Service Conditions

These failures relate to improper operation of the component or part. Abnormally severe conditions of speed, loading, temperature, and chemical environment or operation without scheduled maintenance, inspection, and monitoring are often major contributors to service failures. The startup or shutdown procedures of complex equipment must be established and heeded in order to get optimum use of the equipment. In the event of failures with the procedures established, the procedures and maintenance schedule must be reevaluated and reestablished.

11.6 Information Sources and Specifications

There are literally thousands and thousands of materials available in the market now and the number continues to increase because new materials are continually being developed. It is, therefore, an enormous task to gather information, especially at the conceptual stage of design, on all kinds of materials. At this stage the best sources of information are handbooks on materials, physics, and chemistry. There are a number of these that cover the different classes of materials, such as metals, ceramics, plastics, and polymers, and a good start is a series of handbooks by the ASM International. As a rule, there are professional societies for each of the classes of materials that can be initially contacted for information. Most of the information is now in electronic format (computer diskettes or CD-ROM) and it makes information searches much easier. The materials charts described in this chapter have also been compiled in electronic format by the University of Cambridge.

At the embodiment and detailed design stages when a class of materials has been selected, we can obtain information directly from the materials suppliers (vendors) or from the associations that represent them. Examples of these are the American Iron and Steel Institute (AISI), The Iron Castings Society or the American Foundrymen's Society, the Copper Development Association (CDA), the Americal Aluminum Association (AAA), and the Titanium Development Association. These associations compile standard products and practices such as sizes, shapes, and availability on each of the materials. In order to minimize cost, the designer (team) needs to be familiar with these. In general, if the sizes and shapes are different from what are offered by the suppliers, we are looking at added costs and perhaps even delay in delivery of materials. In the end, the designer will need to work with at least a couple of suppliers who can furnish the materials. This will avoid any future problems in case one supplier has difficulty meeting the order due to equipment breakdown or labor problems.

In addition to the above, the designer has to consult the codes or specifications about the component or structure being designed. Some of the groups that write the codes are listed in Table 11-6. We note that the majority of those in Table 11-6 represent consumer groups. Thus, the materials specifications are for their own particular materials' procurement needs. The most notable of these are the military (MIL) specifications for the Department of Defense. The SAE and the AMS are the specifying groups for the automotive and the aerospace industries. Most of these specifications are, however, very close to those of the products and processes from the different suppliers or from the independent specifying groups, which are the ASTM, ANSI, and ASME. They are given other numbers to give the consumer some confidentiality in the materials being ordered. The CSA is an example of another country's standards. Other countries also have their respective standards. Because many companies have manufacturing plants in different countries, the materials standards and specifications of different countries, including the United States, are frequently interchangeable. In the United States, the most prominent are the ASTM specifications.

Table 11-6 Some specification-writing groups.

Group Name	Designation
Association of American Railroads	AAR
American Association of State Highway Officials	AASHTO
American Bureau of Shipping	ABS
American Gear Manufacturers Association	AGMA
American National Standards Institute, Inc.	ANSI
American Petroleum Institute	API
American Railway Engineering Association	AREA
American Society of Mechanical Engineers	ASME
American Society for Testing and Materials	ASTM
American Water Works Association, Inc.	AWWA
Canadian Standards Association	CSA
Manufacturers Standardization Society of the Valve and Fittings Industry	MSS
Society for Automotive Engineers	SAE
Aerospace Materials Specifications	AMS
The United States Government	
Department of Defense	MIL and JAN
General Services Administration	FED

Summary

We have discussed the different phases and the various activities that we have to go through in engineering design. At each of these phases, we should be simultaneously making materials selection while ascertaining compliance to the various technical, commercial, and governmental constraints that must be satisfied. Optimum selection involves the consideration of various interrelated constraints among which are physical and mechanical factors, processing, and fabricability, life performance factors, cost and availability, and other factors such as codes and statutory requirements. The criteria for materials selection are based on the property, processing, and environmental profiles of the materials. One of the property profiles is the performance or efficiency index that turns out to be a combination of two or more properties and depends on the type of loading. We have learned that two of these indices are the specific strength and the specific moduli that are sought for in materials destined for the aerospace industry. Based on these indices, materials selection at the conceptual stage can be made with materials charts that will indicate equal performance-index materials. The performance index depends on the external shape, as well as, the internal structure of the material. Examples of external shapes are rectangular and I-beam sections. An example of an internal structure is the honey-combed structure. Processing profile of a material indicates its weldability, formability, and machinability and the ease and total cost of manufacture of the component or structure. Environmental profile relates to usage of energy of the structure or component and the recyclability or reuse of the material. Designing for reuse of a component or recyclability is now an important factor in the selection of materials. To support the design activities, we look at materials and component failure studies for engineering experiences and various sources of information and specifications from specifying bodies or organizations.

References

1. N. Cross, *Engineering Design Methods,* John Wiley & Sons, Ltd., 1989.
2. C. A. Abel, K. L. Edwards, and M. F. Ashby, "Materials, Processing, and the Environment in Engineering Design, Materials & Design," vol. 15, November, 1994, pp. 179–193. (This article is a summary of the "Workshop on Materials, Processing, and the Environment," at the 9th Int. Conf. on Engineering Design (ICED '93), held at The Hague, Netherlands, August 17–19, 1993.)
3. M. F. Ashby, "On the Engineering Properties of Materials," *Acta Met., 37,* 1273–1293 1989.
4. M. F. Ashby, "Materials and Shape," *Acta Met., 39,* 1025–1039, 1991.
5. M. F. Ashby, *Materials Selection in Mechanical Design,* Pergamon Press, 1992.
6. *Selection of Materials for Component Design,* an ASM Sourcebook, ASM, 1986.
7. W. Beitz, *Designing for Ease of Recycling, J. of Eng. Design,* vol. 4, no.1, pp. 11–23, 1993.
8. W. Beitz, "Designing for Ease of Recycling- General Approach and Industrial Application," pp.731–738, in Proceedings of the ICED '93 (see Ref. 2) held at The Hague, August 17–19, 1993.
9. J. G. Parkhouse, "Structuring: A Process of Material Dilution, pp. 367–374, in Proc. of the 3d Int. Conf. on Space Structures held at the U. Of Surrey, UK, Edited by H. Nooshin, 1984.
10. Failure Analysis and Prevention, Vol. 10, Metals Handbook, 8th Ed., ASM, 1975.

Terms and Concepts

Clarification of need
Conceptual stage
Design stages
Detail stage
Efficiency of materials
Embodiment stage
Life-cycle analysis
Materials charts
Materials failure analyses
Materials profiles and selection
Modes of loading
Performance of materials
Recylability
Shape factor—external
Shape factor—internal

Questions and Exercise Problems

11.1 The efficiency or performance indices, specific modulus = E/ρ, and specific strength = σ/ρ, are erroneously given the units of either length in inch or stress in psi. Using SI units show that the correct units are in $(\text{velocity})^2$. For the specific modulus, the velocity is the velocity of the propagation of longitudinal sound through the material.

11.2 The indices can have units of length. This is particularly true for the specific strength because the square of the velocity does not have a physical meaning. To get the length unit, the index has to be divided by another factor. The resulting length represents the length of a fiber of the material of uniform cross section that can be suspended in the Earth's gravitational field, assuming no wind effects, before it deforms (yield strength) or breaks (tensile strength) due to its own weight. What is the factor?

11.3 Calculate the specific moduli and strengths of the materials given in Table P11.3 and indicate the three materials with the highest values.

Table P11.3 Densities, tensile moduli, and strengths of materials—specific moduli and strengths.

Material	Density, ρ, kg/m^3	Elastic Modulus, E, GPa	Yield or Tensile Strength, σ, Mpa	$\frac{E}{\rho}$, (m/s)^2	$\frac{\sigma}{\rho g}$, m
Aluminum, 1050-H18	2,705	69	145		
Aluminum Alloy, 3004-H38	2,720	70	250		
Aluminum alloy, 2024-T861	2,770	72.4	455		
Aluminum alloy, 6061-T6	2,700	68.9	276		
Aluminum alloy, 7075-T6	2,800	71	503		
Copper, C10200-H08	8,940	117	345		
Copper-Beryllium, C17200	8,250	128	1345		
Brass, Cu-30Zn, C26000	8,520	110	359		
Bronze, copper-Al, C63020	7,450	124	793		
Titanium, CP -Grade 2	4,507	105	140		
Titanium alloy, Ti-6Al-4V	4,430	110.5	850		
Nickel 200 (Pure)	8,890	204	148		
Nickel alloy 718	8,190	211	1036		
Magnesium alloy AZ80A-T5	1,761	45	275		
304 stainless steel	8,000	193	205		
PH 17-4 stainless steel	7,800	196	1170		
Maraging steel	7,800	190	1750		
4340 low-alloy steel	7,850	205	1500		
Wood, ash	670	15.8	116		
Wood, mahogany	530	13.5	70		
Composite, 50%Gr-Epoxy	1,500	189	1050		
Composite, 50%Gr-Polyester	2,000	48	1240		
Composite, 60%Kevlar-Epoxy	1,400	76	1240		
Polymer, TS epoxy	1,200	3.45	89.6		
Polymer, TS polyester	1,220	5.40	59		

12

Selection of Ferrous Materials

12.1 Introduction

Metallic materials are by any measure the predominant materials selected for engineering applications because of their attributes. These attributes are their ability to match the engineering requirements of the component or part with their physical and mechanical properties, which are due primarily to their atomic bonding and arrangement (crystal structure). The fact that they can be thermally processed to attain desirable mechanical properties is also a significant advantage. They are also easily deformed and fabricated into the shape of the final product at the lowest possible total cost. Fabrication may involve welding, machining, and forming processes.

Metallic materials are generally classified as ferrous and nonferrous. Ferrous materials consist of steels and cast irons, while nonferrous materials consists of the rest of the metals and alloys. The materials from each group are further classified and are given certain designations according to the American Society for Testing and Materials (ASTM) Standard E 527—The Unified Numbering System (UNS) of Metals and Alloys. The primary series of E 527 is given in Table 12-1. A secondary series, which is referred to as Table 2, in Table 12-1 but not included here, gives the coding of the different metals and alloys (indicated as items, for instance, 18 items in E00001-E99999) belonging to that group. We see in Table 12-1 that the ferrous metals and alloys have as many designations as the rest of the nonferrous metals and alloys. This reflects the number and predominant usage of ferrous metals and alloys in commerce, in construction, and in engineering industries. Thus, it is natural to discuss these groups of materials, ferrous and nonferrous, separately, and we start with the ferrous materials in this chapter.

Ferrous metals and alloys are basically irons with carbon added to them. Alloys with less than 2%C are classified as steels, while those with more than 2%C are called cast irons. As the name implies, cast irons

Table 12-1 Primary series of ASTM E 527, the unified numbering system (UNS) of metals and alloys. (Reference to Table 2 in the table is a secondary series in the ASTM E527 specification.)

Nonferrous Metals and Alloys	
A00001–A99999	aluminum and aluminum alloys
C00001–C99999	copper and copper alloys
E00001–E99999	rare earth and rare earthlike metals and alloys (18 items: see Table 2)
L00001–L99999	low melting metals and alloys (15 items: see Table 2)
M00001–M99999	miscellaneous nonferrous metals and alloys (12 items: see Table 2)
N00001–N99999	nickel and nickel alloys
P00001–P99999	precious metals and alloys (8 items: see Table 2)
R00001–R99999	reactive and refractory metals and alloys (14 items: see Table 2)
Z00001–Z99999	zinc and zinc alloys
Ferrous Metals and Alloys	
D00001–D99999	specified mechanical properties steels
F00001–F99999	cast irons and cast steels
G00001–G99999	AISI and SAE carbon and alloy steels
H00001–H99999	AISI H–steels
J00001–J99999	cast steels (except tool steels)
K00001–K99999	miscellaneous steels and ferrous alloys
S00001–S99999	heat and corrosion resistant (stainless) steels
T00001–T99999	tool steels
Specialized Metals and Alloys	
W00001–W99999	welding filler metals, covered and tubular electrodes, classified by weld deposit composition (see Table 2)

are predominantly produced as castings. In contrast, steels are predominantly produced as wrought products (i.e., deformed and shaped after casting), although steel castings may also be produced. In cast iron, the preferred form of carbon is elemental graphite, while the form of carbon in steels is normally in combined formed such as iron carbide or cementite. In order to induce the graphite formation in cast irons, 1 to 3 percent silicon is necessary in the cast iron compositions. Because steels have to be hot-rolled or hot-forged, manganese is added to steel compositions. When added, the residual sulfur content in the steel combines with the manganese rather than with the iron. The manganese sulfide that forms is solid at the hot-rolling temperatures for steel and will allow the deformation to proceed without splitting or cracking. If it is not added, iron sulfide forms (which is liquid at the hot-rolling temperatures) and will induce the steel to split or crack during hot deformation. We shall now discuss separately the various classifications of steel and cast iron alloys.

12.2 Classification, Designations, and Specifications for Steels

Steels are classified or grouped according to some common characteristics. The most common classification is by their composition and then by their yield or tensile strength. They are also classified by their final processing or finishing methods as well as by their sizes and shapes. In order to procure steels, we also need to specify the quality of the product. We shall now discuss these classifications in turn and indicate how the selection of ferrous materials is made.

12.2.1 Classification According to Composition

According to composition, the classification is broadly done with the carbon content and the alloy content, as shown in Table 12-2. According to carbon content, steels are commonly classified as low carbon, medium carbon, and high carbon, with the ranges in composition as those shown in Table 12-2. According to the alloy content, they are plain carbon steels when no alloying element is added, except manganese which may be added up to 1.65%, silicon up to 0.60%, and copper up to 0.60%. Silicon and copper in the composition are not usually intentionally added but are the result of the recycling of scrap iron, such as an automobile scrap with a copper radiator. When other alloying elements are added, such as manganese in excess of 1.65%, nickel, chromium, and molybdenum, they are called low-alloy steels when the total alloy content is less than 5%, and they are called high-alloy steels when their total alloy content is more than 5%. In the latter category, we have the tool steels and the stainless steels. Thus, with no alloy addition except manganese up to 1.65%, steels are either low-carbon, medium-carbon, or high-carbon, plain carbon steels. However, with a total alloy content of less than 5%, steels are categorized as either low-carbon, medium carbon, or high carbon low-alloy steels. The tool steels and stainless steels have their own nomenclature that we shall discuss separately.

The plain carbon and low-alloy steels are coded according to the AISI-SAE system of designation shown in Table 12-3. This system of designation is an example of both the suppliers (members of AISI—American Iron and Steel Institute) and the customers (SAE—Society for Automotive Engineering) coming to a common understanding about the coding or designation of alloys. From the suppliers' side, a designated steel means that it is produced in tonnages, while from the customers' side, it means that the steel is used in appreciable quantities by at least two users. However, it does not imply that other grades are unavailable, nor does it imply that any particular steel producer makes all the listed grades.

The first two digits of each designation signify the type of alloy additions made to iron. "10" is given to plain carbon steels with manganese (Mn) up to 1.00%, while "15" signifies plain carbon steels with Mn up to 1.65%. As indicated above, these steels may also contain up to 0.60%Si, and 0.60%Cu. The prefixed two-digit codes for different alloy additions are indicated in the table. The last two or three digits, XX or XXX when divided by 100, represent the nominal weight percent of carbon. [The only exception is the XX in 9XX, the various SAE grades of high-strength low-alloy (HSLA) steels. The XX in these HSLA steels represents the required minimum yield strength of the steels in kips per sq. in (ksi).] Letters between the first two digits and the last two XX or three XXX represent special additions. For example, 43BVXX is a nickel-chromium-molybdenum steel with special additions of boron (B) and vanadium (V). Leaded steels are represented by the letter L between the first two and the last two or three digits. The letter H is added at the end of a designation code when the steel has to comply with hardenability requirements, and the steels are referred to as H-steels. In addition to the main alloying elements, residual concentrations of phosphorus, sulfur, and silicon are usually present in the steel. Costwise, the low-carbon 10XX steels, which are commonly called mild steels, are the cheapest, and form the basis for cost comparison.

The compositions of the AISI-SAE steels are specified in ranges and limits, examples are shown in

Table 12-2 Classification of steels according to composition.

Carbon Content	Alloy Content
Low Carbon—less than 0.25%	Plain Carbon—no alloying element except Mn up to 1.65%
Medium Carbon—0.25–0.55%	Low Alloy—total alloy content <5%
High Carbon—greater than 0.55% tool steels and stainless steels	High Alloy—total alloy content >5%

Table 12-3 AISI-SAE system of designations for carbon and low alloy steels.

Numerals and digits	Type of steel and nominal alloy content	Numerals and digits	Type of steel and nominal alloy content	Numerals and digits	Type of steel and nominal alloy content
Carbon Steels		*Nickel-Chromium-Molybdenum Steels*		*Chromium Steels*	
10XX(a)	Plain carbon (Mn 1.00% max)	43XX	Ni 1.82; Cr 0.50 and 0.80; Mo 0.25	50XXX	Cr 0.50
11XX	Resulfurized			51XXX	Cr 1.02 } C 1.00 min
12XX	Resulfurized and rephosporized	43BVXX	Ni 1.82; Cr 0.50; Mo 0.12 and 0.25; V 0.03 min	52XXX	Cr 1.45
15XX	Plain carbon (max Mn range-1.00 to 1.65%)	47XX	Ni 1.05; Cr 0.45; Mo 0.20 and 0.35	*Chromium-Vanadium Steels*	
				61XX	Cr 0.60, 0.80 and 0.95; V 0.10 and 0.15 min
		81XX	Ni 0.30; Cr 0.40; Mo 0.12		
Manganese Steels		86XX	Ni 0.55; Cr 0.50; Mo 0.20		
13XX	Mn 1.75	87XX	Ni 0.55; Cr 0.50; Mo 0.25	*Tungsten-Chromium Steel*	
		88XX	Ni 0.55; Cr 0.50; Mo 0.35	72XX	W 1.75; Cr 0.75
Nickel Steels		93XX	Ni 3.25; Cr 1.20; Mo 0.12		
23XX	Ni 3.50	94XX	Ni 0.45; Cr 0.40; Mo 0.12	*Silicon-Manganese Steels*	
25XX	Ni 5.00	97XX	Ni 0.55; Cr 0.20; Mo 0.20	92XX	Si 1.40 and 2.00; Mn 0.65, 0.82 and 0.85; Cr 0.00 and 0.65
Nickel-Chromium Steels		98XX	Ni 1.00; Cr 0.80; Mo 0.25		
31XX	Ni 1.25; Cr 0.65 and 0.80	*Nickel-Molybdenum Steels*		*High-Strength Low-Alloy Steels*	
32XX	Ni 1.75; Cr 1.07	46XX	Ni 0.85 and 1.82; Mo 0.20 and 0.25	9XX	Various SAE grades
33XX	Ni 3.50; Cr 1.50 and 1.57	48XX	Ni 3.50; Mo 0.25		
34XX	Ni 3.00; Cr 0.77			*Boron Steels*	
				XXBXX	B denotes boron steel
Molybdenum Steels		*Chromium Steels*		*Leaded Steels*	
40XX	Mo 0.20 and 0.25	50XX	Cr 0.27, 0.40, 0.50 and 0.65	XXLXX	L denotes leaded steel
44XX	Mo 0.40 and 0.52	51XX	Cr 0.80, 0.87, 0.92, 0.95, 1.00 and 1.05		
Chromium-Molybdenum Steels					
41XX	Cr 0.50, 0.80 and 0.95; Mo 0.12, 0.20, 0.25 and 0.30				

(a) XX or XXX the last two or three digits of these designations indicates the percent carbon when divided by 100, except in 9XX for the SAE grades. The XX in the SAE grades designates the mininum yield strength in ksi.

Tables 12-4, and 12-5 for selected standard plain carbon, and carbon H-steels, and Tables 12-6 and 12-7 for selected alloy steel plates and alloy H-steels. These tables also incorporate the UNS designations according to Table 12-1. The range in composition is the "window of processing" for each steel. For example, a 1020 steel in Table 12-4 means it has a nominal 0.20%C and the steel producer can supply the steel in the range 0.17–0.23%C. We must note that the AISI designations indicate only standard practices, but are not specifications. For the SAE, which is a consumer organization, each of the alloys listed in Table 12-3 is just a part of a specification under various SAE standards that are published in the SAE Handbook.

For procurement purposes, a specification has to be included. The most comprehensive and widely used specifications are the ASTM specifications that have designations different from those of the AISI-SAE designations. These specifications are predominantly targeted for a specific product and are generally oriented toward the performance of the fabricated end product, with considerable latitude given to the chemical composition of the steel. An ASTM designation is an arbitrarily chosen number prefixed by the letter "A", which is designated for ferrous materials. The ASTM specifications or designations are also adopted by the ASME and are prefixed by the letter "S". Thus, an AXYZ code by ASTM becomes SAXYZ code by the ASME, where XYZ is an arbitrary number. Examples of ASTM specifications for structural quality carbon steel plates are listed in Table 12-8.

12.2.2 Classification According to Strength

The classification of the steels according to strength starts with the properties of structural quality plain carbon steels that have low yield strengths generally less than 40 ksi. These are the most common materials used for construction. The term structural quality in Table 12-8 refers to the hot-rolled plates, bars, and shapes with mechanical property requirements that

Table 12-4 Composition ranges and limits of selected AISE-SAE standard plain carbon steels—structural shapes, plate, strip, sheet and welded tubing.

AISE-SAE Designation	UNS Designation	Heat Composition ranges/limits	
		%C	%Mn
1020	G10200	0.17–0.23	0.30–0.60
1038	G10380	0.34–0.42	0.60–0.90
1045	G10450	0.42–0.50	0.60–0.90
1524	G15240	0.18–0.25	1.30–1.65
1541	G15410	0.36–0.41	1.30–1.65

Table 12-5 Composition ranges and limits for selected AISE-SAE standard carbon H-steels.

AISE-SAE Designation	UNS Designation	Heat Composition ranges and limits wt%		
		C	Mn	Si
1038H	H10380	0.34–0.43	0.50–1.00	0.15–0.30
1045H	H10450	0.42–0.51	0.50–1.00	0.15–0.30
1524H	H15240	0.18–0.26	1.25–1.75	0.15–0.30
1541H	H15410	0.35–0.45	1.25–1.75	0.15–0.30

Table 12-6 Composition ranges and limits for selected AISI-SAE alloy steel plates.

AISE-SAE Designation	UNS Designation	Heat Composition ranges and limits, wt%					
		C	Mn	Si	Cr	Ni	Mo
1345	G13450	0.41–0.49	1.50–1.90	0.15–0.30	...	...	...
4145	G41450	0.41–0.49	0.70–1.00	0.15–0.30	0.80–1.15	...	0.15–0.25
4340	G43400	0.36–0.44	0.55–0.80	0.15–0.30	0.60–0.90	1.65–2.00	0.20–0.30
4620	G46200	0.16–0.22	0.40–0.65	0.15–0.30	...	1.65–2.00	0.20–0.30
8620	G86200	0.17–0.23	0.60–0.90	0.15–0.30	0.35–0.60	0.40–0.70	0.15–0.25

Table 12-7 Composition ranges and limits for selected AISE-SAE alloy H-steels.

AISE-SAE Designation	UNS Designation	Heat Composition ranges and limits, wt%					
		C	Mn	Si	Cr	Ni	Mo
1345H	H13450	0.42–0.49	1.45–2.05	0.15–0.30	...	...	...
4145H	H41450	0.42–0.49	0.65–1.10	0.15–0.30	0.75–1.20	...	0.15–0.25
4340H	H43400	0.37–0.44	0.55–0.90	0.15–0.30	0.65–0.95	1.55–2.00	0.20–0.30
4620H	H46200	0.17–0.23	0.35–0.75	0.15–0.30	...	1.55–2.00	0.20–0.30
8620H	H86200	0.17–0.23	0.60–0.95	0.15–0.30	0.35–0.65	0.35–0.75	0.15–0.25

Table 12-8 ASTM Specifications for structural quality carbon steel plates.

Specification	Steel type and condition
Carbon Steel	
A36	Carbon steel plates, bars and shapes
A131	Carbon and HSLA steel plates, bars, shapes and rivets for ships
A283	Carbon steel plates of low or intermediate tensile strength
A284	Carbon-silicon steel plates of low or intermediate tensile strength for machine parts and general construction
A440	Carbon steel plates, bars and shapes of high tensile strength
A529	Carbon steel plates, bars, shapes and sheet piling with 290 MPa (42 ksi) minimum yield point
A573	Carbon steel plates for applications requiring toughness at atmospheric temperatures
A678	Quenched and tempered carbon steel plates
A709	Carbon, alloy and HSLA steel plates, bars and shapes for bridges

are used in the construction of structures such as bridges, buildings, and ships. Table 12-9 lists the mechanical properties of the selected ASTM structural quality carbon steel plates in Table 12-8. We see that the yield strengths are generally below 40 ksi, except for A678 that are quenched and tempered, and A440 and A 529 that require high tensile and minimum yield strengths as indicated in Table 12-8. High-strength structural steels are those with yield strengths generally between 40 ksi and 120 ksi, while ultra-high strength structural steels are those with extremely high-yield strengths, with a minimum 200 ksi.

The high-strength structural steels up to 80 ksi minimum yield strength may be achieved in the hot-rolled condition with ferrite-pearlite or reduced pearlite microstructures. The steels may be either with low alloy additions as solid solution strengtheners that are called HSLA (high-strength, low-alloy)

Table 12-9 Mechanical properties of structural quality carbon steel plate.

ASTM specification	Material grade or type	Tensile strength MPa	ksi	Yield Strength MPa	ksi	Elongation in 50 mm or 2 in.,%
A36	...	400–550	58–80	220–250	32–36	23
A131	A, B, D, E, CS, DS	400–490	58–71	220	32	24
A283	A	310–380	45–55	165	24	28
	B	345–415	50–60	185	27	26
	C	380–450	55–65	205	30	23
	D	415–515	60–75	230	33	21
A284	A	345	50	170	25	25
	B	380	55	190	27	25
	C	415	60	205	30	23
	D	415	60	230	33	22
A440	...	435	63	290	42	19
A529	...	414–586	60–85	290	42	19
A573	58	400–490	58–71	220	32	...
	65	450–530	65–77	240	35	...
	70	480–620	70–90	290	42	...
A678	A	483–621	70–90	345	50	20
	B	552–689	80–100	414	60	20
	C	586–793	85–115	448	65	17
A709	36	400–550	58–80	220	32	21

steels, or with microalloy additions of columbium (niobium) [Cb (Nb)], vanadium (V), and/or titanium (Ti) that are called microalloyed HSLA steels. The latter are hot-rolled with controlled temperatures and deformation, while the former are hot-rolled with no controls. The HSLA steels depend on some reduction in ferrite grain size and solid solution strengthening to attain the hot-rolled strength. The microalloyed HSLA steels have, in general, carbon contents less than 0.10%C and undergo thermal mechanical control processing (TMCP) such as that described in Sec. 8.6.2. Some of these processes will also be discussed shortly in relation to Fig. 12-1. The strength from the microalloyed HSLA steels stems predominantly from their very fine, ferrite grain size with some contribution from precipitation hardening. The compositions of selected SAE 9XX designated HSLA steels in Table 12-3 and described in SAE specification J410c are shown in Table 12-10 and their mechanical properties in Table 12-11. Typical compositions of selected high-strength structural as-hot-rolled HSLA, and hot-rolled microalloyed HSLA steels are given in Table 12-12. A summary of the characteristics and intended uses of HSLA steels as described by ASTM specifications is given in Table 12-13 and a summary of their mechanical properties in Table 12-14.

Yield strengths in excess of 80 ksi are hard to obtain in the as-hot-rolled condition with ferrite-pearlite structures, even with the controlled thermal-mechanical processing of the microalloyed HSLA steels. To attain this strength level, the steels have to be processed to obtain the very fine structures of

Table 12-10 Compositions, of selected SAE 9XX HSLA steels described in SAE J410c.

Grade(c)	Heat composition limits, % C, max	Mn, max	P, max	Other elements
945	0.22	1.35	0.04	Nb, V
950	0.23	1.35	0.04	Nb, V
955	0.25	1.35	0.04	Nb, V, N
960	0.26	1.45	0.04	Nb, V, N
965	0.26	1.45	0.04	Nb, V, N
970	0.26	1.65	0.04	Nb, V, N
980	0.26	1.65	0.04	Nb, V, N

Table 12-11 Mechanical properties of selected SAE 9XX HSLA steels according to SAE J410c.

	Min tensile strength		Min yield strength		Min elongation, %	
Grade	**MPa**	**ksi**	**MPa**	**ksi**	**In 200 mm or 8 in.**	**In 50 mm or 2 in.**
945	415	60	310	45	19	22 to 25
950	450	65	345	50	18	22
955	483	70	380	55	17	20
960	520	75	415	60	16	18
965	550	80	450	65	15	16
970	590	85	485	70	14	14
980	655	95	550	80	10	12

Table 12-12 Typical compositions of selected as-rolled high strength structural and micro-alloyed HSLA steels.

	Composition, %										
Steel	**C**	**Mn**	**P**	**S**	**Si**	**Cr**	**Ni**	**Mo**	**Cu**	**V**	**Others**
As-Rolled High-Strength Structural Steels											
1	0.22	1.35	0.025	···	0.25	···	···	···	0.25	···	···
2	0.20	1.10	0.025	···	0.25	···	···	···	0.25	0.04	0.01 Ti
3	0.15	0.70	0.040	···	0.75	0.55	···	0.15	···	···	0.04 Zr
4	0.15	0.75	0.055	···	0.10	···	0.75	···	0.65	···	···
5	0.10	0.75	0.100	···	0.60	0.75	0.85	···	0.85	···	0.06 Zr
6	0.10	0.35	0.120	···	0.50	0.80	0.50	···	0.50	···	
7	0.12	0.75	0.025	···	0.25	0.25	0.75	0.13	0.75	···	···
Microalloyed High-Strength Low-Alloy Steels											
8	0.09	1.25	0.010	0.012	0.30	···	···	···	···	0.09	0.05 Al; 0.005 N; 0.09 Nb
9	0.14	1.25	0.03	0.03	0.30	···	···	···	···	0.02	0.01 Nb
10	0.16	1.40	0.03	0.03	0.30	···	···	···	···	0.02	0.01 Nb
11	0.06	0.45	0.01	0.02	0.10	···	···	···	···	···	0.02 Nb
12	0.06	0.75	0.01	0.02	0.10	···	···	···	···	···	0.04 Nb
13	0.06	0.95	0.01	0.02	0.10	···	···	···	···	···	0.10 Nb

lower bainite, and/or quenched and tempered martensite. The conventional process is to finish hot-roll, air-cool to room temperature, then austenirize, quench and temper the steels. Recent more advanced processing of microalloyed steels involves either accelerated cooling or direct quenching immediately after the steels pass the last hot-rolls to produce bainitic and/or martensitic structures and then tempering. These processes are schematically shown in Fig. 12-1. Examples of high-strength steels that are conventionally quenched and tempered (QT) with minimum 100 ksi yield strengths are ASTM specifications A514/ A517 and modified A710. An example of a direct quenched and tempered steel is HT-80 (Japanese code). The U.S. Navy recently developed two steels, HSLA 80 and HSLA 100, according to MIL-S-24645A. These are low carbon Cu-Ni precipitation-hardened steels based on ASTM A710 that replaced quenched and tempered steels for ease of processing and cost savings. The properties of these steels are shown in Fig. 12-2.

The ultrahigh strength steels are all quenched and tempered medium carbon, low-alloy steels or aged high-alloy steels and low-carbon maraging steels with minimum 200 ksi yield strength, used primarily as aerospace and defense industry structural materials.

Table 12-13 Summary of Characteristics and Applications of Structured HSLA Steels Described in ASTM Specifications.

ASTM specification	Title	Alloying elements	Available mill forms	Special characteristics	Intended uses
A242	High-stength low-alloy structural steel	Cr, Cu, N, Ni, Si, Ti, V, Zr	Plates, bars and shapes up to 102 mm (4 in.) thick	Atmospheric corrosion resistance four times that of carbon steel	Structural members, in welded, bolted or riveted constructions
A440	High-strength structural steel	Cu, Si	Plates, bars and shapes up to 102 mm (4 in.) thick	Atmospheric corrosion resistance twice that of carbon steel	Structural members, primarily in bolted or riveted constructions
A441	High-strength low-alloy structural manganese-vanadium steel	V, Cu, Si	Plates, bars and shapes up to 203 mm (8 in.) thick	Atmospheric corrosion resistance twice that of carbon steel	Welded, bolted or riveted structures, but primarily welded bridges and buildings
A572	High-strength low-alloy niobium-vanadium steels of structural quality	Nb, V, N	Plates, bars, shapes and sheet piling up to 152 mm (6 in.) thick	Yield strengths of 290 to 450 MPa (42 to 65 ksi), in six grades	Welded, bolted or riveted structures, but mainly bolted or riveted bridges and buildings
A588	High-strength low-alloy structural steel with 50 ksi minimum yield point to 4 in. thick	Nb, V, Cr, Ni, Mo, Cu, Si, Ti, Zr	Plates, bars and shapes up to 203 mm (8 in.) thick	Atmospheric corrosion resistance four times that of carbon steel; nine grades of similar strength	Welded, bolted or riveted structures, but primarily welded bridges and buildings where weight savings or added durability is important
A618	Hot-formed welded and seamless high-strength low-alloy structural tubing	Nb, V, Si, Cu	Square, rectangular, round and special shape structural tubing, welded or seamless	Three grades of similar yield strength; may be purchased with atmospheric corrosion resistance twice that of carbon steel	General structural purposes, including welded, bolted or riveted bridges and buildings
A633	Normalized high-strength low-alloy structural steel	Nb, V, Cr, Ni, Mo, Cu, N, Si	Plates, bars and shapes up to 152 mm (6 in.) thick	Enhanced notch toughness; yield strengths of 290 to 415 MPa (42 to 60 ksi) in five grades	Welded, bolted or riveted structures for service at temperatures down to −45°C (−50°F)
A656	High-strength low-alloy hot rolled structural vanadium-aluminum-nitrogen and titanium-aluminum steels	V, Al, N, Ti, Si	Plates, normally up to 15.9 mm (5/8 in.) thick	Yield strength of 552 MPa (80 ksi)	Truck frames, brackets, crane booms, rail cars and other applications where savings is important

Table 12-14 Ranges of mechanical properties of HSLA steels according to ASTM specifications.

ASTM specification	Types or Grades	Min. Yield Strengths MPa	Min. Yield Strengths Ksi	Min. Tensile Strengths MPa	Min. Tensile Strengths Ksi	Min. Elongation in 50 mm (2 in)
A242	Types 1 & 2	290 to 345	42 to 50	435 to 480	63 to 70	21%
A440		290 to 345	42 to 50	435 to 485	63 to 70	21%
A441		275 to 345	40 to 50	415 to 485	60 to 70	21%
A572	Grades 42, 45, 50, 55, 60, 65	290 to 450	42, 45, 50, 55, 60, 65	415 to 550	60, 60, 65, 70, 75, 80	24 to 17%
A588	Grades A to J	290 to 345	42 to 50	432 to 485	63 to 70	21%
A618	Grades I, II, III	345	50	448 to 483	70, 70, 65	22%, 22%, 20%
A633	Grades A to E	290 to 415	42 to 60	430 to 690	63 to 100	23%
A656	Grades 1 and 2	552	80	655 to 793	95 to 115	12% in 200 mm

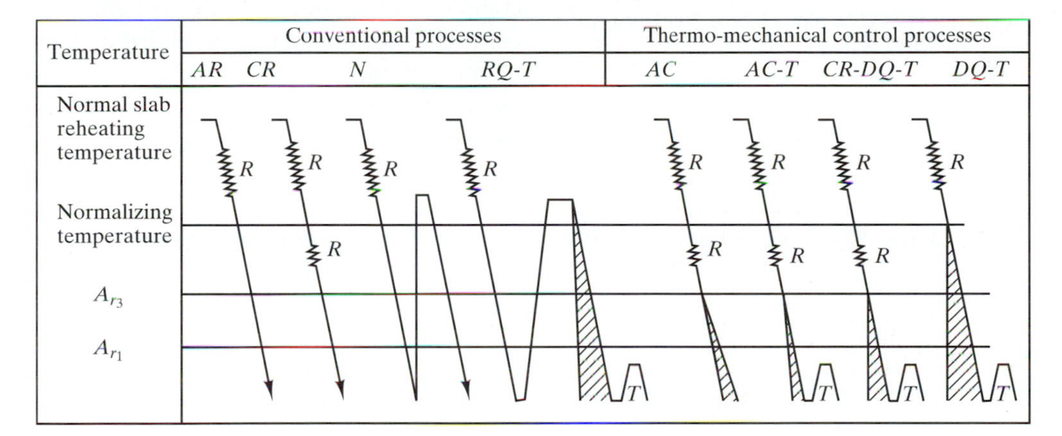

Fig. 12-1 Schematic of conventional and advanced thermo-mechanical control rolling processes of plates. AR–as-rolled; CR–controlled-rolled; N–rolled and normalized; RQ-T–rolled quenched and tempered; AC –controlled rolled, accelerated cooling; AC-T - AC plus tempering; CR-DQ -T - CR plus direct-quenched and tempered; DQ - T - rolled, direct-quenched and tempered.

The most common of these are the medium carbon AISI/SAE 4130, 4140, 4340 and modifications of 4340 such as 300M, D-6a, and D-6ac. Typical mechanical properties of 4340 are given in Table 12-15 and those of D-6a steel are given in Table 12-16. The 4340 may also be ausformed and then quenched and tempered, as indicated in Sec. 9.7.2g, to attain properties much higher than those that may be obtained with the conventional quench and temper process. The compositions and mechanical properties of some high alloy ultra-high strength steel plates, according to ASTM specifications, are shown in Tables 12-17, and 12-18, respectively. A538 and A590 are maraging steels, and A605 is basically the proprietary HP 9%Ni-4%Co steel originally developed by the former Republic Steel Corporation.

12.2.3 Classification According to Product Shape, Finish Processing, and Quality Descriptors

Typical product classification of flat hot-rolled carbon and low-alloy steels is done according to the thickness and width. This is shown in Table 12-19 which shows the ranges of thickness and width of flat hot-rolled bars, plates, strip, and sheet products. In addition to the bars shown in Table 12-19, the term

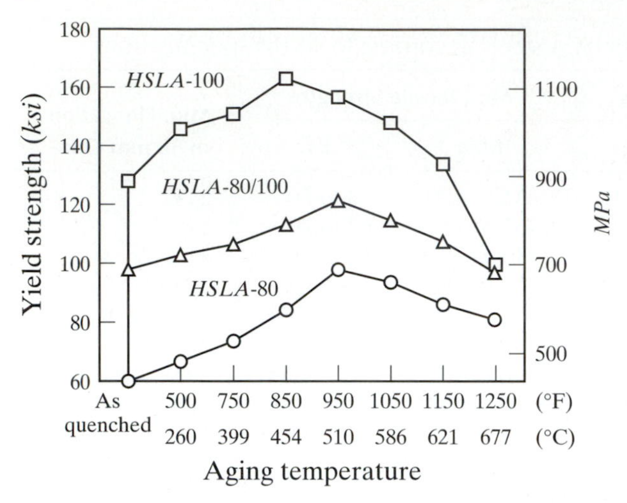

Fig. 12-2 Yield strengths of U.S. Navy HSLA Steels–AC plus aging $1\frac{1}{4}$ in. (32 mm) thick plate.

"bar" also includes (1) rounds, squares, hexagons, and similar cross sections 9.52 mm ($\frac{3}{8}$ in.) and greater across, (2) small angles, channels, tees, and other standard shapes less than 76 mm (3 in.) across, and (3) concrete-reinforcing bars, commonly called rebars.

In addition to the product forms in Table 12-19, there are also "shape" products that include structural shapes such as I-beams and special shapes. Structural shapes are flanged, are 76 mm (3 in.) or greater in at least one cross-sectional dimension, and are used in structures such as bridges, buildings, ships, and railroad cars. Special shapes are those designed by users for specific applications.

The finishing process classification signifies the last processing the steel has undergone. The most common, of course, are hot-rolled, cold-rolled or cold-finished, annealed, normalized, quenched and tempered, and coating processes such as porcelain enameling, hot-dip galvanizing, and electrolytic galvanizing (zinc-coating).

Table 12-15 Mechanical properties of quenched and tempered 4340 steel(a) at various tempering temperatures.

Tempering temperature		Tensile strength		Yield strength		Elongation in 50 mm or 2 in., %	Reduction in area, %
°C	°F	MPa	ksi	MPa	ksi		
205	400	1980	287	1860	270	11	39
425	800	1500	217	1365	198	14	48
650	1200	1020	148	860	125	20	60

(a) Oil quenched from 845°C (1550°F) and tempered at various temperatures.

Table 12-16 Mechanical properties of quenched and tempered D-6a steel(a).

Tempering temperature		Tensile strength		Yield strength		Elongation in 50 mm or 2 in., %	Reduction in area, %
°C	°F	Mpa	ksi	Mpa	ksi		
205	400	2000	290	1620	235	8.9	25.7
425	800	1630	236	1570	228	9.6	36.8
650	1200	1030	150	970	141	18.4	60.8

(a) Normalized at 900°C (1650°F). oil quenched from 845°C (1550°F) and tempered at various temperatures.

Table 12-17 ASTM Compositions of pressure vessel quality high alloy ultra-high strength steel plates.

Specification	Material Grade	Composition, % (a)						
		C	Cr	Ni	Mo	Co	Ti	Others
A538	A	0.03	···	17.0-19.0	4.0-4.5	7.0-8.5	0.10-0.25	0.05-0.15 Al
	B	0.03	···	17.0-19.0	4.6-5.1	7.0-8.5	0.30-0.50	0.05-0.15 Al
	C	0.03	···	18.0-19.0	4.6-5.2	8.0-9.5	0.55-0.80	0.05-0.15 Al
A590	···	0.03	4.50-5.50	11.50-12.50	2.75-3.25		0.20-0.35	0.40 Al
A605	···	0.16-0.23	0.65-0.85	8.50-9.50	0.90-1.10	4.25-5.00	···	0.06-0.12 V

(a) When a single value is shown, it is a maximum limit.

Table 12-18 Mechanical Properties of Pressure Quality ASTM Ultra-high strength steels.

ASTM Specification	Material grade or type	Tensile strength MPa	Tensile strength ksi	Yield strength MPa	Yield strength ksi	Elongation in 50 mm or 2 in., %
A538	A	1450	210	1380–1620	200–235	8
	B	1650	240	1580–1790	230–260	6
	C	1930	280	1900–2100	275–305	6
A590	···	1300	188	1240–1450	180–210	14
A605	···	1310-1520	190-220	1170 (b)	170 (b)	14

(a) Where a single value is shown, it is a minimum. (b) Minimum and/or maximum values depend on the thickness of the plate. (c) As-rolled class 1 plate is limited to 25 mm (1 in.) maximum thickness. (d) As-rolled and aged class 2 plate is limited to 25 mm (1 in.) maximum thickness.

Table 12-19 Product classification of flat-rolled carbon and low-alloy steels according to size, width, and thickness. [3]

Thickness (in.)	Specified Width (in.)					
	To 3.5	More than 3.5 to 6	More than 6 to 8	More than 8 to 12	More than 12 to 28	More than 48
0.230 and above	Bar[a]	Bar[a]	Bar[a]	Plate[a]	Plate[a]	Plate[a]
0.229 to 0.204	Bar[a]	Bar[a]	Strip	Strip	Sheet	Plate[a]
0.203 to 0.180	Strip	Strip	Strip	Strip	Sheet	Plate[a]
0.179 and below	Strip	Strip	Strip	Strip	Sheet[b]	Sheet[b]

[a] Subject to certain conditions, these dimensions are sold as carbon sheet or strip as well as bars or plate.

[b] These product classifications for hot-rolled sheet are based on the median point of the minimum thickness ordered plus full published thickness tolerances.

In addition to the finishing process, there are quality descriptors that indicate the suitability of products for certain applications or fabrication processes. The quality descriptors for some of the product forms shown in Table 12-19 for carbon and low-alloy steels are given in Table 12-20. The quality descriptors imply certain mechanical and physical attributes of the product that may be due to the control of one or more of the following factors during manufacture: (1) degree of internal soundness; (2) relative uniformity of chemical composition; (3) relative freedom from surface imperfections; (4) relative size of discard cropped from the

Table 12-20 Quality descriptors for steel products.

Product Form	Finishing Process	Quality Descriptors: Carbon Steels	Quality Descriptors: Alloy Steels
Plate	Hot-rolled	Regular quality Structural quality Cold-drawing quality Cold-pressing quality Cold-flanging quality Pressure vessel quality Marine quality	Regular quality Drawing quality Pressure vessel quality Structure quality Aircraft quality Aircraft physical quality
Bar	Hot-Rolled	Merchant quality Special quality Special hardenability Special internal soundness Nonmetallic inclusion requirement Special surface quality Scrapless nut quality Axle shaft quality Cold extrusion quality Cold heading and cold forging quality	Regular quality Aircraft structural or steel subject to magnetic particle inspection Axle shaft quality Cold-Heading quality Special cold-heading quality Rifle barrel quality, gun quality, shell or AP shot quality
	Cold-Finished	Standard quality Special hardenability Special internal soundness Nonmetallic inclusion requirement Special surface Cold heading and cold forging quality Cold extrusion quality	Regular quality Aircraft structural or steel subject to magnetic particle inspection Axle shaft quality Cold-heading quality Special cold-heading quality Rifle barrel quality, gun quality, shell or AP shot quality
Strip	Hot-rolled	Commercial quality (CQ) Drawing quality (DQ) DQ special killed (DQSK) Structural quality	
Sheet	Hot-Rolled Cold-Rolled Long terne Galvanized Electrolytic Zn-coated	Commercial quality (CQ) Drawing quality (DQ) DQ special killed (DQSK) Structural quality	

ingot; (5) extensive testing during manufacturing; (6) number, size, shape, and distribution of nonmetallic inclusions (i.e., cleanness or inclusion rating of steel); and (7) relative hardenability requirements.

Most of the quality descriptors in Table 12-20 are self-explanatory with the exception of "merchant quality" for hot-rolled carbon steel bars. This descriptor is for bars that are made for noncritical applications requiring modest strength and mild bending or forming, but not for forging or heat-treating purposes. A descriptor for a particular product is not necessarily carried over to another product; for example, standard-quality cold-finished bars are made from special-quality hot rolled bars.

Sheets and Strips. Low-carbon, plain carbon steel sheets and strips form the majority of steel products. They are used primarily for consumer goods and are selected for stamping applications, such as for automobile bodies and appliances. Typical compositions for these unalloyed steels are 0.05–0.10%C, 0.25–0.50%Mn, 0.035%P max, and 0.04%S max. The commercial quality (CQ) product is suitable for simple bending (bend flat on itself in any direction in standard room-temperature bend test) and moderate forming. CQ does not require mechanical properties tests and uniformity in either composition or mechanical properties. However, the hardness of a cold-rolled CQ sheet is less than 60 HRB when it is shipped and it will strain aged after shipping. When greater ductility or more uniform properties than CQ are required, drawing quality (DQ) is specified. DQ material is suitable for deep-drawn parts and other parts requiring severe deformation. For the most severe deformation and for resistance to stretcher strains (Lüder bands), DQ special-killed (DQSK) is specified. Special-killed steel is usually aluminum-killed, but other deoxidizers may be used to obtain desired characteristics. For either DQ or DQSK, the supplier usually guarantees that the material is capable of being formed into a part within an established breakage allowance. Structural quality (SQ), formerly called physical quality (PQ), is applicable when specified strength and elongation values are required in addition to bend tests. Typical mechanical properties and bend tests of structural low-carbon sheets and strips according to ASTM specifications are shown in Table 12-21.

Plates. Steel plate is predominantly used in the hot-finished (hot-rolled) condition. Its strength and toughness properties are improved by controlled hot-rolling and by quenching and tempering. It is used mainly in the construction of buildings, bridges, ships, railroad cars, storage tanks, pressure vessels, pipes, large machines, and other heavy structures, where good formability, weldability, and machinability are required. Because the latter properties are

Table 12-21 Minimum mechanical properties of structural quality low-carbon sheet and strip steels according to ASTM specifications.

ASTM specification No.	Grade	Tensile strength		Yield strength		Elongation in 50 mm or 2 in.,
		MPa	ksi	MPa	ksi	%
Hot-rolled sheet and strip						
A570	A	310	45	170	25	23-27
	B	340	49	205	30	21-25
	C	360	52	230	33	18-23
	D	380	55	275	40	15-21
	E	400	58	290	42	13-19
Cold-rolled sheet						
A611	A	290	42	170	25	26
	B	310	45	205	30	24
	C	330	48	230	33	22
	D	360	52	275	40	20
	E	570	82	550	80	...

impaired by high-carbon contents, most of steel plate compositions are low-carbon and medium-carbon, with the low-carbon unalloyed steel grades predominating. Alloy steel plates are also produced but sometimes they are heat-treated to achieve mechanical properties superior to the hot-finished product.

Regular quality is the most common quality of plain, carbon steel plate with a maximum of 0.33%C. Regular quality is usually ordered to standard composition ranges, is not ordinarily produced to mechanical property requirements, and is analogous to merchant quality for bars because there are no restrictions on deoxidation, grain size, check analysis, chemical uniformity, internal soundness, or freedom from surface imperfections. Applications for this quality plate are for mild cold-bending, mild hot-forming, punching, and welding for noncritical parts of machinery.

Structural and pressure vessel quality steel plates are produced to meet standard specifications prepared by specification-writing groups, of which ASTM is the most widely used in the United States. ASTM A6 covers the general requirements for the structural quality plate products described in a particular ASTM specification, while ASTM A20 discusses the general requirements for the pressure vessel quality products. Examples of structural quality ASTM specifications were cited earlier in Table 12-8. ASTM specifications for pressure vessel quality steel plates are given in Table 12-22.

Bars and Shapes. Structural bars and shapes, such as I-beams, channels, tees, and others, are used con-

Table 12-22 ASTM specifications for pressure vessel quality steel plates(a).

Specification	Steel type and condition	Specification	Steel type and condition
Carbon Steel		*Carbon Steel (continued)*	
A285(b)(c)	Carbon steel plates of low or intermediate tensile strength	A724	Quenched and tempered carbon steel plates for layered pressure vessels not subject to postweld heat treatment
A299(b)(c)	Carbon-manganese-silicon-steel plates	A738	Heat treated carbon-manganese-silicon steel plates for moderate-and lower-temperature service
A442(b)(c)	Carbon steel plates for applications requiring low transition temperature.	*Alloy Steel*	
A455(b)	Carbon-manganese steel plates of high tensile strength	A202(b)	Cr-Mn-Si alloy steel plates
A515(b)(c)	Carbon-silicon steel plates for intermediate-and higher-temperature service	A203(b)	Nickel alloy steel plates
A516(b)(c)	Carbon steel plates for moderate-and lower-temperature service	A204(b)(c)	Molybdenum alloy steel plates
A537(b)(c)	Heat-treated carbon-manganese-silicon steel plates	A225(b)	Mn-V alloy steel plates
A562	Titanium-bearing carbon steel plates for glass or diffused metallic coatings	A302(b)(c)	Mn-Mo and Mn-Mo-Ni alloy steel plates
A612(b)(c)	Carbon steel plates of high tensile strength for moderate- and lower-temperature service	A353(b)	Double normalized and tempered 9% nickel alloy steel plates for cryogenic services
A662(b)	Carbon-manganese steel plates for moderate-and lower-temperature service	A387(b)(c)	Cr-Mo alloy steel plates for elevated-temperature service
		A517(b)	Quenched and tempered alloy steel plates of high tensile strength
		A533(b)(c)	Quenched and tempered Mn-Mo and Mn-Mo-Ni alloy steel plates

Table 12-22 ASTM specifications for pressure vessel quality steel plates(a). (Continued)

Specification	Steel type and condition	Specification	Steel type and condition
Alloy Steel		*Alloy Steel (continued)*	
A542(b)(c)	Quenched and tempered Cr-Mo alloy steel plates	A736	Age-hardening low-carbon Ni-Cu-Cr-Mo-Nb alloy steel plates
A543(c)	Quenched and tempered Ni-Cr-Mo alloy steel plates		
		High-Strength Low-Alloy Steel	
A553(b)	Quenched and tempered 8% and 9% nickel alloy steel plates	A734	(See above under "Alloy Steel")
A645(b)	Specially heat treated 5% nickel alloy steel plates for low-or croygenic-temperature service	A737	HSLA steel plates for applications requiring high strength and toughness
		Ultrahigh-Strength Steel	
A658(b)	36% nickel alloy steel plates with very low coefficient of thermal expansion	A538	Precipitation-hardening (maraging) 18% nickel alloy steel plates
A734	Quenched and tempered alloy and HSLA steel plates for low-temperature service	A590	Precipitation-hardening (maraging) 12% nickel alloy steel plates
A735	Low-carbon Mn-Mo-Nb alloy steel plates for moderate- and lower-temperature service	A605	Quenched and tempered 9% nickel, 4% cobalt, Mo-Cr alloy steel plates

(a) Covered in A20. (b) This ASTM specification is also published by ASME, which adds an "S" in front of the "A" (for example. SA285). (c) Also covered in A647, which specifies the special requirements for pressure vessel quality steel plate for nuclear applications.

siderably in the construction industry. *Merchant quality bars* are selected for noncritical parts of bridges, buildings, ships, agricultural implements, road-building equipment, railway equipment, and general machinery. These are applications that require only mild cold-bending, mild hot-forming, punching, and welding. Although they should be free from visible shrinkage pipe, they may contain pronounced chemical segregation, internal porosity, surface seams, and other surface irregularities. Thus, merchant quality bars are not suitable for forging, heat-treating, or other applications in which internal soundness or relative freedom from surface imperfections is a significant factor.

Special quality bars are selected when end use, method of fabrication, or subsequent processing require restrictions of more uniform composition and much less (minimum) surface defects than merchant quality bars. These applications, which include many structural uses, may require hot-forging, heat-treating, cold-drawing, cold-forming, or machining. Some end use may require one or more of the following special requirements: special hardenability, internal soundness, nonmetallic inclusion rating, and surface condition. The letter "A" is used with the quality descriptor when one or more of these are required, for example, cold-forging quality A. When a single restriction, other than the four special ones mentioned here, the letter "B" is added to the descriptor. Cold extrusion quality B, for example, is used for hot-rolled bars used to produce solid or hollow shapes by severe cold extrusion. For very severe cold-plastic deformation, the letter "D" is used.

Structural quality bars are materials with certain mechanical property requirements set by specifications. As with hot-rolled plates, ASTM A6 describes these requirements for each of the specifications in Table 12-22.

Concrete reinforcing bars are classified separately from merchant quality and special quality bars. They are available as either plain rounds or deformed rounds. Deformed reinforcing bars are used almost exclusively in the construction industry to provide tensile and compressive strengths to concrete structures. The surface of deformed bars has lugs or protrusions, which are hot-formed in the final roll pass between rolls (with patterns cut in them), to inhibit movement relative to the surrounding concrete. Plain reinforcing bars are used more often for dowels, spirals, and structural ties and supports than as substitutes for deformed bars.

Concrete reinforcing bars are produced according to ASTM specifications A615, A706, or A722. A615 is for plain carbon steel that is adequate for most reinforcing bar applications. A706 is for a high strength low-alloy steel that is intended for applications requiring a combination of strength, weldability, ductility, and improved bending properties. A722 covers deformed and plain uncoated high-strength steel bars for prestressing concrete. Concrete reinforcing bars are also available rolled from railroad rails (ASTM A616) and from axles for railroad cars (A617).

Regardless of product category, the tensile properties of as-hot-rolled carbon steel bars are influenced mainly by chemical composition, thickness or cross-sectional area, and variables in roll mill design and practice. For carbon and HSLA steels, carbon content is the dominant factor; carbon and manganese are used to produce steels with pearlitic microstructures and the required tensile strength. The carbon content is increased to obtain high strength and the minimum expected yield. The tensile strengths of commonly used grades of hot-rolled carbon steel bars are shown in Fig. 12-3.

The problem with carbon and HSLA steels that depend on carbon and manganese to attain strength is the accompanying deterioration in toughness properties and weldability with increasing carbon content, as shown in Fig. 12-4 for impact toughness. Here is where the new *microalloyed HSLA steels* outshine the carbon and HSLA steels because their increased strength is attained through simultaneous improvement in toughness and weldability. For this reason, they are referred to as *high-performance structural steels*. The high performance attributes of microalloyed HSLA

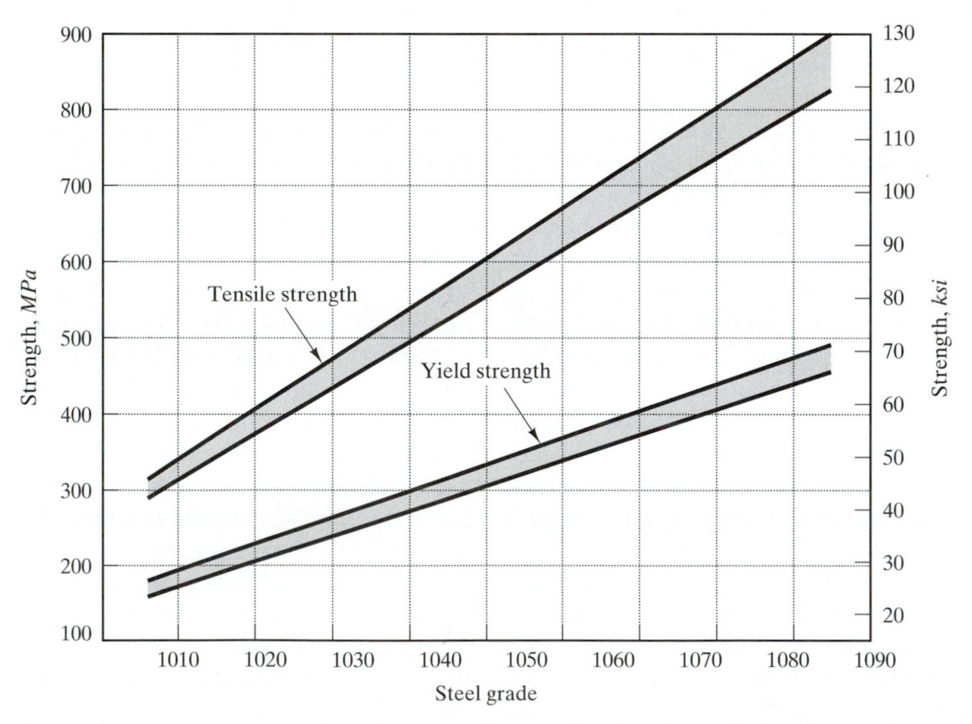

Fig. 12-3 Minimum expected yield and tensile strengths of as–hot-rolled carbon steel bars.

steels arise from their low carbon content (generally $< 0.10\%C$), and very fine ferrite grain size.

Cold-Finished Bars. Any grade of carbon or alloy steels that can be hot-rolled can also be cold-finished. Cold-finished bars (round, square, hexagonal, flat, or special shapes) are cold-finished either by machining (turning or grinding) or by cold-drawing through a die. Cold-finished bars fall into four categories: 1) cold-drawn bars; 2) turned and polished bars; 3) cold-drawn, ground, and polished; and 4) turned, ground, and polished. The differences in finishing by machining versus cold-drawing may involve applicability, surface characteristics, cost, and mechanical properties.

For applicability, it is obvious that turning and centerless grinding are limited to rounds. We see, therefore, that cold-drawing is a much broader process applicable to all product forms. However, steels with greater than 0.55%C are seldom cold-drawn because they must be annealed first before cold-drawing, which increases cost; their mechanical properties, after annealing and cold-drawing are somewhat lower than those of lower carbon grades cold-drawn from hot-rolled stock; and there is little advantage in machinability of the high-carbon annealed and cold-drawn steel over the lower, carbon hot-rolled and cold-drawn steel. Furthermore, only bars up to a certain size can be drawn and larger bars must be cold-finished by machining.

There is also a difference in the number and severity of the surface imperfections that may be present

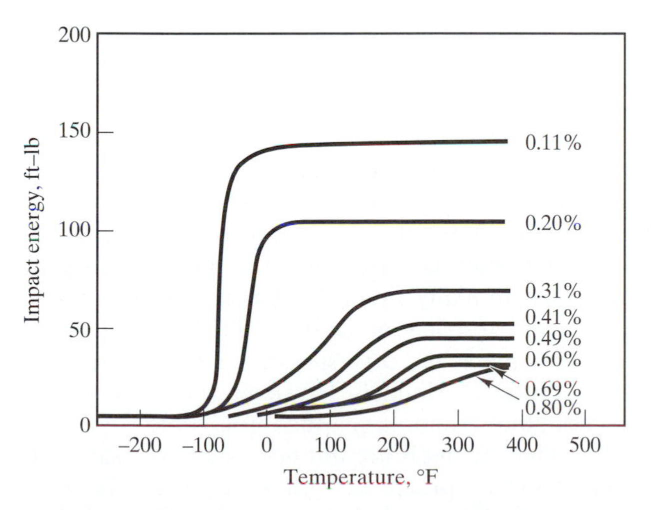

Fig. 12-4 Effect of carbon content on Charpy V-notch impact properties of normalized ferrite-pearlite steels.

in machined and cold-drawn bars. Because stock is removed in turning and grinding, shallow imperfections and decarburization after hot-rolling may be removed after machining. When a material is drawn, stock is only displaced, and surface imperfections are only reduced in depth. However, the length of these imperfections may be increased after cold-drawing.

The cost for turning and grinding operations is more than that for cold-drawing. For example, a typical price extra (in 1978) for turning, grinding, and polishing round bars of a nonresulfurized bar of the same size is $10.50 per 100 lbs., compare with $7.50 for cold-drawing, grinding, and polishing. Thus, we see that cold-drawing has a cost advantage over machining.

The yield and tensile strength properties of cold-drawn finished bars are greater than the machined finish bars. The machined finished bars retain the tensile properties of the hot-rolled bars. Within the framework of commercial drafts, the yield and tensile strengths of cold-drawn bars increase dramatically during the first 15 percent reduction in area (cold-work) as shown in Fig. 12-5 for bars less than 25 mm (1 in.) in cross section with less than 690 MPa (100 ksi) tensile strength before cold-drawing. The yield strength increases more than the tensile strength; therefore, the yield strength to tensile strength ratio increases after cold-drawing.

12.3 Procurement and Specifications

A materials procurer usually writes a summary of the company's needs to the material supplier or vendor by writing a specification. A *specification* is a written document of the technical and other requirements that a product must meet. When a purchaser orders a material according to a specification, and the vendor or supplier accepts the specification, the document becomes a part of the contract to purchase (buyer) and to sell (supplier or vendor) and becomes binding. An adequate specification may contain the following items:

1. The *scope* may cover the product classification, including the size range when necessary, the condition, and any comments on product processing that are deemed helpful to either the supplier or purchaser. The classification of the product

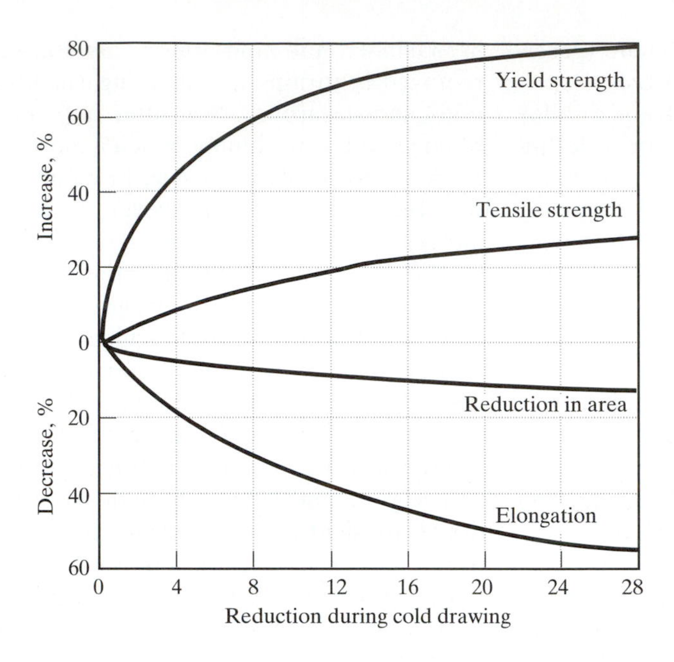

Fig. 12-5 Effect of cold-drawing on the tensile properties of steel bars (originally less than 25 mm cross section and tensile strength less than 690 MPa).

according to one of those discussed in Sec. 12.2 will be very helpful.

2. The *chemical composition* may be specifically mentioned or it may be indicated by a well-recognized designation (e.g., AISI/SAE) or specification (e.g., ASTM) based on chemical composition.
3. The *quality* of the product being sought may be indicated by one of the quality descriptors indicated in Table 12-20 along with whatever necessary additional requirements, which may include the type of processing.
4. *Quantitative* requirements may be indicated such as the allowable ranges of the composition and all the necessary physical and mechanical properties that characterize the material. The necessary test methods to determine these properties may have to be specified by reference to standard test methods.
5. Any other requirement may become part of a specification. Some examples are: special tolerances, surface preparation, edge and finish on flat-rolled products, packaging, loading, and transport conditions.

Because the specification becomes a binding document, the materials procurer must consult with the materials supplier before it is finalized. The obvious reason is cost. If a specification is very much different from the suppliers' normal processing practices, the materials specification can only be met if the purchaser is willing to pay for changes in the normal processing practices. In addition to cost, delivery time for the product might be delayed. Thus, for convenience and economy, the customers' specifications are finalized with assurance from the materials' supplier so that they can be met without much penalty in cost and time. For this reason, it is not unusual that the standard products of the suppliers are the same as the customers, with the latter giving their own codes or number. On occasion, the customer and vendor groups agree to use the same materials' specification. An example of this is the SAE-AISI steel composition designation that we have discussed.

We must also realize that the ASTM codes are specifications, while the AISI-SAE designation system is just a part of the specification. The ASTM specifications represent the consensus among producers, specifiers, fabricators, and users of steel products and may be used for the procurement of materials. In many cases, the dimensions, tolerances, limits, and restrictions in the ASTM specifications are the same as the corresponding items in the AISI Steel Products Manuals. However, there are differences in the terminology—the terms "type", "grade", and "class" of steels are not necessarily the same. In steel industry practices, "grade" denotes chemical composition; "type" indicates the deoxidation practice, for instances, killed, semikilled, or rimmed; and "class" indicates other attributes, such as strength or surface finish. In ASTM specifications, however, the terms are used interchangeably. In ASTM A533, for example, "type" denotes chemical composition, while "class" indicates the strength level. In ASTM A515, "grade" signifies the strength level, with the permissible maximum carbon content being dependent on both the strength level and the plate thickness. In ASTM A302, "grade" connotes the requirements for both the chemical composition and the mechanical properties. A materials' procurer must be aware of these differences to communicate precisely what a company needs.

12.4 Selection of Carbon and Low-Alloy Structural Steels [4a]

Steels used for civil engineering structures are predominantly the low-carbon plain carbon and low-carbon low alloy steels. The alloy additions in the latter steels are made with the specific purpose of increasing corrosion resistance or strength. The low-carbon plain carbon steels are commonly called *mild steels.* This class of steels are the materials used in the construction industries for roads, buildings, bridges, pipelines, and ships. They may be in the form of *bars, plates and strips, and various structural shapes.* For this reason, they are referred to as *structural steels* and are predominantly *hot-rolled steels.* In addition, low-carbon, cold-rolled and annealed *sheet steels* are used in the *automotive, appliance, transportation, and food industries.* The total steel tonnages used in the above applications are estimated to be upwards to 80 percent of all steel shipments. Procurement of these steels are done usually, but not exclusively, by referring to the ASTM specifications; some of these specifications have already been discussed in Sec. 12.2.

The safety, reliability, and economy of a structure depends on its design, its fabrication, its inspection, its maintenance and on materials selection. The most efficient and economical structures are based on the optimized trade-offs between these parameters. The use of the best material in combination with a deficient design or fabrication or both, or the use of the best design in combination with deficient materials or fabrication or both, does not yield a safe, reliable, or economical structure. It is estimated in 1995 that a one percent improvement in the performance and durability of the nation's (U.S.) infrastructures (i.e, roads and bridges) can result in about $30 billion in savings. [4a]

The attributes of a material that are important in the design of structures are: its modulus, yield strength, yield strength to tensile strength ratio (yield-tensile ratio), ductility, fracture toughness, fatigue resistance, corrosion resistance, fabricability that may involve forming, welding and machining, and availability and cost. The importance of one property or a combination of these properties to the safety and reliability of the structure depends on the design and the loading conditions. In static applications, such as storage tanks, pipes, pressure vessels, and buildings, the yield strength and the fracture toughness are important considerations but fatigue is rarely considered in the design. In contrast, fatigue is the primary consideration in the design of bridges because of cyclic loads and stresses.

The increase in strength of a material may allow the use of thinner gauges with some benefits. A thinner gauge means less weight that makes a component easier to transport. When used in a transport vehicle, it saves energy and becomes more environmentally friendly. Automobiles built since the early 1970s are much lighter and have much lower gasoline consumption than their predecessors because of high-strength steels. The thinner gauge also allows easier fabrication by welding and induces better properties of the weld joint. The thinner gauge allows the weld joint to cool slower and in terms of thickness constraint, the thinner gauge is less conducive to plane strain or brittle fracture of the weld joint. This assumes, of course, that the steels are equally weldable and that the quality of weld is the same. In some of these structures, such as storage tanks, welded water pipes, and similar structures fabricated from thin (usually less than 0.75 in. thick) plates, strips, or sheets, steels with yield strengths higher than 50 ksi are, however, not advisable. The use of thinner gauge steels, with higher yield strengths than the minimum 50 ksi or less steels that are currently being used, may cause unacceptable distortion and buckling.

One of the important current requirements of structural steels for bridges and structures subjected to large earthquake forces is a low, yield strength-to-tensile strength ratio, commonly called the yield-tensile ratio. The current AASHTO (American Association of State Highway and Transportion

Officials) requirement is for the yield-tensile ratio to be less than 0.85. Civil engineers relate this requirement to what is called the structural ductility of structural members. The structural ductility of structural members is defined as the ability to absorb energy and deform plastically (inelastically) while maintaining the load-carrying capacity. The ability to absorb energy is very important to the safe and economical design of structures subjected to large earthquake forces. A structurally ductile member provides sufficient warning of impending collapse under gravity or lateral forces. Also for plastic design of steel members in bridges, structural ductility is required for the complete plastic hinge mechanism to form.

The performance of welded components may wipe out the advantage that may arise from another property. For example, the maximum allowable design stress in fatigue may be increased in proportion to an increase in the tensile strength of the base material when using high-strength steels. However, extensive fatigue tests on welded components in bridge structures show that steel properties do not affect their fatigue life. The fatigue properties of welded components were found to depend only on the stress fluctuation (i.e., the stress range), on the size and shape of weld imperfections, and on the geometry of the welded component (see Fig. 16-4). Thus, high-strength steels have no particular advantage over low-strength steels because the stress range cannot be increased.

From the foregoing discussion, we can conclude that the selection of structural materials depends not only on the base material but also on the configuration and properties of the welded or fabricated component. The properties of a welded component depend to a large extent on the weldability of the steel. In addition to weldability, formability and machinability are also considered when a material is selected because it might have to be formed or machined when a structure or a component is fabricated. Thus, we need to know the attributes of a material that will determine the fabricability and the subsequent performance of the component or structure.

12.4.1 Weldability, Formability, and Machinability of Carbon and Low-Alloy Steels

Weldability. The weldability of a material is a measure of the ease of forming a sound, strong weld joint. A sound joint means that the weld does not have defects or imperfections (such as cracks) in it and a strong joint means that the weld must be at least as strong as the base material.

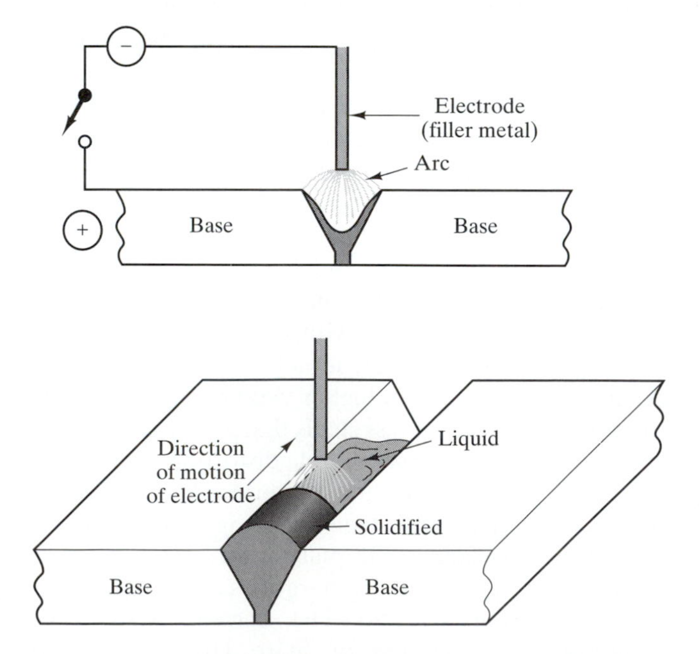

Fig. 12-6 Schematic of fusion welding process—electric arc melts base and filler metal; base materials are fused together when melt solidifies.

The *predominant joining process is fusion welding* as schematically illustrated in Fig. 12-6 (previous page). The edges of the base materials to be joined are machined in such a way as to provide a trough for the liquid metal puddle that forms during the process. The machined edges form a trough (a "V" groove) when the base materials are brought close together. Fusion of the base materials is accomplished by the electric arc produced when the negative electrode (filler metal) is brought close to the workpieces (base materials). The electrode is also called the filler metal because it provides the material to fill the V-groove. The base materials and the filler metal melt to form the liquid puddle or pool that subsequently solidifies to fuse the base materials. The electrode is moved along the length of the "V"-groove at a uniform velocity and the base materials in front of the electrode melt, as the liquid pool behind the electrode solidifies.

It should be obvious that the temperature produced by the electric arc is sufficiently high to melt the base material and the filler metal. The heat is eventually transferred from the melt or liquid pool to the base material and a temperature profile, illustrated in Fig. 12-7a, is produced in the base material. The result of this temperature profile in the base material is to produce different microstructures after cooling, which are schematically illustrated in Fig. 12-7c, corresponding to the different ranges of temperatures in the profile and to their location in the accompanying equilibrium diagram, Fig. 12-7b. Figure 12-7b uses a plain carbon steel with 0.15%C composition as an example and the vertical dotted line shows the ranges of temperature where different

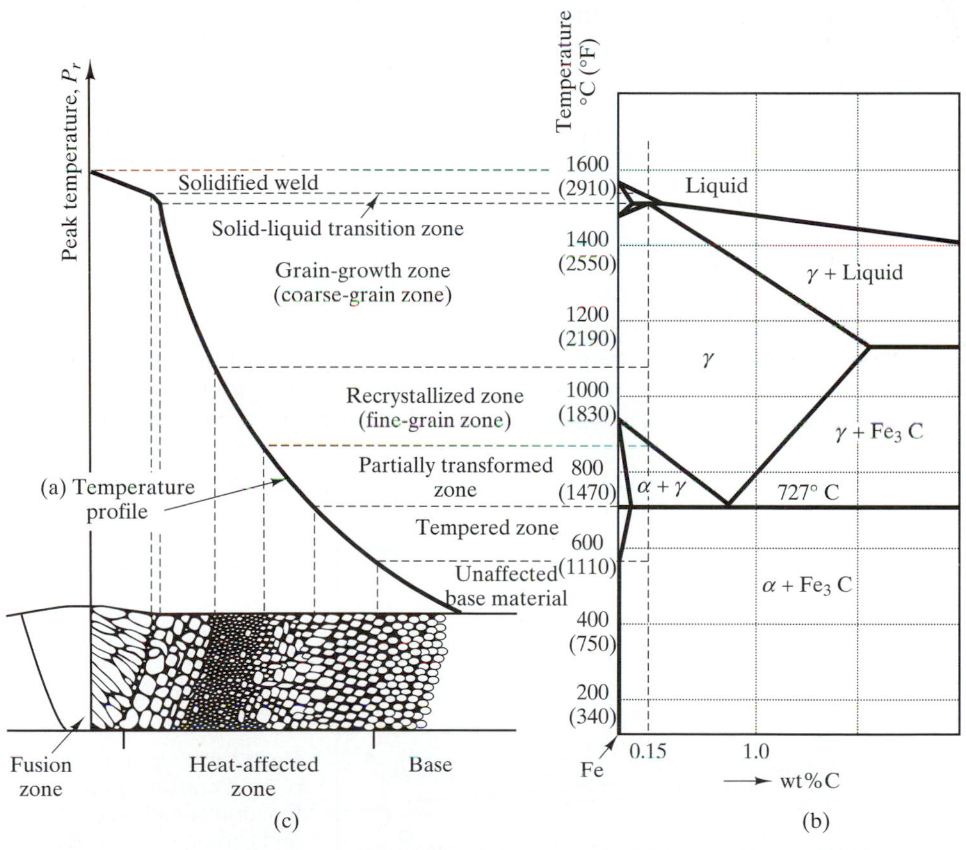

Fig. 12-7 (a) Schematic temperature profile in base material showing the ranges of temperature where different phases were obtained [from phase diagram (b)] and transformed to the different microstructures in (c) for a 0.15%C steel.

phases are located. With Fig. 12-7b, the temperature ranges in the temperature profile show the phases at high temperatures that transform to the phases shown in Fig. 12-7c at room temperature, depending on the cooling rate. Figure 12-7c illustrates the three zones typical in a weld joint: (1) the fusion or weld pool area, (2) the heat-affected zone (HAZ) that consists of areas in the material that attained temperatures below the liquidus temperature in Fig. 12-7b to the subcritical temperatures, where annealing, aging, or tempering may occur, and (3) the unaffected base material whose temperature is below the stress relief temperature.

The weldability of steels is primarily influenced by the composition, the heat input, and the rate of cooling; all three factors relate to the hardenability of the HAZ region, the most critical region. The hardenability depends on the amount of carbon and alloying elements that are brought into solution in austenite during the heating of the base material. The composition is commonly expressed as carbon equivalent, C.E., such as

$$CE = C + \frac{Mn}{6} + \frac{Cr + Mo + V + Cb}{5} + \frac{Ni + Cu}{15} \tag{12-1}$$

The symbols on the right side of Eq. 12-1 signify the compositions in weight percent of the elements. The ease or difficulty of forming a sound weld is indicated in Fig. 12-8 which shows the three regions for the susceptibility of cracking in the HAZ in a carbon equivalent versus carbon content plot. The susceptibility of HAZ cracking relates to the hardness of the martensite that forms during the cooling of the plate. As we learned in Chap. 9, the carbon content of martensite (same as austenite) is the sole contributor to the hardness of martensite and Fig. 12-9 shows the effect of carbon content on the sensitivity of the weld to crack because of the brittle martensite that forms.

The formation of martensite is a function of both composition and cooling rate. High carbon content or carbon equivalent makes it easier to form martensite as illustrated in Fig. 12-9. In welding, the cooling rate depends on the heat input and the section size (thickness) of the base steels. The heat input is given by

$$H = \frac{E \times I \times 60}{S} \tag{12-2}$$

where H is the heat input in Joules per inch, E is the arc voltage in volts, I is the arc current in amperes, and S is the arc (electrode) travel speed in inches per minute. A high-energy heat input means the cooling rate of the weld will be slower for the same thickness of steel. For the same heat input, the thicker material represents a greater mass of material to extract the heat and the cooling rate will increase. In order to avoid HAZ cracking, we must attempt to lower the cooling rate of the weld below the critical cooling rate to form the martensite in the steel. Low-carbon and low-alloy steels have very high critical cooling rates and will not form hard brittle martensites at the HAZ. They require no precautions during welding, Zone I in Fig. 12-8, except for the selection of the proper filler metal to pro-

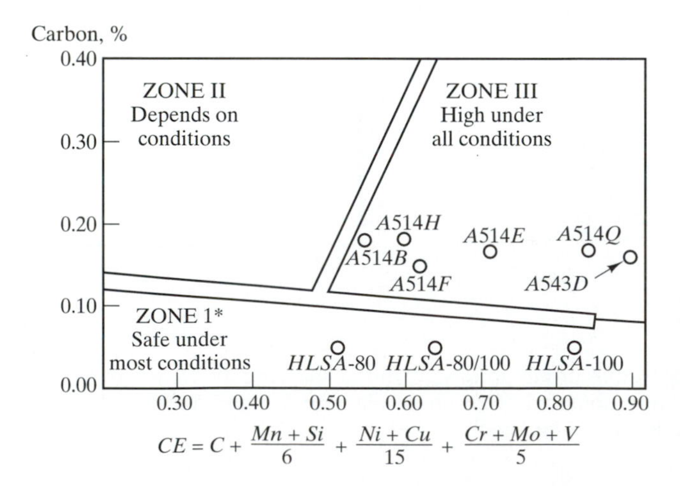

Fig. 12-8 Susceptibility of HAZ cracking as a function of carbon content and carbon equivalent of high-strength steels. (*High-strength filler metals may require additional care). [4b]

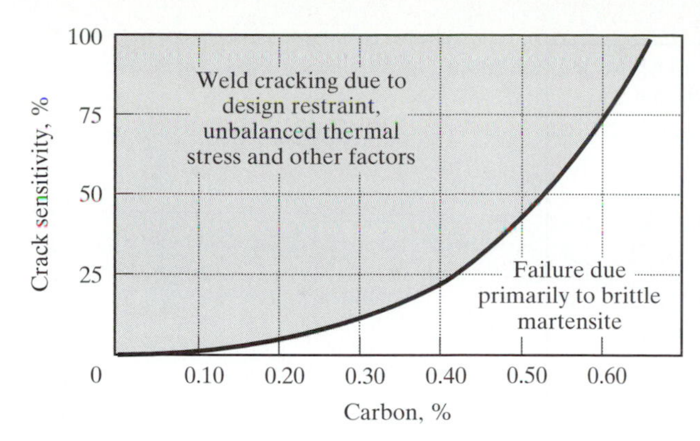

Fig. 12-9 Crack sensitivity increases with carbon content of brittle martensite in steel.

duce a strong weld. For higher carbon content and carbon equivalent, it is easier to form martensite, and the cooling rate of the weld joint must be reduced below the critical cooling rate to avoid forming martensite. The cooling rate can be reduced by preheating the base materials (work-pieces), Zone II in Fig. 12-8. For much higher carbon content and carbon equivalent when martensite formation is unavoidable, the weld section may have to be post-weld heated to reduce the susceptibility of the martensite to crack, Zone III in Fig. 12-8.

It is generally known that the "weak link" in a welded structure is the weld joint. Thus, it is important that the correct materials and process selection be made to produce a sound and strong joint.

Formability. In terms of formability, we consider sheet and bulk formability or workability. Sheet formability is the ability of the sheet steel to be stretched or drawn, as discussed in Chap. 8. In both stretching and drawing, the steel must have good ductility, plastic anisotropy ratio, $r > 1.0$, and a high strain-hardening exponent, n. Bulk formability, or workability, is the relative ease with which metal can be shaped through the deformation processes of forging, extrusion, or rolling. In the broadest sense, the sheet and bulk formability are related only in so far as both processes require good ductility of the material. However, while tensile and shear stresses are predominantly involved in drawing and stretching (sheet formability), compressive stresses are involved in bulk formability. Thus, most materials essentially exhibit similar bulk formability characteristics, except for free machining steels that have a lot of nonmetallic sulfide inclusions purposely added to increase their machinability. In general, the bulk formability of carbon and alloy steels improves as the deformation rate increases. The improvement is attributed to the increased heat generated at high deformation rates.

Selection of sheet steel involves understanding the available grades (quality descriptors) of commercial quality, drawing quality, and drawing quality, special-killed. The choice of a material for forming into a particular part depends on many factors; it is usually done in consultation with the technical staff of the suppliers or steel producers. Choice of sheet steel may also be helped by the forming limit diagram of the material and a circle grid analysis. An example of the forming limit diagram (FLD) is shown in Fig. 12-10 and is usually supplied by the steel producer or supplier for their product. The material's FLD indicates failure when strains are at the critical zone and above and satisfactory performance below the critical zone. In the circle grid analysis that is used to yield the FLD, a blank sheet is gridded with uniform sized and spaced circles before undergoing deformation. The circles become ellipses after the deformation and stretch is represented by positive minor strain, plane strain by zero minor strain, and draw by negative minor strain. The terms major and minor refer to the major and minor axes of the ellipses. A circle grid analysis of the product to be formed will indicate whether better uniform elongation (for stretch) or higher r-value (for draw) is needed when breakage strains occur on the right or left, respectively, of the FLD.

Machinability. The term machinability is used to indicate the ease or difficulty with which a material can be machined to the size, shape, and desired surface finish. The terms used to indicate machinability of steels under specified conditions are the machinability index and machinability ratings. These are qualitative ratings with no clear-cut or unambiguous meanings and no standard or universally accepted method to measure them. Qualitative judgments are based on one or more of the following criteria:

- *Tool Life:* The amount of material that can be removed by a standard cutting tool under standard cutting conditions before tool performance becomes unacceptable or tool wear reaches a specified amount.

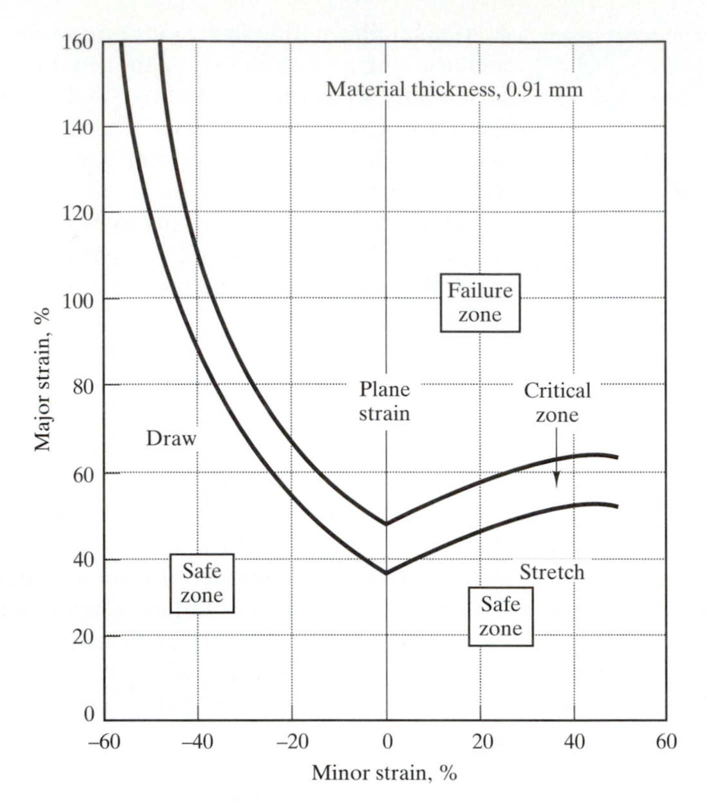

Fig. 12-10 Schematic of forming limit diagram (FLD). Strains in critical zone and above it will result in excessive breakage. Drawing requires high *r* values while stretch requires high *n* values or high uniform elongation. [1, 2]

- *Cutting Speed:* The maximum speed at which a standard tool under standard conditions can continue to provide satisfactory performance for a specified period.
- *Power Consumption:* The power required to remove a unit volume of material under specified machining conditions.
- *Comparisons* with a standard steel based on experience in machine shops.
- *Quality* of surface finish.
- *Feeds* resulting from a constant thrust force.

The most common criteria used are the tool lives and the cutting speeds. Materials with superior machinability increase tool lives at equal cutting speeds or permit higher cutting speeds while maintaining equal tool lives. Either better tool lives or higher cutting speeds increases productivity and lowers machining costs.

Most machinability ratings are based on the performance of steels in one type of operation (usually turning) and with one type of cutting tool. Consequently, it is important to determine if the machinability rating of a material using high-speed tool steels (HSS) correlates with the machinability rating using other tools such as carbide or coated-carbide tools. Two correlations are shown in Fig. 12-11 where the cutting speeds for coated carbide tools, and the cutting speeds for HSS tools are plotted as functions of the cutting speeds of indexable-carbide tools in m/min. The statistical analysis of the two correlations indicates very high correlation coefficients and indicates that a machinability rating of a material using one tool has a similar rating using other cutting tools. The materials (workpieces) used for the correlations are three types of stainless steels, 14 grades of construction steel, and 5 types of cast irons.

There is also a correlation of machinability ratings obtained with different machining processes involving the same 22 types of ferrous materials used in Fig. 12-11. Figure 12-12 shows the correlation between turning and boring using a cut depth of 1.0 mm (0.04 mil) and indicates a one-to-one correlation. The machinability rating is 100 times the ratio of the rec-

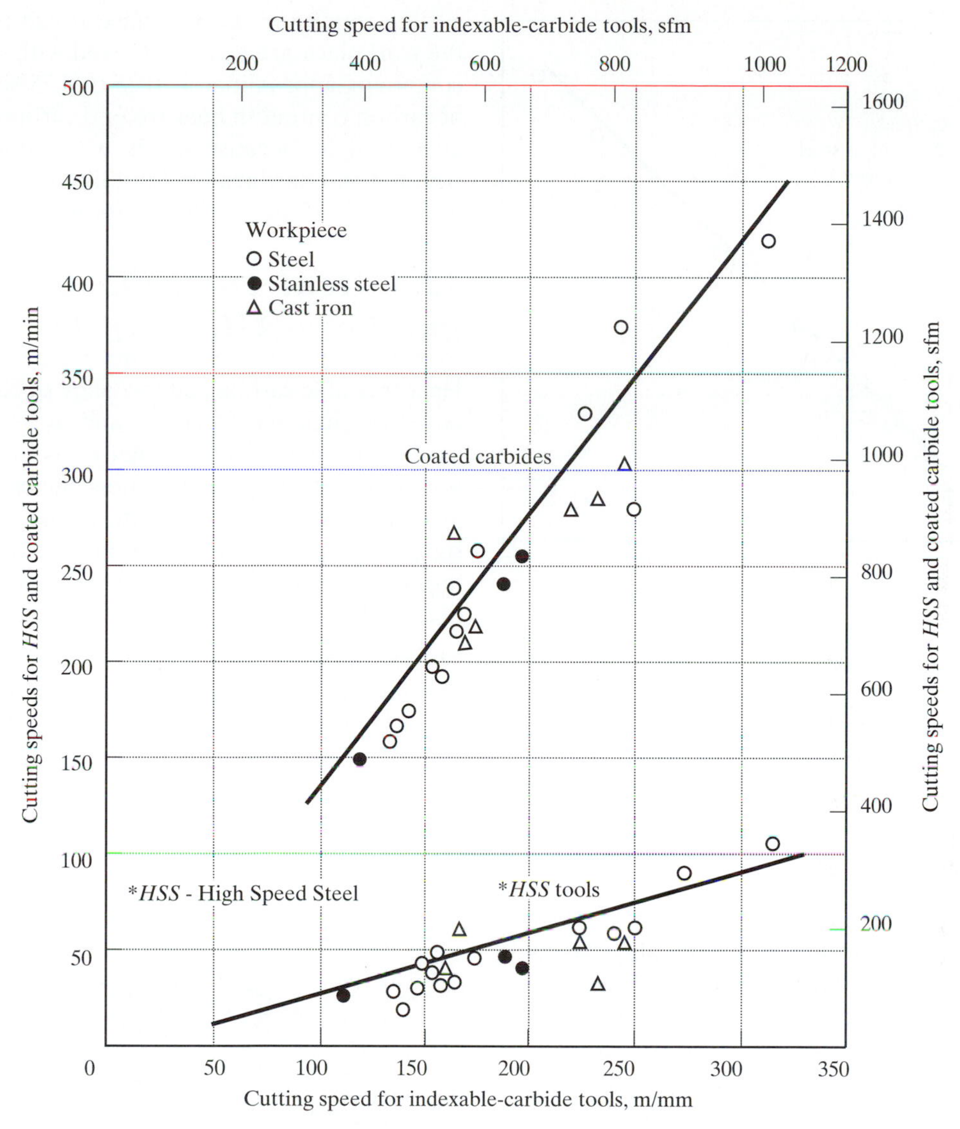

Fig. 12-11 Correlations of cutting speed machinability ratings with different tools on the same workpieces. [1, 2]

ommended cutting speed for the material to that for 1212 steel for comparable cutting conditions—the 1212 steel has a hardness level between 150 to 200 HB (Brinell). There are other data that indicates, that the turning operation correlates well also with reaming, face milling, and end milling, but not very well with drilling.

The machinability of carbon and alloy steels depends on many factors that include the composition, microstructure, the dispersion of second-phase particles, the mechanical properties, such as strength and hardness, and the physical properties, such as thermal conductivity. Almost always, carbon steels of comparable carbon content and hardness have better machinability than alloy steels. The exception to this are alloy steels that are quenched and tempered to hardness levels greater than 300 HB that exhibit superior machinability than carbon steels. Machinability ratings are available and can be consulted in the selection of materials. Again these

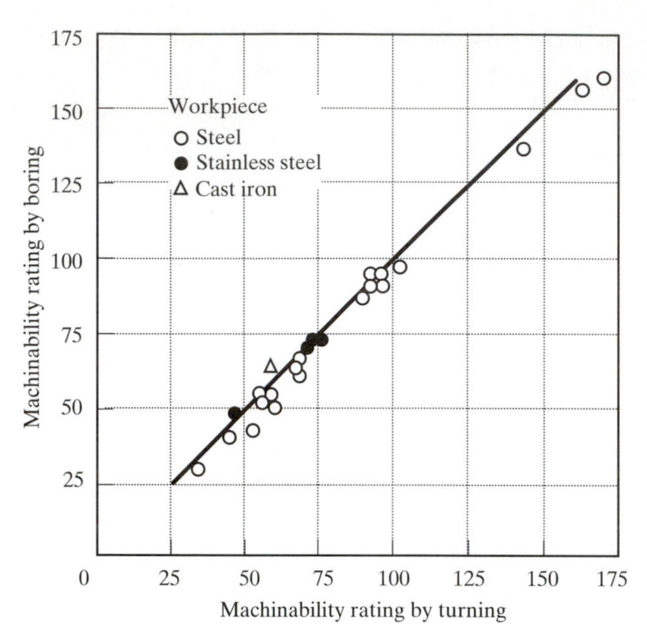

Fig. 12-12 Correlation of machinability ratings between two machining operations. [1, 2]

ratings are based on the percentage of the cutting speed for AISI 1212 steel.

Low-carbon steels containing less than 0.15%C are low in strength in the annealed condition and machine poorly because they are soft and gummy and adhere to the cutting tool. The machinability of these steels can be improved by cold-working to raise the strength and lower the ductility. Steels in the 0.15 to 0.30%C range machine satisfactorily in the as-rolled, as-forged, annealed, or normalized condition with predominantly pearlitic structure. The medium-carbon steels up to 0.55%C machine best when annealed to produce a mixture of lamellar pearlite and spheroidized carbide. If not partially spheroidized, the strength and hardness may be too high for optimum machinability. For steels with higher than 0.55%C, a complete spheroidization treatment is preferred. Quenched and tempered structures are generally not desired for machining.

Both tool life and production rate are adversely affected by increases in carbon content. To minimize tool wear and maximize production rate, selection of steels should be made with the lowest carbon content consistent with the mechanical property requirements. Carbon content also affects surface finish in machining, although its effect can be greatly modified by the nature of the cutting operation or by the cutting conditions. Low-surface roughness values of the workpiece are easily achieved with carbon steels containing approximately 0.25 to 0.35%C. The effect of carbon content in heat-treated carbon steels is relatively slight because steels with about the same strength and hardness will not likely have more than 0.10%C difference in composition.

12.5 Heat-treatable Carbon and Low-Alloy Steels

Heat-treatable carbon and low-alloy steels are selected based on their hardenability and are predominantly used in the quenched (hardened) and tempered condition. In these steels, the hardness, the carbon content, and the percent of martensite in the structure play significant roles in the selection process. The following empirical correlations, which we learned earlier in Sec. 9.7, have been observed for these steels:

1. The hardness of the as-quenched steel depends only on the carbon content and the percent of martensite, as shown in Fig. 12-13. The alloy con-

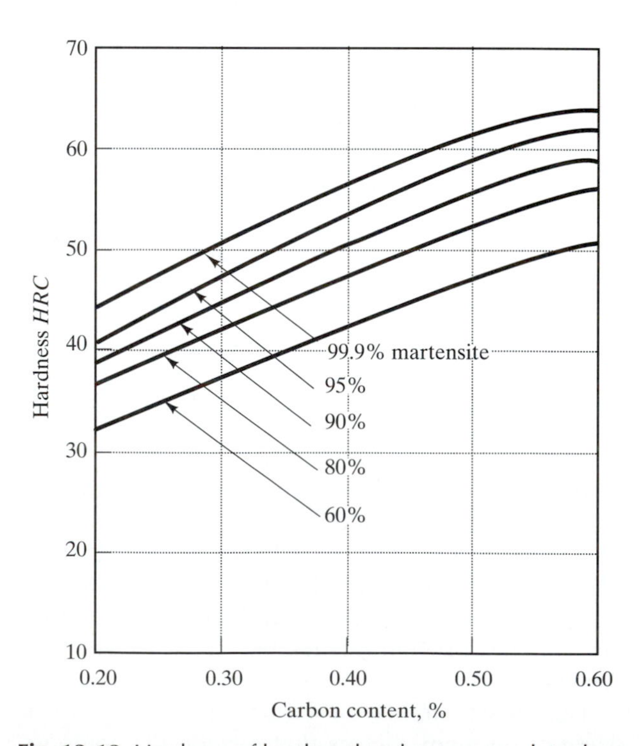

Fig. 12-13 Hardness of hardened and untempered steels as function of carbon content and amount as-quenched martensite. [1,2]

tent of the steel has no influence on the as-quenched hardness.

2. The alloy content increases the hardenability of the steel (i.e., martensite will form or the steel hardens to a greater depth from the surface). In terms of size, the ideal critical diameter, D_I, increases, which means larger diameters can be hardened through. The alloy content also allows the formation of martensite at a much less drastic quench and induces less residual quenching stresses and distortion during the quenching process.
3. The quenched and tempered structure exhibits the best combination of strength and toughness properties. When carbon and low-alloy steels are quenched and tempered to the same hardness up to 500 HB (~50 HRC), they essentially have the same yield and tensile strengths, as shown in Fig. 12-14a and b. However, they exhibit different ductilities in terms of percent reduction in areas, as shown in Fig. 12-15, and the alloy steels have to be tempered at higher temperatures to produce the same hardness as the carbon steels. The quenched and tempered hardness is about 5 to 6 HRC units less than the as-quenched martensite structure, as seen in Fig. 12-16.
4. Because the fatigue limit is about one-half the tensile strength, the fatigue limits of these steels are also related to the hardness—an example of this correlation is shown in Fig. 12-17. The fatigue limit is influenced as well by the martensite content in the microstructure. Figure 12-18 shows that for fatigue applications, a 100 or 99.9% quenched and tempered martensite is the recommended microstructure. Figure 12-19 shows also the effect of the size of inclusions on the fatigue performance of steels. From this result, we can expect that "cleaner" steels will perform even better and, in fact, this is the case. Thus, for fatigue we might have to order desulfurization or vacuum refining of the liquid melt as discussed in Chap. 7.

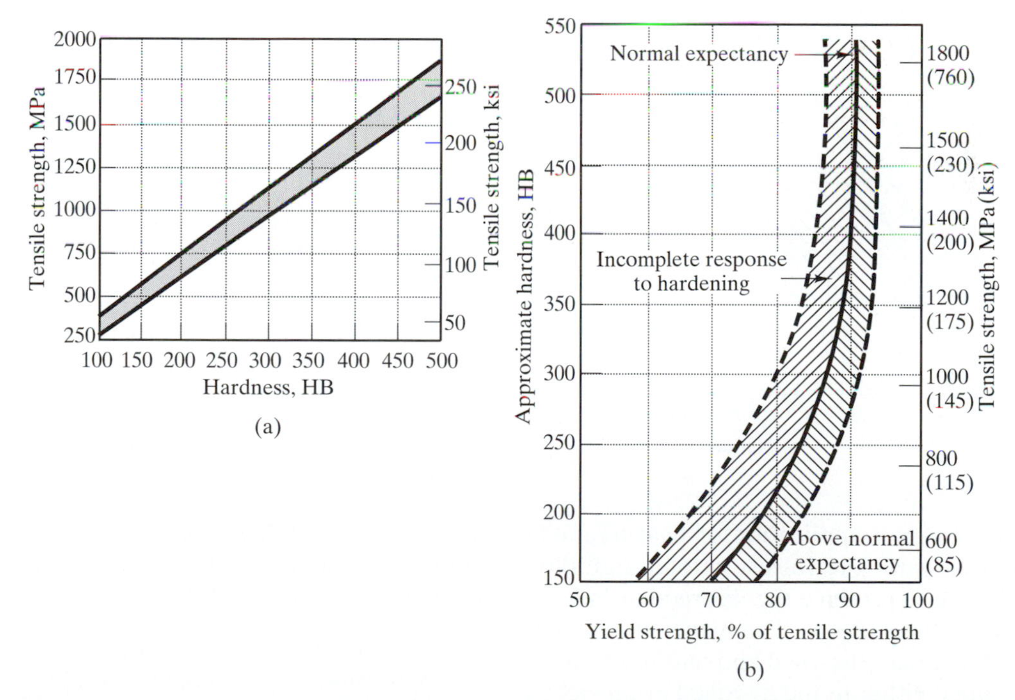

Fig. 12-14 (a) Relation of yield and tensile strengths for quenched and tempered steels. (b) Tensile strength of as-rolled, normalized, or quenched and tempered steels as function of Brinell hardness. [1, 2]

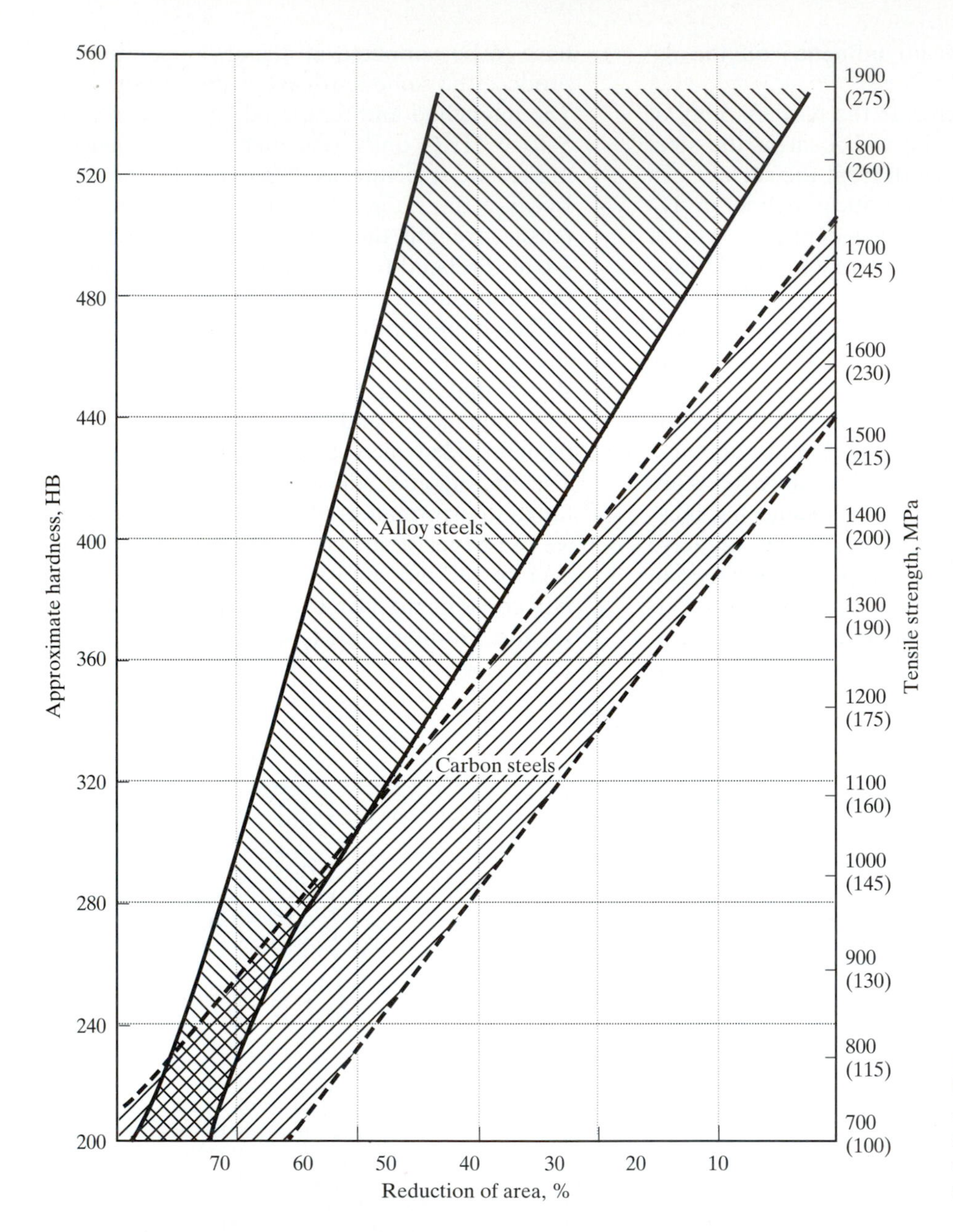

Fig. 12-15 Correlation between tensile strength and reduction in area for quenched and tempered carbon and low-alloy steels. [1, 2]

The choice for any application of heat-treatable steels must start with the lowest carbon and alloy contents possible to produce the desired hardness at the locations in the sections. The higher the carbon content, the higher are the yield and tensile strengths that are obtained either in the as-rolled or quenched and tempered condition. However, higher carbon contents (greater than 0.35%C) have a greater tendency for quench cracking and make fabrication of the parts by welding more difficult. As a rule of thumb, the major uses for various carbon levels in steels are the following: [3]

- up to 0.20%C is very successful for welding applications or for maximum toughness.
- 0.30%C is suitable for a combination of hardness (for wear) and toughness. Scarifier points and cold chisels are two typical applications.

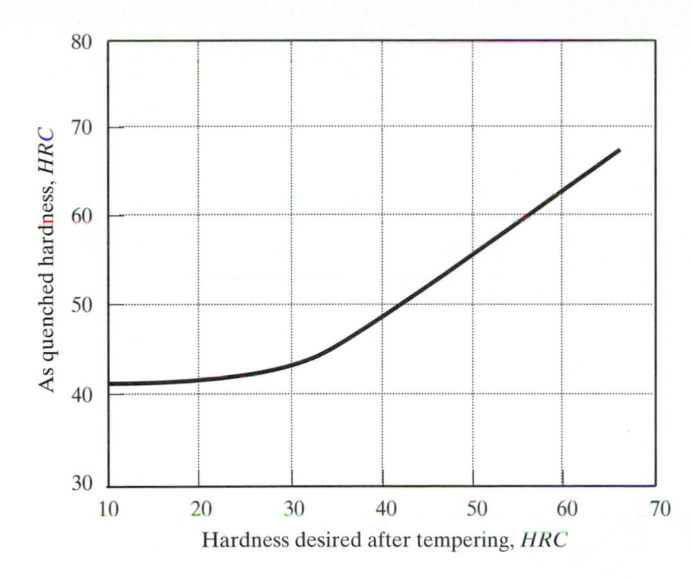

Fig. 12-16 Minimum as-quenched hardness as function of desired final tempered hardness. [1, 2]

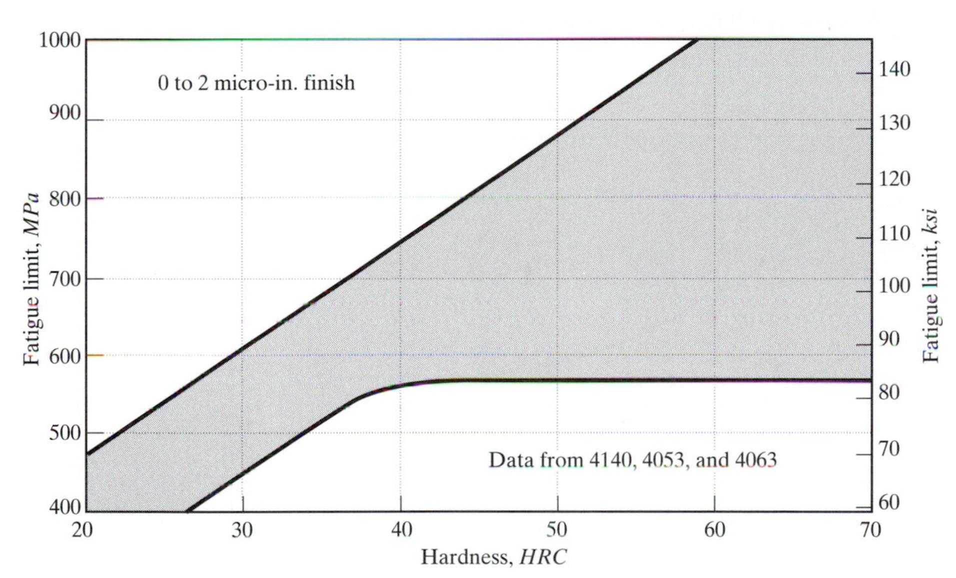

Fig. 12-17 Fatigue limit of through-hardened steels as function of hardness. [1, 2]

- 0.40%C is generally used for high-strength applications.
- 0.50%C is best for high-strength and wear resistant applications.
- 0.60%C is popular for heat-treated springs.
- greater than 0.60%C are for specialized applications such as tools and bearing races.

12.5.1 Carbon Steels

For the plain carbon steels, 10XX and 15XX, the low carbon grades, i.e., less than 0.25%C content, are very shallow hardening steels (hardening with small depths from surface or thin gauges) and are not used for hardenability. However, the steels with 0.16-0.22%C and minimum 0.70%Mn are suitable for carburizing or carbo-nitriding and water quenching. The medium carbon grades containing 0.25-0.55%C are the most used of the heat-treated carbon steels and four of the carbon and carbon-boron H-steels are shown in Table 12-23. Some of the 15XX steels have lower carbon content than 0.25%C but these steels have higher manganese content than the 10XX to compensate for the hardenability. When added, the

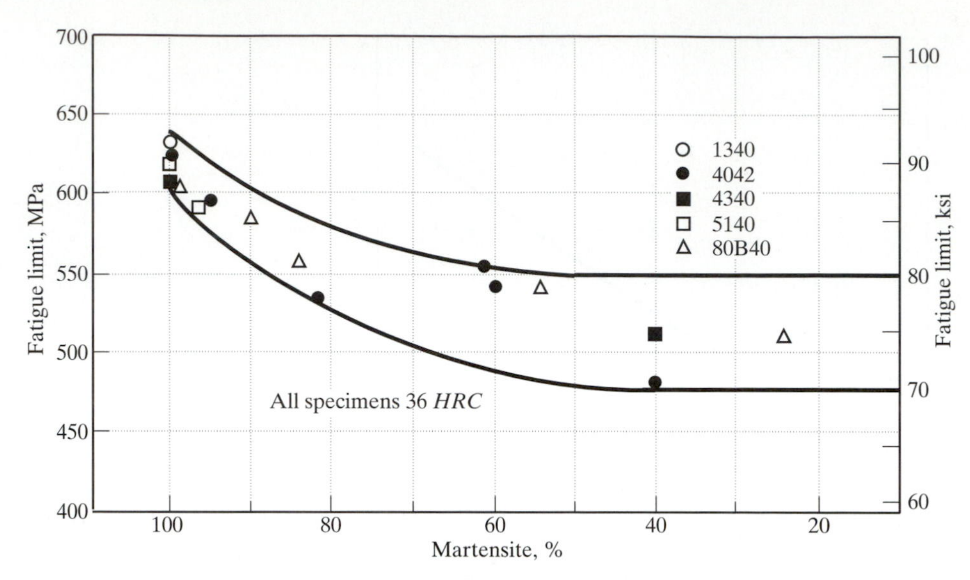

Fig. 12-18 Fatigue limit of through-hardened steels as function of amount of martensite in microstructure. [1, 2]

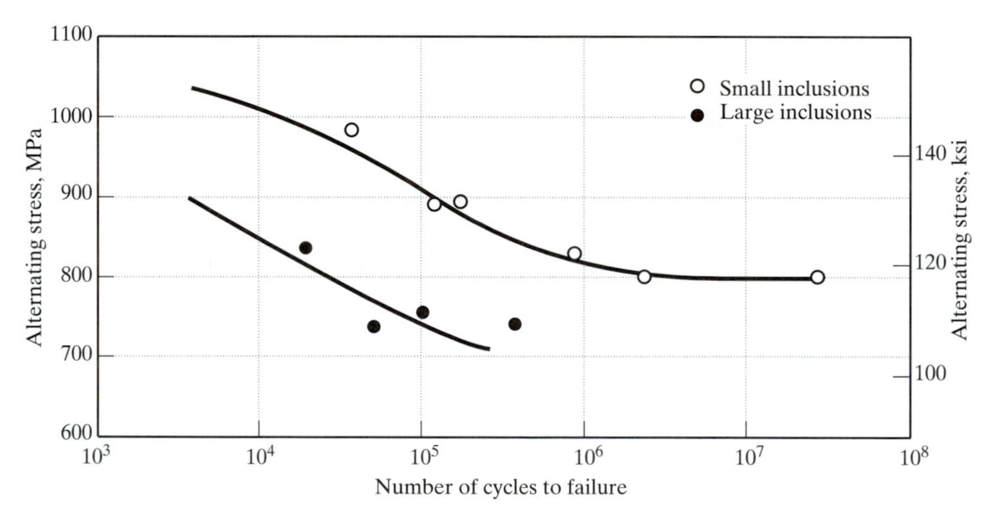

Fig. 12-19 Influence of size of nonmetallic inclusions on fatigue performance of 4340H steels. [1, 2]

boron content is about 0.001–0.003%. Most of these steels are used for crankshafts, couplings, tie rods, and other machinery parts that require hardness values within the range of 229 to 444 HB. Many of the common hand tools, such as pliers, open-end wrenches, screwdrivers, and a few edged tools (such as tin snips and brush knives) are also made from these steels. Also, these steels are very suitable for induction hardening and water quenching of a cylindrical, simple regular shape and uniform cross section, as well as a nonuniform section and irregularly shaped part.

Plain carbon steels containing 0.55–1.00%C are more restricted in application than the medium carbon grades because they have decreased machinability, poor formability, and poor weldability, which

Table 12-23 Composition of Four Carbon and Carbon-boron H steels.

SAE or AISI No.	Ladle chemical composition, w%				
	C	Mn	Si	P, maximum	S, maximum
1045H	0.42/0.51	0.50/1.00	0.15/0.35	0.040	0.050
1541H	0.35/0.45	1.25/1.75	0.15/0.35	0.040	0.050
15B41H	0.35/0.45	1.25/1.75	0.15/0.35	0.040	0.050
15B62H	0.54/0.67	1.00/1.50	0.40/0.60	0.040	0.050

make the fabrication of parts from these steels more difficult and more expensive. The one H steel within this carbon range in Table 12-23 (i.e., 15B62H) is suitable for heavy machinery parts, such as shafts, collars, and the like, and may be used either as normalized and tempered for low strength or as quenched and tempered for moderate strength. Steels such as 1070 and 1095 are very suitable for springs that require resistance to fatigue and permanent set. In the full-hardened condition (55 HRC and higher), these steels are used where abrasion resistance is the primary concern, such as for agricultural tillage tools as plowshares, moldboards, coulters, cultivator shovels, disks for harrows and plows, mower and binder knives, ledger plates, band knives, and knives for cutting grass, hay, or grain. Most of the parts made from these steels are hardened by conventional quenching (i.e., by oil, water, or brine quenching). However, special techniques such as austempering and martempering are often applied to reduce distortion, elimination of breakages, and to induce greater toughness.

The following is a summary of the characteristics of some heat-treatable plain carbon steels, 10XX and 15XX: [3]

- 0.16–0.22%C—suitable for carburizing or carbonitriding as well as water quenching. Steels with 0.70%Mn minimum are preferred.
- 0.33–0.37%C—suitable for induction hardening and water quenching of parts that have a nonuniform section and irregular shape, for instance, rear axle shaft of a passenger car.
- 0.40–0.45%C—suitable for induction hardening and water quenching of parts with a simple regular shape and uniform cross section.
- 0.46–0.50%C—suitable for induction hardening and water quenching of cylindrical parts.
- 0.78–0.95%C—suitable for austempered, martempered, and shallow hardened parts. Austempered and martempered parts are abrasion-resistant tools and cutting devices (saw blades and knives). Shallow-hardening is done with water or brine quenching or caustic quenching. Holes, grooves, and notches induce cracking problems.
- The 15XX grades with higher manganese have poor machinability, tend to produce microsegregation, have poor toughness, and are difficult to heat-treat.

12.5.2 Alloy Steels

The carbon steels, even for the high carbon grades, are shallow hardening. To induce hardening to greater depths from the surface or to through-harden greater sections, alloy additions are made to the steel. The following elements are the common additions to low-alloy steels to increase hardenability in ascending order: nickel, silicon, manganese, chromium, molybdenum, vanadium, and boron. The addition of several alloying elements in small amounts is more effective than the larger addition of one or two alloying elements. For these additions to be effective, they must be in solution in austenite. The alloying elements chromium, molybdenum, and vanadium form alloy carbides that require higher austenitizing temperatures and longer times at temperature to dissolve in austenite. Thus, allowance must be made during austenitization to dissolve these carbides in order to achieve their full effect on hardenability. In addition to increasing hardenability, the alloying elements also retard the softening of the steels during tempering as illustrated in Fig. 12-20. The steels in Fig. 12-20 all have nominal 0.45%C and at a certain tempering

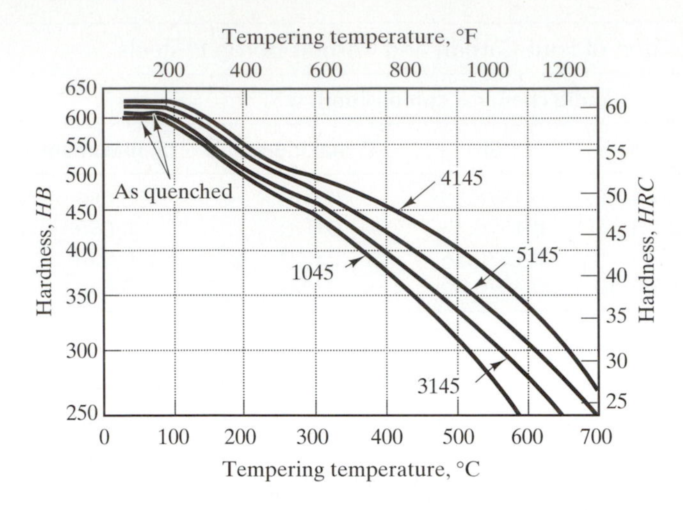

Fig. 12-20 Softening during tempering of a plain carbon and three low-alloy steels with nominal 0.45%C. [1, 2]

temperature the alloy steels have higher hardnesses than the plain carbon steel, 1045.

Selection based on hardenability can be done either by the percent of martensite present at different locations in the sections or by the hardnesses at the different locations in the sections of the steels. Based on the discussion at the beginning of this section, both the percent martensite and the hardness are related to the tensile strength of the steel. The need for either the amount of martensite in the microstructure or the hardness level depends on the mode of loading that the part will be subjected to. For pure tension, shear, and fatigue, the required microstructure is 99.9 percent quenched and tempered martensite or the section must be fully through hardened. However, the practical definition of full hardening by the SAE Iron and Steel Committee is minimum 90 percent martensite structure throughout. For bending or torsion, a 50 percent minimum martensite structure at the center and an 80 percent minimum martensite to a depth of at least 25 percent of the distance from the surface to the center (three-quarter-radius location) of the finished part may be sufficient. To ensure this, the 80 percent martensite hardness is specified at the half-radius location. Alloy steel selection guides based on the presence of at least 80 and 50 percent minimum martensite, for different locations in round sections of varying diameters for oil-quenched and tempered, and water quenched and tempered steels are available.

Although selection of steels can be made based on the amount of martensite formed, a more common method is the selection based on the attainment of a hardness level at a certain location of the section or part. The data needed for this selection process are: (a) the correlation of the Jominy equivalent cooling (J_{ec}) rates in various section sizes for various quenching media, and (b) the minimum hardenability curve of the steel. The relations of J_{ec} and various round diameters for agitation in water and oil are given in Figs. 9-52 and 9-53. Figure 12-21 shows the J_{ec} at various locations of round sections for different quenching media, which includes the data from Figs. 9-52 and 9-53. The various quenching media are given in the table below Fig. 12-21 and are numbered one through seven. Figure 12-21 includes two sets of curves for austenization in (1) a nonscaling protective atmosphere (left figures), and (2) in air (nonprotective atmosphere) that produces scales (right figures).

The procedure for the selection of steels based on hardenability is schematically illustrated in Fig. 12-22. The process usually starts with the need or requirement for a specific hardness (or strength) at a specific location of a certain size (diameter) of a steel. We must keep in mind that this hardness must be in the quenched and tempered condition. We take the following steps to select the required steel:

1. We use Fig. 12-16 to determine the as-quenched hardness (AQ) from the required quenched and tempered (Q + T) hardness.
2. We then determine how we shall austenitize the steel (i.e., whether with atmosphere or in air, to

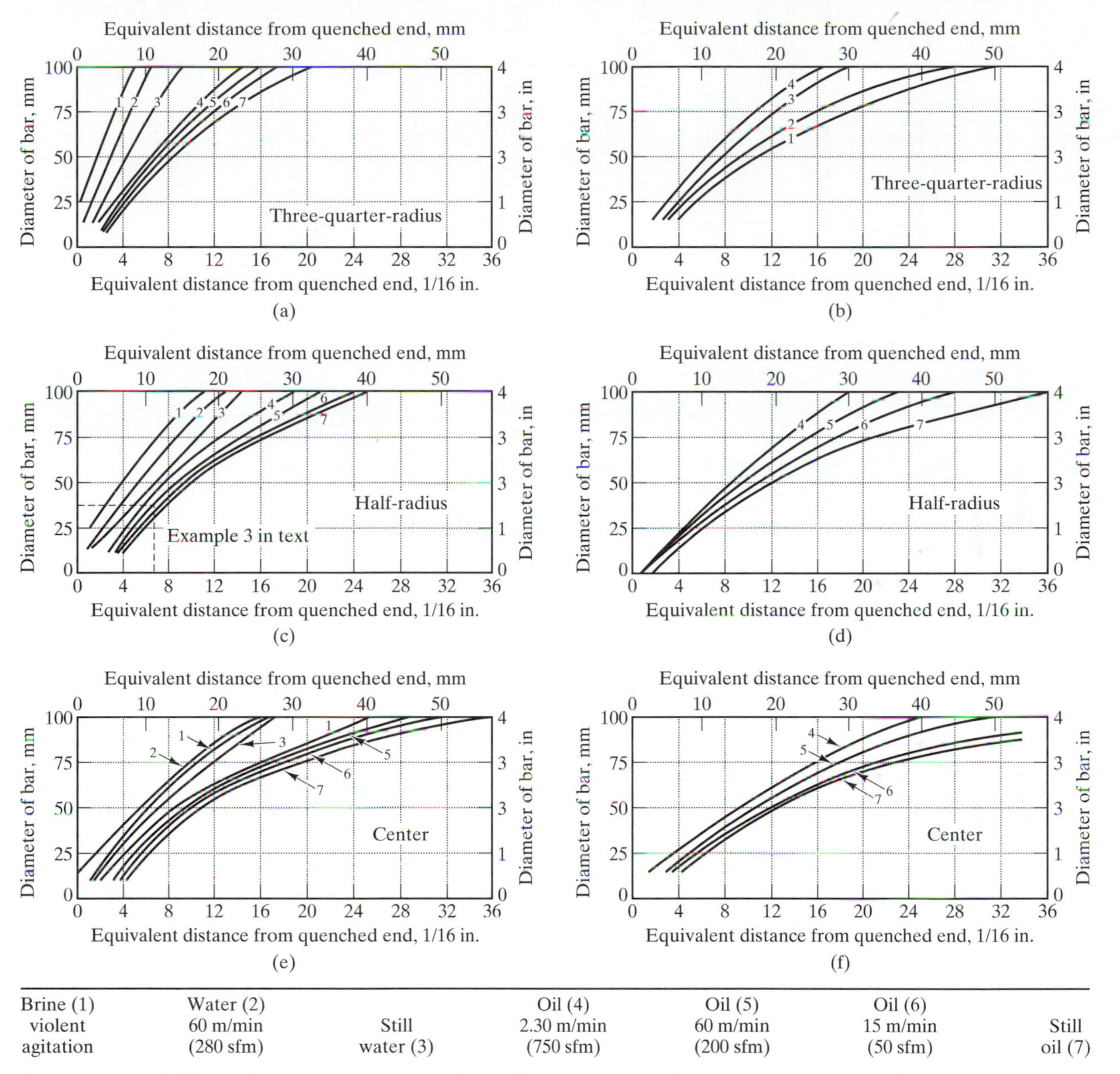

Fig. 12-21 J_{ec} of different locations in sections of round bars of varying diameters quenched in different media (numbers correspond to media in table below figures). (a), (c), and (e)—Austenitized in nonscaling atmosphere; (b), (d) and (f)—Austenitized in air.

use either the nonscaling (left) or scaling (right) figures in Fig. 12-21) and the quenching medium that we have to use.

3. Using the figure for the specific section location in Fig. 12-21, we look for the intersection of the size of the steel (1) (in Fig. 12-22a) with the curve for the quenching medium and determine the J_{ec} from the abscissa, see (2) in Fig. 12-22a.
4. We then look at published minimum hardenability curves for H steels and use the J_{ec} from Step 3 to determine the hardness on the ordinate, see (4) in Fig. 12-22b.

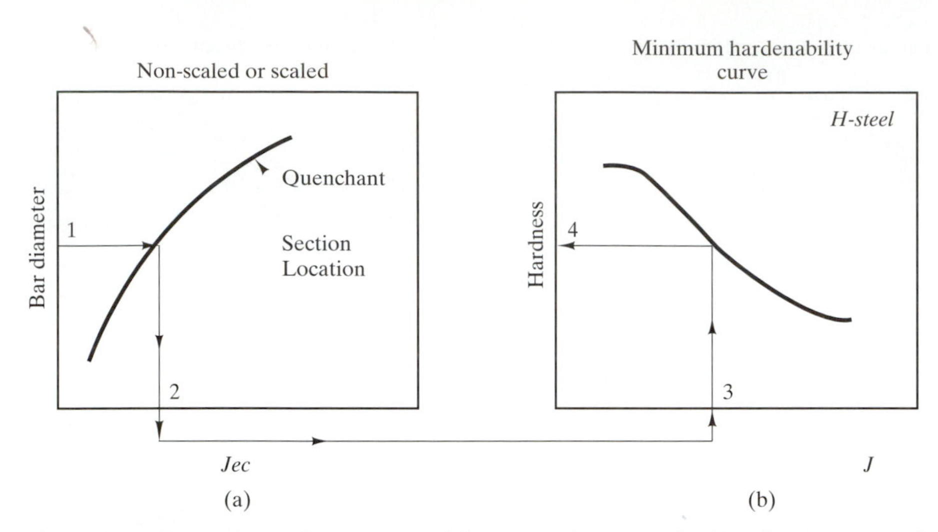

Fig. 12-22 Schematic in selecting H-steel for required as-quenched hardness. (a) one of figures in Fig. 12-21 with location desired and quenchant, (b) minimum hardenability of steel. Starting at point 1, the bar diameter, determine J_{ec} at 2; use J_{ec} at 3 in minimum hardenability curve for steel and determine minimum as-quenched hardness. Hardness at 4 must equal or exceed required as-quenched hardness.

5. We match the hardness obtained in Step 4 with the required AQ hardness in Step 1. If the hardness from Step 4 equals or exceeds the AQ hardness in Step 1, the steel exhibiting the hardenability curve used in Step 4 is a possible choice. In contrast, if the hardness from Step 4 is less than the AQ hardness, the steel with the hardenability curve in Step 4 will not satisfy the requirement.

We should appreciate that there may be more than one steel that will satisfy the requirements and therefore we should look at all hardenability curves to determine other steels that may satisfy the requirements. The search for all the steels that will satisfy the required hardness is facilitated by Table 12-24 which classifies the H steels according to six minimum as-quenched (AQ) hardness levels at various J_{ec} values. The AQ hardness levels are: 30, 35, 40, 45, 50, and 55 HRC. Under each hardness level, the steels corresponding to the J_{ec} value obtained in Step 3 above will give the steels that will satisfy the requirement. Thus, using Table 12-24 we eliminate Steps 4 and 5 above. We use the AQ hardness obtained in Step 1 as one of the six levels in Table 12-24 and then look for the steels using the J_{ec} value obtained in Step 3. Examples of this selection process are given in the case studies in Chap. 16.

If there is more than one candidate steel to select from, we make the selection based on other characteristics of the candidate steels such as machinability, forgeability, weldability, sensitivity to quench-cracking, distortion, availability, and cost. The characteristics of some of the most used low-alloy steels are given below. The italicized steels have established good characteristics. [3]

- 13XX—Difficult steels to process because of poor machinability, wide range in hardenability, and poor fracture toughness. Manganese tends to float in the melt and can create manufacturing problems. In general, these steels are poor selections for a heat-treated part.
- 40XX up to 0.27%C—economical direct-quench carburizing steels for gears; tendency to microcrack is very low.
- 4032 to 4027—More economical grades can be substituted for these grades.
- 41XX—widely used but erratic heat-treating response; hard to produce with acceptable surface and internal cleanliness. They tend to quench-crack when the nominal carbon content exceeds 0.40%; good for parts that are quenched mildly in oil or parts to be nitrided.
- *4320* and *4320* modified (with 0.70 to 0.90%Mn)—excellent steels for heavily loaded

Table 12-24 Classification of H steels according to minimum hardnesses at various Jominy equivalent cooling distances from quenched end.

Distance from quenched end, 1/16th in.	H steels with a minimum hardenability curve that intersects the specified hardness at the indicated distance from the quenched end of the hardenability specimen
30 HRC	
$2\frac{1}{2}$	8617, 118, 4620, 5120, 1038, 1522, 4419
3	4812, 4027, 1042, 1045, 1146, 1050, 1524, 1526, 4028, 6118
$3\frac{1}{2}$	4720, 6120, 8620, 4032
4	4815, 8720, 4621, 8622 1050
$4\frac{1}{2}$	46B12, 4817, 4320, 8625, 5046
5	4037, 1541, 4718, 8822
$5\frac{1}{2}$	94B15, 8627, 4042,1541, 15B35
6	94B17
$6\frac{1}{2}$	4820, 1330, 4130, 8630, 1141
7	9130, 5130, 5132, 4047
$7\frac{1}{2}$	1335, 50B46, 15B37
8	5135
$9\frac{1}{2}$	1340
10	8635, 5140, 4053, 50B40
11	4640
12	8637, 1345, 50B44, 5145 94B30
14	50B50
16	4135, 5147, 8645, 8740
20	4063
22	4068, 50B60, 5155, 86B30, 9260
24	4137, 5160, 6150, 81B45 51B60, 8650
32	4140
35 HRC	
$1\frac{1}{2}$	8617
2	4812, 4118, 4620, 5120, 3038, 1522, 4419, 6118
$2\frac{1}{2}$	4028, 4720, 8620, 4027, 1042, 1045, 1146, 1050, 1524, 1526
3	9130, 46B12, 4320, 6120, 8720, 4621, 8622, 8625, 4032, 4815
$3\frac{1}{2}$	4815, 4817, 94B17, 5046, 1050, 4781, 8822
35 HRC *(Continued)*	
4	8627, 4037
$4\frac{1}{2}$	94B15, 4042, 1541
5	4820, 1330, 4130, 5130, 8630, 5132, 1141, 50B46, 4047, 15B35, 94B17
$5\frac{1}{2}$	1335
6	5135
$6\frac{1}{2}$	15B37
7	8635, 1340, 5140, 4053
8	4063, 1345, 5145
$8\frac{1}{2}$	8637
9	4640, 4068, 50B40
$9\frac{1}{2}$	8640, 50B44, 5150
10	8740, 9260
$10\frac{1}{2}$	4135, 50B50
13	4137
16	4140, 6150, 81B45, 86B30
40HRC	
1	5120, 6120
$1\frac{1}{2}$	4118, 4620, 4320, 4720, 8620, 8720, 1038, 1522, 1526, 4621
2	8622, 8625, 4027, 1045, 1524, 4028, 4718
$2\frac{1}{4}$	1146
$2\frac{1}{2}$	4820, 8627, 4032, 1042, 1050
3	4037, 8822
$3\frac{1}{2}$	4130, 5130, 8630, 5046, 1050, 1541
4	1330, 5132, 4042
$4\frac{1}{2}$	5135, 1141, 4047
5	1335, 50B46, 15B35
$5\frac{1}{2}$	8635, 5140, 4053, 15B37
6	1340, 9260, 4063
$6\frac{1}{2}$	8637, 5145, 1345
7	4640, 4068
$7\frac{1}{2}$	8640, 5150
8	4135, 8740, 50B40
$8\frac{1}{2}$	6145, 9261, 50B44, 5155
9	4137, 8642, 5147, 50B50, 94B30
$9\frac{1}{2}$	8742, 8645, 5160, 9262

(Continued)

Table 12-24 Classification of H steels according to minimum hardnesses at various Jominy equivalent cooling distances from quenched end. (*Continued*)

Distance from quenched end, 1/16th in.	H steels with a minimum hardenability curve that intersects the specified hardness at the indicated distance from the quenched end of the hardenability specimen	Distance from quenched end, 1/16th in.	H steels with a minimum hardenability curve that intersects the specified hardness at the indicated distance from the quenched end of the hardenability specimen
40 HRC *(Continued)*		***45 HRC*** *(Continued)*	
$10\frac{1}{2}$	6150, 50B60	13	8653, 8660
11	4140	14	9840, 4145
$11\frac{1}{2}$	81B45, 8650, 5152	16	85B45, 4147
12	86B30	17	4337
13	51B60	18	4150
14	8655	22	4340
15	4142	26	4161
$15\frac{1}{2}$	8750	30	E4340
18	4145, 8653, 8660	36	9850
19	9840, 86B45	***50HRC***	
20	4147	1	4032, 5132, 1038
24	4337, 4150	$1\frac{1}{2}$	1335, 5135, 8635, 4037, 1042, 1146, 1045
32	4340	2	4135, 1541, 15B35, 15B37
36+	E4340, 9850	$2\frac{1}{4}$	1050
45 HRC		$2\frac{1}{2}$	4042
1	4027, 4028, 8625	3	8637, 5140, 5046, 4047
$1\frac{1}{2}$	8627, 1038	$3\frac{1}{2}$	4137, 1141, 1340
2	4032, 1042, 1146, 1045	4	4640, 5145, 50B46
$2\frac{1}{2}$	4130, 5130, 8630, 4037, 1050, 5132	$4\frac{1}{2}$	8640, 8740, 4053, 9260
3	1330, 5046, 1541	5	8642, 4063, 1345, 50B40
$3\frac{1}{4}$	1050	$5\frac{1}{2}$	8742, 6145, 5150, 4068
$3\frac{1}{2}$	1335, 5135, 4042, 4047	6	4140, 8645
4	8635, 1141	$6\frac{1}{2}$	9261, 50B44, 5155
5	8637, 1340, 5140, 50B46, 4053, 9260, 15B37	7	5147, 6150
$5\frac{1}{2}$	5145, 4063	$7\frac{1}{2}$	5160, 9262, 50B50
6	4135, 4640, 4068, 1345	8	4142, 81B45, 8650
$6\frac{1}{2}$	8640, 8740, 5150, 94B30	$8\frac{1}{2}$	5152, 50B60
7	4137, 8642, 6145, 9261, 50B40	$9\frac{1}{2}$	4337, 8750, 8655
$7\frac{1}{2}$	8742, 50B44, 5155	10	4145, 51B60
8	8645, 5147	$10\frac{1}{2}$	9840
$8\frac{1}{2}$	4140, 6150, 5160, 9262, 50B50	11	8653, 8660
9	50B60	$11\frac{1}{2}$	8645
$9\frac{1}{2}$	81B45, 8650, 86B30	12	85B45
10	5152	13	4340, 4147
11	51B60, 8655	14	4150
$11\frac{1}{2}$	4142	20	E4340
12	8750	22	9850, 4161

(*Continued*)

Table 12-24 Classification of H steels according to minimum hardnesses at various Jominy equivalent cooling distances from quenched end. (*Continued*)

Distance from quenched end, 1/16th in.	H steels with a minimum hardenability curve that intersects the specified hardness at the indicated distance from the quenched end of the hardenability specimen	Distance from quenched end, 1/16th in.	H steels with a minimum hardenability curve that intersects the specified hardness at the indicated distance from the quenched end of the hardenability specimen
55HRC		*55HRC* (*Continued*)	
1	1141, 1042, 4042, 4142, 1045, 1146, 1050, 8642	$5\frac{1}{2}$	8650, 5152, 4068
$1\frac{1}{2}$	50B46	6	50B50
2	8742, 5046, 4047, 5145	$6\frac{1}{2}$	5160, 9262
$2\frac{1}{2}$	6145	7	4147, 8750, 8655
3	4145, 8645, 1345	$7\frac{1}{2}$	50B60
$3\frac{1}{2}$	86B45, 5147, 4053, 9260	9	8653, 51B60, 8660
$4\frac{1}{2}$	5150, 40635	$9\frac{1}{2}$	4150
5	81B45, 6150, 9261, 5155	17	9850

carburized gears with contact stress over 250 ksi; can be made very clean; carburized with great freedom from microcracking; have good short- and long-life fatigue properties; a major disadvantage is cost.

- *4340*—outstanding engineering material, especially in the 2 to 5 inch sections; can be melted clean, heat treats with minimum difficulty, but just like 4320, is very expensive. The nickel content is beneficial for short-cycle fatigue properties.
- 44XX—these are carburizing steels that contain substantial amounts of molybdenum and are prone to microcracking when quenched directly from the furnace.
- 46XX—carburizing grades that machine and heat-treat uniformly with good case hardenability, but the core hardenability is low except for 4626 that contains 0.70 to 1%Ni. However, even 4626 exhibits much lower hardenability than other grades. (In general, 4320 provides better value than 4626, even though it is priced slightly higher.)
- *4815*, *4817*, and *4820*—exhibit exceptionally high-case hardenability and excellent resistance to short- and long-life fatigue; excellent for heavily loaded carburized gears.
- 5015 and 5046—carbon steels with 0.80 to 1.10%Mn and high residuals can be used instead of these grades to produce substantial savings.
- 5115 through 5160, 6118, 6150, and 8115—carbon-mangansese-boron steels with residuals can also be substituted for these steels.
- *8615 through 8627*—excellent carburizing steels for fine- and medium-pitch gears; case carbon should be controlled to 0.95% to prevent microcracking during quench.
- *8630*—excellent for abrasive wear resistance when water-quenched and tempered at 400° F.
- *8637*—suitable for water quenching in sections up to 2 in. and tempering to a machinable hardness.
- *8640 through 8645*—suitable for oil-quenching in thin sections and sections up to 2-inches in diameter or equivalent in flats.
- *8655 through 8660*—excellent steel for hot-wound, quenched, and tempered coil springs and torsion bars to 2-in. diameter and leaf springs to 1.50 inch thick.
- 87XX—the cost of these steels cannot be justified when compared to the less expensive 86XX grades.
- 8822—excellent steel for heavy-duty carburized gearing but will usually microcrack in direct quenching and requires vigorous oil quenches when reheated for hardening to prevent bainite formation.
- 9254, 9255, 9260—these are rather difficult to produce clean and to roll with good surface,

because of the high silicon content. When used for valve coil springs, they have the reputation to develop transverse cracks. Cr-V valve spring wire is preferred. For heavy springs, the 50B60 or 51B60 grades are better value.

- 50B44, 50B46, 50B50—can be replaced by lower-cost C-Mn-B steels.
- *50B60 and 51B60*—outstanding values in steels for heavy springs.
- 81B45—can be replaced with a high-residual C-Mn-B grade
- *86B45*—an excellent steel for heavy shafts and forgings at a substantial savings over 4340.
- *94B17*—has potential as a good carburizing steel.
- 94B30—can be replaced with a high-residual lower-cost C-Mn-B grade.

12.6 Selection of Tool Steels [5,6]

Tool steels are those that are used to make tools specifically for cutting, forming, or shaping a material into a part or component for a specific use. In service, most tools are subjected to very high, rapidly applied loads. In cutting or machining, the tips of the tools are raised to very high-temperatures that may make them red in color. Because of these conditions, the selection of a tool steel for service is made according to: (1) its resistance to softening at elevated temperatures or ability to retain high red hardness or hot hardness; (2) its resistance to wear, deformation, and breakage; and (3) its toughness so as to absorb the sudden application of loads. However, the selection of the proper tool steel is only one of four factors that determine the performance of a tool in service. The other three factors are proper design of the tool, the accuracy with which the tool is made, and the application of the proper heat treatment to the tool steel. A tool can perform successfully in service only when all four factors are satisfied.

The AISI designations, the types, and the major elements in the compositions of selected tool steels are shown in Table 12-25. These steels are produced according to the following ASTM standards:

- ASTM A600—standard requirements for molybdenum and tungsten high-speed steels
- ASTM A681—standard requirements for hot-work, cold-work, shock-resisting, special-purpose, and mold steels
- ASTM A686—standard requirements for water-hardening tool steels.

These specifications may be used for procuring the steels; the latter may also be procured using trade names specific to tool steel producers. The types of tools steels shown in Table 12-25 have also specifications in France, Great Britain, Germany, Japan, and Sweden.

12.6.1 High-Speed Steels

These are steels that are used to make tools for high-speed cutting. The two types of tool steels in this group are the Group M (molybdenum high-speed steels), and the Group T (tungsten high-speed steels). Both groups of steels have equivalent performance, including hardening ability, but the initial cost of the Group M steels is approximately 40 percent lower than the Group T steels. This cost advantage arises from the fact that the same number of atoms of molybdenum and tungsten are needed to combine with the carbon atoms to produce the carbide particles. Because the atomic weight of molybdenum is lower than tungsten, we need a lower weight addition of molybdenum than tungsten and, therefore, lower cost because the additions are priced on a per weight basis. In addition to this cost advantage, the United States has a good domestic source of molybdenum and consequently the Group M steels comprise more than 95 percent of all high-speed steels in the United States.

Typical applications for both groups include cutting tools of all sorts, including drills, reamers, end mills, mill cutters, taps, broaches, and hobs. Some are suitable for cold-work applications, including cold-header die inserts, thread-rolling dies, punches, and blanking dies. For the latter applications, the steels are usually underhardened, which means that they are austenitized at lower temperatures than those that are recommended for cutting applications. This process leaves some of the alloy carbides undissolved in austenite prior to quenching and therefore produces an underhardened (softer) quenched and tempered product but with better toughness properties.

The cutting applications require that the tools remain sharp for a long time and that they must have primary high-hardness (strength) to begin with for wear resistance and must resist softening during the cutting operation. In terms of dislocation concepts, this means that the motion of dislocations is hindered and

Table 12-25 Examples of principal types of tool steels and their compositions. [5, 6]

Designation	C	Mn	Si or Ni	Cr	V	W	Mo
		Water-Hardening Tool Steels (a)					
W1*	0.60 to 1.40 (a)	...	...	...	...	...	...
		Shock-Resisting Tool Steels					
S5*	0.55	0.80	2.00 Si	...	...	...	0.40
		Oil-Hardening Cold Work Tool Steels					
O1*	0.90	1.00	...	0.50	...	0.50	...
		Air-Hardening Medium-Alloy Cold Work Tool Steels					
A2*(b)	1.00	...	...	5.00	...	...	1.00
		High-Carbon High-Chromium Cold Work Tool Steels					
D2*(b)	1.50	...	...	12.00	...	...	1.00
		Chromium Hot Work Tool Steels					
H11	0.35	...	...	5.00	0.40	...	1.50
H12*	0.35	...	...	5.00	0.40	1.50	1.50
		Tungsten Hot Work Tool Steels					
H21*	0.35	...	...	3.50	...	9.50	...
		Molybdenum Hot Work Tool Steels					
H43	0.55	...	...	4.00	2.00	...	8.00
		Tungsten High Speed Tool Steels					
T1*(b)	0.70	...	...	4.00	1.00	18.00	...
		Molybdenum High Speed Tool Steels					
M1*(b)	0.80	...	...	4.00	1.00	1.50	8.50
M2*(b)	0.85	...	...	4.00	2.00	6.25	5.00
		Low-Alloy Special-Purpose Tool Steels					
L3	1.00	...	...	1.50	0.20	...	...
		Low-Carbon Mold Steels					
P6	0.10	...	3.50 Ni	1.50	0.20	...	...
P21	0.20	1.20 Al	4.00 Ni	...	...	...	...

*Stocked in almost every warehousing district and made by most tool steel producers.

(a) Various carbon contents are available in 0.10% ranges. (b) Available as free-cutting grade. (c) Available with vanadium contents of 2.40 or 3.00%. (d) Neither AISI nor SAE has assigned type numbers to these steels.

blocked. In tool steels, this is accomplished by dispersion hardening with a large number of hard, wear-resistant particles of alloy carbides in the microstructure. Thus, the compositions of these steels include carbon in the high-carbon range, and the carbide-forming elements molybdenum, tungsten, chromium, and vanadium. The T steels do not contain molybdenum but the M steels contain tungsten in addition to the molybdenum addition. Cobalt is added to improve the red hardness. The resistance to softening of various tool steels is illustrated in Fig. 12-23 where we see that the M2 grade (curve 4) exhibits secondary hardening during tempering at high temperatures (ca. 550°–600°), instead of softening, as shown by the other three tool steels. The usual maximum hardness that can be developed in both these groups of steels is about 65 HRC. For higher wear resistance, some of the steels can be heat-treated to provide 66-68 HRC.

The M steels readily decarburize and are more sensitive to austenitizing conditions (i.e, temperature and atmosphere), especially the high-molybdenum and low-tungsten. Thus, the austenitizing temperature of the M steels is lower than the T steels; 1175° to 1230°C for the M steels versus 1205°C to 1300°C for the T steels.

12.6.2 Hot-Work Tool Steels—Group H

These steels are intended to withstand the combinations of heat, pressure, and abrasion associated with manufacturing operations involving shearing, punch-

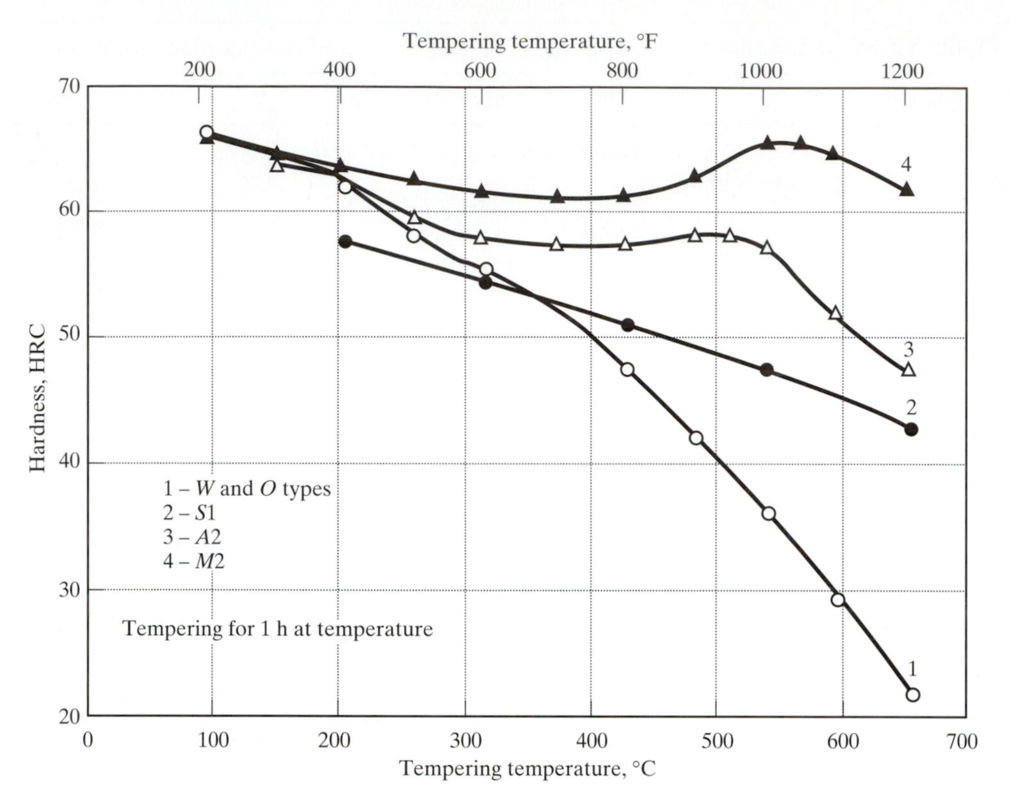

Fig. 12-23 Resistance to softening of four tool steels at various temperatures; 4 has the highest resistance and 1 has the lowest.

ing, or forming of metals at high temperatures. They are used for hot die work of all kinds such as dies for extrusion, mandrels, hot-forging, die-casting, and hot shears. The chromium hot-work steels are used for the lower melting metals of aluminum and magnesium, while the tungsten and the molybdenum hot-work steels are used for the higher melting brass, nickel, and steel alloys.

The chromium hot work steels (H10 to H19) have medium carbon and medium alloy contents to induce toughness at the normal working hardnesses of 40 to 55 HRC and enable them to be water-cooled in service. The most widely used of this group are H11, H12, and H13 that may be air-hardened to full working hardnesses up to 152 mm (6 in.). The air-hardening of these steels induces low distortion. The H11 is also used in the aerospace industry to make highly stressed parts because it has the ability to resist softening during continued exposure up to 540° and at the same time, it provides moderate toughness and ductility at room temperature. Because it resists softening at high temperatures, it can be tempered at high temperatures to completely relieve residual hardening (quenching) stresses.

As with the high-speed steels, the tungsten and molybdenum hot-work steels have very similar characteristics. The higher alloy contents of these steels make them more resistant to high-temperature softening. However, the higher alloy contents and higher carbon contents (for the molybdenum hot-work steels) in these steels make them more prone to brittleness and make them more difficult to water-cool safely in service. Although air-hardenable, the tungsten hot-work steels are usually quenched in oil or hot salt to minimize scaling. As in high-speed steels, the molybdenum steels require greater care in heat-treatment to minimize decarburization.

12.6.3 Cold-work Tool Steels

These steels are not resistant to softening at high temperatures and are restricted to applications that do not require prolonged or repeated heating above 205° to 260°C. There are three groups of cold-work tool steels: Group A—air-hardening steels; Group D—high-carbon high-chromium steels; and Group O—oil-hardening steels.

The principal alloying elements in the Group A steels are manganese, chromium, and molybdenum that allow these steels to be air-hardened to achieve full hardness in sections up to 102 mm (4 in.) in diameter. Air hardening of these steels produces the least distortion and the least tendency to crack during hardening. Typical applications for these A steels include shear knives, punches, blanking and trimming dies, forming dies, and coining dies. The inherent dimensional stability of these steels make them very good gauges and precision measuring tools.

The Group D tool steels contain from 1.5 to 2.35%C and 12%Cr and with the exception of D3, which also contain 1%Mo. These steels with the exception of D3 are air-hardened to produce full hardness; D3 is quenched in oil; therefore, tools made from D3 are more susceptible to distortion and have greater quench-cracking tendency. Group D steels have high resistance to softening at elevated temperatures and exhibit excellent wear resistance, especially D7, which has the highest carbon and vanadium contents. Typical applications of these steels include long-run dies for blanking, forming, thread-rolling, and deep-drawing; dies for cutting laminations; brick molds; gauges; burnishing tools; rolls; and shear and slitter knives.

The Group O steels have high-carbon contents and sufficient alloying elements to allow small-to-medium sized sections to achieve full hardness when oil-quenched from the austenitizing temperature. Because of the high-carbon, they have high wear resistance at normal low temperatures. However, they have poor resistance to softening as indicated in Fig. 12-23. These steels are used extensively in dies and punches for blanking, trimming, drawing, flanging, and forming. They are also used for machinery components (such as cams, bushings, and guides), and for gauges (where good dimensional stability and wear resistance are needed).

12.6.4 Shock Resisting Steels—Group S

The principal alloying elements in this group of steels are manganese, silicon, chromium, tungsten, and molybdenum in varying amounts, and a 0.50%C for all steels. These steels exhibit high strength, high toughness, and low-to-medium wear resistance and are used primarily for chisels, rivet sets, punches, driver bits, hammers, and other applications requiring high toughness and resistance to shock-loading. Types S1 and S7 are also used for hot-punching and shearing, which require some heat resistance. Because of their excellent toughness with high strength, they are also considered for structural applications.

12.6.5 Low-Alloy Special-Purpose Steels—Group L

These steels contain high carbon and small varying amounts of chromium, vanadium, nickel, and molybdenum. These are oil-hardening steels and are generally used for machine parts such as arbors, cams, chucks, and collets and other applications requiring good strength and toughness. L2 and L3 are similar to AISI-SAE 52100 steel and are used for similar applications including bearings, rollers, clutch plates, high-wear springs, and feed fingers.

12.6.6 Mold Tool Steels—Group P

These are basically low-carbon steels, except for P20, with alloying elements of chromium, and varying amounts of nickel, molybdenum, and vanadium, and for P21 some aluminum is added. They are used almost exclusively in low-temperature die-casting dies and in molds for the injection or compression molding of plastics. Because of their low carbon, they are very soft and have a low work-hardening rate in the annealed condition to allow the production of a mold impression by cold hubbing for the P2 to P6 steels. After the impression is formed, the mold is carburized, hardened, and tempered to a surface hardness of about 58 HRC. Types P4 and P6 are deep hardening, with P4 achieving full hardness in the carburized case by air-cooling.

Types P20 and P21 are normally supplied heat-treated to 30 to 36 HRC, which allows them to be machined readily into large, intricate dies and molds.

No subsequent high-temperature heat treatment is required, thereby avoiding distortion and size changes. However, for plastic molds, P20 is sometimes carburized and hardened after the mold impression is machined. With aluminum addition, P21 is a precipitation-hardening steel that is supplied prehardened to 32 to 36 HRC and is preferred for critical-finish molds because of its excellent polishability.

12.6.7 Water Hardening Tool Steels—Group W

These are basically plain carbon steels with nominal carbon contents ranging from 0.60 to 1.40%; the most popular grades contain about 1.00%C. Small amounts of chromium or vanadium are added to most of the steels; chromium to increase hardenability and wear resistance, and vanadium to maintain fine austenite grain size for toughness. Being plain carbon steels, they are shallow hardening and form relatively thin, fully hardened zone even when quenched drastically. Sections more than 13 mm ($^1/_2$ in.) thick will generally have a hard case over a strong, tough, and resilient core. They have low resistance to softening at high temperatures and are suitable for: cold-heading, striking, coining, and embossing tools; woodworking tools; hand metal-cutting tools, such as taps and reamers; wear-resistant machine tool components; and cutlery.

The Group W tool steels are supplied in four grades or qualities. These grades according to the SAE are:

- *Special (Grade 1).* The highest-quality water hardening tool steel. Hardenability is controlled, and composition is held to close limits. Bars are subjected to rigorous testing to ensure maximum uniformity in performance.
- *Extra (Grade 2).* A high-quality water hardening tool steel that is controlled for hardenability and is subjected to tests to ensure good performance in general applications.
- *Standard (Grade 3).* A good-quality water-hardening tool steel that is not controlled for hardenability and that is recommended for applications in which some latitude in uniformity can be tolerated.
- *Commercial (Grade 4).* A commercial quality water-hardening tool steel that is neither controlled for hardenability nor subjected to special tests.

12.6.8 Guide to Selection and Summary of Properties of Tool Steels

Figures 12-24a and b summarize the performance of the tool steels correlating hot-hardness (resistance to softening) ratings with toughness ratings in (a), and wear-resistance ratings and toughness ratings in (b). In the ratings, one is low and nine is high. The ideal tool steel would be one with high toughness, high hot-hardness, and high wear resistance. It is obvious that all the three performance factors are not high for each of the tool steels. Thus, selection of tool steels may have to be based on two performance factors with the third being compromised. Table 12-26 is a reference guide for tool selection for different applications.

12.7 Selection of Stainless Steels [7]

Stainless steels (SS) are characterized by a minimum 10.5 wt%Cr addition to iron. For most SS, the maximum chromium content is about 30% and the minimum iron content is 50%. The carbon is normally present from less than 0.03 percent to a maximum of 1.2 percent in certain martensitic grades. The stainless characteristic arises from the formation of an invisible and very adherent chromium-rich oxide surface film. This puts the steel in a passive state and when the film is breached, it immediately heals when oxygen is present. Other elements added to improve particular characteristics include nickel, molybdenum, copper, titanium, aluminum, silicon, niobium, nitrogen, sulfur, and selenium.

12.7.1 Designations, Classes, and Properties of Stainless Steels (SS)

Figure 12-25 is a summary of the compositional and property linkages in the SS family; Table 12-27 lists the UNS and the AISI designations and the compositions of selected standard and special SS; and Table 12-28 lists the properties of the selected SS compared to other alloys. At the center of Fig. 12-25 is the 304 grade that is the most used SS. Modifications of this grade are indicated in the arrows to produce the other classes and to achieve the specific, desired characteristics of the SS.

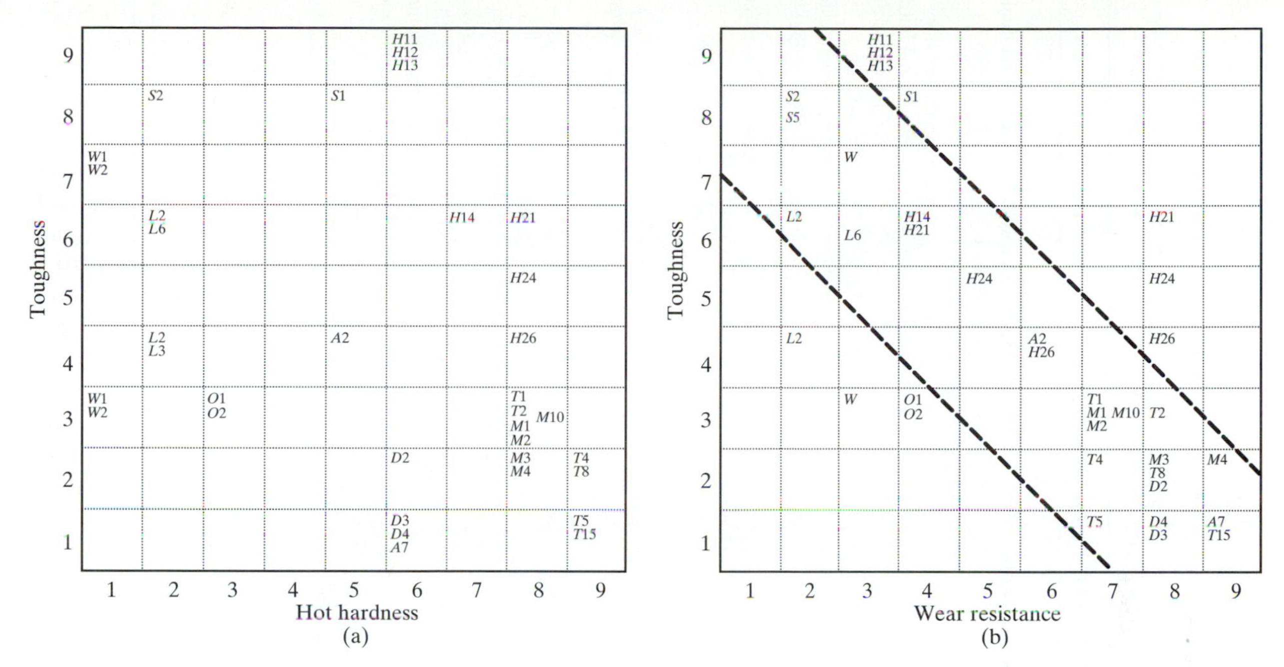

Fig. 12-24 Performances of tool steels (a) matrix toughness as function of hot-hardness, (b) matrix toughness as function of wear resistance. 9 is highest rating and 1 is lowest rating. Toughness, wear, and hot-hardness are the three criteria used for tool steel selection. [6]

In the United States, the wrought grades of SS are designated by AISI (American Iron and Steel Institute) numbers, by the UNS (Uniform Numbering System), or by proprietary names given by specialty steel producers. The *AISI designation* is the older and more widely used and this *consists of a three-digit number*. The 200 and 300 grades are austenitic, while the 400 grades are either ferritic or martensitic. The UNS system is a recent adoption and has a considerably greater number of SS because it incorporates all of the recently developed SS. The *UNS designation consists of the letter **S** followed by a five-digit number*. When the alloy has an AISI number, the first three numbers of the UNS designation correspond to the AISI number. The last two numbers indicate modifications to the grade. "00" means no modification and indicates the basic AISI grade. For *high nickel contents (~25 to 35%Ni)*, the designation starts with *N* followed by five digits, for example, N08020 in the austenitic alloys. In this text, we shall use the AISI designation because this is the more popular designation in the United States.

Table 12-27 lists examples of the five groups of SS. The first four are based on the microstructures of the alloys: *ferritic, martensitic, austenitic, and duplex* (ferritic plus austenitic) and the fifth class is based on *precipitation-hardening (PH)*. In terms of composition, the Schaeffler diagram in Fig. 12-26, which is a plot of chromium equivalent (of all the ferrite stabilizers) versus the nickel equivalent (of all the austenite stabilizers), shows where the different classes are located. The ferrite stabilizers are the elements that widen the ferrite phase field and include silicon and all the carbide formers—chromium, molybdenum, vanadium, niobium, and titanium. The austenite stabilizers are those that enlarge the austenite phase field and include nickel, manganese, carbon, and nitrogen. The mechanical properties of the selected SS in Table 12-27 are shown in Table 12-28 and are compared with those of a 1080 steel, a 6061 aluminum, and a copper alloy.

Ferritic SS have the same BCC (body-centered cubic) structure as that of iron at room temperature as well as at high temperatures. Thus, they cannot be hardened by heat-treatment. Their annealed yield strengths range from 275 to 350 MPa (40 to 50 ksi). Their poor toughness and susceptibility to sensitization limit their fabricability and the usable section

Table 12-26 Summary guide for tool steel selection. [6]

Application areas	Tool steel groups, AISI letter symbols, and typical applications						
	High-speed tool steels, M and T	**Hot-work tool steels, H**	**Cold-work tool steels, D, A, and O**	**Shock-resisting tool steel, S**	**Mold steels, P**	**Special purpose tool steels, L**	**Water-hardening tool steels, W**
Cutting tools Single-point types (lathe, planer, boring) Milling cutters Drills Reamers Taps Threading dies Form cutters	General-purpose production tools: M2, T1 For increased abrasion resistance: M3, M4, M10 Heavy-duty work calling for high hot hardness: T5, T15 Heavy-duty work calling for high abrasion resistance: M42, M44	...	Tools with sharp edges (knives, razors) Tools for operations in which no high speed is involved, yet stability in heat treatment and substantial abrasion resistance are needed	Pipe cutter wheels	...	...	Uses that do not require hot hardness or high abrasion resistance Examples with carbon content of applicable group: Taps (1.05–1.10%C) Reamers (1.10–1.15%C) Twist drills (1.20–1.25%C) Files (1.35–1.40%C)
Hot-forging tools and dies Dies and inserts Forging machine plungers and piercers	For combining hot hardness with high abrasion resistance: M2, T1	Dies for presses and hammers: H20, H21 For severe conditions over extended service periods: H22–H26	Hot trimming dies: D2	Hot-trimming dies Blacksmith tools Hot-swaging dies	...	...	Smith tools (0.65–0.70%C) Hot chisels (0.70–0.75%C) Drop forging dies (0.90–1.00%C) Applications limited to short-run production
Hot extrusion tools and dies Extrusion dies and mandrels Dummy blocks Valve extrusion tools	Brass extrusion dies: T1	Extrusion dies and dummy blocks: H21–H26 For tools that are exposed to less heat: H10–H14, H19	...	Compression molding: S1	...	...	...

(Continued)

Table 12-26 Summary guide for tool steel selection. (*Continued*)

Application areas	Tool steel groups, AISI letter symbols, and typical applications						
	High-speed tool steels, M and T	**Hot-work toll steels, H**	**Cold-work tool steels, D, A, and O**	**Shock-resisting tool steel, S**	**Mold steels, P**	**Special purpose tool steels, L**	**Water-hardening tool steels, W**
Cold-forming dies Bending, forming, drawing, and deep-drawing dies and punches	Burnishing tools: M1, T1	Cold-heading die casings: H13	Drawing dies: O1 Coining tools: O1, D2 Forming and bending dies: A2 Thread rolling dies: D2	Hobbing and short-run applications: S1, S7 Rivet sets and rivet busters	…	Blanking, forming, and trimmer dies when toughness has precedence over abrasion resistance: L6	Cold-heading dies: W1 or W2 (C ~ 1.00%) Bending dies: W1 (C ~ 1.00%)
Shearing tools Dies for piercing, puching, and trimming Shear blades	Special dies for cold and hot work: T1 For work requiring high abrasion resistance: M2, M3	For shearing knives: H11, H12 For severe hot-shearing applications: H21, H25	Dies for medium runs: A2, A6, O1 Dies for long runs: D2, D3 Trimming dies (also for hot trimming): A2	Cold and hot shear blades Hot puching and piercing tools Boilermaker tools	…	Knives for work requiring high toughness: L6	Trimming dies: (0.90–0.95% C) Cold-blanking and punching dies (1.00%C)
Die casting and molding dies	…	For aluminum and lead: H11, H13 For brass: H21	A2, A6, O1	…	Plastic molds: P2-P4, P20	…	…
Structural parts for severe service conditions	Roller bearings for high temperature environment: T1 Lathe centers: M2, T1	For aircraft components (landing gears, arrester hooks, rocket cases): H11	Lathe centers: D2 D3 Arbors: O1 Bushings A4 Gauges: D2	Pawls Clutch parts	…	Spindles and clutch parts (if high toughness is needed): L6	Spring steel (1.10–1.15%C)
Battering tools, hand and power	…	…	…	Pneumatic chisels for cold work: S5 For higher performance: S7	…	…	For intermittent use: W1 (0.80%C)

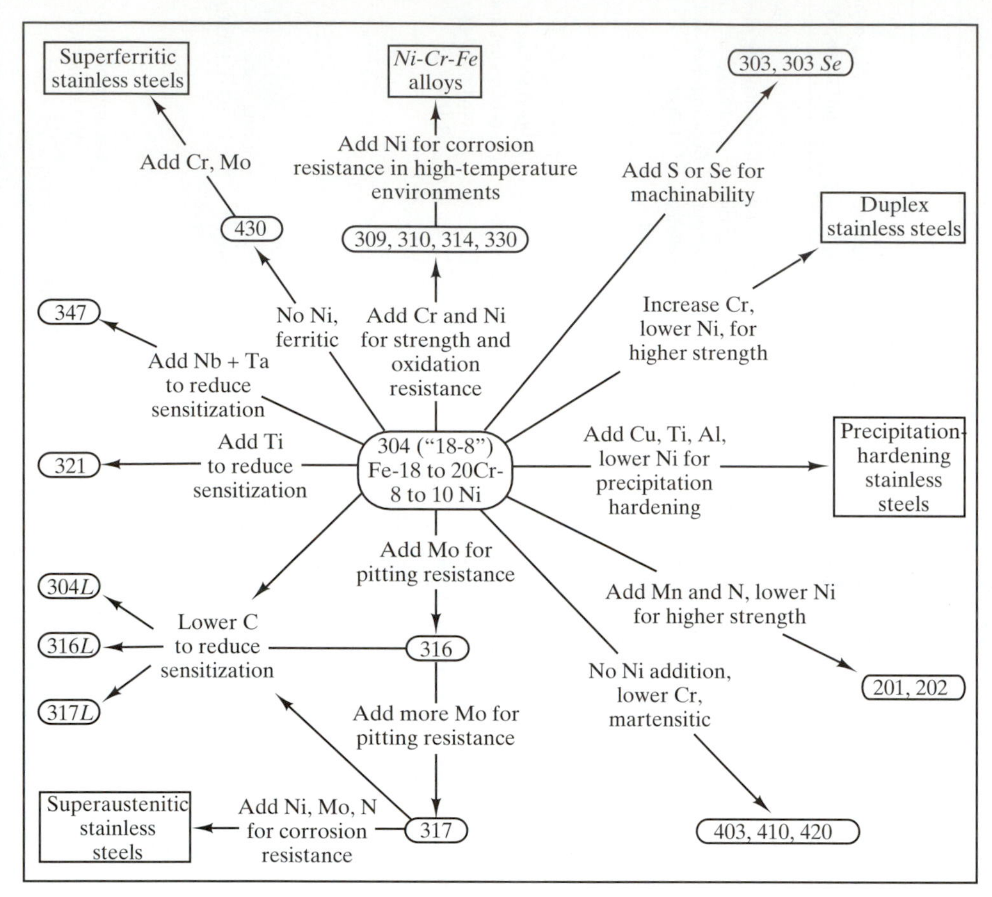

Fig. 12-25 Compositional and property linkages in stainless steel alloys. Starting from 304, improvements in property indicated by arrows lead to appropriate alloys.

size. Their chief advantages are their resistance to chloride stress-corrosion cracking, atmospheric corrosion, and oxidation at a relatively low cost.

The ferritic SS contain between 10.5 to 30%Cr, with small amounts of austenite-forming elements, such as carbon, nitrogen, and nickel. Their usage depends on the chromium content. The low-Cr alloys (−11%Cr) (Type 405 and 409) have fair corrosion and oxidation resistance as well as good fabricability at low cost. They are widely use in automotive exhaust systems and Type 409 is the most widely used ferritic SS. The intermediate-Cr alloys (16 to 18%Cr) (Types 430 and 434) are used for automotive trim and cooking utensils—they have poor toughness and weldability that make them difficult to fabricate. The high-Cr alloys (19 to 30%Cr) (Types 442 and 446) are the superferritics referred to in Fig. 12-25. They are applied where high level of corrosion and/or oxidation resistance is/are required. The superferritics usually contain very low carbon and nitrogen with stabilizing elements titanium and niobium to prevent sensitization and to improve as-welded properties (see Sec. 9.7.3b—Embrittling Phenomena for a review of sensitization and stabilization). They also contain either aluminum or moybdenum. The superferritics exhibit exceptional resistance to localized corrosion induced by exposure to aqueous chlorides and much better than the austenitic grades. The austenitic SS are plagued by localized corrosion such as pitting, crevice corrosion, and stress-corrosion cracking (SCC) in the presence of aqueous chlorides. Therefore, the superferritics are often used in heat exchangers and piping systems for chloride-bearing aqueous solutions and seawater.

The *austenitic SS* constitute the largest class, in terms of both the number of alloys and usage. They

Table 12-27 Composition of selected standard and special stainless steels.

UNS	AISI	Composition, wt% max									
designations	types	C	Ma	Si	P	S	Cr	Ni	Mo	N	Others
Ferritic alloys											
S40500	405	0.08	1.00	1.00	0.040	0.030	11.50–14.50	···	···	···	0.10–0.30 Al
S43000	430	0.12	1.00	1.00	0.040	0.030	16.00–18.00	···	···	···	···
Martensitic alloys											
S41000	410	0.15	1.00	1.00	0.040	0.030	11.50–13.00	···	···	···	···
S42000	420	0.15	1.00	1.00	0.040	0.030	12.00–14.00	···	···	···	···
S44004	440C	0.95-1.20	1.00	1.00	0.040	0.030	16.00–18.00	···	0.75	···	···
Austenitic alloys											
S20161	Gall-Tough	0.15	4.00–6.00	3.00–4.00	0.040	0.040	15.00–18.00	4.00–6.00	···	0.08–0.20	···
S21800	Nitronic 60	0.10	7.00–9.00	3.50–4.50	0.040	0.030	16.00–18.00	7.00–9.00	···	0.08–0.20	···
S30400	304	0.08	2.00	1.00	0.045	0.030	18.00–20.00	8.00–10.50	···	···	···
S30403	304L	0.03	2.00	1.00	0.045	0.030	18.00–20.00	8.00–12.00	···	···	···
N08020	20 Cb-3	0.07	2.00	1.00	0.045	0.035	19.00–21.00	32.00–38.00	2.00–3.00	···	8 x C-1.00 Nb; 3.00–4.00 Cu
Duplex alloys											
S32950	7- Mo Plus	0.03	2.00	0.60	0.035	0.010	26.0–29.0	3.50–5.20	1.00–2.50	0.15–0.35	···
Precipitation-hardenable alloys											
S17400	17-4Ph	0.07	1.00	1.00	0.040	0.030	15.50–17.50	3.00–5.00	...	···	0.15–0.45 Nb; 3.00–5.00 Cu
S45500	Custom 455	0.05	0.50	0.50	0.040	0.030	11.00–12.50	7.50–9.50	0.50	···	0.10–0.50 Nb; 1.50–2.50 Cu 0.80–1.40 Ti

Table 12-28 Properties of selected stainless steels compared to 1080 steel, 6061 aluminum, and a copper alloy.

UNS or AISI type	Condition	Rockwell hardness	Average tensile properties: Yield strength, 0.2% offset, MPa	ksi	Ultimate tensile strength, MPa	ksi	Elongation in 50.8 mm (2.0 in.), %	Reduction of area, %
Austenitic stainless								
Type 304	Annealed	81 HRB	241	35	586	85	60.0	70.0
N08020	Annealed	84 HRB	276	40	621	90	50.0	65.0
S20161	Annealed	93 HRB	365	53	970	140	59.0	64.0
S21800	Annealed	95 HRB	414	60	710	103	64.0	74.0
Ferritic								
Type 405	Annealed	81 HRB	276	40	483	70	30.0	60.0
Type 430	Annealed	82 HRB	310	45	517	75	30.0	65.0
Duplex								
S32950	Annealed	100 HRB	570	82	760	110	38.0	78.0
Martensitic								
Type 410	Annealed	82 HRB	276	40	517	75	35.0	70.0
	Oil quenched from 1010°C (1850°F) and tempered:							
	at 250°C (500°F)	43 HRC	1089	158	1337	193	17.0	62.0
	at 593°C (1100°F)	26 HRC	724	105	827	120	20.0	63.0
Type 420	Annealed	92 HRB	345	50	655	95	25.0	55.0
	Oil quenched from 1038°C (1900°F) and tempered at 316°C(600°F)	52 HRC	1482	215	1724	250	8.0	25.0
Type 440C	Annealed	97 HRB	448	65	758	110	14.0	30.0
	Oil quenched from 1038°C(1900°F) and tempered at 316°C (600°F)	57 HRC	1896	275	1975	285	2.0	10.0
Precipitation hardened								
S45500	Annealed	31 HRC	793	115	1000	145	14.0	70.0
	Water quenched from 1038°C(1900°F)and aged:							
	at482°C (900°F)	49 HRC	1620	235	1689	245	10.0	45.0
	at 566°C (1050°F)	40 HRC	1207	175	1310	190	15.0	55.0
S17400	Annealed	31 HRC	793	115	965	140	12.0	50.0
	Water quenched from 1038°C (1900°F) and aged:							
	at 482°C(900°F)	44 HRC	1262	183	1365	198	15.0	52.0
	at 621°C(1150°F)	33 HRC	869	126	1131	164	17.0	59.0
Carbon steel								
AISI 1080	Annealed	97 HRB	455	66	821	119	15.0	22.0
	Oil quenched from 816 °C (1500 °F) and tempered at 204 °C(400 °F)	42 HRC	980	142	1304	180	12.0	35.0
Aluminum alloy								
Type 6061	Annealed	...	55	8	124	18	25.0	...
	Aged	56 HRB	276	40	311	45	12.0	...
Copper alloy								
Al bronze (95Cu-5 Al)	Annealed	45 HRB	173	25	380	55	65.0	...

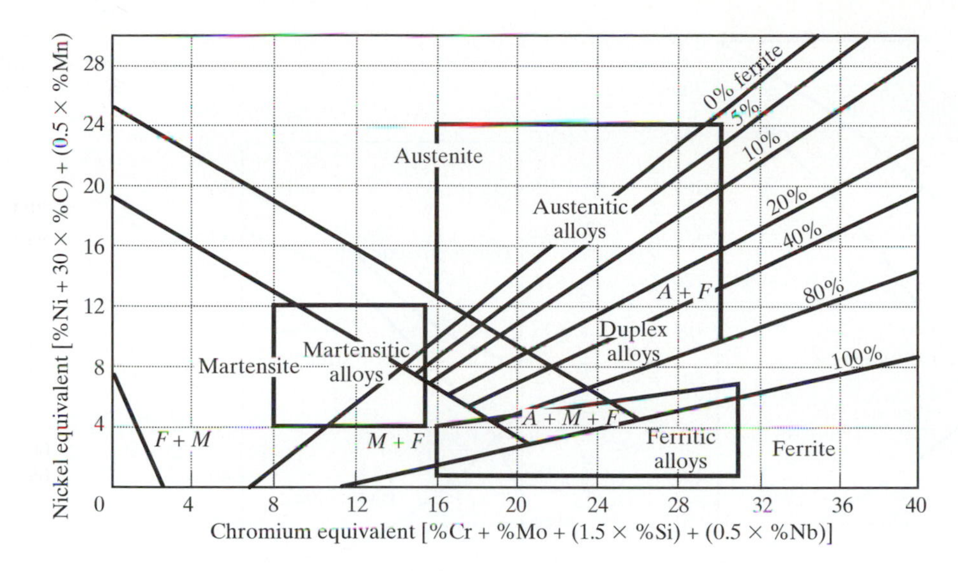

Fig. 12-26 Schaefler constitution diagram for stainless steels.

have the FCC structure at room and high temperatures; therefore, they cannot be heat-treated. They are ordinarily non-magnetic and exhibit excellent ductility, formability, and toughness even at cryogenic temperatures in the annealed condition. Type 304 is the most used stainless steel grade due to its excellent formability, as demonstrated in Table 12-28 by its ductility and adequate corrosion characteristics. The austenitic SS can only be strengthened by cold-work and in a very dramatic way, as indicated in Fig. 12-27 for AISI 301 and 305. With low-alloy content, such as in AISI 301 and 304, heavy cold-work or deformation, such as that in machining or cold-forming operations, induces the austenite to transform to martensite which is magnetic.

The austenitic SS contain generally from 16 to 26 percent chromium, up to 35 percent nickel, and up to 20 percent manganese. Nickel and manganese are the principal austenite formers, although carbon and nitrogen are also used because they dissolve readily in austenite. The 2XX series contain some nitrogen, up to 7%Ni, and require from 5 to 20%Mn to increase the solubility of nitrogen in austenite and to prevent martensite formation. The interstitial nitrogen in solid solution increases the strength of the austenite. The 3XX series contain larger amounts of nickel and up to 2%Mn. Types 301 and 304 are the leanest and are considered the base alloys in the 3XX series. As indicated, 304 is the most used. Annealed 3XX SS have typical tensile yield strengths from 200 to 275 MPa (30 to 40 ksi), whereas the high-nitrogen 200 grades may have yield strengths up to 500 MPa (70 ksi).

A wide range of corrosion resistance can be achieved by balancing the ferrite stabilizers, such as chromium and molybdenum, with the austenite stabilizers, see Fig. 12-25. Molybdenum is added to Type 304 to produce Types 317 and 316 for enhanced pitting corrosion resistance in chloride environments. The high chromium grades, Types 309 and 310, are used in oxidizing environments and high-temperature applications. High-nickel grades such as N08020 are used in severe reducing acid environments. Alloys containing high nickel, molybdenum (~6%), and nitrogen (~0.20%) are sometimes referred as the superaustenitics in Fig. 12-25. Titanium and niobium are added to stabilize the carbon in Types 321 and 347 to prevent intergranular corrosion after elevated temperature exposure. Lower carbon grades (with L or S suffixes after the AISI numbers), such as 304L, also prevent intergranular corrosion. Figure 12-28 shows the time-temperature-carbide precipitation (sensitization) curves of austenitic SS with various carbon contents. The L grade with a maximum 0.03%C has its carbide precipitation curve pushed to the right and therefore allows the austenite to be cooled slow without being sensitized (no carbide precipitation).

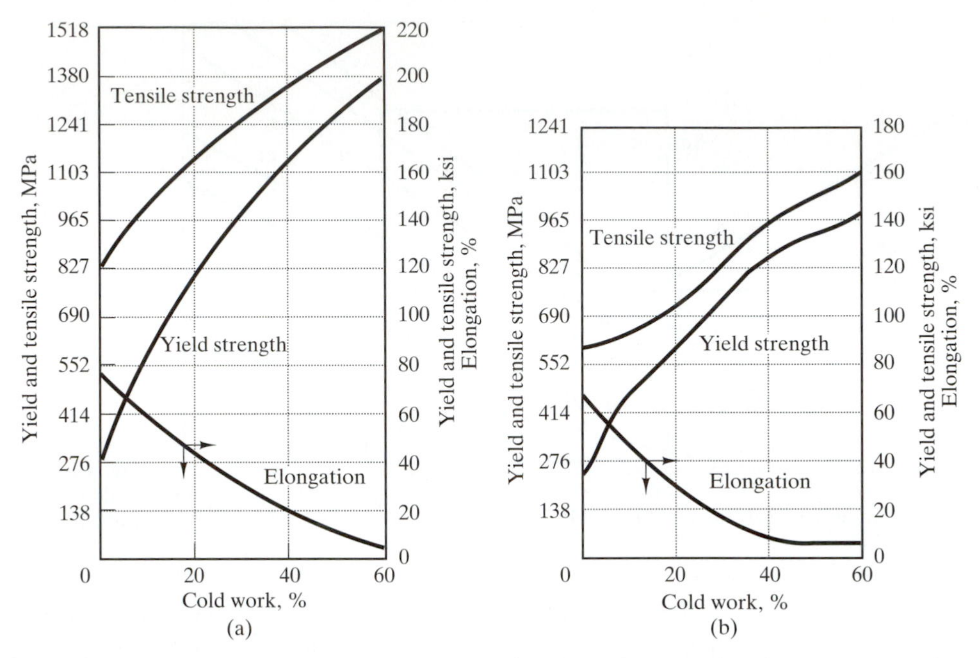

Fig. 12-27 Cold-work significantly increses yeild and tensile strengths of austenitic stainless steels, (a) 301 type, and (b) 305 type.

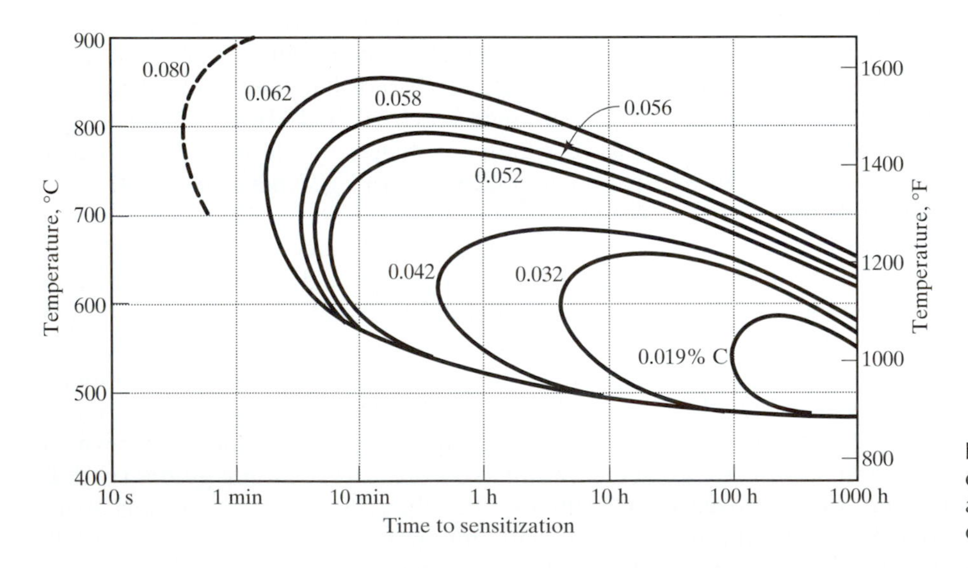

Fig. 12-28 Time-temperature curves for sensitization of autenitic SS for various carbon contents.

Martensitic SS are transformed from the austenite FCC (face-centered cubic) structure at high temperatures to the BCT (body-centered tetragonal) martensite structure at low temperatures when the austenite is rapidly cooled in air or in a liquid. In the annealed condition (i.e., the very slow cooling of the austenite in the furnace), the microstructure consists of the soft ferritic phase. Steels in this group contain usually no more than 14%Cr, except for Types 440A, 440B, and 440C that contain 16 to 18 percent. The carbon content is ordinarily low or medium carbon, except for the 440 types that have high carbon content from 0.60

to 1.20 percent. The chromium and the carbon contents are balanced in order to ensure the formation of martensite. Niobium, silicon, tungsten, and vanadium may be added to modify the tempering characteristics. Small amounts of nickel are added to improve corrosion resistance in some media and to improve toughness.

In order to obtain the desired structure and properties of martensitic SS, they are heat-treated like the low-alloy steels (i.e., they are austenitized, quenched, and tempered). Properties of some annealed and quenched and tempered martensitic SS are shown in Table 12-28. As shown with the low-alloy steels, the hardness and the strength of the as-quenched martensite and the Q + T structure depend on the carbon content. The alloys with 16%Cr and from 0.6 to 1.1%C may exhibit 60 R_C hardness and tensile yield strength of 1900 MPa (280 ksi). The high hardnesses of these steels also make them very wear resistant. Type 440C (1.1%C) have excellent adhesive and abrasive wear resistance, similar to tool steels, whereas Type 410 (0.1%C) have relatively poor wear resistance.

Molybdenum and nickel are added to martensitic SS to improve corrosion and toughness properties. Nickel also acts to balance the effect of high chromium to maintain the desired structure with not much free ferrite. Molybdenum and nickel additions are, however, restricted in order not to produce retained austenite (not fully martensitic) in the quenched structure. This limitation of alloy addition to obtain the 100 percent martensitic structure restricts the corrosion resistance to moderate levels.

Martensitic SS are specified when the application requires good tensile strength, creep, and fatigue strength properties, in combination with moderate corrosion and heat resistance up to approximately 650°C. In the United States, low- and medium-carbon martensitic steels (for example, type 410 and modified versions of it) have been used primarily in steam turbines, jet engines, and gas turbines. Type 420 and similar alloys are used in cutlery, valve parts, gears, shafts, and rollers. Martensitic SS are also used in petroleum and petrochemical equipment. Type 440 grades with high carbon are also used for surgical and dental instruments, scissors, springs, cams, and ball bearings, in addition to the uses of type 420.

Duplex SS exhibit improved properties over either austenitic or ferritic SS because of the presence of about equal amounts (50–50) of the austenite and ferrite phases in the microstructure. They have improved stress-corrosion cracking resistance over the austenitic SS because of the ferrite phase, and have improved toughness and ductility over the ferritic SS because of the austenite phase. The two-phase alloys also exhibit synergism in their yield strengths. They exhibit tensile yield strengths from 550 to 690 MPa (80 to 100 ksi) in the annealed condition, which is about twice the yield strength of either phase alone.

Current commercial grades contain 22-26%Cr, 4-7%Ni, up to 4.5%Mo, about 0.7% Cu and W, and 0.08-0.35%N. They are loosely divided into four generic types: (1) Fe-23Cr-4Ni-0.1N, (2) Fe-22Cr-5.5Ni-3Mo-0.15N; (3) Fe-25Cr-5Ni-2.5Mo-0.17N-Cu; and (4) Fe-25Cr-7Ni-3.5Mo-0.25N-W-Cu. The last type is frequently referred to as the "super" duplex SS. The continual modifications of the composition have improved corrosion resistance, workability, and weldability. In particular, nitrogen additions have improved the pitting corrosion resistance and weldability of these alloys.

Duplex SS are used in a range of industries, in particular, the oil and gas, petrochemical, pulp and paper, and pollution control industries. They are commonly used in aqueous, chloride-containing environments, and as replacements for austenitic SS that have suffered either chloride SCC or pitting during service. The super-duplex grades are resistant to oxygenated or chlorinated seawater. Because of possible embrittlement, the non-welded materials are limited to applications below 280°C, while welded materials are limited below 250°C.

Precipitation-Hardening (PH) SS can be austenitic (such as S66286), semiaustenitic (such as S17700), or martensitic (such as S17400). These steels have very low carbon contents and because of this, the primary hardening is due to precipitation-hardening, even for the martensitic grade. The alloying elements used in PH steels are aluminum, titanium, niobium, and/or copper. The PH grades have generally good ductility and toughness with moderate-to-good corrosion resistance. A better combination of strength and corrosion resistance is achieved with the PH alloys than with the martensitic SS alloys. These properties are due to the greater alloying elements and the restricted carbon content (0.04%C maximum). The latter is critical for good toughness and good ductility, but at the expense of wear resistance.

The PH alloys exhibit high tensile yield strengths, up to 1700 MPa (250 ksi). Cold-working prior to aging can increase the yield strength more. Because of these strength levels, most of the applications for the PH steels are in the aerospace and other high-technology industries. The most popular of the PH steels is S17400.

12.7.2 Selection of SS

The selection of SS is primarily based on corrosion resistance and secondarily on mechanical properties. Other factors such as fabricability, wear, service temperatures, toughness, and physical properties may come into consideration for specific purposes. Based on corrosion, the most serious to guard against are the localized forms that can cause unexpected and sometimes catastrophic failure, while most of the structure remains unaffected. Some of these forms are SCC (stress corrosion cracking), crevice corrosion, pitting, and intergranular attack in sensitized materials such as in weld heat-affected zones (HAZ).

The selection of materials based on corrosion is a very difficult task because corrosion is not easily quantifiable. Corrosion for a specific application is very difficult to assess because of very subtle factors. Even a seemingly minor impurity in the medium in parts per million concentration can dramatically increase the corrosive attack. The presence of stray electrical currents and heat transfer through the steel to and from the corrosive medium can dramatically increase the damage. At elevated temperatures, minor changes in the atmosphere can affect the scaling and sulfidation. Laboratory data can be very misleading in predicting service performance. Even actual service data has limitations because of some of the subtle factors mentioned above. Thus, the designer quite often relies on previous experiences and for these, Tables 12-29 and 12-30 provide the applicability of the different types of SS in various classes of environments, in various acids, bases, organics, and in pharmaceuticals.

As a general rule of thumb, the selection process starts with the basic type 304 (S30400), the most commonly used stainless steel. Type 304 has good corrosion resistance and resists most oxidizing acids, many sterilizing solutions, most organic chemicals and dyes, and a wide range of inorganic chemicals, see Table 12-30. For industrial processes requiring a higher level of corrosion, types 316 (S31600) and 317 (S31700) should be considered. These grades have a molybdenum addition (Fig. 12-25) to increase their pitting resistance against chlorides. Type 316 is generally specified for use with corrosive chemicals used to produce inks, rayons, photographic chemicals, paper, textiles, bleaches, and rubber. It is also used for surgical implants within the hostile environment of the human body. For severe corrosive environments in which chloride SCC is the primary concern, the superaustenitics, duplex SS, or the superferritics should be considered. For less severe corrosive environments, ferritic SS such as type 430 may be adequate. The low-alloy content of this grade makes it less corrosion resistant, but makes it also much cheaper than type 304, 316, and the superalloys. Type 430 resists effectively foods, fresh water, and nonmarine atmospheric corrosion.

12.8 Selection of Cast Irons

Cast irons are alloys with more than 2%C in iron. The type of cast iron produced during solidification depends on the phases that form during the eutectic reaction. We recall that the iron-carbon has two possible systems: (1) the metastable iron-cementite (Fe_3C), and (2) the iron-graphite (C), shown in Fig. 12-29. We see that there are two possible eutectic reactions: one occurring at 1154°C with a liquid composition of 4.26%C for the iron-graphite, and the other, at 1148°C with a liquid composition of 4.30%C. We shall neglect the very small difference in the liquid compositions and say that both eutectic reactions occur at 4.30%C. If we cool the liquid under equilibrium conditions (i.e., infinitely slow), we shall obtain the following reaction at 1154°C,

$$L \rightarrow \gamma + C\ (\text{graphite}) \qquad (12\text{-}3)$$

However, cooling is never done in equilibrium and the liquid temperature is not going to stay infinitely long at 1154°. In practice and in a very short time, the

Table 12-29 Corrosion resistance of standard types of stainless steel to various environments.

Type	Mild atmo-spheric and fresh water	Atmospheric		Salt water	Chemical		
		Industrial	Marine		Mild	Oxidizing	Reducing
Austenitic stainless steels							
201	x	x	x	···	x	x	···
202	x	x	x	···	x	x	···
205	x	x	x	···	x	x	···
301	x	x	x	···	x	x	···
302	x	x	x	···	x	x	···
302B	x	x	x	···	x	x	···
303	x	x	···	···	x	···	···
303Se	x	x	···	···	x	···	···
304	x	x	x	···	x	x	···
304H	x	x	x	···	x	x	···
304L	x	x	x	···	x	x	···
304N	x	x	x	···	x	x	···
S30430	x	x	x	···	x	x	···
305	x	x	x	···	x	x	···
308	x	x	x	···	x	x	···
309	x	x	x	···	x	x	···
309S	x	x	x	···	x	x	···
310	x	x	x	···	x	x	···
310S	x	x	x	···	x	x	···
314	x	x	x	···	x	x	···
316	x	x	x	x	x	x	x
316F	x	x	x	x	x	x	x
316H	x	x	x	x	x	x	x
316L	x	x	x	x	x	x	x
316N	x	x	x	x	x	x	x
317	x	x	x	x	x	x	x
317L	x	x	x	x	x	x	x
321	x	x	x	···	x	x	···
321H	x	x	x	···	x	x	···
329	x	x	x	x	x	x	x
330	x	x	x	x	x	x	x
347	x	x	x	···	x	x	···
347H	x	x	x	···	x	x	···
348	x	x	x	···	x	x	···
348H	x	x	x	···	x	x	···
384	x	x	x	···	x	x	···
Ferritic stainless steels							
405	x	···	···	···	x	···	···
409	x	···	···	···	x	···	···
429	x	x	···	···	x	x	···
430	x	x	···	···	x	x	···
430F	x	x	···	···	x	···	···
430FSe	x	x	···	···	x	···	···
434	x	x	x	···	x	x	···

(Continued)

Table 12-29 Corrosion resistance of standard types of stainless steel to various environments. (*Continued*)

Type	Mild atmospheric and fresh water	Atmospheric		Salt water	Chemical		
		Industrial	Marine		Mild	Oxidizing	Reducing
436	x	x	x	...	x	x	...
442	x	x	...	...	x	x	...
446	x	x	x	...	x	x	...
Martenistic stainless steels							
403	x	...	...	...	x	...	...
410	x	...	...	...	x	...	...
414	x	...	...	...	x	...	...
416	x	...	...	...	...	...	...
416Se	x	...	...	...	...	...	...
420	x	...	...	...	...	...	...
420F	x	...	...	...	...	...	...
422	x	...	...	...	...	...	...
431	x	x	x	...	x	...	...
440A	x	...	...	...	x	...	...
440B	x	...	...	...	...	...	...
440C	x	...	...	...	...	...	...
Precipitation-hardening stailness steels							
Ph 13-8 Mo	x	x	...	...	x	x	...
15-5 PH	x	x	x	...	x	x	...
17-4 PH	x	x	x	...	x	x	...
17-7 PH	x	x	x	...	x	x	...

An x notation indicates that the specific type may be considered for application on the corrosive environment.

liquid will cool below 1148° and the following reaction can occur,

$$L \rightarrow \gamma + Fe_3C \tag{12-4}$$

The temperature separation between the two reactions is only 8°C and since we need to have undercooling of the liquid for solidification to proceed (Sec. 7.4.1), there is not much undercooling for reaction Eq. 12-3 before the liquid is also undercooled with respect to Eq. 12-4. To increase the temperature difference between the two eutectic reactions and to provide greater undercooling for Eq. 12-3, silicon is always added to cast iron to induce the graphite formation. Thus, if we cool the liquid very slowly, the graphite formation will occur. However, in spite of the silicon addition, the separation between the two reactions is not really very large and if we cool the liquid quickly, the cementite formation will be induced.

When silicon is added, the eutectic carbon composition is found to decrease according to the following equation

$$\%C + \frac{\%Si}{3} = 4.3 \tag{12-5}$$

We note, of course, that without silicon, the eutectic carbon composition is 4.3 percent. In a given alloy composition, it is convenient to take the left side of

Table 12-30 Applications of various SS grades in some acids, bases, organics, and pharmaceuticals.

Environment	Grades(a)
Acids	
Hydrochloric acid	Stainless is not generally recommended except when solutions are very dilute and at room temperature (pitting may occur).
Mixed acids	There is usually no appreciable attack on type 304 or 316 as long as sufficient nitric acid is present.
Nitric acid	Type 304L and 430 and some higher-alloy stainless grades have been used.
Phosphoric acid	Type 304 is satisfactory for storing cold phosphoric acid up to 85% and for handling concentrations up to 5% in some unit processes of manufacture. Type 316 is more resistant and is generally used for storing and manufacture if the flourine content is not too high. Type 317 is somewhat more resistant than type 316. At concentrations ≤85%, the metal temperature should not exceed 100°C (212°F) with type 316 and slightly higher with type 317. Oxidizing ions inhibit attack.
Sulfuric acid	Type 304 can be used at room temperature for concentrations <80 to 90%. Type 316 can be used in contact with sufuric acid ≤ 10% at temperatures ≤50°C (120°F) if the solutions are aerated: the attack is greater in air-free solutions. Type 317 may be used at temperatures as high as 65°C (150°F) with ≤5% concentration. The presence of other materials may markedly change the corrosion rate. As little as 500 to 2000 ppm of cupric ions make it possible to use type 304 in hot solutions of moderate concentration. Other additives may have the opposite effect.
Sulfurous acid	Type 304 may be subject to pitting, particularly if some sulfuric acid is present. Type 316 is usable at moderate concentrations and temperatures.
Bases	
Ammonium hydroxide, sodium hydroxide, caustic solutions	Steels in the 300 series generally have good corrosion resistance at virtually all concentrations and temperatures in weak bases, such as ammonium hydroxide. In stronger bases such as sodium hydroxide, there may be some attack, cracking, or etching in more concentrated solutions and/or at higher temperatures. Commercial-purity caustic solutions may contain chlorides, which will accentuate any attack and may cause pitting of type 316, as well as type 304.
Organics	
Acetic acid	Acetic acid is seldom pure in chemical plants but generallly includes numerous and varied minor constituents. Type 304 is used for a wide variety of equipment including stills, base heaters, holding tanks, heat exchangers, pipelines, valves, and pumps for concentrations ≤99% at temperatures < ~50°C (120°F). Type 304 is also satisfactiry—if small amounts of turidity or color pickup can be tolerated—for room temperature storage of glacial acetic acid. Types 316 and 317 have the broadest range of usefulness, especially if formic acid is also present or if solutions are unaerared. Type 316 is used for fractionating equipment, for 30-90% concentrations where type 304 cannot be used, for storage vessels, pumps, and process equipment handling glacial acetic acid, which would be discolored by type 304. Type 316 is likewise applicable for parts having temperatures >50°C (120°F), for dilute vapors, and for high pressures. Type 317 has somewhat greater corrosion resis tance than type 316 under severly corrosive conditions. None of the stainless steels has adequate corrosion resistance to glacial acetic acid at the boiling temperature or at superheated vapor temperatures.
Aldehydes	Type 304 is generally satisfactory.
Amines	Type 316 is usually preferred to type 304.
Cellulose accetate	Type 304 is satisfactory for low temperatures, but type 316 or type 317 is needed for high temperatures.
Formic acid	Type 304 is generally acceptable at moder ate temperatures, but type 316 is resistant to all concentrations at temperatures up to boiling.

(Continued)

Table 12-30 Applications of various SS grades in some acids, bases, organics, and pharmaceuticals. (*Continued*)

Environment	Grades(a)	Environment	Grades(a)
Esters	With regard to corrosion, esters are comparable to organic acids.	Tall oil (pulp and paper industry)	Type 304 has only limited use in tall-oil distillation service. High rosin acid streams can be handled by type 316L, with a minimum molybdenum content of 2.75%. Type 316 can also be used in the more corrosive high fatty acid streams at temperatures <245°C (475°F), but type 317 will probably be required at higher temperatures.
Fatty acids	Type 304 is resistant to fats and fatty acids ≤~150°C (300°F), but type 316 is needed at 150-260°C (300-500°F), and type 317, at higher temperatures.		
Paint vehicles	Type 316 may be needed in exact color and lack of contamination are important.		
Phthalic annydride	Type 316 is usually used for reactors, fractionating columns, traps, baffles, caps, and piping.	Tar	Tar distillation equipment is almost all type 316 because coal tar has a high chloride content: type 304 does not have adequate resistance to pitting.
Soaps	Type 304 is used for parts such as spray towers, but type 316 may be preferred for spray nozzles and flake-drying belts to minimize off-color product.	Urea	Type 3161, is generally required.
Synthetic detergents	Type 316 is use for preheat, piping, pumps, and reactors in catalytic hydrogenation of fatty acids to give salts of sulfonated high-molecular alcohols.	**Pharmaceuticals**	Type 316 is usually selected or all parts in contact with the product because of its inherent corrosion resistance and greater assurance of product purity.

(a) The stainless steels mentioned may be consisdered for use in the indicated environments. Additional information or corrosion expertise may be neccessary prior to use in some environments: for example, some impurities may cause localized corrosion (such as chlorides causing pitting, or stress-corrosion cracking of some grades.

Eq. 12-5 and to define it as the carbon equivalent (C.E.),

$$\text{C.E.} = \%\text{C} + \frac{\%\text{Si}}{3} \tag{12-6}$$

When phosphorus is added, the C.E. is defined by

$$\text{C.E.} = \%\text{C} + \frac{\%\text{Si} + \%\text{P}}{3} \tag{12-7}$$

We say that when the C.E. is 4.3, the alloy is eutectic. Thus, when the C.E. is less than 4.3, the alloy is hypoeutectic; when the C.E. is greater than 4.3, the alloy is hypereutectic. Cast irons may have the same C.E. but differ in carbon and silicon contents.

The carbon and silicon contents of a cast iron, expressed as C.E., establish the solidification temperature range of the alloy as well as its casting, mechanical, and other properties. However, cast irons with the same C.E. but appreciably different carbon and silicon contents will not have similar casting properties. For example, carbon is more than twice as effective as silicon in preventing solidification shrinkage, while silicon is more effective than carbon in keeping thin sections from becoming hard.

12.8.1 Types and Characteristics of Cast Irons [8]

The ranges of compositions for typical unalloyed cast irons are given in Table 12-31 and graphically illustrated in Fig. 12-30. The first two types in Table 12-31 (i.e., white and malleable irons), follow Eq. 12-4 and form cementite, while the other three (gray, ductile, and compacted) follow Eq. 12-3 and form graphite during the eutectic reaction. The types of cast irons are classified according to the form or shape that the carbon or graphite has in the microstructure.

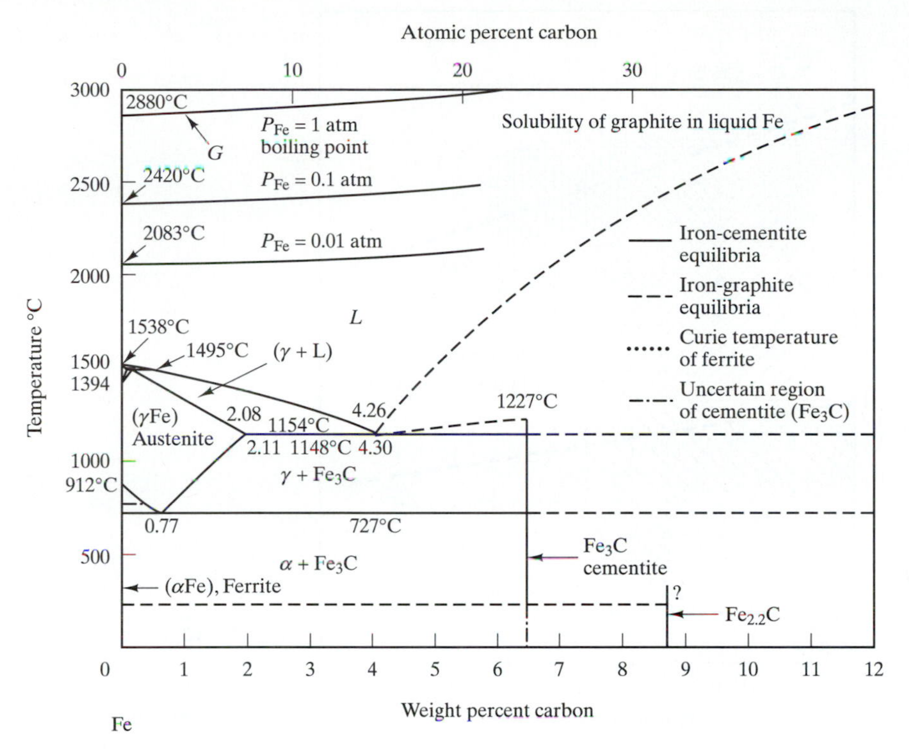

Fig. 12-29 Iron-carbon equilibria: the iron-cementite and iron-graphite systems have very close eutectic temperatures without silicon.

Table 12-31 Range of compositions of unalloyed types of cast iron.

	Percent (%)				
Type of Iron	**Carbon**	**Silicon**	**Manganese**	**Sulfur**	**Phosphorus**
White	1.8-3.6	0.5-1.9	0.25-0.8	0.06-0.2	0.06-0.2
Malleable (Cast White)	2.2-2.9	0.9-1.9	0.15-1.2	0.02-0.2	0.02-0.2
Gray	2.5-4.0	1.0-3.0	0.2-1.0	0.02-0.25	0.02-1.0
Ductile	3.0-4.0	1.8-2.8	0.1-1.0	0.01-0.03	0.01-0.1
Compacted Graphite	2.5-4.0	1.0-3.0	0.2-1.0	0.01-0.03	0.01-0.1

White cast irons have typically low carbon and low silicon, are hypoeutectic (C.E. < 4.3), and solidify with mostly the carbon as cementite, as demonstrated in Fig. 12-31. Cementite or iron carbide is a hard and brittle phase. Subsequently, white cast iron is a hard and brittle material and derives its name from the white crystalline fracture surface it produces when it breaks. It has a very high compressive strength, excellent wear resistance, and retains its hardness for limited periods of time up to red heat. It is therefore excellent for wear and abrasion-resistant applications, but it is normally unmachinable. It is not as easy to cast as the other cast irons since its solidification temperature is generally higher, and the formation of iron carbide leads to a larger solidification shrinkage than most of the irons, in particular, the gray iron.

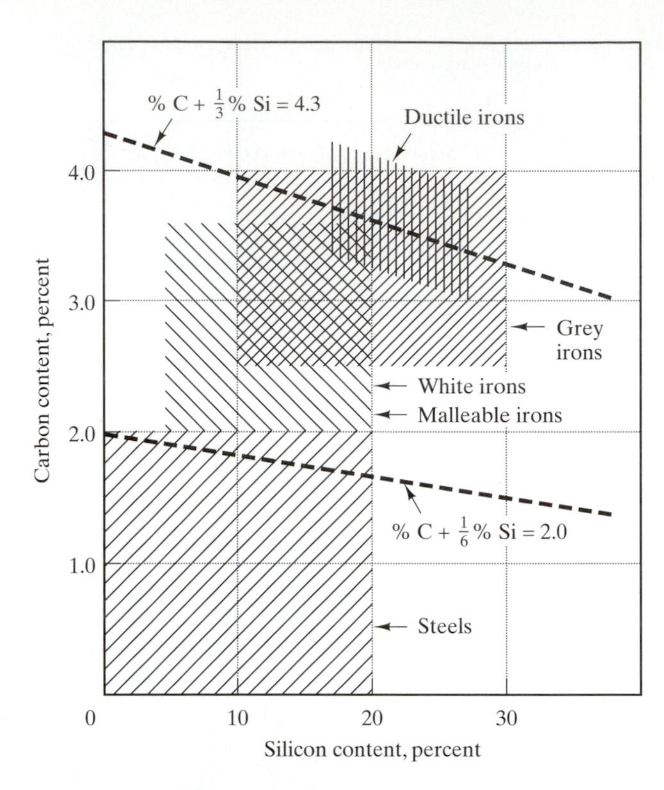

Fig. 12-30 Approximate ranges in carbon and silicon contents in the various types of cast irons and steels.

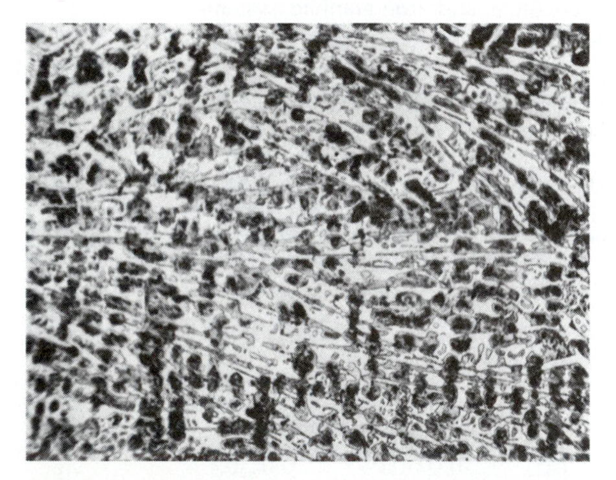

Fig. 12-31 White iron microstructure—white area is cementite (iron carbide) and dark areas are unresolved pearlite.

Malleable cast iron is characterized by the majority of its carbon appearing as irregularly shaped nodules of graphite in the microstructure, as shown in Fig. 12-32. It is first solidified as white cast iron (i.e., the carbon as cementite), and then given an extended tempering treatment in the range of 800° to 970°C. At this temperature range, the treatment causes the cementite to dissociate to austenite and the irregular nodular graphite as in Fig. 12-32. Because of the way it is formed (via a tempering reaction), this form of graphite is called temper carbon. When the iron is cooled very slowly from the tempering temperature, the austenite further decomposes to ferrite and additional graphite to produce the *regular or ferritic malleable iron with the graphite in a ferritic matrix,* as seen in Fig. 12-32. When the iron is cooled faster from the tempering temperature, or when other alloying elements are present, to induce the transformation of the austenite to pearlite, bainite, or martensite, we obtain the higher strength *pearlitic malleable iron with the graphite in either pearlite, bainite, or martensite* matrix.

Gray cast iron is produced when the carbon equivalent is very close to the eutectic (4.30%) and when the cooling rate during solidification is slow. The carbon forms as interconnected graphite flakes in a eutectic cell, as shown in the scanning electron micrograph in Fig. 12-33. The graphite flake is Type VII in Fig. 12-34 as established by ASTM A247; Type VII is further subclassified into five types in Fig. 12-35. Type A flake graphite, which is uniformly distributed with

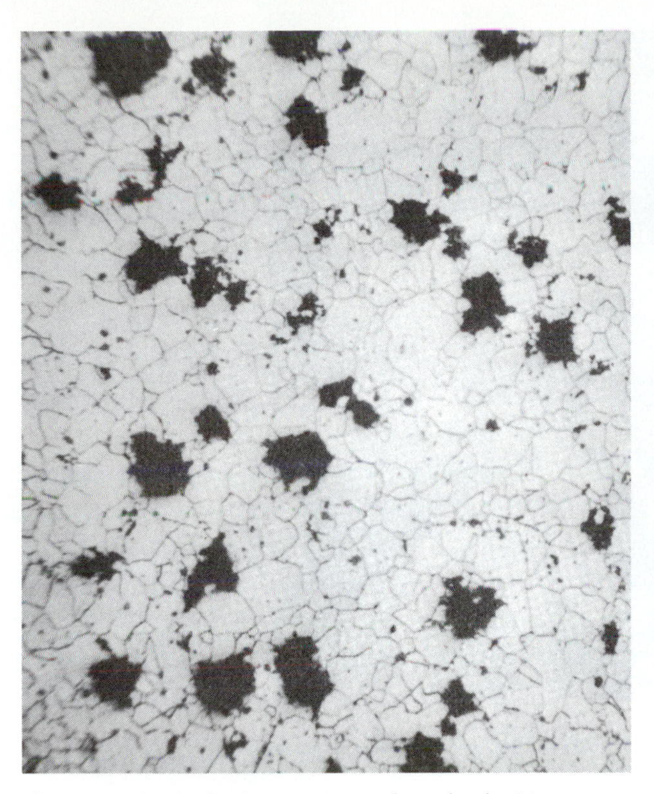

Fig. 12-32 Typical microstructure of regular ferritic malleable cast iron.

random orientation, is the common preferred type for mechanical applications. When gray iron is broken, the fracture occurs mostly along the graphite interface and produces the characteristic gray color of the surface from which the name of the iron is derived. Because the majority of iron castings produced are of the gray type, the generic term cast iron is often improperly used to mean gray cast iron specifically. The presence of the graphite flakes provides the following valuable characteristics of gray cast iron:

1. The ability to produce sound castings economically in complex shapes such as water cooled engine blocks
2. Good machinability even at wear-resisting hardness levels and without burring
3. Dimensional stability under differential heating such as in brake drums and discs
4. High vibration damping capacity as in power transmission cases
5. Borderline lubrication retention as in internal combustion engine cylinders.

It is possible to induce the formation of cementite (white iron) in selected areas of a gray cast iron. This is done by localized rapid solidification of an area to produce the white iron in the gray iron as shown in Fig. 12-36. This white iron is commonly referred to as *chilled iron* and is induced in areas of the casting where wear and abrasion are significant considerations.

Ductile cast iron is sometimes referred to as nodular iron and is called SG or spherulitic graphite iron in Europe since the graphite occurs as nodules and spheroids or spherulites. The most common form is Type I in Fig. 12-34, although Types II, V, and VI may occasionally occur. The compositions and the C.E. of gray, ductile, and compacted irons are very similar, and the change from the flake to the nodular or spheroidal form of graphite is achieved by specific controlled additions of magnesium and/or cerium to the melt before solidification. Magnesium and cerium, which are called nodularizing agents, react with the sulfur and oxygen in the molten iron, and changes the way the graphite is formed. Since the compositions of ductile irons are very similar to gray irons, they are as easily cast as the gray irons and can be cast in a wide range of section sizes—either very thin or very thick.

Compacted graphite iron is a very recent addition to the types of cast irons and its name is derived from the blunted form of the graphite, Type IV in Fig. 12-34. This graphite structure and the resulting properties of the iron are intermediate between those of the gray and the ductile irons. The compacted graphite shape has been observed for a long time before its commercial production and has been called quasi-flake, aggregated flake, seminodular, and vermicular graphite. Its commercial production is very similar to that of ductile iron with the controlled additions of magnesium and/or cerium, but requires also the addition of another element such as titanium to minimize the formation of spherulitic graphite. It retains much of the castability of gray iron, but has higher strength and ductility.

Analogous to plain carbon steels that contain carbon and up to 1%Mn, the foregoing types of cast irons containing only carbon and silicon may also be referred to as plain cast irons or unalloyed cast irons, as Table 12-31 indicates. When other elements are added

Fig. 12-33 Scanning electron micrograph showing interconnected graphite flakes in gray iron.

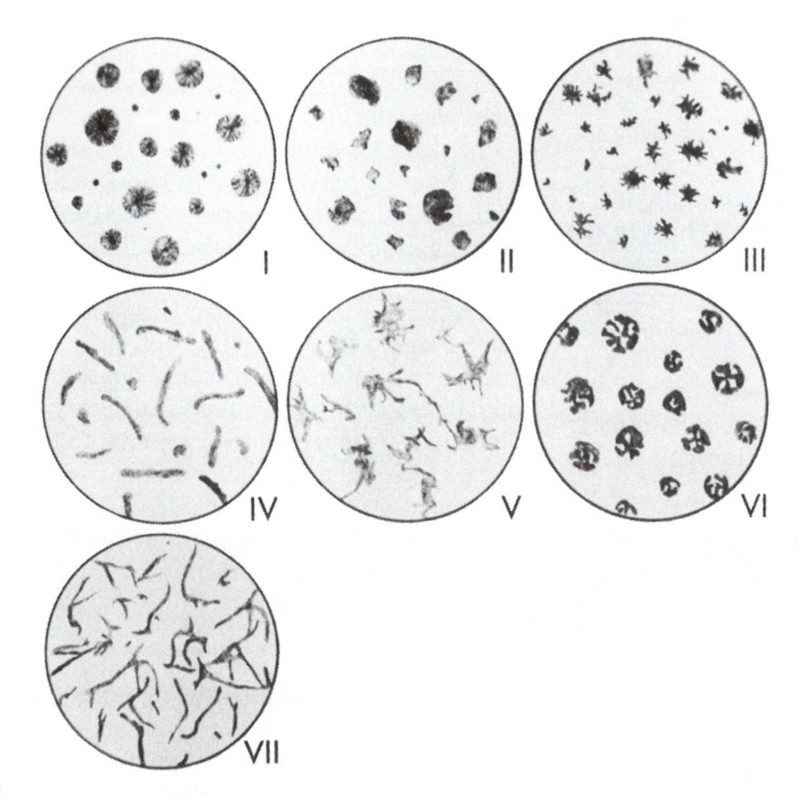

Fig. 12-34 Seven types of graphite according to ASTM A-247.

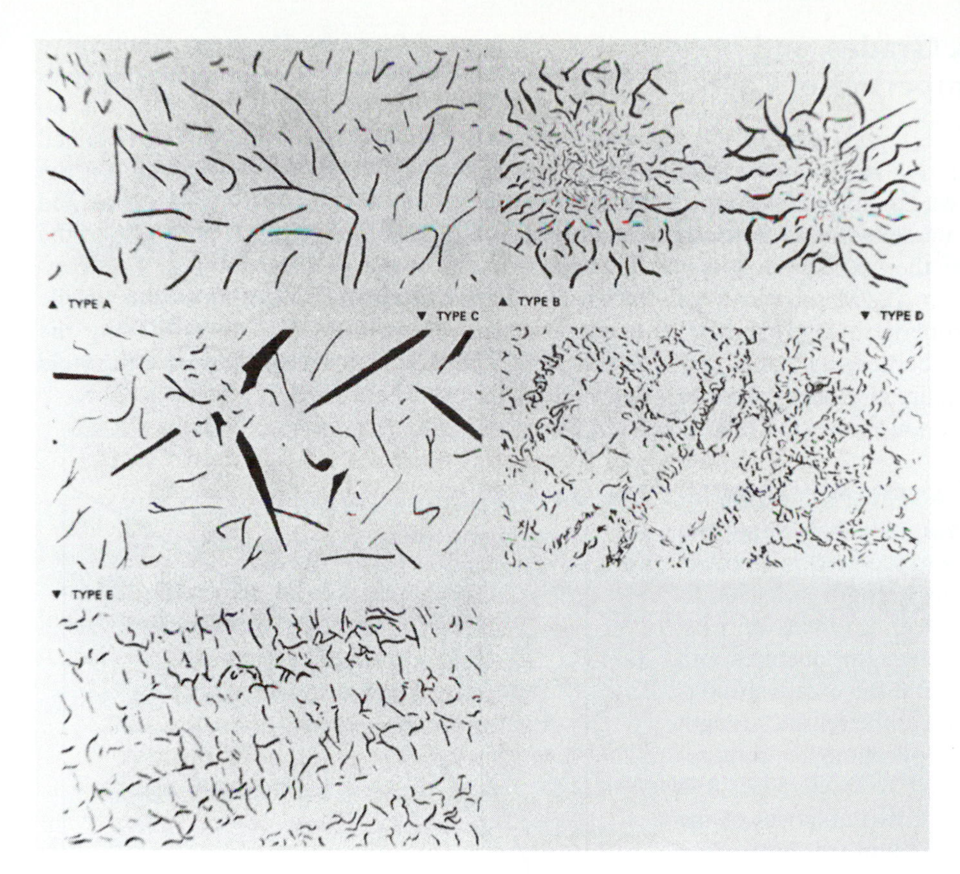

Fig. 12-35 Five subclassifications of Type VII flake graphite in Fig. 12-34.

to cast irons in amounts over 0.1%, they are called alloy additions. *Low alloy cast irons* contain no more than 3 percent total additions that are used to modify the matrix microstructure or to enhance the properties of the base iron or to extend the base iron's application into a casting size in which it would be unsuitable without the alloying additions. The alloy additions may be made in the ladle before the melt is cast.

High-alloy cast irons have from 3 to 40% total alloy addition and are used for either their corrosion resistance, high temperature properties, or wear and abrasion resistance. These alloys retain the castability of the base irons to enable the production of simple or complex shaped castings. High-alloy additions are applied only to white, gray, and ductile cast irons and these high alloy irons are produced by specialized foundries. Alloying additions are not made for malleable irons because they interfere with the formation of the temper carbon. The effects of alloying additions on compacted graphite properties are being investigated.

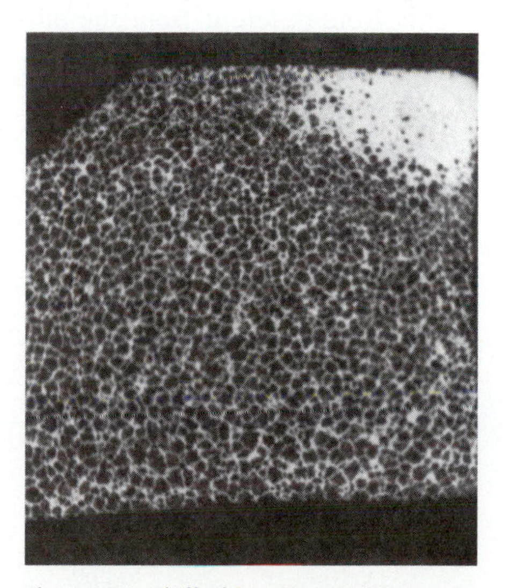

Fig. 12-36 Chilled iron in gray iron.

12.8.2 Specifications, Grades, and Mechanical Properties of Cast Irons

The specifications for the different cast irons are predominantly made by the ASTM. For automotive applications, the specifications are made by the SAE which may or may not be adopted by the ASTM. The grades or classes of irons according to ASTM specifications are based on their tensile properties, while the SAE specifications specify minimum hardnesses. The advantage of the latter is that the required hardnesses can be verified by the customer with a non-destructive test.

12.8.2a Gray Cast Iron. The ASTM specification for castings for general engineering use is A48 and the grades are from 20 to 60. The grade or class number indicates the minimum tensile strength in ksi, (kips per sq. in.), and the grades are in intervals of 5 ksi from 20 to 60 ksi. The yield strength, ductility, and composition are not specified and the composition is left to the foundry to vary to meet the tensile strength requirement. For automotive applications, the requirement for each grade according to SAE J431 is a minimum hardness, but the first two numbers of the grade code implies also the minimum tensile strength in ksi, as seen in Table 12-32. Thus, the ASTM and the SAE specifications are equivalent.

The presence of the graphite flakes in the microstructure yields very unique mechanical properties for gray iron, different from the other types and other metals.

1. *Modulus of Elasticity.* The stress-strain curve of gray iron does not have a linear portion (i.e., does not obey Hooke's law) and therefore there is no unique modulus. Two types of moduli may be obtained as indicated in Fig. 12-37. The tangent modulus is the tangent at the origin of the stress-strain curve and the other is a secant modulus that is the slope of the line connecting the origin and the point on the stress-strain curve representing 25 percent of the tensile strength. This is the reason for not specifying the yield strength because we cannot determine the yield

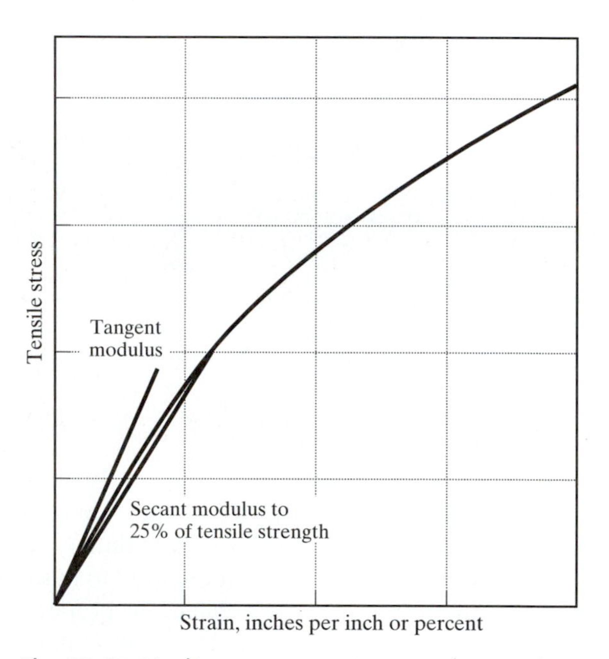

Fig. 12-37 Nonlinear stress-strain curve for gray iron showing how tangent modulus and secant modulus are obtained.

Table 12-32 Mechanical properties of automotive gray iron according to SAE J431.

SAE grade	Hardness, HB	Minimum transverse load		Minimum deflection		Minimum tensile strength	
		kg	lb	mm	in.	MPa	ksi
G1800	187 max	780	1720	3.6	0.14	118	18
G2500	170 to 229	910	2000	4.3	0.17	173	25
G3000	187 to 241	1000	2200	5.1	0.20	207	30
G3500	207 to 255	1110	2450	6.1	0.24	241	35
G4000	217 to 269	1180	2600	6.9	0.27	276	40

(a) Properties determined from as-cast test bar B (1.2-in., or 30.5-mm. diam)

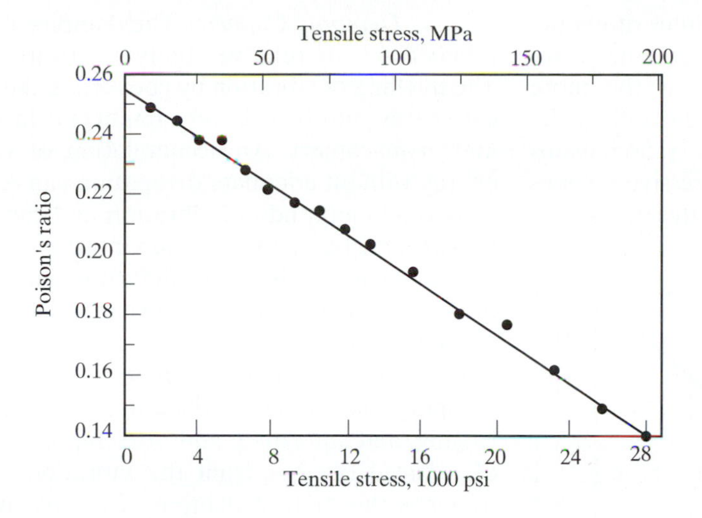

Fig. 12-38 Poisson's ratio of gray iron as a function of tensile stress.

strength in the manner described in Chap. 4. The secant modulus is the more commonly quoted value.

The modulus of elasticity is affected by the structure, composition, and processing of the gray iron. It decreases with increasing graphite content and longer graphite flake length, decreases with increasing carbon equivalent, increases when alloying elements are added and when the strength of the cast iron increase, and decreases with increasing section size and by annealing the iron.

2. *Poisson's Ratio.* For most materials, the Poisson's ratio is a constant. A value of 0.3 is usually used for steels. For most materials, there is a lateral strain to compensate the longitudinal strain when they are pulled in tension because of the constancy in volume (mass). However, for gray iron, there is a volume change in the iron as the strain increases because of the graphite flakes. The change in volume results in an increase in longitudinal strain without much lateral contraction (strain) and therefore, the Poisson's ratio decreases with increasing stress, as shown in Fig. 12-38. When the linear relation in Fig. 12-38 is extrapolated to zero stress, the Poisson's ratio unaffected by the graphite-related increase in volume is 0.26 in this case. This zero-stressed value is affected by the strength level and ranges from 0.24 for the softest irons to 0.27 for the high-strength gray irons.

3. *Compression Properties.* Because the graphite flakes act as notches, gray iron is very brittle and its tensile strength depends on the amount and shape of the flakes. As with any brittle material, it performs better in compression than in tension, and the graphite flakes do not influence the compressive strength. The compressive strength of gray iron is three to four times its tensile strength and depends more on the matrix structure (i.e., whether ferritic, pearlitic, bainitic or martensitic). The relation of compressive strength to tensile strength and hardness are shown in Figs. 12-39 and 12-40. Figure 12-41 shows the influence of the compressive stress on the modulus of

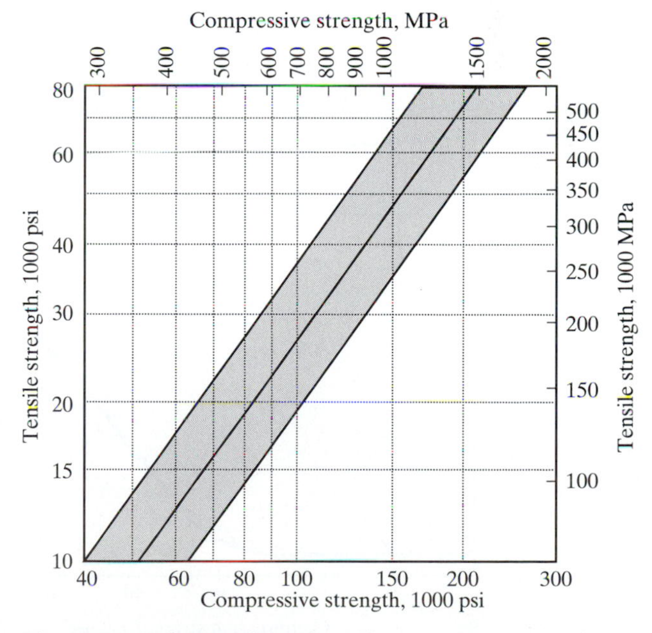

Fig. 12-39 Relation of tensile and compressive strengths in gray iron.

elasticity and Poisson's ratio. The modulus of elasticity is approximately constant with increasing compressive stress; the tangent modulus is the more appropriate in compression and is greater than the secant modulus in tension. The Poisson's ratio remains approximately constant at low-compressive stresses up to about 25 ksi and increases above this stress.

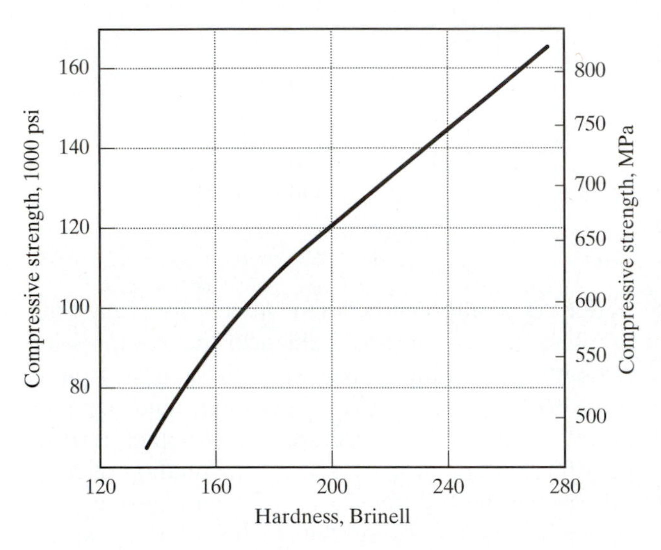

Fig. 12-40 Relationship of compressive strength as function of hardness of gray iron.

4. *Damping Capacity.* The damping capacity of a material is its relative ability to absorb vibration. The quelling of vibration by converting the mechanical energy into heat is very important in structures and in machinery. An accumulation of vibrational energy without adequate dissipation can result in an increasing amplitude of vibration and can lead to a catastrophic failure. Excessive vibration can result in inaccuracy in precision machinery, in excessive wear on gear teeth and bearings, and can induce fretting in mating surfaces.

Components made of materials with a high damping capacity can reduce noise such as chatter, ringing, and squealing, and also minimize the level of induced stresses from the vibration. Gray cast iron has the highest damping capacity among the types of cast iron, steel, and aluminum as manifested in Table 12-33. For this reason, gray iron is ideally suited for machine bases and supports, engine cylinder blocks, and brake components. This behavior is attributed to the graphite flake structure and the nonlinear stress-strain curve.

The specific damping capacity of lower class (strength) gray irons is better than the higher class irons, as illustrated in Fig. 12-42 and as Table 12-33 indicates, coarse graphite flakes are better than fine flakes. Large cast-section thickness increases specific

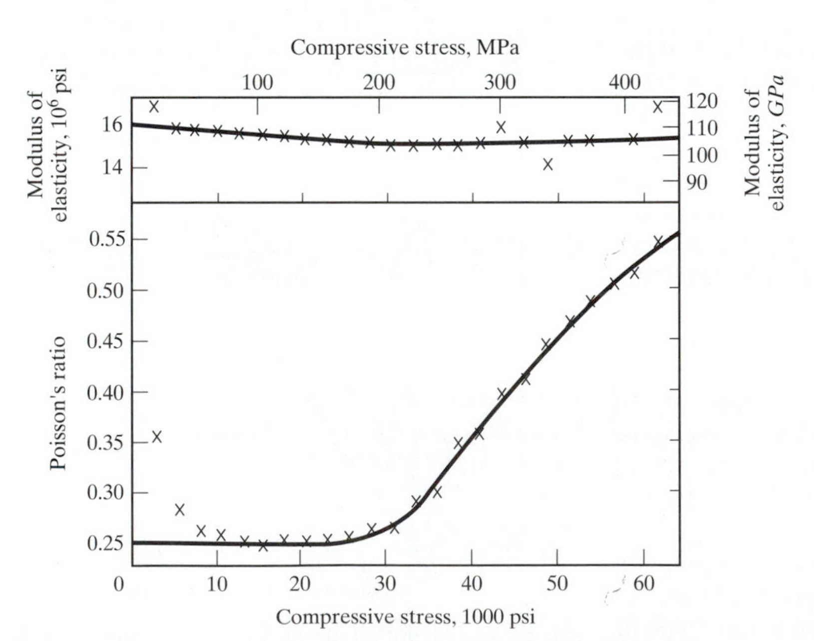

Fig. 12-41 Modulus of elasticity and Poisson's ratio depend on the compressive stress in gray iron.

Table 12-33 Relative damping capacities of some materials.

Material	$\delta \times 10^{-4}$ *
White Iron	2–4
Malleable Iron	8–15
Ductile Iron	5–20
Gray Iron, Fine Flake	20–100
Gray Iron, Coarse Flake	100–500
Eutectoid Steel	4
Armco Iron	5
Aluminum	0.4

* Natural Log of the Ratio of Successive Amplitude

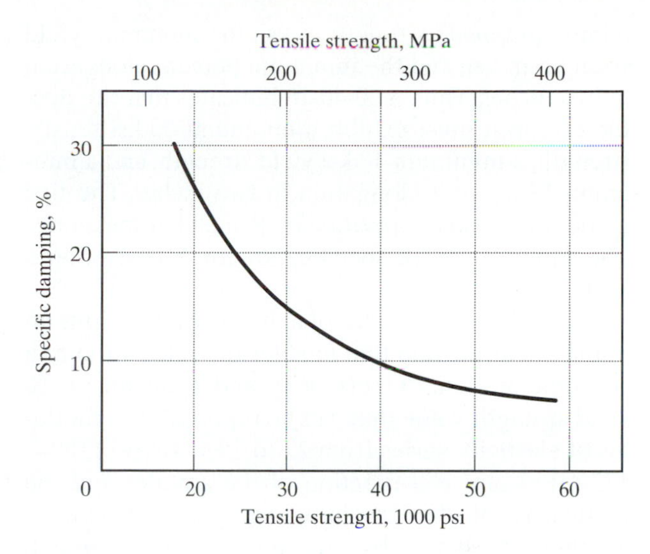

Fig. 12-42 Damping capacity of gray iron as a function of tensile strength.

Table 12-34 Grades of ductile cast iron according to ASTM A-536.

Grade and Heat Treatment	Tensile Strength minimum		Yield Strength Minimum		Percent elongation min. 2″	Typical Brinell Hardness*	Matrix Micro-structure*
	psi	MPa	psi	MPa			
60-40-18 (1)	60,000	414	40,000	276	18	149–187	ferrite
65-45-12 (2)	65,000	448	45,000	310	12	170–207	ferrite & pearlite
80-55-06 (3)	80,000	552	55,000	379	6	187–255	pearlite & ferrite
100-70-03 (4)	100,000	690	70,000	483	3	217–269	pearlite
120-90-02 (5)	120,000	828	90,000	621	2	240–300	tempered martensite

* not specified in A-536
(1) May be annealed after casting
(2) Generally obtained as-cast
(3) An as-cast grade with a higher manganese content
(4) Usually obtained by a normalizing heat treatment
(5) Oil-quenched and tempered to desired hardness

damping capacity. Inoculation and increased phosphorus content decrease the damping capacity. The matrix microstructure influences the damping capacity—ferritic (lowest strength) exhibiting the highest, and a tempered martensitic matrix is better than a pearlitic matrix.

12.8.2b Ductile cast iron. There are five grades of ductile cast irons classified by their tensile properties in ASTM Specification A-536. These are shown in Table 12-34 and are the result of the matrix microstructure surrounding the spheroidal or nodular graphite. We see that the highest grade has a quenched and tempered martensitic matrix microstructure, the lowest grade has a ferritic matrix, and the intermediate grades have ferrite-pearlite and/or pearlite matrix structures. The hyphenated numbers that indicate the grade represent, in the order given, the minimum tensile strength in ksi, the minimum yield strength in ksi, and the minimum percent elongation in two inches. Thus, a 60-40-18 indicates that the ductile cast iron must exhibit a minimum 60 ksi tensile strength, a minimum 40 ksi yield strength, and a minimum 18 percent elongation in two inches. The five grades may also be specified by Brinell hardness only. The equivalent SAE specification for ASTM A536 is SAE J434c.

The stress-strain curve of a ductile iron exhibits an initial linear portion that enables the calculation of a constant modulus of elasticity and a unique 0.2% yield strength value that can be reported. The modulus of elasticity varies from 23 to 25 $\times 10^6$ psi (159—172 GPa) and is a function of the amount and the nodularity of the graphite. Increasing amount of graphite; reflected by the total carbon content, reduces the load-bearing cross section of the iron and thus the modulus. Irregularly shaped nodules also decrease the modulus as shown in Fig. 12-43 where the modulus is calculated from resonant frequency techniques and is called dynamic. The Poisson's ratio for ductile iron is constant and is about 0.28. Because of the initial linearity of the stress-strain curve, the damping capacity of ductile iron is not significantly greater than steel, as seen in Table 12-33 and in Fig. 12-44.

The tensile strength properties are correlatable, as in steels, to both the hardness and the fatigue (endurance) limit. For steels, there is only one correlation of strength and hardness for as-rolled, normalized, and quenched and tempered structures (see Fig. 9-38) and the ratio of the fatigue limit to the tensile strength is about 0.5. However, the correlations for ductile iron depend on the matrix microstructure, as shown in Figs. 12-45 and 12-46 for strength and hardness, and Fig. 12-47 for the fatigue (endurance) limit to tensile strength ratio for the different structures.

12.8.2c Compacted Graphite (CG) Cast Iron. The shape of the graphite in CG cast iron being between those of gray and ductile cast irons reflects the CG properties being intermediate between those of gray and ductile irons. CG iron has greater strength and ductility than gray, but less than those of ductile iron. This is shown in Fig. 12-48 for the tensile strength as a function of the carbon equivalent (C.E.). Since both CG and ductile irons depend on the additions of magnesium and rare-earth, it is not surprising that some nodular graphite will appear in the microstructure of CG iron. The amount of the spherical graphite in the microstructure increases the tensile strength of CG iron, as shown in Fig. 12-49. In addition to the dependency on C.E., the strength

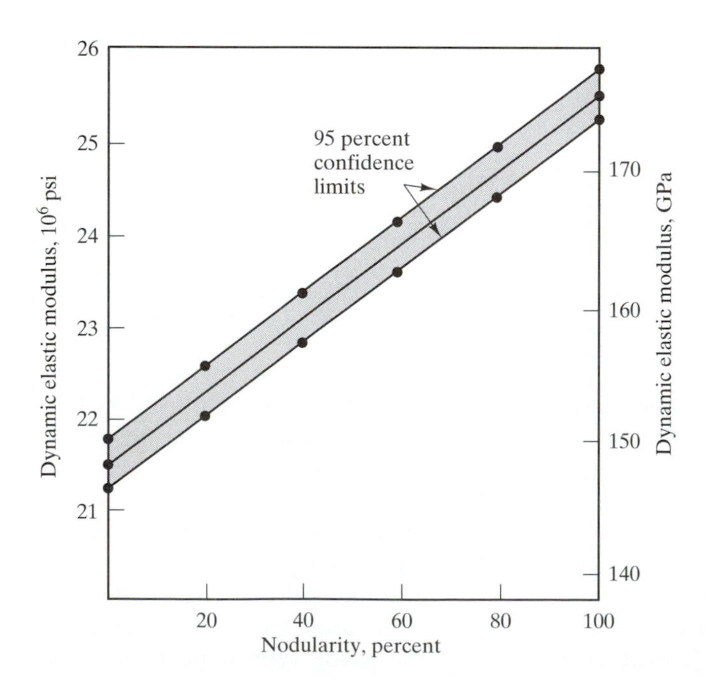

Fig. 12-43 Dynamic modulus of elasticity as a function of the nodularity of graphite in ductile cast iron.

depends also on the matrix microstructure and the section size and this is summarized in Fig. 12-50.

There are five grades of CG irons specified in ASTM specification A842, initially adopted in 1985 and reapproved in 1991. As with the gray irons, the grades are based on the minimum tensile strengths, but this time expressed as MPa or N/mm^2, rather than ksi. This change in units reflects the usage of the SI units when the specification was first adopted. The tensile properties and hardness requirements for these grades are given in Table 12-35.

The stress-strain curve of CG iron also exhibits an initial linearity which means that a particular CG iron has a unique modulus of elasticity and a unique

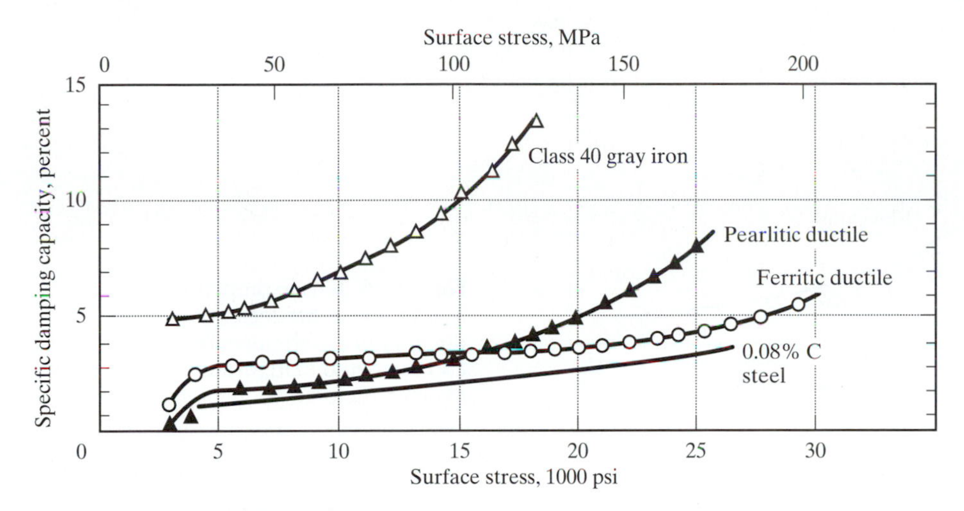

Fig. 12-44 Damping capacity of ductile iron compared with steel and gray iron.

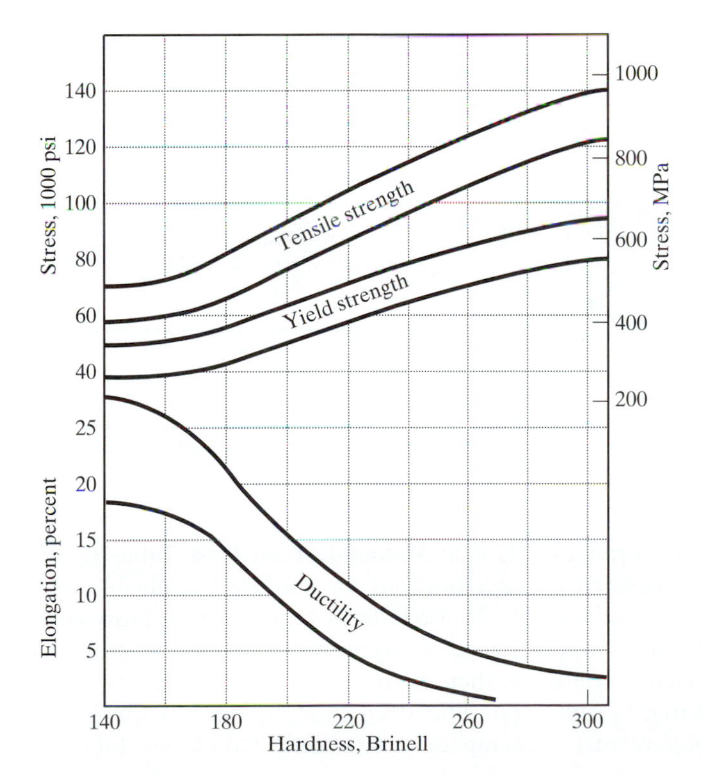

Fig. 12-45 Tensile properties as function of hardness for as-cast, normalized, or annealed ductile iron with ferrite and/or pearlite matrix structure.

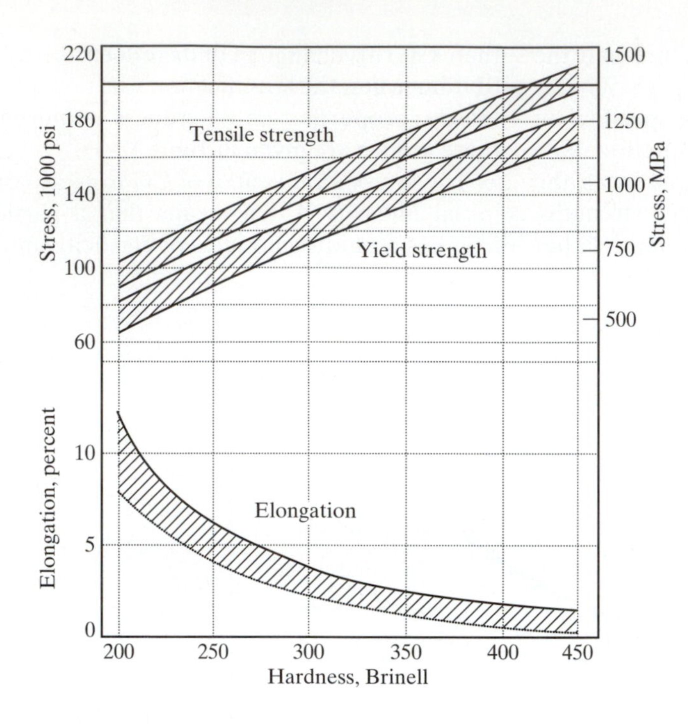

Fig. 12-46 Tensile properties as function of hardness for ductile irons with quenched and tempered martensitic matrix structure.

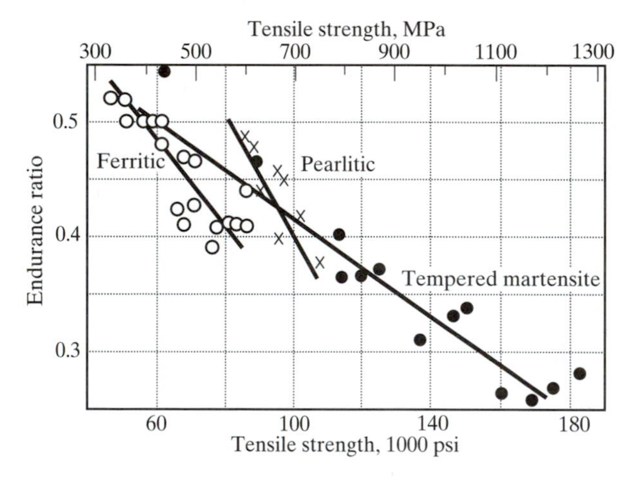

Fig. 12-47 Fatigue limit to tensile strength (endurance ratio) as function of tensile strength in ductile irons.

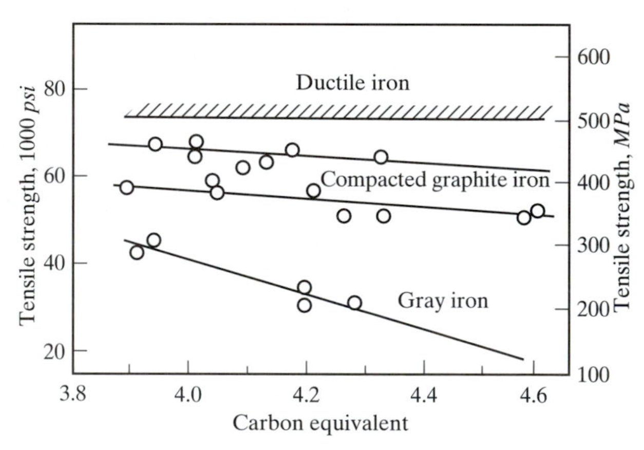

Fig. 12-48 Tensile strength of compacted graphite iron compared with those of gray and ductile irons as functions of carbon equivalent.

0.2 percent offset yield strength. The modulus ranges from 20 to 24 $\times$ 10^6 psi which reflects its dependency on the amount and shape of the graphite, the section size, and the matrix microstructure. The Poisson's ratio averages to be 0.275 and its damping capacity is only 60 percent that of gray iron. (Relative also to gray iron, ductile iron's damping capacity is only 34 percent.)

12.8.2d Malleable Cast Iron Table 12-36 shows the specifications for grades of malleable cast irons. The table also indicates the matrix microstructures and the typical applications of malleable iron. As indicated earlier, the matrices are classified as either ferritic or pearlitic, even though some of the latter may have tempered martensitic structures. Table 12-37 shows

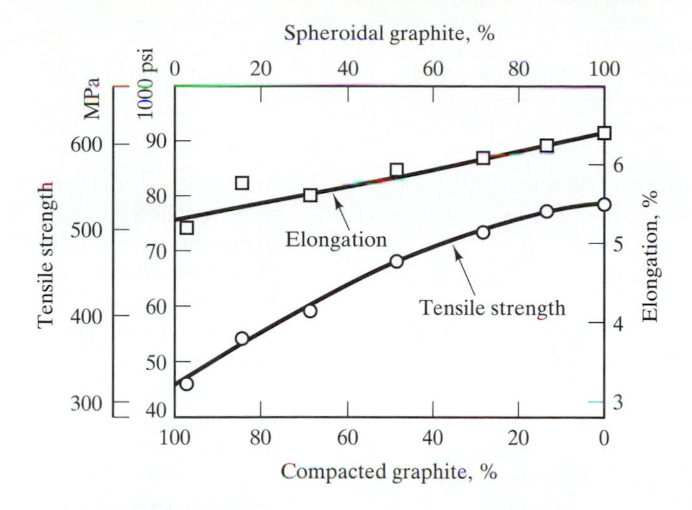

Fig. 12-49 Tensile properties of compacted graphite as function of amount of nodular graphite in the microstructure.

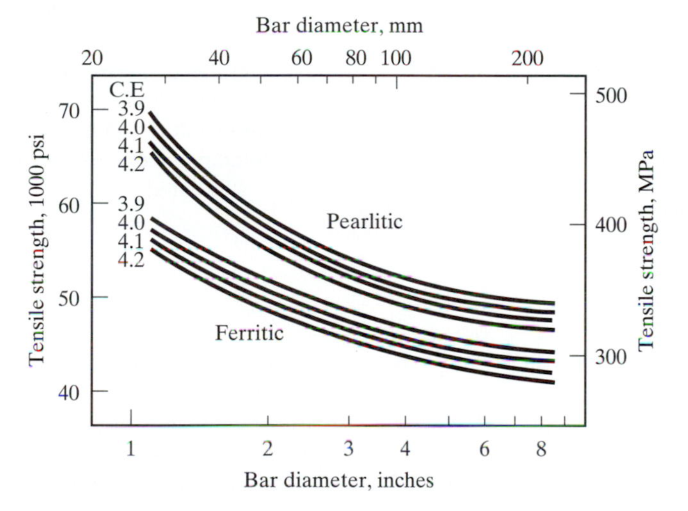

Fig. 12-50 Tensile strength of CG iron as function of carbon equivalent, matrix structure, and section size.

Table 12-35 ASTM A842—tensile and hardness requirements for CG iron.

	MINIMUM TENSILE PROPERTIES REQUIRED			HARDNESS REQUIREMENTS	
GRADE	**Yield Str., MPa**	**Tensile Str., MPa**	**%El., in 50 mm**	**Brinell, HB**	**BID *, in mm**
250	175	250	3.0	179 max	4.50 min
300	210	300	1.5	143-207	5.0 - 4.2
350	245	350	1.0	163-229	4.7 - 4.0
400	280	400	1.0	197-255	4.3 - 3.8
450	315	450	1.0	207-269	4.2 - 3.7

*Brinell Impression Diameter (BID) is the diameter in millimeters of the impression of a 10 mm steel ball using a 3000 kg load.

Table 12-36 Grades of malleable iron according to ASTM, SAE, and other specifications and their microstructure and typical applications. [2]

Specification No.	Class or grade	Microstructure	Typical Applications
Ferritic			
ASTM A47; ANSI G48.1; FED QQ-I-666c	32510 35018	Temper carbon and ferrite	General engineering service at normal and elevated temperature for good machinability and excellent shock resistance.
ASTM A338	32510 35018	Temper carbon and ferrite	Flanges, pipe fittings, and valve parts for railroad, marine, and other heavy duty service up to 345°C (650°F).
ASTM A197; ANSI G49.1		Free of primary graphite	Pipe fittings and valve parts for pressure service.
Pearlitic and Martensitic			
ASTM A220; ANSI G48.2; MIL-I-11444B	40010 45008 45006 50005 60004 70003 80002 90001	Temper carbon in necessary matrix without primary cementite or graphite	General engineering service at normal and elevated temperatures. Dimensional tolerance range for castings is stipulated.
Automotive			
ASTM A602; SAE J158	M3210	Ferritic	For low-stress parts requiring good machinability: steering gear housings, carriers, and mounting brackets.
	M4504	Ferrite and tempered perlite(b)	Compressor crankshafts and hubs.
	M5003	Ferrite and tempered perlite(b)	For selective hardening: planet carriers, transmission gears, differential cases.
	M5503	Tempered martensite	For machinability and improved response to induction hardening.
	M7002	Termpered martensite	For high-strength parts: connecting rods and universal joint yokes.
	M8501	Tempered martensite	For high strength plus good wear resistance: certain gears.

(a) For mechanical properties, see Table 12-37. (b) May be all tempered martnesite for some applications.

Table 12-37 Mechanical properties of the various grades of malleable iron in Table 12-36. [2]

Specification No.	Class or grade	Tensile strength		Yield strength		Hardness HB	Elongation(b), %
		MPa	ksi	MPa	ksi		
Ferritic							
ASTM A47, A338;							
ANSI G48.1;							
FED QQ-I-666c	32510	345	50	224	32	156 max	10
	35018	365	53	241	35	156 max	18
ASTM A197	···	276	40	207	30	156 max	5
Pearlitic and Martensitic							
ASTM A220;							
ANSI G48.2;							
MIL-I-11444B	40010	414	60	276	40	149-197	10
	45008	448	65	310	45	156-197	8
	45006	448	65	310	45	156-207	6
	50005	483	70	345	50	179-229	5
	60004	552	80	414	60	197-241	4
	70003	586	85	483	70	217-269	3
	80002	655	95	552	80	241-285	2
	90001	724	105	621	90	269-321	1
Automotive							
ASTM A602; SAE J158	M3210(c)	345	50	224	32	156 max	10
	M4504(d)	448	65	310	45	163-217	4
	M5003(d)	517	75	345	50	187-241	3
	M5503(e)	517	75	379	55	187-241	3
	M7002(e)	621	90	483	70	229-269	2
	M8501(e)	724	105	586	85	269-302	1

(a) For microstructures and typical applications see Table 12-36. (b) Minimum in 50 mm (2 in.). (c) Annealed. (d) Air quenched and tempered. (e) Liquid quenched and tempered.

the properties of the various grades of malleable irons. The first two digits of the five-digit ASTM code for the different grades reflect the minimum yield strength in ksi, last two digits the minimum ductility, and the third digit the modification. The SAE grades have four-digit codes with the first two and the last two digits reflecting the same properties as the ASTM codes, but minimum hardness levels are specified.

12.8.3 Design of Iron Castings [8]

Since castings are used as-cast, they are the engineering components we design for and the casting process is the equivalent of the forming and fabrication processes we give to wrought materials to form the components. The castings might require some machining and possibly homogenization and heat-treating. Thus, the design of as-cast components or castings involves the consideration of the following interrelated factors: (1) the technical or functional requirements or constraints of the casting or component; (2) the shape and section size of the casting; (3) the material to be used for the casting; and (4) the manufacturing processes that may include the casting or forming process, machining, and maybe some welding.

The technical or functional requirements of a casting may include: the forces or stresses the casting must withstand, and whether it is strength or stiffness limited (see Chap. 11); the accuracy or wear resistance; and the environmental conditions to which the casting will be exposed to, such as the temperature, the atmosphere and whether vibrations will be involved. We should recall that an efficient design for strength is not necessarily a suitable design for minimum deflection (strain) or stiffness. We must also note that we can make castings with complex shapes as easily as those with simple shapes.

Aside from pure tension and pure shear modes of loading on the casting, the shape of the casting will enhance its performance with any loading. Thus, if the component (casting) requirement is for stiffness or minimum deflection (strain), then we might be able to take advantage of the effectiveness of some shapes over the common solid shapes such as rounds, squares, and flats for the same mass of material. An example is the three-point bending mode shown in Fig. 12-51 where the component on the right (B) with the same mass as the flat plate on the left (A) was found to deflect only one-fifth as much as the flat plate. The performance index being proportional to $E^{1/3}/\rho$ in plate bending (see Table 11-2), it was also found that if the softest cast iron shaped as B, with approximately one-half the modulus of the elasticity of steel, were substituted for steel shaped as A, the iron will still be 2.5 times as rigid as the steel plate with the same mass. The additional advantages of the softest gray iron over steel are its machinability, thermal shock resistance, and castability. It is also significant to note that shape B with ribbings or reinforcing members can easily be obtained by any casting process, but would be very difficult and expensive to produce by other fabricating methods.

Design for maximum strength, on the other hand, requires generally a smooth, free-flowing shape with profile changes blended by generous fillets to minimize stress concentrations. Examples of designing for strength are shown in Fig. 12-52 where levers of the same mass but with different cross sections exhibit large differences in strength, both vertically and laterally. Design (a) uses standard wrought sections welded together. In this design, the I-beam section has constant section modulus and moment of inertia, and when loaded as a cantilever beam at the smaller hub (arrow indicates load), the bending moment and the stress are highest at the junction of the beam and the larger hub where stress concentration occurs because of the change in section.

In a casting, a taper of the I-section can be formed, as shown in Design (b), to increase the moment of inertia, I_x, which compensates for the increased bending moment and minimizes the stress concentration with a much gradual change in section. Thus, for the same load as in Design (a), the stress at the junction of the tapered I-beam and the larger hub is considerably reduced. The maximum stress imposed on the tapered section (Design (b)) is only 40 percent that of the stress in Design (a). Such a reduction in stress means that we can substitute a material with lower yield strength for a costlier higher yield strength material, for example, using a 40,000 psi yield strength for a 100,000 yield strength material. For resistance to torsional loading and/or compressive loads that involve buckling, the tubular section in (c) and the inverted U section in (d) exhibit further decrease in stress as summarized in Table 12-38. Components (b), (c), and (d) can be easily formed by casting but are more difficult and expensive by other forming methods.

Although complex castings with non-uniform cross sections are easily casted, not all the types of cast irons perform equally well. Thus, the right type of cast iron needs to be selected. Thin sectioned castings requiring ductility and toughness are best made of malleable iron. Thin sections with any type of cast iron will tend to form chill or white iron and tempering will produce the malleable iron. Chain, sprockets, tool parts, and hardware are made of fer-

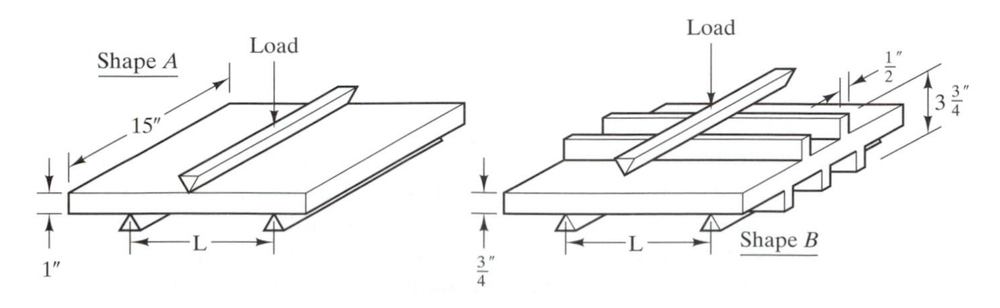

Fig. 12-51 For stiffness design (minimum deflection), the shape of the component increases its performance.

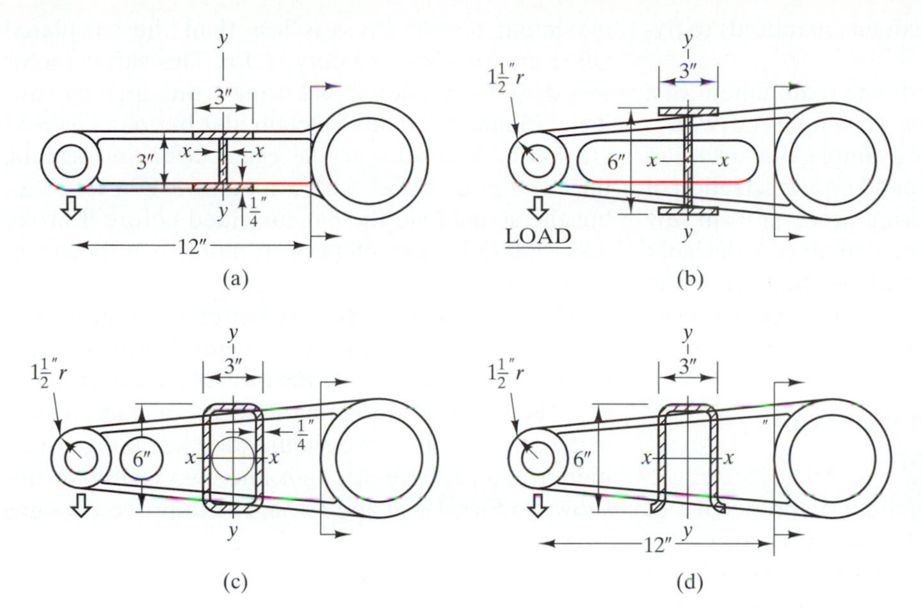

Fig. 12-52 Various designs of levers of the mass but with different cross sections; performance of the various designs are given in Table 12-38.

Table 12-38 Performances as decrease in maximum stress of the various designs of the lever in Fig. 12-52

Section		Moment of Inertia		Section Modulus		Decrease in Stress*	
View	Type	X Axis	Y Axis	X Axis	Y Axis	X Axis	Y Axis
(a)	I	3.1 in.^{4}	1.1 in.^{4}	2.1 in.^{3}	0.75 in.^{3}	–	–
(b)	I	15.8 in.^{4}	1.1 in.^{4}	5.3 in.^{3}	0.75 in.^{3}	60%	nil
(c)	Tube	18.0 in.^{4}	6.0 in.^{4}	6.0 in.^{3}	4.0 in.^{3}	57%	81%
(d)	U	16.0 in.^{4}	7.1 in.^{4}	5.3 in.^{3}	4.1 in.^{3}	53%	84%

*Percent decrease in imposed bending stress as compared to View A due to increased moment of intertia and section modulus.

ritic malleable iron. Pearlitic malleable iron is used for connecting rods, drive train and axle components, and spring suspensions for automobiles and trucks.

The gray irons are the most produced castings because of the versatility and the ease in making castings in a number of grades. The low-strength gray iron is very fluid as molten iron and solidifies with virtually no solidification shrinkage (no change in volume between the liquid and the solid iron) so it can be poured into thin and very complex castings. Typical applications for low strength gray iron are manifolds for internal combustion engines and gas burners with attached venturis. The higher strength gray irons are used for a wide variety of machinery parts such as gear blanks, forming dies, and automobile engine blocks and heads. The hardness and strength of gray iron vary with section size. If a section is too thin, the higher hardness and strength can make the iron unmachinable. If cast in heavy sections, the strength is reduced appreciably.

Ductile iron can also be used in a wide variety of casting size and section thickness. However, thin sections may require annealing, and alloy additions may be necessary to obtain the higher strength grades in heavy sections. Compacted graphite irons are suitable for moderately thin and medium section castings ranging from truck exhaust manifolds to flywheels.

The accuracy or wear-resistance requirement of a component might have to be accomplished with a chill section of the casting, for example, the surface of an engine cam. This might depend on the selection of a casting process (manufacturing process) from several casting processes that are available. A designer will need to consult the staff of the foundries who can help develop a design that is conducive to being manufactured at the least cost.

12.8.4 Special Design Concepts in Using Gray Iron

In the discussion of the mechanical properties of gray iron, Sec. 12.8.2a, we learned that gray iron does not exhibit a linear stress-strain curve, has a greater modulus in compression than in tension, and has at least three times greater compressive strength than tensile strength. Because of the nonlinearity of the stress-strain curve, gray iron does not have a unique yield strength. Thus, when we design for strength in static loading, we use the tensile strength as the criterion but using only one-fourth of its value as the allowable design stress. Based on this practice, the secant modulus at one-fourth the value of the tensile strength is also used and is a more conservative modulus to be used in structural design than the tangent modulus. The tangent modulus is more appropriate in designing precision equipment where the working stress and resulting strain are very close to zero.

The higher modulus in compression of gray iron provides a built-in safety factor in beam design. The structural formulae for beams in three-point bending assume that the moduli in compression and in tension are equal. Thus, in a simple center-loaded beam, the maximum compressive stress at the top surface and the maximum tensile stress at the bottom surface are equal for most materials and the neutral axis is at the center line shown in Fig. 12-53. For gray iron, the nonlinear stress-strain curve and higher modulus induce a higher stress in compression than in tension. To balance the forces, the neutral axis will move above the center-line towards the compression side. Thus, there are more material in tension than in compression. Because of this larger tensile section, the maximum tensile stress is less than the calculated value and provides a safety factor. This safety factor was demonstrated in a test using a one-inch by two-inch (25 mm by 50 mm) rectangular bar of a Class 35 gray iron. According to the classical beam formula, the bar should break with a load of 5180 (2350 kg), but the actual load the bar sustained before it broke was 8700 (3945 kg), 68 percent more than the calculated value.

Thus, gray iron performs better in compression and in conjunction with its excellent damping capacity, it is used frequently as the base of a lot of machinery. Therefore, in designing castings or components using gray cast iron, we should take advantage of its higher compressive strength and two examples are shown in Figs. 12-54 and 12-55. The improved designs

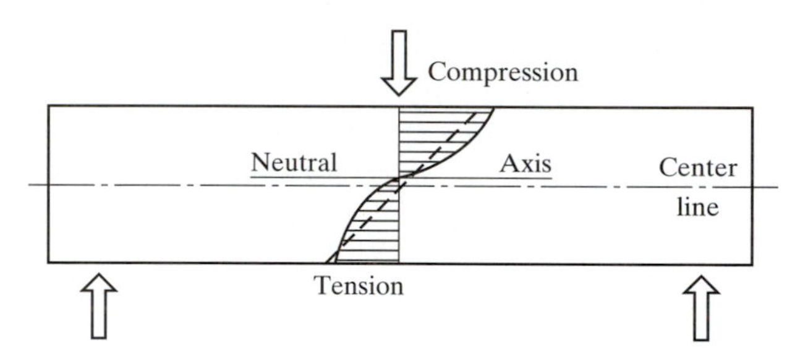

Fig. 12-53 Higher modulus and stress in compression shift the neutral axis in bending towards the compression side and provides a larger beam section subjected to tension.

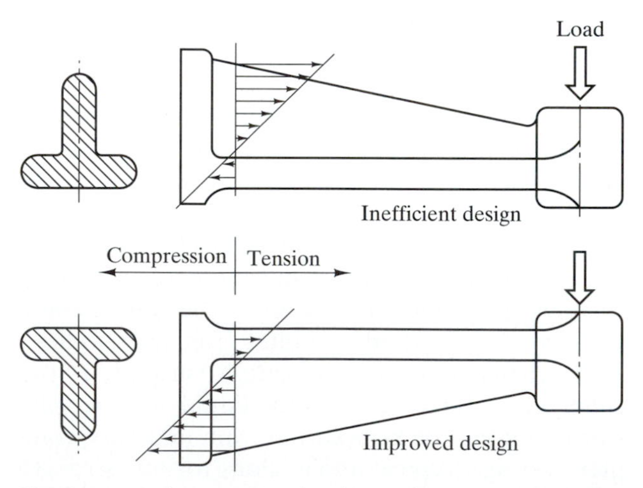

Fig. 12-54 A design example taking advantage of the higher compressive strength of gray iron.

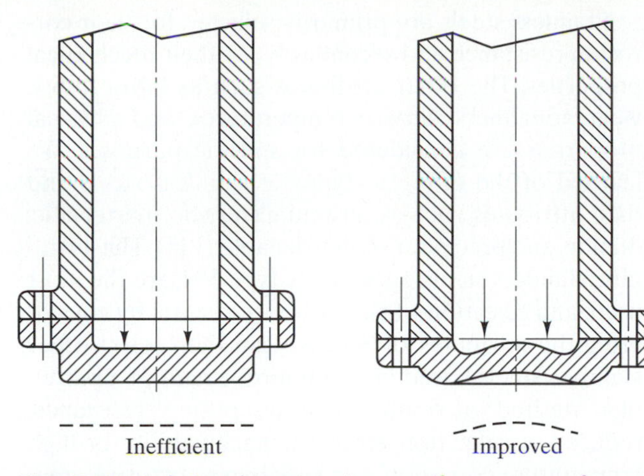

Fig. 12-55 A design that minimizes tensile stresses preforms much better for gray iron.

in these figures have shifted the tensile stresses to areas where there is more material to carry them. Thus, with the same applied load, the maximum tensile stress is lowered or less metal can be used in the casting.

Summary

In this chapter, we learned of the various types of ferrous materials and their standard designations. Selection of these materials depends not only on composition, strength, how they are processed, and their shape, but also on various "quality descriptors" and their fabricability that will reflect on the overall performance and cost of components or structures.

The most used ferrous materials are the structural steels that go into construction of ships, buildings, pipelines, bridges, appliances, automobiles, and many others. They may be procured as sheets, strips, plates, bars, and structural shapes such as angles, channels, and I-beams; as-hot rolled, cold-rolled, annealed, temper-rolled, or coated; and with various quality descriptors. Compositionwise, they may be plain carbon or low alloy steels and selection of steels starts with the lowest carbon content possible of plain carbon steels. According to strength, we have low strength (common variety), high or medium strength, and ultra-high strength structural steels. In the high strength category, we also have conventional and high-performance steels, which are discussed more in Chap. 16. The high performance steels exhibit much easier fabricability, especially in welding and forming, and superior performance in toughness and fatigue, in addition to strength. Procurement of structural steels is usually done with ASTM specifications which are usually cross-referenced by many consumer groups, such as the API (American Petroleum Institute) and the Federal government in MIL specifications, and technical organizations, such as the ASME.

Heat-treated steels depend on the formation of the martensite structure when quenched from an austenitizing temperature. This structure is the hardest microconstituent and is very brittle. Therefore, it has to be tempered to give some ductility and toughness. For the same composition, a properly quenched and tempered steel (see Sec. 9.7.3) exhibits the best combination of hardness, strength, and toughness. High hardness is needed for abrasion resistance and wear. The hardness or strength of as-quenched martensite is a function of carbon content only. Thus, we should consider plain carbon steels first before going to alloy steels, and here again, we should consider the lowest carbon content possible. However, plain carbon steels, even with high-carbon content, have very low hardenabilities (the ease of forming martensite), and they can only be hardened in thin gauges (shallow hardening). The latter when quenched from high temperatures exhibit distortion and dimensional problems because of the need for high severity quenches (see Sec. 9.7.2e). Alloy steels have greater hardenabilities that allow martensite formation with less severe quenches for easier processing and to avoid distortion and residual stress problems. Higher hardenabilities also mean that martensite can form to a greater depth or thickness of the material. Empirically, the hardness of quenched and tempered martensite is correlated to tensile strength and the strengths of carbon and low-alloy steels are the same up to about 500 HB (50 HRC) quenched and tempered hardness.

A quantitative measure of hardenability is obtained with the Jominy test and the resulting hardenability curve. A significant concept with the hardenability curve is called the Jominy equivalent cooling rate. Each Jominy distance in the hardenability curve gives a hardness and a cooling rate to achieve that hardness. Thus, the Jominy equivalent cooling rate says that, if we achieve a cooling rate in a part by whatever quench we use, the hardness of that part can be obtained from the hardenability curve. We can also use the concept in reverse, that is, if we want a certain hardness (strength), we should quench it with the equivalent cooling rate. Selection of heat-treated steels depend on the hardenability and the size of the steel needed, the quench medium, and the characteristics of each steel.

Tool steels are primarily used to cut, form, or shape materials into a part or component. The selection of tool steels depends on: "red hardness"; wear, deformation, or breakage resistance; and toughness. The ability to maintain hardness at temperatures when the tip of the tool turns red in color is due to secondary hardening that was discussed in Sec. 9.7.2b. Selection of tool steels based on the three factors is usually a compromise to get the optimum performance needed. The performance of a tool depends on the design, on the accuracy of how it is made, and on proper heat treatment. We have learned the various attributes of the different groups of tool steels and a guide for the selection process is provided.

Stainless steels are primarily selected for their corrosion resistance and secondarily for their mechanical properties. The other attributes such as fabricability, wear, toughness, service temperatures, and physical properties are considered for specific purposes. We learned of the different types of stainless steels and their attributes, such as, austenitic, ferritic, martensitic, duplex, and precipitation-hardenable (PH). The austenitic stainless steels, specifically type 304, are the most used, and selection of stainless steels starts by considering the attributes of 304. We can then select other stainless steels based on the improvement of a particular desired attribute, such as pitting resistance, reduced sensitization, strength, machinability, or high temperature corrosion. For very harsh corrosive environments, we might have to go for the superaustenitic or superferritic or duplex grades.

The cast irons are primarily produced as castings and are different from steels in that the carbon forms as graphite, rather than as carbides in combination with iron or other alloying elements. They are classified according to the form or shape of the graphite—gray (flakes), ductile or nodular (rounded), compacted (blunted or stubby flakes), white iron (as carbide), and malleable (temper carbon). The most used type is gray iron that exhibits excellent damping capacity and is usually specified according to grades indicating its tensile strengths. Gray iron has very unique properties and there are a few special design concepts we need to be aware of when we use gray iron. We learned the attributes of all the types of cast irons and the specifications to use. Cast irons can also be classified as unalloyed, low-alloyed, and high-alloyed. The properties of cast irons depend also on the matrix microstructures that may be ferritic, pearlitic, bainitic, and tempered martensitic.

References

1. "Properties and Selection: Ferrous Materials," vol 1, *Metals Handbook*, 10th ed., ASM.
2. "Properties and Selection: Irons and Steels," vol 1, *Metals Handbook*, 9th ed., ASM.
3. R. F. Kern and M. E. Suess, *Steel Selection*, John Wiley & Sons, 1979.
4. *High Performance Structural Steels*, 1995 Conference Proceedings, Edited by R. Asfahani, Published by ASM International, 1995. a) J. M. Barsom, "High-Performance Steels and Their Use in Structures," p. 3. b) A. D. Wilson, "Successful Development and Application of High-Performance Plate Steels," p. 99.
5. *Properties and Selection of Tool Materials*, ASM, 1975.
6. "Tool Steels" in Ref. 1.
7. *Stainless Steels*, an ASM specialty Handbook, ASM International, 1994.
8. *Iron Castings Handbook*, Iron Castings Society, Inc., 1981.

Terms and Concepts

AISI—American Iron and Steel Institute
AISI-SAE steel designations
ASME codes
ASTM—American Society for Testing and Materials
Austenitic stainless steels
Bars
Base metal
Cast iron
CE—carbon equivalent for welding
CE—carbon equivalent in cast iron
Chilled iron
Cold work tool steels
Compacted graphite (CG) iron
Controlled thermal mechanical processing
Damping capacity
Design concepts with gray iron
Ductile (cast) iron
Duplex stainless steels
Ferritic stainless steels
Formability
Forming limit diagram - FLD
Fusion zone
Graphite forms in cast iron
Gray (cast) iron
H-steels
HAZ cracking susceptibility
HAZ–heat affected zone
Heat input
Heat-treated carbon steels
Heat-treated low-alloy steels
High-alloy steels
High performance steels
High-speed tool steels
High strength structural steels
Hot work tool steels
Hot-rolled
HSLA (high-strength low-alloy) steels
Low-alloy steels
Machinability
Machinability ratings
Malleable (cast) iron
Martensitic stainless steels
Materials
Merchant quality bars
Microalloyed HSLA steels
Mild steels
Mold tool steels
PH (precipitation hardened) stainless steels
Plain carbon steels
Plates
Quality descriptors
Quenched and tempered steels
Re-bars
SAE–Society for Automotive Engineers
Shape products
Sheets
Shock-resisting tool steels
Stainless steels
Strips
Structural steels
Tool steels
Ultra-high strength steels
UNS—unified numbering system (ASTM)
Water-hardening tool steels
Weldability
White (cast) iron
Workpiece

Questions and Exercise Problems

12.1 What is the difference in composition between steels and cast irons? What are the roles of manganese and silicon in steels and cast irons, respectively?

12.2 How are steels classified according to (a) carbon content and (b) alloy content? In which category is the bulk (majority) of steels produced and give examples of its applications?

12.3 What do the terms bars, plates, sheets, strips, and shapes signify?

12.4 What is the basic structural quality ASTM steel grade? What are ultra-high strength steels?

12.5 What is the difference between the conventional HSLA and the microalloyed HSLA steels in terms of (a) composition, (b) processing, (c) structures, and (d) properties?

12.6 What are the quality descriptors for steel products?

12.7 What is the difference between a merchant bar and a rebar?

12.8 What is meant by cold-finishing? What are the differences in cold-finishing by machining and cold finishing by cold-drawing?

12.9 What is the difference between a specification and the AISI-SAE designations?

12.10 How is the weldability of a steel rated? What is the HAZ in welding? the fusion zone? the base metal? What is the influence of the thickness of the steel plate and the heat input on the HAZ structures and properties? (Note: You will need to use the heat-treating principles in Chap. 9)

12.11 What is the difference between sheet and bulk formability? What properties are used for the selection of steels for sheet and bulk formabilities? What is the forming limit diagram?

12.12 What are some factors that are used to rate the machinability of a steel? What steel is used as a basis for machinability ratings for carbon and alloy steels?

12.13 What is the role of carbon in quenched and tempered low-alloy steels? the role of alloying elements? What is the rule of thumb in regards to carbon content when selecting heat-treated steels?

12.14 What are the applications and relative usage of (a) low carbon, (b) medium carbon, and (c) high-carbon heat-treated, plain carbon steels?

12.15 What are the two methods used to select heat-treated steels according to their hardenabilities?

12.16 What are the influences of inclusions and the amount of martensite on the fatigue performance of heat-treated steels?

12.17 What are tool steels? What is meant by "red hardness" and why is it important? How is red hardness achieved in tool steels and what is it called?

12.18 What are high-speed tool steels and their main applications? Hot-work tool steels and their uses? Cold-work tool steels and applications? Shock resisting tool steels? Water-hardening tool steels and their uses?

12.19 What are stainless steels and the five different types of stainless steels?

12.20 What is the most significant factor used in the selection of stainless steels and the predominant types of stainless steels in the market?

12.21 What is meant by stabilization and its significance in regards to stainless steels? What is another method to reduce the sensitization of stainless steels?

12.22 Without chemical analyses, what would be a quick way to differentiate an austenitic stainless steel from a ferritic, a martensitic, or a duplex stainless steel?

12.23 In addition to corrosion resistance, the 304 stainless steel is very formable. What is the formability of 304 due to?

12.24 What is the predominant type of corrosion that will occur in stainless steels? Which type is better for chloride environments?

12.25 How are cast irons classified? What are the different types?

12.26 In addition to the form of graphite, the cast irons are also classified according to their matrices. How do these matrices form in cast iron?

12.27 What are some unique properties of gray iron that warrant a different design approach from that used for steels and the other cast irons?

12.28 What amounts of alloying elements are present in low-alloy cast iron? high-alloy cast iron?

12.29 What is a chilled iron and where is white cast iron used?

13

Selection of Nonferrous Metals

13.1 *Introduction*

After ferrous metals and alloys, the next two most used structural metals are aluminum and titanium alloys. Their usage arises from their low densities and excellent corrosion resistance. These properties allow them to be substituted for structural steels in structures requiring high performance materials. Both alloys can be sufficiently strengthened by alloying and by heat-treating to produce very high specific strengths and specific moduli. The latter terms were called the performance indices or structural efficiency factors (see Chap. 11) and are the attributes that are sought in materials used in the transport industry, especially the aerospace industry. Thus, most of the structural aluminum and titanium alloys are used in the aerospace industry with aluminum capturing about a 3:1 share over titanium because of cost. In addition, aluminum is also used for its good electrical conductivity and formability. Magnesium is also a very light metal, but has found only limited structural application (mostly as castings) because of its associated formability and corrosion problems.

Copper and nickel have about the same densities as iron and their performance indices will not be much different from those of steels. In addition, their costs are higher than steel and therefore, they cannot economically substitute for steels in structural applications. Their application is not based on their mechanical properties, but rather on other properties such as electrical, corrosion, and high temperature properties. Copper finds its largest application in the building and construction industry where it is used as electrical wirings and as tubings for potable water. Nickel has its greatest application as an alloying element for use in stainless steels and low-alloy steels. As the major element in a material, nickel-based alloys are used for their corrosion resistance and high-temperature performance for which they are called superalloys. The latter are used in jet-aircraft turbine engines that are subjected to severe mechanical stresses.

The metals and alloys mentioned above are the subjects of this chapter because they are the most widely used nonferrous metals in engineering applications. Because of their importance in structural applications, aluminum and titanium will be discussed first.

13.2 Aluminum and Aluminum Alloys [1, 2]

The procurement or selection of aluminum and its alloys is done by indicating the alloy designation first followed by a code indicating how it is processed—the latter is indicated by what it is called the temper designation. Therefore, before anything else, we should learn about the alloy and the temper designations. As the steel designations are handled by the American Iron and Steel Institute (AISI), the aluminum designations are handled by the Aluminum Association (AA) and is covered by the American National Standards Institute (ANSI) standard H35.1.

13.2.1 Alloy Designations

Aluminum and its alloys have designations according to whether they are wrought products or cast products. The designation sytem for the wrought alloys is shown in Table 13-1 and that for the cast alloys is shown in Table 13-2.

13.2.1a Wrought Aluminum and Aluminum Alloys. The four-digit designation system starts with the first digit indicating the major alloy addition to aluminum. The first digit "1" is designated for the unalloyed or essentially pure aluminum and the digits, 2 through 8, are designated for the aluminum alloys with the major alloying element(s) indicated in Table 13-1.

For the essentially pure aluminum 1XXX alloys with minimum 99.00 percent aluminum content, the last two XX in the designation correspond to the two digits after the decimal point in the minimum aluminum content. Thus, the minimum aluminum content will be 99.XX percent, expressed to the nearest 0.01 percent. The first X or the second digit of the 4-digit designation signifies some control of the composition. A "0" means that there is no control of the natural impurity limits. A "1 to 9" second digit, which is assigned consecutively as needed, signifies a special control or addition of one or more individual impurities.

For the 2XXX through 8XXX aluminum alloys, the last two XX have no special meaning, but signify only a certain type of alloy in the group (i.e., the major alloying elements are the same). A "0" for the first X or second digit in the 4-digit designation signifies the first alloy of the same major alloy additions (same last two XX). A "1 to 9" second digit signifies a composition modification of the original alloy. The AA follows explicit rules to determine whether a proposed composition is a modification of a previously registered alloy or is an entirely new alloy.

To illustrate the designation system, we refer to Table 13-3 which shows selected wrought 1XXX aluminum and 2XXX wrought alloys. Examples of the 1XXX designation are 1050, 1150, and 1350. These three essentially-pure aluminum have a minimum of 99.50 percent aluminum content; 1050 specifies no control over the natural limits of the impurities, 1150 imposes a combine maximum of 0.45 percent for (Si + Fe), and 1350 specifies a lower maximum for Si and additions of boron (B), vanadium (V) and titanium (Ti). The latter is used primarily as an

Table 13-1 Wrought aluminum alloy designation.

Designation	Major Alloying Element
1XXX	None, 99.00% min. aluminum
2XXX	Copper (Cu)
3XXX	Manganese (Mn)
4XXX	Silicon (Si)
5XXX	Magnesium (Mg)
6XXX	Magnesium and silicon
7XXX	Zinc (Zn)
8XXX	Other than the above elements
9XXX	Unused

Table 13-2 Cast aluminum alloy designation.

Designation	Major Alloy Addition(s)
1XX.Y	None, 99.00% min., Aluminum
2XX.Y	Copper (Cu)
3XX.Y	Si-Mg, Si-Cu, Si-Cu-Mg
4XX.Y	Silicon (Si)
5XX.Y	Magnesium (Mg)
7XX.Y	Zinc (Zn)
8XX.Y	Tin (Sn)
9XX.Y	Other elements from those above
6XX.Y	Unused

Table 13-3 Examples of alloy and UNS designations for selected wrought unalloyed aluminum and wrought 2XXX alloys.

Grade designation			Composition wt%													
														Unspecified other elements		
Aluminum Association	UNS No.	ISO No. R209	Si	Fe	Cu	Mn	Mg	Cr	Ni	Zn	Ga	V	Specified other elements	Ti	Each	Total	Al, minimum
1050	A91050	Al 99.5	0.25	0.40	0.05	0.03	0.03	···	···	0.05	···	0.05	···	0.03	0.03	···	99.50
1100	A91100	Al 99.0 Cu	0.95 (Si + Fe)		0.05–0.20	0.05	···	···	···	0.10	···	···	···	···	0.05	0.15	99.00
1200	A91200	Al 99.0	1.00 (Si + Fe)		0.05	0.05	···	···	···	0.10	···	···	···	0.05	0.05	0.15	99.00
1150	···	···	0.45 (Si + Fe)		0.05–0.20	0.05	0.05	···	···	0.05	···	···	···	0.03	0.03	···	99.50
1350	A91350	F-Al 99.5	0.10	0.40	0.05	0.01	···	0.01	···	0.05	0.03	···	0.05 B, 0.02(V+Ti)	···	0.03	0.10	99.50
2017	A92017	AlCu4MgSi	0.20–0.8	0.7	3.5–4.5	0.40–1.0	0.40–0.8	0.10	···	0.25	···	···	···	0.15	0.05	0.15	rem
2117	A92117	AlCu2.5Mg	0.20–0.8	0.7	3.5–4.5	0.40–1.0	0.40–1.0	0.10	···	0.25	···	···	0.25 Zr + Ti	···	0.05	0.15	rem
2018	A92018	···	0.9	1.0	3.5–4.5	0.20	0.45–0.9	0.10	1.7–2.3	0.25	···	···	···	···	0.05	0.15	rem
2218	A92218	···	0.9	1.0	3.5–4.5	0.20	1.2–1.8	0.10	1.7–2.3	0.25	···	···	···	···	0.05	0.15	rem
2618	A92618	···	0.10–0.25	0.9–1.3	1.9–2.7	···	1.3–1.8	···	0.9–1.2	0.10	···	···	···	0.04–0.10	0.05	0.15	rem
2219	A92219	AlCu6Mn	0.20	0.30	5.8–6.8	0.20–0.40	0.02	···	···	0.10	...	0.05–0.15	0.10–0.25 Zr	0.02–0.10	0.05	0.15	rem
2319	A92319	···	0.20	0.30	5.8–6.8	0.20–0.40	0.02	···	···	0.10	···	0.05–0.15	0.10–0.25 Zr	0.10–0.20	0.05	0.15	rem
2419	A92419	···	0.15	0.18	5.8–6.8	0.20–0.40	0.02	···	···	0.10	...	0.05–0.15	0.10–0.25 Zr	0.02–0.10	0.05	0.15	rem
2519	A92519	···	0.25	0.30	5.3–6.4	0.10–0.50	0.05–0.40	···	···	0.20	...	0.05–0.15	0.10–0.25 Zr	0.02–0.10	0.05	0.15	rem

electrical conductor. The 1100 aluminum has a minimum 99.00 percent aluminum and an imposed maximum of 0.95 percent for (Si + Fe). The 1200 aluminum also has 99.00 percent minimum but has a maximum of 1.00 percent for (Si + Fe).

Examples of the wrought aluminum alloy designation are 2017 and 2117. These two alloys have essentially the same alloy contents of copper, silicon, manganese, and magnesium—2117 is a modification of 2017 with the addition of 0.25% maximum of zirconium (Zr) + titanium (Ti). Other examples are 2X18, and 2X19 types. Each of these types has alloy modifications (i.e., each type has more than one X integer).

Table 13-3 also lists the available UNS (Unified Numbering System) code for each aluminum or alloy. Each UNS number is preceded by the letter A (for aluminum) as indicated in Table 12-1 (ASTM E 527). The five-digit UNS number consist of the 4-digit Aluminum Association designation preceded by 9. The ISO codes in Table 13-3 refer to those given to selected pure aluminum and aluminum alloys by the International Organization for Standardization (ISO). This system of designation signifies the major and secondary alloying additions with their compositions.

13.2.1b Cast Aluminum and Aluminum Alloys. Cast aluminum and aluminum alloys are designated with three digits, XXX. A fourth digit, Y, separated by a decimal point indicates the composition for either a casting or an ingot. Y = 0 indicates the composition of the alloy for a casting, while Y = 1 or 2 indicates the composition of specific alloys for an ingot that may have to be remelted by foundries.

As in the wrought designations, the first of the XXX three-digit designation indicates the major alloy addition with "1" designated for essentially pure aluminum or no alloy addition. The designations 3, 6, 8, and 9 for cast alloys signify different alloying elements than those in the wrought alloy designation. The 3 in cast alloys indicate silicon addition with copper and/or magnesium, while 3 in wrought alloys indicates manganese addition. The designation 3 in cast alloys also includes silicon-magnesium alloys that were given the 6 designation in the wrought alloys. Thus, the 6XXX designation is unused in cast alloys. The 8 in cast alloys is specifically designated for tin, and 9 is used for other elements.

For essentially pure aluminum castings or ingots, the last two digits in the three-digit code, 1XX, signify the same meaning as in the wrought alloys (i.e., the numerals after the decimal point in the minimum composition of aluminum, as in 99.XX percent, expressed to the nearest 0.01 percent). For alloy castings or ingots, the last two digits in the three-digit 2XX through 8XX are arbitrarily assigned to indicate certain types of alloys in the family or group. Modifications to original alloy designations are indicated by prefixed capital letters to the three-digit code and are indicated sequentially starting with A, B, C, and so on, but with the letters I, O, Q, and X being omitted; X is reserved for experimental alloys.

Examples of the above designations can be found in Table 13-4, which also shows the designated UNS numbers. The latter consists of the prefix A followed by a five-digit number, the middle three digits being the Aluminum Association (AA) designation and the last digit being either 0, 1, or 2 indicating castings or ingots. The first digit is reserved for modification—0 being designated for the original alloy and 1 through 8 for alloy modifications.

13.2.2 Temper Designations

The processing of aluminum and its alloys are coded by letters and numbers and are indicated next to the alloy designations after a hyphen. The letter code with numbers constitutes the *temper designation* and indicates precisely how the material is to be processed. Although some of the designations are not applicable to all alloys, the temper designations are generally applicable to both wrought and cast products.

13.2.2a Basic Temper Designations. The appropriate processings are indicated by the following letter codes:

- *F, As Fabricated. F* is given to products that have been shaped by cold-working, hot-working, or casting processes in which no special control over thermal conditions or strain-hardening was employed. For wrought products, there are no limits on the mechanical properties.
- *O, Annealed. O* is given to wrought products that have been annealed to obtain the lowest strength and to cast products that are annealed to improve

Table 13-4 Examples of alloy and UNS designations for selected unalloyed aluminum and aluminum alloys for castings and ingots.

Grade designation			Composition wt%												
Aluminum Association	UNS No.	Product(a)	Si	Fe	Cu	Mn	Mg	Cr	Ni	Zn	Sn	Ti	Unspecified other elements Each	Total	Al, min(b)
100.1	A01001	Ingot	0.15	0.6–0.8	0.10	···	···	···	···	0.05	···	···	0.03	0.10	99.00
201.0	A02010	S	0.10	0.15	4.0–5.2	0.20–0.50	0.15–0.55	···	···	···	···	0.15–0.35	0.05	0.10	rem
201.2	A02012	Ingot	0.10	0.10	4.0–5.2	0.20–0.50	0.10–0.55	···	···	···	···	0.15–0.35	0.05	0.10	rem
A201.0	A12010	S	0.05	0.10	4.0–5.0	0.20–0.40	0.15–0.35	···	···	···	···	0.15–0.35	0.03	0.10	rem
B201.0	A22010	S	0.05	0.05	4.5–5.0	0.20–0.50	0.25–0.35	···	···	···	···	0.15–0.35	0.05	0.15	rem
305.0	A03050	S, P	4.5–5.5	0.6	1.0–1.5	0.50	0.10	0.25	···	0.35	···	0.25	0.05	0.15	rem
305.2	A03052	Ingot	4.5–5.5	0.14–0.25	1.0–1.5	0.05	···	···	···	0.05	···	0.20	0.05	0.15	rem
A305.0	A13050	S, P	4.5–5.5	0.20	1.0–1.5	0.10	0.10	···	···	0.10	...	0.20	0.05	0.15	rem
A305.2	A13052	Ingot	4.5–5.5	0.13	1.0–1.5	0.05	···	···	···	0.05	···	0.20	0.05	0.15	rem

(a) S = Sand Casting: P = Permanent mold; (b) rem = 100.00% minus sum of all alloy additions.

ductility and dimensional stability. The *O* may be followed by a digit other than zero.

- *H, Strain-Hardened (for wrought products only). H* indicates products that have been strengthened by cold-work or strain-hardening, with or without supplementary thermal treatment to produce some reduction in strength. The *H* is always followed by two or more digits that indicate the processing and the degree of cold-work or strain-hardening. Part of this system was introduced in Sec. 8.4.2 and will be immediately discussed more after the letter codes.
- *W, Solution Heat-Treated. W* indicates an unstable condition applicable only to alloys whose strengths change naturally (spontaneously) at room temperature over a period of months or even years after solution treatment. The designation is specific only when the period of natural aging is indicated, for example, W $^{1}/_{2}$h
- *T, Solution Heat Treated. T* is given to alloys whose strength is stable within a few weeks after it has been solution-treated. The *T* is always followed by one or more digits.

13.2.2b Designation for Degree of Cold-Work or Strain-Hardening. The first digit following *H* indicates the processing and the second digit indicates the degree of cold-work or strain-hardening of the aluminum or alloy based on tensile strength. The designations are:

- *H1x, Strain-Hardened Only.* The first digit, "1" designates products that have been strain-hardened to obtain the desired tensile strength without supplementary thermal treatment. The second digit *x* indicates the degree of cold-work or strain-hardening based on the annealed tensile strength.
- *H2x, Strain-Hardened and Partially Annealed.* The first digit "2" is designated for products that have been strain-hardened more than the desired amount and then reduced in strength to the desired level by partial annealing. The second digit *x* indicates the degree of strengthening (strain-hardening) left after the product has been partially annealed.
- *H3x, Strain-Hardened and Stabilized.* The first digit "3" is designated for products that have been strain-hardened and whose mechanical properties are stabilized by a low-temperature thermal treatment or as a result of heat introduced during fabrication. This designation applies only to those alloys that, unless stabilized, will gradually age-soften at room temperature. The second digit *x* indicates the degree of strain-hardening remaining after stabilization. Stabilization usually improves the ductility.

For alloys that age-soften at room temperature, each H2x temper has the same minimum ultimate strength as the H3x temper with the same second digit. For other alloys, each H2x temper has the same minimum ultimate tensile strength with slightly higher elongation as the H1x with the same second digit.

The second digit *x* in the above designations indicates the degree of strain-hardening and may be a numeral from 1 through 9. Numeral 8 indicates a temper with an ultimate tensile strength that is achieved by 75 percent cold reduction in area following full annealing and is designated the full hard condition. The temperature of cold-reduction must not exceed 50°C or 120°F. Tempers between 0 (annealed) and 8 (full hard) are designated by the numerals 1 through 7. A material having an ultimate tensile strength halfway between that of the 0 and the 8 tempers is designated by the numeral 4, midway between the 0 and the 4 tempers by the numeral 2, and midway between the 4 and the 8 tempers by the numeral 6. Numeral 9 designates a temper whose minimum ultimate tensile strength exceeds that of the 8 temper by about 14 MPa (2 ksi) or more. When the above digit *x* is odd, the ultimate tensile strength of the material is the arithmetic mean of the two adjacent even digits. For example, the ultimate tensile strength of the H15 temper will be the arithmetic mean of the H14 and H16 ultimate tensile strenghts.

For alloys that cannot be sufficiently cold-reduced to establish an ultimate tensile strength applicable to the 8 temper (75 percent cold reduction in area after full annealing), the 6-temper tensile strength may be approximately established by a 55 percent cold reduction following full anneal, or the 4-temper tensile strength by a 35 percent cold reduction approximately, following full anneal.

A variation of the two-digit system following H may be made by adding a third digit (from 1 to 9). The third digit is used when the degree of control of the temper or the mechanical properties is different from but close to those of the two-digit H temper designation to which it is added, or when some other characteristic is significantly affected. The minimum ultimate tensile strength of a three-digit H temper is at least as close to that of the two-digit H temper as it is to either of the adjacent two-digit H tempers. The following three-digit H temper designations have already been assigned for wrought products of all alloys.

- *Hx11* is given to products that incur sufficient strain-hardening after final annealing but qualifies neither as O temper nor as Hx1 temper in that it has not much or consistent amount of strain-hardening.
- *H112* given to products that may acquire some strain-hardening during working at elevated temperature and for which there are no mechanical property limits.
- *Patterned or Embossed Sheet.* Three-digit H tempers are given to embossed sheets according to a standard designation.

13.2.2c Designations for Heat-Treated Conditions. The *W* and *T* are designations given to wrought and cast aluminum alloys that are heat-treatable (i.e., those that can be strengthened by heat-treating or thermal processing). *W* indicates an unstable condition and is not ordinarily used. The *T* designation is followed by a numeral from 1 through 10 that indicates the processing given to the wrought or cast alloy. The temper designations and the brief explanations of the processings are given below:

- *T1, Cooled from an Elevated Temperature Shaping Process and Naturally Aged to a Substantially Stable Condition.* This designation is given to products that are not cold-worked after an elevated-temperature shaping process such as casting or extrusion and for which mechanical properties have been stabilized by room-temperature aging. This designation also applies to products that are flattened or straightened after cooling from the shaping process, but the effects of the cold-work imparted by the flattening or straightening are not accounted for in specific property limits.
- *T2, Cooled from an Elevated-Temperature Shaping Process, Cold-Worked, and Naturally Aged to a Substantially Stable Condition.* This designation is given to products that are cold-worked specifically to improve strength after cooling from a hot-working process such as rolling or extrusion and for which mechanical properties have been stabilized by room-temperature aging. The designation also applies to products in which the effects of cold-work imparted by flattening or straightening are accounted for in specific property limits.
- *T3, Solution Heat-Treated, Cold-Worked, and Naturally Aged to a Substantially Stable Condition.* T3 is given to products that are cold-worked specifically to improve strength after solution heat-treatment and for which mechanical properties have been stabilized by room temperature aging. It applies also to products in which the effects of the cold-work imparted by flattening or straightening are accounted for in specified property limits.
- *T4, Solution Heat-Treated and Naturally Aged to a Substantially Stable Condition.* This is given to products that are not cold-worked after solution heat-treatment and for which the mechanical properties have been stabilized by room-temperature aging. It also applies to products in which the effects of the cold-work imparted by flattening or straightening are not accounted for in specific property limits.
- *T5, Cooled from an Elevated-Temperature Shaping Process and Artificially Aged.* This is given to products that are not cold-worked after cooling from an elevated temperature-shaping process, such as casting or extrusion, and for which mechanical properties have been improved by artificial aging (i.e., precipitation hardening at temperatures higher than room temperature). It applies also to products in which the effects of the cold-work imparted by flattening or straightening are not accounted for in specific property limits.
- *T6, Solution Heat-Treated and Artificially Aged.* This is given to products that are not cold-worked after solution heat-treatment and for which mechanical properties, or dimensional stability, or both, have been substantially improved by artificial aging (i.e., precipitation hardening at

temperatures higher than room temperature). It applies also to products in which the effects of the cold-work imparted by flattening or straightening are not accounted for in specific property limits.

- *T7, Solution Heat-Treated and Overaged or Stabilized.* This is given to wrought products that have been artificially aged after solution heat treatment beyond the peak strength to provide some special characteristic, such as enhanced resistance to stress-corrosion-cracking or exfoliation. It is also given to cast products that are artificially aged after solution heat-treatment to provide dimensional and strength stability.
- *T8, Solution Heat-Treated, Cold-Worked, and Artificially Aged.* This is given to products that are cold-worked, after solution heat-treatment, specifically to improve strength, and for which mechanical properties, or dimensional stability, or both, have been substantially improved by artificial aging. The effects of cold-work, including that imparted by flattening or straightening, are accounted for in specific property limits.
- *T9, Solution Heat-Treated, Artificially Aged, and Cold-Worked.* This is given to products that are cold-worked after artificial aging to specifically improve the strength.
- *T10, Cooled from an Elevated-Temperature Shaping Process, Cold-Worked, and Artificially Aged.* This is given to products that are cold-worked specifically to improve strength after cooling from an elevated-temperature shaping process, such as rolling or extrusion, and for which mechanical properties have been substantially improved by artificial aging. The effects of cold-work, including that for imparted by flattening or straightening, are accounted for in specific property limits.

13.2.2d Additional T Temper Variations. Variations of the above T temper designations can be made by adding more digits. Specific sets have been assigned to wrought products that have been mechanically stress-relieved. For products that are *stress-relieved by stretching* after solution heat-treatment or after cooling from an elevated temperature-shaping process, see the following table with the permanent strains,

Product form	Permanent Strain, %
Plate	1.5–3
Rod, Bar, Shapes, and Extruded Tube	1.0–3
Drawn Tube	0.5–3

- *Tx51.* The additional digits, 51, are given specifically to stretching of plate, of rolled or cold-finished rod and bar, of die or ring forgings, and of rolled rings. The products received no further straightening after stretching.
- *Tx510.* The additional digits, 510, are for stretching of extruded rod, bar, shapes, and tubing, and to drawn tubing. Products in this temper receive no further straightening after stretching.
- *Tx511.* The additional digits are given to products that have undergone minor straightening after the stretching to comply with standard tolerances.

For stress-relieving by compression, Tx52 is given to products that have been mechanically stress-relieved by compression after solution heat-treatment or after cooling from a hot-working process to produce a permanent strain of 1 to 5 percent.

For stress relieving by combined stretching and compression, Tx54 is given to die forgings that are stress-relieved by restriking cold in the finish die.

Temper designations have also been assigned to wrought products that have been heat treated from the O or F temper to demonstrate response to heat treatment.

- T42 is the designation given to a product that has been solution heat-treated from the O or the F temper and then naturally aged to a substantially stable condition.
- T62 is the designation given to a product that has been solution heat-treated from the O or the F temper and then artificially aged.

13.2.2e Annealed Products. In annealed products, a digit following the letter O indicates a product with special characteristics. For example, for heat-treatable alloys, O1 indicates a product that has been annealed at approximately the same time and temperature required for solution heat-treatment, and then air-cooled to room temperature. This specification is used for products that have to be machined by the user prior to solution heat-treatment. There is no mechanical property specification.

13.2.3 Selection and Applications of Wrought Aluminum and Aluminum Alloys [2]

Wrought aluminum alloys are classified into two basic types: nonheat-treatable and heat-treatable alloys. The nonheat-treatable alloys include all the various grades of pure aluminum and all other alloys that derive their strength from solid-solution strengthening and by cold-work or strain hardening from the anneal temper. These include the 1XXX, 3XXX, 4XXX, and the 5XXX alloys, although a few of them belong in the 7XXX and the 8XXX series.

The heat-treatable alloys are those that contain one or more of the elements of copper, magnesium, silicon, and zinc that have the characteristics of increasing solubility in aluminum as the temperature increases and have the generic characteristics for precipitation hardening discussed in Chap. 9. These include the 2XXX, 6XXX, and the 7XXX alloys, although a few of them are also in the 4XXX and the 5XXX series where the combination of the above elements is present.

Tables 13-5 and 13-6 list the typical compositions of the most common nonheat-treatable and heat-treatable alloys, while Table 13-7 lists the typical mechanical properties and applications of both groups of alloys.

13.2.3a Nonheat-Treatable Alloys. These alloys exhibit very low yield and tensile strengths in the annealed (O) state and are, therefore, easily formable (workable) to different shapes.

- *1xxx Alloys.* 99.00 percent or more pure aluminum has many applications, especially in the electrical and chemical industries. These grades are characterized by excellent corrosion resistance, high thermal and electrical conductivities, low mechanical properties, and excellent workability. Typical uses include: chemical equipment, reflectors, heat exchangers, electrical conductors and capacitors, packaging foil, architectural applications, and decorative trim. One of the important applications of pure aluminum is as a cladding material, to improve the corrosion resistance of the heat-treatable alloys, or to improve the finishing characteristics of other nonheat-treatable alloys. 1350 was purposely developed as an electrical conductor.
- *3xxx Alloys.* Manganese is the major alloying element of this group. These alloys have about 20 percent more strength than the 1xxx alloys as a result of dispersion and solid solution hardening. Manganese content in these alloys is limited to about 1.5 percent and a few alloys contain magnesium as an additional element. Alloys 3003, 3004, and 3105 are used as general-purpose alloys for moderate-strength applications requiring good workability. These applications include beverage cans, cooking utensils, heat exchangers, storage tanks, awnings, furniture, highway signs, roofing, siding, and other architectural applications. They constitute the most used aluminum alloys, tonnage wise.
- *4xxx Alloys.* Silicon is the major alloying element and can be added in a substantial amount (up to 12 percent) to cause a dramatic lowering of the melting range without producing brittleness. For this reason, the major use of these aluminum-silicon alloys is as welding wires and brazing alloys in the joining of aluminum alloys. Thus, they are not characterized by their mechanical properties and are omitted in Table 13-7. They are normally nonheat-treatable, but, when used to weld the heat-treatable alloys, will pick up some of the alloying constituents of the heat-treatable alloys and will respond to heat-treatment to a limited extent. The alloys containing appreciable amounts of silicon become dark gray to charcoal when surface-anodized and are thus in great demand for architectural applications. Alloy 4032 is heat-treatable, has a low wear-resistance, and is used to produce forged engine pistons.
- *5xxx Alloys.* Magnesium is the major alloying element with some manganese to produce a moderate to high-strength alloy that can be cold-worked or strain-hardened. Magnesium is considerably more effective as a hardener and can be added in greater amounts. These alloys were developed as "marine alloys" to be highly resistant in salty marine environments and they also exhibit good weldability. For high magnesium alloys, certain restrictions are placed on the amount of cold-work and the operating temperature to avoid susceptibility to SCC (stress-corrosion cracking). Their uses include: architectural, ornamental, and decorative trim; cans and can ends;

Table 13-5 Compositions of the most common nonheat-treatable wrought aluminum and aluminum alloys. [2]

	Alloying elements, wt% (nominal value)						
Alloy	Silicon	Copper	Manganese	Magnesium	Chromium	Other	Aluminum, minimun %
1199	···	···	···	···	···	···	99.99
1180	···	···	···	···	···	···	99.80
1060	···	···	···	···	···	···	99.60
1350	···	···	···	···	···	···	99.50
1145	···	···	···	···	···	···	99.45
1235	···	···	···	···	···	···	99.35
1100	···	0.12	···	···	···	···	99.00
3102	···	···	0.22	···	···	···	rem
3003	···	0.12	1.2	···	···	···	rem
3004	···	···	1.2	1.0	···	···	rem
3104	···	0.15	1.1	1.0	···	···	rem
3005	···	···	1.2	0.40	···	···	rem
3105	···	···	0.6	0.50	···	···	rem
4043	5.2	···	···	···	···	···	rem
4343	7.5	···	···	···	···	···	rem
4643	4.1	···	···	0.20	···	···	rem
4045	10.0	···	···	···	···	···	rem
4145	10.0	4.0	···	···	···	···	rem
4047	12.0	···	···	···	···	···	rem
5005	···	···	···	0.8	···	···	rem
5042	···	···	0.35	3.5	···	···	rem
5050	···	···	···	1.4	···	···	rem
5052	···	···	···	2.5	0.25	···	rem
5252	···	···	···	2.5	···	···	rem
5154	···	···	···	3.5	0.25	···	rem
5454	···	···	0.8	2.7	0.12	···	rem
5654	···	···	···	3.5	0.25	0.10 Ti	rem
5056	···	···	0.12	5.0	0.12	···	rem
5456	···	···	0.8	5.1	0.12	···	rem
5457	···	···	0.30	1.0	···	···	rem
5657	···	···	···	0.8	···	···	rem
5082	···	···	···	4.5	···	···	rem
5182	···	···	0.35	4.5	···	···	rem
5083	···	···	0.7	4.4	0.15	···	rem
5086	···	···	0.45	4.0	0.15	···	rem
7072	···	···	···	···	···	1.0 Zn	rem
8001	···	···	···	···	···	0.6 Fe, 1.1 Ni	rem
8280	1.5	1.0	···	···	···	0.45 Ni, 6.2 Sn	rem
8081	···	1.0	···	···	···	20 Sn	rem

Table 13-6 Composition of some commercial heat-treatable wrought aluminum alloys. [2]

Alloy	Alloying elements, wt% (nominal value)							Aluminum
	Silicon	Copper	Manganese	Magnesium	Chromium	Zinc	Other	
2011	···	5.5	···	···	···	···	0.40 Bi, 0.40 Pb	rem
2014	0.8	4.4	0.8	0.5	···	···	···	rem
2017	0.5	4.0	0.7	0.6	···	···	···	rem
2117	···	2.6	···	0.35	···	···	···	rem
2218	···	4.0	···	1.5	···	···	2.0 Ni	rem
2618	0.18	2.3	···	1.6	···	···	1.1 Fe, 1.0 Ni, 0.07 Ti	rem
2219, 2419	···	6.3	0.30	···	···	···	0.10 V, 0.18 Zr, 0.06 Ti	rem
2024, 2124, 2224	···	4.4	0.6	1.5	···	···	···	rem
2025	0.8	4.4	0.8	···	···	···	···	rem
2036	···	2.6	0.25	0.45	···	···	···	rem
4032	12.2	0.9	···	1.0	···	···	0.9 Ni	rem
6101	0.50	···	···	0.6	···	···	···	rem
6201	0.7	···	···	0.8	···	···	···	rem
6009	0.8	0.37	0.50	0.6	···	···	···	rem
6010	1.0	0.37	0.50	0.8	···	···	···	rem
6151	0.9	···	···	0.6	0.25	···	···	rem
6351	1.0	···	0.6	0.6	···	···	···	rem
6951	0.35	0.28	···	0.6	···	···	···	rem
6053	0.7	···	···	1.2	0.25	···	···	rem
6061	0.6	···	0.28	1.0	0.20	···	···	rem
6262	0.6	0.28	···	1.0	0.09	···	0.6 Bi, 0.6 Pb	rem
6063	0.4	···	···	0.7	···	···	···	rem
6066	1.3	1.0	0.8	1.1	···	···	···	rem
6070	1.3	0.28	0.7	0.8	···	···	···	rem
7001	···	2.1	···	3.0	0.26	7.4	···	rem
7005	···	···	0.45	1.4	0.13	4.5	0.14 Zr, 0.03 Ti	rem
7016	···	0.8	···	1.1	···	4.5	···	rem
7021	···	···	···	1.5	···	5.5	0.13 Zr	rem
7029	···	0.7	···	1.6	···	4.7	···	rem
7049	···	1.6	···	2.4	0.16	7.7	···	rem
7050	···	2.3	···	2.2	···	6.2	0.12 Zr	rem
7150	···	2.2	···	2.4	···	6.4	0.12 Zr	rem
7075, 7175	···	1.6	···	2.5	0.23	5.6	···	rem
7475	···	1.6	···	2.2	0.22	5.7	···	rem
7076	···	0.6	0.50	1.6	···	7.5	···	rem
7178	···	2.0	···	2.7	0.23	6.8	···	rem

Table 13-7 Typical tensile properties—descriptions and selected applications of commonly used wrought aluminum alloys. [1]

Alloy and temper	Ultimate tensile strength		Yield strength		Elongation in 50 mm (2 in.), %		Description and selected applications
	MPa	ksi	MPa	ksi	1.6 mm (1/16 in.) thick specimen	13 mm (1/2 in.) diameter specimen	
1100-O	90	13	34	5	35	45	Commercially pure aluminum, highly resistant to chemical attack and weathering. Low cost, ductile for deep drawing, and easy to weld. Used for high-purity applications such as chemical processing equipment. Also for nameplates, fan blades, flue lining, sheet metal work, spun holloware, and fin stock
1100-H14	124	18	117	17	9	20	
1350-O	83	12	28	4	...	...	Electrical conductors
1350-H19	186	27	165	24	...	...	
2011-T3	379	55	296	43	...	15	Screw machine products. Appliance parts and trim, ordnance, automotive, electronic, fasteners, hardware, machine parts
2011-T8	407	59	310	45	...	12	
2014-O	186	27	97	14	...	18	Truck frames, aircraft structures, automotive, cylinders and pistons, machine parts, structurals
2014-T4, T451	427	62	290	42	...	20	
2014-T6, T651	483	70	414	60	...	13	
2017-T6, T451	427	62	276	40	...	22	Screw machine products, fittings, fasteners, machine parts
2024-O	186	27	76	11	20	22	For high-strength structural applications. Excellent machinability in the T-tempers. Fair workabilitiy and fair corrosion resistance. Alclad 2024 combines the high strength of 2024 with the corrosion resistance of the commercially pure cladding. Used for truck wheels, many structural aircraft apllications, gears for machinery, screw machine products, automotive parts, cylinders and pistons, fasteners, machine parts, ordnance, recreation equipment, screws, and rivets
2024-T3	483	70	345	50	18	...	
2024-T4, T351	469	68	324	47	20	19	
Alclad:							
2024-O	179	26	76	11	20	...	
2024-T3	448	65	310	45	18	...	
2024-T4, T351	441	64	290	42	19	...	
2219-O	172	25	76	11	18	...	Structural uses at high temperature (to 315°C, or 600°F). High-strength weldments
2219-T87	476	69	393	57	10	...	

(*Continued*)

Table 13-7 Typical tensile properties—descriptions and selected applications of commonly used wrought aluminum alloys. (*Continued*)

Alloy and temper	Ultimate tensile strength		Yield strength		Elongation in 50 mm (2 in.), %		Description and selected applications
	MPa	ksi	MPa	ksi	1.6 mm (1/16 in.) thick specimen	13 mm (1/2 in.) diameter specimen	
3003-O	110	16	41	6	30	40	Most popular general-purpose alloy. Stronger than 1100 with same good formability and weldability. For general use including sheet metal work, stampings, fuel tanks, chemical equipment, containers, cabinets, freezer liners, cooking utensils, pressure vessels, builder's hardware, storage tanks, agricultural applications, appliance parts and trim, architectural applications, electronics, fin stock, fan equipment, name plates, recreation vehicles, trucks and trailers. Used in drawing and spinning
3003-H12	131	19	124	18	10	20	
3003-H14	152	22	145	21	8	16	
3003-H16	179	26	172	25	5	14	
3004-O	179	26	69	10	20	25	Sheet metal work, storage tanks, agricultural applications, building products, containers, electronics, furniture, kitchen equipment, recreation vehicles, trucks and trailers
3004-H38	283	41	248	36	5	6	
3105-O	117	17	55	8	24	...	Residential siding, mobile homes, rain-carrying goods, sheet metal work, appliance parts and trim, automotive parts, building products, electronics, fin stock, furniture, hospital and medical equipment, kitchen equipment, recreation vehicles, trucks and trailers
3105-H14	172	25	152	22	5	...	
3105-H18	214	31	193	28	3	...	
3105-H25	179	26	159	23	...	...	
5005-H34	159	23	138	20	8	...	Specified for applications requiring anodizing; anodized coating is cleaner and lighter in color than 3003. Uses include apliances, utensils, architectural applications requiring good electrical conductivity, automotive parts, containers, general sheet metal, hardware, hospital and medical equipment, kitchen equipment, name plates, and marine applications

(*Continued*)

Table 13-7 Typical tensile properties—descriptions and selected applications of commonly used wrought aluminum alloys. (*Continued*)

Alloy and temper	Ultimate tensile strength		Yield strength		Elongation in 50 mm (2 in.), %		Description and selected applications
	MPa	ksi	MPa	ksi	1.6 mm (1/16 in.) thick specimen	13 mm (1/2 in.) diameter specimen	
5052-O	193	28	90	13	25	30	Stronger than 3003 yet readily formable in the intermediate tempers. Good weldability and resistance to corrosion. Uses include pressure vessels, fan blades, tanks, electronic panels, electronic chassis, medium-strength sheet metal parts, hydraulic tube, appliances, agricultural applications, architectural uses, automotive parts, building products, chemical equipment, containers, cooking utensils, fasteners, hardware, highway signs, hospital and medical equipment, kitchen equipment, marine applications, railroad cars, recreation vehicles, trucks and trailers
5052-H112	...	...	...	...	...	...	
5052-H32	228	33	193	28	12	18	
5052-H34	262	38	214	31	10	14	
5056-O	290	42	152	22	...	35	Cable sheathing, rivets for magnesium, screen wire, zippers, automotive applications, fence wire, fasteners
5056-H18	434	63	407	59	...	10	
5083-O	290	42	145	21	...	22	For all types of welding assemblies, marine components, and tanks requiring high weld efficiency and maximum joint strength. Used in pressure vessels up to 65°C (150°F) and in many cryogenic applications, bridges, freight cars, marine components, TV towers, drilling rigs, transportation equipment, missile components, and dump truck bodies. Good corrosion resistance
5083-H321	317	46	228	33	...	16	
5086-H32	290	42	207	30	12	...	Used in generally the same types of appliations as 5083, particularly where resistance to either stress corrosion or atmospheric corrosion is important
5086-H34	324	47	255	37	10	...	
5086-H112	269	39	131	19	14	...	
5454-O	248	36	117	17	22	...	For all types of welded assemblies, tanks, pressure vessels. ASME code approved to 205°C (400°F). Also used in trucking for hot asphalt road tankers and dump bodies; also, for hydrogen peroxide and chemical storage vessels
5454-H32	276	40	207	30	10	...	
5454-H34	303	44	241	35	10	...	
5454-H112	248	36	124	18	18	...	
5456-H321 and -H116	352	51	255	37	...	16	For all types of welded assemblies, storage tanks,

(*Continued*)

Table 13-7 Typical tensile properties—descriptions and selected applications of commonly used wrought aluminum alloys. (*Continued*)

Alloy and temper	Ultimate tensile strength		Yield strength		Elongation in 50 mm (2 in.), %		Description and selected applications
	MPa	ksi	MPa	ksi	1.6 mm (1/16 in.) thick specimen	13 mm (1/2 in.) diameter specimen	
							pressure vessels, and marine components. Used where best weld efficiency and joint strength are required. Restricted to temperatures below 65°C (150°F)
5657-H25	159	23	138	20	12	...	For anodized auto and appliance trim and nameplates
6061-O	124	18	55	8	25	30	Good formability, weldability, corrosion resistance, and strength in the T-tempers. Good general-purpose alloy used for a broad range of structural applications and welded assemblies including truck components, railroad cars, pipelines, marine applications, furniture, agricultural applications, aircraft, architectural applications automotive parts, building products, chemical eqiupment, dump bodies, electrical and electronic applicaitons, fasteners, fence wire, fan blades, general sheet metal, highway signs, hospital and medical equipment, kitchen equipment, machine parts, ordnance, recreation equipment, recreation vehicles, and storage tanks
6061-T4	241	35	145	21	22	25	
6061-T6 and -T651	310	45	276	40	12	17	
6063-T5	186	27	145	21	12	...	Used in pipe railing, furniture, architectural extrusions, appliance parts and trim, automotive parts, building products, electrical and electronic parts, highway signs, hospital and medical equipment, kitchen equipment, marine applications, machine parts, pipe, railroad cars, recreation equipment, recreation vehicles, trucks and trailers
6063-T6	241	35	214	31	12	...	
7050-T7651	552	80	490	71	...	11	High-strength alloy in aircraft and other structures. Also used in ordinance and recreation equipment
7075-T6 and -T651	572	83	503	73	11	11	For aircraft and other applications requiring highest strengths. Alclad 7075 combines the strength advantages of 7075 with the corrosion-resisting properties of commercially pure aluminium-clad surface. Also used in machine parts and ordnance
Alcad:							
7075-O	221	32	97	14	17	...	
7075-T6 and -T651	524	76	462	67	11	...	

street-light standards; boats and ships; cryogenic tanks; crane parts; and automotive structures.

- *8xxx Alloys.* These alloys encompass a variety of compositions, including nonheat-treatable and heat-treatable. Among the nonheat-treatable, 8001 is an aluminum-nickel-iron alloy that is used in nuclear energy applications. It has good corrosion resistance in water at high temperature and pressure, and exhibits similar properties as 3003. Some aluminum-iron-manganese and aluminum-iron-manganese-zinc are produced as thin sheet and foil for applications as fin stock. The aluminum-tin-nickel-copper alloy 8280 is used as bearings; tin provides the antifriction characteristics, while nickel and copper contribute to strengthening. Another bearing alloy, 8081, containing 20%Sn and 1%Cu has bearing characteristics superior to other alloys for automotive use. 8090 alloy is a heat-treatable aluminium-lithium that is being developed for the aerospace industry and will be discussed with other aluminum-lithium alloys in an upcoming section.

Aluminum alloy foil is an extremely important and widely used wrought product. It is generally produced commercially from the above nonheat-treatable alloys. These alloys include 1100, 1145, 1235, 3003, 5052, and 5056. *Aluminum honeycomb core* for aircraft is made of 3003-H19, 5052-H39, or 5056-H39 foil, but alloy 2024 heat-treated to the T81 temper is used when long service at elevated temperature is required.

13.2.3b Heat-Treatable Alloys. These alloys are basically selected for structural applications because of their high strengths and for the inherent lightness and corrosion resistance of aluminum.

- *2xxx Alloys.* Copper is the principal alloying element in these alloys, often with magnesium as a secondary addition. Age-hardening or precipitation-hardening was discovered with alloy 2017, which was the first alloy introduced in 1911 by Wilm in Germany. These alloys do not have as good a corrosion resistance as most other aluminum alloys, and under certain conditions, may be prone to intergranular corrosion. Thus, in the form of sheets, these alloys are often cladded with a high-purity aluminum or with a magnesium-silicon alloy of the 6xxx group to provide galvanic protection. These alloys find widespread use for parts and structures requiring high specific strengths. They are commonly used to make truck and aircraft wheels, truck suspension parts, aircraft fuselage, wing skins, and structural parts, and others requiring good strength at temperatures below 150°C. Except for alloy 2219, these alloys have limited weldability.
- *6xxx Alloys.* These alloys contain silicon and magnesium, in approximately the proportions required to form the compound, Mg_2Si, magnesium silicide, which precipitates out and strengthens the alloys during heat-treating. With medium strength but not as strong as the 2xxx and 7xxx alloys, they have good formability, weldability, machinability, and corrosion resistance. They may be formed to desired shapes in the T4 temper condition and then further strengthened to the T6 temper condition after forming. Uses include architectural applications, bicycle frames, transportation equipment, bridge railings, and welded structures.
- *7xxx Alloys.* These alloys contain zinc, from 1 to 8 percent, and a smaller amount of magnesium to produce moderate to very high-strength alloys. Other elements, such as copper and chromium, are also added in small amounts. High-strength 7xxx alloys exhibit reduced SCC (stress-corrosion cracking) resistance and are used in slightly overaged condition to provide better combinations of strength, corrosion resistance, and fracture toughness. They are used in airframe structures, mobile equipment, and other highly stressed parts.

Aircraft Alloys. Because of their high specific strengths and specific moduli, the *alloys 2024 and 7075 have been the conventional aircraft structural materials,* constituting from 70 to 80 percent of the weight of current aircraft. Figure 13-1 shows the basic materials used in a commercial aircraft and indicates where these two alloys are used. For this application, fracture toughness and fatigue strengths are two important properties and are shown in Tables 13-8 and 13-9. Higher toughness requirements have been met through the high-purity modification of these alloys such as 2124, 2224, 7175, 7475, and

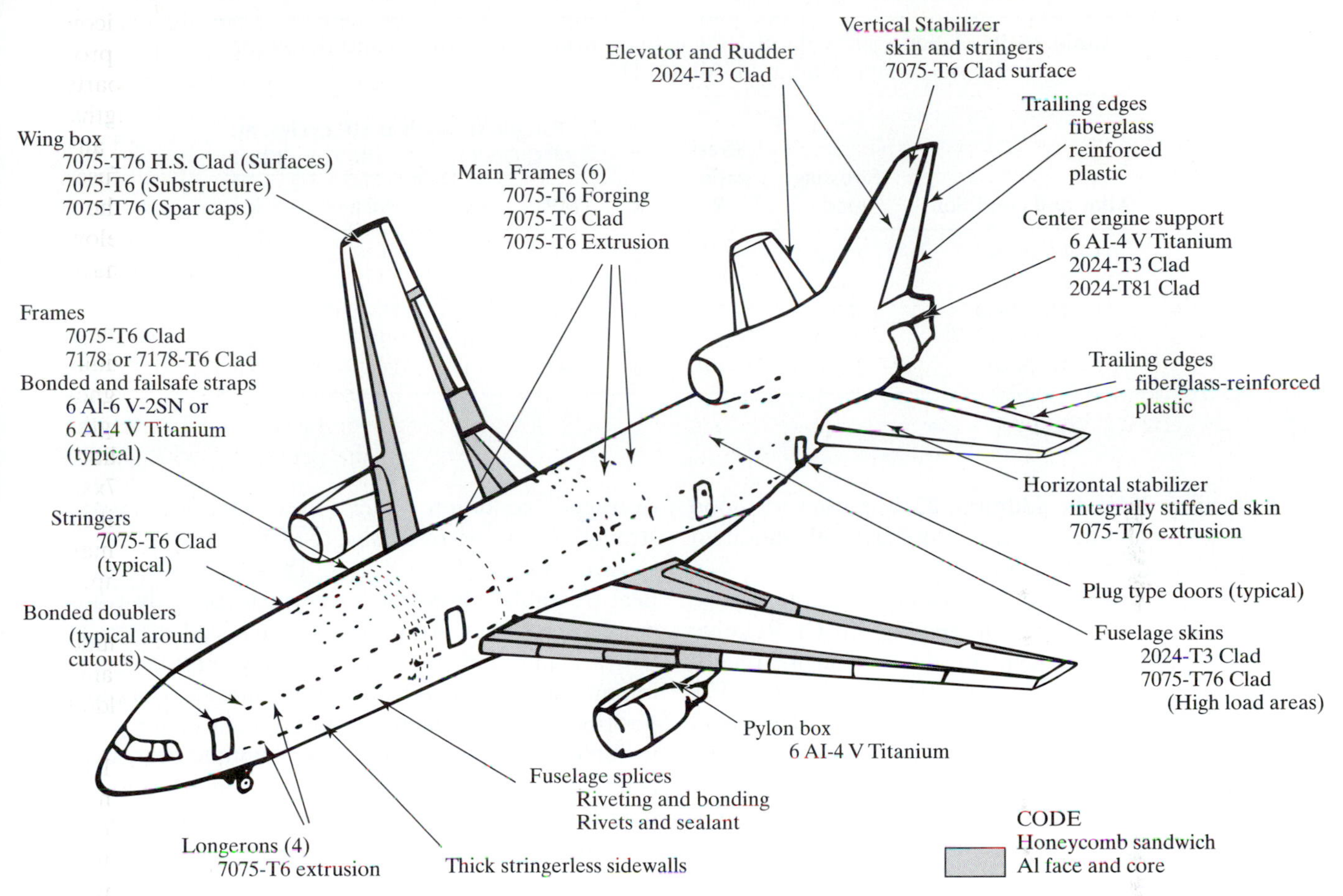

Fig. 13-1 Usage of 7XXX and 2XXX aluminum alloys in commercial aircraft. [3]

Table 13-8 Fracture Toughness of Aerospace Aluminum Alloy Plates at Room Temperature and Below. [1]

Alloy and condition	Room temperature yield strength		Specimen design	Orientation	Fracture toughess, K_{I_c} or $K_{I_c}(J)$ at:			
					24°C(75°F)		−196°C(−320°F)	
	MPa	ksi			MPa$\sqrt{m}$	ksi$\sqrt{in.}$	MPa$\sqrt{m}$	ksi$\sqrt{in.}$
2014-T651	432	62.7	Bend	T-L	23.2	21.2	28.5	26.1
2024-T851	444	64.4	Bend	T-L	22.3	20.3	24.4	22.2
2219-T87	382	55.4	Bend	T-S	39.9	36.3	46.5	42.4
7039-T6	381	55.3	Bend	T-L	32.3	29.4	33.5	30.5
7075-T651	536	77.7	Bend	T-L	22.5	20.5	27.6	25.1
7075-T7351	403	58.5	Bend	T-L	35.9	32.7	32.1	29.2
7075-T7351	392	56.8	Bend	T-L	31.0	28.2	30.9	28.1

Table 13-9 Fatigue Strengths at 10^6 Cycles at Room Temperature and Below or Aerospace Aluminum Alloys. [1]

Alloy and condition	Stressing mode	Stress ratio, R	K_t	Fatigue strength at 10^6 cycles, at: 24°C (75°F) MPa	ksi	−196°C (−320°F) MPa	ksi
2014-T6 sheet	Axial	−1.0	1	115	17	170	25
2219-T62 sheet	Axial	−1.0	1	130	19	15	2.2
2219-T87 sheet	Axial	−1.0	1	150	22	115–170	17–25
7039-T6 Sheet	Axial	−1.0	1	140	20	215	31
7075-T6 sheet	Axial	−1.0	1	96	14	145	21

with alloy 7050. In addition, 2219 is used for good weldability and high-temperature strength, modified as 2419 for higher toughness.

Another property of importance in the aerospace industry is *stress-corrosion cracking* (SCC). Relative SCC ratings for wrought products of high-strength aluminum alloys show that 2024-T8 and 7075-T73 exhibit A ratings—very high resistance to SCC for all forms of product and in all directions relative to the rolling direction, except the ST (short-transverse) for 2024-T8. (see Table 16-11) The T8 temper involves cold-working that helps to avoid SCC and T73 is an overaged condition. The latter sacrifices some strength for better resistance to SCC.

Aluminum-Lithium Alloys. These alloys are also heat-treatable alloys but have yet to be fully commercialized. They are touted as advanced materials and are targeted to replace the conventional aluminum alloys 2024 and 7075 in the aircraft. The compositions of some aluminum-lithium alloys are given in Table 13-10. They have the very unique characteristics of being lighter and yet have higher moduli than the conventional aluminum alloys. These characteristics are shown for three selected alloys, 2090,

Table 13-10 Compositions of Aluminum-Lithium Alloys. [1]

Element	Composition, wt% 2090(a)	2091(a)	8090(a)	Weldalite 049(b)
Silicon	0.10	0.20	0.20	...
Iron	0.12	0.30	0.30	...
Copper	2.4–3.0	1.8–2.5	1.0–1.6	5.4
Manganese	0.05	0.10	0.10	...
Magnesium	0.25	1.1–1.9	0.6–1.3	0.4
Chromium	0.05	0.10	0.10	...
Zinc	0.10	0.25	0.25	...
Lithium	1.9–2.6	1.7–2.3	2.2–2.7	1.3
Zirconium	0.08–0.15	0.04–0.16	0.04–0.16	0.14
Titanium	0.15	0.10	0.10	...
Other, each	0.05	0.05	0.05	(Ag 0.4)
Other, total	0.15	0.15	0.15	...
Aluminum	bal	bal	bal	bal

(a) Registered limits, (b) Nominal composition

2091, and 8090 in Table 13-11 and translate to much higher specific strengths for comparable strengths and higher specific moduli. These alloys also exhibit better toughness and fatigue properties.

Alclad and Clad Aluminum Products. Aluminum products are sometimes coated on one or both surfaces with a metallurgically bonded, thin layer of pure aluminum or aluminum alloy. If the combination of core and cladding alloys is selected so that the cladding is anodic to the core, the product is called alclad. A clad product, on the other hand, is a combination in which the cladding is not intentionally made anodic to the core. The cladding of the Alclad product protects the core at exposed edges as well as at abraded or corroded areas and acts as a sacrificial anode. Clad products are designed to provide improved surface appearance or other characteristics required for special applications. Brazing products are commercial examples of clad products in which a cladding alloy having a melting point appreciably lower than the core is used for subsequent joining of several parts into an assembly.

The corrosion potentials of the cladding and the core must be sufficiently different, with the cladding being anodic, in order to protect the core. The corrosion potentials for different unalloyed and alloyed aluminum are shown in Tables 13-12a and -12b. We see that pure aluminum, 1060, is about 0.15V more anodic than 2024-T3 and -T4 tempers and is used for most Alclad 2XXX products. Alloy 7049 has the most anodic corrosion potential and can be used for Alclad products of most of the aluminum alloys.

13.2.4 Selection of Aluminum and Aluminum Alloy Castings [2]

Aluminum casting alloys are based on the same alloy systems as those of the wrought aluminum alloys, except for the absence of manganese as a major alloying element as discussed earlier. They are also classified as nonheat-treatable and heat-treatable alloys. Because the products are castings, the nonheat-treatable alloys can not be strengthened by cold-working. Typical applications of the aluminum alloy castings are shown in Table 13-13.

The 3XX alloys are the most used for aluminum castings, as the 3XXX alloys are for wrought products. As indicated earlier in Sec. 13.2.1, the difference between these two alloys is in their major alloying element. The wrought alloys contain manganese while the cast alloys have silicon. Silicon forms a eutectic with aluminum as shown in Fig. 13-2 that provides the fluidity and castability of these high-volume (marketwise) silicon-containing aluminum casting alloys. Silicon contents from about 4 percent to the eutectic composition of about 12 percent reduce scrap losses, permit production of much more intricate shapes with greater variations in section

Table 13-11 Physical Properties of Selected Aluminum-Lithium Alloys. [1]

Property	2090	2091	8090
Density, g/cm^3 (lb/in.^{3})	2.59 (0.094)	2.58 (0.0931)	2.55 (0.092)
Melting range, °C (°F)	560–650 (1040–1200)	560–670 (1040–1240)	600–655 (1110–1210)
Electrical conductivity, %IACS	17–19	17–19	17–19
Thermal conductivity, at 25°C (77°F), W/m · K (Btu · in./ft^2 · °F · h)	84–92.3 (580–640)	84 (580)	93.5 (648)
Specific heat at 100°C (212°F), J/kg · K (cal/g · °C)	1203 (0.2875)	860 (0.205)	930 (0.22)
Average coefficient of thermal expansion from 20 to 100°C (68 to 212°F), μm/m · °C (μin./in. · °F)	23.6 (13.1)	23.9 (13.3)	21.4 (11.9)
Solution potential, mV(a)	−740	−745	−742
Elastic modulus, GPa (10^6 psi)	76 (11.0)	75 (10.9)	77 (11.2)
Poisson's ratio	0.34	...	...

(a) Measured per ASTM G 60 using a saturated calomel electrode

Table 13-12a Electrode or solution potentials of non-heat-treatable wrought commercial aluminum alloys. [4]

Values are the same for all tempers of each alloy.

Alloy	Potential(a), V
1060	−0.84
1100	−0.83
3003	−0.83
3004	−0.84
5050	−0.84
5052	−0.85
5154	−0.86
5454	−0.86
5056	−0.87
5456	−0.87
5182	−0.87
5083	−0.87
5086	−0.85
7072	−0.96

(a) Potential versus standard calomel electrode.

Table 13-12b Electrode or solution potentials in heat-treated conditions of commercial wrought aluminum alloys. [4]

Alloy	Temper	Potential(a), V
2041	T4	−0.69(b)
	T6	−0.78
2219	T3	−0.64(b)
	T4	−0.64(b)
	T6	−0.80
	T8	−0.82
2024	T3	−0.69(b)
	T4	−0.69(b)
	T6	−0.81
	T8	−0.82
2036	T4	−0.72
2090	T8E41	−0.83
6009	T4	−0.80
6010	T4	−0.79
6151	T6	−0.83
6351	T5	−0.83
6061	T4	−0.80
	T6	−0.83
6063	T5	−0.83
	T6	−0.83
7005	T6	−0.94
X7016	T6	−0.86
X7021	T6	−0.99
X7029	T6	−0.85
X7146	T6	−1.02
7049	T73	−0.84(c)
	T76	−0.84(c)
7050	T73	−0.84(c)
	T76	−0.84(c)
7075	T6	−0.83(c)
	T73	−0.84(c)
	T76	−0.84(c)
7475	T6	−0.83(c)
	T73	−0.84(c)
	T76	−0.84(c)
7178	T6	−0.83(c)

(a) Potential versus standard calomel electrode. (b) Varies ±0.01 V with quenching rate. (c) Varies ±0.02 V with quenching rate.

thickness, and yield castings with higher surface and internal quality.

Although a large number of aluminum alloys have been developed for casting, the following are the basic types:

1XX.0 Alloys—There is only one important application of commercially pure aluminum casting and it is based on its high electrical conductivity. Aluminum is used as collector rings and conductor bars, which are cast integrally with steel laminations to produce rotors for certain types of electric motors. Most cast aluminum motor rotors are produced for controlled electrical performance with a minimum occurrence of microshrinkage and cracks during casting. Based on these, the carefully controlled pure alloys 100.0 (99.00% min. Al), 150.0 (99.50% min. Al), and 170.0 (99.70% min. Al) are predominantly used. Alloy 100.0 contains a significantly larger amount of iron and other impurities to improve castability and crack resistance, and to lower shrinkage formation. For the same reason, alloy 150.0 has a better casting performance than alloy 170.0. Because of the better castability of 100.0, it is recommended for rotors larger than 152 mm (6 in.).

Although gross casting defects may adversely affect electrical performance, the conductivity of the alloys are almost exclusively controlled by the impurities in solid solution. The influence of impurities in solid solution was earlier shown in Table 3-3 and reproduced in Table 13-14 that indicates that the average increase in resistivity is much larger when the elements are in solution than when out of solu-

Table 13-13 Typical applications of aluminum casting alloys. [1]

Alloy	Representative appliations
100.0	Electrical rotors larger than 152 mm (6 in.) in diameter
201.0	Structural members: cylinder heads and pistons; gear, pump, and aerospace housings
208.0	General-purpose castings; valve bodies, manifolds, and other pressure-tight parts
222.0	Bushings; meter parts; bearings; bearing caps; automotive pistons; cylinder heads
238.0	Sole plates for electric hand irons
242.0	Heavy-duty pistons; air-cooled cylinder heads; aircraft generator housings
A242.0	Diesel and aircraft pistons; air-cooled cylinder heads; aircraft generator housings
B295.0	Gear housing; aircraft fittings; compressor connecting rods; railway car seat frames
308.0	General-purpose permanent mold castings; ornamental grilles and reflectors
319.0	Engine crankcases; gasoline and oil tanks; oil pans; typewriter frames; engine parts
332.0	Automotive and heavy-duty pistons; pulleys, sheaves
333.0	Gas meter and regulator parts; gear blocks; pistons; general automotive castings
354.0	Premium-strength castings for the aerospace industry
355.0	Sand: air compressor pistons; printing press bedplates; water jackets; crankcases. Permanent: impellers; aircraft fittings; timing gears; jet engine compressor cases
356.0	Sand: flywheel castings; automotive transmission cases; oil pans; pump bodies. Permanent: machine tool parts; aircraft wheels; airframe castings; bridge railings
A356.0	Structural parts requiring high strength; machine parts; truck chassis parts
357.0	Corrosion-resistant and pressure-tight applications
359.0	High-strength castings for aerospace industry
360.0	Outboard motor parts; instrument cases; cover plates; marine and aircraft castings
A360.0	Cover plates; instrument cases; irrigation system parts; outboard motor parts; hinges
380.0	Housing for lawn mowers and radio transmitters; air brake castings; gear cases
A380.0	Applications requiring strength at elevated temperature
384.0	Pistons and other severe service applications; automatic transmissions
390.0	Internal combustion engine pistons, blocks, manifolds, and cylinder heads
413.0	Architectural, ornamental, marine, and food and dairy equipment applications
A413.0	Outboard motor pistons; dental equipment; typewriter frames; street lamp housings
443.0	Cookware; pipe fittings; marine fittings; tire molds; carburetor bodies
514.0	Fittings for chemical and sewage use; dairy and food handling equipment; tire molds
A514.0	Permanent mold casting of architectural fittings and ornamental hardware
518.0	Architectural and ornamental castings; conveyor parts; aircraft and marine castings
520.0	Aircraft fittings; railway passenger car frames; truck and bus frame sections
535.0	Instrument parts and other applications where dimensional stability is important
A712.0	General-purpose castings that require subsequent brazing
713.0	Automotive parts; pumps; trailer parts; mining equipment
850.0	Bushings and journal bearings for railroads
A850.0	Rolling mill bearings and similar applications

Source: Compiled from *Aluminum Casting Technology,* American Foundrymen's Society, 1986.

tion. The electrical conductivity increases as the purity of aluminum increases from the 56%IACS (International Annealed Copper Standard) for 100.0 to 60 percent IACS for the 170.0 casting.

2XX.0 Alloys—The value of the copper addition to this alloy is to improve the strength and hardness at elevated temperatures; therefore, the 2XX.0 alloys find applications in high temperature environments. These alloys are heat-treatable when the alloy contains 4 to 5 percent copper. However, they have limited fluidity and marginal castability that requires careful gating and generous riser feeding during solidification to ensure a sound casting. In addition, pressure-tight parts of intricate design are difficult to

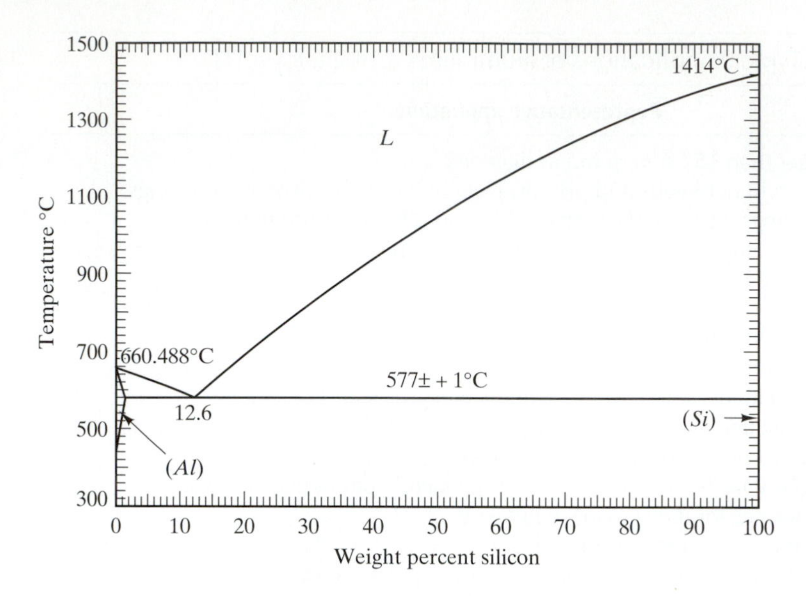

Fig. 13-2 The aluminum-silicon binary equilibrium diagram. [10]

Table 13-14 Effect on resistivity of aluminum on elements in and out of solution [2]

Element	Maximum solubility in Al, %	Average increment(a) in resistivity per wt%, microhm-cm	
		In solution	Out of solution(b)
Chromium	0.77	4.00	0.18
Copper	5.65	0.344	0.030
Iron	0.052	2.56	0.058
Lithium	4.0	3.31	0.68
Magnesium	14.9	0.54(c)	0.22(c)
Manganese	1.82	2.94	0.34
Nickel	0.05	0.81	0.061
Silicon	1.65	1.02	0.088
Titanium	1.0	2.88	0.12
Vanadium	0.5	3.58	0.28
Zinc	82.8	0.094(d)	0.023(d)
Zirconium	0.28	1.74	0.044

(a) Add above increase to the base resistivity for high-purity aluminum, 2.65 microhm-cm at 20°C (68°F) or 2.71 microhm-cm at 25°C (77°F). (b) Limited to about twice the concentration given for the maximum solid solubility, except as noted. (c) Limited to approximately 10%. (d) Limited to approximately 20%.

obtain; their resistance to hot-cracking is relatively poor; and they are susceptible to stress-corrosion cracking. Because of these characteristics, the 3XX alloys are often substituted for the 2XX alloys.

Alloy 295 with 4.5%Cu was the initial alloy used for a lot of aluminum castings. Alloy 296 with the addition of 0.6%Si to Alloy 295 improved the resistance to hot-cracking in permanent mold castings.

However, these alloys have since been replaced by alloys 355 and 356, which have similar mechanical properties, but lower specific gravities, better corrosion resistance, and superior casting characteristics.

Alloy 222, with higher copper content of about 10 percent, was developed for used as pistons for internal combustion engines. However, it has been replaced in this application by alloys 332, 336, and 339, and to a lesser extent by 242 and A242. Alloys 230, 242, A242, and 243 also have high strength and hardness at elevated temperatures to justify their usage as diesel engine pistons and air-cooled cylinder heads for aircraft engines.

Alloys 201, A206, 224, and 249 alloys are recent developments that exhibit significantly higher tensile strengths than those of any previous aluminum casting alloy and are being used to cast premium-quality aerospace parts.

3XX.0 Alloys—Because these alloys contain silicon with copper and/or magnesium, they have good castability and at the same time can be strengthened by heat-treating. For these reasons, these alloys are largely used for aluminum castings.

Si-Cu alloys. Silicon provides excellent casting characteristics while copper imparts moderately high strength and improved machinability but at the expense of reduced ductility and lower corrosion resistance. Alloys 319 (6%Si-3.5%Cu) and 380(8.5%Si-3.5%Cu) are the general-purpose alloys for aluminum castings. Alloy 319 is the preference for sand casting, but may also be used in permanent mold casting, while Alloy 380 is the preference for die casting. Alloys 308, 383, and 384 and their variations that may contain up to 3 percent zinc and other elements are other die-casting alloys. For permanent mold casting, Alloy 333 (9.0%Si-3.5%Cu) is recommended. Many castings of these alloys (319, 380, and 333) are supplied in the F temper, as-cast condition. Some castings are heat-treated to the T5 temper to improve hardness, machinability, and stability of properties. The strength of all the alloys can be substantially improved by a full T6 temper.

Si-Mg alloys. These alloys are hardened by Mg_2Si, magnesium silicide, in the same manner as the 6XXX wrought alloys. The most popular of these is 356 (7%Si-0.3%Mg) and the higher purity A356. They have excellent castability, weldability, pressure tightness, and corrosion resistance. They are heat-treatable to provide various combinations of tensile and physical properties that are attractive for many applications, including aircraft and automotive parts. Alloy 357 is similar to 356 but has a higher magnesium (0.5%Mg) and can be heat-treated to a higher strength level. The high-purity versions of both alloys, A356 and A357, provide higher ductility and are among the premium-quality sand and permanent mold castings specified for military and aircraft applications.

Other alloys of the aluminum-silicon-magnesium silicide group are 359 (9%Si-0.6%Mg), the die-casting alloy 360 (10.5%Si-0.5%Mg), A360 (lower iron than 360), and 364 (8.5%Si-0.3%Mg-0.4%Cr-0.03%Be). Corrosion resistance of these Al-Si-Mg is superior to the general-purpose Al-Si-Cu alloys 319 and 380.

Si-Mg-Cu Alloys. The combination of magnesium and copper provides a greater increase in strength during heat treatment than either copper or magnesium alone. However, the higher strength is achieved with some sacrifice in ductility and corrosion resistance.

Representative sand and permanent mold alloys of this group are 355 (5%Si-1.3%Cu-0.5%Mg), and the higher tensile property version C355 (also a premium-quality casting alloy). Tank engine cooling fans as well as high-speed rotating parts and impellers are made of C355. When premium-casting procedures are applied, even higher tensile properties can be obtained with the heat-treated 354 alloy (9%Si-1.8%Cu-0.5%Mg).

Pistons for automotive internal combustion engines are made of permanent mold cast 332 alloy (9.5%Si-3%Cu-1%Mg). The higher silicon 336 alloy with nickel addition (12%Si, 2.5%Ni-1%Cu-1%Mg) is used for diesel and higher output engines. 336 has improved elevated-temperature properties, a lower expansion coefficient, and exhibits improved wear resistance. The pistons are usually used in the T5 temper or a variation of T5, such as T551, to increase hardness, to improve machinability, and to provide growth stability.

4XX.0 Alloys—These are primarily aluminum-silicon binary alloys and are classified into (a) eutectic and hypoeutectic alloys ($\leq$12%Si), and (b) hypereutectic alloys ($>$12%Si). They exhibit high corrosion resistance, good weldability, and low specific gravity, but are somewhat difficult to machine.

Eutectic and Hypoeutectic Alloys. Alloys 413, 443, and 444 are the important binary aluminum alloys. Alloy 443 (5.3%Si, hypoeutectic alloy) may be used in all casting processes for parts that require good ductility, corrosion resistance, and pressure tightness—deemed more important than strength. Alloy 413 and A413 (12%Si, eutectic alloy) for die casting also exhibit good corrosion resistance, and are superior to alloy 443 in terms of castability and pressure tightness. Alloy A444 (7%Si-0.2%Fe max) has good corrosion and especially high ductility when it is cast in permanent mold and heat-treated to a T4 condition. It is also selected when impact resistance is a primary consideration.

Hypereutectic Alloys. These aluminum-silicon binary alloys have outstanding wear resistance, lower thermal expansion coefficient, and very good casting characteristics. They have been used in Europe for years for heavy-duty internal combustion engine pistons and in similar high-temperature applications. In the United States, alloys such as 393 (22%Si-2.3%Ni-1%Mg-0.1%V) have been used in similar applications for a long time.

In the early 1970s, the first all-aluminum engine block without iron liners or electroplating on the cylinder bores was the die-cast 390 alloy (17%Si-4.5%Cu-0.5%Mg). Since that time, the engines of five premium automobiles in Europe have been designed and manufactured using 390 alloy and using the same bare-bore technology. The usage of Alloy 390 is due in part to its outstanding wear characteristics and strength. Today, 390 is used extensively in small engines (e.g., chain saws), computer disc spacers, pistons for air-conditioning compressors, air compressor bodies, master brake cylinders, pumps, and other components in automatic transmissions.

The hypereutectic alloys are hard to machine because of their hard, large-sized primary silicon phase. The use of polycrystalline diamond cutting tools has alleviated the problems of poor tool life when these alloys are machined and has yielded good surface finish on the workpiece and chip characteristics. To guarantee the best machinability, mechanical properties, and performance of parts cast from these alloys, the primary silicon size must be controlled and refined. This refinement process can be done by treating the melt with a few hundredths of a percent of phosphorus that nucleates primary silicon particles during solidification and refines the primary silicon to 8 to 10 percent of the unrefined primary silicon size.

5XX.0 Alloys—These alloys contain magnesium as the major alloying element and are characterized by their excellent corrosion resistance, good machinability, and attractive appearance. Just like the wrought alloys with magnesium, they are especially good for marine and sea water applications and exhibit moderate-to-high strength and toughness properties. They are suitable for welded assemblies and are often used in architectural and other decorative or building needs because of their attractive appearance when they are anodized (see Table 13-13 for other applications).

Compared with the 3XX and 4XX alloys, the 5XX alloys require more care in gating, in properly locating and sizing risers, and in providing greater chilling to produce sound castings. For best corrosion resistance, these alloys must be prepared with high-quality, low-impurity content metals and handled with great care in the foundry. Controlled melting and pouring practices are needed to compensate for the greater oxidizing tendency of the alloys (magnesium) when molten. This care is accentuated because many of the applications of these alloys require polishing and/or fine surface finishing, where defects caused by oxide inclusions are particularly undesirable. The relatively poor castability of these alloys and the associated special foundry practices needed to produce quality castings increase the cost of the castings of these alloys and make them less desirable, unless absolutely necessary.

7XX.0 Alloys—The major alloying in these alloys is zinc, but as we previously noted, these alloys also contain some magnesium; therefore, they are basically aluminum-zinc-magnesium alloys. These alloys age naturally (precipitation hardening at room temperature) to a reasonable strength level without solution heat-treatment, making them useful for castings with shapes that are difficult to solution heat-treat and quench without cracking and distortion. They have good machinability and high melting points that make them suitable castings to be assembled by brazing. They have, however, the same castability problems and special handling procedures as the 5XX.0 alloys.

8XX.0 Alloys—These tin-containing alloys are selected for bearings and bushings because they have high load-carrying capacity and fatigue strength. They also exhibit superior corrosion resistance to internal combustion engine lubricating oils in comparison to most other metals for bearing composition. All four alloys of this family can be cast in both sand and permanent molds. However, careful control of gating and other casting practices are necessary to produce sound castings and to overcome a marked susceptibility to hot-cracking.

Alloys 850 and 851 are primarily used for connecting rods and crankcase bearings for diesel engines. Alloys 852 and 853 have higher compressive yield strengths and hardnesses and are used for truck roller bearings, large, rolling mill bearings, and other bearings as well as bushings that are made to be heavily loaded.

Selection of Aluminum Casting Alloys. The five major factors that influence alloy selection for casting applications consist of the following:

- Casting process and design considerations—include fluidity, resistance to hot-tearing, solidification range, and die soldering (in die casting)
- Mechanical property requirements—include strength and ductility, heat-treatability, and hardness
- Service requirements—include pressure tightness characteristics, corrosion resistance, surface treatments, dimensional stability, and thermal stability
- Economics—include machinability, weldability, ingot and melting costs, and heat-treatment

Table 13-15 serves as a guide in selecting casting alloys. It shows the classification of alloys into sand casting, permanent mold casting, and die-casting types; their ratings of castability in these three shaping processes; and the ratings on corrosion resistance, machinability, and weldability of their castings.

13.3 Titanium and Titanium Alloys [5]

In its mineral form as rutile, titanium was first discovered over 200 years ago, about 1790, but it was not commercially produced until the early 1950s. Since then, the commercial production of titanium and titanium alloys in the United States has increased from zero to more than 23 million kg/yr (50 million lb/yr). This remarkable growth was ushered by (1) the development of a relatively safe, economical method to produce titanium metal in the late 1930s, and (2) the appreciation of its commercial importance after its mechanical and physical properties were obtained in the late 1940s. The potential of titanium was immediately recognized of titanium as a high-strength-to-weight ratio, and as a higher temperature (than aluminum) material for aircraft application. This potential has been realized with about 20 to 30 percent of the current aircraft's weight being titanium alloys. About 80 percent of all titanium production goes to the aerospace industry. In addition to its high specific strength, titanium's commercial importance comes also from its outstanding corrosion resistance. Table 13-16 lists selected unalloyed or commercially pure titanium grade and commercially available titanium alloys, their descriptions, and some typical applications.

13.3.1 Commercially Pure and Modified Titanium

Commercially pure titanium has been available as mill (as-fabricated by producers) products since 1950 and is used for applications that require moderate strength combined with good formability and corrosion resistance. Initial production was largely due to aerospace demands for a material lighter than steel and more heat resistant than aluminum alloys. However, most of the current applications of commercially pure titanium stem from its excellent corrosion resistance and good weldability.

Titanium is a very reactive metal and forms compounds very quickly with the elements carbon, oxygen, nitrogen, and hydrogen. Incorporation of these elements in the composition usually strengthens, but at the same time embrittles (lowers the ductility) pure titanium as seen in Fig. 13-3. Thus, in the melting and welding of titanium, it is imperative to minimize the exposure of titanium to these elements and quite often the processing is done under vacuum or argon atmosphere. Nevertheless, commercially pure titanium has generally carbon, oxygen, and nitrogen together with iron and silicon as principal impurities

Table 13-15 Ratings of castability, corrosion resistance, machinability, and weldability for aluminum casting alloys; Ratings are 1 for best and 5 for worst. An alloy may have different ratings for each of the casting processes.

Alloy	Resistance to hot cracking(a)	Pressure tightness	Fluidity(b)	Shrinkage tendency(c)	Corrosion resistance(d)	Machin-ability(e)	Weld-ability(f)
Sand casting alloys							
201.0	4	3	3	4	4	1	2
208.0	2	2	2	2	4	3	3
213.0	3	3	2	3	4	2	2
222.0	4	4	3	4	4	1	3
240.0	4	4	3	4	4	3	4
242.0	4	3	4	4	4	2	3
A242.0	4	4	3	4	4	2	3
295.0	4	4	4	3	3	2	2
319.0	2	2	2	2	3	3	2
354.0	1	1	1	1	3	3	2
355.0	1	1	1	1	3	3	2
A356.0	1	1	1	1	2	3	2
357.0	1	1	1	1	2	3	2
359.0	1	1	1	1	2	3	1
A390.0	3	3	3	3	2	4	2
A443.0	1	1	1	1	2	4	4
444.0	1	1	1	1	2	4	1
511.0	4	5	4	5	1	1	4
512.0	3	4	4	4	1	2	4
514.0	4	5	4	5	1	1	4
520.0	2	5	4	5	1	1	5
535.0	4	5	4	5	1	1	3
A535.0	4	5	4	4	1	1	4
B535.0	4	5	4	4	1	1	4
705.0	5	4	4	4	2	1	4
707.0	5	4	4	4	2	1	4
710.0	5	3	4	4	2	1	4
711.0	5	4	5	4	3	1	3
712.0	4	4	3	3	3	1	4
713.0	4	4	3	4	2	1	3
771.0	4	4	3	3	2	1	...
772.0	4	4	3	3	2	1	...
850.0	4	4	4	4	3	1	4
851.0	4	4	4	4	3	1	4
852.0	4	4	4	4	3	1	4
Permanent mold casting alloys							
201.0	4	3	3	4	4	1	2
213.0	3	3	2	3	4	2	2
222.0	4	4	3	4	4	1	3
238.0	2	3	2	2	4	2	3
240.0	4	4	3	4	4	3	4
296.0	4	3	4	3	4	3	4

(*Continued*)

Table 13-15 Ratings of castability, corrosion resistance, machinability, and weldability for aluminum casting alloys; Ratings are 1 for best and 5 for worst. An alloy may have different ratings for each of the casting processes. (*Continued*)

Alloy	Resistance to hot cracking(a)	Pressure tightness	Fluidity(b)	Shrinkage tendency(c)	Corrosion resistance(d)	Machin-ability(e)	Weld-ability(f)
308.0	2	2	2	2	4	3	3
319.0	2	2	2	2	3	3	2
332.0	1	2	1	2	3	4	2
333.0	1	1	2	2	3	3	3
336.0	1	2	2	3	3	4	2
354.0	1	1	1	1	3	3	2
335.0	1	1	1	2	3	3	2
C355.0	1	1	1	2	3	3	2
356.0	1	1	1	1	2	3	2
A356.0	1	1	1	1	2	3	2
357.0	1	1	1	1	2	3	2
A357.0	1	1	1	1	2	3	2
359.0	1	1	1	1	2	3	1
A390.0	2	2	2	3	2	4	2
443.0	1	1	2	1	2	5	1
A444.0	1	1	1	1	2	3	1
512.0	3	4	4	4	1	2	4
513.0	4	5	4	4	1	1	5
711.0	5	4	5	4	3	1	3
771.0	4	4	3	3	2	1	...
772.0	4	4	3	3	2	1	...
850.0	4	4	4	4	3	1	4
851.0	4	4	4	4	3	1	4
852.0	4	4	4	4	3	1	4
Die casting alloys							
360.0	1	1	2	2	3	4	
A360.0	1	1	2	2	3	4	
364.0	2	2	1	3	4	3	
380.0	2	1	2	5	3	4	
A380.0	2	2	2	4	3	4	
384.0	2	2	1	3	3	4	
390.0	2	2	2	2	4	2	
413.0	1	2	1	2	4	4	
C443.0	2	3	3	2	5	4	
515.0	4	5	5	1	2	4	
518.0	5	5	5	1	1	4	

(a) Ability of alloy to withstand stresses from contraction while cooling through hot short or brittle temperature range. (b) Ability of liquid alloy to flow readily in mold and to fill thin sections. (c) Decrease in volume accompanying freezing of alloy and a measure of amount of compensating feed metal required in form of risers. (d) Based on resistance of alloy in standard salts spray test. (e) Composite rating based on ease of cutting, chip characteristics, quality of finish, and tool life. (f) Based on ability of material to be fusion welded with filler rod of same alloy.

Table 13-16 Selected unalloyed titanium and titanium alloys, their descriptions, and applications.

Alloy, UNS Number, and Common Names	General Description	Applications
Unalloyed Titanium		
ASTM Grade 2 UNS: R50400 Grade 2	Grade 2 Ti is the workhorse for industrial applications requiring good ductility and corrosion resistance. The guaranteed minimum yield strength of 275 MPa (40 ksi) for Grade 2 is comparable to those of annealed austenitic stainless steels.	Grade 2 can be used in continuous service up to 425°C (800°F) and in intermittent service up to 540°C (1000°F). Used in the fabrication of components for aircraft, reaction vessels, heat exchangers, and electrochemical processing equipment.
Modified Titanium		
Ti-0.2Pd UNS: R52400 and R52250 ASTM Grade 7 (R52400) ASTM Grade 11 (R52250)	The two Ti-0.2Pd ASTM Grades (7 and 11) have better resistance to crevice corrosion at low pH and high temperature than corresponding unalloyed Grades 1 and 2.	Used for special corrosion applications. Grade 7 is comparable to Grade 2 in strength, while Grade 11 is comparable to Grade 1 in strength.
Alpha and Near-Alpha Alloy		
Ti-3Al-2.5V UNS: R56320 Tubing alloy, ASTM Grade 9, and Half 6-4	Ti-3Al-2.5V has excellent cold formability and 20 to 50% higher strength than unalloyed Ti.	Available as foil, seamless tubing, pipe, forgings, and rolled products. Used mostly in tubular form in various aerospace and nonaerospace applications.
Alpha-Beta Alloy		
Ti-6Al-4V and Ti-6Al-4V ELI UNS: R56400 and R56401 (ELI) Ti 6-4 ASTM Grade 5	Ti-6Al-4V is the most widely used Ti alloy, accounting for 50% of all the titanium tonnage worldwide. The alloy is most commonly used in the annealed. Ti-6Al-4V is also hardenable in sections up to 25 mm (1 in.) with yield strengths as high as 1140 MPa (165 ksi).	The aerospace industry accounts for more than 80% of Ti-6Al-4V usage. The biggest user of Ti-6Al-4V outside of the aerospace is for medical protheses, which accounts for about 3% of the market. The alloy also has a variety of weight reducing applications in high performance automotive and marine equipment.
Beta Alloy		
Ti-3Al-8V-6Cr-4Mo-4Zr UNS: R58640 Beta C™ and 38-6-44	Beta C™ has similar characteristics to Ti-13V-11Cr-3Al, but it is easier to melt. It is cold rollable and drawable, and is used mainly as bar and wire for springs.	Beta C™ is used in fasteners, springs, torsion bars, and as foil for sandwich structures. It is also used for tubulars and casings in oil, gas, and geothermal wells.

in their composition. These are called residual impurities that come from the raw materials used.

The reactivity with oxygen is what gives the very beneficial corrosion-resistant property of titanium and its alloys. This effect is the same as those observed in aluminum alloys and in stainless steels. The titanium surface reacts quickly and forms a very tight oxide on the surface that protects the bulk material from further attack. As indicated in Chap. 10 (on corrosion) this is the process of passivation. Titanium is passivated by the tight oxide that is like a very thin ceramic coating on the surface. When the oxide is breached or scratched, titanium reacts quickly with the oxidizing atmosphere to heal the opened surface.

ASTM classifies commercially pure titanium into four grades—Grades 1, 2, 3, and 4, according to minimum tensile properties. Tables 13-17 and 13-18 indicate the design tensile properties for flat products (sheet, strip, and plate) and for bars and shapes of these four grades. Basically, the oxygen and iron contents of the grades, also shown in Table 13-17, deter-

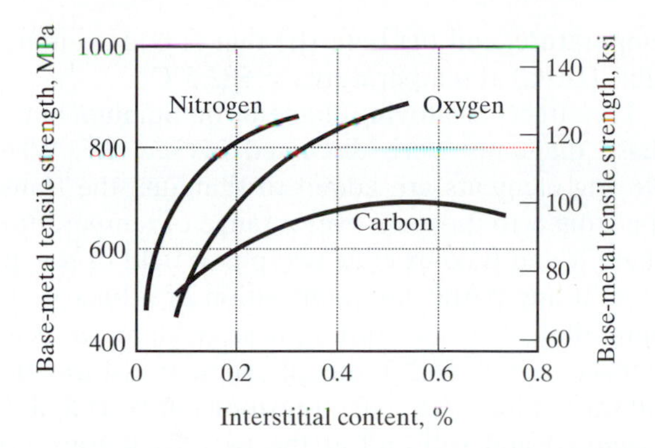

Fig. 13-3 Effect of interstitial element content on the strength of titanium.

mine the strength levels. In the higher strength grades, oxygen and iron may be intentionally added to the residual amounts from the raw material (sponge titanium) to increase the strength. Carbon and nitrogen are kept to minimum levels to avoid loss in ductility.

In addition to the basic grades, there are three other grades, Grades 7, 11, and 12, which are called modified titanium. The design properties of Grades 7 and 11 are the same as those of Grades 2 and 1. These modified grades are used when better corrosion resistance is required.

Grade 2 titanium (99.2% min. Ti) is the most used of the grades because it offers moderate strength, excellent formability, and excellent corrosion resistance. The

Table 13-17 Design tensile properties of unalloyed titanium sheets, strips, and plates. [5]

ASTM grade	Designation(a)	Ultimate tensile strength (L-LT)(b) MPa	Ultimate tensile strength (L-LT)(b) ksi	Tensile yield strength (L-LT)(b) MPa	Tensile yield strength (L-LT)(b) ksi	Elongation(c), % L	Elongation(c), % LT
Grade 1 (0.18 O, 0.2 Fe wt% max)	CP4	240	35	170	25	24	24
Grade 2 (0.25 O, 0.3 Fe wt% max)	CP3	345	50	275	40	20	20
Grade 3 (0.35 O, 0.3 Fe wt% max)	CP2	445	65	380	55	18	18
Grade 4 (0.40 O, 0.5 Fe wt% max)	CP1	550	80	480	70	15	15

(a) Designation of corresponding ASTM grades in MIL-T-9046. (b) S-basis properties in both the longitudinal and long-transverse directions for thicknesses less than 25 mm (1 in.). (c) S-basis elongation for thicknesses of 0.635 to 25 mm (0.025 to 1 in.).
Source: MIL-HDBK-5

Table 13-18 Design tensile properties of annealed chemically pure titanium extruded bars and shapes. [5]

Property	Minimum longitudinal properties for extruded bars and shapes(a) less than 75 mm (3 in.) thick: Grade 1	Grade 2	Grade 3	Grade 4
Ultimate tensile strength, MPa (ksi)	275 (40)	345 (50)	445 (65)	550 (80)(b)(c)
Tensile yield strength, MPa (ksi)	205 (30)	275 (40)	380 (55)	480 (70)(b)(c)
Elongation, % for extrusions:,				
4.77 to 25 mm (0.188 to 1.000 in.) thick	25	20	18	15(b)(c)
25 to 50 mm (1.001 to 2.000 in.) thick	20	18	15	12
50 to 75 mm (2.001 to 3.000 in.) thick	18	15	12	10

(a) Extrusions per MIL-T-81556. (b) For rolled bar per MIL-T-9047, the minimums for ASTM grade 4 apply to sections up to 100 mm (4 in.) thick with a maximum 103 cm^2 (16 $in.^2$) cross section. (c) Minimums also apply to long-transverse direction in rolled rectangular bar 13 to 75 mm (0.5 to 3 in.) thick per MIL-T-9047.
Source: MIL-HDBK-5, 1 Dec, 1991.

strength is comparable to annealed stainless steels as shown in Table 13-19 that compares the hardnesses (strengths) of various condenser tube materials. Its formability and corrosion resistance are derived from its relatively low impurity contents. The corrosion rates of Grade 2 titanium in some selected media are given in Table 13-20.

Grade 2 titanium is used in the aircraft industry for airframe skin and nonstructural components, in addition to the chemical and processing industries. Other aircraft applications include exhaust-pipe shrouds, fireproof bulkheads, gas-turbine bypass ducts, engine cowlings, and formed brackets and skins for hot areas—they are also used for galley equipment, chemical toilets, and floor supports under these areas. In the chemical and processing industries, Grade 2 titanium is used for reaction vessels and heat exchangers because of its corrosion resistance to sea water, moist chlorine, moist metallic chlorides, chlorite and hypochlorite solutions, nitric and chromic acids, organic acids, sulfides, and many industrial gaseous environments. Because of its resistance to deposit, impingement, and crevice corrosion, it is used extensively in tubular and plate-type heat exchangers for condensers, evaporators, and in other components such as for marine vessels, power stations, oil refineries, offshore platforms, and water-purification and desalination plants.

13.3.2 Titanium Alloys

Pure titanium has two crystalline forms (allotropes): (1) alpha (α) that is hexagonal, close-packed (HCP) at temperatures $< 882.5°C$, the allotropic transition temperature; and (2) beta (β) that is body-centered cubic (BCC) at temperatures $> 882.5°C$.

The effects of alloying elements on titanium binary phase diagrams were discussed in Sec. 9.5. When alloying elements are added to titanium, the transition from α to β occurs over a range of temperatures where α and β co-exist (a two-phase field of $\alpha + \beta$). Thus, at any particular composition of a binary titanium alloy, there are three phase fields in the solid state—α, $\alpha + \beta$, and β; the allotropic transformation will start at the $\alpha/(\alpha + \beta)$ transition curve (called the α-*transus*) and will end at the $(\alpha + \beta)/\beta$ transition boundary (called the β-*transus*). Some elements will raise the α-transus to enlarge the α-phase field. These are called α-stabilizers, are predominantly nontransition metals, and are commonly referred to as simple metals (SM). In addition to the simple metals, carbon and nitrogen are also α-stabilizers. Other elements lower the β-transus to enlarge the β-phase field. These are called β-stabilizers and are predominantly the transition metals (TM).

13.3.2a Classifications and Designations of Alloys. Technical, commercial titanium alloys are rarely binary alloys. They are predominantly multicomponent alloys and are broadly classified as "α," "β," or "$\alpha + \beta$" alloys, according to whether they exhibit the phases α (HCP), β (BCC), or $\alpha + \beta$ (duplex structure) in their microstructure at room temperature after air cooling from above the $(\alpha + \beta)/\beta$ transition temperature. Within the $\alpha + \beta$ classification, an alloy that contains much more α than β is called is called a near-α alloy, and an alloy containing a great deal more β is called a near-β alloy. The β

Table 13-19 Typical hardnesses of condenser materials.

Alloy	Hardness(a), HB	Hardness(b),HV
Titanium (ASTM grade 2)	180	145
Admiralty brass (annealed)	75	60
Aluminum brass	82	85
90-10 cupronickel (annealed)	65	75
70-30 cupronickel (annealed)	70	95
18-8 stainless steel (ST)	...	142
Type 304 stainless steel	180	...

(a) "CodeWeld Titanium Tubing," TIMET, Pittsburgh, p 7. (b) "Sumitomo Titanium," Sumitomo Metal Industries, Tokyo, p 14.

Table 13-20 Corrosion rates of Grade 2 titanium in selected media.

Corrosions	Concentration, %	Temperature		Corrosion rate	
		°C	°F	μm/yr	mils/yr
Acetic acid	5, 25, 75	100	212	nil	nil
	50, 99.5	100	212	0.25	0.01
Aluminum chloride, aerated	25	25	77	<2	<0.1
	5, 10	60	140	<2.5	<0.1
	10	100	212	<2.5	<0.1
Ammonium chloride	1, 10, saturated	20–100	68–212	<13	<0.5
Ammonium sulfate	5	25	77	nil	nil
	Saturated + 5% H_2SO_4	25	77	25	1
Aqua regia (3:1)	100	25	77	nil	nil
	100	77	170	890	35
Calcium Chloride	28	Boiling		nil	nil
	5, 10, 30	100	212	<25	<1
Calcium hypochlorite	Saturated	25	77	nil	nil
	2, 6	100	212	1.3	0.05
Chlorine					
Saturated with H_2O	...	25	77	125	5
More than 0.013% H_2O	...	79	175	nil	nil
Dry	...	32	90	Rapid	Rapid
Copper nitrate	Saturated	25	77	nil	nil
Cupric chloride	20, 40	Boiling		nil	nil
Ferric chloride	10, 20	25	77	nil	nil
	5	60	140	nil	nil
	10–40	Boiling		nil	nil
	30	93	200	nil	nil
	10–30	100	212	<13	<0.5
	5 + 10% NaCl	100	212	<13	<0.5
Ferric sulfate	10	25	77	nil	nil
Ferrous sulfate	Saturated	25	77	nil	nil
Hydrochloric acid	5	35	95	<50	<2
	10	35	95	1000	40
	20	35	95	4400	175
Hydrochloric acid plus copper sulfate	10 + 0.05	65	150	<50	<2
	10 + 0.1	65	150	<25	<1
	10 + 0.2, 0.25, or 0.5	65	150	nil	nil
	10 + 1	65	150	<25	<1
Hydrogen sulfide	Saturated water	25	77	<125	<5
Lactic acid	10–85	100	212	<125	<5
	10–100	Boiling		<125	<5
Lead acetate	Saturated	25	77	nil	nil
Magnesium chloride	5–40	Boiling		nil	nil
	5–40	100	212	<125	<5
Nitric acid	5	100	212	<25	<1
	10	100	212	<50	<2
	40–50, 69.5	100	212	<25	<1
	65	175	347	<125	<5

(*Continued*)

Table 13-20 Corrosion rates of Grade 2 titanium in selected media. (*Continued*)

Corrosions	Concentration, %	Temperature		Corrosion rate	
		°C	°F	μm/yr	mils/yr
	40	200	392	<1250	<50
	70	270	518	<1250	<50
	20	290	554	300	12
Phosphoric acid	5–30	25	77	<50	<2
	35–85	25	77	<1250	<50
	85	38	100	1000	40
	5–35	60	140	<1250	<50
	10	79	175	1250	50
	5	100	212	<1250	<50
Seawater	...	25	77	nil	nil
Silver nitrate	50	25	77	nil	nil
Sulfuric acid	15	25	77	nil	nil
	1	60	140	nil	nil
	3	60	140	1.3	0.05
	5	60	140	730	29
Zinc chloride	Saturated	25	77	nil	nil
	10	Boiling		nil	nil
	20	100	212	<125	<5

alloys are actually metastable, and the precipitation or separation of the α phase in the metastable β is a method to strengthen this group of alloys. A listing of some titanium alloys according to α, near α, α + β, near β, and β alloys is given in Table 9–2.

The designations and codes for titanium alloys have yet to be completely formalized. Although there is a Titanium Development Association, it has yet to fully develop a coding and designation system similar to the AISI designations for steel and the Aluminum Association designations for aluminum. In the meantime, alloys are formally referred to by indicating the compositions (in weight percent) of the alloying elements added. Thus, a Ti-8Al-1Mo-1V alloy means it contains 8%Al, 1%Mo, and 1%V. Such a system is quite acceptable. The problem arises for the uninitiated or the novice in titanium alloys when the literature refers to their common codes in industry. Most of the common codes refer to the numbers without the elements. Thus, the example given here is commonly referred to as Ti-811. The most used titanium alloy Ti-6Al-4V is commonly referred to as the 6-4 alloy or Ti-6-4. For the Ti-5Al-2Sn-2Zr-4Mo-4Cr alpha-beta alloy, the common name Ti-17 indicates the total weight percent of alloying present.

Each of the common titanium alloys in each class, except two of the β alloys, have now been designated a UNS (Unified Numbering System) code by the ASTM. Some of the miscellaneous and other alloys have UNS codes, which probably reflect their usage and may be used to select or procure materials.

The characteristics of the titanium alloys arise predominantly from the characteristics of their crystal structures. These are:

- *Alpha Alloys.* These alloys generally have superior creep resistance as compared to the beta alloys, and are therefore preferred for high-temperature applications. They also do not exhibit the ductile-to-brittle transition behavior common to BCC metals (the beta alloys exhibit it as well) and therefore, they are more suitable for cryogenic applications. They are characterized by satisfactory strength, toughness, and weldability, but poorer forgeability than the beta alloys. This results in a greater tendency for forging defects. Because they are single phase, alpha alloys are not heat-treatable and are commonly used in the annealed or recrystallized condition to minimize or eliminate residual stresses caused by working. For the near-alpha alloys, they exhibit variations in microstructures (similar to those of the alpha-beta alloys) due to the presence of the beta phase. These variations are the results of the processing of these alloys in either the β-field (β processed) or in the (α + β)-field (α/β processed) and their influence on the properties are shown in Table 13-21.

- *Beta Alloys.* Because of their BCC structure, these alloys have excellent forgeability over a wider range of forging temperatures than alpha alloys, and a beta alloy sheet is cold-formable in the solution-treated condition. As a class, beta and near-beta alloys have increased fracture toughness over alpha-beta alloys at a given strength level, with the advantage of heavy section heat-treatment capability. In recent years, this characteristic is given more attention because of the increased need for damage tolerance in aerospace structures. Some of the beta alloys containing molybdenum also exhibit good corrosion resistance. They exhibit better room-temperature forming and shaping characteristics than do alpha-beta alloys, higher strength than alpha-beta alloys at temperatures where yield strength (instead of creep strength) is the applicable criterion, and better response to heat treatment (solution treatment, quenching, and aging) in heavier sections than the alpha-beta alloys. However, these advantages are hampered by the close control of processing and fabrication steps needed to achieve the optimal properties. Thus, beta alloys have found limited application so far where very high strength is required, such as springs and fasteners. The high strength comes from the finely dispersed alpha particles in the retained beta microstructure as a result of heat-treatment.
- *Alpha-Beta Alloys.* The compositions of these alloys are such that their microstructures can contain between 10 to 50 percent beta, although most often the beta is about 20 percent. The presence of both the HCP and the BCC phases offers the properties of both phases—good strength, toughness, weldability, and forgeability. In addition, the presence of both phases offers the best superplastic elongation as seen in Table 13-22, much better than the sum of the separate elongations of the alpha (pure titanium) and the beta alloys. Figure 13-4 shows that the optimum beta-content for elongation is between 20 to 30 percent beta. During heat-treating, the presence of nonequilibrium phases, such as the alpha-primed martensite or metastable beta, results in substantial increases in tensile and yield strengths after the aging treatment. The best combination of properties is achieved by solution treating at a temperature close to, but below, the beta transus, followed by quenching and then aging.

Table 13-21 Influence of β and α/β processing on properties of near alpha alloys.

Property	β processed	α/β processed
Tensile strength	Moderate	Good
Creep strength	Good	Poor
Fatigue strength	Moderate	Good
Fracture toughness	Good	Poor
Crack growth rate	Good	Moderate
Grain size	Large	Small

13.3.2b Room Temperature Mechanical Properties. Table 13-23 lists the mechanical properties of Grades 2 and 7, commercially pure titanium, and Ti-6-4 alloys. It includes the tensile, impact, hardness, and the moduli properties. Table 13-24 also shows the dependence of the yield strength and the plane strain fracture toughness of predominantly alpha-beta alloys on the shape of the alpha phase or the processing of the alloys. We see that the fracture toughness varies by as much as a multiple of two or three for a given alloy, depending on the processing.

Ti-6Al-4V Alloy. Ti-6-4 is currently the most widely used titanium alloy, accounting for more than 50 percent of all titanium tonnage in the world. The aerospace industry accounts for more than 80 percent of this usage. The next biggest market for Ti-6-4 is its usage in medical prosthesis, which accounts for 3 percent of the market. The automotive, marine, and the chemical industries also use small amounts of Ti-6-4.

There is an ELI (extra-low interstitial content) grade of the alloy with low oxygen and iron that exhibits high damage tolerance properties, especially at cryogenic temperatures. When palladium is added (about 0.2 wt% Pd), enhanced corrosion is realized.

Table 13-22 Superplastic conditions and elongations of titanium alloys.

Alloy	Test temperature		Strain rate, s^{-1}	Strain rate sensitivity factor, *m*	Elongation, %
	°C	°F			
Commercially pure titanium	850	1560	1.7×10^{-4}	...	115
α-β alloys					
Ti-6Al-4V	840–870	1545–1600	1.3×10^{-4} to 10^{-3}	0.75	750–1170
Ti-6Al-5V	850	1560	8×10^{-4}	0.70	700–1100
Ti-6Al-2Sn-4Zr-2Mo	900	1650	2×10^{-4}	0.67	538
Ti-4.5Al-5Mo-1.5Cr	870	1600	2×10^{-4}	0.63–0.81	>510
Ti-6Al-4V-2Ni	815	1500	2×10^{-4}	0.85	720
Ti-6Al-4V-2Co	815	1500	2×10^{-4}	0.53	670
Ti-6Al-4V-2Fe	815	1500	2×10^{-4}	0.54	650
Ti-5Al-2.5Sn	1000	1830	2×10^{-4}	0.49	420
Near-β and β alloys					
Ti-15V-3Sn-3Cr-3Al	815	1500	2×10^{-4}	0.50	229
Ti-13Cr-11V-3Al	800	1470	...	...	<150
Ti-8Mn	750	1380	...	0.43	150
Ti-15Mo	800	1470	...	0.60	100

Ti-6-4 is available in wrought, cast, and sintered (P/M—powder metallurgy) product forms, with the wrought products accounting for more than 95 percent of the market. The aerospace industry uses all wrought product forms. Forgings are used to fabricate various attachment fittings; sheets and plates are used to fabricate clips, brackets, skins, bulkheads, and so on; extrusions are used for parts such as wing chords and other parts with long, constant cross sections; wires for fasteners on wings; and tubings for components such as torque tubes. In missile and space applications, Ti-6-4 has been used for rocket engine and motor cases, pressure vessels, wings, and in applications where weight is critical.

Although not as corrosion-resistant as commercially pure titanium alloys, Ti-6-4 has, nonetheless, excellent corrosion resistance compared to other alloy systems. It is highly resistant to natural environments and a great many aqueous chemicals up to at least their boiling temperatures. It is resistant to general corrosion by sea water and brine, oxidizing acids, aqueous chloride solutions, wet chlorine gas, and sodium hypochorite at typical product-operating temperatures. It is, however, vulnerable to reducing acids such as hydrofluoric, hydrochloric, sulfuric, oxalic, formic, and phosphoric acids. Table 13-25 shows the general corrosion rates in specific media.

The other titanium alloy that is commonly used is Ti-3Al-2.5V, which is commonly known as "Half 6-4" in the industry. It has an intermediate strength between unalloyed titanium and Ti-6-4, and has excellent cold formability for the production of seamless tubing, strip, and foil. It has a high specific strength and comparable weldability as Ti-6-4 but is more amenable to cold-working than Ti-6-4. It is used primarily as a tubing alloy.

In comparison to both steel and aluminum, titanium alloy usage in applications not requiring corrosion resistance and not imposing weight restrictions is hampered by its high cost. Table 13-26 shows a comparison of the cost of titanium with stainless steel, aluminum, and nickel alloys. The exact price per pound might change from time to time, but the relative cost will not vary much.

13.3.2c Selection of Wrought Titanium Alloys. Just as the selection of stainless steels starts with the most used, type 304, selection of titanium alloys starts also by determining first whether the attributes of Ti-6Al-4V alloy match the requirements of the design. If other attributes are needed from a titanium alloy other than those offered by Ti-6Al-4V, Fig. 13-5 summarizes the direction to go to make the selection of the right alloy. We look for near-beta and beta alloys if we are

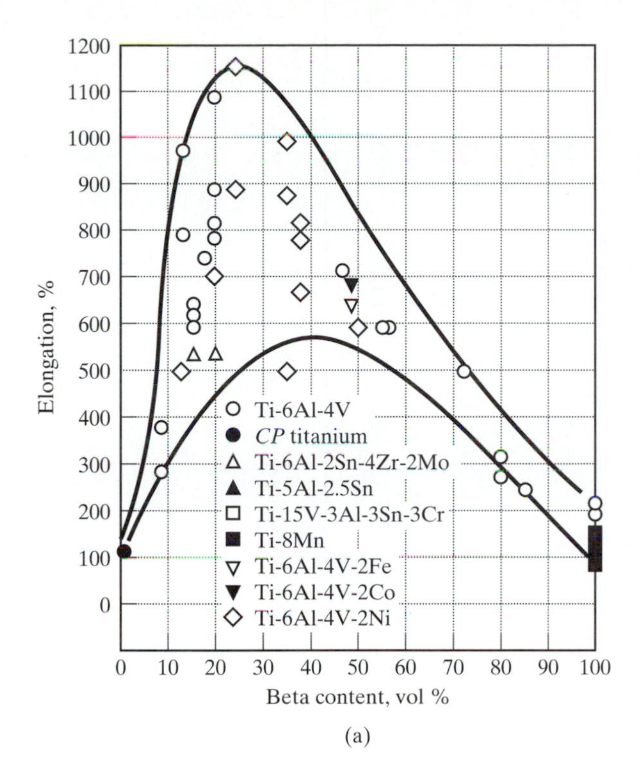

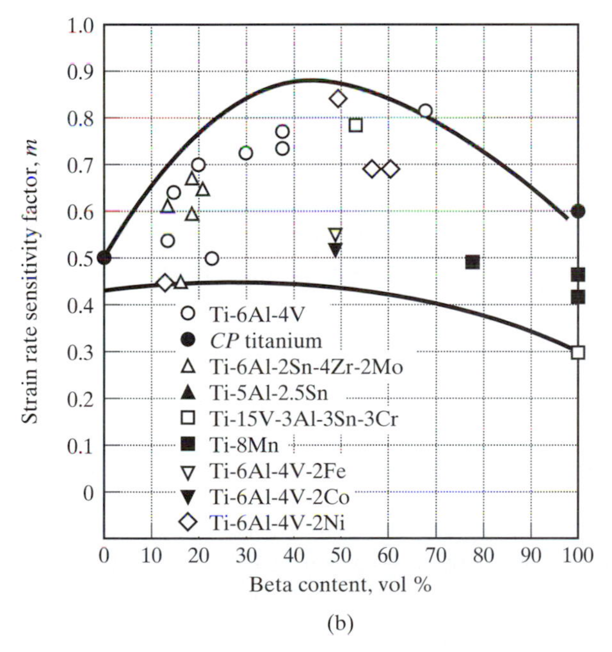

Fig. 13-4 Influence of the beta-phase content on (a) the superplastic elongation, and (b) m, the strain-rate sensitivity of several titanium alloys.

seeking more formability, strain-rate sensitivity, heat-treatment capacity and room-temperature strength than Ti-6Al-4V. On the other hand, if we are looking for greater high-temperature strength, better weldability, higher flow stress, and increasing the beta transus, then we should look for near-alpha and alpha alloys.

13.3.2d Titanium Castings. The lost-wax investment-casting technology has allowed the production of titanium alloy castings for use in the aerospace industry. The compositions of the titanium castings are basically the same as the wrought alloy compositions. Thus, the most widely used titanium casting alloy is also Ti-6Al-4V; Grade 2 is the most commercially pure titanium casting, as seen in Table 13-27.

Aerospace casting applications include the range from major structural components weighing more than 135 kg (300 lb) each to small switch guards weighing less than 30 g (1 oz). Ti-6Al-4V castings are used extensively for large, complex housings for turbine engines. They are also used in a variety of airframe applications, including cargo-handling equipment, flow diverters, torque tubes for brakes, and helicopter hubs. In missile and space applications, they are used for wings, missile bodies, optical sensor housings, and ordnance. Ti-6Al-4V castings are also used to attach the external fuel tanks to the Space Shuttle, as well as the boosters to the external tanks. They are also used in marine environments and as medical implants. If cost is a consideration, Ti-6Al-4V castings are about two to three times the cost of a nickel superalloy casting.

For higher temperature applications, Ti-6Al-2Sn-4Zr-2Mo and Ti-6Al-2Sn-4Zr-6Mo are specified. We see that the first of these two alloys captures about 7 percent of the casting market, seen in Table 13-27. These alloys are specified for service up to 595°C (1100°F) and will be in greater demand because aircraft engine manufacturers are designing for higher operating temperatures to increase the fuel efficiency and the thrust of the engines. Figure 13-6 shows the high temperature properties of the castings of these advanced titanium alloys compared with the properties of two other Ti-6Al-4V casting alloys. Along these lines, advanced titanium aluminides are also being developed for high-temperature applications in turbine engines.

Table 13-23 Minimum and average mechanical properties of selected wrought titanium alloys at RT.

Numerical composition, %	Conditions	Minimum and average tensile properties(a)				Average or typical properties				
		Ultimate tensile strength, MPa (ksi)	0.2% yield strength, MPa (ksi)	Elongation, %	Reduction in area, %	Charpy impact strength, $J/(ft-lb_f)$	Hardness	Modulus of elasticity, GPa (10^6 psi)	Modulus of rigidity, GPa (10^6 psi)	Poisson's ratio
Commercially pure titanium										
99.2 Ti (ASTM grade 2)	Annealed	340–434 (50–63)	280–345 (40–50)	28	50	34–54 (25–40)	200 HB	102.7 (14.9)	38.6 (5.6)	0.34
99.2 Ti (ASTM grade 7)	Annealed	340–434 (50–63)	280–345 (40–50)	28	50	43 (32)	200 HB	102.7 (14.9)	38.6 (5.6)	0.34
α-β alloys										
6 Al-4V	Annealed	900–993 (130–144)	830–924 (120–134)	14	30	14–19 (10–14)	36 HRC	113.8 (16.5)	42.1 (6.1)	0.342
	Solution + Aged	1172 (170)	1103 (160)	10	25	···	41 HRC	···	···	···
6Al-4V (low O_2)	Annealed	830–896 (120–130)	760–827 (110–120)	15	35	24 (18)	35 HRC	113.8 (16.5)	42.1 (6.1)	0.342

Table 13-24 Dependence of yield strength and fracture and toughness on microstructure and/or processing of various titanium alloys.

Alloy	α morphology or processing method	Yield strength		Plane-strain fracture toughness (K_{Ic})	
		MPa	ksi	MPa $\sqrt{m}$	ksi $\sqrt{in.}$
Ti-6Al-4V	Equiaxed	910	130	44–66	40–60
	Transformed	875	125	88–110	80–100
	α − β rolled + mill annealed	1095	159	32	29
Ti-6Al-6V-2Sn	Equiaxed	1085	155	33–35	30–50
	Transformed	980	140	55–77	50–70
Ti-6Al-2Sn-4Zr-6Mo	Equiaxed	1155	165	22–33	20–30
	Transformed	1120	160	33–55	30–50
Ti-6Al-2Sn-4Zr-2Mo forging	α + β forged, solution treated and aged	903	131	81	74
	β forged solution treated and aged	895	130	84	76
Ti-17	α − β processed	1035–1170	150–170	33–50	30–45
	β processed	1035–1170	150–170	53–88	48–80

Table 13-25 General corrosion rates of Ti-6Al-4V in specific media.

Medium	Concentration, wt%	Temperature- °F	Corrosion rate, mils/yr	Medium	Concentrtion, wt%	Temperature °F	Corrosion rate, mils/yr
Acids				*Acids (mixed)*			
Hydrochloric	2	100	nil–1.2	Sulfuric-nitric	10–90	160	19.0–23.0
Hydrochloric	5	100	2.4–7.2	*Acids (vapors)*			
Hydrochloric	10	100	20.0–24.0	Hydrochloric	37	100	328.0–408.0
Hydrochloric	30	100	208.0–253.0	Nitric	70	150	nil
Nitric	65	Boiling	3.0–5.0	Nitric	70	200	20
Nitric (white fuming)	90	180	6.0	Sulfuric	96	100	nil
Phosphoric	10	Room	0.8–2.0	Sulfuric	96	150	nil
Phosphoric	10	170	132.0	Sulfuric	96	200–300	0.4–0.6
Phosphoric	85	Room	16.0–24.0	*Alkalis*			
Sulfuric	2	100	15.6–21.6	Sodium hydroxide	25	Boiling	1.8–2.0
Sulfuric	10	100	38.4–39.6	*Chlorides*			
Acids (inhibited)				Aluminum chloride	25	Boiling	780.0–840.0
5% HCl + 1% $CuSO_4$	...	100	0.6	Barium chloride	25	Boiling	nil
5% HCl + 1% CrO_3	...	100	nil	Calcium chloride	28	208	2.7–2.9
10% HCl + 1% $CuSO_4$	...	150	3.0	Cupric chloride	40	Boiling	0.2
10% HCl + 5% $CuSO_4$	...	150	4.0	Ferric chloride	20	Room	nil
10% HCl + 1% CrO_3	...	150	nil–0.4	Ferric chloride	30	200	nil
10% HCl + 5% CrO_3	...	150	nil–0.2	Magnesium chloride	5	Boiling	nil
Acids (mixed)				Magnesium chloride	20	Boiling	nil
Aqua regia (3 parts	...	Room	2.0	Magnesium chloride	40	Boiling	nil
HCl, 1 part HNO_3				Mercuric chloride	Sat'd	200	4.0
Sulfuric-nitric	90–10	Room	18.0	Nickel chloride	20	200	nil
Sulfuric-nitric	90–10	160	295.0–298.0	Seawater	...	Room	nil
Sulfuric-nitric	70–30	Room	22.0–25.0	Sodium chloride	20	Room	nil
Sulfuric-nitric	70–30	160	269.0	Stannic chloride	100	Molten	nil
Sulfuric-nitric	50–50	Room	20.0–25.0	*Gases*			
Sulfuric-nitric	50–50	160	175.0–179.0	Sulfur dioxide (dry)	...	Room	nil
Sulfuric-nitric	30–70	Room	4.0	*Organic chemicals: acids*			
Sulfuric-nitric	30–70	160	92.0–95.0	Formic	50	Boiling	7.97
Sulfuric-nitric	10–90	Room	nil	Oxalic	1	Room	12.0–13.0

Source: D. W. Stough et al., "The Corrosion of Titanium, "TML Report No.57, Titanium Metallurgical Laboratory, Battelle Memorial Institute, 29 Oct. 1956, pp. 109–111.

Table 13-26 Product forms of Ti-6Al-4V alloy and comparison of price with other materials.

Product	Size and weight ranges	Price comparison(a)
Ingot	3200 to 13,600 kg (7000 to 30,000 lb)	. . .
Billet	Normally 100 mm (4 in.) diam to about 355 mm (14 in.) diam or square. Billets up to 5000 lb have been sold, but this is not necessarily the upper limit.	. . .
Bar	Cross sections up to 0.4 × 0.4 mm(16 × 16 in.)	. . .
Die forging	From <0.5 kg to >1300 kg (<1 lb to >3000 lb)	Ti, $30/lb; Al, $10/lb; stainless steel, $8/lb
Plate	Typical dimensions: Thickness: 5 to 75 mm (0.1875 to 3 in.); Width: 915 and 1220 mm (36 and 48 in.); Length: 1.8, 2.4, and 3 m (72, 96, and 120 in.)	. . .
Sheet	Typical dimensions: Thickness: 0.4 to 4.75 mm (0.016 to 0.187 in); Width: 915 and 1220 mm (36 and 48 in); Length 1.8, 2.4, and 3 m (72, 96, and 120 in.)	Ti, $16/lb; stainless steel, $3/lb; Al $2–4/lb; Inco 718, $10/lb
Tube	Specialty item	. . .
Forged block	Available in a wide range of sizes, with maximum size related to ingot size and the amount of work that can be imparted to the forged block	Ti, $8/lb; stainless steel and Al, $2.50–3/lb
Extrusion	From circle sizes of about 25 to 760 mm (1 to 30 in.) diam. Minimum thickness of about 3 mm (1/8 in.) for small circle sizes, and about 13 mm (1/2 in.) for large circle sizes	Ti, $13–15 lb; 300 series stainless steel, $3–4/lb: 15-5PH, $4–5/lb; 13-8PH, $9–12/lb: Al, $2–4/lb
Wire	Typically manufactured in sizes ranging from 0.28 to 12.2 mm (0.011 to 0.4860 in.) diam	1/4 in. wire: Ti $26/lb; A283, $6/lb; stainless steel, $7.50/lb; 8740, $1/lb; Al 7075 $2.30/lb

(a) Due to its lower density, 1 lb of titanium is approximately 1.7 to 1.8 more material by volume than 1 lb of steel or nickel-base alloy. Absolute dollar costs may not be current. Relative cost comparison should be done between different materials.

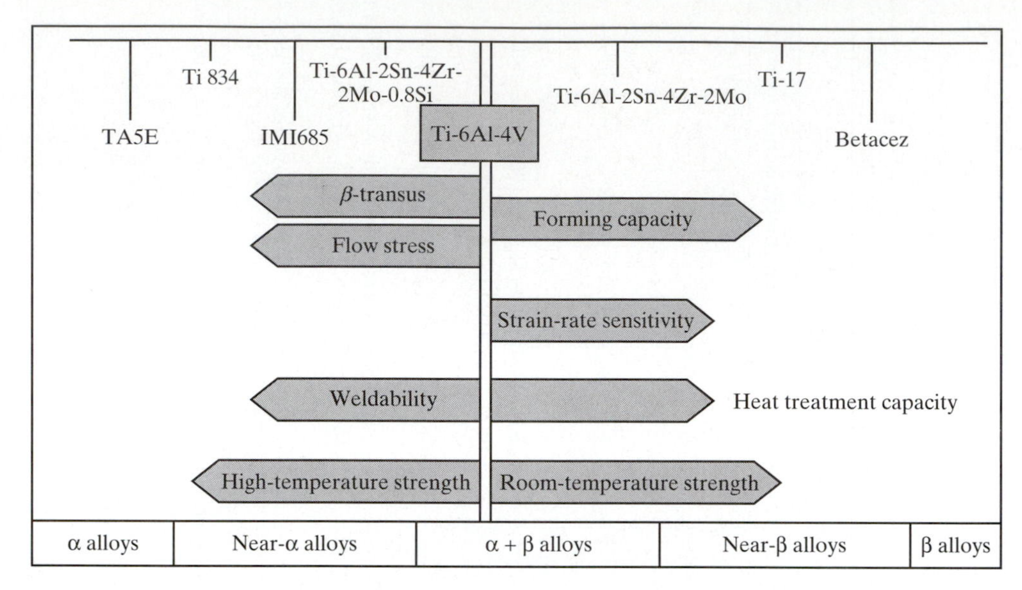

Fig. 13-5 A guide in the selection of titanium alloys, starting with Ti-6Al-4V.

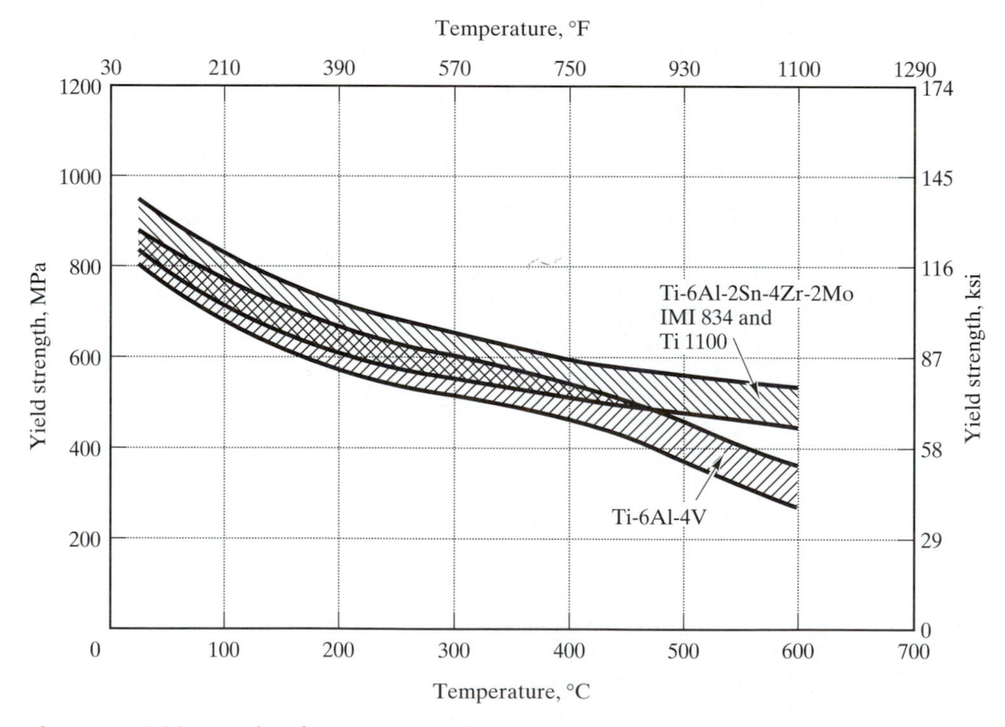

Fig. 13-6 Yield strengths of cast titanium alloys at elevated temperatures.

Table 13-27 Titanium alloys for castings and their relative usage.

Alloy	Estimated relative use of castings	Numerical Composition, wt %													Special properties(a)(b)
		O	N	C	H	Al	Fe	V	Cr	Sn	Mo	Nb	Zr	Si	
Ti-6Al-4V	85%	0.18	0.015	0.04	0.006	6	0.13	4	···	···	···	···	···	···	General purpose
Ti-6Al-4V ELI(c)	1%	0.11	0.010	0.03	0.006	6	0.10	4	···	···	···	···	···	···	Cryogenic toughness
Commercially pure titanium (grade 2)	6%	0.25	0.015	0.03	0.006	···	0.15	···	···	···	···	···	···	···	Corrosion resistance
Ti-6Al-2Sn-4Zr-2Mo	7%	0.10	0.010	0.03	0.006	6	0.15	···	···	2	2	···	4	···	Elevated-temperature creep
Ti-6Al-2Sn-4Zr-6Mo	<1%	0.10	0.010	0.03	0.006	6	0.15	···	···	2	6	···	4	···	Elevated-temperature strength
Ti-5Al-2.5Sn	<1%	0.16	0.0015	0.03	0.006	5	0.2	···	···	2.5	···	···	···	···	Cryogenic toughness
Ti-3Al-8V-6Cr-4Zr-4Mo (Beta-C)	<1%	0.10	0.015	0.03	0.006	3.5	0.2	8.5	6	···	4	···	4	···	RT strength
Ti-15V-3Al-3Cr-3Sn (Ti-15-3)	<1%	0.12	0.015	0.03	0.006	3	0.2	15	3	3	···	···	···	···	RT strength
Ti-1100	<1%	0.07	0.015	0.04	0.006	6.0	0.02	···	···	2.75	0.4	···	4.0	0.45	Elevated-temperature properties
IMI-834	<1%	0.10	0.015	0.06	0.006	5.8	0.02	···	···	4.0	0.5	0.7	3.5	0.35	Elevated-temperature properties
Total	100%														

(a) Relative to Ti-6Al-4V. (b) RT, room temperature. (c) ELI, extra low interstitial

13.4 Magnesium and Magnesium Alloys [6]

For structural applications, magnesium has a great potential. It is much lighter than aluminum and titanium and would, therefore, have potentially high specific strength and specific modulus. Magnesium, being a hexagonal metal, has limited ductility, however; therefore, it is more difficult to deform it to shape, requiring higher temperatures to enhance its deformability. Consequently, for structural applications, we should expect that magnesium will be used predominantly as castings. This is substantiated by Table 13-28 where the tonnages of castings are more than four times that of wrought products in structural applications. Compared to aluminum and titanium, the other major problem of magnesium is its very corrosive nature. As was seen in the galvanic series in Fig. 10-7, it is the most anodic and would therefore require more stringent corrosion control measures.

We see in Table 13-28 that the structural applications account for a little more than 15 percent of all magnesium usage in 1988. The majority of applications for magnesium is seen to be in nonstructural applications such as: an alloying element in aluminum alloys, particularly for aluminum can stock; a desulfurizer in the steel industry; a nodulizer in cast iron; a chemical; and a sacrificial anode to protect buried line pipes and others.

13.4.1 Magnesium Alloys and Their Designations

The alloy code designation for magnesium alloys is done according to ASTM specification B-297 and has a one-letter or two-letter code followed by a one-, two- or three-digit number as indicated in Table 13-29. The first letter(s) represents the major alloying element(s), while the one-, two- or three-digit number represents the composition in weight percent of the alloying element(s), rounded to a whole numbers. The codes for the alloying elements are given in Table 13-29. After

Table 13-28 Structural and nonstructural applications of magnesium.

	Shipments, metric tons					
Applications	**1983**	**1984**	**1985**	**1986**	**1987**	**1988**
Structural						
Die castings	27 900	30 400	29 700	28 800	26 600	28 500
Gravity castings(a)	2 000	1 300	1 200	1 600	1 800	2 100
Wrought products(b)	7 100	6 600	4 800	5 400	8 400	7 400
Total structural	37 000	38 300	35 700	35 800	36 800	38 000
Nonstructural						
Aluminum alloying	110 800	113 500	121 000	122 100	122 100	134 300
Desulfurziation	13 400	17 400	19 100	20 300	21 900	28 600
Nodular iron	8 900	9 800	11 300	12 300	14 200	15 800
Metal reduction	9 200	12 200	10 300	9 600	8 800	10 200
Chemical(c)	8 200	7 800	8 000	8 000	7 200	8 100
Electrochemical(d)	7 600	7 700	9 100	8 300	8 000	8 000
Other(e)	9 300	9 300	10 300	10 000	17 000	8 200
Total Nonstructural	167 400	177 700	189 100	190 600	199 200	213 200
Total all uses	204 400	216 000	224 800	226 400	236 000	251 200

(a) Includes sand, permanent mold, investment, and plaster castings. (b) includes extrusions, forgings, sheet, plate, and photoengraving stock (c) Grignard reaction or pyrotechnic applications. (d) Cast or extruded anodes and batteries. (e) Shipments into the U.S.S.R., the People's Republic of China, and Comecon countries.

Source: International Magnesium Association

Table 13-29 Standard ASTM system for alloy and temper designations for magnesium alloys.

First part	Second part	Third part	Fourth part
Indicates the two principal alloying elements	Indicates the amount of the two principal alloying elements	Distinguishes between different alloys with the same percentages of the two principal alloying elements	Indicates condition (temper)
Consists of two code letters representing the two main alloying elements arranged in order of decreasing percentage (or alphabetically if percentages are equal)	Consists of two numbers corresponding to rounded-off percentages of the two main alloying elements and arranged in same order as alloy designationst in first part	Consists of a letter of the alphabet assigned in order as compositions become standard	Consists of a letter followed by a number (separated from the third part of the designation by a hyphen)
A—aluminum B—bismuth C—copper D—cadmium E—rare earth F—iron G—magnesium H—thorium K—zirconium L—lithium M—manganese N—nickel P—lead Q—silver R—chromium S—silicon T—tin W—yttrium Y—antimony Z—zinc	Whole numbers	Letters of alphabet except I and O	F—as fabricated O—annealed H10 and H11—slightly strain hardened H23, H24, and H26—strain-hardened and partially annealed T4—solution heat-treated T5—artificially aged only T6—solution heat-treated and artificially aged T8—solution heat-treated, cold-worked, and artificially aged

the alloy code, a letter is assigned sequentially from A to Z, except the letters I and O, to indicate the modifications of the alloy, with A being the initial alloy. A temper designation, which is separated by a hyphen, is given as in aluminum alloys after the alloy code.

As examples of the designation system, we take first AZ91C-T6. The letters A and Z represent the major alloying elements aluminum and zinc; the two-digit number 91 indicates nominal 9 wt%Al and 1%Zn; C indicates the third modification of this alloy, and T6

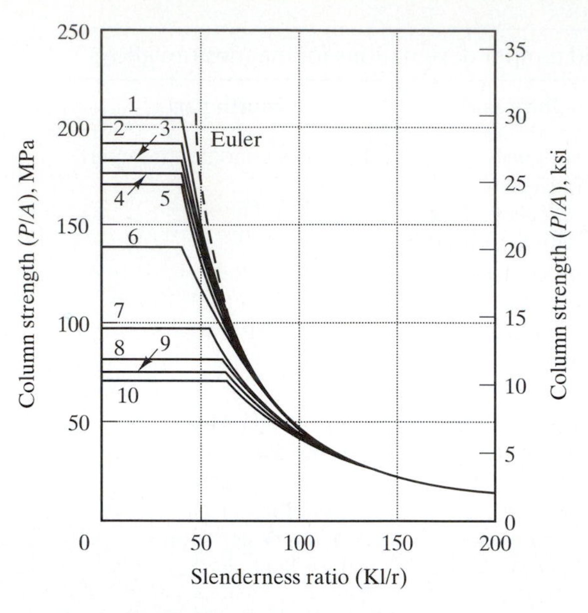

Alloy	Curve number	Least dimension of area range	
ZK60A-T5	1	<1290 mm^2	<2.000 in^2.
	2	1290–1935 mm^2	2.000–2.999 in^2.
	5	1935–3225 mm^2	3.000–4.999 in^2.
AZ80A-T5	2	6.35–38.09 mm	0.250–1.499 in.
	3	38.100–63.49 mm	1.500–2.499 in.
	4	63.50–127.00 mm	2.500–5.000 in.
ZK21A-F	6	<3226 mm^2	<5.000 in^2.
AZ61A-F	7	6.35–127.00 mm	0.250–5.000 in.
AZ31B-F	8	≤6.35 mm	≤0.250 in.
	9	6.35–63.49 mm	0.250–2.499 in.
	10	63.50–127.00 mm	2.500–5.000 in.

Minimum column strength curves for several magnesium extrusion alloys. *P*, ultimate column load; *A*, cross-sectional area; *K*, constant that depends on end conditions; *l*, column length; and *r*, minimum radius of gyration of column cross section.

Fig. 13-7 Column strengths of several magnesisium extrusion alloys. [6]

means it is solution-heat-treated and artificially aged. A second example is AM100A-T61, that has a three-digit code. The first two letters A and M represent aluminum and manganese. The first two digits 10 represents 10 wt%Al and 0 represents 0.1 wt% Mn that was rounded to the nearest number. K1A and M1A are examples of one-letter code alloys that mean they contain only one major alloying element.

The compressive yield strengths of magnesium castings are the same as their tensile yield strengths. For wrought magnesium alloys, however, the compressive yield strength is considerably less than the tensile yield strength. The ratio of the compressive to tensile yield strengths varies from about 0.4 for M1A to an average value of about 0.7 for the other magnesium alloys. This has practical implications in designing magnesium alloy structures that are compressively loaded axially, as in columns. The maximum design stress of a column is taken to be the minimum compressive yield strength of the magnesium alloy. Figure 13-7 shows the column strength curves for extrusions of several magnesium alloys.

Table 13-30 Characteristics of die-casting magnesium alloys.

Alloy(a)	General characteristics
AZ91D	Most commonly used die-casting alloy. Good strength at room temperature, good castability, good atmospheric stability, excellent saltwater corrosion resistance
AM60B	Good elongation and toughness, excellent saltwater corrosion resistance, good yield and tensile properties
AS21X1	Best creep resistance of die-casting alloys, good room-temperature properties, useful in high-temperature applications
AS41XB	Good creep resistance up to 175°C (350°F), good room-temperature properties, excellent saltwater corrosion resistance, useful in high-temperature applications

(a) All alloys are in the as-cast condition.

13.4.2 Applications of Magnesium

We noted that castings are the predominant forms of magnesium structural applications. Furthermore, about 95 percent of all the castings are made by die-casting by which the liquid is forced to the mold or die cavity by an applied pressure. The general purpose die-cast alloy is AZ91D. For castings requiring better ductility (elongation) and toughness, AM60B is selected. This alloy is used in the production of die-cast automobile wheels and some archery and other sports equipment. Table 13-30 shows the characteristics of other die-casting alloys.

Early applications of magnesium in Germany in the 1920s were focused on structural applications for the automotive and aerospace industries. In the 1930s, the German automakers were using magnesium die-castings as oil pump housings, engine cooling fans, blower impellers, and gearboxes. Aircraft manufacturers used magnesium alloys in tail wheels, brake shoes, brake levers, and pistons. In the 1950s, 1960s, and the early 1970s, Volkswagen was a major consumer of magnesium and magnesium alloys. In 1971, Volkswagen consumed 42,000 metric tons of primary magnesium at its casting shops in Germany, Brazil, and Mexico. This dramatically dropped to 9,000 tons in 1975 when the company introduced new, redesigned models.

Up until recently, structural applications of magnesium in cars in the United States have been for small components. Table 13-31 compares the automotive applications of magnesium alloys in the United States, in Europe, and in Japan. We see that the usage in the United States falls into the categories of supports, brackets, housings, and covers. In comparison, Europe (primarily in Germany and secondarily in Sweden), usage includes the larger components that consist of instrument panel support beams, seat frames, and wheels. In the United States, magnesium alloy wheels are specified on performance cars but have not been broadly adapted to the mass market. This difference reflects the experience level of product designers in using magnesium alloys. However, the U.S. auto manufacturers are now seriously looking into using more magnesium alloys because of the federally mandated CAFE (Corporate Average Fuel Economy) and the EPA's (Environmental Protection Agency) emmisions limits. Antipollution equipment will add weight to the vehicle and in order to maintain the required CAFE, the dead (inertia) weight of the vehicle must decrease. The decrease in dead-weight of a vehicle also induces other outcomes, in addition to fuel economy. For example, a one-pound weight reduction in an unsprung rotating mass, such as a wheel or brake rotor, reduces inertial forces, enhances vehicle acceleration and deceleration, and improves the active suspension design features. A weight reduction in reciprocating components is significantly greater than in the case of dead-weight reduction. It has been said that a 10-gram reduction in piston or connecting rod weights can measurably increase fuel efficiency.

Table 13-31 Automotive parts in the U.S., Europe, and Japan made of magnesium alloys. [7]

North America	Europe	Japan
Steering column lock housing	Gear box	Steering column assembly
Clutch housing	Instrument panel support beam	Steering wheel cores
Pedal support brackets	Cylinder head covers	Valve covers
Seat supports	Seat frames	Cylinder head covers
Transfer case	Cam covers	Inmani top cover
Armrest support	Oil pumps	Inmani chamber
Accessory drive brackets	Oil pan bottoms	Motorcycle wheels
Engine accessory brackets	Wheels/systems	
Valve cover	Transmission housing	
Steering column brackets		
Induction system housing		
Cam cover		
Oil filter adaptor		

13.5 Copper and Copper Alloys

13.5.1 Alloy and Temper Designations of Copper and Copper Alloys

In the same manner as in aluminum alloys, the procurement of copper and copper alloys is done by specifying the alloy designation and its temper designation. The generic classification of wrought and cast copper alloys are given in Table 13-32 with their designated UNS numbers and the major alloying elements. The UNS consists of five digits after the letter C (i.e., CXXXXX). As with stainless steels, the first three digits are the alloy numbers given by the Copper Development Association (CDA) and the last two digits are for modifications made to the original alloy. For example, the CDA designation for cartridge brass is CDA 260. In the UNS, the code for the alloy is C26000. Modifications of this alloy may be designated C26001, or C26010, and so on. Wrought alloys are assigned C10000 to C79999; cast alloys are assigned C80000 to C99999. The designation system is administered by the CDA.

Table 13-32 Generic classification and designated UNS number of copper alloys.

Generic name	UNS numbers	Compositions
Wrought alloys		
Coppers	C10100–C15760	>99% Cu
High-copper alloys	C16200–C19600	>96% Cu
Brasses	C20500–C28580	Cu-Zn
Leaded brasses	C31200–C38590	Cu-Zn-Pb
Tin brasses	C40400–C49080	Cu-Zn-Sn-Pb
Phosphor bronzes	C50100–C52400	Cu-Sn-P
Leaded phosphor bronzes	C53200–C54800	Cu-Sn-Pb-P
Copper-phosphorus and copper-silver-phosphorus alloys	C55180–C55284	Cu-P-Ag
Aluminum bronzes	C60600–C64400	Cu-Al-Ni-Fe-Si-Sn
Silicon bronzes	C64700–C66100	Cu-Si-Sn
Other copper-zinc alloys	C66400–C69900	...
Copper-nickels	C70000–C79900	Cu-Ni-Fe
Nickel silvers	C73200–C79900	Cu-Ni-Zn
Cast alloys		
Coppers	C80100–C81100	>99% Cu
High-copper alloys	C81300–C82800	>94% Cu
Red and leaded red brasses	C83300–C85800	Cu-Zn-Sn-Pb (75–89% Cu)
Yellow and leaded yellow brasses	C85200–C85800	Cu-Zn-Sn-Pb (57–74% Cu)
Manganese bronzes and leaded manganese bronzes	C86100–C86800	Cu-Zn-Mn-Fe-Pb
Silicon bronzes, silicon brasses	C87300–C87900	Cu-Zn-Si
Tin bronzes and leaded tin bronzes	C90200–C94500	Cu-Sn-Zn-Pb
Nickel-tin bronzes	C94700–C94900	Cu-Ni-Sn-Zn-Pb
Aluminum bronzes	C95200–C95810	Cu-Al-Fe-Ni
Copper-nickels	C96200–C96800	Cu-Ni-Fe
Nickel silvers	C97300–C97800	Cu-Ni-Zn-Pb-Sn
Leaded coppers	C98200–C98800	Cu-Pb
Miscellaneous alloys	C99300–C99750	...

The *wrought and cast copper alloys* in Table 13-32 may be grouped into *six families:*

- ***Coppers.*** have minimum copper content of 99.0 percent or higher.
- ***High copper alloys.*** For wrought products, these are alloys with less than 99.0 percent but more than 96 percent copper content and do not fall into any other copper alloy groups. For cast alloys, these are alloys with more than 94 percent copper content and to which silver may be added for special properties.
- ***Brasses.*** These alloys contain *zinc* as the principal alloying element with or without other alloying elements such as iron, aluminum, nickel, and silicon. There are *three groups of wrought alloy brasses:* copper-zinc alloys (brasses), copper-zinc-lead alloys (leaded brasses), and copper-zinc-tin alloys (tin brasses). There are *four groups of cast alloy brasses:* copper-tin-zinc alloys (red, semi-red, and yellow brasses), "manganese bronze" alloys (high-strength yellow brasses), leaded "manganese bronze" alloys (leaded high-strength yellow brasses), and copper-zinc-silicon alloys (silicon brasses and bronzes).
- ***Bronzes.*** These are copper alloys in which the major alloying element is not zinc or nickel. The term "bronze" was originally used for copper alloys with tin. Today, the term is used with the name of a major alloying element. There are *four wrought*

bronzes: copper-tin-phosphorus alloys (tin and phosphor bronzes), copper-tin-lead-phosphorus alloys (leaded tin and phosphor bronzes), copper-aluminum alloys (aluminum bronzes), and copper-silicon alloys (silicon bronzes). There are also four cast bronzes: copper-tin alloys (tin bronzes), copper-tin-lead (leaded tin bronzes), copper-tin-nickel alloys (nickel tin bronzes), and copper-aluminum alloys (aluminum bronzes). The "manganese bronzes" mentioned in the brasses have zinc as the principal alloy with additions of tin, manganese, and lead.

- ***Copper Nickels.*** These are also known as *cupronickel alloys* with nickel as the principal alloying element from 10 to 30 percent, with or without other added elements.
- ***Copper-Nickel-Zinc Alloys.*** These are known commonly as "nickel silvers" and contain zinc and nickel as the principal and secondary alloying elements, with or without other designated elements.

The temper designations are covered in ASTM B 601 "Standard Practice for Temper Designations for Copper and Copper Alloys—Wrought and Cast" and some of the more common designations are shown in Table 13-33. If we were to designate the basic tempers to parallel those in the aluminum designations, the basic tempers would be: (1) M, as manufactured; (2) O, annealed; (3) H, hardened by cold-rolling or cold-drawing; (4) T, heat-treated; and (5) W, welded-tube tempers. The last two tempers are not shown in Table 13-33. The H tempers are specified on the basis of cold-reduction imparted by rolling or drawing; these were discussed in Sec. 8.4.2 and shown in Table 8-3. As indicated there, the nominal temper designations are related to the amount of reduction stated in the increase of the Browne & Sharpe (B&S) gauge numbers for rolled sheet (percent reduction in thickness) and drawn wire (percent reduction in diameter).

13.5.2 Properties and Selection Criteria of Copper and Copper Alloys

The nominal compositions, the forms into which they are manufactured, and the mechanical properties of representative wrought copper and copper alloys from the six alloy families are shown in Table 13-34. The selection criteria of these alloys are primarily based on the following factors: (1) electrical and thermal conductivity, (2) corrosion resistance, (3) fabricability, (4) strength or hardness, and (5) color. Copper is the highest electrical conductor of the technical commercial metals. Because of this, conductivities of materials are rated based on an IACS (International Annealed Copper Standard) rating. A 100% IACS rating is assigned to annealed copper with a volume resistivity of 0.017241 ohm-mm^2 per meter at 20°C (68°F). Because electrons also are the thermal energy carriers, copper is also the highest thermally conductive metal. We note as well from the standard electrode potentials in Table 10-2 that copper is the most cathodic of all the technical commercial metals discussed in this text. Therefore, we find that the majority of copper applications utilizes the conductivity and corrosion-resistant properties. The latter properties are complemented by good fabricability and moderate strength. As illustrated by the ductilities (elongations in 50 mm) in Table 13-34, copper and its alloys can be readily formed to any of the products indicated. The addition of lead or bismuth makes them very machinable and they are easily soldered, brazed, and welded. Strength can be achieved by either solid-solution, cold-worked, or heat-treating in some alloys. Copper alloys yield different pleasing colors that permit them to be used for decorative purposes only, or when a particular combination of color and finish is desired in combination with strength and corrosion properties. Table 13-35 lists the different colors of copper and copper alloys.

Coppers. These materials are primarily chosen for their electrical conductivities. Since impurities impair the electrical conductivity, these are kept to a minimum. Nonetheless, some impurities are intentionally added to achieve certain characteristics.

The most common metal of this group is C11000, electrolytic tough pitch (ETP) copper. Its normal processing leaves about 0.04%O in combination with copper as CuO. This level of impurity does not measurably lower the conductivity (it still exhibits a conductivity of 101% IACS), but it lowers the ductility a little. Oxygen-free coppers (C10100 and C10200) can be obtained by melting under nonoxidizing atmospheres or by adding deoxidizers such as phosphorus to the melt. These coppers are required for applications requiring high-conductivity coupled with ductility, low-gas permeability, freedom from hydrogen

Table 13-33 Partial temper designation codes for copper alloys according to ASTM B 601.

Temper designation	Temper name or material condition	Temper designation	Temper name or material condition
Cold-worked tempers		*Annealed tempers*	
H00	$^{1}/_{2}$ hard	O10	Cast and annealed (homogenized)
H01	$^{1}/_{4}$ hard	O11	As-cast and precipitation heat treated
H02	$^{1}/_{2}$ hard	O20	Hot forged and annealed
H03	$^{1}/_{3}$ hard	O25	Hot rolled and annealed
H04	Hard	O30	Hot extruded and annealed
H06	Extra hard	O31	Extruded and precipitation heat treated
H08	Spring		
H10	Extra spring	O40	Hot pierced and annealed
H12	Special spring	O50	Light annealed
H13	Ultra spring	O60	Soft annealed
H14	Super spring	O61	Annnealed
		O65	Drawing annnealed
Cold-worked and stress-relieved tempers		O68	Deep-drawing annnealed
HR01	H01 and stress relieved	O70	Dead-soft annnealed
HR02	H02 and stress relieved	O80	Annealed to temper, $^{1}/_{8}$ hard
HR04	H04 and stress relieved	O81	Annealed to temper, $^{1}/_{4}$ hard
HR08	H08 and stress relieved	O82	Annealed to temper, $^{1}/_{2}$ hard
HR10	H10 and stress relieved		
HR20	As-finned	*Annealed tempers*	
HR50	Drawn and stress relieved	OS005	Average grain size, 0.005 mm
		OS010	Average grain size, 0.010 mm
As-manufactured tempers		OS015	Average grain size, 0.015 mm
M01	As-sand cast	OS025	Average grain size, 0.025 mm
M02	As-centrifugal cast	OS035	Average grain size, 0.035 mm
M03	As-plaster cast	OS050	Average grain size, 0.050 mm
M04	As-pressure die cast	OS060	Average grain size, 0.060 mm
M05	As-permanent mold cast	OS070	Average grain size, 0.070 mm
M06	As-investment cast	OS100	Average grain size, 0.100 mm
M07	As-continuous cast	OS120	Average grain size, 0.120 mm
M10	As-hot forged and air cooled	OS150	Average grain size, 0.150 mm
M11	As-forged and quenched	OS200	Average grain size, 0.200 mm
M20	As-hot rolled		
M30	As-hot extruded		
M40	As-hot pierced		
M45	As-hot pierced and rerolled		

embrittlement, or low out-gassing tendency. To ensure complete deoxidation, excess phosphorus is added and retained in the composition.

High purity coppers are relatively soft in the annealed condition and are usually used in a hardened H temper condition for better handling. Cold-work slightly decreases the electrical conductivity from the 101%IACS in the annealed temper (O temper) to 97 percent in the spring-rolled temper (H08). When used at elevated temperatures (which are encountered in soldering and semiconductor packaging operations), these metals soften as seen in Fig. 13-8. To retard softening, silver-bearing (C11300-C11600), cadmium-bearing (C14300 and C14310), and zirconium-bearing (C15000-C15100) coppers are selected. The cadmium- and zirconium-bearing coppers attain higher strengths when cold-worked or strain-hardened and exhibit better softening resistance than the silver-bearing and

Table 13-34 Composition, commercial forms, and properties of representative wrought coppers and copper alloys.

Alloy number (and name)	Nominal composition, %	Commercial forms(a)	Mechanical properties(b) Tensile strength MPa	Tensile strength ksi	Yield strength MPa	Yield strength ksi	Elongation in 50 mm (2 in.), %(b)	Machinability rating, %(c)
C10100 (oxygen-free electronic copper)	99.99 Cu	F, R, W, T, P, S	221–445	32–66	69–365	10–53	55–4	20
C11000 (electrolytic tough pitch copper)	99.90 Cu, 0.04 OF	R, R, W, T, P, S	221–455	32–66	69–365	10–53	55–4	20
C14300	99.9 Cu, 0.1 Cd	F	221–400	32–58	76–386	11–56	42–1	20
C15000 (zirconium-copper)	99.8 Cu, 0.15 Zr	R, W	200–524	29–76	41–496	6–72	54–1.5	20
C17200 (beryllium-copper)	99.5 Cu, 1.9 Be, 0.20 Co	F, R, W, T, P, S	469–1462	68–212	172–1344	25–195	48–1	20
C17500 (copper-cobalt-beryllium alloy)	99.5 Cu, 2.5 Cu, 0.6 Be	F, R	310–793	45–115	172–758	25–110	28–5	...
C26000 (cartridge brass, 70%)	70.0 Cu, 30.0 Zn	F, R, W, T	303–896	44–130	76–448	11–65	66–3	30
C26800, C27000 (yellow brass)	65.0 Cu, 35.0 Zn	F, R, W	317–883	46–128	97–427	14–62	65–3	30
C28000 (Muntz metal)	60.0 Cu, 40.0 Zn	F, R, T	372–510	54–74	145–379	21–55	52–10	40
C36000 (free-cutting brass)	61.5 Cu, 3.0 Pb, 35.5 Zn	F, R, S	338–469	49–58	124–310	18–45	53–18	100
C44300, C44400, C44500 inhibited admiralty	71.0 Cu, 28.0 Zn, 1.0 Sn	F, W, T	331–379	48–55	124–152	18–22	65–60	30
C51100	95.6 Cu, 4.2 Sn, 0.2 P	F	317–710	46–103	345–552	50–80	48–2	20
C60800 (aluminum bronze, 5%)	95.0 Cu, 5.0 Al	T	414	60	186	27	55	20
C65500 (high-silicon bronze, A)	97.0 Cu, 3.0 Si	F, R, W, T	386–1000	56–145	145–483	21–70	63–3	30
C67500 (manganese bronze, A)	58.5 Cu, 1.4 Fe, 39.0 Zn, 1.0 Sn, 0.1 Mn	R, S	448–579	65–84	207–414	30–60	33–19	30
C70600 (copper-nickel, 10%)	88.7 Cu, 1.3 Pc, 10.0 Ni	F, T	303–414	44–60	110–393	16–57	42–10	20
C71500 (copper-nickel, 30%)	70.0 Cu, 30.0 Ni	F, R, T	372–517	54–75	138–483	20–70	45–15	20
C75200 (nickel silver, 65-18)	65.0 Cu, 17.0 Zn, 18.0 Ni	F, R, W	386–710	56–103	172–621	25–90	45–3	20

(a) F, flat products; R, rod; W, wire; T, tube; P, pipe; S, shapes. (b) Ranges are from softest to hardest commercial forms. The strength of the standard copper alloys depends on the temper (annealed grain size or degree of cold work) and the selection thickness of the mill product. Ranges cover standard tempers for each alloy. (c) Based on 100% for C36000.

Source: Copper Development Association Inc.

Table 13-35 Standard color controlled wrought copper alloys.

UNS number	Common name	Color description
C11000	Electrolytic tough pitch copper	Soft pink
C21000	Gilding, 95%	Red-brown
C22000	Commercial bronze, 90%	Bronze-gold
C23000	Red brass, 85%	Tan-gold
C26000	Cartridge brass, 70%	Green-gold
C28000	Muntz metal, 60%	Light brown-gold
C63800	Aluminum bronze	Gold
C65500	High-silicon bronze, A	Lavender-brown
C70600	Copper-nickel, 10%	Soft lavender
C74500	Nickel silver, 65-10	Gray-white
C75200	Nickel silver, 65-18	Silver

the electrolytic-tough-pitched (ETP) coppers when strain-hardened, as seen in Figs. 13-9, and 13-10.

Cold-rolled silver-bearing copper is used extensively for automobile radiator fins. As seen in Fig. 13-10, moderate cold-rolling (about 21 percent reduction in area) will produce a harder material after soldering or baking than severe cold-rolling (90 percent reduction in area). Radiator cores are usually baked for three minutes at 345°C and we see that the strength of the 90 percent reduced silver-bearing copper after the bake operation would be only about 235 MPa compared to about 305 MPa for the 21 percent reduced material after the three-minute-bake at 345°C. We should recall that the precipitous drop from about 410 (0 anneal time—as cold-worked) to about 225 MPa (at 345°C) for the 90 percent reduced material is due to recrystallization. A severely cold-worked material will recrystallize at a lower temperature than a lesser cold-worked material (see laws of recrystallization—Sec. 8.5.2). As Fig. 13-10 suggests, a cadmium-bearing copper can be severely cold-worked

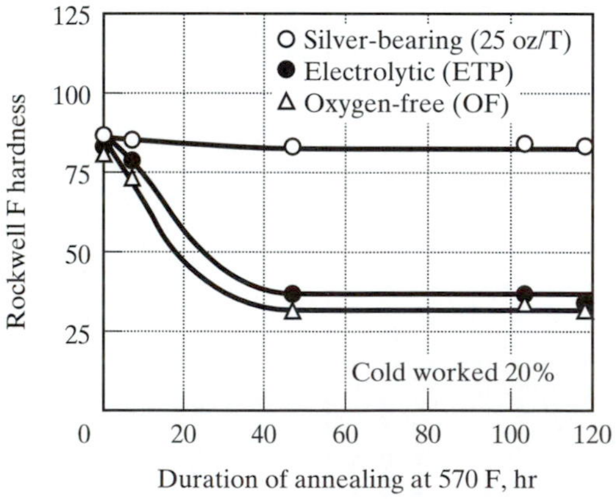

Fig. 13-8 Silver retards the softening of copper elevated temperatures.

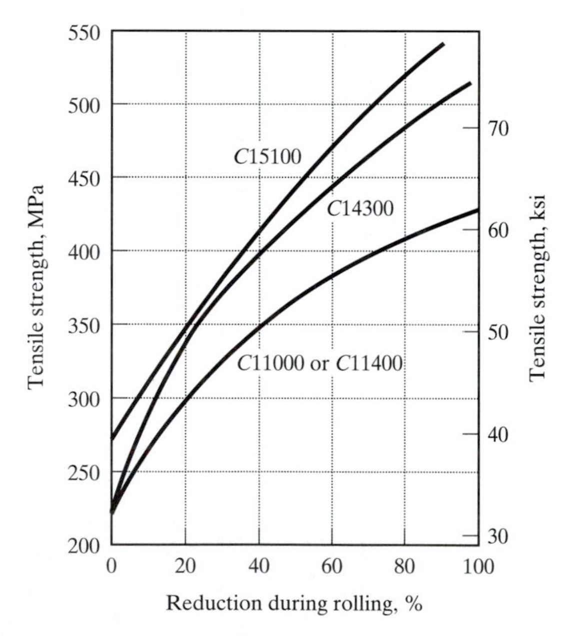

Fig. 13-9 Zirconium- and cadmium-bearing coppers strain-hardened more than coppers or silver-bearing copper.

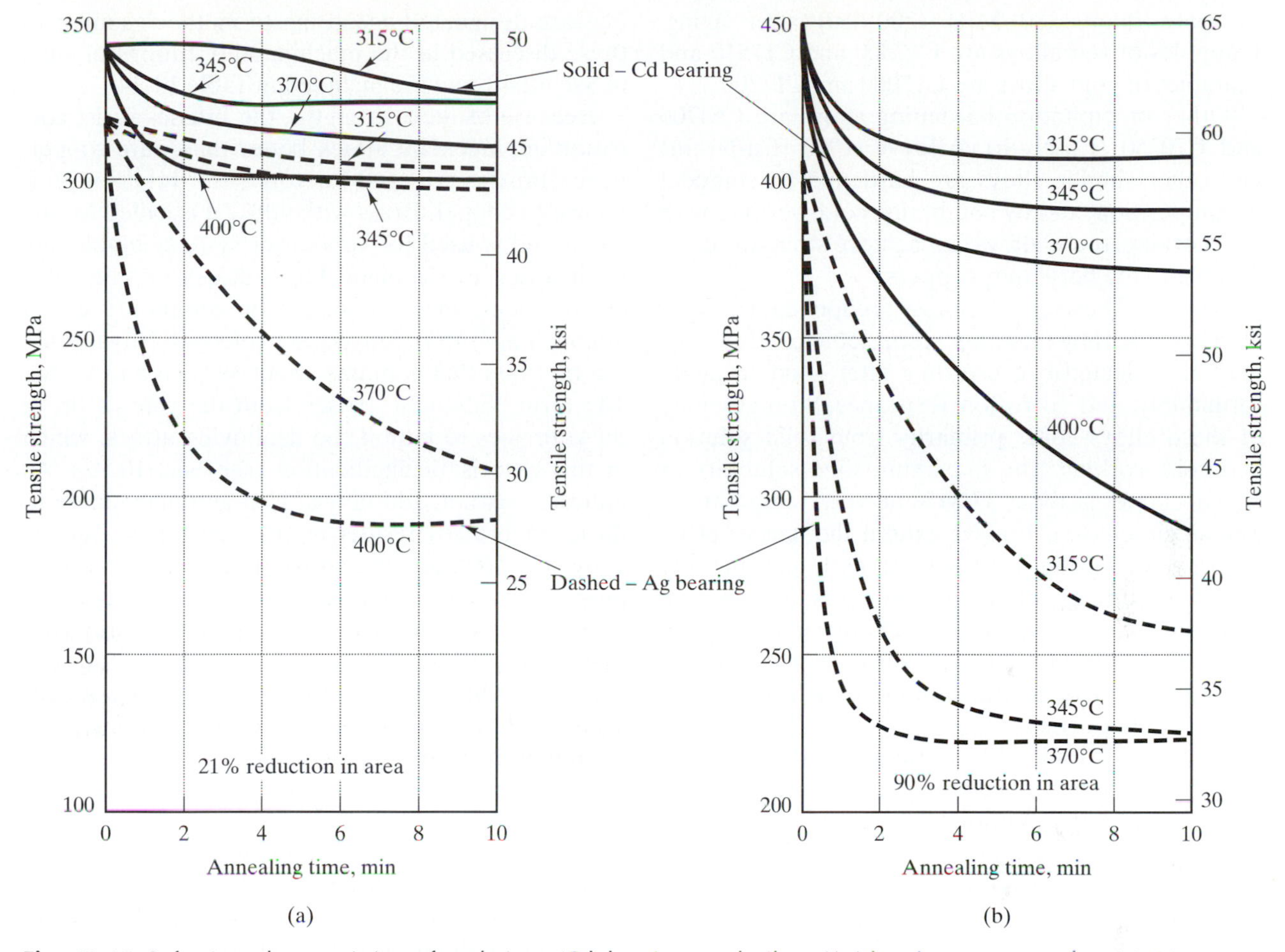

Fig. 13-10 Softening characteristics of cadmium (Cd)-bearing, and silver (Ag)-bearing coppers after (a) 21 percent reduction in area (thickness) from 0.1 to 0.075 mm in thickness, and (b) 90 percent reduction in area (thickness) from 0.75 to 0.75 mm in thckness.

without it being susceptible to drastic softening at 345°C during soldering and baking.

Another application where softening resistance is of paramount importance is in lead frames for electronic devices, such as plastic dual-in-line packages. During packaging and assembly, lead frames may be subjected to temperatures up to 350°C for several minutes or up to 500°C for several seconds. Leads must maintain strength because they are pressed into socket connectors; softened leads collapse, causing rejects. Alloys C15100 (Cu-Zr), C15500 (Cu-Ag-Mg-P), C19400 (Cu-Fe-P-Zn), and C19500 (Cu-Fe-Co-Sn-P) are popular for this application because they have good conductivity, strength, and softening resistance.

High Coppers and Precipitation Hardening Alloys. This group consists predominantly of heat-treatable precipitation-hardening alloys that include Cu-Be, Cu-Cr, and Cu-Ni-P alloys. The Cu-Be alloys are grouped into red and gold alloys that contain from 0.2 to 0.7 wt%, and from 1.6 to 2.0 wt% Be, respectively. The red alloys can have yield strengths from 170 to 550 MPa (25 to 80 ksi) with no heat treatment to greater than 895 MPa (130 ksi) after aging. The gold alloys can develop yield strengths from about 205 to 690 MPa (30 to 100 ksi) with no heat treatment to more than 1380 MPa (200 ksi) after aging. Examples of red alloys are C17500 and C17510 and examples of gold alloys are C17000 and C17200.

Other precipitation hardening alloys are C64700 and C70250 (Cu-Ni-Si). Alloy C71900 (Cu-Ni-tin) and other similar alloys are hardened by spinodal decomposition, and by combining cold-working with hot-working, these alloys can achieve strengths comparable to the beryllium-coppers.

Brasses. These are the most predominant copper alloys. While they maintain respectable conductivity, they are selected because they offer good strength, formability, and corrosion resistance. Strengthening of these alloys come primarily from solid solution and cold-working. The maximum solid solubility of zinc in copper is about 37 wt% at room temperature. The single-phase α-brasses exhibit the unique characteristic of having increased strength and ductility at the same time, as the zinc content is increased up to about 35 wt%, see Fig. 13-11. Because of these properties, C26000 (30%Zn alloy) is used in the drawing of cartridges for ammunitions and is commonly called Cartridge brass.

When lead is added to these alloys, they become very machinable. The free cutting brass alloy C36000 with 3.0 wt%Pb addition is given a 100 percent machinability rating and the ratings of all the other copper alloys are based on it. The machinability ratings are designated according to methods similar to those discussed in the machinability ratings of steel in Sec. 12.4.1 and are included in Table 13-34.

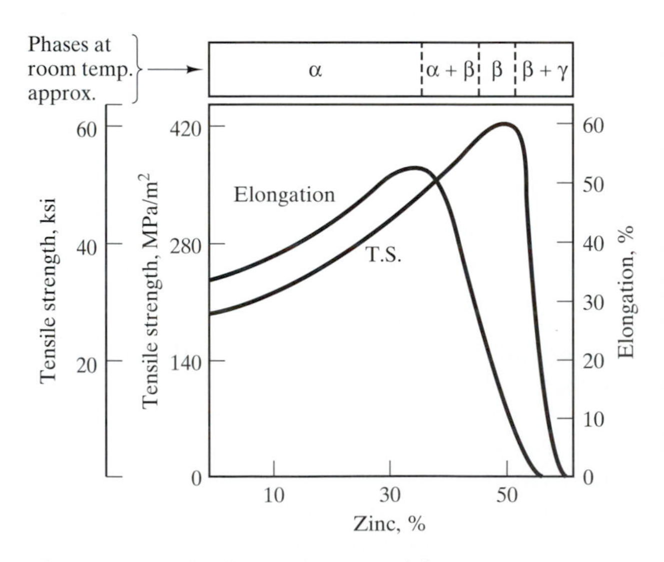

Fig. 13-11 Single phase α-brasses exhibit a unique increase in both strength and ductility simultaneously. [9]

Because of their strengths, the brasses resist corrosion impingement attack better than pure copper; hence, brasses are used for condenser tubes in preference to copper. Brass with 40%Zn is called Muntz metal and is used for condenser systems employing fresh water as a coolant. For brackish or salt-water environment, the *tin brasses* or commonly called *inhibited admiralty brasses* are selected. Naval brass is a tin brass that contains about 39%Zn and 1%Sn. The term "inhibited" arises from the role of tin in these brasses to inhibit the dealloying attack, which in this case is, dezincification (see Sec. 10.3.2). As indicated earlier, addition of iron and manganese to these tin brasses results in the manganese bronze alloy (C67500 and cast alloys) that is used for ship propellers among other uses.

Bronzes. The term bronze was initially referred to copper-tin alloys. It is now used to include all copper alloys whose major alloying addition is not zinc. Thus, we have now tin and phosphor bronzes, aluminum bronzes, and silicon bronzes.

Copper-tin alloys are called *tin bronzes.* Their strength increases by increasing the tin content and peaks at about 20%Sn, after which the strength decreases sharply with additional tin. The maximum solubility of tin in copper is about 15 wt% and tin bronzes containing from 9 to 11 wt% are called gun metals and exhibit excellent strength and toughness, with moderate ductility. Tin bronzes have better corrosion resistance than brasses and are less susceptible to stress-corrosion cracking (SCC).

The production of tin bronzes requires procedures to prevent the oxidation of tin to SnO_2. When it forms, SnO_2 disperses in the melt and appears as patches on the surface that weaken and render the material brittle. The formation of SnO_2 can be avoided by deoxidizing the melt with phosphorus, as in electrolytic copper. To ensure the complete absence of SnO_2, a residual content of phosphorus is left in the composition that also serves to increase the strength. These alloys are called *phosphor bronzes* and those containing 2%P and higher have strengths approaching those of steel. Phosphor bronze-6 (4.5%Sn and 0.2%P) has been shown to successfully replace titanium and stainless steels in handling chlorinated lime slurry and calcium hypochlorite with 65 percent chlorine.

Aluminum bronzes contain from about 3%Al to about 13%Al with single or combined additions of iron, nickel, silicon and cobalt. The alloys above 9%Al can be hardened by quenching from above a critical temperature. The hardening process is a martensitic-type process, similar to the martensitic transformation in steels. The strengthening arises from the subsequent temper annealing after the quenched. Aluminum bronzes with nickel and zinc form a class of materials, called *shape memory alloys,* which utiltize the reversible martensitic transformation.

The corrosion resistance of aluminum bronzes is better than that of phosphor bronzes, presumably because of a protective oxide that forms due to the aluminum addition. They resist corrosion from chloride and potash solutions, nonoxidizing mineral acids, and many organic acids. They are used extensively in the pulp and paper industry because of their improved resistance to alkaline environments. Aluminum bronze tubing, pipes, pumps, valves, tanks, shafts, and so on, are used in marine environments, in the pulp and paper industry, and in autoclaves for fatty acids.

Silicon bronzes contain up to 4%Si to improve machinability, formability, and corrosion resistance and are suitable for all types of welding. Their corrosion resistance is similar to the aluminum bronzes and their formability equals that of pure copper. Applications of silicon bronzes include heat-exchanger tubes, hydraulic-pressure fittings, bearing plates, piston rings, rivets, bolts, screws, clamps, and similar hardware.

Copper Nickels. These are commonly called cupro-nickels and contain between 3 to 30%Ni. These alloys are designed for marine service and have been used extensively as sea-water condenser tubes. In order to avoid pitting corrosion, the limit of nickel content is placed at 30 percent. The range of composition from 10 to 30 percent are used for condenser tubes in brackish and sea-water environment and 10%Ni is quite popular because it also prevents biofouling, as Fig. 13-12 indicates. Cupro-nickels are especially preferred in polluted water environment.

Nickel Silvers. These are alloys with zinc additions from 10 to 29 percent and nickel additions from

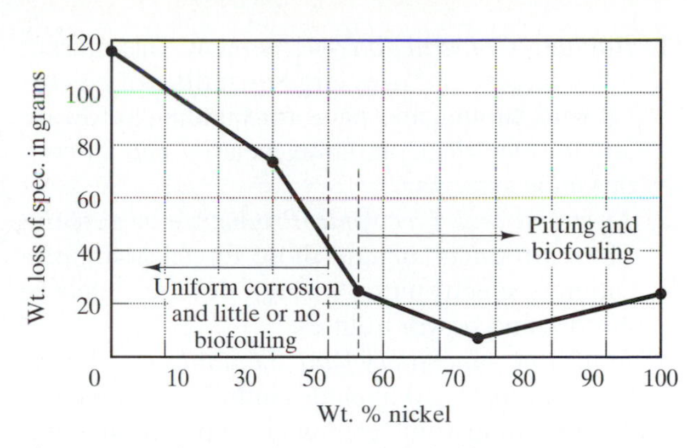

Fig. 13-12 Behavior of copper-nickel alloys in sea water.

10 to 18 percent. They have high strength and good ductility which make them easy to form by which stamping, rolling, or drawing. Because of their silvery appearance and corrosion resistance, they are used as silver-plated flatware, hollow ware, and musical instruments. The high zinc alloys are used in many spring applications because of their strength and stiffness.

Machinability. In order to improve the machinability of any of the copper and copper alloys discussed above, insoluble elements such as lead, tellurium, selenium, and bismuth are added. In addition to these, sulfur addition also increases machinability in the same manner as in the resulfurized steel grades. Thus, any of the alloys with one of these elements in their composition has improved machinability. However, the alloys with the insoluble elements require special attention during hot-rolling and hot-forming because they form a liquid phase that tends to segregate at grain boundaries and will induce splitting and fracture. The leaded high-zinc brasses avoid this problem because they transform to the beta phase at high temperatures. The beta phase can dissolve lead and thus avoids the liquid phase. Because of OSHA regulations, however, the usage of lead is being phased out.

13.5.3 Applications of Copper and Copper Alloys

The five major markets with their percent share of copper and copper alloys are the following:

1. *Building Construction*—41 percent. This market is the largest and uses large quantities of electrical wire, tubing, and parts for building hardware and for electrical, plumbing, heating, and air conditioning systems.
2. *Electrical and Electronic Products*—24 percent. These products include those for telecommunications, electronics, wiring devices, electric motors, and power utilities.
3. *Industrial Machinery and Equipment*—14 percent. This category includes industrial valves and fittings; industrial, chemical, and marine heat exchangers; and various types of heavy equipment, off-road vehicles, and machine tools.
4. *Transportation*—11 percent. Applications in this sector include road vehicles, railroad equipment, and aircraft parts; automobile radiators and wiring harnesses are the most important products in this category.
5. *Consumer and General Products*—10 percent. These include electrical appliances, fasteners, coinage, and jewelry.

Table 13-36 End-use applications of copper and copper alloys and their relative shares in the United States in 1989.

Application	% of total 1989	1980
Building wiring	16.9	10.7
Plumbing and heating	14.9	13.4
Autos, trucks, and buses	9.8	8.7
Telecommunications	8.1	13.1
Power utilites	7.7	7.4
Air conditioning and commercial refrigeration	7.1	6.3
In-plant equiptment	7.1	8.4
Electronics	5.7	4.2
Industrial valves and fittings	3.4	3.5
Appliances and extension cords	2.7	2.9
Coinage	0.9	2.7
Other	15.7	18.7
Total	**100.0**	**100.0**

Source: Copper Development Association Inc.

The percent share of each of the above markets may fluctuate with times, but the relative shares of the markets will most likely remain for sometime. Table 13-36 is a more detailed breakdown of the market shares of the end users from the above categories. Building wiring and plumbing constituted the two largest markets in 1989. The third largest market for copper is the automotive industry and this is also due mostly to the wiring harnesses.

The above markets explain the relative usage of wrought copper and copper alloys in 1989 in the United States shown in Table 13-37. In this table, the coppers and high coppers from the wire and brass mills constitute about 75 percent of the usage. The next largest group of alloys are the brasses—both plain and leaded—that account for about 20.4 percent of the usage. The rest of the alloys account for less than 5 percent.

13.6 Nickel and Nickel Alloys [6]

Although changes in our market-driven economy occur, the relative usage of nickel is predominantly as an alloying element in stainless and low-alloy steels. As indicated in Table 13-38, about 57 percent is used in stainless steels and an additional 9.5 percent is used in the heat-treatable low-alloy steels.

Nickel-based alloys, which account for about 13 percent of the total nickel usage, are used primarily for their corrosion-resistant and/or heat-resistant properties. As with aluminum, titanium, and copper, nickel also can be used commercially pure. The nickel alloys are extensive solid solutions with copper, chromium, iron, molybdenum, tungsten, and tantalum and retain the face-centered cubic (FCC) structure. Except for copper, the other solid solution elements are also carbide formers. Thus, in addition to strengthening by solid solution, they also form submicroscopic carbide particles with the residual carbon to produce a dispersion-strengthening effect. The alloys can be further hardened by precipitation of coherent particles, the most prominent being γ' and γ'' by the addition of titanium, aluminum, and/or niobium. The coherent particles are very effective in rendering nickel-based alloys extremely heat-resistant. In addition to these coherent particles, the heat-resistant and creep properties are further enhanced by oxide dispersions, such as yttria (Y_2O_3). The roles of alloying elements in nickel-based alloys are shown in Table 13-39, which also includes their roles in iron-base alloys for comparison.

Table 13-37 Relative useage of copper and copper alloys in 1989.

Copper or copper alloy group	Designation	Approximate U.S. shipments in 1989 Mg × 10^3	lb × 10^6	Remarks
Wire mill products				
Coppers	C10000–C15900	1522	3356	C11000 is the predominant material.
Brass mill products				
Copper	C10000–C15900 C16000–C16900 C18000–C18900	570	1257	Includes modified copper, cadmium copper, and chromium copper
Common brasses	C20000–C29900	219	482	Of this amount, 90% is strip, sheet, and plate.
Leaded brasses	C30000–C39900	347	766	Of this amount, 96% is rod.
Tin bronzes (phosphor bronzes)	C50000–C53900	14	31	Unleaded only
Aluminum bonzes, silicon bronzes, and manganese bronzes	C60000–C68400	13	28	
Copper-nickels	C70000–C72900	32	71	
Nickel silvers	C73000–C79900	4	9	
Others	C17000–C17900 C19000–C19900 C40000–C49900 C54000–C54900 C68500–C69900	49	107	Includes beryllium-coppers, copper-iron alloys, tin brasses, leaded tin bronzes, aluminum brasses and silicon brasses.
Total		2770	6107	

Source: Copper Development Association Inc.

Table 13-38 Relative usage of nickel.

Product	% Usage
Stainless Steel	57.0
Alloy Steel	9.5
Nickel-based Alloys	13.0
Copper-based Alloys	2.3
Plating	10.4
Foundry	4.4
Other	3.3

13.6.1 Alloys—Their Designations and Characteristics

The designation system for nickel and nickel-based alloys is shown in Table 13-40 and consists of the letter *N* followed by a five-digit code, that is, NXXXXX according to the Unified Numbering System (UNS). The first two digits signify the group of alloys, according to the major alloying elements; commercially pure nickels are assigned 02 and 03, with the latter being precipitation or dispersion-hardened. With the major alloying elements for each group, the commercial names of available alloys are also indicated in parenthesis. When the alloy is indicated to be precipitation-hardened, it is given a new code. The last three digits in the five-digit UNS code are the same as the common alloy number, or an assigned number, in the case of an alloy with a letter code. Examples of UNS designation for selected nickel alloys can be gleaned from Table 13-41, which is a list of some common nickel alloys and their compositions that are used for corrosion purposes.

Commercially Pure and Low-Alloy Nickels (N02XXX and N03XXX). The range of nickel con-

Table 13-39 Roles of alloying elements in iron- and nickel-based alloys.

Effect	Iron base	Nickel base
Solid-solution strengtheners	Cr, Mo	Co, Cr, Fe, Mo, W, Ta
FCC matrix stabilizers	C, Mn, Ni	...
Carbide form		
MC type	Ti	W, Ta, Ti, Mo, Nb
M_7C_3 type	...	Cr
$M_{23}C_6$ type	Cr	Cr, Mo, W
M_6C type	Mo	Mo, W
Carbonitrides		
M(CN) type	C, N	C, N
Forms γ' Ni_3 (Al, Ti)	Al, Ni, Ti	Al, Ti
Retards formation of hexagonal η (Ni_3Ti)	Al, Zr	...
Raises solvus temperature of γ'	...	Co
Hardening precipitates and/or intermetallics	Al, Ti, Nb	Al, Ti, Nb
Forms γ'' (Ni_3Nb)	...	Nb
Oxidation resistance	Cr	Al, Cr
Improves hot corrosion resistance	La, Y	La, Th
Sulfidation resistance	Cr	Cr
Increases rupture ductility	B	B, Zr
Causes grain-boundary segregation	...	B, C, Zr

tent in these materials is from 94 to 99.5 percent minimum. Alloy 200 contains 99.5 percent nickel minimum, while Alloy 301 contains from 4.00 to 4.75% Al and from 0.25 to 1.00% Ti.

Alloy 200 (with 0.10%C max.) and its low carbon variety Alloy 201 (with 0.02%C max) are used extensively in the food, chemical, and paper and pulp processing industries. They offer excellent corrosion resistance in reducing environments with reasonable heat-transfer characteristics and are used for caustic liquor evaporation and in alkaline solutions of varying concentration. As a general rule, oxidizing environments promote corrosion, while reducing conditions retard corrosion of nickel.

Alloy 200 was initially used for food processing, kitchen hardware, and roofing. It continues to be selected for use as a coinage material. It also finds applications in the electronics industry as plated pins for printed circuit board interconnects. With its thermal coefficient of expansion being nearly equal to that of steel, it is very compatible with steel and is used as a cladding material to fabricate nickel-lined vessels.

Nickel-copper Alloys (Monels) (N04XXX AND N05XXX). In addition to their corrosion resistance in reducing chemical environments, these alloys have also been found to exhibit excellent corrosion resistance in sea water, where they manifest excellent performance in nuclear submarines and various surface vessels. Alloy 400 (Monel - 66%Ni and 33%Cu) is the base alloy in this series and can be magnetic depending upon composition and previous processing. Alloy R-405 has controlled sulfur addition for machinability and alloy K-500 (K-Monel or Monel

Table 13-40 UNS designation of nickel and nickel alloys.

Designation	Major Alloying Element(s)
N02XXX	None, but impurity limits are specified
N03XXX	None, precipitation- or dispersion-hardened
N04XXX	Copper (Monels)
N05XXX	Copper (precipitation-hardened)
N06XXX	Chromium with other elements (Nichrome, Inconel, Hastelloy, Eatonite)
N07XXX	Chromium (precipitation-hardened)
N08XXX	Iron and Chromium (Incoloy, Haynes, Armco, and Carpenter)
N09XXX	Iron and Chromium (precipitation-hardened)
N10XXX	Molybdenum (Hastelloys)
N13XXX	Cobalt (with Cr, Mo, W; or precipitation-hardened)
N14XXX	Iron
N19XXX	Iron and cobalt (precipitation-hardened)
N99XXX	Brazing filler metals

K-500) is a precipitation hardening variety with aluminum and titanium additions and is totally nonmagnetic.

Being homogeneous solid solutions of nickel and copper, they exhibit, in general, better corrosion resistance than either nickel or copper alone. With exceptions, they are more resistant than nickel in reducing environments and more resistant than copper in oxidizing environments. Their corrosion resistance in alkaline solutions is similar to Alloy 200 (pure nickel), but much less in concentrated caustic solutions and are, therefore, not recommended for caustic evaporators and concentrators. They can withstand hydrofluoric acid at most concentrations up to 92 percent and temperatures up to 110°C. Monel 400, the most common alloy in this group, can withstand sulfuric acid up to 80 percent and hydrochloric up to 20 percent.

These alloys find applications as fasteners and liquid handlers in the marine, aerospace, and chemical processing industries. Because of its nonmagnetic characteristics, Alloy K-500 is used for gyroscope application, as anchor cable aboard minesweepers, and for propeller shafts on a variety of vessels. It is used as valves and pumps to handle organic acids, caustic, and dry chlorine in chemical process industries. It is also used as sucker rods and for associated Christmas tree well-head applications, especially in sour gas (with sulfur) environments, in the oil and gas industry.

Nickel-Chromium and Nickel-Chromium-Iron Alloys (N06XXX and NO7XXX). Nickel alloys with at least 15 percent chromium provide both oxidation and carburization resistance at temperatures exceeding 760°C (1400°F). The chromium promotes the formation of a protective surface oxide, and the nickel exhibits good retention of the protective coating, especially during cyclic exposure (on-off service) to high temperatures. In atmospheres that are oxidizing to chromium but reducing to nickel, nickel-chromium alloys may be subject to internal oxidation. To reduce the susceptibility to internal oxidation, iron is added. Figure 13-13 shows some of these alloys (numbers and letters) nearest their nickel content shown in the vertical scale. These may be grouped into (1) Ni-Cr and Ni-Cr-Fe alloys that contain from about 45 to 80 wt% Ni, and (2) Fe-Ni-Cr alloys that contain from 30 to 45 wt% Ni. The austenitic stainless steels containing nickel up to 20 wt% are grouped as Fe-Cr-Ni alloys and are shown only for comparison. In these alloys, the sequence of the symbols of the elements indicates their relative amounts in the composition; the first having the highest amount and the third having the least amount. The chromium content in most of the alloys range from 15 to 25 wt.%.

Table 13-41 Selected nickel and nickel-based alloys for corrosion purposes and their UNS designations. [9]

Common alloy designation	UNS designation	Chemical composition, %										
		C(max)	Nb	Cr	Cu	Fe	Mo	Ni	Si	Ti	W	Other
Nickel (Commercially Pure)												
200	N02200	0.1	...	...	0.25 max	0.4 max	...	99.2 min	0.15	0.1 max	...	...
Nickel-copper (Monel)												
400	N04400	0.15	...	...	31.5	1.25	...	bal	0.5	...	...	...
Nickel-molybdenum (Hastelloy)												
B	N10001	0.05	...	1.0 max	...	5.0	28	bal	1.0	...	...	...
Nickel-chromium-iron												
600	N06600	0.08	...	16.0	0.5 max	8.0	...	bal	0.5	0.3 max	...	...
800	N08800	0.1	...	21.0	0.75 max	44.0	...	32.5	1.0	0.38	...	...
Nickel-chromium-iron-molybdenum												
825	N08825	0.05	...	21.5	2.0	29.0	3.0	42	0.5	1.0	...	...
G	N06007	0.05	2.0	22.0	2.0	19.5	6.5	43	1.0	...	1.0 max	...
Nickel-chromium-molybdenum-tungsten												
690	N06690	0.02	...	29.0	...	10.0	...	61	...	0.3	...	...
C-276	N10276	0.01	...	15.5	...	5.5	16.0	57	0.08	...	4.0	...
Precipitation-hardening												
K-500	N05500	0.25	...	...	29.0	2.0 max	...	63	0.5 max	0.6	...	2.7Al
718	N07718	0.05	5.0	18.0	...	19	3.0	53	...	0.4 max	...	...
X-750	N07750	...	0.9	15.5	...	7.0	...	bal	...	2.5	...	...

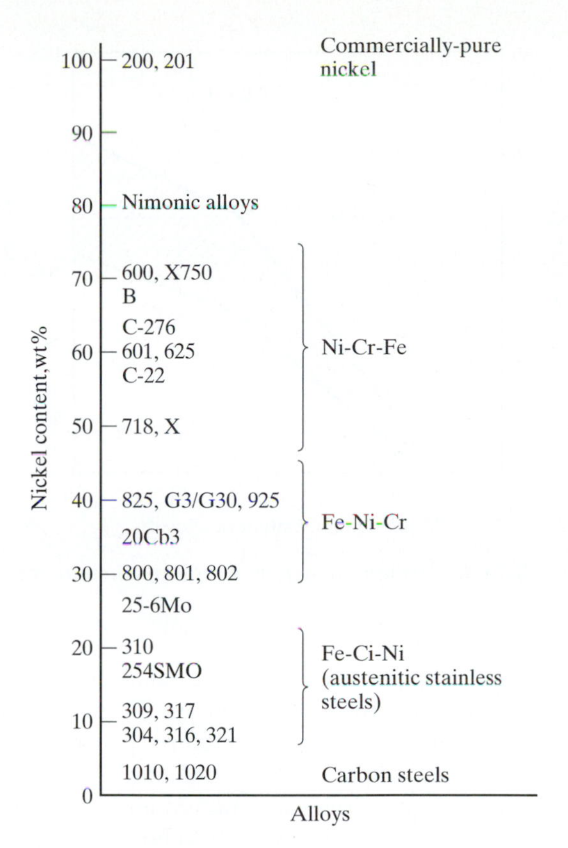

Fig. 13-13 Nickel-base alloy chart showing some alloys and their groupings according to nickel content.

High nickel content alloys are used as electrical resistance materials for heating applications in furnaces and appliances. Some of the alloys that are commercially important are:

- 80Ni-20Cr (with 1.5Si) — commonly called Nichrome alloy. This alloy has very high electrical resistivity characteristics and can be used at temperatures up to 1150° to 1175°C (2100° to 2150°F)
- 60Ni-24Fe-16Cr—This alloy is suitable for less exacting applications and can be used at temperatures up to 950°C (1750°F); for example, as clothes dryer heating elements
- 45Fe-35Ni-20Cr—This alloy is a cost-effective consideration for operations at temperatures up to 1065°C (1950°F). However, it exhibits a large temperature coefficient of resistance that must be considered when designing an element

A common application for the above alloys, for example, is as a spiral-wound electrical heating element contained within a metal sheath, for use as an appliance heating element. The alloys are also used for heating in domestic fan heaters and in thermal storage units.

When not used as heating elements, the Ni-Cr and Ni-Cr-Fe alloys are used as heat-resistant and corrosion-resistant materials. The initial applications in the United States as heat-resistant alloys were in thermal process equipment and in the chemical process industry, where carburizing environments and elevated temperatures limited the performance of stainless steels. They are now being used in commercial and military jet engines. The two earliest alloys of this group are Alloy 600 (76Ni-15Cr-8Fe) and the Nimonic alloy (80Ni-20Cr +Ti/Al). Alloy 600 is the basic alloy in the Ni-Cr-Fe group and has been the preferred alloy for all steam generator tubing in nuclear power plants. With the recognition that Alloy 600 might exhibit stress-corrosion cracking in superheated pure waters, Alloy 690 (corrosion-resistant alloy variant discussed below) is now being used as the replacement for alloy 600 in existing generators as well as in new designs. The Nimonic alloy and its variations are considered superalloys and are suitable for many of the hot sections in aircraft gas turbine engines—blades, turbine rings, fasteners, and the like. Some heat-resistant variants of these alloys are:

- Alloy 601. This alloy has a lower nickel content (61 percent) with aluminum and silicon additions for improved oxidation and nitriding resistance
- Alloy X750. This alloy contains aluminum and titanium for precipitation-hardening; originally used for skin of the Bell X-1 experimental aircraft
- Alloy 718. This alloy contains titanium and niobium to overcome strain-aging cracking problems during welding and weld repair
- Alloy X (48Ni-22Cr-18Fe-9Mo + W). This is a high-temperature flat-rolled product for aerospace applications
- Waspalloy (60Ni-19Cr-4Mo- 3Ti-1.3Al). Proprietary alloy for jet engines

If the design requires higher strength, oxide dispersion-strengthened alloys can be used. Mechanically alloyed (MA) nickel alloys with about 1 vol % oxides (typically, yttria, Y_2O_3) may be selected. Mechanical alloying is the process of mechanically mixing oxide and metal powders in a ball mill. The particles are impacted by the steel balls, are fractured into pieces, and are rewelded back together, and then formed into shape by sintering or powder metallurgy (P/M). The MA nickel alloys are particle composites of dispersed oxides in nickel-based alloys with higher strength and high heat resistance; the oxides hinder the motion of dislocations to retain strength longer at high temperatures.

Corrosion resistant alloys in the Ni-Cr-Fe group are:

- Alloy 625. This contains 9%Mo + 3%Nb and exhibits both high temperature and wet corrosion resistance; it resists pitting and crevice corrosion.
- Alloy G3/G30 (Ni-22Cr-19Fe-7Mo-2Cu). The increased molybdenum content in these alloys provides improved pitting and crevice corrosion resistance.
- Alloy C-22 (Ni-22Cr-6Fe-14Mo-4W). This alloy exhibits superior corrosion resistance in oxidizing acid chlorides, wet chlorine, and other severe corrosive environments.
- Alloy C-276 (17%Mo + 3.7%W). This alloy exhibits good sea-water corrosion resistance and excellent pitting and crevice corrosion resistance.
- Alloy 690 (27%Cr). This alloy exhibits excellent oxidation and nitric acid resistance. It is specified for nuclear waste disposal by the vitreous encapsulation method.

Similar to its function in stainless steels, the molybdenum in the above alloys increases the pitting and crevice corrosion resistance. Figure 13-14 shows that crevice corrosion in a 6 percent ferric chloride solution occurs at a higher temperature when the molybdenum content increases. Hastelloy B and B-2 containing about 28%Mo exhibit the best resistance to hot hydrochloric acid environments of any of the nickel-based alloys. The molybdenum-containing alloys are used for many purposes in chemical processing, pulp and paper production (bleachers and washers), and pollution-control equipment.

Iron-Nickel-Chromium Alloys (N08XXX and N09XXX). The iron content in these alloys can be as high as 46 wt% and the combined nickel and chromium content is more than 50 wt%. Other alloying elements are added to obtain desired specific properties. They are used in applications with less stringent temperature

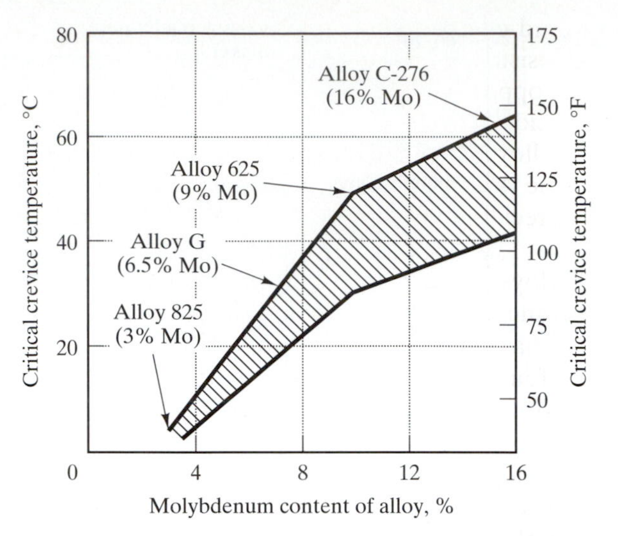

Fig. 13-14 Molybdenum increases crevice corrosion of nickel alloys.

and strength requirements and are relatively cheaper than the Ni-Cr and Ni-Cr-Fe alloys. They are used extensively in the high-temperature petrochemical environments, where sulfur-containing feedstocks (naphtha and heavy oils) are cracked into component distillated parts. They are resistant to both chloride-ion and polythionic acid stress-corrosion cracking. Some of the important commercial alloys in this group are:

- Alloy 800 (Fe-32Ni-21Cr). This is the basic alloy in this group. It is resistant to oxidation and carburization at elevated temperatures.
- Alloy 800H. This a modification of Alloy 800 with controlled carbon (0.05–0.10%) and grain size (>ASTM 5) to optimize stress-rupture properties.
- Alloy 800HT. Similar to 800H with combined addition of titanium and aluminum from 0.85 to 1.2 percent to ensure optimum high-temperature properties.
- Alloy 801. This contains higher titanium (0.75 to 1.5 percent) and exhibits exceptional stress-corrosion cracking (SCC) resistance to polythionic acid.
- Alloy 802. This is a higher carbon version (0.2 to 0.5 percent) that exhibits improved strength at high temperatures.
- Alloy 825 (Fe-42Ni-21.5Cr-2Cu). This is stabilized with titanium addition (0.6 to 1.2 percent)

and contains about 3%Mo for increased pitting resistance in aqueous corrosion applications. Copper content provides resistance to sulfuric acid.

- Alloy 925. Titanium and aluminum are added to the Alloy 825 composition for strengthening via precipitation-hardening.

Alloys 825 and 925 are also used for corrosion resistance because of their molybdenum content. They have excellent resistance to sulfuric acid and have found extensive use as tubing and plate for production of fertilizer and associated products. Alloy 925 is also used for downhole components in sour gas wells and has been forged to block master valves for well-head applications, as well as associated Christmas tree well-head components. Alloy 825 is also used for downhole tubular components where hydrogen sulfide, carbon dioxide, and sodium chloride (salt) are at high temperatures.

Nickel-Molybdenum Alloys (N10XXX). The basic alloy in this group is Hastelloy B and its low carbon variant B-2. Because they have relatively small amounts of chromium (1% max), their corrosion resistance is limited to reducing environments or media. Hastelloy B was specifically developed to withstand hydrochloric acid at all concentrations and temperatures. It is also resistant to sulfuric (except boiling sulfuric acid beyond 60 percent concentration), phosphoric, and hydrofluoric acids and their salts, as well as organic acids and their salts at all concentrations and high temperatures. It is attacked by oxidizing salts such as ferric chloride and cupric chloride.

Because of its high carbon content (1%C) and molybdenum being a former carbide, Hastelloy B exhibits the same sensitization phenomenon observed in stainless steels. During welding, molybdenum carbides form at the grain boundaries in the heat-affected zone and make the alloy susceptible to intergranular corrosion. As in stainless steels, the area around the grain boundaries have less molybdenum content because of the carbide formation, and galvanic cells are developed due to the difference in concentrations within the grain and around the grain boundaries. Knife-line cracks sometimes develop also at the fusion line—the boundary between the fusion and heat-affected zones. To minimize the localized intergranular corrosion, Hastelloy B must be solution heat-treated after welding, in the same manner as done with stainless steels.

To avoid this sensitization problem, Hastelloy B-2 with 0.1%C maximum and some silicon content should be selected. This alloy can be used in the as-welded condition because molydenum carbide does not form. To further ensure no carbide formation, titanium may be specified in the composition to stabilize the carbon, as in stainless steels.

Nickel-based Superalloys. The term superalloy is given to some nickel-based alloys because of their inherent high strength, in addition to their heat and corrosion resistance properties. These superalloys can be used in load-bearing applications at temperatures in excess of $0.8T_m$, where T_m is the absolute melting point of the alloy. The 0.8 fraction of the melting point is higher than any other class of engineering alloys. There are also iron-based and cobalt-based superalloys, but nickel-based superalloys are the most widely used for the hottest parts of the aircraft gas turbine engines. They constitute currently over 50 percent of the weight of advanced aircraft engines. Although these alloys are available in wrought form, their inherent strength at high temperatures makes it very difficult to shape them by deformation methods. Most of the products of these alloys are produced by investment casting. In the early 1970s, directional solidification technology was developed to enable production of columnar or single-crystal jet engine turbine blades. The single crystal minimizes creep deformation at high temperatures. Table 13-42 lists some selected nickel-based superalloys with their compositions, and their temperatures of application. We see that the single crystal products can be used up to 1100°C.

Table 13-42 Nominal compositions of selected nickel-based superalloys and their operating temperatures. [9]

Alloy	Composition, wt%												Approximate year of introduction	Temperature capability,(a)	
	C	Cr	Co	Al	Ti	Mo	W	Nb	Ta	Zr	B	Other		°C	°F
Conventionally Cast															
IN-718	0.05	19	...	0.5	1.0	3.0	...	5.0	...	0.01	0.005	18F	1965	700	1290
IN-6201(b)	0.03	20	20	2.4	3.6	0.5	2.3	1.0	1.5	0.05	0.8	...	1978	1010	1850
MAR-M 246	0.15	9.0	10	5.5	1.5	2.5	10	...	1.5	0.05	0.015	1.5Hf	1966	1025	1885
Directionally Solidified															
MAR-M 002 DS	0.15	9.0	10	5.5	1.5	...	10	...	2.5	0.05	0.015	1.5Hf	1975	1045	1915
IN-6203(b)	0.15	22	19	2.3	3.5	...	2.0	0.8	1.1	0.1	0.01	0.75Hf	1981	1020	1870
Single Crystal															
PW 1480	...	10	5.0	5.0	1.5	...	4.0	...	12	...	...	...	1980	1060	1940
PWA 1484	...	5.0	10	5.6	...	2.0	6.0	...	8.7	...	...	3Re 0.1Hf	1986	1100	2010
CM SX-4G	...	6.2	9.5	5.5	1.0	0.6	6.5	6.5(c)	...	...	...	2.9Re 0.1Hf	1986	1110	2030

(a) 100 hr to rupture at 140 MPa. (b) High-chromium alloy suitable for land and marine-based gas turbines. (c) Combined Nb + Ta.

Summary

The most prominent nonferrous metals and alloys used for engineering purposes are aluminum, titanium, magnesium, copper, and nickel alloys. As will be justified in Sec. 16.3, aluminum is the most widely used of the nonferrous metals and alloys because of its structural performance or efficiency index, corrosion resistance, and cost. Titanium ranks second in structural performance and it is more expensive than aluminum. Magnesium is a potentially promising material for structural purposes, if its deformability and corrosion resistance can be improved. The principal use of copper is for its electrical conductivity. Structurally, copper, brasses, and bronzes are used as pipes and tubings because of their corrosion resistance. The big market for nickel is for alloying of stainless and low alloy steels. Nickel-based alloys are predominantly used for their corrosion performance in severe corrosive and high-temperature environments and find wide applications in the chemical process industries.

Aluminum alloys are classified as heat-treatable or nonheat-treatable alloys. The most used wrought alloys are the nonheat-treatable 3XXX alloys because they exhibit higher strengths than the 1XXX (commercially pure aluminum) alloys, sufficient formability, and good weldability. The other nonheat-treatable 5XXX alloys were developed for marine applications, while the 4XXX alloys are used predominantly as welding and brazing alloys. Of the heat-treatable alloys, the 2XXX and the 7XXX alloys are used principally for aircraft structures, while the 6XXX are used for general-purpose high-strength applications. For castings, the 3XX alloys are also used the most. We must note, however, that the digit "3" in wrought alloys signifies manganese as the major alloying element, while "3" in cast alloys is primarily silicon addition with copper or magnesium or both. Procurement of aluminum alloys are done by specifying the temper designations.

The majority, about 80 percent, of titanium's application goes to the aerospace industry and more than 80 percent of this is accounted by the Ti-6Al-4V alloy. A guide for selection of titanium alloys is provided and starts with the Ti-6Al-4V alloy. The other major use of titanium is due to its outstanding corrosion resistance. The alloy most used for this purpose is Grade 2 unalloyed titanium that has comparable strength properties as annealed austenitic stainless steels. Ti-6Al-4V and Grade 2 Ti are used the most, in both the wrought and cast alloys.

Structural applications of magnesium account for only about 15 percent of magnesium and of these more than 80 percent are castings. Magnesium is making inroads in automotive applications.

There are six families of copper and copper alloys: coppers, high-copper alloys, brasses, bronzes, copper-nickels, and copper-nickel-zinc alloys. The coppers and high coppers constitute about 75 percent of the market that reflects their usage for electrical applications. Then the next most widely used are the plain and leaded brasses that account for about 20.4 percent of the market.

The most significant use for nickel is to improve the corrosion of stainless steels. For nickel-based alloys, they are used for the most severe corrosive environments as well as heat-resistant alloys.

References

1. *Aluminum and Aluminum Alloys,* ASM Specialty Handbook, ASM International, 1993.
2. *Aluminum: Properties and Physical Metallurgy,* J. E. Hatch, Editor, ASM, 1984.
3. F. A. A. Crane and J. A. Charles, *Selection and Use of Engineering Materials,* Butterworths, 1984.
4. *Corrosion,* vol. 13, ASM Handbook, 9th ed. Metals Handbook, ASM International, 1987.
5. *Titanium Alloys,* Materials Properties Handbook, ASM International, 1994.
6. *Nonferrous Materials,* vol. 2, Metals Handbook, 10th ed., ASM International.
7. "Magnesium on the Move," Proceedings of Conference held in Chicago, IL during the 49th Annual World Conference, May 12–15, 1992, International Magnesium Association, ISSN-0161-5769, 1992.
8. H. H. Uhlig and R. Winston Revie, *Corrosion and Corrosion Control,* 3d ed., John Wiley, 1985.
9. S. L. Chawla, and R. K. Gupta, *Materials Selection for Corrosion Control,* ASM International, 1993.
10. Binary Alloy Phase Diagrams, Vol. 1, ASM, 1986.

Terms and Concepts

AA—aluminum association
AA designations for aluminum alloys
Aircraft alloys—7XXX and 2XXX
Alclad and cladded aluminum
Alpha + Beta Ti-6-4 (Ti-6Al-4V) alloy
Alpha transus
Aluminum foils
Aluminum honeycomb
Aluminum-lithium alloys
ANSI—American National Standards Institute
ASTM designations for magnesium alloys
Bearing strength
Beta transus
Brass - copper-zinc alloys
Bronze
CAFE—Corporate Average Fuel Economy
Cast aluminum alloys
CDA—Copper Development Association
CDA designations for copper alloys
Colors of wrought-copper alloys
Cupro-nickel alloys
ETP—Electrolytic Tough Pitch Copper
Heat treatable aluminum alloys
IACS—International Annealed Copper Standard
ISO—International Organization for Standardization
Nickel Silver
Nonheat-treatable aluminum alloys
Superalloys
Superplastic forming of Ti-6-4 alloys
Temper designations
Titanium - CP Grade 2
Transus
UNS designation for nickel alloys
Wrought-aluminum alloys

Questions

13.1 What are heat-treatable and nonheat-treatable aluminum alloys? How do the two types of alloys obtain their strengthening?

13.2 Which alloy designations and their major alloying elements are nonheat-treatable? heat-treatable?

13.3 What is the difference in the wrought and cast aluminum alloy designations? What two digits are used in the last X (after the decimal point) for the cast designations and what do they signify?

13.4 Indicate the similarities and differences in the following aluminum alloys: 1035, 1135, 1235, 1335, and 1435. What is the major alloying addition? Verify your answer using Table 13-3.

13.5 What do the following aluminum alloys indicate: 2018, 2218, 2618, 2024, and 2124? What is the major alloying element?

13.6 What is the difference between a 1035 aluminum and a 135.1 aluminum? Can we convert one to the other? How?

13.7 What are the differences between the channels (structural shape) with 7075-F, 7075-O, and 7075-T6 designations?

13.8 Using Fig. P8.15, indicate the yield strengths of 3105-H12, 3105-H22, and 3105-H32. What are the differences in these alloys?

13.9 What are the allotropes of titanium? On cooling, is the allotropic transformation a contraction or is it an expansion process? Explain.

13.10 What is the most widely used unalloyed titanium and some of its most common structural applications?

13.11 What are near-alpha, alpha + beta, and near-beta titanium alloys?

13.12 What is the most widely used titanium alloy? Cite some of its attributes. For superplastic forming, what is the optimum amount of the beta phase?

13.13 What precaution must be exercised in the processing of titanium and titanium alloys?

13.14 What are some advantages of magnesium and magnesium alloys? What does bearing strength mean?

13.15 Explain the relative usage of magnesium given in Table 13-28.

13.16 What is the difference between a brass and a bronze material?

13.17 Using Fig. 8-18, how would you specify the 70Cu-30Zn brass strip with a yield strength of (a) 10,000 psi, and (b) 22,000 psi?

13.18 What is a Muntz metal? Admiralty or naval brass? Free cutting brass?

13.19 Which industry uses copper and copper alloys the most and for what purpose?

13.20 Which industry uses nickel the most?

13.21 What is the main application of nickel alloys? What is a superalloy?

14

Inorganic Materials—Ceramics And Glasses

14.1 Introduction

Ceramics are ionically or covalently bonded inorganic and nonmetallic compounds of carbon, oxygen, nitrogen, boron, and silicon. Because of the ionic and covalent bonding, they have relatively high melting points, high elastic moduli, high hardnesses and strengths, low electrical and thermal conductivities, and are very brittle. Thus, most engineering uses of ceramics are as high-temperature materials, such as refractories in metal- and heat-treating industries, and as coatings and materials in aircraft and land-based gas turbine engines. They are used as abrasives and for wear-resistant applications because of their hardness. In addition, they are used to contain chemicals (acids and bases) because of their corrosion-resistant properties.

There is a lot of ongoing research and development on ceramics that has resulted in their classification as traditional and advanced ceramics. Traditional ceramics are those that are derived and processed from clay or nonclay minerals and are predominantly oxides. The traditional ceramic products are refractories, cement, whitewares, porcelain enamel, and structural clay products. On the other hand, advanced ceramics, also referred to as engineering or technical ceramics, are typically synthesized with very high purities. As a result, they exhibit better mechanical properties, corrosion/oxidation resistance, or electrical, optical, and/or magnetic properties than the traditional ceramics. Some of these ceramics include the carbides (SiC and BC), nitrides (AlN, SiN, SiAlON, and BN), borides, pure oxides (alumina, zirconia, thoria, beryllia, magnesia, spinel, and forsterite), magnetic ceramics, ferroelectric ceramics, piezoelectric ceramics, and the recently discovered superconducting ceramics.

In the same manner, glasses are also loosely classified as traditional and advanced. In the traditional classification, the glasses form a subset of ceramics because they are predominantly derived from silicates. The difference between glasses and ceramics is in their crystal structures. Ceramics exhibit long-range order, are crystalline, and have crystal structures. Glasses, on the other hand, have only short-range order, are amorphous, and have no crystal structures.

Traditionally, the manufacture of various sodium-calcium-silicate glass products (such as windows and bottles) forms the largest activity. Advanced glasses are also predominantly synthesized and exhibit unique characteristics such as optical transparency, electrical, corrosion resistance, and/or magnetic properties. These include high-strength high-purity silica fibers, nonsilicate glasses such as the halides, chalcogenides, metallic glasses, and oxy-halide glasses. Some glasses can be induced to nucleate and grow crystalline areas within their volume and the products are called glass-ceramics.

We shall now discuss ceramics and glasses separately.

14.2 Ceramics

14.2.1 Crystal Structures

Some of the structures of ionically and covalently bonded ceramics were briefly described in Chap. 2. The crystal structures depend on the ratio of the ionic radii and on the maintenance of electrical neutrality. These ionic radii determine the packing of positive and negative ions and the packing is such as to maximize electrostatic attraction and to minimize electrostatic repulsion between ions. For this to occur, the positive ions (cations) are normally surrounded by negative ions (anions) and vice versa. The normal description of the resulting structure assumes that the larger ions (anions) form the lattice structure and the smaller ions (cations) fit in the interstices of this lattice structure. In general, the packing of the ions follow Pauling's rules which are:

1. The anions surrounding the cations form a coordination polyhedron. The number of anions surrounding a cation is called the cation coordination number (CN) and the *cation coordination polyhedron* (CCP) is formed by the planes connecting the centers of the anions. The *CN is determined by the ratio of the radius of the cation to that of the anion.* The ranges of the ratio for the different CCPs are shown in Table 2-2. The CCP is the building block of an ionic structure.
2. In a stable structure, the CCP units are arranged in three-dimensions to optimize attractive and repulsive interactions between ions. *The stable structure must be electrically neutral, both macroscopically and at the atomic level.*
3. In a stable structure, instead of the edges and the faces, the corners of the CCPs tend to be shared. Sharing the corners separates the cations (within each CCP) the farthest to minimize the repulsive forces between them.
4. The CCPs formed about cations of low CN and high charge tend to be linked at the corners.
5. The number of different constituents in a structure tends to be small. This comes from the difficulty in efficiently packing ions and CCPs of different sizes into a single structure.

Based on Pauling's rules, we can understand why most of the oxide ceramics have crystal structures based on nearly close-packed oxygen ions. The two close-packed structures are the FCC (face-centered cubic) and the HCP (hexagonal close-packed). If the oxygen ions (anions) occupy the FCC and the HCP lattice sites, two interstitial sites are available where the cations (positive ions) may locate in these structures. These were previously discussed in Sec. 2.4.4 and are illustrated again in Fig. 14-1, where the Ts and the Os indicate the tetrahedral and octahedral sites, in both the FCC and the HCP lattices. The site names come from the fact that if cations locate on them, the CCPs are an octahedron and a tetrahedron, which are outlined also in Fig. 14-1. The coordination number of cations at the octahedral site is six—at the tetrahedral site, four. The cation coordination octahedron is comprised of the eight triangular faces connecting the six O^{2-} ions, while the tetrahedron is comprised of the four triangular faces connecting the four O^{2-} ions, in both lattices. The CCPs in both lattice structures with close-packed O^{2-} have corners from which to link with other CCP units to form stable structures, according to Pauling's rules.

We can now describe ceramic structures in terms of FCC or HCP packing of one specie with the other specie filling either the octahedral or the tetrahedral sites. Examples of these are shown in Fig. 14-2. Thus, the NaCl structure in Fig. 14-2a is an FCC packing of Cl^- ions with the Na^+ ions at all the octahedral holes. The MgO structure in Fig. 14-2b is an FCC packing of O^{2-} anions with one Mg^{2+} cation at each of

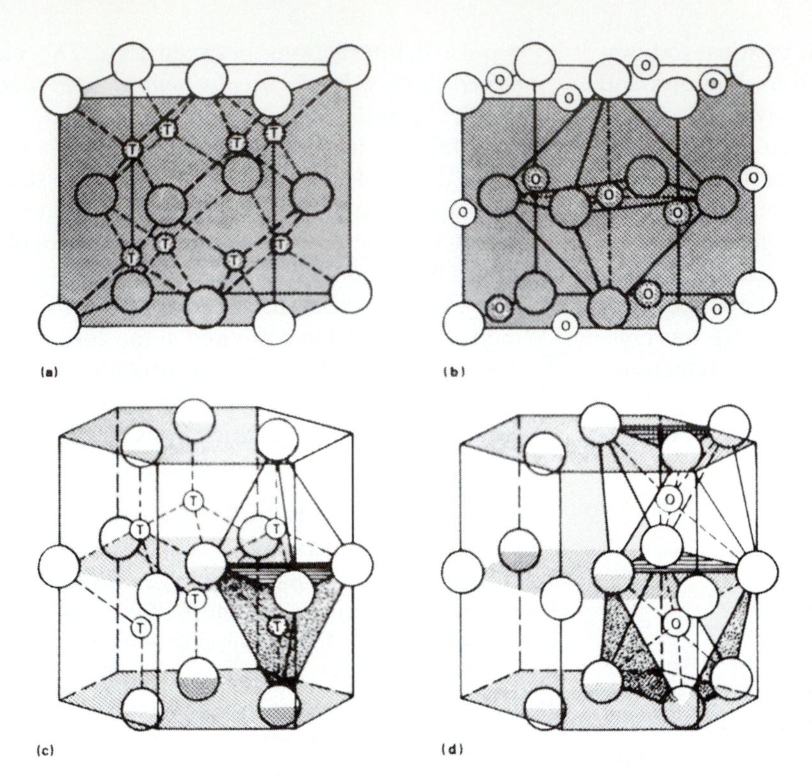

Fig. 14-1 Face-centered cubic (a and b) and hexagonal close-packed (c and d) lattices showing tetrahedral (T) and octahedral (O) positions in the lattices. [1]

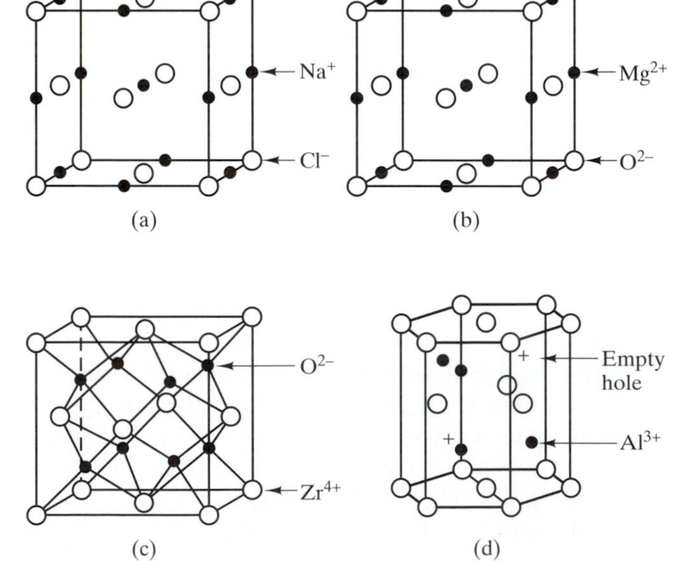

Fig. 14-2 Examples of ceramics with ionic bonding. (a) NaCl, (b) MgO—both have oxygen anions in FCC lattice anc cations in octahedral interstices, (c) cubic zirconia, cations in FCC lattice and anions in tetrahedral interstices, (d) alumina, anions in HCP lattice and cations in 4 of 6 octahedral sites. [4]

the octahedral holes. The ZrO_2 structure in Fig. 14-2c is an FCC packing where the Zr^{2+} cations assume the FCC lattice sites and the oxygen anions occupy all eight tetrahedral sites shown in Fig. 14-1a. This is the calcium fluorite (CaF_2) structure that is one of many other ceramic structures that have not been described in this text to simplify the treatment. The Al_2O_3 structure in Fig. 14-2d is the HCP packing of oxygen ions with four of the six octahedral holes occupied by the aluminum ions in Fig. 14-1d.

In covalent-bonded ceramics, each atom shares its electrons with its neighbors to set forth a fixed number of directional bonds. The ultimate covalent ceramic is diamond that is used industrially for various tooling and wear-resistant applications. As indicated in Chap. 2, diamond also has a face-cubic structure but with four additional atoms at four of the eight tetrahedral sites, Fig. 14-1a. To distinguish this from the simple FCC, this structure is called the diamond cubic structure, shown in Fig. 14-3a. Two simple covalent crystalline ceramics are SiC and the cubic SiO_2. The SiC structure in Fig. 14-3b is essentially that of a diamond cubic except that the carbon atoms at the tetrahedral sites are replaced by Si. The cubic silica (cristobalite) is the high-temperature allotrope (see Secs. 2.6.4 and 2.8) of quartz and has a structure in which all of the carbon atoms in the diamond cubic structure are replaced by silicon atoms; then each Si-Si bond is broken and an oxygen ion is inserted (i.e., changing the Si-Si bonds into -Si-O-Si-, as shown in Fig. 14-3c). At room temperature, the stable crystalline form of silica is a more complicated three-dimensional silicate framework in which all the oxygen atoms bridge the silicon atoms. The different allotropic or polymorphic forms of silica at atmospheric pressure are shown in Table 14-1.

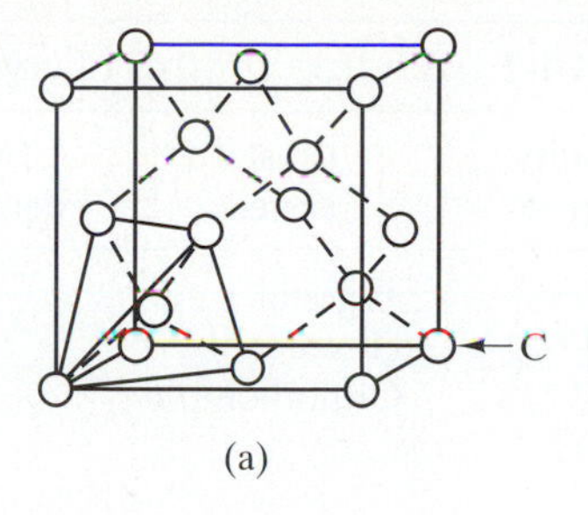

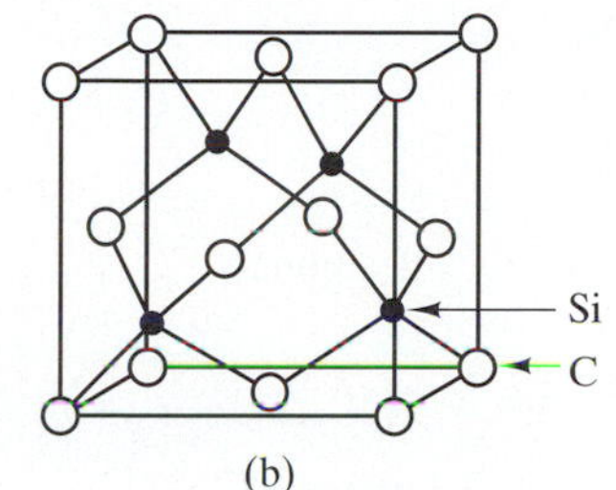

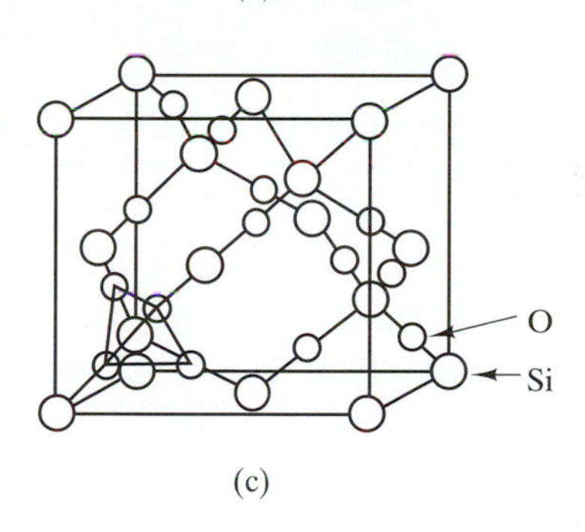

Fig. 14-3 Covalent Bonded Materials (a) Diamond cubic (DC) structure, (b) SiC with zinc-blende structure—similar to DC with the silicon atoms replacing the carbon atoms inside the lattice, (c) cubic silica (cristobalite) with the silicon atoms at the DC positions and the oxygen atom bridging two silicon atoms.

Imperfections in Ceramic Crystal Structures. As in metals, actual ceramic structures contain point defects, line defects (dislocations), and surface defects such as grain boundaries. The point defects may be vacancies, substitutional, interstitial ions, Frenkel, and Schottky defects. All of these have the same meaning as those in metals, and the last two defects were described in Chap. 2.

There are also dislocations in ceramic structures. However, the dislocations in ceramics are inherently much harder to move due to of the localized ionic or covalent bonding that needs to be broken and then reformed when the dislocation passes through the structure. Thus, ceramics are characteristically stronger and harder but more brittle than metals.

Grain boundaries are also present in crystalline ceramics. An example is shown in Fig. 14-4 that illustrates the polycrystallinity (many grains) of a ceramic material. The effect of the grain boundaries in ceramics is the same as that in metals. At low temperatures, they increase the strength and toughness, while at high temperatures, they induce higher creep rate.

The influences of the above imperfections on mechanical properties are overshadowed, however, by the larger macroscopic defects that are produced during processing and handling. The processing of most ceramics involves the consolidation of particles that are initially compacted at low temperatures and are heated (fired) at high temperatures to bond the particles together. The consolidation (sintering) process does not completely densify the ceramic body

Table 14-1 Allotropic forms of crystalline silica at atmospheric pressure.

Classifications of phases	Stability range, K	Crystal system	Theoretical density, g/cm^3	Melting point, K	Remarks
Quartz, low	<845	Trigonal	2.65	···	Most common crystalline phase
Quartz, high	845–1140	Hexagonal	2.52	1743	High-temperature
Tridymite S	1140–1743	Orthorhombic	2.32	1933	Characteristic phase in refractories
S-I	<337				Polymorphic modification
S-II	337–391				Polymorphic modification
S-III	391–436				Polymorphic modification
S-IV	436–483				Polymorphic modification
S-V	483–748				Polymorphic modification
S-VI	748–1743				Polymorphic modification
Tridymite M	···	Hexagonal	2.30	···	Unstable, converts to Tridymite S
M-I	<391				Polymorphic modification
M-II	391–436				Polymorphic modification
M-III	>436				Polymorphic modification
Cristobalite, low	<545	Tetragonal	2.32	···	Low-temperature phase
Cristobalite, high	545–1998	Cubic	2.27	1998	Appears first when above phases are heated to 1672 K

to 100 percent density, therefore, porosities invariably remain within the body. The amount, the size, the distribution, and the continuity of these voids or porosities dictate the mechanical properties of the ceramic material. In addition, macroscopic defects or cracks may be introduced during cooling from high temperatures and/or during final finishing or machining or handling of the sintered ceramic part. At any time, there will be a distribution of these macro- or microcracks that influence the mechanical properties.

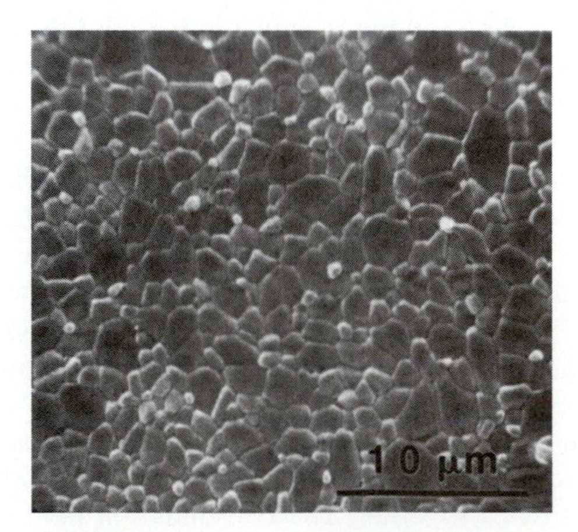

Fig. 14-4 Microstructure of polycrystalline Al_2O_3.

14.2.2 Ceramic Phase Diagrams

Phase diagrams for ceramic materials are also available to be used as guides in the processing of these materials. For the traditional ceramics, the components of the phase diagrams are predominantly oxides or silicates. In general, ceramic phase diagrams are much more complicated than the metallic binary phase diagrams discussed in Chap. 5. Although they are more complex, ceramic binary diagrams also show the same characteristics exhibited by metals when they are alloyed or mixed together. Some examples of ceramic binary phase diagrams are shown in Figs. 14-5 to 14-8. Figure 14-5 shows complete solubility of magnesium and ferrous silicates in the solid, as well as, in the liquid; Fig. 14-6 shows the formation of a very low melting point eutectic when sodium oxide (soda) is mixed with silica. This phenomenon is "fluxing" whereby a solid can be molten at a much lower temperature, or a liquid with high viscosity becomes less viscous and

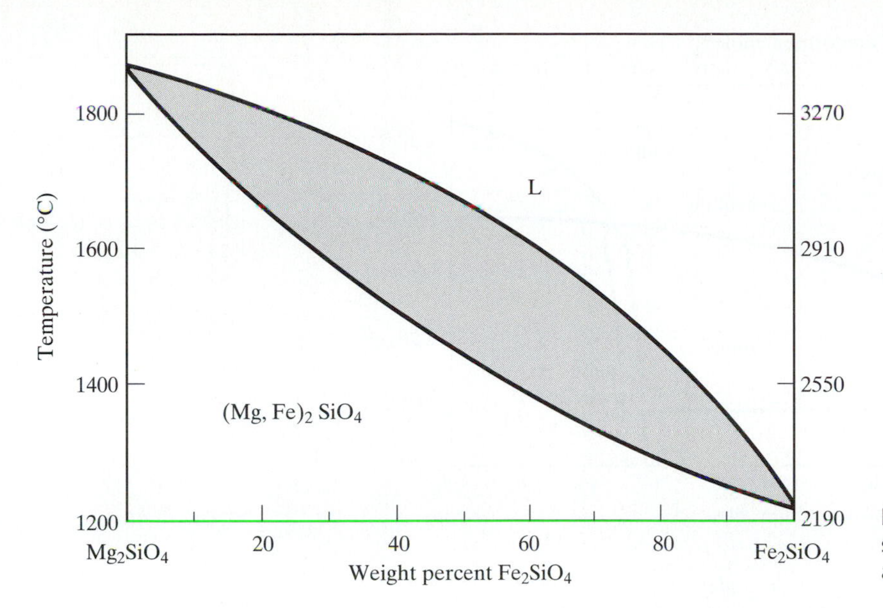

Fig. 14-5 Illustration of complete solid solubility of magnesium and ferrous silicate in each other.

more fluid in order to make processing easier. This is the appropriate diagram that is used in the production of the common glass. This diagram suggests also that pure silica can easily degrade or deteriorate when an alkali or a basic oxide (sodium oxide) is present in the liquid as in slags. We also see in Fig. 14-6 that sodium oxide and silica are completely insoluble in each other, but they react and form stoichiometric compounds (Sec. 5.9), $Na_2O \cdot 2\,SiO_2$ and $NaO \cdot SiO_2$.

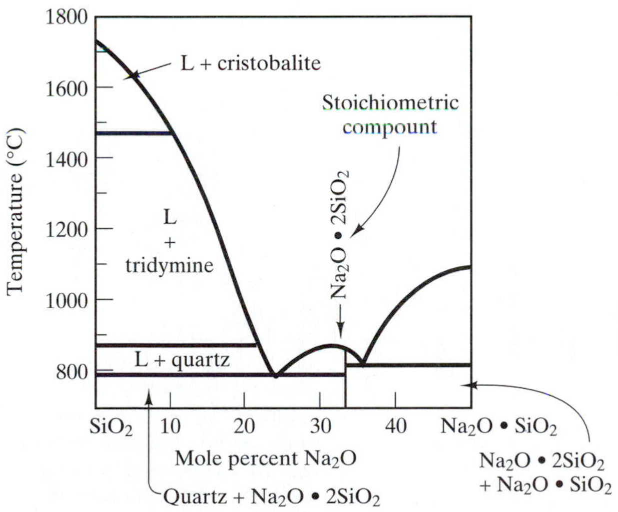

Fig. 14-6 Eutectic system in ceramics and illustrates the fluxing of silica by sodium oxide. Fluxes lower the melting points of ceramics to allow them to be processed at manageable temperatures.

Figure 14-7 shows the binary diagram of silica and alumina that shows mullite as the only stable secondary solid solution phase or a nonstoichiometric compound between silica and alumina. In spite of this, mullite is usually represented by $3\,Al_2O_3 \cdot 2\,SiO_2$ that indicates a definite stoichiometric composition; however, it is not. Mullite is a rare naturally occurring mineral and its most important place of occurrence is in the island of Mull, in the UK. It is one of the most widely used refractory materials because it has a low thermal coefficient of expansion that results in better thermal shock resistance than pure alumina. Figure 14-7 shows that the mullite increases its melting point as the alumina content in solid solution increases.

Figure 14-8 is the partial binary diagram of zirconia and calcium oxide. The M, T, and C in this diagram stand for the allotropic crystal structures of zirconia (i.e., monoclinic, tetragonal, and cubic). When CaO is alloyed with ZrO_2, partial solid solutions of CaO in ZrO_2 indicated by the subscripts "ss" of M, T, and C are observed. This diagram shows also the formation of stoichiometric compounds, the presence of eutectoid reactions, and the partial solubility of CaO in M_{ss} increasing as the temperature increases.

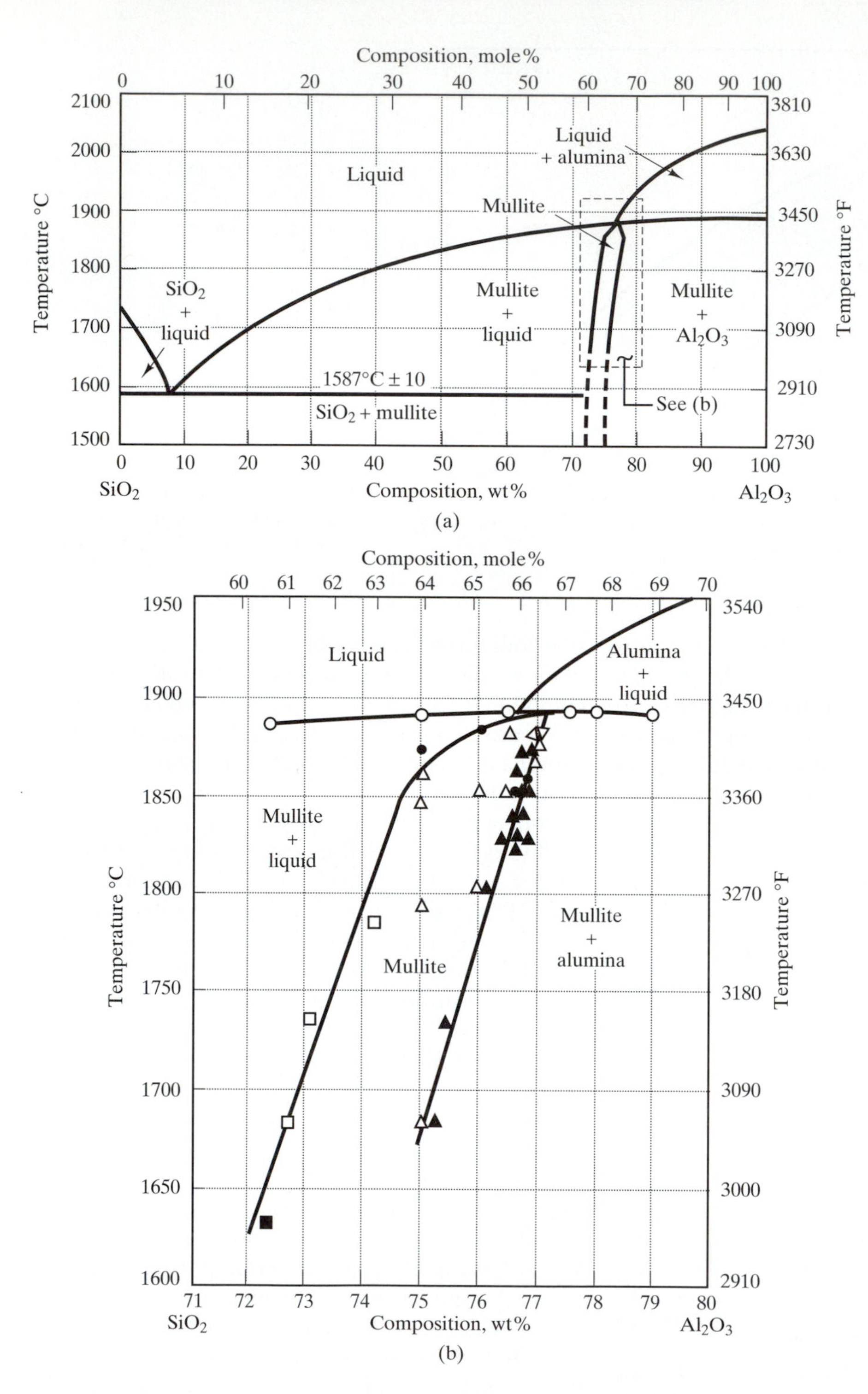

Fig. 14-7 (a) The silica-alumina binary diagram showing the mullite solid solution. (b) Expanded view of dashed rectangular area in Fig. 14-7a showing varying melting point of mullite as alumina increases.

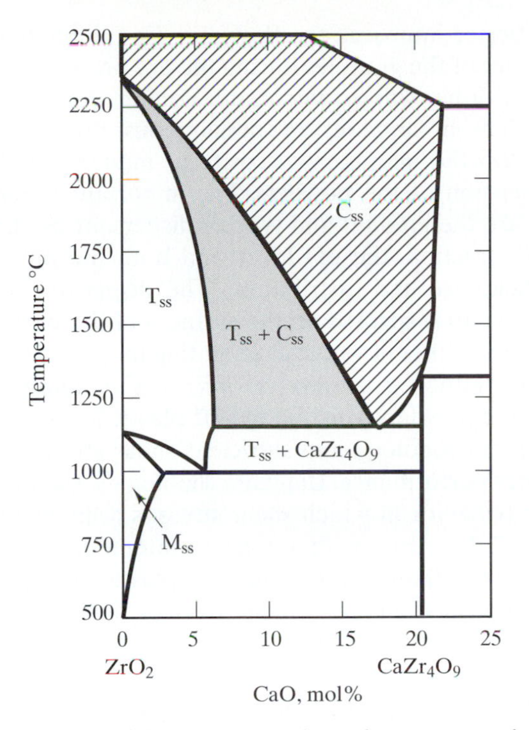

Fig. 14-8 Part of the ZrO_2-CaO phase diagram. C_{ss} refers to the cubid solid solution phase, T_{ss} to the tetragonal solid solution phase, and M_{ss} to the monoclinic solid solution phase.

EXAMPLE 14-1

In the binary diagram of SiO_2-Na_2O, calculate the weight percent of Na_2O in the stoichiometric compound $Na_2O \cdot 2\ SiO_2$.

SOLUTION: The stoichiometric compound contains one mole of Na_2O and two moles of SiO_2. In terms of the stoichiometric compound notation given in Sec. 5.9, A_xB_y, $A = Na_2O$ and $B = SiO_2$; $x = 1$ and $y = 2$. We calculate the weight percent in the same manner as illustrated in Example 5-4.

$$\text{Weight of } Na_2O = 1 \times \text{MW of } Na_2O = 2 \times 23.2 + (1 \times 16) = 62.4$$

$$\text{Weight of } SiO_2 = 2 \times \text{MW of } SiO_2 = 2[1 \times 28.1 + (2 \times 16)] = 120.2$$

$$\text{Total formula weight of } Na_2O \cdot 2\ SiO_2 = 182.6$$

$$\text{wt\% } Na_2O = 62.4/182.6 \times 100 = 34.2\%$$
$$\text{wt\% } SiO_2 = 120.2/182.6 \times 100 = 65.8\%$$

EXAMPLE 14-2

The formula of mullite is given as a stoichmetric compound $3Al_2O_3 \cdot 2SiO_2$. (a) What is the mole percent of silica? (b) If a batch is prepared from pure oxides to produce mullite, what is the minimum amount of alumina needed per 1000 kg (long ton)?

SOLUTION

(a) The compound contains 3 moles of alumina and 2 moles of silica and the total moles is, therefore, 5 in the compound. The mole percent of silica is $2/5 \times 100 = 40$ mol%.

(b)

$$\text{total weight of } Al_2O_3 = 3 \times \text{MW of } Al_2O_3 = 3[(2 \times 27) + (3 \times 16)] = 306$$

$$\text{total weight of } SiO_2 = 2 \times \text{MW of } SiO_2 = 2\,[28.1 + (2 \times 16)] = 120.2$$

$$\text{Total formula weight in } 3Al_2O_3 \cdot 2SiO_2 = 426.2$$

$$\text{Minimum weight of } Al_2O_3 \text{ per 1000 kg of batch} = 306/426.2 \times 1000 = 718 \text{ kg.}$$

14.2.3 Processing of Ceramics

We see from the phase diagrams that the melting points of the constituent oxides of ceramics are rather high. The oxides are never molten and cast as we do for metals. To form ceramic bodies or shapes, we induce particles of the oxides to bind together. The process is called sintering or powder metallurgy. Thus, in order to obtain the right alloy or chemistry of the ceramics, we have to mix powders from the different raw materials, according to the phase diagram, if it is available. For example, if we want to produce mullite from alumina and silica, we have to mix approximately three molecular weights of alumina with two molecular weights of silica.

In very general terms, the manufacture of ceramic bodies or shapes consists of the following steps:

14.2.3a Batching and Preparation of Powders. Raw materials for ceramic manufacture must be in powder form in order to be prepared for forming and mixing. For traditional ceramics, the raw materials are often beneficiated from minerals to remove undesired impurties and then pulverized and sized by passing through sieves or screens. The most common raw materials are *silica, clay, fluxes, and refractory* *materials.* Silica is obtained from quartz rock deposits or from pure quartz sands. The two most common clay materials are kaolin ($Al_2O_3 \cdot 2SiO_2 \cdot 2H_2O$) and talc ($3MgO \cdot 4SiO_2 \cdot H_2O$). Clays are used in traditional ceramic compositions because they enable the ceramic to be plastic when mixed with water. Fluxes are minerals, such as feldspar, that contain alkali oxides to promote the fusion of silica and alumina particles during firing. As seen in Fig. 14-6, the alkali oxides induce the formation of a liquid or a glassy phase on the surfaces of the particles that weld them together on cooling. The formation of this liquid or glassy phase allows the processing of ceramics at practical manageable temperatures. Refractory materials consist of oxides, carbides, nitrides, and other materials that withstand high temperatures.

For advanced ceramics, the raw materials are usually chemically prepared powders of high purity. Examples of powders for advanced ceramics are aluminum oxide, zirconium oxide, silicon carbide, silicon nitride, and aluminum nitride.

For proper densification during firing, a mix of particle sizes will induce densification with less void fraction for a shorter firing (sintering) time. Fine, single-sized particles of about 100 nm (4 μ-in.) result in a highly reactive ordered dense packing that promotes uniformly fine-grained ceramic bodies with little tendency for exaggerated grain growth. However, the packing still leaves about 30 percent interstitial void volume (compare the volume packing fraction of FCC and HCP found in Chap. 2) that requires a great amount of sintering and shrinkage to achieve full densification. Alternatively, a distribution of particle sizes from approximately 20 μm (800 μ-in.) to less than 50 nm (2 μ-in.) allows highly efficient packing that leaves an interstitial pore volume of less than 5 percent. This mix requires less firing (sintering) time for densification, but the larger particles tend to grow excessively, resulting in a nonideal microstructure. Thus, an intermediate size between the two particle sizes will likely produce the optimum result.

Before the powders of raw materials can be shaped or formed, they are prepared into states that are compatible with the forming process. For dry pressing, the powders undergo spray drying. The process of *spray drying* consists of atomizing a water-based slurry (a liquid suspension) containing binders and plasticizers into a drying chamber with heated air. The spray-dried particles are granulated and the plasticizer improves the flow and die-filling characteristics of the granules during the dry pressing.

For either *slip casting* or *tape casting,* the slurry of powders in aqueous or nonaqueous media must develop the necessary viscosity to induce a colloid suspension of the powders. The important parameters are the amount and type of dispersant used, the solids fraction, the pH level, and both the powder-particle size and distribution. The dispersant is an additive that adsorbs on the surfaces of the particles to prevent them from agglomerating in the colloid.

For either *extrusion* or *injection molding,* the ceramic powders must be in stiff plastic masses with Bingham rheology and sufficient shear strength for proper performance. Bingham rheology is the plastic flow behavior in which shear stress is a linear function of shear rate with a nonzero intercept, or shear strength. Mixing of batches for plastic forming is accomplished in high shear devices such as "pug mills" or sigma blade mixers.

14.2.3b Forming. The principal forming processes for ceramic materials are those listed in Table 14-2 that indicates the preferred applicabilities of the processes to produce different sizes and shapes of ceramic bodies. As indicated above, *slip and tape castings* use powders that are in a slurry or colloidal suspension and are therefore called *slurry casting processes. Extrusion and injection molding* are called *plastic processes* because the powders are in stiff plastic masses. *Dry pressing* is the pressing of the spray-dried granulated powders.

The *slip casting* process is illustrated in Fig. 14-9 for drain casting and solid casting. The process involves pouring a low-viscosity slurry into a porous plaster mold that draws the water from the slurry that is in contact with the wall of the plaster mold. In drain casting, the excess slurry is drained out after a layer (from a few mm to more than 10 mm or 0.4 in. for heavy sanitary ware) of relatively dried solids builds up on the mold wall. The ceramic shell is allowed to dry some more in the mold to develop greater strength before the mold is removed. Edges and mold seams on the casting are trimmed and smoothed with a knife and a wet sponge after the mold is removed. In solid casting, the process is similar to casting of metals in that the entire slurry filling the section of the mold is allowed to dry in place. To fill up the shrinkage of the slurry in the solid section,

Table 14-2 Applicabilities of principal forming processes of ceramic materials.

Forming process	Component size	Component shape	Production volume
Slip casting	Large	Complex	Low
Tape casting	Thin sheets	Simple	High
Extrusion	Wide range	Constant cross section	High
Injection molding	Small	Complex	High
Dry pressing	Small to medium	Simple, low aspect ratio	High

Note: Labels in table reflect the most preferred condition for the stated process. In practice most forming processes are used over a range of conditions.

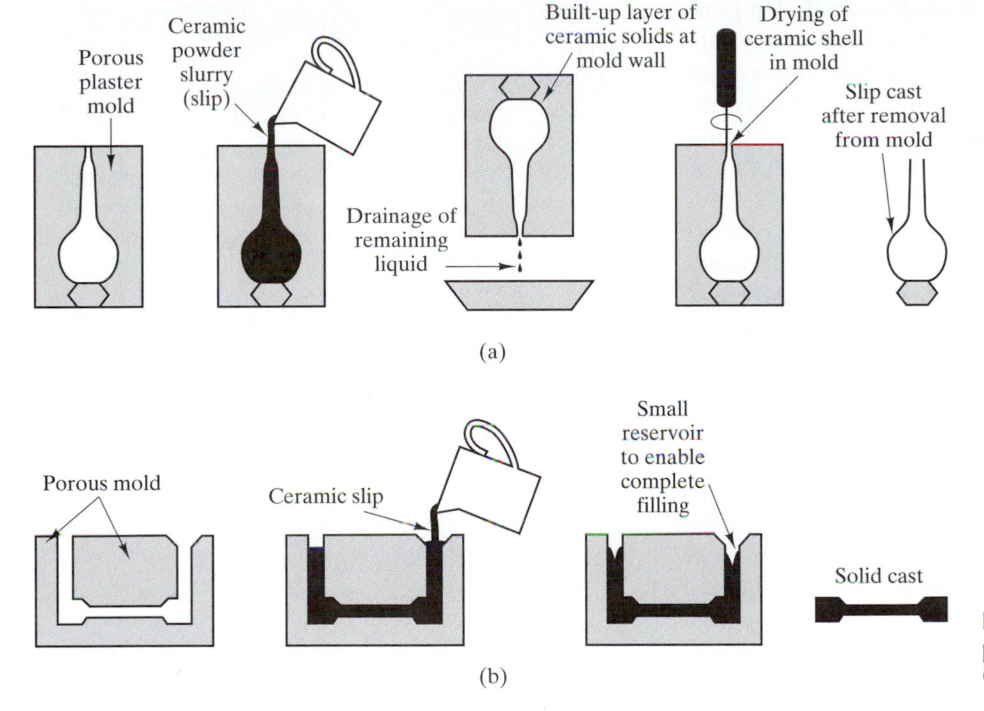

Fig. 14-9 Ceramic slip casting process (a) drain casting, and (b) solid casting.

there is also a reservoir of slurry in a "riser-like" cavity that fills the shrinkage volume.

In *tape casting,* ceramic slurries containing substantial amounts of binder and plasticizer are poured into thin layers on either a glass plate or an impervious polymer film and allowed to dry. This process produces thin flexible ceramic tapes. Tape casting is widely used for the production of ceramic substrates in the electronic industry.

Plastic processes have traditionally been the method to produce bodies of clay-based ceramics due to their highly plastic nature. These methods are also applied to advanced ceramic bodies (aluminum oxide and silicon nitride compositions) by the addition of organic plasticizers to impart the required plasticity. The principal processes are extrusion, injection molding, plastic pressing, and jiggering.

In *extrusion,* the stiff plastic ceramic mass is forced through a rigid die to produce a column of uniform section. The resulting shape can be cut in appropriate lengths as the final product, or as the blank or slug for other forming operations. The extrusion process can be either a batch or a continuous process depending on the equipment available. The ram or piston extruder is a batch process that can use high pressures from hydraulic pumps to push a single batch of stiff plastic ceramic mass through the die. Continuous extrusion is used for plastic masses requiring less pressure than the piston extruder and can be accomplished with the pug mill and the screw-fed plasticating extruders, both of which involve auger extrusion. The auger extruder consists of a cylinder, a feed screw, and a die. The pug mill can be thought of as a mixer and a non-heated extruder in one unit; a schematic of it is shown in Fig. 14-10. The screw-fed plasticating extruders are based on the equipment used to extrude polymers (plastics) that generally have externally heated cylinders, and dies that can be heated and/or cooled, shown in Fig. 14-11.

Injection molding is the process of heating a granular ceramic-binder mix until it becomes plastic (soft), and then forcing the plastic mix into a mold cavity, where it cools and resolidifies to produce the desired shape of the ceramic body. Figure 14-12 illustrates the screw-type and plunger-type machines that are very similar to those used for polymers.

Plastic pressing and *jiggering* are batch processes. In plastic pressing, a preform or extruded slug is pressed under relatively low pressure to conform to the shape of a die. The die material may be an impervious metal or a porous plaster mold. In jiggering, an extruded slug is placed on a revolving plaster form and a template tool is brought in contact with the slug, which is also pressed onto the plaster form while the template tool cuts away excess body. This leaves a shape with one surface conforming to the shape of the plaster mold, and the other surface, conforming to the shape of the template tool. Jiggered shapes must have circular cross sections—good examples of these products are dinner plates and coffee mugs.

Dry-pressing is the process whereby nearly dry, free-flowing powders are consolidated into a predetermined shape of a metal die under uniaxial high pressure. The process is sometimes called dust-pressing, die-pressing, or uniaxial compaction. The compaction force is generally applied in the vertical direction by opposite-acting mechanical or hydraulic rams. Components produced by dry-pressing include such diverse products as cutting tools, grinding wheels, refractories, insulators, seal rings, and nozzles.

14.2.3c Drying and Firing. For structural ceramics and refractories, the "green" ceramic form or compacted powder is heated to dry the ceramic, to burn the added binders and plasticizers, to effect changes in the structure of the minerals, to vitrify or bond the particles, and to densify the body. As an example of some of the effects heating can induce, Table 14-3 shows the changes in clay-based ceramic bodies when

Table 14-3 Reactions that occur during the firing of clay-base ceramic bodies.

Temperature		
°C	°F	Reaction
≤100	≤212	Loss of free moisture
100–200	212–390	Loss of adsorbed water
200–450	390–840	Crystal structure of clay minerals altered by removal of OH groups. Pyrophyllite shows a marked expansion.
400–700	750–1290	Organic matter in the form of lignite is oxidized. Pyrophyllite expands further.
573	1065	Quartz inverts to high-temperature polymorph.
700–950	1290–1740	Pryophyllite reaches maximum expansion: metakaolin converts to spinel in clay.
950–1100	1740–2010	Mica structure is destroyed. Talc decomposes to protoenstatite and glass. Mullite forms from spinel. Pyrophyllite converts to mullite and glass.
1100–1200	2010–2190	Feldspars melt; clay and cristobalite dissolve. Vitrification begins, porosity decreases, shrinkage increases.
>1200	>2190	Protoenstatite from talc converts to clinoenstatite. Mica breaks down into alumina and glass. Glass content increases, mullite needles grow; only closed porosity remain.

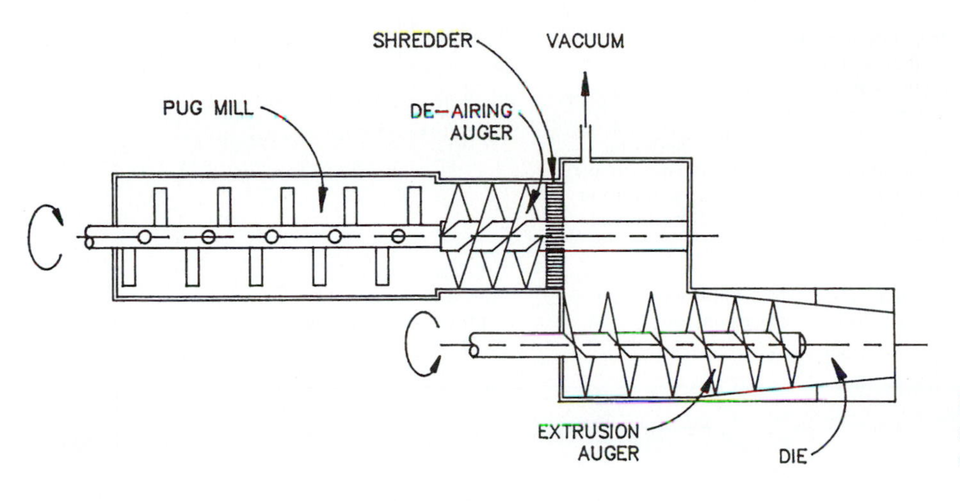

Fig. 14-10 Pug mill with deairing chamber and extrusion auger.

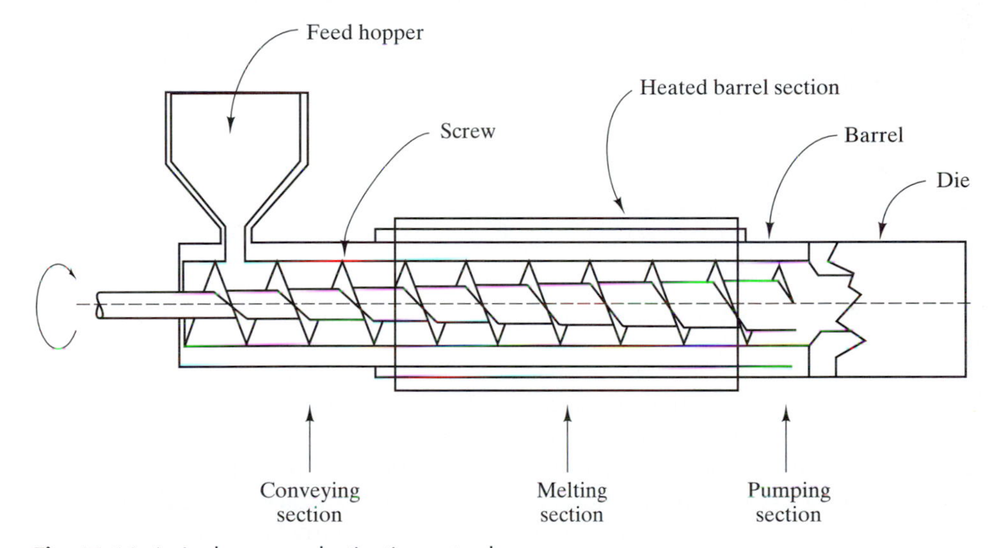

Fig. 14-11 A single screw plasticating extruder.

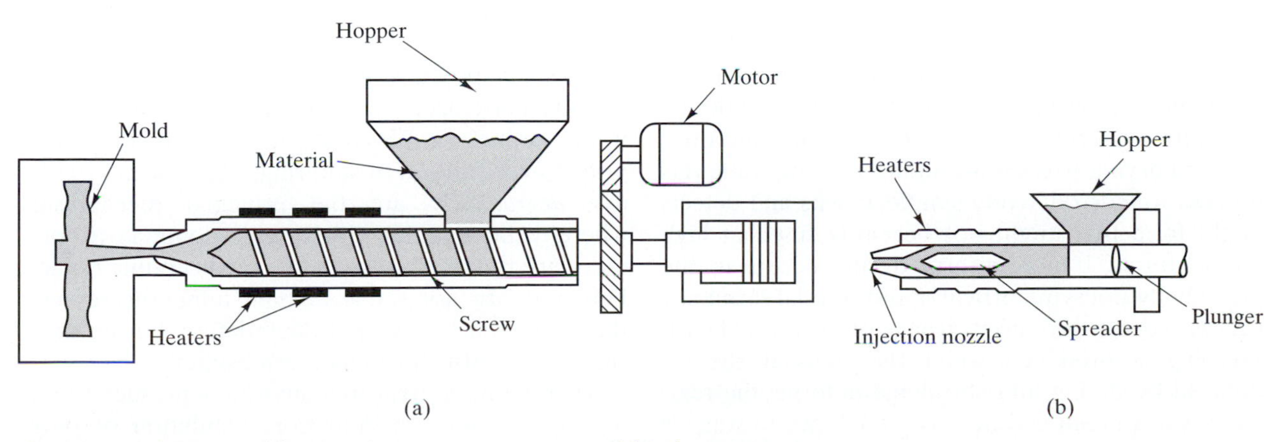

Fig. 14-12 Injection molding machines (a) screw type, and (b) plunger type.

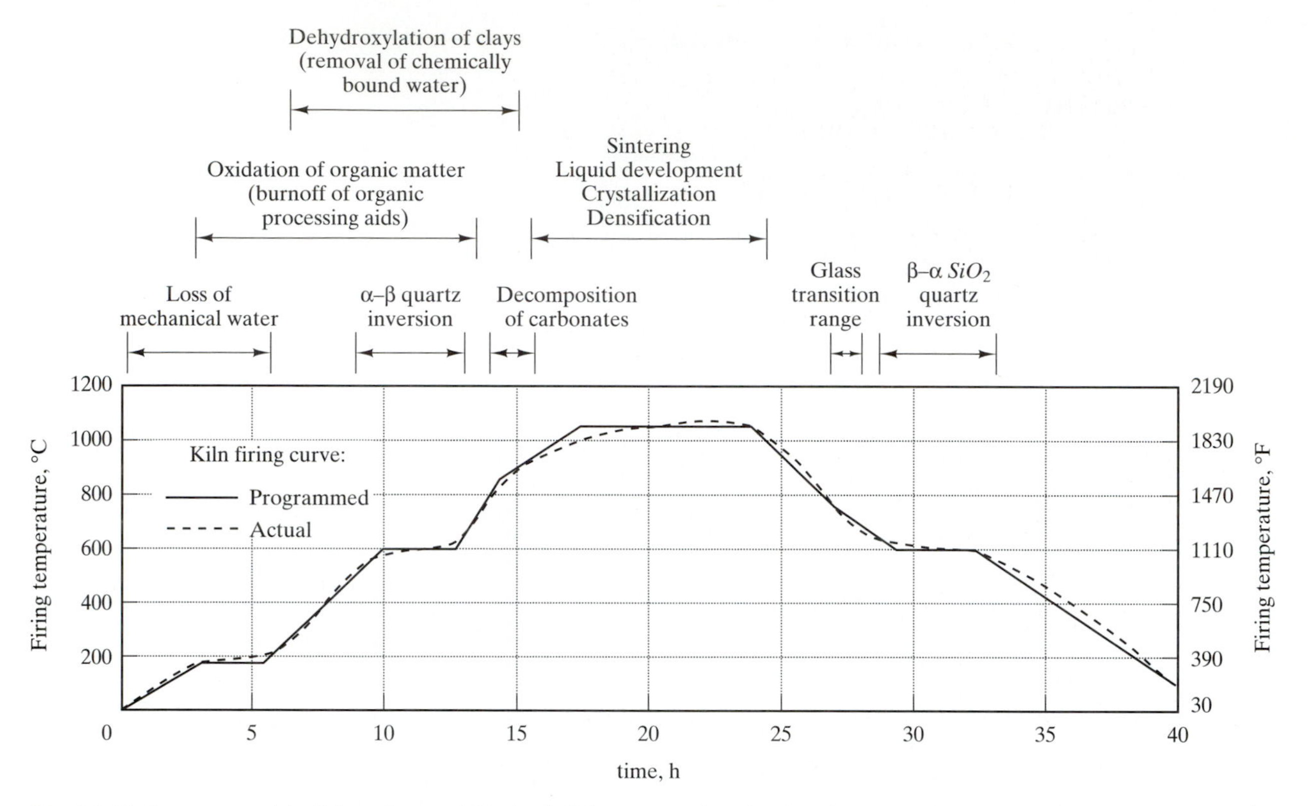

Fig. 14-13 Programmed (solid) and actual (dashed) firing curves for clay-bearing ceramic compositions. Reactions at the ranges of temperatures are indicated. [1]

they are heated. The process of heating the green compact is commonly called firing and it is usually programmed to achieve the effects desired. An example of a programmed firing cycle with the superimposed actual firing curve for a clay-based green body is shown in Fig. 14-13. This figure also shows the effects of heating that we shall now discuss.

Basically, *the drying* (water loss) and *the burning* of organic binders and plasticizers occur at relatively low temperatures up to 400° to 700°C. During drying and burning, shrinkage of the body occurs and thermal and drying stresses are induced. Thus, while this process sounds relatively simple, cracks and defects might form that might not be easily fused at high temperatures. If nonuniform drying occurs in the green body, nonsymmetrical or differential shrinkage occurs and may produce warpage of the formed part. Warping is produced when the thermal stresses induced by drying and shrinkage deforms the relatively weak green ceramic. The tendency to warp is reduced by increasing the uniformity of drying and by reducing the average drying shrinkage of the body. On the other hand, the tendency to warp is increased by nonuniform external films or coatings, particle orientation, or binder migration, which produce a nonuniform surface permeability, and by nonuniformities in the circulation or temperature of the drying air due to setting patterns and supporting hardware. Ideally, drying should occur symmetrically in an isotropic material.

Organic processing aids such as dispersants, binders, plasticizers, and lubricants that were added in the forming process must be removed completely prior to vitrification or sintering. *Thermal and oxidative degradation* are the principal mechanisms whereby the organics, which are predominantly polymers, are removed from the formed ceramic bodies. Polymers are not stable at high temperatures and their removal by thermal degradation can be done simply by heating in an inert atmosphere.

Heating in an oxidizing environment, such as air, induces oxidative degradation. Oxidation of polymers can occur even at room temperature for the unstabilized types and is very pronounced at high

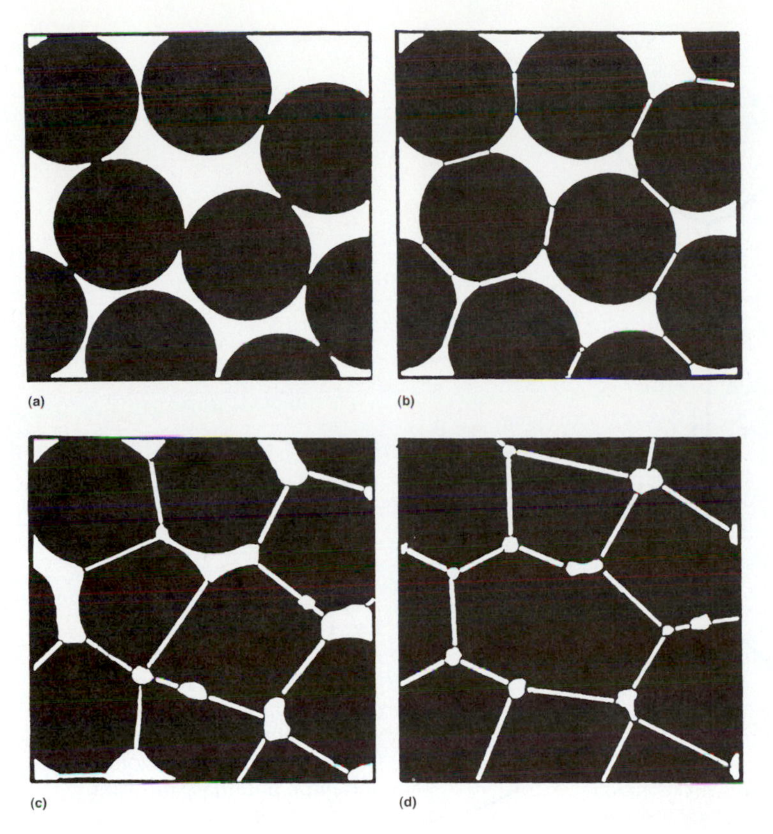

Fig. 14-14 Stages in the vitrification and densification of a ceramic body during sintering; (a) loose powder ("green" compact), (b) initial bonding stage, (c) intermediate stage—grain boundaries form, (d) final stage—densification and elimination of pores along grain boundaries.

temperatures. Oxidation of low molecular weight polyethylene in air is not completed until 500°C and generates more diverse and higher molecular weight species than does thermal degradation. Whereas thermal degradation yields mainly alkanes and alkenes, oxidative degradation forms aldehydes and ketones.

The last step in the firing of ceramic bodies is *sintering,* the process of bonding or vitrifying the particles and densifying the ceramic body. This may occur with or without the formation of a liquid phase at the particle interfaces. Without the liquid, it is called *solid-state sintering* and with the liquid phase, it is called *liquid-phase sintering.* The development of the interparticle bond in solid-state sintering is shown schematically in Fig. 14-14. Initially the powders are loose and some have contact with other particles to start the bond as seen in Fig. 14-14a. This initial bond forms the "neck" between the particles, as shown in an actual scanning electron micrograph in Fig. 14-15. The increase in the neck areas eventually forms the grain boundaries. The process of densification of the

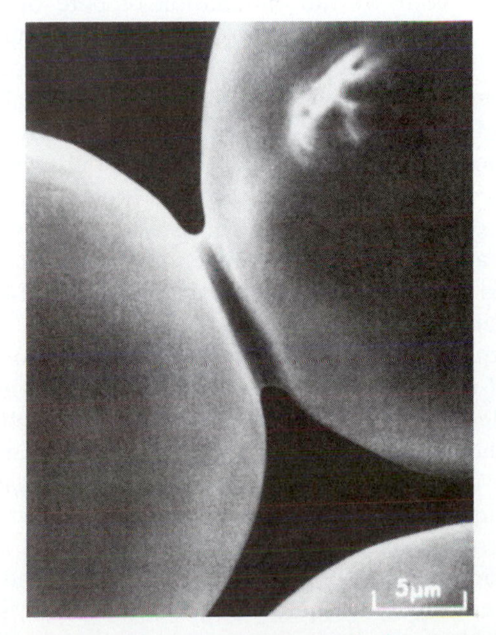

Fig. 14-15 Scanning electron micrograph showing the neck formation between particles at the initial stage of sintering.

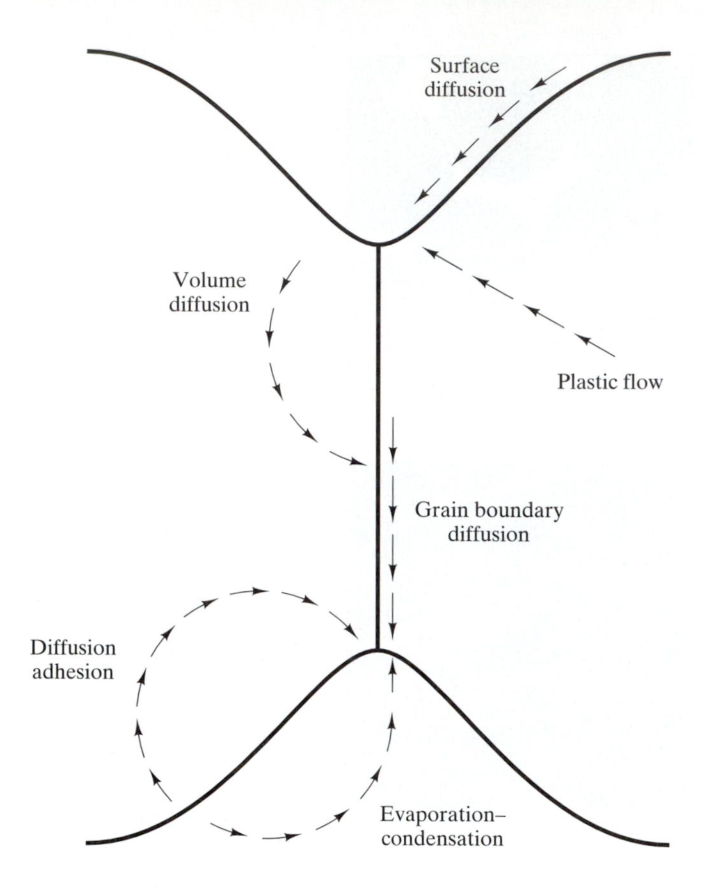

Fig. 14-16 Mass transport mechanism for the growth of the necked region include evaporation-condensation and surface diffusion, and the bulk transport mechanisms volume and grain boundary diffusion and plastic flow.

ceramic body by the elimination of pore or open spaces occur essentially through diffusion (surface, grain, and volume), plastic flow, and evaporation-condensation mechanisms, shown schematically in Fig. 14-16. Figures 14-14c and d schematically show the most efficient way of densification during sintering by keeping the pores confined around the neck areas or grain boundaries. This is also schematically illustrated in Fig. 14-17a. When the pores or open spaces are within the grains, as shown in Fig. 14-17b, full densification cannot be achieved.

Liquid-phase sintering involves the formation of a liquid film that is a result of some additives in the powder mixture that produce low-melting phases. For clay-based minerals, it is due to the feldspar that fluxes some of the solid particles as discussed earlier. The liquid film wets the particles' surfaces, dissolves some of the solid at the surfaces, and provides a fast diffusion medium for the dissolved atoms. The surface tension of the liquid film acts like a low-pressure external stress that induces liquid flow and the rearrangement of the particles that contribute to the rapid densification of the body. In general, liquid-phase sintering results in a more dense (less porosity) ceramic body that exhibits better properties. The resulting sintered microstructure consists of the solid particles (grains) in contact with an interspaced (solidified) liquid matrix that filled and replaced the pores of the green compact. An example of a microstructure as a result of liquid-phase sintering is shown in Fig. 14-18. Most of the advanced ceramics are produced by liquid-phase sintering.

14.2.3d Shaping and Surface Finishing. The fired and sintered ceramic bodies or parts are finished before their actual use to conform to the shape, size, finish or surface, and quality requirements. The extent of finishing depends upon the application. For example, bricks are finished when they are brushed off as they are removed from the kiln (furnace), while ceramic turbine blades require extensive shaping that may cost as much or more than all of the previous

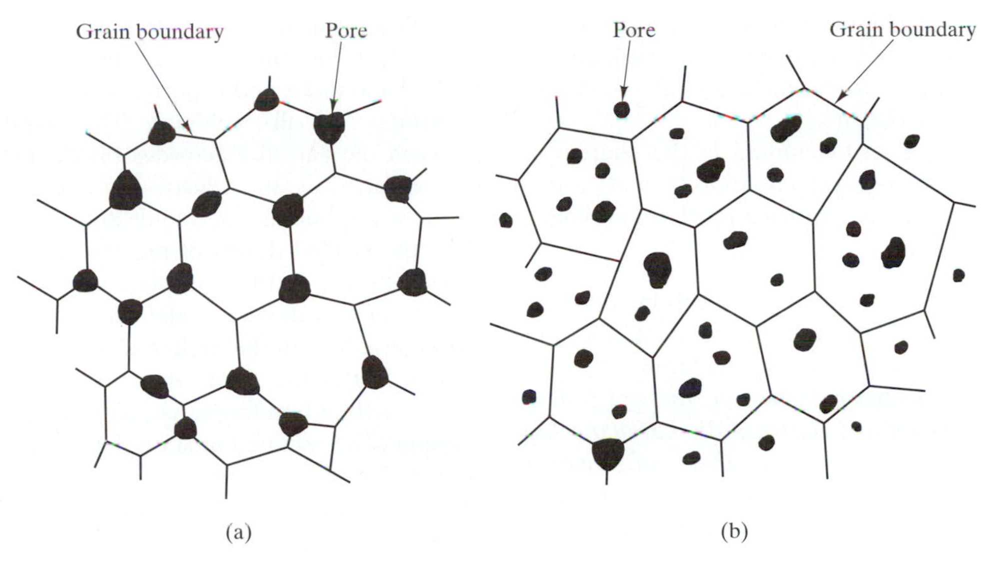

Fig. 14-17 Densification of ceramic occurs when the pores remained along grain boundaries in (a), but does not occur when the pores are within the grains in (b).

Fig. 14-18 Scanning electron micrograph of liquid-phase sintered WC-Co cermet showing angular carbides embedded in solidified liquid cobalt (light areas). [1]

processing steps combined. Most ceramics fall in between these extremes.

Machining is the most common method of shaping. Machining processes utilizing abrasives are called *abrasive machining methods;* those not using abrasives are called *nonabrasive machining methods.* Abrasive machining and surface finishing methods include grinding, lapping, honing, polishing, abrasive fluid jet cutting, and ultrasonic machining. Grinding methods are the most versatile and they generally use diamond abrasives held in a grinding wheel and applied against the work surface in a variety of configurations. Free-abrasive machining is the term used when the machining is accomplished by applying the abrasives to the surface of the workpiece without being fixed in a grinding wheel. Some of these methods are:

- Lapping is a low-pressure, low-speed operation to achieve high geometric accuracy, to correct minor shape errors, to improve surface finish, or to provide tight fits between mating surfaces.
- Honing resembles lapping in principle, but is usually reserved for finishing internal surfaces of cylindrical or spherical parts.
- Polishing is used primarily to improve surface finish and employs fine, loose abrasives with preselected hardness.

- Ultrasonic machining selectively erodes a part by impinging abrasive particles suspended in a liquid medium onto the surface of the workpiece at ultrasonic frequencies.
- Abrasive fluid jet-machining is the shaping of the part with a high-velocity fluid (usually water) stream carrying the abrasive particles impinging onto the surface.

The nonabrasive machining methods include laser beam machining, electrical discharge machining, ion-beam machining, electron-beam machining, chemical machining, and electrochemical machining. Of these methods, *laser beam and electrical discharge machining* are the most used. *Laser beam machining* involves the evaporation of material caused by the focused, intensed laser beam on the surface of the workpiece. The pinpoint accuracy of the focused beam and the speed with which it can be pulsed electronically offers dimensional control and high-volume production to meet demands at affordable costs. As the name implies, *electrical discharge machining* involves electrical discharges and requires three necessary components—that is, two electrodes and a dielectric medium flowing between the two electrodes. One of the electrodes is the workpiece and the other is a die-electrode or wire that makes the machine either die-sinking or wire-cutting. A schematic of a die-sinking machine is shown in Fig. 14-19 and the shaping process involves the melting and evaporation of materials. When the die-electrode approaches the workpiece electrode, an electrical discharge (plasma) is formed and melts and vaporizes the materials of both electrodes (recall welding). The vapor pressure between the gap of electrodes breaks the liquids on the surfaces of the electrodes and the dielectric medium resolidifies the liquids and the gases to produce the eroded debris of materials. In the end, the eroded surface of the workpiece develops the female surface of the die-electrode; the die sinks into the workpiece. Not all the molten electrode materials are removed after the spark due to surface tension and the strength of bonding forces between the liquid and the solid. The retained melt on the surface resolidifies on the electrodes as a recast layer. In order to minimize the recast layer, a pulse-type power supply, rather than a capacitance power supply, is used.

14.2.4 Properties of Structural Ceramics

From the preceding discussion on processing, it must be obvious that ceramic materials contain a certain amount of porosity or voids. The amount, the size, the distribution, and the continuity of these voids affect the mechanical and chemical performance of the ceramic part or component. We shall discuss these properties in relation to structural ceramics and refractories that are the materials of engineering and industrial significance.

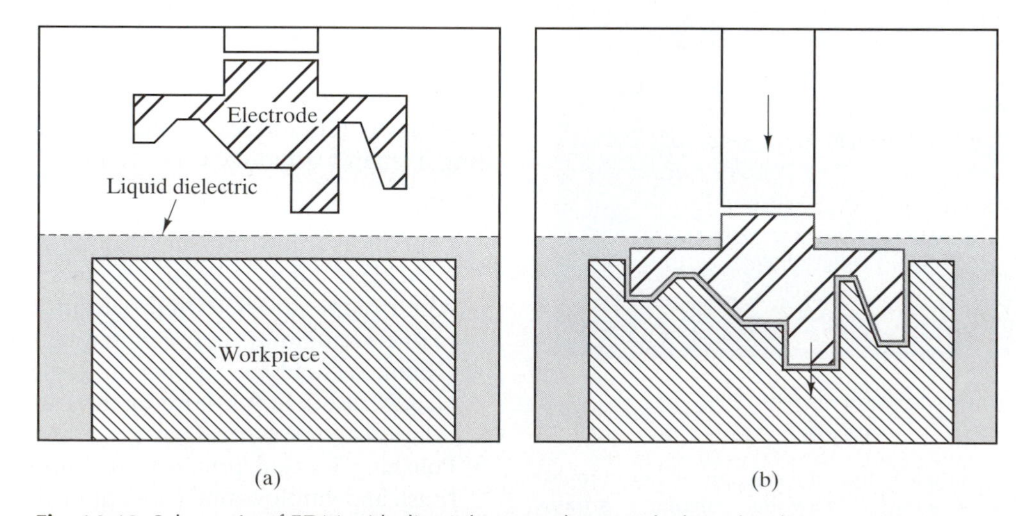

Fig. 14-19 Schematic of EDM with die-sinking machine, (a) before, (b) after.

14.2.4a Porosity or Void Content. The void content may be expressed as *apparent porosity* and as *total porosity*. The apparent porosity is the percent of the total volume that is open pore space; it is a measure of the surface area available for reaction with gases and molten liquid. It is determined by first weighing the ceramic dry (W_d), then reweighing it when it is suspended in water (W_s), and after it is taken out of the water when the pores are filled with water, (W_w). Using units of gram and cm^3, the apparent porosity is

$$\text{Apparent Porosity} = \frac{W_w - W_d}{W_w - W_s} \times 100 \quad (14\text{-}1)$$

The *total porosity* is the percent of the total volume containing voids, in both open and closed pores. Depending on the material, method of manufacture, and the degree of burn (firing), the total porosity may be just slightly more than the apparent porosity or even twice as much. It is given by

$$\text{Total Porosity} = \frac{(\rho_T - \rho_B)}{\rho_T} \times 100 \quad (14\text{-}2)$$

where ρ_T is the true density of the ceramic, and ρ_B is the bulk density, which is given by

$$\rho_B = \frac{W_d}{W_w - W_s} \quad (14\text{-}3)$$

The bulk density is the dry weight of the ceramic divided by its volume.

In general, the porosity is detrimental to the mechanical and chemical properties of ceramics and it needs to be minimized. However, when used as insulators, porosity increases the insulating capacity of ceramics and makes the ceramics much lighter. Both of these characteristics are very desirable for aerospace structures, and porous silica bricks are adhesively bonded to NASA's shuttle as the thermal protection structure (TPS) for the main aluminum frame. The increased porosity in the ceramic may be accomplished by either (a) mixing a bulky combustible substance, like sawdust or ground cork, or a volatile solid, such as naphthalene, with the wet mix; or (b) forcing air into the wet plastic mass; or (c) mixing reagents (into the batch) that will react chemically to form a gas and a product not injurious to the brick.

EXAMPLE 14-3

A sintered alumina component weighed 325 grams dry, 235 grams when suspended in water, and 335 grams when wet. The true density of alumina is 3.97 gram per cm^3. Calculate the bulk density, the total porosity, and fraction of the porosity that is closed?

SOLUTION: We need to use Eqs. 14-1, 14-2, and 14-3. First we note that

$$W_d = 325 \text{ g} \qquad W_s = 235 \text{ g} \qquad W_w = 335 \text{ g}.$$

$$\rho_B = \frac{W_d}{W_w - W_s} = \frac{325}{335 - 235} = 3.25 \text{ g/cm}^3$$

$$\text{Total Porosity} = \frac{\rho_T - \rho_B}{\rho_T} \times 100 = \frac{3.97 - 3.25}{3.97} \times 100 = 18.1 \text{ percent}$$

$$\text{Apparent porosity} = \frac{W_w - W_d}{W_w - W_s} \times 100 = \frac{335 - 325}{335 - 235} \times 100 = 10 \text{ percent}$$

$$\text{Fraction of closed porosity} = \frac{18.1 - 10}{18.1} = 0.45$$

14.2.4b Uniaxial Tensile Strength. Ceramic materials are inherently brittle because of their predominantly covalent and ionic bonding. As such, they do not exhibit much ductility and in a tensile test, a linear stress-strain curve is almost always observed up to fracture, as shown in Fig. 14-20, and the fracture stress is the same as the tensile strength. This linear elastic behavior is the region of applicability of fracture mechanics and the important property that determines the performance of brittle materials is their fracture toughness, K_{I_c}, which was shown in Chap. 4 to be

$$K_{I_c} = Y\sigma\sqrt{\pi a} \quad (14\text{-}4)$$

where Y is a geometric factor, σ is the fracture stress or tensile strength when there is a surface crack of size a or an interior crack of size $2a$. In Eq. 14-4, K_{I_c}

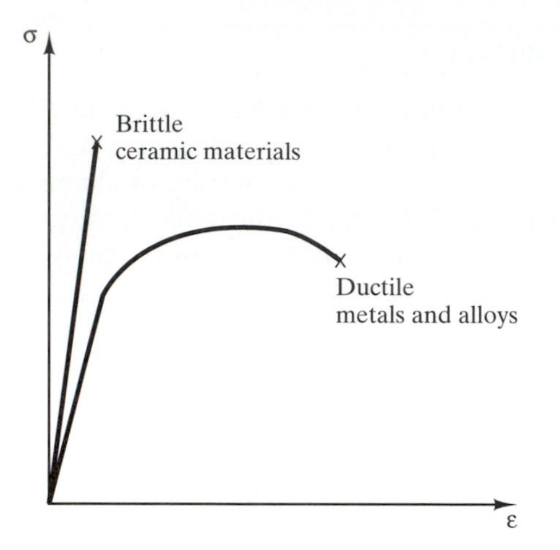

Fig. 14-20 Schematic of stress-strain curve of ceramic materials compared to ductile materials.

and *Y* are constants for a certain geometry of material, but σ will depend on the largest *a,* the crack size, that is oriented perpendicular to the applied stress, σ. Because of the different sizes and the distribution of porosities and cracks as a result of the processing and finishing of a ceramic material, there is an inherent statistical variation of fracture stress or tensile strength when a ceramic material is tested in tension. Thus, we do not have a single tensile strength or fracture stress, but there is a certain definable probability that a given sample will have a given strength.

The statistical variation of strength of brittle materials is explained by Weibull's weakest-link model of the failure of materials, which says that a structure is just as strong as its weakest link. Weibull's model assumes that a structure has a chain of *N* elements, and if any element fails, the structure fails also. If F_i is the probability of failure of the ith element, and the probability of failure of the whole structure is *F,* then the probability of the survival of the structure is:

$$1 - F = (1 - F_1)(1 - F_2)(1 - F_3) \cdots (1 - F_N) \quad (14\text{-}5)$$

Taking the natural log of each side of Eq. 14-5, we get:

$$\ln(1 - F) = \ln(1 - F_1) + \ln(1 - F_2) + \ln(1 - F_3) + \cdots + \ln(1 - F_N) = \sum_{i=1}^{N} \ln(1 - F_i) \quad (14\text{-}6)$$

For the logarithmic terms on the right side of Eq. 14-6, and for small values of F_i,

$$\ln(1 - F_i) \approx -F_i \quad (14\text{-}7)$$

Note that Eq. 14-7 is applicable for each of the ith element, but not for the whole structure, *F.* For the whole structure,

$$\ln(1 - F) = \sum_{i=1}^{N} \ln(1 - F_i) = \sum_{i=1}^{N} -F_i \quad (14\text{-}8)$$

Weibull assumed an arbitrary and simple function of F_i as,

$$F_i = \left(\frac{\sigma - \sigma_u}{\sigma_o}\right)^m \quad (14\text{-}9)$$

where σ is the maximum tensile stress, σ_u is a threshold stress (below which no failure will occur), σ_o is a characteristic strength, and m is the Weibull modulus. Equation 14-9 was chosen because it satisfies the boundary conditions at low and high stresses, but is not founded on any concept of flaw density or severity, or on any fracture mechanics criteria. Substituting Eq. 14-9 into Eq. 14-8, we get

$$\ln(1 - F) = \sum_{i=1}^{N} -F_i = \sum_{i=1}^{N} -\left(\frac{\sigma - \sigma_u}{\sigma_o}\right)^m \quad (14\text{-}10)$$

and in the limit as each element approaches an infinitesimal size, Eq. 14-10 becomes

$$\ln(1 - F) = \int_V -\left(\frac{\sigma - \sigma_u}{\sigma_o}\right)^m dV \quad (14\text{-}11)$$

and

$$F = 1 - \exp \int_V -\left(\frac{\sigma - \sigma_u}{\sigma_o}\right)^m dV \quad (14\text{-}12)$$

For the case of a test specimen experiencing uniform tensile stress across its section, *F* becomes

$$F = 1 - \exp\left[-V\left(\frac{\sigma - \sigma_u}{\sigma_o}\right)^m\right] \quad (14\text{-}13)$$

The threshold stress σ_u is usually set equal to zero, and the characteristic strength, σ_o, is a normalizing strength defined as the characteristic strength at which a unit volume of the material (i.e., $V_o = 1$), will fail in tension. Using these simplifications for a material in uniform tension,

$$F = 1 - \exp\left[-V\left(\frac{\sigma}{\sigma_o}\right)^m\right] \quad (14\text{-}14)$$

and the survival probability, P_s, is

$$P_s = 1 - F = \exp\left[-V\left(\frac{\sigma}{\sigma_o}\right)^m\right] \quad (14\text{-}15)$$

If we are testing identical samples, each of volume V in the gauge length, the survival probability, P_s, is defined as the fraction of the identical samples that will survive a loading to a tensile stress, σ. Equation 14-15 is written for this case as

$$P_s(V) = \exp\left[-\left(\frac{\sigma}{\sigma_o}\right)^m\right] \quad (14\text{-}16)$$

We can now see the validity of the assumed function, Eq. 14-9. When $\sigma = 0$, $P_s(V) = 1$, which means that all samples will obviously survive. As σ increases, an increasing number of samples will fail, and $P_s(V)$ decreases. As $\sigma \rightarrow \infty$, $P_s(V) \rightarrow 0$ (i.e., large stresses will virtually break all samples).

If we set $\sigma = \sigma_o$ in Eq. 14-16, we find that $P_s(V) = 1/e = 0.368$ (~ 0.37). Thus, σ_o is the tensile stress that allows about 37 percent of the sample to survive. The Weibull modulus, *m,* indicates the statistical scatter in the tensile strength of the ceramic material. A low value indicates more scatter in strength than a high value. Traditional ceramics, such as whitewares, bricks, pottery and cement have values from three to five. The more advanced ceramics, for instance, SiC and Si_3N_4 have values of about 10—for these the strength varies less. Metals have much less variation in strength, and for these, $m \approx 100$, which indicates that a material will have a well-defined single failure stress.

The constants σ_o and m for a ceramic material may be determined experimentally. Samples of the same size and shape are tested at varying stresses and the survival probability is determined at each stress level. That is, a batch of samples is tested at stress σ_1, and the fraction of unbroken samples is determined as $P_{S_1}(V)$. A second batch of samples is tested at σ_2, and the fraction of unbroken samples is determined as $P_{S_2}(V)$, and so forth. The constants may be determined in the following manner. If we take the natural logarithms in Eq. 14-16, we obtain

$$\ln\left[\frac{1}{P_s(V)}\right] = \left(\frac{\sigma}{\sigma_o}\right)^m \quad (14\text{-}17)$$

And if we again take logarithms, we obtain

$$\ln\left[\ln\left(\frac{1}{P_s(V)}\right)\right] = m \ln\left(\frac{\sigma}{\sigma_o}\right) \quad (14\text{-}18)$$

We see that Eq. 14-18 is in the form of an equation of a straight line, $y = mx$. Thus, if we plot $\ln(\sigma/\sigma_o)$ against $\ln\{\ln[1/P_s(V)]\}$, we should obtain a straight line with slope *m,* the Weibull's modulus. If we set $P_s = 0.368$, we obtain σ_o. An example of this plot is shown in Fig. 14-21 that shows σ_o as the stress for which $F = 0.632$ and $P_s = 0.368$.

The tensile test is, however, very difficult to perform for brittle materials. The bending or flexure test is the more common test to obtain uniaxial strengths, σ_o, and *m.* In bending, the elements of the material experience varying stress, from maximum tensile at the outer surface to zero at the neutral axis to compressive stress on the inside. To obtain the failure or

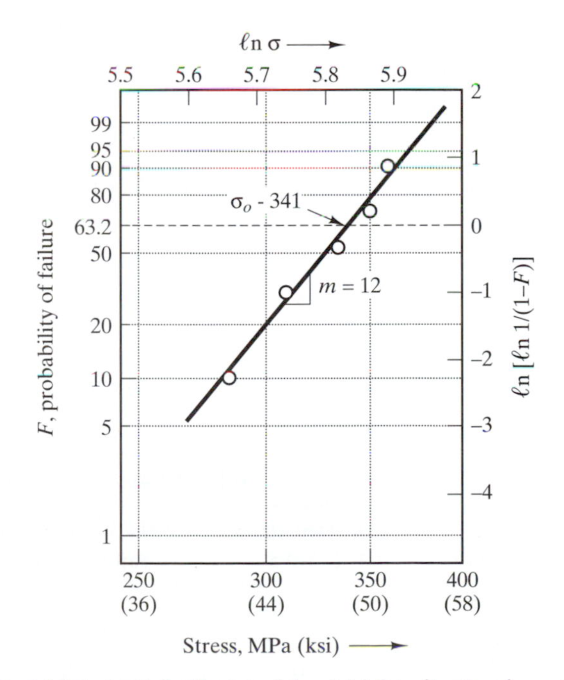

Fig. 14-21 A Weibull plot of Eq. 14-18 indicating how the modulus, m, is obtained and the characteristic strength, σ.

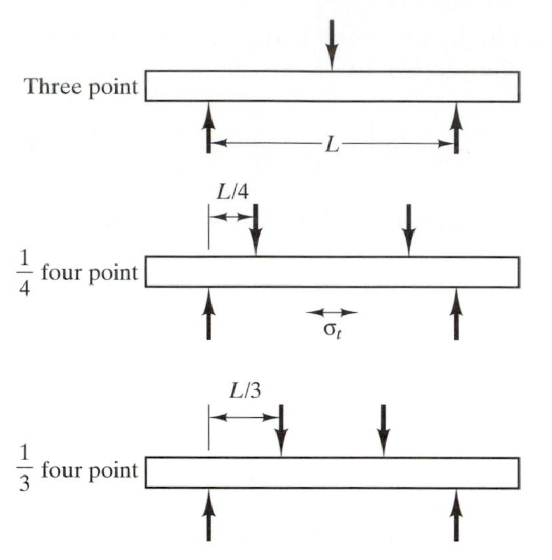

Fig. 14-22 Alternative flexure bend specimens to obtain tensile strengths.

survival probability, we need to perform the stress-volume integration in Eq. 14-12. To avoid this integration, an effective volume is used. In the tensile test, the effective volume of material subjected to a stress σ is the actual volume *V* within the gauge length. For the flexure tests shown in Fig. 14-22, the effective volume of a three-point loaded, rectangular specimen of total volume *V* within the support spans is

$$V_{E_3} = \frac{V}{2(m + 1)^2} \tag{14-19}$$

and that for the $\frac{1}{4}$ four-point loaded, rectangular specimen is,

$$V_{E_4} = \frac{V(m + 2)}{4(m + 1)^2} \tag{14-20}$$

Equations 14-19 and 14-20 indicate that if a bend specimen breaks with a maximum stress called *modulus of rupture (MOR)* at the outer fiber elements, the bend test is equivalent to a tensile test using a specimen with a volume of only V_E that breaks with the same stress as the MOR. We see that the effective volume is a function of the Weibull modulus *m*—for m = 10, $V_{E_3} = V/242$, and $V_{E_4} = (3/121)V$. Thus, we note that the four-point flexure test stresses six times the effective volume of the specimen in the three-point test. For this reason, the four-point test is the more common flexure test because the result comes from a greater volume of the material.

We can compare the relative strengths of two different sized specimens using Eq. 14-15. If we take the same survival probability of different sized specimens, the exponents are equated and the relative strengths are:

$$\frac{\sigma_1}{\sigma_2} = \left(\frac{V_{E_2}}{V_{E_1}}\right)^{1/m} \tag{14-21}$$

The relative strengths of two ceramic fibers of the same diameter, but different lengths, is thus

$$\frac{\sigma_1}{\sigma_2} = \left(\frac{V_2}{V_1}\right)^{1/m} = \left(\frac{L_2}{L_1}\right)^{1/m} \tag{14-22}$$

For example, if m = 10, and $L_2 = 4L_1$, $\sigma_1 = 1.15\ \sigma_2$, or, in other words, the shorter fiber would be 15 percent stronger on the average than the longer fiber.

EXAMPLE 14-4

Identical samples of a ceramic material were tested with the three-point bend test. Each test was conducted at constant stress using 10 samples. The constant stresses used were 50, 60, 70, and 80 MPa. The numbers of specimens that failed at each stress level were 1, 3, 6, and 9 out of 10 samples. Using the Weibull statistics, determine the Weibull modulus, and the characteristic strength.

Table EP14-4 Summary of 3-point bend test.

Test Number	Constant stress, σ, MPa	No. of Specimens Failed out of 10	$P_s = (1 - F)$, survival probability	$\ln[\ln(1/P_s)]$	ln σ
1	50	1	0.9	−2.250	3.912
2	60	3	0.7	−1.031	4.091
3	70	6	0.4	−0.087	4.248
4	80	9	0.1	+0.834	4.382

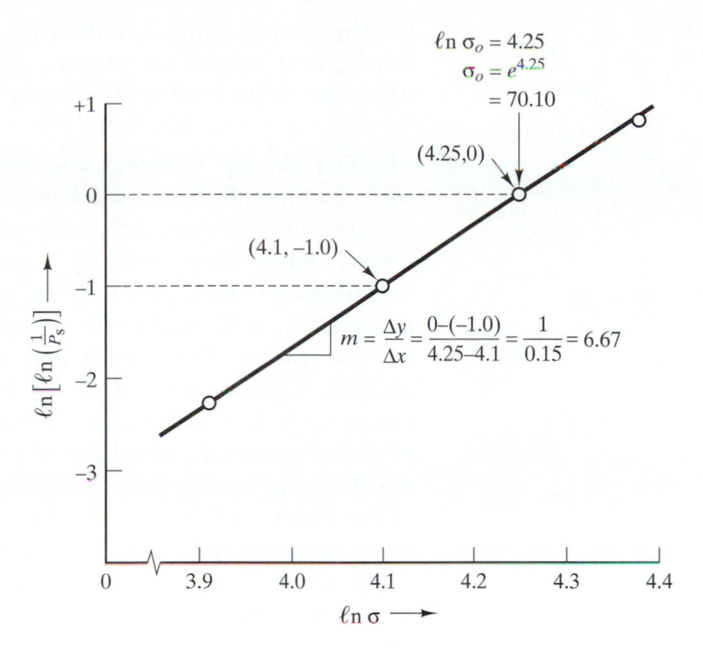

Fig. EP14-4

SOLUTION: Table EP14-4 summarizes the test results. The failure probabilities, Fs, are 0.10, 0.3, 0.6, and 0.9 for the different stress levels and the survival probabilities are $P_s = (1 - F)$. A plot of $x = \ln \sigma$, versus $y = \ln [\ln (1/P_s)]$ is shown in Fig. EP14-4, from which the Weibull modulus, m is found to be 6.67, and the characteristic strength, $\sigma_o = 70.1$ MPa at $y = 0$.

EXAMPLE 14-5

A ceramic material was tested with the 3-point bend test at constant stresses. The stresses were 50, 52, 54, 56, and 58 MPa and the number of specimens tested for each stress was 10. The number of specimens that broke for each of the stress levels is 1, 2, 5, 7, and 9. Determine the Weibull modulus, m, and the characteristic strength.

SOLUTION: Following Example 14-4, the survival probabilities are obtained, and a plot of $x = \ln \sigma$, versus $y = \ln [\ln (1/P_s)]$ is made. The summary of the results is shown in Table EP14-5, and the plot is shown in Fig. EP14-5. The Weibull modulus, $m = 21.3$, and the characteristic strength = 55.5 MPa are indicated in Fig. EP14-5. Comparing the results with those of Example 14-4, we see that the Weibull modulus is higher when the range of properties is narrower.

Table EP 14-5 Summary of test results for Example 14-5.

Test Number	Constant stress, σ, MPa	Failed Specimens out of 10	Survival Probability, $P_s = (1 - F)$	$\ln [\ln(1/P_s)]$	$\ln \sigma$
1	50	1	0.90	−2.250	3.912
2	52	2	0.80	−1.500	3.951
3	54	5	0.50	−0.366	3.989
4	56	7	0.30	0.186	4.025
5	58	9	0.10	0.834	4.060

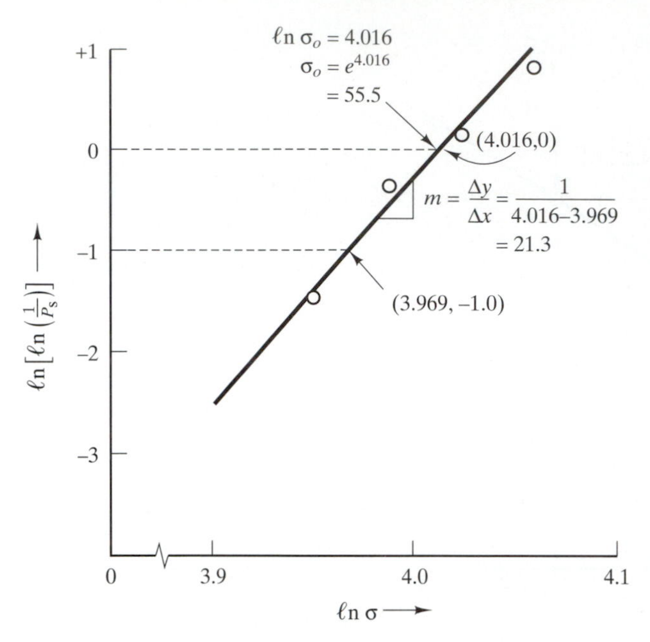

Fig. EP 14-5

EXAMPLE 14-6

Six identical samples of a ceramic material were tested with the 3-point bend test. The flexure strengths obtained were 56, 57, 53, 58, 54.5, and 59 ksi. Determine the Weibull modulus and the characteristic strength of the ceramic material.

SOLUTION: In order to determine the Weibull modulus, we still have to plot ln σ versus ln [ln $(1/P_s)$]. The key is to determine the P_s for each of the flexure strengths. This is done by arranging the flexure strengths from the lowest to the highest strength, and numbering them 1, 2, ..., 6, consecutively. The probability of failure of each strength level is given $F = (i - 0.5)/N$, where i is the ith level and N is the total number of samples, in this case, 6. The results are summarized in Table EP 14-6 and the plot of x = ln σ, versus y = ln [ln $(1/P_s)$], is shown in Fig. EP 14-6.

14.2.4c Uniaxial Compression Strength. Ceramics have traditionally been used more in compression because their compressive strengths are at least several times their uniaxial tensile strengths. In compression, crack surfaces or porosities whose major axes are normal to the compression axis tend to fuse rather than open and the compressive strength does not depend on the largest crack size. Cracks or defects whose axes are oriented not normal to the compressive axis have stress fields at their tips that have tensile stress components. At the worst orientation (−30° to the axial direction), the tensile stress (component) intensity at the tips of the cracks is about one-eighth of the intensity when the specimen is loaded in tension. Thus, using the fracture mechanics equation, Eq. 14-4, the compressive strength in the worst case scenario is eight times the tensile strength. Because of the statistical distribution of cracks, the observed compressive strength is much higher than this. The axial compressive strength is stated to be roughly 15 times larger,

$$\sigma_C \approx 15\,\sigma_{TS} \quad (14\text{-}23)$$

The reason for this is that the tensile stress component of the stress field extends the crack only slightly and then the crack extension stops because the extending crack aligns itself with the compression axis, as shown in Fig. 14-23. When the extension is arrested, the crack is said to have propagated stably. Increasing stress will extend more cracks until the accumulated damage in the specimen causes the specimen to disintegrate into powder. We see there-

Table EP 14-6 Summary of test results for Example 14-6.

Sample number, i	Flexure strength, σ_f	$F = (i - 0.5)/N$	ln [ln$(1/P_s)$], $P_s = 1 - F$	ln σ_f
1	53 ksi	0.0833	−2.250	3.912
2	54.5	0.250	−1.246	3.998
3	56.0	0.417	−0.618	4.025
4	57.0	0.583	−0.133	4.043
5	58.0	0.750	+0.327	4.060
6	59.5	0.917	+0.910	4.086

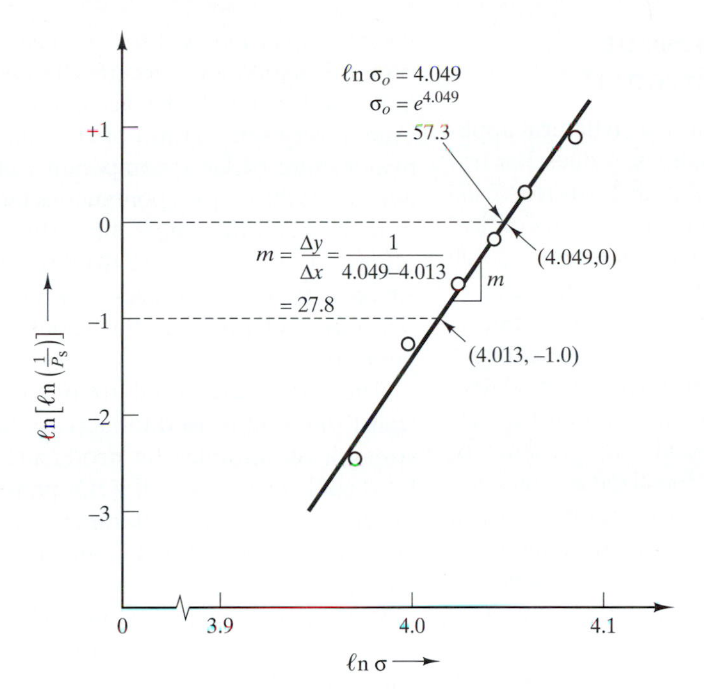

Fig. EP 14-6

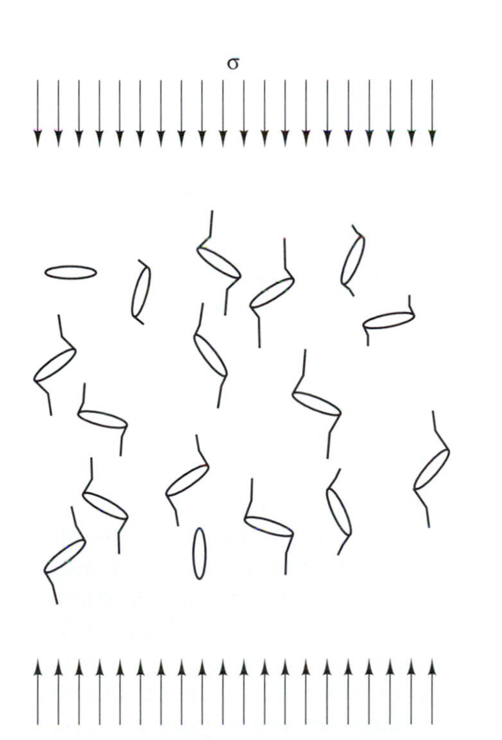

Fig. 14.23 In compression, cracks extend only a short distance and align with the compression axis.

fore that the compressive strength depends not on the largest, worst-oriented, highest-stressed defect, but rather on the entire defect population. Compressive strength may have a dependence on the square root of grain size, since the size of defects may have a relation to the average grain size, or because of microplasticity in the grains. Weibull statistics does not apply because compressive strengths often have very small scatter.

14.2.4d Modulus of Elasticity. The linear stress-strain curve schematically illustrated in Fig. 14-20 enables the determination of the Young's Modulus of Elasticity. As we have seen in Chap. 11, the bending performance of a material is dependent on the Modulus of Elasticity. In ceramics, the porosity decreases their ability to resist deformation under load because the effective modulus of a material decreases with porosity. This decrease can be seen from the moduli of cellular or foamed materials, where

$$E = E_s\left(\frac{\rho}{\rho_s}\right)^2 \tag{14-24}$$

where E and E_s are the moduli of the foam structure and the solid, while ρ and ρ_s are the densities of the foamed material and the solid.

14.2.5 Fracture Toughness of Ceramics and Zirconia*

The biggest drawback of ceramics in structural applications is their low fracture toughness values. For traditional ceramics, 0.5 to 3.0 MPa$\sqrt{m}$ (0.46 to 2.7 ksi $\sqrt{\text{inch}}$) are the usual values and with this, the critical crack size is very small. A fracture strength of 700 MPa (100 ksi) is associated with a defect diameter of only 1 to 30 μm (0.04 to 1.2 mils). Traditional ceramics rarely exceed 350 MPa (50 ksi) fracture strength because of the presence of larger defects than 30 μm.

Studies on ceramics have shown that their fracture toughness may be increased by (1) the presence of microvoids, (2) the presence of localized regions that are in compression, and (3) by fiber reinforcement. At the tip of a crack, there is a lot of stress intensification and the stress field is high. The presence of *microvoids* will relieve this stress field and thus lowers the stress-intensification to propagate the crack. This beneficial effect of microvoids can be taken only within limits, however, since other adverse effects will set in. The presence of *localized regions in compression* will require higher tensile stress levels when the crack propagates through them. *Fiber reinforcement* makes the material a composite. When a crack propagates in fiber-reinforced composite materials, part of the energy of crack propagation is dissipated along the fiber-matrix interface through debonding. Thus, the fiber reinforcement raises the stress level to propagate the crack through the brittle matrix.

14.2.5a ZrO_2 in Another Ceramic Matrix. The first two methods of enhancing the fracture toughness of a ceramic material can be accomplished by incorporating zirconia particles in the ceramic matrix. The methods take advantage of the allotropy of zirconia and the martensitic transformation from the high temperature tetragonal structure to the low temperature monoclinic structure of zirconia. Actually, zirconia exhibits three crystal structures (allotropes) in the solid state: monoclinic, tetragonal, and cubic structures (see Fig. 14-8). The monoclinic is stable up to 1170°C (2140°F) when it transforms to tetragonal; the tetragonal structure is stable up to 2370°C (4300°F) when the cubic structure forms and persists up to the melting point at 2680°C (4855°F). On cooling, the tetragonal to monoclinic transformation occurs martensitically and is accompanied by a relatively large volume expansion of 3 to 5 percent. For brittle materials, this expansion exceeds the elastic and fracture limits and can only be accommodated by cracking, even in very small grains of zirconia (ZrO_2). Thus, the manufacture of large components of pure zirconia is impossible due to the spontaneous failure upon cooling from the sintering temperature. The addition (alloying) of lime (CaO), magnesia (MgO), or yttria (Y_2O_3) to zirconia (ZrO_2) stabilizes the cubic or tetragonal structure and prevents the spontaneous cracking of pure ZrO_2.

The ZrO_2 tetragonal to monoclinic martensitic transformation is used to increase the toughness and strength of ceramics by producing dispersed stabilized cubic or tetragonal ZrO_2 phases in the ceramic matrix, for example, alumina, Al_2O_3. A stabilized cubic or tetragonal ZrO_2 means that the cubic or tetragonal structure is maintained at low temperatures when ordinarily the monoclinic structure is the stable phase. We should note, therefore, that the cubic or tetragonal ZrO_2 are thermodynamically unstable, but kinetically, the transformation to the monoclinic phase does not occur. This is similar to the retained austenite in high-alloy or tool steels at room temperature. As in steels, the martensitic transformation in ZrO_2 can be induced by stress or cold-work. Thus, when the ceramic containing the dispersed stabilized tetragonal ZrO_2 is stressed, the martensitic transformation is induced and produces microcracks in the matrix (Al_2O_3), and localized regions in compression (the dispersed ZrO_2 that transformed from tetragonal to monoclinic). When the tip of a propagating crack meets a transformed ZrO_2 phase, its stress field will be relieved by the cracks and a higher stress is needed also to propagate or extend the crack through the compressive region—the toughness (the crack propagation is hindered) and the strength (higher stress needed to propagate the crack) of the ceramic have increased.

In order to obtain the optimum increases in toughness and strength, the ZrO_2 particle sizes and their total volume fraction must be controlled. The particle size must be large enough in order to transform, yet must be small enough in order that limited microcracking can only occur. In alumina (Al_2O_3) matrix, the optimum size of pure ZrO_2 is about 1 to 2 μm and that of partially stabilized ZrO_2 (PSZ) with lime, magnesia, or yttria (yttrium oxide) is about 2 to 5 μm. The influence of increasing the ZrO_2 volume

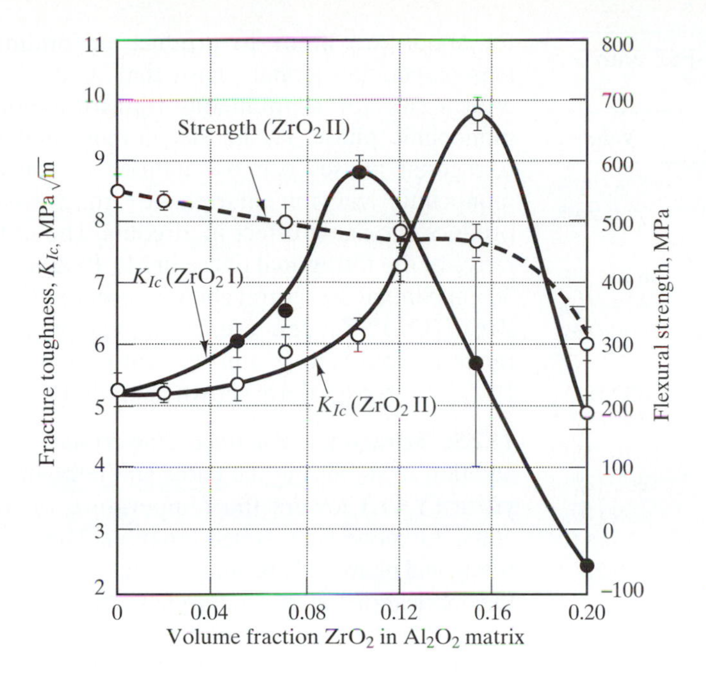

Fig. 14.24 Fracture toughness and flexure strengths of alumina as functions of the volume fraction of zirconia.

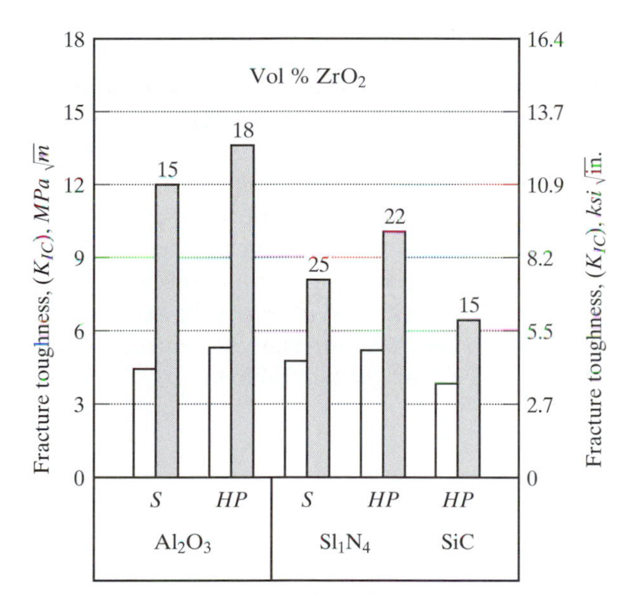

Fig. 14.25 Increases in fracture toughness of alumina, silicon nitride, and silicon carbids. Numbers in shaded bars are vol% zirconia; S stands for sintered and HP for hot-pressed.

fraction in Al_2O_3 is shown in Fig. 14-24 where we see that the toughness and flexure strength (modulus of rupture) increase to some peak values and will decrease, henceforth, since the microcracks that are produced will be very close together and will decrease the strength and toughness. Figure 14-25 shows the increases in toughness of alumina, silicon nitride, and silicon carbide when ZrO_2 is incorporated into their matrices. The number on top of the shaded bars indicates the volume percent of ZrO_2 in the ceramic matrices.

14.2.5b Partially Stabilized Zirconia (PSZ). This is a process of increasing the toughness and strength of ZrO_2 itself by partially stabilizing the cubic ZrO_2 structure with a dopant addition of either MgO, CaO,

Table 14-4 Typical properties of Mg-PSZ with 3 wt% MgO.

Property	Value
Density, g/cm^3	5.75
Hardness, HR45N	74–79
Flexural strength, MPa (ksi)	
at 25°C	634 (92)
at 500°C	414 (60)
at 1000°C	290 (42)
Tensile strength, at 25°C MPa (ksi)	352 (51)
Compressive strength, at 25°C MPa (ksi)	1758 (255)
Young's modulus, GPa (10^6 psi)	200 (29)
Shear modulus, GPa (10^6 psi)	69 (10)
Bulk modulus, GPa (10^6 psi)	373 (54)
Poisson's ratio	0.22
Thermal expansion at 25–1000°C × 10^{-6}/K	10.1
Fracture toughness (K_{Ic}), MPa$\sqrt{m}$ (ksi$\sqrt{in.}$)	8-12 (7–11)
Weibull modules, *m*	20

or Y_2O_3. The production of a PSZ depends on a dopant concentration that does not completely stabilize the cubic phase and on the cubic phase being heat-treatable to produce a two-phase microstructure of cubic and tetragonal and/or monoclinic phases. For example, the cubic phase of a ZrO_2–3 wt% MgO (Mg-PSZ) can be quenched from the cubic single-phase field at temperatures >1750°C (>3180°F) to avoid large grain boundary phase precipitation, followed by isothermal aging at ~1400°C (~2550°F) for about two hours to produce an optimum size, lens-shaped tetragonal phase that is just below the critical size for spontaneous transformation to the monoclinic phase. As in the alumina matrix, these tetragonal phases can be induced to transform to monoclinic via the martensite transformation to produce the same effect in zirconia. The critical size range of the tetragonal phase in Mg-PSZ is from 25 to 30 nm. Similar structural changes occur in CaO doped ZrO_2 (Ca-PSZ); the critical tetragonal phase size range is from 6 to 10 nm. Typical properties of Mg-PSZ (3 wt% MgO) are shown in Table 14-4.

14.2.5c Tetragonal Zirconia Polycrystals (TZP). In addition to stabilizing the cubic and tetragonal phases, yttria (Y_2O_3) lowers the temperature of the tetragonal to monoclinic transformation. Thus, larger-size tetragonal phases can be retained that considerably ease the fabrication of the multi-phase ceramic. It is also found that the toughening of the ceramic increases as the amount of the retained tetragonal phase increases. Therefore, maximum toughness is expected when the ceramic contains 100 percent polycrystalline tetragonal phase—thus the name, TZP, tetragonal zirconia polycrystal. The ceramic that is produced is dense and fine-grained. As in the PSZs, there is a critical grain size above which the tetragonal phase spontaneously transforms to the monoclinic phase and that causes the fracture strength to decrease as shown in Fig. 14-26. This critical grain size increases dramatically when the yttria content is changed from 2 mol% to 3 mol%, as shown in Fig. 14-27, and increases the retention of the tetragonal phase to more than 90 percent. Typical properties of TZP are given in Table 14-5.

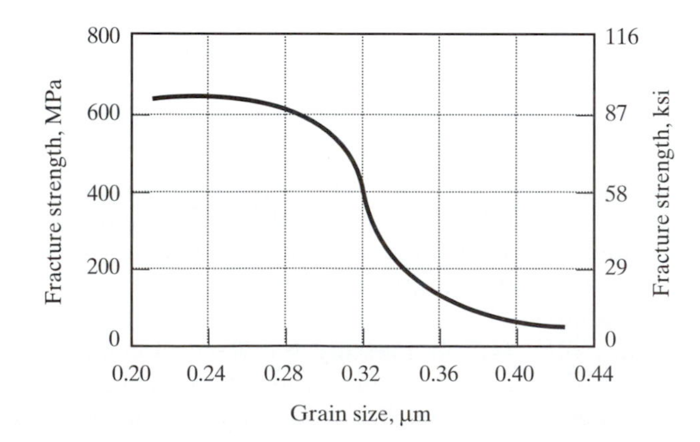

Fig. 14-26 Influence of grain size on fracture strength.

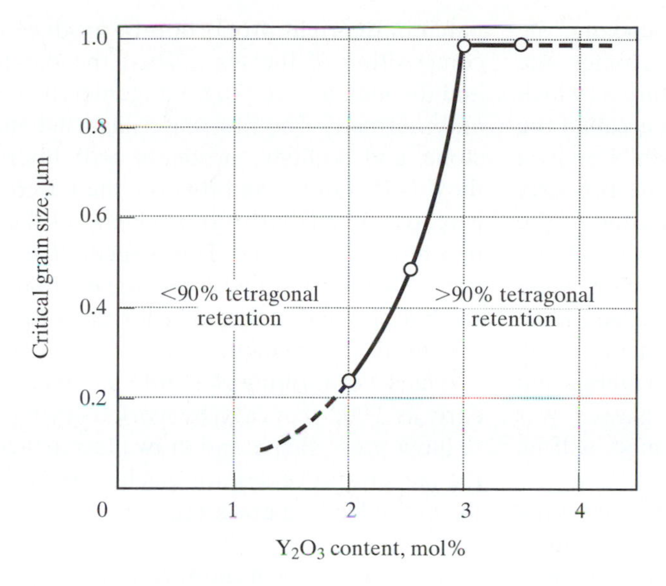

Fig. 14-27 Influence of yttria content on critical grain size.

Table 14-5 Typical Properties of TZP.

Property	Value
Density, g/cm^3	6.05
Hardness, HR45N	83
Modulus of rupture(a), MPa(ksi)	900 (130)
Fracture toughness (K_{Ic}), MPa$\sqrt{m}$ (ksi$\sqrt{in.}$) (b)	14 (12.7)
Elastic modulus, GPa (10^6 psi)	200 (30)
Weibull modulus, *m*	14
Thermal conductivity, W/m · K	2

(a) Four-point bend test. (b) Single-edge notched beam test.
Source: Coors Ceramics

14.3 *Glasses*

As indicated in the Introduction to this chapter, glasses differ from ceramics in that they do not exhibit long range order (crystalline structure) and are therefore amorphous materials. The most widely used commercial glasses are the soda-lime-silicate glasses that utilize the abundance of silica or quartz sand as the raw material. Because of this, glasses have traditionally been treated as a subset of ceramics and have been thought to be silicate glasses. It turns out that other oxides such as boric oxide (B_2O_3), germania (GeO_2), and phosphoric oxide (P_2O_5) can also form glasses. Glasses can also form with nonoxides such as halides, chalcogenides that are compounds of selenium, sulfur, and tellurium, and metallic materials. These developments suggest that we should not tie the term glass only to the silicate glasses. At the same time, we should not lose sight of the fact that the silicate glasses constitute the majority of engineering glasses because the raw material is readily available and inexpensive. Therefore, we shall discuss for the most part the silicate and oxide glasses in this section.

14.3.1 Formation and Primary Processing of Glass

We start with the formation of glass from silica. The various allotropes of crystalline silica were given earlier in Table 14-1 and we noted that the allotrope at the highest temperature before silica melts is the cubic cristobalite whose structure is shown in Fig. 14-3. The liquid that is formed when crystalline silica (cristobalite) melts has an unusually high viscosity. When the liquid is cooled, the very high viscosity of the liquid prevents the silicon and oxygen ions from rearranging and from crystallizing into the equilibrium cristobalite structure. Instead of crystallizing at the melting point to produce an abrupt change in property, the liquid becomes undercooled below its equilibrium melting point. On further cooling, the supercooled liquid solidifies and it becomes the amorphous solid or glass. The temperature at which the supercooled liquid is frozen to solid is called the glass transition temperature, T_g, and is exhibited by a change in the slope of the temperature-property (specific volume) curve. Because the freezing of the liquid involves atomic movements in the liquid, the T_g depends on the cooling rate and is higher when the cooling rate is increased, as schematically illustrated in Fig. 14-28. The phenomenon depicted by Fig. 14-28 is universal and applies also to metals and polymers. The formation of metallic glasses was described in Chap. 7 and that for polymers will be described in the next chapter.

The primary processing of glass is via melting the raw materials and then forming the different products directly from the melt. Since the melting point of pure silica is more than 1700°C (see Fig. 14-6), additions of lime and soda ash are made to lower the melting point in the production of the soda-lime-silica glasses. Both lime (CaO) and soda ash (Na_2O) are fluxes to hasten the melting of silica and to reduce the melting point of silica to more manageable temperatures (see Fig. 14-6). Thus, the first step is to weigh the batch in the right proportion to produce the desired composition of the glass. Then the batch is melted and the melt allowed to homogenize, both chemically and thermally. The raw materials used such as limestone and sodium carbonate will liberate carbon dioxide that mixes and stirs the melt. Before the melt is ready to be formed to products, all these gas bubbles must be removed. This is facilitated by the addition of what are called *fining agents* that enlarge the gas bubbles and induce a faster rise of the bubble to the surface of the melt.

The art of forming glass into products was done as early as 2,000 years ago by workers using their hands to blow, press, shape, and draw glass articles. Many of the current processes are mechanical adaptations of the initial hand operations.

14.3.1a Container Manufacture. The production of containers involves a gob feeder machine, shown in Fig. 14-29, and the automatic container manufacturing operations, shown in Fig. 14-30. The gob feeder delivers precise preshaped glass gobs of a specific weight to individual molds for pressing or blowing operations. Figure 14-30 shows three variations to produce containers including wide-mouth containers.

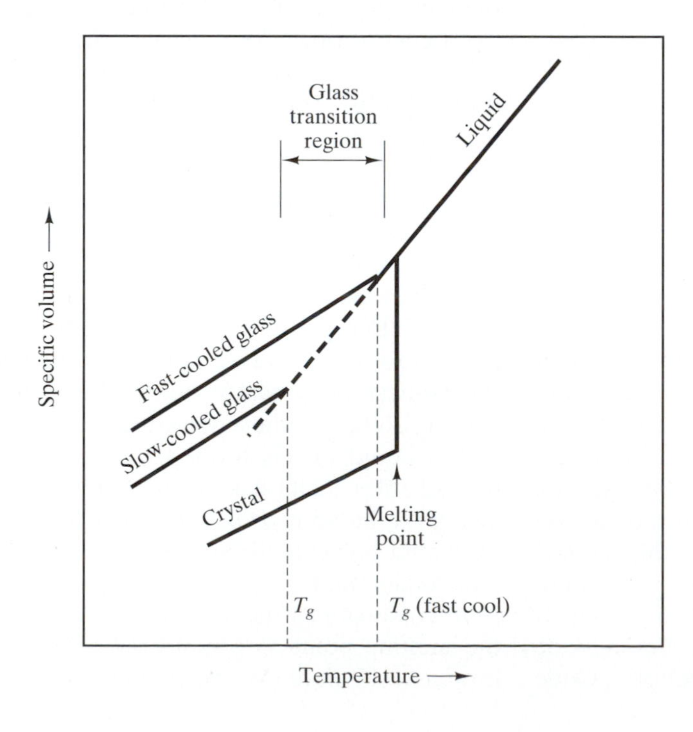

Fig. 14-28 Formation of glass or amorphous structure on cooling from the liquid.

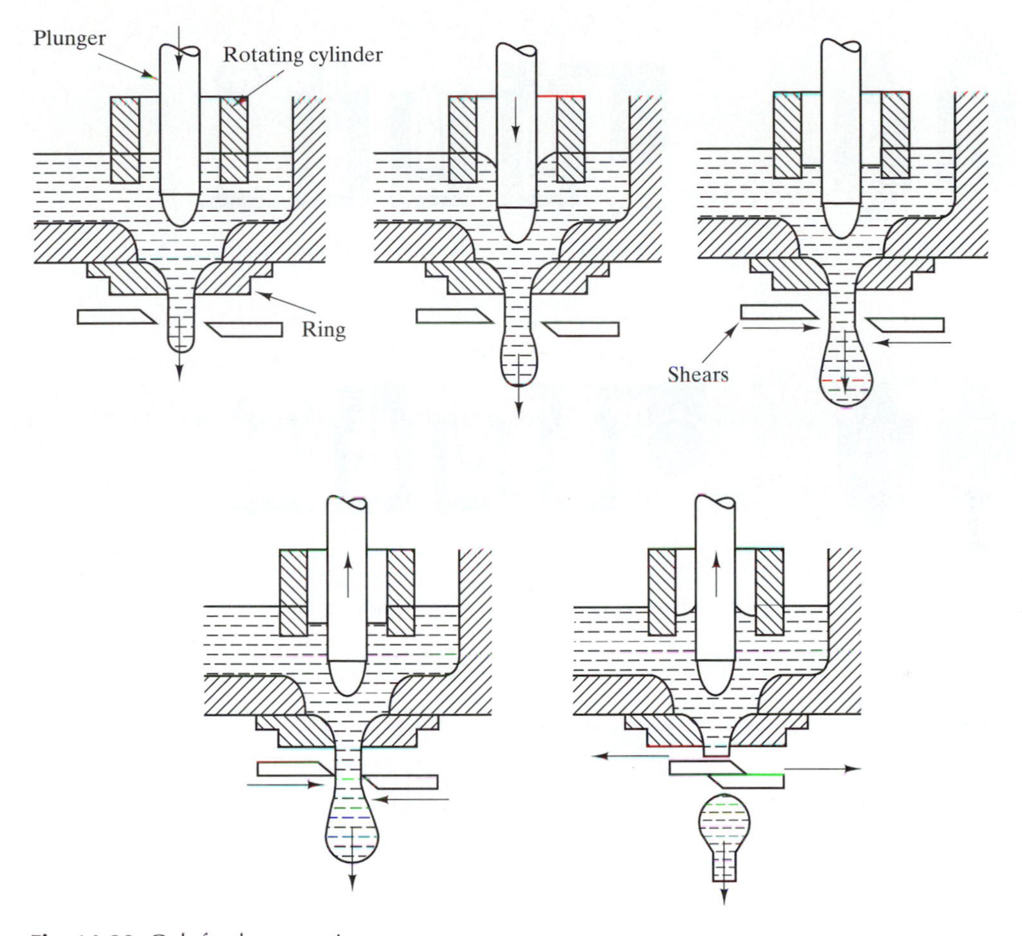

Fig. 14-29 Gob feeder operation.

14.3.1b Flat Glass—Sheet and Plate. The *float glass process* is shown in Fig. 14-31 and consists of floating the molten glass on top of a molten tin bath. The bottom glass surface is supported by the molten tin, while the top surface is exposed to the reducing atmosphere composed of hydrogen and nitrogen (forming gas) above the bath. The glass enters the bath at approximately 1150°C (2100°F) at a viscosity of 1000 Pa · s (10,000 Poise) and exits it at about 650°C (1200°F). Gravity allows the glass to flow out, and surface tension holds it back to produce an equilibrium thickness of 7.1 mm (0.281 in.). Thicknesses from 1.5 to 25 mm (0.06 to 1.0 in.) can be made by either stretching the glass with knurled knobs or compressing the glass with graphite paddles. The tin bath can be 45 m (150 ft) long and can produce a glass sheet 4 m (13 ft 4 in.) wide that can be finished to 3.7 m (12 ft) wide. This process is limited to soda-lime-silica sheet glasses.

The fusion process shown in Fig. 14-32 is a direct-melt process that is used to produce precision sheet glasses with softening points ranging from 660°C (1220°F) to 865°C (1590°F). The softening point is the temperature at which glass will deform under its own weight. (The viscosity of the glass at the softening point is 4×10^6 Pa·s(4×10^7P). In this process, the molten glass flows from the melter into either a trough or an overflow pipe at a viscosity of about 20,000 Pa·s. The glass overflows on each side of the trough and rejoins below the refractory fishtail or root to produce the sheet glass. The thickness of the sheet is dependent on the pull rate and thicknesses as thin as 0.5 to 1mm (0.02 to 0.04 in.) and as thick as 15 mm (0.6 in.)

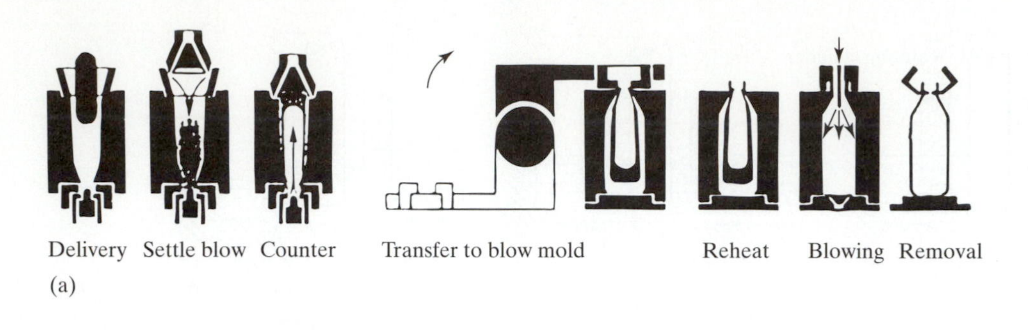

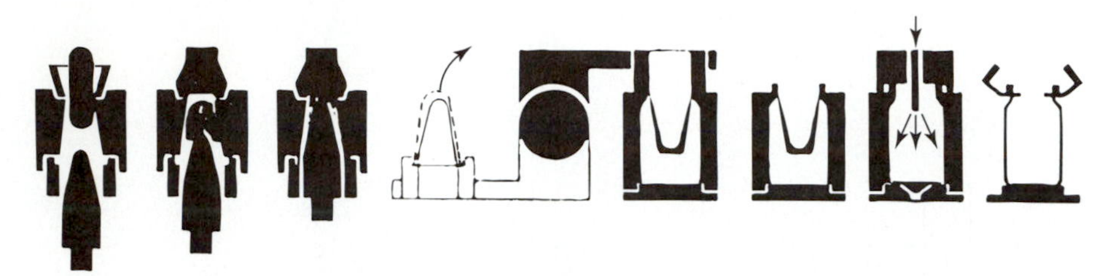

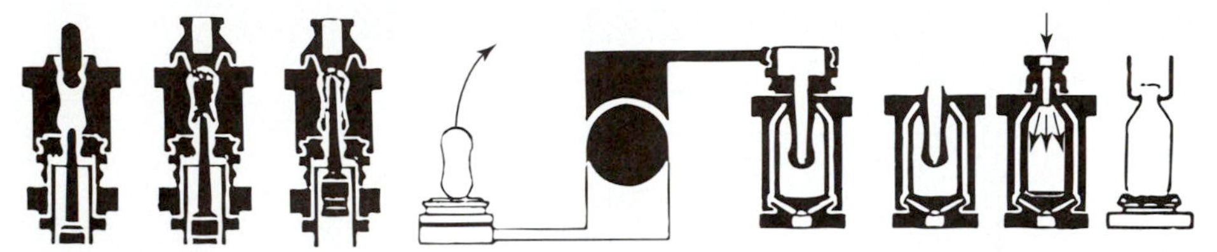

Fig. 14-30 Automatic container manufacturing variations; (a) blow and blow; (b) wide-mouth press and blow, and (c) narrow-necked press and blow.

have been produced. The production widths are half the size of the float process.

14.3.1c Tubings. Figure 14-33 shows the set up in the rotating mandrel Danner process. In this process, the molten glass is allowed to flow out of a tank and coats the rotating refractory mandrel. The glass coating flows down toward the end of the mandrel that is inclined between 15° and 20° from the horizontal and forms a tubing with air blown through the center of the mandrel. The glass tubing is then pulled off the end and out of the furnace.

14.3.1d Glass Fibers. Figure 14-34 shows the process of producing continuous or chopped glass fibers. The melt directly flows through a bushing with many forming orifices and is rapidly cooled to form continuous fibers as the melt streams pass the orifices. A polymeric coating (sizing or coupling agent) is applied to the fibers to prevent damage to the fibers during handling and are then collected in a strand using traditional textile methods. Chopped fibers are produced when the continuous filaments are passed through a chopper instead of being wound in a strand. Continuous and chopped glass fibers are used in polymer matrix composites. Continuous fibers are used when a maximum unidirectional strength is desired, while chopped fibers are used when uniform isotropic strength is desired.

14.3.2 Silicates and Glass Structures

The properties of glasses are also dictated by their structures. To illustrate the terms used in the structures of glasses, the structures of silicates will be illustrated because these have been studied the most.

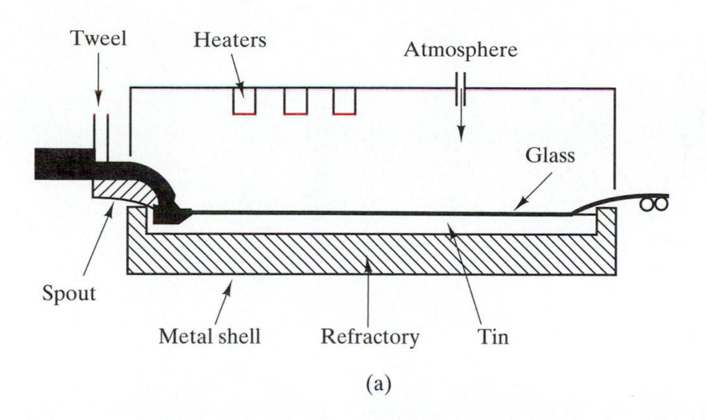

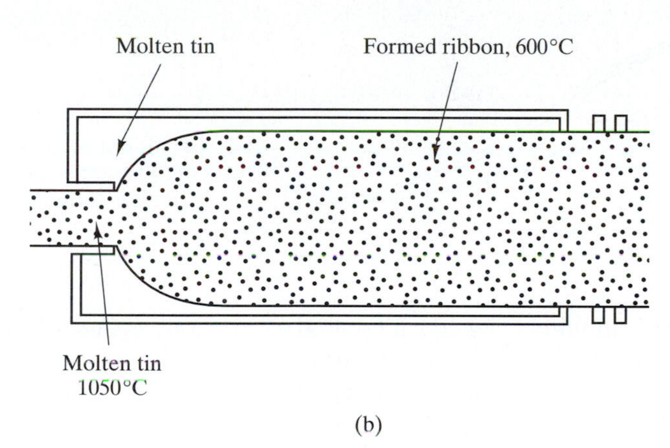

Fig. 14-31 Production of glass plates and sheets by the float-glass process (a) side view, and (b) top view of molten tin bath.

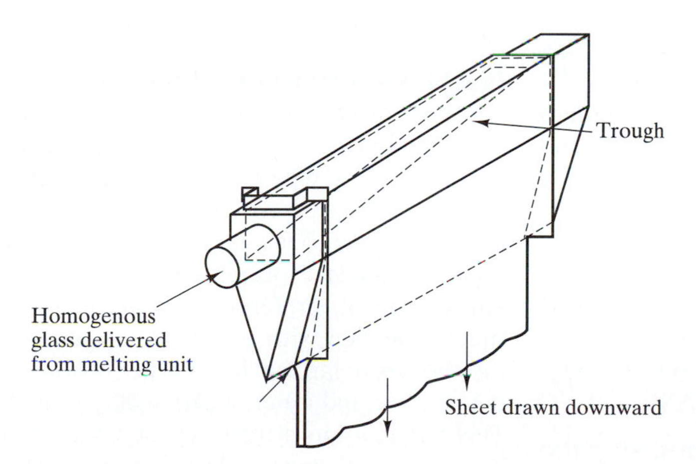

Fig. 14-32 The overflow pipe in the fusion process to produce glass sheets and plates.

The building block for the silicates is the $(SiO_4)^{-4}$ tetrahedron with the silicon ion placed at the center of four symmetrically arranged oxygen ions. The $(SiO_4)^{-4}$ units may be linked to other cations or to oxygen ions and the silicates can be classified by the way the tetrahedra are linked.

14.3.2a Island Structures. These structures are also called the orthosilicates in which the $(SiO_4)^{-4}$ units are linked to each other through other nonsilicon cations. The important materials in this group include zircon ($ZrSiO_4$), phenacite (Be_2SiO_4), willemite (Zn_2SiO_4), the olivines ($Mg_{(2-x)}Fe_xSiO_4$), and the

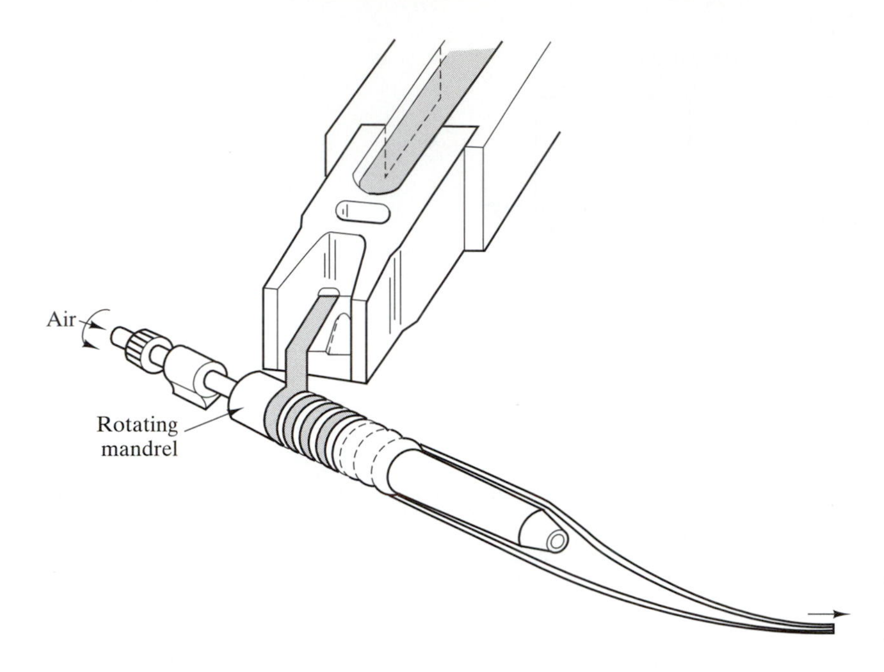

Fig. 14-33 Rotating mandrel used in the Danner process to produce glass tubings.

aluminosilicate group that includes mullite ($Al_6Si_2O_{13}$). For engineering applications, mullite is the most important material in this group.

14.3.2b Isolated Group Structures. These are structures where two or more $(SiO_4)^{-4}$ tetrahedra are linked through a common corner. These can result in the $(Si_2O_7)^{6-}$ grouping, as well as the ring groupings $(Si_3O_9)^{6-}$, $(Si_4O_{12})^{8-}$, and $(Si_6)_{18})^{12-}$, shown in Fig. 14-35. In all these cases, the groupings are isolated from each other. Charge balance and linkage between these groups are maintained by additional cations. Structures with the $(Si_4O_{12})^{8-}$ rings include the zeolites that are used in large volume as catalyst materials in the petrochemical industry. A mineral with the $(Si_6O_{18})^{12-}$ rings is beryl that is closely related to the widely used ceramic cordierite (ideally $Mg_2Al_4Si_5O_{18} = 2MgO \cdot 2Al_2O_3 \cdot 5SiO_2$, magnesium aliminum silicate, MAS).

14.3.2c Chain Structures. These are formed when two oxygen atoms of every $(SiO_4)^{-4}$ tetrahedron are shared and extended parallel chains are formed as shown in Fig. 14-36a. This yields the general formula $n(SiO_3)^{2-}$, with charge balance again maintained by adjacent cations that can also link the parallel chains in three dimensions. These single-chain compounds are called pyroxenes. Alternatively, the chains can be linked with shared oxygen ions as shown in Fig. 14-36b. The general formula is now $n(Si_4O_{11})^{6-}$. Some important materials of this class are the pyroxenes enstatite ($MgSiO_3$), wollastonite ($CaSiO_3$), and spodumene $LiAl(SiO_3)_2$ = LAS. LAS has a low coefficient of thermal expansion and is an important constituent of commercial glass-ceramics.

14.3.2d Sheet Structures. These are formed by sharing three oxygen ions of each $(SiO_4)^{-4}$ tetrahedron as shown in Fig. 14-37. This results in a layer composition represented by $n^2(Si_2O_5)^{2-}$. Most clay materials, such as kaolinite, have these sheet structures, which affect both their behavior in aqueous suspension and their mechanical properties. An additional atom attaches itself preferentially to one side of the sheet—the side with the spare oxygen. As a result, the sheet is polarized: it has a net positive charge on one surface and a net negative charge on the other. This polarization attracts water easily between the sheets and makes clays plastic—the sheets of silicates slide readily over each other. The bonding in the plane of the sheets is very strong, but they cleave or split easily between sheets, such as in mica and talc.

14.3.2e Framework or Network Structures. These are formed when all four oxygen atoms of the $(SiO_4)^{-4}$ tetrahedron are shared with other units.

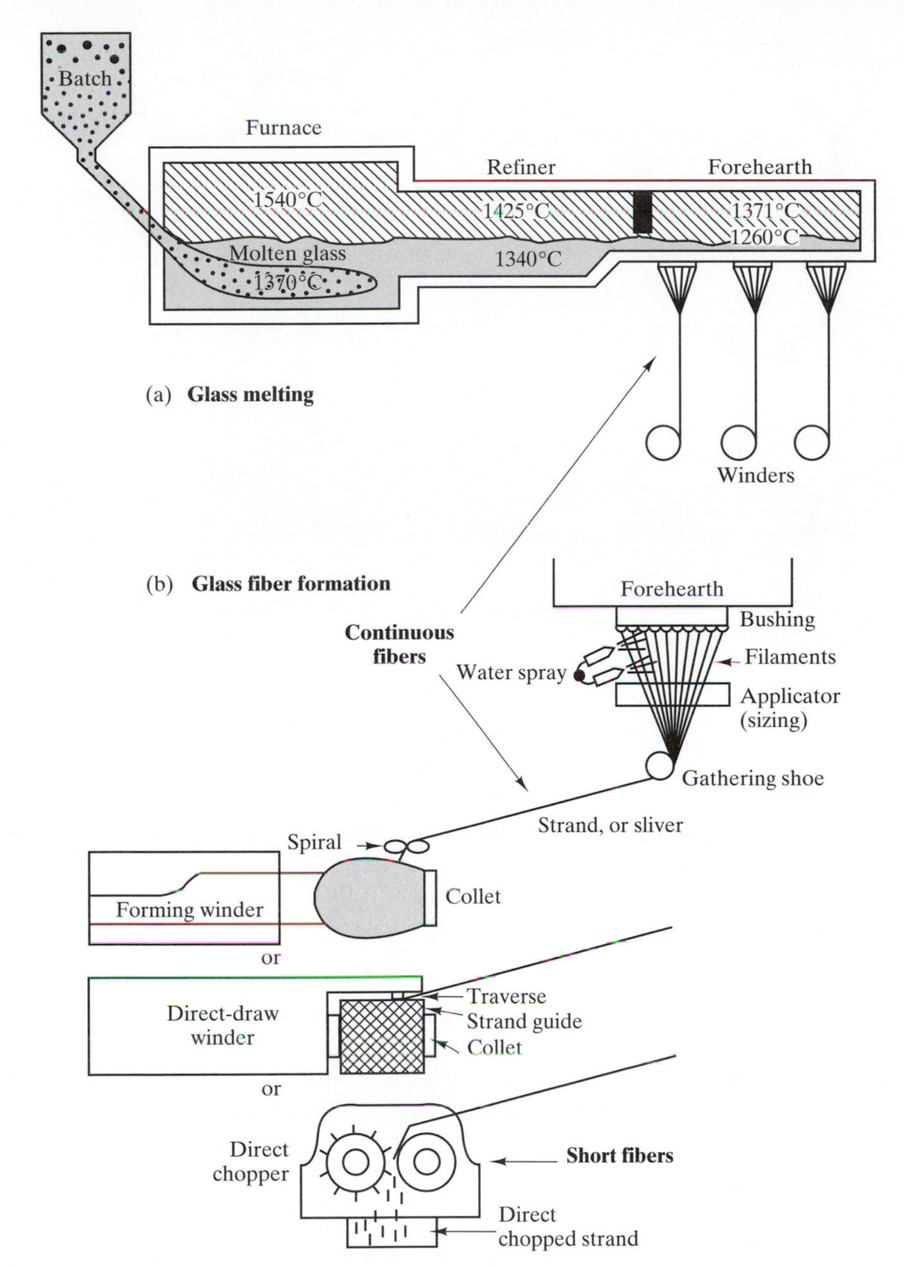

Fig. 14-34 (a) Glass melting, and (b) production of glass fibers. [2]

The ideal examples are the various structural modifications of silica, shown in Table 14-1 and in Fig. 14-3. Included in this group are silicates in which some silicon atoms are replaced by aluminum and with alkali or alkaline earth cations entering the lattice to maintain a charge balance. The most important mineral in this group are the feldspars, major constituents of igneous rock.

14.3.2f Glass or Amorphous Structures. The $(SiO_4)^{-4}$ units in the various structural polymorphs in Table 14-1 are arranged in a regular, ordered manner, as illustrated in Fig. 14-38a. On the other hand, the glassy or amorphous silica structure arises when the $(SiO_4)^{-4}$ units are randomly arranged as schematically shown in Fig 14-38b. The three-dimensional random network shows that the bonding requirements are also satisfied.

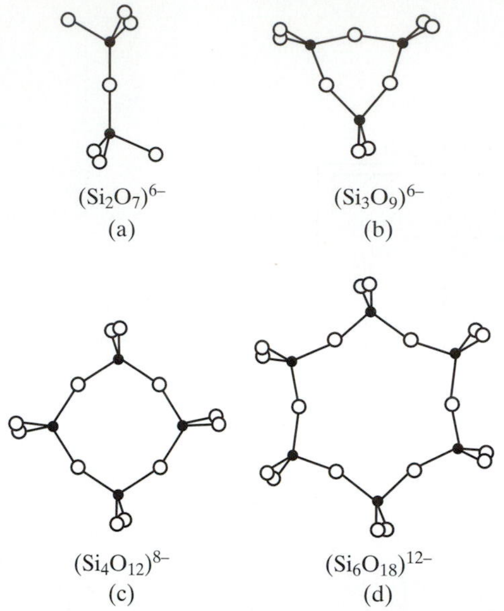

$(Si_2O_7)^{6-}$
(a)

$(Si_3O_9)^{6-}$
(b)

$(Si_4O_{12})^{8-}$
(c)

$(Si_6O_{18})^{12-}$
(d)

Fig. 14-35 Silicate group structures (a) two linked units, (b) three-unit ring, (c) four-unit ring, and (d) six unit ring.

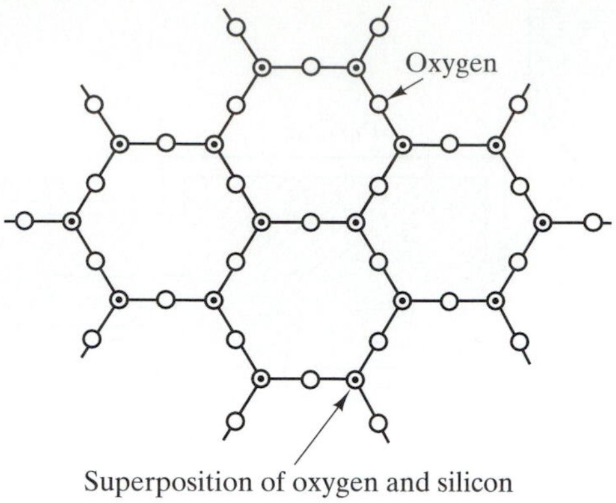

Oxygen

Superposition of oxygen and silicon

Fig. 14-37 Silicate sheet structure.

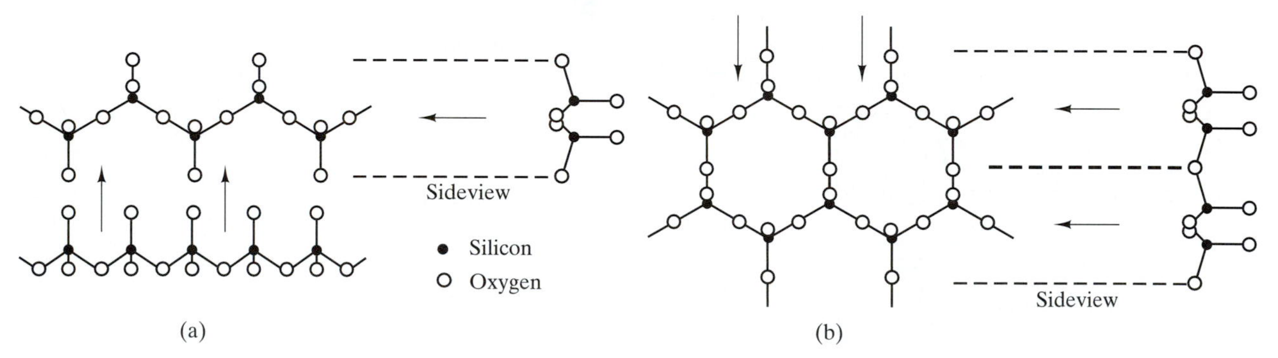

Fig. 14-36 Silicate chain structures. (a) the pyroxene single chain, and (b) the amphibole double-chain.

In general, each of the common oxide glass formers produces a random network. Some of the oxide glass formers are listed in Table 14-6. From this list, germania also forms a random 3-D network; boric oxide forms a random 2-D network; and phosphoric oxide forms randomly arranged linear polymeric chains.

The *oxide glass formers* are those that can produce glassy structures on their own. In general the melts from these oxides have very high viscosities that will induce the glass formation, as was illustrated for silica in Fig. 14-28. This high viscosity comes from the strong bonding between the short-range order units (characteristic of amorphous structure) and makes the shaping of the glass quite difficult. Thus, other oxide additives are added to the pure oxide glasses, primarily to improve their formability and processability. These additives are designated as either *intermediate* or *modifier* oxides. A modifier is an oxide that is incapable of forming a glass on its own, but is frequently added in large amounts to the glass formers to alter or modify the structures and properties of the glass. An intermediate oxide is defined as one that is incapable of forming a glass by itself, but is capable of substituting for a glass former in the glassy amorphous network; in other words, its behavior is "intermediate" between those of a glass

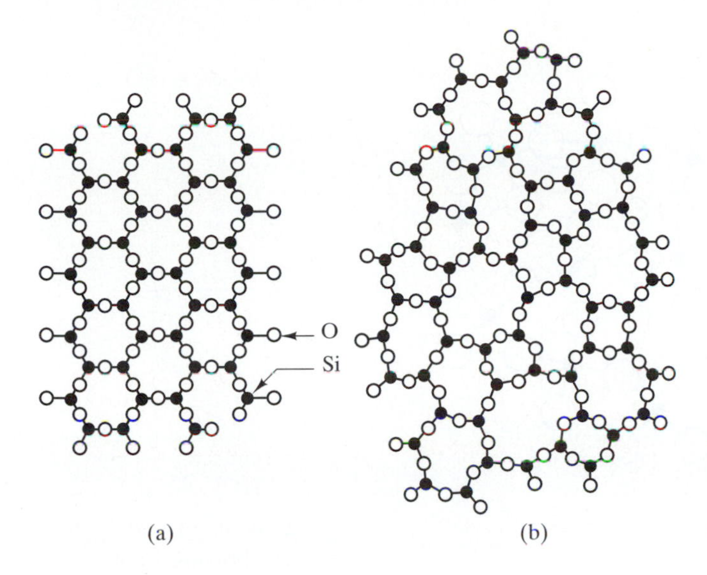

Fig. 14-38 (a) regular 3-D crystalline network of silicates, (b) random network of silicates in glassy structure.

Table 14-6 Classification of oxides into glass formers, intermediates, and modifiers. [3]

Glass Formers	Intermediates	Modifiers
B_2O_3, SiO_2, GeO_2, Al_2O_3, V_2O_5, P_2O_5, Sb_2O_5, ZrO_2	TiO_2, ZnO, PbO, Al_2O_3, CdO, BeO, ZrO_2, ThO_2	PbO_2, MgO, PbO, ZnO, BaO, CaO, Na_2O, Li_2O, K_2O, CdO, SnO_2, Ga_2O_3, La_2O_3

former and a modifier. A list of typical intermediate and modifier oxides are also given in Table 14-6.

When added to glass formers, the modifiers can alter their random network structure by either forming nonbridging ions and thus decreasing the connectivity, such as in alkali-silicate glasses with an example shown in Fig. 14-39, or by changing the coordination number of the glass former. Examples of the latter are: (1) the formation of $(BO_4)^{4-}$ tetrahedra units (coordination number, CN 4) in addition to the regular $(BO_3)^{3-}$ triangular units (CN 3) in alkali-borates; and (2) the formation of $(GeO_6)^{6-}$ octahedral units (CN 6) in addition to the regular $(GEO_4)^{4-}$ tetrahedra units (CN 4) in alkali-germanate glasses. These changes affect the bonding and will be reflected in the properties of the glasses.

The intermediate oxide formers usually have the same coordination number as the glass formers. However, this is not a necessary condition. The necessary condition is for the intermediates to increase the connectivity of the amorphous network. Thus the effects of the intermediate oxides are opposite to those of modifiers; that is, either reduce the concentration of nonbridging ions or reconvert of the CN of the glass-former ion into its original CN. It follows that the changes in properties induced by the intermediates are opposed to those induced by the modifiers.

14.3.3 Physical and Chemical Properties of Glasses

As indicated in Chap. 4, these are the properties that are controlled by the bonding and arrangement of atoms in the material and are called the microstructure insensitive properties.

14.3.3a Viscosity—Melting and Glass Transition Temperatures. As indicated earlier, the primary purpose of additives to the glass formers is to improve their processability and formability. The addition of modifiers is made to glass formers to increase the fluidity of the melt and to make it easier for the viscous fluid to be blown or pressed into shape.

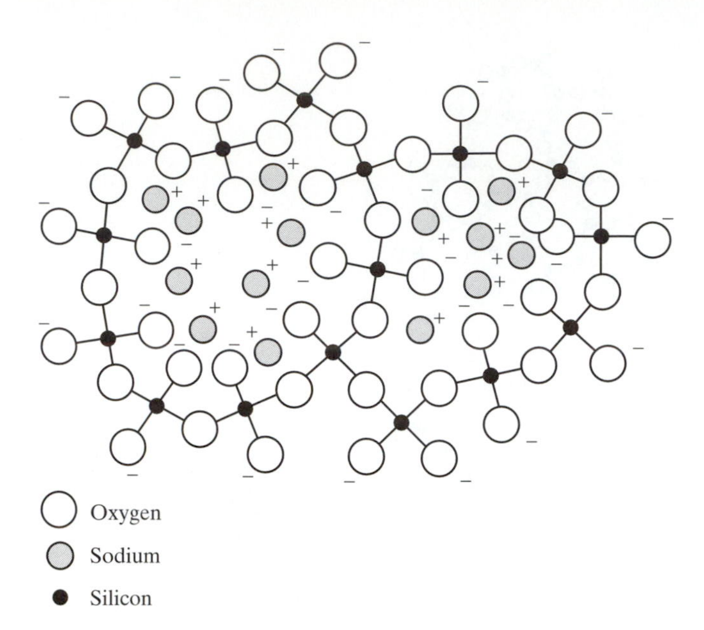

Fig. 14-39 $Na_2O - SiO_2$ glass structure showing the non-bridging oxygen anions.

The definition of viscosity was given earlier in Eq. 4-85 as

$$\eta = \frac{\tau}{\dot{\gamma}} \qquad (14\text{-}25)$$

η is the viscosity, τ is the shear stress, and $\dot{\gamma}$ is the shear strain rate. The unit of viscosity in the SI system is Pa·s and in the English system is Poise (P). The relationship is 1 Pa·s = 10 P. In the processing of glasses, there are certain points or temperatures that are characterized by distinct viscosities. These are:

- The *melting point* is the temperature at which the glass has a viscosity of 10 Pa·s (100 P) and is commonly used as a reference point when comparing differences in meltability between two glasses.
- The *working (forming) point* is the temperature at which the glass flows sufficiently for the glass to be formed or sealed and the viscosity of glass is typically 10^3 Pa·s (10^4 P).
- The *softening point* is the temperature at which glass will deform under its own weight and the viscosity is typically 4×10^6 Pa·s (4×10^{17}P).
- The *annealing point* is the temperature at which the internal strains are reduced to an acceptable commercial limit in 15 min and the viscosity is typically 10^{12} Pa·s (10^{13} P).
- The *strain point* is the temperature at which the internal stresses are reduced to low values after four hours. The viscosity of the glass is 3.2×10^{13} Pa·s (3.2×10^{14} P) which makes the glass substantially solid.

The primary factors that control the viscosity are the connectivity of the network, the strength of the bonds, and the rate of change of the melt structure with temperature. Vitreous silica, which exhibits a strongly bonded 3-D network that slowly breaks down with increasing temperature, exhibits a viscosity-temperature curve with a very shallow slope. This curve ranges from a glass transition temperature, T_g, of about 1100°C (2012°F) to melting temperatures in excess of 2000°C (3632°F). The addition of small amounts of alkali oxide dramatically decreases its viscosity and increases the slope of the viscosity-temperature curve. With as little as 10 mol% alkali oxide, the silica melt can be "fined" (removal of gaseous bubbles) at temperatures below 1600°C (2912°F). Further additions of alkali oxide rapidly decrease the temperatures required for forming a usable melt, with processing temperatures dropping below 1300°C (2372°F). The T_g of alkali-silicate glasses decreases also in a similar manner. Additions of intermediate oxides that increase the network connectivity increase the viscosity of the melts and the temperature required to form a satisfactory glass.

The effect of adding alkali oxides on the viscosity of boric oxide is quite different from the effect on

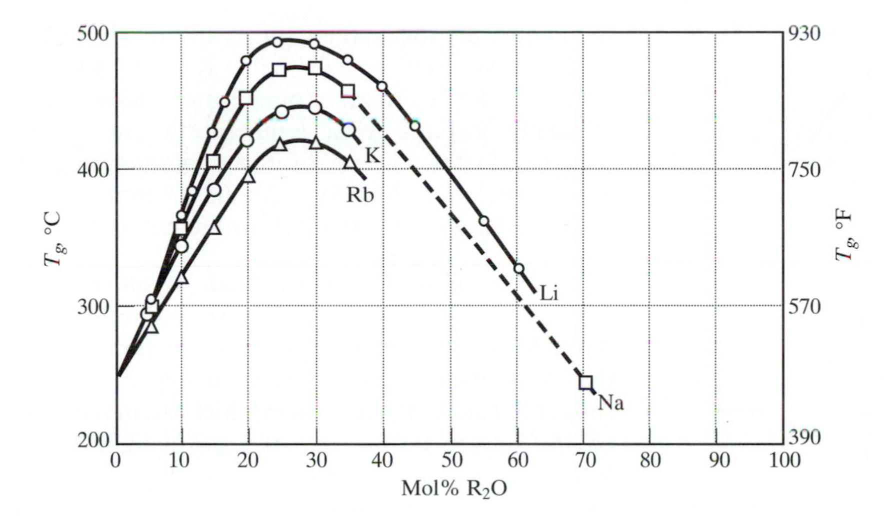

Fig. 14-40 Effect of alkali oxide concentration on the glass transition temperature of alkali-borate glasses.

the viscosity of silica. Because of a change in the coordination number from 3 to 4, boric oxide changes from a planar to a random 3-D network that increases its viscosity. The viscosity peaks at around 25 to 30 mol% alkali oxide and then subsequently decreases. The latter is attributed to the formation of nonbridging oxygen ions at high alkali concentrations. The viscosity behavior correlates directly with the T_g behavior that is shown in Fig. 14-40.

14.3.3b Density The density of glasses is dependent on the void-space of the amorphous network as well as on the masses of the ions present (discussed earlier in Chaps. 2 and 4). In general, the addition of species that fill the voids or interstices of the amorphous network will tend to increase the density by reducing the void space. The addition of ions into the structure will change the density in proportion to the mass of the added ions relative to that of the ions present in the glass.

For silicate glasses, the variation of density with composition of the alkali ion exemplifies the effect of the placement of other ions into the interstices of a glassy structure. The density of amorphous silica increases monotonically as the concentration of the alkali ion increases. However, the increase in density is not necessarily proportional to the atomic mass of the added ion. Although sodium-silicate glasses are significantly denser than the corresponding lithium-silicate glasses, potassium-silicate glasses have almost identical densities as the sodium-silicate glasses.

The densities of the alkali borates and germanates exhibit similar trends as do the alkali silicates, but with some differences. The increase in density is not monotonic with the increase in the alkali ion concentration, but peaks and then decreases, as shown in Fig. 14-41. We also see in Fig. 14-41 that a higher atomic weight ion yields a higher density, with the exception of the sodium ion (lower atomic weight) producing a higher density glass than potassium (higher atomic weight). The difference between alkali-silicates and that of the borates and germanates is attributed to the difference in the changes in the structure induced by the alkali ion discussed in Sec. 14.3.2.

14.3.3c Thermal Expansion. The thermal expansion behavior of glasses are categorized according to temperature regions. At temperatures below the glass transition region (see Fig. 14-28), the material is truly an amorphous solid glass, is brittle, and behaves in a linear elastic manner. Above the glass transition region, the material behaves as a viscous fluid and is called a melt. At the glass transition region, the material exhibits visco-elastic behavior and is still commonly referred to as a glass, in spite of the fact that it is in a state of transition between solid and liquid phases.

The thermal expansion of solid glass is primarily controlled by the expansion of the bond length (as discussed in Chap. 4), that results from an increased thermal energy. Structural changes that increase the number of nonbridging bonds will increase the

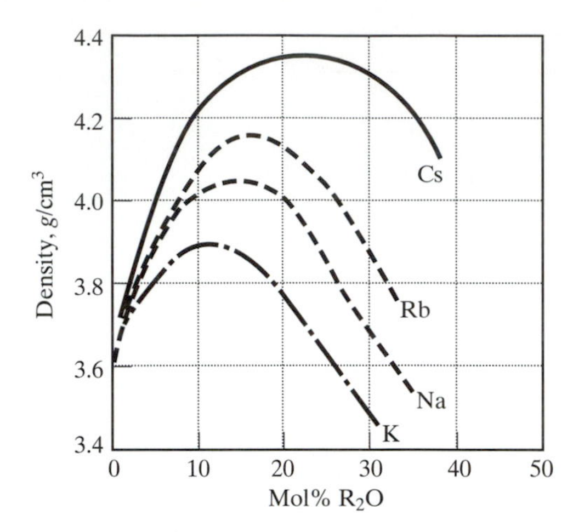

Fig. 14-41 Densities of alkali-germanates as functions of alkali-oxide concentration.

coefficient of thermal expansion, whereas changes in the coordination number may either increase or decrease the linear thermal expansion coefficient, α. Thus, for alkali silicates, α increases for a given alkali concentration (mol%) in increasing order as Li < Na < K < Rb < Cs. The reduction of the non-bridging oxygen ions by the addition of an intermediate oxide reduces the α of the alkali-silicates.

The addition of alkali oxides (modifiers) to borates and germanates produces an anomalous effect, with the effect being more pronounced in borates than in germanates. For the borates, the unmodified two-dimensional sheetlike structure consisting of the $(BO_3)^{3-}$ triangles (CN 3) has a large α. At low alkali ion concentrations, the addition changes the CN from 3 to 4. This change forms the $(BO_4)^{4-}$ tetrahedra units and induces 3-dimensional network bonding at all four corners that results in the decrease of the α. As the alkali concentration increases, the alkali ion begins to create nonbridging oxygen ions, the connectivity of the network decreases, and the α increases. Thus, the anomalous behavior shows an initial decrease to produce a minimum in α before it increases as the alkali concentration increases. For germanates, the initial decrease is not as pronounced since the change is from an initial three-dimensional random network with a CN of 4 to another three-dimensional network with a CN of 6.

14.3.3d Thermal Conductivity and Diffusivity. These properties are weak functions of composition. In general, the thermal conductivities of glasses are in the range 0.149 to 2.09 W/m · K (0.001 to 0.005 cal/cm · s · K) at room temperature. These values gradually increase with temperature but remain within a factor of two of the room-temperature value. Thermal diffusivity is about 0.5 mm^2/s (0.02 ft^2/hr) and does not vary much with temperature.

14.3.3e Chemical Durability and Weathering of Glasses. Chemical attack of glass surfaces occurs through either dealkalization or dissolution. Glasses that have mobile sodium and potassium ions are subject to leaching processes in which the mobile ions on the surface diffuse into the surrounding liquid. When the glass is not immersed in a liquid, but just exposed to the atmosphere, the leaching process is called *weathering*. The leaching process can occur either under high humidity conditions or under alternating cycles of moisture condensation and evaporation. The alkali leached from the surface reacts with water to form hydroxides and thus creates a very high pH environment on the surface of the glass. The high pH environment causes the silicate network to dissolve and the formation of visible deposits on the surface eventually reduces the transparency of the material. The weathering resistance of several types of commercial glasses is summarized in Fig. 14-42. In general, those that have good water resistance also exhibit good weatherability. Because the leaching of alkali ions initiates the weathering process, glasses that are low in alkali generally exhibit good weathering resistance as well.

14.3.3f Young's Modulus, E and Poisson's Ratio. On a macroscopic scale, the stress on stretching a glass is the combined response of the individual bonds of the material. Thus, the stress-strain curve of a glass and its slope (the Young's Modulus, E) depend on the bond type, concentration of each type, and the network structure. Glasses with linear polymeric chain structure have generally low moduli. Although the primary intramolecular chain bonds are strong, the intermolecular chain bonds of the van der Waals type are weak and dictate the stress-strain response. Layered or sheet structures have also lower moduli than the three-dimensional network structures because of the weak bonding between layers. In three-dimensional network structures, the concentration of nonbridging oxygen ions will affect the moduli.

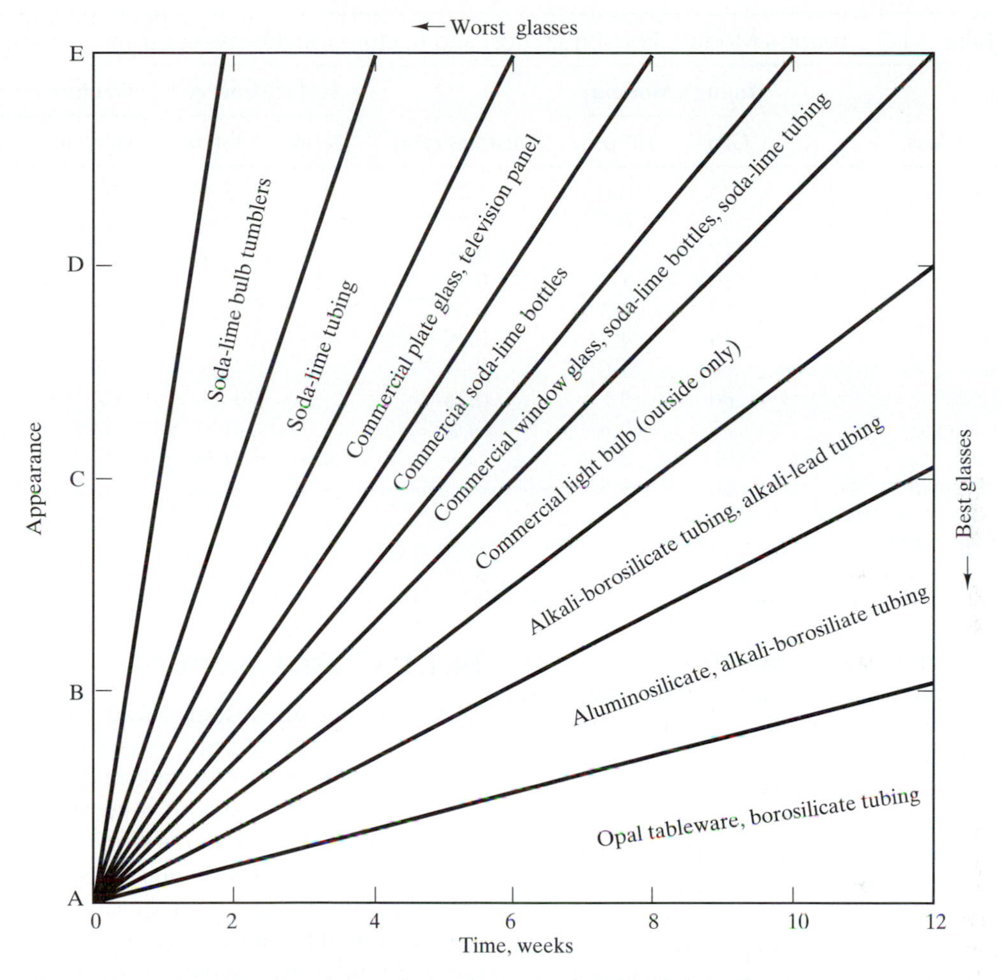

Fig. 14-42 Average weatherability of various glasses under high humidity.

The Young's Moduli of selected glasses are given in Table 14-7. Boric oxide glass has a planar sheet structure and therefore manifests a much lower modulus than silica and germania. The Young's moduli of most commercial silicate glasses range from 50 GPa ($\sim 7 \times 10^6$ psi) for low-T_g glasses to about 75 GPa (11×10^6 psi) for fused silica and E-glass fibers. The high moduli glasses in Table 14-7 have other oxides and/or silicon nitride present in the structure.

The Poisson's ratio, ν, of pure vitreous silica is 0.17. Most of the multicomponent silicate glasses have Poisson's ratios in the range 0.2 to 0.3. Because glasses are isotropic, the shear and bulk moduli can be calculated using the relationships given in Eqs. 4-23a, b, and c.

14.3.4 Mechanical Properties

Being brittle materials, the strength of glasses are again dependent on the macroscopic flaw sizes, the volume of the material, the environmental conditions, and the fracture toughness, as in ceramics. The fracture toughness of some of the glasses are also given in Table 14-7 and we note the very low values of K_{I_c}. The observed strengths of silicate glasses depend on the volume of material (diameter), flaw sizes (surface condition), and the environmental conditions (temperature). For pure silica, all in pristine condition and tested at room temperature, the strength of a 0.03 mm (0.001 in.) diameter fiber is higher than a 1 mm (0.04 in.) diameter fiber. This reflects the depen-

Table 14-7 Young's Moduli, Poisson's ratios, and fracture toughnesses of some glasses.

Glass	Young's Modulus		Poisson's ratio	Surface energy		Fracture toughness	
	GPa	10^6 psi		N/m	lbf/ft	MPa$\sqrt{m}$	ksi$\sqrt{in.}$
SiO_2	73	10.6	0.17	4.4	0.30	0.79	0.72
GeO_2	43	6.2	0.21	...	...	...	...
B_2O_3	17	2.5	0.26	...	...	...	...
Se	10	1.5	0.33	...	...	...	...
As_2Se_3	17	2.5	0.29	...	...	...	...
$Ge_3As_4Se_3$	29	4.2	0.26	...	...	...	...
Ge-As-Se	18	2.6	...	...	...	0.25	0.23
PbO-$4B_2O_3$	60	8.7	0.26	...	...	...	...
Aluminosilicate	75	10.8	...	4.0	0.27	0.91	0.83
Borosilicate	60	8.7	...	4.6	0.31	0.77	0.70
Soda-lime-silica	66	9.6	...	3.9	0.26	0.75	0.68
$20La_2O_3$-$30Al_2O_3$-$50SiO_2$	100	14.5	...	...	...	...	...
$37.5Y_20_3$-$8.6Al_2O_4$-$37.5Si_3N_4$-$6.4SiO_2$	183	26.5	0.28	...	...	...	...

dence of the Weibull probability of failure on the volume of the material and that larger cracks are likely to occur in a larger volume. Between two 0.03 mm diameter fibers, the strength is higher in vacuum than in air, which has some relative humidity. In general, glasses exhibit lesser strength in the presence of water or air with high humidity. The strength of glasses with abraded surfaces will be lower due to the surface flaws introduced.

The hardness values of some glasses are shown in Table 14-8. Because of the manner in which hardnesses are obtained by indentation, the values for silicate glasses do not vary over a wide range. Pure silica has a microindentation hardness of about 6.0 GPa and the values for most silicate glasses are within about 30 percent of this value. Again, glasses containing some alumina and silicon nitride exhibit higher hardnesses, while nonoxide glasses are softer.

14.4 Glass-Ceramics and Secondary Processing

Glass-ceramics are glasses that have been formed into shape and then given a special secondary processing in order to achieve the desired effect. We shall describe this unique, secondary processing along with the other secondary processes of annealing, tempering, and ion-exchange.

14.4.1 Glass-Ceramics

This term signifies that these materials contain ceramics (crystalline) and glasses (amorphous). The ceramic phase is polycrystalline (many crystalline grains) and is the result of a controlled nucleation and growth process from a glass matrix containing nucleating agents. These nucleating agents are the equivalent in the glass matrix to the inoculants in liquid metal melts to induce heterogeneous nucleation. Glass-ceramics are by definition at least 50 percent crystalline (ceramics) by volume and are generally more than 90 percent crystalline. Because the volume of the ceramic phase can be controlled by varying the glass composition, glass-ceramics represent one of the true engineered materials.

The most common method of producing glass-ceramics is to melt the batch of the constituents, hence the shape is fabricated while it is in the glassy state by using most of the methods described in Sec. 14.3.1. Then the formed glass is given the thermal processing schematically illustrated in Fig. 14-43 (i.e., a nucleation process at a low temperature and then growth of the crystalline nuclei at a higher temperature).

14.4.1a Nucleation. The heterogeneous nucleation of crystalline nuclei on the surfaces of the nucleating agents in the glass matrix is accomplished near the annealing point of the host glass matrix. The nucleating agents can be dispersed phases of a metal,

Table 14-8 Hardnesses of glasses.

Glass	Hardness	
	GPa	10^6 psi
SiO_2	6.2(a)	0.90(a)
	4.7(b)	0.68(b)
GeO_2	2.4(b)	0.35(b)
B_2O_3	2.0(a)	0.29(a)
Soda-lime-silica	4.5(b)	0.65(b)
$12.5Na_2O$-$17.5CaO_2$-$70SiO_2$	5.5(b)	0.80(b)
$37MgO$-$13Al_2O_2$-$50SiO_2$	6.6(a)	0.95(a)
$18Y_2O_3$-$24Al_2O_3$-$58SiO_2$	8.1(a)	1.17(a)
$37.5Y_2O_3$-$18.6Al_2O_3$-$37.5Si_3N_4$-$6.4SiO_2$	11.4(b)	1.65(b)
$30Na_2O$-$70B_2O_3$	4.7(a)	0.68(a)
Sodium borosilicate	4.1(b)	0.59(b)
$10BaF_2$-$30ZnF_2$-$30YF_3$-$30ThF_4$	3.0(a)	0.44(a)
$57ZrF_4$-$36BaF_2$-$3LaF_3$-$4AlF_3$	2.5(a)	0.36(a)
Se	0.3(a)	0.04(a)
As_2Se_3	1.3(a)	0.19(a)
$Ge_{25}Se_{75}$	1.9(a)	0.28(a)
$Ge_{40}As_{15}S_{45}$	2.6(a)	0.38(a)
$Ge_{30}Sb_{10}Se_{60}$	1.9(a)	0.28(a)

(a) Vickers (b) Knoop

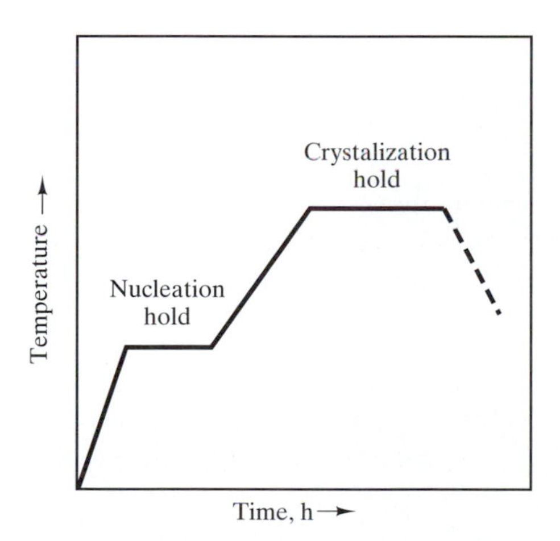

Fig. 14-43 Heat treatment process to nucleate and grow ceramic crystals in glass.

titanate, zirconate, or flouride, which are structurally incompatible with the matrix and are highly unstable as glasses. The crystalline nuclei are formed epitaxially on the surfaces of the nucleating agents.

14.4.1b Crystal Growth. The crystalline nuclei are allowed to grow at a higher temperature to allow the formation of the polycrystalline structure; the same growth process as used in the solid nuclei in a melt to produce a solid polycrystalline structure. Because of the initial high density of the small crystalline nuclei, the resulting microstructure is a highly uniform, fine-grained polycrystalline ceramic in a matrix of residual glass. Depending on the composition and heat treatment, grain sizes can range from <0.1 μm ($<$ 4μ-in.) to >10 μm (>400 μ-in.).

The process that was just described to produce glass-ceramics is commonly known as *ceramming* and because the nucleation and growth process occurs at a relatively high viscosity, the shapes of the components in the glass state are preserved and undergo little or no shrinkage (1 to 3 percent) or deformation. Unlike ceramics that are formed via sintering, glass ceramics are fully densified, free of porosity, and their uniform microstructure ensures that their physical properties are reproducible. Thus, they are stronger than their parent glasses and can be designed to meet a wide range of thermal expansion requirements with excellent thermal shock resistance, dielectric properties, chemical durability, or high strength and toughness.

The glass-ceramics are limited to those compositions or constituents that can form glasses and whose nucleation and growth of the crystalline phase can be controlled. The most used glass-ceramics are the 3-D network structures of aluminosilicate tetrahedra comprised of compositions from the LAS-lithium aluminum silicates (Li_2O-Al_2O_3-SiO_2)and the MAS-magnesium aluminum silicates (MgO-Al_2O_3-SiO_2) systems. Because the silicate structures have very low-bulk thermal expansions, their glass-ceramics also exhibit low thermal expansion with excellent thermal stability and thermal shock resistance.

The glass-ceramics of the LAS (spodumene) system have great commercial importance because of their very low thermal expansion and excellent chemical durability. The crystalline ceramics that form in this system are essentially a single solid solution of either β-quartz or β-spodumene and the glass-ceramics contain only a residual glass or accessory phase. The glass-ceramics containing either β-quartz or β-spodumene are produced from the same glass composition but with different ceramming temperatures. The nucleating agent for the β-quartz solid solution is a mixture

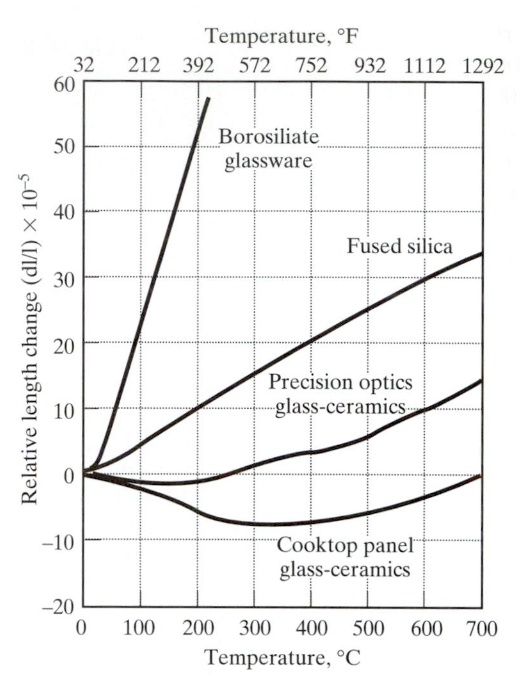

Fig. 14-44 Relative changes in lengths versus temperature for two β-quartz glass-ceramics compared to those of fused silica and borosilicate glassware.

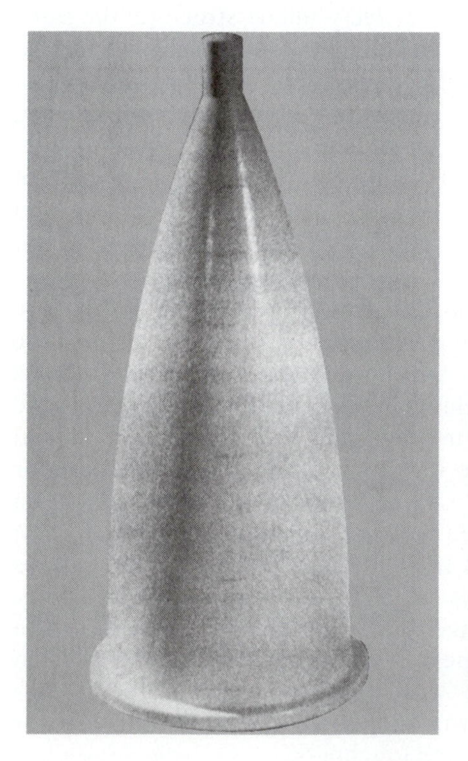

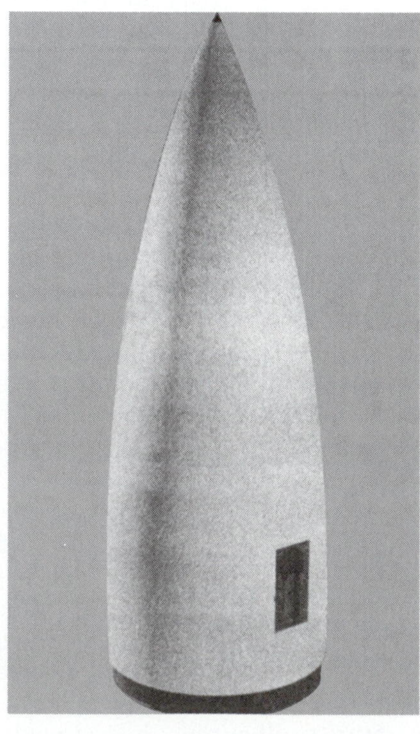

Fig. 14-45 (Left) As-cast radome blank made of pyroceram prior to finishing, (Right) finished pyroceram radome including guidance section shell.

of ZrO_2 and TiO_2 and the ceramming temperature is $\leq$ 900°C ($\leq$ 1652°F). This produces very small crystals (<100 nm or 1000 Å) and when these are coupled with the low birefringence (refraction of light into two slightly different rays) of β-quartz phase and the closely matched refractive indices of the crystals and the residual glass, a transparent, yet highly crystalline glass-ceramic body is obtained. The glass-ceramics containing the β-spodumene solid solution is cerammed at temperatures > 1000°C (>1832°F). This involves the transformation of the β-quartz to β-spodumene between 900° and 1000°C (between 1652° and 1832°F) that is accompanied by a five- to ten-fold increase in grain size and results in an opaque glass-ceramic.

Figure 14-44 illustrates the near-zero or negative thermal expansions of two β-quartz glass ceramics for precision optics and cook-top panels compared with those of fused silica and a borosilicate glass. The near-zero thermal expansion behavior combined with transparency, optical polishability, excellent chemical durability, and strength greater than that of glass has made the β-quartz glass ceramics well-suited for telescopic mirror blanks, infrared transmitting range tops, stove windows, and transparent cookware. The β-spodumene glass-ceramics have found wide application as cookware, architectural sheets, benchtops, hot-plate tops, heat exchangers, regenerators, and as matrices for fiber-reinforced composites.

An example of a glass-ceramic of the MAS (cordierite) system is Pyroceram 9606 (a Corning tradename) that is a silica-rich cordierite-based (see silicate isolated group structure in Sec. 14.3.2) glass-ceramic containing a mixture of phases that are mostly cordierite, with some cristobalite, rutile, and magnesium titanate. The cristobalite phase increases the overall thermal expansion of the glass-ceramics, but it provides a unique strengthening method after the ceramming process. When the surface of the glass-ceramic is treated with a hot alkali solution, it leaches the cristobalite phase and leaves a porous skin, about 0.38 mm (0.015 in.) thick, which serves as an abrasion-resistant layer that inhibits the initiation of flaws and increases the strength from 120 to 240 MPa (17 to 35 ksi). This material, with its high transparency to radar, is the standard glass-ceramic used for missile radomes. A radome (missile nose cone—dome over the radar) is an aerodynamic fairing over the radar antenna of high-speed, radar-guided, air-to-air, and surface-to-air missiles. An as-cast and a finished Pyroceram radome are shown in Figs. 14-45. The composition, the crystalline phases, and some properties of Pyroceram 9606 are given in Table 14-9.

Table 14-9 Composition, properties, and application of cordierite-based Pyroceram glass-ceramic.

Composition, wt%(a)		
SiO_2	56.1	
Al_2O_3	19.8	
MgO	14.7	
CaO	0.1	
TiO_2	8.9	
Ab_2O_3	0.3	
Fe_2O_3	0.1	
Properties:		
Crystalline phases		Cordierite Cristobalite Rutile Mg-dititanate
Coefficient of thermal expansion (10–700°C or 32–1290°F). $K^{-1} \times 10^{-6}$		4.5
Modulus of rupture. MPa (ksi)		250 (36)
Fracture toughness, $MPa\sqrt{m}$ ($ksi\sqrt{in.}$ in.)		2.2 (2.0)
Young's modulus, GPa (10^6 psi)		120 (17.4)
Thermal conductivity, W/m · K (cal/cm · s · °C)		38 (0.09)
Hardness, HK_{100}		700
Dielectric constant at 8.6 GHz		5.5
Loss tangent at 8.6 GHz		0.0003
Softening temperature, °C (°F)		>1300 (>2370)
Commercial applications		
Radomes		

(a) Corning Glass Works specifications

14.4.2 Other Secondary Processing

As in the processing of metals, residual stresses can occur after the processing of glass and can be very detrimental, if these are tensile in nature. If they are tensile, they need to be relieved and this is done by *annealing*, as in metals. The process is simply to heat the glass to the annealing point.

The plate or sheet glass that is produced by the float glass method (Sec. 14.3.1) is in the annealed condition, which means that the glass has no residual stresses. Of the 4.45×10^9 ft^2 U.S. production of float-glass in 1989, about 55 percent is used in the annealed condition and the other 45 percent is used as safety glass. Annealed glass can be cut, ground, drilled, and beveled as needed. However, glass in the annealed state is not very strong and can easily be broken by thermal gradients, wind loads, impact, and so on. Moreover, when annealed glass breaks, it forms typically dagger shape shards radiating from the origin of failure, as shown in Fig. 14-46a. Because of concerns over personal injuries and possible deaths resulting from accidents involving annealed glass, laws were passed requiring safety glass in buildings and vehicles in the United States and elsewhere.

Safety glasses maybe laminated or tempered. Of the 45 percent float glass that was used in 1989 for safety glasses, about 42 percent was laminated and 58 percent was tempered. A laminated glass consists of two pieces of annealed glass sandwiching a central layer of polyvinyl butyral (PVB) plastic. When a laminated glass breaks, the shards that are produced adhere to the tough PVB interlayer, instead of falling or flying, as the case may be, as shown in Fig. 14-46b. Tempered glass is produced by tempering (to be described next) the annealed glass to produce a residual compressive stress on the surface, and because of this compressive stress, is about five times stronger than the annealed glass when loaded in a bending mode. When a tempered glass breaks, small rock-saltlike particles are produced that are incapable of causing major injuries because of their size, blunt shape, and dull edges, as shown in Fig. 14.46c.

14.4.2a Tempering Process. The tempering process involves heating the glass to a tempering range and then rapidly cooling or quenching the surface. The temperature profiles of the surface and the midplane of the plate or sheet during the heating and rapid cooling steps are schematically shown in Fig. 14-47. The tempering range is shown in Fig. 14-48 to be between 620° and 640°C and the viscosity of the glass is between $10^{10.2}$ and $10^{9.5}$ P. During the rapid cooling, the surface is first solidified and cooled below the strain point while the midplane section is still above the strain point. The quenched outer fibers would like to contract but they are constrained from contracting by the midplane glassy material. The outer fibers are initially pulled in by the inside section and are initially in tension while the midsection is in

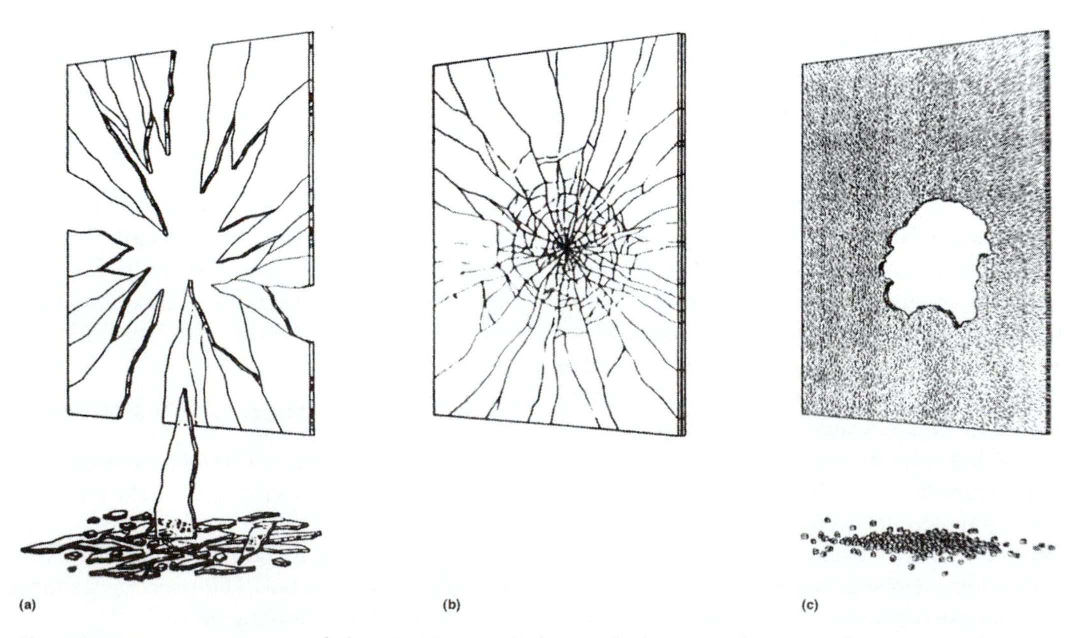

Fig. 14-46 Fracture patterns of glass in (a) annealed state, (b) laminated state, and (c) tempered state.

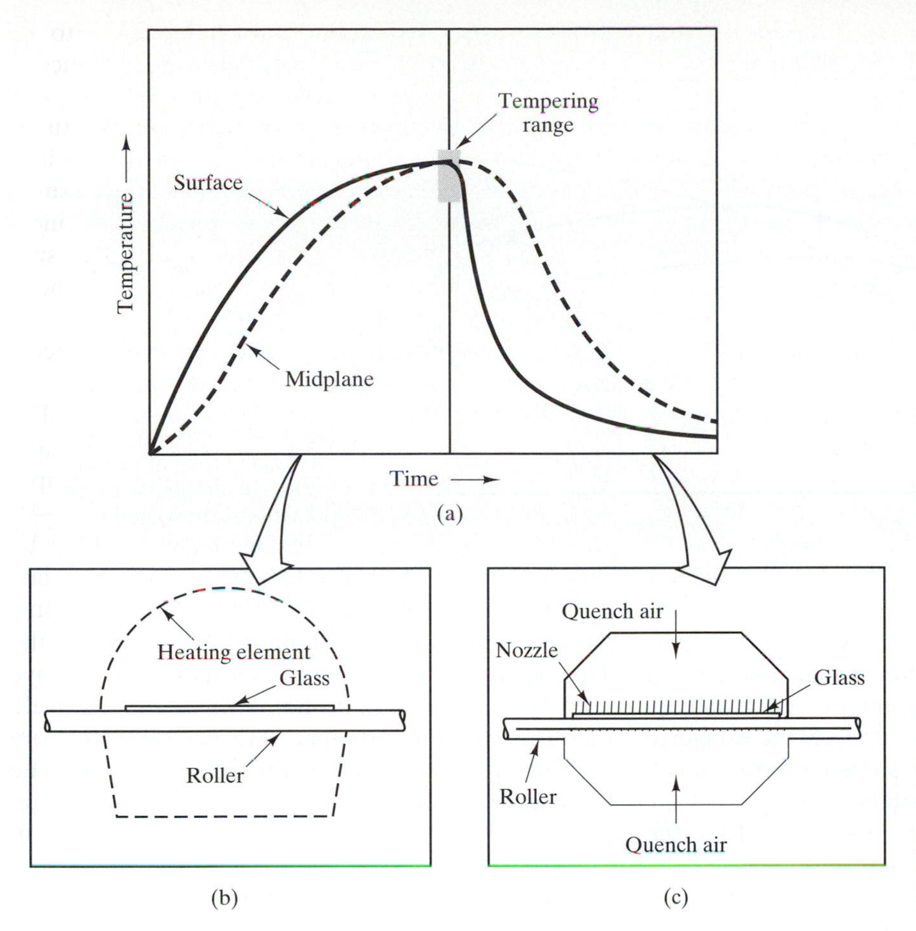

Fig. 14-47 Tempering Process and Equipment. (a) Surface and midplane temperature profiles during heating and cooling, (b) heating portion of tempering furnace, and (c) quenching section of tempering furnace.

compression. As cooling continues, the midsection material contracts, but it is restrained by the already solid surface layer; therefore, the midsection will be in tension. Thus, there is a reversal in the stress-states of the surface and midsection material during the initial cooling and at the final cooling stage. At the end, a compressive stress develops at the surface. When the tempered glass is bent, this compressive stress must first be over-ridden before a tensile state of stress develops on the surface. Therefore, the modulus of rupture (rupture strength) of a tempered glass is considerably higher than that of annealed glass.

14.4.2b Ion-Exchange. A surface compressive layer can also be developed in glasses by ion-exchange. This is referred to as chemical tempering, chemical strengthening, or "stuffing." Basically, a larger ion is exchanged for a smaller ion and is stuffed into the same site occupied by the smaller ion in the glass structure. The stuffed surface layer is put into a

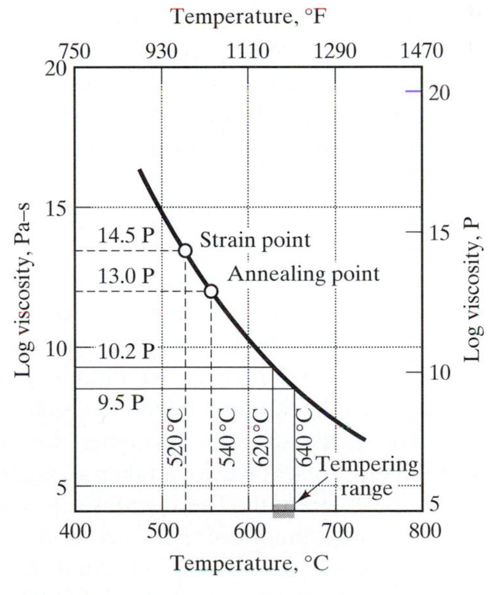

Fig. 14-48 Temperatures and viscosities of glass in the tempering range.

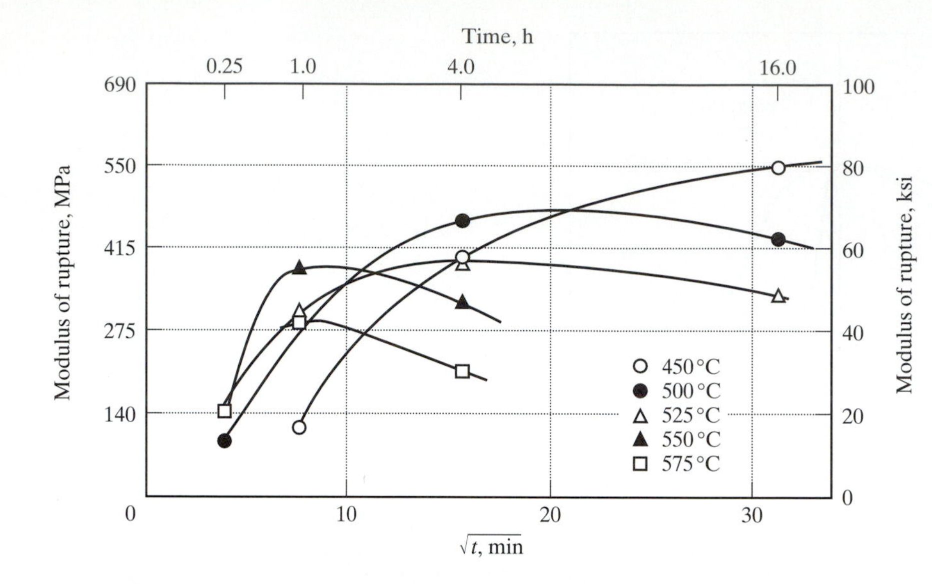

Fig. 14-49 Strengthening as a function of time at various ion-exchange temperatures for potassium-exchanged $1.1Na_2O \cdot 1Al_2O_3 \cdot 4SiO_2$.

compressive state. The exchange is done at a temperature below the glass transition temperature, T_g, in order not to give the glass network the opportunity to relax in a short time. The farther below the exchange temperature is from T_g, the less stress relaxation will occur, but the exchange rate will be slower.

The ion exchange pairs most widely used are the sodium ion (Na^+) for the lithium ion (Li^+) and the potassium ion (K^+) for the sodium ion (Na^+). The ion-exchange is accomplished at temperatures in the range of from above 200°C (392°F) to above the annealing point of the glass. The most practical ion-exchange is done by immersing the glass into molten sodium nitrate (melting point of 310°C, or 590°F) or potassium nitrate (melting point of 337°C, or 639°F). The sodium nitrate bath can be used up to 450°C (842°F), while potassium nitrate can be used up to 525°C (977°F).

Because of stress-relaxation, the strengthening of stuffed ion-exchanged glasses as measured by the modulus of rupture is very temperature-dependent. The response is very similar to precipitation-hardening in metallic alloys, whereby a peak strength is achieved with time and then the strength decreases, as shown in Fig. 14-49. The peak strength is higher for lower-exchange temperatures; however, it takes a longer time to attain the peak strength. The decrease in strength after the peak is attributed to stress-relaxation.

Not all glasses can be strengthened by ion-exchange. The compositions that yield the best strength results are the alumina-silica compositions. The borosilicate glasses cannot build up enough stresses because they do not contain enough alkali ions. The common soda-lime glasses exhibit very high rates of stress relaxation and therefore cannot retain the stresses developed by the ion-exchange

14.5 Design and Selection of Ceramics and Glasses

14.5.1 Design Methodology

Ceramics and glasses are brittle materials and the design methodology for both materials are therefore the same. In general, brittle materials are used as much as possible in compressive loading since they have much better properties in compression than in tension. Nonetheless, tensile stresses may arise in certain applications and they need to be accounted for during the design stage. For example, glasses are commonly used as containers or pressure vessels where tensile hoop stress can arise.

The three general design techniques available are: *empirical, deterministic,* and *probabilistic design.* Empirical design is a trial-and-error approach to match the material to a particular component and the environmental conditions to which the component is exposed. If the prototype prepared from the selected material fails, a new material is used. This process is repeated until a suitable material is found. Deterministic design relies on measured average properties of materials, combined with a safety factor, to produce successful results. This is the common method used in design using metallic materials, for example steels, because the scatter of the yield and tensile strengths in metals is very small.

The principal difference between brittle materials (ceramics and glasses) and metallic materials is the much broader scatter in the strength properties of ceramics and glasses. The strength in brittle materials is controlled by the distribution of volumetric and surface flaws (defects) as the result of their (ceramics and glasses) processing, handling, and use. Because of the wide variation in the defects, the tensile strength exhibits a very broad scatter. Thus, the design technique best suited for brittle materials is the probabilistic design that consists of determining the design stress for a certain size of the material and for a certain risk or probability of failure.

The most widely used probability design technique is the Weibull probability or statistics that was discussed earlier in relation to the determination of uniaxial tensile strength. The same Weibull equations are used in the design process for brittle materials. The design is based on Eqs. 14-14 and 14-15 that give the probability of failure, or the probability of survival when the brittle material is subjected to a uniaxial tensile stress of σ. We see from these two equations that if we were to use σ as the design uniaxial tensile stress, σ will depend on the selected probability of failure or survival, the Weibull modulus, and the size of the component. The dependence of the tensile design stress, σ_{design}, on these parameters is summarized in Fig. 14-50. The MOR_o (modulus of rupture) in this graph is the maximum outer-fiber stress in a four-point bend test specimen that causes failure 63.2 percent of the time, F is the probability of failure, and V are the sizes of the materials. We see that the allowable design stress is very sensitive to the Weibull modulus, m, and the probability of failure. Since the probability of failure is usually chosen and fixed for the design, we see that the best material from a group of brittle ceramic materials is that which has the highest Weibull modulus, m, assuming that the MOR_o values are not greatly different.

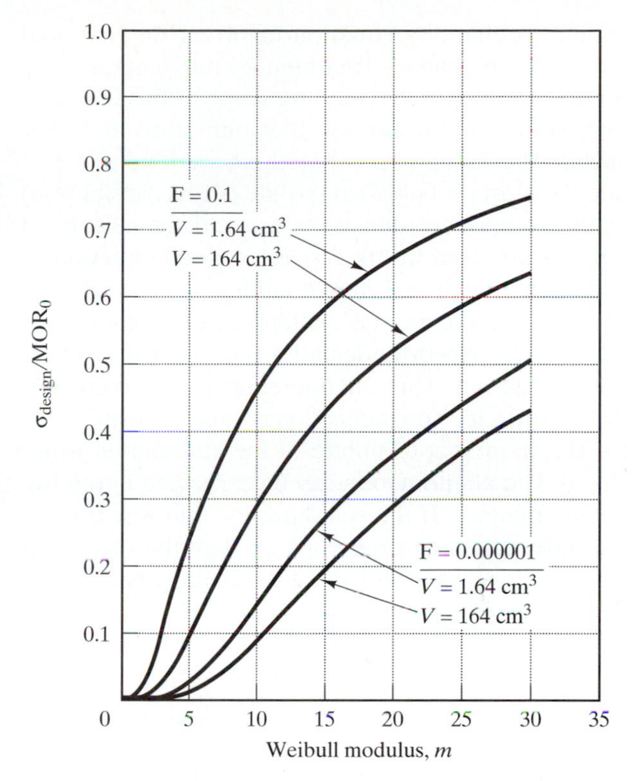

Fig. 14-50 Ratio of design stresses to characteristic strength for a specific test bar size as a function of the Weibull modulus, probability of failure, and volumes stressed in uniaxial tension.

The choice of a probability of failure for a component depends on the consequence of failure. If the failure of a component will result in the possible loss of life, factory downtime, loss of customers, or damage to the company's image, then we should choose a very small probability of failure, such as one in one million (F = 0.000001) or less. When the component can be easily replaced and causes little inconvenience to the user, the probability of failure of one in one thousand (F = 0.001) might be acceptable. When evaluating experimentally the benefits of a prototype component, a failure rate or probability of one in ten (F = 0.1) could be acceptable.

Figure 14-50 is for the special case of uniaxial tension of ceramic components (i.e. only one principal stress, σ_1). In general, however, the stress state in a component under stress is tri-axial and the other principal stresses, σ_2, and σ_3, can affect the probability of failure of the components. This leads to a more general probability equation involving the principal stresses σ_1, σ_2, and σ_3. Treatment of this general probability is, however, beyond the scope of this text. It is only sufficient to mention that numerical methods, such as the finite-element method, are used to calculate the most probable stress-state (stress distribution) in the component during its use and the calculated stresses are used in the general Weibull equation to determine the probability of failure.

Most oxide ceramics exhibit the phenomenon of slow crack growth under load that must be considered in design. This phenomenon is observed predominantly in "toughened" glass and is responsible for the spontaneous failure of the automobile windshield. The phenomenon can be expressed in the following manner. If the standard test that was used to measure σ_{TS} takes a time t (test), then the stress that the sample will support safely for a time t (service) is

$$\left(\frac{\sigma}{\sigma_{TS}}\right)^n = \frac{t\ (\text{test})}{t\ (\text{service})} \tag{14-26}$$

where n is the slow crack growth exponent. Its value for oxides is between 10 and 20 at room temperature. For carbides and nitrides, n can be as large as 100.

EXAMPLE 14-7

Using the ceramic material in Example 14-6, determine the maximum design tensile stress the material must be subjected to for (a) a one in ten probability of failure, and (b) a one in a million probability of failure. The volume of the material is 100 cm^3.

SOLUTION The design tensile stress will be obtained by using Fig. 14-50. We need to have the Weibull modulus and the characteristic strength of the material; m = 27.8 and $\sigma_0 = (MOR)_0 = 57.3$ ksi from Example 14-6.

Now from Fig. 14-50, and using V = 164 and m = 27.8,

(a) for F = 0.1, (one in ten), σ_2 tensile design stress = 0.62 σ_0

(b) for F = 0.000001, (one in a million), σ_2 tensile design stress = 0.41 σ_0.

If σ_1 is the maximum design tensile for a 100 cm^3 the material, using Eq. 14-22,

$$\sigma_1 = \sigma_2 x \left(\frac{V_2}{V_1}\right)^{1/m} = \sigma_2 x \left(\frac{164}{100}\right)^{1/27.8} = 1.018\ \sigma_2$$

Thus, for

(a) F = 0.1, $\sigma_1 = 0.62 \times 57.3 \times 1.018 = 36.1$ ksi

(b) F = 0.000001, $\sigma_1 = 0.41 \times 57.3 \times 1.018$ = 23.9 ksi.

14.5.2 Selection of Structural Ceramics

The process of selecting a structural ceramic material for application in a component involves the consideration of the best combination of properties that include the strength, the maximum operating temperature capability, the Young's modulus, the coefficient of thermal expansion, the thermal conductivity, the thermal shock resistance, the manufacturability, and the cost.

Table 14-10 lists the major types of structural ceramics and their typical properties. As a broad generalization, oxide materials are less expensive and are easier to process, whereas the non-oxides offer better mechanical properties and greater thermal stability. As a design strategy, alumina (Al_2O_3) is considered the default (the first to choose) material. Alumina offers very good, but not outstanding, performance with respect to strength, toughness, wear resistance, thermal stability, and thermal shock resistance. It is excellent in terms of high-thermal conductivity, and it is available at relatively low cost. If the design requires high strength, then silicon nitride and silicon carbide can be considered. Figure 14-51 shows the flexure strength of alumina at various temperatures together with those of silicon nitride, silicon carbide, and partially stabilized zirconia, and indicates the range of operating temperatures for these materials. If toughness is paramount at low temperatures, the more expensive partially or fully stabilized zirconia can be considered. If catastrophic failure protection is required at high temperatures and if cost is not an obstacle, a broad range of selection possibilities exists among ceramic matrix fiber-reinforced composites.

Structural ceramics are currently finding applications in automotive engines since current federal government regulations on mileage and emissions from cars can be better met by increasing the operating temperature of the internal combustion engines and exhaust gases. The material most suited for automotive gas and diesel engines is silicon nitride and the properties of three types of silicon nitride are

Table 14-10 Major types of structural ceramics and their properties. [1]*

Material	Crystal structure	Theoretical Density g/cm^3	Knoop or Vickers hardness		Transreverse rupture strength		Fracture toughness		Young's Modulous		Poisson's ratio	Thermal expansion, $10^{-6}/K$	Thermal conductivity $W/m \cdot K$
			GPa	10^4 psi	MPa	ksi	$MPa\sqrt{m}$	$ksi\sqrt{in.}$	GPa	10^6 psi			
Glass-ceramics	Variable	2.4–5.9	6–7	0.9–1.0	70–350	10–51	2.4	2.2	83–138	12–20	0.24	5–17	2.0–5.4(a) 2.7–3.0(b)
Pyrex glass	Amorphous	2.52	5	0.7	69	10	0.75	0.7	70	10	0.2	4.6	1.3(a) 1.7(c)
TiO_2	Rutile tetragonal	4.25	7–11	1.0–1.6	69–103	10–15	2.5	2.3	283	41	0.28	9.4	8.8(a)
Al_2O_3	Hexagonal	3.97	18–23	2.6–3.3	276–1034	40–150	2.7–4.2	2.5–3.8	380	55	0.26	7.2–8.6	27.2(a) 5.8(d)
Cr_2O_3	Hexagonal	5.21	29	4.2	> 262	> 38	3.9	3.5	> 103	> 15		7.5	10–33(e)
Mullite	Orthorhombic	2.8	···	···	185	27	2.2	2.0	145	21	0.25	5.7	5.2(a) 3.3(d)
Partially stablized ZrO_2	Cubic monoclinic, tetragonal	5.70–5.75	10–11	1.5–1.6	600–700	87–102	(f)	(f)	205	30	0.23	8.9–10.6	1.8–2.2
Fully stablized ZrO_2	Cubic	5.56–6.1	10–15	1.5–2.2	245	36	2.8	2.5	97–207	14–30	0.23–0.32	13.5	1.7(a) 1.9(g)
Plasma-sprayed ZrO_2	Cubic	5.6–5.7	···	···	6–80	0.9–12	1.3–3.2	1.2–2.9	48(h)	7	0.25	7.6–10.5	0.69–2.4
CeO_2	Cubic	7.28	···	···	···	···	···	···	172	25	0.27–0.31	13	9.6(a) 1.2(d)
TiB_2	Hexagonal	4.5–4.54	15–45	1.5–6.5	700–1000	102–145	6–8	5.5–7.3	514–574	75–83	0.09–0.13	8.1	65–120(i) 33–80(j)
TiC	Cubic	4.92	28–35	4.0–5.1	241–276	35–40	···	···	430	62	0.19	7.4–8.6	33(a) 43(d)
TaC	Cubic	14.4–14.5	16–24	2.3–3.5	97–290	14–42	···	···	285	41	0.24	6.7	32(a) 40(d)
Cr_3C_2	Orthorhombic	6.70	10–18	1.5–2.6	49	7.1	···	···	373	54		9.8	19

(*Continued*)

Table 14-10 Major types of structural ceramics and their properties. (*Continued*)

Material	Crystal structure	Theoretical Density g/cm^3	Knoop or Vickers hardness		Transreverse rupture strength		Fracture toughness		Young's Modulous		Poisson's ratio	Thermal expansion, $10^{-6}/K$	Thermal conductivity $W/m \cdot K$
			GPa	10^4 psi	MPa	ksi	$MPa\sqrt{m}$	$ksi\sqrt{in.}$	GPa	10^6 psi			
Cemented carbides.	Variable	5.8–15.2	8–20	1.2–2.9	758–3275	110–475	5–18	4.6–16.4	396–654	57–95	0.2–0.29	4.0–8.3	16.3–119
SiC	α. hexagonal	3.21	20–30	2.9–4.4	(i)	(i)	(m)	(m)	207–483	30–70	0.19	4.3–5.6	63–155(a) 21–33(d)
	β. cubic	3.21	...	...	...	...	...	...	...	...	...	...	...
SiC (CVD)	β. cubic	3.21	28–44	4.1–6.4	(n)	(n)	5–7	4.6–6.4	415–441	60–64		5.5	121(a) 34.6(g)
Si_3N_4	α. hexagonal	3.18	8–19	1.2–2.8	(o)	(o)	(p)	(p)	304	44	0.24	3.0	9–30(a)
	β. hexagonal	3.19	...	...	...	...	...	...	...	...	...	...	...
TiN	Cubic	5.43–5.44	16–20	2.3–2.9	...	...	...	...	251	36	...	8.0	24(s) 67.8(q) 56.9(r)

*Letters in parenthesis refer to the footnotes. Please refer to Ref. 1 at the end of the Chapter.

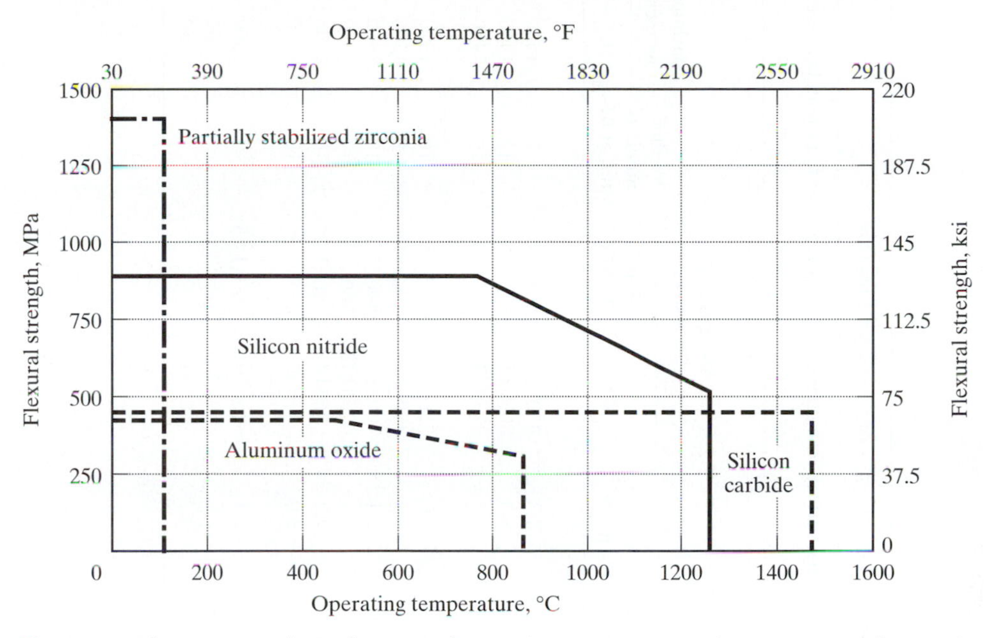

Fig. 14-51 Flexure strengths and ranges of operating temperatures of some structural ceramic materials.

given in Table 14-11, along with those of silicon carbide, three oxide ceramics, and the three common metals used in automotive industry—cast iron, steel, and aluminum. The following is a summary of the key features of the ceramic materials in Table 14-11:

- Hot-pressed silicon nitride (HPSN) exhibits the highest specific strength (strength/density) at 600°C (1112°F) of all the materials in Table 14-11 and has excellent thermal shock resistance.
- Sintered silicon nitride (SSN) has a high strength and can be formed into complex shapes.
- Reaction-bonded silicon nitride (RBSN) can be formed into complex shapes with no firing shrinkage.
- Hot-pressed silicon carbide (HPSC) is the strongest of the silicon carbide family and maintains strength to very high temperatures (1500°C, or 2732°F).
- Sintered silicon carbide (SSC) has high-temperature capability and can be formed into complex shapes.
- Reaction-bonded silicon carbide (RBSC) can be formed into complex shapes and has high thermal capability.
- Partially stabilized zirconia (PSZ) is a good insulator and has high strength and toughness. It has a thermal expansion close to iron, facilitating shrink fit attachments.
- Lithium-aluminum-silicate (LAS) is a good insulator and has very low thermal expansion.
- Aluminum titanate is a good insulator with relatively low thermal expansion.

14.5.3 Selection of Refractories

Refractories are ceramic materials that can withstand high temperatures and can maintain their physical and mechanical properties at these high temperatures. They are very important materials in the metallurgical industries, especially the iron and steel industry, which consumes about 63 percent of the refractories.

The most common refractories are made from natural materials and are composed of metal oxides, or carbon, graphite, or silicon carbide. Newer synthetic carbides, nitrides, borides, silicides, and high-purity oxides are used for higher service temperatures when the expense is warranted. Two-thirds of all refractories used by the industry are preformed bricks and other fired shapes. The remainder take the form of monolithic materials, such as castables, plastics, and gunning or ramming mixes. Refractory insulating materials are used to reduce heat losses and conserve

Table 14-11 Properties of silicon nitride compared with other structural ceramics and metals.

These properties are typical of these classes of materials, but in many cases, large variations may exist between various formulations. Strengths are for four-point bending spans of 19.05/9.525 mm (0.745/0.375 in.) and bar-cross sections of 6.35 × 3.175 mm (0.25 × 0.125 in.).

Material	Young's Modulus		Poisson's ratio	Thermal conductivity W/m . K		Thermal expansion, $\times 10^{-6}$/K	Specific Heat J/g . °C	Density, g/cm^3	Strength, MPa (ksi)		Weibull modulus (m), at RT	Maximum use temperature	
	GPa	10^6 psi		RT	600°C (1130°F)				RT	600°C (1130°F)		°C	°F
Silicon nitride													
Hot-pressed (HPSN)	290	42.1	0.3	29	22	2.7	0.75	3.3	830 (120)	805 (117)	7	1400	2550
Sintered (SSN)	290	42.1	0.28	33	18	3.1	1.1	3.3	800 (116)	725 (105)	13	1400	2550
Reactions-bonded (RBSN)	200	29.0	0.22	10	10	3.1	0.87	2.7	295 (43)	295 (43)	10	1400	2550
Silicon carbide													
Hot-pressed (HPSC)	430	62.4	0.17	80	51	4.6	0.67	3.3	550 (80)	520 (75)	10	1500	2730
Sintered (SSC)	390	56.6	0.16	71	48	4.2	0.59	3.2	490 (71)	490 (71)	9	1500	2730
Reaction-bonded (RBSC)	413	59.9	0.24	225	70	4.3	1.0	3.1	390 (57)	390 (57)	10	1300	2370
Partially stabilized zirconia (PSZ)	205	29.7	0.30	2.9	2.9	10.5	0.5	5.9	1020 (148)	580 (84)	14	950	1740
Lithium-aluminum-silicate	68	9.9	0.27	1.4	1.9	0.5(a)	0.78	2.3	96 (14)	96 (14)	10	1200	2190
Aluminum-titanate	11	1.6	0.22–0.26	2	5	1.0	0.88	3.0	41 (6)	...	15	1200	2190
Common metals (Reference)													
Cast iron	170	24.7	0.28	49	40	12	0.45	7.1	620 (90)	100 (14.5)	...	500	930
Steel	200	29.0	0.28	38		14	0.45	7.8	1550 (218)	140 (20)	...	600	1110
Aluminum	70	10.2	0.33	160		22.4	0.96	2.7	370 (54)	0	...	350	660

(a) Maximum excursion from 350 to 800°C (1600 to 1470°F). Initial expansion is negative.

energy and can be made in the form of bricks, fiber board or blanket, or special vacuum cast shapes.

Refractories are categorized as either clay or nonclay products and as acid or basic refractories. Acid refractories are those containing SiO_2 or ZrO_2, while basic refractories are those that contain MgO, CaO, or CrO_3. The acid-base classification signifies the reactivity of the refractories with acidic or basic materials. Acid refractories will react with and will be degraded by basic materials, while basic refractories will react with and be degraded by acid materials. This classification comes to fore when selecting refractories to be in contact with the slags of molten metals. Clay refractories are fireclay and high-alumina, while nonclay refractories are those designated as basic (magnesia, dolomite, chromite, and their combinations), extra-high alumina, mullite, silica, silicon carbide, and zircon.

Fireclay generally consists of the mineral kaolinite ($Al_2O_3 \cdot 2SiO_2 \cdot 2H_2O$) with minor amounts of other clay minerals and impurities. Fireclay refractories have alumina content ranging from 25 to 45 percent and are classified as low-, medium-, high, or super-duty based on their resistance to high temperature or on their refractoriness. Their refractoriness increases with increasing alumina content and decreasing amounts of impurities, such as alkalis and iron oxide. They offer moderate to high resistance to thermal stress, low thermal expansion, and adequate thermal insulation. They are essentially acidic and offer some resistance to attack by acidic materials but fail rapidly at high temperatures when exposed to chemically basic materials.

High-alumina refractories are generally composed of bauxite or other raw materials that contain from 50 to 87.5 percent alumina. Refractories with compositions greater than these are considered nonclay refractories. Depending on its composition and the impurities present, a high-alumina refractory is generally multipurpose, and offers fair-to-excellent resistance to chipping and somewhat higher volume stability than most other clay refractories.

Magnesia brick consists mainly of the mineral periclase (MgO) and is available in chemically-bonded, pitch-bonded, burned or fired, and burned and pitch-impregnated forms. Historically, the source of MgO is from calcined natural magnesite ($MgCO_3$). Higher purity MgO up to 98 percent can be obtained from sea water or underground brines. *Dolomite* brick is made from a highly calcined natural mineral of the composition $CaCO_3 \cdot MgCO_3$. Because the lime component hydrates readily, unfired bricks made with dolomite grain are usually tar- or pitch-bonded; burned dolomite brick is often impregnated with tar or pitch to extend its service life. *Chrome* brick is from naturally occurring chrome ore, which is a complex solid solution series of spinel-type minerals among which are picrochromite ($MgO \cdot Cr_2O_3$) and ferrous chromite ($FeO \cdot Cr_2O_3$).

High-purity magnesite and chrome-free compositions of high-purity sea water or brine-well magnesia provide maximum refractoriness and resistance to oxides. These compositions, as well as dolomite and magnesite-dolomite, can be coal-tar bonded and used in basic oxygen furnace (BOF) applications. Chrome-magnesite bricks, which have a larger proportion of chrome than magnesite, exhibit less thermal expansion than the high-magnesia composition. Magnesite-carbon bricks are used in the most severe areas of the BOF and electric arc furnace and offer outstanding resistance to corrosion by steel-making slags.

Mullite ($3Al_2O_3 \cdot 2SiO_2$) bricks are made from kyanite, sillimanite, andalusite, bauxite, or mixtures of aluminum silicate minerals, with an overall composition of alumina of about 70 percent. This is the most stable of the alumina-silica combination and offers excellent resistance to loading at high temperatures.

Silica brick is made from quartzites, silica gravel, or novaculite by bonding the particles with the addition of 3 to 3.5 percent of CaO, which forms a small amount of glass upon firing. It is characterized by a high coefficient of thermal expansion between room temperature and 500°C (932°F) and, therefore, must be heated and cooled extremely slowly through this temperature range. Silica brick is available in three grades: super-duty—has a very low alumina and alkali contents; regular, or conventional, duty; and coke-oven quality.

Semi-silica brick is made from siliceous clays and contains mainly cristobalite bonded with a glassy phase. It contains typically 18 to 25 percent alumina and 72 to 80 percent silica. This brick has excellent load-bearing quality to 1300°C (2372°F), but like regular silica brick, has a relatively high thermal expansion from room temperature to 500°C (932°F).

Refractory selection is made with the ultimate goal of achieving minimum overall operating costs per ton of product (e.g., steel). The objective is to obtain the optimum service performances of the refractory linings in the heating and melting furnaces and in the ladles in order to prolong their replacement intervals. In general, the refractories are required to withstand the following service conditions:

- All ranges of temperatures up to 1760°C (3200°F) or higher.
- Sudden changes in temperature—"thermal shock."
- Stresses, mainly compressive, at both high and low temperatures.
- The action of slags, ranging from acid to basic in character.
- The action of molten metals, always at high temperatures and capable of exerting great pressures and buoyant forces.
- The action of gases, such as SO_2, CO, Cl_2, CH_4, H_2O, and volatile oxides and salts of metals. All are capable of penetrating and reacting with the brick.
- The action of dusts in gases, which may be fluxing or nonfluxing, and of acids or basics.
- Impact and abrasive forces at both high and low temperatures, as those in stacks of furnaces.

At the same time a refractory is being subjected to one or more of the above conditions, it may be required to function as (a) a storehouse for heat, as in checkers, (b) as a conductor of heat, as in walls of furnaces, or (c) as an insulator.

The *physical and mechanical properties* of some fired refractory bricks are shown in Table 14-12. The bulk density and porosity were earlier described in Sec. 14.2.4, Eqs. 14-1, 14-2, 14-3. The other pertinent properties to be considered in selecting refractories are the following:

Cold Crushing Strength. The cold crushing strength of ceramics is the ability of a ceramic to withstand handling/shipping and impact/abrasion at low tempertatures. It reflects the processing of the ceramics through their porosity, bulk density, and other attributes, and is different from the compressive strength. For example, soft-burned clay brick exhibits compressive crushing strength above 1,000 psi, but may be very friable, particularly if dry-pressed. It does not indicate strength at service temperature.

Softening Temperature. A standard test for determining the softening behavior of alumina-silica bricks is done with the *Pyrometric Cone Equivalent* or *P.C.E. Test.* In this test, the softening behavior of a small-cone sample is compared with that of standard pyrometric cones of known time-temperature softening behavior. The P.C.E. is reported as the number of that standard cone whose tip behaves in the same manner as the tip of the refractory cone being tested. Figure 14-52 shows the P.C.E. of various alumina-silica refractories when heated under the test conditions. The results are only used as guides. Normally, the maximum service temperatures of fireclay brick are considerably below the P.C.E. temperatures.

Silica and basic bricks do not conform to the P.C.E. test. The melting point of pure silica about 1727°C (3140°F) is not impressively high, and yet a silica brick, containing 95-98% SiO_2, can be brought up to temperatures exceeding 1649°C (3000°F). The strong, interlocking crystalline structure of silica makes it approach the sharp melting behavior of a pure compound. Silica can remain rigid to within a few degrees of its melting point.

On the other hand, the main constituents of basic bricks (magnesite, chrome, chrome-magnesite, and forsterite) all melt at temperatures far above that of silica. However, the raw materials of basic bricks contain other minerals, considerably lower in refractoriness than the main constituents. These low-refractory minerals can form liquids at temperatures as low as 1315°C (2400°F) and they largely dictate the melting behavior of the brick as a whole. The detrimental effects of these low-refractory minerals can be minimized or even eliminated by corrective additions, special firing treatments, or more directly, by using a higher purity raw material.

Strength and Behavior under Load at High Temperatures. The softening temperature correlates with the strength of refractories at high temperatures. At the softening temperature, the glassy or vitrified ceramic bond gradually becomes a viscous liquid and the refractory starts exhibiting plastic flow. Under load, the refractory behavior will depend on the amount of load, the duration and rate of loading, the amount and viscosity of the liquid, and on the crystalline structure of the solids. Loaded heavily and rapidly as in normal compression testing, all refractories will fail in shear until temperatures are

Table 14-12 Some physical and mechanical properties of some fired refractory bricks.

	Magnesia (95% MgO)	Chrome (30% Cr_2O_3)	90% alumina	70% alumina	Zircon	Fireclay (Missouri superduty)	Silicon carbide	Silica (superduty)
Bulk density, g/cm^3	2.80–2.95	3.06–3.14	2.90–2.96	2.53–2.60	3.60–3.72	2.31–2.37	2.56–2.66	1.78–1.87
Potosity, %	15–19	16–20	14–18	17.5–21.5	19–23	11–14	11–15	20–24
Cold crushing strength, MPa	48–70	35–55	62–95	27–48	48–76	12–21	69–83	27–41
(ksi)	(7–10)	(5–8)	(9–14)	(4–7)	(7–11)	(1.7–3)	(10–12)	(4–6)
Modulus of rupture, MPa	17–24	14–21	17–21	7.6–11	15–23	4.8–6.9	21–24	4.1–6.9
(ksi)	(2.5–3.5)	(2–3)	(2.5–3)	(1.1–1.6)	(2.2–3.3)	(0.7–1)	(3–3.5)	(0.6–1)
Reheat test, % permanent linear change after heating to:								
1600°C (2910°F)	...	...	...	+3.5 to 6.0	...	0.0 to 0.9	...	...
1650°C (3000°F)	...	...	...	...	0.0	...	...	...
1725°C (3135°F)	−0.2 to 1.0	...	+0.1 to 1.0	...	...	...	−0.1 to 0.1	...
Load test at 170 KPa (25 ksi); withstands load to temperature, °C (°F)	1620 (2950)	1400 (2550)	1760 (3200)	1450 (2640)	1600 (2910)	1450 (2640)	1650 (3000)	1680 (3055)

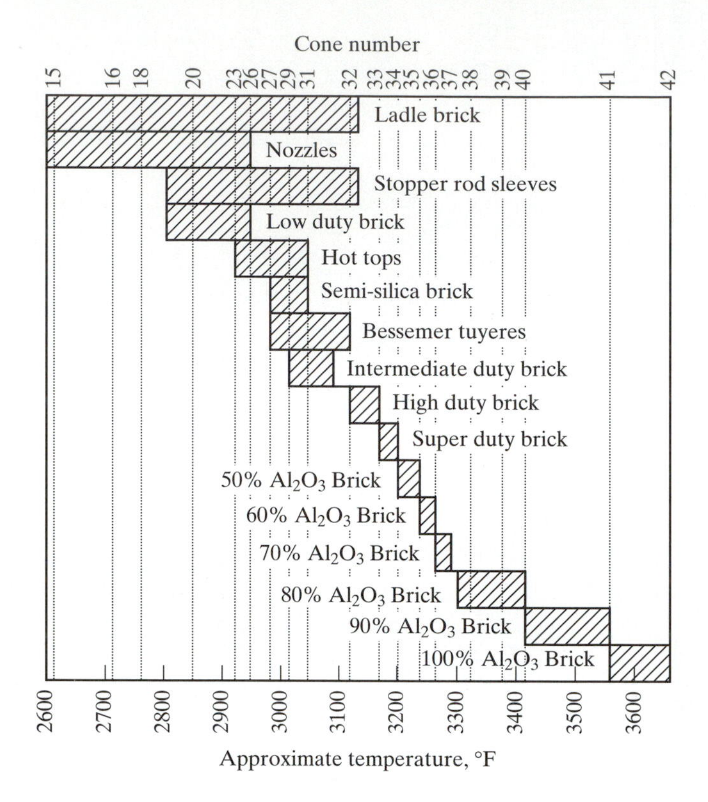

Fig. 14-52 Pyrometric cone equivalent (P.C.E.) alumina-silica refractories

reached at which they contain so much liquid of such low viscosity that almost instant deformation occurs. Under such conditions, the compressive strength or the modulus of rupture may not change materially below the temperatures of initial liquid formation, but will decrease rapidly as the temperatures deviate higher from the temperatures of initial liquid formation.

Under light, sustained loads, comparable to those encountered in service, the refractory will creep, the same phenomenon as in metals. Fireclay refractories have the widest softening temperature range and are most susceptible to creep. The load for the standard test is 170 kPa (25 psi) at a temperature held for $1^1/_2$ hours. The temperature of initial deformation and the total amount of deformation will depend on the flux content, the degree of burn, and the porosity. Impurities will lower the temperature of initial deformation because of fluxing (tends to form eutectics) and the viscosity of the liquid. Hard-burned bricks will creep less than soft-burned bricks since the harder firing develops a more refractory glassy phase that produces higher viscosity liquids at test temperatures.

Clay-bonded, high alumina bricks behave in a similar manner as the fireclay bricks, but the temperature range of plastic deformation is higher because of the greater alumina content. With purer raw materials, denser and superior bonded high alumina bricks can be produced with much improved resistance to deformation under load. In fact, the glassy bond, and not the alumina content, reflects the hot-load behavior.

High-purity silica bricks show little evidence of creep under 170 kPa (25 psi) or even 340 kPa (50 psi), remaining rigid until they fail in shear at temperatures from 1649° to 1704°C (3000° to 3100°F). Because of their strong, crystalline structure, and the viscosity of silica melt, these bricks can contain up to 30 percent liquid before they fail.

The creep of basic bricks under load at high temperatures varies widely depending on the bond structure. The bond structure may vary from gangue silicates in the raw material producing forsterite, $2MgO \cdot SiO_2$, to spinel bonds formed from additions of Al_2O_3 or Cr_2O_3 with MgO, or the direct MgO-MgO bond, or the MgO-chrome spinel bonds. As a result of these variations in bond structure, the creep under load at high temperature will vary from as little as 2 percent strain after 24 hours at 1593°C (2900°F) to failure after only one hour at temperatures as low as 1510°C (2750°F).

Carbon refractories show little or no loss in strength at temperatures up to 1760°C (3200°F). Such refractories are finding increased applications where reducing atmospheres prevail.

Thermal Expansion and Volume Changes. Since they are subjected to temperature changes, linear and volume changes of the refractories are of primary concern because of their effects on thermal stresses and spalling (breaking up into pieces). Figure 14-53 shows the linear expansion curves of some refractories. Those with a uniform expansion rate generally present the least difficulties when temperatures fluctuate widely. Of these, those with the lowest total expansion, as a general rule, are less prone to thermal spalling. Thus, fused and fireclay bricks are very favorable materials. Silica bricks exhibit large linear expansion at temperatures up to 582°C (1080°F), which is due in part to the allotropic changes of quartz to tridymite and then to cristobalite. However, if care is taken in heating and cooling below a dull, red-heat temperature, silica bricks behave admirably at higher temperatures. In fact, one of the problems in using magnesite bricks in roofs of open-hearth furnaces arises from their continuing linear expansion at high temperatures. Temperature changes (at high temperatures) induce thermal stresses on roofs with magnesite bricks but will have no effect on silica roofs.

Permanent volume changes occur during the drying and the firing of the brick. In service, further volume changes may occur when the brick is used at temperatures above that at which it was fired. Volume changes also occur when the brick is fluxed with dusts or slags and subsequent vitrification ensues. *Secondary expansion* in fireclay bricks may occur due to the exfoliation of the clay grains. *"Bloating"* is due to gas evolution and expansion in an overheated and hence pyro-plastic clay body. Bloating usually indicates misapplication of the refractories. However, it is desired in ladle lining

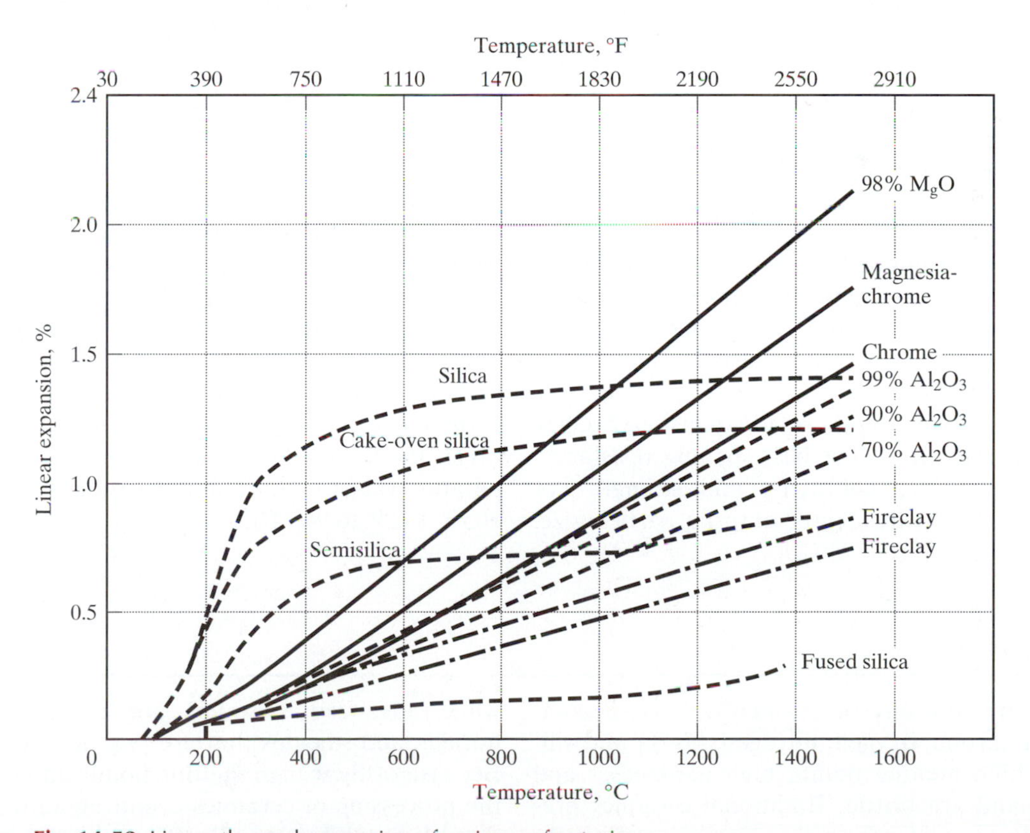

Fig. 14-53 Linear thermal expansion of various refractories.

because it produces a monolithic, liquid-tight lining. Permanent expansion may also occur due to reactions with the environment. Fireclay brick expands about 12 percent when it reacts with alkalies to form alkali-aluminum silicates. This leads to surface peeling of the refractories. With some bricks, expansion or "bursting" occurs due to the absorption of iron oxide.

Resistance to Thermal Shock (Spalling). Spalling is the breaking away of pieces of refractory from the hot face, thus exposing fresh surfaces. There are *three types of spalling: thermal, structural,* and *mechanical.* Thermal spalling results from too rapid an expansion or contraction of the hot face of the refractory due to sudden temperature changes. Structural spalling is that brought about by flux absorption or vitrification that sets up zones in the brick. These zones may result in different expansions and sensitivity to thermal shock from the original brick or may result in shrinkage to induce separation of pieces. Mechanical spalling may be caused by mechanical abuse, as in the removal of slag accumulations from refractory surfaces or by shifting loads and stresses, particularly in arches, where refractories are pinched and crack off.

The thermal shock resistance of ceramics is related to their tensile strength and fracture toughness and is measured by the highest drop in temperature during cooling, ΔT, before the material fractures. This is related to the tensile strength in the following manner.

$$\sigma_{TS} = E\alpha(\mathbf{\Delta T}) \qquad (14\text{-}27)$$

where E is the Young's modulus, α is the coefficient of linear thermal expansion that has units of strain per degree change in temperature, and ΔT is the maximum drop in temperature before surface cracks are observed.

Because of their low thermal expansion, fireclay bricks, as a class, have the best spalling resistance. However, since their elasticity and strength are strongly affected by the degree of firing, particle size, and porosity, there is a wide variation in performance. Basic bricks generally do not have the best resistance to thermal shock because of their high coefficients of thermal expansion. However, many special basic brick compositions, such as the spinel-bonded magnesite brick, have spalling resistance equivalent to most fireclay or high alumina bricks. As indicated earlier, the spalling resistance of silica brick varies from very poor to excellent, depending on whether they are used in the temperature range below or above 593°C (1100°F).

Abrasion Resistance. The resistance of refractories to abrasion and erosion is primarily a function of the strength. Since strength is dependent on the degree of burn, it is also related to porosity and bulk density. Resistance to abrasion does not change much below 1093°C (2000°F), but decreases at higher temperatures.

Permeability. Permeability is a measure of the rate of diffusion of liquids and gases through the refractory and is governed by the size and number of connected pores or channels that are continuous from one side of the refractory to the other. Thus, it has some relation to the porosity. Permeability to liquids increases with increasing temperature because the viscosities of liquids decrease with temperature—the opposite is true for gases, whose viscosities increase with temperature. While permeability is not commonly determined, it is important because the liquids may contain impurities or substances that can react with the brick.

Carbon Monoxide Resistance. In a refractory that contains unreacted iron oxide and is exposed to carbon monoxide for an extended period of time, solid carbon may be catalytically deposited. The deposited carbon may develop sufficient internal stresses to break the bonds of the refractory. The critical temperature range for carbon deposition is from 495° to 505°C (925° to 940°F).

Summary

Ceramics are ionically or covalently bonded compounds of carbon, oxygen, nitrogen, boron and silicon with high melting points, high hardnesses and strengths, and are brittle. Traditional ceramics predominantly are oxides from clay or nonclay minerals, while advanced ceramics are carbides, nitrides, borides, and silicides that are primarily synthesized. Because of their high melting points and brittleness, the processing of ceramics is entirely different from that of metallic materials. Processing of traditional ceramics involves the mixing of the powders, making the powders plastic in an aqueous or nonaqeous media in order to shape them, drying the powders, and firing or sintering the powders to induce bonding between the particles. Because of this processing, traditional ceramics have porosity, the degree of which depends on the method of sintering. It is therefore not surprising that ceramics exhibit a lot of variability in tensile strength. In this regard, ceramics do not have a tensile strength, but rather the most probable tensile strength is based on Weibull statistics. The spread of tensile strengths is dictated by the Weibull modulus; a high Weibull modulus means that the tensile strengths have a narrow distribution, while a low modulus indicates that the tensile strengths have broader distribution. The advanced structural ceramics may have a Weibull modulus of about 20—in comparison, metals have a Weibull modulus of about 100. The property that dictates the tensile strength of ceramics is the fracture toughness that may be improved by alloying with zirconia. Because of the inherent defects, ceramics have much higher compressive strengths that do not have much scatter.

Glasses are formed when the liquid does not crystallize at the melting temperature during cooling. The most common glasses are the silicates that are based on silica. The formation of glass from silica is due to the very high viscosity of the pure silica melt that prevents the silicon and oxygen ions from crystallizing to the cristobalite structure during cooling. This makes silica a glass former. Boric oxide, germania, and phosphoric oxide are also glass formers. We can also prevent the formation of the crystalline structure by breaking or modifying the Si-O bond by adding what are called modifiers. When the latter are added, the viscosity of the melt decreases and the melting temperature is lowered considerably to allow the melting of glasses, in the same way metals are molten. The shaping and processing of glass products are therefore done at high enough temperatures that depend on the viscosity. Glasses have strain, annealing, softening, and working points. The latter is determined when the glass is shaped and formed into products and the viscosity is 10^4 poise—the melting point of glass is when the viscosity is 100 poise.

Glass-ceramics consist of very fine ceramic particles or grains in residual glass (matrix). It is the product of a secondary thermal processing of glass containing nucleating agents. The two-step process is called ceramming that induces a large number of small nuclei to form at a low temperature, and then the nuclei are induced to grow at a higher temperature. Glass ceramics are by definition at least 50 percent crystalline by volume and are generally more than 90 percent crystalline (ceramics).

Glasses undergo other secondary processes such as annealing, tempering, and ion-exchange. The purpose of annealing is to relieve residual stresses such as in metals. Safety glasses are produced by laminating annealed glasses with polyvinyl butyral (PVB) or by tempering. Tempering induces residual compressive stress on the surface when the glass surface is rapidly cooled or quenched from a tempering temperature. Surface compressive stress can also be developed by ion-exchange. This is referred to as chemical tempering, chemical strengthening, or stuffing. The surface is put into compression when larger ions are exchanged with and occupy the sites of smaller ions in the glass.

Design with ceramics and glasses involves probabilistic design methods that essentially uses the Weibull statistics used in determining the probable tensile strength of ceramics. Selection of structural ceramics starts by first evaluating the characteristics of alumina and from there goes to the other advanced structural ceramics. A case study is given in Chap. 16. A lot of the traditional ceramics are used as refractories in the metal industries and their selection was also discussed.

References

1. *Ceramics and Glasses,* vol. 4, Engineered Materials Handbook, ASM International, 1991.
2. *Composites,* vol. 1, Engineered Materials Handbook, ASM International, 1987.
3. W. D. Kingery, H. K. Bowen, and D. R. Ulmann, *Introduction to Ceramics,* 2d ed., John Wiley, 1976.
4. M. F. Ashby and D. R. H. Jones, *Engineering Materials 2,* 1st ed., Pergamon Press, 1986
5. *The Making, Shaping, and Treating of Steel,* U.S. Steel.

Terms and Concepts

Advanced ceramics
Advanced glasses
Allotropes of silica—quartz, tridymite, cristobalite
Annealing of glass
Annealing point
Apparent porosity
Batching and preparation of powders
Bulk density
Cation coordination polyhedron
Ceramming
Clays
Cold crushing strength of refractories
Creep of ceramics
Deterministic design
Dry pressing
Empirical design
Extrusion
Fining
Firing
Float-glass process
Fluxes
Fracture toughness of ceramics
Glass ceramics
Glass container manufacture
Glass fibers
HPSC—hot pressed silicon carbide
HPSN—hot pressed silicon nitride
Imperfections in ceramics
Injection molding
Intermediate oxide
Ion-exchange in glass
Laminated glass
LAS—lithium aluminum silicate
Layered or sheet structures
Liquid phase sintering
Martensitic transformation of zirconia
Melting point of glass
Modulus of rupture—flexure strength
Modifier oxide
Oxide glass formers
Pauling's rules
Permeability of refractories
Phase diagrams of ceramic materials
Plastic forming processes
Plate and sheet glass manufacture
Porosity or void content
Probabilistic design
PSZ—partially stabilized zirconia
Pug mill
Pyroceram
Pyrometric cone equivalent
Radome
RBSC—reaction bonded silicon carbide
RBSN—reaction bonded silicon nitride
Refractories
Silicate structures
Sintering
Slip casting
Slurry casting
Softening point
Solid-state sintering
Spalling of refractories
Spray drying
SSC—sintered silicon carbide
SSN—sintered silicon nitride
Strain point
Structural ceramics
Survival probability
Tape casting
Tempering of glass
Thermal expansion of refractories
Three-dimensional network structures
Total porosity
Traditional ceramics
Traditional glasses
True density
Tubing manufacture
TZP—tetragonal zirconia polycrystals
Viscosity of glass
Weakest-link-model
Weathering of glasses
Weibull modulus, m
Weibull's statistics
Working (forming) point

Questions and Exercise Problems

14.1 In Fig. 14-8, the stoichiometric compound at 20 mol% CaO can be represented by $CaO \cdot 4ZrO_2$. Determine the minimum amount of CaO per kilogram that is needed to form the compound.

14.2 In Fig. 14-8, what are the allotropic transformations of pure zirconia and their temperatures of transformation?

14.3 Indicate the type(s) of and write the three-phase reactions in Fig. 14-8.

14.4 Using Fig. 14-8, what is the maximum solubility in weight percent of calcium oxide in monoclinic zirconia?

14.5 What is the difference in slurry casting and plastic processes in forming ceramics?

14.6 What is the process of sintering? How is it done?

14.7 A sintered silica brick weighs 1 kg when dry, 1.1 kg when saturated with water, and 0.60 kg when suspended in water. Determine the bulk density, the total porosity, and the porosity due to closed pores. The true density of silica is 2,650 kg per m^3.

14.8 A sintered magnesia brick weighs 1.250 kg when dry, 1.350 kg when saturated with water, and displaces 0.5 kg of water. Determine the bulk density, the total porosity, and the fraction of the closed pores. The true density of magnesia is 3,590 kg per m^3.

14.9 Conventional ceramics have fracture toughness of about 5 $MPa\sqrt{m}$, while advanced ceramics may have 20 $MPa\sqrt{m}$. (a) For the same applied load, how large of a surface edge crack can the advanced ceramics tolerate relative to the conventional ceramics, and (b) for the same surface crack size, how much more stress can the advanced ceramic tolerate before fracture?

14.10 What is the Weibull statistics and what does the Weibull modulus signify?

14.11 Seven samples, each with 164 cm^3 volume, of zirconia exhibited flexure strengths of 605, 595, 650, 630, 640, 615, and 620 MPa. (a) Determine the characteristic strength and the Weibull modulus, and (b) the maximum tensile design stress for a 1 in a million probability of failure.

14.12 If the characteristic strength of a brittle material is 150 MPa and its Weibull modulus is 10, determine the maximum tensile design stress for a 100 cm^3 of the material for a one in ten failure probability.

14.13 Why is it possible to increase the fracture toughness of an oxide, such as alumina, when zirconia is alloyed with it? What are the sources of the increased fracture toughness?

14.14 What is the difference between PSZ and TZP?

14.15 Why does glass form easily with silica?

14.16 What are modifier oxides? Intermediate oxides?

14.17 What is viscosity? What is the difference between the working and strain points of glasses?

14.18 What are glass-ceramics? How are they produced?

14.19 What is the tempering process in glass? What does ion-exchange accomplish in some glasses?

14.20 What is spalling? Differentiate the three types of spalling.

15

Plastics and Reinforced Plastics

15.1 Introduction

In this chapter, we shall learn how to design with and select plastics (polymeric materials) for engineering applications. To accomplish this goal, we must have an understanding of the characteristics, the properties, and the processing of plastics. Composites are an entirely different group of materials, but the most common of these is the reinforced plastics. Reinforced plastics constitute the most significant application of plastics (polymers) for engineering structures, and their performance is limited by the properties of the plastics. Thus, it is appropriate to discuss reinforced plastics with plastics in as much as we are discussing the characteristics and properties of plastics—the matrix material and the "weak link" in reinforced plastics.

The terms *plastics, resins,* and *polymers* are usually taken as synonymous. However, there are technical distinctions. A polymer is a pure unadulterated material that is usually taken as the long chain macromolecular (large molecule) product of the polymerization process. The chains may contain various combinations of carbon, hydrogen, oxygen, nitrogen, chlorine, flourine, and sulfur. They are liquid during manufacture and are thus easily moldable into their finished product state as solid. Pure polymers, like pure metals, are seldom used, however. When additives are used with polymers, the "alloys or blends" are technically referred to as plastics or resins. Thus, we shall be using the term plastics or resins interchangeably. We see, therefore, that plastics comprise a large, diverse class of materials, numbering more than 15,000 and still increasing, that exhibit a broad range of properties and processing characteristics. The predominant bonding in this class of materials is covalent.

Composite materials may be defined as the macroscopic combinations of metallic, ceramic, and/or polymeric materials, having an identifiable interface between them. The key words are macroscopic and identifiable interface. One of the materials is called the matrix and the other is called the filler or reinforcement material. The most common of these composite materials is reinforced plastics in which the matrix is plastics and the filler can be either a metal, a ceramic, or a polymer in the form of either particles, short fibers, or long continuous fibers.

The properties and processability of the plastics depend on the structure and chemical composition of the polymers. It is therefore appropriate to start this chapter to briefly describe how the polymers are produced (the process of polymerization) and the characteristics of their structures. Then we shall describe the types of polymers, their properties, their processing, and how to design with and select plastics.

15.2 Polymerization Processes

While there are natural polymeric materials (for example, wood and cotton), the most commercially important engineering polymers are synthesized. Basically, all polymers are formed by the creation of chemical linkages between relatively small molecules or monomers to form very large molecules (macromolecules) or polymers. The linkages are formed by either one or a combination of two types of reactions: *addition* and *condensation* reactions.

15.2.1 Addition Polymerization

The addition polymerization process is characterized by the simple combination of molecules without the generation of any by-products as a result of the combination. The original molecules do not decompose to produce reaction debris that needs to be removed from the reaction. When units of a single monomer are hooked together, the resulting product is a *homopolymer,* such as polyethylene, that is made from the ethylene monomer. When two or more monomers are used in the process, such as ethylene and propylene, the product is a *co-polymer*.

The formation of a homopolymer is analogous to the linking of identical boxcars of a railroad train. Each of the boxcars is analogous to one *small molecule called the mer* that is linked to other identical mers to *form a chain with* many *mers called* poly*mers.* Just as in the boxcars, the mer must have a way to link or bond with other mers at each end (i.e., it must be at least bi-functional).

The actual polymerization process involves the formation of not only one train (a macromolecule), but many trains (macromolecules); each of the macromolecule compete for the available mers (box-cars in the train depot). Thus, at the end of the polymerization process, the product consists of many macromolecules having a variety of lengths, composed of the same kind of mers.

The addition polymerization process may occur in one of three ways. The first addition reaction involves the external *chemical activation* of molecules that causes them to combine in a chain-reaction-type as illustrated in Fig. 15-1a for the polymerization of ethylene. The initial reaction of this process involves the breaking of the double-bond in ethylene, followed by the propagation reaction, and the termination reaction. The bonding of the atoms occur directly within the reacting molecule. The second way for an addition polymerization to occur is through a *rearrangement of atoms* within both reacting molecules as shown in Fig. 15-1b for the polyurethane polymerization. A third way for an addition polymerization to occur is

$$2CH_2=CH_2+R{-}R \rightarrow 2R{+}CH_2{-}CH_2{+}$$

Initial reaction

$$R{+}CH_2-CH_2{+} + (n-1)CH_2=CH_2 \rightarrow$$

$$R{+}CH_2-CH_2{+}_n$$

Propagation reaction

$$R{+}CH_2-CH_2{+}_n + R- \rightarrow$$

$$R{+}CH_2-CH_2{+}_n R$$

Termination reaction (combination)

(a) Catalyst activated bond opening (ethylene polymerization)

$$nO=C=N-R-N=C=O+nH-O-R-OH \rightarrow$$

$$O=C=N{+}R-\overset{H}{N}-\overset{O}{\overset{\|}{C}}-O-R{+}_n OH$$

(b) Rearrangement mechanism (polurethane polymerization)

$$nH-N-(CH_2)_5-C=O \rightarrow$$

$$\left[\overset{H}{N}-(CH_2)_5-\overset{O}{\overset{\|}{C}}\right]_n$$

(c) Ring-opening mechanism (nylon 6 from caprolactam)

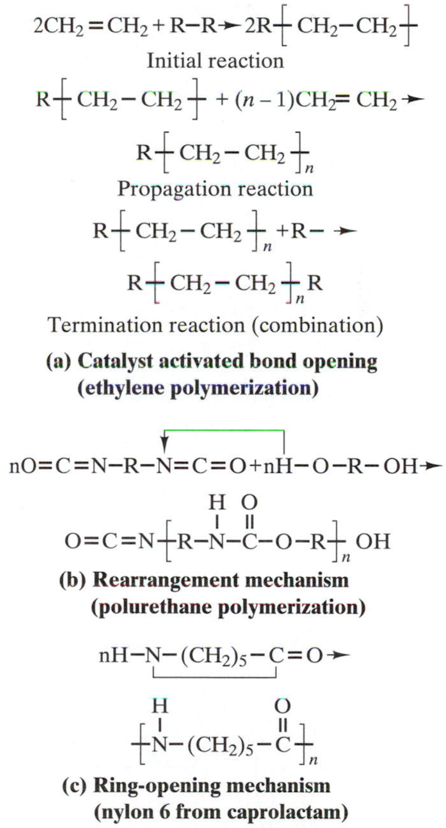

Fig. 15-1 Typical addition polymerizations (no by-products). [1]

through ring opening by which a molecule, composed of a ring of atoms, opens up and connects with other ringlike molecules being opened up under the influence of catalytic activators, Fig. 15-1c. In all the three ways of addition polymerization, the characteristic of an addition polymerization reaction is maintained (i.e., there is neither a by-product nor any loss of atoms from the reacting molecules).

15.2.2 Condensation Polymerization

In this process, the chemical union of two molecules can be achieved only by the *formation of a by-product molecule* (usually small) with atoms from the two molecules to create the link (of the molecules) for the polymerization to continue. In these reactions, the by-product debris normally is removed immediately from the polymer since it may inhibit further polymerization or it may remain as an undesirable impurity in the finished products. Examples of condensation polymerization reactions are illustrated in Fig. 15-2.

15.2.3 Combination Polymerization

It is also possible for both of the above types of reactions to occur sequentially. This is called *combination polymerization* and is the mechanism in the formation of polyesters and polyurethanes. In this process, the *condensation reaction usually occurs first* to form a relatively small polymer, which is then capable of undergoing *further reaction by the addition polymerization with a third ingredient* to form larger polymer molecules. This is illustrated in Fig. 15-3. After Step 1, the condensation step, the still active polyester reacts with the styrene to form what is essentially a polyester-styrene co-polymer.

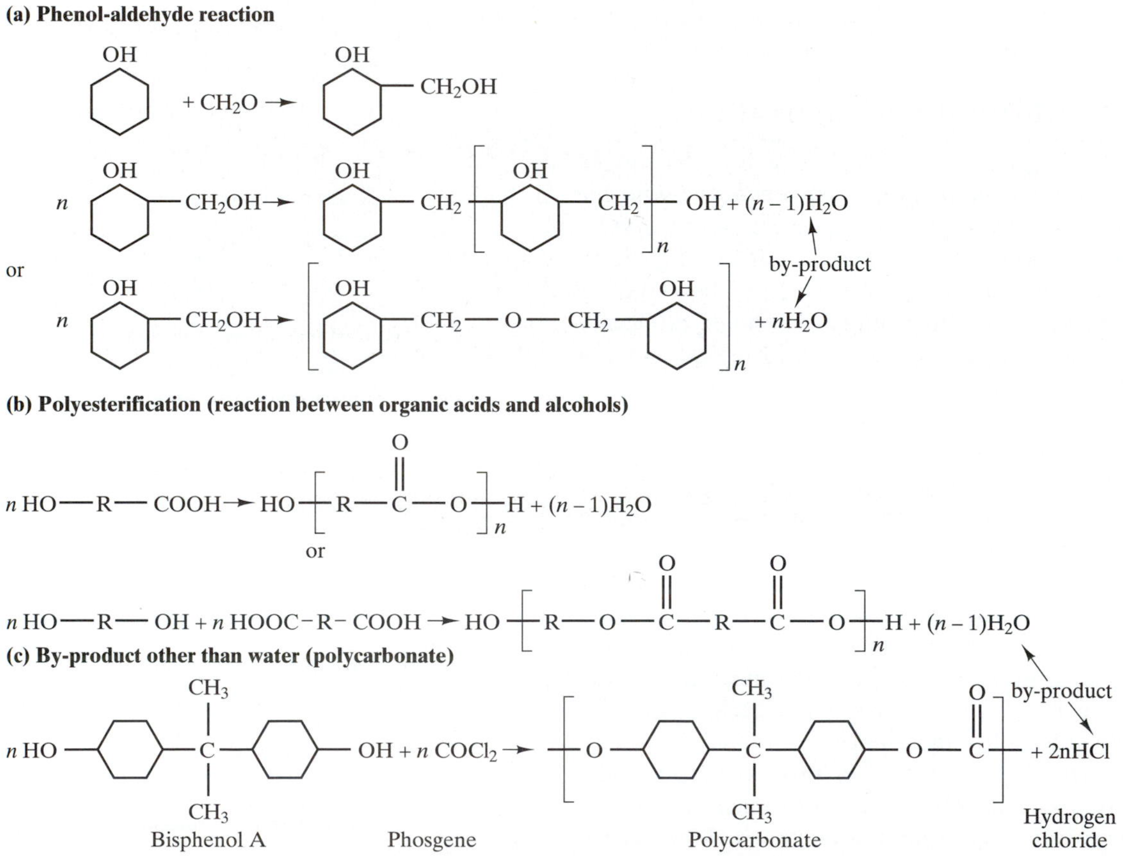

(a) Phenol-aldehyde reaction

OH, + CH_2O → OH, CH_2OH

n OH, CH_2OH → OH, CH_2 [OH, CH_2]$_n$ OH + $(n-1)H_2O$ (by-product)

or

n OH, CH_2OH → [OH, CH_2—O—CH_2, OH]$_n$ + nH_2O

(b) Polyesterification (reaction between organic acids and alcohols)

$n\,HO{-}R{-}COOH \rightarrow HO{-}[R{-}C(=O){-}O]_n{-}H + (n-1)H_2O$

or

$n\,HO{-}R{-}OH + n\,HOOC{-}R{-}COOH \rightarrow HO{-}[R{-}O{-}C(=O){-}R{-}C(=O){-}O]_n{-}H + (n-1)H_2O$

(c) By-product other than water (polycarbonate)

n HO—(bisphenol, C(CH_3)$_2$)—OH + $n\,COCl_2$ → [O—(C(CH_3)$_2$)—O—C(=O)] + 2nHCl (by-product)

Bisphenol A, Phosgene, Polycarbonate, Hydrogen chloride

Fig. 15-2 Examples of condensation polymerization with by-products of water in (a) and (b), and hydrogen chloride in (c). [1]

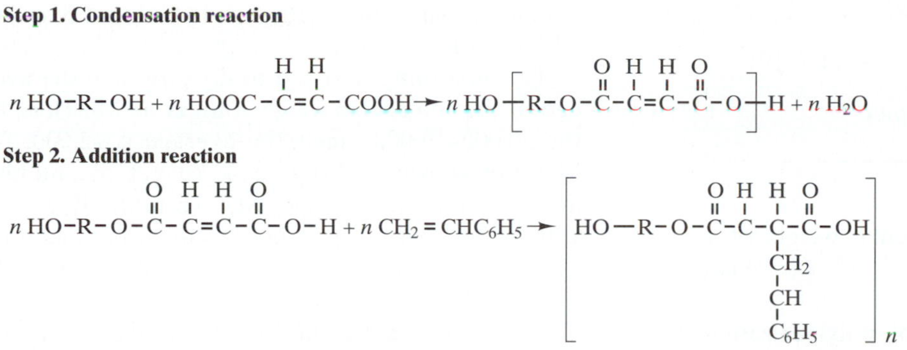

Step 1. Condensation reaction

$$n\,\text{HO-R-OH} + n\,\text{HOOC-}\overset{\text{H}}{\text{C}}\text{=}\overset{\text{H}}{\text{C}}\text{-COOH} \rightarrow n\,\text{HO}\left[\text{R-O-}\overset{\text{O}}{\overset{\|}{\text{C}}}\text{-}\overset{\text{H}}{\text{C}}\text{=}\overset{\text{H}}{\text{C}}\text{-}\overset{\text{O}}{\overset{\|}{\text{C}}}\text{-O}\right]\text{H} + n\,\text{H}_2\text{O}$$

Step 2. Addition reaction

$$n\,\text{HO-R-O-}\overset{\text{O}}{\overset{\|}{\text{C}}}\text{-}\overset{\text{H}}{\text{C}}\text{=}\overset{\text{H}}{\text{C}}\text{-}\overset{\text{O}}{\overset{\|}{\text{C}}}\text{-O-H} + n\,\text{CH}_2\text{=CHC}_6\text{H}_5 \rightarrow \left[\text{HO-R-O-}\overset{\text{O}}{\overset{\|}{\text{C}}}\text{-}\overset{\text{H}}{\text{C}}\text{-}\underset{\underset{\underset{\text{C}_6\text{H}_5}{|}}{\underset{\text{CH}}{|}}}{\underset{\text{CH}_2}{\overset{\text{H}}{\text{C}}}}\text{-}\overset{\text{O}}{\overset{\|}{\text{C}}}\text{-OH}\right]_n$$

Fig. 15-3 Curing of polyester—an example of combination polarization. [1]

15.3 *Structural Features of Polymers and Basic Properties*

The resulting properties of plastics depend on the following structural characteristics.

15.3.1 Molecular Weight (MW)

The *MW refers to the average weight* of the molecules in the mixture of different sizes of molecules that make up the polymer. It is expressed *either as number average MW or as weight average MW*. The number average MW, M_n, is based on the sum of the number fractions of the weight of each specie or size of the molecule present. The expression for the number average MW, M_n, is

$$M_n = \sum x_i M_i \tag{15-1}$$

where x_i is the number fraction of molecules with average molecular weight, M_i and is equal to

$$x_i = n_i / N_T \tag{15-2}$$

where n_i is the number of molecules with average MW of M_i, and $N_T = \sum n_i$ is the total number of molecules.

The weight average MW, M_w, is based on the weight fractions of each specie or size of molecule present in the polymer and is expressed as

$$M_w = \sum f_i M_i \tag{15-3}$$

where f_i is the weight fraction of molecules with average MW M_i, and is equal to

$$f_i = w_i / W_T \tag{15-4}$$

where w_i is the weight of molecules with average MW M_i, and $W_T = \sum w_i$ is the total weight of the polymer.

Another expression relating to MW is the *degree of polymerization,* DP. The DP refers to the number of monomer molecules that combine to form a single polymer molecule. Thus, in terms of the polymer, the DP is estimated by dividing the number average MW by the mer weight,

$$DP = M_n / w_m \tag{15-5}$$

where w_m is the mer formula weight.

High MW polymers tend to be tougher and more chemically resistant, whereas low MW polymers tend to be weaker and more brittle. For example, in the polyethylene (PE) family, low MW PE is almost wax-like, whereas ultra-high MW PE exhibits outstanding chemical resistance and toughness. However, higher MW polymers require more energy in the form of temperature and pressure to polymerize them.

EXAMPLE 15-1

The mers for some polymers based on the ethylene molecule are shown in Fig. 15-9. What are the mer weights of polyethylene, polystyrene, polyvinyl chloride, and polytetrafluoroethylene?

SOLUTIONS: For the mer weights,

Polyethylene:

$$2C + 4H = (2 \times 12) + (4 \times 1) = 28\text{g/mer}$$

Polystyrene:

$$8C + 8H = (8 \times 12) + (8 \times 1) = 104\text{g/mer}$$

Polyvinyl chloride:

$$2C + 3H + 1C1 = (2 \times 12) + (3 \times 1) + (1 \times 35.45)$$
$$= 62.45\text{g/mer}$$

Polytetrafluoroethylene:

$$2C + 4F = (2 \times 12) + (4 \times 19)$$
$$= 100 g/mer$$

EXAMPLE 15-2

The average number molecular weight of a polytetrafluoroethylene (Teflon) is 62,500. What is the degree of polymerization?

SOLUTION: The degree of polymerization = Ave. MW/mer weight = 62,500/100 = 625.

EXAMPLE 15-3.

During the polymerization of polytetrafluoroethylene (PTFE), the number of polymer chains with their respective ranges in molecular weights are shown in the first two columns of Table EP15-3 below. From this data, determine the (a) number average molecular weight, (b) weight average molecular weight, and (c) the degree of polymerization.

SOLUTION:

(a) The number average molecular weight, M_n, is given by Eq. 15-1. We need to determine x_i, the number fraction of polymers (chains) with molecular weight, M_i. To get x_i, we add all the n_i, number of chains in column 1, to get the total number of polymeric chains, N_T = 49,000 chains. Then, the x_i for each range of molecular weights is the number of chains, n_i divided by N_T. Thus, for the 1,000 chains with 10,000–15,000 MW range, the x_i = 1,000/49,000 = 0.020, which is shown in Column 3. The sum of the x_i's is Column 3 should add up to 1.

The next thing we need to do is to take the average of the molecular weight ranges as M_i. Thus, for the 10,000–15,000 range, the average is 12,500. The M_i's are shown in Column 4. To get the number average molecular weight, M_n, we multiply each x_i with its M_i, and the product is shown in Column 5. The sum of all the products in Column 5 is the M_n = 32,096 g/g-mole.

(b) To get the weight average molecular weight, we use Eq. 15-3. The M_i's are already shown in Column 4. We need to determine the f_is, weight fraction of molecules. To get the f_is, we get $w_i = n_i M_i$ (i.e, we multiply Column 1 by Column 4) and each w_i is shown in Column 6. The sum of all the w_i's is W_T, and each f_i is obtained by dividing w_i by W_T, and is shown in Column 7. Multiplying the numbers in each row of Columns 4 and 7 results in Column 8, and the sum of the values in Column 8 is the weight average of molecular weight, M_w = 33,671g/g-mole.

(c) The degree of polymerization PTFE is M_n divided by the mer weight (100 from Example 15-1) ~321.

EXAMPLE 15-4

The polymerization of polyethylene (PE) is initiated and terminated by hydrogen peroxide (H_2O_2). The resultant reaction, according to Fig 15-la, is

$$n[CH_2 = CH_2] + H_2O_2 \rightarrow$$
$$HO - [CH_2 - CH_2]_n - OH$$

Table EP 15-3 Polymerization data and analyses.

1	2	3	4	5	6	7	8
n_i, no. of chains,	**MW range, $\times 10^{-3}$**	**X_i, number fraction**	**M_i, ave. MW $\times 10^{-3}$**	**$X_i M_i$, $\times 10^{-3}$**	**$W_i = n_i M_i$, $\times 10^{-6}$**	**f_i, weight fraction**	**$f_i M_i$, $\times 10^{-3}$**
1,000	10–15	0.020	12.5	0.250	12.5	0.008	0.100
2,000	15–20	0.041	17.5	0.718	35.0	0.022	0.385
4,000	20–25	0.082	22.5	1.845	90.0	0.057	1.282
10,000	25–30	0.204	27.5	5.610	275.0	0.175	4.812
15,000	30–35	0.306	32.5	9.945	487.5	0.310	10.075
12,000	35–40	0.245	37.5	9.188	450.0	0.286	10.725
3,000	40–45	0.061	42.5	2.592	127.5	0.081	3.442
2,000	45–50	0.041	47.5	1.948	95.0	0.060	2.850
Σ = 49,000		Σ = 1.000		Σ = 32.096	Σ = 1572.5	Σ = 0.999	Σ = 33.671

If the density of PE is 900 kg/m^3, determine (a) the volume of a 1 ton (1,000 kg) PE, and (b) the minimum amount (weight) of hydrogen peroxide needed to produce the 1 ton PE with an average number degree of polymerization of 2,000.

SOLUTION:

(a) The volume of the 1 ton PE = 1,000/900 = 1.111 m^3

(b) The minimum amount of hydrogen peroxide needed is the amount for 100 percent efficiency in the reaction. We see in the reaction that one mole of hydrogen peroxide is needed per mole of the polymer. Therefore, when we obtain the number of moles of the polymer, we multiply it by the molecular weight of hydrogen peroxide to get its weight. To get the number of moles of the polymer, we divide its weight (1,000 kg) by its MW. The average MW is

$$\text{average MW} = \text{DP} \times \text{mer weight}$$
$$= 2{,}000 \times 28 = 56{,}000 \text{ g/g-mole}$$

Number of moles of polymer

$$= \frac{1{,}000 \text{ kg} \times 1{,}000 \text{ g/kg}}{56{,}000 \text{g/g-mole}}$$
$$= \text{moles of } H_2O_2$$

Weight of H_2O_2

$$= \frac{1{,}000}{56} \times 34 = 607.1 \text{ g of } H_2O_2.$$

We need only 0.60 kg of H_2O_2 to produce 1 ton of PE with a DP of 2,000.

EXAMPLE 15-5.

The polymerization of PE using hydrogen peroxide (H_2O_2) is only 80 percent efficient. Determine how much PE is produced per 10 grams of H_2O_2 added when the degree of polymerization is 2,000.

SOLUTION: Again the key here is that the number of moles of H_2O_2 reacted is equal to the number of moles of PE produced. The weight of PE produced is obtained by multiplying the number of moles by the average MW.

moles of H_2O_2 reacted

$$= \frac{10 \times 0.8}{34} = 0.2353 = \text{moles of polymer}$$

Weight of PE

$$= 0.2353 \times 2{,}000 \times 28 = 13{,}777\text{g} = 13.177 \text{ kg}$$

15.3.2 Molecular Weight Distribution

The molecular weight distribution (MWD) of molecules refers to the histogram of the weights or sizes of the molecules in the polymer. If the polymer is made of chains very close to the average weight and length, then the MWD is narrow; if the polymer is made up of a wider range of weights and lengths, then the MWD is broad, as shown in Fig. 15-4a. In general, narrow-range MWD resins have better mechanical properties and can be processed easier, and with better control than the broad-range MWD resins. Figure 15-4b shows the flow characteristics of narrow and wide MWD polymers in terms of viscosity as a function of shear rate, and Fig. 15-4c indicates other factors that influence the viscosity as a function of flow rate.

15.3.3 Linear and Branched Molecules

Earlier, the analogy of connecting boxcars to form a train was made in reference to the formation of polymer macromolecules. This analogy implies that the polymer molecules align themselves into long chains without any side protrusions or branches. Indeed, some molecules do form in this manner and are called *linear molecules,* Fig. 15-5a. The term linear does not imply, however, that the molecules are straight. The open circles in Fig. 15-5a represent the element (which is carbon most of the time) that forms the backbone of the structure, and the small dots signify the hydrogen atom. We see that the linear molecules can curl, twist, or fold unto themselves. In addition to the linear molecules, the polymerization process may produce more complex structures. Polymers may form *branched structures* very similar to the clumps of tree branches, as shown in Fig. 15-5b. These *branches* or *protrusions* in an otherwise linear molecule are referred to as *steric hindrances to close approach.* Thus, linear molecules with no steric hindrances resemble spaghetti noodles that can be densely packed and can slide easier over each other than branched molecules. Linear molecules have higher densities, for example, high-density PE (polyethylene), because of the absence of steric hindrances. In addition, linear molecules have higher tensile strengths, higher stiffness, and higher softening temperatures. Branched molecules, on the other

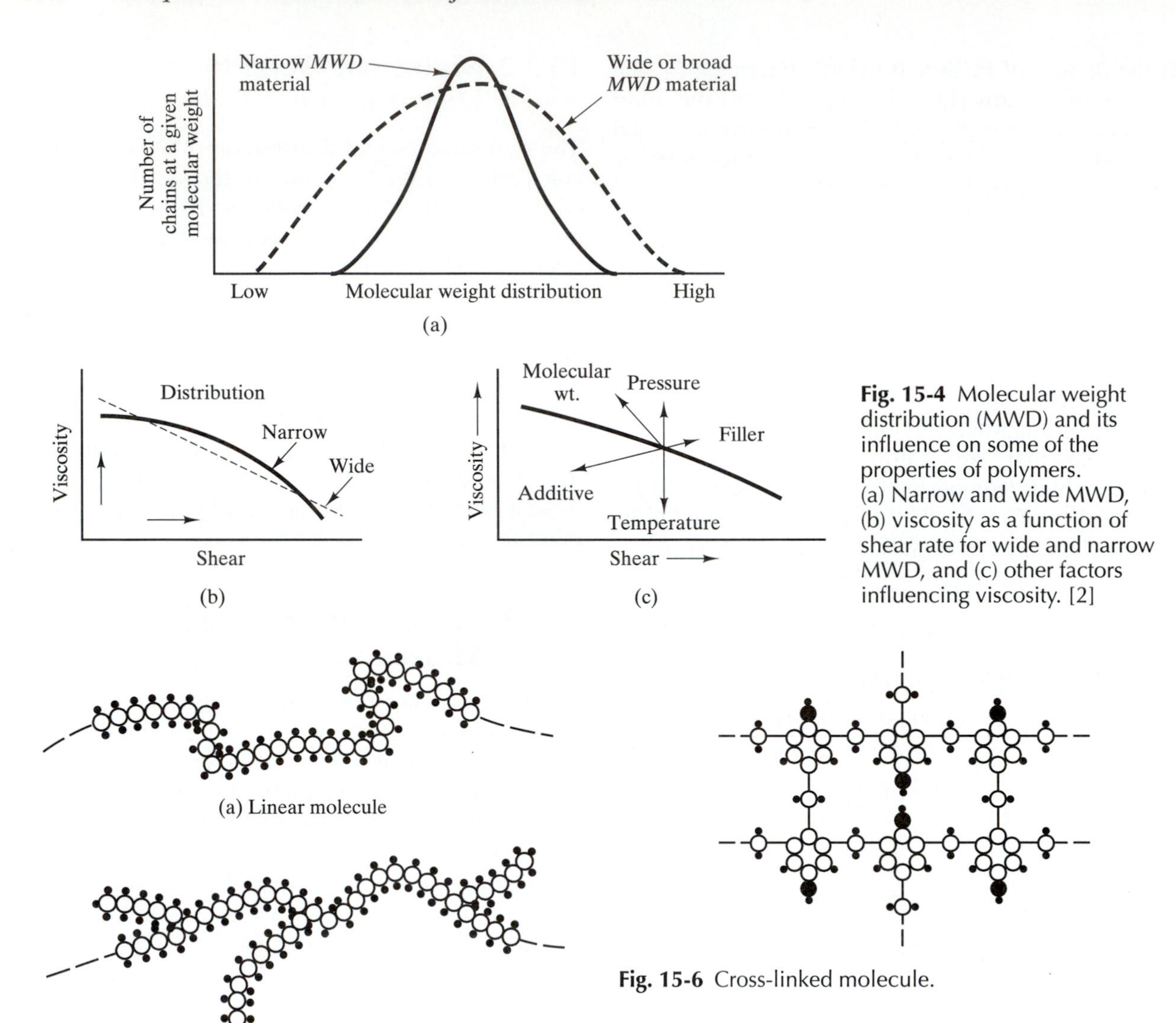

Fig. 15-4 Molecular weight distribution (MWD) and its influence on some of the properties of polymers. (a) Narrow and wide MWD, (b) viscosity as a function of shear rate for wide and narrow MWD, and (c) other factors influencing viscosity. [2]

Fig. 15-6 Cross-linked molecule.

Fig. 15-5 Linear and branched molecules.

hand, have more voids, lesser density, are more flexible, and are more permeable to gases and solvents than linear molecules.

15.3.4 Cross-Linked Molecules

The individual molecular chains described in the linear and branch molecules are not bonded to each other. Depending on the monomeric molecules used to form the polymer, it is possible for the individual macromolecular chains to be strongly bound together. The bonding between two chains is called *cross-linking,* illustrated in Fig. 15-6. This usually occurs when the monomers have more than one double bond in their structure. One double bond is used to create the linking of monomers to form the individual chains and the other remaining double bond can be broken to create the cross-linkages between macromolecular chains. These cross-linkages or cross-bonds make the sliding of the polymer molecules very difficult; the polymer becomes very stiff and is very hard to deform. Fully cross-linked polymers are thermoset (see types of polymers that are discussed shortly) and do not exhibit creep or relaxation, are usually brittle, and do not deform with heat. They decompose at high temperatures and are fairly resistant to solvent attack; solvents may swell such polymers, but will seldom cause complete rupture or dissolution.

15.3.5 Crystalline and Amorphous Polymers

Because of the absence of steric hindrances, linear molecules can be packed and aligned much easier than are branched or cross-linked molecules. The spaghettilike linear structure can be neatly folded and can easily produce alignment of the molecules in some regular pattern. Such alignment creates order and exhibits diffraction, a characteristic of a crystalline structure. On the other hand, cross-linked and branched molecules cannot produce regular patterns and are, therefore, noncrystalline or amorphous. Amorphous polymers do not have melting points but have, instead, glass transition temperatures, T_g, in which they exhibit gradual softening as the temperature is raised. The formation of the amorphous polymer solid proceeds as shown in Fig. 15-7 in the same manner as the formation of glasses and amorphous metals that were discussed in Secs. 14.3.1 and 7.6.

Figure 15-8 illustrates how a unit cell may exist in a chain-folded linear polymer crystal in (a), and a predominantly crystalline polymer in (b). Below the T_g, there is a weak intermolecular van der Waals bonding between the amorphous polymer molecules. As the polymer is heated close to the T_g, these secondary bonds gradually break accounting for the range of softening temperature, as opposed to a sharp melting point of crystalline polymers. Table 15-1 shows some of the differences in properties of crystalline and amorphous polymers. We must appreciate that polymers do not form 100 percent crystalline structures, as metals and ceramics do. They are at best semicrystalline, but nonetheless, the plastics industry calls them crystalline. For example, crystalline heavy-density polyethylene (HDPE) exhibits only about 80 percent crystalline.

A crystalline structure of a polymer may be transformed to an amorphous structure. When a crystalline polymer is heated above its melting point, an amorphous polymer may form when the molten polymer is cooled very rapidly—the same process to produce amorphous metals. Thus, the as-quenched amorphous polymer will have an entirely different set of properties than the crystalline polymer it came from. If molecular movements are sufficiently high, the as-quenched structure may approach its "most-preferred" configuration as time passes. This is similar to an aging process or an isothermal transformation where the unstable phase reverts to the stable phase. If the temperature is raised, the time to attain the stable structure will be shorter.

Polymers are unique in that large ordered and/or aligned domains may exist in the liquid state. These polymers are then called *liquid crystalline polymers* (LCP) and are a special class of thermoplastics (see

Table 15-1 Characteristics of crystalline and amorphous polymers.

Crystalline	Amorphous
Sharp melting point	Broad softening range
Usually opaque	Usually transparent
High shrinkage	Low shrinkage
Solvent resistant	Solvent sensitive
Fatigue/wear resistant	Poor fatigue/wear

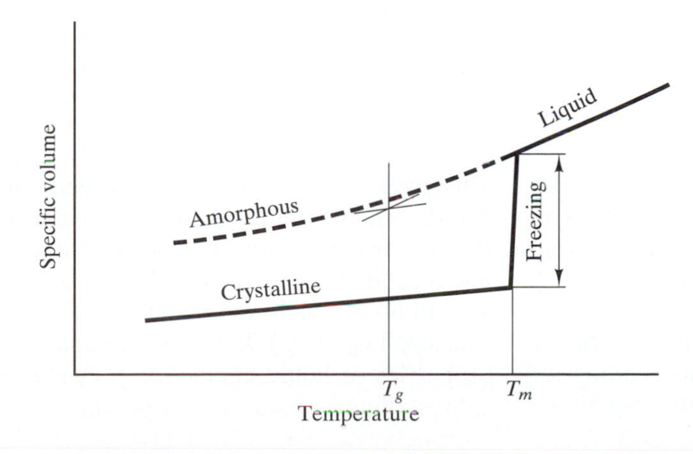

Fig. 15-7 Solidification of amorphous and crystalline thermoplastics.

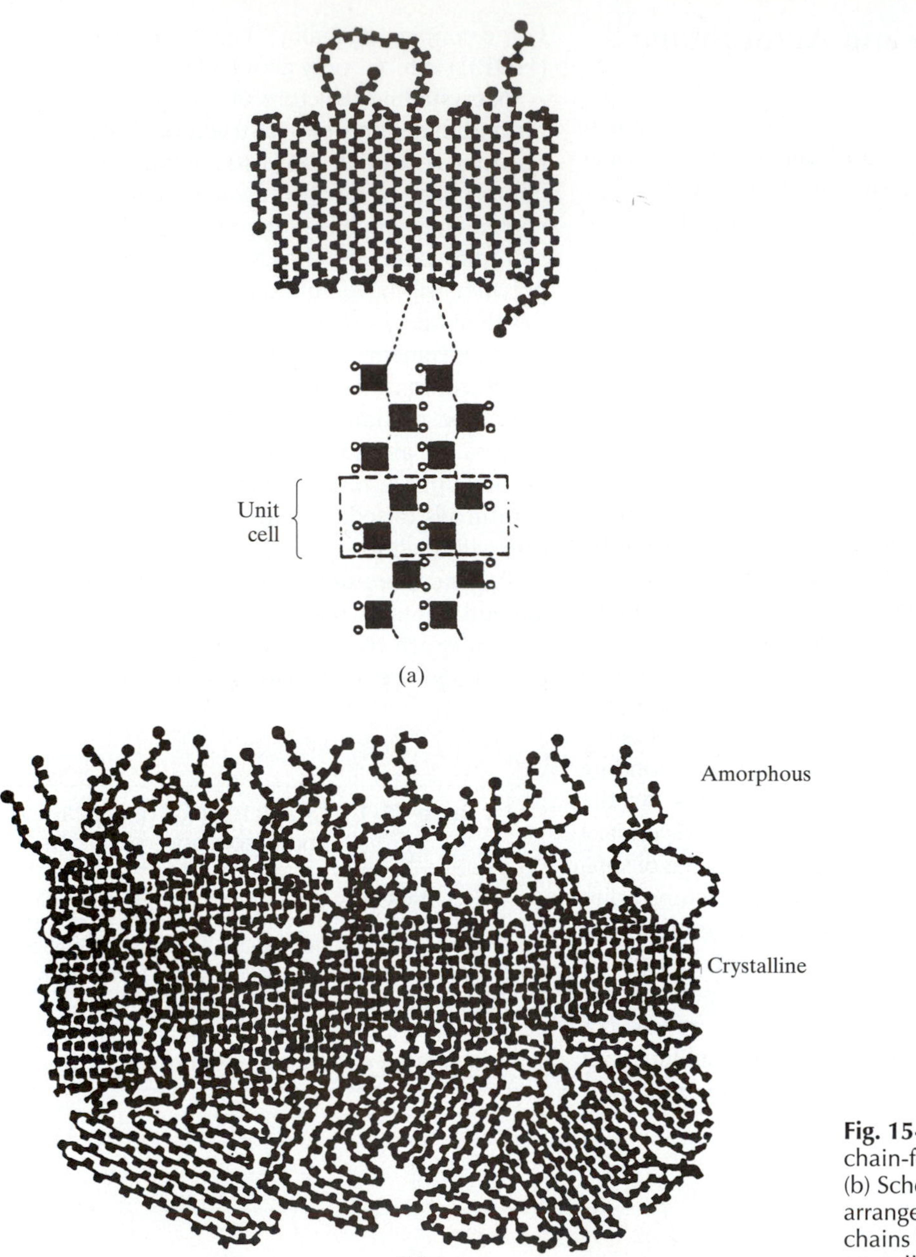

Fig. 15-8 (a) Unit cell in chain-folded linear polymer. (b) Schematic showing the arrangement of folded linear chains in predominantly crystalline polymer. [3]

types of plastics). Their molecules are stiff and rod-like and are organized in large parallel arrays or domains that exhibit low-melt viscosity and are thus processed easier than other high MW polymers. They have the lowest warpage and shrinkage of all the thermoplastics. When injection-molded or extruded, the molecules align into long rigid chains along the flow direction and thus act like reinforcing fibers. This gives the LCPs both high strength and stiffness. As the melt solidifies, the molecular orientation freezes into place. An example of a liquid crystalline polymer is Kevlar (a Du Pont trade name), an aramid fiber, which was mentioned earlier in Chap. 2 (see also Fig. 15-23). Table 15-2 compares the general properties of liquid crystalline polymers with those of crystalline and amorphous polymers.

Table 15-2 Relative properties of crystalline, amorphous, and LCPs. [2]

Property	Crystalline	Amorphous	Liquid crystalline
Specific gravity	Higher	Lower	Higher
Tensile strength	Higher	Lower	Highest
Tensile modulus	Higher	Lower	Highest
Ductility, elongation	Lower	Higher	Lowest
Resistance to creep	Higher	Lower	High
Max. usage temperature	Higher	Lower	High
Shrinkage and warpage	Higher	Lower	Lowest
Flow	Higher	Lower	Highest
Chemical resistance	Higher	Lower	Highest

15.3.6 Intramolecular Chemical Composition

This is the makeup or composition of the units making the mer. For example, in polyethylene, as shown in Fig. 15-9, the mer unit is the ethylene molecule that is entirely made up of carbon and hydrogen. If one (or more) of the hydrogen atoms is replaced by other groups or elements, the steric hindrances created by the substitutions can drastically change the properties of the polymer. For example, whereas polyethylene (PE) is translucent, flexible, and crystalline, polystyrene is transparent, brittle, and amorphous since the styrene mer consists of the benzyl radical replacing one hydrogen atom in the ethylene molecule; it is much harder to align the polymeric molecules in polystyrene than in polyethylene. Since the polymers in Fig. 15-9 are based on the ethylene mer, they are sometimes called olefin-based or vinyl-based polymers. An olefin is a hydrocarbon compound with at least one double bond. The term "vinyl" is the name of the radical when a hydrogen atom is removed from the ethylene molecule and some of the names of the polymers reflect this. For example, the mer unit is vinyl chloride when chlorine is substituted for a hydrogen atom and the polymer is polyvinyl chloride (PVC). Some of the polymeric names reflect the substitutions made for hydrogen; the mer unit is chlorotrifluroethylene when one chlorine and three fluorine atoms are substituted and the polymer is polychlorotrifluoroethylene (PCTFE). When the fluorine atom substitutes for all the hydrogen, the mer unit is tetrafluoroethylene and the polymer is polytetrafluoroethylene (PTFE), whose common name is Teflon. Compared to polyethylene, the examples just cited are opaque.

15.3.7 Co-polymerization

In addition to making changes by substitution in the basic repeating mer unit, as illustrated above for the ethylene mer, it is possible to change the chemical composition and hence the shape, structure, and the properties of a polymer by mixing different types of structural groups or basic repeating units within the chain of a polymer. This is done by a process called *co-polymerization*. Examples of the mers and products of co-polymerization are illustrated in Fig. 15-10. In addition to changing the types of repeating units, the relative amounts of each monomer used in the reaction may also be varied to produce literally an unlimited number of possible combinations of properties. This concept gives the possibility of "tailor-making" or "engineering" the plastic material desired. The x, y, z subscripts represent the relative proportions of the different monomers in the polymer. They do not imply, however, that y moles of styrene, for example, will form a chain connected with a chain of x moles of acrylonitrile in an acrylonitrile-styrene copolymer.

EXAMPLE 15-6

Butadiene and styrene are co-polymerized with equal numbers of moles of butadiene (B) and styrene (S) mers. These mers are arranged alternately in the polymer chain, as shown below. Determine the

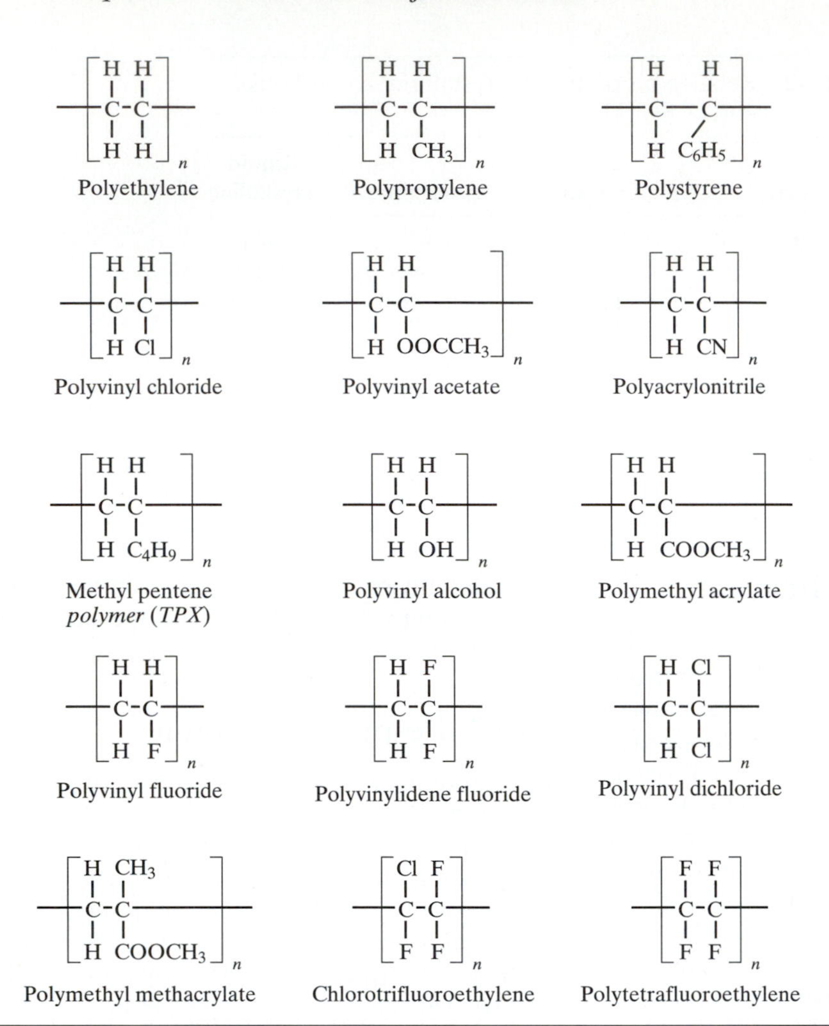

Fig. 15-9 Polymers based on the ethylene molecule. [1]

number average molecular weight of the co-polymer when the degree of polymerization (DP) is 2,500.

$$nB + nS \rightarrow (B\text{-}S\text{-}B\text{-}S\text{-} \cdots B\text{-}S)_n$$

SOLUTION: To get the MW, we multiply the DP by the mer weight. The mer weight of a copolymer is the weighted average of the mer weights of the constituents of the co-polymers, as in

$$\text{Weighted mer weight} = \Sigma\, x_i m_i$$

where x_i is the mole fraction, and m_i is the mer weight of constituent i. In this case, since there is an equal number of moles of butadiene and styrene, the $x_i = 0.5$ for both butadiene and styrene. The formulas for butadiene and styrene are shown in Fig. 15-10.

Mer weight of butadiene

$$= 4C + 6H = (4 \times 12) + (6 \times 1) = 54\text{g/mer B}$$

Mer weight of styrene

$$= 8C + 8H = (8 \times 12) + (8 \times 1) = 104\text{g/mer S}$$

the weighted mer weight

$$= 0.5 \times 54 + (0.5 \times 104) = 79\text{g/mer B-S}$$

Number average MW

$$= 79 \times 2{,}500 = 197{,}500\text{g/g-mole B-S.}$$

EXAMPLE 15-7

If the ratio of B:S in the B-S co-polymer is 1:4, what would be the average molecular weight when the degree of polymerization is 2,500?

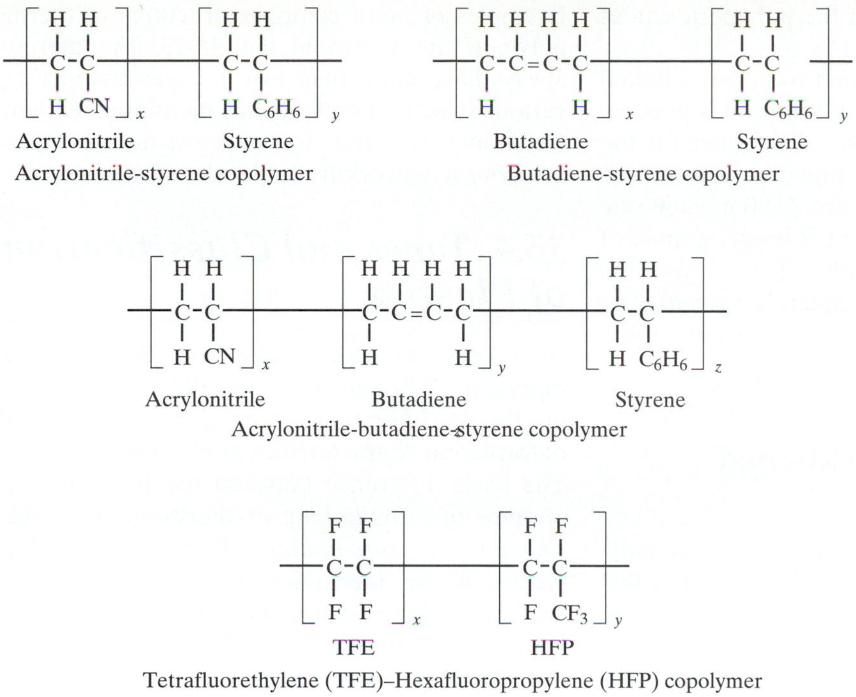

Acrylonitrile Styrene
Acrylonitrile-styrene copolymer

Butadiene Styrene
Butadiene-styrene copolymer

Acrylonitrile Butadiene Styrene
Acrylonitrile-butadiene-styrene copolymer

TFE HFP
Tetrafluorethylene (TFE)–Hexafluoropropylene (HFP) copolymer

Fig. 15-10 Examples of co-polymers with carbon backbone. [1]

SOLUTION: With B:S = 1:4, the mole fraction of B = 1/5 = 0.2, and mole fraction of S is 0.8. Therefore, the weighted mer weight is

Weighted mer weight

$$= 0.20 \times 54 + (0.80 \times 104)$$

$$= 94\text{g/g-mole B-S, and}$$

Number average MW

$$= 2{,}5000 \times 94 = 235{,}000\text{g/g-mole of B-S.}$$

EXAMPLE 15-8

Determine the molar ratio of B:S in a co-polymer with a molecular weight of 148,000 and a degree of polymerization of 2,000. What is the weight percent of B?

SOLUTION: Here we are given the MW and the DP, from which we obtain the weighted mer weight and the mole fraction of B and S.

Weighted mer weight

$$= 148{,}000/2{,}000 = 74\text{g/mer B-S}$$

If x = mole fraction of B, then (1 − x) = mole fraction of S. Then

$$(x)54 + (1 - x)104 = 74$$

From which

$$x = 0.60 \text{ of B} \quad \text{and } 1 - x = 0.40 \text{ of S.}$$

Thus, the molar ratio B:S is 60/40 or 3/2.

$$\text{Weight \%B} = \frac{(3 \times 54)}{(3 \times 54) + (2 \times 104)} \times 100$$

$$= 43.8\% \text{ butadiene}$$

EXAMPLE 15-9

The B-S co-polymers in Examples 15-6, 15-7, and 15-8 are thermoplastic polymers. They can be converted to thermoset polymers by cross-linking the polymeric chains. An example of a cross-linking process is vulcanization, which is the process of cross-linking with sulfur. In cross-linking the B-S, the double bond in the butadiene is broken and a sulfur atom bridges the polymeric chains. Using the B-S in Example 15-8, determine the amount of sulfur needed to completely cross-link all the butadienes in the polymeric chains per mole of the B-S.

SOLUTION: For a butadiene mer to be cross-linked with another butadinene, one sulfur atom is needed. Therefore, the number of moles of butadiene is the number of moles of sulfur. For one mer of B-S there is 0.60 moles of B. Since there are 2,500 mers in one g-mole of B-S, the total moles of B in one g-mole of B-S is $0.6 \times 2{,}500 = 1{,}500$ moles. This is also the moles of sulfur needed to completely vulcanize or cross-link all the butadiene mers, and therefore the

$$\text{Grams S} = 1{,}500 \times 32 = 48{,}000 \text{ g.}$$

15.3.8 Non-Carbon Backboned Polymers

In the examples discussed up to this point, the polymers have the -C-C-linkages (carbon backbone) that arose by breaking double bonds between adjacent carbon atoms. When polymers are produced by rearrangement or by condensation reactions, *polymers containing oxygen (O) or nitrogen (N) in the backbone of the molecules* can be produced. Examples of these are illustrated in Figs. 15-11 and 15-12. Examples of more complex structures of thermoset polymers are shown in Fig. 15-13. The thermosets always have more than two linkages connecting the various structural units and as mentioned earlier, are commonly referred to as cross-linked materials involving a networklike molecular structure.

15.4 Types and Classification of Plastics

In the previous section, there were references to *thermoplastic (TP)* and *thermoset (TS)* polymers. These are the two basic types of plastics according to their deformation characteristics at elevated tempertures. The basic difference between the two types is the *absence of cross-linking* in thermoplastics and the *presence of cross-linking* of the macromolecular chains in the thermoset polymers. As the name implies, a *thermoplastic material will deform with temperature* while *thermoset* has a set form and *will not deform with temperature*.

Thermoplastics consist of long, linear molecules, each of which may have side-chains or groups (i.e., branched molecules, but are not cross-linked). The

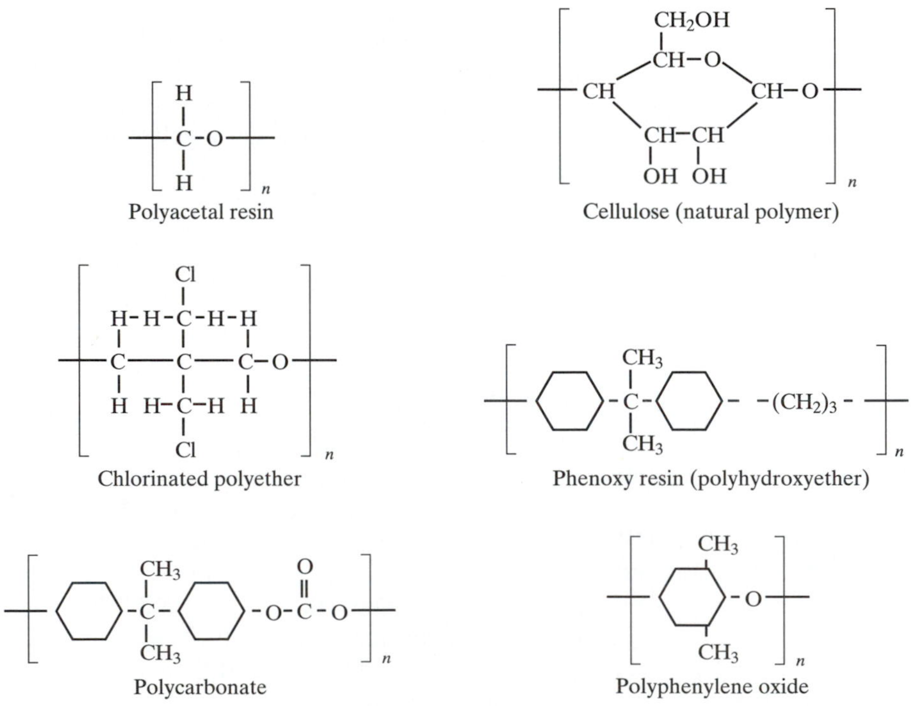

Polyacetal resin

Cellulose (natural polymer)

Chlorinated polyether

Phenoxy resin (polyhydroxyether)

Polycarbonate

Polyphenylene oxide

Fig. 15-11 Polymers with oxygen in the chain. [1]

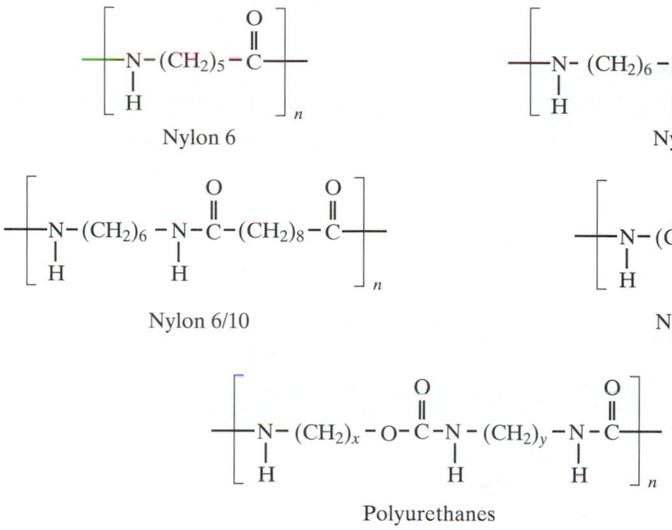

$$\left[-\underset{\mathrm{H}}{\mathrm{N}}-(\mathrm{CH_2})_5-\overset{\mathrm{O}}{\overset{\|}{\mathrm{C}}}- \right]_n$$

Nylon 6

$$\left[-\underset{\mathrm{H}}{\mathrm{N}}-(\mathrm{CH_2})_6-\underset{\mathrm{H}}{\mathrm{N}}-\overset{\mathrm{O}}{\overset{\|}{\mathrm{C}}}-(\mathrm{CH_2})_4-\overset{\mathrm{O}}{\overset{\|}{\mathrm{C}}}- \right]_n$$

Nylon 6/6

$$\left[-\underset{\mathrm{H}}{\mathrm{N}}-(\mathrm{CH_2})_6-\underset{\mathrm{H}}{\mathrm{N}}-\overset{\mathrm{O}}{\overset{\|}{\mathrm{C}}}-(\mathrm{CH_2})_8-\overset{\mathrm{O}}{\overset{\|}{\mathrm{C}}}- \right]_n$$

Nylon 6/10

$$\left[-\underset{\mathrm{H}}{\mathrm{N}}-(\mathrm{CH_2})_{10}-\overset{\mathrm{O}}{\overset{\|}{\mathrm{C}}}- \right]_n$$

Nylon 11

$$\left[-\underset{\mathrm{H}}{\mathrm{N}}-(\mathrm{CH_2})_x-\mathrm{O}-\overset{\mathrm{O}}{\overset{\|}{\mathrm{C}}}-\underset{\mathrm{H}}{\mathrm{N}}-(\mathrm{CH_2})_y-\underset{\mathrm{H}}{\mathrm{N}}-\overset{\mathrm{O}}{\overset{\|}{\mathrm{C}}}- \right]_n$$

Polyurethanes

Fig. 15-12 Polymers with nitrogen in the chain—nylons and polyurethanes. [1]

chains can be thought of as independent, intertwined strings resembling spaghetti. When heated, the individual chains slip, causing plastic flow. Thus, they can be repeatedly melted and reshaped by heating and cooling so that any scrap generated can be reused. No chemical change occurs during the deformation, but the thermoplastic may burn to some degree. Consequently, care must be taken to avoid degrading, decomposing, or igniting these materials. Their softening temperatures vary with the polymer type and grade. They are soluble in specific solvents. TPs generally offer higher impact strength, easier processing, and adaptability to complex designs than do TSs.

Thermosets are resins that undergo chemical change, called curing during processing, to form cross-linked structures and become permanently insoluble and infusible. Thus, they cannot be remelted and reprocessed. Thermoset scrap must either be discarded or used as low-cost filler in other products. In some cases, the scrap may be pyrolyzed to recover some of the calorific value from the resin and some of the reinforcements in reinforced plastics, such as glass, may be reused. Thermosets may be supplied in the liquid form or as a partially polymerized solid molding powder. In their uncured condition, thermosets can be formed to the finished product shape with or without using pressure and then are cured or polymerized by using chemicals (curing agents) or heat.

The distinction between thermoplastics and thermosets is not always clearly drawn, however. For example, we can transform a normally thermoplastic polyethylene (PE) into a thermoset, cross-linked PE by high-energy radiation. There are also plastics that have both types within their generic class: for instance, there are both thermoplastic and thermoset polyester resins.

Elastomers are sometimes portrayed as another type of polymer, but they really are not. These are generally low-modulus flexible materials that can be stretched repeatedly to at least twice their original length, but will return to their roughly original length when the stress is released. We have known about natural and synthetic rubbers (elastomers), such as latex, nitrile, and neoprene, which attain their properties through the process of vulcanization. Vulcanization is the cross-linking or bonding of the molecular chains of the thermoset elastomers (TSE) to make them very hard and extremely difficult to stretch. For this reason, these TSE materials are always required for vehicle tires. However, the TS natural and synthetic rubbers are now being replaced by thermoplastic elastomers (TPEs). Thus, we see that elastomers are similar to polyesters, which may be either thermoset or thermoplastic; the very flexible stretchable elastomers are the TPEs.

The TPEs are predominantly synthetic rubbers that exhibit large strains with minimal stress. They

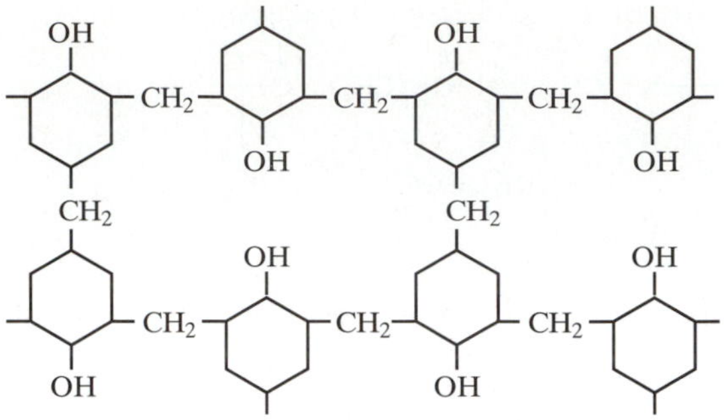

Phenol-formaldehyde resin

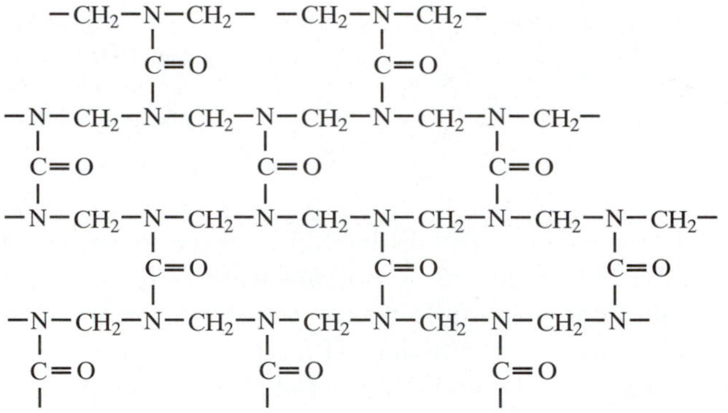

Urea-formaldehyde

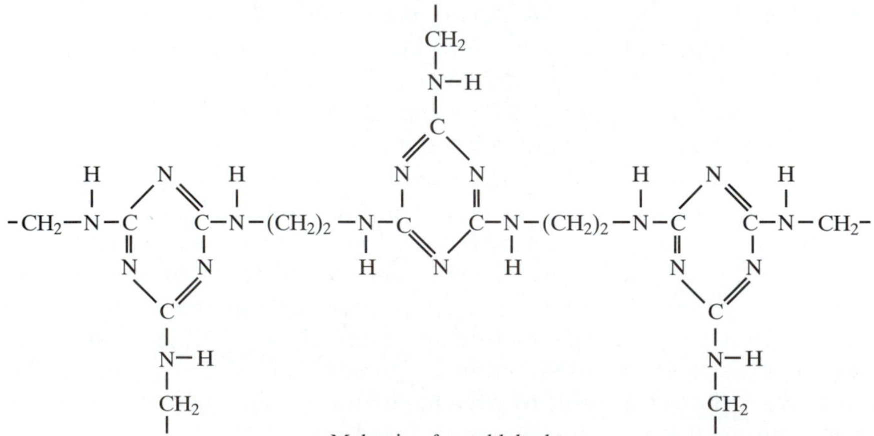

Melamine-formaldehyde

Fig. 15-13 Complex structures of thermoset polymers. [1]

have two specific characteristics: (1) their glass transition temperature (T_g) is below that at which they are applied, and (2) their molecules are highly kinked and coiled. When stress is applied, the molecular chain uncoils and the end-to-end length can be several hundred percent that of the unstressed coiled length. The initial modulus is very low, but when the molecules are extended, the modulus increases dramatically. If cross-linked, they become harder; indeed, they can get very hard if heavily cross-linked, such as in tires for automobiles. When this occurs, they become thermosets—just like a TP polyethylene being converted to a TS polyethylene.

There are at least 30 to 40 different families of thermoplastics now commercially available and about 10 different basic families of thermosets. These are just the common types of polymers. With copolymerization, mixing or blending, and alloying, the total number of polymeric materials can be endless. Yes, just as in metals, alloying and blending are possible. In an effort to classify plastics and to establish a system for ordering or procuring plastics, ASTM D 4000 was established. In this system, the plastics are arranged into broad generic families, each of which is assigned a set of letters. A part of ASTM D 4000 will be discussed in Sec. 15.14.

15.4.1 Classification of Thermoplastics

These materials may be classified into *commodity* and *engineering* thermoplastics. The commodity thermoplastics are generally used in no-load or very low-load applications, while the engineering resins can be designed to carry loads for a long period of time. The commodity thermoplastics account for the majority of plastic materials consumed. Most of these plastics come from the olefinic group, shown in Fig. 15-9. The low-density polyethylene (LDPE) formulations account for about 25 percent of the tonnage, followed by the high-density polyethylenes (HDPE), then polypropylene, polyvinyl chloride, and polystyrene. Together, these materials account for about two-thirds of all plastics consumed. About 90 percent of all plastics (weightwise) consumed are the commodity resins and the rest is accounted for by the engineering resins that include both TPs and TSs.

15.4.1a Commodity Thermoplastics

These include the polyolefins, the styrenes, and the vinyls. The acrylics and the cellulosics belong also in this group.

Polyolefins. The two prominent members of this family that we shall discuss here are polyethylene (PE) and polypropylene (PP). The other members are polybutylene (PB), polymethylpentene (PMP), ethylene-vinyl acetate (EVA), and the ionomers.

Polyethylenes comprise the largest-volume plastics worldwide and are available in many varieties with an equally wide range of properties. Some are flexible, others rigid; some have low impact strength, others are nearly unbreakable; some are clear, others are opaque, and so on. In general, they are semicrystalline and are characterized by toughness, near-zero moisture absorption, excellent chemical resistance, excellent insulating properties, low coefficient of friction, ease of processing, chemically resistant to both acids and bases at room temperature, and excellent dielectric strength. As indicated earlier, PEs can be cross-linked (XLPE) to form infusible thermosets with high heat and crack resistance. Applications are in wire and cable coatings, foams, and rotationally molded products.

PEs are classified by density as follows: (a) ultra- or very low density—0.880 to 0.915 g/cm^3, (b) low density—0.910 to 0.925 g/cm^3, (c) medium density—0.926 to 0.940 g/cm^3, and (d) high density—0.941 to 0.965 g/cm^3. The primary differences among the types are in rigidity, heat resistance, chemical resistance, and ability to sustain loads. In general, as density increases, hardness, strength, heat resistance, stiffness, and resistance to permeability increase as well. The low-density polyethylenes (LDPE) have highly branched structures with moderate crystallinity (50 to 65 percent). They are very flexible and have very low heat resistance (maximum recommended service temperature is 60° to 80°C). The traditional markets for LDPE are in packaging films, wire and cable coating, injection molding, and in pipe and tubing. Linear LDPE (LLDPE) has gradually replaced LDPE in some of these markets. As the name implies, LLDPE has very little branching in its long chain molecules. They exhibit much higher elongation than LDPE, higher tear, tensile, and impact strength, along with improved resistance to environmental stress cracking. These properties allow much stronger, downgauged (thinner) products to be produced with much less material. Ultra- and very low-density polyethylenes (ULDPE and VLDPE) are synonymous designations for linear PEs with densities down to 0.880 g/cm^3.

High-density polyethylenes (HDPE) are highly crystalline, tough materials that can be formed by most processing methods. Within their density range, the stiffness, tensile strength, melting point, and chemical resistance all improve, but the stress-crack resistance and the low-temperature impact strength are lower at the high end. High molecular weight HDPE (HMW-HDPE) are a special class of linear resins with weight average molecular weights in the range 0.5 to 2 million. These are made into blown film for packaging, extruded into pressure pipe and sheets for truck bed and pond liners, and blow-molded into large shipping containers. There is also an ultra-high molecular weight-HDPE (UHMW-HDPE) with weight average molecular weights in excess of 2 million. High-strength, chemical resistance, and lubricity make UHMW-HDPE ideal for gears, slides, rollers, and other industrial parts. It is also used to make artificial hip joints. In fiber form, UHMW-HDPE exhibits liquid crystal properties that can be used as reinforcements and as lightweight ultra-strong fabrics. Figure 15-14 shows the ranges of molecular weights for the HDPEs.

Polypropylenes (PPs) are semitranslucent and milky-white in color, with excellent colorability. Their method of production induces the molecules to crystallize into compact bundles, that makes them stronger than other members of the polyolefins. They are easily formed to a variety of products and are available in many grades as well as co-polymers. PPs exhibit a low density of 0.90 and a good balance of moderate cost, strength, and stiffness, as well as excellent fatigue, chemical resistance, and thermal and electrical properties. Their excellent fatigue resistance allows them to be used as integral "living hinges" in many applications. Like all other polyolefins, PPs have excellent resistance to water and water solutions, such as salt and acid solutions that are destructive to metals. They are also resistant to organic solvents and alkalis.

The greatest commercial uses of homopolymer PP are in fibers and filaments. PP fibers are woven into fabrics and carpets, and they are also used to produce nonwoven fabrics for disposables. PP is also made into nonoriented and oriented films for packaging, which have largely replaced cellophane and glassine.

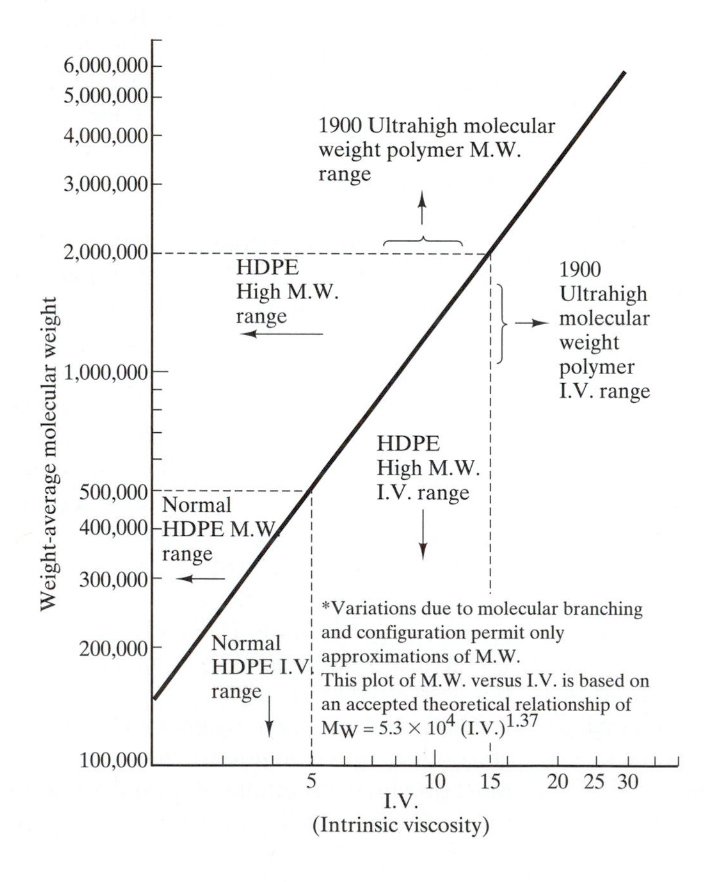

Fig. 15-14 Molecular weights of polyethylene and their linear relationship with viscosity. [2]

PP is also injection molded into caps and closures, appliance components, and auto parts.

PP is able to carry light loads for long periods but does not have outstanding creep resistance. Its strength, rigidity, heat resistance, and dimensional stability can be increased significantly with glass-fiber reinforcement. PP has poor impact toughness but this can be improved by co-polymerization.

Ionomers are characterized by random ionic bonds between long-chain polymer molecules to produce properties that are associated with high-molecular-weight materials. At normal processing temperatures, however, their ionic bonding diminishes, allowing them to be processed in conventional extruders and injection-molding machines. They are extremely tough and having high tensile strength and elongation in the range of 300 to 500 percent. They can be alloyed to produce stiff and heat-resistant grades, while retaining the excellent impact toughness. They are used in the automotive industry as foam injection-molded parts, replacing the heavy rubber and metal bumper guards. Because of their ionic character, they are used as membranes to hold the electrolyte (water) in fuel cells; the latter are then called polymer electrolyte membrane (PEM) fuel cells.

Styrenic Plastics. In general, this family of plastics is characterized by ease of processing, hardness, and excellent dielectric properties, but have limited heat resistance and are attacked by aliphatic and aromatic hydrocarbon solvents. Aside from the homopolymer polystyrene, the other common members in this family are the co-polymers of styrene-acrylonitrile (SAN), styrene-butadiene (SB), and acrylonitrile-butadiene-styrene (ABS).

Polystyrene (PS) is amorphous, the lowest-cost member of the family, and is used for its hardness, rigidity, optical clarity, dimensional stability, and excellent processability. Applications include packaging for food, cosmetics, pharmaceuticals, audiocassette cases, dust covers, disposable drinking cups, and cutlery. Oriented crystalline PS sheet is thermoformed into trays and blister packaging, and foamed crystalline PS is used in a wide variety of insulating and packaging applications. PS is available in a wide range of grades for all types of processes. Modified grades include high heat and various degrees of impact resistance. Clarity and gloss are reduced, however, in the impact grades.

Styrene-acrylonitrile (SAN) co-polymers are transparent, amorphous, and have higher heat and chemical resistance than crystalline PS has. Because they are polar in nature, SAN resins are hygroscopic and require drying before processing. They are injection-molded into such products as dishwasher-safe housewares, refrigerator shelves, medical devices, oven caps, connectors for PVC tubing, and lenses.

Styrene-butadiene (SB) co-polymers combine the transparency of styrene and the elastomeric properties of butadiene to yield a transparent, high-impact strong material. They are not hygroscopic and require no drying before processing; therefore, they are used for packaging and items such as cups, lids, trays, and clamshells for the fast food industry.

ABS (acrylonitrile-butadiene-styrene) co-polymers are alloys of the three constituents in various proportions to yield a wide range of plastics with properties reflecting the constituents' properties. They exhibit a balanced combination of high impact toughness, wide service temperature range, good dimensional stability, chemical resistance, electrical insulating proper ties, and ease of processability. The applications of ABS fall between those of commodity and engineering plastics. In the automotive market, ABS is used to injection-mold interior panels, trim, grilles, wheel covers, and mirror housings. They are also used to mold computer disk housings, telephones, calculators and business machines, and appliances.

Vinyl Plastics. These are predominantly polymers and co-polymers based on the vinyl mers that were discussed Sec. 15.3.6. Although there are literally thousands of compositions with varying properties in this family, there are certain general characteristics that are common to nearly all these plastics. In general, they can be plasticized to give a wide range of products ranging from thin, flexible free films to rigid molded pieces. They exhibit excellent water and chemical resistance, strength, abrasion resistance, and self-extinguishability. In their elastomeric form, vinyls exhibit properties superior to natural rubber in their flex life, resistance to acids, alcohols, sunlight, and wear and aging. They are nontoxic, tasteless, odorless, and are suitable for use as packaging materials that will come in contact with foods and drugs. Some of the more important plastics in this family include polyvinyl chloride (PVC), chlorinated PVC (CPVC), and polyvinylidene chloride.

Polyvinyl chloride (PVC) is the most commercially important plastic in this family and accounts for a major volume of all plastics consumed worldwide. Its acceptance comes from its versatility; it may be utilized in rigid or flexible form. Rigid PVC is sometimes called the poor man's engineering plastic and normally contains a heat stabilizer along with an impact modifier such as ABS or chlorinated PE. Flexible PVC contains plasticizers to soften the resin and to improve the processing. Almost 75 percent of the rigid PVC goes into building and construction applications. Most are processed into extruded products such as pipe, siding, and window profiles. PVC is also used for packaging, blown into bottles, and made into sheets for thermoforming boxes and blister packs. Flexible PVCs are made into food wrap, wire and cable coating, flooring, garden hose, and toys.

PVC is self-extinguishing. However, in a fire it produces hydrochloric acid and other toxic and corrosive chemicals, though it can be burned in a properly designed incinerator without releasing any of these chemicals into the atmosphere.

Chlorinated PVC (CPVC) is produced by the chlorination of PVC and has a higher heat distortion temperature and more combustion resistance. CPVC is used to make pipe and fittings for potable water and industrial chemicals, dark-color window frames, and housings for appliances and business equipment.

Polyvinylidene chloride (PVDC) is a co-polymer of vinylidene chloride and vinyl chloride. It exhibits high strength, abrasion resistance, strong welds, dimensional stability, toughness, durability, and exceptional barrier resistance to oxygen, carbon dioxide, water, and many organic solvents. It can be molded into fittings and parts for use in chemical industries—for example, PVDC pipes are superior to iron pipes for the disposal of acids. PVDC is made into monolayer films for food warp (Saran) and medical packaging, and into co-extruded film and sheet structures as a barrier layer.

Other Commodity Thermoplastics. Some plastics in this category are the acrylics, the acrylonitrile based, and the cellulosics. The *acrylics* are used for their excellent optical transparency and resistance to long-term exposure to sunlight and the weather. Their applications include outdoor signs, glazing, aircraft canopies, skylights, auto tail-lights, dials, buttons, lighting applications, knobs, and machine covers. The most widely used of the acrylics is *polymethylmethacrylate (PMMA)* commonly known as *Plexiglas*. The *acrylonitrile-based resins* are sometimes called *barrier resins* and are primarily used in packaging, where they do not transmit the gas, aroma, or flavor of a package's contents.

The *cellulosics* are not synthetic plastics but rather are made from the naturally occuring polymer, cellulose, which is obtained from wood pulp and cotton linters. Cellulose can be made into a film as cellophane or a fiber, rayon, but it must be chemically modified to produce thermoplastics. Because cellulose can be compounded with many different plasticizers in varying concentrations, its property range is broad and is usually specified by the flow characteristics, according to ASTM D 569. The cellulosics include cellulose acetate (acetates), cellulose acetate butyrate (butyrates), cellulose acetate propionate (propionates), ethyl cellulose, and cellulose nitrate (nitrates). The nitrates were the first plastics to be developed in 1868, originally to make billiard balls. Because of its flammability, its use today is relatively minimal.

15.4.1b Engineering Thermoplastics

These materials exhibit properties that enable them to compete with metals, whereas the heretofore described commodity thermoplastics compete with glass, paper, and wood. The families of *engineering thermoplastics* include the acetals, the fluoroplastics, polyamides (nylon), polyamide-imide, polyarylates, polycarbonates, thermoplastic polyesters, polyetherimide, polyketones, polyphenylene oxide, polyphenylene sulfide, and the sulfone polymers. The properties of some thermoplastics, including commodity and engineering types, are presented in Table 15-3, while Table 15-4 lists the trade names and the manufacturers of some of the engineering thermoplastics, and where they are used.

The *acetals,* formally called polyoxymethylenes (POM), are polymers of formaldehyde, are among the strongest and stiffest thermoplastics, and are characterized by excellent fatigue life, dimensional stability, low friction coefficients, exceptional solvent resistance, and high heat resistance for extended use to 104°C (220°F). The acetals may be either homopolymers or co-polymers. The homopolymers

Table 15-3 Typical property ranges for thermoplastics.[1]*

Thermoplastics		Specific Gravity	Trans-parency	Tensile Strength 10^3 psi	Tensile Modulus, 10^5 psi	Impact Strength Izod ft. 1b/inch	Max use Temp, F (no load)	DTUL at 66 psi	DTUL at 264 psi
ABS	GP	1.05–1.07	No	5.9	3.1	6	160–200	210–225	109–206
	Hi. imp.	1.01–1.06	No	4.8	2.4	7.5	140–210	210–225	188–211
	Ht. res.	1.06–1.08	No	7.4	3.9	2.2	190–230	225–252	226–240
	Trans.	1.07	Yes	5.6	2.9	5.3	130	180	165
Acetals	Homo	1.42	No	10	5.2	1.4	195	338	255
	Copoly	1.41	No	8.8	4.1	1.2–1.6	212	316	230
Acrylics	GP	1.11–1.19	Yes	5.6–11.0	2.25–4.65	0.3–2.3	130–230	175–225	165–210
	Hi. imp.	1.12–1.16	No	5.8–8.0	2.3–3.3	0.8–2.3	140–195	180–205	165–190
		1.21–1.28	No	8.0–12.5	3.5–4.8	0.3–0.4	125–200	170–200	155–205
	Cast	1.18–1.28	Yes	9.0–12.5	3.7–5.0	0.4–1.5	140–220	165–235	160–215
	Multi polymer	1.09–1.14	Yes	6–8	3.1–4.3	1–3	165–175	—	185–195
Cellulosics	Acetate	1.23–1.34	Yes	3.0–8.0	1.05–2.55	1.1–6.8	140–220	120–209	111–195
	Butyrate	1.15–1.22	Yes	3.0–6.9	0.7–1.8	3.0–10.0	140–220	130–227	113–202
	Ethylcellulose	1.10–1.17	Yes	3–8	0.5–3.5	1.7–7.0	115–185	—	115–190
	Nitrate	1.35–1.40	Yes	7–8	1.9–2.2	5–7	140	—	140–160
	Propionate	1.19–1.22	Yes	4.0–6.5	1.1–1.8	1.7–9.4	155–220	147–250	111–228
Eth. copolymers	EEA	0.93	Yes	2.0	0.05	NB	190	—	—
	EVA	0.94	Yes	3.6	0.02–0.12	NB	—	140–147	93
Fluoropolymers	FEP	2.14–2.17	No	2.5–3.9	0.5–0.7	NB	400	158	—
	PTFE	2.1–2.3	No	1.4	0.38–0.65	2.5–4.0	550	250	—
	CTFE	2.10–2.15	Yes–No	4.6–5.7	1.8–2.0	3.5–3.6	350–390	258	—
	PVF_2	1.77	No	7.2	1.7	3.8	300	300	195
	ETFE & ECTFE	1.68–170	No	6.5–7.0	2–2.5	NB	300	220	160
Nylons	6/6	1.13–1.15	No	9–12	3.85	2.0	180–300	360–470	150–220
	6	1.14	No	12.5	—	1.2	180–250	300–365	140–155
	6/10	1.07	No	7.1	2.8	1.6	180	300	—
	8	1.09	No	3.9	—	> 16	—	—	—
	12	1.01	No	6.5–8.5	1.7–2.1	1.2–4.2	175–260	—	120–130
	Copolymers	1.08–1.14	No	7.5–11.0	—	1.5–19	180–250	—	130–350
Polyarylate		1.2	Yes	9.5	2.9	4.2	—	—	345
Polyarylsufone		1.36	No	13	3.7	2	500	—	525
Polybutylene		0.910	No	3.8	0.26	NB	225	215	130
Polycarbonate		1.2	Yes	9	3.45	12–16	250	270–290	265–285
PC/ABS alloy		1.14	No	8.2	3.7	10	220	235	220

(Continued)

Table 15-3 Typical property ranges for thermoplastics. (Continued)

Thermoplastics		Specific Gravity	Trans-parency	Tensile Strength 10^3 psi	Tensile Modulus, 10^5 psi	Impact Strength Izod ft. 1b/inch	Max use Temp, F (no load)	DTUL at 66 psi	DTUL at 264 psi
Polyesters	PET	1.37	No	10.4	—	0.8	175	240	185
	PBT	1.31	No	8.0–8.2	3.6	1.2–1.3	280	310	130
	PTMT	1.31	No	8.2	—	1.0	270	302	122
	Copol.	1.2	Yes	7.3	—	1.0	—	—	154
Polyetheretherketone		1.3	No	13	—	1.6	480	—	320
Polyetherimide		1.27	Yes	15	4.3	1.1	338	405	390
Polyethersulfone		1.37	Yes	12	3.5	1.4–2.2	356	—	397
Polyethylenes	LD	0.91–0.93	No	0.9–2.5	0.20–0.27	NB	180–212	100–120	90–105
	HD	0.95–0.96	No	2.9–5.4		0.4–14	175–250	140–190	110–130
	HMW	0.945	No	2.5	1	HB	—	155–180	105–180
Ionomer		0.94–0.95	Yes	3.4–4.5	0.3–0.7	6–NB	160–180	110	100–120
Polymethylpentene		0.83	Yes	3.3–3.6	1.3–1.9	0.95–3.8	275	—	—
Polyphenylene oxide based mtls.		1.06–1.10	No	7.8–9.6	3.5–3.8	5.0	175–220	230–280	212–265
Polyphenylene sulfide		1.34	No	10	4.8	0.3	500	—	278
Polyimide		1.43	No	5–7.5	5.4	5–7	500	—	680
Polypropylenes	GP	0.90–0.91	No	4.5–6.0	1.6–3.0	0.4–1.2	225–300	200–230	125–140
	Impact copolymer	0.88–0.91	No	3.5–5.0	1.3	2–NB	200–250	160–200	120–135
	Random copolymer	0.89–0.91	No	4.0–5.5	1.0–1.7	1.1–12	190–240	185–230	125–140
Polystyrenes	GP	1.04–1.07	Yes	6.07–7.3	4.5	0.3	150–170	—	180–220
	Hi. impact	1.04–1.07	No	2.8–4.6	2.9–4.0	0.7–1.0	140–175	—	175–210
Polysulfone		1.24	No	10.2	3.6	1.2	300	360	345
Polyurethanes		1.11–1.25	No	4.5–8.4	0.1–3.5	NB	190	—	—
PVC Rigid		1.3–1.5	No-Yes	5–8	3–5	0.5–20	150–175	135–180	130–175
PVC Flexible		1.2–1.7	Yes & No	1.4	—	0.5–20	140–175	—	—
Rigid CPVC		1.49–1.58	Yes-No	7.5–9.0	3.6–4.7	1.0–5.6	230	215–245	200–235
PVC/acrylic		1.30–1.35	No	5.5–6.5	2.75–3.35	15	—	180	170
PVC/ABS		1.10–1.21	No	2.6–6.0	0.8–3.4	10–15	—	—	—
SAN		1.08	Yes	10–12	5.0–5.6	0.4–0.5	140–200	—	190–220
SMA		1.05–1.15	Yes-No	5–9	2.7–4.4	0.5–12	200	—	205–260

[a]All values at room temperature unless otherwise listed. [b]Per ASTM. [c]Notched samples. [d]Deflection temperature underload. *Number in bracket is Reference at end of chapter.

Table 15-4 Trade names, suppliers, and some characteristics of some neat (unreinforced) thermoplastics.[2]

Resin	Trade Name	Manufacturer	HDT[1]	°F[2]	Major Property Advantages	Applications
Polyphenylene sulfide (PPS)	Ryton Supec Fortron	Phillips 66 General Electric Hoechst Celanese	500 (260°C)	425 (218°C) 464 (240)	Heat resistance, good electrical properties, chemical resistance	Automotive, electrical, industrial, consumer
Polyamideimide (PAI)	AT-110 Torlon	Amoco Amoco	525 (274)	450 (232)	Superior mechanical properties, wide temperature range (requires postmold treatments)	Aerospace motor parts, industrial seals, bearings
Polyetherether-ketone (PEEK)	Victrex Stabar	 ICI	430(221)	 600 (316)	Outstanding thermals, nonflammable, good chemical resistance, radiation resistance	Wire and cable insulation, automotive, PCBs
Polytherimide (PEI)	Ultem Upac	General Electric American Cyanamid	345 (174)	350 (177)	High strength and rigidity at elevated temperatures	Electrical,industrial, appliances
Polyarylsuflone (PAS)	Radel	Amoco	340 (171)	350 (177)	Low melt processing temperatures, hydrolytically stable	Electrical, electronics, automotive, transportation
Polyethersulfone (PES)	Victrex Ultrason E	ICI BASF	430 (221)	356 (180)	Transparent, creep resistant at room temperature, high temperature resistance in air and water, low cost	Electrical appliances, industrial applications
Polyarylate (PA)	Arylon Bexloy M Ardel Durel	Du Pont Du Pont Amoco Hoechst Celanese	310–345 (154–174)	250 (121)	Thermal expansion close to metal, excellent dimensional stability, flame resistance, warp resistance, low water absorption	Automotive, appliances, industrial, electrical, electronics
Polyetherketone (PEK)	Victrex Ultrapek	ICI BASF	400 (204)	500 (260)	Good thermal properties, easy processing, excellent dielectrics	Advanced composites, seals, bearings, PCBs, chemical components
Aromatic copolyesters (based on hydroxybenzoic acid)	Xydar Vectra Grantur Victrex SRP	Amoco Hoechst Celanese Granmont ICI	445 (229) 590 (310)	390 (199) 464 (240)	High temperature resistance, high mechanical properties, very good processability, excellent chemical resistance, extremely low coefficient of expansion	Electrical, electronics, automotive underhood parts, fiber-optic devices, aerospace, Tupperware
Polyimides (PI)	Kapton Vespel Kinel Kermid Matrimid Upilex	Du Pont Du Pont Rhone Poulenc Rhone Poulenc Ciba-Geigy Ciba-Geigy	680 (360)	550 (288)	Heat resistance, radiation resistance, good dielectrics, low coefficient of thermal expansion	High-temperature films, electrical insulation and parts, mechanical parts, seals, bearings, aircraft, aerospace

[1]Heat distortion temperature at 820 kPa (264 psi) per ASTM.
[2]Continuous-use temperature per ASTM.

are somewhat tougher and harder than the co-polymers, but they are unstable during processing. The acetals are highly crystalline that accounts for their excellent properties and long-term creep resistance, but their apparent moduli fall off monotonically with long-term loading. Other applications involve replacement of metals where the higher strength of metals is not required, and costly finishing and assembly steps can be eliminated. Typical parts include gears, rollers, bearings, conveyor chains, auto window lift mechanisms and cranks, door handles, plumbing components, and pump parts.

Fluoroplastics are fluorocarbon polymers exhibiting inertness to most chemicals, resistance to high temperature, waxy feel, extremely low coefficients of friction, and excellent dielectric properties that are relatively insensitive to temperature and power frequency. The most common member of this family of plastics is polytetrafluoroethylene (PTFE). Other members are polychlorotrifluoroethylene (PCTFE), fluorinated ethylene propylene (FEP), ethylene chlorotrifluoroethylene (ECTFE), ethylene tetrafluoroethylene (ETFE), polyvinylidene fluoride (PVDF), polyvinylfluoride (PVF), and perfluoroalkoxy (PFA) resin. PTFE, commonly known as Teflon, is extremely heat-resistant (up to 500°F), has outstanding chemical resistance (being inert to most chemicals), the lowest coefficient of friction among the plastics, outstanding low temperature characteristics, and remains usable even at cryogenic temperatures; PTFE is difficult to process via melt extrusion and molding.

Polyamides are commonly called nylons and are identified by the number of carbon atoms in the monomers. When two monomers are involved, the polymer will carry two numbers, for instance, nylon 6/6. Crystalline nylons have high tensile strength, flexure modulus, impact strength, and abrasion resistance. They resist nonpolar solvents, including aromatic hydrocarbons, esters, and essential oils, but they are softened by and absorb polar materials such as alcohols, glycols, and water. Moisture pickup causes dimensional changes and reduced mechanical properties, but overly dry material can cause processing problems, and therefore, moisture control must be undertaken.

The two most widely used nylons are the Nylon 6/6 (a co-polymer of hexamethylene diamine and adipic acid) and Nylon 6 (polycaprolactam) (see Fig. 15-12). Other nylon types are: 6/9, 6/10, 6/12, 11, 12, 4/6, and 12/12.

Polyamide-imides (PAI) are among the highest temperature amorphous thermoplastics, exhibiting a useful service temperature range from cryogenic to almost 260°C (500°F). Their heat resistance approaches that of polyimides, but their mechanical properties are distinctly better. PAI is inherently flame retardant, burns with very low smoke, and has passed FAA standards for aircraft interior use. PAI is chemically resistant, unaffected by aliphatic and aromatic hydrocarbons, acids, bases, and halogenated solvents, but is attacked at high temperatures by steam and strong acids and bases.

Polyarylates (PAR) are relatively new materials and are amorphous, aromatic TP polyesters. They are polycondensation products, with both components being aromatic (ring type) molecules. They exhibit properties between those of polycarbonates and polysulfones, with a deflection temperature under load (DTUL) at 1,820 kPa (264 psi) of about 171°C (340°F). They exhibit good electrical characteristics, high UV and weather resistance, at least 85 percent light-transmission, and remain unaffected by long, outdoor exposure. Automotive lighting has recently become a prime market as designers seek unreinforced transparent plastics that can handle the heat load from the higher-intensity bulbs operating in small spaces.

Polycarbonates (PC) are among the strongest, toughest, and most rigid thermoplastics, with ductilities normally associated with softer, lower-moduli thermoplastics. They are transparent and resistant to a variety of chemicals, but are attacked by a number of organic solvents including carbon tetrachloride. They exhibit good creep resistance, high impact strength, and low moisture absorption, but are adversely affected by weathering (slight color change and embrittlement can occur with exposure to ultraviolet radiation). Among their many applications, the PCs are used as cams and gears, interior aircraft components, automotive instrument panels, headlights, lenses, boat propellers and housings for hand-held power tools as well as small appliances.

Polyesters are the polymeric products of alcohols and organic acids. The two dominant thermoplastics in this family are polyethylene terephthalate (PET), and polybutylene terephthalate (PBT). These have similar properties as the types 6 and 6/6 nylons but have lower water absorption and higher dimensional stability than the nylons. For the PET to attain its maximum properties, its crystallinity must be increased via processing or orientation of its molecules. Orientation increases the tensile strength by 300 to 500 percent and reduces permeability. PET is a water-white polymer and is made into fibers, films, and sheets, is blow-molded and thermoformed into containers for soft drinks and foods. PBT, on the other hand, crystallizes readily, even in chilled molds and is sold in the form of filled and reinforced compounds for engineering applications. Its uses include appliances, automotive, electrical/electronic, materials handling, and consumer products.

Liquid crystalline polymers (LCP), which also belong in the family of thermoplastic polyesters, are aromatic co-polyesters with a tightly ordered, stiff rodlike structure that is self-reinforcing. LCPs are resistant to most organic solvents and acids, inherently flame-resistant, meet federal standards for aircraft interior use, and can withstand temperatures over 540°C (1000°F) before decomposing. They have found applications in aviation, electronics, automotive under hood parts, and chemical processing.

Polyetherimide (PEI) is an amorphous, transparent amber polymer with high temperature resistance, rigidity, impact strength, and creep resistance. PEI has a glass transition temperature of 215°C (420°F) and a DTUL of 200°C (390°) at 1821 kPa (264 psi) and meets FAA standards for aircraft interiors. PEI is used in the medical field because of its heat and radiation resistance, hydrolytic stability, and transparency; in the electronic field, to make burn-in sockets, bobbins, and printed circuit substrates; and in the automotive field, as lamp sockets and under hood temperature sensors.

Polyimides (PI) are linear, aromatic polymers with outstanding properties involving severe environments such as high temperature and radiation, heavy load, and high rubbing velocities. In air, the continuous service temperature of PI is on the order of 260°C (500°F). At high temperatures, PI parts and components retain an unusually high degree of strength and stiffness, but prolonged exposure at high temperatures can cause a gradual reduction in tensile strength. Unfilled PI parts have an unusual resistance to ionizing radiation and have good electrical properties. PIs are used to mold high-performance bearings for jet aircraft, compressors, and appliances and in making printed circuit boards. In the form of a film, it is used for electric motor insulation and in flexible wiring.

Polyketones are crystalline polymers of aromatic ketones that exhibit exceptional high-temperature performance. *Polyetheretherketone (PEEK)* is tough and rigid, with a continuous use temperature rating of 246°C (475°F), is self-extinguishing, and gives off very low smoke emission on burning, and is resistant to most solvents, and is hydrolytically stable due to its crystallinity. *Polyether ketone (PEK)* polymers are similar to PEEK but have slightly higher [about 5.5°C (10°F)] heat resistance. Because of their high price, uses for polyketones are in the highest-performance applications such as metal replacement parts for aircraft and aerospace structures, electrical power plants, printed circuits, oven parts, and industrial filters.

Polyphenylene ether (PPE) or polyphenylene oxide (PPO) is alloyed or combined with other polymers to produce useful materials. PPE is blended with PS (usually the high-impact type) over a range of ratios to yield products with DTUL ratings from 80° to 177°C (175° to 350°F). Because both PPE and PS are hydrophobic, the alloys exhibit very low-water absorption rates, high-dimensional stabilty, and excellent dielectric properties over a wide range of frequencies and temperatures. PPE/PS alloys are used to mold housings for appliances and business machines, automotive instrument panels and seat backs, and fluid-handling equipment. PPE can also be alloyed with nylon to provide increased resistance to organic chemicals and better high-temperature performance. PPE/nylon blends have been used to injection-mold automobile fenders.

Polyphenylene sulfide (PPS) is characterized by its excellent high-temperature performance, inherent flame retardance, and good chemical resistance. Its crystalline structure is best achieved at high mold temperatures from 120° to 150°C (250° to 300°F). Most PPS is sold in the form of filled and reinforced compounds for injection molding. Because of its heat and flame resistance, it has been used to replace thermoset phenolics in electrical/electronic applications in molding connectors and sockets; PPS withstands contact with metals at temperatures up to 260°C (500°F).

The material is gaining wide use in molding surface-mounted components, circuit boards, and the like. Other applications include industrial pump housing, valves, and downhole parts for the oil industry, along with molded parts for heaters, dryers, microwave ovens, and irons, and in automotive under-the-hood components.

The sulfones are amorphous engineering TPs noted for their high DTUL temperatures, outstanding dimensional stability and electrical properties, excellent chemical resistance, and for being biologically inert, rigid, strong, and easily processed by different methods. These strong, rigid plastics are the only types that will remain transparent at service temperatures as high as 200°C (392°F). The basic types are the polysulfones (PSUs) that are polymers with SO_2 groups in their backbone, polyarylsulfones (PASU), polyethersulfones (PES), and polyphenylsulfones.

Thermoplastic Elastomers (TPEs). The basic characteristics of TPEs were discussed earlier at the beginning of Sec. 15.4. These materials do not belong to a single family of materials but have a common set of properties and are used for shock absorption, noise and vibration control, sealing, corrosion protection, abrasion and friction resistance, electrical and thermal insulation, waterproofing, and all types of load-bearing products. Elastomers are differentiated on the bases of how long it will take to return to their original size after they are deformed and the deforming force removed. The ASTM standard D 1566 details the tests for elastomeric materials. *Basically, an elastomer must be capable of retracting, within one minute, to less than 1.5 times its original length, after being stretched at room temperature to twice the original length and held for one minute prior to release of the load.* The ASTM D 2000 and SAE J 200 standards designate rubbers or elastomeric materials according to types and class as indicated in Table 15-5. The types are designated according to their maximum service temperatures, from 21° to 135° C (70° to 275° F) for rubbers, and from 71° to 274° C (160° to 525° F) for elastomers, using the letters A through J. The class designation is given based on the maximum volume of swelling upon immersion in a prescribed ASTM #3 oil test using the letters A through K for the 10 classes of volume swell. The type and class designations are written together

Table 15-5 Types and classes of elastomers (according to ASTM D-2000 and SAE J200).[2]

Type	Temperature °C (°F)	Class	Volume Swell, Max %
A	70 (158)	A	Not required
B	100 (212)	B	140
C	125 (257)	C	120
D	150 (302)	D	100
E	175 (347)	E	80
F	200 (392)	F	60
G	225 (437)	G	40
H	250 (482)	H	30
J	275 (527)	J	20
		K	10

to classify the many different elastomers or rubbers, as illustrated in Table 15-6. Of the TPEs, the styrenic co-polymers are the most used. Their structure consists normally of a block of rigid styrene on each end with a rubbery phase in the center, such as styrene/butadiene/styrene (SBS), styrene/isoprene/ styrene (SIS), styrene/ethylene-butylene/styrene (SEBS), and styrene/ethylene-propylene/styrene (SEPS).

Alloys and Blends. In addition to the variations of polymers that are created due to co-polymerization, finished thermoplastics can also be combined to create new materials. Today the range of materials that can be produced by alloying and blending is almost limitless, and this has become the fastest growing segment of the engineering thermoplastics. Further discussion will be deferred to Sec. 15.5. on compounding.

15.4.2 Thermosets

These materials offer one or more of the following: (1) high thermal stability, (2) resistance to creep and deformation under load, (3) high dimensional stability, and (4) high rigidity and hardness. These advantages are in addition to the light weight and excellent insulating electrical properties common to all plastics. They are amenable to low cost, mechanized forming or shaping through compression and transfer molding processes, and the recently developed thermoset injection molding techniques.

Table 15-6 Some popularly used elastomer (per ASTM D2000 and SAE J200).[2]

Type ↓ ↓ Class	Typical Rubber
AA	Natural rubber, styrene butadiene, butyl, ethylene propylene, polybutadiene, Polyisoprene
AK	Polysulfide
BA	Ethylene propylene, styrene butadiene (high temperature) Butyl
BC	Chloroprene, Chlorinated polyethylene
BE	Chloroprene, Chlorinated polyethylene
BF	Nitrile
BG	Nitrile, urethane
BK	Polysulfide, nitrile
CA	Ethylene propylene
CE	Chlorosulfonated polyethylene, chlorinated polyethylene
CH	Nitirle, epichlorohydrin Ethylene/acrylic
DA	Ethylene propylene
DE	Chlorinated polyethylene, Chlorosulfonated polyethylene
DF	Polyacrylate (butyl-acrylate type)
DH	Polyacrylate
FC	Silicone (high strength)
FE	Silicone
FK	Fluorinated silicone
GE	Silicone
HK	Fluorinated rubbers

The thermoset molding compounds are predominantly used in what are called advanced composite materials that are comprised of two major ingredients: (1) a resin system that generally contains such components as curing agents, hardeners, inhibitors, and plasticizers, and (2) fillers and/or reinforcements, which may consist of mineral or organic particles, inorganic, organic, or metallic fibers, and/or inorganic or organic chopped cloth or paper. The resin system provides the dimensional stability, electrical qualities, heat resistance, chemical resistance, decorative, and flammability qualities. The fillers and reinforcements provide the strength and toughness and sometimes the electrical qualities.

Some of the thermoset plastics and their properties are listed in Table 15-7. The materials shown in this table are mostly filled or with reinforcements, except for some of the epoxies and the urethane, and thus cannot be compared directly with the unreinforced thermoplastics in Table 15-3. However, as indicated earlier, these generally have higher strength and modulus. Table 15-8 shows the major thermosets used in composites, and of these, the polyesters and the epoxies are the dominant materials. Other generic families of thermoset molding compounds are the alkyds, allylics, bismaleimides (BMI), melamines, phenolics, polyimides, silicones, ureas, and polyurethanes.

The *thermoset polyesters* are unsaturated (i.e., they have at least two double bonds in the monomers) and can be cured to be brittle and hard, tough and resilient, or soft and flexible. The cross-linking of unsaturated polyesters is initiated by a peroxide catalyst that is selected based on requirements for curing temperature and rate. Among the members of this family are orthophthalic (lowest cost) and isophthalic polyesters, vinyl esters, and blends of various types that result in various materials with a very wide range of properties. The primary use of thermoset polyesters is as the matrix of what may be called the commodity composites that go to commercial, industrial, and land and sea transportation applications. These applications include chemically resistant piping and

Table 15-7 Properties of some thermoset plastics. [1]

Thermosets	Specific Gravity	Trans-parency	Tensile Strength, 10^3 psi	Tensile Modulus 10^5 psi	Impact Strength Izod, ft lb/inch	Max Use Temp. F (no load)	DTUL at 264 psi
Alkyds							
Glass filled	2.12–2.15	No	4–9.5	20–28	0.6–10	450	400–500
Mineral filled	1.60–2.30	No	3–9	5–30	0.3–0.5	300–450	350–500
Syn. fiber filled	1.24–2.10	No	4.5–7	20	0.5–4.5	300–430	245–430
Diallyl phthalates							
Glass filled	1.61–1.87	No	6–11	14–22	0.4–15	300–400	330–540
Mineral filled	1.65–1.80	No	5–9	12–22	0.3–0.5	300–400	320–540
Epoxies (bis A)							
No filler	1.06–1.40	Yes	4–13	2.15–5.2	0.2–1.0	250–500	115–500
Graphite fiber reinf.	1.37–1.38	No	185–200	118–120	—	—	—
Mineral filled	1.6–2.0	No	5–15	—	0.3–0.4	300–500	250–500
Glass filled	1.7–2.0	No	10–30	30	10–30	300–500	250–500
Epoxies (novolac)							
No filler	1.12–1.24	No	5–11	2.15–5.2	0.3–0.7	400–500	450–500
Epoxies (cycloaliphatic)							
No filler	1.12–1.18	Yes	10–17.5	5–7	—	480–550	500–550
Melamines							
Cellulose filled	1.45–1.52	No	5–9	11	0.2–0.4	250	270
Flock filled	1.50–1.55	No	7–9	—	0.4–0.5	250	270
Fabric filled	1.5	No	8–11	14–16	0.6–1.0	250	310
Glass filled	1.8–2.0	No	5–10	24	0.6–18	300–400	400
Phenolics							
Woodflour filled	1.34–1.45	No	5–9	8–17	0.2–0.6	300–350	300–370
Mica filled	1.65–1.92	No	5.5–7	25–50	0.3–0.4	250–300	300–350
Glass filled	1.69–1.95	No	5–18	19–33	0.3–18	350–550	300–600
Fabric filled	1.36–1.43	No	3–9	9–14	0.8–8	220–250	250–330
Polyesters							
Glass filled BMC	1.7–2.3	No	4–10	16–25	1.5–16	300–350	400–450
Glass filled SMC	1.7–2.1	No	8–20	16–25	8–22	300–350	400–450
Glass cloth reinf.	1.3–2.1	No	25–50	19–45	5–30	300–350	400–450
Silicones							
Glass filled	1.7–2.0	No	4–6.5	10–15	3–15	600	600
Mineral filled	1.8–2.8	No	4–6	13–18	0.3–0.4	600	600
Ureas							
Cellulose filled	1.47–1.52	No	5.5–13	10–15	0.2–0.4	170	260–290
Urethanes							
No filler	1.1–1.5	No & Yes	0.2–10	1–10	5–NB	190–250	—

reactors, truck and automotive cabs and bodies, appliances, bathtubs and showers, automobile hoods, decks, and doors, and in building materials, where they are filled with gypsum or alumina trihydrate to make them flame retardant. Polyesters are also used in "cultured marble", a highly-filled material that resembles the natural materials for sinks and vanities. Polyester concretes are used to patch highways and bridges and polyester patching compounds are used to repair auto-body damage.

Epoxies are the result of polymeric reactions with epichlorohydrin. The two most important plas-

Table 15-8 Characteristics of major thermosets and the processes used in the manufacture of reinforced plastics/composite parts or components.[2]

Thermosets	Properties	Processes
Polyesters	Simplest, most versatile, economical, and most widely used family of resins; good electrical properties, good chemical resistance, especially to acids	Compression molding, filament winding, hand layup, mass molding, pressure bag molding, continuous pultrusion, injection molding, spray-up, centrifugal casting, cold molding, encapsulation
Epoxies	Excellent mechanical properties, dimensional stability, chemical resistance (especially to alkalis), low water absorption, self-extinguishing (when halogenated), low shrinkage, good abrasion resistance, excellent adhesion properties	Compression molding, filament winding, hand layup, continuous pultrusion, encapsulation, centrifugal casting
Phenolic resins	Good acid resistance, good electrical properties (except arc resistance), high heat resistance	Compression molding, continuous lamination
Silicones	Highest heat resistance, low water absorption, excellent dielectric properties, high arc resistance	Compression molding, injection molding, encapsulation
Melamines	Good heat resistance, high impact strength	Compression molding
Daillyl o-phthalate	Good electrical insulation, low water absorption	Compression molding

tics in this family are (1) the diglycidyl ether of bisphenol A (DGEBPA) and its homologs, and (2) the glycidyl ethers of various novalac resins. The DGEBPA is the product of the reaction with bisphenol A (BPA) in the presence of an alkali and the glycidyl ethers are the products of reactions with the various phenolic novolacs. The reaction to produce DGEBPA is shown in Fig. 15-15 while Fig. 15-16 illustrates the glycidel ethers of three novalacs. The DGEBPA is the most widely used and it can range from a viscous liquid to a high-molecular weight solid. Higher molecular weight homologs are produced by controlling the ratio of epichlorohydrin (glycidyl chloride) to BPA; the higher the BPA concentration during the reaction, the higher is the molecular weight. The glycidyl ethers of novolac resins provide higher temperature performance and improved chemical resistance than the BPA-based epoxies.

The epoxies are more expensive than the equivalent TS polyesters, but they exhibit much better performance than the polyesters. For this reason, the epoxies are currently used far more often than all the other matrices in advanced composites, for example, graphite fiber epoxy, for structural applications in commercial and military aircraft. Their general properties include: toughness, less shrinkage during curing, good weatherability, low moisture absorption, curing without the evolution of by-products, good wetting, and adhesion to a wide variety of surfaces. Other excellent qualities are: good mechanical properties and thermal capabilities, excellent fatigue resistance, outstanding electrical properties from low to high temperatures, exceptional water resistance, practically resistant to fungus, and general corrosion resistance. The adhesive property makes epoxies excellent for potting and encapsulating electronic and other components, for coatings to most materials

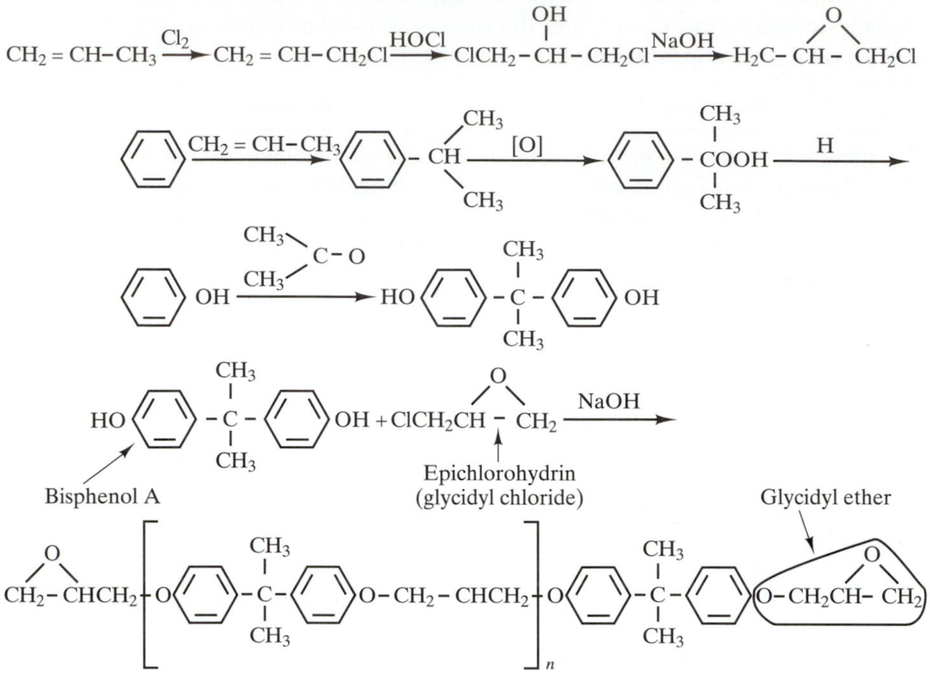

$$CH_2=CH{-}CH_3 \xrightarrow{Cl_2} CH_2=CH{-}CH_2Cl \xrightarrow{HOCl} ClCH_2{-}CH(OH){-}CH_2Cl \xrightarrow{NaOH} H_2C{-}CH{-}CH_2Cl \text{ (epoxide, O bridging } H_2C \text{ and } CH)$$

$$C_6H_6 \xrightarrow{CH_2=CH-CH_3} C_6H_5{-}CH(CH_3)_2 \xrightarrow{[O]} C_6H_5{-}C(CH_3)_2{-}COOH \xrightarrow{H}$$

$$C_6H_5{-}OH \xrightarrow{(CH_3)_2C{=}O} HO{-}C_6H_4{-}C(CH_3)_2{-}C_6H_4{-}OH$$

$$HO{-}C_6H_4{-}C(CH_3)_2{-}C_6H_4{-}OH + ClCH_2CH{-}CH_2 \text{ (epoxide)} \xrightarrow{NaOH}$$

Bisphenol A

Epichlorohydrin (glycidyl chloride)

Glycidyl ether

$$CH_2{-}CHCH_2 \text{ (epoxide)} {-}\left[O{-}C_6H_4{-}C(CH_3)_2{-}C_6H_4{-}O{-}CH_2{-}CH(OH)CH_2\right]_n{-}O{-}C_6H_4{-}C(CH_3)_2{-}C_6H_4{-}O{-}CH_2CH{-}CH_2 \text{ (epoxide)}$$

Fig. 15-15 Synthesis of di-Glycidyl ether of bisphenol A (DGEBPA). [4]

such as steel, aluminum, plastic, and other materials, and for adhesives to join similar or different materials of plastic, steel, aluminum, wood, or glass.

Alkyds thermoset compounds are also based on unsaturated polyester resins and are therefore similar to the TS polyesters, but have lower levels of monomers. They are part of a group that includes the bulk-molding compounds (BMCs) and sheet-molding compounds (SMCs) that have fast molding cycles at low pressure to make them much easier to process than other TSs by vacuum, compression, transfer, or injection molding.

The principal *allylic* resins are the diallyl phthalates (DAP) and the diallyl isophthalates (DAIP), with DAP being used predominantly. The DAIP exhibits a higher continuous service temperature range [177° to 232°C (350° to 450°F)] than DAP [149° to 177°C (300° to 350°F)]. In general, the allyls are more expensive and for this reason are used in very few consumer products. The major use of DAPs is in electrical connectors in communications, computer, and aerospace systems.

Bismaleimides (BMIs) are very high temperature-resistant condensation polymers made from maleic anhydride and a diamine, such as methylene diamine (MDA). BMIs exhibit the same desirable features as epoxies but with higher continuous service temperature ranges—205° to 232°C (400° to 450°F). As such they are used as the matrix for composites used in military aircraft and in aeropace. They are also used in the manufacture of printed circuit boards and as heat-resistant coatings.

Melamines are amino thermoset resins, two of which are melamine formaldehyde (MF) and urea formaldehyde (UF). Melamines are best known for their extreme hardness, excellent and permanent colorability, and self-extinguishing flame resistance. Because they do not impart taste or odor to solid or liquid foods they are used for dishware and household goods. Various kinds of fillers are used with melamines to meet varying requirements. For dishes and kitchen ware, the filler is alpha-cellulose, mineral-filled for electrical resistance, and fabric or glass-fiber reinforcement for impact and tensile strengths. Melamines have relatively good chemical resistance, although they are attacked by acids and alkalis. They are also used as adhesives, coatings, laminating resins, and for electrical applications.

Phenolics (phenol-formaldehyde) continue to be low-cost general purpose thermosets to meet a multi-

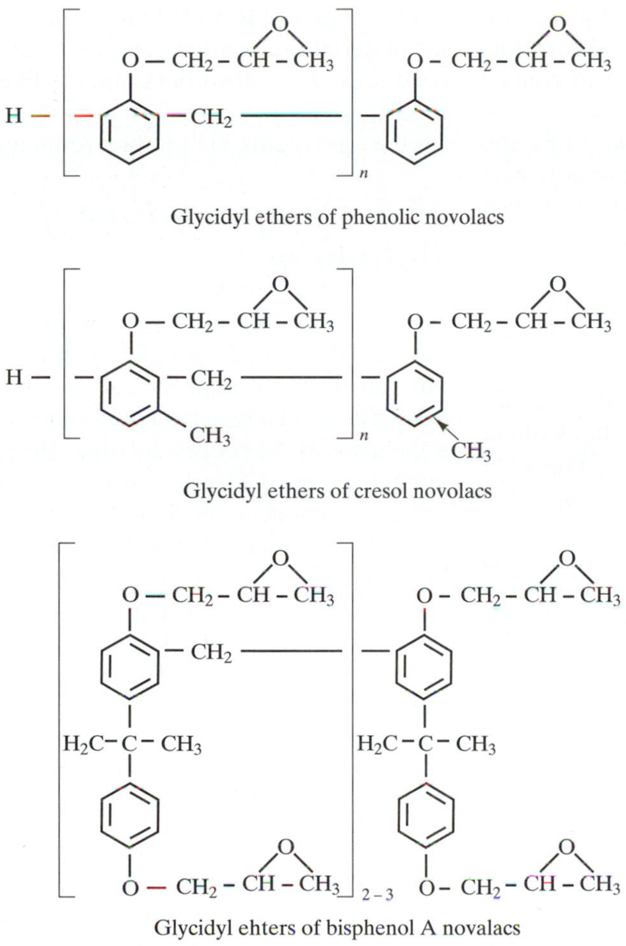

Glycidyl ethers of phenolic novolacs

Glycidyl ethers of cresol novolacs

Glycidyl ehters of bisphenol A novalacs

Fig. 15-16 Glycidyl ethers (epoxies) of novalacs. [4]

tude of applications. They are formulated with a one- or two-stage phenolic curing system. In general, the one-stage resins are slightly more critical to process. They have properties that are somewhat inferior to those of the more expensive TSs, but they are usually easier to mold. They exhibit superior heat and flame resistance, a high DTUL, good electrical properties, excellent moldability and dimensional stability, and good water and chemical resistance. The heat-resistant types are used in motor housings for appliances, handles for pots and pans, and other such products.

Urea formaldehyde (UF) is the other member of the amino family of TS plastics. The UFs are available in a wide range of colors, from translucent, colorless and white through all the colors to a lustrous black. These are nonflammable (self-extinguishing), odorless, and tasteless materials and char at about 200°C (392°F). Temperatures from −21° to +80°C (−70° to +175°F) have no effect, but higher temperatures over prolonged periods will cause fading and blistering. The applications of the UFs include sanitary wares such as toilet seats, knobs, closures, buttons, electrical accesories such as housings and switches, laminates, and so on.

The ureas, the melamines, and the phenolics are materials that compete for the same applications. Each of the three has characteristics that designers can use. Phenolics have by far the largest market, due to their performance and cost advantages. The advantages of the melamines over the ureas include better retention of electrical properties at elevated temperatures, and in the presence of moisture, superior heat resistance, improved surface hardness and gloss, better resistance to staining, and lower-water absorption, and in general, are better products when in contact with foods and drinks.

Polyimides (PIs) are available as TSs and TPs, and are some of the best heat- and fire-resistant polymers known. They can retain a significant portion of their room-temperature mechanical properties in the range of −240° to 315°C (−400° to +600°F). Moldings and laminates are generally based on the TSs. They have good wear resistance, low coefficient of friction, outstanding electrical properties over wide temperature and humidity ranges, are unaffected by dilute acids, aromatic and aliphatic hydrocarbons, esters, alcohols, hydraulic fluids, JP-4 fuel, and kerosene. However they are attacked by dilute alkalis and concentrated inorganic acids.

Silicones (SIs) differ from most other plastics in that they contain only silicon and oxygen (a siloxane bond) and no carbon in their main polymer chain. This structure gives the SIs a wider temperature capability, with better moisture and oxidation resistance, than most of the carbon chain polymers. A wide-range of flexibility can be obtained in SIs with variations of the side groups and cross-linking, but their tensile strengths are generally inferior to those of the carbon polymers. These high-cost molding compounds with fillers and reinforcements are used in applications requiring retention of physical, electrical, and dimensional properties after long-term exposure, and good corrosion resistance.

A complete series of silicone products is theoretically possible from true liquids through the TP stage to a rubbery stage, and finally to a complex three-dimensional TS structure characterized by extreme heat resistance, inertness, and hardness. Examples of some silicone products are illustrated in Fig. 15-17.

Polyurethanes (PURs) are produced by the reaction of polyisocyanates with polyester- or polyether-based resins and can be either TPs or TSs. They are available in three forms: rigid foam, flexible foam, and as an elastomer. (The elastomers were the earliest TPEs that were introduced in the market in the 1950s.) They are characterized by high strength and good chemical and abrasion resistance, with superior resistance to ozone, oil, gasoline, and many solvents. The rigid foam product is widely used as an insulation material in buildings, appliances, and similar type applications. The flexible foam is an excellent cushioning material for furniture, and the elastomer is used in solid tires, shock absorbers, and the like. The T_g of the flexible foams is well below room temperature—and for the rigid foams, higher than room temperature.

15.5 Compounding of Polymers

Just like in metals, very few polymers are used in their pure, unadulterated form; hence, they are mixed with other materials to improve and enhance their properties for usage in a variety of applications. *Compounding* is the name of the process by which the constituents are intimately mixed together in the molten state into as nearly a homogeneous mass as is possible. This is the general term given to (1) the process of alloying or blending polymers; (2) using additives and fillers such as colorants, flame retardants, antistatic agents, plasticizers, and others; and (3) adding reinforcement—or a combination of all three. The product of the last one is usually referred to as reinforced plastics or polymer matrix composites.

15.5.1 Alloying and Blending

Alloys and blends are combinations of polymers that are mechanically mixed. They do not depend on chemical bonding, but often require "compatibilizers" to keep the constituents from segregating. In alloys and blends, the best characteristics of each constituent is usually retained and quite often the objective is to find two or more constituents whose mixture yields synergistic property improvements beyond those that are purely additive in effect, as schematically illustrated in Fig. 15-18. Most often, the desired property improvements are in the areas of impact strength, weather resistance, improved low-temperature performance, and flame retardation.

When the combination of polymers produces a product that has a single glass transition temperature and that exhibits a synergistic effect in properties, the product is called an *alloy*. When the combination produces a product with multiple glass transition temperatures and with properties that are the average of the contributions of the individual constituent polymers, the product is called a *blend*. To keep the constituents of a polymer blend from separating,

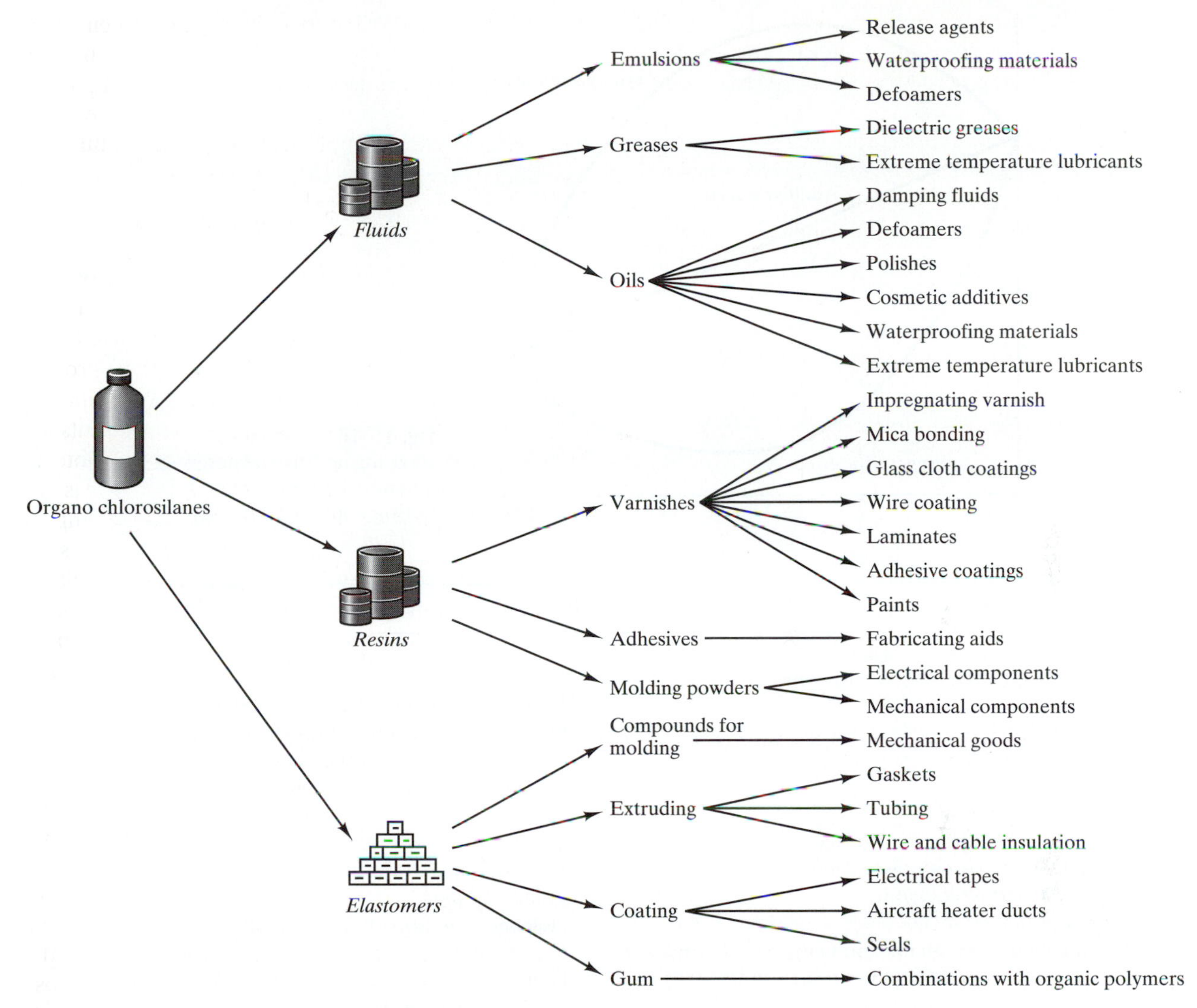

Fig. 15-17 Examples of silicone products. [2]

especially when they are chemically incompatible, compatibilizers are added to the mixture. The compatibilizer is a material that has an affinity to both polymers to be mixed. Examples of plastic alloys and their properties are shown in Table 15-9 and the trade names and producers of some alloys are given in Table 15-10.

15.5.2 Additives and Fillers [1]

These materials are compounded with plastics to change and improve their physical, mechanical, and processing properties. Some common fillers and their uses are given in Table 15-11 that also includes some reinforcements. The types of *additives and fillers* and their effects are:

- *Fillers* may be inorganic, organic, mineral, natural, or synthethic. They are more commonly used with the thermosetting resins such as the phenolics, ureas, and melamines, although they are also used for some thermoplastics. Large amount of fillers are called *extenders* since they allow the production of a large volume of the plastic with a small amount of resin, thereby reducing the cost of the plastic. Because the

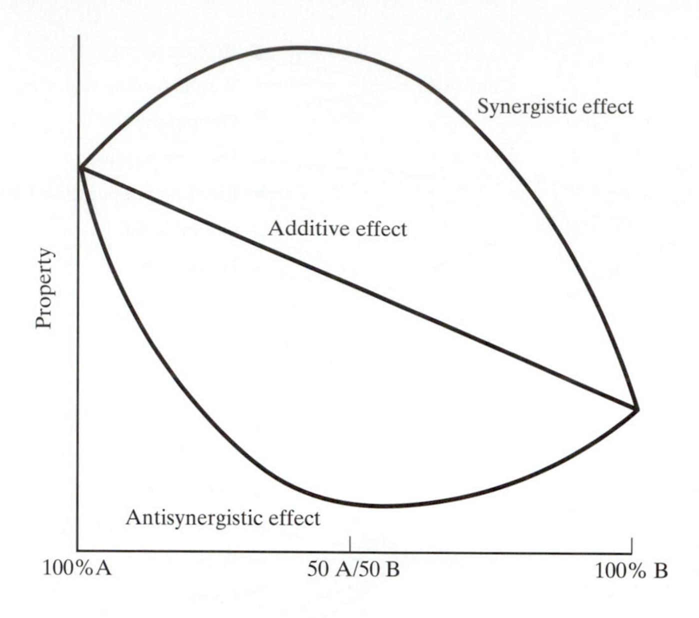

Fig. 15-18 Schematics showing additive, synergistic, and antisynergistic effects during compounding. [2]

Table 15-9 Some commercial plastic alloys and their outstanding properties.[2]

Alloy	Properties
PVC/acrylic	Flame, impact, and chemical resistance
PVC/ABS	Flame resistance, impact resistance, processability
Polycarbonate/ABS	Notched impact resistance, hardness, heat-distortion temperature
ABS/polysulfone	Lower cost
Polypropylene/ethylene-propylene-diene	Low-temperature impact resistance and flexibility
Polyphenylene oxide/polystyrene	Processability, lower cost
Styrene acrylonitrile/olefin	Weatherability
Nylon/elastomer	Notched Izod impact resistance
Polybutylene terephthalate/polyethylene terephthalate	Lower cost
Polyphenylene sulfide/nylon	Lubricity
Acrylic/polybutylene rubber	Clarity, impact resistance

properties of an extended plastic often decrease, its use is limited to less critical applications. One of the most widely used filler is wood flour. Examples of mineral fillers are calcium carbonate, silica, and clay.

- *Plasticizers* are used generally to enhance flexibility, resiliency, and melt flow of the plastics. A plasticizer is not chemically linked to the plastic, but acts like an internal lubricant by reducing the van der Waals forces between the polymer chains, and by separating the chains to prevent them from intermeshing. Without the plasticizers, it would be impossible to make plastic sheets, tubes, films, and other flexible forms. As a general rule, the higher the ratio of plasticizer to plastics (resin), the greater is the flexibility achieved. The most popular general-purpose plasticizers are the phthalates, although epoxies, phosphates, adipate diesters, and polyesters are also used.
- *Heat stabilizers* prevent the degradation of resins during processing when melts are subjected to high temperatures, or they extend the life of end products of which they become a part. PVC is

Table 15-10 Some plastic alloys with their producers (trade names) and properties.[2]

Materials	Producers	Properties
PPO/PS	GE (Noryl)	Polyphenylene oxide (PPO) has high strength and high heat resistance but oxidizes at temperatures required for processing: adding polystyrene (PS) makes it possible to process.
ABS/PC	Mobay (Bayblend), Fiberite	Acrylonitrile-butadiene-styrene (ABS) improves processability of polycarbonates; PC contributes toughness and heat resistance
PC/PET PC/PBT	GE (Xenoy)	PC, though tough and able to withstand very high temperatures, lacks good resistance to chemicals; polyethylene terephithalate (PET) and polybutylene terephthalate (PBT) make up for this lack
PET/PBT	GAF (Gafite), Hoechst Celanse (Celanex), GE (Valox)	Alloying with PET lowers PBT's impact resistance but brings down its cost
PVC/ABS	General Tire & Rubber, GE, Cycoloy, Cycovin, various compounders	Polyvinyl chloride (PVC) adds flame retardance and rigidity to ABS, a more easily processed resin
PP/elastomer	Reichhold, Hoechst Celanese Montedison	Polypropylene (PP) contributes good heat resistance and processability; elastomers add impact resistance

particularly conducive to degradation during processing, and is thus a prime consumer of heat stabilizers. Some of the stabilizers are organotin and barium. Generally, liquid stabilizers are used for flexible PVC and barium/zinc liquids are the most used. Other resins requiring stabilization are chlorinated PE and blends of ABS and PVC.

- *Antioxidants* protect materials from deterioration through oxidation brought on by heat, light, or chemically induced mechanisms. Deterioration is manifested by embrittlement, melt-flow instability, loss of tensile properties, and discoloration. The three main preventive mechanisms to control deterioration of polymers are by: (1) absorbing or screening ultraviolet light, (2) deactivating metal ions, and (3) decomposing hydro-peroxides to nonradical products.
- *Ultraviolet (UV) light absorbers* stabilize the color and lengthen the life of the product because virtually every plastic degrades in sunlight in a number of ways, the most common being discoloration and the loss of physical properties. The polymers that are particularly susceptible to this type of degradation are the polyolefins, PS, PVC, ABS, the polyesters, and the polyurethanes. The most effective UV absorber is black, in any form, whether carbon black, black paint, or black dye. However, it cannot be used universally. Hence, a variety of chemicals are used, some of which are benzophenones, nickel complex of an alkylated hindered phenol, benzotriazoles, and hindered amine light stabilizers (HALS).
- *Antistatic agents*, sometimes called destaticizers, are used to reduce the buildup of electrostatic charges on the surface of the plastics due to their inherent poor electrical conduction. They are used to attract moisture to increase surface conductivity and to reduce the likelihood of a spark or a discharge. The space suits of the astronauts have antistatic agents added to them to prevent electrical discharge that may damage the delicate electronic components in the shuttle. The plastics that are particularly susceptible to accumulation of

Table 15-11 Common fillers or reinforcements and their uses.[1]

Filler/reinforcement	Uses
Alumina trihydrate	Extender: flame retardant: smoke suppressant
Barium sulfate	Used as a filler and white pigment: increases specific gravity, frictional resistance, chemical resistance
Boron fibers	High tensile strength and compressive load-bearing capacity; expensive
Calcium carbonate	Most widely used extender/pigment or filler for plastics
Calcium sulfate	Extender, also enhances physical properties, increases impact, tensile, compressive strength
Carbon black	Filler; used as a pigment, antistatic agent, or to aid in crosslinking; conductive
Carbon/graphite fibers	Reinforcement: high modulus and strength: low density: low coefficient of expansion: low coefficient of friction: conductive
Ceramic fibers	Reinforcement: very high temperature resistance: expensive
Feldspar and nepheline syenite	Specialty filler, easily wet and dispersed: enables transparency and translucency; weather and chemical resistance
Glass reinforcement (fiber, cloth, etc.)	Largest volume reinforcement: high strength, dimensional stability, heat resistance, chemical resistance
Kaolin	Second-largest volume extender/pigment, with largest use in wire and cable, SMC and BMC, vinyl flooring
Metal fillers, filaments	Used to impart conductivity (thermal and electrical) or magnetic properties or to reduce friction: expensive
Mica	Flake reinforcement: improves dielectric, thermal, mechanical properties: low in cost
Microspheres, hollow	Reduces weight of plastic systems; improves stiffness, impact resistance
Microspheres, solid	Improves flow properties, stress distribution
Organic fillers	Extenders/fillers, like wood flour, nutshell, corncobs, rice, peanut hulls
Polymeric fibers	Reinforcement; lightweight
Silica	Fillers/extenders/reinforcements; functions in thickening liquid systems and making them thixotropic, in helping avoid plate-out with PVC, in acting as a flatting agents
Talc	Extenders/reinforcements/fillers; higher stiffness, tensile strength and resistance to creep
Wollastonite	High loadings possible: can improve strength, lower moisture absorption, elevate heat and dimensional stability, improve electricals

electrostatic charge are polyethylenes, polypropylenes, polystyrenes, nylons, polyesters, urethanes, cellulosics, acrylics, and acrylonitriles. The most common antistatic agents are amines, quarternary ammonium compounds, phosphate esters, and polyethylene glycol esters.

- *Coupling or sizing agents* are added to improve the bonding of plastic to its inorganic filler materials, such as glass fibers. Silanes and titanates are used for this purpose.
- *Flame retardants* are added to reduce the flammability of plastics because they either insulate the plastic, create an endothermic cooling reaction, coat the plastic thereby excluding oxygen, or actually influence the combustion through a reaction with materials that have different physical properties. The retardants can be inorganics such as alumina trihydrate (ATH), antimony oxide, or zinc borate, or organics such as phosphate esters, and halogenated compounds of various types. ATH is the most widely used flame retardant and is very effective in thermosets and in some thermoplastics. Bromine and chlorine compounds are used in a variety of thermosets

and many thermoplastics from PE, PP, PVC through ABS, and into the engineering plastics, but toxicity concerns may eventually lead to their replacement by other chemicals. The phosphorus-based retardants are potential replacements of the halogenated compounds.

- *Blowing agents* are used singly or in combination with other substances to produce cellular structure (foam) in a plastic mass. They cover a wide variety of products and techniques, but compounders are limited to chemical agents that decompose or react under the influence of heat to form a gas. Chemical blowing agents range from simple salts such as ammonium or sodium bicarbonate to complex nitrogen-releasing agents. The nitrogen-releasing compounds are the most used and examples are the azo compounds, such as azodicarbonamide, the N-nitroso compounds, and the sulfonyl hydrazides.
- *Lubricants* are used to enhance the resins' processability and the appearance of the final products. To be effective, the lubricants must be compatible with the resins in which they are used, do not adversely affect the properties of the end-products, and are easily combined. The five types of lubricants are: metallic stearates, fatty acid amides and esters, fatty acids, hydrocarbon waxes, and low molecular weight polyethylenes.
- *Colorants* must provide colorfastness when exposed to light, temperature, humidity, chemicals, and so on, but without reducing other desirable properties such as flow during processing, resistance to chalking and crazing, and impact strength. Pigments and dyes may be used. However, dyes yield colors with poorer lightfastness, heat stability, and the tendency to bleed and migrate in the plastic system. Thus, dyes are used much less than pigments. Pigments may be organic or inorganic. Organic ones provide stronger, more transparent colors and are higher priced. Inorganic ones are denser and are usually of a larger particle size. Carbon blacks are also used, as a colorant, to protect the plastics from thermal UV degradation, and as a reinforcing fiber. There are various special colorants such as metallics, fluorescents, phosphorescents, and pearlescent colorings.

15.6 Composites and Reinforced Plastics

A composite material, as defined earlier, is a macroscopic combination of two or more distinct materials with a recognizable interface between them. However, because composites are generally used for their structural properties, the definition is restricted to include only those materials that exhibit greatly enhanced mechanical properties as the result of the combination of a reinforcement (such as fibers or particles) supported by a binder (matrix) material (the continuous phase in the composite). The systems of composite materials are indicated in Table 15-12 and we note that all the major classes of materials, that is, metals, ceramics, glasses, and polymeric materials (plastics), may serve both as the matrix and as the filler or reinforcement material. Organic reinforcements are, however, not used with the metals or engineering ceramics.

The purpose of the reinforcements is different for the different matrices. For metals, it is usually to improve their high temperature creep properties and to improve their hardnesses. For polymers, it is for the improvement of their stiffness and strength, and occasionally for toughness. For ceramics, it is usually to improve their toughness, as indicated in Chap. 14. Although the term matrix has the connotation of being the greater amount of material, it may in fact have the lesser volume fraction, particularly in polymer composite materials or reinforced plastics. In fact, the matrix needs only to wet and form a continuous phase around the reinforcement. If we consider a continuous fiber composite, the maximum ideal fiber volume fraction is 90.7 percent, which makes the matrix volume fraction only 9.3 percent. The fiber volume in polymer matrix composites do not approach the ideal maximum, but 60 to 75 percent fiber volume fraction is not unusual. In cemented carbides, the particle volume fraction approaches 80 percent or more.

15.6.1 Classifications of Composites

Composites are classified firstly *according to the matrix*. Thus, we have *metal matrix, polymer matrix, and ceramic matrix composites*. In addition, they are classified *according to the form of the reinforcement. Accordingly, there are particulate, fiber, and laminar*

Table 15-12 Different matrices, reinforcements, and properties of composite materials.[2]

Matrix Material	Reinforcement Material	Properties Modified
Metal	Metal, ceramic, carbon, glass fibers	Elevated temperature strength Electrical resistance Thermal stability
Ceramic	Metallic and ceramic particles and fibers	Elevated temperature strength Chemical resistance Thermal resistance
Glass	Ceramic fibers and particles	Mechanical strength Temperature resistance Chemical resistance Thermal stability
Organics, Thermosets, Thermoplastics	Carbon, glass, organic fibers, glass beads, flakes, ceramic particles, metal wires	Mechanical strength Elevated temperature strength Chemical resistance Antistatic Electrical resistance EMF shielding Flexibility Wear resistance Energy absorption Thermal stability

composites. For most applications, the reinforcement has a higher strength and modulus than the matrix. In some instances, such as rubber-modified polymers, the filler is more compliant and more ductile than the matrix to yield improved toughness in the composite. This is also true for ceramic composites, (i.e., more compliant materials are added to improve the toughness). We shall now discuss the composites based on the form of the reinforcement.

A *particulate composite* is one in which the "particle" reinforcement has roughly equal dimensions in all directions. Thus, particles may be rods, spheres, flakes, and many other shapes whose aspect ratios are about 1.0. The aspect ratio is the ratio of the length to the cross-sectional dimension. *In these composites, the size, shape, and distribution, in addition to the amount and the modulus of the particles affect the composite properties*. Smaller sizes and also more rounded shapes are usually better. Uniform distribution of the particles should be strived for; in metals and crystalline ceramics, the particles must not be along the grain boundaries.

A *fiber composite* is one in which the reinforcements have aspect ratios much greater than 1.0 (i.e., the length is much greater than the cross-sectional dimension). *Fiber composites* are further classified according to (1) *discontinuous or short fiber composites,* and (2) *continuous fiber reinforced composites*. In short or *discontinuous fiber composites,* the *properties vary with the length of the fiber*. When the length of the fiber is such that any further increase in length no longer increases the modulus of the composite, then the composite is *continuous fiber reinforced*. Most continuous fiber composites have, in fact, fiber lengths comparable to the overall dimension of the component or structure.

Laminar composites are composed of at least two layers of materials (one for each constituent), whose two dimensions are much larger than the third. Examples of laminar composites are cladded materials for corrosion protection (ALCLAD, and stainless steel cladded materials), plywood, the U.S. coins, and most recently, ARALL, which is a made of layers of aramid polymer and aluminum sheets. The plywood is a laminated composite that is made by bonding together several sheets of discontinuous or continuous fiber composites.

15.6.2. Reinforced Plastics

By far the most widely used composites are the polymer matrix composites or reinforced plastics. Products, such as tennis rackets, golf clubs, pleasure boats, and many more, abound in our daily lives. When we talk about composites, the average person immediately thinks of these as the polymer matrix materials. The reasons for this are the ease with working or shaping polymeric materials due to their low melting points (glass transition temperatures) and the excellent corrosion resistance of polymers. Also most of the commercial products, are the low cost "chopped" fiber composites.

For engineering applications, continuous fiber polymer matrix composites are generally used. However, their use is *limited to relatively low temperatures because of the low polymer glass transition temperatures*. For the high-speed civil transport (HSCT) plane, matrix resins are being developed that can be used continuously at 425°C. High speeds result in higher drag forces and high "skin" temperatures, which is why resins with higher continuous use temperature are being developed. *For high temperature applications, metal matrix composites* have to be used. *For still much higher temperatures, ceramic matrix composites* are being developed. The driving force for all *these advanced composites* is the need for better fuel efficiency and higher speeds of aircrafts. Better fuel efficiency requires light weight and high-combustion temperatures, which result in higher thrust for jet engines.

From here on, our discussion will be predominantly on continuous-fiber reinforced plastics. In fiber-reinforced plastics (FRP), most of the load is sustained by the fibers. The function of the matrix is to surround and to form a continuous phase around the fibers and to bind or adhere with the fibers in order to be able to transfer load to and between fibers, and to protect the fibers from the environment and from damage during handling. The matrix also keeps the fibers in proper orientation and position to carry the intended loads, distributes the loads more or less evenly among the fibers, provides resistance to crack propagation if the fibers crack, and provides all of the interlaminar shear strength of the composite. The matrix (the polymer or plastic) is usually the weak-link in the composite because there is no available resin that can withstand very high loads. Thus, the matrix is generally the first to form microcracks (crazing) that will grow into larger cracks through coalescence, debonds from the fiber surface, and fractures at much lower strains than desired. Furthermore, the matrix generally dictates the overall service temperature and environmental resistance of the composite. Thus, the selection of the polymer matrix becomes very important. We shall now discuss some of the most widely used fibers and matrices in fiber-reinforced composites.

15.6.2a Fibers. Currently, the three most widely used fibers are glass, carbon/graphite, and the aramid or aromatic polyamide, known as Kevlar.

As noted previously, *glass* is amorphous silica that is derived by adding oxide modifiers to lower the melting point and the viscosity of the liquid. The four glass compositions that are used commercially to produce continuous glass fiber components are shown in Table 15-13. The *A-glass* is the high-alkali

Table 15-13 Types and compositions of glass fibers.[4]

	Material, wt%							
Glass type	Silica	Alumina	Calcium oxide	Magnesia	Boron oxide	Soda	Calcium Fluoride	Total minor oxides
E-glass	54	14	20.5	0.5	8	1	1	1
A-glass	72	1	8	4	···	14	···	1
ECR-glass	61	11	22	3	···	0.6	···	2.4
S-glass	64	25	···	10	···	0.3	···	0.7

grade that is essentially soda-lime-silica and is used in applications requiring good chemical resistance. The *E-glass* is the electrical grade, which is essentially calcium alumino-borosilicate, that offers good electrical properties and durability. *E-glass is the most widely used reinforcing glass fiber*. The *ECR glass fibers* are used in applications requiring good electrical properties, coupled with better chemical resistance. The S-glass is magnesium-alumino silicate with no boron oxide and the fibers are used in applications requiring higher tensile strength and higher thermal stability. S-glass has better high-temperature property retention than any of the other glasses. However, it is also expensive and is therefore used only in applications where cost-performance can be justified. The inherent properties of the four glass fibers are given in Table 15-14.

S-glass fiber is predominantly used in the aircraft/aerospace industry. ECR and S-glasses exhibit excellent resistance to moisture over a period of time, compared to A-glass. E-glass is predominantly used in printed circuit boards because of its excellent electrical properties, superior dimensional stability, good moisture resistance, and lower cost. These are the same attributes that make E-glass the most widely used reinforcement in plastics.

In addition to the above four grades, C-glass and R-glass have been developed that are used sparingly. The C-glass is chemically resistant and the R-glass is a high-strength, high modulus grade.

In the production of glass fibers described in Sec. 14.3.1, the fibers pass through a sizing applicator bath to completely *coat the fibers with a sizing or coupling agent*. The sizing agent protects the fiber from handling damage and imparts lubricity, resin (matrix) compatibility, and adhesion properties to the final product; that is, the fiber properties are tailored to the specific end-use requirement. Thus, in addition to the differences in the grades of glasses, *the sizing agent also differentiates the varying fiberglass products.*

Carbon/graphite fibers are used predominantly for advanced composites for the aerospace and aircraft industries. Substantial broadening of the application base has taken place to include recreational sports equipment as well as industrial and commercial products. This is due to the dramatic drop in price and the increase in mechanical properties of the carbon/graphite fibers. In the early 1970s, the cost of carbon fibers was more than \$220/kg. (\$100/lb). In the 1980s, the price went down as low as \$9/kg (\$4/lb).

The allotrope of carbon that is predominantly used is *graphite. Graphite* has a hexagonal structure with very strong covalent bonding of the carbon atoms in the hexagonal layer (basal) planes but weak dispersive bonding between the planes. *Graphene* is the term used to describe the hexagonal layer, as prescribed by the International Committee for the Characterization and terminology of Carbon. The covalent bonding in the graphenes generates the highest absolute modulus, the highest specific modulus, and the highest theoretical tensile strength of all known materials. The weak dispersive bonding between the graphenes produces a low-shear modulus and cross-plane Young's modulus that is detrimental to fiber properties. Table 15-15 gives the theoretical properties of graphite compared to those of diamond, while Fig. 15-19 shows the stress-strain curves of some actual fibers, among which is GY-70, one of the commercially available carbon/graphite fibers, as shown in Table 15-16. The latter also indicates the manufacturers of graphite fibers.

Table 15-14 Properties of the different types of glass fibers.[4]

	Specific gravity	Tensile strength		Tensile modulus		Coefficient of thermal expansion, 10^{-6} K	Dielectric constant (a)	Liquidus temperature	
		MPa	ksi	GPa	10^6 psi			°C	°F
E-glass	2.58	3450	500	72.5	10.5	5.0	6.3	1065	1950
A-glass	2.50	3040	440	69.0	10.0	8.6	6.9	996	1825
ECR-glass	2.62	3625	525	72.5	10.5	5.0	6.5	1204	2200
S-glass	2.48	4590	665	86.0	12.5	5.6	5.1	1454	2650

(a) At 20 °C (72 °F) and 1 MHz.

Table 15-15 Theoretical properties of graphite and diamond.[4]

	Tensile modulus		Tensile strength		Density, g/cm^3	Specific modulus	Specific strength
	GPa	10^6 psi	GPa	10^6 psi			
Graphite(a)	1020	150	150	20	2.26	451	66
Diamond	20	3	90	15	3.51	177	25

(a) Parallel to basal plane.

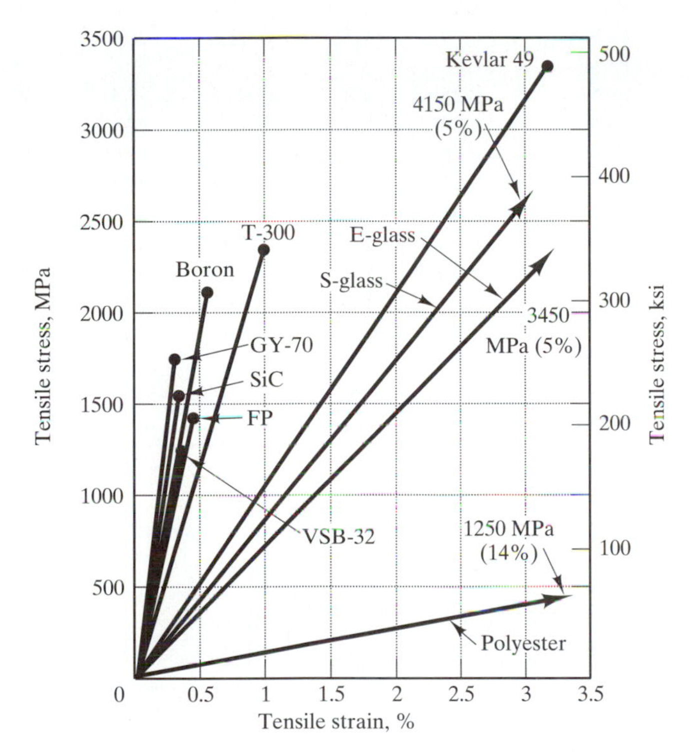

Fig. 15-19 Stress-strain curves for various fibers. [4]

All continuous carbon/graphite fibers produced to date come from organic precursors. The three different precursor materials are rayon, polyacrylonitrile (PAN), and isotropic and liquid crystalline pitches. Rayon and isotropic pitch precursors are used to produce low-modulus carbon fibers (≤ 50 GPa, or 7×10^6 psi). Higher modulus carbon fibers are made from either PAN or liquid crystalline (mesophase) pitch and the steps for each process are schematically shown in Fig. 15-20. From either the PAN or the mesophase pitch precursor, an oriented precursor fiber is spun, slightly oxidized to thermoset the fibers, and then carbonized between 800 and 1200°C to produce carbon fibers. At this stage, the fibers may be stretched to increase their preferred orientation, that is, the alignment of the graphenes to the length of the fiber. Graphitization is the heat treatment of the fibers at temperatures in excess of 2,000°C in inert atmospheres. This step also reduces the level of impurities and stimulates crystal growth. With both precursors, the higher the process temperatures that are used in the carbonization and graphitization, the higher are the moduli of the resultant fibers. The properties of carbon/graphite fibers produced from PAN and mesophase pitch precursor fibers are shown in Table 15-17. The effect of the alignment of the graphenes (preferred orientation) on modulus is shown in Fig. 15-21.

Table 15-16 Typical mechanical properties of commercially available carbon fibers.[4]

Product Name	Manufacturer	Precursor type	Density g/cm^3	Tensile strength GPa	Tensile strength 10^6 psi	Tensile modules GPa	Tensile modules 10^6 psi
AS-4	Hercules, Inc.	PAN	1.78	4.0	0.580	231	33.5
AS-6	Hercules, Inc.	PAN	1.82	4.5	0.652	245	35.5
IM-6	Hercules, Inc.	PAN	1.74	4.8	0.696	296	42.9
T-300	Union Carbide/Toray	PAN	1.75	3.31	0.480	228	32.1
T-500	Union Carbide/Toray	PAN	1.78	3.65	0.530	234	33.6
T-700	Toray	PAN	1.80	4.48	0.650	248	36.0
T-40	Toray	PAN	1.74	4.50	0.652	296	42.9
Celion	Celanese/ToHo	PAN	1.77	3.55	0.515	234	34.0
Celion ST	Celanese/ToHo	PAN	1.78	4.34	0.650	234	39.0
XAS	Grafil/Hysol	PAN	1.84	3.45	0.500	234	34.0
HMS-4	Hercules, Inc.	PAN	1.78	3.10	0.450	338	49.0
PAN 50	Toray	PAN	1.81	2.41	0.365	393	57.0
HMS	Grafil/Hysol	PAN	1.91	1.52	0.220	341	49.4
G-50	Celanese/ToHo	PAN	1.78	2.48	0.360	359	52.0
GY-70	Celanese	PAN	1.96	1.52	0.230	483	70.0
P-55	Union Carbide	Pitch	2.0	1.73	0.260	379	55.0
P-75	Union Carbide	Pitch	2.0	2.07	0.300	517	75.0
P-100	Union Carbide	Pitch	2.15	2.24	0.325	724	100
HMG-50	Hitco/OCF	Rayon	1.9	2.07	0.300	345	50.0
Thornel 75	Union Carbide	Rayon	1.9	2.52	0.365	517	75.0

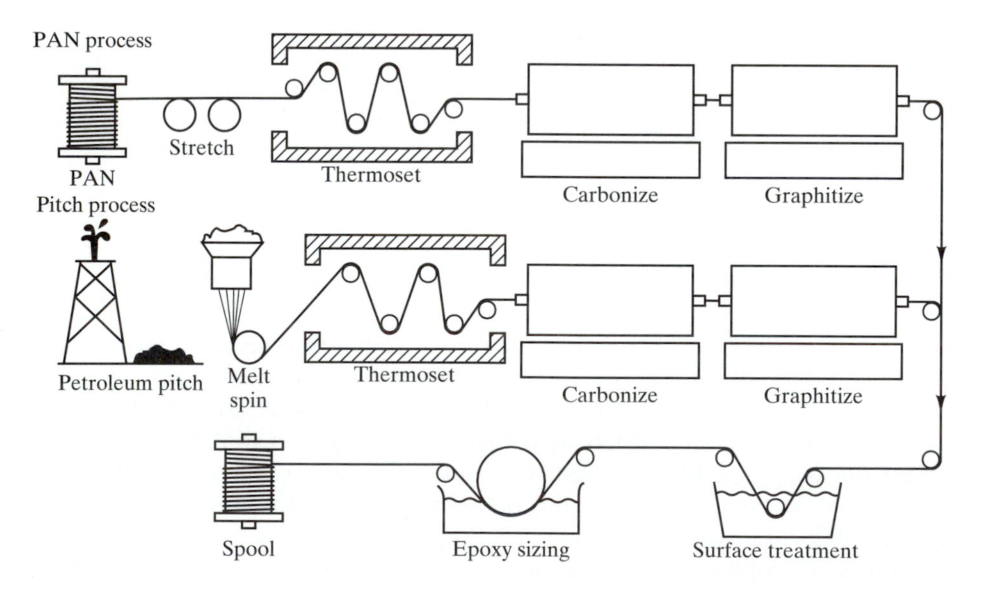

Fig. 15-20 Production of carbon/graphite fibers. Highly oriented fibers are obtained with PAN by hot-stretching, but are the natural consequences of the mesophase (liquid crystalline) order. [4]

In addition to the mechanical properties, carbon/graphite fibers also have very attractive physical properties. One very significant property is their negative coefficient of thermal expansion (CTE). The negative CTE is due to the bending of the graphenes. When this is combined with the positive CTE of the matrix, it produces a composite with a near-zero CTE over a temperature range of several hundred degrees. The

Table 15-17 Properties of standard graphite fibers from PAN and mesophase pitch precursors.[4]

Properties	Standard grades from PAN				Standard grades from mesophase pitch					
	Low modulus		High modulus		Low modulus		High modulus		Very high modulus	
Axial										
Tensile modulus, GPa (10^6 psi)	230	(30)	390	(55)	160	(25)	380	(55)	725	(110)
Tensile strength, GPa (10^6 psi)	3.3	(0.48)	2.4	(0.35)	1.4	(0.20)	1.7	(0.25)	2.2	(0.32)
Elongation at break, %	1.4	···	0.6	···	0.9	···	0.4	···	0.3	···
Thermal conductivity, W/m . K (Btu . in./h . ft^2 . °F)	8.5	(59)	70	(490)	···	···	100	(690)	520	(3600)
Electrical resistivity, $\mu\Omega \cdot$ m ($\mu\Omega \cdot$ cm)	18	(1800)	9.5	(950)	13	(1300)	7.5	(750)	2.5	(1250)
Coefficient of thermal expansion at 21°C, (70°F), 10^{-6}/K	−0.7	···	−0.5	···	···	···	−0.9	···	−0.6	···
Transverse										
Tensile modulus, GPa (10^6 psi)	40	(6)	21	(3)	···	···	21	(3)	···	···
Coefficient of thermal expansion at 50°C, (120°F), 10^{-6}/K	10	(2)	7	(1)	···	···	7.8	···	···	···
Bulk										
Density, g/cm^3	1.76	···	1.9	···	1.9	···	2.0	···	2.15	···
Filament diameter, μm(μin.)	7.8	(280–310)	7	(280)	11	(430)	10	(390)	10	(390)
Carbon assay, %	92–97	···	100	···	97+	···	99+	···	99+	···

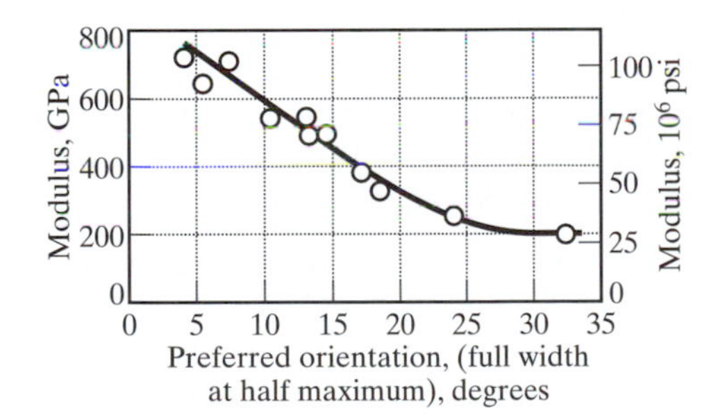

Fig. 15-21 Modulus of carbon/graphite fiber from mesophase pitch as a function of preferred orientation. [4]

near-zero CTE induces low thermal fatigue stresses and long service life. Another property is the high thermal conductivity of graphite that may equal or exceed that of copper. This property is being exploited in the semiconductor industry as a heat transfer medium to cool the integrated circuits. Carbon fibers also have a relatively high electrical conductivity, about 1/50 or less that of copper. Because of this, free-floating fibers produced during composite manufacture must be avoided and removed because they can cause electrical shorts of electrical equipment.

The most widely used organic fiber is *para-aramid Kevlar,* which is poly para-phenylene-terephthalamide. The fiber is also known as *PPD-T* because it is the product of the condensation reaction of para-phenylene diamine and terephthaloyl chloride, as shown in Fig. 15-22. The aromatic ring contributes to thermal stability, while the para configuration leads to stiff, rigid molecules that induce the polymer chains to aggregate and to form ordered domains. They are in the class of *liquid crystalline polymers.* When PPD-T solutions are extruded through a spinneret and drawn during fiber manufacture, the liquid crystalline domains orient and align in the direction of flow, as shown in Fig. 15-23. Thus, we can understand why Kevlar has a very high modulus and a high strength in the longitudinal fiber axis. The stress-strain curve of Kevlar is also shown in Fig. 15-19 where we see Kevlar 49 exhibiting higher modulus and strength than the E- and S-glass fibers.

The mechanical properties of the three types of Kevlar fibers are shown in Table 15-18. Kevlar 49 is

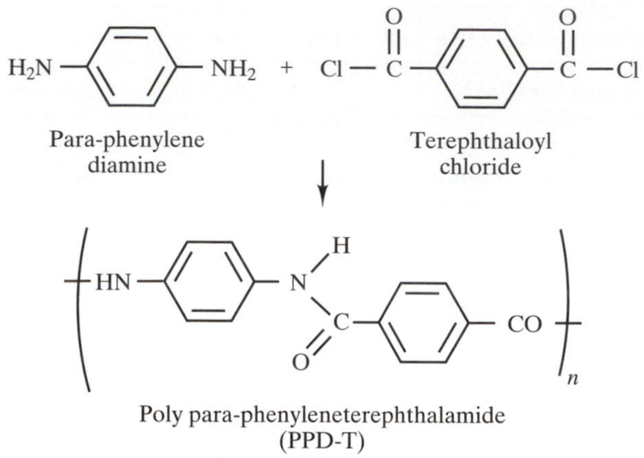

$H_2N-C_6H_4-NH_2 + Cl-CO-C_6H_4-CO-Cl$

Para-phenylene diamine

Terephthaloyl chloride

$\downarrow$

$[-HN-C_6H_4-NH-CO-C_6H_4-CO-]_n$

Poly para-phenyleneterephthalamide (PPD-T)

Fig. 15-22 Chemical structure of para-aramid Kevlar fiber. [4]

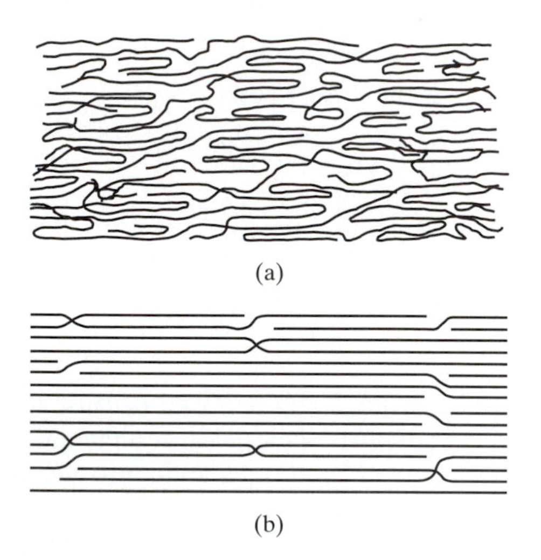

Fig. 15-23 Liquid kevlar (a) before, and (b) after it passes the spinneret. [4]

the dominant form used in structural composites due to its higher modulus. Kevlar 29 is used when higher toughness, damage tolerance, or ballistic stopping performance is desired, such as in bullet-proof vests, and Kevlar 149 is the ultra-high modulus form whose modulus approaches the theoretical value predicted for p-aramid fibers.

There are *other fibers used in fiber-composites*. Natural fibers such as flax, jute, hemp, cotton, and silk can be used as reinforcements but they impart relatively low strength and low modulus. Continuous monofilament boron and silicon carbide fibers are produced by chemical vapor deposition on substrate wires drawn through glass reactor tubes. Boron is used in the aerospace industry and was the dominant fiber used until carbon/graphite supplanted it in polymer matrix composites. Both boron and silicon carbide (SiC) are used as reinforcements in metal matrix composites, notably aluminum and titanium. Continuous SiC-based fibers known as Nicalon are also produced by controlled pyrolysis of spun polymeric polymers. Alumina and silicon nitride fibers are also available. Fiber FP shown in Fig. 15-19 is the only almost pure alumina, continuous polycrystalline fiber commercially available—the others are essentially alumina-silica and alumina-silica-boria ceramic oxides. Silicon nitride fibers are currently being developed. The potential properties of continuous Si_3N_4 are very attractive.

15.6.2b Matrices. *Thermoset resins* are the predominant matrices for thermal and dimensional stability. These were discussed in Sec. 15.4.2; therefore, it is only necessary to mention that the two major thermosets used are the *polyesters* and the *epoxies*. The polyesters are used primarily in commercial, industrial, and transportation applications. For this reason, they account for a large portion of the polymer matrices used in composites. The widespread fabrication of glass fiber reinforced polyesters and vinyl esters is due to silane, the coupling (sizing) agent used with glass fiber to produce its (fiber) adhesive properties. The polyester resins do not adhere very well, however, with graphite and Kevlar fibers, therefore, their use with graphite and Kevlar fibers are practically nonexistent. The polyesters have relatively large cure shrinkage that restricts their use to low-performance composites.

Table 15-18 Properties of the three types of Kevlar fibers.[4]

Material	Density g/cm^3	Filement diameter		Tensile modulus (a)		Tensile strengths (a)		Tensile elongation, %	Available yarn count, No. filaments
		μm	μm	GPa	10^6 psi	GPa	10^6 psi		
Kevlar 29 (high toughness)	1.44	12	470	83	12	3.6 2.8(b)	0.525 0.400(b)	4.0	134–10,000
Kevlar 49 (high modulus)	1.44	12	470	131	19	3.6–4.1	0.525–0.600	2.8	134–5000
Kevlar 149 (ultra-high modulus)	1.47	12	470	186	27	3.4	0.500	2.0	134–1000

(a) ASTM D 2343 impregnated strand. (b) ASTM D 885, unimpregnated strand.

Epoxies, on the other hand, are used extensively for a variety of demanding structural applications, such as those in the aerospace industry. The polyesters cost less, but the epoxies have a broad range of physical and mechanical properties coupled with ease of processing that make them very invaluable. The uncured resins have a variety of physical forms, ranging from low viscosity liquids to nontacky solids, which when combined with a large selection of curing agents, offers the composite fabricator a wide range of processing conditions. However, the epoxies are very susceptible to moisture absorption that decreases the glass transition temperature (T_g) of an epoxy resin. To avoid subjecting the composite to temperatures higher than this wet T_g, the maximum service temperature is limited to 120°C, for heavily loaded applications, and 80° to 105°C, for toughened epoxy resins. In general, this restraint has helped to avoid serious thermal-performance difficulties.

When higher service temperature is required, a polyimide matrix is used. These resins are available with a maximum hot/wet in-service temperature of 260°C. However, the condensation polymerization of these resins releases volatiles during cure that produce voids in the resulting composite. Nonetheless, they can produce good-quality, low-void content composite parts, which, unfortunately, are quite brittle.

The dual goal of improving hot/wet properties and impact resistance of composite matrices have led to the *development, but of limited use, of new high-temperature thermoplastic resin matrices.* These resins are the engineering thermoplastics that were discussed in Sec. 15.4.1; some of these are polyetheretherketone (PEEK), polyphynelene sulfide (PPS), polyetherimide (PEI), and polyamideimide (PAI).

15.7 Plastics Properties Used in Design

Designing components and structures with plastics require a good understanding of the properties of plastics, particularly, the behavior of plastics under stress.

15.7.1 Density

One of the major advantages of plastics in engineering is their low density. As indicated in Chap. 4, their low density arises primarily from the light masses of the atoms that constitute them. In the form of reinforced plastics, these materials offer very high specific strengths and specific moduli that are used in the aerospace industry. The density of polymers depends on their crystallinity and on their structure, as discussed in Sec. 15.3.

15.7.2 Water Absorption and Water Transmission

Water absorption is the percentage increase in weight of a plastic due to its absorption of water. These tests are, of course, specified by the ASTM and are done for different lengths of time at varying temperatures, as well as varying solutions. Water absorption affects the mechanical and electrical properties as well as the dimensions of parts made of plastics because they tend to swell. Plastics with very low water-absorption tend to have better dimensional stability. In composites, the influence of water is considered in terms of induced hygro-thermal stresses that could lead to debonding of the fiber-matrix interfaces.

Water transmission is the rate at which water vapor permeates a plastic. Plastics may have different transmission rates for water vapor and for other gases. For example, PE is a good barrier for moisture or water vapor, but other gases permeate it rather easily. Nylon, on the other hand, is a poor barrier to water vapor but a good barrier to other vapors. The effectiveness of a vapor barrier is rated in term of *perms*. A rating of one perm means that one sq. ft. of the barrier is penetrated by one gram of water vapor per hour under a pressure differential of one inch of mercury. One inch of mercury equals about 3.45 kPa (0.5 psi).

The problem of water transmission is also important in vehicle tires and certain blow-molded bottles, which must be virtually impermeable to air and other gases. The most impermeable of the rubbers is butyl rubber; however, the carcass of a tire is not made of this rubber. Because of its impermeability to gases, butyl rubber is also used as roof coating. With plastic bottles, different layers of both coinjected and coextruded plastics can be used to fabricate the bottle to make it impermeable to different vapors and gases.

15.7.3 Thermal Properties

Examples of thermal properties of TPs are given in Table 15-19. In terms of design-involving stresses, the two most significant properties to be noted are the *glass transition temperature, T_g*, and the *coefficient of thermal expansion, CTE.* The T_g is the point below which plastic behaves as glass does—it is very strong and rigid, but brittle. Above T_g, it is neither as strong nor rigid as glass, but it is also not as brittle. The T_g is unique to amorphous TPs and as indicated earlier, there is no unique melting point for these materials. Designers should be aware that the modulus of elasticity of the plastics are reduced drastically in the vicinity of the T_g, by a factor of as much as 1,000, which is schematically illustrated in Fig. 15-24. Therefore, the environmental temperature of an amorphous TP must be kept below its T_g. Figure 15-24 also illustrates what was earlier said about elastomers or rubbers—rubbers are TPs with T_g's below their application temperatures. For example, the T_g of natural rubber is −75°C (−103°F). We compare this to PMMA (Plexiglas) with a T_g of 100°C, which makes it brittle and glassy at room temperature.

Very closely related to the T_g is the *deflection temperature under load (DTUL),* also called the *heat deflection temperature (HDT)*. It is the result of the ASTM D648 test in which a 12.7 mm ($\frac{1}{2}$-in.)-deep plastic test bar with a span of 101.6 mm (4 in.) is bent under load of either 455 kPa (66 psi) or 1821 kPa (264 psi). The test is conducted in an oil bath, and the temperature of the bath is increased at a constant rate of 2°C per minute. The DTUL is the temperature at which the sample shows a deflection of 0.254 mm (0.010 in.). Since the stress and deflection are given, the test indicates the temperature at which the flexural modulus decreases to values of 240 MPa (35 ksi) at 455 kPa (66 psi) stress and 971 MPa (141 ksi) at 1821 kPa (264 psi) stress. Tables 15-3 and 15-7 show the DTULs of some thermoplastic and thermoset polymers.

The *coefficient of thermal expansion (CTE)* is an important consideration if dissimilar materials are assembled together, as in composite materials where it is a signifigant problem. Unmatched CTEs of the constituent materials in composites lead to thermal stresses that could lead to cracking of the brittle constituents and debonding at the interfaces. The thermal stress is given by

$$\sigma = E_p(\alpha_p - \alpha_2)(\Delta T) \qquad (15\text{-}6)$$

where σ is the stress on the plastic, E_p is the modulus of the plastic, α_p is the CTE of the plastic, α_2 is the CTE of the other material, and ΔT is the temperature change. When the part or component is cycled from low to high temperatures and back to low temperatures, the stress subjects the part to thermal fatigue, thereby inducing debonding at the interfaces.

15.7.4 Mechanical Properties

These properties refer to the response of materials to loading or deformation. Thermoset plastics are almost always brittle due to their cross-linked structures and, as indicated earlier, they do not deform with temperature. Thus, we shall primarily discuss the

Table 15-19 Thermal properties of some thermoplastics compared with other common materials.[2]

Plastics (morphology)		Density g/cm^3 ($lb./ft.^3$)	Melt temperature, T_m°C(°F)	Glass transition temperature T_g°C(°F)	Thermal Conductivity (10^{-4} cal/s · cm°C (BTU/lb.°F)	Heat Capacity cal/g°C (BTU/lb.°F)	Thermal Diffusivity $10^{-4}cm^2/s$ $10^{-3}ft^2/hr.$)	Thermal Expansion 10^{-6}cm/cm°C (10^{-6}in./in°F)
PP	(C)	0.9 (56)	168 (334)	5 (41)	2.8 (0.068)	0.9 (0.004)	3.5 (1.36)	81 (45)
HDPE	(C)	0.96 (60)	134 (273)	−110 (−166)	12 (0.290)	0.9 (0.004)	13.9 (5.4)	59 (33)
PTFE	(C)	2.2 (137)	330 (626)	−115 (−175)	6 (0.145)	0.3 (0.001)	9.1 (3.53)	70 (39)
PA	(C)	1.13 (71)	260 (500)	50 (122)	5.8 (0.140)	0.075 (0.003)	6.8 (2.64)	80 (44)
PET	(C)	1.35 (84)	250 (490)	70 (158)	3.6 (0.087)	0.45 (0.002)	5.9 (2.29)	65 (36)
ABS	(A)	1.05 (66)	105 (221)	102 (215)	3 (0.073)	0.5 (0.002)	3.8 (1.47)	60 (33)
PS	(A)	1.05 (66)	100 (212)	90 (194)	3 (0.073)	0.5 (0.002)	5.7 (2.2)	50 (28)
PMMA	(A)	1.20 (75)	95 (203)	100 (212)	6 (0.145)	0.56 (0.002)	8.9 (3.45)	50 (28)
PC	(A)	1.20 (75)	266 (510)	150 (300)	4.7 (0.114)	0.5 (0.002)	7.8 (3.0)	68 (38)
PVC	(A)	1.35 (84)	199 (390)	90 (194)	5 (0.121)	0.6 (0.002)	6.2 (2.4)	50 (128)
Aluminum		2.68 (167)	660 (1220)		3000 (72.5)	0.23	4900 (1900)	19 (10.6)
Copper		8.8 (594)	1083 (1981)		4500 (109)	0.09	5700 (2200)	18 (10)
Steel		7.9 (493)	1536 (2797)		800 (21.3)	0.11	1000 (338)	11 (6.1)
Maple wood		0.45 (28.1)			3 (0.073)	0.25	27 (10.5)	60 (33)
Zinc		6.7 (418)	420 (788)		2500 (60.4)	0.10	3700 (1430)	27 (15)

C = Crystalline resin. A = Amorphous resin.

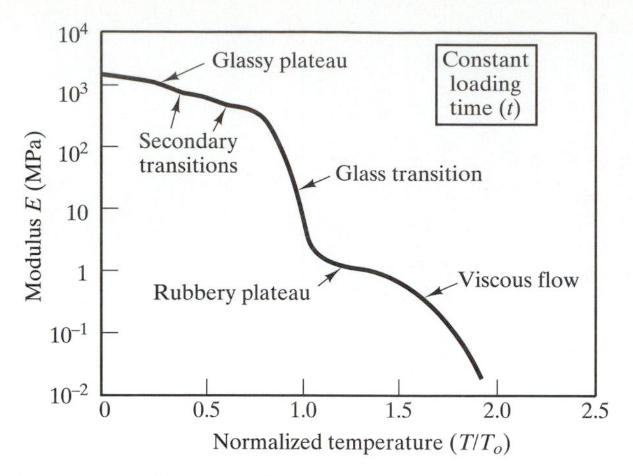

Fig. 15-24 Schematic of modulus of a thermoplastic as a function of normalized temperature. [3]

mechanical properties of thermoplastics that are quite different from those of metals.

The difference in mechanical properties of plastics from those of metals is largely due to their relatively low glass-transition temperatures, T_gs. As seen in the previous section, the glass transition temperature of plastics has the same significance as the melting points of metals and ceramics. Metals and ceramics have very high melting points while the glass transition temperatures of polymers are comparatively very low. The relative temperature of T/T_g or T/T_m, where T is the application temperature, determines the movement of atoms or molecules and the creep behavior. For metals and ceramics, T/T_m is very low at room temperature, thus the assumption of constant stiffness and strength is quite valid. On the other hand, T/T_g at room temperature for plastics is predominantly greater than 0.5 and thus creep of plastics usually occurs at room temperature. Thus, properties of TPs are dominated by their viscoelastic behavior relative to the applied stress. Their properties vary with time, and the design should incorporate whether the stress is applied for a short-term or for a long-term.

15.7.4a Short-Term Properties. The short-term behavior and mechanical properties of plastics are a result of their responses to loads usually lasting only a few seconds or minutes, up to a maximum of 15 minutes. These short-term responses have been used to define the basic or reference design and engineering properties of conventional metallic materials. The properties, which include tensile strength, compressive strength, flexural strength, shear strength, and the associated moduli, are also obtained for plastics for quality control purposes, to ensure that constant properties of the plastics are obtained during production. These properties are also provided by materials' suppliers on data sheets and in computerized data banks. For many engineering plastics that are considered linearly elastic, homogeneous, and isotropic, their tensile and compressive properties are observed to be identical. This eliminates the need to measure their compressive properties. Furthermore, if their tension and compression properties are equal, standard beam-bending theory suggests there is no need to measure bending properties. However, as a concession to the nonlinear, anisotropic behavior of most plastics, the flexural properties are also obtained and presented in marketing data.

The tensile test for plastics is conducted according to ASTM D 638 and is basically the same as that for metals described in Chap. 4. Tensile stress-strain curves for polycarbonates are shown in Fig. 15-25. Figure 15-25a shows the stress-strain curve for PCs compared to steel, and Fig. 15-25b is the stress-strain curve on an expanded scale. We see that the PC has similar tensile characteristics as those of metals. The *yield strength* is obtained also with the *0.2% offset strain* method. The *tensile strength* is the maximum tensile stress. If the maximum occurs at the yield point, it is designated as the tensile strength at yield. If the maximum stress occurs at fracture, it is called the tensile strength at fracture. The non-linear behavior of the resins requires the determination of the *tangent and the secant moduli* as shown in Fig. 15-26 in the same manner as was done for gray cast iron. The influences of the strain rate and temperature are indicated in Fig. 15-27. The *area under the stress-strain curve is also indicative of the impact toughness.*

Linear-amorphous polymers, such as PMMA and PS, show five regimes in which the Young's modulus may vary from 3–3,000 MPa. These regimes are illustrated in Fig. 15-24 depending on the normalized temperature, T/T_g, and are

1. the glassy regime or plateau, with a relatively large modulus, about 3GPa
2. the glass transition regime, where the modulus drops steeply from about 3 GPa to about 3 MPa

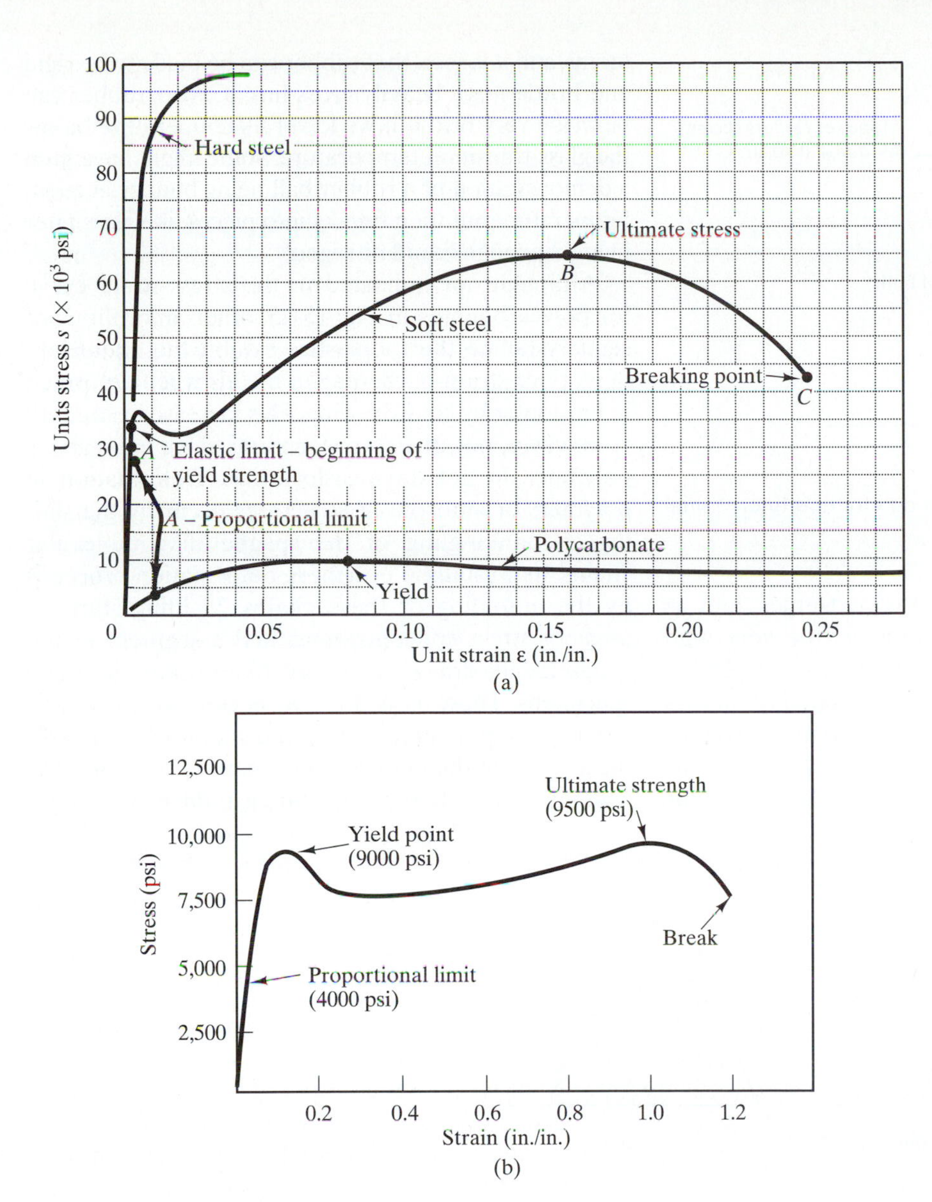

Fig. 15-25 Stress-curves of polycarbonates. (a) compared with hard and soft steels, and (b) on an expanded scale, indicating tensile properties. [2]

3. the rubbery regime, with a low modulus of about 3 MPa
4. the viscous regime, when the polymer starts to flow, as in liquids
5. the decomposition regime in which the chemical breakdown starts.

Well below the T_g, the polymer molecules pack tightly together, either in an amorphous tangle, or in poorly semicrystalline structured material. At these temperatures, the polymer behaves like a glass, is brittle, and its tensile strength depends on its fracture toughness, K_{I_c}, and the largest crack size present in the material, shown schematically in Fig. 15-28.

As in metals, the modulus or stiffness of the polymer is the average of the stiffnesses of its bonds, which are covalent along the chain and secondary bonds between chains. Along the covalent chain, the modulus approaches that of diamond that is 10^3 GPa. For the secondary bonds between chains, (von der Waals) the modulus is that of a simple hydrocarbon such as paraffin wax that is 1 GPa. The moduli of

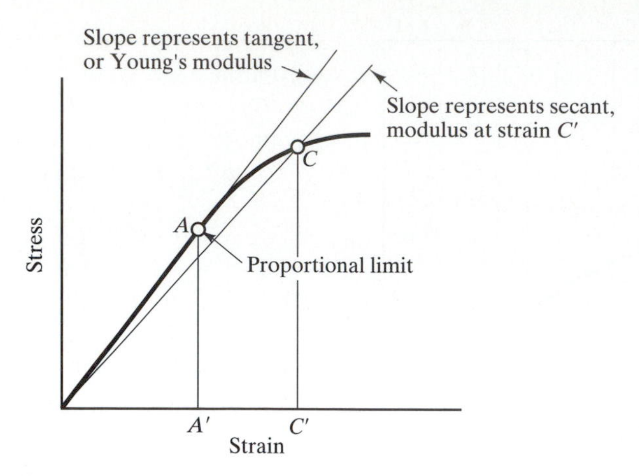

Fig. 15-26 Tangent and secant moduli for non-linear initial stress-strain curve.

polymers are between these two extremes, and as shown in Fig. 15-29, the modulus can be very high when the molecules are aligned as in drawn fibers.

The effect of cross-linking on the moduli of plastics is illustrated for elastomers in Fig. 15-30 in the *vulcanization* of polyisoprene. *Vulcanization,* as described earlier, is the cross-linking of rubber molecules with sulfur atoms, and it is seen that rubber can be made to be rigid and brittle if we heavily cross-link it. Also, rubber can be made very brittle if we lower its temperature below the glass-transition temperature. Some might have seen a demonstration of a rubber ball being bouncy at room temperature but then breaks into pieces like glass after being dipped in liquid nitrogen!

Although not indicated in the stress-strain curve for polycarbonates in Fig. 15-25, other amorphous or semicrystalline thermoplastics exhibit the equivalent of work or strain-hardening in metals when deformed in tension or by cold-drawing. This increase in strength is, however, not due to dislocations since amorphous polymers do not have dislocations. Deformation in polymers or amorphous structures occur by the sliding and untangling of the spaghettilike molecular chains. In crystalline polymers, deformation proceeds by the unfolding of these chains. Yielding starts at areas of stress concentrations and a segment of the gauge length draws down, just like necking in metal specimens. There is a difference, however, between metals and polymers. While the metal continues to draw down in the necked region until it fractures, the necking of a polymeric specimen induces the align-

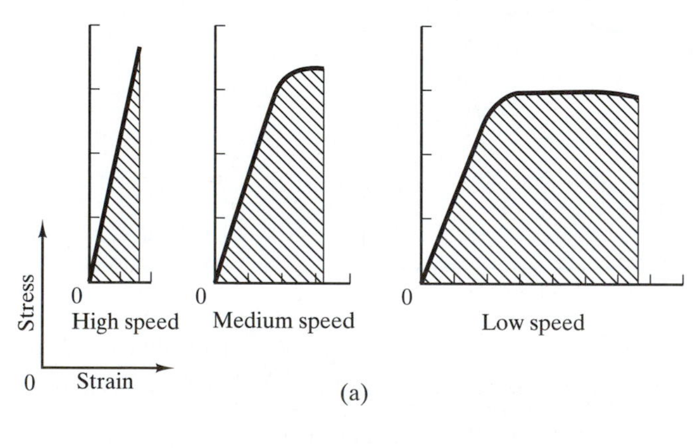

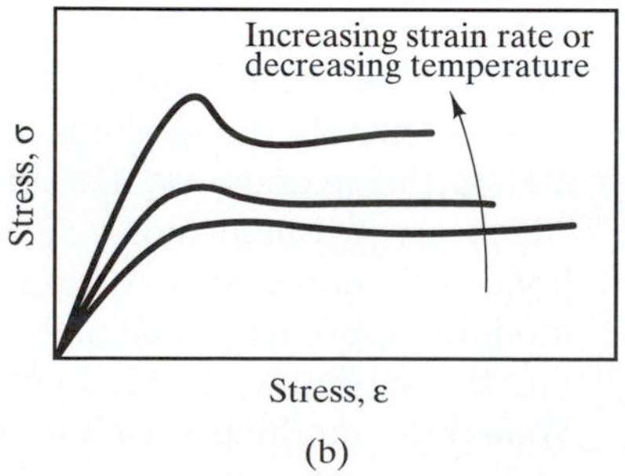

Fig. 15-27 Schematic figures showing the influences of strain-rate (speed of test) and temperature on shapes of stress-strain curves. (a) a ductile plastic can show a linear elastic behavior at very high speeds; areas under curves are indicative of absorbed impact energy, (b) temperature effect. [2]

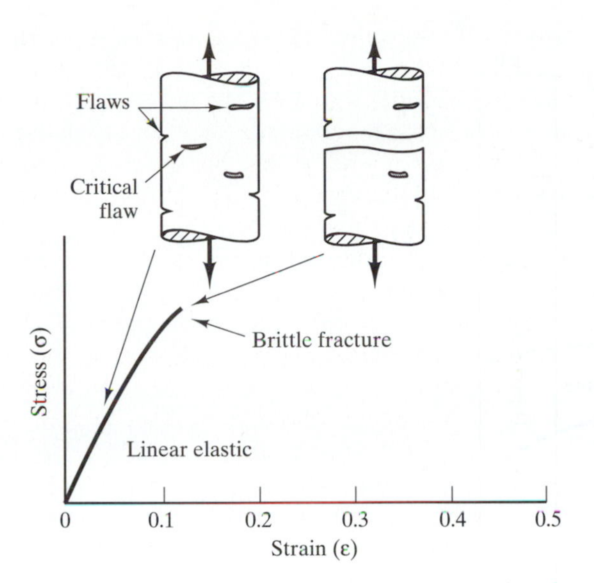

Fig. 15-28 The tensile strength of polymers in the glassy regime depends on their fracture toughness and the largest crack size present. [3]

ment of the molecular chains that makes the neck region stronger than the surrounding areas. Thus, it will take a higher stress to deform this drawn and aligned chain segment while the surrounding unaligned chain segments will take lesser stress to deform. In this manner the alignment of the rest of the chain will proceed, just like the inhomogeneous Lüder's band propagation in annealed steels. The stress remains essentially constant until all the chains are aligned and then the stress increases dramatically until the polymer fractures. The increase in strength is due to the force to break the main bonding of the polymer chain and one can think of this as work-hardening in polymers. This phenomenon is illustrated in Fig. 15-31.

We note in Fig. 15-31 that the thermoplastics undergo a lot of plasticity due to the alignment of molecular chains. This is the origin of the name "plastics". We should also appreciate that the molecular chains in cross-linked molecules cannot be straightened and this accounts for their relative strength and brittleness (no ductility). Furthermore, it is noted

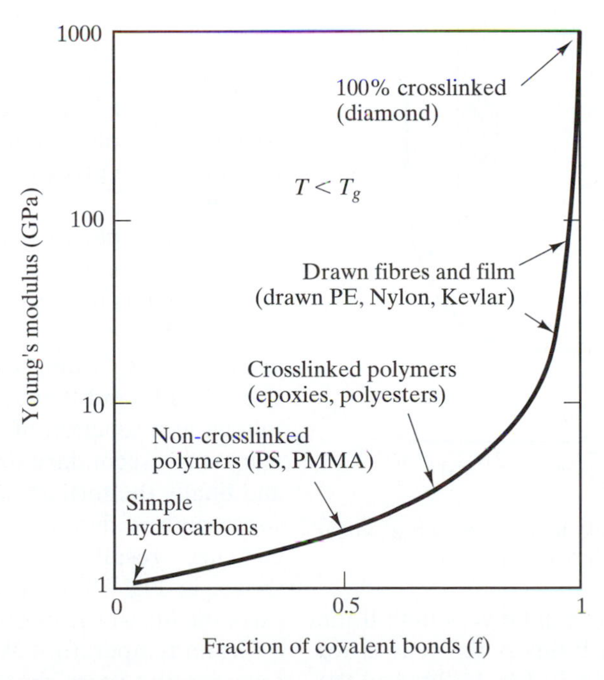

Fig. 15-29 Influence of covalent bonding and molecular orientation on the modulus.

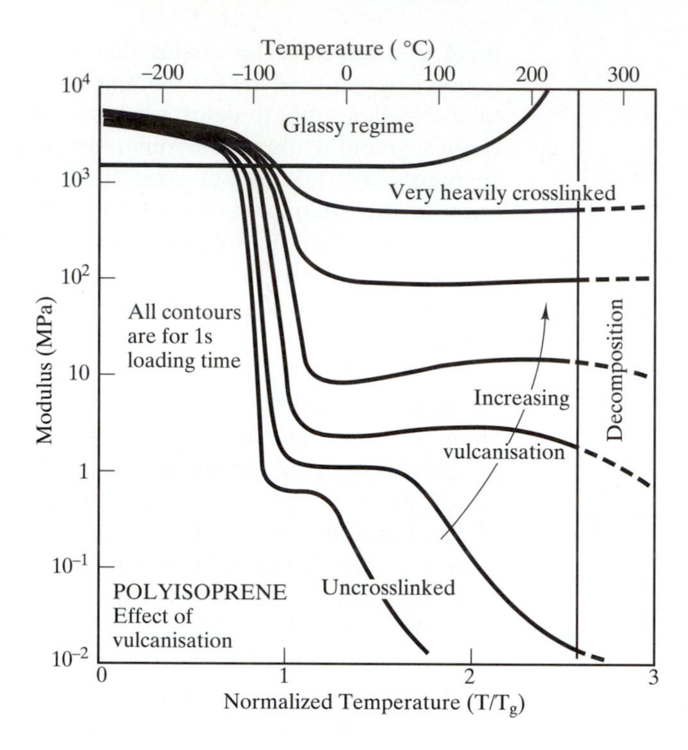

Fig. 15-30 Influence of cross-linking (vulcanization) on the modulus. [3]

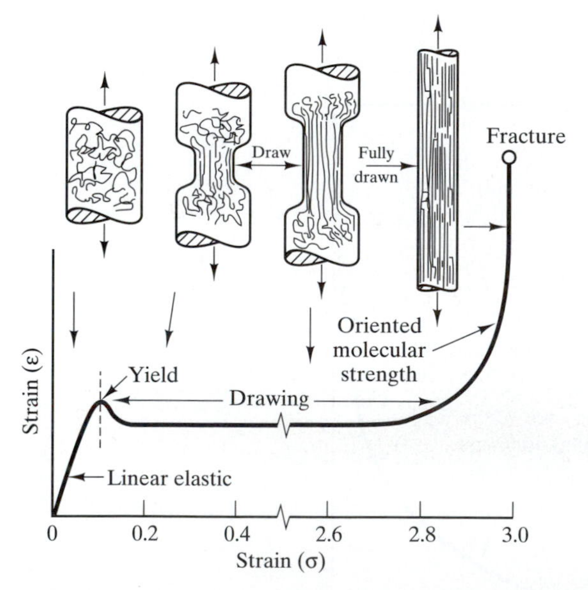

Fig. 15-31 Plasticity and "work hardening" of a linear structured polymer during tensile test. [3]

that the strength of polymers can be very high if the molecular chains are aligned; this is the reason for the high moduli (drawn fibers in Fig. 15-29) and the high strengths of some organic fibers such as Nylon and Kevlar.

15.7.4b Long-Term Properties. While the short-term mechanical properties are important for quality control, the long-term behavior and mechanical properties are used in design when the plastics are used in stressed structures. This is due to their relatively low glass transition temperatures that make them viscoelastic at room temperature. Thus, the three relevant long-term properties are *creep, stress relaxation,* and *fatigue*.

The *creep phenomenon* is for constant load or stress (less than the yield strength) and is the same as that for metals that was described in Chap. 4, shown again in Fig. 15-32. As in metals, we observe an instantaneous strain response due to the applied stress (load), and then a viscous strain (creep strain) follows as a function of time in three stages: the primary stage, secondary stage with constant creep rate, and finally the tertiary stage until rupture. When the stress-rupture data is plotted in log-log scales, a linear curve results for each of the temperatures, as shown in Fig. 15-33. The log-log linear curve indicates the life of the plastics under a constant stress at a certain temperature. When creep is interrupted by releasing the stress, the strain is gradually recovered and the plastics will tend to return to their previous shape before the creep test. This phenomenon is

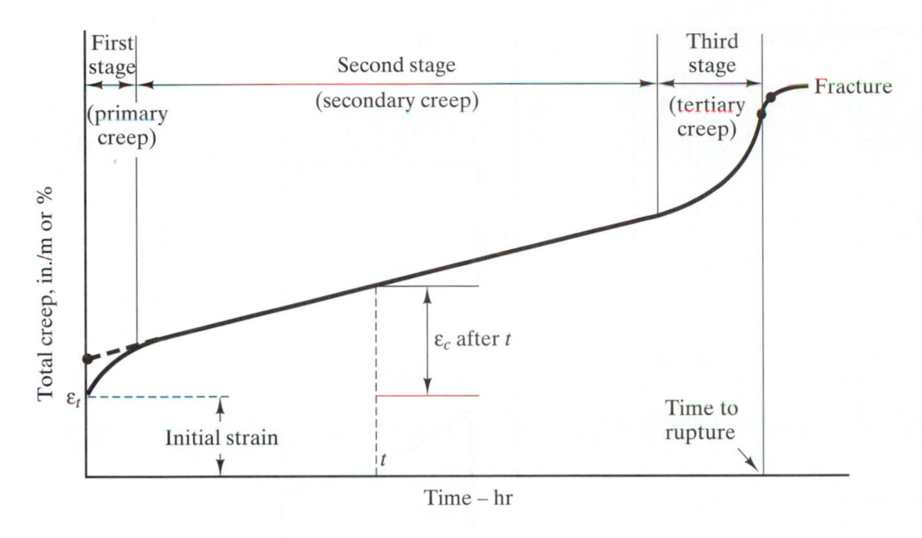

Fig. 15-32 The three stages in creep under constant load at one temperature.

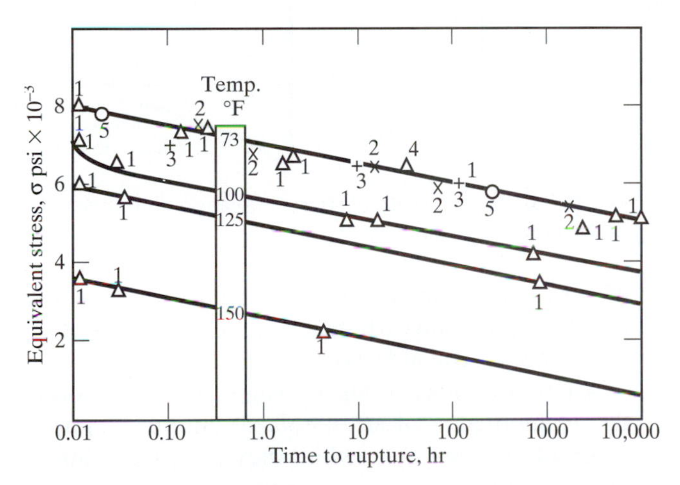

Fig. 15-33 Stress-life (rupture) data for a rigid 2-in. diameter PVC pipe for various temperatures. [2]

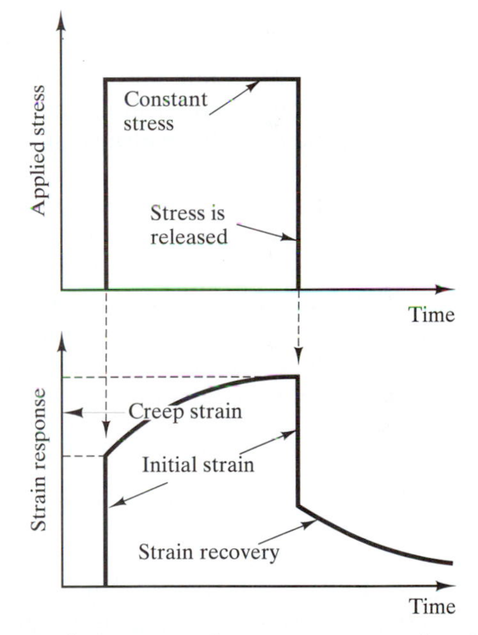

Fig. 15-34 Phenomenon of strain recovery or relaxation. The initial elastic strain is recovered immediately on releasing the elastic load. Creep strain is recovered after a period of time.

called strain relaxation and is schematically shown in Fig. 15-34.

In order to account for the increasing strain with time during creep, the concept of *apparent creep modulus, E_C,* is introduced by

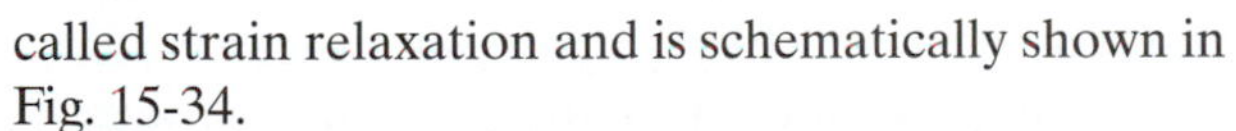

$$E_C = \frac{\text{Applied stress}}{\text{Initial strain} + \text{Creep strain}} \qquad (15\text{-}7)$$

where the applied stress is constant and held below the elastic limit of the material. When plotted on a log-log scale, the log (time) versus log (apparent modulus) plot is usually a straight line after the initial 100 hours. Thus, we can perform creep tests up to 1,000 hrs and the linearized creep data between 100 to 1,000 hours in a log-log plot can be extrapolated to longer times to determine an allowable stress on the plastics for a designated life. An example of this is the creep data of ABS at 23°C (73°F) for 1,000 psi constant stress as shown in Fig. 15-35. The actual creep data is from 100 to 1,000 hours in the upper linear curve. The dashed portion of the upper line and the lower line are extrapolations. The end of useful

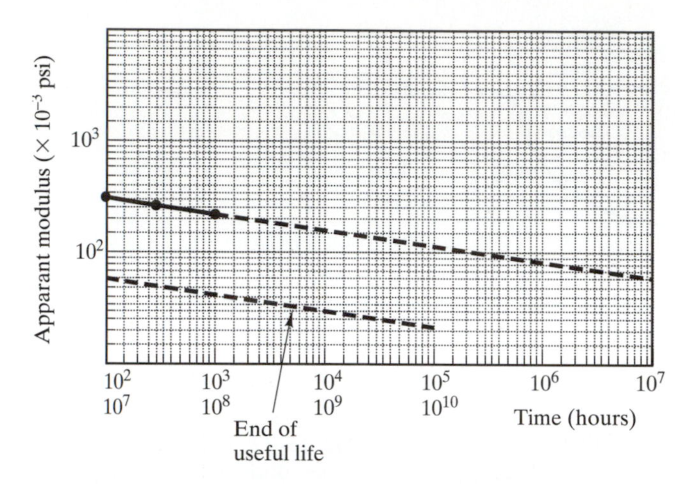

Fig. 15-35 Creep modulus data for 100 to 1,000 hrs. for ABS can be extrapolated to longer times (design requirement) in a log-log plot to determine the apparent creep modulus for that time. [2]

life indicated at 5×10^8 hours corresponds to the total strain (initial strain + creep strain) at the yield strength of this particular ABS, which is taken as the criterion for failure.

The phenomenon of *stress-relaxation* occurs under constant strain and is schematically shown in Fig. 15-36. The plastic is deformed (strained) a fixed amount and the stress at this constant strain decreases (relaxes) as a function of time. As in creep, the constant strain imposes an immediate initial stress on the plastic according to the short-term behavior,

$$\sigma_o = E\,\epsilon_o \qquad (15\text{-}8)$$

where E is the tensile modulus and ϵ_o is the constant strain. As time passes, the stress decreases gradually, but it may take years for the stress to decrease near a value of zero. The stress-relaxation behavior or the rate of decrease in stress is extremely temperature dependent, especially in the region of the glass transition. At a given temperature in this transition region, the stress changes rapidly with time. Because the stress decreases, the modulus of the plastics, now called *relaxation modulus,* E_R, decreases also,

$$E_R = \frac{\sigma(t)}{\epsilon_o} \qquad (15\text{-}9)$$

The data as schematically illustrated in Fig. 15-36 may be converted to isochronous (constant time) curves, seen in Fig. 15-37 for polycarbonates. The stress relaxation modulus curves at constant strains as a function

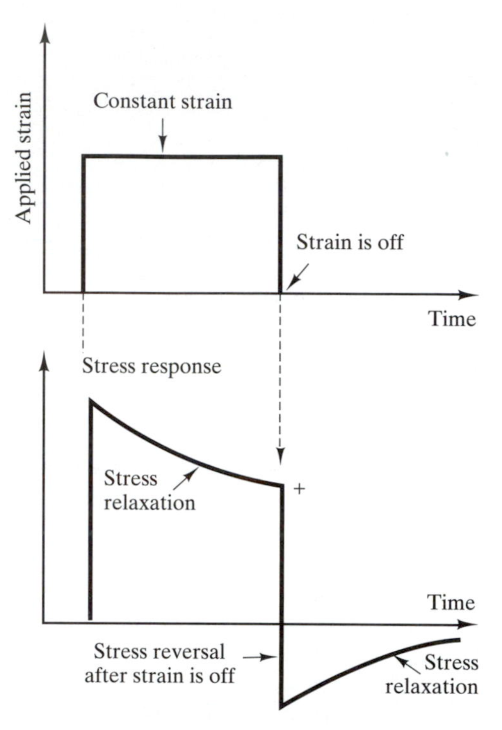

Fig. 15-36 Phenomenon of stress relaxation with time under constant strain.

of time are obtained from these curves, as shown in Fig. 15-38, for polycarbonates. In general, relaxation data is not as available as creep data.

There are many product designs that incorporate the phenomenon of stress-relaxation. For example, when plastics are assembled, they are placed into a permanently deflected (strained) condition, such as that in press fits, bolted assemblies, and in some plastic springs. Stress-relaxation data provides practical information on the stress needed to hold a metal insert in Fig. 15-39. Plastics parts with excessive fixed strains imposed on them for extended periods of time may fail. For developing initial design concepts, a strain limit of 20 percent of the strain at the yield point or of the yield strength is suggested for high-elongation plastics. For low-elongation brittle materials, 20 percent of the elongation at fracture is suggested as an initial guide line. Prototype parts should then be thoroughly tested at end-use conditions to confirm the design.

Fatigue. Fatigue failure of plastics under repeated cyclic load occurs basically in the same manner as in metals that was described in Chap. 4. Fatigue cracks initiate and then propagate and fracture occurs when

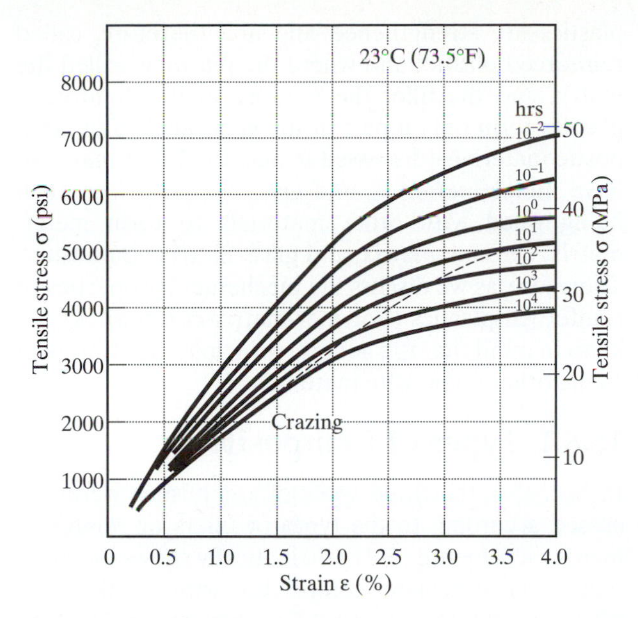

Fig. 15-37 Isochronous (constant time) stress-strain curves for polycarbonates from stress relaxation data. [2]

Fig. 15-38 Relaxation modulus for polycarbonates as a function of time for varying constant strains. [2]

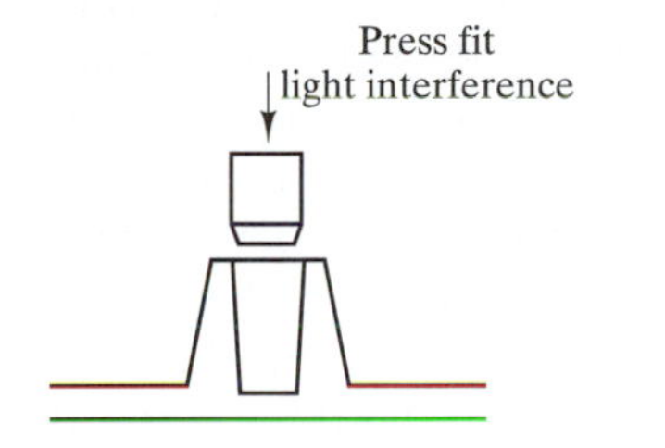

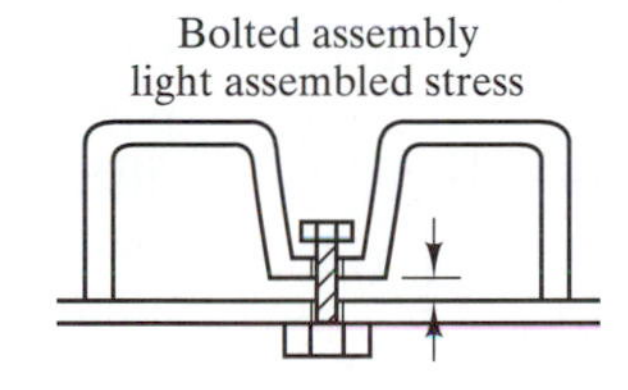

Fig. 15-39 Examples of constant strain applications. [2]

the remaining material can no longer support the stress. However in plastics, failure may be defined before the actual fracture, in accordance with ASTM D 671; fatigue failure may be considered to have occurred also when the elastic modulus has decreased to 70 percent of its original value. In addition, because of their viscoelastic behavior and low thermal conductivity, plastics are prone to heating if the cyclic stress or cyclic rate is high. Thus, practical failure may include thermal softening, partial melting of the material, excessive change in dimensions, warping, crazing, cracking, or the formation of internal voids or deformation markings.

Fatigue behavior of plastics is obtained using the same tests as in metals—the most common is the constant stress amplitude test, and of the three possible tests, the bending fatigue test is the most commonly used. Typical room temperature test results for various TPs and TSs are illustrated in Figs. 15-40 a, b, and c. The results are for complete reversal, zero-mean stress, and cantilever bending in accordance to ASTM D 671. In general, the results indicate that a plastic exhibits an endurance limit (Fig. 15-40a), which is the stress below which the plastic is less susceptible to fatigue failure. The exceptions are nylon in Figs. 15-40a and b, and urea in Fig. 15-40c.

15.8 Properties of Reinforced Plastics and Composite Materials

So far we have discussed the properties of unfilled plastics and for the most part, their properties are isotropic, except when the molecular chains are highly

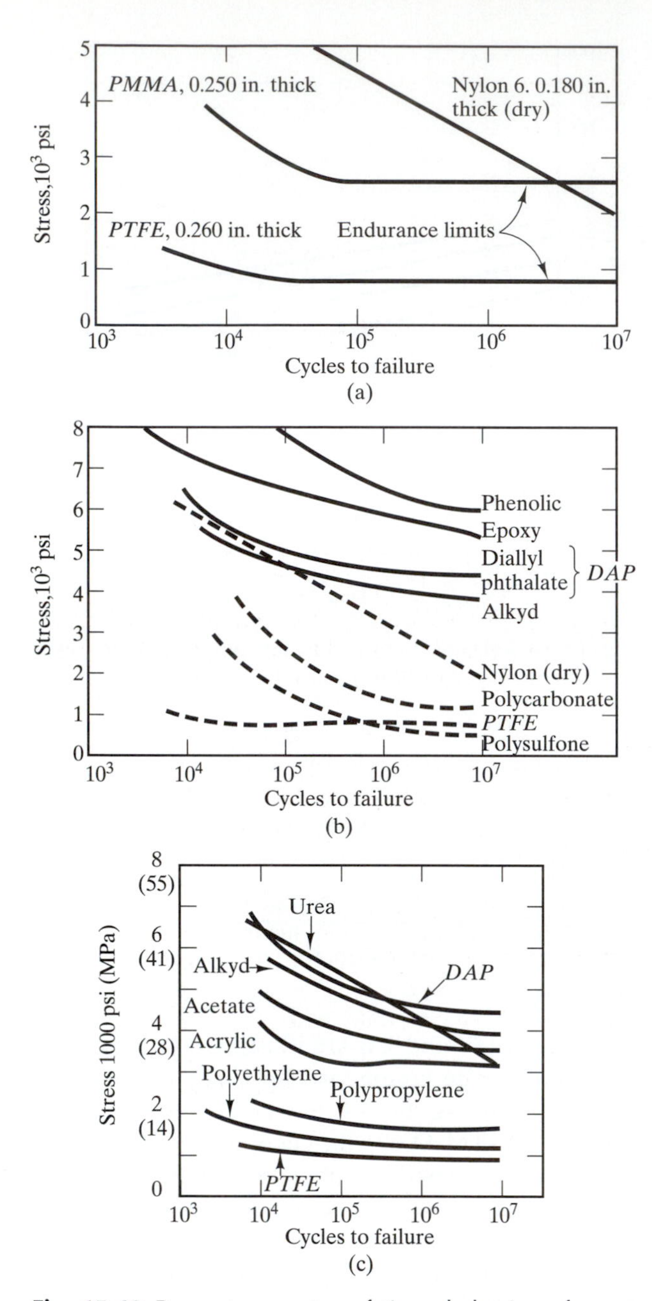

Fig. 15-40 Room temperature fatigue behavior of some thermoplastics and thermosets. Fatigue testing is per ASTM D 671—completely reversed cantilever bending at constant load and a frequency of 1800 cycles/min. [2]

oriented as in fibers. When the plastics are filled with other materials, their properties are improved (discussed in Sec. 15.5 on compounding). In this section we shall limit the discussion to the cases when the plastics are strengthened and are, therefore, called reinforced plastics and where the plastic is called the matrix and the filler the reinforcement. Reinforced plastics form only a part of the general class of composite materials (discussed in Sec. 15.6) that may also have metals, ceramics, and glasses as matrices, each being filled with other materials to form metal-matrix, ceramic-matrix, and glass-matrix composites. Therefore, as we discuss the mechanical properties of reinforced plastics (polymer-matrix composites), we keep in mind that the general principles apply as well to the other composite materials.

15.8.1 Types of Composites

In Sec. 15.6, the three types of composites were discussed according to the types of fillers or reinforcements used. Figure 15-41 schematically shows the three types: (1) particulate composites with particles as fillers or reinforcements; (2) fiber composites in which the reinforcing fibers may be either as long as the dimensions of the component, called continuous-fiber composites, or short and chopped called discontinuous-fiber composites; and (3) laminar composites in which the reinforcement is planar in form. In plastics, particles of other materials are added essentially as fillers to improve the properties or lower the cost. (In metals and ceramics, particles are added primarily to improve their high temperature strength and toughness.) Reinforced plastics are mostly fiber composites, continuous- or discontinuous-fiber, and laminar composites called laminates that are produced by stacking the layers or plies of continuous-fiber reinforced plastics. Because of the obvious directional properties of these plies, laminates can be designed to suit the specific strength and mode of loading of the component—a truly engineered material. Thus, we shall limit our discussion to fiber composites and show how we can "engineer" a laminate.

15.8.2 Properties of Fiber-Reinforced Plastics

We have discussed in Sec. 15.6.2 the functions of the fibers and the matrices and we shall now discuss the resulting properties of the combination. The macroscopic combination of the fibers and the matrix results in properties that are unattainable by either of the constituent materials. For example, fibers

Structure	Reinforcement	Composite	Properties
	Particle	Particulate	Isotropic
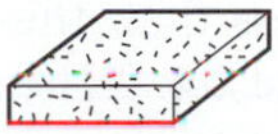	Short/chopped fibers	Discontinuous fiber	Random – isotropic aligned – anisotropic
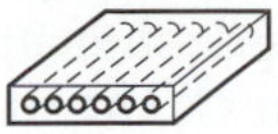	Continuous fiber	Continuous fiber	Orthotropic
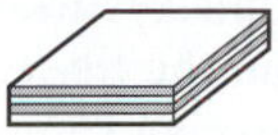	Laminae or layers	Laminar laminates	Through-thickness anisotropy; planar isotropy/anisotropy

Fig. 15-41 Structures, reinforcements, types, and properties of composites.

have high tensile strengths and moduli, but sustain zero compressive or flexural stress, while plastics have low moduli and strengths. The combination, a fiber-reinforced plastic, has very attractive properties of high strength and stiffness (high moduli) and can sustain compressive or flexural stresses. In order to obtain these desirable properties, the fibers and matrix must have some form of bonding between them at the interface to make the composite. Otherwise, the combination is not a composite and the composite properties are not achieved.

Composite properties may be estimated by the rule of mixtures, which was referred to earlier in Chap. 8 regarding dispersion strengthening. Mathematically, the rule states that

$$P_C = \sum_{i=1}^{n} V_i P_i \qquad (15\text{-}10)$$

and

$$\sum_{i=1}^{n} V_i = 1 \qquad (15\text{-}11)$$

where P_C is the composite property and the subscript i refers to i^{th} constituent, V is its volume fraction, and P is its property. *We must note that V is the volume fraction and not the mass or weight fraction by which most solid solutions or mixtures are expressed. For a two constituent system,* the rule of mixture is

$$P_C = V_1 P_1 + V_2 P_2 \qquad (15\text{-}12)$$

and,

$$V_1 + V_2 = 1 \qquad (15\text{-}13)$$

The composite property may be isotropic or anisotropic. An example of an isotropic property is the density. Density does not depend on the direction and the above equation applies for all types of composites. Other properties will be isotropic or anisotropic depending on the type of composite, as indicated in Fig. 15-41. Properties are isotropic for particulate and randomly oriented short-fiber composites, but are anisotropic in continuous-fiber as well as in oriented or aligned short-fiber composites.

15.8.2a Stressing Unidirectional Fiber Composite in Longitudinal Direction. The derivation of the mechanical properties of composites, using the rule of mixtures, depends on the following assumptions: (1) both the fiber and the matrix behave as linear elastic materials, that is, both exhibit a linear elastic stress-strain curve; (2) their Poisson's ratios are about equal; and (3) there is perfect bonding between the fiber and the matrix.

When the unidirectional fiber composite is stressed along the longitudinal direction of the fiber, the strains of the composite, the matrix, and the fiber are all equal,

$$\epsilon_C = \epsilon_f = \epsilon_m \qquad (15\text{-}14)$$

where the subscripts c, m, and f, refer to the composite, matrix, and fiber. For this case, the rule of mixtures for the modulus of the composite is

$$\begin{aligned} E_C &= V_m E_m + V_f E_f \\ &= (1 - V_f) E_m + V_f E_f \end{aligned} \qquad (15\text{-}15)$$

Because $E_f \gg E_m$, we see that the modulus of the composite in the longitudinal direction depends on the modulus of the fiber. Multiplying both sides of Eq. 15-15 with the equal strains from Eq. 15-14, we obtain

$$E_C \epsilon_C = (1 - V_f) E_m \epsilon_m + V_f E_f \epsilon_f \qquad (15\text{-}16)$$

From which we obtain the strength of the composite as,

$$\sigma_C = (1 - V_f)\sigma_m + V_f\sigma_f \quad (15\text{-}17)$$

We see that the stress on, or the strength of, the composite also follows the rule of mixtures.

15.8.2b Stressing the Composite along the Tranverse Fiber Direction. When a unidirectional composite is stressed in the *transverse direction,* the stress on the composite is the same as that on the matrix as well as on the fiber,

$$\sigma_C = \sigma_m = \sigma_f \quad (15\text{-}18)$$

In this case, the strain of the composite follows the rule of mixtures, thus

$$\epsilon_C = (1 - V_f)\epsilon_m + V_f\epsilon_f \quad (15\text{-}19)$$

Combining Eqs. 15-18 and 15-19, we obtain the following relation for the modulus of the composite in the transverse direction.

$$\frac{1}{E_C} = \frac{(1 - V_f)}{E_m} + \frac{V_f}{E_f} \quad (15\text{-}20)$$

Since $E_f \gg E_m$, $V_f/E_f < (1 - V_f)/E_m$, and we see that in the transverse direction, the modulus of elasticity of a unidirectional polymer matrix composite is controlled by the modulus of the matrix. Figure 15-42 is a schematic of a layer or lamina or ply of a unidirectional composite showing the longitudinal and the transverse directions.

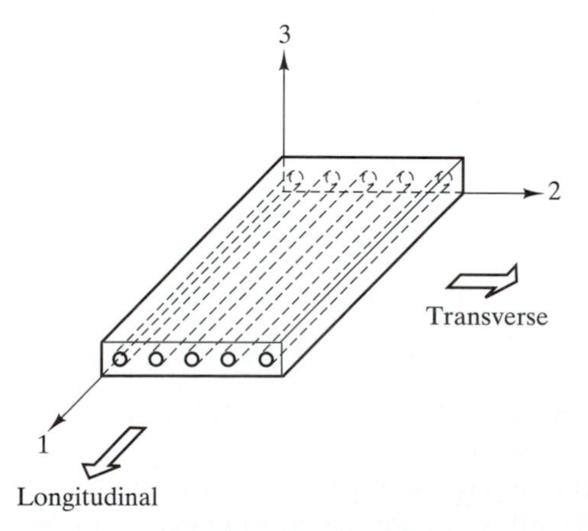

Fig. 15-42 Schematic of a unidirectional showing longitudinal and tranverse directions.

15.8.2c Critical Fiber Volume Fraction. In a unidirectional composite, there is a critical volume fraction of fibers to make an effective composite. This is shown in Fig. 15-43 with the assumption that the fibers have failure strains less than the matrix, and that all fibers fail at the same time. The stress-strain curves of the fiber and the matrix are also shown in the inset of Fig. 15-46. The tensile strength of the fiber is σ_{fu} and its fracture strain is ϵ_f^*. The stress-strain of the matrix suggests that the matrix work-hardens and has an elongation greater than the fiber and that the matrix shall be capable of carrying some load. At ϵ_f^*, the matrix stress is (σ_m) and the tensile strength of the composite is (applying Eq. 15-17).

$$\sigma_{Cu} = \sigma_{fu}V_f + (\sigma_m)_{\epsilon_f} \cdot (1 - V_f) \quad (15\text{-}21)$$

Equation 15-21 indicates the strength of the composite when there is sufficient fiber to control the strength. When extended to the extreme values of fiber-volume fractions, we see from this equation that when $V_f = 0$, $\sigma_{cu} = (\sigma_m)_{\epsilon_f^*}$, and when $V_f = 1$, $\sigma_{cu} = \sigma_{fu}$. These are indicated in Fig. 15-43.

However, when the fiber volume fraction is small, that is below V_{min} in Fig. 15-43, the matrix sustains most of the stress on the composite. Starting from $V_f = 0$, (i.e., at $\sigma_{cu} = \sigma_{mu}$), and neglecting the fiber contribution, Eq. 15-21 becomes

$$\sigma_{Cu} = \sigma_{mu}(1 - V_f) \quad (15\text{-}22)$$

as indicated in Fig. 15-43. We see from Fig. 15-43 that the straight lines represented by Eqs. 15-21 and 15-22 intersect at $(V_f)_{min}$, which is defined as the minimum fiber-volume fraction that ensures fiber-controlled composite failure. We obtain $(V_f)_{min}$ by equating Eqs. 15-21 and 15-22,

$$(V_f)_{min} = \frac{\sigma_{mu} - (\sigma_m)_{\epsilon_f^*}}{\sigma_{fu} + \sigma_{mu} - (\sigma_m)_{\epsilon_f^*}} \quad (15\text{-}23)$$

The critical fiber-volume fraction to be an effective composite is given by the condition that $\sigma_{Cu} \geq \sigma_{mu}$, and when we substitute $\sigma_{Cu} = \sigma_{mu}$ in Eq. 15-21, we obtain

$$(V_f)_{crit} = \frac{\sigma_{mu} - (\sigma_m)_{\epsilon_f^*}}{\sigma_{fu} - (\sigma_m)_{\epsilon_f^*}} \quad (15\text{-}24)$$

$(V_f)_{crit}$ is a more important property than $(V_f)_{min}$.

There is also a maximum effective fiber-volume fraction. If we stack the fibers in contact with each

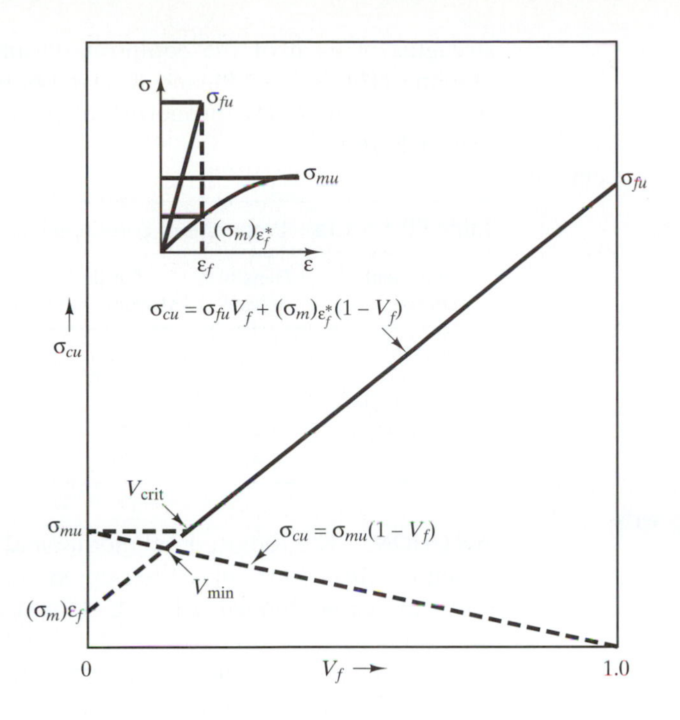

Fig. 15-43 Schematic of the strength of the composite as a function of fiber volume fraction. A critical fiber volume fraction is needed for the fiber to make an effective composite. [5]

other, the maximum fiber-volume fraction is 0.907, previously discussed in Sec. 15.6. However, the matrix cannot enclose all the longitudinal surfaces of the fibers. Because the matrix needs to enclose and bind with the fiber surfaces, the highest fiber volume fraction used is much lower than 0.90 and the commonly used effective fiber-volume fraction is about 0.60 to 0.70.

The discussion up to this point is for only one fiber in a unidirectional composite. There is no reason why we cannot use more than one fiber in a composite. When we use two or more fibers with a single matrix, we produce a *hybrid composite*.

EXAMPLE 15-10

Silica flour (ground silica) is used as an extender (filler) for acetal. Using the rule of mixtures, estimate the density of the mixture when equal weights of silica and acetal are used. The density of silica is 2.65 g/cm^3 and that of acetal is 1.42 g/cm^3.

SOLUTION: If we use as a basis of calculation, 100 gms of the mixture, there will be 50 grams of silica and 50 grams of acetal.

Volume of silica

$$= \frac{50}{2.65} = 18.87 \text{ cm}^3$$

Volume of acetal

$$= \frac{50}{1.42} = 35.21 \text{ cm}^3$$

Total volume

$$= 18.87 + 35.21 = 54.08 \text{ cm}^3$$

density

$$= \frac{100}{54.08} = 1.85 \text{ g/cm}^3$$

or using the rule of mixtures,

$$\rho = \frac{18.87}{54.08} \times 2.65 + \frac{35.21}{54.08} \times 1.42 = 1.85 \text{ g/cm}^3$$

EXAMPLE 15-11

As in the previous example, what would be the weight per cent of silica in acetal for the density of the mixture to be 2.0 g/cm^3?

SOLUTION: To solve the problem, we shall use the rule of mixtures to get the volume fraction of silica and then convert the volume fraction to weight fraction (per cent). If we let V= volume fraction of silica, then

$$2.0 = \text{V} \times 2.65 + (1 - \text{V}) \times 1.42$$

solving for

$$V = 0.4715; (1 - V) = 0.5285$$

To convert to weight fraction, we assume 100 cm^3 of mixture with 47.15 cm^3 of silica and 52.85 cm^3 of acetal.

Weight of silica

$$= 47.15 \times 2.65 = 124.95 \text{ g}$$

Weight of acetal

$$= 52.85 \times 1.42 = 75.05 \text{ g}$$

weight per cent of silica

$$= \frac{124.95}{124.95 + 75.05} \times 100 = 62.5 \text{ wt\%}$$

EXAMPLE 15-12

A cast polyester is filled with 50 percent by weight of chopped E-glass. Estimate the density and the modulus of elasticity of the material. Densities of polyester and E-glass are 1.22 and 2.58 g/cm^3. Their moduli are 5.4 and 72.4 GPa.

SOLUTION: We use as a basis, 100 grams of the E-glass filled polyester. The volume of each constituent,

$$\text{Volume of polyester} = 50/1.22 = 40.98 \text{ cm}^3$$

$$\text{Volume of E-glass} = 50/2.58 = 19.38 \text{ cm}^3$$

$$\text{Total volume} = 60.36 \text{ cm}^3$$

$$V_{\text{polyester}} = 40.98/60.36 = 0.6789$$

$$V_{\text{E-glass}} = 19.38/60.36 = 0.3211$$

(a) Density

$$= 100/60.36 = 1.66 \text{ g/cm}^3$$

or

$$= 0.6789 \times 1.22 + 0.3211 \times 2.58 = 1.66 \text{ g/cm}^3.$$

(b) Modulus of Elasticity

$$= 0.3211 \times 72.4 + 0.6789 \times 5.4 = 26.9 \text{ GPa}$$

EXAMPLE 15-13

Unidirectional fiber-epoxy-matrix composites are made with equal volumes of fiber and epoxy. Estimate the longitudinal and transverse moduli and strengths of each of the composites using E-glass, graphite-HM, Kevlar 49, and Kevlar 149 fibers with the epoxy matrix. The properties of the constituent materials are:

Table EP 15-13a Properties of constituent materials.

Constituent Material	Density, g/cm^3	Tensile Modulus, GPa	Tensile Strength, MPa
Epoxy (matrix)	1.25	3.5	70
E-Glass (fiber)	2.58	72.5	3,450
Graphite-HM (fiber)	1.90	390	2,400
Kevlar-49 (fiber)	1.44	131	3,700
Kevlar-149 (fiber)	1.47	186	3,400

SOLUTION: The longitudinal moduli and strengths of unidirectional 50 vol% fiber-epoxy matrix composite are estimated using Eqs. 15-15 and 15-17. The transverse moduli are estimated using Eq. 19-20. The transverse strengths of the composites are dictated by the matrix properties, assuming the interfacial bonding between the matrix and the fiber is ideal and have infinite strength. Otherwise, the transverse strength is limited by the interfacial bonding strength.

Application of Eqs. 15-15 and 15-17 also assumes ideal bonding between matrix and fiber in order for the stress to be transferred from matrix to fiber. The estimates of the longitudinal moduli and strengths are:

for the E-glass-epoxy:

$$E_C = 0.50 \times 3.5 + 0.50 \times 72.5 = 38 \text{ GPa}$$

$$\sigma_C = 0.50 \times 70 + 0.50 \times 3450 = 1735 \text{ MPa}$$

for the Graphite-epoxy:

$$E_C = 0.50 \times 3.5 + 0.50 \times 390 = 196.8 \text{ GPa}$$

$$\sigma_C = 0.50 \times 70 + 0.50 \times 2400 = 1235 \text{ MPa}$$

for the Kevlar 49-epoxy:

$$E_C = 0.50 \times 3.5 + 0.50 \times 1313 = 67.2 \text{ GPa}$$

$$\sigma_C = 0.50 \times 70 + 0.50 \times 3700 = 1885 \text{ MPa}$$

for the Kevlar 149-epoxy:

$$E_C = 0.50 \times 3.5 + 0.50 \times 186 = 94.2 \text{ GPa}$$

$$\sigma_C = 0.50 \times 70 + 0.50 \times 3400 = 1735 \text{ MPa}$$

for the Transverse Moduli Properties:

Table EP 15-13b Longitudinal and transverse properties of 50 vol% fiber epoxy matrix composites.

Composite (50 vol% fiber)	Longitudinal Properties		Transverse Properties	
	Modulus, GPa	Strength, MPa	Modulus, GPa	Strength, MPa
E-glass-epoxy	38	1760	6.68	70 (matrix)
Graphite HM-epoxy	196.8	1235	6.94	70
Kevlar 49-epoxy	67.2	1885	6.82	70
Kevlar 149-epoxy	94.8	1735	6.87	70

for the E-glass-epoxy:

$$\frac{1}{E_C} = \frac{0.50}{3.5} + \frac{0.50}{72.5}, \quad E_C = 6.68 \text{ GPa}$$

The other transverse moduli can be calculated in a similar manner. The results are summarized in Table EP 15-13b above.

We should note that the longitudinal properties of a unidirectional composite are dictated primarily by the fiber properties, while the transverse properties depend on the matrix properties.

EXAMPLE 15-14

A hybrid unidirectional fiber composite consists of 30 vol% graphite-HM and 30 vol% Kevlar 49 fibers in an epoxy matrix (40 vol %). Estimate the longitudinal specific modulus and specific strength of the composite.

SOLUTION: Since the specific modulus is E/ρ and specific strength is σ/ρ, we need to determine the modulus, the strength, and the density of the composite. Using the data in Example Table 15-13a.

$$E_C = 0.40 \times 3.5 + 0.30 \times 390 + 0.30 \times 131$$
$$= 157.7 \text{ GPa}$$
$$\sigma_C = 0.4 \times 70 + 0.3 \times 2{,}400 + 0.3 \times 3{,}700$$
$$= 1858 \text{ MPa}$$
$$\rho_C = 0.4 \times 1.25 + 0.3 \times 1.9 + 0.3 \times 1.44$$
$$= 1.50 \text{ g/cm}^3, \text{ or } 1{,}500 \text{ kg/m}^3.$$

The specific modulus,

$$\frac{E_C}{\rho_C} = \frac{157.7 \times 10^9 \text{ Pa(N/m}^2)}{1{,}500 \text{ (kg}_m/\text{m}^3)} =$$

$$105 \times 10^6 \frac{[\text{kg}_m \times (\text{m/sec}^2)]/\text{m}^2}{\text{kg}_m/\text{m}^3} = 105 \times 10^6 \frac{\text{m}^2}{\text{sec}^2}$$

As indicated in Problem 11-1, the square root of E/ρ is the velocity of sound in the material. In this case the velocity of sound in the longitudinal direction is 10,250 m/sec.

The specific strength,

$$\frac{\sigma_C}{\rho_C} = \frac{1.858 \times 10^9 \text{ Pa}}{1{,}500(\text{kg}_m/\text{m}^3)} = 1.24 \times 10^2 \frac{\text{m}^2}{\text{sec}^2}$$

As in E/ρ, the unit of σ/ρ is the square of velocity. However, as indicated in Problem 11-2, this will have more physical significance, if we divide it by the gravitational acceleration of 9.80 m/sec^2 to give the specific length of a material that can be suspended without breaking, assuming no wind effects and constant acceleration. Thus,

$$\frac{1.24 \times 10^6 \text{ (m}^2/\text{sec}^2)}{9.80 \text{ (m/sec}^2)} = 126{,}000 \text{ m}$$

15.9 Laminates

When we stack layers or laminae or plies of the unidirectional composite shown in Fig. 15-42, we produce a laminate. Depending on how we stack the layers relative to each other, we can produce different kinds of laminates with entirely different properties; thus, we can "engineer" laminates as we desire.

15.9.1 Fiber Orientation

We start by defining the orientation relationship of the fiber with respect to the laminate as shown in Fig. 15-44. The longer dimension of the laminate is designated as the x-axis and the width is the y-axis. The orientation of the fiber as we stack each unidirectional ply in Fig. 15-42 with respect to the longitudinal laminate x-axis is denoted with an angle θ. According

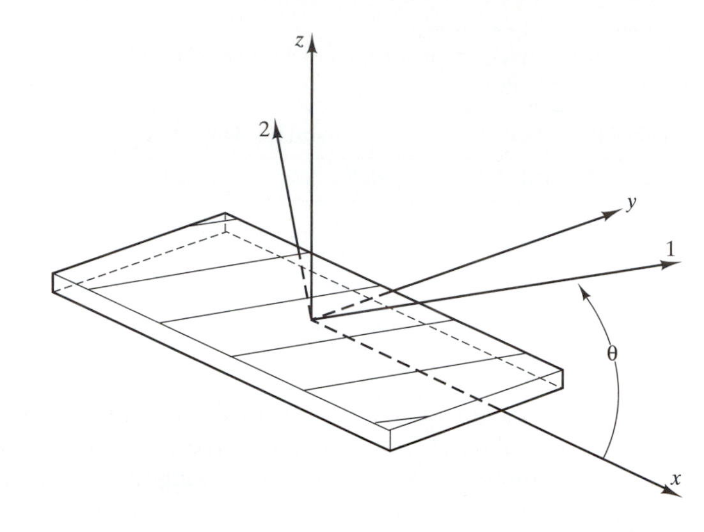

Fig. 15-44 Definition of laminate orthogonal axes (x and y) in relation to the fiber axes (1 and 2).

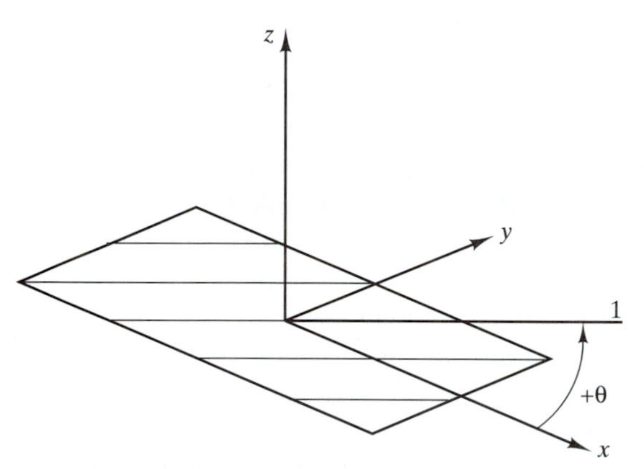

Fig. 15-45 Orientation convention of fiber axis, 1 relative to laminate axis, x. Counterclockwise rotation is $+\theta$; clockwise rotation is $-\theta$.

to convention, a counterclockwise angle with respect to the x-axis is denoted $+\theta$, and a clockwise angle is denoted $-\theta$, as shown in Fig. 15-45. Thus, a ply with the fiber parallel to the x-axis has a 0° orientation and a ply with the fiber perpendicular to the x-axis (i.e., parallel to the y-axis), has a 90° orientation. A laminate with all the plies having 0° orientation is called a *unidirectional laminate.* A laminate with alternate layers of 0° and 90° orientations is called a *cross-ply laminate.* A laminate containing a layer(s) or a ply(ies) with $\pm\theta$ orientation is called an *angle-ply laminate.*

15.9.2. Laminate Codes [5]

We can stack any number of plies with different orientations in a laminate, as required by the design to yield the unique properties and characteristics. The stacking of the laminae or plies is identified by using the *laminate orientation code.* In this code, (1) each ply is designated a number that represents its fiber orientation angle, (2) adjacent plies are separated by a slash if their orientation angles are different, (3) the plies are listed in sequence from one laminate face to the other, starting with the top lamina, with brackets indicating the beginning and end of the code, and (4) adjacent laminae or plies of the same orientation are denoted by a numerical subscript. Examples of laminate orientation codes are shown in Fig. 15-46.

Laminates possessing a symmetry in laminae orientations about the geometric midplane are indicated by coding only half of the stacking sequence. *A symmetric laminate with an even number of laminae* is indicated by a code, starting from the top lamina and stopping at the plane of symmetry. A subscript S to the bracket notation indicates that only one-half of the laminate is shown, see Fig. 15-47a. *A symmetric laminate with an odd number of laminae* is coded the same way as the even numbered symmetric laminate, except that the center lamina, which contains the plane of symmetry, is listed last and a bar placed over it, to indicate that half of it lies on either side of the plane of symmetry, Fig. 15-47b. Repeating sequences of laminae are called *sets* and are enclosed in parenthesis. A set is coded in the same way. Examples of sets are shown in Fig. 15-48. Codes for *hybrid laminates* are illustrated in Fig. 15-49. Each lamina or ply of the hybrid laminate is coded as before but

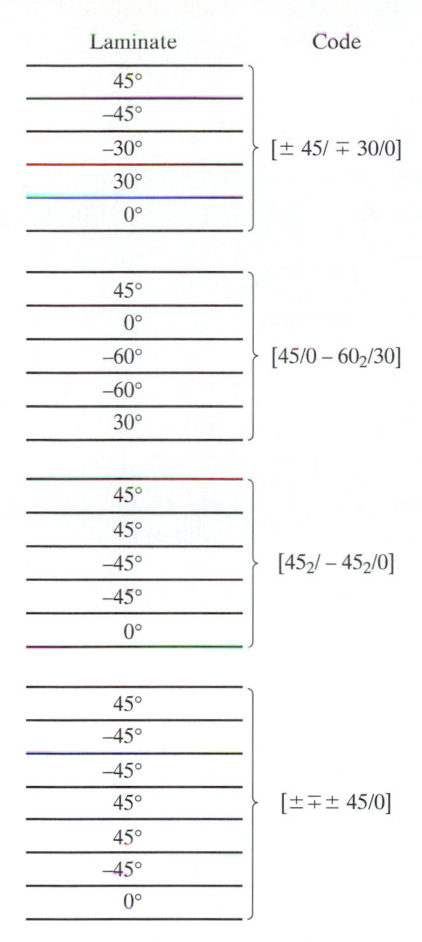

Fig. 15-46 Examples of laminates and their codes.

subscripts are added to indicate the fiber used in the lamina.

15.9.3 Properties of Laminates

Figure 15-50 illustrates the properties of different laminates in polar coordinates. A unidirectional laminate is orthotropic and its properties are given mathematically by the equations in Sec. 15.8.2. We see that the strength is highest for a unidirectional composite along the longitudinal axis and lowest along transverse axis. This reflects the fact that the tensile strength of the fiber dictates the longitudinal strength of the laminate and the strength of the matrix (plastics) dictates the transverse strength.

A cross-ply laminate with an equal number of 0° and 90° has equal strengths in the longitudinal and transverse directions, but still exhibits the lowest strengths along the ±45° directions. The laminate with isotropic properties (larger circle in Fig. 15-50) is a quasi-isotropic laminate and is represented by the laminate code in Fig. 15-47. It is called quasi-isotropic because it is only isotropic in the plane of the laminate and exhibits different properties along the thickness of the laminate.

15.10 Fabrication of Plastics

The various processes to be briefly discussed here are those that are used to fabricate all types and shapes of plastic products, ranging from common household packages to structural reinforced plastics that are used in the aircraft. Table 15-20 shows the

Laminate	Code
0°	
+45°	
−45°	
90°	$[0/\pm45/90]_s$
90°	
−45°	
+45°	
0°	

(a)

Laminate	Code
0°	
45°	$[0/45/\overline{90}]_s$
90°	
45°	
0°	

(b)

Fig. 15-47 Coding of symmetric laminates. (a) even numbered plies, (b) odd-numbered plies. Symmetric laminate (a) is quasi-isotropic laminate.

Laminate	Code
45°	
0°	
90°	
45°	
0°	
90°	$[(45/0/90)_2]_s$ or $[45/0/90]_{2s}$
90°	
0°	
45°	
90°	
0°	
45°	

Laminate	Code
45°	
0°	
90°	
45°	
0°	
90°	$[(45/0/90)_4]$ or $[45/0/90]_4$
45°	
0°	
90°	
45°	
0°	
90°	

Fig. 15-48 Laminate codes with sets.

Laminate*		Code
0°	B/Ep	
45°	Gr/Ep	$[0_B/\pm 45_{Gr}/90_{Gr}]_s$
–45°	Gr/Ep	
90°	Gr/Ep	
90°	Gr/Ep	
–45°	Gr/Ep	
45°	Gr/Ep	
0°	B/Ep	

Laminate*		Code
0°	B/Ep	
0°	B/Ep	
45°	Gr/Ep	$[0_{2B}/45_{Gr}/\overline{90}_{Gr}]_s$
90°	Gr/Ep	
45°	Gr/Ep	
0°	B/Ep	
0°	B/Ep	

*B = boron . Gr = graphite . Ep = epoxy .

Fig. 15-49 Coding hybrid laminates.

Table 15-20 Relative production of plastics by different processes.

Material	% by Weight
Extrusion	36
Injection	32
Blowing	10
Calendering	6
Coasting	5
Compression	3
Powder	2
Others	6

relative amounts of fabricated plastics from the different processes, and we see that the two most prominent processes are extrusion and injection. The principles of these two processes are very similar to the same processes used in forming the 'green' ceramic compacts described in Chap. 14. Blowing, the third most popular process in Table 15-20, is a similar process to the blowing of glass containers also described in Chap. 14.

Polymers are usually available in the form of granules, powders, pellets, and liquids. The common features of all the fabrication processes are: (1) mixing, melting, and plasticizing the raw materials; (2) trans-

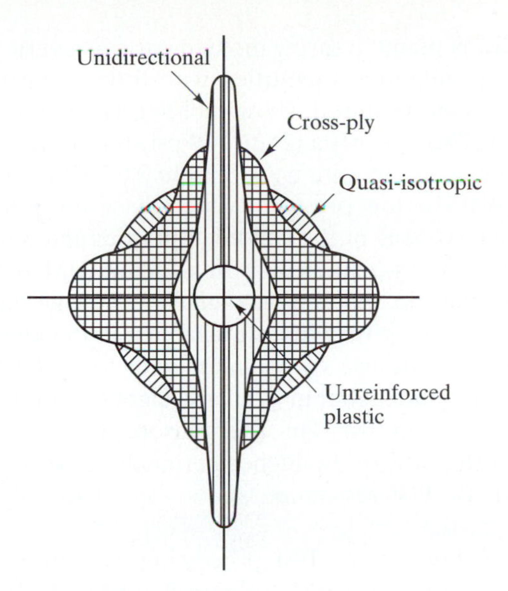

Fig. 15-50 Properties of different laminates in polar coordinates. Longitudinal (0°) is vertical; transverse direction (90°) is horizontal.

porting and shaping the melt; (3) drawing and blowing; and (4) finishing. Mixing, melting, and plasticizing are usually done by introducing the mix through a hopper into an externally heated cylinder with an auger (screw), which mixes and pushes the mixture forward to the heated zone to be plasticized, and then pushes the melt through a die or into a mold. The drawing and blowing processes orient the polymeric molecules in the production of different shapes (by blow-molding, thermoforming, etc.). The finishing involves the solidification of the melt, and may include coating and surface modification, cutting, and other in-line post-forming operations.

15.10.1 Extrusion

This process owes its prominence to its versatility, and its relatively low cost, continuous operation. As described in the compacting of ceramic green, extrusion and injection molding (IM) have very similar equipment. A major advantage of extrusion over IM is its comparatively low pressure along with its continuous operation to fabricate plastic components. The pressure in extrusion ranges usually from 1.4 to 10.4 MPa (200 to 1,500 psi) but can go up to 34.5 to 69 MPa (5,000 to 10,000 psi). In IM, the pressures go from 14 to 210 MPa (2,000 to 30,000 psi), but its major disadvantage is that it is a discontinuous melting and molding cyclic process and requires more stringent controls.

In both extrusion and IM, the material mix passes from a hopper into an externally heated cylinder with an auger (screw). The auger mixes and compresses the material before it is molten and then homogenizes the melt, as shown in Fig. 15-51. Then the auger pushes the melt to the end of the cylinder and forces the melt through a die that gives the desired shape of the extrudate with no break in continuity. On leaving the extruder, the product is drawn by a pulling device, at which stage it may be subject to cooling, usually by water or blown air. The draw aligns or orientates the molecules of the melt (extrudate) in the longitudinal direction of the product and induces a higher longitudinal strength than the tranverse strength. Depending on the product's use, this may or may not be favorable. The degree of orientation can be controlled. To date, thermoplastics are the only materials processed through extruders; markets have yet to be developed for extruded thermosets. The products that may be obtained with extrusion

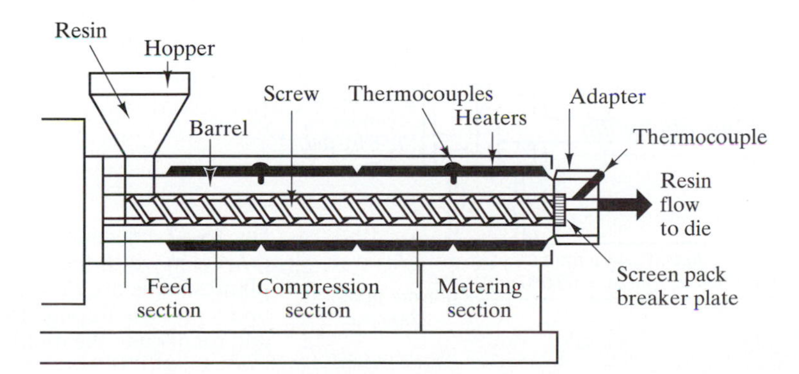

Fig. 15-51 A cross-section of a single screw (auger) extruder. [2]

when coupled with other in-line post-forming operations are films, sheets, tapes, filaments, pipes, rods, as well as many others.

15.10.2 Injection Molding

This process is the most suitable for the manufacture of components with 3-dimensional and complex shapes since it can be more accurately controlled and predicted. Figures 15-52 illustrates an injection molding equipment. The plastic is fed from the hopper, then is mixed and pushed by the auger inside the cylinder into a heated portion where it melts, and the melt fed into the shot chamber (front of the screw). When a preset limit switch or a position transducer is reached, the required shot (melt) size is in the chamber and the auger stops rotating. Then the auger acts as a ram to push or inject the melt into the mold at a pressure that may reach as high as 210 MPa (30,000 psi) in the nozzle. The melt pressure within the mold cavity may range from 1 to 15 tons/sq. in., depending on the plastic's rheology/flow behavior.

Most of the plastics processed by injection molding are thermoplastics, but some thermosets are also processed such as TS polyesters, phenolics, epoxy, and so on. The TPs reach their maximum temperatures prior to entering the "cool" mold cavities, whereas the TSs reach their maximum temperatures in the "hot" mold.

15.10.3 Blow Molding

This is the third most popular process in the manufacture of plastic parts and it offers the advantage of economically manufacturing molded parts in a variety of sizes, in unlimited quantities, with little or virtually no finishing required. Blow molding (BM) requires only 0.17 to 1.03 MPa (25 to 150 psi) but some resins (plastics) may require up to 1.38 to 2.07 MPa (200 to 300 psi). The low pressures generally result in lower internal stresses in the solidified plastics and a more uniform stress distribution. In addition, BM is very appropriate in molding irregular (reentrant) curves and offers the possibility of variable wall thicknesses, along with the use of polymers with high chemical resistance and higher molecular weight (MW) than is permissible in IM. For this reason, items can be blown that utilize the higher permeability, oxidation resistance, UV resistance, and so on, of the higher-MW plastics.

The three major BM processing categories are: (1) extrusion BM (EBM); (2) injection BM (IBM); and (3) stretch BM, for either EBM or IBM, to obtain bi-oriented products that provide significantly improved cost-to-performance advantages. Almost 75 percent of the BM processes are EBM, a little less than 25 percent for IBM, and less than 1 percent for other BM techniques. About 75 percent of the IBM products are bi-oriented.

Extrusion Blow Molding (EBM). This process is schematically shown for a continuous operation in Fig. 15-53. A parison (a tubular extrudate) through a die-head of a vertical extruder is fed into a split mold cavity. The extruder may have more than one die-head to produce multiple parisons. At a preset parison length, the split mold cavity closes around it and

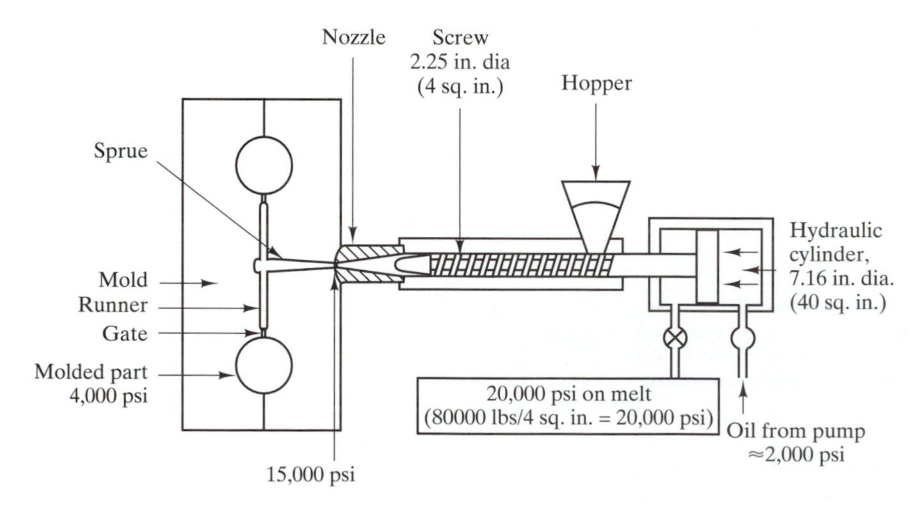

Fig. 15-52 Injection molding process in which the molten plastics in the nozzle is rammed under pressure through the sprue and runner into the mold. [2]

pinches one end. Usually a blow pin is located opposite the pinched end of the "tube" and compressed air inflates the tube against the pinched end and against the female cavity of the mold surfaces. On contact with the relatively cool mold surface, the blown parison cools and solidifies to the part shape. Next the mold opens, ejects the part, then repeats the cycle (i.e., closing around and pinching the parison, shaping it, and ejecting the shaped part).

Injection Blow Molding (IBM). The three stages in IBM are schematically illustrated in Fig. 15-54. In the first stage, an exact amount of molten resin is injected (using an IM equipment such as that shown in Fig. 15-52) around core pins that are placed in one or more preformed cavities in a mold. The two-part mold opens, and the core pins carry the hot plastic to the second-stage blow mold where the plastic is blown with compressed air introduced via the core pin. The blown plastic contacts the cooled cavity surface and solidifies. The third stage is the separation and ejection of the blown part from the mold and the core pin.

15.10.4 Forming Processes

These processes provide a great variety of formed or shaped plastics in a variety of sizes. These are basically divided into hot- and cold-forming as the analogous processes for the hot- and cold-deformation processes of metals in Chap. 8. Thermoforming (hot)

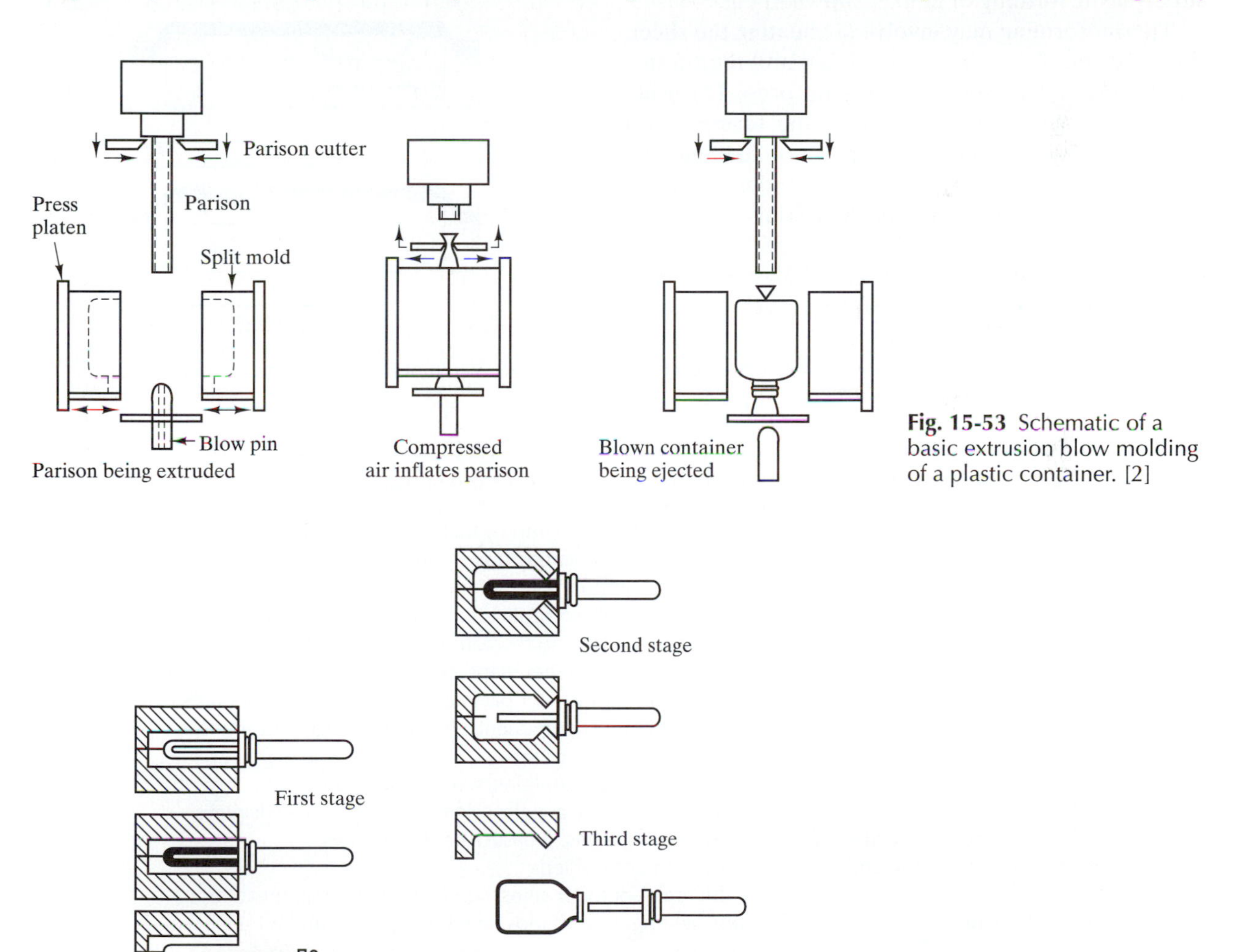

Fig. 15-53 Schematic of a basic extrusion blow molding of a plastic container. [2]

Fig. 15-54 Stages in a basic injection blow molding process. [2]

is the most productive and most diversified. The other techniques require less heat and may be applicable only to certain types of plastics.

Thermoforming consists of heating extruded thermoplastic sheet, film, and profile to its softening temperature, then forcing the hot, flexible material against the contours of a mold by pneumatic means, mechanical means, or a combination of pneumatic and mechanical means. Pneumatic means consist of either creating a pressure differential between the plastic and the mold by pulling a vacuum, or compressed air is used to force the material against the mold. The basic concept of vacuum thermoforming, schematically illustrated in Fig. 15-55, was adapted in superplastic forming of metals, shown in Fig. 8-55.

Thermoforming may involve (1) heating the sheet, film, and the like, in a separate oven and then transferring the hot plastic to a forming press, (2) using equipment that combines heating and forming in a single unit, or (3) a continuous operation feeding off a roll of plastic or directly from the extruder die (postforming). Practically all the materials used are extruded TPs, and very few TSs are used that may be either unreinforced or reinforced. Polystyrene (PS) and other plastics with high melt viscosities are particularly desirable. Table 15-21 lists some of these plastics and the products into which they can be formed.

Other forming processes require less heat or sometimes do not require it at all and they can use modified metalworking tools. They are classified as; (1) cold-forming (performed at room temperature with unheated tools); (2) solid-phase forming (plastic is heated below its melting point and then formed); and (3) compression molding of reinforced composite sheets (heat is used). Other methods used are forging (open- and close-die forming), stamping, rubber pad or diaphragm forming, fluid forming, spinning, explosive forming, and others. These processes are especially

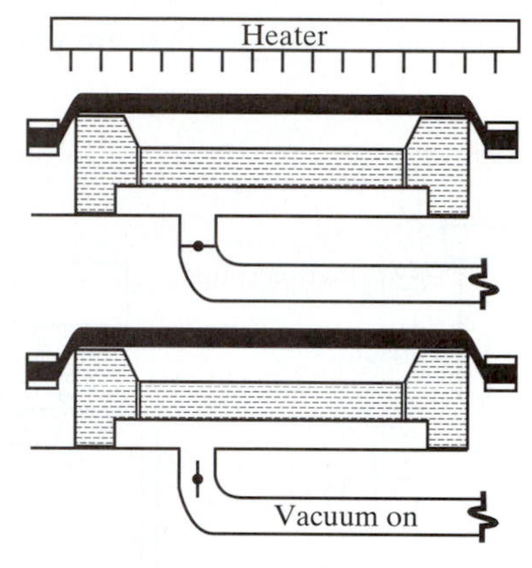

Fig. 15-55 Vacuum thermoforming of thermoplastics. [2]

Table 15-21 Various thermoformed products from some thermoplastics.

Polystyrene	Various packaging applications, including transparent meat trays, trays for cookie and candy boxes, blister packages
Polystyrene foam	Meat trays, egg cartons, take-out food containers
Acrylic	Signs and other outdoor applications like motorcycle windshields, snowmobile hoods, and recreational-vehicle bubble tops
Rigid vinyls	Lighting panels, signs, relief maps, bus-interior panels, dishes and trays for chemicals, blister packages, automobile dashboards
Acrylonitrite butadiene styrene (ABS)	Recreational-vehicle components, luggage, refrigerator liners, business-machine housings
Cellulose acetate	Blister packages, rigid containers, machine guards
Cellulose propionate	Machine covers, safety goggles, signs, shipping trays, displays
Cellulose acetate butyrate	Skylights, outdoor signs, pleasure boat tops, toys
High-density polyethylene	Camper tops, canoes, sleds
Nylon	Reuseable trays, outdoor signs, surgical equipment, meat trays
Polycarbonate	Outdoor lighting, face shields, machine guards, aircraft panels and ducts, signs
Polypropylene	Truck-fender liners, drinking cups, juice and dairy-product containers and lids, test-tube racks

suitable for ultrahigh molecular weight, high-density plastics that are difficult or impossible to process by other methods. Thermoplastic composites can be stamped to produce high-performance parts.

15.10.5 Other Processes

Most of these fabrication processes are especially suited for TS plastics and they include compression molding and transfer molding, (CM and TM), reaction injection molding (RIM), liquid injection (LIM), rotational molding, casting, calendering, coining, and the fusible-core technique.

Compression and transfer molding (CM and TM) are the two main methods to produce molded parts from thermoset (TS) resins. In these processes, the material is compressed into the desired shape and cured with heat and pressure. They are not generally used for TPs. The schematics of these processes are shown in Fig. 15-56. CM was the major fabrication process for plastics during the first half of the 20th century because of extensive use of the phenolics (TS) and it accounted for 75 percent by weight of the fabricated plastics. The development of lower-cost fabricated TPs have reduced the amount of plastics processed by CM and TM to only about 3 percent (see Table 15-20). They are still used particularly in the production of certain low-cost, as well as heat-resistant and dimensionally precise parts. CM and TM are classified as high pressure processes, requiring from 13.8–69 MPa (2,000–10,000 psi) molding pressures.

Reaction injection molding (RIM) is the high pressure mixing of two or more reactive liquid components and then injection of the mixture into a closed mold at low pressures. Large and thick parts can be molded using fast cycles with relatively low-cost materials. It is a very attractive process because it requires low energy with relatively low investment costs. Different materials can be used such as nylon, polyester (TS), and epoxy, but TS polyurethane (PUR) is predominantly used.

Liquid injection molding (LIM) is essentially the same as RIM, (with LIM being used longer), and is used in the automated low-pressure processing of TS resins. It offers fast cycles, low labor cost, low capital investment, energy, and space savings. LIM is very competitive to potting, encapsulating, compression transfer, and injection molding, particularly when insert molding is required.

Rotational molding is basically like the centrifugal casting of cast iron where a plastic is introduced into a rotating mold, is melted, builds up a layer of molten plastic on the inside surface of the rotating

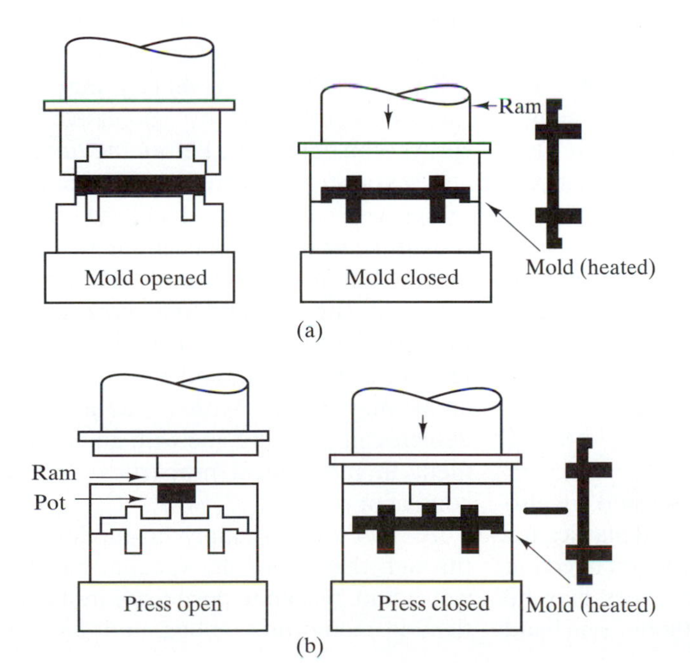

Fig. 15-56 Schematics of (a) compression molding, and (b) transfer molding.

mold, and is cooled and solidified while the mold is rotating by cooling the outside mold surface with air or water. The solidified plastic is removed from the mold and the cycle is repeated. It is capable of producing large as well as small hollow items with uniform wall thicknesses with certain plastics. Its production rate is low, but the total cost of equipment and the production for moderate-sized, and especially large parts is also low.

Casting is the fabrication of a plastic component by pouring a fluid monomer-polymer solution into an open mold where it completes its polymerization. It generally does not involve pressure or vacuum casting, although certain materials and complex parts may require one or the other.

Calendering is the process used in the production of plastic films and sheets by melting the plastic and passing the pastelike mass through nips of a series of heated and rotating speed-controlled rolls into webs of specific thichness and width.

Coining is the process that combines the best of injection molding and compression molding. It is sometimes called injection compression or injection stamping. The process involves using an IM machine to melt the plastic (unreinforced or reinforced TP or TS) and injects a fixed amount of the melt into an open compression mold. After injection is completed, the mold is closed with the male plug fitting into the female mold and the pressure on the melt made uniform.

The *fusible-core technique* is a takeoff of cored investment metal casting and is used to make simple to very complex hollow structural products. The plastic melt coats and covers the entire core surface. Once the plastic solidifies, the core is molten (fused) and the core melt is removed via an existing opening or through a hole drilled through the plastic part to the core.

15.11 Fabrication of Reinforced Plastics

Some of the processes we have just discussed are also applicable in the fabrication of reinforced plastics, for instance, compression molding and reaction injection molding. The various fabrication processes may be grouped into: (1) contact molding methods (as in hand lay-up, spray-up, vacuum bag, pressure bag, autoclave, et cetera); (2) matched mold methods (as in compression molding, transfer molding, resin transfer molding, injection molding, compression-injection molding, stamping, etc.); and (3) other methods (as in filament winding, cold-press molding, pultrusion, continuous lamination, centrifugal casting, encapsulation, rotational molding, reaction injection molding, etc.)

15.11.1 Contact Molding Methods

Hand lay-up is the oldest and in many ways the simplest and most versatile process, but it is very slow and labor intensive. It consists essentially of the hand-tailoring and placing of layers of (usually glass fiber) mat, fabric, or both on a one-piece mold and simultaneously saturating the layers with a liquid TS resin (commonly polyester). The assemblage is cured, with or without heat, and usually with no pressure. Alternatively, a preimpregnated sheet (a pre-preg) with partially cured dry material such as sheet-molding compound (SMC) may be used, but in this case heat is applied and with a good probability, low pressures will be applied.

Spray-up consists of the simultaneous spraying of chopped fibers (usually glass fibers) and a catalyzed resin in a random pattern on the surface of the mold with an air spray gun. The resulting, rather fluffy, mass is consolidated with serrated rollers to squeeze out air and reduce or eliminate voids. As in the hand layup, a first layer of a gel coat may be applied over the mold, followed by successive passes of the sprayed-on composite before the final gel is applied.

Vacuum Bag. A molded part made by hand layup or sprayup can be cured without the application of pressure. For many applications, this approach is sufficient, but maximum consolidation is usually not achieved. There will be some porosity, fibers may not fit closely into internal corners with sharp radii and tend to spring back, and resin-rich or resin-starved areas may occur. With moderate pressure, these defects can be overcome with an attendant improvement in mechanical properties and better quality control of parts.

A method to apply a moderate pressure is through the use of the vacuum bag technique. The wet-liquid, resin composite and mold is enclosed in a flexible membrane or bag, and a vacuum is drawn inside the enclosure. Atmospheric pressure on the outside of the bag then presses the bag or membrane uniformly against the wet composite. Pressure ranges from 69 to 383 kPa (10 to 14 psi).

Pressure bag. If more pressure is required than what the vacuum bag system offers, a second envelope can be placed around the whole assemblage and air pressure admitted between the inner bag and the outer envelope. The latter technique is called the vacuum-pressure bag process.

Autoclave. Higher pressures can be obtained by placing the vacuum assemblage in an autoclave. Air or stream pressures of 690 to 1380 kPa (100 to 200 psi) are commonly achieved. In these processes, an initial vacuum may or may not be employed. The bag must be well sealed to prevent infiltration of high-pressure air, steam, or water into the molded part.

Pressure bag molding. A preform that is made by a sprayup, a mat, or a combination of materials (using a perforated screen to produce the preform) is used. Only enough resin is included in the preform to hold the fibers in place. (The resin must be compatible with the matrix resin and is about 0.5 weight percent in the preform.) An inflatable elastic pressure bag is placed inside the preform and the assembly is put into a closed mold. The matrix resin is injected into the preform and then the pressure bag is inflated to about 345 kPa (50 psi). Heat is applied and the part is cured with the mold. After curing, the bag is deflated and pulled through an opening at the end of the mold, and the part is removed.

15.11.2 Matched Mold Methods

Compression and transfer molding have been described in Sec. 15.10.5.

Resin-Transfer Molding (RTM). This is a closed mold, low-pressure process in which a dry reinforcement preform is placed in the mold, the mold closed, and then the preform is impregnated with a liquid resin (usually polyesters, although epoxies and phenolics may be used) by an injection or tranfer process, through an opening in the center of a mold. The resin is a two-component system (includes the catalyst, hardener, etc.) that was previously mixed in a static mixer and then metered into the mold through a runner system. The resin displaces the air in the mold, escaping through vents provided. When the resin fills up the mold, the vents and resin inlets are closed and the resin is allowed to cure and the part is removed.

Reaction Injection Molding (RIM). This time the process is called reinforced RIM (RRIM). In RRIM, a dry reinforcement preform is also placed in a closed mold. Next, a reactive resin system is mixed under high pressure in a specially designed mix head. After mixing, the reacting liquid flows at low pressure through the runner system to fill the mold cavity and impregnates the reinforcement. When the mold cavity is filled, the resin quickly completes its reaction. The cycle time to produce a molded part can be as little as one minute. Therefore, we see that RRIM is very similar to RTM. However, RRIM uses less complex preforms and lower reinforcement content than RTM.

15.11.3 Other Methods

These methods are primarily used with continuous fiber reinforced plastics.

Pultrusion. In pultrusion and in contrast to extrusion, a combination of liquid resin and continuous fibers is pulled through a heated die of the shape required for continuous profiles or shapes. Shapes include structural I-beams, L-channels, tubes, angles, rods, sheets, and so on. The resins commonly used are polyesters with fillers, although epoxies and urethanes are used whenever their properties are required. Longitudinal fibers are generally continuous rovings. Mat or woven fibers may be used for cross-ply properties.

Filament Winding. In filament winding (FW), continuous filaments are wound onto a mandrel after passing a resin bath, unless preimpregnated (prepreg) filaments or tapes are used in which case, the resin bath is not necessary. The shape of the mandrel is the internal shape of the finished part and the configuration of the winding depends upon the relative speed of rotation of the mandrel and the rate of travel of the reinforcement-dispensing mechanism. Different configurations can be employed on successive passes (layers) and the orientation of the filaments can be tailored (engineered) to the stresses set up in the part. The three most common winding configurations are (1) helical winding, shown in Fig. 15-57, in which the filaments are at a significant angle with the axis of the mandrel; (2) circumferential winding, in which the filaments are wound like a thread on a spool; and (3) polar winding, in which the filaments are nearly parallel to the axis of the mandrel, passing over its ends on each pass.

Because the filaments are continuous and tightly packed, a high filament to resin volume ratio is achieved and results in products with very high specific strengths (strength-to-weight ratio) and specific moduli (Young's modulus to weight ratio). The most common fiber used is glass fibers, but graphite and other fibers can also be used. Care must be taken that the integrity of the fibers is maintained by avoiding surface abrasions or scratches.

15.12 Designing with Plastics and Reinforced Plastics

Designing with plastics exemplifies the adage that designers must be thoroughly familiar with the properties, processing, and manufacturability of materials when they are selected for the design of a component. It is also getting more common due to the constraints placed on the performance of a component or equipment that takes advantage of the lightweight and formability characteristics of plastics. Transport vehicles, such as the automobile and the airplane, are examples of where the lightweight properties of plastics are advantageous. While there is no federal mandate imposed on airplanes, materials that are used for their construction have always required high specific strengths and specific moduli in order to increase fuel efficiency. Because of federal mandates on fuel efficiency and on pollution, car manufacturers are always looking for components where plastics can be utilized in place of the traditional steel or metallic components to lower the gross inertia weight.

The big problem when designing with new materials such as plastics is the lack of experience or engineering data that designers have with them. The reticence in using new materials is obviously based on the unknown performance and risks when they are used in place of materials that have proven to perform very well. In addition, when the design analysis is made, the methods and the engineering formulas used are those available in engineering handbooks or textbooks that are based on the linear elastic behavior of metals and on standard shapes or sections. Thus, the elastic materials properties such as moduli, yield strengths, and Poisson's ratio are used in the design formulas, sheets, charts, or in the finite-element analysis software to determine the performance of the material.

The problem in using the available methods and techniques is the assumption that plastics behave elastically as metals at room temperature. Most structural metals such as steel have very high melting points and behave elastically at room temperature. In contrast, most plastics have relatively low glass transition or melting points, such that at room temperature most engineering plastics are viscoelastic. Thus, designing with plastics must account for their viscous property (flow of matter under stress) in

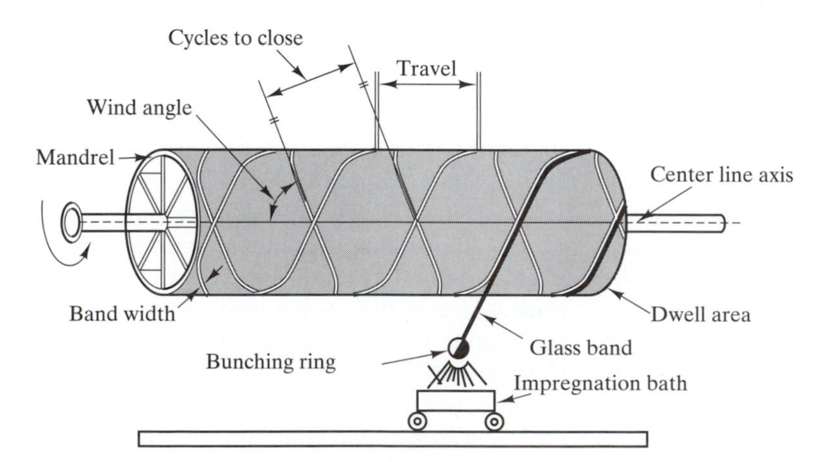

Fig. 15-57 Helical winding of continuous filaments (fibers). [2]

addition to their elastic property. The properties of importance are, therefore, creep, stress-relaxation, and fatigue that involve the interaction of stress, time, temperature, and other environmental conditions. For example, in bending, the important material property is the Young's modulus. The modulus obtained with the short-term tensile test has no relevance when the plastic will be subjected to long-term loading, even when the load is below its initial short-term yield strength. In its place, the apparent creep or stress-relaxation modulus is used for the duration the plastic is expected to be under load or stressed.

The other problem is the size and shape of the component or structure. Quite often, the very first attempt of designers when considering the replacement of a metallic structure with other materials such as plastics is to use the same size and shape as the existing structure. This approach may not be appropriate and some modifications may have to be adopted to achieve the design goals. We shall illustrate this point with an example.

We consider the weight-saving objective for automobiles and consider the body panels that are in sheet form and that need to be both stiff and strong in bending. The materials considered are shown in Table 15-22 with their properties. Panels from these materials with the same thickness (or volume) are first analyzed. In bending, the flexural rigidity is EI, where E is the tensile modulus and I is the moment of inertia. With the thicknesses of the panels being equal, I is the same for all materials, and the flexural stiffness and strengths of the panels depend directly on the materials properties. Thus, the relative panel properties in Fig. 15-58 are identical to the relative material properties in Table 15-22. We see that the metal panels are superior to those of the plastics.

If the objective is to save weight, however, the lower densities of plastics allow them to be used in thicker sections than metals. If we now assume the four panels to have equal weight, they will have different thicknesses. Then, the panels' stiffnesses depend on (Et^3) and their strengths on (σt^2), where E and σ are the material's modulus and strength. In terms of density, the relative stiffnesses of the panels are proportional to (E/ρ^3) and their relative strengths to (σ/ρ^2), where ρ is the material's density. These are shown in Fig. 15-59 for equal weight panels, and we see that the panels with plastic materials are now much better than steel.

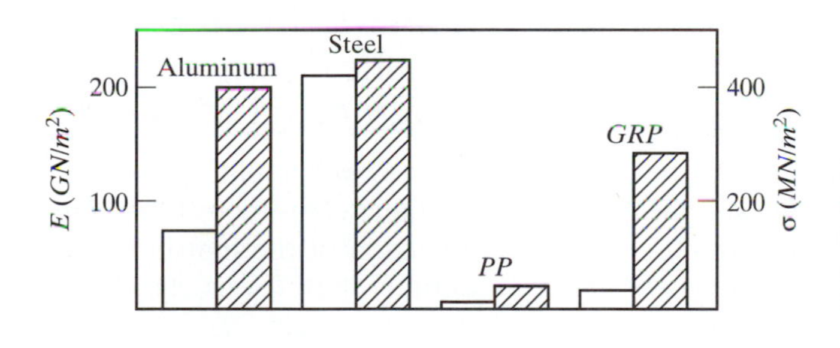

Fig. 15-58 Relative stiffnesses (open bars) and strengths (shaded bars) in flexure of equal thickness sheets from the materials. [2]

Table 15-22 Materials and their properties considered for automotive body panel. [2]

Property	Aluminum	Mild Steel	Polypropylene (PP)	Glass-fiber Reinforced Plastics (GRP)
Tensile modulus (E) GPa (10^6 psi)	70 (10)	210 (30)	1.5 (0.21)	15 (2.2)
Tensile strength (σ) MPa (ksi)	400 (58)	450 (65)	40 (5.8)	280 (40.5)
Specific gravity (S)	2.7	7.8	0.9	1.6

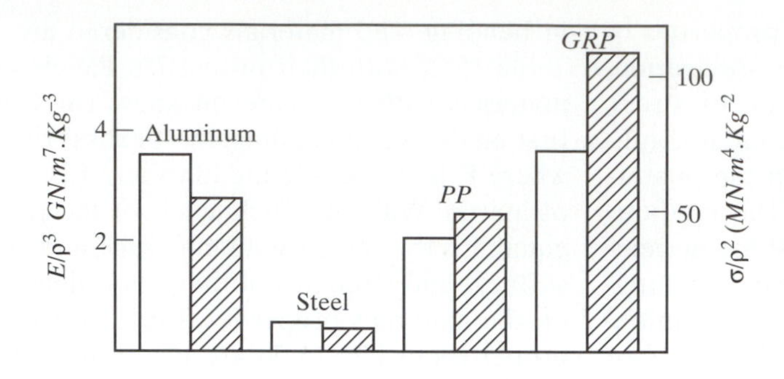

Fig. 15-59 Relative stiffnesses (open bars) and relative strengths (shaded bars) in flexure for sheets of equal weights from the materials. E = modulus, σ = strength, s = density. [2]

We note from the above example that panels made of plastics can be made stiffer and stronger, in spite of the fact that the plastics have inherently lower tensile modulus and tensile strength than steel. The point to be learned is that the properties of the product need to be evaluated, not the properties of the materials making the products. In addition, metals and plastics do not have to compete under the extreme conditions of either equal volume or equal weight. Their effectiveness between these extremes will depend on the requirements of the particular application. For the automotive body panels, it is clear that a thicker plastic panel is better than steel with the same weight. However, with the objective of saving weight, the plastic panel stiffness and strength can be adjusted in order to save weight. For example, for the materials data in Table 15-22, we can consider a GRP (graphite reinforced plastic) panel to replace a steel panel loaded in bending with the same flexural stiffness. If we do this, we find that the thickness of the GRP panel must be about 2.5 times the thickness of a steel panel and the weight of the GRP is only about half the weight of a steel panel; thus, we save 50 percent of the steel panel weight. We also find that the flexural strength of the GRP panel is about 3.5 times that of a steel panel.

We should quickly note, however, that the relative merits of the materials in the cited example are based only on the factor of saving weight. When we consider the substitution of one material for another, in this case the GRP panels for steel auto body panels, we must also consider other factors such as cost, fabricability, repairability, and others. Thus, relative merits of materials must be viewed in the entirety of the design, including recyclability and disposal of the used panel. If we do this, we might immediately find that the cost of and other factors relating to the GRP might be prohibitive for general substitution, and may only be appropriate for special cases.

Nonetheless, the example illustrates a significant conclusion—that is, that plastic products are often stiffness deficient, whereas metal products are strength deficient. Thus, metal components are made thicker in order to avoid failure by overloads. This action also makes the metal components much stiffer than required by their service requirements. On the other hand, plastic parts are made thicker or shaped differently to provide adequate stiffness and at the same time, are made much stronger than required.

15.12.1 Design of Continuous Fiber-Reinforced Plastics/Composites—Engineered Materials

At this juncture, we should appreciate that we can create or engineer continuous fiber composite materials. The selection of the matrix and the reinforcement is entirely left to us. For most composite applications, we usually select high-strength and high-modulus fibers, depending on cost. We should also note that we are not limited to one kind of fiber. We can mix and match to form hybrid composites. For high-strength composites, thermosetting resins are used and the epoxies appear to exhibit generally acceptable properties. *Specific tensile moduli and specific tensile strengths* of 65 percent unidirectional fiber-volume reinforced epoxy matrix composites are shown in Fig. 15-60, which also includes the same properties for steel and aluminum for comparison. The different fibers used with the epoxy matrix are indicated in the figure near the resulting composite properties (points). Recall that *the two coordinates of*

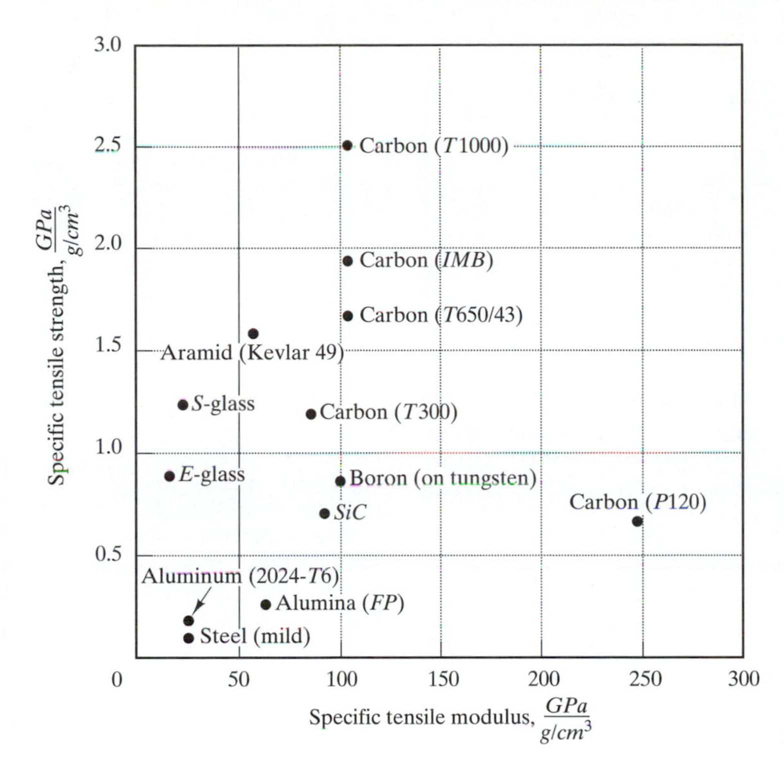

Fig. 15-60 Specific tensile moduli versus specific tensile strengths of various unidirectional epoxy-matrix-65 vol% fiber composites. Fibers used are indicated near points. Same properties for steel and aluminum are included for comparison.

the figure are the performance indices of materials with the least weight for stiffness (modulus) and for failure (strength) of the materials. Thus, we see the advantage of composite materials in lightweight structures, such as airplanes, compared to metals.

Having chosen the constituent materials, we can also engineer a composite in order that the length of the fibers are oriented along the maximum stress of the loading mode. The design or the engineering of a continuous fiber composite involves the *mechanics of laminate composite materials,* which cannot be covered in this text. It is sufficient only to illustrate simple examples. Thus, if a structure is stressed in pure tension, then the alignment of all the fiber along the length of the structure is sufficient. If the structure is stressed in torsion, the alignment of fibers ±45° from the length (axis) of the structure is required. If the structure is stressed transversely, the fibers must be aligned 90° to the axis of the structure.

The performances of continuous fiber-reinforced plastics/composites have been illustrated in Fig. 15-50. The results from Fig. 15-50 stresses the importance of knowing the loading mode of a component during the design stage in order to engineer the composite. However, in the absence of the knowledge of the loading mode, we can safely use a quasi-isotropic structure. However, the quasi-isotropic configuration (see Fig. 15-47) has lower performance in the 0° and 90° directions but higher in the ±45° directions. Figure 15-61 is a plot of the same performance indices, as in Fig. 15-60, for the quasi-isotropic fiber reinforced epoxy matrix composites. The results from Figs. 15-50, 15-60, and 15-61 indicate that the quasi-isotropic laminates lose much of the dramatic increases in the performance indices over the traditional metals when the fibers are aligned according to the loading mode, for example, in the 0° direction for tensile stress.

When designing with laminate composites, we must also realize that they behave differently under stress than metals and other monolithic materials. The quasi-isotropic laminates respond to in-plane stress loading like a monolithic material, but the out-of-plane (through thickness) and often the shear properties are still poor. If not corrected and/or accounted for, these factors can cause delamination under compressive loading or inadequate out-of-plane load-carrying capabilities. Other laminates

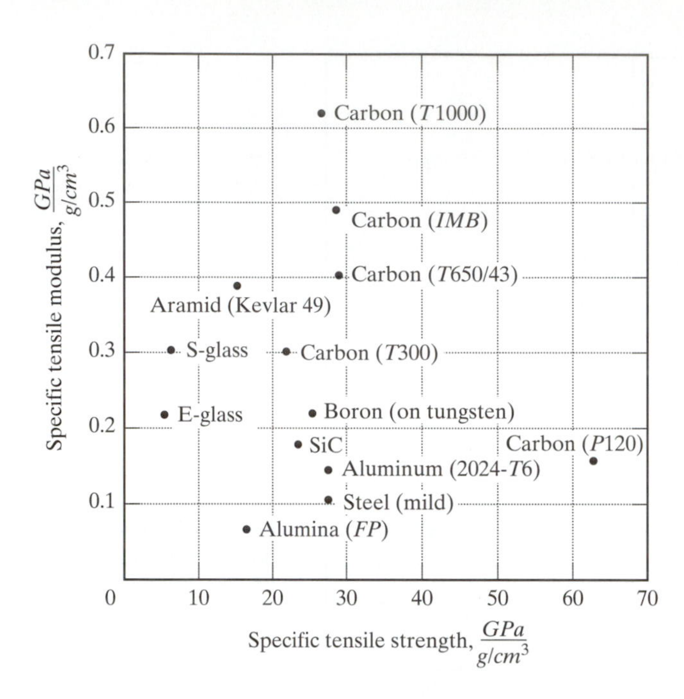

Fig. 15-61 Specific tensile modulus versus specific tensile strength for various quasi-isotropic epoxy matrix-65 vol% fiber composites. Properties of steel and aluminum included.

may respond differently. For example, a simple tensile force on an angle laminate may induce unexpected shear, bending, and/or twisting of the laminate depending on the laminae orientation. These topics are all covered in the mechanics of laminate composites and should be referred to for proper design. In addition, the performance of the composite depends a great deal on the selected matrix, which will be discussed in the following section.

This brief discussion on design with plastics and reinforced plastics points the need for using design methods appropriate for plastics and composites. The designer needs to be receptive to the different methods of handling them. It is important to keep in mind that when designing with plastics or composites that the exact replica of the metal part need not be used.

15.13 Selection of Plastics

The selection of plastics for an engineering application is a formidable task, since there are more than 15,000 known (and growing) plastic materials. While Sec. 15.4 described the various types of plastics, each of the types or families of plastics has numerous members and the final selection eventually comes down to the specific properties of each of the members. For this, the designer will eventually have to rely on the supplier or the manufacturer to provide the data.

15.13.1 Guides to Selecting Plastics

Some of the relevant properties of thermoplastics and thermosets are given in Tables 15-3 and 15-7. The data from these tables and from the supplier is needed at the detailed design. However, at the conceptual stage, we need general overviews on the properties of plastics to guide us in the selection. Some of these overviews are given in Figs. 15-62, 15-63, 15-64, and 15-65 on mechanical properties, operating temperatures, chemical properties, and other miscellaneous properties of the different plastic materials.

15.13.2 Matrix Selection for Reinforced Plastics/Composites

Selection of the constituents of the composite material must consider all the available matrices (resins, metal, ceramics) and reinforcements. In considering these constituents, their compatibility with each other must also be considered, in addition to the service requirements and environments. First, the matrix and the fiber must bond. Second, the coefficients of thermal expansion of both matrix and reinforcement

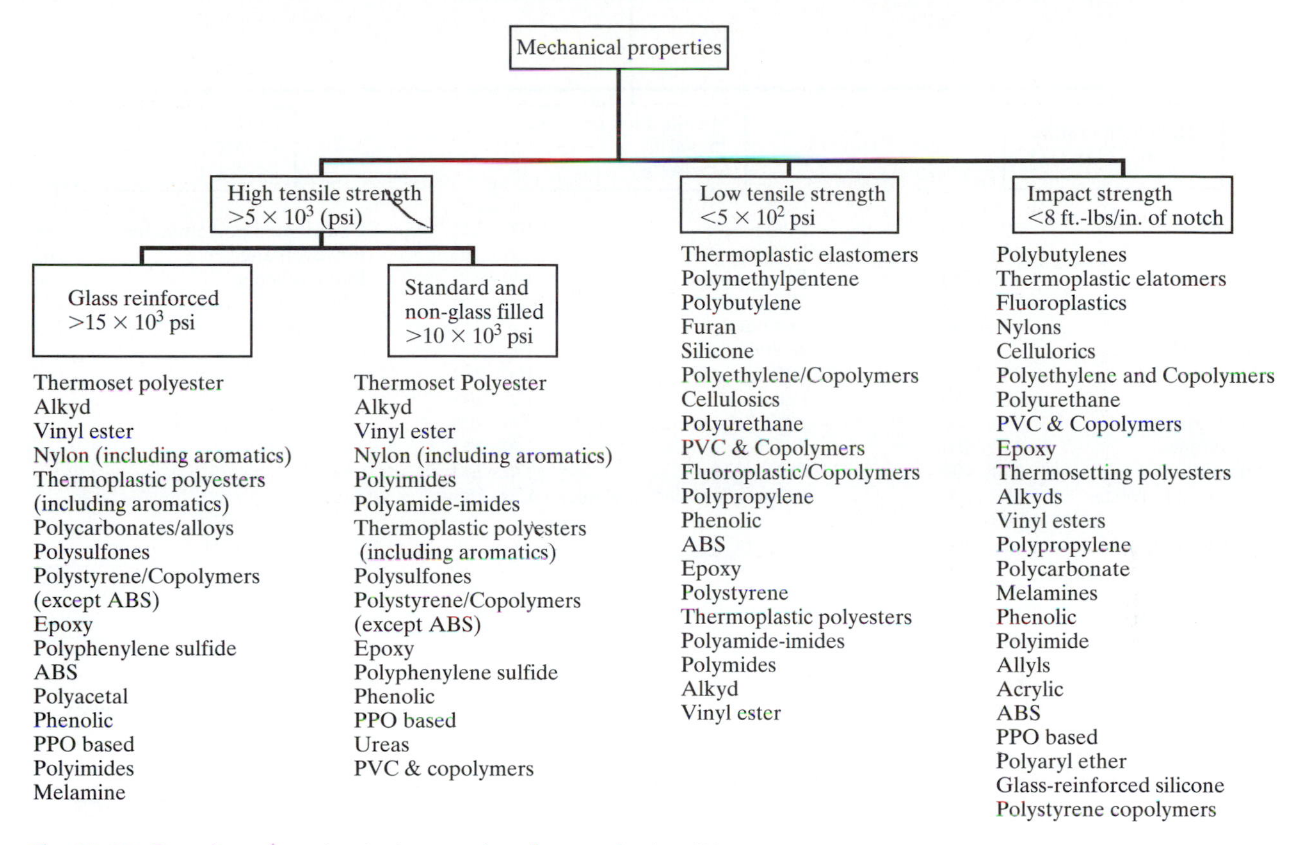

Fig. 15-62 Overview of mechanical properties of some plastics. [2]

must be very close to each other to avoid high thermal stresses that may lead to debonding at the fiber-matrix interface. Third, although the matrix and the fiber must bond, they shall not react with each other in order to maintain the integrity of each constituent. Fourth, for metal matrix composites, the possibility of producing a galvanic cell between the two constituents needs to be factored in.

The overriding consideration in material selection is the final composite material requirements (the physical, mechanical, thermal, chemical, and electrical). These requirements may automatically eliminate some materials and may point to the material family best suited for the design. In reinforced plastics/composites, the matrix resin, as far as the composite strength is concerned, provides only the interlaminar shear strength and is the weak link. For the most part, the matrix holds and aligns the fibers in place in each lamina or ply. Nonetheless, it is still an important constituent that must be selected judiciously since without it or with poor choice, there is either no composite or the composite performance will be poor. A wrong choice of matrix can lead also to processing problems.

Although thermoplastics are making inroads as matrices in reinforced plastics/composites, the dominant matrices are the thermosets for dimensional stability and high-temperature applications. Polyester thermosets are the most widely used in commercial non-critical applications, but in more critical applications, epoxies are used. Aside from being compatible with the fiber, some considerations for matrix (epoxy resin) selection are:

- *No By-Products During Curing.* The curing reaction of the matrix and the curing agent must not give off a by-product that can create defects in the composite.
- *Fully Cured.* The resin-curing agent combination must be capable of developing a fully cured

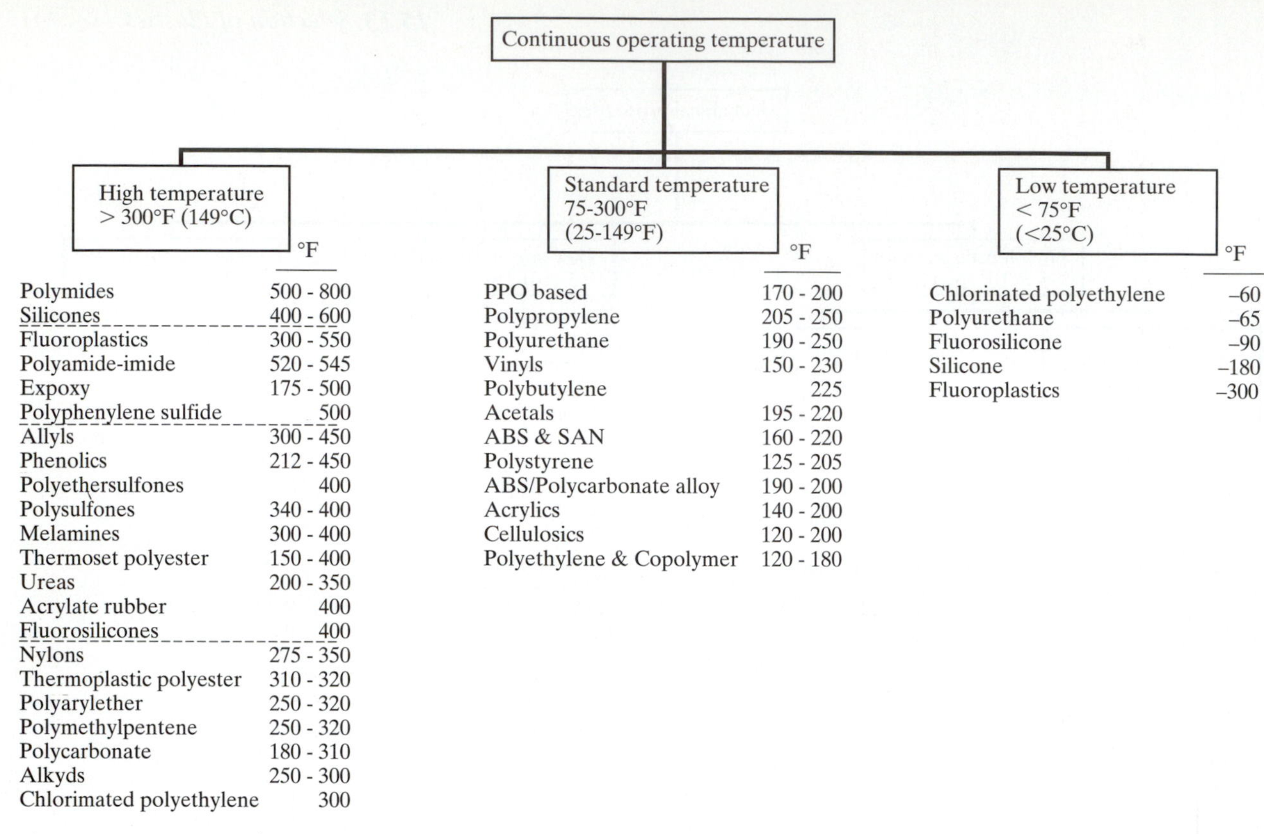

Fig. 15-63 Overview of temperature properties of some plastics. [2]

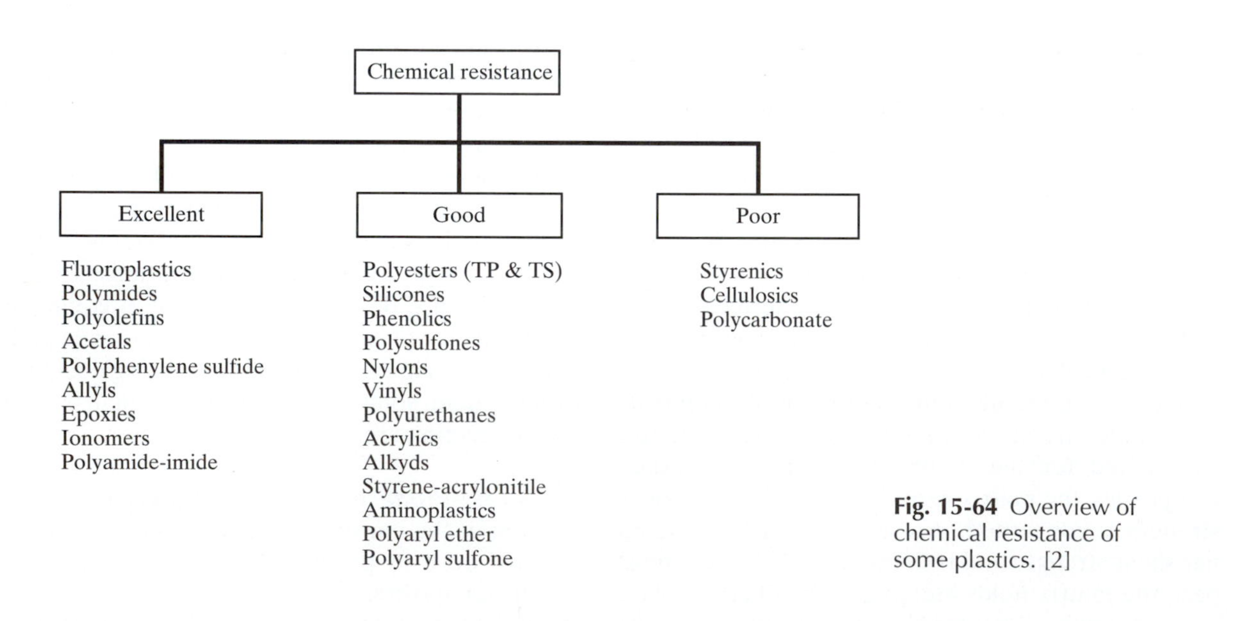

Fig. 15-64 Overview of chemical resistance of some plastics. [2]

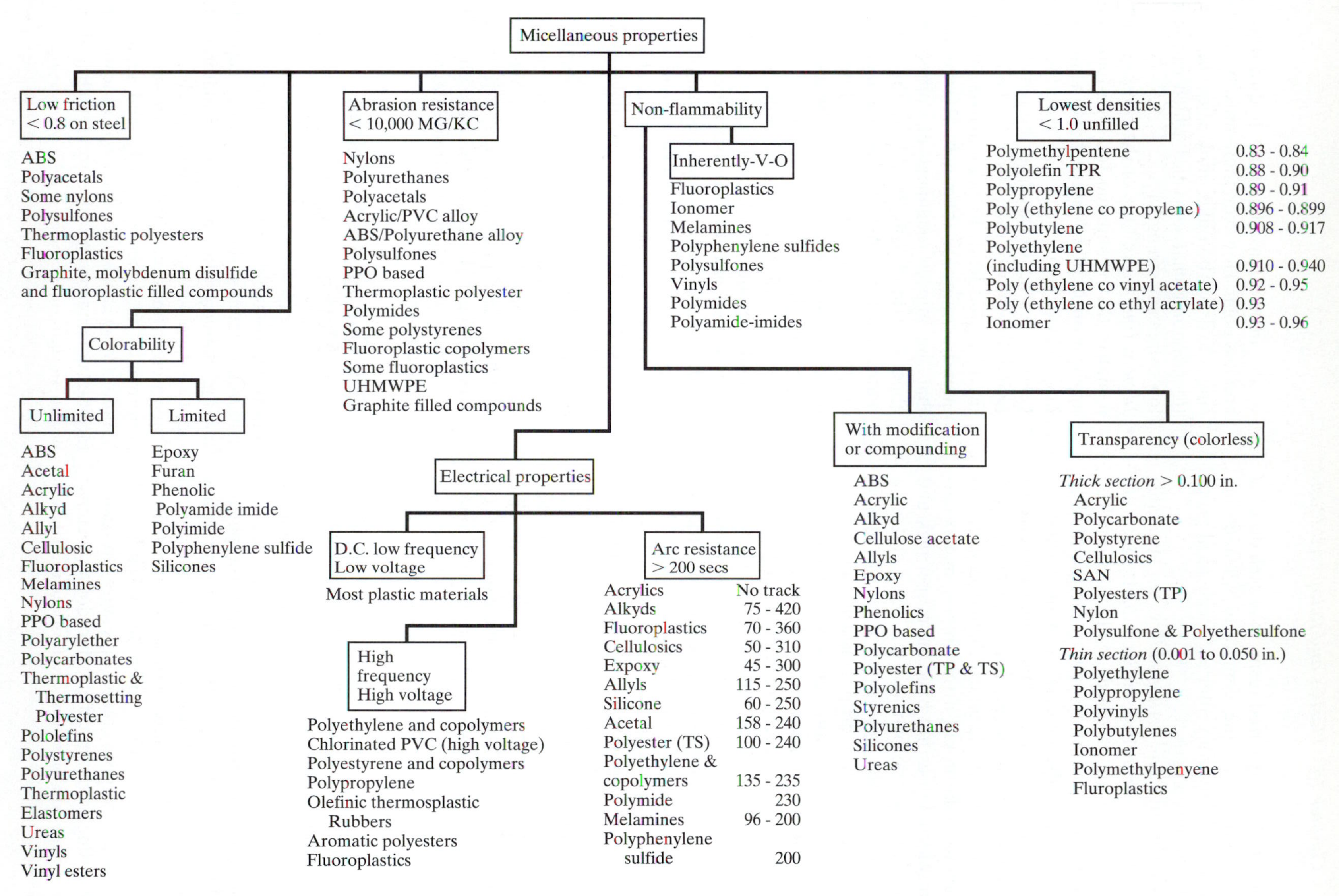

Fig. 15-65 Overview of miscellaneous properties of some plastics [2].

product. Partially cured resins may spell disaster during service of the composite when curing might ensue and may alter the mechanical properties of the composite.

- *Processability and Adaptability to Available Tooling.* As described in Sec. 15.11, there are many laminating processes to choose from. The matrix must be compatible with the tooling available and the processing of the composite. For example, a design engineer may conceive an outstanding component, and a manufacturing engineer can provide an excellent equipment on which to make it, but if the prepreg matrix system selected does not have the proper tack and drape on the tool surface, this often creates a nightmare in the production floor. The matrix viscosity must not drop drastically during the cure which might leave starved areas (no resin or void) in the cured composite. For example, the cure rate and the viscosity of the resin as a function of temperature are critical in autoclave processing. The viscosity of the resin must be sufficiently high when heated in order for the resin not to drain from the fibers, but it must not cure too quickly before autoclave pressure can be applied to the component before it gels and sets.
- *Glass Transition Temperature, T_g.* For elevated temperature application, T_g must be a criterion for selection. Selection of a much higher T_g resin than the service temperature, just to be on the safe side, is not advisable. This impacts the curing process by requiring higher cure temperature and longer cure time. The higher temperature can cause thermal stresses and strains that might be sufficient to induce microcracks in high-temperature, brittle matrix systems on cooling. These cracks impair the strength of the composite and may also introduce variability in strengths. Therefore, it is best to select materials on actual service temperature needs.
- *Moisture or Hygro-thermal Effects.* The permeability and the absorptivity of water in the matrix is an important criterion in selecting the matrix. Absorbed water tends to segregate at the fiber-matrix interface of the composite. When heated, the water vapor creates pressure to cause debonding of the interface. In order to decrease the absorption of water, the molecular structure of the matrix must be more hydrocarbon-like and less polar. However, there must be a balance since hydrocarbon-like molecules have poor fiber-bonding characteristics, which can lead to low composite strengths.
- *Impact Toughness.* One of the consequences of a matrix that is suitable for high-temperature application is that it must have a high cross-link density in its structure and, therefore, it becomes brittle. In this case, the tensile elongation or ductility of the matrix can be lower than the fiber and leads to poor matrix and, subsequently, composite toughness. Results indicate that the impact toughness of composites are about 20 J/m (5 ft-lb_f/ft). Above this value, the matrix fracture energy must be increased to improve the composite toughness. Impact toughness of the matrix can therefore be used as a criterion for selection.
- *Long Term Effects.* Long-term effects of environmental exposure on the matrix dictate the long term performance of the composite. Materials' vendors data such as heat resistance, fatigue, creep, thermal shock, abrasion, and electrical properties; resistance to solvents and corrosive chemicals, stain, and ultraviolet light; and other properties must be consulted.
- *Shelf Life.* On the eventual processability of the matrix, it is also important to determine the effect of storage on the shop floor. Will the tack and drape of a prepreg change? Will the curing process remain the same? There are physical and chemical tests to help determine the answers to these questions.

15.13.3 Composite Selection Methodology

Selecting a composite material to use for a component of an assembly starts by matching the component requirements, the manufacturing capabilities, and the characteristics of available materials from suppliers. Component requirements include the size, shape, weight, finish, cost, volume, environment, loads, performance, fit, function, quality, and repairability. Manufacturing capabilities focus on fabrication space, facilities, equipment, and labor force; technical expertise in design, analysis, test, and processing; quality control, materials management, specifications, and procurement policies. Suppliers' materials characteristics include properties, dimensions, and performance of the most advanced materials; lead time and cost; quality level, and general manufacturing processes. The design engineer must blend all the above into a material system design that will produce a component with the desired performance, quality, repairability, cost, and weight within a competitive time period.

Because the performance of the reinforced plastics/composite is limited by the performance of the matrix, selection of the composite material essentially comes down to the choice of the best plastic matrix. The guides in the last section may be used in the conceptual and embodiment stages. In the detail stage, we need to rate the various materials according to the design requirements and according to the products that they produce. We shall discuss two methods of evaluating the materials—one based simply on the design or service requirements and the other, based on a more elaborate method that includes among others the manufacturability, availability, and visual appearance of the product or composite to allow for a more objective decision making.

Selection based on Service Requirements. An example of the manner in which the final material selection is made starts from a table of the relative ratings of the properties of the matrices of the fiberglass reinforced plastics shown in Table 15-23. The relative ratings in Table 15-23 are usually supplied by material suppliers and those in Table 15-23 are from ICI-LNP. Some of the plastic matrices are listed in the left column and each of their properties such as strength and stiffness, toughness, and so on, that is used as a design criterion is rated. The large numbers refer to the ratings between the families or classes of plastics such as styrenics, olefinics, nylons, et cetera and the small numbers refer to the ratings between the members within a family. The most desirable rating is one and the least desirable is six.

Then using a blank table such as that shown in Table 15-24 as the worksheet, the selection proceeds in the following manner:

1. In the "Design Criteria" row, we cross the material properties (columns) that are required of the part or component being made.
2. For each of the properties (columns) selected, transfer the large bold numbers from the ratings table, Table 15-23, to the worksheet. For example, if strength and stiffness is required, then the large bold numbers 3, 5, 1, 3, 2, and 6, from Table 15-23 to Table 15-24 under the column "strength and stiffness," top to bottom.
3. Add the large numbers for each row (i.e., for each group of plastics, to find the group with the lowest-point subtotal, which will be the best for the given application on the basis of performance.
4. Add the cost factor and find the group with the lowest total points, which will be the best for the application on a cost-performance basis.
5. Once the group is selected, the procedure is repeated for the plastics within the group to find the material with lowest point total as the best for the application on a cost-performance basis.

Examples of the selection process are given in the Case Studies in Chapter 16.

More Elaborate Method of Selection. A more elaborate method of selecting the plastic matrix is a two-step process that is illustrated with Tables 15-25 and 15-26.

With Table 15-25, the first step is to evaluate each material based in its properties listed in the first column. The properties (selection parameters) of each candidate material is given a rating from 1 to 10, with 10 the highest and the ideal rating. A zero rating means the property is unacceptable. In contrast to the simple model, all the properties are rated even though they may not be one of the major design requirements. An ideal material will have a total of 100 points in the "evaluation rating" column for the ten properties listed. The importance of each of the properties for a particular application is given a priority rating, also from 1 to 10. An average of the priority ratings is obtained and for the example used in Table 15-25, this is 8.0. This average priority rating times the 100 total for the ideal material yields the maximum points obtainable for the application. The total points of each material is obtained by multiplying the rating of each property by its priority rating and then summing all the products to give the total points. The sum is divided by the maximum total ranking to get a percentage ranking.

In most selection processes, at least 10 materials are evaluated simultaneously. The best material is

Table 15-23 Ratings of design properties of some thermoplastic plastics (from ICI-LNP).[2]
(1-most desirable; 6-least desirable; large numbers for families of plastics; small numbers for plastics within group)

G/R Resin Groups \ Design Criteria	Strength and Stiffness	Toughness	Short-Term Heat Resistance	Long-Term Heat Resistance	Environ-mental Resistance	Dimensional Accuracy in Molding	Dimensional Stability	Wear and Frictional Properties	Cost
Styrenics	**3**	**6**	**6**	**6**	**6**	**1**	**5**	**6**	**2**
ABS	2	1	1	1	1	3	2	3	3
SAN	1	2	2	2	2	1	1	1	2
Polystyrene	3	3	3	3	3	2	3	2	1
Olefins	**5**	**4**	**4**	**5**	**3**	**5**	**5**	**3**	**1**
Polyethylene	2	2	2	2	2	1	1	2	1
Polypropylene	1	1	1	1	1	1	1	1	2
Other Crystalline Resins	**1**	**1**	**2**	**4**	**4**	**4**	**4**	**2**	**3**
Nylons									
6	2	2	2	2	5	1	4	3	1
6/6	1	3	1	1	4	2	3	2	2
6/10, 6/12	3	1	3	3	3	2	2	3	4
Polyester	4	4	2	1	2	2	1	4	1
Polyacetal	5	5	5	2	1	3	2	1	1
Arylates	**3**	**2**	**3**	**3**	**5**	**1**	**2**	**4**	**4**
Modified PPO	4	3	4	4	3	4	4	4	1
Polycarbonate	2	1	3	3	4	1	3	3	2
Polysulfone	2	2	2	2	2	2	2	1	3
Polyethersulfone	1	3	1	1	1	3	1	2	4
High Temp Resins	**2**	**4**	**1**	**1**	**2**	**4**	**1**	**4**	**5**
PPS	1	2	2	2	1	1	2	2	1
Polyamide-imide	2	1	1	1	2	2	1	1	2
Flourocarbons	**6**	**2**	**2**	**1**	**1**	**6**	**6**	**1**	**6**
FEP	2	1	2	1	1	2	2	1	2
ETPE	1	2	1	2	2	1	1	2	1

Table 15-24 Blank selection worksheet for the selection of a thermoplastic matrix resin.[2]

Material Characteristics	Strength and Stiffness	Toughness	Short-term Heat Resistance	Long-term Heat Resistance	Environ-mental Resistance	Dimensional Accuracy in Molding	Dimensional Stability	Wear and Frictional Properties	Point Subtotal	Cost	Point Total
Design Criteria / G/R Resin Groups											
Styrenics ABS SAN Polystyrene											
Olefins Polyethylene Polypropylene											
Other Crystaline Resins Nylons 6 6/6 6/10, 6/12 Polyester Polyacetal											
Arylates Modified PPO Polycarbonate Polysulfone Polyethersulfone											
High Temp. Resins PPS Polyamide-imide											
Flourocarbons EEP ETEE											

Table 15-25 Evaluation of material based on its properties (see text for discussion).[4]

Selection parameters (Material A)	Evaluation rating (1-10)	×	Priority factor (1-10)	=	Total ranking
Tensile strength at room temperature(a)	0(b)		10		0
Compressive strength at temperature	2(c)		10		20
Elastic, tensile, and compressive moduli	2(c)		10		20
Density(a)	10(c)		10		100
Thermal conductivity(a)	5		5		25
Coefficient of thermal expansion(a)	10		5		50
Specific heat	10		5		50
Service temperature(a)	0(b)		10		0
Volume reisistivity	5		5		25
Dielectric constant at temperature(a)	10(c)		10		100
Scoring	$\frac{\text{Material A}}{\text{Ideal material}}$ 100 max···		8.0 average max		$\frac{\text{390 (48.7\% of ideal material)(d)}}{\text{800 max ranking}}$

(a) Minimum required value of property must be available to be acceptable. (b) Unacceptable. (c) Acceptable. (d) Material unacceptable because tensile strength and service temperature do not meet the minimum required values.

Table 15-26 Evaluation of a material based on its manufacturability and the quality of product produced.[4]

Selection parameters (Material B)	Evaluation rating (1-10)	×	Priority factor (1-10)	=	Total ranking
Good visual apperance	9		6		54
Compatibility in material assembly	5		7		35
Ease of fabrication/good production rate	9		8		72
Low weight/compact shape	10		10		100
Low cost	10		10		100
Good quality	10		8		80
Good maintainability/excellent performance	5		8		40
Good properties at temperature (mechanical, thermal, physical, and electrical)	6		8		48
Availability of process facilities	0		6		0
Material availability	8		7		56
Scoring	$\frac{\text{Material B}}{\text{Ideal material}}$ 100 max		7.8 average max		$\frac{\text{585 (75\% of ideal material)(a)}}{\text{780 max ranking}}$

(a) Other materials need to be considered to meet component requirements.

the one that has the highest number of points for the 10 properties and meets all the minimum requirements. However, a good goal for selection is for the material to achieve 90 percent of the ideal material evaluation rating and of the total maximum ranking and to show no zero rating in the table.

In the example illustrated in Table 15-25, four of the properties—tensile strength, density, coefficient of thermal expansion, and dielectric constant—had minimum required values that must be attained. We see that the tensile strength and the service temperature requirements are not met and are therefore given zero ratings; these ratings lead to an automatic rejection for this material.

Candidate materials with the highest points and percentages after evaluation of their properties are

Table 15-27 ASTM D 4000 Line Callout.

1	2	3	4	5	6	7
Broad Generic Type	Specific Grp Cl Gr xx x x	Reinforcement	% Reinforcement	Table	Cell Requirements xxxxx	Suffix

further evaluated according to Table 15-26. The selection parameters in Table 15-26 form a part of the management decision making in material selection based on the fabricability and the performance of the component. Fabricability factors include availability of facilities and equipment, availability of the candidate materials to meet production quantities and schedules, ease of fabrication at a production rate according to a sales program plan, and meeting quality control requirements including visual appearances with low reject rates. Design and management objectives toward a successful and long-term production program are based on the performance of the component that includes low weight/compact shape, compatibility in assembly, low net cost (including tooling and facilities setup charges), good maintainability with long mean time between failure (MTBF), and excellent performance in customer's application.

As with Table 15-25, Table 15-26 is applied to each candidate material, which is evaluated (rated from 1 to 10) in terms of each of the selection parameters and with each selection parameter given a priority ranking. The selection criteria are the same as those applied for Table 15-25 and we see that the example material rated in Table 15-26 will be rejected because it has a zero rating and both the evaluation and the total ranking percent are less than 90 percent. Thus, we need to consider other materials.

15.14 ASTM D 4000 Line Callout For Plastic Materials

Once the plastics are selected, the specifications of the plastics are given in the ASTM D 4000 Line Callout system that is shown in Table 15-27. For full details, please refer to ASTM D 400 standard. The line callout specification consists of seven groups of letters and/or numbers shown Table 15-27. These groups are:

1— This consists of two or more letters identifying the broad generic class of plastics according to the standard terminology for plastics in ASTM D 1600, shown in Table 1 of ASTM D 4000.

2— This consists of four xxxx digits, the *first two digits* specifies the *group* (grp), the *third digit* specifies the *class* (cl), and the *fourth digit* specifies the *grade* (gr).These digits are found in the specific standard for the plastic. For those plastics with no standards, the four digits are 0000.

3— For reinforced plastics, a letter identifies the reinforcement or filler according to Table 2 of ASTM D 4000. C is for carbon and graphite fiber; G is for glass; L is for lubricants, for example, PTFE, graphite, silicon, and molybdenum disulfide; M is for mineral fillers; and R is for combinations of reinforcements and fillers.

4— Two digits specifies the volume percent of reinforcement and/or filler.

5— One letter refers to a Cell Table that lists number codes of the requirements needed and are indicated as the five digits in 6 in the designation order given. Tables A and B are included in the standards for those materials with standards. Cell Tables C through H are included in ASTM D 4000 for those materials with no standards. The specific table to be used is indicated in Table 1 of ASTM D 4000.

6— Five digits specify the range of properties required and are taken from the cell table indicated in 5.

7— A suffix is used to supersede or supplement the property or cell table requirements. This consists of two letters and three digits. The first letter indicates the special requirement; the second leter indicates the condition or test method or both; and the three digits indicate the specific requirement.Table 3 of ASTM D 4000 illustrates the suffix letters that can be used, as well as how the three-digit code is obtained.

Summary

Polymers are large macromolecular chains that are formed by joining a large number of small units called mers. There will be a number of these chains with different lengths and molecular weights and polymers, therefore, have an average molecular weight. The distribution of the molecular weights and the structure of the polymeric chains determine the properties of polymers. The chains may be linear, branched, and cross-linked. They may contain only one mer, or two or more mers. The "backbone" may consists of C-C bonds only or may contain O, N, and Si. Just like metals, polymers are seldom used pure. When they are mixed or alloyed, they are called plastics or resins.

There are two types of plastics, according to their thermal deformation characteristics. Thermoplastics are plastics that can be thermally deformed repeatedly and reformed into different shapes. Thermosets are plastics that cannot be thermally deformed repeatedly or reformed into different shape after they are formed into a shape. (Thus, thermosets are nonrecyclable and are either disposed as solid waste or burned to recover some of the energy.) This difference in deformation behavior arises from the absence or presence of cross-linking of the polymeric chains. The chains in thermoplastics are generally not cross-linked or may be very lightly cross-linked; the chains in thermosets are heavily cross-linked. During deformation, the chains in thermoplastics can slide past each other while the chains in the thermosets cannot move. Because of this, thermoplastics are relatively weak and they creep, while thermosets can be hard, rigid, resistant to creep and deformation, and have dimensional stability.

Thermoplastics are further classified as commodity or engineering thermoplastics. The commodity thermoplastics are generally used in no-load or very low-load applications, while the engineering thermoplastics can carry loads for a long period of time and can compete with metals in some structural applications. About 90 percent by weight of all plastics are commodity thermoplastics, and only 10 percent are engineering resins that include both thermoplastics and thermosets. Only a few of the thermoplastics are used as engineering resins, but most of the thermosets can be used as engineering resins.

Compounding is the general term used in mixing two or more pure polymers or plastics and leads to literally thousands of alloys or blends of plastics. Additives and fillers are compounded with plastics to change and to improve their physical, mechanical, and processing properties. These may be utilized as extenders, plasticizers, heat stabilizers, antioxidants, UV light absorbers, antistatic agents, coupling or sizing agents, flame retardants, blowing agents, lubricants, and colorants.

When the fillers are particles or fibers, the resulting material is a composite; in this case, it is called reinforced plastics. Reinforced plastics is just one of the many composites that include metal matrix composites and ceramic matrix composites. Reinforced plastics is polymer matrix and constitutes the widely used composite. The most significant reinforced plastics are the fiber composites, which may be discontinuous or continuous fiber composites. The discontinuous fiber composites have short-chopped fibers as fillers, while the continuous fiber have fibers as long as the part. The discontinuous fiber composites are commonly called commodity composites, while the continuous fiber composites are called advanced composites. The most commonly used fibers are glass fibers. For advanced composites, the two commonly used ones are carbon/graphite and Kevlar fibers. The long fiber composites can be laid as laminates and can be engineered to obtain strength properties in the direction(s) desired. The orientation of the fiber with respect to the length and width of the laminate is given in angles and the number of laminae in this orientation is indicated as a subscript. The resulting laminate is given a code that includes all the ply (lamina) orientations and numbers of plies (laminae) in each orientation.

Selection of the matrix material for reinforced plastics is very important because the matrix is the weak link. The most common matrix for the chopped fiber composites are polyesters, while that of the continuous fiber are epoxies. Summary overviews on mechanical properties, operating temperatures, chemical properties, and other miscellaneous properties for different plastics can be used as guides. Two methods for the selection of the matrix material are given and two case studies are given in Chap. 16. These methods consider evaluation ratings of many factors, including ease of manufacture, storage, cost, and others, in addition to the mechanical and physical properties. The candidate materials with the highest total evaluation ratings are selected.

In addition to the physical and mechanical properties used for design of metallic and ceramic components, we have also to consider water absorption and water transmission for plastic components. This is especially true for composite materials where hygrothermal stresses can cause delamination at the interfaces. In addition, for composites, the coefficient of linear thermal expansion of the polymer matrix must be carefully matched with the fibers to minimize delamination due to thermal fatigue. If we select engineering thermoplastics for structural applications, the laboratory short-term data given by suppliers or in handbooks may not be applicable. The viscoelastic properties of creep and stress relaxation must be considered for long-term performance under load.

References

1. *Plastics Engineering Handbook of the Society of Plastics Industry (SPI), Inc.,* Edited by Michael L. Berins, Fifth Edition, Van Nostrand Reinhold, New York, 1991.
2. D. V. Rosato, D. P. Di Mattia, and D. V. Rosato, *Designing with Plastics and Composites: A Handbook,* Van Nostrand Reinhold, New York, 1991.
3. M. F. Ashby and D. R. H. Jones, *Engineering Materials 2,* 1st ed., Pergamon Press, 1986.
4. *Engineered Materials Handbook.* Vol. 1, Composites, ASM International, 1987.
5. B. D. Agrowal and L. J. Broutman, Analysis and Performance of Fiber Comoposites, Second Edition, John Wiley, N. Y., 1990.

Terms and Concepts

Addition polymerization
Additives and fillers
Amorphous polymers
Angle ply laminate
Antisynergism
Apparent creep modulus
Branched molecules
Ceramic matrix composite
Co-polymerization
Commodity thermoplastics
Composites
Compounding
Condensation polymerization
Creep of plastics
Critical fiber-volume fraction
Cross-linked molecules
Cross-ply laminate
Crystalline polymers
Curing
Degree of polymerization
DTUL — deflection temperature under load
Elastomers
Engineered composites
Engineering thermoplastics
Epoxies
Fabrication of plastics
Fabrication of reinforced plastics
Fatigue of polymers
Fiber composite
Glass fiber
Graphene
Graphite fiber
HDPE — high density polyethylene
Hybrid composite
Hygro-thermal effects
Ionomers
Kevlar fiber
Laminar composite
Laminate codes
Laminates
LDPE—low-density polyethylene
Line callout for plastics—ASTM D 4000
Linear molecules
Liquid crystalline polymers
Longitudinal properties
Long-term mechanical properties
Matrix
Mer
Metal matrix composite
Molecular weight distribution
Number average molecular weight
Particulate composite
Plastics
Ply
Polyesters
Polymer alloys

Polymer blends
Polymerization
Polymer matrix composite
Polymers
Prepregs
Quasi-isotropic laminate
Rearrangement mechanism
Reinforced plastics
Resins
Ring-opening mechanism
Rule of mixtures
Selection methodologies for plastics
Short term mechanical properties
Strain relaxation
Stress relaxation
Symmetric laminate
Synergism
Thermoplastic polymers
Thermoset polymers
Transverse properties
Unidirectional fiber composite
Vulcanization
Water absorption
Water transmission
Weight average molecular weight

Questions and Exercise Problems

15.1 Differentiate the terms polymers, plastics, and resins.

15.2 What is polymerization? Differentiate addition from condensation polymerization.

15.3 What are the three mechanisms of addition polymerization?

15.4 Fig. 1-15 in Chap. 1 shows that the angle between the C-C covalent bonds is 109.5°. Determine the length of a polyethylene polymer chain stretched end to end when the degree of polymerization is 10,000. Note: You need to use data from Table 1-4.

15.5 Determine the length of a polypropylene polymer chain stretched end to end when the degree of polymerization is 10,000. What is the average molecular weight of the polypropylene?

15.6 Ordinarily, the C-C backbone in a polymer chain is not arranged linearly, as indicated in the two previous problems. Instead, the backbone coils randomly and the average distance of a randomly coiled chain end to end is given by

$$\text{Random end-to-end distance} = R_o\sqrt{N}$$

where R_o is the interatomic C-C bond and N is the total number of bonds. Determine the random end-to-end distance for the polyethylene molecule in Prob. 15.4.

15.7 Referring to Fig. 15-9, determine the mer weight of polymethyl methacrylate (PMMA).

15.8 What would be the number average degree of polymerization if the number average molecular weight of PMMA is 125,000 g/g-mole?

15.9 For a degree of polymerization of 750, how much hydrogen peroxide is needed to polymerize 1 kg of methyl methacrylate? How much is needed if the degree of polymerization is 3,000 for the same weight of methyl methacrylate? Assume 100 percent efficiency in reaction.

15.10 The polymerization of polyethylene produced the results shown in Table P15.10. From this data, determine (a) the number average molecular weight, (b) the weight average molecular weight, and (c) the number average degree of polymerization.

15.11 The polymerization of polymethylmethacrylate (PMMA) produced the results in Table P15.11. From these results, determine (a) the number average molecular weight, (b) the weight average molecular weight, and (c) the number average degree of polymerization.

15.12 SAN (styrene-acrylonitrile) is the co-polymerization product of acrylonitrile and styrene. If 5 kg of each were polymerized, determine (a) the average mer

Table P15.10 Number and range of molecular weights of polyethylene molecules.

Number of molecules	Range of Mol. Weights	Number of molecules	Range of Mol. Weights
1,000	0–5,000	22,000	25,000–30,000
2,000	5,000–10,000	30,000	30,000–35,000
3,000	10,000–15,000	15,000	35,000–40,000
4,000	15,000–20,000	5,000	40,000–45,000
15,000	20,000–25,000	3,000	45,000–50,000

Table P15.11 Number and range of molecular weights of PMMA molecules.

Number of molecules	Range of Molecular Wts.	Number of molecules	Range of Molecular Wts.
5,000	40,000–50,000	10,000	70,000–80,000
10,000	50,000–60,000	8,000	80,000–90,000
15,000	60,000–70,000	2,000	90,000–100,000

weight, and (b) the number average molecular weight of the polymer when 10 grams of the hydrogen peroxide initiator/terminator were consumed.

15.13 If 2 kg of acrylonitrile and 4 kg of styrene were polymerized, how much hydrogen peroxide initiator-terminator is required to produce a number average molecular weight of 80,000 and the efficiency of the hydrogen peroxide is 90 percent?

15.14 If a 10 kg of the SAN (styrene acrylonitrile) co-polymer is desired with a number average molecular weight 146,800 and a degree of polymerization of 2,000, determine (a) the amounts of styrene and acrylonitrile needed, and (b) the amount of hydrogen peroxide initiator-terminator required.

15.15 If a 10 kg of a B-S (butadiene styrene) co-polymer is produced with a number average molecular weight of 212,000 and a degree of polymerization is 3,000, determine (a) the amounts of butadiene and styrene required, and (b) the amount of sulfur needed to attain 50 percent vulcanization.

15.16 A sand mold for metal casting can be thought of as a packing of sand particles impregnated with a polymeric material. Consider a sand mold with a bulk density of 1.45 g/cm^3 and the true density of the sand is 2.20 g/cm^3. Assume that the pores in the sand mold are open and are fully impregnated with a polycarbonate with a density of 1.2 g/cm^3. Determine (a) the porosity in the sand mold, and (b) the density of the sand mold after impregnation.

15.17 Clay with a density of 2.4 g/cm^3 is used as an extender in a polyester matrix. What is the density of a composite with 70 weight percent clay in polyester? The density of polyester is 1.22 g/cm^3.

15.18 Prove that the maximum fiber volume fraction in a unidirectional fiber composite is 0.907.

15.19 A unidirectional reinforced plastics contains 70 weight percent E-glass fibers with average diameter of 50 μm. Determine the average distance between the fibers in the composite, assuming they are arranged in a hexagonal pattern.

15.20 A graphite-HM fiber epoxy matrix composite is being considered to replace a 7075-T6 aluminum alloy with a tensile strength of 585 MPa. Considering both the modulus and the strength properties, determine the minimum number of 10-μm graphite-HM fibers per cm^3 of the epoxy matrix composite that must be used. How much weight savings can be obtained?

15.21 Consider Problem 15.20 again, but now the material to be replaced is a quenched and tempered 4340 steel with a tensile strength of 1500 MPa. The density of 4340 steel is 7.85 g/cm^3 and its modulus is 205 GPa

15.22 Boron fiber reinforcements are produced by chemical vapor deposition of boron on tungsten wire substrates. If the tungsten wire is 5 μm and the final boron wire is 15 μm, determine the volume and weight percents of boron in the wire. The properties of boron and tungsten are:

	Density, g/cm^3	Tensile Modulus, GPa	Tensile Strength, MPa
Boron	2.36	379	3,450
Tungsten	19.4	407	4,000

15.23 If the 15-μm boron fiber described in the previous problem is used as the reinforcement for an epoxy matrix composite, determine the specific modulus and specific strength of the composite.

15.24 Differentiate the following polymers: (a) thermoplastics versus thermosets, (b) crystalline versus amorphous, (c) alloys versus blends, and (d) commodity versus engineering polymers.

15.25 What is the difference between graphene and graphite?

15.26 Describe the processing of graphite from PAN and mesophase pitch precursors.

15.27 Describe and differentiate the phenomena of stress relaxation and strain relaxation in plastics.

15.28 Differentiate between unidirectional, cross-ply, and angle-ply laminates.

15.29 What is a symmetric laminate? A quasi-isotropic laminate?

16
Materials Selection Case Studies

16.1 Introduction

The overall objective in engineering design is to produce a structure or component that will perform the desired function efficiently and safely with the minimum cost. The latter includes the material cost, and the fabrication, operation, and the maintenance of the component or structure. In the final analysis, we design structures or components to prevent them from failing during service in order to avoid cost in lives, primarily, and in downtime of the operation. As we noted in Sec. 11.5, the causes of failure may be deficiencies in design, in materials selection, in processing of the material, in fabrication, or in the assembly of the structure or component. Clearly, we see that in order to obtain the optimum performance of the structure or component, it entails doing all of the above factors correctly and of these factors, correct design and materials selection are paramount. The best design with a wrong material or the best material with a wrong design will not produce the desired optimum performance.

We shall limit our discussion to the optimum selection of materials. To make the optimum selection of materials, we must know the detailed characteristics of the materials in terms of their properties (mechanical, physical, and chemical), their fabricability (weldability, formability, and machinability), and their cost. Most of the characteristics of various materials have been discussed in the previous chapters. We shall now discuss in greater detail some case studies to illustrate materials selection. For the most part, the cases were actual design projects to demonstrate the thought processes that went into the materials selection. There are examples for steels, aluminum, structural ceramics, and reinforced plastics. We shall also illustrate how concurrent engineering and life-cycle analysis or assessment are conducted. A final case illustrates how a poor material selection can be disastrous.

16.2 Steels

Steels are by far the predominant materials used for a lot of engineering structures and components. We shall first discuss constructional and structural steels, and then heat-treated steels.

16.2.1 Structural Steels

Within this class are the conventional and the high-performance structural steels. These steels constitute the materials used in the construction of buildings, roads and bridges, storage and pressure vessels, pipelines, automobiles, and the like. Because of the tonnages involved in these applications, there is an opportunity to save a lot of money with the proper selection of the steel.

A list of ASTM specified conventional structural steels, their grades, applications, compositions, and typical properties, is given in Table 16-1. These steels are processed either as hot-rolled, normalized, or quenched and tempered. They basically contain carbon and manganese and are hot-rolled, with no controls of temperature and deformation, as indicated by the AR process in Fig. 12-1[1a][1]. The structural steel specified for general application is ASTM A36 with a minimum 250 MPa (36 ksi) yield strength in the as-rolled condition. As seen in Table 16-1, to attain this minimum yield strength for thicker gauges of A36, or to increase the strength of steels in the as-hot-rolled condition, the carbon or manganese content, or both, are increased with or without small additions of some elements such as nickel, chromium, and vanadium. This is illustrated, for example, in the compositions of A36 and A588 in Table 16-2[1a]. Minimum yield strengths up to 345 MPa (50 ksi) may be obtained in this manner in the as-rolled or normalized condition without much adverse influence on the fracture toughness and weldability of the steel. However, increasing the strength of these hot-rolled conventional steels to a 485 MPa (70 ksi) minimum yield strength by increasing the carbon and alloy contents will adversely affect their fracture toughness, ductility, and weldability, as indicated in Sec. 12.2.2 and 12.4.1. Thus, the hot-rolled low-carbon and low-manganese steels with low-strength after hot-rolling are quenched and tempered, shown as the RQ-T process in Fig. 12-1, to attain strengths greater than 345 MPa (50 ksi). This is illustrated in Table 16-3[1a] for a 485 Mpa (70 ksi) minimum yield strength ASTM A852 steel that is quenched and tempered and has the same composition as the as-rolled A588, a 345 MPa (50 ksi) minumum yield strength. ASTM A514-F in Table 16-2 is a 100 ksi minimum quenched and tempered steel with a higher alloy content. The conventional steels with minimum yield strengths from 50 to 100 ksi are called high-strength low-alloy (HSLA) steels, which we shall compare with the high-performance microalloyed HSLA steels.

In an executive report titled, "High-performance Construction Materials and Systems: An Essential Program for America and Its Infrastructure" [cited in Ref. 1b] and prepared by the Civil Engineering Research Foundation (CERF), "a high-performance steel is one with the following desired properties:

- Strength, ductility, fracture toughness, fatigue resistance, corrosion resistance, fire resistance
- Formability, weldability, and improved weld joint performance
- Uniformity in properties through 'clean steel' technology and process control

It is anticipated that these high-performance steels can reduce the amount of material needed in construction and to reduce the fabrication costs by providing more efficient manufacturing processes."

The high-performance structural steels are the microalloyed high-strength low-alloy (HSLA) steels with lower carbon contents, usually less than 0.10%C, (some typical compositions are indicated in Table 12-12). The strength of these steels is achieved via controlled thermo-mechanical processsing as described in Sec. 8.6.2 and in Fig. 12-1 to produce either very fine ferrite grain sizes during air cooling or bainitic and/or martensitic structures after direct quenching or accelerated cooling after hot-rolling. Tempering or coiling of these steels also may induce some precipitation hardening when the microalloyed carbides or precipitate particles from other elements, such as copper, are formed. The direct quenching and tempering processes after controlled rolling, shown in Fig. 12-1, enable the direct production of high-strength steels up to 100 ksi minimum yield strength after hot-rolling, thus eliminating the extra heating (austenitizing) and quenching of the RQ-T process and thereby making the steels cheaper. To increase ductility, formability, and impact toughness in these steels, the sulfur content is reduced to less than 10 ppm by weight in composition and rare-earth or

[1]The number in brackets refer to the references at the end of the chapter.

Table 16-1 Specifications, grades and compositions of common structural steels.

ASTM-SPEC	Grade	Specification For	Chemical Composition, wt % C, max	Mn	Others	Y.S., Ksi min (MPa)	T.S., Ksi min (MPa)	Y.S./T.S., Ratio	% Elongation in 2″ (8″)
A36/ A36M-93a	Plates to 3/4″	General Applications	0.25	<1.0	0.40% Si, max 0.20% Cu	36 (250)	58–80 (400–550)	0.45–0.62	23(20)
	3/4″–1 1/2″		0.25	0.80–1.20	0.40% Si, max 0.20% Cu				
	over 4″		0.29	0.85–1.20	0.15–40% Si 0.20% Cu				
A131/ A131M-93a	A	Ships	0.23	more than 3 × % C	0.35% Si max	34 (235)	58 to 71 (400–490)	0.48–0.59	23(21)
	B		0.21	0.80–1.10	0.10–0.35% Si				
	CS (N)		0.16	1.00–1.35	0.10–0.35% Si				
	E(N)	N=normalized	0.18	0.70–1.35	0.10–0.35% Si				
	H.S. Grades		0.18	0.90–1.60	0.10–0.50% Si	46 (315)	68 to 85 (470–585)	0.54–0.68	22(19)
					0.50% Ni max 0.25% Cr, max 0.08% Mo, max	51 (360)	71 to 90 (409–620)	0.57–0.72	22(19)
					0.35% Cu, 1% V 0.059, Cb(Nb)	57 (390)	74 to 94 (510–650)	0.61–0.77	22(19)
A242/ A242M-93		Welded, Riveted or bolted construction for general purposes	0.15%	1.00% max	0.20% Cu, min	50(to 3/4″) (345)	70 (480)	0.71	21(18)
						46(3/4″–1 1/2″) (315)	67 (460)	0.69	21(18)
						42(1 1/2″–4″) (290)	63 (435)	0.67	21(18)
A529/ A529M-93	42	Riveted, bolted, or Welded Construction	0.27%	1.20%	0.20% Cu	42 (290)	60–85 (415–585)	0.49–0.70	22(19)
	50		0.27%	1.35%		50 (345)	70–100 485–690)	0.50–0.71	21(18)

(Continued)

Table 16-1 Specifications, grades and compositions of common structural steels. (*Continued*)

ASTM-SPEC	Grade	Specification For	Chemical Composition, wt % C, max	Mn	Others	Y.S., Ksi min (MPa)	T.S., Ksi min (MPa)	Y.S./T.S., Ratio	% Elongation in 2″ (8″)
A572/ A572M-93	42	Bridges, buildings and other	0.21	1.35%	0.40% Si max 0.05% Cb 0.02–0.15%V	42 (290)	60 (415)	0.7	24(20
	50	structures with good notch toughness	0.23	1.35%		50 (345)	65 (450)	0.77	21(18)
	60		0.26	1.35%		60 (415)	75 (520)	0.80	18(16)
	65		0.23 0.26	1.65%		65 (450)	80 (550)	0.81	17(15)
A573/ A573M-93	58	Structural Quality	0.23	0.60–0.90	0.10–0.35% Si	32 (220)	58–71 (400–490)	0.45–0.55	24(21)
	65		0.24–0.26	0.85–1.20	0.15–0.40% Si	35 (240)	65–77 (450–530)	0.45–0.54	23(20)
	70		0.27–0.28	0.85–1.20	0.15–0.40% Si	42 (290)	70–90 (485–620)	0.47–0.60	21(18)
A588/ A588M-93	A	Welded construction of bridges and	0.19% max	0.80–1.25	0.15–0.65% Si 0.25–0.40% Ni, 0.30–0.70% Cr,	50(<4″) (345)	70 (485)	0.71	21(18)
	B	buildings, corrosion substantially	0.20	0.75–1.35	0.20–0.50% Cu, 0.01–0.10% V	46(4″–5″) (315)	67 (460)	0.69	21
	C	better than A36	0.15	0.80–1.35		42(5″–8″) (290)	63 (435)	0.67	21
A633/ A633M-93	A	Steel plates for welded, riveted, or bolted	0.18	1.00–1.35	0.15–0.50% Si 0.05% Cb max	42 (290)	63–83 (430–570)	0.51–0.67	23(18)
	C	construction (Normalized condition)	0.20	1.15–1.50	0.15–0.50% Si 0.01–0.05% Cb	50 (<2.5″) (345)	70–90 (485–620)	0.56–0.71	23(18)
	D		0.20	0.70–1.60	0.15–0.50% Si 0.35% Cu 0.25% Ni, 0.25% Cr 0.08% Mo	46(2.5″–4″) (315)	65–85 (450–590)	0.54–0.71	23(18)

(*Continued*)

Table 16-1 Specifications, grades and compositions of common structural steels. (*Continued*)

ASTM-SPEC	Grade	Specification For	Chemical Composition, wt % C, max	Chemical Composition, wt % Mn	Chemical Composition, wt % Others	Y.S., Ksi min (MPa)	T.S., Ksi min (MPa)	Y.S./T.S., Ratio	% Elongation in 2″ (8″)
	E		0.22	1.15–1.50	0.15%–0.50% Si 0.04–0.11% V 0.01–0.03% N	60(<4″) (415)	80–100 (550–690)	0.60–0.75	23(18)
A656/ A656M-93	50	Plates for truck frames, brackets crane booms,	0.18	1.65	0.60% Si max 0.08–0.15% V 0.020% N	50 (345)	60 (415)	0.83	23(20)
	60	railcars, and similar applications (As-Hot-Rolled)	0.18	1.65	0.005–0.15%	60 (415)	70 (485)	0.86	20(17)
	70		0.18	1.65		70 (485)	80 (550)	0.875	17(14)
	80		0.18	1.65		80 (550)	90 (620)	0.89	15(12)
A709/ A709M-93	36	Quenched and tempered alloy steel, shapes,	0.25	<1.0	0.40% Si, max 0.20% Cu	36 (250)	58–80 (400–550)	0.45–0.62	23(20)
	50	plates, and bars. Structural use in bridges.	0.23	1.35	0.40% Si, max	50 (345)	65 min (450)	0.77	21(18)
	50W	Grade W atmospheric corrosion better than structural grade, A36	0.15-0.20	0.80–1.35	0.40% Si, 0.40 Ni, 0.45% Cr, 0.35% Cu, 0.08% V	50 (345)	70 min (485)	0.71	21(18)
	70W		0.19	0.08–1.35	0.40% Si, 0.30% Cu, 0.40 Ni, 0.55% Cr, 0.08% V	70 (485)	90–110 (620–760)	0.70–0.78	19
	100		0.21-0.21	0.70–1.30	Ni, Cr, Mo, V, B additions of varying concentrations,	100<21/2″) (690)	110–130 (760–895)	0.77–0.91	18

(*Continued*)

Table 16-1 Specifications, grades and compositions of common structural steels. (*Continued*)

ASTM-SPEC	Grade	Specification For	Chemical Composition, wt %			Y.S., Ksi min (MPa)	T.S., Ksi min (MPa)	Y.S.,/T.S., Ratio	% Elongation in 2″ (8″)
			C, max	Mn	Others				
	100W					90(2.5″–4″) (620)	100–130 (690–895)	0.69–0.90	16
A808/ A808M-93		High Strength Low-C, Mn, Cb, V steel	0.12 max	1.65	Si–0.50% 0.02–0.10% Cb	50(<1 1/2″) (345)	65 (450)	0.77	22(18)
					0.10% V max 0.15%–(Cb + V) max	46(1.5″–2″) (315)	65 (450)	0.71	22(18)
						42(2″–2.5″) (290)	60 (415)	0.70	22(18)

Table 16-2 Compositions of some 2-in. -thick ASTM specified steel plates[Reference 1a].

	A36	A572*	A588-A	A514-F
Yield Strength Min. (Ksi)	36	50	50	100
Processing**	AR	AR	AR	Q&T
Carbon	0.26 max	0.23 max	0.19 max	0.10–0.20
Manganese	0.80–1.20	1.35 max	0.80–1.25	0.60–1.00
Phosphorus	0.04 max	0.04 max	0.04 max	0.035 max
Sulfur	0.05 max	0.05 max	0.05 max	0.035 max
Silicon	0.15–0.40	0.15–0.40	0.30–0.65	0.15–0.35
Nickel	–	–	0.40 max	0.70–1.00
Chromium	–	–	0.40–0.65	0.40–0.65
Molybdenum	–	–	–	0.40–0.60
Vanadium	–	0.01–0.15	0.02–0.10	0.03–0.08
Copper	–	–	0.25–0.40	0.15–0.50
Boron	–	–	–	0.0005–0.006

* Type 2
** AR = As-rolled.
Q&T = Quenched and tempered.

Table 16-3 Strength of Weathering Steels with the same composition increases with Q&T.

	A588	A852
Yield Strength Min. (Ksi)	50	70
Processing*	AR	Q&T
Carbon	0.19 max	0.19 max
Manganese	0.80–1.25	0.80–1.35
Phosphorus	0.04–max	0.035 max
Sulfur	0.05 max	0.04 max
Silicon	0.30–0.65	0.20–0.65
Nickel	0.40 max	0.50 max
Chromium	0.40–0.65	0.40–0.70
Vanadium	0.02–0.10	0.02–0.10
Copper	0.25–0.40	0.20–0.40

* AR = As-rolled
Q&T = Quenched and tempered.

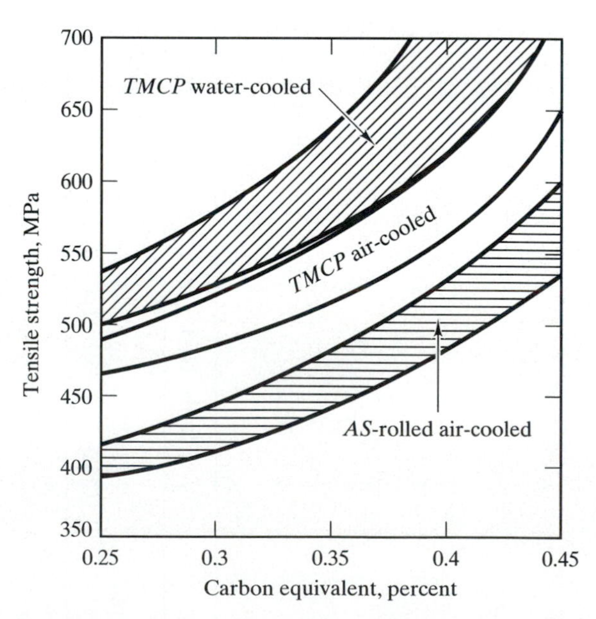

Fig. 16-1 Tensile strength as a function of carbon equivalent of as-rolled (conventional) and TMCP steels[1a].

calcium is added to control the sulfide shape (see Chap. 7).

In general, the microalloyed thermo-mechanically controlled processed (TMCP) HSLA steels exhibit higher yield and tensile strengths, much better toughness, weldability, and improved weld joint performance than do the as-rolled conventional HSLA steels. They also exhibit excellent formability and ductility. These steels, however, exhibit much higher yield-to-tensile strength ratios than do the conventional as-rolled and normalized HSLA steels. Figure 16-1 shows the relationship of carbon equivalent and tensile strengths for as-rolled conventional HSLA steels and TMCP air-cooled and water-cooled microalloyed HSLA steels. For a given desired strength, the TMCP microalloyed steels require lower carbon equivalents. The lower carbon equivalents will induce a much lesser difference in the fracture toughness of the base metal and the HAZ—an improved weld joint performance. Since the carbon equivalent relates directly to the hardenability of the HAZ, lower carbon equivalents require no preheat in thin plates or less pre-heat when needed in thick plates and make the welding process easier. These are shown in the results of the modeling of the required preheat of conventional and high-performance 70 and 100 ksi minimum yield strength bridge steels in Figs. 16-2 and 16-3.

The development of the high-strength high-performance structural steels (microalloyed HSLA steels) provides the opportunity to substitute them for the conventional lower-strength structural steels. This substitution usually involves using a thinner gauge (less weight) of the higher-strength microalloyed steels for the same function performed by the lower-strength conventional steels. Depending upon the cost-differential of the two materials, an immediate savings in material cost may or may not be achieved. (Material costs are usually per pound basis.) Unequivocally, there are definitely cost advantages for the microalloyed steels in the fabrication, maintenance, and operational costs of the structure[1a]. *In fabrication using welding*, a thinner gauge material requires lesser filler metal volume and decreases the number of passes and welding time required. During cooling, a thinner gauge cools slower than a thicker gauge material and therefore reduces further the tendency to form martensite in the HAZ and to form cold cracks. The low sulfur content of the microalloyed steels reduces or entirely eliminates lamellar tearing. Depending on the composition and thickness, microalloyed steels may or may not require preheat and/or post-weld stress-relief treatment. When needed, the thinner plates will require less time and energy for both treatments. All these translate to ease and increased efficiency along with lower costs of fabrication. In addition, when used in high-rise buildings, the use of thinner gauge high-strength steels reduces the structure weight and the cost of the foundation, and increases

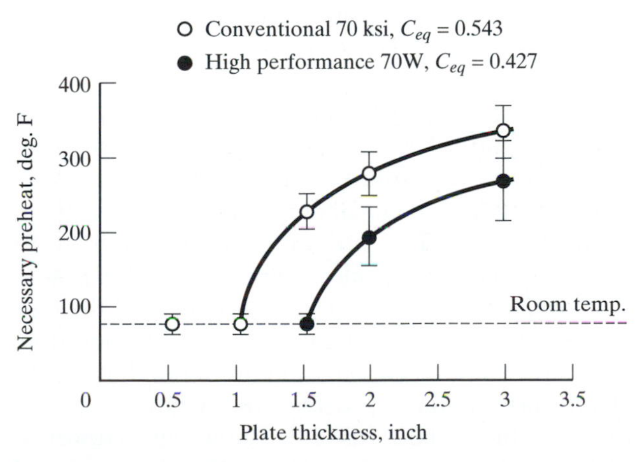

Fig. 16-2 Predicted preheat temperatures for butt joints of 70-ksi plates at 50 kJ/in.[1a].

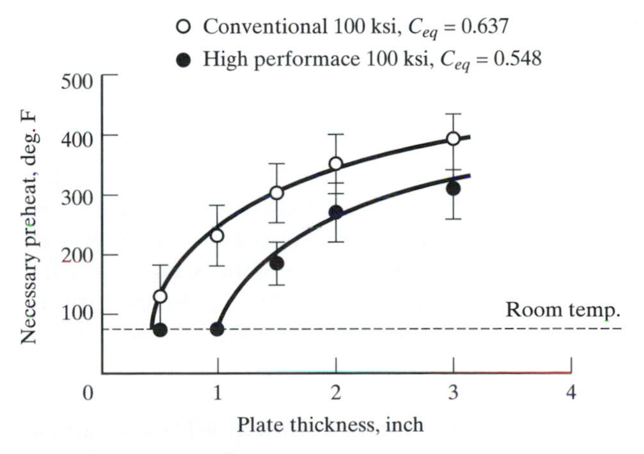

Fig. 16-3 Predicted preheat temperatures for butt joints of 100-ksi plates at 50 kJ/in.[1a].

the interior space of the building. *In maintenance*, it takes less time and money to nondestructively test thinner gauge plates. If repairs are needed, again it will take less time and money to do weld repairs. *During operation*, the lower weight of the structure means less fuel consumption (the main factor used to increase the Corporate Average Fuel Efficiency (CAFE) in miles per gallon of cars) and increase pay-loads for freight trucks, ships, off-highway and construction equipment. Most of the heavy off-highway and construction equipment have to be transported from one job site to another. A lower weight means less transport costs in terms of fuel consumption and the equipment to transport them. In addition, the better weld-joint performance of the microalloyed steels means less maintenance and a longer operating life. In a commercial office building, the increased office space means more income since office rentals are on a per square foot basis.

The substitution of the high-performance microalloyed steels for the same strength and thickness as the 70 and 100 ksi minimum yield strength conventional steels also provides cost savings in the thermal processing. Because the latter have to be quenched and tempered, the length of the conventional quenched and tempered steel plates is limited by the length of the heating and quenching facilities, which is about 50 feet[1a]. This constraint may be eliminated in the TMCP microalloyed steels for 50 mm (2 in.) or less thick plates. The elimination of the heating and quenching reduces the cost of the material, and the longer length of the TMCP plates reduces the welding cost in a long span structure such as a bridge.

We shall now consider cases in the design of structures[1a] where these materials may be used. The cases are classified as either statically or dynamically loaded structures because different properties of the material are required for each case. Fatigue is important for dynamically loaded structures but is inconsequential in statically loaded structures. In either case, strength, ductility, fracture toughness, and fabricability are important. Fabricability includes forming, machining, welding, and weld-joint performance, as discussed in Sec. 12.4.1.

16.2.1a Statically Loaded Structures[1a]. Examples of these structures are storage tanks, pipes, and buildings. Small storage tanks, water pipes, and similar structures are commonly fabricated from thin, usually less than 19-mm (0.75 in.)-thick plates or sheet steels with yield strength equal to or less than 345 MPa (50 ksi). The mode of loading in these structures precludes the use of thinner, higher-strength steels than are presently used because unacceptable distortion and/or buckling may occur. This is particularly true for buried water pipes. In addition, welding of the low-strength conventional steels in thin sections presents no difficulties and problems. Nonetheless, cracks are observed to propagate from corners of improperly designed and unstrengthened large openings in large storage tank walls. Although the microalloyed steels exhibit better HAZ and weld-performance properties, they cannot compensate for the damage or fracture caused by design deficiencies. Thus, high-performance steels may not be necessary for these applications.

Statically loaded large diameter gas and oil transmission pipelines constitute another case. These are critical structures whose failure may cause loss of life and severe property and environmental damage. They are usually fabricated from thin steel plates [less than 25 mm (1-in.) thick] with yield strengths of $\leq$483 MPa (70 ksi). The handling of these pipes with equipment during installation and the low operating temperatures to which some of these pipes are exposed, require high-fracture toughness and low-impact transition temperatures. In addition, pipe sections are girth-welded in the field to form the pipeline. This require steels that need no preheat or postheat-treat. The critical nature of these structures and their materials' requirements make the microalloyed steels the choice for this application. The Trans-Alaska pipeline is built of these steels.

High-rise buildings, unlike storage tanks, water pipe, gas and oil transmission pipe lines, are commonly fabricated from very thick plates of conventional steels with 345 MPa (50 ksi) or less minimum yield strengths. The use of higher-strength, $\geq$483 MPa (70 ksi), thinner microalloyed steels can potentially reduce the weight of the building and the cost of the foundation, as discussed earlier. In addition, these steels are amenable to being welded using the electroslag welding process—the routine process for the thick plates used in buildings. However, such a potential is yet to be realized because of another requirement in high rise buildings, especially in earthquake-prone areas. This requirement is included in the Load and Resistance Factor Design (LRFD) by the American Institute of Steel Construction (AISC), and allows for a rotation capacity of a compact section of flexural members prior to fracture or collapse when they are plastically deformed. Although the flexural members in these structures may include the weldments and their properties and the connections between components, the requirement is imposed on the base properties of the steel. The requirement is for the steel to have a yield-tensile strength ratio of less than 0.85[1a,c,d]. In Japan, this is reduced to 0.80[1e]. This requirement is also adapted by the American Association of State Highway and Transportation Officials (AASHTO) in regards to bridges, which we shall now discuss.

16.2.1b Dynamically Loaded Structures[1a,c]. Bridges are examples of dynamically loaded structures. Fluctuating loads are imposed on the static load (weight) of the structure and may initiate and propagate fatigue cracks from weld imperfections and/or stress concentration areas such as those at weld terminations and weld toes. The fatigue crack propagates until it reaches the critical crack size that causes the fracture. The critical crack size depends on the applied load and the fracture toughness of the material (see Chap. 4). Thus, the fatigue of the weldment and the fracture toughness of the steel are very important for bridges. However, extensive studies on welded components in bridges reveal that the base steel properties do not influence the fatigue life. The factors that influence the fatigue life of welded components are the stress fluctuations (range), the size and shape of the weld imperfections, and the geometry of the welded component. These studies produce the AASHTO design stress range curves shown in Fig. 16-4[2] for different categories of weldments in bridges. These curves represent the lower 95 percent confidence limit for 95 percent survival for a particular weldment.

Presently, bridges may be fabricated from plates up to 102-mm (4 in.) thick having 36-, 50-, 70- or 90/100 ksi minimum yield strength. About 80 percent of the steels are of the 50 ksi minimum yield strength[1a]. The use of the most recently added A852 Q & T steel with 70 ksi yield strength in bridges is increasing. However, the 90/100 ksi minimum yield strength materials have limited applications due to the special care required in welding to avoid hydrogen cracking. In addition, the requirements for bridge construction, which are live load deflection, vibration, buckling strength, and fatigue resistance, are all independent of the strength of the steel. They are governed more by the modulus of elasticity and are adversely affected by thinner gauges of high-strength materials, especially in the I-girder[1c] configuration prevalent in the United States.

Recent developments have shown that 80- and 100-ksi minimum yield strength high-performance

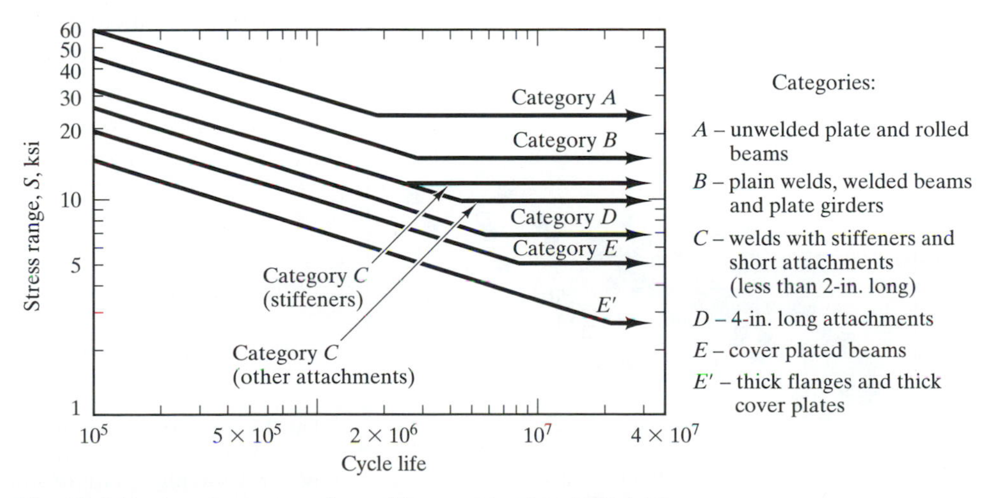

Fig. 16-4 Design stress range for weld categories A to E′[Ref. 2].

steels can be produced in thicknesses that are used for bridges. However, these steels exhibit yield-to-tensile strength ratios greater than 0.85, as shown in Fig. 16-5[1a]. Thus, based on the current AASHTO requirement, the use of these steels in bridges is stymied. The weldability and the ease of fabrication with microalloyed steels, resulting in more economical bridges, cannot be taken advantage of.

This discussion illustrates the difficulty in using new structural materials, as exemplified in the design with plastics in Chap. 15. In the first place, the suppliers, fabricators, and users must have a consensus of opinion about the characteristics of the new material, in this case, the high-performance steels. A lot of materials engineers feel that the restriction on the yield-tensile ratio is unrealistic and therefore this requirement needs further study and discussion. In addition, as indicated with plastics, we simply cannot use the same design as that used for the conventional steels. Thus, innovative designs using the new materials are needed[1a,1c]. These are two objectives of the Federal Highway Administration (FHWA) Project DTFH61-93-R-00007, "Innovative Bridge Designs Using Enhanced Performance Steels."

16.2.1c Use of Microalloyed Steels in Automobiles and Off-highway Vehicles and Equipment. The use of microalloyed HSLA steels in automobiles is now standard because of the need to reduce the inertia weight of the vehicles. Gone are the gas guzzlers with gross inertia weight of 2 tons or more. In addition to fuel efficiency, lighter weight increases the acceleration performance of the car. They are used for frames or chassis, bumpers, body panels, and others. While the dominant position of steel in the automobile is constantly being challenged by aluminum and plastics, its cost advantage, performance, ease of fabrication, repairability, and recyclability makes it the material for the affordable automobile for the mass customers for some time to come. Selection of automotive materials will be discussed in greater detail in Sec. 16.4.

Off-highway vehicles and equipment are those that are used in the construction, mining, and agricultural machinery. Typical structural steels used are ASTM A572 and A656. The largest usage category is in the manufacture of the vehicle frames that includes the main frames, undercarriage frames, engine frames, and loader frames[1b]. The frames constitute the backbone structure of the vehicle and carry loads over multiple paths. They provide the mounting and the alignment for all the other vehicle components, including the power train, various implements, and the operator station. They are normally welded assemblies of castings, forgings, plates, and special rolled sections. Some plate forming is necessary. Thus, weldability is a process criterion for selection. However, their dominant performance criterion is fatigue, in particular weld fatigue, although maximum loading may apply at points where other mounted structures transfer their loads. Weld fatigue is usually the limiting performance and this is the reason for using moderate strength structural steels.

16.2.1d Benefits in Adopting High-performance Steels—an Example. In general, to take advantage of the benefits of new materials, such as the high-performance steels, requires a close cooperation between the fabricator and the material supplier. This

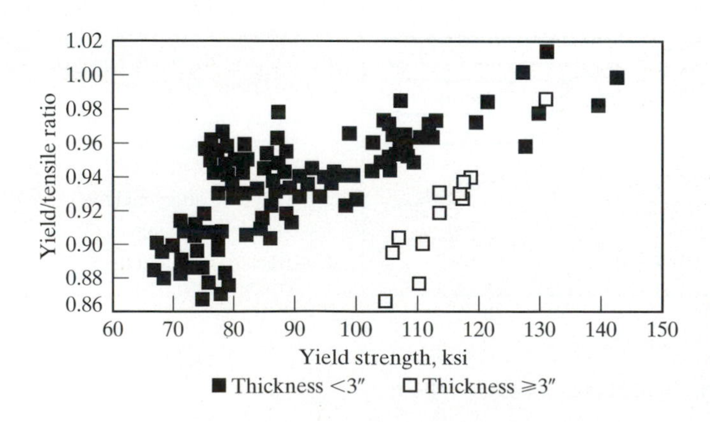

Fig. 16-5 Yield strength-tensile strength ratios of grades 65, 80, and 100 high-performance steels. [1a]

usually means that the fabricator or manufacturer is willing to change the material it is currently using. It also implies that the material is available from a supplier. When this cooperation is successful, both the fabricator and material supplier reap economic benefits. Such a cooperation is especially beneficial for fabricators with no research and development department and results in usage of a material even before it becomes a part of the standards.

An example of this cooperative effort is reported by Pike[1f] of Grove Worldwide, a manufacturer of mobile hydraulic cranes and aerial work platforms. Grove has been using ASTM A514 quenched and tempered 690 MPa (100 ksi) minimum yield strength structural plate. When a 690 MPa (100 ksi) minimum yield strength TMCP microalloyed steel became available, it worked with suppliers to determine if it could replace thinner sections of the A514 steel. Due to the nature of the processing, the maximum thickness of the TMCP microalloyed steel at this strength level is about 9.5 mm (⅜-in.) thick plate. Grove immediately recognized the economic benefits because A514, being a quenched and tempered steel, was more expensive than the TMCP high-performance steel. In addition, the visual appearance of A514 that resulted from the quenching and tempering process left something to be desired. The quench and tempering process was necessary for the A514 in order to attain the desired mechanical properties. Aside from adding cost, the quench and temper process produces some waviness (distortion due to the quenching) on the surface of the plate. During storage or transport, the low parts of the wavy-surfaced plate collect moisture and induce corrosion and pitting that did not become evident until before preparation for painting. Painting did not completely hide these pits and the result was poor visual appearance. Grove considered visual appearance as important, especially in high visibility areas such as booms, outrigger boxes, and the frame side plates of cranes.

The opportunity to replace the A514 with a cheaper and aesthetically better material arose. Grove took advantage of it to produce the same quality product performance-wise, and it is cheaper! In the course of the fabrication trials, the new material exhibited the high-performance properties of toughness, weldability, and formability. The Charpy V-notch impact toughness requirement of 20 J (27 ft-lbs) absorbed energy at −40°C (−40°F) was easily met. The material was readily weldable with minimal risk of cracking due to the low carbon equivalents with very good weld joint performance. Enhanced cold-forming was observed due to the low carbon content and inclusion shape control. Bend radii of twice the thickness (2T) are regularly achieved and smaller radii are occasionally used.

Pike described two other successful high-performance steel substitutions for a tubing and a bar material and concluded in his paper, "The net combined savings for the described material improvements is very significant. Further cost savings and performance improvements are built into new designs that utilize these materials. Such use of high-performance steels permits Grove to deliver more added value to customers. The benefits of these new materials far outweigh the difficulties of pursuing and using them."

From this example, we should learn that designers and fabricators can take advantage of the benefits of new materials even before these materials become part of a standard. In order to do this, fabricators must have a knowledgeable materials person who can undertake the risks in making the change. Such a change can be facilitated by a counterpart materials person with the supplier.

Summary: To summarize this section, we saw a lot of advantages and economic benefits in using the new high-performance structural steels over the conventional structural steels. However, our discussion showed that these new materials are not used in every design. Materials selection in design must be according to the "fitness for purpose" design philosophy by which the structure or component is designed to provide the necessary safety and performance at the least cost. We see also that the structures or components using these new materials must undergo innovative designs, for example, in bridges, in order to take advantage of the full benefits. Their application also depends on convincing the AASHTO and AISC that their higher yield strength to tensile strength ratio is not a detriment. If this hurdle is overcome, there is substantial savings in fabrication costs, as has been illustrated in the example where the high-performance steel replaced a conventional Q&T steel.

16.2.2 Stainless Steel as a Viable Structural Material[1g]

The initial price ratio of stainless steel to the conventional structural steel we have been discussing is about eight to one on a per weight basis. To the uninitiated, this is sufficient to eliminate stainless steel. However, if we consider more seriously what stainless steel has to offer, it becomes a very viable structural material. In the first place, the austenitic stainless steels can be cold-worked to attain comparable mechanical strengths as the carbon structural steels, as illustrated in Fig. 12-27. Because of its corrosion resistance, not much corrosion allowance in thickness is provided over the design thickness, it does not have to be painted, requires no maintenance, and lasts longer than the conventional structural steel. In contrast, corrosion allowance has to be added to the design thickness of the structural steel that may be two or three times the design thickness (increases the weight) and then has to be given corrosion protection. The corrosion allowance cancels part of the cost advantage. If we add operational, maintenance, and eventual replacement costs of the structural steel in the life of the structure, we come to realize that stainless steel can be a very smart choice.

The above considerations led to the use of stainless steel in some of United Kingdom's offshore oil and gas platforms for cladding for accommodation modules, fire, and blast walls, and components such as cable trays and ladders. It was shown that the installed cost of stainless steel offshore is only about 15 percent greater than that for the structural steel, although the initial material cost ratio is about eight to one.

The durability of stainless steel is its most important asset. However, under severe conditions, even some stainless steels will corrode and it is therefore important that the most appropriate alloy is selected for the working conditions and environment. In general, as indicated in Chap. 12, we start the selection process with the austenitic stainless steels, which are actually specified for building and construction. As indicated earlier, these steels will significantly increase in strength when warm- or cold-worked. As a result, the American Society of Civil Engineers (ASCE) has published the following guide for using stainless steel: *Specification for the Design of Cold-Formed Stainless Steel Structural Members*, ANSI/ASCE, 8–90. The guide emphasizes the modern Load and Resistance Factor Design (LRFD) principles but includes the allowable stress design method in the appendix.

European designers, in particular, in the U.K. with the Steel Construction Institute (SCI), have adopted the U.S. design and issued the "Design Manual for Structural Stainless Steel." To simplify the selection of stainless steel grades, the Design Manual is based on the three stainless steels—304L, 316L, and Duplex 2205. These steels are projected to be used offshore, as well as onshore, and the "L" (low-carbon) varieties will greatly reduce the occurrence of intergranular corrosion due to sensitization when welding is used for fabrication. The typical properties of these steels are given in Table 16-4. The inclusion of Duplex 2205 in the Design Manual provides the range of strength for stainless steel structural applications.

16.2.2a Case Study: Stainless Steels for Reinforcement of Concrete in Highway Bridges. Corrosion of reinforcing bars (rebars) in concrete is a ubiquitous problem and is a serious concern especially in bridges with reinforced concrete decks. Bridges are supposed to be designed to have a 120-year life, but repairs or remedial treatment might have to be done after only 20 years due to corrosion.

Concrete is a particulate composite of coarse and fine aggregates that are bound together by portland cement. The cement is a synthetic mineral mixture which, when ground into a powder and mixed with water, forms a stonelike mass. This mass results from a series of chemical reactions whereby the constituents hydrate to form a material of high hardness that is extremely resistant to compressive loading. During hydration, the cement forms a paste that has excellent adhesive properties and binds the aggregates and the steel bars together. In terms of compos-

Table 16-4 Properties of stainless steels in design manual.

Stainless Steel Type	0.2% Offset Yield Strength, MPa
304L	250–280
316	275–300
Duplex 2205	520–600

and binds the aggregates. Going a step further, a reinforced concrete is a "hybrid" cement matrix composite with the coarse and fine aggregates and steel bars as the reinforcements.

The four main compounds that constitute portland cement are: (1) tricalcium silicate, $3CaO.SiO_2$ (C_3S); (2) dicalcium silicate, $2CaO.SiO_2$ (C_2S); (3) tri-calcium aluminate, $3CaO.Al_2O_3$ (C_3A) ; and (4) tetracalciumaluminoferrite, $4CaO.Al_2O_3.Fe_2O_3$ (C_4AF). Because of the predominance of calcium oxide in the constituents of cement, the pH (~ 10) of the stonelike mass that forms is basic. If the pH is maintained at this level for the rest of the life of the reinforced concrete, the steel can have a very long life. The problem is that the pH around the steel bars is not maintained at this level and corrosion ensues.

The inherent problem of concrete is its porosity. With the different sizes of aggregates, there are a lot of voids left in the stonelike mass that is not filled. Water penetrates the body of the concrete and in time will diminish the basicity of the cement. In addition, when the water freezes in cold weather, it cracks the concrete and eventually causes a separation between the concrete and the steel bar. There is also the difference in thermal expansion of the steel and the concrete that can also lead to interfacial separation. When this occurs, water seeps through the interface and the steel bar is now exposed to water, and not the basic cement, and corrosion ensues. When corrosion starts, the oxide (the rust) that forms has a higher volume than the metal and therefore stresses and further opens the concrete-steel bar interface. The corrosion is enhanced further when salts and de-icing agents are applied to the road surface to melt the snow. When corrosion of the bars is excessive, the bars have to be replaced or the road surface (or bridge deck as the case may be) reconstructed.

A study was made to determine if stainless steel can be used as reinforcing bars. The results shown in Fig. 16-6[1g] indicate that type 316 can be used. In this figure, an unshaded box indicates that there was no observable corrosion up to a period of 10 years when

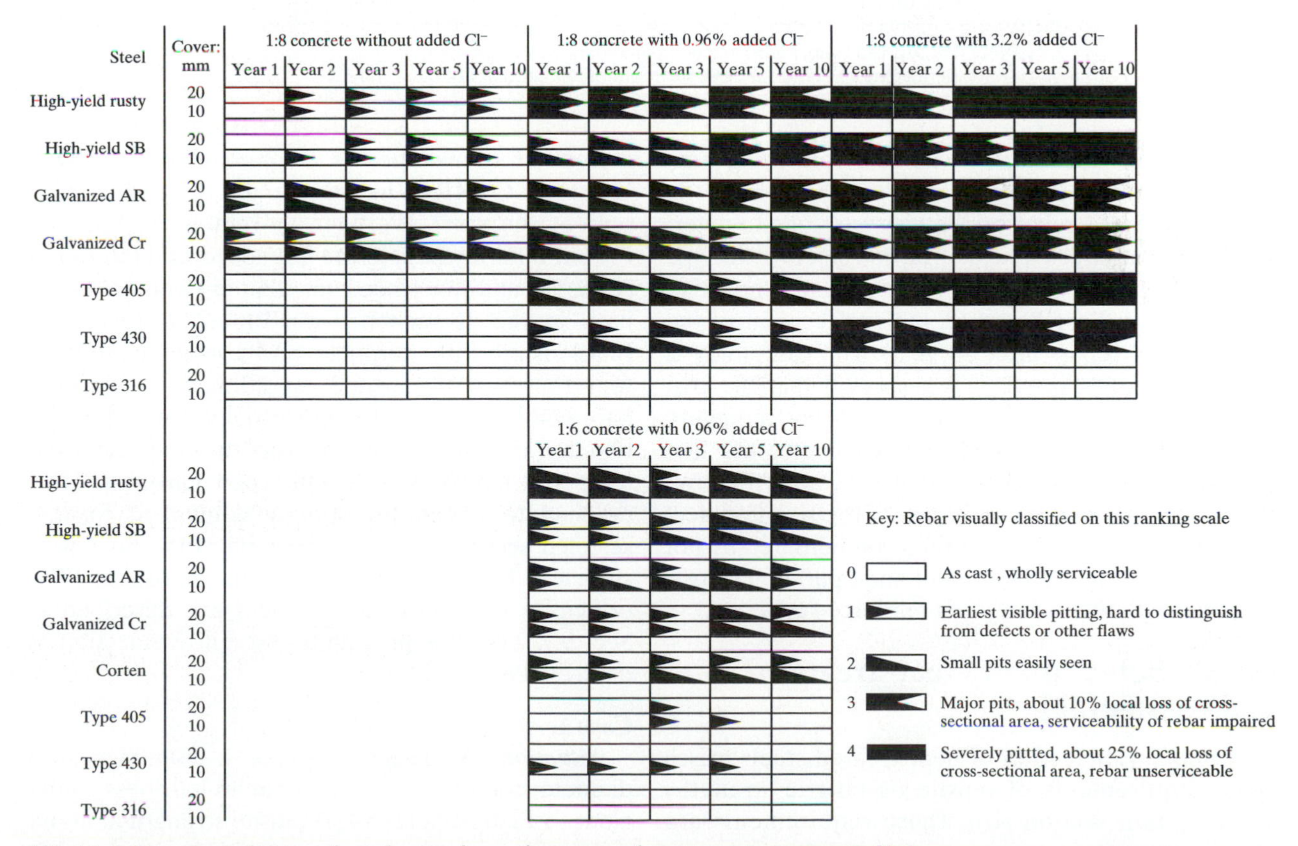

Fig. 16-6 Visual conditions of reinforcing bars after years of exposure in concrete. [1g]

Table 16-5 Life-cycle cost of a rebar-bridge pier[Ref. 1g].

Cost of capital		8.50%
Inflation rate		2.50%
Real interest rate		5.85%
Desired life-cycle duration		120.0 years
Downtime per maint/replace event		51.0 days
Value of lost production		20,000 Mu/day
	Carbon ST.	*ST.ST*
Material costs	20,530	130,000
Fabrication costs	13,856	13,856
Other installation costs	0	0
Total Initial Costs (*Mu*)	34,386	143,850
Maintenance costs	0	0
Replacement costs	61,709	0
Lost Production	224,797	0
Material-related costs	0	0
Total Operating Cost (*Mu*)	286,506	0
Total I.C. Cost (*Mu*)	320,892	143,856

Note: For this example, carbon steel replacement was taken at 30-year intervals. Replacement of the stainless steel reinforcement will not be required.

type 316 was embedded in concrete containing various amounts of chlorides. Thus, a very simple life-cycle cost analysis of a rebar bridge pier was done to determine the cost-effectiveness of carbon steel versus type 316 stainless steel rebar. The results are shown in Table 16-5 and are based on the replacement of the carbon steel rebars after 30 years. The case study is given here as reported. It is suspected that the cost comparisons were made on present-day monetary unit (Mu), and the results show that after the first replacement, the stainless steel can be cost-effective. The larger initial installation cost of the stainless steel reinforced concrete is more than compensated by not incurring the large replacement and operational costs incurred with the carbon steel reinforced concrete.

16.2.3 Selection of Heat-Treated Steel [4]

Summary: Selection of heat-treated steels arises from requirements of tensile or fatigue strengths for a certain section size. These requirements are converted to hardnesses, in particular, Rockwell Hardness C (HRC) in order to get the carbon content of the steel. We have to remember that the required hardness is in the quenched and tempered (Q & T) condition since heat-treated steels are never used in the as-quenched condition. However, the relationship with hardness and carbon content is found in the as-quenched condition, such as in the hardenability curves; therefore, to determine the carbon content we need to convert the quenched and tempered hardness to as-quenched hardness. Then proceed to obtain the Jominy equivalent for the selected section size and determine the alloys with sufficient hardenability to yield the hardness. (The principles of steel heat-treatment are described in Sec. 9.7, and the procedure for alloy selection is described in Sec. 12.5.)

Case 1.

Situation: A design requires a 50-mm (2-in.) diameter round bar with a minimum hardness equivalent to 1250 MPa (185 ksi) tensile strength at about the ¾-radius position in its cross section. The steel will be heat-treated in a reducing atmosphere (non-scaling atmosphere) and will be quenched in an agitated oil bath at an equivalent velocity of 200 ft/min. We need to determine the optimum steel that can be used.

Solution: To convert the tensile strength requirement to hardness, we use either Fig. 9-35 or Fig. 12-14a. The minimum 1250 MPa tensile strength is equivalent to 390 HB (Brinell Hardness). We need to convert this to HRC using Fig. 9-45 to get 42 HRC in the quenched and tempered condition. The latter in turn is converted to the as-quenched hardness using Fig. 9-44 or Fig. 12-16 to get 50 HRC as-quenched hardness.

Thus, the requirement is equivalent to an as-quenched hardness of 50 HRC at the ¾-radius position of a 50-mm (2-in.) diameter bar that is quenched in agitated oil. We proceed by using either Fig. 12-21a (Number 5 quenchant) or Fig. 9-53 to get the Jominy equivalent cooling at the ¾-radius of a 50-mm diameter bar. We arrive at 7.5 J_{ec}, which means that the 50 HRC must be obtained 7.5 Jominy units (i.e., $7.5 \times 1/16$-in.) from the quenched end. We refer to Table 12-24, and using the 50 HRC, we obtain three possible steels—5160, 9262, and 50B50—that will give 50 HRC at 7.5 J from the quenched end. To increase the number of steels from which to select, we can also use the steels that will give 50 HRC at the 8J position, since if 50 HRC is obtained at 8J, then the hardness at 7.5 J is greater than 50 HRC. We also get three steels—4142, 81B45, and 8650.

Judging from the characteristics of the six alloy steels given at the end of Sec. 12.5, the correct choice will be between 81B45 and 8650 because they do not appear to present any problems during processing. On paper, 81B45 looks the better of the two because it has a lower total alloy content (see Table 12-3), a lower carbon content, and will be cheaper due to the lower alloy content and hardenability is obtained with the addition of boron. Final choice should be accomplished, however, by using the two alloys to produce the actual intended part and determine their actual performance.

Case 2.

Situation: A 45-mm (1.75-in.) diameter shaft is required to sustain a maximum torsional stress of 170 MPa (25 ksi) and a maximum bending stress of 550 MPa (80 ksi). A company uses 4140H steel for other parts or components that it produces and wants to find out if 4140H will yield the required performance for this problem as well. The company has an agitated oil quenching facility; furthermore, it is known from other similar parts that at least an 80 percent martensite structure should be present at the ¾-radius position in the shaft.

Solution 1: The maximum bending stress is a tensile stress at the outer surface of the shaft. Since the maximum torsional (shear) stress is one-half the tensile stress (See Chap. 4), then the maximum torsional stress is really 275 MPa (40 ksi) from the bending stress and exceeds the stated required 170 MPa (25 ksi) torsional stress. Thus, we concern ourselves only with satisfying the bending (tensile) stress requirement.

Furthermore, since the shaft rotates, the outer surface elements of the shaft experience alternating tensile and compressive stress. Thus, the probable mode of failure will be bending fatigue and we need to design for the bending fatigue performance of the shaft. The required maximum bending stress of 550 MPa (80 ksi) is taken as the fatigue limit. When we use Fig. 12-17, we see that the shaft must have a minimum quenched and tempered hardness of 35 HRC. Using Fig. 12-16, the required minimum as-quenched hardness is 45 HRC. The carbon content of a steel that will yield 45 HRC with 80 percent martensite is 0.37 percent (from Fig. 12-13). This carbon content is the lower limit of the range of carbon content of 4140H (i.e., 0.37 to 0.44). Thus, composition-wise, we see that 4140H is more than adequate because when we order 4140H, we will mostly likely get an average 0.40 percent carbon content.

The ultimate question to be answered is: Will a 45-mm (1.75-in.) diameter 4140H steel produce the required as-quenched 45 HRC hardness at the ¾-radius position? To resolve this, we use Fig. 12-21a, Curve 5, or the ¾-radius curve of Fig. 9-53 first, to determine the Jominy equivalent cooling rate of the 45-mm bar diameter or the distance from the quenched end of the hardenability curve, and proceed then, to the minimum hardenability curve of 4140H to determine the hardness at the Jominy distance. This procedure is schematically shown in Fig. 16-7[4] from where we note that the Jominy equivalent distance is 6.5 and the minimum as-quenched hardness from the hardenability curve is

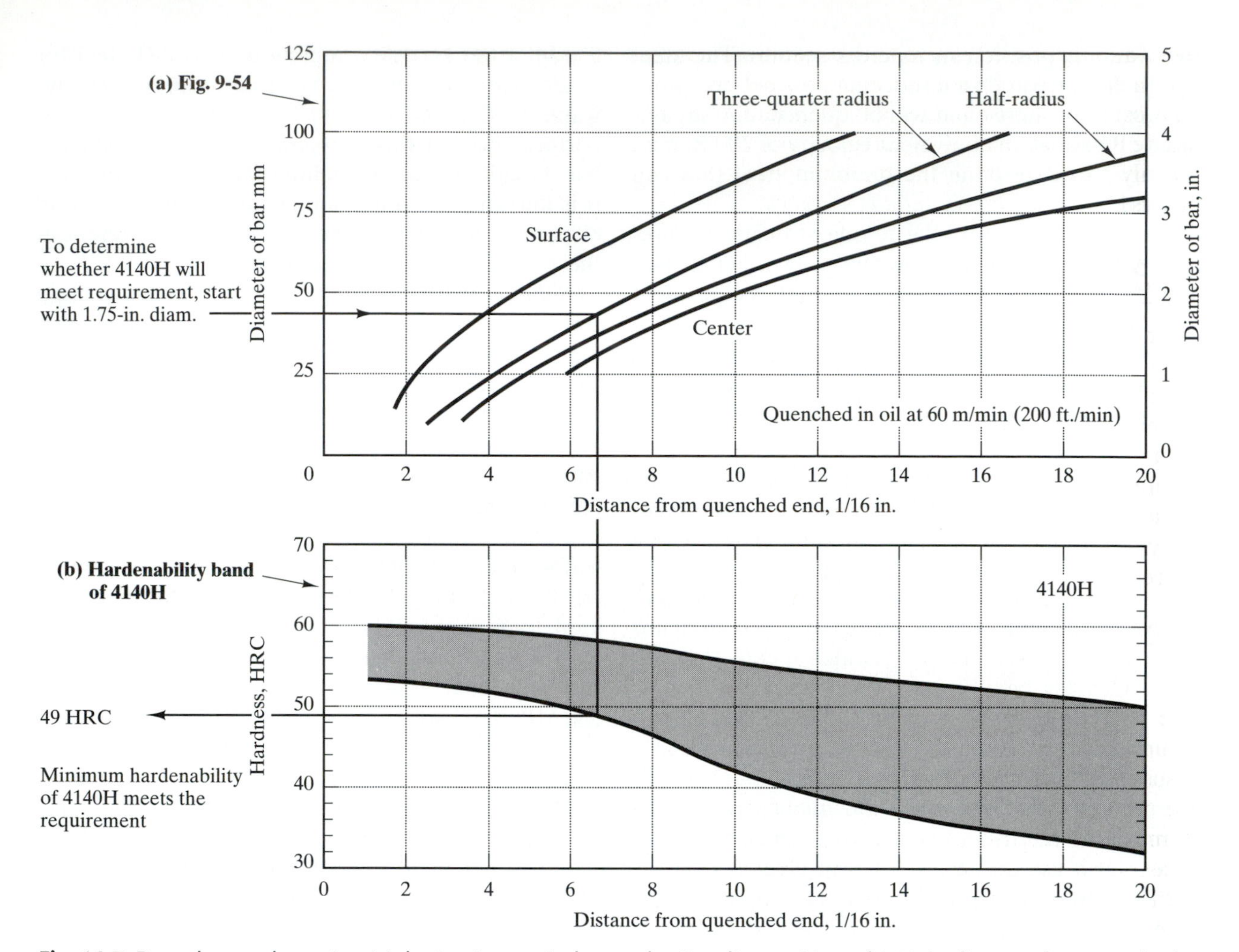

Fig. 16-7 Procedure to determine (a) the Jominy equivalent at the ¾-radius position of 1.75 in diameter bar quenched in oil, and (b) the hardness at that position with the hardenability curve (Reproduced with permission from ASM).

49 HRC, which is more than the required 45HRC. Thus, we see that 4140H is more than adequate to meet the requirements.

Solution 2: If we were to determine the possible alloys that can meet the design requirements, we should use Table 12-24. The question to be answered is: What alloys can produce as -quenched 45 HRC at the ¾-radius position in a 45-mm (1.75-in.) diameter bar when quenched in an agitated oil? We see from Fig. 16-7, that the equivalent Jominy distance from the quenched end is 6.5. From Table 12-24, we find that the alloys that will produce 45 HRC 6.5 Jominy distance from the quenched end are 8640, 8740, 5150, and 94B30. If we include the alloys that will produce 45 HRC at 7 Jominy distance, we have the additional alloys of 4137, 8642, 6145, 9261, and 50B40. From these alloys, we determine that 4137 (0.37 percent average carbon content) is among the possible alloys and this method also confirms that 4140H should be more than adequate to meet the design requirements.

Added Note: We see from the last solution that 4140H may not be the optimum alloy for the application and may actually cost more or be harder to process than the others. However, from the position of the company who has experienced processing the 4140H steel for other parts, the choice of 4140H eliminates stocking another grade of steel. An added steel grade in the inventory can lead to mixups when the steels are not properly identified. In addition, depending on the quantity needed, the other steel may actually cost more. Consolidating all orders into one grade can be more economical by ordering larger quantities of only one steel from the steel supplier. Small lot orders from steel service centers is costly.

16.3 Aluminum Alloys

As indicated in the introduction of Chap. 13, the next most widely used alloys for structural purposes are the aluminum alloys. Most of the structural uses of aluminum are in the aerospace industry where their light weight comes to fore. We shall first discuss the justification of the use of aluminum alloys and then present a case study for the selection of the aluminum alloys for the supersonic Concorde and NASA's space shuttle.

16.3.1. Justification of Aluminum Use in Skins of Aircraft[5]

The demand on the performance of structures and components is probably the greatest in the aerospace industry and that performance, in turn, depends on the materials selected. The demand for greater performance of aircraft structures comes from the more catastrophic consequences of failure of the structures in the air, which almost always includes loss in lives.

As in any transport system, the aircraft consists of three components: (1) the means of propulsion or the power-plant (engine) and the fuel; (2) the carrier, which includes the fuselage, the airframes (wings), landing gears and other structures, the control systems and the crew; and (3) the payload, consisting of cargo and/or passengers. The design of an aircraft always attempts to maximize the payload, which has to be balanced by safety. There is very little room for overdesign because it always leads to a reduction in the payload and increased operational cost. In addition, the design of a commercial aircraft is quite different from the design of a military aircraft. In general, the demands on a military aircraft structure are greater than on a commercial aircraft. Thus, the design lives of military aircraft are about an order of magnitude shorter than those of commercial aircraft—slightly over 6,000 hours for military jets versus 50,000 to 100,000 hours for commercial jets.

The life of an aircraft depends on a repeated cycle of operations consisting of the following: (1) parking, (2) takeoff and climb to altitude, (3) cruise or operational, (4) descent and landing. The most demanding phase for a commercial jet is the takeoff when it requires the engines to be operated at almost full power and the angle of climb be sufficient to clear all obstructions near the airport. In contrast, the cruise phase is relatively undemanding. Short-range aircrafts spend much less time in the cruise mode than do long-range aircrafts, compared to the length of the time for their respective cycle of operations. On the other hand, military jets must not only be capable of a fast takeoff, but they must also be prepared to execute demanding high-speed maneuvers to evade or to pursue enemy jets; there is rarely a cruise or steady flight phase.

To illustrate the justification of the use of aluminum alloys in aircraft, we consider the stresses on the parts of the structure during the different phases of the aircraft operation. We shall consider the two main structures—the wings and the fuselage or the hull.

Wings. The wings are subjected to variable complex stresses. In the park phase on the ground, the wings act like a cantilever with a force acting on it, consisting of its own weight, the weight of the fuel inside it, and the weight of the engines, if these are wing-mounted. The upper wing surface is in tension and the lower surface is in compression. This stress pattern continues while the plane is taxiing. The largest forces on the wing occur when the aircraft is airborne because the wings support the entire aircraft and cargo. Because of the lift, there is a reversal of the loading modes experienced during parking and taxiing. The wing bends upward and the upper surface is now in compression with the lower surface in tension. Since the wings support the entire structure in flight, the stresses are relatively high. We see, therefore, that the wing structure is subjected to stress reversals and that fatigue performance must be one of the properties that the wing structure must have. Relative to fatigue performance, cracks will initiate and then grow to a critical size, before a catastrophic failure occurs. Thus, the material in the structure must exhibit a slow crack growth rate and high-fracture toughness, to allow for the largest critical crack size. During steady flight, the upper surface is under compression. For this, the wing must have sufficient stiffness and must resist buckling.

In Chap. 11, we have seen that the efficiency or performance indices of a plate (panel) externally loaded or by its self-weight in bending are $E^{1/3}/\rho$ and $\sigma_{ys}^{1/2}/\rho$ for stiffness and strength criterion, respectively. Table 16-6 compares these indices for 17-4 PH stainless steel, Ti-6Al-4V, and 7075-T6 aluminum alloy. We see that the 7075-T6 alloy exhibits the highest indices and justifies the usage of this alloy in the wing skins and the fuselage (see Fig. 13-1). In addition to this, titanium is costlier than aluminum and therefore aluminum will be the dominant material for aircraft structures for sometime to come.

It can be shown that the buckling stress of a panel is

$$\frac{\sigma}{\rho} = 1.5355 \frac{E^{1/3}}{\rho} \left(\frac{P}{b^2}\right)^{2/3} \tag{16-1}$$

where σ is the buckling strength, σ/ρ is the specific buckling strength, and P/b^2 is the structural loading index. A plot of P/b^2 vs. σ/ρ is shown in Fig. 16-8 for four different materials: CPTi is commercially pure titanium; FV520 is a precipitation hardening (PH) stainless steel; Ti-6Al-4V, and 7075-T6 aluminum alloy. We see that the specific buckling strength for 7075-T6 is the highest of all the materials at low loading index. At about a loading index of 50 MN/m², the Ti-6Al-4V alloy exhibits the higher specific buckling strength and becomes competitive. Thus, in order for the titanium alloy to be competitive with the aluminum alloy for wing skins, the loading index must exceed 50 MN/m². In order to achieve this, we can either decrease the width of the titanium panel, *b*, compared to that of an aluminum panel, or the titanium panel must be subjected to a higher loading intensity than the aluminum panel. Reducing the width the panel allows titanium to be used in thinner gauges and the panel becomes lighter. However, as seen in Fig. 16-8, *b* is the distance between stiffeners in the panel. Thus, the decrease in weight of the titanium skin will increase the number of stiffeners (which will increase the weight) in the panel and thus counteract the effect of titanium thickness reduction. Therefore, there is a limit to the benefits of decreasing the weight by thickness reduction. Without going into the details, it has been shown that aircraft frames with titanium are not as practical and economical as the aluminum frames.

We should note, however, that the titanium panel compared above has the same shape as the aluminum

Table 16-6 Efficiency indices of materials as wing panels in bending.

	Young's Modulus		Yield Strength		Density		$E^{1/3}/\rho$	$\sigma_{ys}^{1/2}/\rho$
Materials	**GPa**	**psi × 10⁶**	**MPa**	**ksi**	**Mg/m³**	**lbm/in³**	**$kg^{-2/3} \cdot m^{7/3}$**	**$kg^{-1/2} \cdot m^2$**
17-4 PH S.S.	200	29	1262	183	7.83	0.283	1.60	14.21
Ti-6Al-4V	110	16	870	126	4.43	0.160	2.31	20.85
7075-T6 Al	72	10.4	480	70	2.80	0.101	3.18	24.51

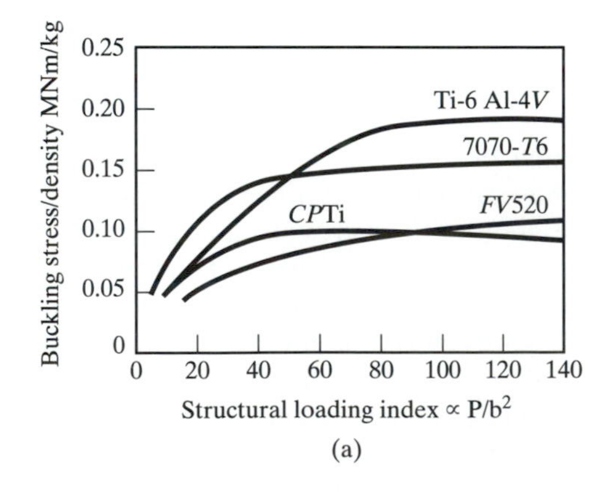

(a)

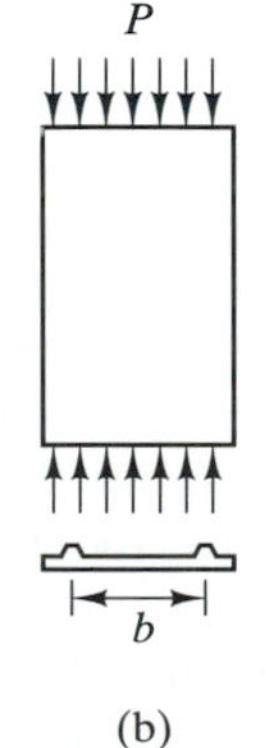

(b)

Fig. 16-8 (a) Specific buckling strength of aluminum, titanium, and stainless steel as a function of loading index of panels in (b) [Ref. 5].

panel. This brings out once again a common thread when considering new materials for applications as a replacement for older materials. It is seen that a new material needs to have some element of redesign in order to compete. This was indicated in the discussion of plastics and with the new microalloyed steels discussed earlier with structural steels. It turns out that a titanium panel with a sandwich honeycomb structure will be structurally very competitive. The titanium honeycomb structure can be manufactured by superplastic forming and diffusion bonding. However, these added processings add to the cost factor and because of this, it is unlikely that titanium will ever replace aluminum for the skins of an aircraft in the foreseeable future.

Table 16-7 Skin temperatures as function of speed. [Ref. 5]

Speed, Mach Numbers	Skin Temperatures °C	Skin Temperatures °F
2.0	100	212
2.5	150	302
3.0	200	392
3.5	300	572
4.0	370	698

Fuselage or Hull. The fuselage or hull is the long, almost cylindrical shell, closed at its ends, that carries the whole payload in the aircraft. The hull is supported by the wings at about its midlength position. During flight, the upper fuselage surface is subjected to tensile stress, while the lower fuselage surface is under compressive stress, especially in the area under the wings. In addition, when the aircraft is at a high altitude, the cabin must be pressurized, which makes the cabin a pressure vessel with cylindrical hoop and longitudinal stresses superimposed on the payload stresses. Because it is a pressure vessel, the outside cabin materials must exhibit high static strength and high fracture toughness. Furthermore, because of pressurization and depressurization, the fuselage materials must also exhibit low-cycle fatigue performance. In addition to these, if the aircraft flies at very high speeds, drag and frictional effects induce high-temperatures on the skin surfaces, so that materials must have resistance to heat-softening. Table 16-7 shows the projected skin temperatures from Mach 2 to Mach 4 (a Mach number indicates the speed in multiples of the speed of sound), while Fig. 16-9 shows the specific strengths of some alloys as a function of temperature. Some aluminum alloys reduce their strengths rapidly in the range of 100° to 50°C and therefore, are not used above these temperatures. This is the

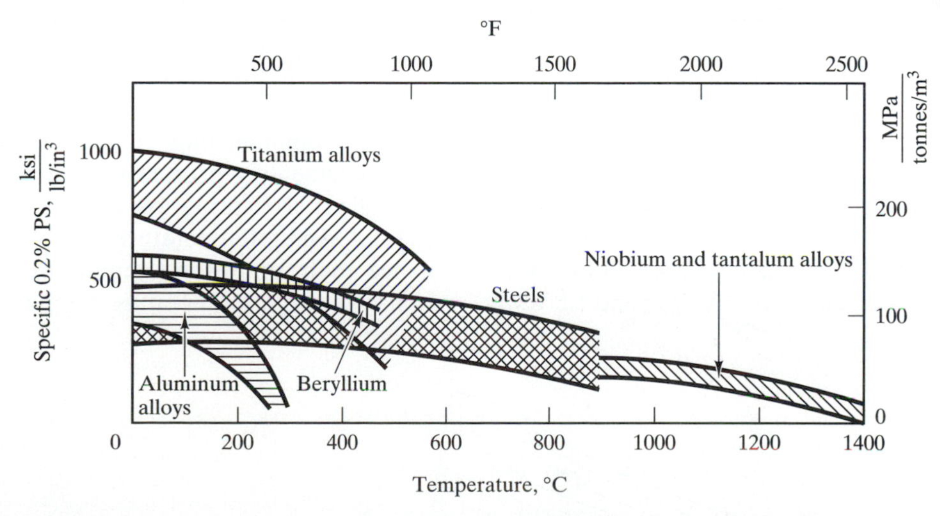

Fig. 16-9 Specific 0.2% offset yield strength (PS) of different alloys as function of temperature [Ref. 5].

reason why the Concorde, the trans-Atlantic supersonic plane, has a maximum Mach 2.2 speed. We see that the projected skin temperatures for the high speed civil transport (HSCT) plane (projected speed is Mach 3-5) being developed by NASA is around 400°C, titanium is being considered for this case.

16.3.1a Case Study 1—Selection and Development of an Aluminum Alloy for the Concorde [5]. The Concorde is a supersonic, commercial airplane that was designed to enable business persons from either side of the Atlantic, mainly the United States and Europe, to fly to a destination, conduct short business meetings, and return on the same day. Its maximum Mach 2.2 speed induces skin heating, as indicated in Table 16-7, and Fig. 16-10 indicates the skin temperatures due to drag effects. We see that the highest temperature is 128°C at the nose. The temperature at the leading edge of the wing is 105°C, while that over most of the fuselage is over 90°C but less than 100°C. These drag temperatures were considered to increase locally due to radiation from engine parts or jet emission and therefore temperatures up to 175°C were incorporated into the design. It was also projected that forgings, sheets, plates, and extrusions of the selected material would be needed.

At the time when the materials were selected, titanium was not considered because of the insufficient expertise prevailing in the United Kingdom relating to its processing and application to confidently and successfully build a supersonic transport of this material. On the other hand, there was considerable expertise and confidence that an aluminum alloy could be developed. Early on, the 7xxx alloys were rejected because they were known to lose their mechanical properties rapidly at temperatures slightly in excess of 100°C. This restricted the consideration to the 2xxx-type alloys. Four alloys were considered and evaluated: 2014, 2024, X2020, and RR58. 2014 and 2024 are very similar alloys that contain 0.5%Mg and 1.5%Mg, respectively, in addition to about 4.4%Cu. 2014 also contains 0.8% silicon (see Table 13-6). X2020 was an experimental alloy containing copper, cadmium, and lithium, and RR58 was a British engine alloy. The evaluation was done by extrapolating the results to obtain the tensile strength expected after soaking at temperatures for 30,000 hours. The results indicate that the X2020 alloy was outstanding at 130°C while the RR58 was best at 175°C. Between these two alloys, the X2020 alloy was found to have an inferior notched fatigue strength. Thus, the RR58 was selected and a special processing was developed to produce the different forms of products.

The applicability of the RR58 alloy was first tested for forged impellers in an engine that required good creep resistance at 175° to 250°C. The Concorde requirement was for a much longer life at a lower temperature. Normally, the strength at room temperature is favored by finer grain sizes, but the creep resistance decreases with a finer grain size. This was shown to be true for a sheet material of RR58, with about a 15-μm grain size, exhibiting a lower creep resistance than the forgings, having about a 200-μm grain size. Mechanical working (deformation) procedures were therefore modified with control of the amounts of cold work at each stage to optimize the grain size.

The heat treatment procedures were also optimized. The optimum heat treatment for sheet and plate was solution heat-treatment at 530°C for about 45 minutes, followed by quenching into cold water. Sheet forming is normally carried out immediately after solution treatment, but the required formability to produce some of the Concorde parts was not obtained from the alloy in this condition. Thus, a recovery annealing process was developed to avoid coarse grain size formation when normal annealing was used. Aging temperature was also optimized to balance room or lower temperature strength compared to the high-temperature creep resistance. Low-

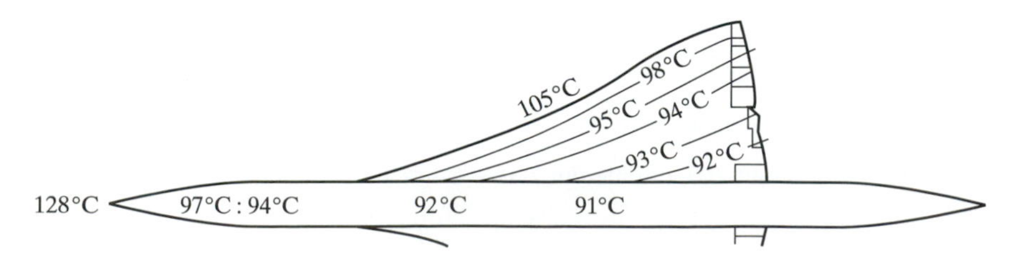

Fig. 16-10 Body and wing temperatures of Concorde[Ref. 5].

temperature aging produced higher low-temperature tensile strength, while high-temperature creep resistance is better with high-temperature aging. The optimized aging temperature was 20 hours at 190°C.

Special attention was also given to the residual stress and the susceptibility to stress corrosion of the forgings. In die-forgings with complex shapes, rough machining was carried out in the as-forged condition, then solution-heat treatment, and followed by quenching into boiling water. After artificial aging and final machining, the residual stresses were found to be less than 77 MPa (11 ksi). Stress-corrosion resistance is further improved by overaging at 215°C. In forgings of a more regular form, stress relieving was carried out by a cold-compression process, and there was no need for the overaging treatment at 215°C. Aging was done at 200°C.

The chemical composition of RR58 was also optimized. It was found that the silicon content must be held within the range 0.18 to 0.25% to avoid inferior properties. It was also necessary to balance the nickel and iron contents to about 1.1 percent each; an excess of iron forms the phase Al_3CuFe, while an excess of nickel forms the phase AlCuNi. In either case, the copper will be depleted and the hardening effect when the phase Al_2CuMg forms on aging is reduced.

The result of all the special processing and controls was to successfully confer the required creep resistance of the RR58 to place it ahead of its competitors. This is shown in Fig. 16-11 as RR58 SST (supersonic transport) processing.

16.3.1b Case Study 2—Materials Selection for NASA's Space Shuttle [6]. The National Aeronautics and Space Administration's (NASA) Space Transportation System (STS), commonly known as the Space Shuttle Orbiter, was designed to provide a system for transporting personnel, equipment, and supplies into low earth orbit. The system, shown in Fig. 16-12, consists of four major elements: the solid rocket boosters, the space shuttle main engines, the external tank, and the space shuttle orbiter. The orbiter houses the astronauts, the main engines, and the payloads and is launched into space in a piggy-back

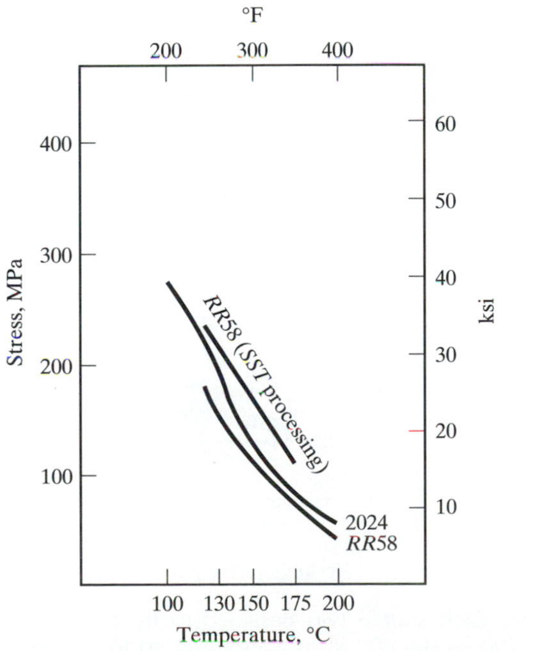

Fig. 16-11 Estimated stresses to produce 0.1% total plastic strain in 1000 hrs[Ref. 5].

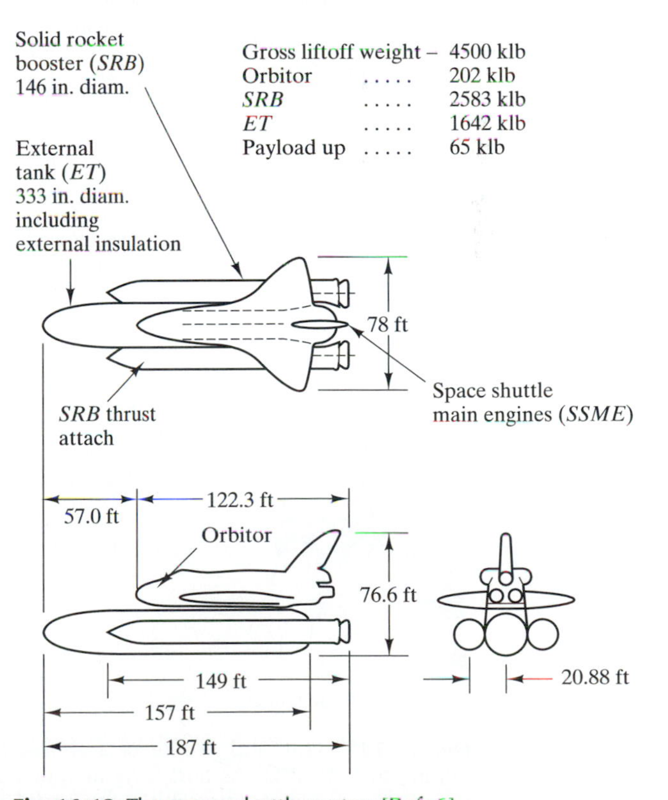

Fig. 16-12 The space shuttle system[Ref. 6].

position attached to the solid rocket boosters and the external tank.

A typical mission profile is shown in Fig. 16-13. About 89 percent of the thrust for the launch is provided by the solid rocket boosters. The remaining 11 percent is provided by the shuttle main engines, which burn hydrogen with oxygen supplied from the external tank. Both the rocket boosters and the external tank are separated from the orbiter a few minutes into the flight and are recovered from the ocean, cleaned, and reused. During reentry into the earth's atmosphere from space orbit, the shuttle orbiter experiences temperatures exceeding 1260°C (2300°F) on its lower fuselage and in excess of 1455°C (2650°F) along the leading edges and the nose cap. Maximum heating occurs at an altitude of about 45 km (150,000 ft) when the orbiter has slowed down to about Mach 8. At 15 km (50,000 ft), the orbiter is maneuvered aerodynamically to land as a glider.

The materials requirements for the design of the shuttle system included: (1) minimum weight (68,000 kg, or 150,000 lb, dry weight) and cost;

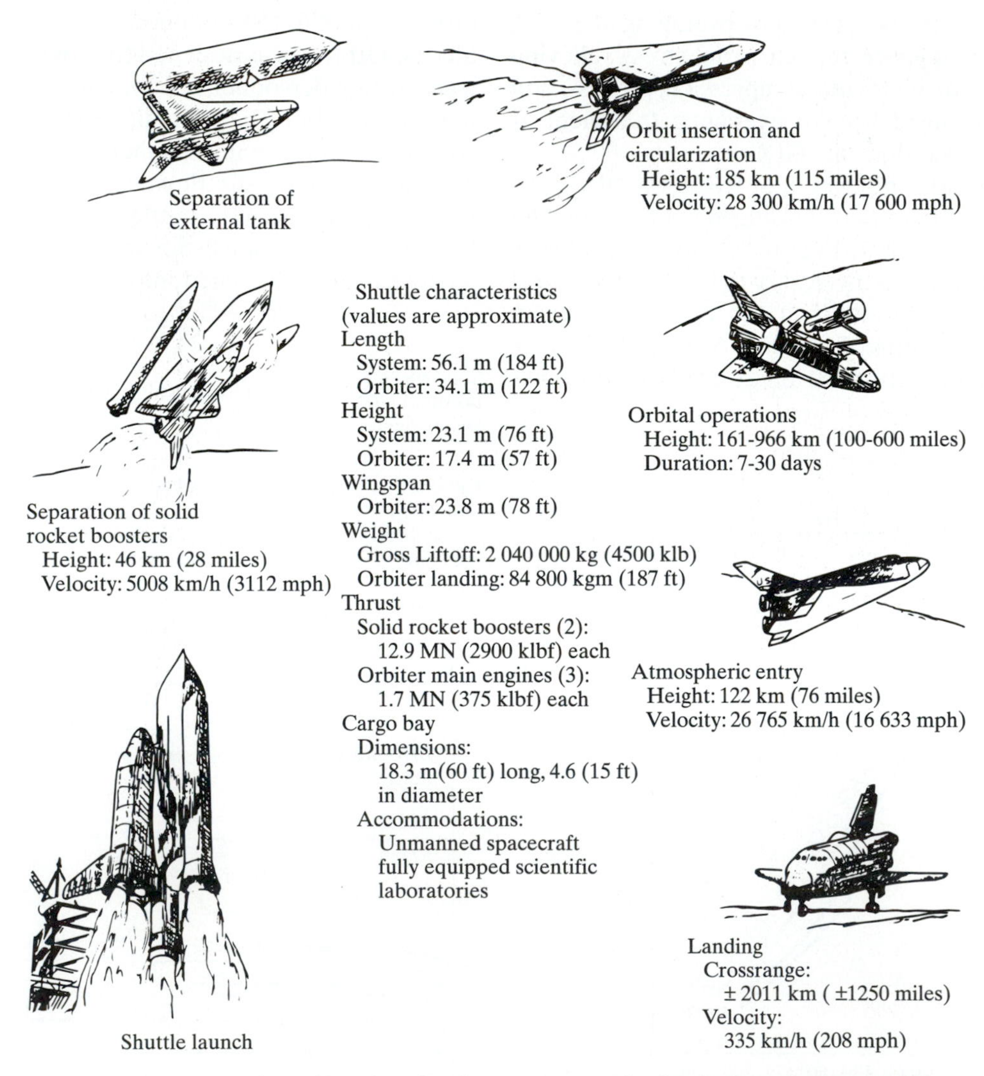

Fig. 16-13 Events and profile of a shuttle mission. Each shuttle was designed to fly a minimum 100 missions and carry as much as 29,500 kg (65,000 lb) of cargo and up to seven crew members into orbit. It can return 14,500 kg (32,000 lb) of cargo to earth. [6]

(2) exposure to extreme temperatures from −253°C (−423°F) for liquid hydrogen tanks to approximately 1455°C (2650°F) for the nose cap, leading edges and metallic pressure ports; (3) the shuttle must be electrically grounded to insure lightning protection and control of electromagnetic interference; (4) exposure to one of the most corrosive environments at the Kennedy Space Center due to heat, high humidity, salt air, and the daily condensation of dew onto the structures. The shuttle was originally designed to avoid exhibiting corrosion for a 10-year life, and it was anticipated that it would fulfill its 100-mission launch capability during this period.

The minimum weight requirement was an overriding requirement and it meant selecting materials for their inherent structural efficiencies, such as specific strengths, σ/ρ, over their corrosion resistance. Early studies indicated that the savings per pound of weight in the orbiter vehicle structure or systems amounted to over $30,000 during the mission life. Based on the added minimum cost requirement, the material choice for the major structural parts of the orbiter, such as the wings, tail, fuselage, and cabin is aluminum alloys. As in the commercial aircraft, the two significant aluminum alloys for this application are the 2xxx and the 7xxx alloys and because of the possible high-temperature exposure up to 175°C, the 2xxx alloys are used more than the 7xxx. This is the same reason for the choice of the 2xxx alloys for the Concorde. The typical materials used for the construction of the shuttle orbiter are shown in Fig. 16-14, where we see that the usage of the 2xxx alloys is more than the 7xxx alloys. This is a reversal of the relative usage of these two alloys in commercial aircraft, as shown in Fig. 13-1. The speed of commercial aircrafts is subsonic and therefore the skin temperatures are less than 100°C, which makes the 7xxx alloys better than the 2xxx alloys.

It was anticipated that stress-corrosion was a major problem for the structural materials exposed to the corrosive environment. As indicated in Chap. 10, stress-corrosion of a material will occur only when all the following three factors are present: (1) a susceptible alloy, (2) a specific environment, and (3) tensile stress. If any one of the three is eliminated, stress-corrosion will not occur. The approach used on the orbiter was to eliminate or minimize all the three factors to the fullest practical extent.

The selection of aluminum alloys was based on the susceptibility to stress-corrosion cracking (SCC) of the different tempers of the 2xxx and 7xxx alloys, especially in the short-transverse direction, as shown in Table 16-8. An added requirement was for the stress-corrosion threshold to be at least 170 MPa (25 ksi). (The stress-corrosion threshold is based on the $\mathbf{K}_{I_{SCC}}$ which is the plane-strain stress-intensity-factor threshold for a given material, environment, and temperature below which preexisting fatigue cracks do not propagate under static loads. The threshold value was obtained with standard tests in a salt spray environment for 30 days for a given crack size.) These requirements eliminated the T3, T4, and the T6XX tempers of nearly all the 2xxx and the 7xxx alloys since their SCC ratings in Table 16-8 are poor, and/or their stress-corrosion thresholds are lower than the minimum 170 MPa (25 ksi) required; the stress-corrosion thresholds could be as low as 48 MPa (7 ksi). For the 2xxx alloys, the T8XX tempers were used predominantly, although in a few cases, the T6 or T62 tempers were used for the 2124 and 2219 alloys. For the 7xxx alloys, the preferred tempers were the T73 and the T76.

In an effort to reduce the tensile stress factor in stress-corrosion, the alloys were ordered in the stress-relieved temper from the aluminum alloy producers (for example, T651 or T851) whenever possible. This temper condition also reduces machining distortion. Interference fits were limited to stress levels below 67 percent of the stress-corrosion thresholds. Tables were prepared to allow the materials and process engineers to ascertain stress levels resulting from interference fit pins and bushings into various size lugs. Residual stresses in assembly were minimized by shimming. Forming by bending, which puts the short-transverse direction in tension, was not permitted.

The elimination of the environment signified painting the structure. The requirements for the paint system were: (1) it must provide the corrosion protection for aluminum for a minimum exposure of 10 years to the seacoast environment without touch up; because (2) it must also serve as the base for the adhesively-bonded thermal protection system (TPS) ceramic tiles to protect the aluminum skins from the shuttle reentry (to the earth's atmosphere) temperatures indicated earlier; (3) the paint system must also endure a maximum temperature of 175°C (350°F), the maximum application temperature of the 2xxx

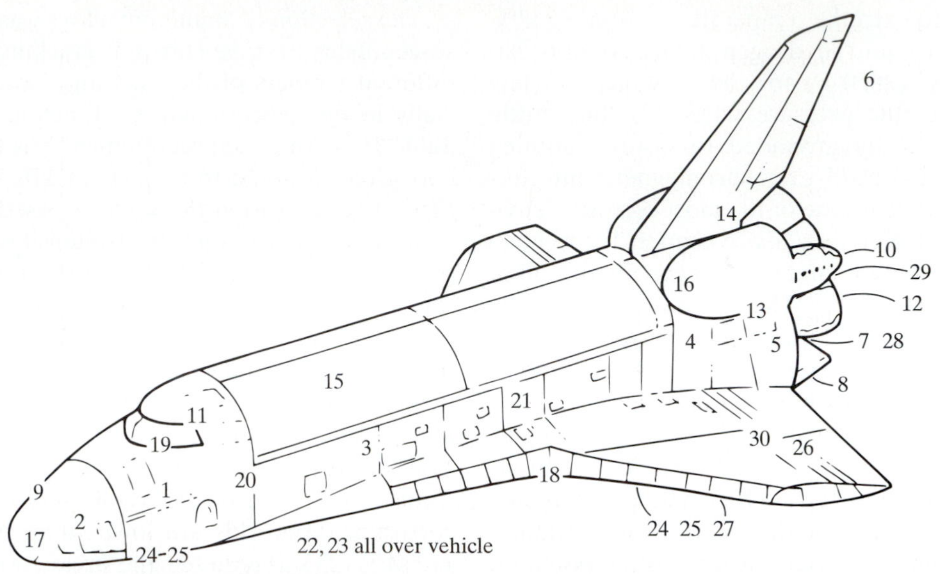

1. Cabin: aluminum alloy 2219
2. Forward fuselage: aluminum alloy 2024
3. Mid fuselage: aluminum alloys 2124, 7075
4. Aft fuselage: aluminum alloy 2124
5. Aft heatshield: aluminum alloy 2124
6. Tail: aluminum alloys 2124 and 2024
7. Engine mounted heat shield: Inconel alloy 625 sandwich
8. Body flap: aluminum alloy 2024
9. ACS engine nozzles: niobium alloy C103
10. OMS engine nozzles : nobium alloy FS-85
11. Window frames: S-65 Beryllium
12. Thrust structure: Ti-6AI-4V (diffusion bonded), Ti-6AI-4V and boron epoxy
13. APU: René 41 turbine, Hastelloy alloy B, Hastelloy alloy X, René 95
14. Tail conical seal: Inconel alloy 718 sandwich
15. Cargo bay doors: Graphite/epoxy
16. OMS pod structure: Graphite /epoxy
17. Nose cone: RCC
18. Wing leading edges:RCC
19. Windows: Quartz (outer), alumosilicate (inner)
20. Pressure vessels: Kevlar-Ti-8AI-4V, Kevlar-Inconel alloy 718, Ti-6AI-4V, aluminum alloy 2219
21. Mid fuselage support tubes: Boron aluminum, diffusion bonded Ti-6AI-4V
22. Plumbing systems: Type 304 and 21-6-9 stainless steels, Ti-3AI-2.5V, Inconel alloy 718
23. High-strength fasteners: A-286, MP35N. Inconel alloy 718
24. Landing gear: 300M steel, aluminum alloy 7075
25. Brakes: Beryllium, carbon-carbon
26. Flipper doors/rub panels: Ti-6AI-4V, Inconel alloy 625
27. ET door: Beryllium S-65
28. High-temperature fasteners: Udimet 500
29. Main engine nozzle: Inconel alloy 718, A-286 tubes
30. Elevon cove seal: Niobium alloy C103

Fig. 16-14 All the major structural and corrosion resistant materials make the construction of the space shuttle[Ref. 6].

Table 16-8 Relative SCC resistance ratings of heat-treatable high-strength wrought aluminum alloys.

Resistance ratings are as follows: A, very high; B, high; C, intermediate; D, low.

Alloy and temper(a)	Test direction(b)	Rolled plate	Rod and bar(c)	Extruded shapes	Forgings
2011-T3.-T4	L	(d)	B	(d)	(d)
	LT	(d)	D	(d)	(d)
	ST	(d)	D	(d)	(d)
2011-T8	L	(d)	A	(d)	(d)
	LT	(d)	A	(d)	(d)
	ST	(d)	A	(d)	(d)
2014-T6	L	A	A	A	B
	LT	B(e)	D	B(e)	B(e)
	ST	D	D	D	D
2024-T3, -T4	L	A	A	A	(d)
	LT	B(e)	D	B(e)	(d)
	ST	D	D	D	(d)
2024-T6	L	(d)	A	(d)	A
	LT	(d)	B	(d)	A(e)
	ST	(d)	B	(d)	D
2024-T8	L	A	A	A	A
	LT	A	A	A	A
	ST	B	A	B	C
2048-T851	L	A	(d)	(d)	(d)
	LT	A	(d)	(d)	(d)
	ST	B	(d)	(d)	(d)
2124-T851	L	A	(d)	(d)	(d)
	LT	A	(d)	(d)	(d)
	ST	B	(d)	(d)	(d)
2219-T3, -T37	L	A	(d)	A	(d)
	LT	B	(d)	B	(d)
	ST	D	(d)	D	(d)
2219-T6, -T8	L	A	A	A	A
	LT	A	A	A	A
	ST	A	A	A	A
6061-T6	L	A	A	A	A
	LT	A	A	A	A
	ST	A	A	A	A
7005-T53.-T63	L	(d)	(d)	A	A
	LT	(d)	(d)	A(e)	A(e)
	ST	(d)	(d)	D	D
7039-T63.-T64	L	A	(d)	A	(d)
	LT	A(e)	(d)	A(e)	(d)
	ST	D	(d)	D	(d)
7049-T73	L	A	(d)	A	A
	LT	A	(d)	A	A
	ST	A	(d)	B	A
7049-T76	L	(d)	(d)	A	(d)
	LT	(d)	(d)	A	(d)
	ST	(d)	(d)	C	(d)
7149-T73	L	(d)	(d)	A	A
	LT	(d)	(d)	A	A
	ST	(d)	(d)	B	A
7050-T736	L	A	(d)	A	A
	LT	A	(d)	A	A
	ST	B	(d)	B	B
7050-T76	L	A	A	A	(d)
	LT	A	B	A	(d)
	ST	C	B	C	(d)
7075-T6	L	A	A	A	A
	LT	B(e)	D	B(e)	B(e)
	ST	D	D	D	D
7075-T73	L	A	A	A	A
	LT	A	A	A	A
	ST	A	A	A	A
7075-T736	L	(d)	(d)	(d)	A
	LT	(d)	(d)	(d)	A
	ST	(d)	(d)	(d)	B
7075-T76	L	A	(d)	A	(d)
	LT	A	(d)	A	(d)
	ST	C	(d)	C	(d)
7175-T736	L	(d)	(d)	(d)	A
	LT	(d)	(d)	(d)	A
	ST	(d)	(d)	(d)	B
7475-T6	L	A	(d)	(d)	(d)
	LT	B(e)	(d)	(d)	(d)
	ST	D	(d)	(d)	(d)
7475-T73	L	A	(d)	(d)	(d)
	LT	A	(d)	(d)	(d)
	ST	A	(d)	(d)	(d)
7475-T76	L	A	(d)	(d)	(d)
	LT	A	(d)	(d)	(d)
	ST	C	(d)	(d)	(d)
7178-T6	L	A	(d)	A	(d)
	LT	B(e)	(d)	B(e)	(d)
	ST	D	(d)	D	(d)
7178-T76	L	A	(d)	A	(d)
	LT	A	(d)	A	(d)
	ST	C	(d)	C	(d)
7079-T6	L	A	(d)	A	A
	LT	B(e)	(d)	B(e)	B(e)
	ST	D	(d)	D	D

(a) Ratings apply to standard mill products in the types of tempers indicated and also in Tx5x and Tx5xxx (stress-relieved) tempers and may invalidated in some cases by use of nonstandard thermal treatments or mechanical deformation at room temperature, by the user.
(b) Test direction refers to orientation of direction in which stress is applied relative to the directional grain structure typical of wrought alloys, which for extrusion and forgings may not be predictable on the basis of the cross-sectional shape of the product: L, longitudinal: LT, long transverse: ST, short transverse,
(c) Sections with width-to-thickness ratios equal to or less than two, for which there is no distinction between LT and ST properties.
(d) Rating not established because product not offered commercially,
(e) Rating is one class lower for thicker sections: extrusions, 25mm (1 in.) and thicker: plate and forgings, 38 mm (1.5 in.) and thicker.

alloys that might arise when the reentry temperatures might be transferred by conduction to the surface. It must also be capable of withstanding space vacuum and low-temperatures (−155°C or −250°F) without degradation, and (4) it must exhibit minimum off-gassing to avoid giving off toxic fumes (hazard for crew members) inside the cabin or the condensation of volatile material on windows or optical (thermal) control surfaces.

The paint system chosen was a chromate inhibited, epoxy polyamine primer that was found to be tough, abrasion resistant, and durable. Surfaces to be painted were either anodized according to MIL-A-8625 type II, class I, or chemically filmed according to MIL-C-5541, class IA. Each coat of paint was 0.015 to 0.023 mm (0.6 to 0.9 mil) thick and demonstrated corrosion protection to aluminum after 1500 hours of salt spray exposure, even in areas scratched through to bare aluminum. Surfaces, to which the TPS tiles were bonded, achieved additional corrosion protection from the 0.13 to 0.23 mm (5 to 9 mils) thick room-temperature vulcanized (RTV) adhesive layer used to bond the tiles.

The finish paint system for the shuttle, Discovery and Atlantis, followed the scheme shown in Table 16-9. The original plan to put two coats of primer in the crew compartment was replaced with an anodized layer and one primer. This substitution realized substantial weight savings. Each coat of the chromated epoxy polyamine primer to cover 8157 m^2 (88,000 ft^2) of surface weighed more than 500 kg (1100 kg).

Table 16-9 Finish paint system for the shuttles, Discovery and Atlantis.

Area	Coating
Exterior TPS surface	Anodize + 1 coat chromated epoxy polyamine primer + 1 coat RTV adhesive
Exterior non-TPS surfaces	Anodize + 1 coat chromated epoxy polyamine primer
Interior surface	Anodize + either 1 coat chromated epoxy polyamine primer or 1 coat polyurethane
Crew compartment	Chemical film or anodize + 1 coat chromated epoxy polyamine primer + 1 coat polyurethane or anodize only

Based on the above considerations, the forward fuselage of the shuttle was fabricated as a sheet metal skin-stringer using the 2024-T6 alloy. Suspended inside the forward fuselage was an all-welded aluminum pressurized cabin made from 2219-T6 and -T8 alloys. The cabin is approximately conical in shape, about 5 m (17 ft) long and tapering from 5 to 2.4 m (17 to 8 ft) in diameter at its forward end. The mid and aft fuselage structures were machined from 2124-T851 alloy plate. Major frames were made of 7075-T76 or -T73 alloys.

Aluminum honeycomb sandwich structure was extensively used in the wing and body flap areas. The honeycomb core is usually made of thin (0.025 to 0.075 mm, or 1 to 3 mil) 5056-H39 alloy. The face sheet skin of this sandwich structure could not be alclad because corrosion, if it occurs, will proceed in the plane of the sheet and will delaminate the bond line. To prevent corrosion, the honeycomb cores were protected with conversion coatings and were nonperforated; the face sheets used corrosion-resistant adhesive primers, and the sandwich assemblies were sealed at the edges to prevent water entry.

Summary: The high-strength 2xxx and 7xxx aluminum alloys are the dominant materials for aircraft structures because of their high specific moduli and strengths (over titanium and steel) and because of cost (over titanium). The 7xxx alloys exhibit higher-strengths at low-temperatures but the strengths decrease rapidly above 100°C. For this reason, the 7xxx alloys are used predominantly for structural frames inside the aircraft (spacecraft) and for the skin of a commercial aircraft that flies at subsonic speed where the skin temperature remains below 100°C. At supersonic speeds, when the skin temperature may exceed 100°C, the 2xxx alloys are used. Different temper conditions of these alloys exhibit varying degrees of susceptibility to stress corrosion cracking (SCC) and final selection of processing (temper condition) must take this into account. At higher skin temperatures (higher Mach numbers) and higher loading index, titanium alloys are used.

16.4 Materials Selection for the Automotive Industry

The automotive industry has always been known to be very competitive, as far as materials usage is concerned. Knowledgeable persons know that a one-cent per pound cost differential in materials can be translated into huge profits or savings for the automotive companies. Materials usage is also determined by the ease in the processing and fabricability of components for autos. Ever since the early 1970s, this competition for materials usage has become fiercer and it has not let down. The following events explain this situation:

- Legislation to minimize repairs by imposing a threshold of 5 mph (miles per hour) speed before damage to cars can occur—this led to front-and rear-energy absorbing bumpers.
- The Clean Air Act in the 1970s and an even cleaner air act (emissions control) in the 1990s.
- The energy crisis in the early 1970s led to efforts to curb the consumption of fossil fuels (predominantly imported oil). This has led to the legislation of the 55 mph speed limit and the mandate that automotive companies must have a fleet Corporate Average Fuel Efficiency (CAFE) of 28 mpg (miles per gallon) by the mid-1980s.
- Recycling of materials used in the automobile.
- The 1993 Partnership for the Next Generation of Vehicles (PNGV) whose goal is to build an 80 mpg car—with the performance and roominess of today's sedans.

The above events have imposed on the automotive companies to minimize: (1) fuel consumption, (2) pollutants given off by the combustion of the fuel, and (3) solid waste. Minimization of fuel consumption was done primarily through the reduction in the gross inertia vehicle weight. For steels, the traditional materials, it meant reducing the thickness and the development of the *high-performance, high-strength steels*. It also meant that some steels have to be replaced by lighter materials such as *aluminum* and *plastics*. Most of these plastics were originally the thermoset varieties. It was realized in the late 1980s and early 1990s that they were accelerating the exhaustion of available landfills. Up to that time, there was steady growth in the usage of thermoset plastics and this realization reversed the trend in the early 1990s. The thermoset plastics are being replaced by the *engineering thermoplastics* (ETPs) that are reformable and are recyclable.

In addition to using lightweight materials, a more efficient combustion of the fuel using higher engine operating temperatures was undertaken; this led to the use of structural ceramics in engines. Cars were also designed to minimize aerodynamic drag and structural designs were incorporated to increase the performance efficiencies of the materials. Further pressure by lower limits of maximum emission of pollutants requires the installation of emission control equipment that increases the weight of the vehicle. This increases the pressure to further reduce weight and has led to the use of *diecast magnesium* (the lightest structural metal) components in cars.

Iron and steel comprise about 55 percent of the gross weight of today's autos. This is threatened however by the PNGV, an alliance between the government and the Council for Automotive Research (CAR) in the United States. At first glance, aluminum appears to have (by default) the driver's seat in this project because of its inherent lightweight and high specific strength (i.e., strength-to-weight ratio). Using an all aluminum body construction, it was shown that as much as 170 kg (380 lb) of the weight of the same body made of steel can be reduced [7a]. It remains to be seen whether it can replace steel as the dominant material for autos due to some processing and manufacturing difficulties, for instance, welding and its cost relative to steel. From the steel side, global steel producers formed a consortium to accept the challenge of the PNGV. The consortium commissioned Porsche Engineering Services, a North American unit of Porsche AG, to design the lightest possible steel-bodied automobile. This project has produced the UltraLight Steel Auto Body (ULSAB) that involved innovative design, fewer parts, and more efficient manufacturing. The ULSAB project used holistic (simultaneous or concurrent) engineering, high-strength steels (more than two-thirds of the weight), and advanced technologies to obtain comparable weight savings as the aluminum body car[7b].

Among the competing materials, which include steel, aluminum, magnesium, plastics, and reinforced plastics, materials selection for high volume-production structures is based on product performance characteristics. Among these qualities are strength, stiffness, fracture toughness, creep, fatigue limit, corrosion resistance, and factors such as weight, cost, and fabricability including forming, welding, and machining. The product performance characteristics have to match the performance requirements—part geometry, the envelope (the space allowed by the design for the part to function), and assembly. For example, light-truck frames are generally made of steel because the design calls for a platform frame in which the section height of the beam-in-bending is limited. For this structure, the product of EI (E = modulus of the material, and I = moment of inertia of the section) controls the bending performance. Since I is proportional to d^3 (d is the depth or height of the beam section), a limited section height requires a material with a high modulus. Consequently, the space restriction and part-durability criteria for the truck platform frame requires the selection of steel [7c].

The engine cradle is generally constructed as a platform that limits the bending-strength section height to under 100 mm (4 in.). This section height allows aluminum extrusion shapes, such as boxes, and channels, with the required thickness to achieve the same bending strength and stiffness as the current stamped-steel design. In applications where new materials are used, finite element analysis (FEA) models help to tailor the section to the envelope (space) and bending constraints. A recent redesign of a midsize sedan engine cradle using aluminum extrusions doubled lateral stiffness, and reduced the weight by 25 percent. The aluminum-intensive Audi A8 has formed extrusions coupled together for its space frame structure [7a]. If magnesium extrusions are available with sufficient corrosion resistance, further savings in weight may be realized, as indicated by Table 16-10[8] that shows weight savings of magnesium relative to aluminum and steel for equal bending strength and equal stiffness.

In components requiring high tensile strength, fiber-reinforced plastics or composites can save weight up to 50 percent and increase the life of the components. Examples are fiberglass springs, drive shafts, and natural-gas fuel tanks for automobiles.

16.4.1 Structural Ceramics for Engine Components

The requirements of higher fuel efficiency and cleaner emissions from combustion engines, both gasoline and diesel, can be partiallly met by higher engine operating temperatures. To operate at higher temperatures means that ceramic engine components must be used in place of traditional metallic materials, predominantly the nickel alloys. In addition to increasing fuel efficiency and cleaner combustion emissions, higher operating temperature increases the engine's power output. When ceramics are used, the other properties of light weight, hardness, and corrosion resistance also induce beneficial effects in faster responses of turbomachinery, better wear resistance, and longer life. Some ceramic engine components that are being used are listed in Table 16-11[9]. We shall describe here the swirl chamber and the turbocharger rotor, both of which traditionally use nickel alloys.

16.4.1a The Swirl Chamber[9b, 10]. This is the precombustion chamber in indirect-injection diesel engines, which is typical of high-speed diesel engines used for passenger cars and light trucks. It is basically a cup, shown in Fig. 16-15[10], which contains the initial combustion; the combustion gases exit through an angled hole in the bottom of the cup into the cylinder. The substitution of a ceramic pre-combustion chamber for the traditional high-temperature nickel alloy saves cost and weight. Furthermore, the better insulating characteristics of ceramics promotes hotter and faster combustion, which in turn improves noise and exhaust emissions. Clearly, the precombustion material must withstand the initial ignition and the continuous thermal up and down-shocks and must therefore exhibit excellent thermal shock resistance.

The design objectives and the methods successfully used to incorporate the "swirl-chamber" in the Toyota Crown 2L-THE diesel engine will now be described. The objective was to increase the maximum 930°C operating temperature observed with the nickel alloy chamber to about 970°C with the ceramic chamber. This change in operating temperature was expected to increase the power output of the engine from 96 to 106 PS, as shown in Fig. 16-16[10]. This power increase represents an increment over the nor-

Table 16-10 Potential weight savings when magnesium is used to achieve the same bending strength and stiffness. [8]

Material	Thickness	Bending strength	Stiffness	Weight
For equal thickness				
1025 steel	100	100.0	100.0	100.0
6061-T6 aluminum sheet and extrusions	100	97.2	34.5	34.5
AZ31B magnesium extrusions	100	47.2	22.4	22.5
ZK60A-T5 magnesium extrusions	100	88.9	22.4	22.5
AZ31B-H24 magnesium sheet	100	73.4	22.4	22.5
For equal bending strength				
1025 steel	100	100	100.0	100.0
6061-T6 aluminum sheet and extrusions	101	100	35.8	34.8
AZ31B magnesium extrusions	146	100	69.2	32.9
ZK60A-T5 magnesium extrusions	106	100	26.7	23.9
AZ31B-H24 magnesium sheet	117	100	35.6	26.3
For equal stiffness				
1025 steel	100	100	100	100.0
6061-T6 aluminum sheet and extrusions	143	199	100	49.2
AZ31B magnesium extrusions	165	129	100	37.2
ZK60A-T5 magnesium extrusions	165	242	100	37.2
AZ31B-H24 magnesium sheet	165	200	100	37.2
For equal weight				
1025 steel	100	100	100	100
6061-T6 aluminum sheet and extrusions	290	817	841	100
AZ31B magnesium extrusions	444	930	1962	100
ZK60A-T5 magnesium extrusions	444	1753	1962	100
AZ31B-H24 magnesium sheet	444	1451	1962	100

Note: Comparison made at room temperature for rectangular beams of constant width with the following minimum yield strengths: 1025 steel, 250 MPa (36 ksi); 6061-T6 aluminum, 240 MPa (35 ksi); magnesium alloys, average of minimum tensile yield and compresive yield strengths. All comparisons expressed in percent.

mal linear increase due to engine displacement, as shown in Fig. 16-17[10].

In order to use ceramics, it was necessary to incorporate the brittle behavior of ceramics in the design, and to reduce stress concentration and tensile stresses on the part. Ceramics cannot tolerate many flaws or defects because of their inherent low fracture toughness and therefore must be processed to minimize the defects. To reduce the stress concentration on the part, it was necessary to redesign the part as shown in Fig. 16-18[10]. Finite element analysis (FEA) models indicated that large tensile thermal stresses occurred at the upper and lower outer surfaces, as shown in Fig. 16-19[10]. In order to reduce these external surface stresses, the outer metal ring shown in Fig. 16-18 was shrink-fitted to the chamber. During engine operation, the cylinder head tends to deform elliptically due to the temperature gradient, as shown in Fig. 16-20[10]. To accommodate this deformation, a clearance on the side was provided as indicated in Fig. 16-15. There is also a restraining force between the swirl chamber and the cylinder head gasket. To reduce this stress, another clearance was provided between

Table 16-11 Commercial ceramic engine components in some cars[Ref. 10a].

Components	Production start	Engine	Manufacturer
Glow plug	1981	Isuzu	Kyocera
Swirl chamber	1983	Isuzu	Kyocera
Glow plug	1983	Mitsubishi	Kyocera
Intake heater	1983	Isuzu	Kyocera
Swirl chamber	1984	Toyota	Toyota
Rocker arm tip	1984	Mitsubishi	NGK
Glow plug	1985	Mazda	Kyocera
Glow plug	1985	Nissan	NTK
Turbocharger rotor	1985	Nissan	NTK
Rocker arm tip	1987	Nissan	NTK
Turbocharger rotor	1988	Isuzu	Kyocera
Link injector	1989	Cummins	Toshiba
Turbocharger rotor	1989	Toyota	Toyota
Turbocharger rotor	1990	Toyota	Kyocera
Rocker arm tip	1990	Mazda	Kyocera

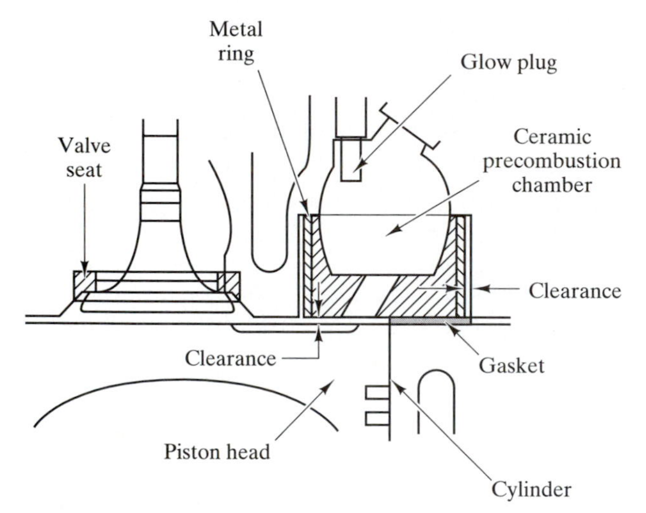

Fig. 16-15 Ceramic precombustion (swirl) chamber for the indirect-injection diesel engine. [10]

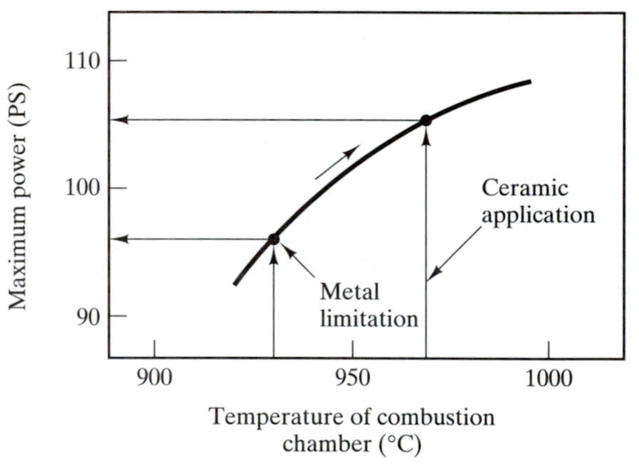

Fig. 16-16 Maximum power as function of combustion temperature[Ref. 10].

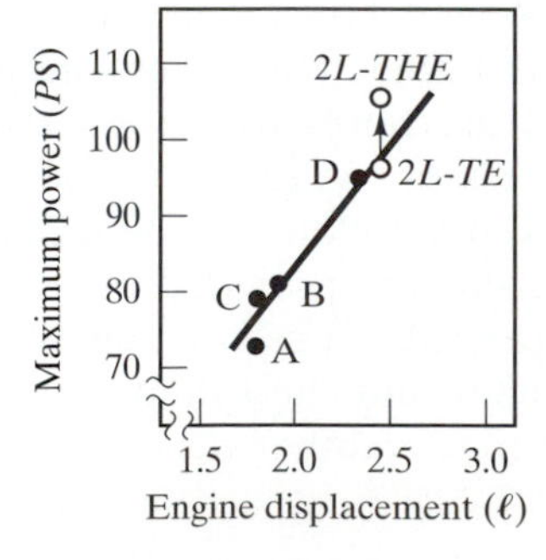

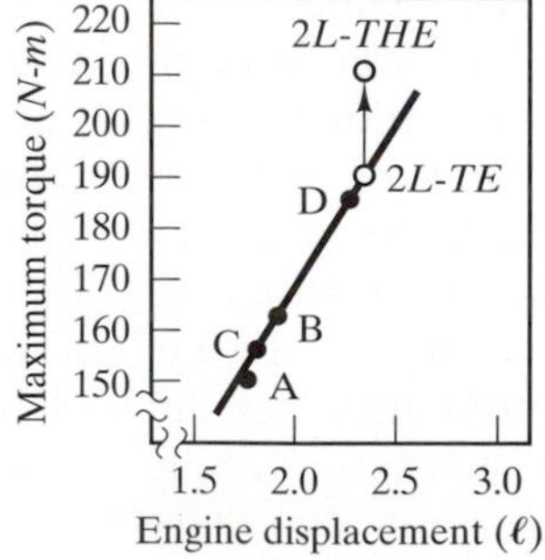

Fig. 16-17 Incremental increase over the linear relations of engine power and torque as function of engine displacement due to increased temperature. [10]

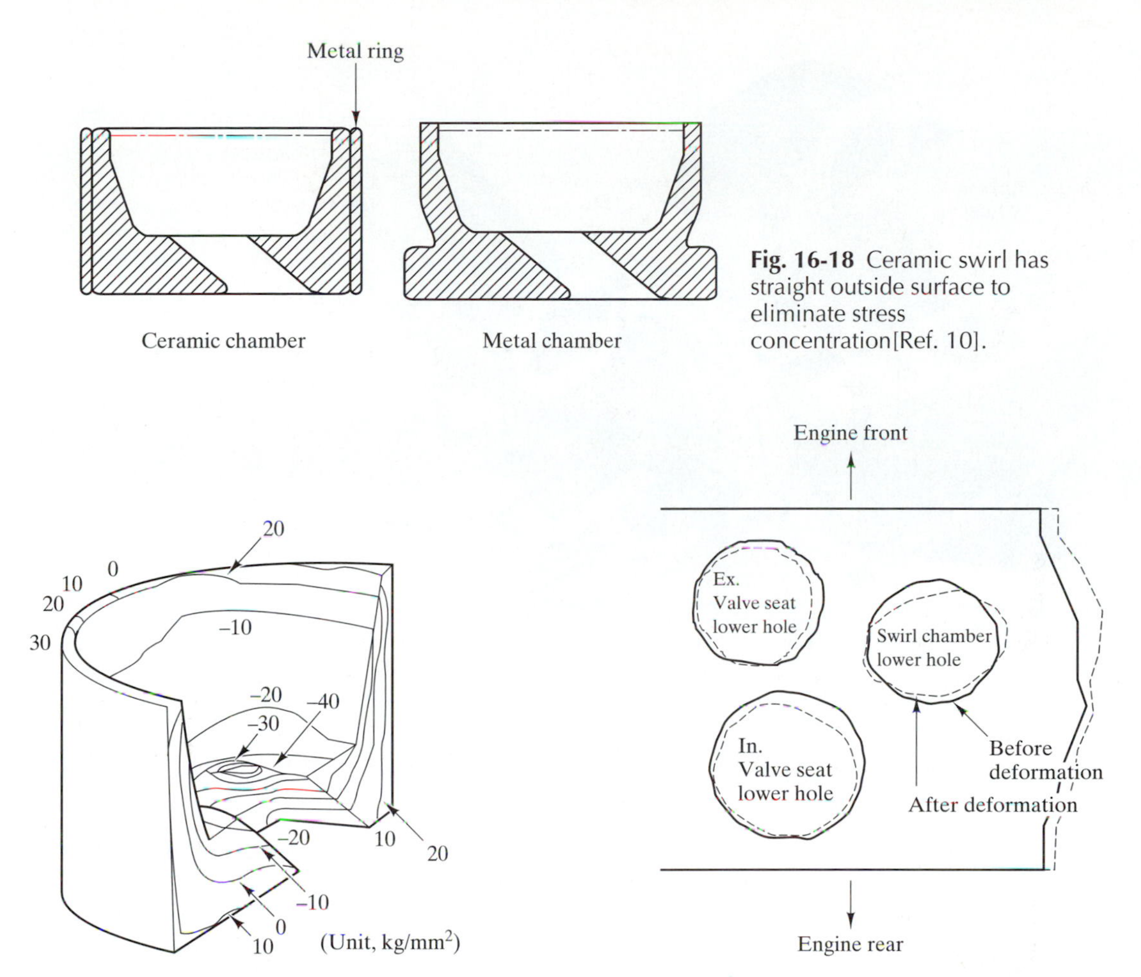

Fig. 16-18 Ceramic swirl has straight outside surface to eliminate stress concentration[Ref. 10].

Fig. 16-19 Predicted thermal stress profile by finite element modeling [Ref. 10].

Fig. 16-20 Deformation of cylinder head due to thermal stresses[Ref. 10].

the cylinder head and the chamber, as shown in Fig. 16-15.

16.4.1b Turbocharger Rotor[9b,c,d]. An automotive turbocharger, shown in Fig. 16-21[9d], is a device that takes the hot exhaust gas from the engine to drive a turbine wheel on one end of the shaft of the turbocharger. On the opposite end of the shaft is a compressor wheel that is driven by the turbine wheel and compresses the intake air. This allows a larger quantity of fuel to be burned in the engine per unit time and increases the power output relative to that of a nonturbocharged engine. When the system as a whole is properly designed, it can improve either the specific fuel consumption, the available torque, or reduce pollutants.

Depending on the specific application, turbine wheel temperatures can exceed 1000°C (1830°F). Rotational speeds can range from 90,000 rpm in larger units used on large-displacement diesel engines to over 200,000 rpm in smaller turbochargers used for passenger cars. At such speeds, the centrifugal force can reach peak stresses of 260 MPa (38 ksi). These speeds are necessary for the compressor wheel to provide air at rates ranging from 10 to 40 kg/min (22 to 88 lb/min) at pressures of 100 to 300 kPa (15 to 45 psi).

The replacement of the high-temperature nickel alloy turbine wheel with a ceramic turbine can produce weight savings of about 60 percent. This translates to approximately 50 percent weight and inertia saving for the rotating assembly (i.e, the turbine, the shaft, and the compressor). Figure 16-22[9b] shows the test

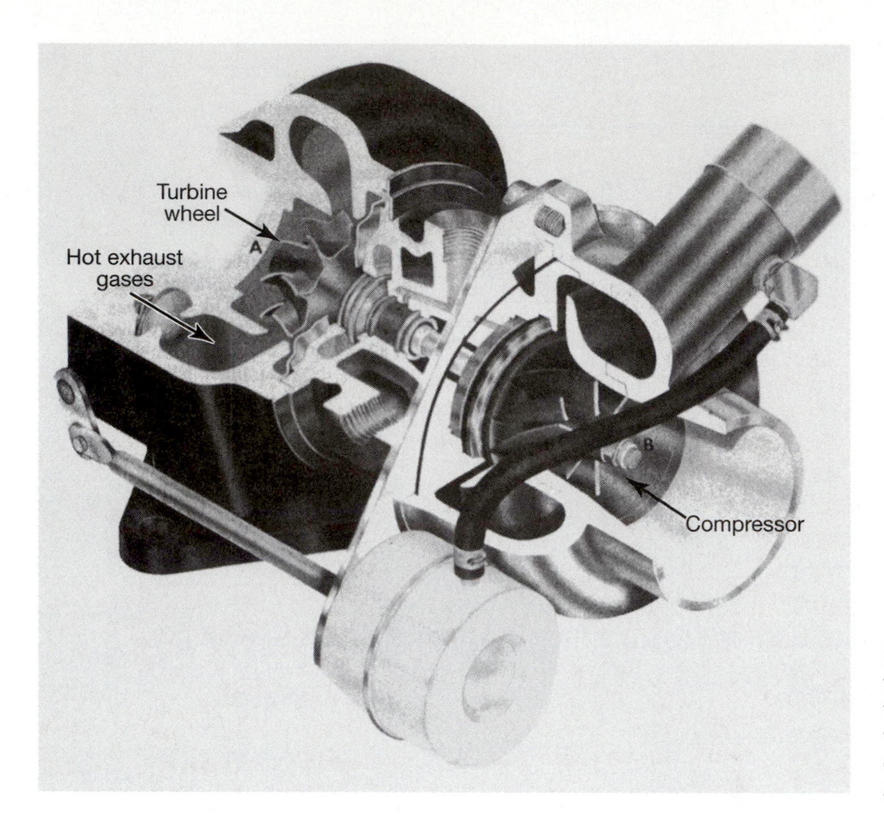

Fig. 16-21 Cut-out photo of a turbocharger. Hot exhaust gas from engine drives the turbine wheel which in turn drives the compressor for the intake air[Ref. 9d].

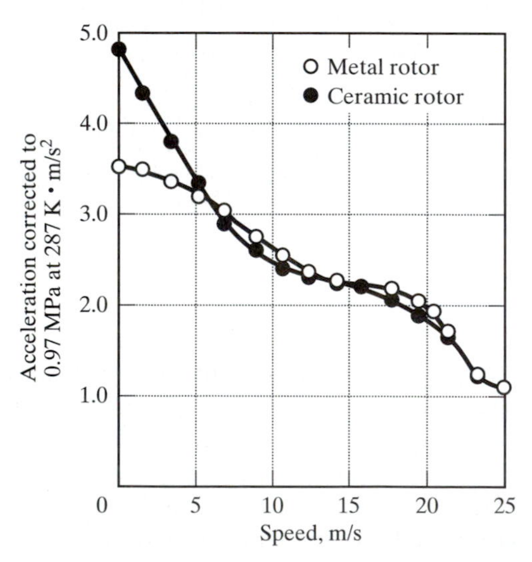

Fig. 16-22 Relative wide open throttle acceleration of ceramic and metal turbocharger rotors in 2.3L engine. [9b]

data comparing initial vehicle acceleration using a ceramic turbocharger versus a nickel-base alloy turbocharger; the turbocharger response time from idle to maximum is halved and this significantly improves the acceleration feel of the 1360 kg (3000 lb) test car in which it was installed.

It is clear that the ceramic material for the rotors must have low density, high-temperature strength, oxidation resistance, toughness, and thermal shock resistance to withstand cold-start conditions. In addition, cold-starts, shutdowns, accelerations, and decelerations result in transient thermal gradients that can induce severe stresses. The ceramic materials for the blades must have sufficient as-fired surface and volumetric strength at temperature to survive at the stress levels induced by centrifugal forces and thermal gradients.

16.4.1c Selection of the Ceramic Material. The common structural ceramics considered for engineering design are either oxides or nonoxides. The oxides are mainly based on alumina (Al_2O_3) and zirconia (ZrO_2), while the nonoxides are silicon carbide (SiC) and silicon nitride (Si_3N_4). The properties of these materials are given in Table 16-12.

Based on the above properties and the requirements of the applications stated earlier, the nonoxide ceramics, silicon nitride and silicon carbide, have better high-temperature strength, are lighter, and have

Table 16-12 Properties of oxide and nonoxide engineering ceramics.

⇓ Properties/Ceramics ⇒	Al_2O_3	ZrO_2	Si_3N_4 SSN*	Si_3N_4 RBSN*	SiC
Bend Strength, MPa, @RT	440	1020	900	350	500
@1000°C	340	450@800°C	600	350	500
Fracture Toughness, K_{Ic}, $MPa\sqrt{m}$	4.5	8.5	7	6	4.5
Thermal Shock Resistance, ΔT°C	200	350	900	600	370
Density, gm/cm^3	3.97	5.68–6.10	3.20	3.12	3.18

*SSN is sintered silicon nitride, while RBSN is reaction-bonded silicon nitride.

very good thermal shock resistance. Between silicon nitride and silicon carbide, silicon nitride is the better material because of its toughness and thermal shock resistance and is in fact the material chosen for the applications given in Table 16-11. The higher toughness allows the material to withstand foreign object damage better, while the thermal shock resistance allows it to withstand sudden changes in temperature, as in cold-starts.

Summary: Structural ceramics are used in combustion engines because of their inherent high-temperature and corrosion-resistant properties. Because they are light materials, they induce faster acceleration and fuel efficiency. However, the inherent brittleness and variability of the properties of these materials must be incorporated in the design.

16.4.2 Application of Concurrent Engineering in Product Development [11]

Concurrent engineering is also referred to as simultaneous or holistic engineering because it considers, at the design stage of the product or component, all the requirements from the material, the manufacturability of the product, and the expected use, performance, and life of the product. We shall follow an example of this approach in the actual successful design of a new automotive structural cross member that results in considerable savings in time and cost.

Case Need: Over more than 25 years, the automotive cross member shown in Fig. 16-23 has been used. Production problems were experienced in that some splits occurred during the forming operation. It was decided to redesign a new cross member and the concurrent engineering approach was utilized.

Concurrent Engineering Approach: This approach, indicated as *B* in Fig. 16-24, is compared with the sequential or "compartmentalized" design approach, indicated as *A* in Fig. 16-24. In the old approach, the product was designed by the designer based on the customer's needs, without inputs from either the materials engineer, the fabricating engineer, or the tool engineer. Once it was designed, the product testing and development stage was conducted that consisted of sequentially designing the tool (soft), building the tool (hard), and then trying out materials and different processes for the product. Then the blank development was completed that consisted of trying the blank die and conducting pilot runs. If there were any problems in this process, it went back to the designer, the sequential process was repeated, and there was a significant loss of time and effort.

In the concurrent engineering approach, all the problems that may arise are preempted at the outset by bringing together the designer, the materials engineer, the manufacturing or fabricating engineer, and the tool engineer to discuss and to set the design parameters from the requirements of the product (customer specification). The five different indicated tasks are performed simultaneously by analyzing and synthesizing the different trade-offs of the requirements to obtain an optimized product in terms of materials, ease of fabrication, tooling, and overall cost. The process of optimization in this concurrent approach is facilitated by the availability of computers and softwares. These model the processes and can make predictions on: the selection of materials; forming and nesting; the stresses and equipment needs

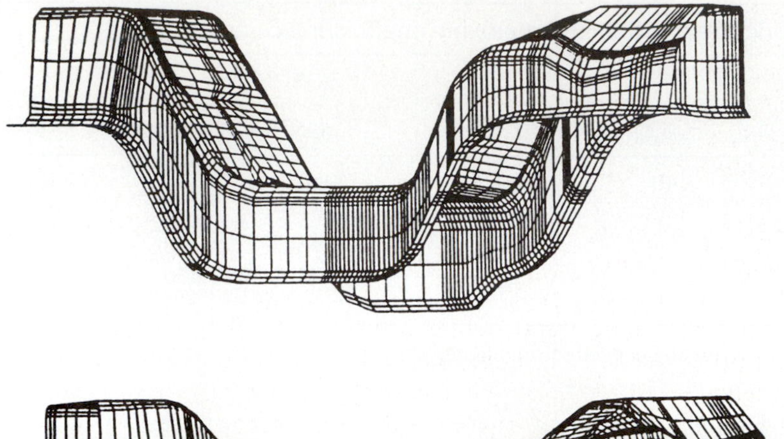

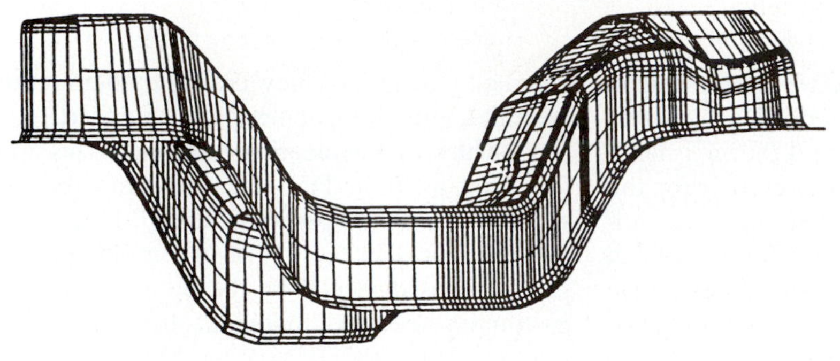

Fig. 16-23 Production automotive structural cross member. Grids or mesh of lines is for the numerical finite element modeling. [Ref. 11].

A. Sequential or compartmentalized design

Start → End

Product design	Soft-tooling (design)	Hard-tooling (build)	Product development	Process development	Die tryout	Pilot
					(Blank development)	

Product testing & development

B. Concurrent or simultaneous engineering design

Start → End ←← Save 33% of time →→

Computer aided product and process design & development	Build hard tooling (blank die)	Die tryout	Pilot

Product testing & development

Tasks →

(1)	(2)	(3)	(4)	(5)
Forming and nesting, predict blank shape, severity, thin-out, residual stresses	Determine tooling costs and number of operations required	Determine required press capacity in terms of forces and energies	Determine final part performance using deformed material properties	Predict blank die shape and spotting

Fig. 16-24 (A) Sequential or compartmentalized design procedure, (B) concurrent engineering design.

(press) during the fabrication; and on the tooling required. The need for laboratory mockups, hard prototypes, and physical testing is not eliminated, but modeling significantly reduces the amount of time required. For the case considered here, as much as one-third of the time used in the sequential approach was saved, as indicated in Fig. 16-24.

Tasks 1 and 4 relate to the material's properties and performance and will be discussed here. Forming and nesting evaluate the structure to ensure that it is formable. Inputs required include material type, thickness, and forming press characteristics. The forming simulation (modeling) predicts thinouts, residual stresses, and fitups with other components. Task 4 evaluates and predicts the performance of the product or structure based on the properties of the material after it has been formed into the product.

Validation of Softwares (Model) Using Old Product. In any modeling, we need to make sure that the software can predict the performance of the product. This is done by validating it with a known product that has considerable design and manufacturing information—the old product. Mathematically, this translates to solving a particular problem (equation) with known boundary conditions and a solution and to determine if the software predicts the correct solution. In terms of the old product, some of the boundary conditions are the material that was used and its thickness as well as the type of forming operation. The thickness was 4.8 mm and the properties of the materials to be used in Eq. 4-53 or Eq. 8-8 are shown in Table 16-13.

The software program that was used predicts the shape of the blank, the forming severity and thinout areas where splitting may occur, including the thickness distribution in the part. The modeling starts by using the print (drawing) of the part. The only required input information for the software is the meshed geometry (the grids of lines shown in Fig. 16-23 that form the elements in finite element analysis) of the part, the materials properties, and the boundary conditions. The displacement boundary conditions emulate the effects of tooling, process, and the constraints of the tooling (such as hole locations).

Table 16-13 Materials properties used in old product.

Yield Strength, Mpa	262
n, strain-hardening exponent	0.18
K, strength coefficient	551
r-value, normal anisotropy	1.0

The predicted strains during forming of the old product are superimposed on the forming limit diagram (FLD) of the material, shown in Fig. 16-25. The predicted strains (dots) include values in the critical and failure zones of the FLD. (see explanation of FLD in Sec. 12.4.1, Fig. 12-10). Thus, the model predicted the breakage or splitting problem encountered during the production of the old part. In addition, the model correctly predicted the locations where the old part might break, as shown in Fig. 16-26. These are exactly the same locations where splits occurred during production over the years. The model also correctly predicted the thickness variations in the formed part where thinning and thickening from the original thickness of 4.8 mm occurred, as shown in Fig.16-27. The only wide discrepancy being the top area of concern where the actual thickness is 4.06 mm versus the predicted 4.83 mm. Figure 16-28 also shows that the model correctly predicted the actual blank size.

New Cross Member. Based on the confidence derived from the validation of the software to predict the behavior of the material during forming according to the old design, the model was used to redesign a new cross member. The latter is shown in Fig.16-29, and Fig.16-30 shows the predicted strains superimposed on the FLD of the material. Only a couple of strains encroached in the critical zone; therefore, the results predict that the forming of this new cross member is easier than the old member.

The results in Fig. 16-27 also suggest that most of the areas show thickening. In addition, because of the deformation process, the yield or flow strength of the material changed from the original material properties in Table 16-13. To accomplish Task 4 in Fig. 16-24, these changes in strength and thickness can be fed to a standard finite element analysis (FEA) model (software) to predict the stiffness, fatigue life, and crash worthiness of the part. This resulted in a potential reduction in part thickness (weight) of 15 percent relative to thickness derived when the original yield strength was used.

Summary: This case indicates the methodology and advantages of concurrent engineering. It also

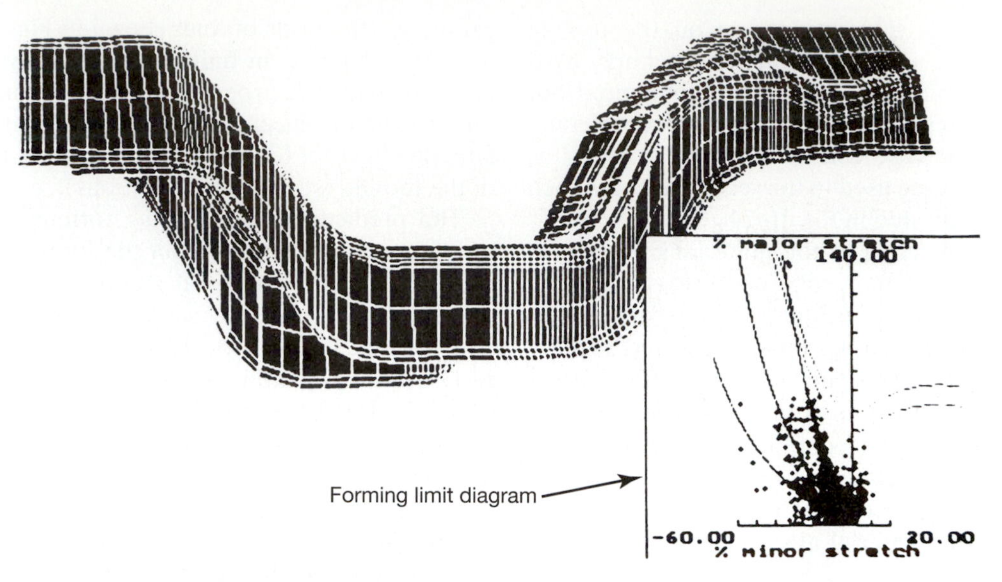

Fig. 16-25 Predicted strains from modeling in forming limit diagram (FLD) to indicate possible problems during the production of old cross member by forming.

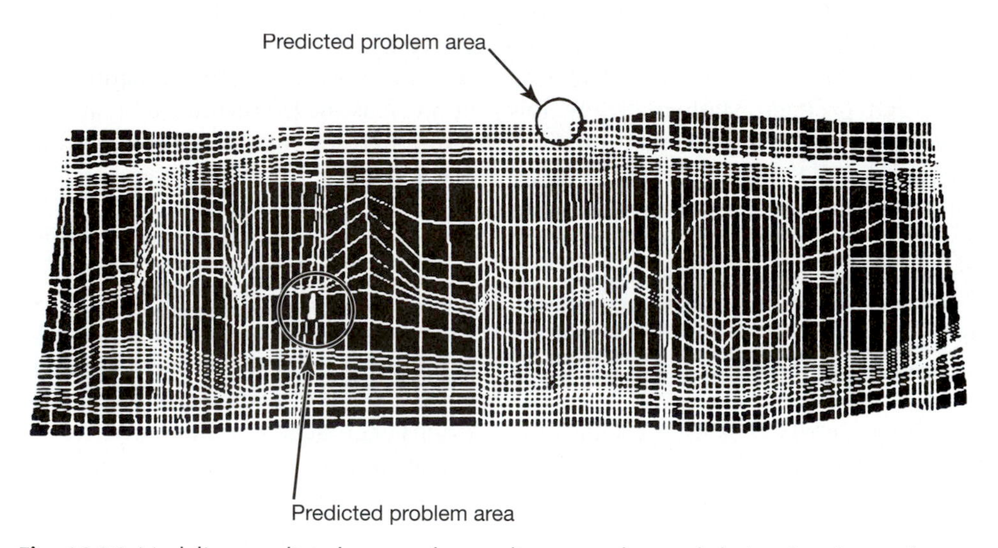

Fig. 16-26 Modeling predicted areas where splits were observed during forming and production of old cross member.

indicates that computational tools are available and should be used at the very outset. Finally, it provides an example of the significance during fabrication of the materials characteristics, such as the forming limit diagram, and the tensile properties, n and K, in Table 16-13. Using the materials properties of the final product (as deformed) in design can save weight and cost, relative to using the materials properties in the as-received (annealed) condition.

16.5 Selection Based on Life-Cycle Analysis (LCA) [12, 13]

The "dust to dust" life cycle of a material is schematically shown in Fig. 16-31. An engineering material undergoes many processes from its raw form from the earth to its usable product form. In each of the intermediate processes, energy is spent and the envi-

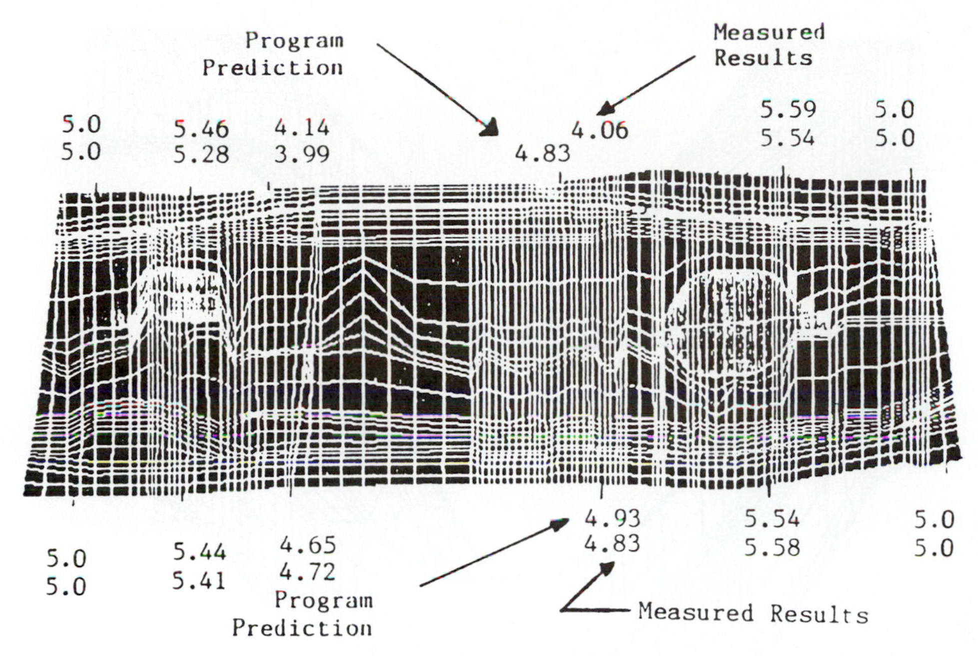

Fig. 16-27 Predicted and actual measured thickness of production cross-member showing thickening from original 4.80 mm thickness, except in problem areas.

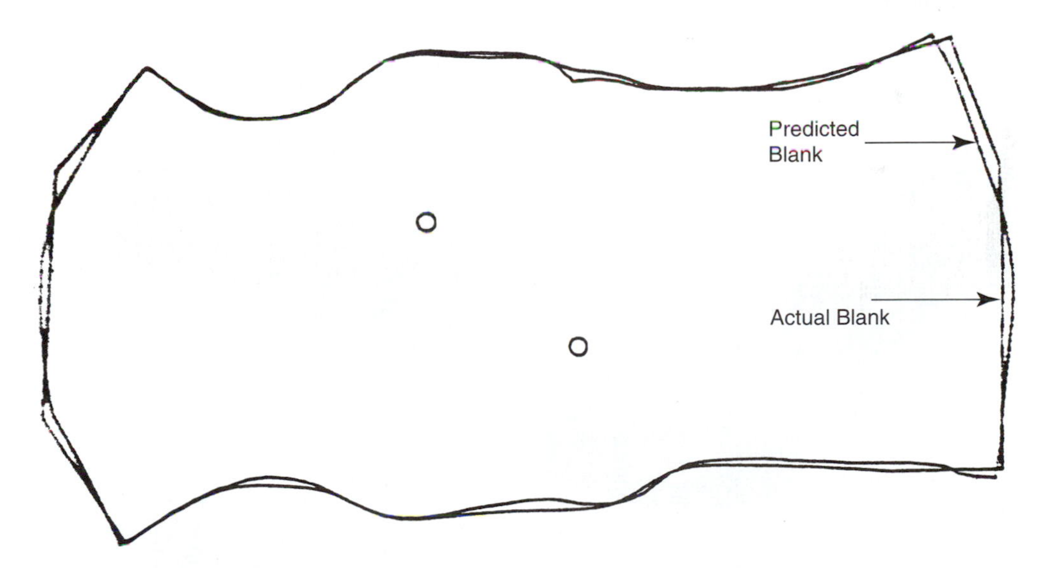

Fig. 16-28 Predicted and actual blank size of production cross member.

ronment (land, water, and air) is affected by whatever emissions and waste are produced from the combustion of fuel, chemical treatment, and/or solid waste. After its use as a product, it can be either recycled or reprocessed to form a useful product or it can be disposed back to the earth. If it is recycled, part of its energy content is again put to use. Before its disposal back to earth, some of its energy may be recovered by combustion.

Life-cycle analysis (LCA) is the concurrent consideration during the design stage of the environmental impact of a material usage for a product, in

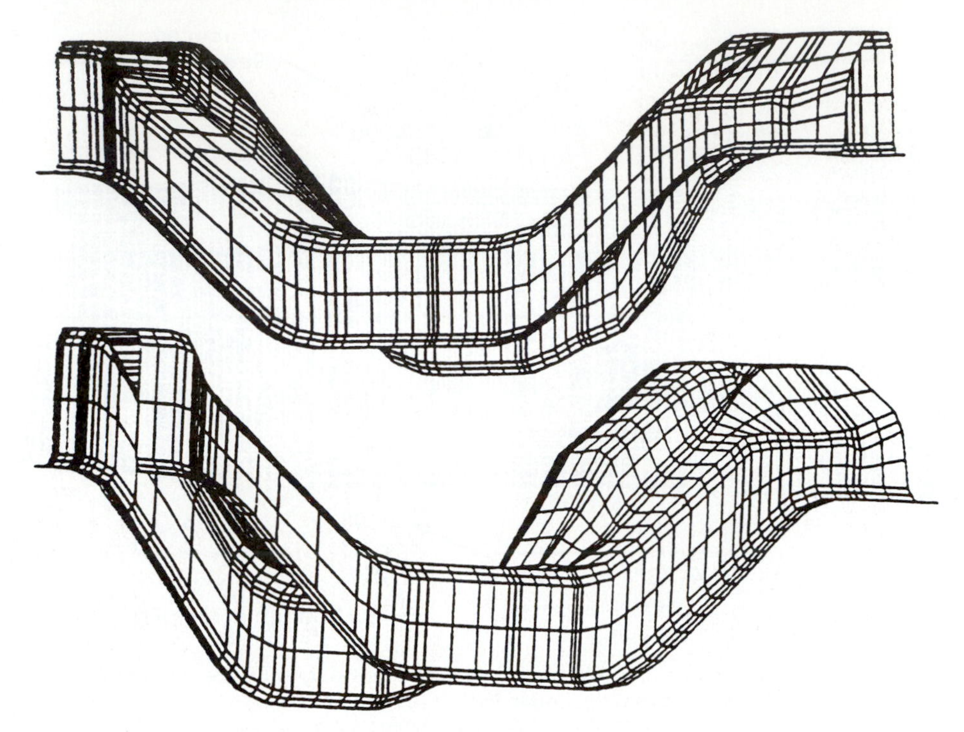

Fig. 16-29 New design of the cross member with a depression at midwidth.

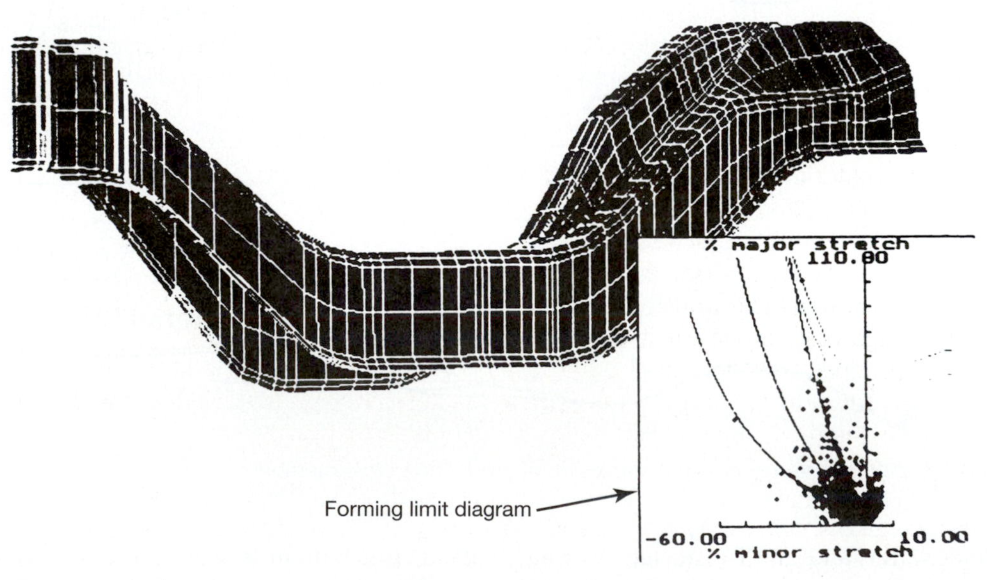

Fig. 16-30 Predicted strains in FLD indicate new design forms better.

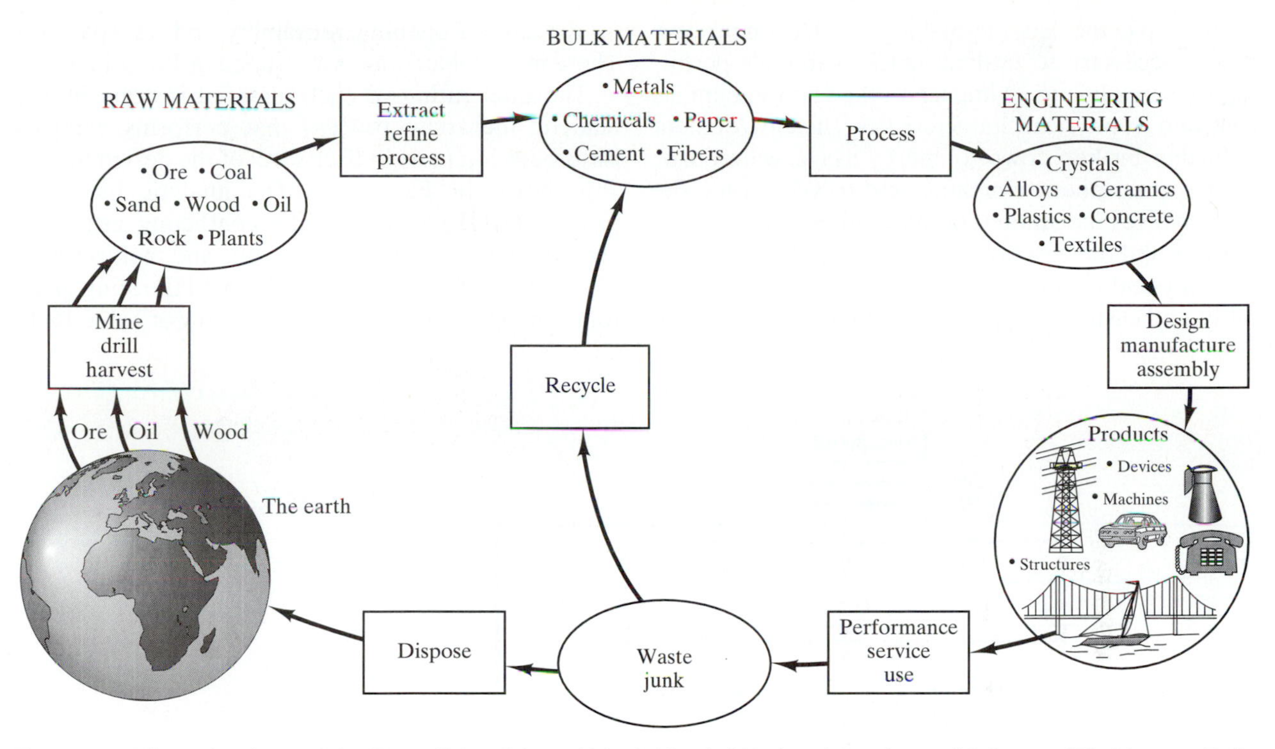

Fig. 16-31 Life cycle of materials. (From "Materials and Man's Needs," National Academy of Sciences, Washington, D.C., 1974).

addition to the technical design and manufacturing issues, as well as economic factors. Among the many factors that can be used to assess the environmental impact, energy usage is commonly used. Energy usage, especially if it comes from fossil fuels, indicates the extent the material facilitates the depletion of these fossil fuels. In addition, because the fossil fuels are combusted, it indicates as well the extent the processing of the material contributes to the pollution of the air and water (acid rain for example). Because energy usage costs money, it indicates the initial cost of the product, considering all other factors being equal. In addition to energy, water is also used in the processes and it also can be polluted.

Selection of materials using LCA is therefore the determination of the relative environmental impact of various materials when they are used for a product. When entirely different materials are being considered for one application, the comparison is done for the entire life cycle of each of the materials. When the same material, but different designs, is used for the same application, the LCA may just involve the concurrent design for manufacturability, an example being the automotive cross member discussed in the last section. When the material is used in transport vehicles, such as aircrafts and automobiles, energy usage during the life of the product is also evaluated.

16.5.1 Material Selection for Beverage Containers [12]

Figure 16-32 shows the life cycle of a beverage container and the materials considered for this case study—plastics, glass, and steel. For these three different materials, the life cycle begins with the mining and refining of raw materials. During the refining processes and the manufacture of the container from each of the materials, energy and water are used, while emissions and wastes are generated. After the useful life of the beverage container, it is either recycled or disposed to landfills or other sites.

To help in the decision-making or selection process, it was necessary to evaluate each material's performance according to its impact on the environment, its cost, and its physical characteristics. The environment is further broken down to energy usage, waterborne wastes, atmospheric emissions, and recycling rate. The cost involves the amount of material needed, the cost per ton, and the cost of the material per 1,000 gallons. The physical characteristics include its weight, durability, and customer appeal. Customer appeal embodies the ease of opening, reusability, and transparency. These material attributes are stated in Table 16-14.

The final rating of each material is done by the analytic hierarchy process that performs comparisons between pairs in each level of the design hierarchy shown in Fig. 16-33. The analytic hierarchy process (AHP) considers the conflicting attributes according to the design constraints and makes tradeoffs between them. The result is an AHP rating given for each material shown in the last row of Table 16-14

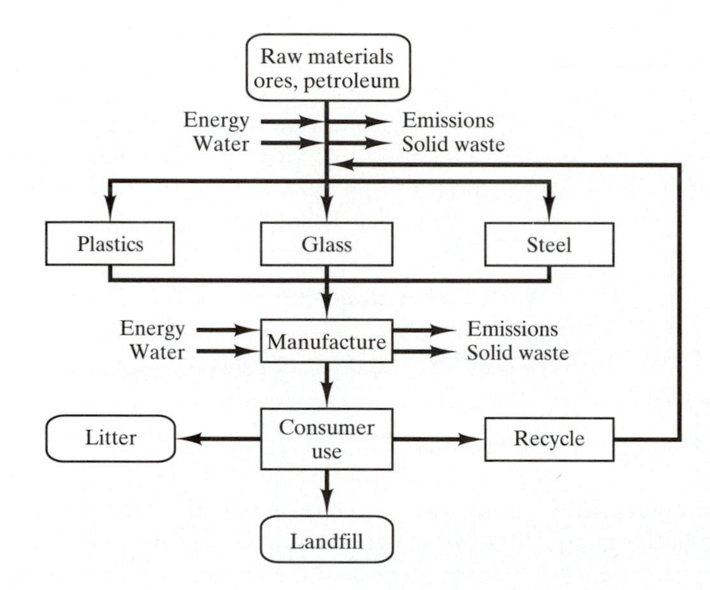

Fig. 16-32 Life cycle of a beverage container [Ref. 12].

Table 16-14 Materials attributes (based on 1000 gallons in 12 ounce containers)[12].

	Plastic	Glass	Steel
Production Energy (million BTU)	63	64	54
Waterborne Wastes (lbs)	69	56	34
Atmospheric Emissions (lbs)	241	261	222
Recycling Rate	6%	15%	75%
Virgin Raw Material (lbs)	988	7541	2746
Virgin Material cost ($/ton)	150	45	80
Material Cost per 1000 gallons ($)	74.1	170	109.8
Weight Savings (0=worst 10=best)	9	1	4
Durability (0=worst 10=best)	8	1	7
Customer Appeal (0=worst 10=best)	9	5	2
Final AHP rating	0.46	0.20	0.34

and illustrated in a bar graph in Fig. 16-34. According to this method, plastics is the best selection.

16.5.2 Material Selection for Automobile Fenders [13]

This case study was conducted to compare four different fender designs for an average compact class automobile. It provides a performance comparison of materials in terms of resource use, impact on global climate, and recyclability.

The four materials considered for the fender were: (1) steel sheet, (2) aluminum sheet, (3) an injection-molded polymer blend (PPO/PA) of polyphenylene oxide (PPO)and nylon (PA - polyamide), and (4) a sheet-molding compound, a glass-reinforced unsaturated polyester resin. The fender made from each material was designed in order to yield the same mechanical performance as the other three, for example, equal stiffness, bending strength, dent resistance, et cetera. In other words, all fenders had the same functional performance. The thicknesses and weights of the fender designs for the different materials are listed in Table 16-15.

Again, from among many parameters, the energy requirement of each material was selected as the basis of comparison. The results are shown in Fig. 16-35 which shows the total energy requirement from the incipient use of the fender through the first life of the automobile, a recycling process, and then the second application of the fender. The energy value at the zero kilometer state represents the energy requirement of each material to mine and refine the material and to manufacture the fender. The energy requirement dur-

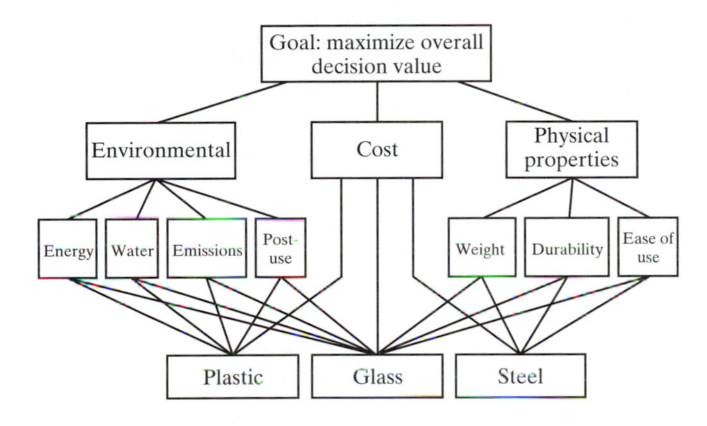

Fig. 16-33 Design hierarchy in life-cycle analysis[12].

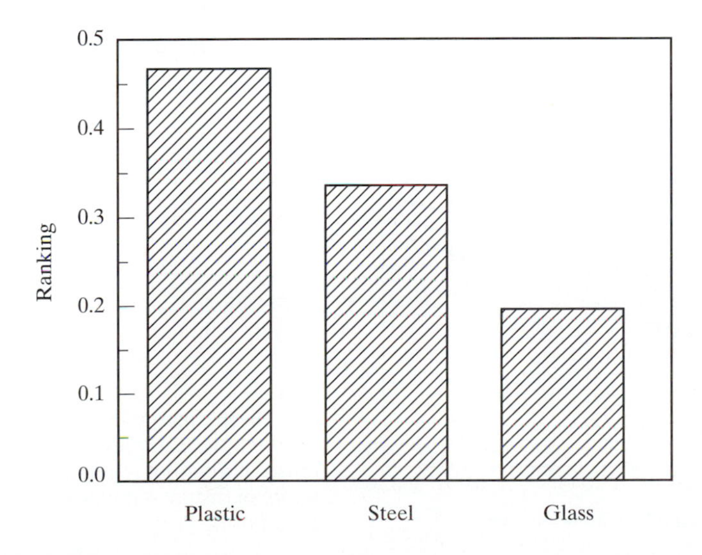

Fig. 16-34 Concurrent multiattribute design evaluation scores[12].

ing use is that due to each material's weight; the step or break in the lines represents the recycling stage.

We see that aluminum has the highest energy requirement of all the four materials. This requirement reflects the cost of the product and comes primarily from the electrolytic process used in the extraction of aluminum from alumina and from the production of alumina. SMC has the lowest energy requirement with about 29 percent, then steel with about 34 percent, and lastly the PPO/PA blend with 66 percent of the energy usage of aluminum. The relatively high PPO/PA energy value is due to its inherent energy content, as well as from the materials used in the polymer production. The relatively low value for the SMC compared to the PPO/PA blend arises from the fact that the SMC uses cheap mineral extenders (filler materials) and therefore contains presumably less than half of the "neat" pure SMC resin.

The slopes of the lines emanating from the zero-kilometer energy value reflects the energy consumption per kilometer of each material. Thus, we see that steel exhibits the highest energy requirement per kilometer because it is the heaviest. During the first life of the automobile, the energy usage is about 900 MJ, about 2.5 times that used to produce the steel fender. For the same life of the automobile selected, aluminum uses only about 300 MJ. This accentuates the advantage of light weight in the transport industry. Nevertheless, the total energy demand for the aluminum up to the first life of the automobile remains the highest. The steel now comes second, the PPO/PA blend comes third, and the SMC remains the lowest. We note, however, that the selected life of the automobile of less than 150,000 km is relatively low. Presumably, if the life goes beyond 160,000 km (100,000 mi.), the total demand for the steel fender can cross that for the aluminum fender during its first life cycle. We can see this crossover by extending the lines of aluminum and steel during the first life cycle.

After the first life cycle, the material is recycled into a new part. Not all the material can be recycled. For the plastic fenders, the mix is 70 percent virgin materials and 30 percent recycled. For the metals, there will be losses during remelting, casting, and so forth. The highest energy demand comes from the PPO/PA blend and then steel and aluminum demands are about the same and SMC remains the lowest. Through the second use, we see that energy demand is the highest for steel, then the PPO/PA, the aluminum, and the SMC remains the lowest.

We should realize that the LCA using the energy parameter is just one of many parameters to help in the decision making to select the appropriate material. There are also factors that were considered, such as the emissions and the effect on global warming,

Table 16-15 Materials, thicknesses, and weights for fender designs. [13]

Material	Thickness, mm	Weight, kg
Steel	0.7	5.60
SMC	2.5	4.97
PPO/PA	3.2	3.35
Aluminum	1.1	2.80

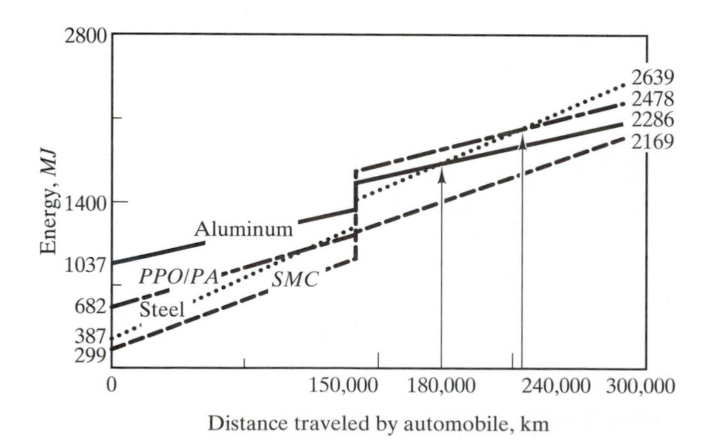

Fig. 16-35 Energy usage for the production, use, recycling, and reuse of different fenders in terms of distance traveled by the automobile [13].

that impact the environment. While these have some relation to energy usage, we cannot discuss them all here. Based on energy alone, it appears that the SMC material is the best material for the fender through the first life, a recycling phase, and through a second application. However, final selection has to consider other factors such as cost, fabricability, fitup with other components, repairability, et cetera.

16.6 Selection of Glass-Reinforced Plastics [14]

The selection process was described in Sec. 15.13.3 and two examples are given here for the simple method in which the design service requirements are only evaluated. The evaluation adopts the property ratings of the glass-reinforced plastics shown in Table 15-27 and the blank worksheet shown in Table 15-28. Using a blank worksheet, the method of evaluation was described in Sec. 15.13.3 and is repeated below for completeness. This is the longhand procedure of computerized databases.

1. In the Design Criteria row, we cross the material properties (columns) that are required of the part or component being made.
2. For each of the properties (columns) selected, transfer the large bold numbers from the ratings table, Table 15-27, to the worksheet. For example, if strength and stiffness are required, then transfer the large bold numbers 3, 5, 1, 3, 2, and 6, from Table 15-27 to Table 15-28 under the column strength and stiffness, top to bottom.
3. Add the large numbers for each row (i.e., for each group of plastics) to find the group with the lowest-point subtotal, which will be the best for the given application on the basis of performance.
4. Add the cost factor and find the group with the lowest total points, which will be the best for the application on a cost-performance basis.
5. Once the group is selected, the procedure is repeated for the plastics within the group to find the material with the lowest point total as the best for the application on a cost-performance basis.

The results of the two examples are given in Tables 16-16 and 16-17, for a gasoline-powered chain saw housing, and an impeller for a chemical-handling pump. For the chain saw housing, the service requirements (design criteria) of importance are the strength and stiffness, toughness, short-term heat resistance, and environmental resistance. The toughness requirement comes from the possibility of mishandling the saw and the saw drops. Short-term heat resistance is due to the intermittent use of the chain saw. The bold numbers in Table 16-16 are the same numbers under each corresponding property for each of the materials. Based on these, the group of plastics showing the overall most desirable rating (lowest point total) is the nylon (polyamide) family. Within the nylon family, nylon 6/6 exhibits the lowest point total of 11 and should, therefore, be the first choice.

For the impeller, the service requirements are strength and stiffness, short-term heat resistance, long-term heat resistance, and environmental resistance. The toughness is not included because the impeller will not be exposed to impact loading. The long-term heat resistance is added because of the continuous operation of pumps. The results of the evaluation in Table 16-17 show the high-temperature resins exhibit the lowest point total and polyphenylene sulfide (PPS) is better than polyamide-imide.

16.7 Application of Fracture Mechanics to Materials Selection

We shall discuss the application of fracture mechanics to materials selection via the failure analysis of the catastrophic failure of a 260-in. diameter motor case shown in Fig. 16-36[15]. The lessons to be learned from this case are the following:

- Neglect or the ignorance of the application of fracture mechanics in design can be disastrous.
- Classical design (allowable stress method) based on a fraction of yield strength of the material is not sufficient. The fracture toughness of the material must be considered and design must be made to prevent brittle fracture, especially for thick, high-strength materials that are prone to brittle fracture.
- The material properties of significance to the structure or component are those in the as-fabricated condition, not the as-received base materials properties given by the supplier.

Table 16-16 Selection of nylon 6/6 for gasoline-powered chain-saw housing [14].

Material Characteristics	Strength and Stiffness	Toughness	Short-Term Heat Resistance	Long-Term Heat Resistane	Environ-mental Resistance	Dimensional Accuracy in Molding	Dimensional Stability	Wear and Frictional Properties	Point Subtotal	Cost	Point Total
G/R Resin Groups ⇓ / Design Criteria ⇒	X	X	X		X						
Styrenics (ABS, SAN, Polystyrene)	**3**	**6**	**6**		**6**				**21**	**2**	**23**
Olefins (Polyethylene, Polypropylene)	**5**	**4**	**4**		**3**				**16**	**1**	**17**
Other Crystalline Resins											
Nylons	**1**	**1**	**2**		**4**				**8**	**3**	**11**
6	2	2	2		5				11	1	12
6/6	1	3	1		4				9	2	11
6/10, 6/12	3	1	3		3				10	4	14
Polyester	4	4	2		2				12	1	13
Polyacetal	5	5	5		1				16	1	17
Arylates (Modified PPO, Polycarbonate, Polysulfone, Polyethersulfone)	**3**	**2**	**3**		**5**				**13**	**4**	**17**
High Temp. Resin (PPS, Polyamide-imide)	**2**	**4**	**1**		**2**				**9**	**5**	**14**
Fluorocarbons (FEP, ETFE)	**6**	**2**	**2**		**1**				**11**	**6**	**17**

Table 16-17 Selection of nylon 6/6 for gasoline-powered chain-saw housing [14].

Material Characteristics	Strength and Stiffness	Toughness	Short-Term Heat Resistance	Long-Term Heat Resistane	Environ mental Resistance	-Dimensional Accuracy in Molding	Dimensional Stability	Wear and Frictional Properties	Point Subtotal	Cost	Point Total
G/R Resin Groups ⇓ Design Criteria ⇒	X		X	X	X						
Styrenics											
ABS	3		6	6	6				21	2	23
SAN											
Polystyrene											
Olefins											
Polyethylene	5		4	5	3				17	1	18
Polypropylene											
Other Crystalline Resins											
Nylons											
6	1		2	4	4				11	3	14
6/6											
6/10, 6/12											
Polyester											
Polyacetal											
Arylates											
Modified PPO	3		3	3	5				14	4	18
Polycarbonate											
Polysulfone											
Polyethersulfone											
High Temp. Resin											
PPS	2 1		2 1	2 1	1 2				6 6	1 5	7 11
Polyamide-imide	2	1	1	2				6	2	8	
Fluorocarbons											
FEP	6		2	1	1				10	6	16
ETFE											

Fig. 16-36 Pieces of failed motor case laid out in approximately the proper relation to each other[15].

- Weld repairs on structures should be approached with a lot of care and thought.

The Case. The motor case was a 260-in. diameter, 65-feet high chamber. It was constructed of a 0.73-in. thick plate of 250 Grade maraging steel with a 240 ksi yield strength. The design stress was 160 ksi (which is ⅔ of the yield strength) and was designed to withstand proof pressures of 960 psi. The motor case was fabricated by joining the steel plates by automatic submerged arc welding. The motor case failed during hydrotest at a pressure of 542 psi, only 56 percent of design-proof pressure.

The failure analysis report of this case is as follows: Although this motor case was designed to withstand proof pressures of 960 psi, it failed during hydrotest at a pressure of 542. In this 240 ksi yield strength material, the failure occurred at a very low membrane (structure) stress of 100 ksi. The fracture was both premature and brittle with crack velocities approaching 5,000 ft/sec. The result of this 65-ft-high chamber literally flying apart is shown in Fig. 16-36. *Postfailure examination revealed that the fracture had originated in an area of two defects that had probably been produced by manual gas-tungsten arc weld repair—a lesson.* After the crack initiated, it branched into multiple cracks as shown in Fig. 16-37, leading to complete catastrophic failure of the motor case.

The real lesson in this failure is the lack of design knowledge that went into the material selection. First, the chamber was 0.73 in. thick, which puts it into the plane-strain regime for the high-strength material considered. Grade 250 maraging steel with a yield strength of about 240 ksi was chosen. For plane-strain conditions, this was not a particularly good choice since the base metal had a plane-strain fracture

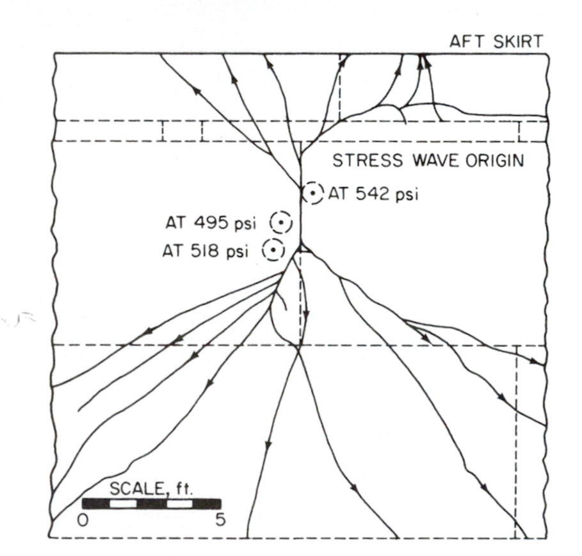

Fig. 16-37 Fracture paths about failure origin; dashed lines indicate welds[15].

toughness, K_{I_c}, of only 79.6 ksi$\sqrt{in.}$ At the design stress of 160 ksi, a critical defect only 0.08 in. in depth could have caused catastrophic failure. Still, the chamber manufacturer thought that defects of this size could be detected. In retrospect, this was not a very judicious decision. Furthermore, the error was compounded by the fact that welding this material produced an even lower K_{I_c} value (*a lesson*) ranging from 39.4 to 78.0 ksi$\sqrt{in.}$ the toughness level depending on the location of the flaw in the weld.

A postfailure analysis was run on several types of flaw configurations and weld positions. The results of the analysis indicated the K_{I_c} values ranged from 38.8 to 83.1 ksi$\sqrt{in.}$ with the average being 55.0 ksi$\sqrt{in.}$ As the exact value of the fracture toughness at the failure origin in the chamber is not known, this average value is used as an estimate of K_{I_c}. Postfailure examination of the fracture origin indicated the responsible defect could best be approximated by an internal ellipse that was 0.22 in. in depth and 1.4 in. long. The critical dimension was the in-depth value of 0.22 in. since the crack would propagate through the thickness from this dimension. To calculate the critical defect size from fracture mechanics, the simple equation

$$K_{I_c} = \sigma\sqrt{\pi a} \qquad (16\text{-}2)$$

was used, which yielded results that are only 4 percent off the more rigorous solution. (In the equation, σ is the applied stress and a is one-half the critical defect size). Based on the fracture toughness of 55.0 Ksi$\sqrt{in.}$ a plot of the membrane stress versus defect size for crack instability is shown in Fig. 16-38. At the observed 100-ksi failure stress for the chamber, a critical defect size, 2a, of 0.2 in. was predicted, which is very close to the observed value, also indicated in Fig. 16-38. Besides the Grade 250 data, there is also shown in Fig. 16-38 a curve for Grade 200 maraging steel. Although this material has a yield strength that is about 10 percent lower than Grade 250, its toughness is about triple the average K_{I_c} value for the Grade 250, with K_{I_c} being about 150 ksi$\sqrt{in.}$ for a member 0.7 in. thick. The curve for Grade 200 in Fig. 16-38 is the plot of Eq. 16-2 using the fracture toughness of 150 K_{I_c}. Significantly, the defect that led to the failure of the Grade 250 chamber would not have failed a Grade 200 chamber, and in fact, yield stresses could have been reached without failure. At the failure stress of 100 ksi, it would have taken a flaw 1.42 in. in depth to burst a Grade 200 chamber. This would not have been possible since the thickness was only 0.73 in; A leak would have been detected—leak-before-break criterion. Even at the design stress of 160 ksi, it would take a

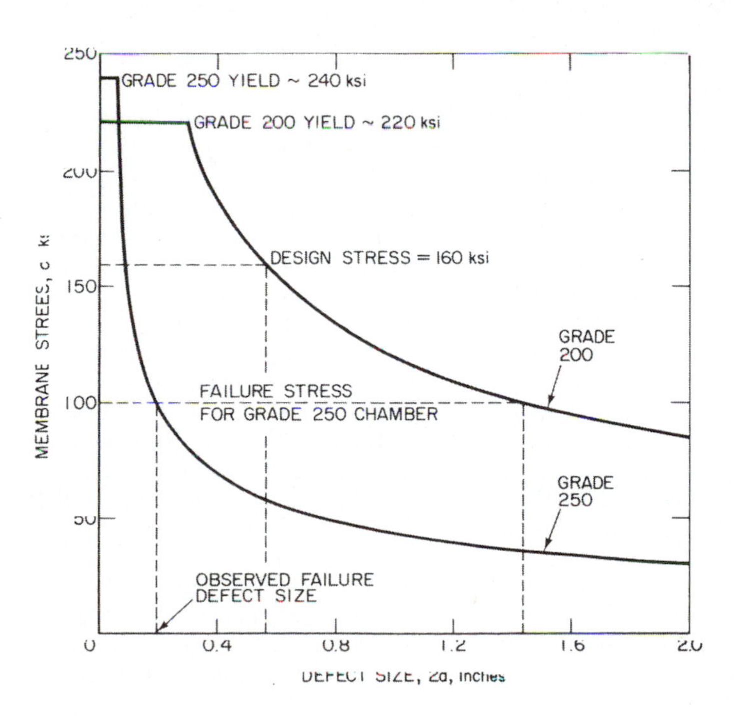

Fig. 16-38 Design curves for 260-in motor case to determine critical defect sizes in materials with different fracture toughness[15].

flaw 0.56 in. in depth to cause plane-strain fracture, and this is a very large flaw. In all probability, a flaw this large in this type of material (Grade 200) would be arrested under plane-stress conditions as soon as it grew through the thickness.

Thus, the Grade 200 maraging steel would have been a more reliable material for this chamber than the Grade 250 maraging steel. The proof of this is that a competitor for this job for NASA has successfully built the 260-in. diameter chamber with the Grade 200 maraging steel. Not only were there two successful proof tests of Grade 200 chambers, there were also two firings that developed thrusts of more than 6 million lb_f.

In summary, using a lower-strength level steel with higher toughness, even at a design stress that was a higher percentage of the yield strength, would have been better.

References

1. *Proceedings of the 1995 Conference on High-Performance Structural Steels* in Cleveland, Ohio, R. Asfahani, Editor, ASM International, Metals Park, Ohio.
   (a) J. M. Barsom, "High-Performance Steels and Their Use in Structures," p. 3.
   (b) D. L. Blunier, et al, "Requirements of High-Performance Steels for Heavy Fabrications," p. 45.
   (c) R. Sause and J. W. Fisher, "Application of High-Performance Steel in Highway Bridges," p. 13.
   (d) T. K. Sooi, et al, "Stress-Strain Properties of High-Performance Steel and the Implication for Civil Structure Design," p. 35.
   (e) K. Ichise, et al, "Development of High-Performance Steels for Structures," p. 23.
   (f) G. S. Pike, "Application of High-Performance Steel in Mobile Hydraulic Cranes and Aerial Work Platforms," p. 53.
   (g) D. J. Cochrane, "European Developments in the Application of Structural Austenitic and Duplex Stainless Steels," p. 299.
2. J. M. Barsom and S. T. Rolfe, *Fracture & Fatigue Control in Structures*, 2d ed; Prentice Hall 1897.
3. R. A. White and E. F. Ehmke, *Materials Selection for Refineries and Associated Facilities, National Association of Corrosion Engineers*. 1991.
4. E. R. Kuch, "Hardenable Carbon and Low-Alloy Steels," in vol. 1, *Metals Handbook*, 10th ed.; ASM International, 1990.
5. F. A. A. Crane and J. A. Charles, *Selection and Uses of Engineering Materials* Butterworths & Co., Ltd., 1984.
6. L. J. Corb, "Corrosion of Manned Spacecraft" as part of *Corrosion in the Aerospace Industry* Corrosion–Metals Handbook, vol. 13, p. 1058, ASM International, 1987.
7. *Advanced Materials & Processes, 149*, (5) 1996.
   (a) G. Lucas, 'Aluminum Structural Applications', p. 29.
   (b) D. C. Martin, 'Trends in Automotive Steel', p. 27
   (c) D. Tice, et al, 'Materials Selection for Auto, p. 802.
8. S. Housch, B. Mikuchi, and A. Stevenson, *Selection and Application of Magnesium and Magnesium Alloys,* vol. 2, Metals Handbook, 10th ed.; ASM International, Metals Park, OH 1990.
9. *Ceramic and Glasses,* vol. 4, Engineered Materials Handbook, ASM International, 1987.
   (a) K. Okajima and R. Matsuda, "Design Practices for Structural Ceramics in Gasoline Engines," p. 728.
   (b) A. F. McLean, "An Overview of the Ceramic Design Process," p. 676.
   (c) L. C. Lindgren, et al, "Design Practices for Structural Ceramics in Gas Turbine Engines," p. 716.
   (d) C. C. Baker and D. E. Baker, "Design Practices for Structural Ceramics in Automotive Turbocharger Wheels," p. 722.
10. S. Kamiya, et al, *Silicon Nitride Swirl Lower-Chamber for High-Power Turbocharged Diesel Engines,* SAE Paper 850523, 1985.
11. J. Beckman, et al, *A Simultaneous Engineering Approach for Automotive Structural Cross Member,* SAE Paper 930524, 1993.
12. D. L. Thurston and A. Blair, *A Method for Integrating Environmental Impacts into Product Design,* p. 765, in Proceedings of International Conference on Engineering Design, The Hague, Netherlands, Aug. 17–19, 1993.
13. M. Harsch, et al, *Life-Cycle Assessment, Adv. Materials & Processes,* 149, (6), p. 43, 1996.
14. D. V. Rosato, D. P. DiMattia, D. V. Rosato, *Designing with Plastics and Composites,* Van Nostrand Reinhold, 1991.
15. W. W. Gerberich, "Fracture Mechanics Approach to Design Application," cited reference in Chap. 6, of Ref. 2 above.

Appendix
Metric Conversion Guide

This section is intended as a guide for expressing weights and measures in the Systéme International d'Unités (SI). The purpose of SI units, developed and maintained by the General Conference of Weights and Measures, is to provide a basis for worldwide standardization of units and measures. For more information on metric conversions, the reader should consult the following references:

- "Standard for Metric Practice," E 380, *Annual Book of ASTM Standards,* vol 14.02., 1987, American Society for Testing and Materials, 1916 Race Street, Philadelphia, PA 19103.
- "Metric Practice," *ANSI/IEEE* 268, 1982, American National Standards Institute, 1430 Broadway, New York, NY 10018.
- *Metric Practice Guide—Units and Conversion Factors for the Steel Industry,* 1978, American Iron and Steel Institute, 1000 16th Street NW, Washington, DC 20036
- *The International System of Units,* SP 330, 1986, National Bureau of Standards. Order from Superintendent of Documents, U.S. Government Printing Office, Washington, DC 20402-9325
- *Metric Editorial Guide,* 4th ed. (revised), 1985, American National Metric Council, 1010 Vermont Avenue NW, Suite 320, Washington, DC 20005-4960
- *ASME Orientation and Guide for Use of SI (Metric) Units,* ASME Guide SI 1, 9th ed., 1982, The American Society of Mechanical Engineers, 345 East 47th Street, New York, NY 10017

Base, supplementary, and derived SI units

Measure	Units	Symbol
Base units		
Amount of substance	mole	mol
Electric current	ampere	A
Length	meter	m
Luminous intensity	candela	cd
Mass	kilogram	kg
Thermodynamic temperature	kelvin	K
Time	second	s
Supplementary units		
Plane angle	radian	rad
Solid angle	steradian	sr
Derived units		
Absorbed dose	gray	Gy
Acceleration	meter per second squared	m/s^2
Activity (of radionuclides)	becquerel	Bq
Angular acceleration	radian per second squared	rad/s^2
Angular velocity	radian per second	rad/s
Area	square meter	m^2
Capacitance	farad	F
Concentration (of amount of substance)	mole per cubic meter	mol/m^3
Conductance	siemens	S
Current density	ampere per square meter	A/m^2
Density, mass	kilogram per cubic meter	kg/m^3
Electric charge density	coulomb per cubic meter	C/m^3
Electric field strength	volt per meter	V/m
Electric flux density	coulomb per square meter	C/m^2
Electric potential, potential difference, electromotive force	volt	V
Electric resistance	ohm	Ω
Energy, work, quantity of heat	joule	J
Energy density	joule per cubic meter	J/m^3
Entropy	joule per kelvin	J/K
Force	newton	N
Frequency	hertz	Hz
Heat capacity	joule per kelvin	J/K
Heat flux density	watt per square meter	W/m^2
Illuminance	lux	lx
Inductance	henry	H
Irradiance	watt per square meter	W/m^2
Luminance	candela per square meter	cd/m^2
Luminous flux	lumen	lm
Magnetic field strength	ampere per meter	A/m
Magnetic flux	weber	Wb
Magnetic flux density	tesla	T
Molar energy	joule per mole	J/mol
Molar entropy	joule per mole kelvin	J/mol · K
Molar heat capacity	joule per mole kelvin	J/mol · K
Moment of force	newton meter	N · m
Permeability	henry per meter	H/m
Permittivity	farad per meter	F/m
Power, radiant flux	watt	W
Pressure, stress	pascal	Pa
Quanitity of electricity, electric charge	coulomb	C
Radiance	watt per square meter steradian	W/m^2 · sr
Radiant intensity	watt per steradian	W/sr
Specific heat capacity	joule per kilogram kelvin	J/kg · K
Specific energy	joule per kilogram	J/kg
Specific entropy	joule per kilogram kelvin	J/kg · K
Specific volume	cubic meter per kilogram	m^3/kg
Surface tension	newton per meter	N/m
Thermal conductivity	watt per meter kelvin	W/m · K
Velocity	meter per second	m/s
Viscosity, dynamic	pascal second	Pa · s
Viscosity, kinematic	square meter per second	m^2/s
Volume	cubic meter	m^3
Wavenumber	1 per meter	1/m

Conversion factors

To convert from	to	multiply by
Area		
in.^{2}	mm^2	6.451 600 E + 02
in.^{2}	cm^2	6.451 600 E + 00
in.^{2}	m^2	6.451 600 E − 04
ft^2	m^2	9.290 304 E − 02
Bending moment or torque		
lbf · in.	N · m	1.129 848 E − 01
lbf · ft	N · m	1.355 818 E + 00
kgf · m	N · m	9.806 650 E + 00
ozf. · in.	N · m	7.061 552 E − 03
Bending moment or torque per unit length		
lbf · in./in.	N · m/m	4.448 222 E + 00
lbf · ft/in.	N · m/m	5.337 866 E + 01
Current density		
A/in.^{2}	A/cm^2	1.550 003 E − 01
A/in.^{2}	A/mm^2	1.550 003 E − 03
A/ft^2	A/m^2	1.076 400 E + 01
Electric field strength		
V/mil	kV/m	3.937 008 E + 01
Electricity and magnetism		
gauss	T	1.000 000 E − 04
maxwell	μWb	1.000 000 E − 02
mho	S	1.000 000 E + 00
Oersted	A/m	7.957 700 E + 01
Ω · cm	Ω · m	1.000 000 E − 02
Ω circular-mil/ft	μΩ · m	1.662 426 E − 03
Energy (impact, other)		
ft · lbf	J	1.355 818 E + 00
Btu (thermochemical)	J	1.054 350 E + 03
cal (thermochemical)	J	4.184 000 E + 00
kW · h	J	3.600 000 E + 06
W · h	J	3.600 000 E + 03
Flow rate		
ft^3/h	L/min	4.719 475 E − 01
ft^3/min	L/min	2.831 000 E + 01
gal/h	L/min	6.309 020 E − 02
gal/min	L/min	3.785 412 E + 00
Force		
lbf	N	4.448 222 E + 00
kip (1000 lbf)	N	4.448 222 E + 03
tonf	kN	8.896 443 E + 00
kgf	N	9.806 650 E + 00
Force per unit length		
lbf/ft	N/m	1.459 390 E + 01
lbf/in.	N/m	1.751 268 E + 02
kip/in.	N/m	1.751 268 E + 05
Fracture toughness		
ksi$\sqrt{\text{in.}}$	MPa$\sqrt{\text{m}}$	1.098 800 E + 00
Heat content		
Btu/lb	kJ/kg	2.326 000 E + 00
cal/g	kJ/kg	4.186 800 E + 00
Heat input		
J/in.	J/m	3.937 008 E + 01
kJ/in.	kJ/m	3.937 008 E + 01
Impact energy per unit area		
ft · lbf/ft^2	J/m^2	1.459 002 E + 01
Length		
Å	nm	1.000 000 E − 01
μin.	μm	2.540 000 E − 02
mil	μm	2.540 000 E + 01
in.	mm	2.540 000 E + 01
in.	cm	2.540 000 E + 00
ft.	m	3.048 000 E − 01
yd	m	9.144 000 E − 01
mile	km	1.609 300 E + 00
Length per unit mass		
in./lb	m/kg	5.599 740 E − 02
yd/lb	m/kg	2.015 907 E + 00
Mass		
oz	kg	2.834 952 E − 02
lb	kg	4.535 924 E − 01
ton (short, 2000 lb)	kg	9.071 847 E + 02
ton (short, 2000 lb)	kg · 10^3(a)	9.071 847 E − 01
ton (short, 2240 lb)	kg	1.016 047 E + 03

(Continued)

(a) kg × 10^3 = 1 metric ton

Conversion factors (*Continued*)

To convert from	to	multiply by
Mass per unit area		
oz/in.^{2}	kg/m^2	4.395 000 E + 01
oz/ft^2	kg/m^2	3.051 517 E − 01
oz/yd^2	kg/m^2	3.390 575 E − 02
lb/ft^2	kg/m^2	4.882 428 E + 00
Mass per unit length		
lb/ft	kg/m	1.488 164 E + 00
lb/in.	kg/m	1.785 797 E + 01
denier	kg/m	1.111 111 E − 07
tex	kg/m	1.000 000 E − 06
Mass per unit time		
lb/h	kg/s	1.259 979 E − 04
lb/min	kg/s	7.559 873 E − 03
lb/s	kg/s	4.535 924 E − 01
Mass per unit volume (includes density)		
g/cm^3	kg/m^3	1.000 000 E + 03
lb/ft^3	g/cm^3	1.601 846 E − 02
lb/ft^3	kg/m^3	1.601 846 E + 01
lb/in.^{3}	g/cm^3	2.767 990 E + 01
lb/in.^{3}	kg/m^3	2.767 990 E + 04
oz/in.^{3}	kg/m^3	1.729 994 E + 03
Power		
Btu/s	kW	1.055 056 E + 00
Btu/min	kW	1.758 426 E − 02
Btu/h	W	2.928 751 E − 01
erg/s	W	1.000 000 E − 07
ft · lbf/s	W	1.355 818 E + 00
ft · lbf/min	W	2.259 697 E − 02
ft · lbf/h	W	3.766 161 E − 04
hp (550 ft · lbf/s)	kW	7.456 999 E − 01
hp (electric)	kW	7.460 000 E − 01
Power density		
W/in.^{2}	W/m^2	1.550 003 E + 03
Pressure (fluid)		
atm (standard)	Pa	1.013 250 E + 05
bar	Pa	1.000 000 E + 05
in. Hg (32°F)	Pa	3.386 380 E + 03
in. Hg (60°F)	Pa	3.376 850 E + 03
lbf/in^2(psi)	Pa	6.894 757 E + 03
torr (mm Hg, 0°C)	Pa	1.333 220 E + 02

To convert from	to	multiply by
Specific area		
ft^2/lb	m^2/kg	2.048 161 E − 01
Specific energy		
cal/g	J/g	4.186 800 E + 00
Btu/lb	kJ/kg	2.326 000 E + 00
Specific heat capacity		
Btu/lb · °F	J/kg · K	4.186 800 E + 03
cal/g · °C	J/kg · K	4.186 800 E + 03
Stress (force per unit area)		
tonf/in.^{2} (tsi)	MPa	1.378 951 E + 01
kgf/mm^2	MPa	9.806 650 E + 00
ksi	MPa	6.894 757 E + 00
lbf/in.^{2} (psi)	MPa	6.894 757 E − 03
MN/m^2	MPa	1.000 000 E + 00
Temperature		
°F	°C	5/9 · (°F − 32)
°R	K	5/9
°F	K	5/9 · (°F + 459.67)
°C	K	°C + 273.15
Temperature interval		
°F	°C	5/9
Thermal conductivity		
Btu · in./s · ft^2 · °F	W/m · K	5.192 204 E + 02
Btu/ft · h · °F	W/m · K	1.730 735 E + 00
Btu · in./h · ft^2 · °F	W/m · K	1.442 279 E − 01
cal/cm · s · °C	W/m · K	4.148 000 E + 02
Thermal expansion		
μin./in. · °C	10^{-6}/K	1.000 000 E + 00
μin./in. · °F	10^{-6}/K	1.800 000 E + 00
Velocity		
ft/h	m/s	8.466 667 E − 05
ft/min	m/s	5.080 000 E − 03
ft/s	m/s	3.048 000 E − 01
in./s	m/s	2.540 000 E − 02
km/h	m/s	2.777 778 E − 01
mph	km/h	1.609 344 E + 00

(*Continued*)

Conversion factors (*Continued*)					
To convert from	**to**	**multiply by**	**To convert from**	**to**	**multiply by**
Viscosity (dynamic and Kinematic)			**Volume per unit time**		
poise (P)	Pa · s	1.000 000 E − 01	ft^3/min	m^3/s	4.719 474 E − 04
cP	Pa · s	1.000 000 E − 03	ft^3/s	m^3/s	2.831 685 E − 02
lbf · s/in.^{2}	Pa · s	6.894 757 E + 03	in.^{3}/min	m^3/s	2.731 177 E − 07
ft^2/s	m^2/s	9.290 304 E − 02			
in.^{2}/s	mm^2/s	6.451 600 E + 02	**Wavelength**		
			Å	nm	1.000 000 E − 01
Volume					
in.^{3}	m^3	1.638 706 E − 05			
ft^3	m^3	2.831 685 E − 02			
fluid oz	m^3	2.957 353 E − 05			
gal (U.S. liquid)	m^3	3.785 412 E − 03			

SI prefixes-names and symbols

Exponential expression	Multiplication factor	Prefix	Symbol
10^{18}	1 000 000 000 000 000 000	exa	E
10^{15}	1 000 000 000 000 000	peta	P
10^{12}	1 000 000 000 000	tera	T
10^{9}	1 000 000 000	giga	G
10^{6}	1 000 000	mega	M
10^{3}	1 000	kilo	k
10^{2}	100	hecto(a)	h
10^{1}	10	deka(a)	da
10^{0}	1	BASE UNIT	
10^{-1}	0.1	deci(a)	d
10^{-2}	0.01	centi(a)	c
10^{-3}	0.001	milli	m
10^{-6}	0.000 001	micro	μ
10^{-9}	0.000 000 001	nano	n
10^{-12}	0.000 000 000 001	pico	p
10^{-15}	0.000 000 000 000 001	femto	f
10^{-18}	0.000 000 000 000 000 001	atto	a

(a) Nonpreferred. Prefixes should be selected in steps of 10^3 so that the resultant number before the prefix is between 0.1 and 1,000. These prefixes should not be used for units of linear measurement, but may be used for higher order units. For example, the linear measurement, decimeter is nonpreferred, but square decimeter is acceptable.

Physical Properties of Selected Elements

Element	Symbol	Atomic Mass, g/mol	Density, g/cm^3	Electrical Resistivity, $(\Omega\text{-m})10^{-8}$@293°K	Crystal Structure	Lattice Constant/ Angle, Å/deg.	Melting Point, °C
Aluminum	Al	26.91	2.699	2.655	FCC, A1	4.40496	660.2
Antimony	Sb	121.757	6.62	39@0°C	Rhombo, A7	4.5067	630.5
						α = 57.11°	
Arsenic	As	74.921	5.72	33.3	Rhombo, A7	4.1319	817°C
						α = 54.12	@28 atm
Beryllium	α−Be	9.012	1.848	4	HCP, A3	a = 2.2859	1277
						c = 3.3845	
	β−Be (>1260°C to M.Pt)				BCC, A2	2.5515 (@1270°C)	
Bismuth	Bi	208.980	9.80	106.8@0°C	Rhombo, A7	a = 4.7460	271.3
						α = 57.23	
Boron	B	10.811	2.30		Rhombo, A7	a = 9.45	2300
						α = 23.8	
Cadmium	Cd	112.411	8.65@20°C	6.83@0°C	HCP, A3	a = 2.9793	320.9
						c = 5.6198	
Calcium	Ca	40.078	1.55	3.91@0°C	FCC, A1	5.5884	838
Carbon	C (graphite)	12.011	1.50–1.80		Hex, A9	a = 2.4612	...
						c = 6.7090	
	C (diamond)		3.51		Dia Cubic, A4	3.5669	
Cerium	α-Ce(<−177°C)	140.115	8.23	75@25°C	FCC, A1	4.85	
	β-Ce(RT)		6.66		Hex, A31	a = 3.6810	
						c = 11.857	
	γ-Ce(>61°C)		6.77		FCC, A1	5.1610	
	σ-Ce(>726°C)		6.77		BCC, A2	4.12	
Cesium	Cs	132.905	1.903	20	BCC, A2	6.141	28.7
Chromium	Cr	51.996	7.19	12.9@0°C	BCC, A2	3.8848	1875
Cobalt	ϵ-Co	58.933	8.85	6.24	HCP, A3	a = 2.5071	1495
						c = 4.0686	
	α-Co(>422°C)				FCC, A1	3.5447	
Columbium	Cb	92.906	8.57	~15	BCC, A2	3.3004	2468
(Niobium)	(Nb)						
Copper	Cu	63.546	8.96	1.678	FCC, A1	3.6146	1083
Gallium	Ga	69.723	5.907		Orthorhombic, A11	a = 4.5186	29.78
						b = 7.6570	
						c = 4.5258	
Germanium	Ge	72.61	5.323	46	Diamond Cubic, A4	5.6574	937.4
Gold	Au	196.967	19.32	2.21	FCC, A1	4.0782	1063

(*Continued*)

Physical Properties of Selected Elements (*Continued*)

Element	Symbol	Atomic Mass, g/mol	Density, g/cm^3	Electrical Resistivity, (Ω-m)10^{-8}@293°K	Crystal Structure	Lattice Constant/ Angle, Å/deg.	Melting Point, °C
Hafnium	Hf	178.49	13.09		HCP, A3	a = 3.1946	2222
						c = 5.0510	
Indium	In	114.82	7.31	8.37	FC Tetr., A6	a = 3.2530	156.2
						c = 4.9470	
Iron	α-Fe	55.847	7.874	9.71	BCC, A2	2.8665(@25°C)	1536.5
	γ-Fe				FCC, A1	3.6467(>912°C)	
	δ-Fe				BCC, A2	2.9315(>1394°C)	
Lanthanum	α-La	138.906	6.19	57	Hex, A3	a = 3.7740	920
						c = 12.171	
	β-La		6.18		FCC, A1	5.303(>310°C)	
	γ-La		5.97		BCC, A2	4.26	
Lead	Pb	207.2	11.34	20.65	FCC, A1	4.9502	325.6
Lithium	αLi	6.941	0.534	8.55	HCP, A3	a = 3.111(<-193°C)	180.54
						c = 5.093	
	βLi				BCC, A2	3.5093(25°C)	
Magnesium	Mg	24.305	1.738	4.45	HCP, A3	a = 3.2094	650
						c = 5.2107	
Manganese	α-Mn	54.938	7.43 (α)	185	Cubic, A12	8.9126 (α)	1245
	β-Mn		7.29 (β)	91	Cubic, A13	6.3152(>710°C)(β)	
	γ-Mn		7.1(γ)	45.1	FCC, A1	3.860(γ)	
	δ-Mn				BCC, A2	3.080(>1143°C)	
Mercury	αHg	200.59	13.546	98.4(@50°C)	Rhomb., A10	a = 3.005	−38.9
						α = 70.53(<−38.9°C)	
Molybdenum	Mo	95.94	10.22	5.2(@0°C)	BCC, A2	3.147	2610
Nickel	Ni	58.693	8.902	6.84	FCC, A1	3.524	1453
Osmium	Os	190.2	22.583	−9.5	HCP, A3	a = 2.7341	2700
						c = 4.3918	
Palladium	Pd	106.42	12.02	10.8	FCC, A1	3.8903	1552
Platinum	Pt	195.08	21.45	10.6	FCC, A1	3.9236	1769
Potassium	K	39.098	5.344		BCC, A2	5.321	63.7
Rhenium	Re	186.207	21.04	19.3	HCP, A3	a = 2.7609	3180
						c = 4.458	
Rhodium	Rh	102.905	12.44	4.51	FCC, A1	3.8032	1966
Rubidium	Rb	85.468	1.532		BCC, A2	5.705	38.9
Ruthenium	Ru	101.07	12.2	7.6(@0°C)	HCP, A3	a = 2.7058	~2500
						c = 4.2816	

(*Continued*)

Physical Properties of Selected Elements *(Continued)*

Element	Symbol	Atomic Mass, g/mol	Density, g/cm^3	Electrical Resistivity, $(\Omega\text{-m})10^{-8}$@293°K	Crystal Structure	Lattice Constant/ Angle, Å/deg.	Melting Point, °C
Selenium	Se	78.96	4.79	12(@0°C)	Hex., A8	a = 4.3659 c = 4.9537	217
Silicon	Si	28.085	2.33	2.3×10^5@0°C	Dia. Cubic, A4	5.4306	1410
Silver	Ag	107.868	10.49	1.59	FCC, A1	4.0857	960.8
Sodium	Na	22.990	0.97	4.2	BCC, A2	4.2906	97.8
Tantalum	Ta	180.948	16.6	12.45	BCC, A2	3.3030	2996
Tellurium	Te	127.60	6.24	4.36×10^5	Hex, A8	a = 4.4566 c = 5.9264	449.5
Thallium	α-Tl	204.383	11.85	18(@0°C)	HCP, A3	a = 3.4566 c = 5.5248	303
	βTl (>230°C)				BCC, A2	3.879(>230°C)	
Thorium	αTh	232.038	11.72	13(@0°C)	FCC, A1 (25°C)	5.0842	1750°C
	βTh				BCC, A2	4.11(>1360°C)	
Tin	αSn(gray)	118.710	5.765		Dia. Cub., A4	6.4892(<13°C)	231.9
	βSn(white)		7.298	11.0	Tetragonal, A5	a = 5.8318 c = 3.1818	
Titanium	αTi	47.88	4.507	42.0	HCP, A3	a = 2.9508 c = 4.6835	1668
	βTi (>882.°C)				BCC, A2	3.3065	
Tungsten	W	183.85	19.3	5.65	BCC, A2	3.1652	3410
Vanadium	V	50.941	6.1	24.8 to 26	BCC, A2	3.0240	1900
Yttrium	Y	88.906	4.47	60.8	HCP, A3	a = 3.6482 c = 5.7318	1495
Zinc	Zn	65.39	7.133	5.916	HCP, A3	a = 2.6650 c = 4.9470	420
Zirconium	αZr	91.224	6.489–6.574	40	HCP, A3	a = 3.2316 c = 5.1475	1852
	βZr (>863)°C		6.046(@979°C)		BCC, A2	3.6090	

Photo 1 Stacking faults in stainless steels appear as fringes in transmission electron microscopy. (From Hirsch et al, Electron Microscopy of thin crystals, Plenum Press, N. Y., 1967.)

Photo 2 Twins in annealed 70Cu–30Zn cartridge brass. Dark strips (bands) are twins to adjacent light areas. Nominal tensile strength is 43ksi. Mag.~75x.

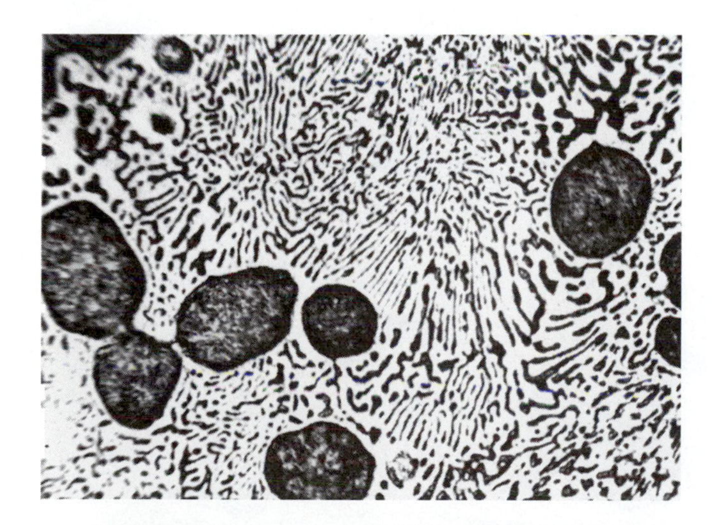

Photo 3 Microstructure of 50Pb–50Sn (half-and-half solder). As slowly solidified consists of lamellar eutectic structure and primary or pro-eutectic lead-rich phase. A hypo-eutectic microstructure. Mag~400x

Photo 4 Microstructure of eutectoid (1080) steel—pearlite. Mag.~2500x.

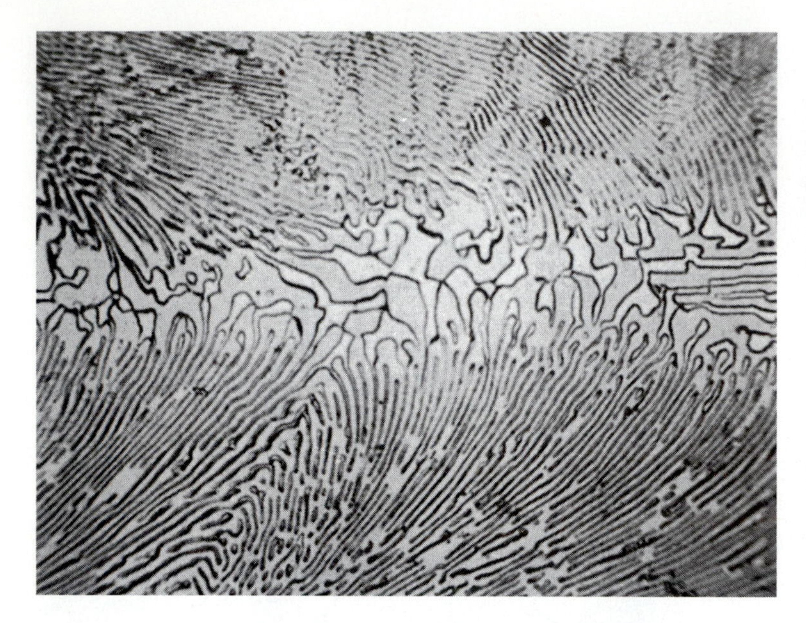

Photo 5 Microstructure of eutectic Cu–83P alloy. Mag~500x

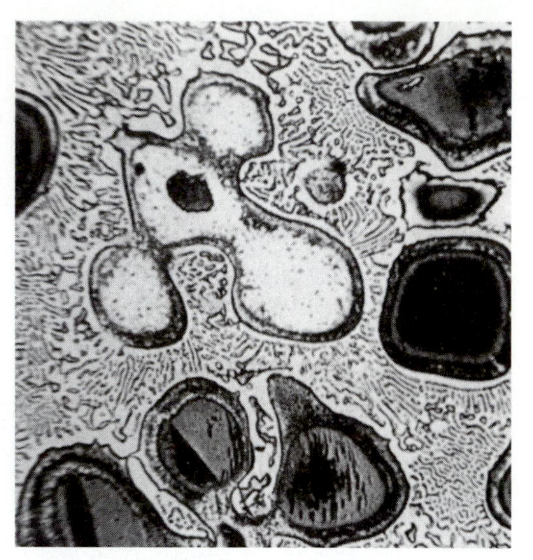

Photo 6 Microstructure of hypo-eutectic Cu–P alloy. Primary phase is copper-rich phase. Mag.~400x

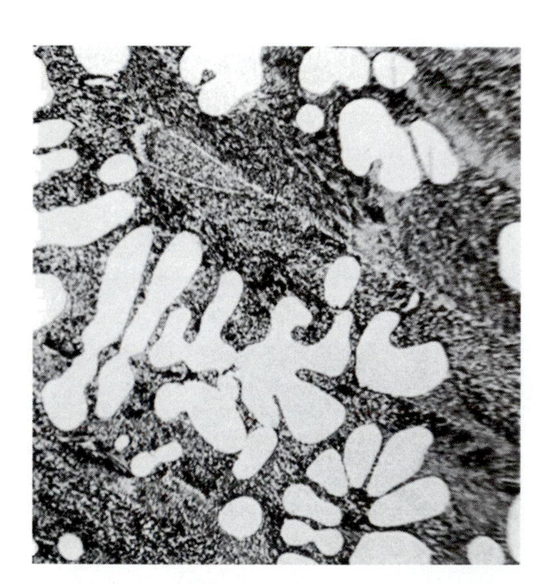

Photo 7 Microstructure of hyper-eutectic Cu–P alloy. Primary phase is Cu_3P. Mag.~400x

Photo 8 Microstructure of half-hard temper 70Cu–30Zn cartridge brass. Mag.~75x. Nominal tensile strength, 62ksi. Annealed structure is photo 2.

Photo 9 Microstructure of full-hard temper 70Cu–30Zn cartridge brass. Mag.~75x. Nominal strength 76ksi. Annealed structure is photo 2.

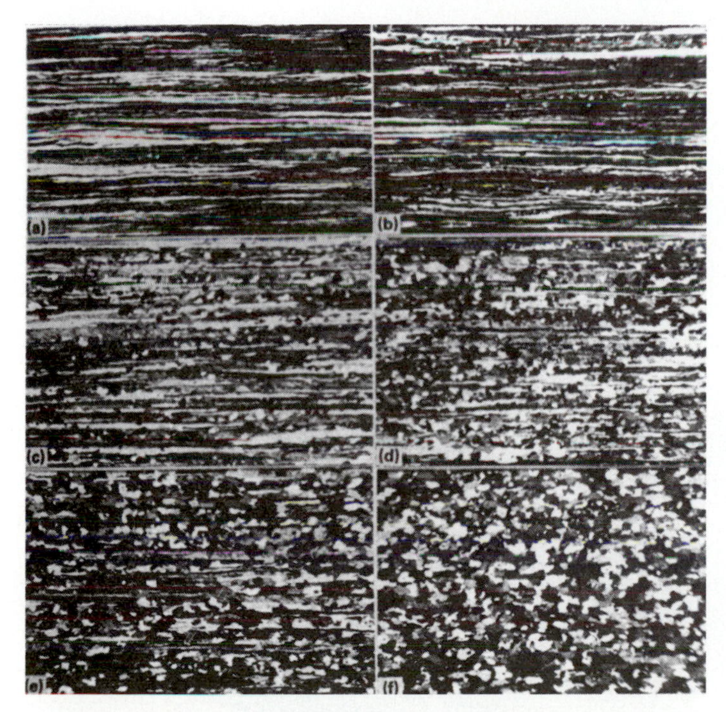

Photo 10 Polarized light micrographs showing progress of recrystallization in 5182–H19 at 245°C (470°F) (a) as-rolled; (b) after 1h at 245°C; (c) after 2h; (d) after 3h; (e) after 4h, and (f) after 7h. Mag.~100x. (From, "Aluminum, properties and physical metallurgy," John E. Hatch, Editor, Published by ASM, 1984.)

Photo 11 Microstructure of austenite in 1080 euctectoid-steel. Mag.~400x

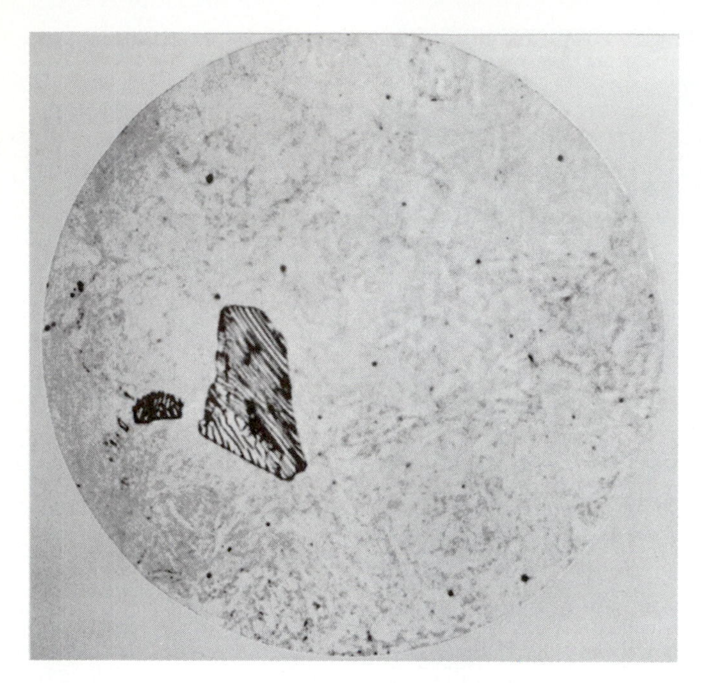

Photo 12 Start of transformation to pearlite of 1080 steel on slow cooling; transformation was interrupted—light areas are martensite that formed during quench of untransformed austenite. Mag.~1000x

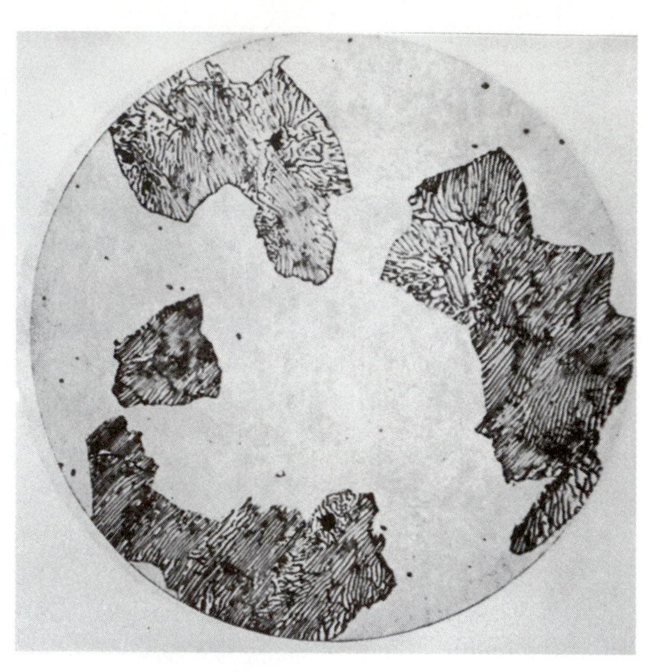

Photo 13 Interrupted transformation to 25% pearlite. Mag.~1000x

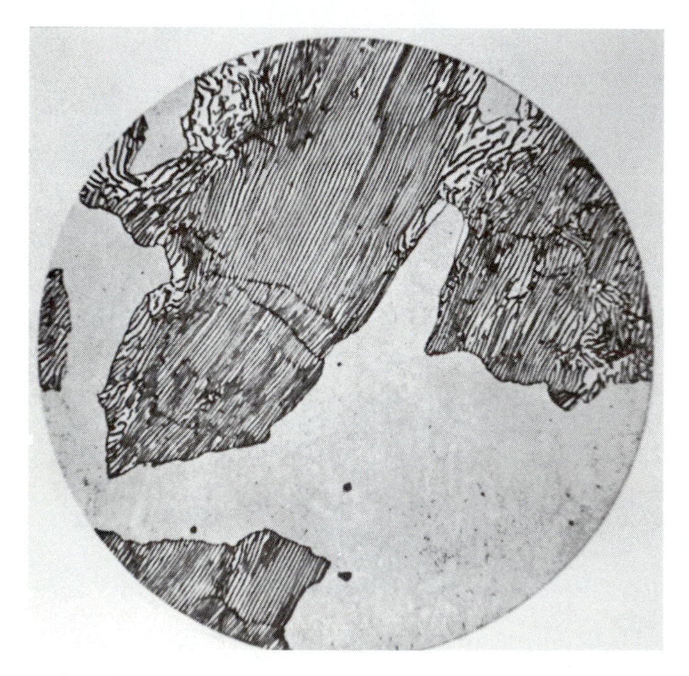

Photo 14 Interrupted transformation to 50% pearlite. Mag.~1000x

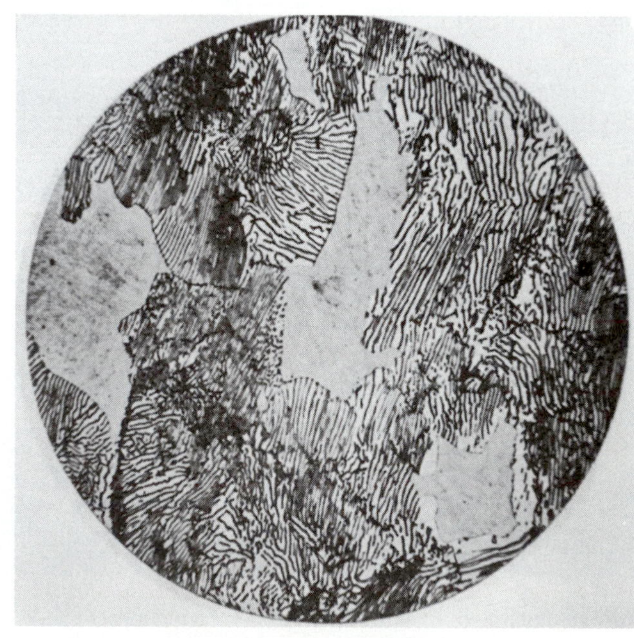

Photo 15 Interrupted transformation to 75% pearlite. Mag.~1000x

Photo 16 Transformation to 100% pearlite. Mag.~1000x

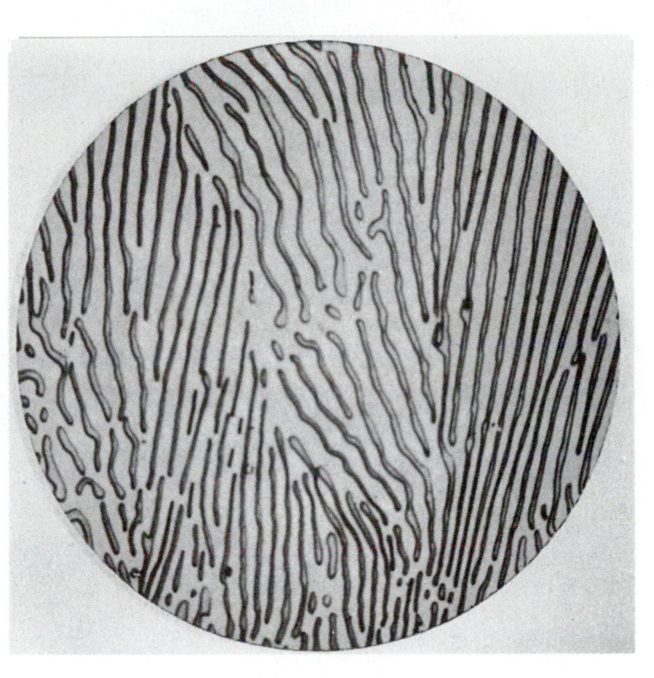

Photo 17 Coarse pearlite formed at about 720°C (~1325°F) in 1080. Mag.~2500x. Hardness~5HRC

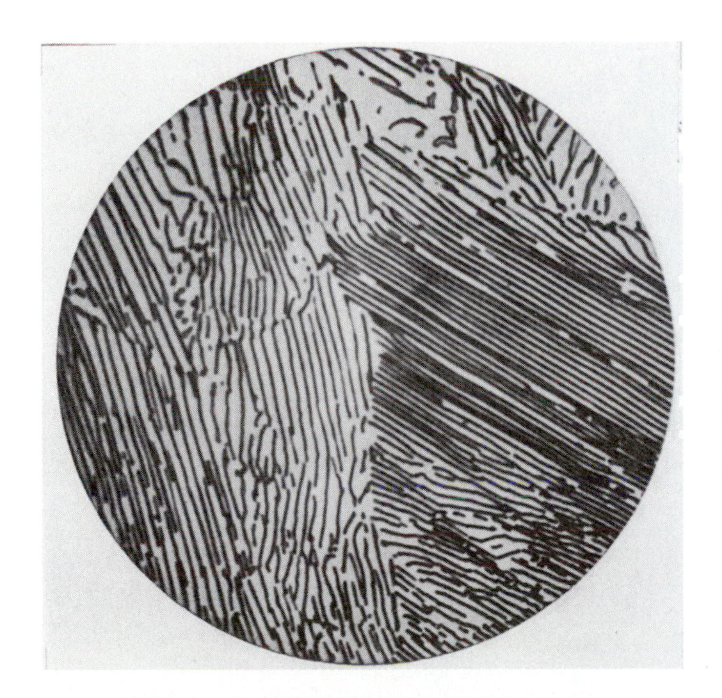

Photo 18 Coarse pearlite formed at about 705°C (~1300°F) in 1080. Mag.~2500x. Hardness~15HRC

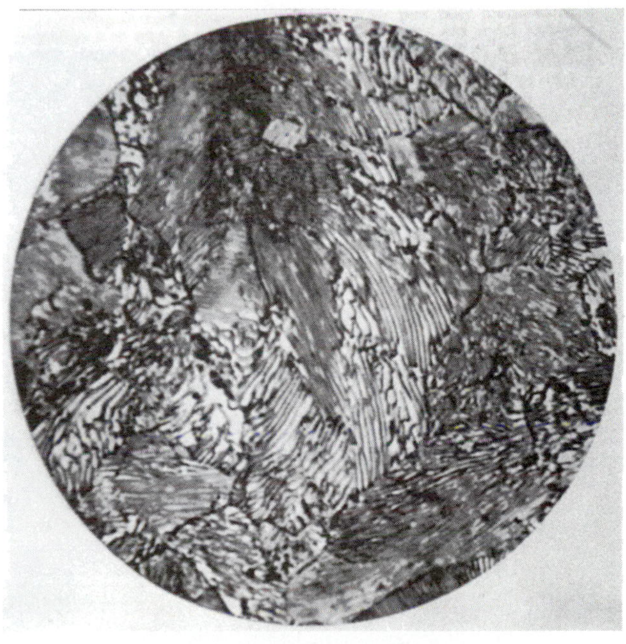

Photo 19 Fine pearlite formed at about 595°C (~1100°F) in 1080. Mag.~3000x. Hardness~40HRC

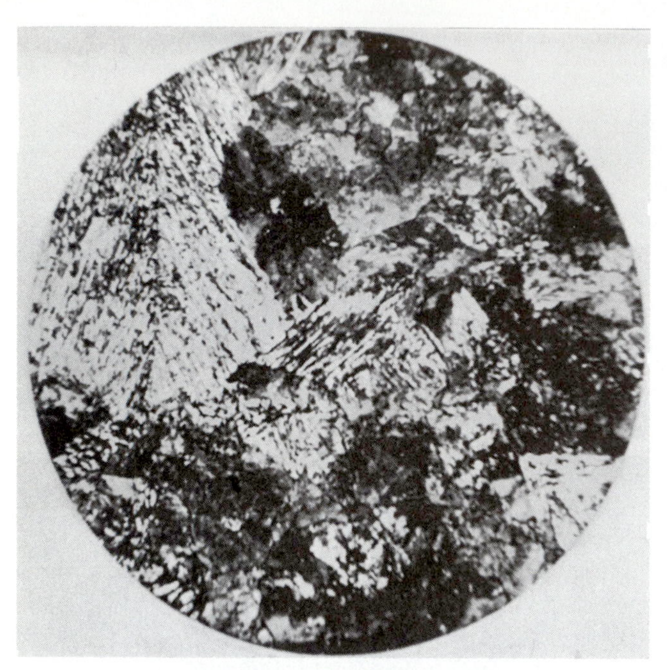

Photo 20 Upper bainitic microstructure formed in 1080 at 495°C (~925°F), just below nose of C-curve. Structure consists of "feathery" bainite and patches of very fine pearlite (unresolved). Mag.~2500x

Photo 21 Lower bainitie structure formed in 1080 at about 290°C (~550°F). Structure is acicular and is very similar to the martensite structure in photo 22. Mag.~2500x

Photo 22 Martensite structure in 1080. Mag.~2000x.

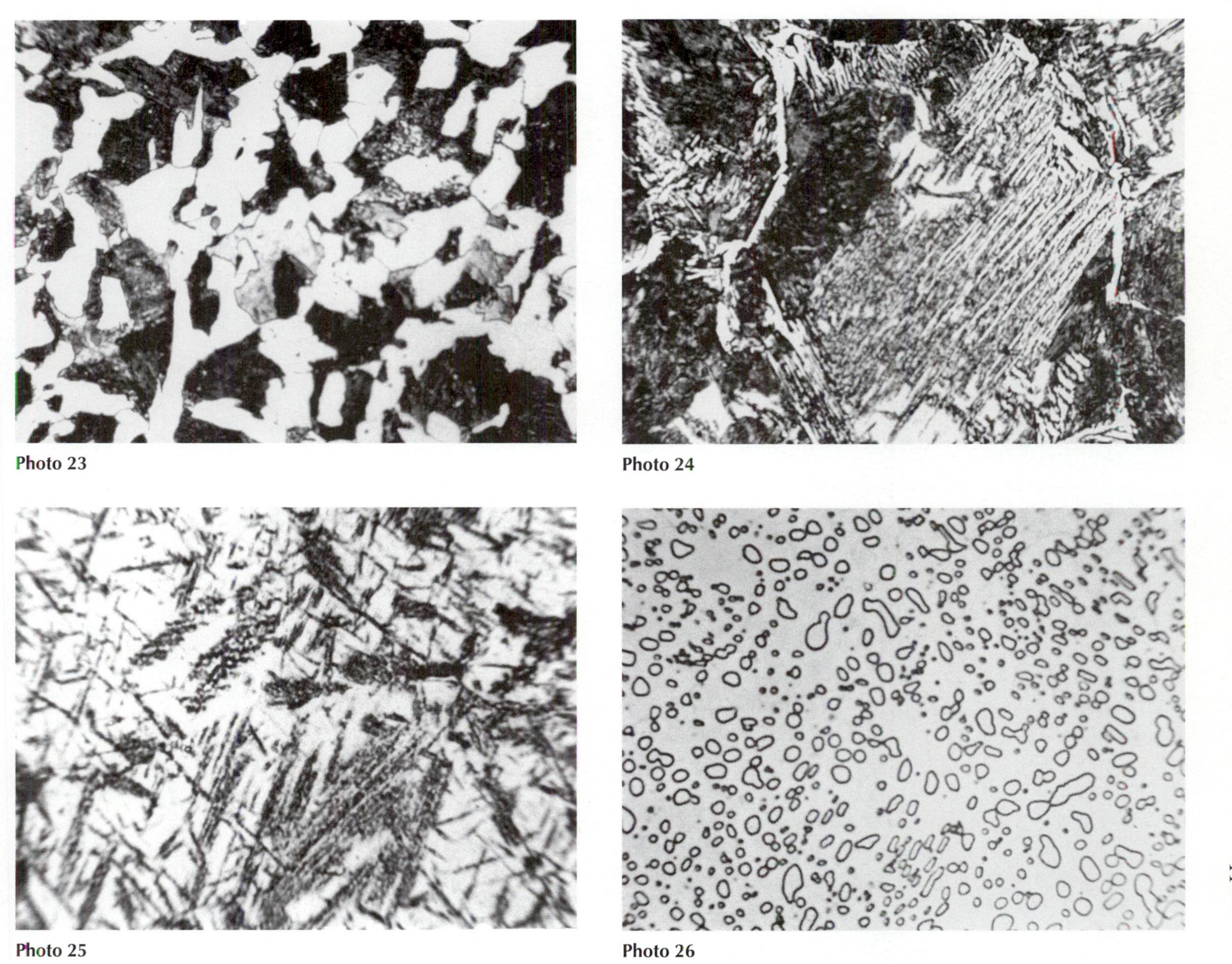

Photo 23

Photo 24

Photo 25

Photo 26

Photo 23, 24, and 25 From 1045 steel, austenitized at 843°C (1550°F); 23—normalized (air cooled) and tempered at 482°C (900°F) for 2h, structure is ferrite–pearlite (very fine) Mag.~500x; 24—oil-quenched 15sec, air cooled 5 min, oil-quenched to R.T.—structure consists of pro-eutectoid ferrite, upper bainite, and very fine pearlite; 25—water quenched 4sec, air cooled 3min, water quenched to R. T., structure consists of feathery and acicular bainite plus martensite, 26—spheroidized structure of 52100.

Index

A

Acetals, 684–88
Acrylics, 684
Activation energy, 110
  mechanisms of, 237–38
Actual critical diameter, 365
Addition polymerization, 667–68
Allotrophy, 45–46
Allowable design stress, 150–51
Alloys, 20, 91–96
Alloy steels, 491–98
Aluminum/aluminum alloys, 332–35, 540–63, 769–79
  cast aluminum/aluminum alloys, 542
  castings, selection of, 557–63
  temper designations, 542–46
  wrought aluminum/aluminum alloys, 540–42
    selection/applications of, 547–57
Amorphous metals, 264–65
Amorphous (noncrystalline) materials, 29–30
Amorphous polymers, 673–75
Amphoteric dopants, 113
Annealing, 270, 296–302, 319–21
  grain growth stage, 301–2
  recovery stage, 297
  recrystallization stage, 298–301
Anodic protection, 422–24
Aqueous corrosion rates, factors affecting, 414–15
As-quenched martensite, hardness of, 347–51
Atomic diameter, 2
Atomic mass, 3–4
Atomic number, 3
Atoms:
  arrangement of electrons in, 5–10
  macroscopic movements of, 220–36
  mass of, 3
  microscopic movements of, 236–39
  per unit cell, 42
  sizes of, 22–25
  structure, 2–3
Automobile fenders, material selection for, 793–95
Automotive industry, materials selection for, 779–86
  structural ceramics for engine components, 781–86

B

Basal planes, 34–35
Beam electrons, 14
Beverage containers, material selection for, 792–93
Blow molding, 730–31
  extrusion blow molding (EBM), 730–31
  injection blow molding (IBM), 731
Body centered cubic (BCC) structure, 32–33
Bonding, 15–22
  covalent, 17–18
  electronegativity, 16–17, 21–22
  hydrogen, 22
  intermolecular attraction, 17
  intramolecular bond, 17
  ionic, 16–17
  metallic, 17, 19–20
  mixed ionic and covalent bonding, 21–22
  principal mechanisms of, 16–20
  saturation, 15, 17–19
  tetrahedral, 18
  unsaturated, 18–20
  van der Waal's secondary bond, 17, 21–22
Bragg's law, 71–72, 74
Bravais lattices, 31–32
Brinnell hardness number (BHN), 174
Brinnell test, 174
Brittle materials, 134, 139–40

C

Carbon and low-alloy structural steels, 479–86
Carbon steels, 489–91

Casting(s), 258–64
  cast structure/segregation, 260–63
  continuous casting, 263–64
  macroscopic defects in, 258–60
Cast irons, 512–35
  compacted graphite (CG), 526–28
  design of iron castings, 531–34
  ductile, 525–26
  gray, 522–25
    special design concepts, 534
  malleable, 528–31
  specifications/grades/mechanical properties of, 522–31
  types/characteristics of, 516–21
Cathodic protection, 421–22
Cellulosics, 684
Ceramics, 604–31
  crystal structure, 605–8
  design/selection of, 650–63
  drying/firing, 614–18
  forming, 612–14
  fracture toughness of, 628–31
  phase diagrams, 608–11
  powders, batching/preparation of, 611–12
  processing of, 611–20
  refractories, selection of, 655–62
  shaping and surface finishing, 618–20
  structural, properties of, 620–27
  See also Glasses; Structural ceramics
Characteristic X-ray emissions, and chemical analysis, 13–14
Charpy V-notch (CVN) test, 171, 172–73
Coble mechanism (creep), 166, 168
Coefficient of linear thermal expansion (CLTE), 124–25
Cold and warm working, 278–85
  cold-work:
    plastic work and stored energy of, 292–96
    strengthening with, 279–81
  cold-work condition, specifying, 281–85
  deformation (crystallographic) texture, 285–86
  residual stresses, 291–92
  sheet forming, 286–91
Cold-work hardening, 136
Cold-work tool steels, 501
Combination polymerization, 668–69
Commodity thermoplastics, 681–84
  acrylics, 684
  cellulosics, 684
  polyolefins, 681–83
  styrenic plastics, 683
  vinyl plastics, 683–84
Common metallic structures, 32–35
Compacted graphite (CG) cast iron, 526–28
Composites, 134, 701–3
  continuous, design of, 738–40
  matrix selection for, 740–44
  selection methodology, 744–49
  types of, 720
Compression, 128
Compression and transfer molding (CM and TM), 733
Compressive stress, 127
Concurrent engineering, in product development, 786–91
Condensation polymerization, 668
Conjugated structure, 87
Constant strain amplitude test, 158–64
Constant stress amplitude test, 156–58
Constructive interference, 70–71
Contact potential, 101–2
Continuous casting, 263–64
Continuous cooling transformation (CCT) diagrams, 345–47
Coordination number, 20, 40, 42
Co-polymerization, 675–78
Copper/copper alloys, 583–92
  alloy and temper designations of, 583–84
  applications of, 591–92
  properties/selection criteria of, 585–91
Corrosion, 390–428
  amount/rate of, 409–12
  aqueous corrosion rates, factors affecting, 414–15
  corrosion current, 412–14
  designing to control, 418–25
  effect on mechanical properties, 416–18
  electrochemical aqueous corrosion, 391–409
  flowing electrolyte, effect of, 415–16
  forms of, 391
  minimizing, through design details and assembly of components, 424–26
  See also Electrochemical aqueous corrosion
Coulombic attractive force, 13–14
Covalent bonding, 17–18
Creep, 124, 164–71
  design involving, 167–71
  diffusional, 166
  dislocation, 166–67
  Larson-Miller parameter, 169–71
  mechanisms of, 165–67
  Sherby-Dorn parameter, 168–69
  steady-state (secondary), 164
  temperature-compensated time (TCT) parameter, 168
  time-temperature parameters, 167–71
  transient (primary), 164
  unstable (tertiary), 164–65
  viscous, 165–66
Cross-linked molecules, 672
Crystalline polymers, 673–75
Crystalline structures, 29–81
  amorphous (noncrystalline) materials, 29–30
  common metallic structures, 32–35
  creep in, 166
  diamond cubic (DC) structure, 38–40
  imperfections in, 63–70
  interstices/holes in crystal structures, 35–38
  ionic crystal structures, 40–44
  materials, 29–30
  packing fraction/density, 47–49
  point space lattices, 30–31
  polymers, 44–45
  slip directions, 61
  slip planes, 60–61
  unit cell, 30
  and X-ray diffraction, 70–77
Crystals:
  directions in, 49–52
  HCP planes/directions, 57–60
  Miller indices, 49–60
  planes in, 52–57
Crystal structure, ceramics, 605–8
CsCi structure, 41
Current rectifiers, 114–17

D

Deflection temperature under load (DTUL), 164
Density, 123
Design:
  activities, 431
  engineering design issues, 451–52
  information sources/specifications, 456–57
  materials and component failures, 452–56
  materials selection, 431–51
  methodology, 429–31
  phases of, 430–31
  See also Engineering design issues; Materials selection
Destructive interference, 70–71
Diamond cubic (DC) structure, 38–40
Diffusion, 219–42
  along defects/on surfaces, 238–39
  diffusion coefficient, 221
  diffusion coefficient (D), nature of, 236–37
  Fick's laws of diffusion, 220–36
  mechanisms of, 236–39
Diffusional creep, 166
Dislocation creep, 166–67
Dislocations, 67
  density of, 136
Dislocation strengthening, 136

Dispersed phase strengthening, 136–39
Dopants, 102, 113
Doping, 86–87, 89, 108
  of pure silicon, 231–33
Drop-weight tear test (DWTT), 172
Ductile cast iron, 525–26
Ductility, 139–40
  measures of, 144
Dynamic properties, 156
Dynamic tear (DT) test, 172

E

Edge dislocation, 67–69
Elastomers, 134
Electrical conductivity, 89–91
  in metals/alloys, 91–96
  superconductivity, 96–101
Electrically conductive materials, 82
Electrically nonconductive materials, 82
Electrochemical aqueous corrosion, 391–409
  cell voltage and electrode potential, 393–94
  electrochemical corrosion cells, types of, 397–94
  electrochemical reactions at the anode/cathode, 392–93
  Nernst equation, 396–97
  Pourbaix diagram, 404–9
  thermodynamics and cell voltage, 394–96
Electrochemical corrosion cells, 397–94
  differential cold-work/stress cells, 401–4
  differential temperature cells, 401
  electrolyte concentration cells, 397–401
  Galvanic cells, 397
Electrodes:
  avoiding contact of, 420
  controlling, 420–21
  elimination of, 418–19
Electron configuration, 6–10
Electron diffraction, 70
Electronegativity, 16–17, 21–22
Electronic conduction, 91
Electrons, 2–4
  beam, 14
  energy levels of, 5–6
  free, 89
  probability density, 2
  quantum states of, 6
Electron velocity, 126
Embrittling phenomena, 373–74
Engineering design issues, 451–52
  fluoroplastics, 684, 688
  performance/cost, 452
  polyamide-imide, 684, 688
  polyamides, 684, 688
  polyarylates, 684, 688
  polycarbonates (PCs), 684, 688
  polyetherimide (PEI), 684, 689
  polyketones, 684, 689
  polyphenylene oxide (PPO), 684, 689
  polyphenylene sulfide (PPS), 684, 689–90
  risk, 451–52
  sulfone polymers, 684, 690
  thermoplastic elastomers (TPEs), 690
  thermoplastic polyesters, 684, 688–89
Engineering stress-strain curve, 140–43
Engineering thermoplastics, 681, 684–90
  acetals, 684–88
Equilibrium phase diagrams, 184–218
  Al-Mn diagram, 208–11
  basic concepts, 185–86
  complete solubility in the solid state, 186–88
  eutectic systems, 192–99
  Gibbs Phase Rule, 188–89
  iron-carbon system, 199–207
  Lever Rule, 188, 189–91
  liquidus curves of, 211–13
  solidus curves, 213–14
  terminology, 185–86
  three-phase reactions, 191–92
Eutectic systems, 192–99
  complete solubility in the solid state, 192–93
  development/control of microstructures in, 195–99
  partial solubilities in both components, 193–95
  partial solubility in one component, 193
Extrinsic semiconductors, 102, 108–14
Extrusion, 729–30
Extrusion blow molding (EBM), 730–31
Extrusion and wire drawing, 295–96

F

Face centered cubic (FCC) structure, 33–34
Factor of safety, 150
Failures of materials and components, 452–56
  arising from processing/fabrication, 455–56
  design deficiencies, 452–53
  due to handling/assembly errors, 456
  due to service conditions, 456
  material imperfections, 455
  material selection deficiencies, 453–55
Fatigue properties/testing, 156–64
  constant strain amplitude test, 158–64
  constant stress amplitude test, 156–58
Fe-Fe_3C metastable system, 200–206
  eutectoid system in diagram, 202–6
Fermi-Dirac distribution function, 103–5
Ferrous materials, 460–538
  alloy steels, 491–98
  carbon and low-alloy structural steels, 479–86
  carbon steels, 489–91
  cast irons, 512–35
  heat-treatable carbon and low-alloy structural steels, 486–89
  procurement/specifications, 477–79
  stainless steels, 502–12
  steels,
    composition/designations/specifications for, 461–77
  tool steels, 498–502
  See also Cast irons; Nonferrous metals; Stainless steels; Tool steels
Fiber-reinforced plastics, 720–25
  continuous, design of, 738–40
Fick's first law, 220–21
Fick's laws of diffusion, 220–36
Fick's second law, 221–22
  and carburizing and nitriding processes, 223–27
  and decarburization, 227–31
  and doping of pure silicon, 231–33
  and homogenization, 222–23
Filament winding, 735–36
Flowing electrolyte, and corrosion, 415–16
Fluoroplastics, 684, 688
Forward bias, 116–17
Fracture mechanics, 151
  and materials selection, 796–800
Fracture toughness, 151–53
  design using, 154–56
  testing, 153–54
Free electrons, 89
Free electron theory of metals, 20
Frenkel defect, 67

G

Gibbs Phase Rule, 188–89
Glass-ceramics, 644–50
  crystal growth, 645–47
  ion-exchange, 649–50
  nucleation, 644–45
  secondary processing, 644–50
  tempering, 648–49
Glasses, 631–44
  chain structures, 636
  chemical durability/weathering of, 642
  container manufacture, 632
  density, 641
  design/selection of, 650–63
  float glass process, 633–34
  formation/primary processing of, 631–34

framework/network structures, 636–37
glass/amorphous structures, 637–39
glass fibers, 634
island structures, 635–36
isolated group structures, 636
mechanical properties, 643–44
sheet structures, 636
silicates/glass structures, 634–39
thermal conductivity/diffusivity, 642
thermal expansion, 641–42
tubings, 634
viscosity, 639–41
Young's modulus and Poisson's ratio, 642–43
See also Ceramics
Glass-reinforce plastics, 795–96
Glass transition temperature, 124
Grain boundaries, 64–66
Grain size, 64–66
strengthening, 135–36
Gray cast iron, 522–25

H

Hardenability, 355–58
Hardenability bands, 358
Hardening (heat treating), 323–31
due to spinoidal and order transformations, 339–42
martensite transformation, 328–31
precipitation hardening, 331–35
of steels, 342–84
titanium alloys, 335–39
transformation C-curves, 325–28
Hardness, 173–75
Brinnell test, 174
Mohs hardness scale, 174
Rockwell hardness test, 175
Scleroscope hardness tester, 174
scratch test, 174
Vickers test, 174–75
Heat-treatable carbon and low-alloy structural steels, 486–89
Heat-treated steels, 767–6
Heat treating, *See* Hardening (heat treating)
Heterogeneous nucleation, 254–56
Hexagonal close-packed (HCP) structure, 34–35
High-speed steels, 498–99
High temperature superconducting (HTS) materials, 98–100
theory of, 101
Homogeneous austenite, 370–73
Homogeneous nucleation, 252–54
Homogenization, 266
and Fick's second law, 222–23
Hot working, 302–8
controlled thermal mechanical processing, 306–8
dynamic recovery/recrystallization, 303–4
static recrystallization, 304–6
Hot-work tool steels (Group H), 499–500
Hybridization, 10
Hydrogen, atomic size for, 3
Hydrogen bonding, 22
Hydrostatic pressure, 128

I

Ideal critical diameter, 365
Impact properties/testing, 171–73
Ingots, macroscopic defects in, 258–60
Injection blow molding (IBM), 731
Injection molding, 730
Inorganic materials, *See* Ceramics; Glasses
Insulators, 82
Intensity of diffracted beam, 72–73
Interatomic bonding, 131–33
Interplanar spacing (d), 72
Interstices/holes in crystal structures, 35–38
Interstitial defect, 66–67
Interstitial diffusion, 237–38
Intrinsic semiconductors, 102–8
Ionic bonding, 16–17
Ionic conduction, 91
Ionic crystal structures—ceramic materials, 40–44
CsCi structure, 41
NaCI structure, 41–42
silica structure, 43–44
zinc blende structure, 42–43
Ions, sizes of, 22–25
Iron-carbon system, 199–207
$Fe\text{-}Fe_3C$ metastable system, 200–206
stable Fe-C (graphite) system, 206–7
Isothermal transformation (I-T) diagrams:
special processes using, 367–69
of steels, 343–45
Isotopes, 3, 100

J

Jominy hardenability test, 355–58

K

Kilovolts (keV), 13
Ksi (kips per square inch), 130

L

Ladle metallurgy, 246–49
Laminates, 725–27
fiber orientation, 725–26
laminate codes, 726–27
properties of, 727
Larson-Miller parameter, 169–71
Lateral strain, 128–29
Lattice constants (parameters), 30
Lattice vibrations, 126
Lever Rule, 188, 189–91
Life-cycle analysis (LCA), selection based on, 791–95
Linear/branched molecules, 671–72
Line defects, 67
Liquid injection molding (LIM), 733
Liquid state, characteristics of, 250–51
Liquidus curves of phase diagram, 211–13
Live-load stress range, 162
Longitudinal (tensile) strain, 128–29
Lorentz force, 99
Low-alloy special-purpose steels (Group L), 501
Lower-yield stress, 143
Low temperature superconducting (LTS) materials, 98–100
motion/interaction of electrons in, 100–101

M

Macromolecule, 18
Magnesium/magnesium alloys, 580–83
applications of, 582–83
designations, 580–82
Magnetic Resonance Imaging (MRI), 98
Malleable cast iron, 528–31
Martensite transformation, 328–31
Materials, 63–70
electrical properties of, 82–121
imperfections within a grain, 66–70
edge dislocation, 67–69
Frenkel defect, 67
interstitial defect, 66–67
line defects (dislocations), 67
stacking faults, 69–70
subgrain boundary, 69
substitutional defect, 67
twins, 69
vacancy, 66
mechanical properties, 133–51
allowable design stress, 150–51
creep, 164–71
design using fracture toughness, 154–56
ductility, 139–40
factor of safety, 150–51
fatigue, 156–64
fracture mechanics/toughness, 151–53
fracture toughness testing, 153–54
hardness, 173–75
impact, 171–73

Materials *(cont.)*
strength, 134
strengthening of metallic materials, 134–39
tensile test (ATSM E-8), 140–46
tensile test for plastics (ATSM D 68), 146–49
physical properties:
coefficient of linear thermal expansion (CLTE), 124–25
density, 123
glass transition temperature, 124
melting point, 123–24
microstructure insensitive, 123–27
thermal conductivity, 125–27
strain, 128–31
stress, 127–28
surface structures, 63–66
Young's modulus, 131–33
Materials selection, 431–51
criteria/tools for, 434–37
internal regular shaped features, influence of, 450–51
interrelated constraints, 431–34
materials charts, 439–45
performance/efficiency of materials, 437–39
shape factors, 445–48
shape sections, efficiency of, 448–50
Materials selection case studies, 754–80
aluminum alloys, 769–7
automotive industry, 779–86
concurrent engineering in product development, 786–94
fracture mechanics, 796–800
glass-reinforced plastics, selection of, 795–96
life-cycle analysis, selection based on, 791–95
steels, 754–69
Maxwell-Boltzmann distribution, 104
Melting, 244–50
primary processing, 244–46
remelting, 249–50
secondary processing (ladle metallurgy), 246–49
Melting point, 123–24
Mer, 19
Metallic bonding, 17, 19–20
Metallic glasses, 264–65
Metals, 82, 134
electrical conductivity in, 91–96
nonmetal to metal transition, 85–89
relationship to bonding/energy bands, 82–85
Metric conversion guide, A-1 to A-15
Microelectronic circuits, processing of, 233–36
Miller-Bravais indices, 58–59
Miller indices, 49–60, 74, 77
Mixed ionic and covalent bonding, 21
Modulus of elasticity, *See* Young's modulus
Mohs hardness scale, 174
Mold tool steels (Group P), 501–2

N

Nabarro-Herring mechanism (creep), 166, 168
NaCI structure, 41–42
Negative repulsive force, 15
Neutron diffraction, 70
Neutrons, 3, 4
mass of, 4
Nickel/nickel alloys, 592–600
designations/characteristics of, 593–99
Nil ductility test (NDT), 172
Nominal stress-strain curve, 140–43
Non-carbon backboned polymers, 678
Nonequilibrium processing diagrams, 343–37
Nonferrous metals, 539–603
aluminum/aluminum alloys, 540–63
copper/copper alloys, 583–92
magnesium/magnesium alloys, 580–83
nickel/nickel alloys, 592–600
titanium/titanium alloys, 563–79
Nonhardening thermal processes, 316–19
Nonmetals (insulators), 82–85
Nonstoichiometric compounds, 113
Non-uniform deformation of test specimen, 143–44
Normalizing, 321–23
Notch-impact testing, 171
N-type semiconductors, 108, 110–12
Nuclei, growth of, 256–58
Nucleons, 3–4

O

Octahedral interstice, 36, 37–38
Ohm's Law, 90
Order transformation, hardening due to, 339–42

P

Parallelipiped unit cell, 30
Partially stabilized zirconia (PSZ), 629–30
Particles, 2
fundamental, 4
Particle strengthening, 136–39
Pascal (Pa), 127–29
Pauli Extension Principle, 6, 10, 83, 103
Periodic Table, 2, 3, 6, 10–15
Phonons, 126
Planes, 52–57
Plastic deformation, 270–315
annealing, 270, 296–302
ceramics/glass/plastics/reinforce plastics, forming of, 310–11
cold/warm working, 278–85
deformation processes, 271–73
hot working, 302–8
rotation of crystal planes during, 277–78
of single crystals, 273–75
superplasticity/superplastic forming, 308–10
Plastics, 666–753
ASTM D 4000 line callout for, 749
blow molding, 730–31
commodity thermoplastics, 681–84
compression and transfer molding (CM and TM), 733
density, 709
designing with, 736–40
engineering thermoplastics, 681, 684–90
extrusion, 729–30
fabrication of, 727–34
forming processes, 731–33
injection molding, 730
liquid injection molding (LIM), 733
long-term properties, 716–19
mechanical properties, 710–19
polymerization processes, 667–69
reaction injection molding (RIM), 733
rotational molding, 733–34
selection of, 740–49
short-term properties, 712–16
thermal properties, 710
thermoforming, 732
thermosets, 690–96
types/classification of, 678–90
water absorption/water transmission, 709–10
See also Composites; Laminates; Polymerization processes; Polymers; Reinforced plastics
Point lattices, 32
Point space lattices, 30–31
Polyamide-imide, 684, 688
Polyamides, 684, 688
Polyarylates, 684, 688
Polycarbonates (PCs), 684, 688
Polyesters, 684, 688–89
Polyetherimide (PEI), 684, 689
Polyketones, 684, 689
Polymerization processes, 667–69
addition polymerization, 667–68
combination polymerization, 668–69
condensation polymerization, 668
Polymers, 19, 44–45, 133, 134
additives/fillers, 697–700
allotropy/polymorphism, 45–46

alloys/blends, 696–97
compounding of, 696–701
conducting, 87
conjugated, 87–88
co-polymerization, 675–78
cross-linked molecules, 672
crystalline and amorphous polymers, 673–75
crystallinity in, 44–45
intramolecular chemical composition, 675
linear/branched molecules, 671–72
molecular weight distribution, 671
molecular weight (MW), 669–71
non-carbon backboned polymers, 678
structural features of, 669–78
thermoplastic, 134
Polymorphism, 45–46
Polyolefins, 681–83
Polyphenylene oxide (PPO), 684, 689
Polyphenylene sulfide (PPS), 684, 689–90
Positive electrostatic (Coulombic) attractive force, 15
Pourbaix diagram, 404–9
Precipitation hardening, 331–35
aluminum alloys, 332–35
Primary creep, 164
Principal quantum number, 5–6
Prismatic planes, 34
Probability density, 2, 17
Protons, 3, 4
mass of, 4
Psi (pounds-force per square inch), 130
p-type semiconductors, 108–10
extrinsic, 109
Pultrusion, 735
Pure shear, 128

Q

Quantum numbers, 6
Quench, severity of, 358–65

R

Reaction injection molding (RIM), 733, 735
Rectifier equation, 117
Rectifiers, 114–17
Reflectivity, 20
Refractories, 655–62
Reinforced plastics, 703–9
contact molding methods, 734–35
designing with, 736–40
fabrication of, 734–36
fiber-reinforce plastics, properties of, 720–25
fibers, 703–8
filament winding, 735–36
matched mold methods, 735
matrices, 708–9
matrix selection for, 740–44
pultrusion, 735
reaction injection molding (RIM), 735
resin-transfer molding (RTM), 735
Residual stresses, thermal stress-relief of, 316–19
Resin-transfer molding (RTM), 735
Retained austenite, 370–73
Reverse bias, 116
Rockwell hardness test, 175
Rotational molding, 733–34

S

Saturation bonding, 15, 17–19
Schmid's Critical Resolved Shear Stress (CRSS) Law, 275–77
Schrodinger (wave) equation, 5
Scleroscope hardness tester, 174
Scratch test, 174
Secant modulus, 147–48
Secondary creep, 164
Semiconductivity, 102–17
Semiconductors, 82
density of states, 103–5
devices based on p-n junctions, 114–17
extrinsic, 102, 108–14
Fermi-Dirac function, 103–5
intrinsic, 102–8
Shear stress, 127, 129–30
Sheet forming, 286–91
actual forming operations, 291
good drawability, drawing/properties for, 286–89
stretchability, stretching/properties for, 289–91
Shells, 5–6
Sherby-Dorn parameter, 168–69
Shock resisting steels (Group S), 501
Silica structure, 43–44
Single crystals, 64, 265–66
dislocations, existence/evidence of, 273–75
plastic deformation of, 273–75
Slip directions, 61
Slip planes, 60–61
Slip systems, number of, and deformability, 61–63
Solidification, 251–58
casting, 258–64
heterogeneous nucleation, 254–56
homogeneous nucleation, 252–54
nuclei, growth of, 256–58
Solids, diffusion in, 219–42
Solid solution strengthening, 134–35
Solidus curves of phase diagram, 213–14
Spinodal transformation, hardening due to, 339–42
Stable Fe-C (graphite) system, 206–7
Stainless steels, 502–12, 766–67
designations/classes/properties of, 502–12
selection of, 512
Steady-state (secondary) creep, 164
Steels, 461–77
composition, classification by, 462–64
hardening of, 342–84
nonequilibrium processing diagrams, 343–37
precautions, 369–84
heat-treated, 767–69
product shape/finish processing/quality descriptors, classification by, 469–67
stainless, 766–67
strength, classification by, 464–69
structural, 755–65
Strain, 128–31
lateral, 128–29
longitudinal (tensile), 128–29
Strain-hardening, 136
Strengthening of metallic materials, 134–39
dislocation strengthening, 136
dispersed phase/particle strengthening, 136–39
grain size strengthening, 135–36
solid solution strengthening, 134–35
Stress, 127–28
compressive, 127
lower-yield, 143
maximum, 157
minimum, 157
shear, 127, 129–30
tensile, 127
Stress-strain curves, 140
Structural ceramics, 620–27
modulus of elasticity, 627
porosity/void content, 621
selection of, 652–55
uniaxial compression strength, 626–27
uniaxial tensile strength, 621–26
Structural steel, 755–65
dynamically loaded structures, 763–64
high-performance steels, benefits of adopting, 764–65
microalloyed steels, 764
statically loaded structures, 762–63
Styrenic plastics, 683
Subshells, 6
Substitutional defect, 67
Sulfone polymers, 684, 690
Superconductivity, 96–101
definition of, 96–97
Superconductors, 82
critical properties of, 97–100
high temperature superconducting (HTS) materials, 98–100
theory of, 101

Superconductors *(cont.)*
  low temperature superconducting (LTS)
    materials, 98–100
    motion/interaction of electrons in,
      100–101
  pinning force, 99
Surface hardening process, 374–84
  processes with change in surface
    composition, 377–84
  processes with no change in surface
    composition, 374–77
Surface structures, 63–66

T

Tangent modulus, 147
Tear-strength, 134
Temperature-compensated time (TCT)
  parameter, 168
Tempering, 351–55
Tensile failure strength, 134
Tensile strain, 128
Tensile stress, 127
Tensile test (ATSM E-8), 140–46
Tensile test for plastics (ATSM D 68),
  146–49
Tertiary creep, 164–65
Tetragonal zirconia polycrystals (TZP),
  630–31
Tetrahedral bonding, 18
Tetrahedral interstice, 36, 37–38
Thermal conductivity, 125–27
Thermal lattice vibration, 99–100
Thermally activated process, 166
Thermal processes:
  annealing, 319–21
  hardening (heat treating), 323–31
  heat-treating of steels, 342–84
  martensite transformation, 328–31
  nonhardening, 316–19
  normalizing, 321–23
  precipitation hardening, 331–35
  spinodal and order transformations,
    339–42
  surface hardening, 374–84
  titanium alloys, 335–39
  transformation C-curves, 325–28
Thermal stress-relief of residual stresses,
  316–19
Thermal vibrations, 92–94
Thermoforming, 732
Thermoplastic elastomers (TPEs), 690
Thermoplastic polyesters, 684, 688–89
Thermoplastic polymers, 134
Thermosets, 690–96
Three-phase reactions, 191–92
Through hardening, 347–69
  precautions, 369–84
Titanium/titanium alloys, 563–79
  alloys:
    classifications/designations of, 568–71
    room temperature mechanical
      properties, 571–72
    selection of wrought titanium alloys,
      572–73
    titanium casting, 573
  commercially pure and modified
    titanium, 563–68
  heat treating, 335–39
Tool steels, 498–502
  cold-work tool steels, 501
  high-speed steels, 498–99
  hot-work tool steels (Group H), 499–500
  low-alloy special-purpose steels
    (Group L), 501
  mold tool steels (Group P), 501–2
  shock resisting steels (Group S), 501
  water hardening tools steels (Group W),
    502
Transformation C-curves, 325–28
Transient (primary) creep, 164
Transistors, 117
True stress-true strain curves, 140, 144–46

U

Unified atomic mass unit (u), 3
Uniform deformation of test specimen,
  143–44
Unit cell, 30
  number of atoms per, 42
Unsaturated bonding, 18–20
Unstable (tertiary) creep, 164–65

V

Vacancy, 66
Vacancy diffusion, 237
Vacant electron site, 109
van der Waal's secondary bond, 17, 21–22
Vickers hardness number (VHN), 175
Vickers test, 174–75
Vinyl plastics, 683–84
Visco-elastic (visco-plastic) deformation,
  165
Viscous creep, 165–66
Voltage amplifiers, 114, 117
Volume packing fraction (VPF), 47–49

W

Water hardening tools steels (Group W),
  502
Work function, 101–2
Wrought aluminum/aluminum alloys,
  540–42, 547–57
  heat-treatable alloys, 554–57
  nonheat-treatable alloys, 547–54

X

X-ray diffraction, 70–77
  Bragg's law, 71–72
  constructive/destructive interference,
    70–71
  determining crystal structure/lattice
    parameter, 74–75
  diffraction angle/direction, 72
  intensity of diffracted beam, 72–73
  interplanar spacing (d), 72
  structure factors for cubic structures,
    73–74

Y

Young's modulus, 17, 126–27, 131–33, 160
  See Modulus of elasticity

Z

Zinc blende structure, 42–43
Zirconia:
  fracture toughness of, 628–31
  partially stabilized zirconia (PSZ), 629–30
  tetragonal zirconia polycrystals (TZP),
    630–31
Zone-refining process, silicon, 102–3